T0350650

THE PLANTS OF

SUDAN AND SOUTH SUDAN

An annotated checklist

EDITED AND COMPILED BY

Iain Darbyshire, Maha Kordofani, Imadeldin Farag, Ruba Candiga and Helen Pickering

Kew Publishing
Royal Botanic Gardens, Kew

ROYAL BOTANIC GARDENS

First published in 2015 by
Royal Botanic Gardens, Kew,
Richmond, Surrey, TW9 3AB, UK
www.kew.org

Distributed on behalf of the Royal Botanic Gardens, Kew in North America by the
University of Chicago Press, 1427 East 60th Street, Chicago, IL 60637, USA

ISBN 978-1-84246-473-1
eISBN 978-1-84246-517-2

British Library Cataloguing in Publication Data
A catalogue record for this book is available from the British Library

Design and layout by Christine Beard
Kew Publishing
Royal Botanic Gardens, Kew

Front cover images: *upper* – Arbaat Valley, Red Sea Hills, Sudan, April 2011 (photo.: Imadeldin Farag);
lower – Boma Plateau near Towot, South Sudan, April 2012 (photo.: Aaron Davis, RBG Kew).
Back cover map of Sudan and South Sudan: base map retrieved from
http://www.arcgis.com/home/item.html?id=c4ec722a1cd34cf0a23904aadf8923a0 November 2014

Printed in the UK by Marston Book Services Ltd
Printed in the USA by The University of Chicago Press

For information or to purchase all Kew titles please visit
www.kewbooks.com or email publishing@kew.org

Kew's mission is to inspire and deliver science-based plant conservation worldwide, enhancing
the quality of life.

Kew receives half of its running costs from Government through the Department for
Environment, Food and Rural Affairs (Defra). All other funding needed to support Kew's vital
work comes from members, foundations, donors and commercial activities including book sales.

Contents

List of figures

List of tables

Foreword

by Professor Essam Ibrahim Warrag

It is a great pleasure to see the publication of this checklist for the flora of Sudan and South Sudan and to write the foreword for Sudan. The flora of Sudan represents part of our wealth and heritage, it is a source of pride that should be documented, managed and conserved for future generations. Sudan covers a vast area of tropical northeast Africa, ranging from desert in the north to savannah woodland in the south and contains a diverse flora. The Sudan flora has always been important for people as a source for livestock grazing, fodder, construction material, firewood, food and traditional medicine. Furthermore, it has vast environmental value in sustaining and determining ecosystems and biodiversity.

This complete checklist of the flora that covers all of the higher plants known in Sudan is an outstanding contribution that provides a baseline reference for future botanical and conservation measures in Sudan. The comprehensive listing of species with accepted scientific names, relevant synonymy, habit and habitat notes will have many uses: aids in identification and correct naming of species are essential resources for biodiversity estimates and biogeographic studies. The preliminary list of threatened plant species will provide conservationists, land management agencies and relevant government departments with key information on potential conservation priorities.

I welcome this important addition to our knowledge of plant biodiversity in Sudan, and at the same time I want to congratulate Dr Iain Darbyshire, Dr Helen Pickering, Dr Maha Kordofani, Ruba Candiga, Imadeldin Ahmed Abdalla Farag and the many individuals and institutions, especially the Royal Botanic Gardens, Kew, who made it possible.

Prof. Essam Ibrahim Warrag
Director, Institute of Environmental Studies
University of Khartoum
Khartoum
Sudan

Foreword

by Professor Sebsebe Demissew

With the completion of the *Flora of Djibouti* in 1994, the *Flora of Somalia* in 2006, the *Flora of Ethiopia and Eritrea* in 2009 and the *Flora of Tropical East Africa* in 2012, many of us were looking for a modern source of botanical information on the floras of Sudan and South Sudan to complete the knowledge gap in the flora of Northeastern and Eastern Africa. Thus the completion of this checklist is a most welcome development.

The checklist of the plants of Sudan and South Sudan covers all taxa of seed plants: the flowering plants (angiosperms) and gymnosperms occurring in the two countries. It includes 183 families with 1,351 genera and 3,969 species (4096 taxa). The system of the family circumscription for the angiosperms follows the latest classification of the Angiosperm Phylogeny Group III produced in 2009 which is also being regularly updated. The new treatment shows that the 10 most diverse plant families in order of rank in the two countries combined are Leguminosae (Fabaceae), Poaceae, Asteraceae, Cyperaceae, Rubiaceae, Malvaceae, Acanthaceae, Euphorbiaceae and Lamiaceae.

Although, as expected, the checklist does not include descriptions of each species, the life-form (tree, shrub, climber and herb) is documented. In addition, relevant synonyms together with the relevant literature are provided, including reference to the completed Floras of adjacent regions, the *Flora Ethiopia & Eritrea* (FEE) and the *Flora of Tropical East Africa* (FTEA), which can provide more detailed information on each species including descriptions.

The checklist also provides the distribution of each of the species within Sudan and South Sudan where nine regions are identified. The first six: NS (Northern Sudan), RS (Red Sea), DAR (Darfur), KOR (Kordofan), CS (Central Sudan) and ES (Eastern Sudan) occur in the Sudan and the remaining three: BAG (Bahr el Ghazal), UN ((Greater) Upper Nile) and EQU (Equatoria) occur in South Sudan. In addition, the distribution of each species in tropical and southern Africa and beyond is recorded.

Within the checklist, species of high conservation concern are also highlighted and their IUCN status assigned using the IUCN (2001) Categories as DD – Data Deficient; NT – Near Threatened; VU – Vulnerable; EN – Endangered and CR – Critically Endangered, or assigning an 'RD' Category for potential globally threatened species.

As exemplified in other modern floras, there is a specimen citation which is important in helping to ascertain the identity of each species and infraspecific taxon. In cases where the species occurs in both Sudan and South Sudan, two specimens are cited, one for each country.

It is hoped that the publication of this checklist will stimulate further botanical research and documentation both in the botanical community at large and within the two countries. The checklist will also help development projects in the two countries and regional organizations such as the Intergovernmental Authority on Development (IGAD) in Eastern Africa, created in 1996, who coordinate activities that affect the seven countries of IGAD (Djibouti, Eritrea, Ethiopia, Kenya, Somalia, Sudan and Uganda), politically and scientifically.

The checklist is a product of a joint collaboration between scientists working in Sudan (University of Khartoum), South Sudan (University of Juba) and the Royal Botanic Gardens, Kew, which will undoubtedly strengthen the relationship between these institutions and promote knowledge transfer and future collaboration.

I would like to express my strong appreciation to the authors for their dedication to providing this comprehensive checklist.

Prof. Sebsebe Demissew
Keeper, National Herbarium and Leader of the Ethiopian Flora Project (1996–2010)
Department of Plant Biology and Biodiversity Management
College of Natural Sciences
Addis Ababa University
Ethiopia

Acknowledgements

Completion of this checklist has only been possible thanks to the considerable support from the staff and facilities of the Herbarium at the Royal Botanic Gardens, Kew, for which we would particularly like to thank Professor David Mabberley, the Keeper of the Herbarium at the commencement of the project, and Professor David Simpson who saw its completion. The continuous support and encouragement throughout the project of Dr Abdelazim Ali Ahmed, head of the Department of Botany at the University of Khartoum, is highly appreciated. We acknowledge the support of the College of Natural Resources and Environmental Studies at the University of Juba.

The concept of the checklist was initiated by Dr Loutfy Boulos from Cairo University and Dr Henk Beentje at Kew. We thank them both for allowing us to take their ideas forward and to build on some of the early drafts of the family accounts they produced. Dr Shahina Ghazanfar at Kew is also thanked for her very useful involvement in the early stages of development of the checklist. We are hugely grateful to George Gosline for his considerable input through the creation of the database and his unstinting help with many of the design features of the publication. We are grateful to the volunteer support of Robert Vanderstricht who also contributed to aspects of the database design as well as writing several family accounts.

We are indebted to Sebsebe Demissew of Addis Ababa University, Ib Friis of the University of Copenhagen and Kaj Vollesen, Mike Lock and Roger Polhill, all retired staff members from the Royal Botanic Gardens, Kew, for their wide range of helpful information and support throughout the project.

We would like to thank the many botanists who provided specialist advice on plant family treatments and identifications. From the Kew Herbarium, we particularly thank: Henk Beentje (Asteraceae), Gill Challen (Euphorbiaceae and Phyllanthaceae), Martin Cheek (various families), Aaron Davis (Rubiaceae), David Goyder (Apocynaceae), Timothy Harris (Gentianaceae), Nicholas Hind (Asteraceae/Compositae), Gwil Lewis and Brian Schrire (Leguminosae), Maria Vorontsova (Poaceae), Odile Weber (Aloe in Xanthorrhoeaceae) and Paul Wilkin (Dioscoreaceae and other monocots). From the University of Khartoum, we particularly wish to thank Haytham Hashim Gibreel for his major contribution to several family accounts (Aizoaceae, Amaranthaceae, Clusiaceae, Gisekiaceae, Hypericaceae, Limeaceae, Lophiocarpaceae, Molluginaceae, Nyctaginaceae and Phytolaccaceae). In addition, we are grateful to Zachary Rogers from Missouri Botanical Garden (Thymelaeaceae); Eberhard Fisher from University of Koblenz (Scrophulariaceae and allies); Olivier Lachenaud from Université Libre de Bruxelles (Psychotria in Rubiaceae); David Johnson from Ohio Wesleyan University (Xylopia in Annonaceae) and Marc Sosef from the National Botanic Garden of Belgium (Ochnaceae) for their expert advice.

We are grateful to Juliet Williamson for producing the four excellent new illustrations of endemic plants for this book and Victoria Gordon Friis for her kind permission to re-use the plates of *Dorstenia annua* and *Brachystephanus sudanicus* that were first published in *Kew Bulletin* in 1982.

ACKNOWLEDGEMENTS

The support, patience and design help we received from the Kew Publishing, especially Gina Fullerlove, Lydia White, Christine Beard and Georgina Smith, are greatly appreciated. Similarly we thank the staff in the Library and Archives who assisted in finding obscure biographical information on some of the collectors.

This checklist would not have been possible without the generous financial support from the following donors: (1) the World Collections Programme, funded by the U.K. Department for Culture Media and Sport, which supported the initiation of the project; (2) the Bentham-Moxon Trust for funding research visits to Kew by staff from the University of Khartoum; (3) the B.A. Krukoff Fund for the Study of African Botany who supported Dr Iain Darbyshire's participation in this project; (4) the University of Cambridge Student Conference on Conservation Science and associated internship which supported Imadeldin Farag's research into the conservation priority species in Sudan; (5) the U.K. Department for International Development at the British Embassy in Juba for funding the participation of Ruba Candiga as a collaborator in this project; and (6) finally, the Joseph Banks Society who assisted with securing funding for the publication of this volume.

1. Introduction

In this chapter we first discuss how and why the current checklist came to be written, before giving a brief overview of the geography, climate and vegetation of Sudan and South Sudan to provide some context for the plant checklist, then we end with a review of the current extent of knowledge of the Sudanese flora and the need for further exploration in the region. When referring to the two countries together within this work, we refer to the 'Sudan region'. The regional subunits of the two countries that are referred to in this and subsequent chapters are discussed in more detail in chapter 4, and the reader can refer to Figure 4.1 for their locations.

The evolution of the checklist

Recent decades have seen major advances in the documentation of the tropical and subtropical flora of eastern Africa, with the completion of the *Flora of Tropical East Africa* (1952–2012), the *Flora of Ethiopia and Eritrea* (1989–2009), the *Flora of Somalia* (1993–2006), the *Flora of Egypt* (1999–2005) and all the related monographic work on African flowering plants. There therefore exists a near-complete and rigorous regional baseline on which applied botanical work can be developed, for example the identification of conservation priority regions, habitats and species. However, a very notable and sizable gap within this regional framework is the flora of Sudan and South Sudan which remains one of the great floristic 'black holes' in Africa and so hinders a truly regional approach to plant conservation efforts in eastern Africa.

The most recent country-wide account of the Sudan region's plant diversity is the *Flowering Plants of the (Anglo-Egyptian) Sudan* by F.W. Andrews (1950–56) which itself built on the earlier *Flora of the Sudan* by Broun & Massey (1929). Whilst Andrews' work provides a very useful starting point for a modern checklist, the content is both incomplete and outdated. Crucially, Andrews' flora was written prior to the major period of modern botanical documentation in tropical eastern Africa; at that time the *Flora of Tropical East Africa* had only just begun and none of the large families had yet been treated, whilst the other modern regional or national Floras had not even been contemplated. There were some subsequent updates to Andrews' work, notably by Wickens (1968) and Drar (1970). More recently, there have been excellent botanical accounts published focussing upon specific localities, including Jebel Gourgeil and northwest Darfur (Quézel 1969), the Jebel Marra massif in Darfur (Wickens 1976), Jebel Uweinat on the Sudan-Libya-Chad border (Léonard 1997; 1999a; 1999b; 2000) and the mountains of the Sudan-Uganda border area east of the Nile (Friis & Vollesen 1998; 2005). The woody flora has also received some further attention, most notably through El Amin's (1990) guide to the trees and shrubs of Sudan (but see also, for example, Sahni 1968; Sommerlatte & Sommerlatte 1990; Braun *et al.* 1991; Vogt 1995). However, large areas of the two countries and large parts of the flora remain uncovered by any modern treatment. It is therefore clear that an updated, accurate and comprehensive checklist is sorely needed to provide a baseline reference for future botanical and conservation work in the Sudan region and to fill the gap in our regional knowledge.

It was with this in mind that the idea for a new checklist for Sudan, based largely on the existing knowledge held in relevant herbaria, was first mooted in the mid-2000s by Drs. Loutfy Boulos of the University of Cairo and Henk Beentje of the Royal Botanic Gardens, Kew (further, Kew). Whilst Dr. Boulos drafted a number of early accounts for families in the Brassicales and Caryophyllales, the project was not taken further. In 2008, one of the current authors, Prof. Maha Kordofani, contacted Dr. Beentje and Dr. Shahina Ghazanfar at Kew in order to resurrect the idea of a country-wide checklist. Dr. Iain Darbyshire agreed to lead the project from the Kew side, securing some small funding from the U.K. government's World Collections Programme to cover the costs of a visit to the U.K. by Prof. Kordofani to attend a planning workshop. Around the same time, Dr. Helen Pickering agreed to assist in the project on a voluntary basis. A suitable format and research methods were agreed for the checklist and the compilation of the data got underway. In the following years, we have managed to secure small grants from several sources in order to fund several further research trips to the U.K. by Prof. Kordofani and her colleagues at the University of Khartoum, including Imadeldin Farag who co-authors this work. Imadeldin has also spent some considerable time in the field during this period and was able to make plant collections for incorporation into the checklist, adding valuable records from the Red Sea region and Central Sudan.

Following a referendum on independence in 2011, the largely Christian south of Sudan seceded from the largely Islamic north, so splitting Sudan in two and creating the newest African nation, that of South Sudan. Hence, what started out as a botanical checklist for a single country changed in mid-flow to a checklist for two independent nations. Fortunately, the regional breakdown of Sudan that we had adopted from the outset allowed for easy separation of the two countries. We were also able to add a South Sudanese collaborator to the project, Dr. Ruba Candiga of the University of Juba, who visited Kew under a British Government, Department for International Development (DfID) grant in 2012.

The checklist has been compiled through the combination of literature searches using library and web-based resources, and through reference to herbarium material from the Sudan region. Indeed, one of the key objectives of this work, beyond providing an accurate baseline for plant-related work in the region, was to demonstrate how useful the accumulated knowledge held in herbaria can be in providing detailed information on a region's plant diversity. Supplementing this historical information, we have been able to draw on the current in-country botanical expertise of the Sudanese and South Sudanese co-authors of this work and their colleagues.

The resultant work that we present here, and also maintain electronically as a database which we aim to regularly update and augment in the future, is much more than simply a list of names. It is more a foundation for a complete documentation of the plants of the Sudan region. We hope that this work will become a widely consulted reference source for all scientists, land managers and conservation practitioners with an interest in the region and its plants.

Geography

The former united nation of Sudan, which gained independence from joint British-Egyptian rule in 1956, was the largest country in Africa, at over 2.5 million km², covering a vast area of dry tropical northeast Africa. Following South Sudanese independence in 2011, the Republic of Sudan now occupies an area of 1,886,068 km² between c. 8.7° and 22°N latitude and 21.8 and 38.6° E longitude. It is currently the third largest country in Africa behind Algeria and the Democratic Republic of the Congo. It borders Egypt to the north, Libya to the northwest, Chad to the west, the Central African Republic to the southwest, South Sudan to the south and Eritrea and Ethiopia to the east. North of the Eritrean border, Sudan has an 850 km stretch of coastline on the Red Sea, on which lies the major port city of Port Sudan. Estimates based on the 2008 census place the population of Sudan at over 30 million people. The Republic of South Sudan is considerably smaller, with an area of 619,745 km^2 between c. 3.6° and 12.2° N latitude and 23.6° and 35.9° E longitude. South Sudan is landlocked, bordering Sudan to the north, the Central African Republic to the west, the Democratic Republic of the Congo and Uganda to the south, Kenya to the southeast and Ethiopia to the east. The population in 2013 was estimated at just over 11 million people.

The economies of the two countries are largely rural, with extensive pastoralism and with both smallholder and commercial arable agriculture, particularly intensive along the fertile alluvial plains of the Nile and its tributaries, with major irrigation schemes in some regions. The two countries also share considerable oil wealth in the border region north of the Bahr el Ghazal River, and disputes over these oil reserves were a major factor behind the 2012 war between the two newly separated countries.

The Red Sea coastal plain averages 10–30 km wide but can extend to 50 km wide in some areas. Inland, much of the land in the Sudan region is low-lying, with extensive flat plains averaging between 500 m and 1000 m elevation. Highland areas are scattered and generally small in area, the most important being, in Sudan: (1) Jebel Uweinat in the extreme northwest on the Sudan-Libya-Chad border; (2) the Jebel Marra massif in the central Darfur region which includes the highest peak in Sudan, the Deriba Caldera at just over 3000 m elevation; (3) the northern outlier of the Marra massif, Jebel Gourgeil; (3) Jebel Maydub (Meidob) and the Teiga Plateau, two smaller hilly areas in north Darfur; (4) the Red Sea Hills, an interrupted mountain chain running parallel to the Red Sea coast from southeast Egypt, through the Hala'ib Triangle and Sudan and through much of northern Eritrea; and (5) the Nuba Mountains which cover a large though rather diffuse area of south-central Sudan in Kordofan region. And in South Sudan: (1) the Imatong-Lafit-Dongotona-Didinga series of mountain chains in south-central Equatoria and continuing into Uganda, the Imatong range including the highest peak in South Sudan, Mt Kinyeti at over 3100 m; and (2) The Boma Plateau on the border between Upper Nile and Equatoria regions in the east of the country and extending into Ethiopia.

The Sahara Desert occupies a vast area of the north of Sudan, where it is divided primarily into the Libyan Desert to the northwest and the Nubian Desert to the northeast, separated by the Nile valley. The Sahara is the great southern barrier between the Palearctic and Afrotropic Biogeographical Realms.

Perhaps the single-most important geographic feature of the Sudan region, and uniting the two countries, is the Nile Basin, since the White and Blue Niles and their tributaries are the lifeblood of this extensively dryland region. The White Nile, in South Sudan known locally as the Bahr el Jebel, flows north from its headwaters in the Lake Victoria Basin, entering South Sudan from Uganda near Nimule and continues north, via a major eastern curve beyond Lake No at the confluence of the Bahr el Ghazal river, then gradually turns north again beyond the city of Malakal. It continues north into Sudan, where it is known locally as the Bahr al Abyad, eventually meeting the Blue Nile at Khartoum, Sudan's capital; historically and strategically one of the most important cities in Africa. The White Nile has several major tributaries in South Sudan, most notably the Bahr el Ghazal – Jur – Bahr el Arab system which flows from the west, the Yei River flowing from the southwest and the Sobat flowing from the southeast. All these rivers have associated areas of seasonally or permanently flooded alluvial plains. The most famous and extensive of these is the Sudd region along the White Nile which is one of the world's largest tropical wetlands at c. 5.7 million ha and is internationally recognised under the Ramsar Convention on Wetlands (www.ramsar.org). However, the Sudd region was seriously threatened in the past by the Jonglei Canal scheme which aimed to by-pass these swamps in order to increase water resources downstream by reducing evaporation and transpiration losses in the Sudd region. This scheme was started in 1978 but halted in 1984 due to political instability in the region, though it was still very much a part of Khartoum's and Egypt's long-term vision for the Nile prior to the South's secession.

The Blue Nile, locally the Bahr el Azraq, flows northwest into Sudan from Ethiopia where it has its source in Lake Tana, and converges with the White Nile at Khartoum. The Roseires Dam, completed in 1966, was constructed on the Sudanese section of the Blue Nile near the town of Er Roseires for irrigation and, later, hydroelectric purposes. Further downstream is the earlier Sennar Dam, constructed in 1925 for irrigation schemes. The Blue Nile has several major tributaries in Sudan, most notably the Dinder and Rahad Rivers which flow from the Ethiopian highlands to the north of the Blue Nile. These rivers have extensive associated wetlands near the Ethiopian border, which are protected within the Dinder National Park which was Sudan's first Ramsar site and is also a UNESCO Biosphere Reserve (www.ramsar.org). The area to the south of Khartoum between the two Niles is subject to intensive irrigated agriculture on the fertile alluvial plains.

North of Khartoum the united Nile flows through the Sahara in a large sigmoid curve before flowing north into Egypt. In this northern section there are a series of cataracts (named the 2nd to the 6th Cataracts, the 1st being in Egypt) which caused navigation problems for some of the early expeditions along the Nile. The northernmost portion of the Nile in Sudan is affected by the Aswan Dam in Egypt and the resultant formation of Lake Nasser, the Sudan portion of which is called Lake Nubia. It is also dammed several times in Sudan, including the huge Merowe Dam, built at the 4th Cataract in the northern desert, which was completed in 2009.

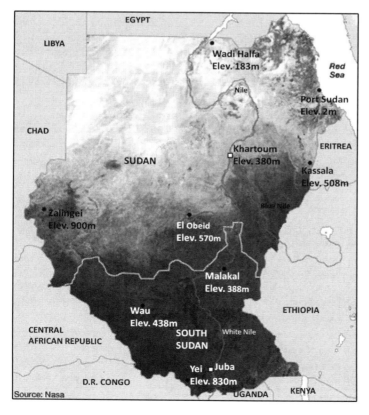

Figure 1.1: satellite image of Sudan and South Sudan showing the location of the nine towns and cities for which weather data are presented in Figure 1.2 (Satellite image source: NASA Earth Observatory – http://earthobservatory.nasa.gov/ImageUse/).

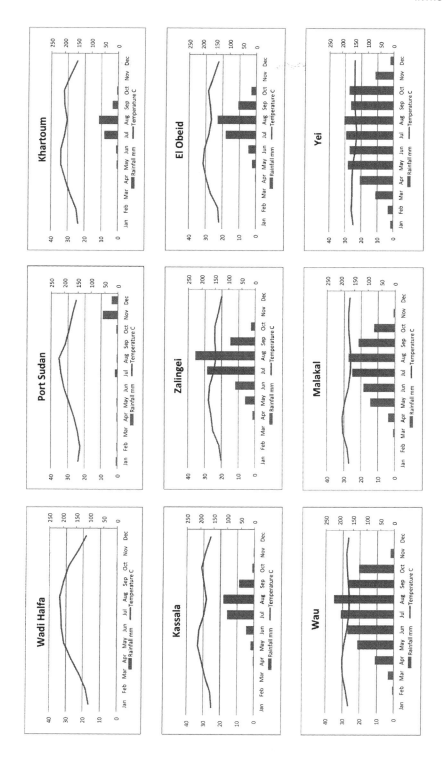

Figure 1.2: mean monthly temperature (line, lefthand scale) and monthly rainfall (bars, righthand scale) for nine towns and cities across Sudan and South Sudan; see figure 1.1 for their locations (data from FAO 1984).

Climate

The climate of the Sudan region varies considerably, most markedly along a north-south gradient which is well represented in Figure 1.2 (data from FAO 1984). Annual rainfall is extremely low and unreliable in the northern deserts with, for example, an average of only 3 mm per year at Wadi Halfa in the extreme north. Moving south, rainfall volume and reliability gradually increases until, in the far south of South Sudan, it can exceed 1300 mm per year in the lowlands and can be considerably higher in the montane regions. Rainfall is highly seasonal, with a single wet season, peaking in all inland regions in the boreal summer, typically in July and August. The wet season becomes more prolonged in the south. The major exception to this rainfall pattern is along the Red Sea coast which receives winter rainfall peaking in November and December, rather akin to an arid Mediterranean rainfall pattern.

Temperatures are characteristically high throughout the year and daytime temperatures can regularly exceed 40°C in the lowland interior. In the hyper-arid north, temperatures peak in the boreal summer but in areas of higher rainfall the summer wet season results in a small temperature dip at that time.

Vegetation

In botanical terms, the Sudan region is truly where north meets south and where east meets west in Africa. The sparse desert flora of the Sahara that dominates the northern portion of Sudan gradually gives way to the south, first to dry scrubby grasslands and then to an increasingly woody vegetation, dominated by vegetation zones that extend all the way to Mauritania and Senegal on the west African coast and that almost reach their eastern-most extent in the Sudan region. Further south still, wet tropical forest elements are found, essentially the northeastern-most extent of the greater Congo Basin flora. And in the southeastern part of the region dry semi-desert thornbush and grasslands represent the northwestern-most extent of a flora that dominates much of the Horn of Africa. Super-imposed on this zonation is the Nile Basin and its associated extensive seasonal and permanent wetlands. Each of the highland regions have their own characteristic vegetation types, with affinities to variously the Saharan massifs and to the equatorial African mountains.

The Sudan region is therefore very rich in phytogeographical terms, and contains no fewer than seven of White's (1983) 22 phytochoria of Africa and Madagascar. Broadly running from north to south, these are:

- **XVII Sahara regional transition zone** comprising the desert flora of the Libyan and Nubian deserts in northern Sudan and extending to the Red Sea lowlands (White's mapping units 67, 68b, 69 and 71). This phytochorion dominates North Africa from Morocco and Mauritania in the west to Egypt and Sudan in the east.

- **XXII Mangrove (azonal vegetation)** comprising the mangrove colonies dominated by *Avicennia marina* along the Red Sea Coast (White's mapping unit 77 which was not recorded in Sudan on White's map due to issues of scale). Mangroves are widespread elsewhere along the tropical and subtropical African coastline.

- **XXII Fresh water swamp and aquatic (azonal vegetation)** wetlands are fairly widespread in parts of the Sudan region, particularly along the Niles and their tributaries, where the swamps are sometimes extensive (see 'Geography' above).

- **XVI Sahel regional transition zone** comprising the semi-deserts and sparse *Acacia* wooded grasslands of Darfur, Kordofan, Central Sudan and the Red Sea region (White's mapping units 43 and 54a). The Sahel region is a narrow zone along the entire southern fringe of the Sahara, extending to Mauritania and Senegal in the west. This phytochorion also contains White's 'Sahelomontane vegetation' (Mapping unit 19b) found only on Jebel Marra and Jebel Gourgeil in Darfur, although both also have some phytogeographical affinities with the major Saharan massifs such as the Tibesti and Ennedi highlands in Chad.

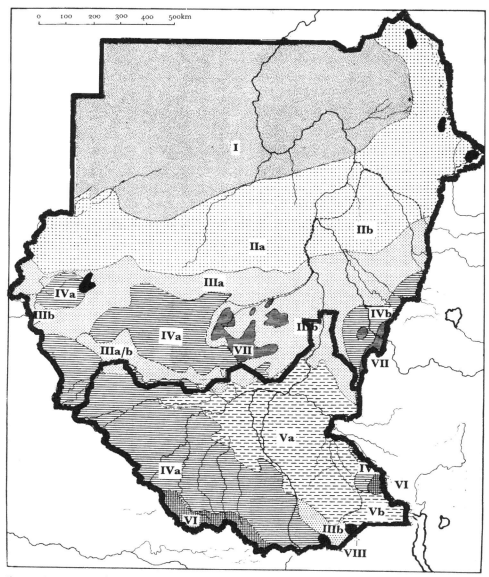

Figure 1.3: a summary of the vegetation of Sudan (adapted from Wickens 1976; reproduced from Kew Bulletin Additional Series V, with permission of the Trustees of the Royal Botanic Gardens, Kew). See text for an explanation of each of the numbered vegetation units.

- **III Sudanian regional centre of endemism** comprising the fire-prone woodlands and wooded grasslands that dominate much of southern Darfur, southern Kordofan, Bahr el Ghazal and parts of Upper Nile and Equatoria (White's mapping units 27, 29a and 35b). This phytochorion forms a narrow zone south of the Sahel from Senegal in the west and gradually broadening eastwards where it extends to western Ethiopia and northwestern-most Uganda. White also includes within this phytochorion the complex mosaic of edaphic grasslands and woodlands associated with the Nile Basin that dominate much of eastern South Sudan and extend north into Central Sudan (White's mapping units 61, 62, 63 and 64).

- **XI Guinea-Congolia/Sudania regional transition zone** comprising lowland and riverine, mainly semi-evergreen Guineo-Congolian forest patches within a matrix of Sudanian vegetation elements, found in southern Equatoria region (White's mapping unit 11a). This transitional phytochorion runs in a thin band from Senegal to Uganda whilst the core Guineo-Congolian regional centre of endemism covers mainly the Guinea coast of west Africa and much of the Congo Basin.

- **IV Somalia-Masai regional centre of endemism** comprising the arid bushlands and thickets and associated arid grasslands that just extend into eastern-most Equatoria (White's mapping unit 42). This vegetation type dominates most of the dry lowlands of northeast Africa from Eritrea to Somalia and south to central Tanzania.

- **VIII Afromontane archipelago-like regional centre of endemism** comprising the montane forests and grasslands of the Red Sea Hills in the north and the Imatong and Didinga Mountains in southern Equatoria (White's mapping unit 19a). As the name suggests, the Afromontane region is a series of isolated montane 'islands' of vegetation in tropical and southern Africa, mainly in a discontinuous chain along the east of the continent but also with isolated areas in the Cameroon Highlands and parts of the Angolan escarpment. On the Boma Plateau in eastern South Sudan, White records cultivation and secondary grassland having replaced upland forest (his mapping unit 17). The transition from Afromontane vegetation to the surrounding lower-altitudinal elements is represented by transitional evergreen and semi-evergreen bushlands and thickets (White's mapping unit 38).

Harrison & Jackson (1958) provided an excellent ecological classification of the vegetation of the Sudan Region. They considered annual rainfall and soil types to be the two most important factors in determining the vegetation in the region and so incorporated these into their classification. Harrison & Jackson's map was later modified by Wickens (1976) and it is his simplified vegetation map that is presented in Figure 1.3. The vegetation units are mainly in marked latitudinal zones, with the major exception of the flood zone associated with the White Nile and its tributaries, and with the minor exception of the 'island-like' montane vegetation zones. This latitudinal zonation is a reflection of the strong north-south changes in annual rainfall in the Sudan region. The vegetation units are summarised below:

I. **Desert**: annual rainfall less than 75 mm, highly unreliable. Vegetation is largely absent, mainly restricted to seasonal watercourses. Ephemeral herbs and grasses occur after the infrequent rain events. This region falls within the Sahara phytochorion of White (1983).

II. **Semi-desert scrub and grassland**: annual rainfall 75–400 mm, unreliable and mainly falling in July–August, except in the Red Sea lowlands where the rainfall is later, November–January. A mixture of annual and perennial grasses dominate the discontinuous herb layer whilst the woody species are scattered or absent, usually occurring as bushes to 2 m tall, becoming more dense along drainage lines. This region falls within the Sahel phytochorion of White (1983). Harrison & Jackson (1958) divided this zone into five subdivisions but Wickens (1976) simplifies this to two:

a. **Semi-desert on lithosols**: important woody species (where present) include *Acacia tortilis*, *Leptadenia pyrotechnica*, *Maerua crassifolia* and *Salvadora persica*. In areas of saline soils, *Tamarix* species can dominate. Important grasses include a number *Aristida* spp., with the perennial *Panicum turgidum* on sandy soils.

b. **Semi-desert on clay soils**: important woody species here include *Acacia mellifera*, *Balanites aegyptiaca*, *Capparis decidua* and *Ziziphus spina-christi*. Common grasses on clays include *Cymbopogon* spp., *Schoenefeldia gracilis* and *Sehima ischaemoides*. The spiny herb *Blepharis edulis* can form pure stands in some areas.

III. **Thorn savanna and scrub**: annual rainfall 280–800 mm. This area is characterised by a mosaic of thickets and wooded grasslands with the dominant tree species being thorny and low in stature, most commonly *Acacia* species. The herb layer is often dominated by few or single grass species, these varying depending on local soils, drainage etc. Wickens (1976) divided this zone into two main subdivisions based on soil types:

a. **Thorn savanna and scrub on sandy soils**: *Acacia senegal* is the dominant tree over large areas; Harrison & Jackson (1958) concluded that much of the *A. senegal* woodlands may be secondary since they often follow cultivation; dominant grasses include *Aristida sieberiana*, *Eragrostis tremula* and, in disturbed areas, *Cenchrus biflorus*.

b. **Thorn savanna and scrub on clay soils**: the woody component is often dominated by *Acacia mellifera* thickets, sometimes in association with e.g. *Adansonia digitata*, *Boscia senegalensis* and *Commiphora africana*. *Acacia seyal* and *Balanites aegyptiaca* can dominate in higher rainfall areas. Locally dominant, grasses include *Cymbopogon* spp., *Echinochloa rotundiflora*, *Hyparrhenia anthistirioides*, *Schoenefeldia gracilis*, *Sehima ischaemoides* and *Sorghum purpureosericeum*.

IV. **Deciduous savanna woodland**: annual rainfall 450–1300 m. This is an area dominated by a low diversity of non-spiny broadleaf woody species with an understorey of tussocky perennial grasses and is subject to regular dry season burning. It falls within White's (1983) Sudanian phytochorion. Wickens (1976) divides this zone into two subdivisions based on soil types:

a. **Deciduous savanna woodland on latosols**: the dominant trees of this zone include *Albizia amara* subsp. *sericocephala*, *Anogeissus leiocarpa*, *Combretum* spp., *Isoberlinia doka* and *Khaya senegalensis*. The *Khaya* is often the most frequent large tree in the wetter parts of this zone or associated with termite mounds in drier parts. Other large tree species in wetter parts include *Daniellia oliveri* and *Parkia biglobosa*. The *Isoberlinia* can be single-dominant on lighter soils or can form mixed woodland with *Burkea africana* and *Terminalia mollis*. *Vitellaria paradoxa* can be common in areas of derived open woodland. Amongst the most important grasses are a number of *Aristida*, *Eragrostis*, *Hyparrhenia* and *Pennisetum* spp. In the driest parts of this region, a woodland of *Anogeissus leiocarpa*, *Prosopis africana*, *Sclerocarya birrea* and *Terminalia laxiflora* can dominate and in these areas the grass layer can contain a more significant annual element.

b. **Deciduous savanna woodland on clay soils**: these woodlands are dominated by *Anogeissus leiocarpa* and *Combretum* spp. including the near-endemic *C. hartmannianum*, together with some *Acacia seyal*. Dominant grasses include *Andropogon gayanus*, *Hyparrhenia* spp. and *Setaria incrassata*.

V. **Flood region**: annual rainfall 800–1000 m or more. This area includes areas that are either occasionally, seasonally or permanently flooded; Wickens (1976) divided it into two subzones:

    a. **Swamp and wetland savanna**: includes (1) the permanent swamps of the Sudd region, dominated by *Cyperus papyrus* stands together with swamp grass species including *Echinochloa pyramidalis*, *E. stagnina*, *Oryza* spp., *Phragmites mauritianus* and *Vossia cuspidata*; (2) semi-permanent swamps with a more mixed grass, sedge and *Typha* community; and (3) the seasonally flooded wooded grasslands with important tree species including *Acacia seyal*, *A. sieberiana*, *Balanites aegyptiaca*, *Borassus aethiopum* and *Hyphaene thebaica*.

    b. **'Toposa' Grassland**: this area is a transition between the flood region proper and the semi-desert grasslands and thickets further southeast in Kenya. It holds a mixed grassland with e.g. *Bothriochloa insculpta*, *Chrysopogon plumulosus*, *Hyparrhenia rufa*, *Setaria incrassata* etc. and some thickets of *Acacia mellifera*. This area falls within White's (1983) Somalia-Masai phytochorion.

VI. **Lowland forest**: annual rainfall over 1300 m. Small relict patches of rainforest e.g. at Azza, Aloma Plateau, Laboni, Lotti and Talanga remain within a matrix of woodland, some of which is derived from deforested areas. The dominant trees of the intact forest are varied but include *Celtis zenkeri*, *Chrysophyllum albidum*, *Holoptelea grandis* and *Mildbraediodendron excelsum*. Riverine fringing forests are also fairly widespread in this region, with important species including *Cola gigantea*, *Erythrophloeum suaveolens*, *Fleroya stipulosa*, *Khaya grandifoliola* and *Syzygium guineense* subsp. *guineense*. Large trees in the woodland areas derived from forest include *Albizia zygia*, *Terminalia schimperiana* and *Vitex doniana*. This region falls within White's (1983) Guinea-Congolia/ Sudania transition zone with the forests being Guineo-Congolian.

VII. **Hill vegetation**: included as a subdivision of Harrison & Jackson's zone IIIA (zone III here), this is a vegetation type occurring on isolated inselbergs ('jebels') where the vegetation is more luxuriant than the surrounding plains; the flora is mainly woody with a mixed species composition, often including e.g. *Anogeissus leiocarpa*, *Combretum hartmannianum*, *Ficus* spp., *Sterculia* spp., *Stereospermum kunthianum*, *Terminalia brownii* etc. The seasonal watercourses (khors/wadis) that runoff from these inselbergs can also have a differing vegetation to the surrounding plains.

VIII. **Montane vegetation**: defined as upland areas where the vegetation differs markedly from that of the surrounding lowlands due to the combined influence of altitude (and so temperature) and, in some cases, associated increases in precipitation. The main 'montane' areas otherwise have varying amounts in common floristically and so should be subdivided into the four main blocks; the first three fall within White's (1983) Afromontane phytochorion:

    a. **Imatong-Dongotona-Lafit Mts**: with a relatively moist montane forest of *Olea capensis* subsp. *macrocarpa*, *Podocarpus milanjianus* and *Syzygium guineense* subsp. *afromontana* as the main climax species and e.g. *Hagenia abyssinica* at higher altitudes. Above the tree-line is a well-developed Ericaceous bushland zone and areas of montane grassland.

    b. **Didinga Mts**: similar to Imatong-Dongotona-Lafit but with the forest at lower altitudes dominated by *Juniperus procera* and *Olea europaea* subsp. *cuspidata*.

c. **Red Sea Hills:** which support drier montane forest dominated by *Juniperus procera* and/or *Olea europaea* subsp. *cuspidata*, although much of this forest has now been removed. In the driest areas, succulent tree species such as *Dracaena ombet* and *Euphorbia* spp. can dominate though they have again been heavily deforested.

d. **Jebel Marra:** supports patchy dry forest with *Olea europaea* subsp. *laperrinei* fairly widespread at higher altitudes together with *Faidherbia albida* on younger soils. Valleys can have a forest vegetation, often with *Ficus palmata*. Elsewhere, grasslands are widespread. The lower slopes essentially support 'Hill Vegetation' (see VII. above).

Botanical coverage in Sudan and South Sudan

The Sudan region has a long history of biological exploration and was one of the first tropical African regions to be extensively botanised by European scientists and explorers in the early to late 1800s, largely owing to the focus of exploration activity along the Nile valley, as well as the ease of access via Red Sea trading routes. The history of botanical exploration and the principal collectors in Sudan and South Sudan are covered in more detail in chapter 2 which documents how botanical exploration continued in Sudan throughout the colonial and post-colonial periods. However, despite this rich history of exploration, our knowledge of the flora of the Sudan region remains very patchy, with botanists having tended to focus on small areas within this vast region. Indeed, Frodin (2001: 478) notes "much of the country and especially the long-troubled and remote south...remains poorly collected and documented". Very few parts of Sudan and South Sudan can be said to be at all well botanised and documented, the main exceptions being: (1) the Erkowit region of the Red Sea Hills; (2) the Red Sea lowlands in the vicinity of Port Sudan; (3) the Jebel Marra massif in Darfur; (4) the Khartoum region; (5) the Gezira region of the Blue Nile valley; (6) some small sections of the White Nile valley, such as the vicinity of the southern portion of the Jonglei Canal project in Upper Nile; and (7) the Imatong and Didinga Mountains of Equatoria. Whilst several of these represent some of the most important localities for plant diversity in the two countries and so understandably have been the focus of attention for botanists, there are still numerous potentially important botanical areas that remain under-explored. Wickens (1976: 16) remarks, for example, that "the Nuba Mountains [in southern Kordofan] and the Boma Plateau [in eastern Upper Nile/ Equatoria] are still virtually unknown botanically". Elsewhere, for example, the eastern border with Ethiopia has seen very little botanising since Schweinfurth's expeditions in the 1860s despite his work having demonstrated this area to be of high botanical interest.

It is very much hoped that this current work will stimulate a fresh interest in the flora of these two countries and that this may result in more extensive exploration in some of the many poorly documented regions. Sadly, the post-independence history of Sudan has been blighted by protracted periods of civil war and regional conflicts, some of which continue to the present day. Political unrest both within and between Sudan and South Sudan continues to hamper our ability to explore, document and conserve the rich biodiversity of these two countries, with a number of extensive areas being largely inaccessible. However, it is hoped that with future stability across the region will come a greater focus upon the biological wealth of the country and the need to document, research and conserve its important species and habitats.

2. Notes on some of the principal collectors in the Sudan Region and their historical setting

Introduction

Early plant collection in Sudan was closely related to the political history of exploration and colonialism. Prior to the 1884 Berlin Conference, most of Africa was seen as an open opportunity for European powers to make treaties with local rulers and exploit their lands. In Sudan this 'Scramble for Africa' was exacerbated by the desire to explore and control the Nile, a theme which continued well into the 20th Century. In most of sub-Saharan Africa the cessation of inter-European rivalries largely allowed the Imperial Power to explore, map and start to exploit the territories which they had been allocated. In Sudan this process was somewhat delayed by the Mahdist revolution which was not defeated until 1898, at which point Sudan was brought under joint British and Egyptian rule as the Anglo-Egyptian Sudan.

Exploration (1830s–1898)

The earliest Sudanese specimens held in the Kew herbarium date from the 1830s, when a small number of explorers and traders roamed the country. Some collectors were involved in government efforts or military expeditions to suppress slavery, while others operated with, or under the protection of, Arab slave and ivory traders. When the Nile route was closed to them, slave traders began to use a route north through Darfur to Egypt, where slavery was still legal. Two of the three main collectors during the mid-19th Century, Georg Schweinfurth and Johannes Pfund, were associated with either slave traders or the attempt to suppress them, while Theodor Kotschy was part of a European-led scientific expedition employed by the Egyptian government. British collectors at this time were principally involved in the search for the source of the Nile.

J.T. Bent

Theodore Bent was principally an archaeologist. From 1875 until shortly before his death in 1897, together with his wife, he travelled widely in the Mediterranean and Middle East visiting archaeological sites. He made botanical collections in Sudan, Arabia and Socotra. His Sudan collections are recorded from '21° lat. Sea Coast to between 3,000 and 4,000 ft.'. On his final journey to Arabia he contracted malaria and died soon after returning to England, aged 55.

F. Cailliaud

Frédéric Cailliaud was a French naturalist, mineralogist and conchologist. On behalf of his patron, Viceroy Muhammad Ali, he collected minerals in Egypt, Nubia and Ethiopia and was part of a military expedition to conquer the Kingdom of Sennar (in modern-day Sudan). Cailliaud searched for gold, unsuccessfully, while the commander Ismail, son of Muhammad Ali, raided the area for slaves. Cailliaud made a detailed survey, including a plant collection, of the mountainous area along the modern Sudan-Ethiopia border. On his return to France, in 1827, he was appointed Conservator of the Museum of Natural History in Nantes. His specimens are in the Herbarium of Montpellier University.

T.M. Kotschy

Theodor M. Kotschy was born in Austria in 1813. As a teenager he undertook several local botanical expeditions, which enabled him to reject a planned ecclesiastical career. At the age of 23 he joined J. Von Russegger's 1836–1838 expedition, as a botanist and zoologist, to examine the mineral resources of North Africa on behalf of the Viceroy of Egypt. In the spring of 1836 he made collections in the Mediterranean and around Cairo, sending 80,000 specimens to Europe. In December the party left Cairo for Sudan, taking 10 weeks to reach Khartoum, and on to El Obeid and Tira, making numerous side trips and collections along the way. The party encountered considerable hardship including skirmishes with people searching for gold, loss of camels and sickness. In 1838 they arrived back in Khartoum, where Kotschy recovered his health. The main group then returned to Europe, taking Kotschy's extensive collection of specimens with them. The interest aroused by these collections resulted in the Esslingen Natural History Society raising a subscription of 30–60 florins from its members to support Kotschy's subsequent 1839 expedition to Kordofan. In October, after several months in the field, a storm destroyed his tent and inundated most of his equipment and collections; it took months to retrieve the more than 10,000 specimens, including 70 species new to science. Many of these species are named in honour of Kotschy, as is the legume genus *Kotschya*. In December he was ordered to return to Cairo, by which time he was seriously short of money and had lost most of his equipment. He arrived in Khartoum on 4 March 1840 almost destitute and received almost no help from the Europeans resident there. Eventually he reached Alexandria from where he dispatched his collections to Europe.

Subsequently he made important collections in Persia and Asia Minor (modern-day Iran and Turkey). On his return to Austria he had to sell many of his collections to various institutions in order to clear his debts. Having no scientific degrees, he was forced to take a low paid assistant job in a provincial museum while continuing to catalogue his collections. Eventually his abilities and work received recognition and he was given a prestigious scientific post in Vienna, where he remained until his death in 1866.

J. Petherick

James Petherick was a Welsh mining engineer, born in 1813, who worked from 1845–1848 for Muhammad Ali Pasha of Egypt, looking for coal in the Red Sea Hills and Upper Egypt. This being unsuccessful, he moved to Kordofan trading gum arabic and later ivory. In 1848 he became the British Vice-Consul in Khartoum. Eleven years later, he took his first leave to England, where the Royal Geographical Society raised a subscription for him to help Speke and Grant by meeting them with provisions at Gondokoro (near present-day Juba). His failure to arrive on time resulted in Speke giving very negative reports to the authorities in Khartoum, Cairo and London, which seriously undermined Petherick's reputation. After the death of Speke and with the testimony of many people from Khartoum, he was finally able to clear his name.

J.D.C. Pfund

Johannes Daniel Christian Pfund was born in 1813 in Hamburg where he studied medicine, though there is no record that he graduated. From 1842–1847 he worked as an assistant in the Prague Museum, dealing with requests from the public and making local collections. He was dismissed in 1848 for embezzlement. He then went to Alexandria where he set up as a doctor, but this enterprise failed when a new hospital was built. Subsequently he worked as a teacher, having lost a legacy from his mother. He did however continue with his interest in local flora and in 1873 exhibited his Egyptian collection in Vienna. This led to him being invited to join a military, scientific and anti-slavery expedition to Darfur and Kordofan, led by two American civil war veterans **Raleigh Edward Colston-Bey** and **Erastus Sparrow Purdy**.

In January 1875 the expedition left Wadi Halfa and, after reaching Dongola, split into two groups; Purdy leading one to Darfur, and Colston-Bey the other to Kordofan, with Pfund as the botanist on the latter. Pfund made several side trips, collecting botanical specimens whenever they were forced by various disasters – sickness, dying camels and deserting porters – to make unscheduled and often prolonged halts. After several months in and around El Obeid he was ordered to accompany Colston-Bey, who was seriously unwell, back to Cairo. On his return to El Obeid, after leaving Colston-Bey in Berber, he planned to join Purdy's expedition in Darfur but died, aged 63, at a base camp in El Fasher before doing so.

Pfund's specimens were sent to Cairo where, in 1879, Dr J.H. Zarb catalogued 625 species. The war office gave the collection to the Geographical Society, who donated it to the Entomological Society. In 1950 it passed to the Desert Institute who then gave it to the Cairo University Herbarium where it remains. Duplicates were sent to Kew and Paris. His seed collection was planted at Giza, but no record exists of what happened to it. His personal collection was given to Schweinfurth, who in turn donated it to the Berlin herbarium.

G. Schweinfurth

Georg Schweinfurth was born in 1836 in Latvia, where he took an early interest in natural history and trained himself to walk very long distances in pursuit of plants. He was a polymath who studied botany, chemistry and palaeontology in Heidelberg, Munich and Berlin.

Schweinfurth's journeys in Sudan are excellently documented by Wickens (1971). His first trip to Africa in 1864 lasted three months and took him along the Red Sea coast to Suakin. He repeated this trip in 1865 exploring the same localities and then went on to Kassala, Gedaref and Gallabat, on the Ethiopian border. For five months he made extensive collections before travelling west to Khartoum, from whence he returned to Suakin and sailed back to Germany.

For his third and most famous trip to Sudan (1868–1871) he was funded by the Berlin Academy of Sciences and Berlin Geographical Society to explore Bahr-el-Ghazal and the Nile region from Khartoum south into what is now the Democratic Republic of the Congo. During this period much of the area was controlled by Arab slave traders who had permanent trading posts scattered throughout the region. Schweinfurth travelled from Khartoum under the protection of a Coptic slave trader named Ghattas and made his headquarters in Ghattas' camp on the Jur river. His extensive exploration in southern Sudan and Congo was made with caravans belonging to Ghattas or other traders. The names of these trading posts are generally on the specimen labels. In early December 1870 a fire destroyed the camp in Jur and with it, all Schweinfurth's manuscripts and journals. Fortunately, his plant specimens had already been sent to Europe. He subsequently moved his headquarters to Khalil's camp at Kuchuk Ali, from which he continued circular expeditions to the west. In June 1871 he returned to Khartoum and Europe, retracing his original steps.

Specimens collected before the fire are numbered sequentially from 1–4296. Those collected after the fire have a prefix Ser. II (not always consistently applied), and don't always appear to be sequential. This is probably due to errors of transcription on to the duplicates from the original Berlin specimens. There is also a 'Ser. III', though it is not clear why a third number series was used (Wickens 1971).

In 1873 he was awarded the Royal Geographical Society Gold Medal after publishing *The Heart of Africa* (Schweinfurth 1874) which demonstrated his very wide range of studies and subjects, including being the first European to describe the Acca tribe in the Congo. In 1874 he returned to Cairo where he founded the Cairo Geographical Society and in 1876 made further studies along the Nile. Between 1880 and 1889 he was Director of the Cairo museum and first president of Khedivial Geographical Society. During this period he made brief visits to Lebanon and Socotra and in the early 1890s to Eritrea. He finally left Egypt in 1914 and worked on his collection in Berlin until his death, aged 89. His very numerous collections and generous donations of specimens to a number of museums ensured his position as one of the great plant collectors of Africa. Many of his specimens have been designated as type material and he is commemorated by the genus *Schweinfurthia* in Plantaginaceae.

J. H. Speke and J. A. Grant

John Hanning Speke and James Augustus Grant were two of the best known 19th century seekers of the source of the Nile. They had both been in the Indian army and fought in the first Sikh war of 1846. Grant remained in India and fought in the Indian mutiny in 1857, returning to England in 1858. Speke returned to England earlier and made his first Africa trip with **Richard Burton** in 1854, which ended in disaster. He then fought in the Crimea before making his second trip to Africa with Burton in 1856. They discovered Lake Tanganyika, but at that point Burton was too ill to continue and Speke went on alone to discover and name Lake Victoria, claiming it as the source of the Nile. He returned to England in 1858, a few weeks earlier than Burton and publicised his finding, resulting in a serious dispute with Burton which was never resolved. In 1860 The Royal Geographical Society funded Speke and Grant to confirm the source of the Nile. On reaching Lake Victoria they continued west around the lake to Ripon Falls where the Nile exits the lake and runs north. They followed the river north to Gondokoro in South Sudan, where they were due to meet **James Petherick**, the British Consul in Sudan at the time. Due to very considerable difficulties with slave traders, Petherick arrived late so Speke and Grant continued under the auspices of **Samuel Baker**. Grant wrote an account of the plants he collected which is published in the Transactions of the Linnean Society (Grant *et al.* 1872–1875), together with meteorological observations and drawings of many of the plants, in particular those that were too big to carry with them. The 702 plants, including 113 species new to science, are also listed in the appendix of Speke's (1863) account *Journey of Discovery of the Source of the Nile*. The labels say 'Speke and Grant Expedition' but were in fact collected and ordered by Grant. In 1864 he was awarded the Patron's Medal of the Royal Geographical Society and in 1866 the Companionship of the Bath.

Less well known European collectors in Sudan during the 19th century included **A. Figari**, an Italian who, in 1844, made collections in Italy, Egypt, Ethiopia and Sudan and is commemorated in several species epithets including *Abutilon eufigarii* Chiov. and *Hyparrhenia figariana* (Chiov.) Clayton; and **L.S. Cienkowsky**, a Russian engineer who collected along the Ethiopian/Sudanese border in 1848–1849. Many of Cienkowsky's collections were destroyed in the bombing of the Berlin herbarium during World War II. He is commemorated by several species including *Fadogia cienkowskii* Schweinf. and *Peponium cienkowskii* (Schweinf.) Engl.

Colonial period (1898–1956)

Sudan was administered under joint British and Egyptian rule between 1898 and 1936. From 1936 until Sudanese independence in 1956 Sudan was ruled directly as a British colony.

Collectors during this period were mainly British government servants, most, but not all of whom, were employed as agricultural or forestry officers, often based in-country for many years. Others were more amateur collectors, being employed in various administrative positions or were wives of government staff.

The agricultural potential of the Nile valley and the higher rainfall areas of South Sudan were seen as the main source of future development and revenue. Early resources were focussed on the Gezira region, between the Blue and White Niles, where a series of gravity-fed canals flowed from the Blue Nile and, from 1919, a dam at Sennar irrigated an area of 3,400 square miles on which cotton was the principle crop. A number of forestry research stations and agricultural farms were created, the largest and most permanent of which was the *Gezira Research Station* at Wad Medani, which contained two herbaria as well as economic botany and plant pathology sections. Although cotton research in the irrigated Gezira area was its main focus, it also carried out a wide range of training, taxonomic and botanical studies and maintained a very considerable correspondence with staff at the Kew herbarium.

In 1946 a hugely ambitious project was raised to build the *Jonglei Canal* in Upper Nile Province to partially by-pass the swamps of the Sudd region in Southern Sudan and take the water to rejoin the White Nile near Malakal, where it could be used for irrigation. Although plans were drawn up in 1954–1959, construction did not start until the post-colonial period in 1978.

The *Tozi Research Farm* was set up in 1952 in the Central Highlands. It focussed on research into rain-fed mechanised production of subsistence crops including sorghum, soya beans, sesame and groundnuts. Initially **J.D. Lea** was the officer in charge.

Yambio Experimental Farm in Equatoria province was created in 1948, primarily to grow food crops for local consumption. In 1950 a decision was made to resettle 60,000 Zande people into permanent villages around Yambio to promote both subsistence agriculture and cotton as a cash crop.

F.W. Andrews

Between 1931 and 1938 Frederick William Andrews collected widely throughout Sudan and South Sudan, with the exception of Darfur and the northern area. In 1939 he collected extensively in Equatoria province, while in subsequent years he collected mainly in the Red Sea Hills with minor collections around Khartoum, in Kordofan and South Sudan. Many of the specimens under his number sequence may have been collected by his assistant, Mohammed Ismail; the notes and labels on his specimens in the Wad Medani herbarium are in Arabic script and were subsequently translated into English, with modifications, before being sent to Kew. From 1940–1953, Andrews was the chief economic botanist to the Sudan government, based at Wad Medani. Between 1950 and 1956 he published the three volume *The Flowering Plants of the (Anglo-Egyptian) Sudan*, which has, to date, been the standard work on Sudanese flora. During his time in Sudan he worked closely with Sudanese assistants and corresponded at length with the Kew Herbarium, who did not always answer very promptly; he had to shelve a proposed publication of the distribution of wild sorghum owing to Kew taking ten years to respond to his request for identifications! After leaving Sudan, in 1953 he worked at the West Malling Research Station for two years before he retired, due to ill health.

G. Aylmer

Before the First World War, Guy Aylmer worked in forestry in Malaya. After the war he spent two years in Sierra Leone, before becoming Conservator of Forests in Sudan in 1920, working mostly in Blue Nile Province and South Sudan until 1930, when he contracted blackwater fever. He was then based in Khartoum, in charge of the nursery and arboretum at the Department of Agriculture and Forestry, from where he made several local collections and two trips to the Red Sea Hills in 1932 and 1936, before his retirement in 1937. He was a keen falconer and made collections of African birds, plants and big cat skins. Guy Aylmer should not be confused with Gerald Percival Vincent Aylmer who collected in Somalia.

P.R.O. Bally

Peter Rene Oscar Bally was born in Switzerland in 1895. He worked for the League of Nations after the First World War until 1928, during which period he went to Albania and India to introduce an antidote for malaria developed by his father's pharmaceutical firm. He subsequently worked for an oil company (Mobil) and collected plants in Tanzania for Hoffman-La Roche, developing an interest in succulents. From 1938 to 1958 he was the botanical curator of the Coryndon National Museum in Nairobi, Kenya (subsequently the National Museum, and later still a part of the National Museums of Kenya). During this time he collected in the Red Sea Hills in Sudan. In 1964 he returned to Switzerland as the head of the Geneva Botanical Garden until retiring to Kenya in 1969, where he died in 1980. His collections are at Kew and the East African herbarium. His particular interests were aloes, on which he worked and published with G.W. Reynolds from South Africa, and succulent asclepiads. He is commemorated by the genera *Ballya* (Commelinaceae) and *Ballyanthus* (Apocynaceae).

Babakir Beshir

Collections by 'Beshir' were made in August and September 1951 in Equatoria and Kassala provinces. The labels of these specimens are known to have been translated from Arabic in the Wad Medani herbarium in the 1950s. The labels on the majority of those collected in Equatoria show the collector to be Beshir eff. (an honorific title) and those in Kassala to have been collected by Beshir Babakir. It is not clear whether these are the same person. Given that the dates on the labels are very close it is possible that there were two Beshirs, since Equatoria and Kassala are far apart geographically. Babakir Beshir is known to have done some collecting while at the School of Agriculture in Khartoum and then worked closely with Andrews at Wad Medani, before taking on the role of Keeper of the Khartoum and Wad Medani herbaria in 1954. In 1956/57 he received further training at Kew and the Commonwealth Mycology Institute. In the early 1960s he moved into administration. We do not have any additional information on Beshir eff. (if he is indeed a different collector).

A.F. Broun

Alfred Forbes Broun was Conservator of Forests in Ceylon from 1899 to 1901, when he became Director of Woods and Forests to the Sudan Government until his retirement in 1910. Together with his wife, he collected throughout Sudan and South Sudan, with the exception of Darfur, the Northern and Kassala Provinces. In 1906 he wrote a catalogue of plants in the Sudan for local use only, which was added to by **H.B. MacMichael** from the Sudan Civil Service and **Cyril Crossland,** a marine biologist and Director of the Sudan Pearl Fishery 1905–1922. Mrs Broun sketched many of the plants, some of which were later published, together with those of **Grace Crowfoot**, wife of J.W. Crowfoot, the Principal of Gordon College and Director of Education in Sudan, 1914–1926. In 1929 Broun published the first *Flora of Sudan*, together with **R.E. Massey**, which was the standard work until Andrews' three volume Flora.

T. Cartwright
Tomas Cartwright worked in Kew from 1905–1908 when he went to Sudan as an agricultural and forestry officer. He was based in Kagelu, near the D.R. Congo border, where he worked on rubber and coffee plantations. In 1919 he and **Frederick Sillitoe** collected together in the Yei valley, making the first collections there since Schweinfurth in the late 1860s.

T.F. Chipp
Thomas Ford Chipp worked for several years at Kew, first as a gardener and later in the Herbarium before going to Ghana in 1910 as Assistant Conservator of Forests. In 1919 he became the Assistant Director of Singapore Botanic gardens for two years. He then moved back to Ghana as Deputy Conservator of Forests before returning to Kew in 1922 as Assistant Director until 1931, during which period he visited **Frederick Sillitoe** in Port Sudan in 1928, and spent part of 1929 plant collecting in South Sudan.

J.E. Dandy
In 1925 James Edgar Dandy was employed as a temporary assistant to John Hutchinson at Kew where he wrote a monograph on magnolias. In 1927 he moved to the British Museum as the Assistant Keeper of Botany, becoming Keeper in 1956 until his retirement ten years later. During his time at the British Museum he catalogued the Sloane Herbarium, worked on the Watsonian vice-county system and made a considerable contribution to the nomenclature of the British flora. In 1934 he spent three months in Sudan where he made collections in Darfur and Kordofan in Sudan and Bahr el Ghazal, Upper Nile and Equatoria provinces in South Sudan. These journeys are thoroughly documented in the book *That Hard Hot Land* by Mary Keenan (2011); his collections are mainly housed at the Natural History Museum, London.

M. Drar
Mohammed Drar worked for the Horticultural Department of the Egyptian Ministry of Agriculture and later became Director of the Orman and Qubba Botanic Gardens in Cairo. He participated in the foundation of the Agricultural Museum of Cairo and the Desert Institute. He also worked for UNESCO. He was a co-author of two volumes of a *Flora of Egypt* during the 1950s. Between February and June 1938 he completed a botanical expedition to Sudan, collecting 2,548 specimens, travelling from Zalingei in the west, via El-Obeid and Khartoum to Erkowit on the Red Sea coast. He returned via Kassala to the Nile and then went south to Juba and west as far as Tambura before retracing his steps to El-Obeid and down the Nile to Cairo, with a short side trip to Merowe. His specimens are held in the Cairo University Herbarium. **Gerald Wickens** helped with some of the determinations while he was based at Kew but Drar's checklist was published too late for his collections to be included in Wickens' Ph.D. thesis on the *Flora of Jebel Marra*. His Sudan expedition was synthesised posthumously by Vivi Täckholm (Drar 1970).

E.E. Evans-Pritchard

Edward Evan Evans-Pritchard was a social anthropologist who spent two years (1926 and 1936) carrying out anthropological field work in Upper Nile Province, South Sudan, during which he made a small number of plant collections. In 1931, after completing a Ph.D. at The London School of Economics, he went to Cairo for three years as Professor of Social Anthropology at Fuad University, before returning to Oxford in 1934. Between 1940 and 1945 he spent his military service in Sudan and in 1946 he became Professor of Sociology in Oxford until his retirement in 1970. During his fieldwork in Sudan, he took over 2500 photographs, currently in the Pitt Rivers Museum in Oxford. His main anthropological publications were on the Azande and Nuer peoples.

M.N. Harrison

Harrison was head of Pasture Research in the Department of Animal Production in Sudan from 1947 to the early 1950s. His collections, mainly of grasses, were made between 1947 and 1953 from throughout Sudan, with the exception of Kassala Province. Together with **J.K. Jackson**, he published the most widely used vegetation map of Sudan (Harrison & Jackson 1958). He also made some collections in Eritrea in 1952. In the 1960s he was an advisor for FAO.

J.K. Jackson

Kenneth Jackson was a British silviculturalist based in the Forestry Department, Khartoum, between 1947 and 1964. He made numerous plant collections throughout the country which are now housed in Kew (herbs and trees), the British Museum (herbs) and the Forestry Herbarium in Oxford (trees). His collections are one of the foremost resources for the study of Sudanese plants. In the late 1960s he worked for the FAO as a forestry advisor to Sudan and Nigeria.

B. Kennedy-Cooke

Brian Kennedy-Cooke joined the Sudan Political Service in 1920, initially working in Kordofan and the Red Sea Hills. In 1928 he settled in Kassala where he was the governor from 1935–1941. His collections are from Northern Province, Red Sea Hills and Kassala – some are only labelled as 'B.K.C.'. In 1941 he served with the army in Eritrea before returning to the Sudan government in 1942. In 1944 he published a book on the trees of Kassala province. He collected specimens of trees between 1931 and 1938. From 1943–1956 he worked for the British Council, six years of which were spent in Italy. In 1936 he received the Order of the Nile (3rd Class), in 1937 and 1953 the Coronation Medal, and in 1946 the Officer Legion of Merit U.S.A.

O. Khalid el Kheir

Omar Khalid el Kheir was a Sudanese collector who worked with **Harrison** at the Institute of Pasture Research collecting grasses, mainly in Darfur and the Red Sea Hills. He also collected ferns in the Imatong Mountains of South Sudan.

J.D. Lea

Lea was head of the Tozi Research Farm in the early 1950s. His collections are all from the Tozi area in Central Sudan during 1952 and 1953.

H. Lynes

Hubert Lynes was an amateur botanist who spent most of his career in the navy (1887–1922). In 1920, while still in the navy, he made the first plant collections in Darfur since the Pfund/Purdy expedition, and was the first person to collect on Jebel Marra, writing a description of the natural history of the volcano (Lynes 1921). Many of his collections are in the Natural History Museum, London. After leaving the navy he made collections in Morocco, Guinea Bissau and the D.R. Congo. He was a keen ornithologist who finally settled in Tanzania.

R.E. Massey
Reginald Ernest Massey, a British Colonial administrator, served as the Sudan government botanist from 1912 until the end of the 1920s. He published an updated and expanded version of **Alfred Broun's** plant catalogue which included local names and many drawings. He also published on grasses and he co-authored the first *Flora of Sudan* with Broun (Broun & Massey 1929). In 1915 he made some collections in Ethiopia.

J.G. Myers
John Golding Myers was one of the principle collectors in South Sudan. He fought with the New Zealand forces in the First War. He then returned to New Zealand and worked at the Entomology Institute on cattle ticks. In 1924 he won an entomological scholarship to Harvard, later studying midges in France, fruit pests in Australia and parasites in Surinam. Ten years later, in 1934, he joined the staff at The Imperial College of Tropical Agriculture in Trinidad. In 1937 he became the Economic Botanist in Sudan, where he was employed to carry out a survey of Equatoria Province for agricultural development. He was killed in a road accident in Amadi in 1942. He made over 10,000 collections of plant, rock, soil, insect and mollusc specimens in South Sudan. His plant specimens are mainly housed at Kew and Khartoum (KHU).

H. Padwa
Henry (or Hanery) Padwa was a Kenyan national, born in 1931. He worked as a Laboratory Assistant at the East African Herbarium from 1952 to 1977. He accompanied Jan Gillett on the Ethiopian Boundary Commission expedition of 1952–1953, collecting in the Illemi triangle in South Sudan near the Kenya border in 1953. He also collected in Kenya and southern Ethiopia. With his experience, enthusiasm and reliability he was a popular choice to accompany other collectors in Kenya in succeeding years, notably Peter Bally, Peter Greenway, Bernard Verdcourt and John Williams. In 1956 he returned to South Sudan to collect with John Williams.

A.W. Peers
Peers was a British ecologist who worked with M.N. Harrison in pasture research, mostly on the Ethiopian border and in the Didinga Mountains in 1953.

M.I. Sherif
Majoub Sherif was based in Wad Medani under the leadership of Wilson-Jones in the early 1950s. During 1951 he collected in Upper Nile Province.

F.S. Sillitoe
Frederick Sampson Sillitoe trained in horticulture and worked at the Royal Exotic Nursery in Chelsea before joining Kew in 1901. He worked in Sudan for 27 years, from 1903 until 1930, initially in Khartoum, where he set up public parks and gardens while being responsible for the eight acre Palace garden. In 1919, together with **Tom Cartwright**, he spent three months collecting in the Yei valley along the Sudan/D.R. Congo border. In 1927 he moved to Port Sudan, where he made a small number of collections while again setting up public parks and gardens. In 1919 he was awarded the Order of the Nile by the Sultan of Egypt and in 1927 he received an M.B.E. for services to Sudan. He left Sudan in 1930 spending 18 months in Malta before retiring in 1932. He died in 1957, aged 80.

N.D. Simpson

Norman Douglas Simpson was a country gentleman who combined a passion for cars and motorbikes with an interest in plants. He collected a herbarium of British plants and was a founder member of the Thirsk Botanical Exchange which later became the Botanical Society of the British Isles. After completing the Cambridge diploma of Agriculture he went to Cairo in 1921 to work on cotton research, where he collected a private herbarium of over 8,000 specimens. In 1926 he moved to Sudan and worked for the irrigation services, where in 1929/30 he made an extensive collection of plants, returning to England in 1930. In 1931 he went to the Ceylon Peradeniya Botanic Garden for a year and then retired to Bournemouth.

A.S. Thomas

Arthur Thomas studied agriculture in Cambridge and tropical agriculture in Trinidad before becoming Assistant Superintendent of agriculture in Ghana in 1930. In 1932 he moved to Uganda, firstly as Assistant Economic Botanist and then, until 1946, Economic Botanist in the Agricultural Department based in Kampala. His Uganda collections are cited extensively in the *Flora of Tropical East Africa*. In late 1935 he collected on both the Ugandan and Sudanese sides of the Imatong Mountains. After his retirement he travelled widely in Asia, the Mediterranean and the Americas for his book on tropical gardening, published in 1965.

J.W.G. Wyld

Major Jasper William George Wyld was the Assistant District Commissioner in Upper Nile from 1926–1931; initially in the Dinka/Nuer area and later in Bor. In 1932 he was appointed District Commissioner in Zande District of Equatoria Province. In 1950 he became responsible for resettling the local population into villages around the Yambio Experimental Farm for the cultivation of subsistence and economic crops. His collections are mainly from the area surrounding Yambio; the majority are housed at the Natural History Museum, London.

Post-Colonial Collectors – 1956 onwards

During the immediate post-colonial period the majority of collectors continued to be either Sudanese or British, before becoming more international from the 1970s onwards. But non-Sudanese collectors were no longer resident in Sudan for long periods of time. Instead, they had narrower short-term projects funded by a wide variety of multi-lateral and bi-lateral organisations. Some of these collectors had previously been involved in colonial development and moved into the new, short-term contract world of 'International Development'.

Both development projects and scientific exploration were disrupted by long periods of civil unrest and war. For example, the *Jonglei Canal* project, planned in the colonial period, was started in 1978 but then abandoned in 1984, mainly due to the disruption caused by the Sudan People's Liberation Army (SPLA) independence movement.

T. Ahti

Teuvo Ahti, his wife Leena Hamet-Ahti and Bror Petterssen were members of the 1962 Finnish Biological Expedition to Nubia that was part of the UNESCO programme to document the archaeological and biological material of the area to be covered by the Nasser dam reservoir (Lake Nubia). They collected 345 species of vascular plants (29 of which were cultivated) in Northern Province and around Kassala. This expedition and its collections are well documented by Ahti *et al.* (1973).

I.J. Blair

Blair worked for FAO from 1958–1970 as a livestock and pasture advisor. In 1963–1964 he collected grasses in Darfur as part of a UN Special Fund for the Jebel Marra Project.

S. Carter (Holmes)

Susan Carter is a specialist in succulent plant groups including succulent Euphorbiaceae, having discovered and catalogued over 200 species. She joined Kew in 1957, working on monocot families for the *Flora of Tropical East Africa* until 1962. In 1959 she met Peter Bally and developed her interest in Aloes and Euphorbiaceae, going on several field trips to East Africa with him and other Kew staff. In November 1987 she went to the Red Sea Hills, as the botanical adviser to an international team on behalf of the Cactus and Succulent Society of America. The party focussed on the Red Sea coastline from Erkowit to Jebel Elba in the Hala'ib Triangle, largely following the footsteps of Schweinfurth's expeditions of the 1860s and 70s. There were therefore few species new to science. She did, however, make 252 collections for Kew, including many *Aloe* and *Euphorbia* specimens.

W.J.J.O. & J.J.F.E. de Wilde & B.E.E. de Wilde-Duyfjes

A family group: Willem de Wilde, his wife Brigitta Duyfjes and his brother Jan de Wilde, from the Wageningen Herbarium in the Netherlands. They collected throughout Africa, many European countries and Indonesia, including extensive collections from Ethiopia and Eritrea in the 1960s and 1970s. In Sudan they collected in the Khartoum region and Darfur in early 1965.

R.B. Drummond

Robert Baily (Bob) Drummond worked in Kew on the *Flora of Tropical East Africa* from 1949. In 1953 he went with **J.H. Hemsley** to East Africa, including Sudan, to re-collect specimens that had been destroyed in Berlin. He returned to Kew with 4,800 specimens, mostly from Kenya and Tanzania. In 1955 he moved to Harare where he became Keeper of the National Herbarium and Botanical Garden and was involved in many publications on the Zimbabwean flora, including *Flora Zambesiaca*.

I. Friis

Ib Friis is a Danish botanist who is currently Professor at the Botanical Museum, Copenhagen. One of his main taxonomic interests has been in the Urticaceae, for which he wrote the *Flora of Tropical East Africa* and the *Flora of Ethiopia and Eritrea* accounts in 1989, the *Flora Zambesiaca* account in 1991, the *Flora of Somalia* account in 1999, and, with K.L. Immelman, the *Flora of Southern Africa* account in 2001. From 1970 he developed a working relationship with the Carlsberg Foundation for botanical exploration in Tropical Africa. He was an honorary lecturer at Makerere University in Uganda from 1994 to 2004. He was involved in the *Flora of Ethiopia* project from 1980 and a member of the Editorial Board to the conclusion of the project in 2009. He is one of the most prolific plant collectors in Ethiopia and continues to run expeditions there in collaboration with Addis Ababa University and Kew. In 1980 and 1982 together with **Kaj Vollesen** he made 1,283 collections from the Imatong Mountains and surrounding lowlands. These specimens are held in the Copenhagen herbarium with duplicates in the Kew, East African and Khartoum Agricultural Herbaria. He and Dr. Vollesen together produced the *Flora of the Sudan-Uganda border area east of the Nile* in 2 volumes, published in 1998 and 2005 by the Royal Danish Academy of Sciences and Letters, of which Ib Friis has been a member since 1990.

W.J. Howard

William Howard was a senior forestry advisor with the Land Resources Division of the British Overseas Development Administration in the 1970s. He was project leader for the Imatong Mountains Forestry Project 1977–1980. His specimens were identified in Kew by Gerald Wickens and returned to the Forestry Research herbarium in Soba, outside Khartoum. He later worked in the Land Resources Division in Tolworth.

A.A. Kamil

In 1966/67 AbuBakr Abdelrahman Kamil collected with K.C. Sahni in Central Sudan, Equatoria and the Red Sea Hills. In May 1968 he collected in Darfur, near Jebel Marra.

M.A. Kassas

Mohamed Abdelfatah Kassas was Professor of Botany at Khartoum University. In 1954 he collected in Kordofan, Central Sudan and the Red Sea Hills. In 1965–1966, together with Dr Pettet, he collected in the Red Sea Hills, Darfur, Central and Eastern Sudan. His collections are mainly in the Universities of Khartoum and Cairo.

J. Kielland-Lund

Jan Kielland-Lund is a Norwegian agronomist associated with the Agricultural University of Norway at Aas, near Oslo. He collected in Equatoria between November 1983 and June 1984; his specimens are widely cited by Friis & Vollesen (1998, 2005).

P. Kosper

Phyllis Kosper, an anthropologist, collected in Equatoria, including the Imatong Mountains, in 1982–1983 and later made a few collections in the Nuba Mountains of south Kordofan.

J. Léonard

Jean Léonard from the Jardin Botanique National de Belgique et Université Libre de Bruxelles, was joint leader, with Loutfy Boulos, of a three month interdisciplinary expedition through Libya to Jebel Uweinat on the border of Libya, Sudan and Egypt, between November 1968 and January 1969. The botanical findings were published in a four volume series (Léonard 1997–2000).

J.M. Lock

John Michael Lock trained as an ecologist (Ph.D. Thesis: *The Effects of Hippopotamus Grazing on Grasslands*) but became more taxonomically inclined with time. He worked in Uganda, 1963–1970, and at the University of Ghana 1970–1977. Between 1979 and 1983 he worked in Sudan as the lead botanist on the Development Studies in the Jonglei Canal area, financed by the EU. He was then a Botanical Consultant working in Venezuela, Nigeria, Tanzania, Uganda, Guinea-Bissau and Gambia, before joining Kew as a botanist and Editor of Kew Bulletin in 1990, retiring in 2002. He has prepared three families for *Flora of Tropical East Africa* (Zingiberaceae, Musaceae, and Xyridaceae), and has also contributed to *Flora Zambesiaca*, *Flore du Gabon* and *Flore du Benin*. He continues to work on the Cyperaceae account for *Flora Zambesiaca* and maintains an active interest in the genus *Xyris*.

S. Miehe (neé Klug)

Miehe collected in Darfur, mainly Jebel Marra, between May and September 1982. About half her labels say 'S. Klug/Hamburg BFH' and the remainder S. Miehe (often with neé Klug in brackets). Some specimens collected on the same day are labelled under different names.

'BFH' is the Federal Research Centre for Forestry and Forest Products at the University of Hamburg.

L.J. Musselman

Lytton Musselman was a Senior Fulbright Researcher in the Department of Agriculture and Botany at Khartoum University between 1982 and 1984. His main research interest was parasitic angiosperms which form part of his collections from the environs of Khartoum and Kordofan, and he published on Sudan's parasitic plants (Musselman 1984; Musselman & Hepper 1986). He visited the Imatong Mountains when teaching a taxonomy course in Juba University in 1983, but did not make any collections. He has worked in many sub-Saharan African countries, throughout the Middle East and in several countries in tropical Asia. Currently he is based at Old Dominion University in Virginia, USA.

A. Pettet

Together with his wife Sally, Tony Pettet was a lecturer in the Botany Department in Khartoum University 1962–1967. In 1963 he sent collections from Erkowit, Jebel Marra, Wadi Halfa and Khartoum to Kew for identification. In 1965–1966 he collected in Darfur with **M.A. Kassas**. After leaving Khartoum, he lectured at the University of Ibadan, Nigeria and then Ahmadu Bello University in Zaria, Nigeria, before returning to England where he taught in schools for several years before retirement.

V.C. Robertson

Vernon Robertson, who died in 2012, was an ecologist and Director of Hunting Technical Services from 1952–1977. He was the team leader of the initial feasibility studies of Jebel Marra for agricultural development in 1957–1958.

K.C. Sahni

Professor Kailash Chandra Sahni was educated in the Punjab and at the Commonwealth Research Institute in Oxford. From 1966–1968 he was the FAO/UNDP forestry advisor for Sudan and collected in Central Sudan, Red Sea Hills, Eastern Province and Equatoria, mostly together with **A.A. Kamil**. His book on *Important Trees of the Northern Sudan* was published by FAO in 1968. He then worked for 28 years at the Dehra Dun Research Institute in India where he was the Editor of the Botanical Section of the Encyclopaedia of Indian Natural History. In 1994 he received the World Environment Day Award and in 1998 the Seth Memorial Award.

K. Vollesen

Kaj Vollesen is a Danish botanist with an extensive knowledge of African plants and a specialist interest in the Acanthaceae. His first work in Africa was in Ethiopia with a team from Copenhagen University in 1972–1973, following which he completed an MSc in Copenhagen. He worked in Tanzania 1975–1978 as a plant ecologist in the Selous Game Reserve, on behalf of the Ministry of Natural Resources and Tourism, and subsequently completed a Ph.D. in 1980 at the University of Copenhagen based on his studies of the Selous. In 1980 and 1982, together with **Ib Friis**, he made two trips of about three months each to the Imatong Mountains, leading to the publication of the *Flora of the Sudan-Uganda border area east of the Nile* in two volumes, 1998 and 2005. From 1981–1987 he was based in Kew working on the *Flora of Ethiopia* project. From 1987–2006 he was a Principal Scientific Officer at Kew, leading numerous trips to Tanzania and other parts of East Africa. His collections are at Kew and Copenhagen, with duplicates at the East African and Khartoum Agricultural Herbaria.

G.E. Wickens

After leaving school in 1945, Gerald Wickens spent three years in the Black Watch. He then worked for the Ministry of Agriculture in Nigeria from 1953–1955, where he was responsible for introducing cocoa. Between 1956 and 1960 he was a Conservation and Extension officer in the Federation of Rhodesia and Nyasaland before returning to England and working on a mixed farming project. From 1962–1966 he was the field manager for the Hunting Technical Services project, funded by UNDP, initially in Kordofan and subsequently in Darfur, carrying out vegetation and land use surveys, eventually leading to the publication of *The Flora of Jebel Marra* in 1976. In 1966 he returned to the UK to do an MSc at Aberystwyth and then joined Kew as a Principal Scientific Officer, later working on the conservation of plants in arid and semi-arid regions, during which period he completed a PhD at Reading University and was responsible for the determinations of specimens sent from Sudan. His collections are from 1961 to 1971 and are housed mainly at Kew.

Note on the sources used

Bibliographical information was obtained from a wide range of sources including: bibliographical dictionaries, published records from the Colonial Office, the *Journal of the Kew Guild*, the *Kew Record of Taxonomic Literature*, handbooks and bibliographies of Sudan (e.g. Hill 1967; Hopkins 2007), published obituaries and internet searches – a particularly useful source is the Sudan Archive at Durham University (https://www.dur.ac.uk/library/asc/sudan/), whilst JSTOR Global Plants (http://plants.jstor.org/) contains a wealth of useful information on plant collectors.

3. Conservation priority species in Sudan and South Sudan

Introduction

With much basic information lacking on plant species and their distributions in Sudan and South Sudan, it is not surprising that there is currently no formal Red List of globally or nationally threatened plant species for either country. Here we present an initial step towards this process, through (i) recording the Sudanese species that are already formally assessed as threatened or potentially so on the *IUCN Red List* (http://www.iucnredlist.org); (ii) documenting for the first time a complete list of endemic and near-endemic species in the two countries, and (iii) listing those non-endemic species that are so highly range-restricted, scarce or occur in such threatened habitats that they are potentially globally threatened or near-threatened. Together, these three groups cover the plant priority species for further investigation including targeted fieldwork, full and up-to-date conservation assessments and consideration in conservation planning.

Information on the current distribution of these rare species is often lacking for both countries, and threats to their survival are often unknown, though in some cases these can be inferred from our knowledge of threats to the habitats and localities in which they occur. With these limitations in mind, it would be premature to attempt formal conservation assessments of many of these species at this stage. The aim here is to provide a working list on which to prioritise the information-gathering stage. South Sudan has recently been added to the remit of the IUCN East African Plants Red List Authority, co-ordinated by Quentin Luke in Nairobi, and one of the authors of this checklist (R.C.) has been invited as a member. This is one of the most active red listing authorities at present and it is hoped that South Sudanese conservation priority species will be included within their work in the future. Sudan does not currently fall under the remit of any regional red list authority or working group.

For each species of potential or known conservation concern, we have recorded its regional distribution in Sudan and/or South Sudan (see chapter 4 for further information on these regions) and also its phytogeographic association, with the phytochoria following those of White (1983) and described in chapter 1.

Currently assessed plant species in Sudan and South Sudan

According to the most recent update of the *IUCN Red List* (http://www.iucnredlist.org, accessed January 2014), only 277 plant species from Sudan and South Sudan have had their conservation status formally assessed and evaluated using the categories and criteria of IUCN (2001). Of these, 243 are assessed as of Least Concern (LC), i.e. they are not considered threatened and so of low conservation priority. The remaining 34 species either have a higher assessment of threat: Near Threatened (6 spp.), 'Conservation Dependent' (1 sp. – this is no longer a category), Vulnerable (15 spp.), Endangered (5 spp.) or Critically Endangered (1 sp.), or are considered Data Deficient (5 spp.); these species are listed in Table 3.1. Of the Near Threatened species, *Irvingia gabonensis* is listed on the *IUCN Red list* as occurring in Sudan but this was based on a misidentification of *I. wombolu* and it is therefore omitted from Table 3.1.

Several points are immediately apparent from this list:

a) The vast majority of plant species (c. 93%) in the two countries have not had their conservation status formally assessed and documented.

b) For those that have been assessed, a significant number of conservation priority species are the commercial timber species which were prioritised by several working groups, such as the World Conservation Monitoring Centre. Many of these species have a large distribution range, usually extending into west Africa, and so would not be threatened were it not for their commercial value and resultant over-exploitation.

c) The majority of species assessed were done so under the old version 2.3 of the IUCN categories and criteria (1994) and so are in need of re-assessment under the current version 3.1 (2001). The new criteria and requirements for documentation are considerably more rigorous, such that several of the species currently assessed as threatened would most likely be down-graded were a re-assessment to be carried out. A good example is *Fleroya stipulosa* which is listed as Vulnerable on the Red list (under *Hallea stipulosa*). Whilst exploited commercially in some parts of its range, this species is often a common pioneer species of swampy forests and, whilst it has certainly experienced some declines through habitat loss and over-logging of mature individuals, it is unlikely to be truly threatened across its very broad range.

d) Very few of the range-restricted or endemic species that occur in Sudan and South Sudan have been formally assessed (see section 2 below). A notable exception is the *Aloe* species, which have all been assessed by O. Weber & S. Carter at Kew (with the exception of the newly described *A. ithya*). The other range-restricted species on the Red List are *Baphia abyssinica*, *Beilschmiedia ugandensis*, *Combretum hartmannianum*, *C. rochetianum*, *Cordyla richardii*, *Dracaena ombet*, *Encephalartos meridionalis*, *E. septentrionalis*, *Medemia argun*, *Ottelia brachyphylla*, *O. scabra*, *Sterculia cinerea* and *Suddia sagittifolia*.

Endemic plant species and taxa in Sudan and South Sudan

Table 3.2 records 86 taxa that are believed to be endemic to Sudan and/or South Sudan. Many of these species are poorly known, often from only one or two historical collections. Indeed, we record 27 as having an uncertain status – either they are currently undescribed and/or they require further study to confirm their status as good species. It is very likely that some of these will ultimately prove to be synonymous with more widespread species following further investigation. Of the 86 taxa, 34 occur in Sudan and 59 in South Sudan with seven taxa occurring in both countries.

Table 3.2 also lists a further 22 taxa that are near-endemic to Sudan and/or South Sudan, i.e. they have the majority of their range in those countries but extend into one neighbouring country. Of these, 11 occur in Sudan and 14 in South Sudan with 3 taxa occurring in both countries. Only one of these is considered to be of uncertain status.

For the two countries combined, the percentage plant endemism is 2% or, if near-endemics are included, 2.6%. For Sudan alone, the equivalent figures are 1.2% and 2%, whilst for South Sudan they are 1.7% and 2.4%. In comparison to many countries of tropical Africa, these rates of endemism are strikingly low. For example, Gabon has 508 strict endemic and 97 near-endemic species, comprising 10.8% and 13% of the flora respectively (Sosef *et al.* 2006). The low rates of endemism are perhaps a reflection of the fact that the Sudan region is the meeting point of several floristic regions in Africa (see chapter 1), rather than being the centre of diversity or endemism for any one of the phytochoria.

We would consider all the endemic and near-endemic plants to be conservation priority species for the Sudan region. However, of the 107 taxa listed, only 10 have had their conservation status formally assessed, nearly half of which are *Aloe* species as discussed above (see Table 3.2). Only three other endemic species have so far been assessed under IUCN categories and criteria: *Encephalartos mackenziei*, *Hyperthelia edulis* and *Ottelia brachyphylla* and we feel that the assessment of the *Hyperthelia* needs a reappraisal. The gathering of further information on these species must be considered an absolute priority for plant conservation efforts in Sudan and South Sudan. The first step will be to georeference the existing specimens and gather all existing literature and herbarium data on each of these taxa, a process that has already been started at Kew. However, it is also essential that these species are studied in the field to determine current distribution, population sizes and threats so that species-specific management plans can be drawn up and a formal IUCN conservation assessment can be published for each.

Some of the more striking endemic and near-endemic species in the region are illustrated in plates 3.1–3.9 in order to aid identification and to draw attention to these poorly known but important species.

Other potentially threatened plant species and taxa in Sudan and South Sudan

Table 3.3 records a further 144 taxa that are highly range restricted and/or globally scarce, such that they are considered to be potentially threatened on a global scale. We consider these taxa to be priorities for further investigation and, ideally, a formal assessment of their conservation status following the same procedures as listed for the endemic species above.

A combined total of 285 potential conservation priority taxa are listed in Tables 3.1–3.3. This represents just 7% of the total flora of the region. Even if all these species prove to be threatened using the categories & criteria of IUCN (2001), this percentage figure is much lower than that estimated worldwide by the *IUCN Sampled Red List Index (SRLI) for Plants* which recorded over 20% of plant species as threatened globally (see http://www.kew.org/ucm/groups/public/documents/document/kppcont_027694.pdf). We should note, however, that (i) the SRLI study included all plant groups and not just gymnosperms and angiosperms as treated in the current checklist, and (ii) we may not have captured every potentially threatened species from Sudan and South Sudan within Tables 3.1–3.3; for example, there may be a number of more widespread species that are, nevertheless, threatened by over-exploitation and/or widespread habitat loss. That said, it is unlikely that the percentage of globally threatened species in Sudan and South Sudan would rise to over 10%. This is most likely a further reflection of the fact that the Sudan region is a meeting point for different floristic regions, with mainly the more widespread species of each phytochorion reaching their range limits in this region, rather than being a centre of endemism in any one phytochorion.

Table 3.1: species occurring in Sudan and South Sudan that are listed on the *IUCN Red List* (www.iucnredlist.org) in any category other than Least Concern (LC). The phytochoria follow White (1983): III: Sudanian; IV: Somalia-Masai; VIII: Afromontane; XI: Guinea-Congolia/Sudania transition (the species listed within this phytochorion are essentially Guineo-Congolian), XVI: Sahel, XVII: Sahara; see chapter 1 for details on these phytochoria.

Species	Family	IUCN assessment	Year of Last Assessment	Requires updating?	Principal threat(s)	Distribution in Sudan	Distribution in South Sudan	Phytochorion
Afzelia africana	Leguminosae: Caesalpinioideae	Vulnerable VU A1d	1998	Yes	Timber exploitation	-	BAG EQU	III
Aldrovanda vesiculosa	Droseraceae	Endangered EN B2ab(iii,v)	2012	No	Wetland degradation & pollution	-	BAG	Azonal (swamp)
Aloe canarina	Xanthorrhoeaceae	Data Deficient DD	2010	No	Over-grazing, habitat loss; DD due to uncertainty over status of Sudan plants	-	EQU	IV
Aloe crassipes	Xanthorrhoeaceae	Data Deficient DD	2010	No	Not known	-	EQU	III
Aloe erensii	Xanthorrhoeaceae	Endangered EN B1ab(iii)	2010	No	Habitat loss	-	EQU	IV
Aloe macleayi	Xanthorrhoeaceae	Endangered EN B1ab(iii)	2010	Yes*	Habitat loss (but may benefit from some disturbance – *there is some confusion over the assessment of this species)	-	EQU	VIII
Aloe sinkatana	Xanthorrhoeaceae	Endangered EN B1ab(iii)	2010	No	Habitat loss, medicinal use, climate change	RS	-	XVI
Ansellia africana	Orchidaceae	Vulnerable VU A2cd+3cd+4cd	2013	No	Over-harvesting	DAR	?EQU	III
Baphia abyssinica	Leguminosae	Vulnerable VU A1c	1998	Yes	Habitat loss	-	UN	III / VIII transition
Belischmiedia ugandensis	Lauraceae	Vulnerable VU A2d	1998	Yes	Over-exploitation & habitat loss	-	EQU	XI
Combretum hartmannianum	Combretaceae	Vulnerable VU A1c+2c	1998	Yes	Habitat loss	DAR KOR CS ES	?BAG ?UN ?EQU	III
Combretum rochetianum	Combretaceae	Vulnerable VU A1c+2c	1998	Yes	Habitat loss	CS ES	EQU	III
Cordyla richardii	Leguminosae: Caesalpinioideae	Vulnerable VU B1+2c	1998	Yes	None listed	-	BAG UN EQU	III
Dalbergia melanoxylon	Leguminosae: Papilionoideae	Lower Risk/ near threatened	1998	Yes	Timber exploitation	DAR KOR ES	UN EQU	III
Dracaena ombet	Asparagaceae	Endangered EN A1cd	1998	Yes	Habitat loss, over-exploitation, pests	RS	-	XVI

Species	Family	Status	Year		Threat			
Encephalartos mackenziei	Zamiaceae	Near Threatened NT	2010	No	Very scarce but no evidence of threat	-	EQU	III
Encephalartos septentrionalis	Zamiaceae	Near Threatened NT	2010	No	Rare but no evidence of threat	-	BAG EQU	III
Entandrophragma angolense	Meliaceae	Vulnerable VU A1cd	1998	Yes	Timber exploitation	-	EQU	XI
Fleroya (Hallea) stipulosa	Rubiaceae	Vulnerable VU A1cd	1998	Yes	Timber exploitation	-	BAG EQU	XI
Helichrysum formosissimum	Asteraceae	Data Deficient DD	2010	No	None listed	-	EQU	VIII
Khaya grandifoliola	Meliaceae	Vulnerable VU A1cd	1998	Yes	Timber exploitation	-	EQU	XI
Khaya senegalensis	Meliaceae	Vulnerable VU A1cd	1998	Yes	Timber exploitation	DAR	BAG EQU	III
Medemia argun	Arecaceae	Critically Endangered CR B1+2c	1998	Yes	Small and fragmented populations	NS ?RS ?KOR	-	XVII
Milicia excelsa	Moraceae	Lower Risk/near threatened	1998	Yes	Timber exploitation	-	BAG UN EQU	XI
Najas welwitschii	Hydrocharitaceae	Data Deficient DD	2010	No	Possibly water pollution	CS	-	Azonal (swamp)
Ocotea kenyensis	Lauraceae	Vulnerable VU A1cd	1998	Yes	Timber exploitation	-	EQU	VIII
Ottelia brachyphylla	Hydrocharitaceae	Data Deficient DD	2010	No	Extremely restricted range	-	BAG	Azonal (swamp)
Ottelia scabra	Hydrocharitaceae	Near Threatened NT	2010	No	Possibly competition from alien species	DAR	BAG UN	Azonal (swamp)
Pouteria altissima	Sapotaceae	Lower Risk/conservation dependent	1998	Yes	Timber exploitation	-	EQU	XI
Prunus africana	Rosaceae	Vulnerable VU A1cd	1998	Yes	Over-harvesting of bark	-	EQU	VIII
Sterculia cinerea	Malvaceae	Lower Risk/near threatened	1998	Yes	None listed	KOR CS ES	-	III
Suddia sagittifolia	Poaceae	Vulnerable VU D2	2010	No	Drought	-	UN	Azonal (swamp)
Vitellaria paradoxa	Sapotaceae	Vulnerable VU A1cd	1998	Yes	Over-exploitation	DAR	BAG UN EQU	III

Table 3.2: endemic and near-endemic taxa in Sudan and South Sudan. For the 'IUCN assessment' column: NE = Not Evaluated; DD = Data Deficient; LC= Least Concern; NT = Near Threatened; VU = Vulnerable; EN = Endangered; CR = Critically Endangered. The region codes for Sudan and South Sudan are fully explained in chapter 4: table 4.3 and figure 4.1

Species	Family	Status uncertain	IUCN assessment	Distribution in Sudan	Distribution in South Sudan	Phytochorion	Notes
Endemic species							
Abutilon eufigarii	Malvaceae: Malvoideae	x	NE	CS		?III	Known only from the type; status uncertain
Acalypha hochstetteriana	Euphorbiaceae		NE	KOR CS		III / Azonal (swamp)	
Albizia aylmeri	Leguminosae: Mimosoideae		NE	DAR KOR CS		III	
Albuca steudneri	Asparagaceae	x	NE	ES		?	Known only from the type (destroyed); status uncertain
Aloe crassipes	Xanthorrhoeaceae		DD		EQU	III	Known only from Mt Bangenze region near the D.R. Congo border
Aloe dioli	Xanthorrhoeaceae		LC		EQU	VIII	Imatong Mts only
Aloe ithya	Xanthorrhoeaceae		NE		EQU	VIII	Imatong Mts only
Aloe macleayi	Xanthorrhoeaceae		EN		EQU	VIII	Imatong Mts only
Aloe sinkatana	Xanthorrhoeaceae		EN	RS		XVI	Red Sea Hills only
Bacopa punctata	Plantaginaceae	x	NE		BAG	Azonal (swamp)	Known only from the type
Barleria sp. A	Acanthaceae	x	NE		UN	III	An undescribed but distinct species allied to *B. calophylla*
Barleria tetraglochin	Acanthaceae		NE	KOR CS		III	
Bidens chippii	Asteraceae		NE		EQU	VIII	Imatong Mts only
Bidens isostigmatoides	Asteraceae		NE		EQU	VIII	Imatong Mts only
Biscutella didyma var. *elbensis*	Brassicaceae		NE	RS		XVII	Jebel Elba only (Hala'ib Triangle)
Bothriocline imatongensis	Asteraceae		NE		EQU	VIII	Imatong Mts only
Brachystephanus sudanicus	Acanthaceae		NE		EQU	XI	Imatong Mts; known only from the type
Caralluma darfurensis	Apocynaceae		NE	DAR		III	Known only from the type from near Nyala
Caralluma sudanica	Apocynaceae		NE	KOR		XVI	
Chlorophytum superpositum	Asparagaceae		NE		EQU	?	Known only from the type
Cleome niamniamensis	Cleomaceae		NE		EQU	III	Closely allied to *C. polyanthera*
Clerodendrum triflorum	Lamiaceae		NE	CS ES	UN	III	
Coleochloa glabra	Cyperaceae		NE		EQU	III / VIII transition	Known only from Mt Odo
Coleochloa schweinfurththiana	Cyperaceae		NE		EQU	III / VIII transition	Known only from Mt Bangenze on the D.R. Congo border

Commelina sp. aff. velutina	Commelinaceae	×	NE		EQU	?	Known from a single specimen; allied to C. velutina from Cameroon
Cotula kotschyi	Asteraceae	×	NE	?NS ?RS DAR ?KOR CS	BAG	Azonal (swamp)	Closely allied to C. cinerea
Cynoglossum sp. cf. aequinoctiale	Boraginaceae	×	NE		EQU	VIII	Known from a single specimen from the Didinga Mts
Dicliptera lanceolata	Acanthaceae		NE*		BAG EQU	III	*Assessed as Data Deficient by Darbyshire & Kordofani (2012)
Dipcadi fesoghlense	Asparagaceae	×	NE	CS	EQU	?	Known only from the type; status uncertain
Dorstenia annua	Moraceae		NE		EQU	VIII	Imatong Mts only
Ehretia sp. cf. amoena	Boraginaceae	×	NE		EQU	III	Imatong Mts; status uncertain
Encephalartos mackenziei	Zamiaceae		NT		EQU	III	
Festuca sudanensis	Poaceae		NE		EQU	VIII	Imatong Mts only
Fimbristylis falcifolia	Cyperaceae	×	NE	ES	BAG	?	Known only from the type; status uncertain
Floscopa schweinfurthii	Commelinaceae		NE		EQU	Azonal (swamp)	
Fuerstia bartsioides	Lamiaceae		NE		EQU	?	
Harveya sp. nov.	Orobanchaceae	×	NE		EQU	XI	Known only from Laboni; a distinct but undescribed species
Heliotropium sudanicum	Boraginaceae		NE	CS	EQU	XVI	Known only from the Gezira area
Hibiscus mongallaensis	Malvaceae: Malvoideae		NE		EQU	IV	Known only from the type from the Ilemi Triangle
Hibiscus muhamedis	Malvaceae: Malvoideae	×	NE	KOR		?	Known only from the type; status uncertain
Hibiscus schweinfurthii	Malvaceae: Malvoideae		NE		BAG EQU	III	
Humularia sudanica	Leguminosae: Papilionoideae		NE		EQU	III	Known only from the type from Sue River near Yambio
Hygrophila caerulea	Acanthaceae		NE	KOR	UN	III / Azonal (swamp)	
Hyperthelia edulis	Poaceae		LC		EQU	III	Known only from the Abu Satta Hills
Hyperthelia macrolepis	Poaceae	×	NE		BAG	III	Possibly just an annual variant of the widespread H. dissoluta
Indigofera congolensis var. bongensis	Leguminosae: Papilionoideae		NE		BAG	III	May also occur in C.A.R.
Indigofera conjugata var. schweinfurthii	Leguminosae: Papilionoideae		NE		BAG EQU	III	I. conjugata s.l. assessed as Least Concern on the IUCN red list, but varieties not assessed
Indigofera knoblecheri	Leguminosae: Papilionoideae		NE		EQU	III	
Ipomoea convolvulifolia	Convolvulaceae	×	NE		EQU	?III	Known only from the type; status uncertain
Ipomoea curtipes	Convolvulaceae	×	NE		BAG	?III	Known only from the type; status uncertain
Ipomoea eurysepala	Convolvulaceae	×	NE	KOR		?III or XVI	Known only from the type; status uncertain
Jatropha melanosperma	Euphorbiaceae		NE		BAG	?III	

33

Table 3.2 contd.

Species	Family	Status uncertain	IUCN assessment	Distribution in Sudan	Distribution in South Sudan	Phytochorion	Notes
Ledebouria lilacina	Asparagaceae		NE	KOR	UN	?III	Imatong Mts; known only from the type
Ledermanniella ramosissima	Podostemaceae		NE		EQU	Azonal (rivers)	
Lepidagathis medusae	Acanthaceae		NE		BAG	III	
Lindernia sudanica	Linderniaceae		NE		EQU	VIII	Imatong Mts only
Lippia radula	Verbenaceae		NE		BAG	?III	
Loudetia esculenta	Poaceae		NE		EQU	III	
Misopates marraicum	Plantaginaceae		NE	DAR		XVI	Jebel Marra only
Nepeta sudanica	Lamiaceae		NE	RS		XVI	Extends to the Hala'ib Triangle
Nesaea sp. A	Lythraceae	x	NE	DAR	UN	III / Azonal (swamp)	Status uncertain; requires further study
Nesaea sp. aff. heptamera	Lythraceae	x	NE		UN	III	Status uncertain; requires further study
Ochna micrantha	Ochnaceae	x	NE	CS ES		III	
Ottelia brachyphylla	Hydrocharitaceae		DD		BAG	Azonal (swamp)	Type destroyed; extant material only doubtfully identified
Pandiaka elegantissima	Amaranthaceae		NE		BAG	III	
Panicum bambusiculme	Poaceae		NE		BAG	XI	
Pavetta bilineata	Rubiaceae		NE		EQU	XI	Talanga Forest only
Pavetta sp. nov. near abyssinica	Rubiaceae	x	NE		EQU	VIII	
Plectranthus jebel-marrae	Lamiaceae		NE	DAR	EQU	XVI	Jebel Marra and Jebel Gourgeil only
Pseudognaphalium marranum	Asteraceae		NE	DAR		VIII	Jebel Marra only
Psychotria nubica	Rubiaceae	x	NE	CS		?III	Known only from the type; status uncertain
Pulicaria grantii	Asteraceae		NE	CS		Azonal (rivers)	Known only from the type from the Nile N of Khartoum
Rhamphicarpa elongata	Orobanchaceae		NE	KOR		III / Azonal (swamp)	Known only from the type
Rhynchosia splendens	Leguminosae: Papilionoideae		NE	ES		?III	Known only from the type from the Sudan/Ethiopia border region
Salacia ducis-wuertembergiae	Celastraceae	x	NE	CS		?III	Status uncertain; no extant material
Scilla engleri	Asparagaceae		NE		BAG	?III	
Thesium sp.cf. leucanthum	Santalaceae	x	NE		EQU	III	Imatong Mts only; requires further study
Thunbergia hispida	Acanthaceae	x	NE	CS	UN	?III	Incompletely known species
Thunbergia longifolia	Acanthaceae	x	NE		EQU	?III	Known only from the type; status uncertain
Thunbergia schweinfurthii	Acanthaceae		NE		BAG EQU	III	

Species	Family			Status		Zone	Notes
Tragia bongolana	Euphorbiaceae			NE	BAG	?III	Known only from the type; status uncertain
Tragia schweinfurthii	Euphorbiaceae			NE	BAG	?III	Red Sea Hills only; status uncertain
Umbilicus paniculiformis	Crassulaceae		RS	NE		XVI	
Urginea grandiflora	Asparagaceae	x	RS	NE		?XVI	
Verbascum nubicum	Scrophulariaceae	x	RS	NE		XVI	
Verbascum sudanicum	Scrophulariaceae		DAR	NE		XVI	Jebel Marra only
Near-endemic species							
Abutilon erythraeum	Malvaceae: Malvoideae		RS CS	NE		XVI	Also Eritrea
Afrosciadium dispersum	Apiaceae			NE	EQU	VIII	Also Tanzania
Albizia schimperiana var. *tephrocalyx*	Leguminosae: Mimosoideae			NE	EQU	XI	Also Uganda
Aloe canarina	Xanthorrhoeaceae			DD	EQU	IV	Also Uganda
Bothriocline congesta	Asteraceae			NE	EQU	VIII	Imatong Mts; also Uganda (Mt Kadam)
Emilia emilioides	Asteraceae		KOR CS ES	NE	UN EQU	III	Widespread in Sudan & South Sudan; possibly also C.A.R.
Erodium oreophilum	Geraniaceae		DAR	NE		XVI	Jebel Marra; Tibesti (Chad)
Euphorbia depauperata var. *laevicarpa*	Euphorbiaceae			NE	EQU	VIII	Also Uganda; known only from Imatong Mts and adjacent highlands
Huernia sudanensis	Apocynaceae	x	RS	NE		XVI	Also Ethiopia
Hyparrhenia confinis var. *nudiglumis*	Poaceae		DAR KOR CS	NE		III	Also Ethiopia
Hyparrhenia confinis var. *pellita*	Poaceae		CS ES	NE	UN	III	Includes Ethiopian border area (Metamma)
Jatropha aceroides	Euphorbiaceae		RS	NE		XVI	Also Eritrea
Kickxia aegyptiaca subsp. *virgata*	Plantaginaceae		DAR	NE		XVI	Possibly also Chad
Lavandula antineae subsp. *marrana*	Lamiaceae		DAR	NE		XVI	Also Chad
Lindernia niamniamensis	Linderniaceae			NE	EQU	VIII	Also Uganda
Medemia argun	Arecaceae		NS ?RS ?KOR	CR	UN	XVII	Nubian Desert of Sudan and Egypt
Pennisetum schweinfurthii	Poaceae		KOR ES	NE		III	Also just extending into Ethiopia
Hannoa schweinfurthii	Simaroubaceae			NE	BAG EQU	III	Also D.R. Congo
Rhynchosia teramnoides	Leguminosae: Papilionoideae			NE	EQU	?III	Also Chad
Rhynchosia tricuspidata subsp. *imatongensis*	Leguminosae: Papilionoideae			NE	EQU	VIII	Also Uganda; Imatong Mts only
Spermacoce tenuissima	Rubiaceae			NE	BAG UN	?III	Also C.A.R.
Suddia sagittifolia	Poaceae			VU	UN	Azonal (swamp)	Sudd marshes; also Uganda

35

Table 3.3: additional potential conservation priority species in Sudan and South Sudan that are not endemic and have not so far been assessed under the categories and criteria of IUCN (2001).

Species	Family	Distribution in Sudan	Distribution in South Sudan	Habitat
Acacia venosa	Leguminosae: Mimosoideae	?ES		III
Acanthus seretii	Acanthaceae		EQU	VIII
Achyrospermum axillare	Lamiaceae		EQU	XI
Acridocarpus ugandensis	Malpighiaceae		EQU	III
Aeollanthus ambustus	Lamiaceae		BAG	III
Aloe camperi	Xanthorrhoeaceae	RS		XVI
Aloe labworana	Xanthorrhoeaceae		EQU	III
Aloe vituensis	Xanthorrhoeaceae		EQU	IV
Aloe wrefordii	Xanthorrhoeaceae		EQU	III
Ampelocissus sarcocephala	Vitaceae		BAG	III
Ampelocissus schimperiana	Vitaceae	ES	EQU	III / XI
Anthephora laevis	Poaceae	RS		XVI
Anthospermum pachyrrhizum	Rubiaceae	DAR		XVI
Ascolepis eriocauloides	Cyperaceae	KOR		Azonal
Asystasia africana	Acanthaceae		EQU	XI
Barleria calophylla	Acanthaceae		BAG EQU	III
Barleria lanceata	Acanthaceae	RS		XVI
Bidens ternata var. *vatkei*	Asteraceae	?NS/RS		XVI
Bothriocline monticola	Asteraceae		EQU	VIII
Brandella erythraea	Boraginaceae	RS		XVI
Bulbostylis sphaerocarpa	Cyperaceae	ES		III
Cadaba gillettii	Capparaceae		EQU	III
Cadaba kassasii	Capparaceae	CS		?III
Carex mannii subsp. *thomasii*	Cyperaceae		EQU	VIII
Casearia runssorica	Salicaceae		BAG EQU	XI
Ceropegia melanops	Apocynaceae		BAG	III
Clitoriopsis mollis	Leguminosae: Papilionoideae		EQU	III
Cnestis mildbraedii	Connaraceae		EQU	XI
Coccinia subsessiliflora	Cucurbitaceae		EQU	XI
Coffea arabica	Rubiaceae		UN	III / VIII transition
Coffea neoleroyi	Rubiaceae		UN	III
Combretum schweinfurthii	Combretaceae		BAG EQU	III
Combretum umbricola	Combretaceae		EQU	XI
Commicarpus montanus	Nyctaginaceae	DAR		XVI
Conomitra linearis	Apocynaceae	KOR		III
Convolvulus hamphilahensis	Convolvulaceae	RS		XVI
Crinipes longifolius	Poaceae		EQU	III
Crinum bambusetum	Amaryllidaceae	RS	EQU	III
Crotalaria intonsa	Leguminosae: Papilionoideae		EQU	VIII
Cynoglossopsis latifolia	Boraginaceae	RS		XVI

Table 3.3 contd.

Species	Family	Distribution in Sudan	Distribution in South Sudan	Habitat
Cyperus commixtus	Cyperaceae		UN	Azonal
Cyperus microbolbos	Cyperaceae	RS		XVII
Cyphostemma adenanthum	Vitaceae	ES		III
Cyphostemma alnifolium	Vitaceae		EQU	III
Cyphostemma crinitum	Vitaceae	DAR	BAG EQU	III
Cyphostemma dembianense	Vitaceae	RS ES	EQU	III / VIII
Dialium excelsum	Leguminosae: Caesalpinioideae		?EQU	XI
Dicliptera latibracteata	Acanthaceae		EQU	VIII
Digitaria xanthotricha	Poaceae		BAG UN	III
Drimia sudanica	Asparagaceae		EQU	III
Drimiopsis spicata	Asparagaceae	DAR	BAG EQU	III
Echidnopsis cereiformis	Apocynaceae	RS		XVI
Eriosema schweinfurthii	Leguminosae: Papilionoideae		BAG	III
Eulophia montis-elgonis	Orchidaceae		EQU	VIII
Eulophia stenoplectra	Orchidaceae		EQU	III
Euphorbia collenetteae	Euphorbiaceae	RS		XVI
Euphorbia magnicapsula var. *lacertosa*	Euphorbiaceae		EQU	III
Euphorbia rivae	Euphorbiaceae		EQU	IV
Euphorbia venenifica	Euphorbiaceae	KOR	BAG EQU	III
Fadogia leucophloea	Rubiaceae		BAG	III
Festuca elgonensis	Poaceae		EQU	VIII
Gardenia tinneae	Rubiaceae		BAG	III
Geophila obvallata subsp. *involucrata*	Rubiaceae		EQU	XI
Geranium biuncinatum	Geraniaceae	RS		XVI
Gladiolus sudanicus	Iridaceae	KOR		XVI
Gossypium longicalyx	Malvaceae		UN	III
Guizotia arborescens	Asteraceae		EQU	VIII
Habenaria antennifera	Orchidaceae	DAR		VIII
Heterotis cinerascens	Melastomataceae		EQU	III / Azonal
Hibiscus eriospermus	Malvaceae: Malvoideae	RS		XVI
Hibiscus sudanensis	Malvaceae: Malvoideae		?EQU	III
Holothrix tridentata	Orchidaceae	DAR		III
Hyparrhenia multiplex	Poaceae	DAR		XVI
Hyperthelia cornucopiae	Poaceae		BAG UN EQU	III
Hypoestes strobilifera var. *strobilifera*	Acanthaceae		BAG EQU	III
Impatiens meruensis subsp. *septentrionalis*	Balsaminaceae		EQU	XI
Indigofera achyranthoides	Leguminosae: Papilionoideae		BAG EQU	III
Indigofera andrewsiana	Leguminosae: Papilionoideae		UN EQU	III

Table 3.3 contd.

Species	Family	Distribution in Sudan	Distribution in South Sudan	Habitat
Indigofera biglandulosa	Leguminosae: Papilionoideae	KOR		III (plus II Zambesian)
Indigofera letestui	Leguminosae: Papilionoideae		EQU	III
Indigofera lotononoides	Leguminosae: Papilionoideae	DAR	BAG	III
Indigofera oubanguiensis	Leguminosae: Papilionoideae		BAG	III
Indigofera pseudosubulata	Leguminosae: Papilionoideae		EQU	III
Isoglossa membranacea subsp. *septentrionalis*	Acanthaceae		EQU	XI
Jatropha aethiopica	Euphorbiaceae	CS		III
Jatropha afrotuberosa	Euphorbiaceae		BAG	III
Jatropha gallabatensis	Euphorbiaceae	ES		III
Jatropha neriifolia	Euphorbiaceae	DAR		III / XVI
Jatropha schweinfurthii subsp. *schweinfurthii*	Euphorbiaceae		BAG EQU	III
Justicia afromontana	Acanthaceae		EQU	VIII
Kleinia picticaulis	Compositae		EQU	IV
Kniphofia nubigena	Xanthorrhoeaceae	RS		XVI
Lavandula saharica	Lamiaceae	NS		XVII
Ledebouria maesta	Asparagaceae	DAR KOR	UN EQU	III (plus II Zambesian)
Lepidagathis peniculifera	Acanthaceae		EQU	III
Leptonychia chrysocarpa	Malvaceae: Byttnerioideae		BAG EQU	XI
Lobelia dissecta	Campanulaceae		EQU	VIII
Lotus hebecarpus	Leguminosae: Papilionoideae	RS		XVI
Lotus hebranicus	Leguminosae: Papilionoideae	RS		XVII
Lotus nubicus	Leguminosae: Papilionoideae	RS CS		XVI
Lychnodiscus cerospermus	Sapindaceae		EQU	XI
Merremia gallabatensis	Convolvulaceae	ES		III
Nesaea icosandra	Lythraceae		BAG	III / Azonal
Ochna leucophloeos subsp. *leucophloeos*	Ochnaceae	CS ES		III
Ochrocephala imatongensis	Asteraceae		EQU	III
Olea europaea subsp. *laperrinei*	Oleaceae	DAR		XVI
Orbea laticorona	Apocynaceae	NS RS		XVI
Orbea sprengeri subsp. *sprengeri*	Apocynaceae	RS		IV
Pavetta schweinfurthii	Rubiaceae		BAG	XI
Pennisetum gracilescens	Poaceae	DAR		XVI
Pennisetum pirottae	Poaceae	CS	UN	III
Pentas purseglovei	Rubiaceae		EQU	VIII

Table 3.3 contd.

Species	Family	Distribution in Sudan	Distribution in South Sudan	Habitat
Peponium cienkowskii	Cucurbitaceae		EQU	VIII
Phagnalon schweinfurthii var. *schweinfurthii*	Asteraceae	RS		XVI
Phyllanthus chevalieri	Phyllanthaceae		UN	III
Phyllanthus limmuensis	Phyllanthaceae		EQU	XI
Phyllanthus trichotepalus	Phyllanthaceae		EQU	XI
Plectranthus grandicalyx	Lamiaceae		EQU	VIII
Polygala septentrionalis	Polygalaceae		EQU	III
Portulaca erythraeae	Portulacaceae	?RS		XVI
Psophocarpus obovalis	Leguminosae: Papilionoideae		EQU	III
Psychotria fertitensis	Rubiaceae		BAG EQU	III
Pycnostachys niamniamensis	Lamiaceae		EQU	III
Satanocrater fellatensis	Acanthaceae	ES		III
Scilla chlorantha	Asparagaceae		BAG	?III
Scilla schweinfurthii	Asparagaceae		BAG UN	III
Scutellaria schweinfurthii subsp. *schweinfurthii*	Lamiaceae		BAG EQU	III
Silene lynesii	Caryophyllaceae	DAR		XVI
Spermacoce phyteuma	Rubiaceae		BAG EQU	III
Stachys schimperi	Lamiaceae	RS		XVI
Stathmostelma angustatum subsp. *vomeriforme*	Apocynaceae		UN	III (plus II Zambesian)
Stathmostelma welwitschii	Apocynaceae		EQU	III (plus II var. *bagshawei* Zambesian)
Sterculia cinerea	Malvaceae: Sterculioideae	KOR CS ES		III
Stipa tigrensis	Poaceae	?DAR		XVI
Tacazzea venosa	Apocynaceae	ES		III
Tephrosia cordatistipula	Leguminosae: Papilionoideae		EQU	III
Tephrosia fulvinervis	Leguminosae: Papilionoideae	?KOR		III / XVI
Tephrosia lebrunii	Leguminosae: Papilionoideae		EQU	VIII
Tragia mitis	Euphorbiaceae	ES		III
Tricalysia niamniamensis var. *djurensis*	Rubiaceae		BAG EQU	XI
Trigonella occulta	Leguminosae: Papilionoideae	NS CS		XVI / XVII
Uvaria schweinfurthii	Annonaceae		EQU	XI
Vernonia plumbaginifolia var. *plumbaginifolia*	Asteraceae	CS	BAG	?III
Vernonia unionis	Asteraceae	RS ES		XVI

Potential important plant areas in Sudan and South Sudan

Any assessment of national conservation priorities must identify the regions and habitats of highest diversity and containing the highest number of range-restricted and/or globally threatened species. This process has already been conducted in the Sudan region for some faunal groups, notably for birds through BirdLife International's Important Bird Area Programme (Robertson 2001), which recognises 23 IPAs across the Sudan region, 11 in Sudan and 12 in South Sudan. However, in view of the paucity of detailed botanical information for large parts of the Sudan region, we are a considerable way off being in a position to do the same for Important Plant Areas at present (using the criteria of Plantlife International – see http://www.plantlife.org.uk/uploads/documents/Guide_to_Implementing_IPAs_2004.pdf). That said, during the compilation of the current checklist, we have been able to identify a number of key localities and regions for plant diversity and endemism which we list below. This list is intended only as a starting point, and we are sure that it will be added to and refined as further exploration and documentation is carried out in the many under-studied regions of the two countries.

SUDAN

- **Jebel Uweinat** – this isolated mountain range on the Libya-Egypt-Sudan border is the easternmost of the Saharan massifs. The flora has been well documented by Léonard (1997–2000); species diversity is low but includes the Saharan mountain endemics *Lavandula saharica* and *Stipagrostis rigidifolia* and is the southern limit for several north African and Afro-Asian species including *Cornulaca monacantha* and *Helianthemum lippii*.
- **Oases of the Nubian Desert** – important primarily for the populations of the Critically Endangered palm *Medemia argun*.
- **Jebel Elba – Jebel Asoteriba region** –the northern section of Sudan's Red Sea Hills, extending into the disputed Hala'ib Triangle (see chapter 4). Jebel Elba itself has been reasonably well botanised and the flora has been documented by Drar (1936) as well as within Boulos's (1999–2005) *Flora of Egypt*. The larger mountains still hold patches of dry forest including rare species such as *Dracaena ombet*, whilst the many wadis contain an interesting semi-desert flora. Some of the Red Sea Hills endemics such as *Nepeta sudanica* reach their northern limit here, whilst a number of north African species reach their southeastern limit, for example *Rumex simpliciflorus* and *Silene villosa*. It also has one endemic taxon, *Biscutella didyma* var. *elbensis*.
- **Erkowit region** – in the central Sudanese Red Sea Hills not far from the main Kassala-Port Sudan road near Sinkat and so reasonably accessible and well botanised. Kassas (1956a) described the vegetation of this area and Hassan (1974) produced an illustrated guide to some of its plants. It receives more moisture than the majority of the Red Sea Hills, mainly in the form of mist. It therefore supports a more luxuriant vegetation than in the surrounding hills, including in the past extensive stands of tree *Euphorbia* spp. and *Dracaena ombet* though these are now much diminished and the *Dracaena* is all but extinct here. There are a number of Red Sea Hills endemics here including *Aloe sinkatana* (globally endangered), *Huernia sudanensis* and *Umbilicus paniculiformis*.
- **Karora region** – this is the southeastern section of Sudan's Red Sea Hills, separated from the central section by the Tokar Delta but contiguous with the extensive Eritrean Red Sea Hills. It is rather isolated and so currently understudied. It holds some relict areas of *Juniperus procera* – *Olea europaea* subsp. *cuspidata* forest, though these are much diminished. More intensive botanising in this region is likely to yield records of species previously thought restricted to Eritrea and northern Ethiopia.

- **Jebel Marra** – this isolated volcanic massif in Darfur's Sahel zone has a rich and varied flora including important areas of *Olea europaea* subsp. *lapperinei* forest. Its flora has been well documented by Wickens (1976). It has a number of endemic and near-endemic species including *Misopates marraicum*, *Plectranthus jebel-marrae* (also on Jebel Gourgeil, below), *Pseudognaphalium marranum* and *Verbascum sudanicum*.
- **Jebel Gourgeil** – the northern outlier of the Jebel Marra massif, largely with a similar, though less diverse, flora but also containing a drier North African element. It contains a number of species otherwise unknown from Sudan including the Saharan massif endemics *Commicarpus montanus* and *Glossocardia bosvallia*.
- **Nuba Mountains** – this diffuse mountain range with isolated peaks is poorly studied botanically but is known to be the northernmost point for some dry tropical forest species. It also contains a small number of range-restricted species such as *Barleria tretraglochin* and *Gladiolus sudanicus*.
- **Gallabat region** – of Eastern Sudan on the Ethiopian border. The early collecting in this region by Schweinfurth (see chapters 2 and 4) indicated that it is an area of high botanical interest with several endemic and range-restricted species such as *Jatropha gallabatensis* and *Rhynchosia splendens*. The area has hardly been botanised since Schweinfurth's time and the **Dinder National Park** to the south is very poorly known and may prove to be of equal interest.

SOUTH SUDAN

- **Woodlands over ironstone in Bahr el Ghazal and western Equatoria** – this area is poorly defined at present, with identification of key specific localities hindered by the poor botanical coverage across the region. However, there appear to be a number of endemic species in the Sudanian woodlands of this region, particularly in the herb flora such as *Lepidagathis medusae* and *Pandiaka elegantissima*. The area along the border with Central African Republic northwest of Tambura and up to Sa'id Bundas and Deim Zubeir are worthy of further exploration as is the extensive Southern National Park for which the flora is largely undocumented.
- **Sudd Marshes** – these extensive swamps associated with the White Nile are of huge international importance for their fauna. Plant diversity is rather low, but a number of scarce species occur including the near-endemic and globally vulnerable grass *Suddia sagittifolia* and the globally endangered aquatic *Aldrovanda vesiculosa*. The seasonally flooded woodlands and grasslands of the surrounding Flood Region have a more diverse flora.
- **Boma Plateau** – this understudied region of Upper Nile and Equatoria bordering Ethiopia is particularly important for its remnant patches of wet forest which are an extension of the forest zone of southwest Ethiopia. The forests contain the only wild populations of *Coffea arabica* in South Sudan, as well as a number of other rare species such as *Baphia abyssinica*. The forests are highly threatened by man. The surrounding plateau and inselbergs hold extensive woodlands with some interesting species including the recent discovery of the rare *Coffea neoleroyi*.
- **Mt Bangenze** – this inselberg straddling the D.R. Congo border east of Yambio and SW of Maridi is noted for the endemic *Coleochloa schweinfurthiana* whilst the rocky woodland in the vicinity holds the endemic *Aloe crassipes*.
- **Imatong-Dongotona-Lafit-Didinga Mts** – the submontane and montane forests and montane grasslands of these small mountain ranges hold the highest number of endemic and range-restricted species in the Sudan region. The endemics are mainly herbaceous,

such as two species of *Aloe* (one only described in 2014), two species of *Bidens* and *Lindernia sudanica*. The extant montane forests are the most extensive in the Sudan region and it is the only area in which *Podocarpus milanjianus* forest occurs.

- **Lowland forests of Equatoria** – the most important extant forest patches are Azza Forest in Meridi region; Laboni, Lotti and Talanga Forests in the foothills of the Imatong Mts; and isolated forest patches on the Aloma Plateau. These forests hold some of the highest plant diversity in the Sudan region and include many of the threatened timber species listed in Table 3.1, for example *Khaya grandifoliola* and *Pouteria altissima*. They also contain a wealth of species found nowhere else in the Sudan region. The gallery forests along some of the major river systems in the region, such as the Yei, are also of importance, though are usually less diverse than those above.
- **Ilemi Triangle** – this area contains some of the best examples of dry Somalia-Masai bushland and grassland in South Sudan and contains a number of species found nowhere else in the Sudan region including the globally endangered *Aloe canarina* and *A. erensii*.

Figure 3.1. *Panicum bambusiculme*: **A** habit; **B** spikelet; **C** dissected spikelet; **D** lemma and palea (bracts) of mature spikelet. Drawn by B. Johnsen. Reproduced from Kew Bulletin 33: 422 (1979) with permission of the Trustees of the Royal Botanic Gardens, Kew.

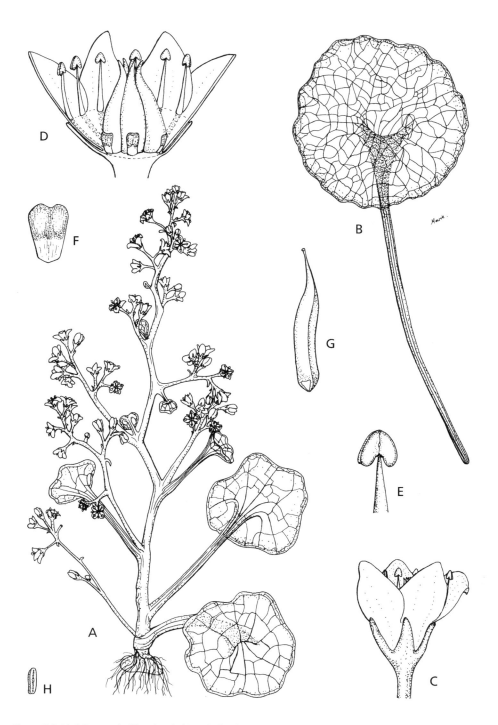

Figure 3.2. *Umbilicus paniculiformis*: **A** habit; **B** leaf; **C** flower; **D** opened flower; **E** anther; **F** nectary; **G** carpel; **H** seed. Drawn by M. Bywater. Reproduced from Kew Bulletin 37: 476 (1982) with permission of the Trustees of the Royal Botanic Gardens, Kew.

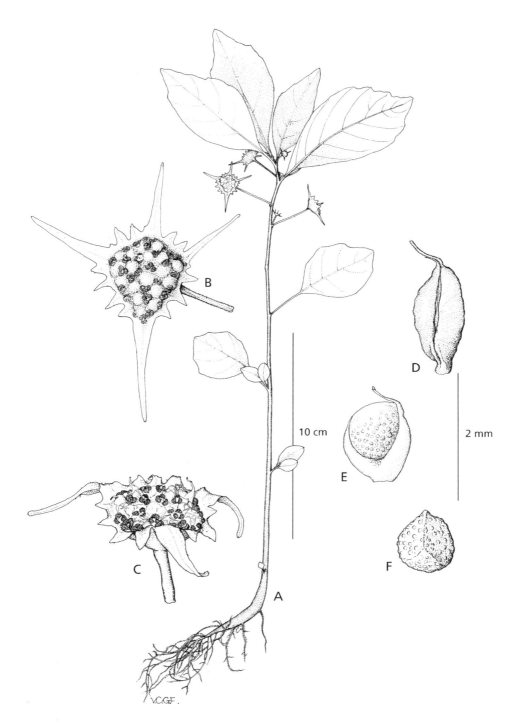

10 cm

2 mm

Figure 3.3. *Dorstenia annua*: **A** habit; **B** flowering receptacle; **C** fruiting receptacle; **D** empty fruit (exocarp); **E** mature fruit with endocarp and exocarp; **F** endocarp. Drawn by V. Gordon Friis. Reproduced from Kew Bulletin 37: 474 (1982) with permission of the artist and the Trustees of the Royal Botanic Gardens, Kew.

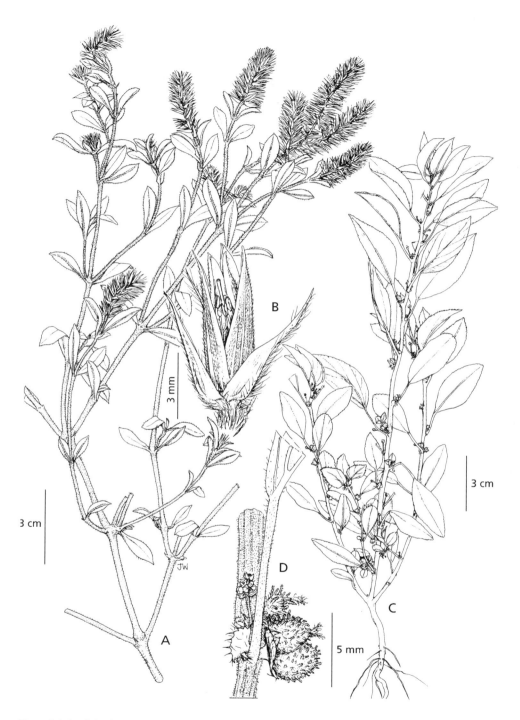

Figure 3.4. *Pandiaka elegantissima*: **A** habit; **B** detail of flower. *Acalypha hochstetteriana*: **C** habit; **D** detail of male and female inflorescences. Drawn by Juliet Williamson.

Figure 3.5. *Heliotropium sudanicum*: **A** habit; **B** detail of fruits. *Nepeta sudanica*: **C** habit; **D** detail of flower. Drawn by Juliet Williamson.

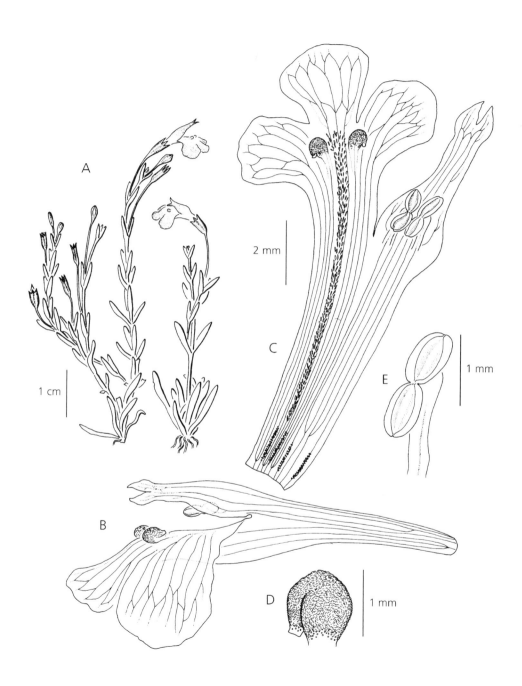

Figure 3.6. *Lindernia sudanica*: **A** habit; **B** corolla; **C** dissected corolla with androecium; **D** staminode; **E** stamen. Drawn by E. Fischer. Reproduced from Kew Bulletin 46: 531 (1991) with permission of the artist and the Trustees of the Royal Botanic Gardens, Kew.

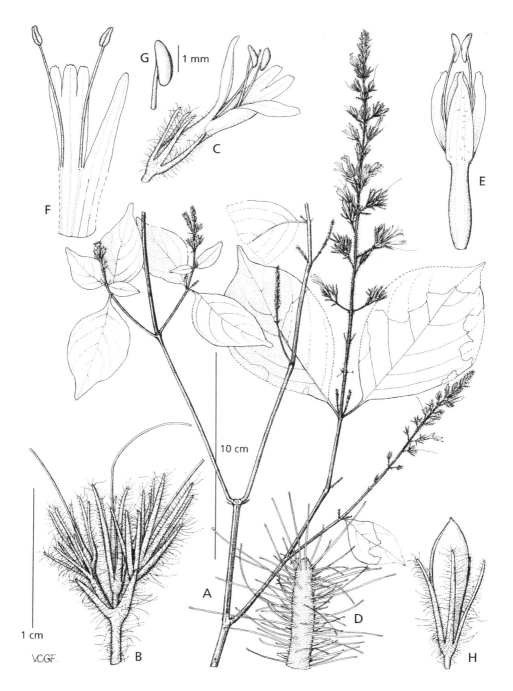

Figure 3.7. *Brachystephanus sudanicus*: **A** habit; **B** detail of partial inflorescence; **C** flower in situ; **D** apical portion of calyx lobe with detail of indumentum; **E** corolla with stamens, dorsal view; **F** dissected corolla with stamens; **G** detail of anther; **H** immature capsule within calyx. Drawn by V. Gordon Friis. Reproduced from Kew Bulletin 37: 466 (1982) with permission of the artist and the Trustees of the Royal Botanic Gardens, Kew.

Figure 3.8. *Lepidagathis medusae*: **A** habit; **B** detail of flower. *Dicliptera lanceolata*: **C** habit; **D** detail of flower. Drawn by Juliet Williamson.

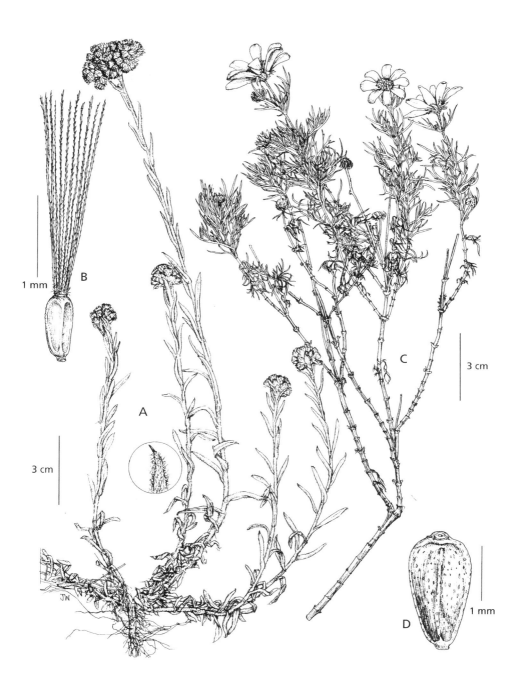

Figure 3.9. *Pseudognaphalium marranum*: **A** habit; **B** detail of achene. *Bidens chippii*: **C** habit; **D** detail of achene. Drawn by Juliet Williamson.

4. The Checklist: explanatory notes

Before using this checklist, we would advise reading the following explanatory notes on the scope, format and conventions used.

Checklist scope and sequence

The checklist aims to provide exhaustive coverage of the taxa of angiosperms and gymnosperms occurring in Sudan and South Sudan, either as natives or naturalised (or potentially naturalised) aliens. Some of the more commonly introduced woody species that have not naturalised are mentioned but not included within the checklist proper. Pteridophytes are not included in the current work, though they may be treated in a subsequent volume – at present for these groups we would refer the reader to the Africa-wide synopsis by Roux (2009) which includes country records for each species, though Sudan and South Sudan are not treated as separate countries.

The family circumscription for the angiosperms follows the latest classification of the Angiosperm Phylogeny Group (APG 2009; with subsequent changes available at http://www.mobot.org/MOBOT/research/APweb/). The sequence of families follows that applied in the Kew herbarium which is largely based on the sequence proposed by Haston *et al.* (2009). The APG family circumscription is not without its controversies and there are some areas where the authors have only reluctantly accepted the current classification. However, for the time being, it seems sensible to adopt a standard classification and APG III is the most widely accepted option. The Leguminosae (Fabaceae) are divided into the three commonly recognised subfamilies. Likewise, we separate the Malvaceae into the seven subfamilies that occur in our region since they are so distinct and may, in future, be separated at the family level (the approach adopted in Heywood *et al.* 2007); the alternative family names are given in brackets.

Within each family, taxa are arranged alphabetically by genus and then by species. Infraspecific taxa are treated separately so, for example, if there are two or more infraspecific taxa of a species in Sudan / South Sudan they are listed separately rather than within an account of the species *sensu lato*.

Each family account is individually authored to give due recognition to those who contributed materially to the compilation of the checklist. In most cases, the accounts were compiled by one or more of the editors of this volume, but we did enlist specialist help where available.

Accepted names

No single source is followed universally for the application of accepted names. Instead, we have attempted to use the most recent taxonomic works available and, where conflicts exist between modern treatments, we have made judgements based on our own observations of the taxa. In general, we have followed the accepted names and synonymy in the *African Plants Database* (http://www.ville-ge.ch/musinfo/bd/cjb/africa/recherche.php) and the related

Tropical African Flowering Plants – Ecology & Distribution hardcopy series (Lebrun & Stork 2003, 2006, 2008a, 2008b, 2010, 2011 and 2012). These are incredibly useful sources of accurate information on African plants and cannot be recommended highly enough. The only areas where there is significant disagreement with these resources are for families and genera for which a very recent African Flora treatment and/or revision is available which supersedes Lebrun & Stork's latest updates (and it is usually not long before all such treatments are incorporated into the African Plants Database). For families of monocotyledonous plants, we have generally followed the accepted names and synonymy in eMonocot (http://www.e-monocot.org/), and for families which are treated under Kew's World Checklist series (http://apps.kew.org/wcsp/home.do), we have consulted that resource widely.

Since the scope of this work is to provide a baseline on which to build further studies of the Sudanese flora, rather than to provide a critical Flora treatment for the two countries, we have in most cases accepted the determinations on the Sudan specimens in the herbaria consulted. We have only attempted to re-identify material where the original identification is clearly incorrect or where the herbarium collections have not been re-curated according to the most recent taxonomic treatment available. Our primary data source is the herbarium at Kew, where the African collections have been widely consulted and often formed the basis for the major regional Flora projects on the continent. Indeed, the *Flora of Tropical East Africa*, *Flora of West Tropical Africa* and *Flora Zambesiaca* have all been co-ordinated from Kew, and the Kew collections were also widely consulted by botanists writing the *Flora of Ethiopia & Eritrea*. Most authors also consulted and re-determined the Sudanese material during their preparatory studies, such that the Sudanese collections at Kew are, on the whole, authoritatively named. It is therefore quite reasonable to rely on these collections as a basis for a modern checklist.

Where uncertainties occur either in the taxonomic status or in the nomenclature, we have noted these in the 'Notes' section for that species. Likewise, we have also noted uncertainties in the identification of the Sudanese material; in these cases we have used the qualifier 'cf'. within the taxon name.

Synonymy

All relevant synonyms are listed, together with the relevant literature in which they were used as accepted names. The starting point of the checklist is F.W. Andrews' *The Flowering Plants of the (Anglo-Egyptian) Sudan* and we have attempted to account for all of the names applied to Sudanese plants both in those volumes and in the major subsequent works relating to Sudan (see 'Literature references', Table 4.1). Any names that were already in synonymy prior to or within Andrews' work are not recorded here unless, of course, they have since been resurrected.

Two types of synonymy are recorded: true synonyms and misapplied names ('*sensu* synonyms'). In view of the fact that most of the major African Floras have been written after *The Flowering Plants of the Sudan*, it is not surprising that many of the names used by Andrews are now in synonymy or were misapplied, either widely or specifically by him.

Synonyms are listed in alphabetical order, but true synonyms are listed before misapplied names.

One of the great difficulties with accounting for all Andrews' taxa is that he did not cite specimens and often only recorded very vague localities, e.g. 'Equatoria' or 'Northern and Central Sudan'. We know that Andrews consulted the collections of Sudanese plants at Kew and the Natural History Museum, London and so with a little detective work we have been able to trace most of the misapplied names with some certainty. That said, there are a number of records in Andrews' work which are clearly incorrect but for which we have been unable to work out how the name has been misapplied, for example *Meiocarpidium lepidotum* (Annonaceae) and *Indigofera oxalidea* and *I. alternans* (Leguminosae: Papilionoideae). The subject of name verification is discussed further under 'Specimen citations'.

Literature citations

As for the synonymy, the starting point for literature citations is taken as Andrews (1950–56); all literature predating that is excluded except where it provides useful information on poorly known taxa, additional to that recorded in Andrews and beyond.

The primary literature sources cited are listed in Table 4.1. For each taxon, our primary aims are (i) to record the major literature relevant to Sudan in which the name is applied as accepted, and (ii) to provide a reference to one or more modern account(s) of the taxon in which the reader can quickly access additional information, for example a full description, more detailed habitat information and more exhaustive literature citations. The two texts we have used as the standards for (ii) are the *Flora of Ethiopia & Eritrea* (F.E.E.) and the *Flora of Tropical East Africa* (F.T.E.A.), since these are both complete, modern Floras which contain detailed information on each taxon treated and which, between them, have a large overlap with the taxa in Sudan and South Sudan.

For those species not covered by either F.E.E. or F.T.E.A., we have cited one or more from a 'secondary' list of literature sources. These are listed in Table 4.2; this table also explains the hierarchy of these sources and in which circumstances they are cited. Some taxa, such as some range-restricted or alien species, are not covered by any of these literature sources, in which case we have cited, where available, scientific papers such as a taxonomic revision in which the taxon is treated.

We have additionally consulted a wide range of taxonomic papers and books during the compilation of the checklist which, for the sake of brevity, are not always listed. For some groups where a publication is particularly critical, the authors have chosen to cite these in addition to the standard texts. For example, in Loranthaceae, *Mistletoes of Africa* by Polhill & Wiens (1998) is an essential reference and so is cited for each species.

Table 4.1: the primary literature sources for the checklist, cited where relevant for all accepted names and synonymy.

Abbreviation	Full citation
Biol. Skr. 51	*Flora of the Sudan-Uganda border area east of the Nile* by I. Friis & K. Vollesen, published in 2 parts, 1998 and 2005, within the Biologiske Skrifter series.
F.E.E.	*Flora of Ethiopia and Eritrea*, published in 8 volumes, 1989–2009 (Vol. 3, 1989, as *Flora of Ethiopia*).
F.P.S.	*The Flowering Plants of the (Anglo-Egyptian) Sudan* by F.W. Andrews, published in 3 volumes, 1950, 1952 and 1956.
F.J.M.	*The Flora of Jebel Marra (Sudan Republic) and its geographical affinities* by G.E. Wickens, published 1976.
F.J.U.	*Flore et vegetation du Jebel Uweinat* by J. Léonard, published in 4 parts (with taxonomy in parts 1–3): Part 1: *Bull. Jard. Bot. Nat. Belg.* 66: 223–340 (1997) Part 2: *Bull. Jard. Bot. Nat. Belg.* 67: 123–216 (1999) Part 3: *Syst. Geogr. Pl.* 69: 215–264 (1999) Part 4: *Syst. Geogr. Pl.* 70: 3–73 (2000)
F.T.E.A.	*Flora of Tropical East Africa*, published in 263 fascicles, 1952–2012.
T.S.S.	*Trees & Shrubs of the Sudan* by H.M. El Amin, published posthumously in 1990

Table 4.2: additional notable literature sources cited in the checklist, with notes on when they are cited.

Abbreviation	Full citation
Bot. Exp. Sud.	*A botanic expedition to the Sudan in 1938* by M. Drar, edited and annotated by V. Täckholm, 1970. (Cited wherever it adds useful information on a species in Sudan and in all cases where the names used by Drar are now in synonymy. Note that we have only seen a few of Drar's Sudanese specimens).
F. Darfur Nord-Occ. & J. Gourgeil	*Flore et végétation des plateaux du Darfur Nord-occidental et du jebel Gourgeil* by P. Quézel, 1969. (Cited wherever it adds useful information on a species in Sudan including additional synonymy. Note that we have not seen the large majority of Quézel's Sudanese specimens).
F.Egypt	*Flora of Egypt* by L. Boulos, published in 4 volumes, 1999–2005. (Cited where species occur in Egypt but not in the F.E.E., F.T.E.A. or F.W.T.A. regions).
F.Som.	*Flora of Somalia* by M. Thulin (ed.), published in 4 volumes, 1993–2006. (Only rarely cited where a species is not treated in any of the other relevant Floras).
F.W.T.A.	*Flora of West Tropical Africa*, Second Edition published in 3 volumes in 5 fascicles, 1954–1972. (Cited where species occur in West Africa but not in the F.E.E. or F.T.E.A. regions).
F.Z.	*Flora Zambesiaca*, to date over 80% completed, published in 13 volumes, 1960– (Only rarely cited, in cases where a species has a disjunct distribution, occurring in Sudan/South Sudan and in southern tropical Africa but not in the F.E.E., F.T.E.A., F.Egypt or F.W.T.A. regions).
F.T.A.	*Flora of Tropical Africa*, published in 9 volumes, 1868–1913. (Only rarely cited, where the species are not covered by any modern Flora accounts and where no other modern descriptions are available).
Trop. Afr. Fl. Pl.	*Tropical African Flowering Plants. Ecology and Distribution* by J.-P. Lebrun & A. Stork, with 7 volumes published to date, 2003–2012. (Cited where the species are not covered by any modern Flora accounts).

Life-forms

Whilst the current work is not intended as a Flora and so does not include descriptions of the species treated, it is useful to have a record of the life-form of each species. Information on life-form, together with the in-country distribution and habitat information, may help with the identification of many of the species within Sudan and South Sudan, or at least narrow the options down to a small number, so long as the reader already has a good idea of the genus they are looking at.

We have not attempted to standardise the life-forms recorded, instead we provide summary information that we think will be useful. The main categories used are: *tree* (with a size qualification if useful), *shrub*, *climber* (then: woody or herbaceous) and *herb* (then: annual, biennial or perennial – if perennial, then the form of perennation is recorded if useful, e.g. rhizomatous, bulbous etc.). For some families, additional life-form information is given if deemed useful, for example noting of succulent species. For grasses, we have omitted 'herb' from the life-form since there seemed little point in repeating this for every species.

Habitat

A short habitat description is provided for each species. Since the majority of herbarium material available for Sudan and South Sudan is historical, the label data are often scant and habitat information is very limited. Therefore, we have often had to rely on Flora accounts from neighbouring countries to provide species habitat information.

The main purpose of providing habitat information is to assist with species identification (see note on life forms above); the habitat notes are brief and are not intended to be exhaustive. At present, we have not attempted to standardise the habitat information. One of the great difficulties of applying standardised habitats in the checklist region is that large areas of the two countries are covered by a complex mosaic of woodlands, bushlands and grasslands that could together fall under a broadly defined 'savanna' biome which is difficult to adequately subdivide based on herbarium data (see chapter 1).

Distribution

(i) Distribution in Sudan and South Sudan

In order to provide a broad overview of the distribution of species within Sudan and South Sudan, we have subdivided the two countries into nine regions, six in Sudan and three in South Sudan – see Table 4.3 and Figure 4.1. The nine regions largely correspond to the Provinces of the former Anglo-Egyptian Sudan, with two exceptions: (1) our Central Sudan comprises both Blue Nile and Khartoum Provs., since the latter was very small and so we considered it best to amalgamate the two; (2) Kassala Prov. is subdivided into our Eastern Sudan and Red Sea, since the flora of the Red Sea region (mainly comprising the Red Sea Hills and the coastal lowlands) is quite distinct from that of the eastern region. These nine regions are now further subdivided into the 27 States of the two countries: 17 in Sudan and 10 in South Sudan (see Table 4.3). We did not adopt these 27 States since it would have been too difficult and time-consuming to assign each specimen to such narrow geographical ranges in the absence of georeference data for the vast majority of specimens. Whilst the nine regions used were originally drawn on political grounds, they do have some phytogeographical value albeit at a crude scale. For example, Northern Sudan captures the majority of the Sahara (Libyan and Nubian) desert in Sudan, whilst Equatoria contains the large majority of the humid tropical vegetation in South Sudan, and the Red Sea region captures unique vegetation types as discussed above.

For each species under 'Distr' (Distribution), we first list the code(s) for each of the regions in Sudan and South Sudan from which it is known. The regions are listed in a geographic sequence starting in the northwest and ending in the south: NS, RS, DAR, KOR, CS, ES, BAG, UN, EQU (see Figure 4.1).

The within-country distribution for each species has been derived mainly from specimen data, principally from the collections at Kew and University of Khartoum. Literature records have also been used where they are deemed reliable. Of particular note here is the work of El Amin (1990) on the woody flora; El Amin travelled widely in Sudan and South Sudan during his studies and made many additional species records in under-collected regions, for example in Bahr el Ghazal where El Amin's work added 86 species records for that region. Quézel's (1969) species list from his expedition to NW Darfur was also an important source of information, adding 96 extra taxa for the Darfur region.

Localities have been checked against online and published gazetteers; we have found Fuzzy Gazetteer (http://dma.jrc.it/services/fuzzyg/) and GeoNames (http://www.geonames.org/) to be particularly useful.

The regional breakdown for each taxon should be considered a first iteration exercise, for a number of reasons. Firstly, regional coverage is still very patchy in the two countries and it is likely that many species are more widespread than currently known (see chapter 1). Secondly, we have not incorporated all the locality records from specimens at other herbaria, most notably the Natural History Museum, London, where the combined facts that tropical Africa is not subdivided into different regions in the collections and that the African angiosperms in many cases require re-curation since many taxa are still filed under synonyms, would have made it too time-consuming to check records for every species. Thirdly, we have not been able to place every specimen locality with certainty since many of the specimens are historical and the localities may have been subject to name changes.

Table 4.3: the nine regions of Sudan and South Sudan applied in the checklist and their relation to the historical and current subdivisions of the two countries.

Region Code	Name	Anglo-Egyptian Sudan Province(s)	Country	Current States
NS	Northern Sudan	Northern	Sudan	Northern (Ash Shamaliyah) River Nile (Nahr an Nil)
RS	Red Sea	Kassala (in part)	Sudan	Red Sea (Al Bahr al Amar)
DAR	Darfur	Darfur	Sudan	North Darfur (Shamal Darfur) West Darfur (Gharb Darfur) Central Darfur (Zalingei) East Darfur (Sharq Darfur) South Darfur (Janub Darfur)
KOR	Kordofan	Kordofan	Sudan	North Kurdufan (Shamal Kurdufan) South Kurdufan (Janub Kurdufan)
CS	Central Sudan	Khartoum, Blue Nile	Sudan	Khartoum (Al Khartum) Al Jazirah White Nile (An Nil al Abyad) Sennar Blue Nile (An Nil al Azraq)
ES	Eastern Sudan	Kassala (in part)	Sudan	Kassala Al Qadarif
BAG	Bahr el Ghazal	Equatoria (post-1948: Bahr el Ghazal)	South Sudan	Northern Bahr el Ghazal Western Bahr el Ghazal Warrap Lakes
UN	(Greater) Upper Nile	Upper Nile	South Sudan	Upper Nile Unity Jonglei
EQU	Equatoria	Equatoria	South Sudan	Western Equatoria Central Equatoria Eastern Equatoria

There are a number of complications regarding both historical and contemporary geographical locations and collecting localities in the Sudan region which are worthy of note here:

(a) Localities of Georg Schweinfurth (1865–71)

Andrews (1950–56) apparently considered all the plant specimens collected during Schweinfurth's journeys in the region in 1865–71 (discussed in more detail in chapter 2) to have been collected within the borders of Sudan / South Sudan. These journeys have since been thoroughly documented by Wickens (1971) and, whilst Andrews' assertion was correct for the large majority of collections, there are two exceptions:

(1) In May–October 1865, Schweinfurth made extensive collections along the current Sudan-Ethiopia border region between Gallabat in easternmost Sudan and Metamma just over the border in Ethiopia. Many of his specimen labels from this part of the journey read 'Flora von Gallabat. Umgegend von Matamma' (*Flora of Gallabat. Neighbourhood of Matamma*; see Wickens 1971: plate 2). These collections may well have been collected on the Ethiopian side of the border. Indeed, they are often treated as such in the *Flora of Ethiopia & Eritrea*, though there is considerable inconsistency between authors, and these specimens are nearly always housed in the Sudan species folders at Kew. Wickens (1971: 130) noted that 'this Gallabat-Metamma region is one of Schweinfurth's important collecting areas. An account of his travels therein is only very briefly documented and there is no accompanying sketch map;

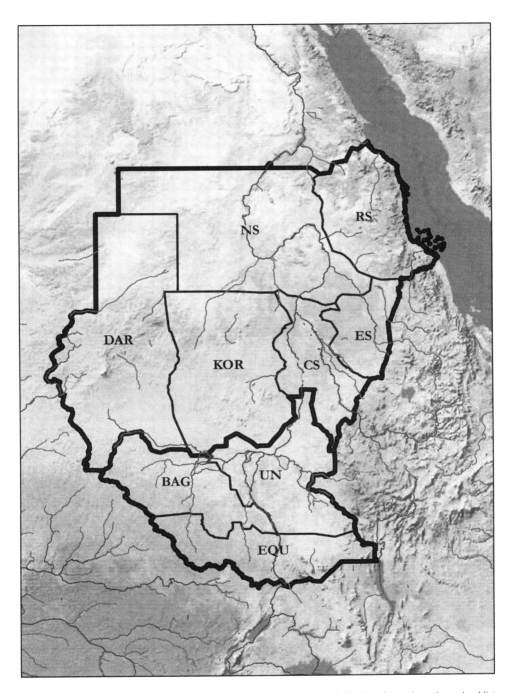

Figure 4.1: map of Sudan and South Sudan showing the subdivision into the nine checklist regions (see Table 4.3 for details). Base map retrieved from http://www.arcgis.com/home/item. html?id=c4ec722a1cd34cf0a23904aadf8923a0 November 2014.

the herbarium labels do not give any additional information. This is extremely unfortunate, for he spent over five months in that area'. The Gallabat area of Eastern Sudan has hardly been studied botanically since Schweinfurth's time and so documentation of the plants on the Sudan side of the border remains very incomplete – it is highly likely that most of the species recorded from Metamma by Schweinfurth will also occur around Gallabat. In view of this fact, together with the lack of precise locality information on the Schweinfurth collections, we have included these specimens within Eastern Sudan (ES) in the current checklist.

(2) Between 1st March and 29th April 1870, Schweinfurth crossed the Nile watershed and made collections in modern-day Democratic Republic of the Congo. This period of Schweinfurth's journey is much better documented than the Gallabat period and the specimen labels have more detailed locality information, such that Wickens (1971) was able to accurately reconstruct this part of his journey. These specimens, many of which are unfortunately still housed in the (South) Sudan species folders at Kew, are excluded from the current work since they clearly originate from beyond our geographical scope. Further, they were collected from the other side of the Nile watershed which is potentially an ecological barrier. That said, the large majority of species collected in that part of the journey have also been found in South Sudan, though there are a very few species for which Andrews' record appears to be entirely based on Schweinfurth's Congolese collections, e.g. *Bertiera aethiopica* Hiern (Rubiaceae). It is quite possible that these exceptions will also turn up in South Sudan following more extensive collecting in the extreme south. All such species are noted in bracketed text or within the 'Notes' sections in the checklist.

(b) Disputed territories

(1) The Ilemi Triangle, an area of 10,300–14,000 km² on the border of southeast South Sudan and northwest Kenya, is claimed by both countries. We have included it within the current checklist since the specimens are usually treated in herbaria as originating from Sudan, and the region was not included in the *Flora of Tropical East Africa*. It is of some phytogeographical significance since it contains some of the most extensive areas of Somalia-Masai *Acacia-Commiphora* bushland in the checklist region (see chapter 1). Several plant collectors have worked in the triangle, most notably H. Padwa who worked on the Ethiopian Boundary Commission surveys (see chapter 2). The contiguous areas to the north and east in South Sudan proper have been much less well studied but are likely to contain the same species assemblages.

(2) The Hala'ib (Halayeb) Triangle, an area of c. 20,500 km² on the Red Sea coast between northeast Sudan and southeast Egypt is claimed by both countries. At present, effective administration is in the hands of the Egyptians, and it is included (under the area GE – 'Gebel Elba and surroundings') in the *Flora of Egypt*. It contains some important historical collecting localities including Jebel Elba and Jebel Shellal. These mountains are contiguous with the chain of Red Sea Hills on the undisputed Sudan side of the border, where they are much less well studied (the only well studied mountains in the Sudanese Red Sea Hills being much further south), but they are likely to contain a similar assemblage of species. We therefore include the Hala'ib Triangle within the current checklist.

(3) The Abyei region, an area of c. 10,550 km² between Kordofan and Bahr el Ghazal, is a disputed territory with special administrative status; it is claimed by South Sudan but currently controlled by the Sudanese government and is an area of much conflict at present. Very few botanical collections have been made from this region; the few collections seen from Abyei have been included within Kordofan, since this region historically fell within the boundary of 'West Kordofan'.

(c) The delimitation of Bahr el Ghazal and Equatoria

Bahr el Ghazal Province was only formally separated from Equatoria Province in 1948 and so locality details for pre-1948 specimens labelled 'Equatoria' have to be checked carefully; whilst we have attempted to do so we cannot guarantee that we have always correctly assigned such specimens. An added complication is that, although Andrews' work was published post-1948, he did not separate out Bahr el Ghazal and so when he recorded a species' distribution as 'Equatoria' (which he did on many occasions), this could refer to either Equatoria proper or Bahr el Ghazal.

(ii) Global distribution

In order to provide a global context, we also record the full distribution of each taxon. This immediately follows the Sudanese distribution, separated by a semi-colon. For localised species, each country within its distribution is recorded, e.g. 'Uganda, Kenya, Tanzania'. For more widespread species a distribution range is recorded, first within tropical and southern Africa and working from west to east, then north to south. Hence, the distribution of a typical widespread tropical African species, for example, may be recorded as: 'Senegal to Somalia, S to Angola, Zimbabwe and Mozambique'. If the species also extends beyond tropical Africa then this part of the distribution is separated by a semi-colon, e.g. 'Senegal to Somalia; N Africa, Arabia'. For very widespread species, a more general distribution is used, e.g. 'palaeotropical' for those species occurring more or less throughout the Old World tropics, 'pantropical' for those species occurring more or less throughout the tropics and 'cosmopolitan' for species with a more or less global distribution. Where a species is not native to the checklist area, its place of origin is noted (if known), together with its current distribution.

Several abbreviations or non-standard country names are used for the sake of clarity or brevity:

C.A.R. – Central African Republic
Congo (Brazzaville) – Republic of Congo
D.R. Congo – Democratic Republic of the Congo

North, South, East and West are all abbreviated to N, S, E and W respectively unless they form part of a country's name e.g. South Africa.

Conservation status

Species of high conservation concern are highlighted within the checklist. For those species that have been formally assessed against the categories and criteria of IUCN (2001) and have been assigned to a category other than Least Concern (LC), the two-letter category code is recorded within the species treatment under the heading 'IUCN'. These codes are as follows:

DD – Data Deficient
NT – Near Threatened
VU – Vulnerable
EN – Endangered
CR – Critically Endangered

Further details, including a listing of the full IUCN assessment criteria, are given in chapter 3 for all the formally assessed species.

In addition, we have highlighted all the species occurring in Sudan and/or South Sudan which are potentially of high conservation priority. This is usually as a result of their restricted distribution and/or scarcity, with the Sudanese populations comprising an important portion of the species global range and/or population. These are recorded under the heading 'IUCN' as:

RD – 'Red Data' (i.e. of potential global conservation concern and likely to be included in a future 'Red List' for the plants of the two countries).

All species recorded as RD are listed and discussed in chapter 3.

We have not attempted to identify species that are of only national conservation concern, i.e. more widespread species globally that happen to be rare in the Sudan region.

Specimen citations

Wherever possible, each species or infraspecific taxon treatment is accompanied by a specimen citation so that the record can be easily verified in the future. The lack of specimen citations in, for example, Andrews (1950–56) and El Amin (1990) and the resultant difficulty in ascertaining the true identity of mis-applied names (see 'Synonymy' above) has clearly demonstrated just how important verifiable records are.

In cases where the species occurs in both Sudan and South Sudan, two specimens are cited, one for each country – in such cases, the Sudan specimen is listed first. The citations are kept to a minimum: we record only the region code (see Table 4.3), collector(s), collection number and date (month and year) of collection if known. In addition, some authors have also recorded phenological information from the specimens (fl: flowering; fr: fruiting). The Kew herbarium was by far the main source referenced for suitable specimens, though many of the specimens cited will also be duplicated in other herbaria both in Sudan and in Europe. The other main herbaria from which specimens are cited are the University of Khartoum (KHU), the Natural History Museum, London (BM), the University of Cairo (CAI) and the East African Herbarium (EA). The only other important herbarium in Sudan is that of the Forest Research and Education Institute, Khartoum (KHF), though we did not consult this collection widely for the current work. Further details on these herbaria can be found on Index Herbariorum (http://sciweb.nybg.org/science2/IndexHerbariorum.asp). The holding herbarium of each specimen is not recorded in the checklist but we have this information, along with specimen locality details, held in the Access Database from which the checklist is generated.

In cases where we have not seen the cited specimen(s), we record this as '(n.v.)' and if there is any doubt over the identity of these specimens, we record this in the 'Note'. This is typically in cases where there are no specimens held at Kew or in Khartoum, and the specimen details are instead derived from the literature, e.g. from taxonomic treatments or from Flora volumes. It should be noted that we have not seen any of the specimens of J. Kielland-Lund or M. Shigeta. These records are derived from Friis & Vollesen (1998, 2005) who, in most cases, verified the identifications and in view of these two authors' extensive knowledge of the eastern African flora, these identifications are deemed likely to be correct.

In some cases, we have been unable to locate a specimen with which to verify the record. The alleged presence of the species in Sudan / South Sudan in these cases is derived either from the literature or from the authors' own knowledge of the countries' floras. Where the record is literature-derived, we always quote the source of the record in the 'Note' section for that species, and we comment on the degree of uncertainty of the record – we also use a '?' before the regional distribution to highlight this uncertainty.

Whilst the checklist aims to be exhaustive in its coverage, it is very likely that some species will have been missed. There will also be future taxonomic changes that will impact upon the list, e.g. lumping or splitting of species, changes to family circumscriptions etc. This checklist is being maintained as an Access Database at Kew and in Khartoum and Juba, and it is our intention that this will (a) be regularly updated and (b) be made available online for researchers interested in the flora of the Sudan region to search and query. The authors would therefore welcome any additions or corrections; these can be sent to Iain Darbyshire at **i.darbyshire@kew.org**.

5. The Checklist: statistical summary

In this chapter we present the summary statistics for the plant diversity and distribution in Sudan and South Sudan and compare these figures to neighbouring countries and Flora regions.

In Table 5.1 the total numbers of angiosperm and gymnosperm families, genera, species and taxa (species + infraspecific taxa) for Sudan and South Sudan are presented. In the region as a whole, **3969 species** are recorded, of which **2117** occur in Sudan and **2961** occur in South Sudan. Andrews (1950–56) documented a total of 3137 species in his *Flowering Plants of the (Anglo-Egyptian) Sudan* (total figure *fide* Frodin 2001) and so our findings increase the known species number by 832, or 27%. It is also a considerably increase (13%) on the later estimate of 3500 species provided by Wickens (*fide* Frodin 2001).

Table 5.1: summary statistics for the angiosperm and gymnosperm richness in Sudan and South Sudan. 'Taxa' refers to species + infraspecific taxa.

Taxonomic Rank	Sudan + South Sudan	Sudan	South Sudan	% overlap
Families	183	135	170	67%
Genera	1351	843	1084	43%
Species	3969	2117	2961	28%
Taxa	4096	2175	3041	27%
Introduced species		140	101	
Introduced taxa		141	103	
Native species		1977	2860	
Native taxa		2034	2938	

The overlap between the two countries at the species level (28%) and even the generic level (43%) is remarkably low. This is largely a reflection of the marked latitudinal zonation of vegetation and phytochoria in the Sudan region (see chapter 1). Hence the vegetation types and species make-up in the northern part of Sudan, dominated by the Nubian and Libyan deserts in the interior and the dry coastal lowlands along the Red Sea, are very different to those of the southern part of South Sudan where many humid tropical species reach their northernmost limits in the forests and woodlands of that region. The main areas of overlap in species composition between the two countries are: (i) in the extensive dry Sudanian woodlands and grasslands that dominate in the border region between the two countries; (ii) along the Nile corridor which acts as a north-south intrusion into the largely latitudinal vegetation zonation and shares many species throughout its Sudan/South Sudan reach; (iii) the isolated montane massifs of the Imatong Mts in South Sudan and Jebel Marra and the Red Sea Hills in Sudan which, whilst very different from one another in some respects, have sufficient altitudinal ranges and associated climatic and vegetation zonation that there is considerable overlap in species make-up between the three; and (iv) the significant number of shared widespread (tropical African to pantropical) weedy species that are adaptable to regional changes in climate and other environmental factors.

The species richness in Sudan and South Sudan is compared to neighbouring countries and Flora regions in Table 5.2. Taking into account its vast area, the flora of Sudan is rather depauperate. It has almost an identical species number to that of Egypt yet is over 85% larger in size (though we acknowledge that there is not a linear relationship between area and species number). Whilst these two countries share a largely arid (Saharan) flora together with the higher diversity associated with the Nile Valley, Egypt's species richness is boosted by the relatively diverse flora along its Mediterranean coastline, which is very different to the Afrotropical elements in the extreme south. Chad also shares the arid (Saharan and Sahelian) flora with Sudan but it has a significantly larger humid zone in the south, largely accounting for its higher total species richness. However, it is also possible that the very limited botanical exploration in large areas of Sudan has resulted in an under-representation of its true floral diversity.

Table 5.2: a comparison of the total angiosperm and gymnosperm richness in Sudan, South Sudan and neighbouring Floras. Data for Egypt: Boulos (1999–2005, summary figures provided by H. Beentje, pers. comm.); Chad: Brundu & Camarda (2013); *Flora of Tropical East Africa* - F.T.E.A.: H. Beentje (pers. comm.); *Flora of Ethiopia and Eritrea* – F.E.E.: Sebsebe Demissew and Ensermu Kelbessa (pers. comm).

Country	Area (km²)	Total species (excluding Pteridophytes)
Sudan	1,886,068	2117
South Sudan	619,745	2961
Egypt	1,001,450	2058
Chad	1,284,000	2256
F.E.E.	1,244,727	5833
F.T.E.A.	1,763,893	11668

The species richness in South Sudan is understandably higher than that of Sudan, Chad and Egypt, since it has a significant humid tropical element to its flora, including species-rich lowland and riverine rainforest.

A regional breakdown of the angiosperm and gymnosperm richness in Sudan and South Sudan is presented in Table 5.3. Easily the highest species richness is attained in the Equatoria (EQU) region of South Sudan. Equatoria contains significant areas of species-rich tropical moist forest, both at lower altitudes (often riverine) and also at higher altitudes, notably in the Imatong and Didinga Mts. At the same time, it also contains a wide range of other habitat types, including Sudanian woodlands in the north, dry *Acacia-Commiphora* bushland and woodland in the extreme southeast, areas of montane grassland in the mountains and wetlands associated with the Nile. The Darfur (DAR) region of Sudan is the only other region to contain over 1000 taxa. However, the large majority of Darfur's plant diversity is restricted to the Jebel Marra and Jebel Gourgeil massifs, areas that are both botanical hotspots and also particularly well botanised (see Wickens 1976). Beyond these Jebels, Darfur has both a rather depauperate and an under-studied flora, such that the region as a whole has a rather low species richness relative to its size. At the other end of the scale, Northern Sudan

(NS) is clearly the least diverse region, reflecting the fact that it is dominated by the Libyan and Nubian deserts; only the Nile Valley and Jebel Uweinat in the extreme northwest have any significant diversity within this region. Elsewhere in the two countries, under-collection is likely to have impacted upon apparent species richness for some regions. We consider Eastern Sudan (ES), Bahr el Ghazal (BAG) and Upper Nile (UN) to be particularly under-represented in terms of their likely species richness since these three regions are poorly botanised in general. As noted in chapter 4, the regional species distributions recorded in the current checklist should be considered a first iteration and it is likely that further field surveys, together with study of herbarium collections not consulted for the current work and more exhaustive literature reviews, will increase the accuracy of these data and fill in some of the gaps in our knowledge.

Table 5.3 plant taxon (species + infrapecific taxa) species richness per region of Sudan and South Sudan; for explanation of the region codes, see chapter 4 and figure 4.1.

Region	Approx. area (km²)	Taxa
NS	492,428	364
RS	216,546	861
DAR	510,214	1257
KOR	380,556	855
CS	162,840	809
ES	109,673	478
BAG	200,149	966
UN	235,929	788
EQU	195,698	2678

Since the vast majority of specimen data for the two countries are not yet geo-referenced, we cannot map species richness and distribution patterns on a finer scale (e.g. using a ¼ degree square grid) as has been done in, for example, Gabon (Sosef *et al.* 2006) or Cameroon (Onana 2011). This would be a particularly useful next step in looking more closely at plant diversity patterns within Sudan and South Sudan.

In Tables 5.4 and 5.5 we present the top ten most species-rich plant families in Sudan and South Sudan respectively. The degree of overlap is considerable, with seven families common to both lists, and with the top three families being the same, albeit in a slightly different order. The Leguminosae comprise nearly 13% of the total flora in both countries which is typical of its importance in tropical Africa as a whole (in the F.T.E.A. region, for example, legumes comprise c. 10% of the total flora). At nearly 16%, the proportion of the flora contributed by the Poaceae is notably high in Sudan, and considerably higher than in South Sudan which reflects the importance of grasses in the extensive sparsely wooded regions of Sudan.

A comparison of the top 10 families in Sudan and South Sudan to those in neighbouring regions is complicated by the fact that all African Floras use a pre-APG III plant family classification, and there have been some marked changes in the delimitation of some tropical plant families following the adoption of APG III. Some families have been lumped together (e.g. the Bombacaceae, Sterculiaceae and Tiliaceae into a greater Malvaceae) whilst others have been split up (e.g. the old Euphorbiaceae and Scrophulariaceae *sensu lato*). For the aid of comparison, in Table 5.6 we have therefore converted the Sudan and South Sudan data to the pre-APG family circumscriptions used in the *Flora of Egypt,* the *Flora of Ethiopia and Eritrea* and the *Flora of Tropical East Africa* (F.T.E.A.). When using this modified family circumscription, the top ten families in South Sudan and F.T.E.A. are the same although in a slightly different sequence, whilst only seven of the ten are the same for Sudan. That said, the overlap between Sudan and these tropical regions is higher than the overlap with Egypt to the north, where the more temperate elements of the flora are well represented by the Brassicaceae, Caryophyllaceae and Chenopodiaceae (= Amaranthaceae in APG III) all of which feature in Egypt's top ten families.

Table 5.4: the ten most species-rich plant families in Sudan.

Rank	Family	Total species Sudan	Total taxa Sudan	% of Sudan flora (species)
1	Poaceae (Gramineae)	332	339	15.7%
2	Leguminosae (Fabaceae)	257	277	12.1%
3	Asteraceae (Compositae)	153	152	7.2%
4	Malvaceae	94	98	4.4%
5	Cyperaceae	84	85	4.0%
6	Convolvulaceae	60	60	2.8%
7	Euphorbiaceae	58	58	2.7%
8	Lamiaceae	57	58	2.7%
9	Acanthaceae	51	51	2.4%
10	Rubiaceae	47	50	2.2%
	TOTAL	**1193**	**1228**	**56.2%**

Table 5.5: the ten most species-rich plant families in South Sudan.

Rank	Family	Total species South Sudan	Total taxa South Sudan	% of South Sudan flora
1	Leguminosae (Fabaceae)	370	394	12.5%
2	Poaceae (Gramineae)	319	321	10.8%
3	Asteraceae (Compositae)	178	183	6.0%
4	Cyperaceae	157	164	5.3%
5	Rubiaceae	153	161	5.2%
6	Acanthaceae	119	120	4.0%
7	Malvaceae	102	103	3.4%
8	Orchidaceae	78	78	2.6%
9	Apocynaceae	77	78	2.6%
10	Lamiaceae	76	79	2.6%
	TOTAL	**1629**	**1681**	**55.0 %**

Table 5.6: a comparison of the ten most species-rich families for selected Floras of northeast and east Africa. Note that the Sudanese families have been converted to the pre-APG III circumscriptions to allow direct comparison with the *Flora of Egypt* (Boulos 1999–2005), the *Flora of Tropical East Africa* (figures courtesy of H. Beentje) and *the Flora of Ethiopia and Eritrea* (figures courtesy of Sebsebe Demissew and Ensermu Kelbessa) The following changes to the family circumscriptions apply:

Acanthaceae = Acanthaceae sensu APG III excl. *Avicennia*.
Amaranthaceae = split into Amaranthaceae s.s. and Chenopodiaceae.
Apocynaceae = Apocynaceae s.s. + Asclepiadaceae (the two were separated in F.E.E. and F.Egypt).
Euphorbiaceae s.l. = [Euphorbiaceae + Peraceae + Phyllanthaceae + Putranjivaceae].
Lamiaceae: for F.T.E.A., we have added *Clerodendrum*, *Karomia*, *Premna* and *Vitex* (total 89 spp.) which were treated
    within Verbenaceae in F.T.E.A.
Malvaceae s.s. = Malvaceae: Malvoideae.
Scrophulariaceae s.l. = [Linderniaceae + Orobanchaceae sensu APG III (excl. *Orobanche* & *Cistanche*) + Plantaginaceae
    sensu APG III (excl. *Callitriche* & *Plantago*) + Scrophulariaceae sensu APG III (excl. *Buddleja*) + Stilbaceae sensu APG
    III (excl. *Nuxia*)].

Rank	Egypt	Sudan	South Sudan	F.T.E.A.	F.E.E.
1	Poaceae 284	Poaceae 332 (incl. 44 cult.)	Leguminosae 370	Leguminosae 1217	Leguminosae 678
2	Asteraceae 228	Leguminosae 257	Poaceae 319	Poaceae 865	Poaceae 609
3	Leguminosae 228	Asteraceae 153	Asteraceae 178	Asteraceae 813	Asteraceae 472
4	Brassicaceae 102	Cyperaceae 84	Cyperaceae 157	Rubiaceae 725	Euphorbiaceae s.l. 242
5	Caryophyllaceae 87	Euphorbiaceae s.l. 70	Rubiaceae 153	Orchidaceae 676	Cyperaceae 231
6	Chenopodiaceae 76	Convolvulaceae 60	Acanthaceae 119	Acanthaceae 597	Acanthaceae 215
7	Boraginaceae 58	Lamiaceae 57	Euphorbiaceae s.l. 111	Euphorbiaceae s.l. 524	Lamiaceae 202
8	Scrophulariaceae s.l. 58	Malvaceae s.s. 57	Orchidaceae 78	Cyperaceae 483	Apocynaceae 179
9	Euphorbiaceae s.l. 55	Acanthaceae 50	Apocynaceae 77	Apocynaceae 406	Orchidaceae 170
10	Lamiaceae 55	Rubiaceae 47	Lamiaceae 76	Lamiaceae 405	Malvaceae s.s. 141

In Table 5.7 we present the top ten most species-rich genera in Sudan and South Sudan. In both countries, all the genera featured are large genera distributed throughout tropical Africa. With this in mind, it is understandable that the overlap between the two countries is considerable, with six genera featuring in both lists. All but one of the genera are from the most species-rich plant families in each of the countries; the exception is *Ficus* (Moraceae) which is diverse in the forests and woodlands of South Sudan. The Leguminosae are particularly well represented, with *Acacia*, *Crotalaria* and *Indigofera* appearing in the top seven in both lists. The Poaceae are well represented in the Sudan list, with *Brachiaria*, *Eragrostis* and *Hyparrhenia* all featuring, which reflects the dominance of this plant family in Sudan. However, only one grass genus, *Panicum*, features in the South Sudan list despite Poaceae being the second largest family in that country, and the Asteraceae, the third largest family in both countries, is poorly represented in the generic lists with only *Vernonia* appearing and then only in the South Sudan list. *Cyperus* is the largest genus in both countries and easily so in South Sudan, a reflection of the fact that this genus is diverse in the permanent marshes associated with the Nile and its tributaries and also in the seasonally wet grasslands and woodlands that are widespread in the region.

Table 5.7: the ten most species-rich plant genera in Sudan, South Sudan, the *Flora of Tropical East Africa* region and the *Flora of Ethiopia & Eritrea* (F.T.E.A. figures courtesy of H. Beentje; F.E.E. figures courtesy of Sebsebe Demissew and Ensermu Kelbessa – infraspecific taxon numbers for F.T.E.A. have not been collated whilst the F.E.E. figures include species and subspecies). Note that in Ethiopia and Eritrea, the genus Eucalyptus is represented by 55 species and so could be considered the eighth most species-rich genus there but since none of these species are native, it is omitted here.

Rank	Sudan			South Sudan			F.T.E.A.		F.E.E.	
	Genus	Spp.	Taxa	Genus	Spp.	Taxa	Genus	Spp.	Genus	Taxa
1	*Cyperus*	39	40	*Cyperus*	64	67	*Crotalaria*	199	*Cyperus*	115
2	*Indigofera*	36	39	*Crotalaria*	45	47	*Euphorbia*	160	*Euphorbia*	109
3	*Euphorbia*	30	30	*Indigofera*	44	47	*Cyperus*	146	*Crotalaria*	88
4	*Crotalaria*	26	27	*Ipomoea*	33	33	*Indigofera*	145	*Indigofera*	83
5	*Ipomoea*	26	26	*Vernonia*	31	31	*Vernonia*	134	*Ipomoea*	61
6	*Hyparrhenia*	25	26	*Acacia*	30	35	*Habenaria*	113	*Acacia*	59
7	*Hibiscus*	20	20	*Panicum*	29	29	*Justicia*	105	*Commiphora*	56
8	*Acacia*	19	27	*Eulophia*	23	23	*Plectranthus*	105	*Hibiscus*	50
9	*Eragrostis*	19	19	*Ficus*	23	23	*Cyphostemma*	93	*Aloe*	48
10	*Brachiaria*	18	18	*Hibiscus*	23	23	*Ipomoea*	89	*Vernonia*	48

Also included in Table 5.7 are the equivalent figures for the F.T.E.A. and F.E.E. regions for comparison. The overlap is again quite considerable – six out of the ten most species-rich genera in F.T.E.A. and eight of the ten in F.E.E. also feature in either the Sudan and/or the South Sudan lists. Of those that do not feature in the top ten in F.T.E.A., only the three grass genera *Brachiaria*, *Eragrostis* and *Hyparrhenia* and the orchid genus *Eulophia* do not have 50 or more species in the F.T.E.A. region and so fall outside the top 30 most species-rich genera there.

6. The Checklist of Gymnosperms and Angiosperms in Sudan and South Sudan

CYCADALES

Zamiaceae

R. Candiga & I. Darbyshire

Encephalartos mackenziei L.E.Newton

Bot. J. Linn. Soc. 140: 187 (2002).
Multi-stemmed cycad. Rocky terrain, roadsides.
Distr: EQU; South Sudan endemic.
IUCN: NT
EQU: Mackenzie s.n. (fl) 4/1995.
Note: a very local species but not considered immediately threatened since it is tolerant of some disturbance.

Encephalartos septentrionalis Schweinf. ex Eichler

F.P.S. 1: 1 (1950); F.T.E.A. Gymnosp.: 6 (1958); T.S.S.: 1 (1990); Biol. Skr. 51: 61 (1998).
Single- or few-stemmed cycad. Rocky hillslopes, amongst grassland & bushland.
Distr: BAG, EQU; C.A.R., D.R. Congo, Uganda.
IUCN: NT
EQU: Anthony A.700 (fl) 11/1952.

EPHEDRALES

Ephedraceae

R. Candiga & I. Darbyshire

Ephedra foliata Boiss. ex C.A.Mey.

F.E.E. 1: 185 (2009).
Syn: *Ephedra aphylla* sensu auctt., non Forssk. – F.P.S. 1: 2 (1950); T.S.S.: 1 (1990).
Climbing or scrambling shrub. *Acacia* woodland & bushland.
Distr: RS; Ethiopia, Djibouti & Somalia; N Africa, Arabia to India.
RS: Farag & Zubair 50 (fl) 2/2012.

PINALES

Podocarpaceae

R. Candiga & I. Darbyshire

Podocarpus milanjianus Rendle

F.P.S. 1: 1 (1950); F.T.E.A. Gymnosp.: 11 (1958); T.S.S.: 3 (1990).

Syn: *Podocarpus gracilior* sensu auctt., non Pilg. – F.P.S. 1: 1 (1950); T.S.S.: 3 (1990).
Syn: *Podocarpus latifolius* sensu Friis & Vollesen, non (Thunb.) R.Br. ex Mirb. – Biol. Skr. 51: 61 (1998).
Tree, often large. Montane forest, often a dominant tree.
Distr: EQU; Nigeria to Kenya, S to Angola & Zimbabwe.
EQU: Howard I.M.79 (fl) 3/1976.

Cupressaceae

R. Candiga & I. Darbyshire

Cupressus lusitanica Mill.

F.E.E. 1: 195 (2009).
Tree. Planted for timber, sometimes naturalised.
Distr: EQU; native of Central America, widely cultivated in the tropics & subtopics.
Note: plantations of this species have been established in the Imatong Mts and it sometimes becomes locally naturalised.

Juniperus procera Hochst. ex Endl.

F.P.S. 1: 2 (1950); F.T.E.A. Gymnosp.: 16 (1958); T.S.S.: 1 (1990); Biol. Skr. 51: 62 (1998); F.E.E. 1: 193 (2009).
Tree. Dry montane forest, often dominant.
Distr: RS, EQU; Eritrea to Somalia, S to Zimbabwe; Egypt & Arabia.
RS: Kennedy-Cooke 259 (fl) 3/1938; **EQU:** Jackson 1303 (fl) 3/1950.

Pinaceae

R. Candiga & I. Darbyshire

[Pinus]

Biol. Skr. 51: 67 (1998).
Note: of the commercially planted *Pinus* species in the Imatong Mts, the most commonly grown are *P. patula* Schiede ex Schltdl. & Cham. and *P. radiata* D.Don. Whether or not they have naturalised is unconfirmed.

NYMPHAEALES

Nymphaeaceae

I. Farag & I. Darbyshire

Nymphaea lotus L.

F.P.S. 1: 14 (1950); F.J.M.: 84 (1976); F.T.E.A. Nymph.: 3 (1989); Biol. Skr. 51: 70 (1998); F.E.E. 2(1): 38 (2000).
Aquatic perennial herb. Pools, swamps, slow-flowing rivers.

Distr: DAR, ?CS, BAG, UN, EQU; palaeotropical, introduced to Europe & the Americas.
DAR: <u>Wickens 2946</u> (fl) 4/1965; **UN:** <u>J.M. Lock 81/4</u> (fl) 2/1981.
Note: Andrews recorded this species from the Nile and tributaries S of Khartoum, so it is likely to occur in Central Sudan, though no specimens have been seen.

Nymphaea maculata Schumach. & Thonn.

F.J.M.: 84 (1976); F.T.E.A. Nymph.: 5 (1989).
Aquatic perennial herb. Pools, slow-flowing rivers.
Distr: DAR, UN, EQU; Senegal to C.A.R., S to Zambia.
DAR: <u>Wickens 1361</u> (fl) 3/1964; **UN:** <u>Simpson 7469</u> (fl) 2/1930.
Note: the name *N. maculata* has been applied in Sudan to plants with white flowers, often smaller than in *N. nouchali*. However, other differences are negligable and it is likely that this is merely another variety of *nouchali*.

Nymphaea micrantha Guill. & Perr.

F.P.S. 1: 14 (1950); Bot. Exp. Sud.: 89 (1970); F.T.E.A. Nymph.: 2 (1989).
Aquatic perennial herb. Pools, slow-flowing rivers.
Distr: ?CS, ?UN, ?EQU; Senegal to Nigeria.
EQU: <u>Drar 1795b</u> 4/1938 (n.v.).
Note: the occurrence of this species in Sudan and South Sudan is unconfirmed. Andrews listed it from "the Nile Rivers and tributaries south of Khartoum" and in his description he refers to the cluster of bulbils at the top of the petioles, which is diagnostic for *N. micrantha*. However, we have seen no material from either country with leaf bulbils. The identification of the cited specimen follows Drar (1970).

Nymphaea nouchali Burm.f. var. *caerulea* (Savigny) Verdc.

F.T.E.A. Nymph.: 7 (1989); F.E.E. 2(1): 40 (2000).
Syn: *Nymphaea caerulea* Savigny – F.P.S. 1: 14 (1950); Bot. Exp. Sud.: 88 (1970); F.J.M.: 84 (1976).
Aquatic perennial herb. Pools, swamps, slow-flowing rivers.
Distr: DAR, KOR, BAG, UN; Senegal to Somalia, S to Angola & South Africa; Madagascar, Egypt.
DAR: <u>Wickens 2959</u> (fl) 5/1965; **BAG:** <u>Schweinfurth 2329</u> (fl) 9/1869.
Note: some of the herbarium material from South Sudan has been named *N. rufescens* but it appears inseparable from *N. nouchali* and so is included here.

PIPERALES

Piperaceae

H. Pickering & I. Darbyshire

Peperomia abyssinica Miq.

F.T.E.A. Piper.: 15 (1996); Biol. Skr. 51: 73 (1998); F.E.E. 2(1): 64 (2000).
Syn: *Peperomia abyssinica* Miq. var. *stuhlmannii* (DC.) Düll – F.T.E.A. Piper.: 15 (1996).
Epiphytic perennial herb. Montane forest.
Distr: EQU; Eritrea to Malawi, ?Mozambique.
EQU: <u>Thomas 1581</u> 12/1935.

Peperomia bangroana C.DC.

F.P.S. 1: 23, as '*bagroana*' (1950); Candollea 61: 356 (2006).
Syn: *Peperomia rotundifolia* sensu auctt., non (L.) Kunth – F.T.E.A. Piper.: 12 (1996); F.E.E. 2(1): 63 (2000).
Mat-forming ?annual or perennial herb. Forest, plantations.
Distr: ?EQU; Sierra Leone to Ethiopia, S to South Africa; Madagascar, Comoros.
Note: no material has been seen from South Sudan; the record is based on Andrews who noted it from near Iwatoka, Equatoria.

Peperomia fernandopoiana C.DC. var. *fernandopoiana*

F.T.E.A. Piper.: 19 (1996); Biol. Skr. 51: 73 (1998); F.E.E. 2(1): 63 (2000).
Syn: *Peperomia holstii* sensu Andrews, non C.DC. – F.P.S. 1: 23 (1950).
Epiphytic perennial herb. Forest.
Distr: EQU; Sierra Leone to Ethiopia, S to Tanzania.
EQU: <u>Andrews 1872</u> 6/1939.

Peperomia molleri C.DC. subsp. *molleri*

F.T.E.A. Piper.: 16 (1996); Biol. Skr. 51: 73 (1998); F.E.E. 2(1): 63 (2000).
Perennial herb, often terrestrial. Forest, often amongst rocks.
Distr: EQU; Liberia to Ethiopia, S to Angola, D.R. Congo & Tanzania; Madagascar.
EQU: <u>Friis & Vollesen 600</u> 12/1980.

Peperomia pellucida (L.) Kunth

F.T.E.A. Piper.: 11 (1996); F.E.E. 2(1): 61 (2000).
Syn: *Peperomia knoblecheriana* Schott – F.P.S. 1: 23 (1950).
Terrestrial annual herb. Wooded grassland, rocks in bushland, riverbanks.
Distr: ?EQU; pantropical.
Note: no material has been seen from South Sudan; the record is based on Andrews who recorded it (under *P. knoblecheriana*) from "banks of the Bahr el Jebel (5–7°N.)".

Peperomia tetraphylla (G.Forst.) Hook. & Arn.

F.T.E.A. Piper.: 12 (1996); Biol. Skr. 51: 74 (1998); F.E.E. 2(1): 60 (2000).
Syn: *Peperomia reflexa* (L.f.) A.Dietr. – F.P.S. 1: 23 (1950).
Epiphytic perennial herb. Montane forest.
Distr: EQU; pantropical.
EQU: <u>Friis & Vollesen 438</u> 11/1980.

Piper capense L.f. var. *capense*

F.P.S. 1: 24 (1950); T.S.S.: 20 (1990); F.T.E.A. Piper.: 5 (1996); Biol. Skr. 51: 74 (1998); F.E.E. 2(1): 59 (2000).
Subshrub. Forest.
Distr: BAG, EQU; Sierra Leone to Ethiopia, S to South Africa; Madagascar, Comoros.
EQU: <u>Myers 11746</u> 8/1939.

Piper guineense Schumach. & Thonn.

F.P.S. 1: 26 (1950); T.S.S.: 20 (1990); F.T.E.A. Piper.: 4 (1996); Biol. Skr. 51: 74 (1998); F.E.E. 2(1): 60 (2000).

Hemi-epiphytic climber. Forest.
Distr: EQU; Guinea Bissau to Ethiopia, S to Angola &
Zambia.
EQU: Friis & Vollesen 446 11/1980.

Piper umbellatum L.

F.P.S. 1: 23 (1950); T.S.S.: 21 (1990); F.T.E.A. Piper.: 8
(1996); Biol. Skr. 51: 74 (1998); F.E.E. 2(1): 60 (2000).
Shrubby perennial herb. Forest.
Distr: BAG, EQU; pantropical.
EQU: Andrews 1748 6/1939.

Hydnoraceae

I. Darbyshire

Hydnora abyssinica A.Br.

F.P.S. 1: 22 (1950); F.T.E.A. Hydnor.: 1 (2002).
Syn: *Hydnora johannis* Becc. – F.E.E. 2(1): 56 (2000).
Subterranean root parasite. Woodland & wooded
grassland, often parasitic on roots of *Acacia*.
Distr: DAR, CS; Eritrea & Somalia, S to Angola, Namibia
& South Africa.
CS: Musselman 6129 (fl fr) 8/1982.

Aristolochiaceae

I. Darbyshire

Aristolochia albida Duch.

F.T.E.A. Aristoloch.: 8 (1986).
Syn: *Aristolochia bongoensis* Engl. – F.P.S. 1: 21
(1950).
Perennial climbing herb, sometimes subshrubby. Forest
margins & clearings, riverine forest.
Distr: BAG, EQU; Senegal to Kenya, S to Angola &
Zimbabwe; Comoros, Mascarenes.
EQU: Andrews 914 (fl) 4/1939.

Aristolochia bracteolata Lam.

F.P.S. 1: 21 (1950); F.T.E.A. Aristoloch.: 6 (1986); Biol. Skr.
51: 73 (1998); F.E.E. 2(1): 54 (2000).
Perennial herb, prostrate or more rarely climbing. Dry
bushland & grassland, dry sandy riverbeds, waste
ground, weed of cultivation.
Distr: RS, DAR, KOR, CS, BAG, UN, EQU; Mali to
Somalia, S to Tanzania; Egypt, Arabia, Pakistan, India &
Sri Lanka.
CS: de Wilde et al. 5680 (fl fr) 2/1965; **UN:** Evans-
Pritchard 37 (fl fr) 11/1936.

Pararistolochia triactina (Hook.f.) Hutch. &
Dalziel

F.P.S. 1: 21 (1950); Adansonia, sér. 2, 17: 479 (1978);
F.T.E.A. Aristoloch.: 2 (1986).
Large woody climber. Riverine forest.
Distr: EQU; Benin to Chad & Uganda, S to Angola & D.R.
Congo.
EQU: Sillitoe 354 (fl fr) 1919.
Note: the identification of the cited specimen is based on
Poncy in Adansonia (l.c.).

MAGNOLIALES

Myristicaceae

I. Darbyshire

Coelocaryon preussii Warb.

Fl. Cameroun 18: 98 (1974).
Large tree. Forest.
Distr: EQU; Nigeria to D.R. Congo; ?introduced in South
Sudan.
EQU: Myers 13409 8/1940.
Note: the single South Sudanese record is "from tree
planted May 1932" at Kagelu Station according to
Myers. The origin of the tree is not stated.

Pycnanthus angolensis (Welw.) Warb. subsp.
schweinfurthii (Warb.) Verdc.

F.P.S. 1: 8 (species only) (1950); T.S.S.: 17 (species only)
(1990); F.T.E.A. Myristic.: 4 (1997); Biol. Skr. 51: 68
(1998).
Tree, often large. Forest, often riverine.
Distr: EQU; D.R. Congo, Uganda, Burundi, Tanzania; full
W African range currently unclear.
EQU: Turner 70 (fr) 5/1931.

Annonaceae

H. Pickering & M. Kordofani

Annona senegalensis Pers. subsp. *senegalensis*

F.P.S. 1: 3 (1950); F.T.E.A. Annon.: 113 (1971); T.S.S.: 5
(1990); Biol. Skr. 51: 63 (1998); F.E.E. 2(1): 11 (2000).
Small tree or shrub. Wooded grassland.
Distr: DAR, KOR, BAG, UN, EQU; Senegal to Ethiopia, S
to South Africa and Angola; Madagascar, Comoros.
KOR: Drar 1887 4/1938 (n.v.); **EQU:** Friis & Vollesen 991
2/1982.

Annona squamosa L.

F.T.E.A. Annon.: 113 (1971); T.S.S.: 5 (1990); F.E.E. 2(1):
10 (2000).
Shrub. Abandoned cultivations.
Distr: DAR; native to the neotropics, now widely
cultivated and occasionally naturalised.
DAR: Kamil 1068 5/1968.
Note: cultivated for its edible fruits, the 'sugar-apple' or
'custard apple'.

Artabotrys monteiroae Oliv.

F.T.E.A. Annon.: 62 (1971); Biol. Skr. 51: 63 (1998); F.E.E.
2(1): 6 (2000).
Woody climber. Riverine forest.
Distr: EQU; Ethiopia to Angola & South Africa;
Madagascar.
EQU: Friis & Vollesen 475 11/1980.

Cleistopholis patens (Benth.) Engl. & Diels

F.T.E.A. Annon.: 31 (1971); T.S.S.: 5 (1990); Biol. Skr. 51:
66 (1998).
Tree. Lowland forest.
Distr: EQU; Sierra Leone to Uganda, S to Angola.
EQU: Jackson 738 5/1949.

Hexalobus crispiflorus A.Rich.
F.P.S. 1: 5 (1950); F.W.T.A. 1(1): 47 (1954); T.S.S.: 7 (1990).
Tree. Forest.
Distr: EQU; Sierra Leone to D.R. Congo, S to Angola.
EQU: <u>Myers 8706</u> 3/1938.

Hexalobus monopetalus (A.Rich.) Engl. & Diels
F.P.S. 1: 5 (1950); F.T.E.A. Annon.: 46 (1971); T.S.S.: 7 (1990); Biol. Skr. 51: 64 (1998).
Shrub or small tree. Wooded grassland.
Distr: DAR, BAG, UN, EQU; Senegal to C.A.R., S to South Africa.
DAR: <u>Wickens 3157</u> 8/1972; **EQU:** <u>Friis & Vollesen 1146</u> 3/1982.

[Meiocarpidium]
Note: Andrews records *Meiocarpidium lepidotum* (Oliv.) Engl. & Diels from Equatoria. The identification must surely be incorrect since this species is restricted to WC Africa, but we have been unable to find the specimen to which the name was applied.

Monanthotaxis buchananii (Engl.) Verdc.
F.T.E.A. Annon.: 97 (1971); Biol. Skr. 51: 64 (1998); T.S.S.: 7 (1990).
Syn: *Popowia buchananii* (Engl.) Engl. & Diels – Bot. Exp. Sud.: 16 (1970).
Syn: *Popowia djurensis* Engl. & Diels – F.P.S. 1: 6 (1950).
Shrub, woody climber or small tree. Riverine forest.
Distr: BAG, EQU; C.A.R to Kenya, S to Mozambique.
EQU: <u>Myers 13280</u> 5/1940.

Monanthotaxis ferruginea (Oliv.) Verdc.
F.T.E.A. Annon.: 106 (1971); T.S.S.: 7 (1990); Biol. Skr. 51: 65 (1998); F.E.E. 2(1): 8 (2000).
Shrub, woody climber or small tree. Forest.
Distr: EQU; Cameroon to Ethiopia, S to Angola & Tanzania.
EQU: <u>Friis & Vollesen 457</u> 11/1980.

Monanthotaxis lucidula (Oliv.) Verdc.
Kew Bull. 25: 27 (1971); T.S.S.: 7 (1990).
Woody climber. Forest.
Distr: EQU; Gabon to D.R. Congo, S to Angola.
EQU: <u>Andrews 1387</u> 5/1939.

Monanthotaxis schweinfurthii (Engl. & Diels) Verdc. var. schweinfurthii
F.T.E.A. Annon.: 93 (1971); T.S.S.: 9 (1990); Biol. Skr. 51: 65 (1998).
Syn: *Enneastemon schweinfurthii* (Engl. & Diels) Robyns & Ghesq. – F.P.S. 1: 4 (1950).
Woody climber. Forest.
Distr: EQU; Gabon to Kenya, S to Angola & Zambia.
EQU: <u>Friis & Vollesen 231</u> 11/1980.

Monodora angolensis Welw.
F.P.S. 1: 6 (1950); F.T.E.A. Annon.: 119 (1971); T.S.S.: 9 (1990); Biol. Skr. 51: 65 (1998).
Syn: *Monodora gibsonii* Bullock ex Burtt Davy – F.P.S. 1: 6 (1950).

Shrub or small tree. Lowland forest.
Distr: EQU; Cameroon to Uganda, S to Angola & Zambia.
EQU: <u>Myers 11383</u> 5/1939.

Monodora myristica (Gaertn.) Dunal
F.P.S. 1: 5 (1950); F.T.E.A. Annon.: 118 (1971); T.S.S.: 11 (1990).
Tree. Forest.
Distr: EQU; Sierra Leone to Kenya, S to Angola & D.R. Congo.
EQU: <u>Sillitoe 281</u> 1919.

Uvaria angolensis Welw. ex Oliv. var. angolensis
F.T.E.A. Annon.: 15 (1971); T.S.S.: 11 (1990); Biol. Skr. 51: 65 (1998); F.E.E. 2(1): 4 (2000).
Syn: *Uvaria bukobensis* Engl. – F.P.S. 1: 6 (1950).
Woody climber or small tree. Forest.
Distr: EQU; Sierra Leone to Ethiopia, S to Angola & Zimbabwe.
EQU: <u>Friis & Vollesen 741</u> 12/1980.

Uvaria chamae P.Beauv.
F.W.T.A. 1(1): 36 (1954); T.S.S.: 11 (1990).
Shrub or small tree. Forest margins.
Distr: BAG; Senegal to D.R. Congo.
BAG: <u>Myers 11505</u> 7/1939.

Uvaria schweinfurthii Engl. & Diels
F.P.S. 1: 7 (1950); F.T.E.A. Annon.: 21 (1971); T.S.S.: 11 (1990); F.E.E. 2(1): 4 (2000).
Tree. Lowland forest.
Distr: BAG, EQU; C.A.R., D.R. Congo, Ethiopia, Uganda.
IUCN: RD
EQU: <u>Andrews 942</u> 4/1939.

Uvariopsis congensis Robyns & Ghesq.
F.T.E.A. Annon.: 71 (1971); T.S.S.: 13 (1990); Biol. Skr. 51: 66 (1998).
Shrub or small tree. Lowland forest.
Distr: BAG, EQU; D.R. Congo to Kenya, S to Angola & Zambia.
EQU: <u>Friis & Vollesen 735</u> 12/1980.

Xylopia acutiflora (Dunal) A.Rich.
F.P.S. 1: 7 (1950); F.W.T.A. 1(1): 42 (1954); T.S.S.: 13 (1990).
Shrub or small tree. Forest.
Distr: ?KOR, BAG, EQU; Sierra Leone to D.R. Congo, S to Angola & Zambia.
EQU: <u>Sillitoe 353</u> 1919.
Note: the record for Kordofan is from El Amin in T.S.S.; we have seen no material with which to verify this.

Xylopia aethiopica (Dunal) A.Rich.
F.T.E.A. Annon.: 76 (1971); T.S.S.: 13 (1990).
Shrub or tree. Swamp forest & moist woodland.
Distr: ?EQU; Senegal to Uganda, S to Zambia & Mozambique.
Note: the record for Sudan is from El Amin in T.S.S. who records it from "high rainfall savanna and swampy forest" without giving a locality. It is likely to occur in Equatoria but this needs confirmation.

Xylopia longipetala De Wild. & T.Durand
Ann. Mus. Congo Belge. Bot., Ser. 2, 1(1): 4 (1899).
Syn: Xylopia parviflora (A.Rich.) Benth. – F.T.E.A.
Annon.: 79 (1971); T.S.S.: 14 (1990); Biol. Skr. 51: 66
(1998); F.E.E. 2(1): 7 (2000).
Syn: Xylopia vallotii Chipp ex Exell – F.P.S. 1: 7 (1950).
Tree. Forest, dense woodland.
Distr: KOR, EQU; Senegal to Ethiopia, S to Angola &
South Africa.
KOR: Simpson 7778 (fl) 4/1930; **EQU:** Myers 10102
11/1938.

Xylopia rubescens Oliv.
F.T.E.A. Annon.: 76 (1971); T.S.S.: 14 (1990).
Tall tree. Riverine & swamp forest.
Distr: EQU; Liberia to Uganda, S to Angola & Zambia.
EQU: Myers 13586 11/1940.

LAURALES

Hernandiaceae

I. Darbyshire

Gyrocarpus americanus Jacq.
F.T.E.A. Hernand.: 7 (1985); F.E.E. 2(1): 33 (2000).
Syn: Gyrocarpus jacquinii Gaertn. – T.S.S.: 17 (1990).
Tree. Rocky hillslopes.
Distr: RS, ES; widespread in the tropics.
Note: this record is derived from El Amin in T.S.S. who
records this species as "growing usually along the Red
Sea coast and in Kassala (J. El Faw, J. Gerein)". We have
seen no Sudanese specimens with which to confirm this
record but this is such a distinctive genus that the records
are likely to be correct.

Monimiaceae

I. Darbyshire

Xymalos monospora (Harv.) Baill.
F.P.S. 1: 7 (1950); F.T.E.A. Monim.: 1 (1968); T.S.S.: 14
(1990); Biol. Skr. 51: 600 (2005).
Shrub or tree. Montane & submontane forest.
Distr: EQU; Bioko & Cameroon, Uganda & Kenya to
South Africa.
EQU: Friis & Vollesen 384 11/1980.

Lauraceae

I. Darbyshire

Beilschmiedia ugandensis Rendle var. ugandensis
T.S.S.: 15 (1990); F.T.E.A. Laur.: 5 (1996).
Syn: Tylostemon ugandensis (Rendle) Stapf – F.P.S. 1:
8 (1950).
Shrub or tree, sometimes large. Riverine & lowland
forest.
Distr: EQU; D.R. Congo, Uganda, Tanzania.
IUCN: VU
EQU: Myers 9157 (fl) 7/1938.
Note: B. Verdcourt originally labelled the South Sudan
material as B. sp. nr. ugandensis, but he includes Sudan

within the distribution of var. ugandensis in F.T.E.A.
without comment; presumably he widened his concept
of that taxon. The whole genus is in great need of
revision – B. ugandensis appears rather local, based upon
current knowledge.

Ocotea kenyensis (Chiov.) Robyns & R.Wilczek
T.S.S.: 15 (1990); F.T.E.A. Laur.: 11 (1996); Biol. Skr. 51:
67 (1998); F.E.E. 2(1): 67 (2000).
Syn: Ocotea viridis Kosterm. – F.P.S. 1: 8 (1950).
Large tree. Montane & submontane forest.
Distr: EQU; Ethiopia to South Africa.
IUCN: VU
EQU: Myers 13427 (fl) 8/1940.

Persea americana Mill.
T.S.S.: 17 (1990); F.T.E.A. Laur.: 1 (1996); Biol. Skr. 51: 67
(1998); F.E.E. 2(1): 16 (2000).
Tree. Farms & naturalised in abandoned cultivations.
Distr: DAR, EQU; native of South America, cultivated
widely in the tropics.
Note: commonly planted for its edible fruits 'avocado';
it has been reported as naturalised in the Imatong Mts
but no specimens have been seen. El Amin in T.S.S. also
reports it being grown on Jebel Marra.

ALISMATALES

Araceae

M. Kordofani, I. Farag & I. Darbyshire

Alocasia macrorrhizos (L.) G.Don
F.T.E.A. Arac.: 4 (1985); Biol. Skr. 51: 495 (2005).
Rhizomatous perennial herb. Abandoned cultivation.
Distr: EQU; native of Malesia to Australia; widely
cultivated in the tropics.
EQU: Shigeta 58 1979 (n.v.).

Amorphophallus abyssinicus (A.Rich.) N.E.Br. subsp. abyssinicus
F.P.S. 3: 280 (1956); F.J.M.: 157 (1976); F.T.E.A. Arac.: 29
(1985); F.E.E. 6: 42 (1997); Biol. Skr. 51: 495 (2005).
Tuberous perennial herb. Woodland & wooded grassland,
rocky areas.
Distr: DAR, BAG, UN, EQU; Ivory Coast to Ethiopia, S to
Namibia & Mozambique.
DAR: Wickens 2005 (fl fr) 7/1964; **BAG:**
Schweinfurth 1806 8/1869.
Note: Andrews also records A. maculatus (now A.
angolensis (Welw. ex Schott.) N.E.Br. subsp. maculatus
(N.E.Br.) Ittenb. ex Govaerts & Frodin), presumed to
be based on Broun 1677 from Bahr el Jebel. Ittenbach
redetermined this sterile collection as Amorphophallus
sp. indet., hence A. angolensis is omitted here.

Anchomanes difformis (Blume) Engl.
F.P.S. 3: 280 (1956); F.T.E.A. Arac.: 25 (1985); Biol. Skr.
51: 495 (2005).
Tuberous perennial herb. Forest.
Distr: BAG, EQU; Sierra Leone to Uganda, S to Angola
& Zambia.
EQU: Friis & Vollesen 1064 3/1982.

Arisaema enneaphyllum Hochst. ex A.Rich.

F.T.E.A. Arac.: 61 (1985); F.E.E. 6: 46 (1997); Biol. Skr. 51: 495 (2005).
Tuberous perennial herb. Open montane forest, ericaceous scrub, roadside banks.
Distr: EQU; Ethiopia, Kenya, Uganda; Yemen.
EQU: Jackson 1413 4/1950.

Arisaema schimperianum Schott

F.T.E.A. Arac.: 63 (1985); F.E.E. 6: 46 (1997); Biol. Skr. 51: 495 (2005).
Syn: *Arisaema* sp. sensu Andrews – F.P.S. 3: 280 (1956).
Tuberous perennial herb. Forest, secondary bushland, roadsides, riverbanks.
Distr: EQU; D.R. Congo, Ethiopia, Uganda.
EQU: Myers 10891 4/1939.

Colocasia esculenta (L.) Schott

F.P.S. 3: 281 (1956); Bot. Exp. Sud.: 17 (1970); F.T.E.A. Arac.: 5 (1985); F.E.E. 6: 38 (1997).
Tuberous perennial herb. Cultivated & naturalised in moist habitats, often near streams.
Distr: CS, EQU; probably native of SE Asia, widely cultivated in the palaeotropics.
EQU: Drar 1780 4/1938 (n.v.).
Note: commonly cultivated as a vegetable, common name 'coco yam'; Andrews notes that it has become semi-naturalised in "Central & Southern Sudan".

Culcasia falcifolia Engl.

F.T.E.A. Arac.: 18 (1985); F.E.E. 6: 35 (1997); Biol. Skr. 51: 496 (2005).
Perennial climbing or trailing herb. Forest including riverine forest.
Distr: BAG, EQU; Ethiopia to Zimbabwe & Mozambique.
EQU: Sillitoe 489 (fr) 1919.

Culcasia scandens P.Beauv.

F.P.S. 3: 281 (1956); F.W.T.A. 3(1): 124 (1968); F.T.E.A. Arac.: 18, in notes (1985).
Perennial epiphytic and/or climbing herb. Forest.
Distr: EQU; Senegal to Angola & D.R. Congo.
EQU: Friis & Vollesen 437 (fl fr) 11/1980.

Lemna aequinoctialis Welw.

F.P.S. 3: 284 (1956); F.T.E.A. Lemn.: 5 (1973); F.E.E. 6: 53 (1997).
Syn: *Lemna perpusilla* sensu auctt., non Torr. – F.P.S. 3: 284 (1956); Bot. Exp. Sud.: 78 (1970); F.J.M.: 157 (1976).
Free-floating aquatic herb. Shallow or seasonal pools.
Distr: DAR, KOR, BAG, UN; widespread in the tropics & subtropics.
KOR: Wickens 783 11/1961; **UN:** J.M. Lock 81/121 8/1981.

Lemna minor L.

F.T.E.A. Lemn.: 4 (1973); F.E.E. 6: 52 (1997).
Free-floating aquatic herb. Pools.
Distr: DAR; cosmopolitan.
DAR: de Wilde et al. 5388 1/1965.

Pistia stratiotes L.

F.P.S. 3: 281 (1956); F.J.M.: 157 (1976); F.T.E.A. Arac.: 66 (1985); F.E.E. 6: 49 (1997).
Free-floating aquatic perennial herb. Open water or exposed on marginal mud.
Distr: DAR, KOR, CS, UN, EQU; pantropical & subtropical.
DAR: Wickens 1633 5/1964; **UN:** J.M. Lock 83/1 1/1983.

Sauromatum venosum (Dryand. ex Aiton) Kunth

F.P.S. 3: 281 (1956); F.T.E.A. Arac.: 58 (1985); F.E.E. 6: 44 (1997).
Tuberous perennial herb. Riverine forest, damp woodland.
Distr: CS; Cameroon to Ethiopia, S to Angola & Zambia; Yemen, India to China.
CS: Cienkowsky s.n. (n.v.).
Note: the only Sudan specimen known is the type of *S. nubicum* Schott, a synonym of *S. venosum*.

Spirodela polyrrhiza (L.) Schleid.

F.T.E.A. Lemn.: 5 (1973); F.E.E. 6: 52 (1997).
Syn: *Lemna polyrrhiza* L. – F.P.S. 3: 283 (1956).
Free-floating aquatic herb. Still water.
Distr: BAG, UN; cosmopolitan.
UN: J.M. Lock 82/24 4/1982.

Stylochaeton hypogeum Lepr.

F.W.T.A. 3(1): 114 (1968).
Syn: *Stylochaeton kerensis* N.E.Br. – F.P.S. 3: 283 (1956); F.E.E. 6: 36 (1997).
Syn: *Stylochaeton borumensis* sensu Wickens, non N.E.Br. – F.J.M.: 157 (1976).
Rhizomatous perennial herb. Floodplains.
Distr: DAR; Mauritania & Senegal to Ethiopia.
DAR: Purdy 17.
Note: the Sudan material seen is sterile and so difficult to place with certainty.

Stylochaeton lancifolius Kotschy & Peyr.

F.P.S. 3: 283 (1956); Biol. Skr. 51: 496 (2005).
Perennial herb. Dry wooded grassland & woodland.
Distr: BAG, UN, EQU; Senegal to C.A.R.
UN: J.M. Lock 81/71 5/1981.

Wolffiella hyalina (Delile) Monod

F.T.E.A. Lemn.: 8 (1973); F.E.E. 6: 53 (1997).
Syn: *Wolffia hyalina* (Delile) Hegelm. – F.P.S. 3: 284 (1956).
Free-floating aquatic perennial herb. Pools, slow-flowing water.
Distr: ?DAR, ?KOR, UN; Mali, Nigeria to Ethiopia, S to Botswana; Egypt.
UN: J.M. Lock 81/122 8/1981.
Note: Andrews records this species from Jebel Barkin in Darfur and Bara in Kordofan but we have not seen the material on which these records are based.

Wolffiella welwitschii (Hegelm.) Monod

F.T.E.A. Lemn.: 7 (1973); F.E.E. 6: 53 (1997).
Syn: *Wolffia welwitschii* Hegelm. – F.P.S. 3: 284 (1956).
Partially submerged aquatic herb. Pools.
Distr: BAG, UN; tropical & southern Africa; tropical America.
UN: J.M. Lock 81/123 8/1981.

Alismataceae

M. Kordofani, I. Farag & I. Darbyshire

Alisma plantago-aquatica L.

F.P.S. 3: 230 (1956); F.T.E.A. Alismat.: 5 (1960); F.E.E. 6: 14 (1997).
Perennial herb. Marshes, muddy margins of ponds & slow-flowing rivers.
Distr: ?KOR; Eritrea to Tanzania, introduced in S Africa; N Africa, Europe & temperate Asia.
Note: Andrews records this species from Rahad in Kordofan; we have not seen the material on which this record was based.

Burnatia enneandra Micheli

F.P.S. 3: 230 (1956); F.T.E.A. Alismat.: 13 (1960); F.E.E. 6: 16 (1997).
Aquatic perennial herb. Shallow water in swamps & marshes.
Distr: BAG, EQU; Senegal to Ethiopia, S to Botswana & South Africa.
BAG: Schweinfurth 2287 (fl fr) 8/1869.

Butomopsis latifolia (D.Don) Kunth

F.Z. 12(2): 16 (2009).
Syn: *Tenagocharis latifolia* (D.Don) Buchenau – F.P.S. 3: 225 (1956); F.T.E.A. Butomaceae: 1 (1960).
Aquatic perennial herb. Marshes & swamps, often standing in water.
Distr: KOR, CS, BAG; Senegal, Ghana to Uganda, S to Botswana; tropical Asia & Australia.
CS: Lea 220 10/1954; **BAG:** Simpson 7695 3/1930.

Caldesia reniformis (D.Don) Makino

F.P.S. 3: 230 (1956); F.T.E.A. Alismat.: 7 (1960).
Aquatic perennial herb. Swamps, margins of lakes & slow-flowing rivers.
Distr: BAG, UN; scattered from Senegal to Kenya, S to Botswana; Madagascar, N Africa, India to Japan, S to Australia.
BAG: Simpson 7658 3/1930.

Limnophyton angolense Buchenau

F.T.E.A. Alismat.: 11 (1960).
Aquatic perennial herb. Swamps.
Distr: BAG; scattered from Guinea to Uganda, S to Botswana.
BAG: Simpson 7384 (fr) 7/1929.

Limnophyton obtusifolium (L.) Miq.

F.P.S. 3: 230 (1956); F.T.E.A. Alismat.: 9 (1960); F.J.M.: 153 (1976); F.E.E. 6: 15 (1997).
Aquatic perennial herb. Swamps & pools or in slow-flowing water.
Distr: DAR, KOR, CS, BAG, UN, EQU; palaeotropical.
KOR: Pfund 798 7/1875; **UN:** Simpson 7517 (fr) 2/1930.

Ranalisma humile (Rich. ex Kunth) Hutch.

F.P.S. 3: 231 (1956); F.T.E.A. Alismat.: 2 (1960).
Stoloniferous perennial herb. Marshes.
Distr: CS; Senegal to Tanzania, S to Zimbabwe.
CS: Schweinfurth 1023 (fr) 1/1869.

Sagittaria guayanensis Kunth

F.T.E.A. Alismat.: 3, in notes (1960); F.W.T.A. 3(1): 11 (1968).
Syn: *Lophotocarpus guayanensis* (Kunth) T.Durand & Schinz – F.P.S. 3: 231 (1956).
Aquatic perennial herb. Muddy pools.
Distr: KOR; Senegal to C.A.R.; widely distributed in the tropics & subtropics.
KOR: Kotschy 423 (fl fr) 12/1839.

Wiesneria schweinfurthii Hook.f.

F.P.S. 3: 233 (1956); F.T.E.A. Alismat.: 13 (1960); F.E.E. 6: 16 (1997).
Slender aquatic perennial herb. Rivers, shallow lakes, water channels.
Distr: BAG; scattered from Senegal to Ethiopia, S to Angola & Botswana.
BAG: Schweinfurth 2157 7/1869.

Hydrocharitaceae

M. Kordofani, I. Farag & I. Darbyshire

Halophila ovalis (R.Br.) Hook.f. subsp. *ovalis*

F.P.S. 3: 226 (1956); F.T.E.A. Hydrocharit.: 25 (1989); F.E.E. 6: 9 (1997).
Submerged aquatic perennial herb. Marine, intertidal zone.
Distr: RS; coastal Eritrea, Kenya & Tanzania; Persian Gulf to W Pacific.
RS: Schweinfurth 199 9/1868.

Halophila stipulacea (Forssk.) Asch.

F.P.S. 3: 226 (1956); F.T.E.A. Hydrocharit.: 27 (1989); F.E.E. 6: 9 (1997).
Submerged aquatic perennial herb. Shallow marine waters.
Distr: RS; coastal Egypt to Tanzania; coastal Indian Ocean.
RS: Broun 1537 12/1908.

Lagarosiphon cordofanus Casp.

F.P.S. 3: 227 (1956); F.T.E.A. Hydrocharit.: 2 (1989); F.E.E. 6: 5 (1997).
Submerged aquatic perennial herb. Still or slow-flowing freshwater.
Distr: KOR, UN; Ethiopia & Uganda to South Africa.
KOR: Kotschy 170 10/1839; **UN:** J.M. Lock 81/262 9/1981.

Lagarosiphon muscoides Harv.

F.T.E.A. Hydrocharit.: 4 (1989).
Submerged aquatic perennial herb. Still or slow-flowing freshwater.
Distr: BAG; Mali to Uganda, S to South Africa.
BAG: Schweinfurth 2158 (fl) 7/1869.

Najas graminea Delile var. *graminea*

F.P.S. 3: 237 (1956); F.T.E.A. Najad.: 8 (1989).
Submerged aquatic annual herb. Lakes, ponds, ditches.
Distr: KOR, BAG; palaeotropical and subtropical.
KOR: Steudner 213 10/1862; **BAG:** Schweinfurth 1717 6/1869.

Najas horrida A.Braun ex Magnus
F.T.E.A. Najad.: 5 (1989); F.E.E. 6: 31 (1997).
Syn: *Najas pectinata* sensu auctt., non (Parl.) Magnus –
F.P.S. 3: 238 (1956); Bot. Exp. Sud.: 87 (1970).
Submerged aquatic annual herb. Lakes, swamps.
Distr: KOR, CS, BAG, UN; Guinea to Eritrea, S to Zambia;
Egypt, Madagascar.
KOR: Wickens 840 11/1961; **UN:** J.M. Lock 83/15 2/1983.

Najas schweinfurthii Magnus
F.P.S. 3: 237 (1956); F.T.E.A. Najad.: 8 (1989); F.E.E. 6: 31
(1997).
Submerged aquatic annual herb. Pools, marshes.
Distr: BAG, UN; Senegal, Chad, Ethiopia, Tanzania.
BAG: Schweinfurth 2140 8/1869.

Najas testui Rendle
F.T.E.A. Najad.: 7 (1989); F.E.E. 6: 31 (1997).
Syn: *Najas graminea* sensu Wickens, non Delile – F.J.M.:
154 (1976).
Submerged aquatic annual herb. Marshes, ponds, rivers.
Distr: DAR; Nigeria to Ethiopia, S to Angola and Zambia.
DAR: Wickens 2497 9/1964.

Najas welwitschii Rendle
F.E.E. 6: 31 (1997).
Submerged aquatic annual herb. Shallow lakes.
Distr: CS; D.R. Congo, Ethiopia, Angola; India.
IUCN: DD
CS: Martin-Saint-Ange 81 1850.

Nechamandra alternifolia (Roxb. ex Wight)
Thwaites
F.P.S. 3: 227 (1956); Aquatic Bot. 13: 505 (1980).
Submerged aquatic annual or perennial herb. Freshwater
pools, marshes.
Distr: CS; native of Indian Subcontinent, naturalised in
Vietnam & Sudan.
CS: Andrews 89 (fl) 12/1938.

Ottelia alismoides (L.) Pers.
F.P.S. 3: 227 (1956); F.T.E.A. Hydrocharit.: 15 (1989).
Aquatic annual or perennial herb. Still or slow-flowing
freshwater.
Distr: KOR, CS; Tanzania; Egypt, India to E Asia,
Australia, scattered in Europe & N America.
CS: Andrews 135 1/1936.

Ottelia brachyphylla (Gürke) Dandy
F.P.S. 3: 227 (1956).
Aquatic perennial herb. Freshwater.
Distr: BAG; South Sudan endemic.
IUCN: RD
BAG: Schweinfurth 2423 (fl) 1869.

Ottelia scabra Baker
F.P.S. 3: 227 (1956); F.T.E.A. Hydrocharit.: 19 (1989).
Aquatic annual or perennial herb. Still freshwater.
Distr: DAR, BAG, UN; Uganda.
IUCN: NT
DAR: Broun 859 2/1906; **UN:** J.M. Lock 81/1 (fl) 1/1981.

Ottelia ulvifolia (Planch.) Walp.
F.P.S. 3: 227 (1956); F.J.M.: 153 (1976); F.T.E.A. Hydrocharit.:

15 (1989); F.E.E. 6: 2 (1997); Biol. Skr. 51: 474 (2005).
Aquatic annual or perennial herb. Still or slow-flowing
freshwater.
Distr: DAR, BAG, UN, EQU; Senegal to Ethiopia, S to
South Africa.
DAR: Wickens 2960 (fl fr) 5/1965; **UN:** J.M. Lock 81/13
2/1981.

Thalassia hemprichii (Ehrenb. ex Solms) Asch.
F.P.S. 3: 229 (1956); F.T.E.A. Hydrocharit.: 22 (1989);
F.E.E. 6: 6 (1997).
Submerged aquatic perennial herb. Shallow marine waters.
Distr: RS; Red Sea to Mozambique; Madagascar, tropical
Indian Ocean, W Pacific.
RS: Schweinfurth 186 9/1868.

Vallisneria spiralis L.
F.T.E.A. Hydrocharit.: 12 (1989); F.E.E. 6: 2 (1997).
Submerged aquatic perennial herb. Still fresh or brackish
water.
Distr: CS, BAG, UN; tropical Africa; Europe, SE Asia,
Japan, Australia.
CS: Schweinfurth 1017 1/1869; **UN:** J.M. Lock 83/16 (fl
fr) 2/1983.

Aponogetonaceae

M. Kordofani, I. Farag & I. Darbyshire

Aponogeton subconjugatus Schumach. &
Thonn.
F.P.S. 3: 233 (1956); F.T.E.A. Aponogeton.: 6 (1989).
Aquatic perennial herb. Temporary pools & swamps.
Distr: ?UN; Senegal to Uganda.
Note: Andrews records this species from Bor in Upper
Nile but we have seen no material with which to verify
this. Lye in F.T.E.A. records this species as "possibly...in
the Sudan".

Aponogeton vallisnerioides Baker
F.P.S. 3: 234 (1956); F.T.E.A. Aponogeton.: 2 (1989); F.E.E.
6: 11 (1997).
Aquatic perennial herb. Shallow pools on rocks, lake
margins.
Distr: BAG, EQU; Sierra Leone to Ethiopia, S to Zambia.
BAG: Myers 11492 (fl) 6/1939.

Potamogetonaceae

M. Kordofani, I. Farag & I. Darbyshire

Potamogeton × assidens Z.Kaplan, Zalewska-
Gal. & Ronikier
Taxon 59: 562 (2010).
Aquatic perennial herb. Shallow ditches.
Distr: CS; Niger; Madagascar, Poland.
CS: de Wilde et al. 5734 2/1965.
Note: a hybrid of *P. nodosus* & *P. perfoliatus*.

Potamogeton crispus L.
F.P.S. 3: 234 (1956); F.E.E. 6: 20 (1997).
Aquatic perennial herb. Ponds, irrigation canals.
Distr: NS, DAR, KOR, CS; palaeotropical.
CS: de Wilde et al. 5849 3/1965.

Potamogeton nodosus Poir.

F.P.S. 3: 234 (1956); F.J.M.: 153 (1976); F.T.E.A.
Potamogeton.: 13 (2006).
Aquatic perennial herb. Shallow lakes & pools, swamps.
Distr: DAR, CS, BAG, UN; cosmopolitan.
DAR: Wickens 1020 (fl) 1/1964; **BAG:** Simpson 7665 3/1930.

Potamogeton octandrus Poir.

F.P.S. 3: 235 (1956); Biol. Skr. 51: 475 (2005); F.T.E.A.
Potamogeton.: 9 (2006).
Aquatic perennial herb. Slow running water or permanent pools.
Distr: BAG, UN; palaeotropical & subtropical.
UN: J.M. Lock 82/6 1/1982.

Potamogeton perfoliatus L.

F.P.S. 3: 235 (1956); F.E.E. 6: 20 (1997).
Aquatic perennial herb. Shallow lakes & pools.
Distr: NS, RS, CS; Mauritania to Ethiopia; Madagascar, widespread in temperate N Hemisphere.
CS: de Wilde et al. 5724 (fl fr) 2/1965.

Potamogeton pusillus L.

F.P.S. 3: 235 (1956); F.J.M.: 153 (1976); F.E.E. 6: 20 (1997); F.T.E.A. Potamogeton.: 7 (2006).
Submerged aquatic perennial herb. Lakes, ponds, slow-flowing rivers.
Distr: DAR; cosmopolitan.
DAR: Wickens 1171 (fl fr) 2/1964.

Potamogeton schweinfurthii A.Benn.

F.P.S. 3: 234 (1956); F.J.M.: 153 (1976); F.E.E. 6: 23 (1997); F.T.E.A. Potamogeton.:10 (2006).
Aquatic perennial herb. Lakes, ponds, slow-flowing rivers.
Distr: BAG, UN; Senegal to Somalia, S to South Africa; N Africa & Mediterranean to Arabia.
UN: Denny 28 (fl) 4/1978.

Stuckenia pectinata (L.) Börner

F.Z. 12(2): 79 (2009).
Syn: Potamogeton pectinatus L. – F.P.S. 3: 235 (1956); F.E.E. 6: 19 (1997); F.T.E.A. Potamogeton.: 5 (2006).
Submerged aquatic perennial herb. Lakes, rivers, irrigation canals, often forming dense stands.
Distr: NS, RS, CS; cosmopolitan.
CS: de Wilde et al. 5689 (fl fr) 2/1965.

Zannichellia palustris L.

F.P.S. 3: 237 (1956); F.E.E. 6: 25 (1997); F.T.E.A. Zannichell.: 1 (2000).
Submerged aquatic annual herb. Rivers & pools over rocky substrate.
Distr: ?CS; cosmopolitan.
Note: Andrews recorded this species from the Gezira canals in Blue Nile Province; we have seen no material with which to verify this record.

Cymodoceaceae

I. Farag, M. Kordofani & I. Darbyshire

Cymodocea rotundata Asch. & Schweinf.

F.P.S. 3: 236 (1956); F.E.E. 6: 29 (1997); F.T.E.A. Cymodoc.: 7 (2002).
Aquatic rhizomatous perennial herb. Shallow water over coral or sand, sometimes exposed at low tide.
Distr: RS; Egypt to Mozambique; Madagascar to tropical Asia & W Pacific.
RS: Schweinfurth 188 9/1868.

Cymodocea serrulata (R.Br.) Asch. & Magnus

F.P.S. 3: 236 (1956); F.E.E. 6: 29 (1997); F.T.E.A. Cymodoc.: 9 (2002).
Aquatic rhizomatous perennial herb. Tidal pools, shallow coastal waters, never exposed.
Distr: RS; Egypt to Mozambique; Madagascar to W Pacific.
RS: Schweinfurth 197 9/1868.

Halodule uninervis (Forssk.) Boiss.

F.E.E. 6: 27 (1997); F.T.E.A. Cymodoc.: 6 (2002).
Syn: Diplanthera uninervis (Forssk.) Asch. – F.P.S. 3: 236 (1956).
Aquatic rhizomatous perennial herb. Shallow coastal waters, mangrove creeks, sometimes exposed at low tide.
Distr: RS; Egypt to South Africa; Asia, Australia, W Pacific.
RS: Schweinfurth 197A 9/1868.

Syringodium isoetifolium (Asch.) Dandy

F.P.S. 3: 237 (1956); F.E.E. 6: 26 (1997); F.T.E.A. Cymodoc.: 5 (2002).
Aquatic rhizomatous perennial herb. Tidal pools, shallow coastal waters, often exposed at low tide.
Distr: RS; Egypt to Mozambique; Madagascar to W Pacific.
RS: Schweinfurth 198 9/1868.

Thalassodendron ciliatum (Forssk.) Hartog

F.E.E. 6: 29 (1997); F.T.E.A. Cymodoc.: 2 (2002).
Syn: Cymodocea ciliata (Forssk.) Ehrenb. ex Asch. – F.P.S. 3: 236 (1956).
Aquatic rhizomatous perennial herb. On coral or mud over coral, at most partially exposed at high tide.
Distr: RS; Egypt to South Africa; Arabia, Madagascar through Indian Ocean to Malesia & Australia.
RS: Broun 1540 1908.

DIOSCOREALES

Dioscoreaceae

H. Pickering

Dioscorea abyssinica Hochst. ex Kunth

F.P.S. 3: 297 (1956); F.E.E. 6: 61 (1997).
Herbaceous climber. Wooded grassland.
Distr: ES, BAG; Senegal to Eritrea & Ethiopia.
ES: Schweinfurth 587 7/1865; **BAG:** Schweinfurth 2032 7/1869.
Note: both D. abyssinica and D. sagittifolia (see below) are closely allied to D. praehensilis Benth. and they may together form a single biological species; the three are found together in SW Ethiopia (P. Wilkin, pers. comm.). They are the wild relatives of the cultivated 'guinea yam' complex (D. cayennensis Lam. s.l.).

Dioscorea bulbifera L.

F.P.S. 3: 295 (1956); F.T.E.A. Dioscor.: 9 (1975); F.J.M.:
159 (1976); F.E.E. 6: 59 (1997).
Herbaceous climber. Lowland forest & secondary
bushland.
Distr: DAR, EQU; Senegal to Ethiopia, S to Angola
& Zimbabwe; tropical Asia & Polynesia, introduced
elsewhere.
DAR: <u>Wickens 2193</u> 8/1964; **EQU:** <u>Myers 11887</u>
10/1938.
Note: Andrews records this species as being widely
cultivated.

Dioscorea dumetorum (Kunth) Pax

F.P.S. 3: 297 (1956); F.T.E.A. Dioscor.: 21 (1975); F.E.E. 6:
58 (1997); Biol. Skr. 51: 501 (2005).
Herbaceous climber. Lowland forest, woodland &
wooded grassland.
Distr: RS, KOR, EQU; Senegal to Ethiopia, S to Angola &
South Africa.
KOR: <u>Patel & Kheir 55</u> 11/1964; **EQU:** <u>Myers 9158</u>
7/1938.

Dioscorea preussii Pax subsp. *preussii*

F.P.S. 3: 295 (1956); F.T.E.A. Dioscor.: 20 (1975); Biol. Skr.
51: 501 (2005).
Herbaceous climber. Lowland forest, bushland &
woodland.
Distr: EQU; Senegal to Kenya, S to Angola & Tanzania.
EQU: <u>Friis & Vollesen 179</u> 11/1980.

Dioscorea quartiniana A.Rich.

F.P.S. 3: 297 (1956); F.T.E.A. Dioscor.: 23 (1975); F.J.M.:
159 (1976); F.E.E. 6: 56 (1997); Biol. Skr. 51: 502 (2005).
Syn: *Dioscorea quartiniana* A.Rich. var. *phaseoloides*
Pax – F.P.S. 3: 297 (1956).
Syn: *Dioscorea quartiniana* A.Rich. var.
schweinfurthiana (Pax) Burkill – F.P.S. 3: 298 (1956).
Herbaceous climber. Forest margins, bushland &
grassland.
Distr: DAR, ES, EQU; Gambia to Ethiopia, S to Namibia &
South Africa; Madagascar.
DAR: <u>Wickens 2346</u> 9/1964; **EQU:** <u>Andrews 1946</u>
6/1939.

Dioscorea sagittifolia Pax var. *sagittifolia*

F.E.E. 6: 61 (1997).
Herbaceous climber. Open forest.
Distr: EQU; Senegal to Ethiopia & Uganda.
EQU: <u>Myers 6959</u> 7/1937.

Dioscorea sansibarensis Pax

F.P.S. 3: 295 (1956); F.T.E.A. Dioscor.: 7 (1975).
Herbaceous climber. Lowland & riverine forest &
woodland, secondary bushland.
Distr: EQU; Guinea to Uganda, S to Angola, Zimbabwe
& Mozambique; Madagascar.
EQU: <u>Myers 9192</u> 8/1938.

Dioscorea schimperiana Hochst. ex Kunth

F.P.S. 3: 297 (1956); F.T.E.A. Dioscor.: 14 (1975); F.E.E. 6:
60 (1997); Biol. Skr. 51: 502 (2005).
Syn: *Dioscorea hirtiflora* sensu Andrews, non Benth. –
F.P.S. 3: 295 (1956).

Herbaceous climber. Forest margins, bushland &
grassland.
Distr: EQU; Nigeria to Ethiopia, S to Zimbabwe &
Mozambique.
EQU: <u>Myers 9502</u> 10/1938.

Tacca leontopetaloides (L.) Kuntze

F.P.S. 3: 308 (1956); F.T.E.A. Tacc.: 1 (1962); F.J.M.: 160
(1976); F.E.E. 6: 63 (1997); Biol. Skr. 51: 507 (2005).
Tuberous perennial herb. Grassland & woodland.
Distr: DAR, ES, UN, EQU; palaeotropical.
DAR: <u>Wickens 1984</u> 7/1964; **EQU:** <u>Sillitoe 340</u> 1919.

PANDANALES

Velloziaceae

H. Pickering

Xerophyta humilis (Baker) T.Durand & Schinz

F.T.E.A. Velloz.: 2 (1975); F.E.E. 6: 312 (1997).
Small perennial herb. Rock crevices, dry bushland.
Distr: ?EQU; Ethiopia, Angola to Mozambique, S to
South Africa.
Note: no material has been seen but this species is
recorded from South Sudan in both F.E.E. and F.T.E.A.

Xerophyta simulans L.B.Sm. & Ayensu

F.T.E.A. Velloz.: 6 (1975); Biol. Skr. 51: 507 (2005).
Syn: *Vellozia* sp. sensu Andrews – F.P.S. 3: 308 (1956).
Shrub. Dry rocky outcrops.
Distr: EQU; Uganda, Tanzania, Zambia, Zimbabwe &
Mozambique.
EQU: <u>Friis & Vollesen 1150</u> 3/1982.

Pandanaceae

I. Darbyshire

Pandanus cf. *chiliocarpus* Stapf

F.T.E.A. Pandan.: 6 (1993).
Syn: *Pandanus* sp. sensu Andrews – F.P.S. 3: 305 (1956).
Syn: *Pandanus* sp. nr. *ugandensis* sensu El Amin –
T.S.S.: 453 (1990).
Tree. Swamps & stream beds, sometimes within forest.
Distr: EQU; (for species) D.R. Congo, Uganda &
Tanzania.
EQU: <u>Myers 6503</u> 5/1937.
Note: Friis & Vollesen (2005: 505) called this taxon *P.
chiliocarpus* but they had seen no specimens; the two
collections at Kew are sterile and cannot be named with
certainty.

LILIALES

Colchicaceae

H. Pickering

Gloriosa superba L.

F.J.M.: 156 (1976); F.E.E. 6: 184 (1997); F.T.E.A. Colchic.:
4 (2005); Biol. Skr. 51: 488 (2005).
Syn: *Gloriosa simplex* L. – F.P.S. 3: 271 (1956).

Perennial herb. Woodland, wooded grassland & forest clearings.
Distr: RS, DAR, CS, BAG, UN, EQU; palaeotropical.
DAR: Wickens 2207 8/1964; **EQU:** Myers 6640 7/1937.

Iphigenia pauciflora Martelli

F.E.E. 6: 187 (1997); F.T.E.A. Colchic.: 7 (2005).
Syn: *Iphigenia ledermannii* Engl. & K.Krause – F.J.M.: 156 (1976).
Perennial herb. Rocky bushland.
Distr: DAR; Senegal to Somalia, S to Malawi.
DAR: Wickens 2084 8/1964.

Smilacaceae

H. Pickering

Smilax anceps Willd.

F.T.E.A. Smilac.: 2 (1989); F.E.E. 6: 65 (1997); Biol. Skr. 51: 494 (2005).
Syn: *Smilax kraussiana* Meisn. – F.P.S. 3: 279 (1956); Bot. Exp. Sud.: 104 (1970); T.S.S.: 455 (1990).
Herbaceous or shrubby climber. Riverine forest, woodland & bushland.
Distr: EQU; Senegal to Ethiopia, S to South Africa; Madagascar, Comoros, Mascarenes.
EQU: Sillitoe 337 1919.

ASPARAGALES

Orchidaceae

H. Pickering & I. Darbyshire

Aerangis kotschyana (Rchb.f.) Schltr.

F.P.S. 3: 312 (1956); F.T.E.A. Orchid. 3: 550 (1989); F.E.E. 6: 296 (1997).
Epiphytic herb. Woodland & wooded grassland.
Distr: CS, EQU; Guinea, Sierra Leone, Nigeria, Ethiopia to South Africa.
CS: Kotschy 449; **EQU:** Myers 6725 9/1937.
Note: *A. brachycarpa* (A.Rich.) T.Durand & Schinz is also likely to occur in South Sudan (EQU) having been recorded from the Ugandan side of the Imatong Mts (Friis & Vollesen 2005: 508).

Angraecum humile Summerh.

F.T.E.A. Orchid. 3: 494 (1989); Biol. Skr. 51: 508 (2005).
Epiphytic herb. Riverine & montane forest.
Distr: EQU; Rwanda, Kenya, Tanzania, Zimbabwe.
EQU: Friis & Vollesen 348 11/1980.

Ansellia africana Lindl.

F.T.E.A. Orchid. 2: 402 (1984).
Syn: *Ansellia gigantea* Rchb.f. var. *nilotica* (Baker) Summerh. – F.P.S. 3: 312 (1956); F.J.M.: 160 (1976).
Epiphytic herb. Woodland & wooded grassland.
Distr: DAR, ?EQU; Guinea Bissau to Kenya, S to Namibia & South Africa.
IUCN: VU
DAR: Wickens 1314 2/1964.
Note: Andrews records this species from Equatoria but we have seen no material with which to verify this record.

Bonatea steudneri (Rchb.f.) T.Durand & Schinz

F.T.E.A. Orchid. 1: 137 (1968); F.E.E. 6: 232 (1997).
Syn: *Bonatea sudanensis* Rolfe – F.P.S. 3: 312 (1956).
Terrestrial herb. Bushland, scrubby grassland, rocky hillslopes.
Distr: RS; Eritrea & Somalia, S to Namibia & South Africa; Arabia.
RS: Sillitoe 1464 1908.

Brachycorythis ovata Lindl. subsp. *schweinfurthii* (Rchb.f.) Summerh.

F.P.S. 3: 313 (1956); F.T.E.A. Orchid. 1: 23 (1968); F.E.E. 6: 205 (1997); Biol. Skr. 51: 508 (2005).
Terrestrial herb. Grassland.
Distr: EQU; Senegal to Ethiopia, S to D.R. Congo & Tanzania.
EQU: Sillitoe 144 1919.

Brachycorythis pleistophylla Rchb.f. var. *pleistophylla*

F.T.E.A. Orchid. 1: 23 (1968); Biol. Skr. 51: 508 (2005).
Terrestrial herb. Grassland, woodland margins.
Distr: EQU; Nigeria to Kenya, S to Angola & South Africa; Madagascar.
EQU: Jackson 3183 5/1954.

Brachycorythis pubescens Harv.

F.P.S. 3: 313 (1956); F.T.E.A. Orchid. 1: 25 (1968); F.E.E. 6: 202 (1997); Biol. Skr. 51: 509 (2005).
Terrestrial herb. Woodland & wooded grassland.
Distr: EQU; Guinea to Ethiopia, S to Angola & South Africa.
EQU: Sillitoe 154 1919.

Bulbophyllum cochleatum Lindl. var. *cochleatum*

F.P.S. 3: 313 (1956); F.T.E.A. Orchid. 2: 315 (1984); Biol. Skr. 51: 509 (2005).
Epiphytic herb. Montane forest.
Distr: EQU; Guinea to Kenya, S to Zambia & Malawi.
EQU: Friis & Vollesen 972 2/1982.

Calyptrochilum christyanum (Rchb.f.) Summerh.

F.P.S. 3: 313 (1956); F.T.E.A. Orchid. 3: 491 (1989); F.E.E. 6: 285 (1997).
Epiphytic herb. Forest, wooded grassland.
Distr: EQU; Gambia to Eritrea, S to Angola, Zimbabwe & Mozambique.
EQU: Myers 7920 6/1937.

Corymborkis corymbis Thouars

F.T.E.A. Orchid. 2: 243 (1984); F.E.E. 6: 198 (1997); Biol. Skr. 51: 509 (2005).
Syn: *Corymborkis welwitschii* (Rchb.f.) Kuntze – F.P.S. 3: 313 (1956).
Terrestrial herb. Forest.
Distr: EQU; Guinea to Ethiopia, S to Angola & South Africa; Madagascar, Mascarenes.
EQU: Friis & Vollesen 654 12/1980.

Diaphananthe fragrantissima (Rchb.f.) Summerh.

F.P.S. 3: 314 (1956); F.T.E.A. Orchid. 3: 528 (1989); F.E.E. 6: 293 (1997).
Epiphytic herb. Forest, bushland.
Distr: EQU; Cameroon to Ethiopia, S to Angola, Zimbabwe & Mozambique.
EQU: Myers 9380 8/1939.

Diaphananthe lorifolia Summerh.

F.T.E.A. Orchid. 3: 526 (1989); Biol. Skr. 51: 510 (2005).
Epiphytic herb. Montane & riverine forest.
Distr: EQU; Rwanda, Ethiopia, Uganda, Kenya, Tanzania.
EQU: Friis & Vollesen 75 11/1980.

Disa aconitoides Sond. subsp. *concinna* (N.E.Br.) H.P.Linder

F.E.E. 6: 240 (1997); Biol. Skr. 51: 510 (2005).
Syn: *Disa concinna* N.E.Br. – F.T.E.A. Orchid. 1: 173 (1968).
Terrestrial herb. Montane grassland.
Distr: EQU; Ethiopia to Zimbabwe & Mozambique.
EQU: Synnott 1038 5/1972.
Note: only two specimens are known from our region, both from Mt Lonyili which is on the South Sudan / Uganda border; they may have been collected on the Ugandan side.

Disa erubescens Rendle subsp. *erubescens*

F.P.S. 3: 314 (1956); F.T.E.A. Orchid. 1: 157 (1968); Biol. Skr. 51: 511 (as *erubens*) (2005).
Terrestrial herb. Montane grassland.
Distr: EQU; Nigeria, Cameroon, Uganda & Kenya, S to Angola, Zimbabwe & Mozambique.
EQU: Myers 11574 7/1939.

Disa fragrans Schltr. subsp. *deckenii* (Rchb.f.) H.P.Linder

Biol. Skr. 51: 511 (2005).
Syn: *Disa deckenii* Rchb.f. – F.P.S. 3: 314 (1956); F.T.E.A. Orchid. 1: 165 (1968); F.E.E. 6: 240 (1997).
Terrestrial herb. Montane grassland.
Distr: EQU; Ethiopia, D.R. Congo, Rwanda, Uganda, Kenya, Tanzania.
EQU: Myers 11636 7/1939.

Disa hircicornis Rchb.f.

F.P.S. 3: 316 (1956); F.T.E.A. Orchid. 1: 171 (1968); Biol. Skr. 51: 511 (2005).
Terrestrial herb. Montane grassland.
Distr: EQU; Nigeria, Cameroon, Ethiopia to Angola & South Africa.
EQU: Myers 11664 7/1939.

Disa ochrostachya Rchb.f.

F.T.E.A. Orchid. 1: 164 (1968); Biol. Skr. 51: 511 (2005).
Terrestrial herb. Montane grassland.
Distr: EQU; Cameroon to Kenya, S to Angola, Zimbabwe & Mozambique.
EQU: Synnott 1037 5/1972.
Note: only one specimen is known from our region, which is from Mt Lonyili on the South Sudan / Uganda border; it may have been collected on the Ugandan side.

Disa scutellifera A.Rich.

F.T.E.A. Orchid. 1: 159 (1968); F.E.E. 6: 241 (1997); Biol. Skr. 51: 511 (2005).
Syn: *Disa schimperi* N.E.Br. – F.P.S. 3: 314 (1956).
Terrestrial herb. Montane grassland.
Distr: EQU; Ethiopia, Uganda, Kenya.
EQU: Myers 11596 7/1939.

Disa welwitschii Rchb.f. subsp. *welwitschii*

F.T.E.A. Orchid. 1: 160 (1968); Biol. Skr. 51: 511 (2005).
Terrestrial herb. Montane grassland.
Distr: EQU; Nigeria to Tanzania, S to Angola & South Africa.
EQU: Prowse 315 6/1953.
Note: we have not seen the cited specimen; the subspecific identification follows Friis & Vollesen.

Disperis anthoceros Rchb.f. var. *anthoceros*

F.T.E.A. Orchid. 1: 229 (1968); F.J.M.: 160 (1976); F.E.E. 6: 247 (1997).
Terrestrial herb. Forest.
Distr: DAR; Nigeria, Cameroon, Ethiopia to South Africa.
DAR: Wickens 2520 9/1964.

Epipogium roseum (D.Don) Lindl.

F.T.E.A. Orchid. 2: 238 (1984); Biol. Skr. 51: 511 (2005).
Saprophytic terrestrial herb. Forest.
Distr: EQU; Guinea to Kenya, S to Angola & Malawi; India to Japan, S to Australia.
EQU: Friis & Vollesen 1139 3/1982.

Eulophia alta (L.) Fawc. & Rendle

F.P.S. 3: 318 (1956); F.T.E.A. Orchid. 3: 435 (1989); F.E.E. 6: 278 (1997).
Terrestrial herb. Swamps, wet grassland.
Distr: EQU; Senegal to Ethiopia, S to Angola & Zimbabwe; neotropics.
EQU: Sillitoe 152 1919.

Eulophia angolensis (Rchb.f.) Summerh.

F.T.E.A. Orchid. 3: 432 (1989); F.E.E. 6: 279 (1997).
Syn: *Eulophia lindleyana* (Rchb.f.) Schltr. – F.P.S. 3: 318 (1956).
Terrestrial herb. Swamps, wet grassland.
Distr: EQU; Senegal to Ethiopia, S to Namibia & South Africa.
EQU: Myers 8740 3/1938.
Note: *E. calantha* Schltr. is also likely to occur in South Sudan (EQU), having been recorded from the Ugandan side of the Imatong Mts (Friis & Vollesen 2005: 512).

Eulophia clitellifera (Rchb.f.) Bolus

F.T.E.A. Orchid. 3: 476 (1989); Biol. Skr. 51: 512 (2005).
Terrestrial herb. Grassland & woodland, in frequently burnt areas.
Distr: ?EQU; Ghana, Ethiopia to South Africa; Madagascar.
Note: no South Sudanese material has been seen but it is recorded as occurring in South Sudan by Friis & Vollesen (2005: 512), who also record it from just over the border in the Ugandan Imatong Mts. F.T.E.A. also records this species from Sudan.

Eulophia cristata (Afzel. ex Sw.) Steud.

F.P.S. 3: 318 (1956); F.T.E.A. Orchid. 3: 439 (1989); F.E.E.

6: 280 (1997); Biol. Skr. 51: 512 (2005).
Terrestrial herb. Grassland & woodland.
Distr: BAG, EQU; Senegal to Ethiopia, S to D.R. Congo & Uganda.
EQU: Myers 6351 3/1937.

Eulophia cucullata (Afzel. ex Sw.) Steud.

F.P.S. 3: 319 (1956); F.T.E.A. Orchid. 3: 437 (1989); F.E.E. 6: 279 (1997); Biol. Skr. 51: 512 (2005).
Terrestrial herb. Grassland & woodland.
Distr: BAG, UN, EQU; Sierra Leone to Ethiopia, S to Angola & South Africa; Madagascar.
EQU: Myers 6460 4/1937.

Eulophia flavopurpurea (Rchb.f.) Rolfe

F.P.S. 3: 318 (1956); F.T.E.A. Orchid. 3: 437 (1989).
Terrestrial herb. Grassland & woodland.
Distr: EQU; Guinea to Ethiopia, S to Angola, Zimbabwe & Mozambique.
EQU: Sillitoe 150 1919.

Eulophia galeoloides Kraenzl.

F.T.E.A. Orchid. 3: 459 (1989); Biol. Skr. 51: 513 (2005).
Saprophytic terrestrial herb. Forest.
Distr: EQU; Ghana to Kenya & Tanzania.
EQU: Friis & Vollesen 400 11/1980.

Eulophia guineensis Lindl.

F.T.E.A. Orchid. 3: 428 (1989); F.E.E. 6: 277 (1997); Biol. Skr. 51: 513 (2005).
Syn: *Eulophia quartiniana* A.Rich. – F.P.S. 3: 319 (1956).
Terrestrial herb. Woodland & bushland.
Distr: ?CS, BAG, EQU; Senegal to Eritrea, S to Botswana & Zimbabwe; Arabia.
EQU: Friis & Vollesen 1265 3/1982.
Note: Andrews records this species from both Fung District and Equatoria; we have only seen material from the latter.

Eulophia horsfallii (Bateman) Summerh.

F.T.E.A. Orchid. 3: 428 (1989); F.E.E. 6: 279 (1997).
Syn: *Eulophia porphyroglossa* (Rchb.f.) Bolus – F.P.S. 3: 321 (1956).
Terrestrial herb. Swamps, riverbanks.
Distr: EQU; Senegal to Ethiopia, S to Angola & South Africa.
EQU: Schweinfurth 2863 2/1870.

Eulophia latilabris Summerh.

F.P.S. 3: 321 (1956); F.T.E.A. Orchid. 3: 430 (1989); Biol. Skr. 51: 513 (2005).
Terrestrial herb. Swamps, wet grassland.
Distr: EQU; Nigeria to Kenya, S to Angola, Botswana & Mozambique.
EQU: Andrews 1469 5/1939.

Eulophia livingstoneana (Rchb.f.) Summerh.

F.P.S. 3: 319 (1956); F.T.E.A. Orchid. 3: 439 (1989); F.E.E. 6: 280 (1997); Biol. Skr. 51: 514 (2005).
Terrestrial herb. Grassland & woodland.
Distr: EQU; C.A.R. to Ethiopia, S to Namibia & South Africa; Madagascar, Comoros.
EQU: Friis & Vollesen 1182 3/1982.

Eulophia mechowii (Rchb.f.) T.Durand & Schinz

Syn: *Eulophia zeyheri* Hook.f. – F.T.E.A. Orchid. 3: 454 (1989).
Syn: *Eulophia abyssinica* sensu Friis & Vollesen, non Rchb.f. – Biol. Skr. 51: 512 (2005).
Terrestrial herb. Grassland & wooded grassland.
Distr: EQU; Nigeria, Ethiopia to Angola & South Africa.
EQU: Sillitoe 139 1919.

Eulophia milnei Rchb.f.

F.T.E.A. Orchid. 3: 456 (1989).
Syn: *Eulophia warneckeana* Kraenzl. – F.P.S. 3: 317 (1956).
Terrestrial herb. Swamps, wet grassland.
Distr: EQU; Senegal to Ethiopia, S to Angola & South Africa.
EQU: Myers 6535 5/1937.

Eulophia montis-elgonis Summerh.

F.P.S. 3: 319 (1956); F.T.E.A. Orchid. 3: 456 (1989); Biol. Skr. 51: 514 (2005).
Terrestrial herb. Montane grassland.
Distr: EQU; Uganda, Kenya.
IUCN: RD
EQU: Myers 11667 7/1939.

Eulophia odontoglossa Rchb.f.

F.T.E.A. Orchid. 3: 458 (1989); F.E.E. 6: 283 (1997); Biol. Skr. 51: 514 (2005).
Syn: *Eulophia shupangae* (Rchb.f.) Kraenzl. – F.P.S. 3: 318 (1956).
Terrestrial herb. Grassland & bushland.
Distr: EQU; Guinea to Ethiopia, S to Angola & South Africa.
EQU: Andrews 1904 6/1939.

Eulophia orthoplectra (Rchb.f.) Summerh.

F.P.S. 3: 321 (1956); F.T.E.A. Orchid. 3: 474 (1989); F.E.E. 6: 281 (1997); Biol. Skr. 51: 514 (2005).
Terrestrial herb. Woodland & grassland.
Distr: EQU; Ivory Coast to Ethiopia, S to Angola, Zimbabwe & Mozambique.
EQU: Myers 8213 1/1938.

Eulophia petersii (Rchb.f.) Rchb.f.

F.T.E.A. Orchid. 3: 440 (1989); F.E.E. 6: 278 (1997); Biol. Skr. 51: 515 (2005).
Syn: *Eulophia schimperiana* A.Rich. – F.P.S. 3: 317 (1956).
Terrestrial herb. Bushland, grassland, rocky ground.
Distr: EQU; Eritrea & Somalia, S to South Africa; Arabia.
EQU: Schweinfurth II.24 5/1870.

Eulophia pyrophila (Rchb.f.) Summerh.

F.P.S. 3: 317 (1956); F.T.E.A. Orchid. 3: 479 (1989); F.E.E. 6: 281 (1997).
Terrestrial herb. Grassland often in burnt areas, rocky ground.
Distr: EQU; Ivory Coast to Nigeria, Ethiopia to Zimbabwe.
EQU: Schweinfurth 2795 12/1869.

Eulophia schweinfurthii Kraenzl.

F.P.S. 3: 317 (1956); F.T.E.A. Orchid. 3: 476 (1989); F.E.E. 6: 281 (1997).
Terrestrial herb. Grassland, bushland & woodland.
Distr: BAG; Ivory Coast, Ethiopia to Angola & South Africa.
BAG: Schweinfurth 2671 1869.

Eulophia speciosa (R.Br.) Bolus

F.T.E.A. Orchid. 3: 473 (1989); F.E.E. 6: 280 (1997); Biol. Skr. 51: 515 (2005).
Syn: *Eulophia wakefieldii* (Rchb.f & S.Moore) Summerh. – F.P.S. 3: 321 (1956).
Terrestrial herb. Grassland, bushland & woodland.
Distr: UN, EQU; Eritrea to Namibia & South Africa; Arabia.
UN: Simpson 7087 6/1929.

Eulophia stachyodes Rchb.f.

F.P.S. 3: 319 (1956); F.T.E.A. Orchid. 3: 444 (1989); F.E.E. 6: 279 (1997); Biol. Skr. 51: 515 (2005).
Terrestrial herb. Grassland & woodland.
Distr: EQU; Nigeria to Ethiopia, S to Angola, Zimbabwe & Mozambique.
EQU: Myers 6479 4/1937.

Eulophia stenoplectra Summerh.

F.T.E.A. Orchid. 3: 479 (1989).
Terrestrial herb. Burnt grassland.
Distr: EQU; Ghana, Uganda.
IUCN: RD
EQU: Myers 8474 2/1938.

Eulophia streptopetala Lindl. var. *streptopetala*

F.T.E.A. Orchid. 3: 469 (1989); F.E.E. 6: 278 (1997); Biol. Skr. 51: 515 (2005).
Terrestrial herb. Grassland, forest margins.
Distr: EQU; Eritrea to Namibia & South Africa.
EQU: Jackson 2999 6/1953.
Note: var. *stenophylla* (Summerh.) P.J.Cribb is also likely to occur in South Sudan (EQU), having been recorded on the Ugandan side of the Imatong Mts (Friis & Vollesen 2005: 515).

Habenaria antennifera A.Rich.

F.E.E. 6: 221 (1997).
Syn: *Habenaria humilior* sensu Wickens pro parte, non Rchb.f. – F.J.M.: 160 (1976).
Terrestrial herb. Montane grassland & scrub.
Distr: DAR; Ethiopia; Yemen.
IUCN: RD
DAR: Wickens 2432 9/1964.

Habenaria armatissima Rchb.f.

F.T.E.A. Orchid. 1: 128 (1968); F.J.M.: 160 (1976); F.E.E. 6: 229 (1997).
Terrestrial herb. Marshy grassland, wooded grassland.
Distr: DAR; Mali to Somalia, S to Namibia, Botswana & Mozambique.
DAR: Wickens 1798 7/1964.

Habenaria bongensium Rchb.f.

F.P.S. 3: 323 (1956); F.W.T.A. 3(1): 196 (1968); F.J.M.: 160 (1976).

Terrestrial herb. Montane grassland.
Distr: DAR, BAG; Benin, Nigeria, Cameroon.
DAR: Wickens 2670 9/1964; **BAG:** Schweinfurth 3974 7/1870.

Habenaria bracteosa Hochst. ex A.Rich.

F.T.E.A. Orchid. 1: 53 (1968); F.E.E. 6: 208 (1997); Biol. Skr. 51: 516 (2005).
Syn: *Habenaria filicornis* sensu Andrews, non Lindl. – F.P.S. 3: 322 (1956).
Terrestrial herb. Montane grassland, forest clearings.
Distr: EQU; Bioko, Cameroon, Ethiopia to Tanzania.
EQU: Myers 11668 7/1939.

Habenaria chirensis Rchb.f.

F.T.E.A. Orchid. 1: 88 (1968); F.E.E. 6: 219 (1997); Biol. Skr. 51: 516 (2005).
Syn: *Habenaria humilior* sensu Wickens pro parte, non Rchb.f. – F.J.M.: 160 (1976).
Terrestrial herb. Montane grassland.
Distr: DAR, EQU; Nigeria to Chad, Ethiopia to D.R. Congo & Tanzania.
DAR: Wickens 2171 8/1964; **EQU:** MacLeay 122 6/1947.

Habenaria cirrhata (Lindl.) Rchb.f.

F.P.S. 3: 324 (1956); F.T.E.A. Orchid. 1: 112 (1968); F.J.M.: 160 (1976); F.E.E. 6: 227 (1997).
Terrestrial herb. Wooded grassland.
Distr: ?RS, DAR, ES, ?EQU; Guinea to Ethiopia, S to Zambia & Malawi; Madagascar.
DAR: Blair 390 9/1964.
Note: Andrews records this species from Equatoria but we have seen no material with which to verify this. The record from the Red Sea region is based on a sterile specimen by Drar (1970).

Habenaria distantiflora A.Rich.

F.E.E. 6: 208 pro parte (1997).
Terrestrial herb. Montane grassland.
Distr: RS; Bioko, Cameroon, D.R. Congo, Ethiopia, Uganda, Kenya; Yemen.
RS: Pettet 146 12/1962.

Habenaria filicornis Lindl.

F.T.E.A. Orchid. 1: 54 (1968); F.J.M.: 160 (1976); F.E.E. 6: 208 (1997).
Terrestrial herb. Montane grassland.
Distr: DAR, KOR; Guinea to Eritrea, S to Angola & South Africa.
DAR: Wickens 2607 9/1964.

Habenaria genuflexa Rendle

F.T.E.A. Orchid. 1: 89 (1968).
Terrestrial herb. Swamps, grassland.
Distr: EQU; Senegal to Uganda, S to Angola & Zambia.
EQU: Jackson 3846 9/1957.

Habenaria huillensis Rchb.f.

F.P.S. 3: 323 (1965); F.T.E.A. Orchid. 1: 96 (1968).
Terrestrial herb. Grassland.
Distr: DAR, EQU; Senegal, Ghana to Kenya, S to Angola & Zimbabwe.
DAR: Jackson 4046 9/1960; **EQU:** Myers 7761 9/1937.

Habenaria humilior Rchb.f.

F.T.E.A. Orchid. 1: 91 (1968); F.J.M.: 160 pro parte
(1976); F.E.E. 6: 220 (1997); Biol. Skr. 51: 516 (2005).
Syn: *Habenaria hochstetteriana* Kraenzl. – F.P.S. 3:
323 (1956).
Terrestrial herb. Montane grassland.
Distr: DAR, EQU; Eritrea to South Africa.
DAR: Wickens 2600 9/1964; EQU: Andrews 1838
6/1939.

Habenaria malacophylla Rchb.f.

F.T.E.A. Orchid. 1: 73 (1968); Biol. Skr. 51: 516 (2005).
Terrestrial herb. Montane forest & grassland.
Distr: EQU; Sierra Leone, Nigeria to Eritrea, S to Angola
& South Africa.
EQU: Friis & Vollesen 82 11/1980.

Habenaria njamnjamica Kraenzl.

F.P.S. 3: 322 (1956); F.T.E.A. Orchid. 1: 80 (1968).
Terrestrial herb. Grassland.
Distr: EQU; C.A.R. to Kenya, S to Zambia.
EQU: Schweinfurth 3960 6/1870.

Habenaria papyracea Schltr.

F.T.E.A. Orchid. 1: 79 (1968); Biol. Skr. 51: 516 (2005).
Terrestrial herb. Grassland.
Distr: EQU; Nigeria, D.R. Congo, Tanzania, Malawi,
Zambia, Zimbabwe.
EQU: Friis & Vollesen 393 11/1980.

Habenaria peristyloides A.Rich.

F.T.E.A. Orchid. 1: 61 (1968); F.E.E. 6: 211 (1997); Biol.
Skr. 51: 517 (2005).
Terrestrial herb. Montane grassland & bushland.
Distr: DAR, EQU; Nigeria, Cameroon, Ethiopia to D.R.
Congo & Tanzania; Yemen.
DAR: Wickens 2719 9/1964; EQU: Myers 11606 7/1939.

Habenaria petitiana (A.Rich.) T.Durand & Schinz

F.T.E.A. Orchid. 1: 56 (1968); F.E.E. 6: 209 (1997); Biol.
Skr. 51: 517 (2005).
Terrestrial herb. Montane grassland, woodland & scrub.
Distr: EQU; Cameroon to Ethiopia, S to South Africa.
EQU: Prowse 335 6/1953.

Habenaria quartiniana A.Rich.

F.T.E.A. Orchid. 1: 66 (1968); F.E.E. 6: 216 (1997); Biol.
Skr. 51: 517 (2005).
Terrestrial herb. Montane grassland & forest margins.
Distr: EQU; Eritrea, Ethiopia, Uganda, Kenya.
EQU: Prowse 313 6/1953.

Habenaria schimperiana Hochst. ex A.Rich.

F.P.S. 3: 323 (1956); F.T.E.A. Orchid. 1: 89 (1968); F.E.E. 6:
220 (1997); Biol. Skr. 51: 517 (2005).
Terrestrial herb. Montane grassland.
Distr: EQU; Ivory Coast, C.A.R. to Eritrea, S to Angola &
South Africa; Yemen.
EQU: Myers 11558 7/1939.

Habenaria walleri Rchb.f.

F.P.S. 3: 323 (1956); F.T.E.A. Orchid. 1: 116 (1968).
Terrestrial herb. Swampy grassland.

Distr: EQU; Guinea, Nigeria to Kenya, S to Zambia &
Malawi.
EQU: Sillitoe 149 1919.

Holothrix arachnoidea (A.Rich.) Rchb.f.

F.T.E.A. Orchid. 1: 7 (1968); F.E.E. 6: 200 (1997).
Terrestrial herb. Montane grassland, rocky hillslopes.
Distr: RS; Ethiopia, Somalia, Kenya; Arabia.
RS: Jackson 2933 4/1953.
Note: *H. brongniartiana* Rchb.f. is also likely to occur
in South Sudan (EQU), having been recorded on the
Ugandan side of the Imatong Mts (Friis & Vollesen 2005:
517).

Holothrix squamata (Hochst. ex A.Rich.) Rchb.f.

F.T.E.A. Orchid. 1: 8 (1968); F.E.E. 6: 200 (1997); Biol. Skr.
51: 517 (2005).
Terrestrial herb. Montane grassland.
Distr: EQU; Ethiopia, Uganda.
EQU: Friis & Vollesen 1279 3/1982.
Note: a local species but not considered to be
threatened.

Holothrix tridentata (Hook.f.) Rchb.f.

F.J.M.: 161 (1976); F.E.E. 6: 200 (1997).
Terrestrial herb. Wooded grassland, rocky hillslopes.
Distr: DAR; Cameroon, Ethiopia.
IUCN: RD
DAR: Wickens 2450 9/1964.

Malaxis maclaudii (Finet) Summerh.

F.P.S. 3: 324 (1956); F.W.T.A. 3(1): 211 (1968); Biol. Skr.
51: 517 (2005).
Terrestrial herb. Forest.
Distr: EQU; Guinea to C.A.R., S to Zambia.
EQU: Andrews 1779 6/1939.

Microcoelia globulosa (Hochst.) L.Jonss.

F.E.E. 6: 288 (1997).
Syn: *Microcoelia guyoniana* (Rchb.f.) Summerh. – F.P.S.
3: 324 (1956).
Epiphytic herb. Forest, particularly margins or riverine
strips, secondary regrowth.
Distr: ?CS; Nigeria to Eritrea, S to Angola & Zimbabwe.
Note: Andrews records this species from Fung District;
we have seen no material with which to verify this
record.

Nervilia kotschyi (Rchb.f.) Schltr. var. kotschyi

F.P.S. 3: 324 (1956); F.J.M.: 161 (1976); F.T.E.A. Orchid.
2: 273 (1984); F.E.E. 6: 253 (1997); Biol. Skr. 51: 518
(2005).
Terrestrial herb. Woodland.
Distr: DAR, CS, EQU; Senegal to Eritrea, S to Zimbabwe
& Mozambique; Madagascar, Comoros.
DAR: Wickens 1805 7/1964; EQU: Andrews 478 4/1939.
Note: *N. adolphi* Schltr. var. *seposita* N.Hallé & Toill.-Gen.
is also likely to occur in South Sudan (EQU), having been
recorded from the Ugandan side of the Imatong Mts
(Friis & Vollesen 2005: 518).

Nervilia subintegra Summerh.

Biol. Skr. 51: 518 (2005).
Terrestrial herb. Forest.

Distr: EQU; Guinea to Togo, Cameroon, C.A.R., D.R. Congo.
EQU: Friis & Vollesen 592 12/1980.

Oeceoclades maculata (Lindl.) Lindl.

F.T.E.A. Orchid. 3: 416 (1989).
Syn: *Eulophidium ledienii* (Stein ex N.E.Br.) De Wild. – F.P.S. 3: 321 (1956).
Terrestrial or epiphytic herb. Forest & woodland.
Distr: EQU; Senegal to Kenya, S to Angola & South Africa; Madagascar, Mascarenes, neotropics.
EQU: Myers 9160 8/1938.

Platycoryne crocea (Schweinf. ex Rchb.f.) Rolfe

F.P.S. 3: 324 (1956); F.T.E.A. Orchid. 1: 146 (1968); F.E.E. 6: 234 (1997).
Syn: *Platycoryne crocea* (Schweinf. ex Rchb.f.) Rolfe subsp. *montis-elgon* (Schltr.) Summerh. – F.T.E.A. Orchid. 1: 148 (1968); Biol. Skr. 51: 519 (2005).
Terrestrial herb. Montane grassland.
Distr: BAG; Cameroon to Ethiopia, S to Angola & Zimbabwe.
BAG: Myers 11493 6/1939.

Platylepis glandulosa (Lindl.) Rchb.f.

F.T.E.A. Orchid. 2: 245 (1984); F.E.E. 6: 256 (1997).
Terrestrial herb. Swamps, riverside in forest.
Distr: EQU; Guinea Bissau to Ethiopia, S to Angola & South Africa.
EQU: Myers 13402 8/1940.

Polystachya bennettiana Rchb.f.

F.T.E.A. Orchid. 3: 357 (1984).
Syn: *Polystachya stricta* Rolfe – F.P.S. 3: 325 (1956).
Epiphytic herb. Woodland & forest.
Distr: ?EQU; Ivory Coast, Nigeria, Cameroon, Eritrea to Zambia & Tanzania.
Note: Andrews records this species from Equatoria; we have seen no material with which to verify this, but Friis & Vollesen (2005: 519) record it from the Ugandan side of the Imatong Mts and so it is likely to occur in South Sudan.

Polystachya concreta (Jacq.) Garay & H.R.Sweet

Kew Bull. 42: 724 (1987).
Syn: *Polystachya tessellata* Lindl. – F.P.S. 3: 325 (1956); F.T.E.A. Orchid. 2: 360 (1984); F.E.E. 6: 264 (1997).
Epiphytic herb. Wet forest, often riverine.
Distr: ?EQU; pantropical.
Note: Andrews records this species from Equatoria; no material has been seen with which to verify this record.

Polystachya cultriformis (Thouars) Lindl. ex Spreng.

F.T.E.A. Orchid. 3: 341 (1984); Biol. Skr. 51: 519 (2005).
Epiphytic herb. Forest.
Distr: EQU; Bioko, Cameroon, Gabon, D.R. Congo to Ethiopia, S to South Africa; Madagascar, Mascarenes.
EQU: Friis & Vollesen 74 11/1980.
Note: *P. eurychila* Summerh. is also likely to occur in South Sudan (EQU), having been recorded from the Ugandan side of the Imatong Mts (Friis & Vollesen 2005: 519).

Polystachya steudneri Rchb.f.

F.T.E.A. Orchid. 2: 375 (1984); F.E.E. 6: 265 (1997); Biol. Skr. 51: 519 (2005).
Epiphytic herb. Woodland & dry forest.
Distr: EQU; Nigeria, Cameroon, Eritrea to D.R. Congo & Kenya.
EQU: Andrews 1858 6/1939.

Polystachya transvaalensis Schltr.

F.T.E.A. Orchid. 3: 357 (1989); Biol. Skr. 51: 519 (2005).
Epiphytic herb. Montane forest & grassland.
Distr: EQU; D.R. Congo to Kenya, S to South Africa.
EQU: Kielland-Lund 913 6/1984 (n.v.).

Rhipidoglossum adoxum (F.N.Rasm.) Senghas

Orchideen (Schlechter) 1: 1110 (1986).
Syn: *Diaphananthe adoxa* F.N.Rasm. – F.T.E.A. Orchid. 3: 536 (1989); F.E.E. 6: 290 (1997); Biol. Skr. 51: 509 (2005).
Epiphytic herb. Montane forest.
Distr: EQU; Ethiopia, Uganda, Kenya.
EQU: Friis & Vollesen 1223 3/1982.

Rhipidoglossum schimperianum (A.Rich.) Garay

Bot. Mus. Leafl. 23: 195 (1972).
Syn: *Diaphananthe schimperiana* (A.Rich.) Summerh. – F.T.E.A. Orchid. 3: 531 (1989); F.E.E. 6: 293 (1997); Biol. Skr. 51: 510 (2005).
Epiphytic or lithophytic herb. Montane forest.
Distr: EQU; Ethiopia, Uganda.
EQU: Myers 11671 7/1939.

Satyrium coriophoroides A.Rich.

F.P.S. 3: 325 (1956); F.T.E.A. Orchid. 1: 201 (1968); F.J.M.: 161 (1976).
Syn: *Satyrium sacculatum* (Rendle) Rolfe – F.T.E.A. Orchid. 1: 200 (1968); F.E.E. 6: 245 (1997); Biol. Skr. 51: 520 (2005).
Terrestrial herb. Montane grassland & woodland.
Distr: DAR, EQU; Cameroon to Ethiopia, S to Zimbabwe.
DAR: Wickens 2133 1964 (n.v.); **EQU:** Myers 11555 7/1939.
Notes: (1) Wickens cites several collections from Jebel Marra under this species but we have been unable to trace the specimens at Kew. (2) *S. carsonii* Rolfe may also occur in South Sudan (EQU), having been recorded from the Ugandan side of the Imatong Mts (Friis & Vollesen 2005: 520).

Satyrium fimbriatum Summerh.

F.P.S. 3: 325 (1956); F.T.E.A. Orchid. 1: 187 (1984); Biol. Skr. 51: 520 (2005).
Terrestrial herb. Montane grassland.
Distr: EQU; Uganda, Kenya, Tanzania.
EQU: Myers 11665 7/1939.

Satyrium sceptrum Schltr.

F.T.E.A. Orchid. 1: 197 (1984); Biol. Skr. 51: 520 (2005).
Terrestrial herb. Montane grassland & bushland.
Distr: EQU; D.R. Congo, Uganda, Kenya, Tanzania, Malawi, Zambia, Zimbabwe.
EQU: Myers 11577 7/1939.

Satyrium trinerve Lindl.

Biol. Skr. 51: 521 (2005).
Syn: *Satyrium atherstonei* Rchb.f. – F.T.E.A. Orchid. 1: 209 (1968).
Terrestrial herb. Moist montane grassland.
Distr: EQU; Guinea, Sierra Leone, Nigeria to Uganda, S to Angola & South Africa.
EQU: MacLeay 121 6/1947.

Satyrium volkensii Schltr.

F.T.E.A. Orchid. 1: 202 (1968); Biol. Skr. 51: 521 (2005).
Terrestrial herb. Montane woodland, bushland & grassland.
Distr: EQU; Nigeria, Cameroon, D.R. Congo, Kenya, S to Zimbabwe.
EQU: MacLeay 70 6/1947.

Stolzia repens (Rolfe) Summerh.

F.T.E.A. Orchid. 2: 330 (1984); F.E.E. 6: 268 (1997); Biol. Skr. 51: 521 (2005).
Epiphytic herb. Montane forest.
Distr: EQU; Ghana to Ethiopia, S to Zimbabwe.
EQU: Friis & Vollesen 1291 3/1982.

Tridactyle filifolia (Schltr.) Schltr.

F.T.E.A. Orchid. 3: 610 (1989); F.E.E. 6: 307 (1997).
Syn: *Tridactyle tridentata* sensu Andrews, non (Harv.) Schltr. – F.P.S. 3: 325 (1956).
Epiphytic herb. Forest, including disturbed areas.
Distr: ?EQU; Sierra Leone to Ethiopia, S to Zambia & Malawi.
Note: Andrews records "*Tridactyle tridentata*" from Asugi, Yei River; whilst we have seen no specimen, this is likely to be an accurate generic record. It is difficult to determine whether this refers to *T. tridentata* s.s. or *T. filifolia* (which are sometimes treated as conspecific) in the absence of specimens.

Hypoxidaceae

H. Pickering

Curculigo pilosa (Schumach. & Thonn.) Engl. subsp. major (Baker) Wiland

F.P.S. 3: 306 (species only) (1956); F.J.M.: 159 (species only) (1976); F.E.E. 6: 89 (species only) (1997); Biol. Skr. 51: 506 (species only) (2005); F.T.E.A. Hypoxid.: 20 (2006).
Rhizomatous perennial herb. Grassland, woodland.
Distr: DAR, ES, BAG, UN, EQU; Senegal to Ethiopia, S to Zambia & Mozambique.
DAR: Wickens 1774 6/1964; UN: J.M. Lock 81/76 5/1981.

Hypoxis angustifolia Lam. var. luzuloides (Robyns & Tournay) Wiland

F.P.S. 3: 306 (species only) (1956); F.J.M.: 160 (species only) (1976); F.E.E. 6: 87 (species only) (1997); Biol. Skr. 51: 506 (species only) (2005); F.T.E.A. Hypoxid.: 5 (2006).
Syn: *Hypoxis* sp. sensu Friis & Vollesen – Biol. Skr. 51: 507 (2005).
Cormous perennial herb. Grassland including wet sites, open woodland.

Distr: DAR, UN, EQU; Guinea to Somalia, S to Angola & South Africa; Madagascar, Mascarenes, Arabia.
DAR: Wickens 1945 7/1964; EQU: Myers 11648 7/1939.

Hypoxis urceolata Nel

F.P.S. 3: 306 (1956); F.T.E.A. Hypoxid.: 18 (2006).
Syn: *Hypoxis multiflora* sensu Andrews, non Nel – F.P.S. 3: 306 (1956).
Syn: *Hypoxis villosa* sensu Friis & Vollesen, non L.f. – Biol. Skr. 51: 506 (2005).
Cormous perennial herb. Grassland, open woodland.
Distr: EQU; D.R. Congo, Uganda, Kenya, Tanzania, Rwanda.
EQU: Andrews 1909 6/1939.

Iridaceae

H. Pickering

Aristea abyssinica Pax

F.T.E.A. Irid.: 10 (1996); F.E.E. 6: 165 (1997).
Syn: *Aristea alata* Baker subsp. *abyssinica* (Pax) Weim. – Biol. Skr. 51: 498 (2005).
Syn: *Aristea alata* sensu Andrews, non Baker – F.P.S. 3: 291 (1956).
Rhizomotous perennial herb. Montane grassland & bushland, forest margins.
Distr: EQU; Nigeria, Cameroon, Ethiopia to South Africa.
EQU: Friis & Vollesen 14 11/1980.

Crocosmia aurea (Pappe ex Hook.) Planch. subsp. pauciflora (Milne-Redh.) Goldblatt

F.Z. 12(4): 50 (1993); F.T.E.A. Irid.: 44, in notes (1996).
Syn: *Crocosmia pauciflora* Milne-Redh. – F.P.S. 3: 291 (1956).
Rhizomatous perennial herb. Forest margins.
Distr: EQU; C.A.R. to Angola & Zambia.
EQU: Cartwright s.n. 1/1911.

[Freesia]

Note: *Freesia laxa* (Thunb.) Goldblatt & J.C.Manning is likely to occur in South Sudan, having been recorded on the Ugandan side of the Imatong Mts (see Friis & Vollesen 2005: 499).

Gladiolus boranensis Goldblatt

F.T.E.A. Irid.: 66 (1996); F.E.E. 6: 175 (1997).
Cormous perennial herb. Montane forest & scrub.
Distr: EQU; Ethiopia, Kenya.
EQU: Myers 14186 10/1941.

Gladiolus dalenii Van Geel subsp. dalenii

F.T.E.A. Irid.: 70 (1996); F.E.E. 6: 176 (1997); Biol. Skr. 51: 499 (2005).
Syn: *Gladiolus natalensis* Reinw. ex Hook. var. *natalensis* – F.J.M.: 158 (1976).
Syn: *Gladiolus psittacinus* Hook. – F.P.S. 3: 293 (1956).
Syn: *Gladiolus quartinianus* A.Rich. – F. Darfur Nord-Occ. & J. Gourgeil: 134 (1969).
Cormous perennial herb. Grassland & wooded grassland.
Distr: DAR, UN, EQU; Senegal to Eritrea, S to Namibia & South Africa; Madagascar, Arabia.
DAR: Wickens 2092 8/1964; EQU: Andrews 1312 5/1939.

Gladiolus dalenii Van Geel subsp. *andongensis* (Baker) Goldblatt

F.T.E.A. Irid.: 71 (1996); F.E.E. 6: 177 (1997); Biol. Skr. 51: 499 (2005).
Cormous perennial herb. Upland grassland & wooded grassland.
Distr: EQU; Senegal to Ethiopia, S to Angola & Mozambique.
EQU: Myers 10932 4/1939.

Gladiolus dichrous (Bullock) Goldblatt

F.T.E.A. Irid.: 79 (1996); Biol. Skr. 51: 500 (2005).
Cormous perennial herb. Montane grassland & rock outcrops.
Distr: EQU; Uganda, Kenya.
EQU: Myers 11605 7/1939.
Note: a very local species but not considered to be threatened.

Gladiolus sudanicus Goldblatt

Goldblatt, Gladiolus Trop. Afr.: 277 (1996); F.E.E. 6: 182 (1997).
Cormous perennial herb. Dry grassland & shrubland.
Distr: KOR; Ethiopia.
IUCN: RD
KOR: Musselman 6237 8/1983.

Gladiolus unguiculatus Baker

F.P.S. 3: 293 (1956); F.J.M.: 158 (1976); F.T.E.A. Irid.: 56 (1996).
Cormous perennial herb. Seasonally wet grassland, wet flushes on rocks.
Distr: DAR, BAG; Senegal to C.A.R., S to Angola, Botswana & Zimbabwe.
DAR: Wickens 1967 7/1964; **BAG:** Schweinfurth 1918 6/1869.

Hesperantha petitiana (A.Rich.) Baker

F.T.E.A. Irid.: 33 (1996); F.E.E. 6: 171 (1997); Biol. Skr. 51: 500 (2005).
Cormous perennial herb. Montane grassland & rocky outcrops.
Distr: EQU; Cameroon, Ethiopia to Zimbabwe.
EQU: Myers 11640 7/1939.

Lapeirousia abyssinica (R.Br. ex A.Rich.) Baker

F.P.S. 3: 293 (1956); F.E.E. 6: 168 (1997).
Cormous perennial herb. Rocky sites.
Distr: ES; Eritrea, Ethiopia.
ES: Schweinfurth s.n. 7/1865.
Note: a very local species but probably not threatened.

Lapeirousia schimperi (Asch. & Klatt) Milne-Redh.

F.J.M.: 158 (1976); F.T.E.A. Irid.: 25 (1996); F.E.E. 6: 170 (1997).
Cormous perennial herb. Seasonally wet areas.
Distr: DAR; Eritrea & Somalia, S to Namibia, Botswana & Zimbabwe.
DAR: Wickens 2695 9/1964.

Moraea afro-orientalis Goldblatt

F.T.E.A. Irid.: 16 (1996); Biol. Skr. 51: 500 (2005).
Syn: *Moraea carsonii* sensu Andrews, non Baker – F.P.S. 3: 293 (1956).
Cormous perennial herb. Montane grassland & bushland.
Distr: EQU; Uganda, Kenya, Tanzania.
EQU: Myers 10952 4/1939.

Moraea schimperi (Hochst.) Pic.Serm.

F.P.S. 3: 293 (1956); F.T.E.A. Irid.: 18 (1996); F.E.E. 6: 168 (1997); Biol. Skr. 51: 501 (2005).
Cormous perennial herb. Swampy grassland, streamsides.
Distr: EQU; Nigeria, Cameroon, Ethiopia to Angola & Mozambique.
EQU: Friis & Vollesen 1222 3/1982.
Note: *M. stricta* Baker is also likely to occur in South Sudan, having been recorded from the Ugandan side of the Imatong Mts; indeed, Friis & Vollesen (2005: 501) include South Sudan within this species' distribution but no specimens have been seen to verify this.

Romulea fischeri Pax

F.T.E.A. Irid.: 36 (1996); F.E.E. 6: 171 (1997); Biol. Skr. 51: 501 (2005).
Syn: *Romulea bulbocodium* sensu Andrews, non (L.) Sebast. & Mauri – F.P.S. 3: 294 (1956).
Syn: *Romulea camerooniana* sensu Wickens, non Baker – F.J.M.: 158 (1976).
Cormous perennial herb. Montane grassland & rocky outcrops.
Distr: RS, DAR, ?EQU; Eritrea, Ethiopia, Somalia, Uganda, Kenya; Arabia.
RS: Pettet 170 12/1962; **EQU:** Myers 11643 7/1939.
Notes: (1) Myers 11643 at Kew was identified by de Vos as cf. *fisheri*; the status of this species in South Sudan requires confirmation. (2) The widespread *R. camerooniana* Baker, difficult to distinguish from *R. fischeri*, may also occur in Sudan.

Zygotritonia bongensis (Pax) Mildbr.

F.P.S. 3: 294 (1956); F.W.T.A. 3(1) 144, in notes (1968).
Cormous perennial herb. Grassland & wooded grassland.
Distr: BAG; Guinea to C.A.R. & D.R. Congo; Socotra.
BAG: Schweinfurth 4025 6/1870.

Xanthorrhoeaceae

H. Pickering

Aloe camperi Schweinf.

F.E.E. 6: 130 (1997); Carter et al., Aloes, Defin. Guide: 621 (2011).
Syn: *Aloe eru* A.Berger – F.P.S. 3: 261 pro parte (1956).
Succulent shrub. Rocky hillslopes, sandy plains.
Distr: RS; Eritrea, Ethiopia.
IUCN: RD
RS: Schweinfurth 19 4/1865.
Note: Quézel (1969: 133) records *A. eru* from Jebel Gourgeil in Darfur. This is likely to be a misidentification, though we haven't seen Quézel's specimens. It may be the same species as that named *A. elegans* on Jebel Marra.

Aloe canarina S.Carter

F.T.E.A. Alo.: 41 (1994); Carter et al., Aloes, Defin. Guide: 612 (2011).
Succulent perennial herb. Open dry bushland.
Distr: EQU; Uganda.

IUCN: DD
EQU: Padwa 231 5/1953
Note: there are some question marks over whether the plants from the Ilemi Triangle truly match the Ugandan material. Plowes (1989) previously named these plants *A. marsabitensis* I.Verd. & Christian.

Aloe crassipes Baker

F.P.S. 3: 261 (1956); Carter et al., Aloes, Defin. Guide: 319 (2011).
Succulent perennial herb. Rocky hillslopes, open wooded grassland.
Distr: EQU; South Sudan endemic.
IUCN: DD
EQU: Myers 6657 5/1937.
Note: Andrews records this species from "North-eastern Sudan: Kishi, between Suakin and Berber", an error in the recording of the collecting locality (see Wickens 1971: 135). *A. crassipes* is restricted to the Mt Bangenze region of Equatoria.

Aloe dioli L.E.Newton

Carter et al., Aloes, Defin. Guide: 489 (2011).
Succulent perennial herb. Montane forest.
Distr: EQU; South Sudan endemic.
IUCN: RD
EQU: Powys & Dioli 824 5/1986.

Aloe cf. elegans Tod.

Bot. Exp. Sud.: 78 (1970); F.J.M.: 155 (1976); Carter et al., Aloes, Defin. Guide: 373 (2011).
Succulent perennial herb. Rocky hillslopes.
Distr: DAR; (for *elegans*) Ethiopia, Eritrea.
DAR: Wickens 2626 9/1964.
Note: the Darfur material is rather isolated from the rest of this species' range and was not mentioned by Carter *et al.* (2011) in their monograph of Aloe. It may refer to a related but different species but the Sudan material is insufficient to confirm this at present (Sebsebe Demissew, pers. comm.). See also discussion by Plowes (1989: 65).

Aloe erensii Christian

Reynolds, Aloes Trop. Afr. & Madagascar: 59 (1966); F.T.E.A. Alo.: 38 (1994); Carter et al., Aloes, Defin. Guide: 307 (2011).
Succulent perennial herb. Dry rocky slopes & cliffs with open bushland.
Distr: EQU; Kenya.
IUCN: EN
EQU: Reynolds 7002 (fl) 5/1953.
Note: the record for South Sudan is from Reynolds who grew the plant on from cultivation, originally collected by J.G. Williams from Kamathia in the Ilemi Triangle.

Aloe ithya T.A.McCoy & L.E.Newton

Haseltonia 19: 64 (2014).
Succulent shrub. Cliffs.
Distr: EQU; South Sudan endemic.
IUCN: RD
EQU: Idriss in McCoy 2892 (fl) 2003.
Note: this species has only just been described from the Imatong Mts; the authors note that the type may match *Andrews* 1843 collected in 1939.

Aloe labworana (Reynolds) S.Carter

F.T.E.A. Alo.: 28 (1994); Biol. Skr. 51: 489 (2005); Carter et al., Aloes, Defin. Guide: 463 (2011).
Succulent perennial herb. Rock outcrops.
Distr: EQU; Uganda.
IUCN: RD
EQU: Friis & Vollesen 553 11/1980.

Aloe macleayi Reynolds

Biol. Skr. 51: 489 (2005); Carter et al., Aloes, Defin. Guide: 348 (2011).
Succulent perennial herb. Montane grassland.
Distr: EQU; South Sudan endemic.
IUCN: EN
EQU: Friis & Vollesen 1289 3/1982.

Aloe macrocarpa Tod.

Reynolds, Aloes Trop. Afr. & Madagascar: 91 (1966); F.E.E. 6: 122 (1997); Carter et al., Aloes, Defin. Guide: 173 (2011).
Succulent perennial herb. Grassland with scattered bushes, amongst rocks.
Distr: KOR; Mali to Somalia.
KOR: Reynolds 8550 (fl) 2/1959.
Note: the record for Sudan is from Reynolds who grew the plant on from cultivation, originally collected by K.N.G. MacLeay in the Nuba Mts.

Aloe schweinfurthii Baker

F.P.S. 3: 261 (1956); F.T.E.A. Alo.: 28 (1994); Biol. Skr. 51: 489 (2005); Carter et al., Aloes, Defin. Guide: 456 (2011).
Succulent perennial herb. Grassland with rock outcrops.
Distr: BAG, EQU; Mali to D.R. Congo & Uganda.
EQU: Myers 6634 5/1937.

Aloe sinkatana Reynolds

Bot. Exp. Sud.: 78 (1970); Carter et al., Aloes, Defin. Guide: 361 (2011).
Syn: *Aloe eru* sensu Andrews pro parte, non A.Berger – F.P.S. 3: 261 (1956).
Succulent perennial herb. Dry riverbeds.
Distr: RS; Sudan endemic.
IUCN: EN
RS: Schweinfurth 206 8/1868.

Aloe tweedieae Christian

F.T.E.A. Alo.: 35 (1994); Carter et al., Aloes, Defin. Guide: 350 (2011).
Succulent perennial herb. Rocky hillslopes, open wooded grassland.
Distr: EQU; Uganda, Kenya.
EQU: Reynolds 7490 (fl) 7/1954.
Note: the single South Sudanese specimen was previously identified by Reynolds (Aloes Trop. Afr. & Madagascar: 230) as *A. secundiflora* Engl. The two are certainly close and *tweedieae* is sometimes treated as a variety of *secundiflora*.

Aloe vituensis Baker

F.T.E.A. Alo.: 46 (1994); Carter et al., Aloes, Defin. Guide: 498 (2011).
Succulent shrublet. Rocky hillslopes with grassland & shrubland.

Distr: EQU; Kenya.
IUCN: RD
EQU: <u>Newton & Powys s.n.</u> 1994.

Aloe wrefordii Reynolds

F.T.E.A. Alo.: 29 (1994); Carter et al., Aloes, Defin. Guide: 369 (2011).
Succulent perennial herb. Rocky hillslopes.
Distr: EQU; Uganda, Kenya.
IUCN: RD
EQU: <u>Martin 56</u> (fl) 3/1934.

Asphodelus tenuifolius Cav.

F.P.S. 3: 264 (1956).
Syn: *Asphodelus fistulosus* L. var. *tenuifolius* (Cav.)
Baker – Bot. Exp. Sud.: 79 (1970).
Syn: *Asphodelus fistulosus* sensu Demissew & Nordal,
non L. – F.E.E. 6: 116 (1997).
Annual herb. Wadis.
Distr: NS, RS, ?EQU; Eritrea, Somalia; Mediterranean, N Africa, Arabia to India.
RS: <u>Carter 1944</u> 11/1987; **EQU:** <u>Drar 1286a</u> 4/1938 (n.v.).
Note: the identification of the South Sudan specimen cited is from Drar (1970) and may be incorrect.

Bulbine abyssinica A.Rich.

F.E.E. 6: 111 (1997); F.T.E.A. Asphodel.: 11 (2002); Biol. Skr. 51: 488 (2005).
Syn: *Bulbine asphodeloides* sensu auctt., non (L.)
Spreng. – F.P.S. 3: 266 (1956); Bot. Exp. Sud.: 79 (1970).
Perennial herb. Dry bushland, woodland & grassland.
Distr: ?CS, BAG, UN, EQU; Eritrea & Somalia, S to Angola, Botswana & South Africa; Arabia.
EQU: <u>Andrews 954</u> 5/1939.
Note: Andrews recorded this species from Fung District and Equatoria; we have not seen material from the former.

Kniphofia nubigena Mildbr.

F.P.S. 3: 271 (1956); Kew Bull. 28: 483 (1973).
Perennial herb. Rocky hillslopes.
Distr: RS; Sudan endemic.
IUCN: RD
RS: <u>Aylmer 559</u> 3/1936.

Kniphofia pumila (Aiton) Kunth

F.E.E. 6: 107 (1997); F.T.E.A. Asphodel.: 7 (2002); Biol. Skr. 51: 489 (2005).
Syn: *Kniphofia comosa* Hochst. – F.P.S. 3: 271 (1956).
Rhizomatous perennial herb. Grassland, near streams.
Distr: ?ES, EQU; D.R. Congo, Eritrea, Ethiopia, Uganda, Kenya.
EQU: <u>Myers 11680</u> 7/1939.
Note: Andrews recorded this species from "Kassala: Gallabat", presumably based on a Schweinfurth collection which we have not seen.

Amaryllidaceae

H. Pickering

Allium cf. alibile Steud. ex A.Rich.

F.P.S. 3: 286 (*A. alibile*) (1956); F.E.E. 6: 154 (*A. alibile*) (1997).

Syn: *Allium* sp. – F.J.M.: 158 (1976).
Bulbous perennial herb. Habitat in Sudan unknown; *A. alibile* occurs in dry bushland & montane grassland.
Distr: DAR; (for species) Ethiopia.
IUCN: DD
DAR: <u>Lynes 37e</u>
Note: this taxon, allied to or conspecific with the Ethiopian *A. alibile* Steud. ex A.Rich., is possibly an escapee from cultivation in Darfur.

Allium spathaceum Steud. ex A.Rich.

F.P.S. 3: 286 (1956).
Syn: *Allium subhirsutum* L. subsp. *spathaceum* (Steud. ex A.Rich.) Duyfjes – F.E.E. 6: 151 (1997).
Bulbous perennial herb. Grassland.
Distr: RS; Eritrea, Ethiopia, Djibouti, Somalia.
RS: <u>Aylmer 129</u> 2/1932.

Ammocharis tinneana (Kotschy & Peyr.) Milne-Redh. & Schweick.

F.P.S. 3: 286 (1956); F.J.M.: 158 (1976); F.T.E.A. Amaryllid.: 19 (1982); F.E.E. 6: 161 (1997).
Bulbous perennial herb. Wooded grassland & bushland, overgrazed areas.
Distr: DAR, KOR, BAG, UN, EQU; C.A.R. to Somalia, S to Namibia & Zimbabwe.
DAR: <u>Wickens 1833</u> 7/1964; **BAG:** <u>Myers 6300</u> (fl) 2/1937.

Boophone disticha (L.f.) Herb.

F.P.S. 3: 286 (1956); F.T.E.A. Amaryllid.: 21 (1982).
Bulbous perennial herb. Grassland & wooded grassland.
Distr: EQU; D.R. Congo to Kenya, S to Namibia & South Africa.
EQU: <u>Jackson 1295</u> 3/1950.

Crinum abyssinicum Hochst. ex A.Rich.

F.P.S. 3: 287 (1956); F.E.E. 6: 161 (1997).
Bulbous perennial herb. Wet grassland, swamps, riverbanks.
Distr: ?CS; Eritrea, Ethiopia, Somalia.
Note: Andrews recorded this species from "Fung District"; the only material we have seen from Sudan which has been named *C. abyssinicum* is *Schweinfurth 13* from Gallabat, Eastern Sudan, which was later annotated "probably not *abyssinica*".

Crinum bambusetum Nordal & Sebsebe

Kew Bull. 57: 467 (2002).
Syn: *Crinum purpurascens* sensu auctt., non Herb. – F.P.S. 3: 287 (1956); Bot. Exp. Sud.: 15 (1970).
Bulbous perennial herb. Grassland, woodland & thicket.
Distr: RS, EQU; Ethiopia.
IUCN: RD
RS: <u>Elwes s.n.</u> 6/1910; **EQU:** <u>Myers 8894</u> 4/1938.

Crinum biflorum Rottb.

Syn: *Crinum distichum* Herb. – Biol. Skr. 51: 497 (2005).
Syn: *Crinum pauciflorum* Baker – F.P.S. 3: 287 (1956).
Bulbous perennial herb. Seasonally wet grassland over rocks.
Distr: BAG, EQU; Senegal to C.A.R.
BAG: <u>Schweinfurth 1975</u> 5/1869.

Crinum glaucum A.Chev.

F.T.E.A. Amaryllid.: 17 (1982).
Bulbous perennial herb. Grassland & wooded grassland.
Distr: UN, EQU; Guinea Bissau, Ivory Coast to Cameroon, Uganda.
EQU: Andrews 948 4/1939.

Crinum jagus (J.Thomps.) Dandy

F.P.S. 3: 286 (1956); F.T.E.A. Amaryllid.: 17 (1982).
Bulbous perennial herb. Rivers, wet grassland.
Distr: EQU; Sierra Leone to Uganda, S to Angola & D.R.Congo.
EQU: Andrews 1212 (fl) 5/1939.

Crinum macowanii Baker

F.T.E.A. Amaryllid.: 11 (1982); F.E.E. 6: 160 (1997); Biol. Skr. 51: 497 (2005).
Bulbous perennial herb. Grassland.
Distr: EQU; Eritrea & Somalia, S to Namibia & South Africa.
EQU: Friis & Vollesen 1026 2/1982.

Crinum ornatum Bury

F.P.S. 3: 287 (1956); F.E.E. 6: 160 (1997); Biol. Skr. 51: 497 (2005).
Syn: *Crinum zeylanicum* sensu Nordal pro parte, non (L.) L. – F.T.E.A. Amaryllid.: 15 (1982).
Bulbous perennial herb. Grassland.
Distr: DAR, KOR, BAG, UN, EQU; Senegal to Somalia, S to D.R. Congo.
DAR: Wickens 1761 6/1964; **EQU:** Myers 8704 3/1938.

Cyrtanthus sanguineus (Lindl.) Walp. subsp. minor Nordal

F.T.E.A. Amaryllid.: 25 (1982); Biol. Skr. 51: 497 (2005).
Syn: *Cyrtanthus* sp. sensu Andrews – F.P.S. 3: 287 (1956).
Bulbous perennial herb. Rocky hillslopes, montane grassland.
Distr: EQU; Kenya, Tanzania.
EQU: Friis & Vollesen 912 2/1982.

Pancratium centrale (A.Chev.) Traub

Pl. Life 19: 59 (1963).
Bulbous perennial herb. Grassland & bushland.
Distr: DAR; Cameroon, Chad, C.A.R., Ethiopia.
DAR: Wickens 3649 6/1977.
Note: a widespread but scarce species.

Pancratium maximum Forssk.

F.P.S. 3: 289 (1956).
Bulbous perennial herb. Semi-desert.
Distr: RS; Arabia.
RS: Strickland 1 4/1897.

Pancratium tenuifolium Hochst. ex A.Rich.

F.T.E.A. Amaryllid.: 26 (1982); F.E.E. 6: 161 (1997); Biol. Skr. 51: 497 (2005).
Bulbous perennial herb. Bushland, open wooded grassland, semi-desert.
Distr: CS, EQU; Guinea Bissau to Somalia, S to Namibia & South Africa.
CS: Lea 208 (fl) 6/1954; **EQU:** Myers 8791 3/1938.

Pancratium tortuosum Herb.

F.P.S. 3: 289 (1956); F.Egypt 4: 87 (2005).
Bulbous perennial herb. Sandy desert.
Distr: RS; Egypt, Arabia.
RS: Schweinfurth 273 9/1868.
Note Quézel (1969: 134) records this species from Jebel Gourgeil in Darfur; this is probably in error.

Pancratium trianthum Herb.

F.P.S. 3: 289 (1956); F.W.T.A. 3(1): 136 (1968).
Bulbous perennial herb. Grassland.
Distr: DAR, KOR, BAG; Gambia to Djibouti; N Africa.
KOR: Kotschy 10; **BAG:** Schweinfurth 1865 3/1869.
Note: the identity of the Sudanese / South Sudanese material requires confirmation.

Scadoxus multiflorus (Martyn) Raf. subsp. multiflorus

F.T.E.A. Amaryllid.: 4 (1982); F.E.E. 6: 158 (1997); Biol. Skr. 51: 498 (2005).
Syn: *Haemanthus multiflorus* Martyn – F.P.S. 3: 287 (1956); Bot. Exp. Sud.: 15 (1970); F.J.M.: 158 (1976).
Syn: *Haemanthus rupestris* Baker – F.P.S. 3: 289 (1956).
Bulbous and rhizomatous perennial herb. Grassland, woodland & forest.
Distr: RS, DAR, KOR, BAG, UN, EQU; Senegal to Somalia, S to Namibia & South Africa; Arabia.
DAR: Wickens 1815 7/1964; **BAG:** Simpson 7147 (fl fr) 6/1929.

Scadoxus puniceus (L.) Friis & Nordal

F.T.E.A. Amaryllid.: 5 (1982); F.E.E. 6: 158 (1997); Biol. Skr. 51: 498 (2005).
Bulbous perennial herb. Montane forest, seasonally wet woodland & grassland.
Distr: CS, EQU; Ethiopia, Tanzania to South Africa.
CS: Lea 207 6/1954; **EQU:** Friis & Vollesen 1209 (fl) 3/1982.
Note: the identity of the South Sudanese specimen is not certain – see Friis & Vollesen (2005: 498).

Asparagaceae

H. Pickering

Albuca abyssinica Jacq.

F.P.S. 3: 260 (1956); F.T.E.A. Hyacinth.: 22 (1996); F.E.E. 6: 145 (1997); Biol. Skr. 51: 486 (2005).
Syn: *Albuca purpurascens* Engl. – F.P.S. 3: 261 (1956).
Syn: *Albuca tayloriana* Rendle – F.P.S. 3: 260 (1956).
Syn: *Albuca wakefieldii* Baker – F.P.S. 3: 260 (1956).
Syn: *Urginea petitiana* (A.Rich.) Solms – F.P.S. 3: 275 (1956).
Bulbous perennial herb. Grassland, bushland & woodland.
Distr: RS, CS, ES, BAG, EQU; Nigeria to Somalia, S to Namibia & South Africa; Arabia.
ES: Beshir 97 9/1951; **BAG:** Andrews 743 4/1939.

Albuca nigritana (Baker) Troupin

F.P.S. 3: 260 (1956); F.W.T.A. 3(1): 103 (1968).
Bulbous perennial herb. Wooded grassland.
Distr: EQU; Senegal to C.A.R. & D.R. Congo.
EQU: Schweinfurth 168 7/1870.

Albuca steudneri Schweinf. & Engl.

F.P.S. 3: 261 (1956).
Bulbous perennial herb. Habitat not known.
Distr: ES; Sudan endemic.
IUCN: RD
ES: Steudner 449 (fl fr) 6/1862 (n.v.).
Note: known only from the type which is believed to have been destroyed in the Berlin bombing; the status of this taxon is uncertain.

Albuca sudanica A.Chev.

F.W.T.A. 3(1): 104 (1968); F.J.M.: 155 (1976).
Bulbous perennial herb. Wooded grassland.
Distr: DAR; Guinea Bissau to C.A.R.; Algeria, Libya.
DAR: Wickens 1876 7/1964.

Albuca virens (Lindl.) J.C.Manning & Goldblatt

Taxon 58: 93 (2009).
Syn: *Ornithogalum ecklonii* Schltdl. – F.P.S. 3: 271 (1956).
Syn: *Ornithogalum tenuifolium* F.Delaroche – F.T.E.A. Hyacinth.: 27 (1996); F.E.E. 6: 147 (1997); Biol. Skr. 51: 487 (2005).
Bulbous perennial herb. Bushland & woodland.
Distr: BAG, EQU; Eritrea & Somalia, S to Namibia & South Africa.
BAG: Schweinfurth 1841 6/1869.

Asparagus africanus Lam. var. *africanus*

F.J.M.: 155 (1976); F.E.E. 6: 68 (1997); Biol. Skr. 51: 490 (2005); F.T.E.A. Asparag.: 6 (2006).
Syn: *Asparagus mitis* A.Rich. – F.P.S. 3: 264 (1956).
Shrub. Woodland, bushland, forest margins.
Distr: RS, DAR, ES, BAG, UN, EQU; Ivory Coast to Somalia, S to South Africa; Arabia, India.
DAR: Wickens 1437 4/1964; **UN:** J.M. Lock 81/60 5/1981.

Asparagus buchananii Baker

F.E.E. 6: 71 (1997); Biol. Skr. 51: 490 (2005); F.T.E.A. Asparag.: 14 (2006).
Syn: *Asparagus racemosus* sensu Friis & Vollesen pro parte, non Willd. – Biol. Skr. 51: 490 (2005).
Herbaceous climber. Grassland, bushland & woodland.
Distr: EQU; Ethiopia & Somalia, S to South Africa.
EQU: Friis & Vollesen 1165 3/1982.

Asparagus flagellaris (Kunth) Baker

F.P.S. 3: 263 (1956); F.J.M.: 155 (1976); F.E.E. 6: 60 (1997); Biol. Skr. 51: 490 (2005); F.T.E.A. Asparag.: 10 (2006).
Syn: *Asparagus abyssinicus* Hochst. ex A.Rich. – F.P.S. 3: 264 (1956); Bot. Exp. Sud.: 78 (1970).
Shrub. Woodland, wooded grassland & bushland.
Distr: RS, DAR, KOR, BAG, UN, EQU; Senegal & Mauritania to Eritrea, S to Namibia & Mozambique; Arabia.
DAR: Wickens 2903 4/1965; **EQU:** Friis & Vollesen 1002 2/1982.

Asparagus racemosus Willd.

F.P.S. 3: 264 (1956); F.E.E. 6: 71 (1997); Biol. Skr. 51: 490 pro parte (2005); F.T.E.A. Asparag.: 12 (2006).

Climbing shrub. Forest margins, bushland, grassland.
Distr: RS, UN, ?EQU; Guinea Bissau to Somalia, S to Angola & Mozambique; Arabia, tropical Asia.
RS: Jackson 2845 4/1953; **EQU:** Shigeta 90 1/1979 (n.v.).
Note: Andrews records this species from Central & Southern Sudan but this appears to be in error. We have only seen material from the Red Sea region; the specimens are rather poor and the identification requires confirmation. We have not seen the Shigeta specimen from Equatoria which was identified by Friis & Vollesen as this species along with their 1165 which is now considered to be *A. buchananii*.

Asparagus scaberulus A.Rich.

F.P.S. 3: 264 (1956); Bot. Exp. Sud.: 79 (1970); F.E.E. 6: 68 (1997); F.T.E.A. Asparag.: 8 (2006).
Erect or climbing shrub. Wadis, dry bushland.
Distr: RS; Chad to Somalia, S to Tanzania; Arabia.
RS: Drar 385 3/1938 (n.v.).
Note: Drar (1970) cites three specimens from the Erkowit region, all of which are sterile; we have seen no other Sudanese material. The presence of this species in Sudan therefore requires confirmation.

Chlorophytum affine Baker var. *curviscapum* (Poelln.) Hanid

F.J.M.: 155 (1976); F.T.E.A. Antheric.: 24 (1997).
Syn: *Chlorophytum tordense* Chiov. – F.E.E. 6: 102 (1997).
Rhizomatous perennial herb. Rock crevices.
Distr: DAR; Ghana, Cameroon & Chad to Somalia, S to Zambia.
DAR: Wickens 1953 7/1964.

Chlorophytum andongense Baker

F.P.S. 3: 267 (1956); F.T.E.A. Antheric.: 10 (1997); Biol. Skr. 51: 491 (2005).
Rhizomatous perennial herb. Woodland & wooded grassland.
Distr: UN, EQU; Sierra Leone to Kenya, S to Angola & Mozambique.
EQU: Jackson 4264 6/1961.

Chlorophytum aureum Engl.

F.P.S. 3: 268 (1956); F.W.T.A. 3(1): 99 (1968).
Rhizomatous perennial herb. Wooded grassland.
Distr: BAG, EQU; Ivory Coast, Benin, Nigeria, C.A.R., D.R. Congo.
BAG: Schweinfurth 1504 4/1869.

Chlorophytum blepharophyllum Schweinf. ex Baker

F.P.S. 3: 267 (1956); F.J.M.: 156 (1976); F.E.E. 6: 105 (1997); F.T.E.A. Antheric.: 51 (1997); Biol. Skr. 51: 491 (2005).
Rhizomatous perennial herb. Woodland, bushland & grassland.
Distr: DAR, ES, BAG, EQU; Senegal to Ethiopia, S to Angola & Mozambique.
DAR: Wickens 1968 7/1964; **EQU:** Friis & Vollesen 1248 3/1982.

Chlorophytum cameronii (Baker) Kativu var. *cameronii*

F.T.E.A. Antheric.: 31 (1997); F.E.E. 6: 102 (1997); Biol. Skr. 51: 491 (2005).

Syn: *Anthericum uyuiense* Rendle – F.P.S. 3: 263 (1956). Rhizomatous perennial herb. Woodland, bushland & grassland.
Distr: DAR, KOR, BAG, UN, EQU; D.R. Congo to Ethiopia, S to Angola & Zambia.
DAR: Wickens 3139 8/1972; **EQU:** Myers 7151 7/1937.

Chlorophytum cameronii (Baker) Kativu var. purpuratum (Rendle) Govaerts

World Checkl. Seed Pl. 3: 14 (1999).
Rhizomatous perennial herb. Woodland & grassland.
Distr: EQU; Cameroon to Somalia, S to Angola, Zimbabwe & Mozambique.
EQU: Jackson 2317 7/1952.

Chlorophytum comosum (Thunb.) Jacques

F.E.E. 6: 102 (1997); F.T.E.A. Antheric.: 55 (1997); Biol. Skr. 51: 492 (2005).
Rhizomatous perennial herb. Forest, riverside rocks.
Distr: EQU; Sierra Leone to Cameroon & Gabon, Ethiopia to South Africa.
EQU: Friis & Vollesen 83 11/1980.

Chlorophytum gallabatense Schweinf. ex Baker

F.P.S. 3: 267 (1956); F.J.M.: 156 (1976); F.E.E. 6: 99 (1997); F.T.E.A. Antheric.: 45 (1997); Biol. Skr. 51: 492 (2005).
Rhizomatous perennial herb. Wooded grassland & bushland, disturbed areas.
Distr: DAR, ES, BAG, UN, EQU; Senegal to Somalia, S to Namibia & Zimbabwe.
DAR: Wickens 1827 7/1964; **EQU:** Myers 11031 4/1939.

Chlorophytum geophilum Peter ex Poelln.

F.J.M.: 156 (1976); F.E.E. 6: 93 (1997); F.T.E.A. Antheric.: 54 (1997).
Rhizomatous perennial herb. Woodland & grassland.
Distr: DAR; Guinea to Ethiopia, S to Zambia.
DAR: Wickens 2006 7/1964.
Note: Quézel (1969: 133) records *Anthericum inconspicuum* Baker (= *C. inconspicuum* (Baker) Nordal) from Darfur; we have not seen the specimen on which this record is based. The true *inconspicuum* is restricted to Somalia-Masai woodland in Ethiopia, Somalia, Kenya and possibly Arabia and so is considered unlikely to occur in Darfur.

Chlorophytum lancifolium Welw. ex Baker subsp. cordatum (Engl.) A.D.Poulsen & Nordal

F.T.E.A. Antheric.: 60 (species only) (1997); Biol. Skr. 51: 492 (species only) (2005); Bot. J. Linn. Soc. 148: 16 (2005).
Syn: *Chlorophytum cordatum* Engl. – F.P.S. 3: 267 (1956).
Rhizomatous perennial herb. Forest.
Distr: EQU; D.R. Congo, Uganda.
EQU: Andrews 1754 6/1939.

Chlorophytum longifolium Schweinf. ex Baker

F.P.S. 3: 268 (1956); F.E.E. 6: 94 (1997); F.T.E.A. Antheric.: 17 (1997); Biol. Skr. 51: 492 (2005).
Rhizomatous perennial herb. Woodland & grassland, often on hillslopes.

Distr: ES, ?EQU; Ethiopia to Namibia & Botswana.
ES: Schweinfurth 8 8/1865.
Note: this species probably also occurs in South Sudan (EQU), having been recorded from the Ugandan side of the Imatong Mts. Drar (1970: 79) recorded "*Chlorophytum possibly longifolium*" from Yei.

Chlorophytum micranthum Baker

F.P.S. 3: 268 (1956); F.T.E.A. Antheric.: 46 (1997); F.E.E. 6: 99 (1997); Biol. Skr. 51: 492 (2005).
Rhizomatous perennial herb. Grassland & woodland.
Distr: UN, EQU; Burundi, Ethiopia, Uganda, Kenya.
EQU: Friis & Vollesen 1168 3/1982.
Note: the World Checklist Series record this species as more widespread than noted here or in the other East African literature.

Chlorophytum nubicum (Baker) Kativu

F.E.E. 6: 94 (1997); F.T.E.A. Antheric.: 15 (1997).
Syn: *Anthericum nubicum* Baker – F.P.S. 3: 263 (1956).
Rhizomatous perennial herb. Dry woodland, bushland & grassland.
Distr: BAG, UN, EQU; Guinea to Ethiopia, S to Zambia & Mozambique.
EQU: Myers 8252 1/1938.

Chlorophytum polystachys Baker

F.P.S. 3: 268 (1956); F.T.E.A. Antheric.: 43 (1997).
Rhizomatous perennial herb. Open woodland, grassland, rock outcrops.
Distr: UN; Burkina Faso, Ghana, Cameroon, D.R. Congo to Kenya, S to Zimbabwe.
UN: Simpson 7211 6/1929.

Chlorophytum pusillum Schweinf. ex Baker

F.P.S. 3: 267 (1956); F.T.E.A. Antheric.: 55 (1997).
Perennial herb. Woodland, rock crevices.
Distr: BAG; Senegal to Uganda, S to Zimbabwe.
BAG: Jackson 3065 7/1953.

Chlorophytum stenopetalum Baker

F.T.E.A. Antheric.: 54 (1997); Biol. Skr. 51: 493 (2005).
Syn: *Chlorophytum schweinfurthii* Baker – F.P.S. 3: 267 (1956); F.J.M.: 156 (1976).
Syn: *Chlorophytum macrophyllum* sensu Andrews, non (A.Rich.) Asch. – F.P.S. 3: 266 (1956).
Rhizomatous perennial herb. Open forest, riverine fringes.
Distr: DAR, ES, BAG, EQU; Mali to Uganda, S to Angola & Mozambique.
DAR: Wickens 1838 7/1964; **BAG:** Schweinfurth 1968 6/1869.
Note: we consider all the Sudanese / South Sudanese material seen to be of the same species, but *C. stenopetalum* and *C. macrophyllum* are very similar and it is possible that the latter also occurs.

Chlorophytum subpetiolatum (Baker) Kativu

F.E.E. 6: 100 (1997); F.T.E.A. Antheric.: 36 (1997); Biol. Skr. 51: 493 (2005).
Syn: *Anthericum monophyllum* Baker – F.P.S. 3: 261 (1956).
Syn: *Anthericum subpetiolatum* Baker – F.P.S. 3: 263 (1956).

Rhizomatous perennial herb. Woodland & grassland.
Distr: BAG, UN, EQU; Nigeria to Somalia, S to Angola,
Zimbabwe & Mozambique.
EQU: Myers 6527 5/1937.

Chlorophytum superpositum (Baker) Marais & Reilly

Kew Bull. 32: 662 (1978).
Syn: *Anthericum superpositum* Baker – F.P.S. 3: 263
(1956).
Rhizomatous perennial herb. Habitat not known.
Distr: EQU; South Sudan endemic.
IUCN: RD
EQU: Schweinfurth III.174 (fr) 6/1870.

Chlorophytum tuberosum (Roxb.) Baker

F.P.S. 3: 268 (1956); F.J.M.: 156 (1976); F.E.E. 6: 97
(1997); F.T.E.A. Antheric.: 28 (1997); Biol. Skr. 51: 493
(2005).
Rhizomatous perennial herb. Grassland, bushland &
woodland, often in seasonally damp areas.
Distr: DAR, KOR, CS, BAG, UN, EQU; Nigeria to Somalia,
S to Tanzania; Indian subcontinent, Myanmar.
DAR: Wickens 3601 6/1977; **EQU:** Myers 8961 4/1938.

Dipcadi fesoghlense (Solms) Baker

F.T.A. 7: 519 (1898); F.P.S. 3: 269 (1956).
Bulbous perennial herb. Habitat not known.
Distr: CS; Sudan endemic.
IUCN: RD
CS: Cienkowsky s.n. (n.v.)
Note: known only from the type specimen; the status of
this taxon is uncertain.

Dipcadi viride (L.) Moench

F.J.M.: 156 (1976); F.T.E.A. Hyacinth.: 3 (1996); F.E.E. 6:
139 (1997); Biol. Skr. 51: 486 (2005).
Syn: *Dipcadi filifolium* Baker – F.P.S. 3: 269 (1956).
Syn: *Dipcadi lanceolatum* Baker – F.P.S. 3: 269 (1956).
Syn: *Dipcadi tacazzeanum* (Hochst. ex A.Rich.) Baker –
F.P.S. 3: 269 (1956).
Syn: *Dipcadi unifolium* Baker – F.P.S. 3: 269 (1956).
Bulbous perennial herb. Woodland, bushland & grassland.
Distr: RS, DAR, KOR, CS, ES, UN, EQU; Mauritania to
Somalia, S to Namibia & South Africa; Arabia.
DAR: Wickens 1902 7/1964; **UN:** J.M. Lock 81/75
5/1981.
Note: Quézel's (1969: 133) record of *D. longifolium*
Baker from Darfur is probably referable here.

Dracaena afromontana Mildbr.

F.E.E. 6: 76 (1997); Biol. Skr. 51: 502 (2005); F.T.E.A.
Dracaen.: 6 (2007).
Shrub or small tree. Montane forest.
Distr: EQU; Ethiopia to D.R. Congo & Malawi.
EQU: Friis & Vollesen 396 11/1980.
Note: J.J. Bos in F.E.E. 6: 77 (1997) records *D.
ellenbeckiana* Engl. as possibly also occurring in Sudan
but we have seen no specimens or other evidence to
support this.

Dracaena fragrans (L.) Ker-Gawl.

T.S.S.: 447 (1990); F.E.E. 6: 77 (1997); Biol. Skr. 51: 501
(2005); F.T.E.A. Dracaen. (2007).

Shrub or tree. Forest.
Distr: EQU; Gambia to Ethiopia, S to Angola, Zimbabwe
& Mozambique.
EQU: Friis & Vollesen 1227 3/1982.

Dracaena laxissima Engl.

F.P.S. 3: 300 (1956); T.S.S.: 447 (1990); Biol. Skr. 51: 503
(2005); F.T.E.A. Dracaen.: 3 (2007).
Shrub. Forest.
Distr: EQU; Nigeria to Kenya, S to Angola &
Mozambique.
EQU: Andrews 1978 6/1939.

Dracaena ombet Heuglin ex Kotschy & Peyr. subsp. ombet

F.P.S. 3: 298 (1956); T.S.S.: 447 (1990); F.E.E. 6: 79
(1997).
Tree. Rocky hillslopes.
Distr: RS; Eritrea; Egypt.
IUCN: EN
RS: Wickens 3069 5/1969.
Note: El Amin (1990) notes that this was once an
important zone-forming species in the Red Sea Hills but
is now almost extinct.

Dracaena steudneri Engl.

F.P.S. 3: 298 (1956); T.S.S.: 449 (1990); F.E.E. 6: 79
(1997); Biol. Skr. 51: 503 (2005); F.T.E.A. Dracaen.: 8
(2007).
Tree or shrub. Montane forest & secondary forest.
Distr: EQU; Ethiopia to Angola, Zimbabwe &
Mozambique.
EQU: Myers 8386 1/1938.

Drimia altissima (L.f.) Ker Gawl.

F.T.E.A. Hyacinth.: 19 (1996); F.E.E. 6: 144 (1997); Biol.
Skr. 51: 486 (2005).
Syn: *Urginea altissima* (L.f.) Baker – F.P.S. 3: 274
(1956); Bot. Exp. Sud.: 79 (1970); F.J.M.: 156 (1976).
Robust bulbous perennial herb. Grassland & wooded
grassland.
Distr: RS, DAR, KOR, ES, UN, EQU; Mauritania to
Somalia, S to Namibia & South Africa.
DAR: Wickens 2919 4/1964; **EQU:** Myers 8392 1/1938.

Drimia elata Jacq.

F.T.E.A. Hyacinth.: 17 (1996); Biol. Skr. 51: 487 (2005).
Bulbous perennial herb. Woodland & grassland, rock
outcrops.
Distr: EQU; Uganda & Kenya, S to Angola & South Africa.
EQU: Friis & Vollesen 918 2/1982.

Drimia indica (Roxb.) Jessop

F.T.E.A. Hyacinth.: 18 (1996); F.E.E. 6: 143 (1997); Biol.
Skr. 51: 487 (2005).
Syn: *Urginea indica* (Roxb.) Kunth – F.P.S. 3: 274
(1956); F.J.M.: 157 (1976).
Bulbous perennial herb. Wooded grassland.
Distr: DAR, BAG, UN; scattered in the palaeotropics.
DAR: Wickens 1582 5/1964; **UN:** Simpson 7493 11/1930.

Drimia sudanica Friis & Vollesen

Nordic J. Bot. 19: 210 (1999); Biol. Skr. 51: 487 (2005).
Bulbous perennial herb. Grassland & wooded grassland

with frequent burning.
Distr: EQU; Sierra Leone, Guinea.
IUCN: RD
EQU: <u>Friis & Vollesen 1126</u> 3/1982.

Drimiopsis barteri Baker

F.P.S. 3: 269 (1956); F.J.M.: 156 (1976); F.T.E.A. Hyacinth.: 7 (1996); F.E.E. 6: 139 (1997).
Bulbous perennial herb. Seasonally damp grassland & woodland.
Distr: DAR, ?EQU; Ghana to Somalia, S to Zambia & Tanzania.
DAR: <u>Wickens 1960</u> 7/1964.
Note: Andrews records this species from Equatoria, but we have seen no South Sudan specimens with which to verify this.

Drimiopsis spicata (Baker) Sebsebe & Stedje

Pl. Divers. & Complex. Patterns: 329. (2005).
Syn: Scilla spicata Baker – F.P.S. 3: 273 (1956).
Bulbous perennial herb. Grassland & wooded grassland.
Distr: DAR, BAG, EQU; D.R. Congo, Ethiopia.
IUCN: RD
DAR: <u>Wickens 3642</u> (fl) 6/1977; **BAG:** <u>Myers 8962</u> 4/1938.

Eriospermum abyssinicum Baker

F.P.S. 3: 271 (1956); F.T.E.A. Eriosperm.: 4 (1996); F.E.E. 6: 136 (1997); Biol. Skr. 51: 494 (2005).
Bulbous perennial herb. Grassland & wooded grassland, often in damp areas amongst rocks.
Distr: DAR, KOR, ES, UN, EQU; Senegal to Ethiopia, S to Namibia & South Africa.
KOR: <u>Wickens 64</u> 7/1962; **EQU:** <u>Friis & Vollesen 944</u> 2/1982.
Note: the African Plants Database records this as a synonym of *E. flabelliforme* (Baker) C.J.Manning but the World Checklist Series has it the other way round; we follow the latter.

Ledebouria edulis (Engl.) Stedje

F.T.E.A. Hyacinth.: 14 (1996); F.E.E. 6: 141 (1997).
Syn: Scilla edulis Engl. – F.P.S. 3: 273 (1956).
Bulbous perennial herb. Grassland & wooded grassland.
Distr: BAG; Ethiopia, Uganda, Kenya, Tanzania.
BAG: <u>Simpson 7188</u> 6/1929.

Ledebouria lilacina (Fenzl ex Kunth) Speta

Phyton (Horn) 38: 106 (1998).
Syn: Scilla lilacina (Fenzl ex Kunth) Baker – F.P.S. 3: 274 (1956).
Bulbous perennial herb. Habitat unknown.
Distr: KOR, UN; Sudan / South Sudan endemic.
IUCN: RD
KOR: <u>Wickens 69</u> 7/1962; **UN:** <u>Broun 1691</u> 6/1909.

Ledebouria maesta (Baker) Speta

Phyton (Horn) 38: 106 (1998).
Syn: Scilla maesta Baker – F.P.S. 3: 274 (1956); F.J.M.: 156 (1976).
Bulbous perennial herb. Wooded grassland, rocky areas, alluvial terraces.
Distr: DAR, KOR, BAG, UN, EQU; Malawi, Mozambique.
IUCN: RD
DAR: <u>Wickens 1818</u> 7/1964; **EQU:** <u>Sillitoe 456</u> (fl) 1919.

Note: in the World Checklist Series, this species is recorded as a Sudan endemic; the relationship to the southern African plants requires further investigation.

Ledebouria revoluta (L.f.) Jessop

F.T.E.A. Hyacinth.: 15 (1996); F.E.E. 6: 142 (1997); Biol. Skr. 51: 487 (2005).
Syn: Scilla hyacinthina (Roth) J.F.Macbr. – F.P.S. 3: 273 (1956).
Bulbous perennial herb. Grassland, bushland & woodland.
Distr: ES, EQU; Eritrea & Somalia, S to Malawi, also Angola, Namibia & South Africa; India, Sri Lanka.
ES: <u>Jackson 1916</u> 7/1951; **EQU:** <u>Kielland-Lund 707</u> 5/1984 (n.v.).

Ornithogalum cf. arabicum L.

F. Darfur Nord-Occ. & J. Gourgeil: 133 (1969).
Bulbous perennial herb. Fixed dunes, old cultivations.
Distr: DAR.
Note: this record is from Quézel (1969) who recorded this taxon as fairly common around Kutum in Darfur. We have not seen the specimen on which it is based but it is unlikely to be the true *O. arabicum* which is mainly a Mediterranean species.

Sansevieria ehrenbergii Schweinf. ex Baker

F.P.S. 3: 300 (1956); F.E.E. 6: 82 (1997); Biol. Skr. 51: 503 (2005); F.T.E.A. Dracaen.: 36 (2007).
Rhizomatous perennial xerophyte. Bushland, rocky ground.
Distr: RS, EQU; Eritrea, Ethiopia, Djibouti, Somalia, Uganda, Kenya, Tanzania; Arabia.
RS: <u>Schweinfurth 31</u> (n.v.); **EQU:** <u>Myers 10544</u> 2/1939.

Sansevieria forskaliana (Schult. & Schult.f.) Hepper & J.R.I.Wood

F.E.E. 6: 82 (1997); F.T.E.A. Dracaen.: 82 (2007).
Syn: Sansevieria abyssinica N.E.Br. – F.P.S. 3: 300 (1956); Bot. Exp. Sud.: 79 (1970).
Syn: Sansevieria liberica sensu Wickens pro parte, non Gérôme & Labroy – F.J.M.: 159 (1976).
Rhizomatous perennial herb. Bushland, woodland, rocky ground.
Distr: DAR, CS, BAG, UN, EQU; Eritrea to Somalia, S to D.R. Congo & Tanzania.
DAR: <u>Wickens 1548</u> 5/1964; **EQU:** <u>Andrews 850</u> 4/1939.

Sansevieria liberica Gérôme & Labroy

F.W.T.A. 3(1): 159 (1968); F.J.M.: 156 pro parte (1976).
Rhizomatous perennial herb. Shaded areas by streams & on rock outcrops.
Distr: DAR; Mali & Liberia to C.A.R.
DAR: <u>Wickens 1946</u> 7/1964.

Sansevieria nilotica Baker

F.P.S. 3: 300 (1956); F.E.E. 6: 82 (1997); Biol. Skr. 51: 504 (2005); F.T.E.A. Dracaen.: 15 (2007).
Rhizomatous perennial herb. Woodland, bushland & riverine forest clearings.
Distr: CS, EQU; C.A.R., Ethiopia, Uganda.
CS: <u>Jackson 1664</u> (fl) 1/1951; **EQU:** <u>Friis & Vollesen 1232</u> 3/1982.

Sansevieria powellii N.E.Br.

F.P.S. 3: 301 (1956); Bot. Exp. Sud.: 79 (1970); F.T.E.A. Dracaen.: 39 (2007).
Perennial herb. Dry bushland.
Distr: ?RS, ES, ?EQU; Ethiopia, Somalia, Kenya, Tanzania.
ES: Kennedy-Cooke 251 3/1938; **EQU:** Drar 1852 4/1938 (n.v.).
Note: (1) Drar (1970) records this species from Erkowit, Yei and Karika, all based on sterile material; the status of this species in South Sudan needs confirmation. (2) Mbugua in F.T.E.A. Dracaen.: 21 (2007) records also S. raffillii N.E.Br. from Sudan but we have seen no collections to verify this.

Scilla chlorantha Baker

F.P.S. 3: 274 (1956).
Bulbous perennial herb. Habitat unknown.
Distr: BAG; C.A.R.
IUCN: RD
BAG: Schweinfurth 162 (fl) 5/1869.

Scilla engleri T.Durand & Schinz

F.P.S. 3: 274 (1956).
Bulbous perennial herb. Habitat unknown.
Distr: BAG; South Sudan endemic.
IUCN: RD
BAG: Schweinfurth 1908 5/1869.

Scilla schweinfurthii Engl.

F.P.S. 3: 273 (1956).
Bulbous perennial herb. Seasonally wet grassland.
Distr: BAG, UN; C.A.R., Chad.
IUCN: RD
UN: J.M. Lock 81/53 5/1981.

Urginea grandiflora Baker

F.P.S. 3: 275 (1956).
Bulbous perennial herb. Habitat unknown.
Distr: RS; ?Sudan endemic.
IUCN: RD
RS: Lord s.n. (fl) 2/1869.
Note: the single specimen is depauperate and has been labelled as possibly being an Albuca. Clearly, this taxon requires further investigation.

ARECALES

Arecaceae (Palmae)

I. Darbyshire

Borassus aethiopum Mart.

F.P.S. 3: 302 (1956); F.J.M.: 159 (1976); F.T.E.A. Palm.: 19 (1986); T.S.S.: 449 (1990); F.E.E. 6: 518 (1997); Biol. Skr. 51: 504 (2005).
Syn: *Borassus flabellifer* L. var. *aethiopum* (Mart.) Warb. – Bot. Exp. Sud.: 92 (1970).
Large tree palm. Along water courses & in drier wooded grassland, often colonial.
Distr: DAR, KOR, CS, BAG, UN, EQU; Senegal to Ethiopia, S to South Africa; Madagascar, Comoros.
DAR: Wickens 3704 6/1977; **UN:** Broun s.n. (fl) 4/1904.

Calamus deerratus G.Mann & H.Wendl.

F.P.S. 3: 302 (1956); F.T.E.A. Palm.: 43 (1986); T.S.S.: 450 (1990).
Syn: *Ancistrophyllum secundiflorum* sensu auctt., non (P.Beauv.) G.Mann & H.Wendl. – F.P.S. 3: 302 (1956); T.S.S.: 449 (1990).
Climbing spiny palm. Riverine & swamp forest.
Distr: BAG, EQU; Senegal to Uganda, S to Angola & Zambia.
EQU: Myers 11334 (fl) 5/1939.

Elaeis guineensis Jacq.

Bot. Exp. Sud.: 92 (1970); F.T.E.A. Palm.: 50 (1986); T.S.S.: 450 (1990).
Tree palm. Riverine & swamp forest, also cultivated.
Distr: EQU; Sierra Leone to Kenya, S to Angola & Mozambique; Madagascar, native range much clouded by widespread cultivation.
EQU: Drar s.n. 4/1938 (n.v.).
Note: it is unclear as to whether or not the 'oil palm' is truly native to South Sudan. It is widely cultivated in South Sudan and up into the Blue Nile Region, but El Amin (1990) also considered it native in Equatoria.

Hyphaene thebaica (L.) Mart.

F.P.S. 3: 304 (1956); Bot. Exp. Sud.: 92 (1970); F.J.M.: 159 (1976); T.S.S.: 451 (1990); F.E.E. 6: 522 (1997); Biol. Skr. 51: 504 (2005).
Tree palm with dichotomous branching. Riverine fringes, oases, depressions in woodland.
Distr: NS, RS, DAR, KOR, CS, ES, EQU; Senegal to Somalia; Libya, Egypt, Middle East, Arabia.
DAR: Wickens 3135 (fl) 5/1971.
Note: Friis & Vollesen (2005) note that this species has been recorded from the plains at Dongotona Mts, Didinga Mts and at Kapoeta, though no specimens are available for confirmation. El Amin (1990) notes that it can form pure forests in the drier parts of Equatoria.

Medemia argun (Mart.) Württemb. ex H.Wendl.

F.P.S. 3: 304 (1956); T.S.S.: 451 (1990); F.Egypt 4: 106 (2005); Palms 53: 9-19 (2009).
Syn: *Medemia abiadensis* H.Wendl. – F.P.S. 3: 304 (1956).
Tree palm. Desert oases.
Distr: NS, ?RS, ?KOR; S Egypt (Nubian Desert).
IUCN: CR
NS: Gibbons & Spanner s.n. (fr) 10/1995.
Note: El Amin (1990) claims that this species extends to North Kordofan; this requires confirmation.

Phoenix dactylifera L.

F.P.S. 3: 304 (1956); F.T.E.A. Palm.: 17 (1986); T.S.S.: 451 (1990); F.E.E. 6: 517 (1997); F.J.U. 1: 307 (1997).
Tree palm. Villages, oases, wadis – cultivated.
Distr: NS, RS (perhaps also other regions); widely cultivated in the dry tropics & subtropics, possibly originating in Arabia.
NS: Léonard 4851 11/1968.
Note: the common 'date palm' of cultivation, widespread in the dry north of Sudan and cultivated elsewhere.

Phoenix reclinata Jacq.

F.P.S. 3: 304 (1956); F.J.M.: 159 (1976); F.T.E.A. Palm.: 15 (1986); T.S.S.: 453 (1990); F.E.E. 6: 515 (1997); Biol. Skr. 51: 504 (2005).
Tree palm, often clump-forming. Forest margins & clearings, riverine fringes.
Distr: RS, DAR, ES, BAG, EQU; Senegal to Somalia, S to Namibia & South Africa; Madagascar, Comoros, Arabia.
DAR: Wickens 1058 (fr) 1/1964; **EQU:** Chipp 50 2/1929.

Raphia farinifera (Gaertn.) Hyl.

F.T.E.A. Palm.: 38 (1986); Biol. Skr. 51: 505 (2005).
Tree palm. Swamps, riverine forest.
Distr: ?EQU; Senegal to Kenya (but ?absent from D.R. Congo), S to Angola, Zimbabwe & Mozambique; Madagascar, Mascarenes.
Note: inclusion of this species here is based on sight records by J.K. Jackson from the Ateppi Valley above Issore and Iyedo Valley, which may possibly have been misidentifications of *R. monbuttorum*; see Friis & Vollesen (2005) for further discussion.

Raphia mambillensis Otedoh

J. Niger. Inst. Oil Palm Res. 6(22): 163 (1982).
Trunkless palm. Along water courses.
Distr: BAG; Nigeria, Cameroon, C.A.R.
BAG: Hoyle 491 (n.v.).

Raphia monbuttorum Drude var. *monbuttorum*

F.P.S. 3: 305 (1956); F.T.E.A. Palm.: 40, in notes (1986); T.S.S.: 453 (1990); Biol. Skr. 51: 505, in notes (2005).
Trunkless or short-trunked palm. Swamps.
Distr: BAG, ?EQU; Nigeria to D.R. Congo.
BAG: Schweinfurth 1738 (fl) 5/1869.
Note: the status of this species in South Sudan (and its relationship to *R. farinifera*) is still in doubt – see note in F.T.E.A. for further discussion. The specimen cited is of a leaf portion only.

COMMELINALES

Commelinaceae

H. Pickering

Aneilema aequinoctiale (P.Beauv.) G.Don

F.E.E. 6: 351 (1997); Biol. Skr. 51: 475 (2005); F.T.E.A. Commelin.: 70 (2012).
Perennial herb. Forest & forest margins, riverine fringes.
Distr: EQU; Guinea to Ethiopia, S to Angola & South Africa.
EQU: Friis & Vollesen 634 12/1980.

Aneilema beniniense (P.Beauv.) Kunth

F.P.S. 3: 238 (1956); F.E.E. 6: 357 (1997); Biol. Skr. 51: 475 (2005); F.T.E.A. Commelin.: 105 (2012).
Perennial herb. Forest including riverine & swamp forest.
Distr: EQU; Senegal to Ethiopia, S to Angola & Zambia.
EQU: Friis & Vollesen 635 12/1980.

Aneilema forsskalii Kunth

F.E.E. 6: 355 (1997).
Syn: *Aneilema tacazzeanum* Hochst. ex A.Rich. – F.P.S. 3: 239 (1956); Bot. Exp. Sud.: 32 (1970) – see note.
Annual herb. Bushland on hillslopes, riverine forest.
Distr: RS; Eritrea, Ethiopia, Djibouti; Arabia.
RS: Aylmer 182 (fr) 3/1932.
Note: Drar (1970) records *A. tacazzeanum* from Equatoria and Darfur but both are based on sterile collections and we believe they were probably misidentified.

Aneilema hirtum A.Rich.

F.E.E. 6: 360 (1997); Biol. Skr. 51: 475 (2005); F.T.E.A. Commelin.: 116 (2012).
Annual herb. Woodland, bushland, forest margins.
Distr: EQU; Ethiopia to Zambia & Malawi.
EQU: Friis & Vollesen 24 11/1980.

Aneilema lanceolatum Benth. subsp. *lanceolatum*

F.P.S. 3: 239 (1956); F.J.M.: 154 (1976); Biol. Skr. 51: 476 (2005); F.T.E.A. Commelin.: 107 (2012).
Syn: *Aneilema lanceolatum* Benth. var. *evolutius* sensu Andrews, non C.B.Clarke – F.P.S. 3: 239 (1956).
Annual herb. Bushland, wooded grassland.
Distr: DAR, KOR, BAG, UN, EQU; Mali to Kenya, S to Congo-Brazzaville & D.R. Congo.
KOR: Wickens 3068 6/1969; **EQU:** Andrews 561 4/1939.

Aneilema petersii (Hassk.) C.B.Clarke subsp. *pallidiflorum* Faden

F.E.E. 6: 354, in notes (1997); F.T.E.A. Commelin.: 92 (2012).
Annual herb. Bushland, woodland, rocky hillslopes, roadsides.
Distr: EQU; Somalia, Uganda, Kenya, Tanzania, Zambia.
EQU: J.M. Lock 81/275 9/1981.

Aneilema spekei C.B.Clarke

F.P.S. 3: 239 (1956); F.E.E. 6: 357 (1997); Biol. Skr. 51: 476 (2005); F.T.E.A. Commelin.:113 (2012).
Annual herb. A wide variety of habitats from swamps to bushland & disturbed areas.
Distr: BAG, EQU; D.R. Congo, Rwanda, Burundi, Ethiopia, Uganda, Kenya, Tanzania, Zambia.
EQU: Friis & Vollesen 563 11/1980.

Aneilema umbrosum (Vahl) Kunth subsp. *ovato-oblongum* (P.Beauv.) J.K.Morton

Biol. Skr. 51: 476 (2005); F.T.E.A. Commelin.: 110 (2012).
Perennial herb. Forest including swamp forest.
Distr: EQU; Senegal to Uganda & Tanzania; neotropics.
EQU: Friis & Vollesen 749 12/1980.

Commelina africana L. var. *africana*

F.P.S. 3: 243 pro parte (1956); F.E.E. 6: 362 (1997); Biol. Skr. 51: 476 (2005); F.T.E.A. Commelin.: 148 (2012).
Perennial herb. Grassland, woodland, marshes, disturbed areas.
Distr: RS, EQU; Guinea Bissau to Somalia, S to Namibia & South Africa; Madagascar, Yemen.
RS: Jackson 2897 4/1953; **EQU:** Schweinfurth 3739 5/1870.

Commelina africana L. var. *lancispatha* C.B.Clarke

F.T.E.A. Commelin.: 150 (2012).
Syn: *Commelina involucrosa* A.Rich. – F.P.S. 3: 243 (1956).
Perennial herb. Woodland, grassland, forest margins, rocky hillslopes.
Distr: ?RS; Sierra Leone, Eritrea to Namibia & South Africa.
Note: Andrews lists *C. involucrosa* from Khor Tamanib in Red Sea District; no specimens have been seen with which to verify this and it may have been a misidentification.

Commelina africana L. var. *villosior* (C.B.Clarke) Brenan

F.T.E.A. Commelin.: 149 (2012).
Perennial herb. Grassland, forest margins, disturbed areas.
Distr: RS, UN, EQU; Senegal to Somalia, S to South Africa; Madagascar, Saudi Arabia.
RS: Carter 1823 11/1987; **EQU:** Myers 9170 8/1938.

Commelina albescens Hassk.

F.E.E. 6: 372 (1997); F.T.E.A. Commelin.: 218 (2012).
Rhizomatous perennial herb. Bushland, grassland, rocky outcrops, seasonally damp areas.
Distr: ES; Eritrea & Somalia, S to Tanzania; Arabia, Pakistan, India.
ES: Beshir 34 8/1951.
Note: Quézel (1969: 132) records this species as common in the areas of Darfur he visited. We suspect that this was based on a misidentification but have not seen Quézel's specimens.

[Commelina amplexicaulis Hassk.]

Note: Andrews lists *C. amplexicaulis* Hassk. from Fung District; the status of this name is uncertain and we have seen no Sudanese material with which to determine its placement.

Commelina benghalensis L.

F.P.S. 3: 241 (1956); F.J.M.: 154 (1976); F.E.E. 6: 373 (1997); Biol. Skr. 51: 477 (2005); F.T.E.A. Commelin.: 200 (2012).
Annual or perennial herb. A wide variety of habitats including disturbed areas.
Distr: RS, DAR, ES, BAG, EQU; Mauritania to Somalia, S to South Africa; Arabia, tropical Asia, introduced elsewhere in the tropics.
ES: Beshir 37 8/1951; **EQU:** Friis & Vollesen 968 2/1982.

Commelina bracteosa Hassk. subsp. *bracteosa* var. *bracteosa*

F.P.S. 3: 245 (1956); F.J.M.: 154 (1976); Biol. Skr. 51: 477 (2005); F.T.E.A. Commelin.: 214 (2012).
Syn: *Commelina aethiopica* sensu Andrews, non C.B.Clarke – F.P.S. 3: 245 (1956).
Perennial herb. Wooded grassland, bushland, forest margins.
Distr: DAR, KOR, BAG, UN, EQU; Senegal to Kenya, S to Angola, Zimbabwe & Mozambique.
DAR: Wickens 2001 7/1964; **BAG:** Jackson 3061 3/1952.

Commelina capitata Benth.

F.P.S. 3: 245 (1956); F.T.E.A. Commelin.: 178 (2012).
Perennial herb. Forest & forest margins.
Distr: EQU; Senegal to Kenya, S to Zambia.
EQU: Dandy 528 (fl) 2/1934.

Commelina diffusa Burm.f.

F.P.S. 3: 241 (1956); F.J.M.: 154 (1976); F.E.E. 6: 362 (1997); Biol. Skr. 51: 477 (2005); F.T.E.A. Commelin.: 140 (2012).
Perennial herb. Swamps, riversides, wet grassland, disturbed areas.
Distr: DAR, BAG; EQU; pantropical & warm temperate.
DAR: Wickens 1998 7/1964; **EQU:** Friis & Vollesen 1105 3/1982.
Note: most Sudanese / South Sudanese material appears to be of subsp. *diffusa* but Friis & Vollesen identified their 1105 as subsp. *montana* J.K.Morton – the Kew sheet is rather poor such that it is difficult to confirm this identification.

Commelina erecta L. subsp. *livingstonii* (C.B.Clarke) J.K.Morton

F.E.E. 6: 372 (1997); Biol. Skr. 51: 478 (2005); F.T.E.A. Commelin.: 211 (2012).
Rhizomatous perennial herb. Wooded grassland, bushland, rocky hillslopes.
Distr: UN, EQU; Senegal to Somalia, S to South Africa; Arabia.
UN: Harrison 60 4/1948.

Commelina forskaolii Vahl

F.P.S. 3: 241 (1956); F.E.E. 6: 374 (1997); Biol. Skr. 51: 478 (2005); F.T.E.A. Commelin.: 203 (2012).
Annual or perennial herb. Grassland, bushland, woodland, disturbed areas.
Distr: RS, DAR, KOR, BAG, UN, EQU; Senegal & Mauritania to Somalia, S to South Africa; Madagascar, Algeria, Egypt, Arabia, India, Vietnam.
KOR: Wickens 546 9/1962; **UN:** J.M. Lock 81/233 8/1981.

Commelina imberbis Ehrenb. ex Hassk.

F.P.S. 3: 245 (1956); F.J.M.: 154 (1976); F.E.E. 6: 369 (1997); F.T.E.A. Commelin.: 187 (2012).
Syn: *Commelina latifolia* sensu Andrews pro parte, non Hochst. ex A.Rich. – F.P.S. 3: 245 (1956).
Perennial herb. Grassland, bushland, rocky hillslopes, disturbed areas.
Distr: DAR, KOR, ?CS, BAG, UN, EQU; Mauritania, Nigeria to Somalia, S to Angola and Zimbabwe; Arabia.
KOR: Wickens 1664 5/1964; **UN:** Sherif A.2806 10/1951.
Note: Drar (1970) recorded *C. latifolia* from Sennar in Central Sudan and Quézel (1969) recorded it from Jebel Gourgeil in Darfur. Both these records may be referable here.

Commelina kotschyi Hassk.

F.P.S. 3: 244 (1956); F.J.M.: 154 (1976); F.E.E. 6: 368 (1997); F.T.E.A. Commelin.: 194 (2012).
Annual herb. Seasonally wet grassland, ditches.
Distr: DAR, KOR, CS; Ethiopia to Angola & South Africa; India.
CS: Aylmer 364 5/1934.

Note: Andrews records this species from Central & Southern Sudan but we have seen no material from the South.

Commelina latifolia Hochst. ex A.Rich.
F.P.S. 3: 245 pro parte (1956); F.E.E. 6: 368 (1997); Biol. Skr. 51: 478 (2005); F.T.E.A. Commelin.: 180 (2012).
Syn: Commelina imberbis sensu Friis & Vollesen, non Hassk. – Biol. Skr. 51: 478 (2005).
Perennial herb. Grassland, bushland, disturbed areas.
Distr: EQU; Eritrea & Somalia, S to D.R. Congo & Tanzania; Arabia.
EQU: Andrews 2007 6/1939.

Commelina purpurea C.B.Clarke
Biol. Skr. 51: 478 (2005); F.T.E.A. Commelin.: 168 (2012).
Perennial herb. Grassland & woodland including seasonally wet areas.
Distr: EQU; Congo-Brazzaville, D.R. Congo, Rwanda, Burundi, Uganda, Kenya, Tanzania.
EQU: Jackson 821.

Commelina schweinfurthii C.B.Clarke subsp. schweinfurthii
F.P.S. 3: 240 (1956); F.E.E. 6: 371 (1997); Biol. Skr. 51: 479 (2005); F.T.E.A. Commelin.: 223 (2012).
Perennial herb. Grassland, woodland, damp ground.
Distr: UN, EQU; Sierra Leone to Ethiopia, S to Zambia & Malawi.
EQU: Myers 9605 10/1938.

Commelina subulata Roth
F.P.S. 3: 241 (1956); F.J.M.: 154 (1976); F.E.E. 6: 364 (1997); Biol. Skr. 51: 479 (2005); F.T.E.A. Commelin.: 162 (2012).
Annual herb. Seasonally wet grassland, poolsides, thin soils over rock.
Distr: DAR, KOR, ES, EQU; Senegal to Eritrea, S to Namibia & South Africa; Arabia, India.
DAR: Wickens 2155 8/1964; **EQU:** Friis & Vollesen 414 11/1980.

Commelina zambesica C.B.Clarke
F.E.E. 6: 370 (1997); F.T.E.A. Commelin.: 205 (2012).
Syn: Commelina latifolia sensu Andrews pro parte, non Hochst. ex A.Rich. – F.P.S. 3: 245 (1956).
Perennial herb. Grassland, woodland, forest margins, swamps.
Distr: EQU; Nigeria to Ethiopia, S to Namibia & South Africa.
EQU: Andrews 1392 5/1939.

Commelina sp. aff. velutina Mildbr.
Perennial herb. Pebbly soil.
Distr: EQU.
EQU: Andrews 1638 6/1939.
Note: the single specimen cited has not been matched but is close to C. velutina from Cameroon.

Cyanotis arachnoidea C.B.Clarke
Biol. Skr. 51: 479 (2005); F.T.E.A. Commelin.: 24 (2012).
Perennial herb. Rock outcrops.
Distr: EQU; Liberia to Kenya, S to Angola & Zimbabwe; India to China, Thailand & Vietnam.
EQU: Friis & Vollesen 579 11/1980.

Cyanotis axillaris (L.) D.Don ex Sweet
F.T.E.A. Commelin.: 14 (2012).
Annual herb. Marshes, wet areas in wooded grassland.
Distr: UN, EQU; Cameroon, Chad, Kenya, Tanzania, Malawi; India to Australia.
EQU: Basinski 21 8/1951.

Cyanotis barbata D.Don
F.J.M.: 154 (1976); F.E.E. 6: 341 (1997); Biol. Skr. 51: 479 (2005); F.T.E.A. Commelin.: 20 (2012).
Syn: Cyanotis hirsuta C.A.Mey. – F.P.S. 3: 247 (1956).
Perennial herb. Grassland, shallow soils over rocks.
Distr: RS, DAR, EQU; Ghana to Eritrea, S to Zimbabwe & Mozambique; Yemen, India to China.
DAR: Wickens 2414 9/1964; **EQU:** Friis & Vollesen 803 12/1980.

Cyanotis caespitosa Kotschy & Peyr.
F.P.S. 3: 247 (1956); F.E.E. 6: 340 (1997); Biol. Skr. 51: 479 (2005); F.T.E.A. Commelin.: 15 (2012).
Perennial herb. Grassland, woodland, wet depressions.
Distr: BAG, EQU; Sierra Leone to Ethiopia, S to Angola & Zambia.
EQU: Andrews 655 4/1939.

Cyanotis cucullata (Roth.) Kunth
Enum. Pl. 4: 107. (1843).
Annual herb. Woodland, weed of cultivation.
Distr: CS; India to Vietnam.
CS: Lea 210 8/1954.
Note: the two Sudanese collections are the only African records known of this Asian species. Beshir (s.n., 1956) noted that it was "said to have appeared only four years ago and very quickly invaded the Gezira scheme" at Wad Medani.

Cyanotis foecunda DC. ex Hassk.
F.J.M.: 154 (1976); F.E.E. 6: 342 (1997); F.T.E.A. Commelin.: 26 (2012).
Perennial herb. Bushland, woodland & grassland, rock outcrops.
Distr: DAR; Cameroon to Ethiopia, S to Namibia, Botswana & Mozambique.
DAR: Wickens 2102 8/1964.

Cyanotis lanata Benth.
F.P.S. 3: 247 (1956); F.J.M.: 155 (1976); F.E.E. 6: 342 (1997); Biol. Skr. 51: 480 (2005); F.T.E.A. Commelin.: 22 (2012).
Annual herb. Woodland & wooded grassland, rock outcrops.
Distr: DAR, CS, ES, BAG, UN, EQU; Senegal to Ethiopia, S to Namibia & South Africa; Yemen.
DAR: Wickens 2127 8/1964; **EQU:** Myers 6643 5/1937.

Cyanotis longifolia Benth.
F.P.S. 3: 247 (1956); F.T.E.A. Commelin.: 16 (2012).
Perennial herb. Grassland, woodland & bushland, rock outcrops.
Distr: BAG, UN, EQU; Senegal to Ethiopia, S to Namibia & Zimbabwe.
EQU: Myers 655 7/1937.
Note: Andrews records this species from Jebel Marra in Darfur but this appears to be in error.

Floscopa africana (P.Beauv.) C.B.Clarke subsp. *petrophila* J.K.Morton

F.W.T.A. 3(1): 28 (1968); Biol. Skr. 51: 480 (2005).
Perennial herb. Forest, often near streams.
Distr: EQU; Senegal to Uganda.
EQU: Friis & Vollesen 607 12/1980.

Floscopa flavida C.B.Clarke

F.P.S. 3: 247 (1956); F.T.E.A. Commelin.: 42 (2012).
Annual herb. Wet grassland, seepage areas.
Distr: BAG; Senegal to Tanzania, S to Angola & Botswana.
BAG: Schweinfurth 2537 10/1869.

Floscopa glomerata (Willd. ex Schult. & Schult.f.) Hassk. subsp. *glomerata*

F.E.E. 6: 348 (1997); Biol. Skr. 51: 480 (2005); F.T.E.A. Commelin.: 46 (2012).
Syn: *Floscopa rivularis* (A.Rich.) C.B.Clarke – F.P.S. 3: 248 (1956); Bot. Exp. Sud.: 33 (1970).
Annual or perennial herb. Swamps, wet grassland, forest margins.
Distr: EQU; Senegal to Ethiopia, S to Namibia & South Africa; Madagascar.
EQU: Friis & Vollesen 308 11/1980.

Floscopa schweinfurthii C.B.Clarke

F.P.S. 3: 248 (1956); F.T.E.A. Commelin.: 44, in notes (2012).
Annual herb. ?Marshes.
Distr: BAG; South Sudan endemic.
IUCN: RD
BAG: Schweinfurth 2648 11/1869.

Murdannia simplex (Vahl) Brenan

F.P.S. 3: 250 (1956); F.J.M.: 155 (1976); F.E.E. 6: 346 (1997); Biol. Skr. 51: 480 (2005); F.T.E.A. Commelin.: 51 (2012).
Perennial herb. Grassland, bushland, woodland, marshes.
Distr: DAR, BAG, UN, EQU; Senegal to Somalia, S to Namibia & South Africa; Madagascar, tropical Asia.
DAR: Wickens 1920 7/1964; **EQU:** Andrews 1077 5/1939.

Palisota mannii C.B.Clarke subsp. *megalophylla* (Mildbr.) Faden

F.T.E.A. Commelin.: 8 (2012).
Syn: *Palisota schweinfurthii* sensu Friis & Vollesen pro parte, non C.B.Clarke – Biol. Skr. 51: 481 (2005).
Perennial herb. Forest including swamp forest.
Distr: EQU; Nigeria to Uganda, S to Angola & Tanzania.
EQU: Myers 11786 (fr) 8/1939.

Palisota schweinfurthii C.B.Clarke

F.P.S. 3: 250 (1956); Biol. Skr. 51: 481, pro parte (2005); F.T.E.A. Commelin.: 9 (2012).
Perennial herb. Forest, particularly along streams.
Distr: EQU; Nigeria to Uganda, S to Angola & Zambia.
EQU: Myers 6477 4/1937.

Pollia condensata C.B.Clarke

F.P.S. 3: 250 (1956); F.E.E. 6: 349 (1997); Biol. Skr. 51: 481 (2005); F.T.E.A. Commelin.: 64 (2012).
Stoloniferous perennial herb. Forest & forest clearings.
Distr: EQU; Sierra Leone to Ethiopia, S to Angola & Mozambique.
EQU: Friis & Vollesen 550 11/1980.

Pollia mannii C.B.Clarke

F.E.E. 6: 348 (1997); Biol. Skr. 51: 482 (2005); F.T.E.A. Commelin.: 62 (2012).
Perennial herb. Forest & forest margins.
Distr: EQU; Ivory Coast to Ethiopia, S to Angola & Tanzania.
EQU: Friis & Vollesen 511 11/1980.

Polyspatha paniculata Benth.

Biol. Skr. 51: 481 (2005); F.T.E.A. Commelin.: 58 (2012).
Stoloniferous perennial herb. Forest.
Distr: EQU; Guinea to Uganda, S to D.R. Congo & Tanzania.
EQU: Friis & Vollesen 467 11/1980.

Stanfieldiella imperforata (C.B.Clarke) Brenan var. *glabrisepala* (De Wild.) Brenan

F.E.E. 6: 346 (1997); Biol. Skr. 51: 482 (2005); F.T.E.A. Commelin.: 39 (2012).
Stoloniferous perennial herb. Forest & forest margins.
Distr: EQU; Ghana, Cameroon to Ethiopia, S to Tanzania.
EQU: Friis & Vollesen 469 11/1980.

Pontederiaceae

M. Kordofani, I. Farag & I. Darbyshire

Eichhornia crassipes (Mart.) Solms

F.T.E.A. Ponted.: 4 (1968); F.E.E. 6: 308 (1997).
Floating aquatic herb. Open water in lakes, ponds and slow-flowing rivers, often becoming a troublesome weed.
Distr: UN; native of Brazil, widely naturalised elsewhere in the tropics & subtropics.
UN: J.M. Lock 82/25 (fl) 4/1982.

Eichhornia diversifolia (Vahl) Urb.

F.P.S. 3: 276 (1956).
Syn: *Eichhornia natans* (P.Beauv.) Solms – F.T.E.A. Ponted.: 6 (1968); F.E.E. 6: 310 (1997).
Floating aquatic herb. Swamps, pools.
Distr: CS, BAG, UN, EQU; native of the neotropics, widely naturalised in tropical Africa, Madagascar & India.
CS: Lewis 384 1937; **UN:** Andrews W.N.106 (fl) 12/1938.

Heteranthera callifolia Rchb. ex Kunth

F.P.S. 3: 276 (1956); F.T.E.A. Ponted.: 8 (1968); F.J.M.: 157 (1976).
Perennial aquatic herb. Swamps, pools.
Distr: DAR, KOR, CS, BAG, EQU; Mauritania & Senegal to Kenya, S to Namibia & South Africa.
DAR: Wickens 2775 (fl fr) 10/1964; **BAG:** Schweinfurth 2234 8/1869.

Monochoria africana (Solms) N.E.Br.

F.P.S. 3: 278 (1956); F.T.E.A. Ponted.: 3 (1968).
Aquatic perennial herb. Wetlands.
Distr: BAG; Burkina Faso, Kenya, Angola, Malawi, Mozambique, South Africa.
BAG: Schweinfurth 2296 (fl fr) 8/1869.
Note: a widespread but scarce species.

ZINGIBERALES

Musaceae

H. Pickering

Ensete ventricosum (Welw.) Cheesman

F.P.S. 3: 252 (1956); F.E.E. 6: 317 (1997); Biol. Skr. 51: 483 (2005).
Tree-like perennial. Montane forest, often in disturbed areas.
Distr: EQU; Ethiopia to Angola & South Africa.
EQU: Schweinfurth II.130 5/1870.

[Musa]

Note: *Musa* cultivars – bananas – are widely grown in Sudan and South Sudan; these are of SE Asian origin.

Marantaceae

H. Pickering

Marantochloa conferta (Benth.) A.C.Ley

Syst. Bot. 36: 287 (2011).
Syn: *Ataenidia conferta* (Benth.) Milne-Redh. – F.P.S. 3: 256 (1956).
Perennial herb. Forest clearings.
Distr: EQU; Ivory Coast to Uganda, S to Angola.
EQU: Hope-Simpson 373 3/1939.

Marantochloa leucantha (K.Schum.) Milne-Redh. var. leucantha

F.T.E.A. Marant.: 6 (1952); F.P.S. 3: 257 (1956); Biol. Skr. 51: 485 (2005).
Perennial herb. Forest.
Distr: EQU; Sierra Leone to Ethiopia, S to Angola & Tanzania.
EQU: Friis & Vollesen 655 12/1980.

Marantochloa mannii (Benth.) Milne-Redh.

F.T.E.A. Marant.: 8 (1952); F.P.S. 3: 256 (1956); F.E.E. 6: 337 (1997).
Perennial herb. Forest.
Distr: EQU; Ivory Coast to Ethiopia, S to D.R. Congo & Tanzania.
EQU: Schweinfurth 3045 2/1870.

Marantochloa purpurea (Ridl.) Milne-Redh.

F.T.E.A. Marant.: 6 (1952); F.P.S. 3: 256 (1956); F.E.E. 6: 337 (1997).
Perennial herb. Forest clearings & streamsides.
Distr: EQU; Sierra Leone to Uganda, S to Angola & Tanzania.
EQU: Myers 6469 4/1937.

Megaphrynium macrostachyum (Benth.) Milne-Redh.

F.T.E.A. Marant.: 4 (1952); F.P.S. 3: 257 (1956).
Perennial herb. Forest.
Distr: EQU; Sierra Leone to Uganda, S to D.R. Congo.
EQU: Myers 9248 8/1938.

Sarcophrynium schweinfurthianum (Kuntze) Milne-Redh.

F.T.E.A. Marant.: 5 (1952); F.P.S. 3: 257 (1956).
Perennial herb. Forest.
Distr: EQU; Cameroon to D.R. Congo & Uganda.
EQU: Schweinfurth 3103 5/1870.

Thalia geniculata L.

T.S.S.: 447 (1990); F.Z. 13(4): 143 (2010).
Syn: *Thalia welwitschii* Ridl. – F.P.S. 3: 257 (1956); F.W.T.A. 3(1): 85 (1968); Bot. Exp. Sud.: 83 (1970).
Straggling perennial herb. Forest.
Distr: CS, BAG, UN, EQU; Senegal to C.A.R., S to Angola & Zimbabwe; neotropics.
CS: Lea 127 11/1952; **EQU:** Andrews 1110 5/1939.

Trachyphrynium braunianum (K.Schum.) Baker

F.T.E.A. Marant.: 2 (1952); F.P.S. 3: 257 (1956); Biol. Skr. 51: 485 (2005).
Robust perennial herb. Forests, often by streams.
Distr: EQU; Guinea to D.R. Congo & Uganda.
EQU: Myers 9189 8/1938.

Costaceae

H. Pickering

Costus afer Ker Gawl.

F.P.S. 3: 253 (1956); F.T.E.A. Zingiber.: 9 (1985); F.E.E. 6: 330 (1997); Biol. Skr. 51: 484 (2005).
Syn: *Costus pterometra* K.Schum. – F.P.S. 3: 254 (1956).
Rhizomatous perennial herb. Forest & forest margins in moist places.
Distr: EQU; Senegal to Ethiopia, S to Angola & Zimbabwe.
EQU: Friis & Vollesen 717 12/1980.

Costus dubius (Afzel.) K.Schum.

F.T.E.A. Zingiber.: 6 (1985); Biol. Skr. 51: 484 (2005).
Syn: *Costus trachyphyllus* K.Schum. – F.P.S. 3: 254 (1956); Bot. Exp. Sud.: 113 (1970).
Rhizomatous perennial herb. Forest, usually in open moist places.
Distr: EQU; Guinea Bissau to Uganda, S to Tanzania.
EQU: Friis & Vollesen 1062 3/1982.

Costus lucanusianus J.Braun & K.Schum.

F.T.E.A. Zingiber.: 8 (1985); F.E.E. 6: 331 (1997).
Rhizomatous perennial herb. Wet areas in forest.
Distr: EQU; Sierra Leone to Ethiopia, S to D.R. Congo & Uganda.
EQU: Andrews 1421 5/1939.

Costus spectabilis (Fenzl) K.Schum.

F.P.S. 3: 254 (1956); F.T.E.A. Zingiber.: 4 (1985); F.E.E. 6: 330 (1997).
Rhizomatous perennial herb. Grassland, wooded grassland, often on rocky ground.
Distr: BAG, UN, EQU; Senegal to Ethiopia, S to Angola, Zimabawe & Mozambique.
EQU: Andrews 899 4/1939.
Note: Andrews records this species from Central & Southern Sudan; the 'Central' element is probably based on *Schweinfurth* 1345 which was collected from Gendua just over the border in Ethiopia.

Zingiberaceae

H. Pickering

Aframomum alboviolaceum (Ridl.) K.Schum.

F.T.E.A. Zingiber.: 34 (1985); F.E.E. 6: 325 (1997); Biol. Skr. 51: 483 (2005).
Rhizomatous perennial herb. Wooded grassland.
Distr: EQU; Senegal to Kenya, S to Angola & Mozambique.
EQU: <u>Myers 6553</u> 5/1937.

Aframomum angustifolium (Sonn.) K.Schum.

F.T.E.A. Zingiber.: 27 (1985); Biol. Skr. 51: 483 (2005).
Rhizomatous perennial herb. Forest & forest margins, often in riverine or swamp forest.
Distr: EQU; Guinea to Uganda, S to Angola & Mozambique; Madagascar, Mascarenes.
EQU: <u>Shigeta 61</u> 1/1979 (n.v.).

Aframomum corrorima (A.Braun) P.C.M.Jansen

F.E.E. 6: 325 (1997).
Syn: *Aframomum usambarense* Lock – F.T.E.A. Zingiber.: 29 (1985).
Rhizomatous perennial herb. Open forest in moist places.
Distr: EQU; Ethiopia to Tanzania.
EQU: <u>Myers 10612</u> 3/1939.

Aframomum luteoalbum (K.Schum.) K.Schum.

F.P.S. 3: 253 (1956); F.T.E.A. Zingiber.: 30 (1985); Biol. Skr. 51: 483 (2005).
Rhizomatous perennial herb. Swamp & riverine forest.
Distr: EQU; D.R. Congo, Uganda, Tanzania.
EQU: <u>Shigeta 159</u> 1979 (n.v.).

Aframomum mala (K.Schum. ex Engl.) K.Schum.

F.T.E.A. Zingiber.: 27 (1985).
Rhizomatous perennial herb. Montane forest & forest margins.
Distr: EQU; Uganda, Kenya, Tanzania.
EQU: <u>Schweinfurth 3338</u> 5/1870.
Note: the single specimen seen is depauperate and was determined by J.M. Lock in 1976 as "probably *A. mala*".

Aframomum polyanthum (K.Schum.) K.Schum.

F.P.S. 3: 253 (1956); F.W.T.A. 3(1): 75 (1968).
Perennial herb. Forest.
Distr: EQU; Cameroon to Ethiopia, S to Zambia.
EQU: <u>Myers 11884</u> 11/1938.
Note: the distribution listed for this species follows the World Checklist Series; it is not mentioned in either F.E.E. or F.T.E.A.

Renealmia congolana De Wild. & T.Durand

F.T.E.A. Zingiber.: 13 (1985).
Syn: *Renealmia* sp. – F.P.S. 3: 255 (1956); Biol. Skr. 51: 484 (2005).
Rhizomatous perennial herb. Moist areas in forest.
Distr: EQU; Congo-Brazzaville, D.R. Congo, Rwanda, Uganda, Tanzania.
EQU: <u>Myers 9660</u> (fr) 10/1938.

Siphonochilus aethiopicus (Schweinf.) B.L.Burtt

F.T.E.A. Zingiber.: 20 (1985); F.E.E. 6: 330 (1997); Biol. Skr. 51: 484 (2005).
Syn: *Kaempferia aethiopica* F.P.S. 3: 255 (1956); Bot. Exp. Sud.: 113 (1970).
Rhizomatous perennial herb. Woodland, bushland & grassland.
Distr: DAR, CS, BAG, UN, EQU; Senegal to Ethiopia, S to Angola & South Africa.
DAR: <u>Wickens 3635</u> 6/1977; **UN:** <u>J.M. Lock 81/65</u> 5/1981.

Siphonochilus brachystemon (K.Schum.) B.L.Burtt.

F.T.E.A. Zingiber.: 19 (1985); Biol. Skr. 51: 485 (2005).
Syn: *Kaempferia macrosiphon* Baker – F.P.S. 3: 255 (1956); Bot. Exp. Sud.: 113 (1970).
Rhizomatous perennial herb. Forest.
Distr: EQU; D.R. Congo, Uganda & Kenya, S to Mozambique.
EQU: <u>Myers 6502</u> 5/1937.

Siphonochilus kirkii (Hook.f.) B.L.Burtt

F.J.M.: 155 (1976); F.T.E.A. Zingiber.: 15 (1985); Biol. Skr. 51: 485 (2005).
Syn: *Kaempferia rosea* Schweinf. ex Baker – F.P.S. 3: 255 (1956).
Rhizomatous perennial herb. Woodland, wooded grassland, dry forest.
Distr: DAR, KOR, BAG, UN, EQU; C.A.R. to Kenya, S to Zimbabwe & Mozambique.
DAR: <u>Wickens 1802</u> 7/1964; **EQU:** <u>Andrews 930</u> 4/1939.

POALES

Typhaceae

M. Kordofani, I. Farag & I. Darbyshire

Typha domingensis Pers.

F.T.E.A. Typh.: 2 (1971); F.J.M.: 157 (1976); F.E.E. 6: 385 (1997).
Syn: *Typha angustata* Bory & Chaub. – F.P.S. 3: 284 (1956).
Syn: *Typha australis* Schumach. – F.P.S. 3: 285 (1956); Bot. Exp. Sud.: 108 (1970).
Syn: *Typha angustifolia* sensu Andrews, non L. – F.P.S. 3: 285 (1956).
Rhizomatous perennial herb. Wetlands.
Distr: ?NS, RS, DAR, CS, UN; pantropical & warm temperate.
DAR: <u>Wickens 2975</u> 5/1965; **UN:** <u>J.M. Lock 81/24</u> (fl) 2/1981.
Note: *T. latifolia* L. may also occur in northern Sudan but this is unconfirmed.

Xyridaceae

H. Pickering

Xyris capensis Thunb. var. *capensis*

F.E.E. 6: 375 (1997); F.T.E.A. Xyrid.: 16 (1999); Biol. Skr. 51: 482 (2005).

Annual herb. Damp places in grassland & woodland, marshes.
Distr: DAR, EQU; Senegal to Ethiopia, S to South Africa; tropical Asia.
DAR: Wickens 2988 6/1965; **EQU:** Jackson 1537 6/1950.

Xyris straminea Nilsson
F.E.E. 6: 375 (1997); F.T.E.A. Xyrid.: 21 (1999); Biol. Skr. 51: 482 (2005).
Syn: *Xyris* sp. – F.P.S. 3: 251 (1956).
Annual herb. Wet flushes over rocks, marshes.
Distr: EQU; Guinea Bissau to Ethiopia, S to Angola & South Africa.
EQU: Friis & Vollesen 566 11/1980.

Eriocaulaceae
H. Pickering

Eriocaulon afzelianum Wikstr. ex Körn.
F.P.S. 3: 251 (1956); F.T.E.A. Eriocaul.: 26 (2006).
Annual or short-lived perennial herb. Wet grassland.
Distr: ?EQU; Cape Verde Is., Senegal to C.A.R., Tanzania, Zambia & Malawi.
Note: Andrews records this species from Equatoria; we have seen no material with which to verify this record.

Eriocaulon bongense Engl. & Ruhland
F.P.S. 3: 251 (1956); F.T.E.A. Eriocaul.: 18 (2006).
Annual herb. Wet grassland.
Distr: BAG; Mauritania & Senegal to Chad & C.A.R., Tanzania to Zambia.
BAG: Schweinfurth 2539 10/1869.

Eriocaulon elegantulum Engl.
F.P.S. 3: 251 (1956); F.T.E.A. Eriocaul.: 17 (2006).
Annual herb. Swamps & seasonal pools.
Distr: BAG; Ghana to Kenya, S to Zimbabwe & Mozambique.
BAG: Schweinfurth 223 10/1869.

Eriocaulon schimperi Körn. ex Ruhland
F.E.E. 6: 379 (1997); Biol. Skr. 51: 482 (2005); F.T.E.A. Eriocaul.: 15 (2006).
Rhizomatous perennial herb. Wet montane grassland.
Distr: EQU; Ethiopia to Malawi.
EQU: Jackson 1534 6/1950.

Eriocaulon setaceum L.
F.T.E.A. Eriocaul.: 5 (2006).
Syn: *Eriocaulon bifistulosum* Van Heurck & Müll.Arg. – F.P.S. 3: 251 (1956).
Aquatic herb. Slow-moving & stagnant water.
Distr: BAG; palaeotropical.
BAG: Schweinfurth 2476 10/1869.

Juncaceae
H. Pickering

Juncus bufonius L.
F.T.E.A. Junc.: 2 (1966); F.J.M.: 161 (1976); F.E.E. 6: 386 (1997).
Tufted annual or perennial herb. Wet grassland, streamsides.

Distr: RS, DAR, CS; largely cosmopolitan though more widespread in temperate regions.
DAR: Wickens 1237 9/1964.

Juncus dregeanus Kunth subsp. *bachitii* (Hochst. ex Steud.) Hedberg
F.P.S. 3: 326 (species only) (1956); F.T.E.A. Junc.: 4 (1966); F.J.M.: 161 (1976); Biol. Skr. 51: 521 (2005).
Tufted perennial herb. Swamps, montane grassland, bushland & forest margins.
Distr: DAR, EQU; Cameroon, Ethiopia to South Africa.
DAR: Wickens 2677 9/1964; **EQU:** Friis & Vollesen 1207 3/1982.

Juncus oxycarpus E.Mey. ex Kunth
F.T.E.A. Junc.: 3 (1966); F.E.E. 6: 387 (1997); Biol. Skr. 51: 522 (2005).
Tufted perennial herb. Swamps, montane grassland & forest.
Distr: EQU; D.R. Congo to Somalia, S to South Africa.
EQU: Friis & Vollesen 311 11/1980.

Juncus punctorius L.f.
F.J.M.: 161 (1976); F.E.E. 6: 389 (1997).
Rhizomatous perennial herb. Swamps & streamsides.
Distr: DAR; Eritrea, Ethiopia, Somalia, Angola & Zimbabwe to South Africa; N Africa, Arabia to Pakistan.
DAR: Wickens 1415 4/1964.

Juncus rigidus Desf.
F.J.U. 1: 305 (1997); F.Egypt 4: 97 (2005).
Syn: *Juncus arabicus* (Asch. & Buchenau) Adamson – F.P.S. 3: 326 (1956).
Rhizomatous perennial herb. Saline depressions, margins of watercourses in dry areas.
Distr: NS, RS; Senegal to Somalia, D.R. Congo to South Africa; N Africa & Mediterranean to Pakistan.
NS: Léonard 5017 12/1968.

Cyperaceae
I. Farag, I. Darbyshire & M. Kordofani

Abildgaardia ovata (Burm.f.) Král
F.E.E. 6: 413 (1997); F.T.E.A. Cyper.: 113 (2010).
Syn: *Abildgaardia monostachya* (L.) Vahl – F.P.S. 3: 328 (1956).
Tufted perennial herb. Grassland & wooded grassland, grazed areas.
Distr: EQU; widespread in warm temperate & tropical regions.
EQU: Schweinfurth 3962 (fr) 6/1870.

Ascolepis capensis (Kunth) Ridl.
F.P.S. 3: 328 (1956); F.E.E. 6: 492 (1997).
Tufted perennial herb. Swamps & wet grassland.
Distr: BAG, EQU; Mali & Ivory Coast to Ethiopia, S to South Africa.
EQU: Jackson 2300 (fl) 6/1952.

Ascolepis eriocauloides (Steud.) Nees ex Steud.
F.E.E. 6: 491 (1997).
Tufted perennial herb. Seasonally wet grassland, wet flushes over rocks.

Distr: KOR; Cameroon, Ethiopia.
IUCN: RD
KOR: Gillespie s.n. (fl) 8/1979.

Ascolepis lineariglumis Lye

F.T.E.A. Cyper.: 267 (2010).
Tufted annual or perennial herb. Wet grassland.
Distr: BAG; Nigeria to Kenya, S to Angola &
Mozambique.
BAG: Myers 7327 (fl) 7/1937.

Ascolepis protea Welw. var. *bellidiflora* Welw.

F.T.E.A. Cyper.: 270 (2010).
Syn: *Ascolepis elata* sensu Andrews, non Welw. – F.P.S.
3: 328 (1956).
Syn: *Ascolepis protea* Welw. subsp. *bellidiflora* (Welw.)
Lye – F.E.E. 6: 491 (1997); Biol. Skr. 51: 522 (2005).
Tufted perennial herb. Wet grassland, streamsides.
Distr: BAG, EQU; Nigeria to Ethiopia, S to Angola &
Mozambique.
BAG: Schweinfurth 1919 (fl) 6/1869.

Bolboschoenus maritimus (L.) Palla

F.T.E.A. Cyper.: 23 (2010).
Syn: *Schoenoplectus maritimus* (L.) Lye – F.E.E. 6: 397
(1997).
Syn: *Scirpus maritimus* L. – F.P.S. 3: 367 (1956).
Rhizomatous perennial herb. Seasonal swamps including
saline areas, ricefields, muddy riverbanks.
Distr: NS, CS; widespread in temperate & tropical regions.
CS: de Wilde et al. 5719 1/1965.

Bulbostylis abortiva (Steud.) C.B.Clarke

F.P.S. 3: 329 (1956); F.E.E. 6: 420 (1997); Biol. Skr. 51:
522 (2005); F.T.E.A. Cyper.: 97 (2010).
Tufted annual herb. Shallow soils over rocks, seasonally
damp grassland & wooded grassland.
Distr: ES, BAG, UN, EQU; Guinea to Ethiopia, S to
Zambia & Malawi; Madagascar.
ES: Schweinfurth 2039 (fl) 9/1865; **UN:** J.M. Lock 81/128
(fr) 8/1982.

Bulbostylis atrosanguinea (Boeckeler) C.B.Clarke

F.P.S. 3: 329 (1956); F.T.E.A. Cyper.: 75 (2010).
Syn: *Bulbostylis setifolia* (A.Rich.) Bodard – F.E.E. 6:
415 (1997); Biol. Skr. 51: 524 (2005).
Tufted perennial herb. Montane grassland & ericaceous
scrub.
Distr: EQU; Ethiopia to Angola, Zambia & Malawi;
Yemen.
EQU: Friis & Vollesen 840 12/1980.

Bulbostylis barbata (Rottb.) C.B.Clarke

F.E.E. 6: 421 (1997); F.T.E.A. Cyper.: 104 (2010).
Tufted annual herb. Wooded & bushed grassland, ditches,
rock crevices.
Distr: KOR; palaeotropical.
KOR: Ritchie 17 (fl) 9/1980.

Bulbostylis cioniana (Pi.Savi) Lye

Mitt. Bot. Staatssamml. München 10: 547 (1971).
Syn: *Fimbristylis cioniana* Pi.Savi – F.P.S. 3: 362 (1956).
Tufted annual herb. Sandy grassland, riverbanks.

Distr: BAG; Sierra Leone to C.A.R.; Spain, N Africa.
BAG: Schweinfurth 2122 (fl) 10/1869.

Bulbostylis coleotricha (Hochst. ex A.Rich.) C.B.Clarke var. *coleotricha*

F.P.S. 3: 329 (1956); F.E.E. 6: 420 (1997); Biol. Skr. 51:
522 (2005); F.T.E.A. Cyper.: 97 (2010).
Tufted annual herb. Shallow seaonally damp soils, usually
over rocks or in rock crevices.
Distr: KOR, BAG, UN, EQU; Senegal to Cameroon,
Ethiopia.
KOR: Wickens 499 (fr) 9/1962; **EQU:** Jackson 2066 (fl)
9/1951.

Bulbostylis densa (Wall.) Hand.-Mazz.

F.J.M.: 161 (1976); F.E.E. 6: 419 (1997); Biol. Skr. 51: 523
(2005); F.T.E.A. Cyper.: 98 (2010).
Tufted annual herb. Montane & lower altitude grassland,
clearings in forest.
Distr: DAR, EQU; palaeotropical.
DAR: Wickens 2411 (fl) 9/1964; **EQU:** Friis &
Vollesen 382 (fr) 11/1980.
Note: Our material should perhaps be treated as subsp.
afromontana (Lye) Haines but the distinction seems
unclear; see note in F.T.E.A. Cyper.: 100 (2010).

Bulbostylis hispidula (Vahl) R.W.Haines subsp. *hispidula*

F.E.E. 6: 416 (1997); Biol. Skr. 51: 523 (2005); F.T.E.A.
Cyper.: 82 (2010).
Syn: *Fimbristylis exilis* (Kunth) Roem. & Schult. – F.P.S.
3: 360 (1956); Bot. Exp. Sud.: 46 (1970).
Syn: *Fimbristylis hispidula* (Vahl) Kunth – F.J.M.: 163
(1976).
Tufted annual (rarely perennial) herb. Seasonally damp
grassland & wooded grassland, shallow soils over rocks.
Distr: DAR, KOR, ES, UN, EQU; pantropical.
KOR: Wickens 531 9/1962; **EQU:** Friis & Vollesen 41 (fr)
11/1980.

Bulbostylis cf. oligostachys (Hochst. ex A.Rich.) C.B.Clarke

F.T.E.A. Cyper.: 87 (2010).
Syn: *Fimbristylis* sp. sensu Wickens – F.J.M.: 163 (1976).
Annual herb. Riverbanks.
Distr: DAR.
DAR: Wickens 2621 (fr) 9/1964.

Bulbostylis oritrephes (Ridl.) C.B.Clarke

Biol. Skr. 51: 524 (2005); F.T.E.A. Cyper.: 77 (2010).
Tufted perennial herb. Grassland & wooded grassland,
shallow soils over rocks, roadsides.
Distr: EQU; Guinea to Uganda, S to Tanzania, Angola,
South Africa.
EQU: Friis & Vollesen 942 (fr) 2/1982.

Bulbostylis pusilla (Hochst. ex A.Rich.) C.B.Clarke subsp. *pusilla*

F.P.S. 3: 329 (1956); F.T.E.A. Cyper.: 93 (2010).
Syn: *Bulbostylis pusilla* (Hochst. ex A.Rich.) C.B.Clarke
subsp. *yalingensis* (Cherm.) R.W.Haines – F.E.E. 6: 418
(1997); Biol. Skr. 51: 524 (2005).
Tufted annual herb. Grassland & wooded grassland,
shallow soils over rocks, seasonally wet areas.

Distr: ?KOR, EQU; Mali to Ethiopia, S to Tanzania.
EQU: Andrews 1852 (fr) 6/1939.
Note: Andrews records this species from Kordofan but we have seen no material from Sudan and his record may be based on a misidentification.

Bulbostylis pusilla (De Wild.) R.W.Haines subsp. congolensis (De Wild.) R.W.Haines

F.E.E. 6: 418 (1997); F.T.E.A. Cyper.: 95 (2010).
Tufted annual herb. Grassland & bushland.
Distr: EQU; Sierra Leone to Ethiopia, S to D.R. Congo & Tanzania.
EQU: Jackson 3843 (fr) 9/1957.

Bulbostylis scabricaulis Cherm.

F.T.E.A. Cyper.: 103 (2010).
Syn: Bulbostylis filamentosa sensu Andrews, non (Vahl) C.B.Clarke – F.P.S. 3: 329 (1956).
Tufted perennial herb. Grassland, swamp margins, rock crevices.
Distr: BAG, EQU; Senegal to Uganda, S to Angola & South Africa; Madagascar.
BAG: Myers 7306 (fr) 7/1937.

Bulbostylis sphaerocarpa (Boeckeler) C.B.Clarke

F.P.S. 3: 328 (1956); F.E.E. 6: 417 (1997); F.T.E.A. Cyper.: 92 (2010).
Tufted annual herb. Seasonally damp soils in open woodland.
Distr: ES, Ethiopia, Tanzania, ?Angola.
IUCN: RD
ES: Schweinfurth 2046 (fr) 9/1865.

Carex chlorosaccus C.B.Clarke

F.P.S. 3: 330 (1956); F.E.E. 6: 505 (1997); Biol. Skr. 51: 524 (2005); F.T.E.A. Cyper.: 432 (2010).
Tussock-forming perennial herb. Forest & forest margins, swamp margins.
Distr: EQU; Bioko, Cameroon, Ethiopia to Malawi.
EQU: Beshir 29A (fl) 8/1951.

Carex conferta Hochst. ex A.Rich.

F.E.E. 6: 503 (1997); Biol. Skr. 51: 525 (2005); F.T.E.A. Cyper.: 428 (2010).
Tufted rhizomatous perennial herb. Swamps, montane forest & transition to ericaceous scrub or montane grassland.
Distr: EQU; Ethiopia to D.R. Congo & Tanzania.
EQU: Friis & Vollesen 1296 (fr) 3/1982.

Carex echinochloe Kunze subsp. echinochloe

F.P.S. 3: 330 (1956); F.E.E. 6: 505 (1997); Biol. Skr. 51: 525 (2005); F.T.E.A. Cyper.: 430 (2010).
Tufted rhizomatous perennial herb. Variously grassland & bushland to open montane forest, swamp margins.
Distr: EQU; Guinea, Bioko, Cameroon, Ethiopia to Tanzania.
EQU: Friis & Vollesen 814 (fr) 12/1980.

Carex johnstonii Boeckeler

F.E.E. 6: 507 (1997); Biol. Skr. 51: 525 (2005); F.T.E.A. Cyper.: 436 (2010).
Tufted perennial herb with short rhizome. Montane forest & upper forest margins.
Distr: EQU; Ethiopia to Malawi.
EQU: Friis & Vollesen 855 12/1980.

Carex mannii E.A.Bruce subsp. mannii

F.P.S. 3: 330 (1956); Biol. Skr. 51: 525 (2005); F.T.E.A. Cyper.: 445 (2010).
Tufted rhizomatous perennial herb. Montane forest with swampy areas.
Distr: EQU; Bioko, Cameroon, D.R. Congo, Rwanda, Uganda.
EQU: Jackson 1532 (fl) 6/1950.

Carex mannii E.A.Bruce subsp. thomasii (Nelmes) Luceño & M.Escudero

Pl. Syst. Evol. 279: 187 (2009).
Syn: Carex thomasii Nelmes – F.P.S. 3: 330 (1956); F.E.E. 6: 508 (1997); Biol. Skr. 51: 526 (2005); F.T.E.A. Cyper.: 443 (2010).
Tufted perennial herb. Open montane forest & bushland.
Distr: EQU; Ethiopia.
IUCN: RD
EQU: Thomas 1794 (fr) 12/1935.

Carex cf. neochevalieri Kük.

Syn: Carex sp. near **echinochloe** sensu Wickens, non Kunze – F.J.M.: 161 (1976).
Montane forest swamps.
Distr: DAR; (for species) Sierra Leone to Cameroon.
DAR: Kassas 351.
Note: the specimen cited has not been seen by us; it was cited by Wickens as "a poor specimen obviously close to C. echinchloe and to C. neo-chevalieri, but more especially the latter".

Carex petitiana A.Rich.

F.E.E. 6: 510 (1997); Biol. Skr. 51: 526 (2005); F.T.E.A. Cyper.: 439 (2010).
Syn: Carex sp. cf. **C. petitiana** sensu Friis & Vollesen – Biol. Skr. 51: 526 (2005).
Tussock-forming perennial herb. Montane forest & forest margins, montane grassland, streamsides.
Distr: EQU; Cameroon, Ethiopia to Tanzania.
EQU: Friis & Vollesen 309 (fr) 11/1980.

Carex steudneri Boeckeler

F.P.S. 3: 330 (1956); F.E.E. 6: 506 (1997); Biol. Skr. 51: 526 (2005); F.T.E.A. Cyper.: 435 (2010).
Tufted rhizomatous perennial herb. Montane grassland & bushland, forest margins, streamsides, rock crevices.
Distr: EQU; Ethiopia to Malawi, South Africa.
EQU: Jackson 1528 (fr) 6/1950.

Coleochloa abyssinica (Hochst. ex A.Rich.) Gilly

F.E.E. 6: 500 (1997); F.T.E.A. Cyper.: 374 (2010).
Syn: Coleochloa abyssinica (Hochst. ex A.Rich.) Gilly var. **castanea** (C.B.Clarke) Pic.Serm. – Biol. Skr. 51: 526 (2005).
Densely tufted perennial herb. Wet rocks, montane grassland.
Distr: EQU; Cameroon, D.R. Congo, Eritrea, Ethiopia, Uganda, Tanzania, Angola.
EQU: Friis & Vollesen 264 (fr) 11/1980.

Coleochloa glabra Nelmes

F.P.S. 3: 331 (1956).
Densely tufted perennial herb. Rock crevices on ridge slopes & summit.
Distr: EQU; South Sudan endemic.
IUCN: RD
EQU: Andrews 852 (fr) 4/1939.
Note: apparently restricted to Mt Odo, with two specimens seen.

Coleochloa schweinfurthiana (Boeckeler) Nelmes

F.P.S. 3: 331 (1956).
Tufted perennial herb. Mountain summit, presumably on exposed rock.
Distr: EQU; South Sudan endemic.
IUCN: RD
EQU: Schweinfurth 3820 (fr) 5/1870.
Note: closely allied to *C. abyssinica* (see note in F.E.E. 6: 500), this species is known only from the type from Mt. Baginze on the border between South Sudan and D.R. Congo.

Cyperus ajax C.B.Clarke

Biol. Skr. 51: 527 (2005); F.T.E.A. Cyper.: 202 (2010).
Rhizomatous perennial herb. Open or secondary montane forest thickets, riverbanks.
Distr: EQU; Uganda to Malawi.
EQU: Friis & Vollesen 404 (fl) 11/1980.

Cyperus albosanguineus Kük.

Biol. Skr. 51: 527 (2005); F.T.E.A. Cyper.: 184 (2010).
Tufted perennial herb. Montane wet grassland, rock crevices.
Distr: EQU; D.R. Congo, Uganda, Kenya, Tanzania.
EQU: Friis & Vollesen 981 (fr) 2/1982.

Cyperus alopecuroides Rottb.

F.P.S. 3: 348 (1956); F.J.M.: 161 (1976); F.E.E. 3: 445 (1989); F.T.E.A. Cyper.: 219 (2010).
Tussock-forming perennial herb. Swamps, riverbeds, seasonally wet grassland.
Distr: NS, DAR, CS, BAG, UN, EQU; palaeotropical.
CS: Andrews 123 (fr) 12/1935; **BAG:** Simpson 7116 (fr) 6/1929.

Cyperus amabilis Vahl

F.P.S. 3: 342 (1956); F.E.E. 6: 460 (1997); Biol. Skr. 51: 527 (2005); F.T.E.A. Cyper.: 158 (2010).
Annual herb, often tufted. Lake & swamp margins, seasonally wet areas, wet sand.
Distr: KOR, BAG, EQU; pantropical.
KOR: Gay 124 (fl) 10/1839; **EQU:** Sillitoe 70 (fr) 1919.

Cyperus amauropus Steud.

F.P.S. 3: 352 (1956); F.E.E. 6: 466 (1997); F.T.E.A. Cyper.: 237 (2010).
Perennial herb with pseudobulb & short rhizome/stolons. Grassland & wooded grassland, shallow soils over rock.
Distr: ?RS; Eritrea & Somalia to Zambia.
Note: Andrews records this species from Khor Tamanib in Red Sea District; we have not seen the specimen on which this was based.

Cyperus angolensis Boeckeler

Biol. Skr. 51: 528 (2005); F.T.E.A. Cyper.: 167 (2010).
Stoloniferous perennial herb. Dry grassland or lightly wooded grassland, often with frequent burning.
Distr: EQU; Ghana to Uganda, S to Angola & South Africa.
EQU: Friis & Vollesen 946 (fl) 2/1982.

Cyperus articulatus L.

F.P.S. 3: 338 (1956); F.E.E. 6: 446 (1997); F.T.E.A. Cyper.: 208 (2010).
Stoloniferous perennial herb. Swamps, pools, lake margins, wet grassland.
Distr: RS, DAR, CS, BAG, UN, EQU; pantropical.
DAR: Broun 1459 1907; **UN:** Harrison 1947 (fr) 6/1947.

Cyperus blepharoleptos Steud.

Pl. Ecol. Evol. 144: 350 (2011).
Syn: *Oxycaryum cubensis* (Poepp. & Kunth) Palla – F.E.E. 6: 428 (1997); F.T.E.A. Cyper.: 126 (2010).
Syn: *Scirpus cubensis* Poepp. & Kunth – F.P.S. 3: 367 (1956).
Aquatic perennial herb. Lakes, pools & swamps, sometimes forming floating mats.
Distr: BAG, UN, EQU; tropical Africa & America.
UN: J.M. Lock 83/17 (fl) 2/1983.

Cyperus boreochrysocephalus Lye

Biol. Skr. 51: 528 (2005); F.T.E.A. Cyper.: 149 (2010).
Slender perennial herb. Grassland & open woodland.
Distr: EQU; Uganda, Kenya, Tanzania.
EQU: Kielland-Lund 927 6/1984 (n.v.).
Note: it is possible that this taxon is conspecific with *C. remotus* (see note in F.T.E.A.).

Cyperus bulbosus Vahl

F.P.S. 3: 345 (1956); F.E.E. 6: 451 (1997); F.T.E.A. Cyper.: 221 (2010).
Perennial herb with basal bulb. Seasonally wet grassland and sand.
Distr: RS, DAR, KOR, CS, UN; palaeotropical.
CS: Jackson 4305 (fl fr) 7/1961; **UN:** Harrison 67 (fl) 6/1947.

Cyperus colymbetes Kotschy & Peyr.

F.P.S. 3: 338 (1956); F.T.E.A. Cyper.: 151 (2010).
Rhizomatous perennial herb. Swampy ground, margins of pools.
Distr: CS, UN, EQU; Uganda & Somalia to Mozambique; Madagascar.
CS: Broun 36 (fl) 12/1903; **EQU:** Myers 8576 (fl) 2/1938.

Cyperus commixtus Kük.

F.Som. 4: 117 (1995).
Slender rhizomatous perennial herb. Pools, lake margins.
Distr: UN; Somalia.
IUCN: RD
UN: Broun 49 (fl fr) 2/1903.
Note: the identity of the single Sudan specimen must be in question but we have seen no Somali material with which to compare it.

Cyperus compressus L.

F.P.S. 3: 347 (1956); Bot. Exp. Sud.: 44 (1970); F.E.E. 6: 459 (1997); F.T.E.A. Cyper.: 195 (2010).

Slender to robust annual herb. Pool margins, ditches & other temporarily wet areas.
Distr: ?DAR, UN; widespread in the tropics.
UN: Drar 832 4/1938 (n.v.).
Note: Andrews based his record of this species on a collection from the Libyan Desert; we have not seen this material but it is likely to be an accurate record. The identification of the cited specimen is from Drar (1970).

Cyperus conglomeratus Rottb. subsp. conglomeratus

F.E.E. 6: 463 (1997).
Tussock-forming perennial herb. Sandy soils, often saline.
Distr: NS, RS, ?KOR, CS; Senegal to Somalia; N Africa, Madagascar, Mascarenes, Arabia to India.
RS: Carter 2015 (fl fr) 11/1987.
Note: Andrews lists also var. *minor* Boeckeler from Red Sea District and Kordofan; we have been unable to trace this name.

Cyperus cuspidatus Kunth

F.P.S. 3: 341 (1956); F.J.M.: 161 (1976); F.E.E. 6: 461 (1997); Biol. Skr. 51: 529 (2005); F.T.E.A. Cyper.: 191 (2010).
Slender annual herb. Seasonally damp grassland, rock outcrops.
Distr: DAR, UN, EQU; pantropical.
DAR: Wickens 1172 (fr) 2/1964; **UN:** Simpson 7254 (fr) 7/1929.

Cyperus cyperoides (L.) Kuntze

F.P.S. 3: 352 (1956); F.E.E. 6: 456 (1997); Biol. Skr. 51: 529 (2005); F.T.E.A. Cyper.: 223 (2010).
Syn: *Cyperus cyperoides* (L.) Kuntze subsp. *flavus* Lye – Biol. Skr. 51: 529 (2005).
Syn: *Mariscus sieberianus* Nees ex C.B.Clarke – Bot. Exp. Sud.: 46 (1970).
Shortly rhizomatous perennial herb. Forest clearings, wooded grassland, streamsides, weed of cultivation.
Distr: BAG, EQU; widespread in the tropics & subtropics.
EQU: Andrews 450 (fl) 4/1939.

Cyperus denudatus L.f.

F.E.E. 6: 439 (1997); Biol. Skr. 51: 530 (2005); F.T.E.A. Cyper.: 190 (2010).
Tufted rhizomatous perennial herb. Swamps, ditches & riversides, damp grassland.
Distr: EQU; tropical & southern Africa; Madagascar, India, Vietnam, Australia.
EQU: Friis & Vollesen 35 (fr) 11/1980.

Cyperus derreilema Steud.

F.P.S. 3: 340 (1956); F.E.E. 6: 435 (1997); Biol. Skr. 51: 530 (2005); F.T.E.A. Cyper.: 197 (2010).
Robust rhizomatous perennial herb. Montane forest, often in clearings, streamsides.
Distr: EQU; Ethiopia to Malawi.
EQU: Johnston 1434 (fr) 2/1936.

Cyperus dichrostachyus Hochst. ex A.Rich.

F.E.E. 6: 436 (1997); Biol. Skr. 51: 530 (2005); F.T.E.A. Cyper.: 198 (2010).
Robust stoloniferous perennial herb. Swamps, pools, riverbanks.

Distr: EQU; Cameroon, Ethiopia to South Africa; Madagascar.
EQU: Friis & Vollesen 1031 (fr) 2/1982.

Cyperus difformis L.

F.P.S. 3: 341 (1956); F.E.E. 6: 436 (1997); Biol. Skr. 51: 531 (2005). F.T.E.A. Cyper.: 173 (2010).
Annual herb. Swamps, pool margins, wet grassland, ditches.
Distr: KOR, CS, BAG, UN, EQU; palaeotropical.
KOR: Ritchie 2 (fl) 9/1980; **EQU:** Myers 6924 (fl) 7/1937.

Cyperus digitatus Roxb. subsp. auricomus (Sieber ex Spreng.) Kük.

F.P.S. 3: 343 (1956); F.J.M.: 162 (1976); F.E.E. 6: 443 (1997); F.T.E.A. Cyper.: 244 (2010).
Syn: *Cyperus digitatus* Roxb. subsp. *auricomus* (Sieber ex Spreng.) Kük. var. *minor* (C.B.Clarke) Kük. – F.P.S. 3: 343 (1956).
Robust rhizomatous perennial herb. Swamps and wet grassland.
Distr: DAR, KOR, CS, ES, BAG, EQU; Senegal to Ethiopia, S to South Africa; Egypt.
DAR: Wickens 1988 (fr) 7/1964; **BAG:** Myers 13878 7/1941.

Cyperus dilatatus Schumach.

F.T.E.A. Cyper.: 236 (2010).
Syn: *Cyperus gracilinux* C.B.Clarke – F.P.S. 3: 346 (1956).
Robust stoloniferous perennial herb. Seasonally wet habitats.
Distr: BAG; Senegal to C.A.R., Somalia to Tanzania.
BAG: Myers 7203 (fl fr) 7/1937.

Cyperus distans L.f.

F.P.S. 3: 346 (1956); F.E.E. 6: 455 (1997); Biol. Skr. 51: 531 (2005); F.T.E.A. Cyper.: 249 (2010).
Syn: *Cyperus distans* L.f. var. *crassispiculosus* Gross & Kük. – F.P.S. 3: 346 (1956).
Syn: *Mariscus longibracteatus* Cherm. – Bot. Exp. Sud.: 46 (1970).
Shortly rhizomatous perennial herb. Swamps, streamsides, damp areas in woodland, forest margins.
Distr: BAG, UN, EQU; pantropical.
UN: J.M. Lock 81/229 (fr) 8/1981.

Cyperus diurensis Boeckeler

F.P.S. 3: 353 (1956); F.E.E. 6: 467 (1997); Biol. Skr. 51: 532 (2005); F.T.E.A. Cyper.: 166 (2010).
Stoloniferous perennial herb. Grassland, woodland, rocky outcrops.
Distr: BAG; Ethiopia to Tanzania.
BAG: Schweinfurth 198A (fr) 8/1869.

Cyperus dubius Rottb. var. dubius

F.P.S. 3: 354 (1956); Biol. Skr. 51: 532 (2005); F.T.E.A. Cyper.: 186 (2010).
Syn: *Cyperus dubius* Rottb. var. *coloratus* (Vahl) Kük. – F.P.S. 3: 354 (1956).
Syn: *Cyperus dubius* Rottb. subsp. *coloratus* (Vahl) Lye – Biol. Skr. 51: 532 (2005).
Tufted perennial herb from bulbous base. Bushland & woodland, often on rocky outcrops, forest margins.

Distr: EQU; palaeotropical.
EQU: Friis & Vollesen 659 (fl) 12/1980.

Cyperus dubius Rottb. var. macrocephalus (C.B.Clarke) Kük.

F.P.S. 3: 354 (1956); F.E.E. 6: 468 (1997); F.T.E.A. Cyper.: 186 (2010).
Tufted perennial herb. Grassland & bushland, often on thin soil over rocks, river & lake shores.
Distr: EQU; Uganda, Kenya, Tanzania, ?Ethiopia.
EQU: Andrews 977 (fl) 5/1939.
Note: Lye in F.E.E. records this taxon as much more widespread than in F.T.E.A.; clearly there is some disagreement over its delimitation.

Cyperus esculentus L.

F.P.S. 3: 345 (1956); F.J.M.: 162 (1976); F.E.E. 6: 451 (1997); F.T.E.A. Cyper.: 227 (2010).
Syn: Cyperus fresenii Steud. – F.P.S. 3: 340 (1956).
Stoloniferous perennial herb. Swamps, wet grassland, weed of cultivation.
Distr: DAR, KOR, BAG, UN, EQU; palaeotropical & subtropical, N America.
DAR: Wickens 1922 (fr) 7/1964; **EQU:** Sillitoe 69 (fl) 1919.

Cyperus exaltatus Retz. var. exaltatus

F.P.S. 3: 344 (1956); F.J.M.: 162 (1976); F.E.E. 6: 444 (1997); Biol. Skr. 51: 533 (2005); F.T.E.A. Cyper.: 245 (2010).
Shortly rhizomatous perennial herb. Swamps, margins of open water.
Distr: KOR, UN, EQU; pantropical.
KOR: Pfund 228 (fl) 5/1878; **UN:** Simpson 7278 (fr) 7/1929.

Cyperus exaltatus Retz. var. dives (Delile) C.B.Clarke

F.T.E.A. Cyper.: 245 (2010).
Syn: Cyperus dives Delile – F.P.S. 3: 343 (1956); Bot. Exp. Sud.: 45 (1970); F.E.E. 6: 444 (1997); Biol. Skr. 51: 532 (2005).
Syn: Cyperus immensus (C.B.Clarke) Kük. var. **petherickii** (C.B.Clarke) Kük. – F.P.S. 3: 343 (1956).
Shortly rhizomatous perennial herb. Swamps, riverbanks, shallow open water.
Distr: BAG, EQU; Senegal to Somalia, S to Botswana & Mozambique; Egypt, Middle East, India to Vietnam.
BAG: Schweinfurth 3727 (fr) 7/1870.

Cyperus fischerianus Schimp. ex A.Rich.

F.E.E. 6: 435 (1997); Biol. Skr. 51: 534 (2005); F.T.E.A. Cyper.: 200 (2010).
Shortly rhizomatous perennial herb, tussock forming. Forest & forest margins, woodland.
Distr: RS, EQU; Eritrea to Malawi.
RS: Andrews 3586 (fl); **EQU:** Kielland-Lund 638 1/1984 (n.v.).

Cyperus foliaceus C.B.Clarke

F.E.E. 6: 437 (1997); Biol. Skr. 51: 534 (2005); F.T.E.A. Cyper.: 193 (2010).
Annual herb. Swamps, streamsides, seasonally wet areas in woodland.
Distr: UN; Togo, Ethiopia to Tanzania.
UN: J.M. Lock 81/213 (fl) 8/1981.

Cyperus glaucophyllus Boeckeler

F.T.E.A. Cyper.: 198 (2010).
Rhizomatous perennial herb. Forest including clearings, streamsides.
Distr: EQU; Uganda & Kenya to Malawi, Swaziland, South Africa.
EQU: Andrews 1906 (fl) 6/1939.
Note: the identification of the single South Sudanese specimen seen requires confirmation.

Cyperus haspan L.

F.P.S. 3: 340 (1956); F.E.E. 6: 438 (1997); F.T.E.A. Cyper.: 205 (2010).
Shortly rhizomatous perennial herb. Swamps, seasonally wet sites.
Distr: BAG, EQU; pantropical.
EQU: Myers 6934 (fr) 7/1937.

Cyperus imbricatus Retz.

F.P.S. 3: 344 (1956); F.E.E. 6: 445 (1997); F.T.E.A. Cyper.: 251 (2010).
Shortly rhizomatous perennial herb. Swamps, streamsides, forest margins.
Distr: BAG, UN, EQU; widespread in the tropics.
BAG: Schweinfurth 1579 (fr) 4/1869.

Cyperus impubes Steud.

F.P.S. 3: 351 (1956); F.E.E. 6: 457 (1997); Biol. Skr. 51: 534 (2005); F.T.E.A. Cyper.: 240 (2010).
Rhizomatous perennial herb. Forest clearings, margins & secondary growth after fire, streamsides.
Distr: EQU; Eritrea & Somalia to Tanzania; Socotra.
EQU: Jackson 1366 3/1950.

Cyperus involucratus Rottb.

Biol. Skr. 51: 534 (2005); F.T.E.A. Cyper.: 187 (2010).
Syn: Cyperus alternifolius L. subsp. **flabelliformis** Kük. – F.P.S. 3: 338 (1956); Bot. Exp. Sud.: 44 (1970); F.E.E. 6: 434 (1997).
Robust rhizomatous perennial herb. Swamps, wet grassland, riverbanks; cultivated as an ornamental.
Distr: CS, EQU; Senegal to Somalia, S to Angola & South Africa.
CS: Kotschy 565 (fr) 1888; **EQU:** Myers 11028 (fl) 4/1939.

Cyperus iria L.

F.P.S. 3: 347 (1956); F.E.E. 6: 453 (1997); Biol. Skr. 51: 535 (2005); F.T.E.A. Cyper.: 214 (2010).
Tufted annual herb. Swamps, pool margins, wet grassland, streamsides.
Distr: DAR, KOR, CS, BAG, UN, EQU; palaeotropical.
DAR: Pfund 335 (fl) 5/1878; **UN:** J.M. Lock 81/203 (fr) 8/1981.

Cyperus jeminicus Rottb.

F.E.E. 6: 463 (1997).
Syn: Cyperus conglomeratus Rottb. var. **multiculmis** (Boeckeler) Kük. – F.P.S. 3: 342 (1956).
Tussock-forming perennial herb. Sand dunes, sandy soils.
Distr: NS, RS, KOR, CS, ES; Senegal to Somalia; Madagascar, N Africa, Arabia, India.
KOR: Jackson 4003 (fr) 9/1959.

Cyperus kerstenii Boeckeler

Biol. Skr. 51: 535 (2005); F.T.E.A. Cyper.: 183 (2010).
Tussock-forming perennial herb. Montane grassland &
boggy ground.
Distr: EQU; Uganda, Kenya, Tanzania.
EQU: Friis & Vollesen 1274 (fl) 3/1982.

[Cyperus kerstingii Engl.]

Note: *C. kerstingii* is listed in the World Checklist Series
as an unplaced name, described from Sudan. We have
seen no material.

Cyperus kyllingiella Larridon

Pl. Ecol. Evol. 144: 351 (2011).
Syn: *Kyllingiella microcephala* (Steud.) R.W.Haines &
Lye – F.E.E. 6: 425 (1997); F.T.E.A. Cyper.: 129 (2010).
Syn: *Scirpus microcephalus* (Steud.) Dandy – F.P.S. 3:
366 (1956); F.J.M.: 165 (1976).
Tufted perennial herb. Sandy soils, often over rock with
seasonal seepage.
Distr: DAR, ES, BAG; Senegal to Ethiopia, S to South
Africa; India.
DAR: Wickens 1773 (fl) 6/1964; **BAG:** Schweinfurth 1916
(fl) 6/1869.
Note: *C. acholiensis* Larridon (syn. *Kyllingiella ugandensis*
R.W.Haines & Lye) is likely to occur in South Sudan but
is so far known only from the type specimen in the
Ugandan Imatong Mts.

Cyperus laevigatus L. subsp. laevigatus

F.P.S. 3: 348 (1956); F.J.M.: 162 (1976); F.E.E. 6: 459
(1997); F.T.E.A. Cyper.: 153 (2010).
Rhizomatous perennial herb. Saline soils, often forming
dense swards.
Distr: NS, RS, DAR, CS, ES, EQU; pantropical.
RS: Jackson 2811 (fl) 4/1953; **EQU:** Clarke 976 (fl)
5/1939.

Cyperus laevigatus L. subsp. distachyos (All.) Ball

F.Egypt 4: 393 (2005).
Syn: *Cyperus laevigatus* L. var. *distachyos* (All.) Coss.
& Durieu – F.J.M.: 162 (1976).
Rhizomatous perennial herb. Wet sand, riverbanks,
including saline areas.
Distr: DAR; S Europe & N Africa to central Asia.
DAR: Wickens 1502 (fr) 4/1964.

Cyperus latifolius Poir.

F.P.S. 3: 347 (1956); F.E.E. 6: 446 (1997); F.T.E.A. Cyper.:
233 (2010).
Stoloniferous perennial herb. Swamps, wet grassland,
ditches, streamsides.
Distr: UN, EQU; Benin, Cameroon to Ethiopia, S to
Angola & South Africa; Madagascar.
UN: Andrews 616 (fr) 4/1939.

Cyperus laxus Lam. subsp. buchholzii (Boeckeler) Lye

F.T.E.A. Cyper.: 201 (2010).
Syn: *Cyperus diffusus* Vahl subsp. *buchholzii*
(Boeckeler) Kük. – Biol. Skr. 51: 531 (2005).
Shortly rhizomatous perennial herb. Forest margins,
grassland, swamps, often near streams.

Cyperus longibracteatus (Cherm.) Kük.

F.P.S. 3: 351 (1956); F.E.E. 6: 455 (1997).
Syn: *Cyperus distans* L.f. subsp. *longibracteatus*
(Cherm.) Lye – Biol. Skr. 51: 531 (2005); F.T.E.A. Cyper.:
250 (2010).
Syn: *Cyperus distans* L.f. subsp. *longibracteatus*
(Cherm.) Lye var. *rubrotinctus* (Cherm.) Lye – Biol. Skr.
51: 532 (2005).
Tufted perennial herb with short rhizome. Montane
forest, forest margins & woodland, streamsides.
Distr: EQU; Senegal to Ethiopia, S to Zambia; Madagascar.
EQU: Schweinfurth 3023 (fl) 7/1870.
Note: treated in F.T.E.A. Cyper.: 250 (2010) as a synonym
of *C. distans*.

Cyperus longus L.

F.E.E. 6: 449 (1997); F.T.E.A. Cyper.: 230 (2010).
Syn: *Cyperus fenzelianus* Steud. – F.J.M.: 162 (1976).
Syn: *Cyperus longus* L. var. *pallidus* Boeckeler – F.P.S.
3: 345 (1956); Bot. Exp. Sud.: 45 (1970).
Stoloniferous perennial herb. Lake margins, swamps, wet
depressions in grassland & woodland.
Distr: NS, DAR, CS, UN; widespread in Africa; S Europe,
W Asia.
CS: Jackson 3991 (fr) 3/1954; **UN:** Simpson 7064 (fl)
6/1929.

Cyperus macrocarpus (Kunth) Boeckeler

F.P.S. 3: 352 (1956).
Shortly rhizomatous perennial herb. Forest clearings,
grassland, woodland, abandoned farmland.
Distr: BAG; Benin to Kenya, S to Angola & South Africa;
Madagascar.
BAG: Schweinfurth 1989 (fr) 6/1869.
Note: this species is treated as a synonym of *C.
cyperoides* in F.T.E.A. Cyper.: 225 (2010).

Cyperus cf. macrorrhizus Nees

F.Som. 4: 126 (1995).
Tussock-forming perennial herb. Coastal plains, on sand.
Distr: RS; Somalia; Egypt to India.
RS: Schweinfurth 645 (fl) 4/1868.
Note: the specimen cited was determined by H. Väre in
2003 at Kew; it requires confirmation. Boulos in F.Egypt
4: 395 (2005) includes *C. macrorrhizus* within the more
widespread *C. capitatus* Vand.

Cyperus maculatus Boeckeler subsp. maculatus

F.P.S. 3: 345 (1956); F.E.E. 6: 450 (1997); F.T.E.A. Cyper.:
230 (2010).
Stoloniferous perennial herb. Sand near pools & rivers.
Distr: CS, ES, UN, EQU; Senegal to Somalia, S to Namibia
& South Africa; Egypt, Madagascar, Mascarenes.
ES: Jackson 4135 (fr) 4/1961; **UN:** Andrews 645 (fl) 4/1939.
Note: closely allied to *C. longus*.

Cyperus mapanioides C.B.Clarke

F.T.E.A. Cyper.: 167 (2010).
Syn: *Cyperus mapanioides* C.B.Clarke var. *major*
(Boeckeler) Kük. – F.P.S. 3: 340 (1956).

Rhizomatous perennial herb. Forest & moist woodland, often along clearings or streams.
Distr: EQU; Sierra Leone to Kenya, S to Angola & Tanzania.
EQU: Schweinfurth 3886 (fr) 7/1870.

Cyperus meeboldii Kük.

F. Darfur Nord-Occ. & J. Gourgeil: 142 (1969); F.E.E. 6: 462 (1997); F.T.E.A. Cyper.: 176 (2010).
Perennial herb. Seasonally wet grassland & pools.
Distr: ?DAR; Senegal, Niger to Somalia, S to Tanzania; India.
Note: this record is from Quézel (1969) who recorded it from the Gertanga region of Darfur; we have not seen Quézel's specimen and the identification requires confirmation.

Cyperus michelianus (L.) Delile subsp. pygmaeus (Rottb.) Asch. & Graebn.

F.T.E.A. Cyper.: 156 (2010).
Syn: Cyperus pygmaeus Rottb. – F.P.S. 3: 343 (1956); F.E.E. 6: 462 (1997).
Tussocky annual herb. Pool margins, seasonally wet habitats.
Distr: NS, CS, ES; Ghana, Nigeria, Ethiopia & Somalia to Tanzania, Namibia; Mediterranean to Asia & Australia.
CS: Pettet 122 (fl) 9/1963.

Cyperus microbolbos C.B.Clarke

F.P.S. 3: 345 (1956); F.E.E. 6: 452 (1997).
Perennial herb with minute bulbs. Coastal sands, forming swards.
Distr: RS; Eritrea.
IUCN: RD
RS: Bent s.n. (fr) 1896.

Cyperus mollipes (C.B.Clarke) K.Schum.

F.P.S. 3: 353 (1956); F.E.E. 6: 468 (1997); F.T.E.A. Cyper.: 180 (2010).
Syn: Cyperus amomodorus K.Schum. – Biol. Skr. 51: 528 (2005).
Syn: Cyperus submacropus Kük. – F.P.S. 3: 353 (1956); F.E.E. 6: 466 (1997).
Syn: Mariscus globifer C.B.Clarke – Bot. Exp. Sud.: 46 (1970).
Syn: Mariscus macropus (Boeckeler) C.B.Clarke – Bot. Exp. Sud.: 46 (1970).
Syn: Mariscus mollipes C.B.Clarke – F. Darfur Nord-Occ. & J. Gourgeil: 142 (1969).
Tufted perennial herb. Grassland & open woodland, thin soils over rocks.
Distr: DAR, KOR, BAG, UN, EQU; Ethiopia & Somalia, S to Zambia & Malawi; India, SE Asia.
KOR: Wickens 77 (fl) 7/1962; **EQU:** Sillitoe 56 (fl) 1919.

Cyperus neoschimperi Kük.

F.T.E.A. Cyper.: 172 (2010).
Syn: Cyperus schimperi (Hochst. ex Steud.) K.Schum. – F.P.S. 3: 353 (1956).
Syn: Cyperus cruentus sensu Lye pro parte, non Rottb. – F.E.E. 6: 465 (1997).
Tufted perennial herb. Rocky hillslopes, rock crevices.
Distr: ?RS; Eritrea, Ethiopia, Somalia, Uganda, Kenya, Tanzania.

Note: no material has been seen for Sudan, but it is listed as occurring there both in F.P.S. and in F.T.E.A. The taxonomy is rather confused – in F.E.E. C. neoschimperi is treated as a synonym of C. cruentus from Egypt and Arabia.

Cyperus niveus Retz. var. leucocephalus (Kunth) Fosberg

F.E.E. 6: 464 (1997); Biol. Skr. 51: 536 (2005); F.T.E.A. Cyper.: 170 (2010).
Syn: Cyperus obtusiflorus Vahl – F.P.S. 3: 342 (1956).
Perennial herb with horizontal rhizome. Dry grassland & woodland, thin soils over rocks.
Distr: BAG, EQU; Benin to Somalia, S to Namibia & South Africa; Arabia.
EQU: Friis & Vollesen 1317 (fl) 3/1982.

Cyperus niveus Retz. var. tisserantii (Cherm.) Lye

F.E.E. 6: 464 (1997); Biol. Skr. 51: 536 (2005); F.T.E.A. Cyper.: 168 (2010).
Perennial herb with horizontal rhizome. Dry grassland & wooded grassland with regular burning.
Distr: EQU; Senegal to Ethiopia, S to Tanzania.
EQU: Myers 10944 (fl) 4/1939.

Cyperus nutans Vahl var. eleusinoides (Kunth) Haines

F.E.E. 6: 453 (1997); F.T.E.A. Cyper.: 243 (2010).
Syn: Cyperus eleusinoides Kunth – F.P.S. 3: 346 (1956).
Rhizomatous perennial herb. Wet grassland, streamsides.
Distr: ES; Nigeria, Eritrea to Mozambique; Asia, Australia.
ES: Schweinfurth 2005A (fr) 10/1865.

Cyperus papyrus L. subsp. papyrus

F.P.S. 3: 337 (1956); F.E.E. 6: 441 (1997); F.T.E.A. Cyper.: 209 (2010).
Robust rhizomatous perennial herb. Swamps, lake & river margins, often forming dense stands, sometimes as floating rafts.
Distr: NS, CS, BAG, UN, EQU; Senegal to Ethiopia, S to South Africa; Egypt.
CS: MacLeay 191 (fl fr) 12/1948; **UN:** J.M. Lock 83/6 (fr) 2/1983.

Cyperus pectinatus Vahl

F.E.E. 6: 439 (1997); F.T.E.A. Cyper.: 151 (2010).
Syn: Cyperus nudicaulis Poir. – F.P.S. 3: 338 (1956).
Shortly rhizomatous or stoloniferous perennial herb. Swamps, lake margins, sometimes floating.
Distr: BAG, UN, EQU; Senegal to Ethiopia, S to Angola & South Africa; Madagascar.
UN: J.M. Lock 81/28 (fl) 2/1981.

Cyperus plateilema (Steud.) Kük.

F.E.E. 6: 467 (1997); Biol. Skr. 51: 536 (2005); F.T.E.A. Cyper.: 182 (2010).
Tufted perennial herb. Montane grassland & bushland, streambanks, rocky outcrops.
Distr: UN, EQU; Ethiopia to Tanzania; Yemen.
EQU: Friis & Vollesen 1054 (fl) 2/1982.

Cyperus platycaulis Baker

F.T.E.A. Cyper.: 190 (2010).

Syn: *Cyperus denudatus* L. var. *lucenti-nigricans* (K.Schum.) Kük. – Biol. Skr. 51: 530 (2005).
Tufted perennial herb. Swamps, pool & lake margins.
Distr: BAG, EQU; Chad to Kenya, S to Angola & South Africa; Madagascar.
EQU: <u>Myers 11593</u> (fr) 7/1939.
Note: in F.T.E.A. Cyper. this species is listed as occurring only in East Africa and Madagascar; the omission of the remainder of its African distribution is presumably an error.

Cyperus pluribracteatus (Kük.) Govaerts
F.T.E.A. Cyper.: 223 (2010).
Syn: *Mariscus psilostachys* C.B.Clarke – F.J.M.: 164 (1976).
Tufted perennial herb. Grassland, rocky outcrops.
Distr: DAR; Uganda to Tanzania, Zimbabwe.
DAR: <u>Wickens 2248</u> (fl) 8/1964.

Cyperus podocarpus Boeckeler
F.P.S. 3: 348 (1956); F.W.T.A. 3(2): 289 (1972).
Slender annual herb. Marshes, pools on rock outcrops.
Distr: BAG; Senegal to C.A.R.
BAG: <u>Schweinfurth 2005</u> (fr) 7/1869.

Cyperus procerus Rottb.
F.T.E.A. Cyper.: 234 (2010).
Syn: *Cyperus procerus* Rottb. var. *stenanthus* Kük. – F.P.S. 3: 347 (1956).
Stoloniferous perennial herb. Wet grassland, swamps.
Distr: BAG; palaeotropical.
BAG: <u>Schweinfurth 2017</u> (fr) 7/1869.

Cyperus pulchellus R.Br.
F.P.S. 3: 342 (1956); F.E.E. 6: 440 (1997); F.T.E.A. Cyper.: 163 (2010).
Slender tufted perennial herb. Seasonally wet grassland & bushland, pool margins.
Distr: DAR, UN; palaeotropical.
DAR: <u>Haekstra 171</u> (fl) 10/1972; **UN:** <u>J.M. Lock 81/211</u> (fl) 8/1981.

Cyperus pustulatus Vahl
F.P.S. 3: 347 (1956); F.E.E. 6: 460 (1997); Biol. Skr. 51: 536 (2005); F.T.E.A. Cyper.: 157 (2010).
Slender to robust annual herb. Wet grassland, pool margins, often on thin soils over rock.
Distr: NS, BAG, UN, EQU; Senegal to Ethiopia, S to Angola & Zambia.
EQU: <u>Myers 7764</u> (fr) 9/1937.
Note: the inclusion of this species for Northern Sudan is based on a record by Malterer (2013) in his report on the survey for the Merowe Dam Project.

Cyperus reduncus Hochst. ex Boeckeler
F.P.S. 3: 341 (1956); F.J.M.: 162 (1976); F.E.E. 6: 437 (1997); Biol. Skr. 51: 537 (2005); F.T.E.A. Cyper.: 217 (2010).
Tufted annual herb. Swamps, pools, wet grassland.
Distr: DAR, BAG, EQU; Senegal to Ethiopia, S to Tanzania.
DAR: <u>Wickens 2736</u> (fl) 9/1964; **EQU:** <u>Sillitoe 68</u> (fl) 1919.

Cyperus remotus (C.B.Clarke) Kük.
Biol. Skr. 51: 537 (2005); F.T.E.A. Cyper.: 150 (2010).
Slender perennial herb. Dry grassland, bushland & woodland.
Distr: EQU; D.R. Congo to Zambia.
EQU: <u>Jackson 1360</u> (fl) 3/1950.
Note: *Cyperus boreochrysocephalus* Lye (see F.T.E.A. Cyper.: 149 (2010)) is possibly conspecific with *C. remotus*.

Cyperus renschii Boeckeler
F.P.S. 3: 340 (1956); Biol. Skr. 51: 537 (2005); F.T.E.A. Cyper.: 247 (2010).
Robust rhizomatous perennial herb. Forest, often along streams or in swamp forest.
Distr: EQU; Senegal to Uganda, S to Angola & Tanzania; Comoros.
EQU: <u>Myers 9768</u> (fl fr) 10/1938.

Cyperus rigidifolius Steud.
F.J.M.: 162 (1976); F.E.E. 6: 447 (1997); F.T.E.A. Cyper.: 165 (2010).
Stoloniferous perennial herb. Wet grassland, swamps, bushland.
Distr: DAR; Ethiopia to South Africa; Arabia.
DAR: <u>Lynes 9</u> (fr) 2/1921.

Cyperus rotundus L.
F.P.S. 3: 344 (1956); F.J.M.: 162 (1976); F.E.E. 6: 449 (1997); Biol. Skr. 51: 537 (2005); F.T.E.A. Cyper.: 211 (2010).
Syn: *Cyperus rotundus* L. var. *nubicus* (C.B.Clarke) Kük. – F.P.S. 3: 344 (1956).
Syn: *Cyperus rotundus* L. var. *spadinea* Boeckeler – Biol. Skr. 51: 538 (2005).
Syn: *Cyperus rotundus* L. subsp. *tuberosus* (Rottb.) Kük. – Bot. Exp. Sud.: 45 (1970).
Stoloniferous perennial or ?annual herb. Swamps, ditches, riverbanks, weed of cultivation.
Distr: NS, RS, DAR, KOR, CS, ES, BAG, EQU; widespread in the tropics.
NS: <u>Wickens 1159</u> (fr) 2/1964; **EQU:** <u>Wood 33</u> (fl) 3/1919.
Note: three subspecies have previously been recognised in Sudan: the widespread subsp. *rotundus*, subsp. *retzii* Kük. restricted to the Red Sea region and subsp. *tuberosus* restricted to Equatoria. However, in F.T.E.A. no infraspecific taxa are recognised.

Cyperus rubicundus Vahl
F.E.E. 6: 464 (1997); Biol. Skr. 51: 538 (2005); F.T.E.A. Cyper.: 160 (2010).
Tufted annual herb. Wet grassland & wooded grassland, pools, swamps, wet rocks.
Distr: EQU; Eritrea & Somalia to South Africa; Madagascar, Macaronesia, Arabia, India, N Australia.
EQU: <u>Kielland-Lund 51</u> 11/1983 (n.v.).

Cyperus schimperianus Steud.
F.P.S. 3: 344 (1956); F.J.M.: 162 (1976); F.E.E. 6: 447 (1997); F.T.E.A. Cyper.: 241 (2010).
Robust rhizomatous perennial herb. Riverbanks, swamps.
Distr: NS, DAR, CS, UN; Cameroon to Ethiopia, S to Tanzania; Egypt, Saudi Arabia.
CS: <u>de Wilde et al. 5722</u> (fr) 2/1965; **UN:** <u>Harrison 1009</u> (fl) 10/1950.

Cyperus spiralis Larridon

Pl. Ecol. Evol. 144: 352 (2011).
Syn: *Kyllingiella polyphylla* (A.Rich.) Lye – F.E.E. 6: 427 (1997); F.T.E.A. Cyper.: 128 (2010).
Syn: *Scirpus steudneri* Boeckeler – F. Darfur Nord-Occ. & J. Gourgeil: 143 (1969).
Rhizomatous perennial herb. Seasonally wet areas in grassland & bushland.
Distr: ?DAR; Chad to Eritrea, S to Zimbabwe.
Note: the record of this species is from Quézel (1969) who recorded it from Kutum, Gertanga and Musbat in Darfur. We have not seen Quézel's material and it is just possible that this was a misidentification of *C. kyllingiella*.

Cyperus squarrosus L.

F.P.S. 3: 352 (1956); F.E.E. 6: 461 (1997); Biol. Skr. 51: 538 (2005); F.T.E.A. Cyper.: 215 (2010).
Syn: *Mariscus squarrosus* (L.) C.B.Clarke – F.J.M.: 164 (1976).
Annual herb. Ditches, roadsides, margins of pools.
Distr: DAR, CS, BAG, UN, EQU; pantropical.
DAR: Wickens 2264 (fl) 8/1964; **UN:** J.M. Lock 81/155 (fl) 8/1981.

Cyperus submicrolepis Kük.

F.P.S. 3: 341 (1956); F.T.E.A. Cyper.: 192 (2010).
Slender annual herb. Seasonally wet areas, shallow pools, depressions on rock outcrops.
Distr: BAG, EQU; Senegal to Uganda, S to Angola & Zambia.
EQU: Andrews 1605 (fr) 6/1939.

Cyperus subumbellatus Kük.

F.E.E. 6: 457 (1997).
Syn: *Mariscus alternifolius* sensu Wickens, non Vahl – F.J.M.: 164 (1976).
Shortly rhizomatous perennial herb. Weed of cultivation & waste ground, open disturbed forest.
Distr: DAR, BAG, EQU; Senegal to Eritrea, S to Namibia & South Africa; Madagascar, Comoros, Caribbean & Surinam.
DAR: Wickens 1917 (fl) 7/1964; **BAG:** Jackson 3060 (fl) 7/1953.
Note: F.T.E.A. lists this taxon as a synonym of *C. cyperoides* (L.) Kuntze but we maintain these two taxa as distinct here.

Cyperus tenuiculmis Boeckeler var. *schweinfurthianus* (Boeckeler) S.S.Hooper

F.T.E.A. Cyper.: 242 (2010).
Syn: *Cyperus schweinfurthianus* Boeckeler – F.P.S. 3: 346 (1956).
Rhizomatous perennial herb. Wet grassland, ditches, swamps.
Distr: BAG, EQU; Senegal to Kenya, S to Angola & Tanzania.
EQU: Sillitoe 61 (fl fr) 1919.

Cyperus tenuispica Steud.

F.P.S. 3: 341 (1956); F.E.E. 6: 437 (1997); F.T.E.A. Cyper.: 193 (2010).
Slender annual herb. Swamps, seasonally wet areas.
Distr: UN, EQU; palaeotropical.
EQU: Dandy 638 (fl) 3/1934.

[*Cyperus* sp. ?nov.]

Note: Wickens lists *Cyperus* sp. nov. from Jebel Marra based on *Wickens* 2679 & 2986; we have not located these specimens.

Diplacrum africanum (Benth.) C.B.Clarke

F.P.S. 3: 358 (1956); Biol. Skr. 51: 539 (2005); F.T.E.A. Cyper.: 415 (2010).
Dwarf annual herb. Bare sand & mud in marshes.
Distr: EQU; Senegal to Kenya, S to Zambia.
EQU: Friis & Vollesen 631 (fr) 12/1980.

Eleocharis atropurpurea (Retz.) J.Presl & C.Presl

F.P.S. 3: 358 (1956); F.E.E. 6: 405 (1997); F.T.E.A. Cyper.: 44 (2010).
Dwarf annual herb. Margins of pools, seasonally wet grassland.
Distr: KOR, CS; widespread in the tropics.
CS: Lea 223 (fl) 10/1954.

Eleocharis brainii Svenson

F.P.S. 3: 358 (1956); F.T.E.A. Cyper.: 44 (2010).
Dwarf annual herb. Swampy grassland, lake margins, often partially submerged.
Distr: BAG; Senegal to Uganda, S to Botswana & Mozambique.
BAG: Schweinfurth 2583 (fl) 10/1869.

Eleocharis fistulosa (Poir.) Schult.

F.T.E.A. Cyper.: 39 (2010).
Syn: *Eleocharis acutangula* (Roxb.) Schult. – F.P.S. 3: 359 (1956); F.E.E. 6: 404 (1997).
Stoloniferous perennial herb. Swamps, seasonally flooded grassland, pool & lake margins, ditches.
Distr: BAG; pantropical.
BAG: Simpson 7650 (fl) 9/1930.
Note: our material has been named *E. fistulosa* var. *robusta* Boeckeler, which is sometimes treated as a separate species, but it seems reasonable to treat the Sudan material under *E. fistulosa* s.l.

Eleocharis nigrescens (Nees) Steud.

F.P.S. 3: 358 (1956); F.T.E.A. Cyper.: 46 (2010).
Tufted annual herb. Margins of pools, damp depressions.
Distr: BAG; Senegal to Nigeria, Uganda to Zimbabwe; Madagascar, tropical America.
BAG: Schweinfurth 2576 (fl) 10/1869.

Eleocharis palustris (L.) Roem. & Schult.

Syn: *Eleocharis* sp. aff. *tibestica* sensu Wickens – F.J.M.: 163 (1976).
Tufted perennial herb. Stream margins, wetlands.
Distr: DAR; temperate & subtropical northern hemisphere.
DAR: Wickens 2676 (fl) 9/1964.
Note: the identity of this specimen requires confirmation; if correct, it represents the southernmost limit of *E. palustris* in Africa.

Eleocharis setifolia (A.Rich.) J.Raynal subsp. *setifolia*

F.E.E. 6: 405 (1997); F.T.E.A. Cyper.: 42 (2010).
Tufted annual herb. Pools & seasonal swamps.

Distr: UN; widespread in the tropics.
UN: J.M. Lock 81/171 (fl) 8/1981.

Eleocharis setifolia (A.Rich.) J.Raynal subsp.
schweinfurthiana (Boeckeler) D.A.Simpson
Kew Bull. 43: 428 (1998); F.T.E.A. Cyper.: 42 (2010).
Syn: *Eleocharis schweinfurthiana* Boeckeler – F.P.S. 3:
359 (1956).
Tufted annual herb. Pools & seasonal swamps.
Distr: BAG; Senegal to Nigeria, D.R. Congo.
BAG: Schweinfurth 1949 (fl) 6/1869.

Fimbristylis bisumbellata (Forssk.) Bubani
F.P.S. 3: 360 (1956); F.J.M.: 163 (1976); F.E.E. 6: 412
(1997); F.T.E.A. Cyper.: 58 (2010).
Tufted annual herb. Seasonal pools in woodland, sandy
riverbanks.
Distr: NS, DAR, CS, ES; palaeotropical & subtropical.
CS: de Wilde et al. 5846 (fl) 3/1965.

Fimbristylis complanata (Retz.) Link var.
complanata
F.P.S. 3: 362 (1956); F.J.M.: 163 (1976); F.E.E. 6: 408
(1997); F.T.E.A. Cyper.: 50 (2010).
Syn: *Fimbristylis subaphylla* Boeckeler – F.P.S. 3: 362
(1956).
Tufted perennial herb with short rhizome. Swamps,
marshy grassland, ditches.
Distr: DAR, UN; pantropical.
DAR: Wickens 1423 (fl) 4/1964; **UN:** J.M. Lock 82/46 (fl)
8/1982.

Fimbristylis dichotoma (L.) Vahl subsp.
dichotoma
F.P.S. 3: 360 (1956); F.J.M.: 163 (1976); F.E.E. 6: 412 (1997);
Biol. Skr. 51: 539 (2005); F.T.E.A. Cyper.: 57 (2010).
Tufted annual or shortly rhizomatous perennial herb. Wet
grassland, swamps, riverbanks.
Distr: DAR, CS, BAG, UN, EQU; widespread in the tropics
& warm temperate regions.
DAR: Wickens 2597 (fl) 9/1964; **EQU:** Friis &
Vollesen 1240 (fr) 3/1982.

Fimbristylis falcifolia Boeckeler
Linnaea 37: 25 (1871).
Herb. Habitat not known.
Distr: ES; ?Sudan endemic.
IUCN: DD
ES: Schweinfurth s.n. (n.v.).
Note: known only from the type specimen which we
have not located. The status of this taxon is uncertain.

Fimbristylis ferruginea (L.) Vahl
F.P.S. 3: 360 (1956); F.E.E. 6: 409 (1997); F.T.E.A. Cyper.:
55 (2010).
Tufted perennial herb with a short rhizome. Damp
ground & muddy hollows including saline areas.
Distr: ES; pantropical.
ES: Schweinfurth 1864 (fr) 9/1865.
Note: from the locality, it seems likely that this is subsp.
sieberiana (Kunth) Lye, as subsp. *ferruginea* is restricted
to coastal areas. Andrews lists this species from "central
and southern Sudan"; it is likely that this is due to
misidentification.

Fimbristylis pilosa Vahl
F.E.E. 6: 410 (1997); F.T.E.A. Cyper.: 60 (2010).
Annual or more commonly perennial herb. Wet grassland
& bushland, swamp margins.
Distr: BAG, UN; Senegal to Ethiopia, S to Angola &
Tanzania.
UN: J.M. Lock 81/256 (fr) 9/1981.

Fimbristylis quinquangularis (Vahl) Kunth
subsp. *quinquangularis*
F.T.E.A. Cyper.: 53 (2010).
Syn: *Fimbristylis miliacea* (L.) Vahl – F.E.E. 6: 410 (1997).
Tufted annual herb. Swampy grassland, pool margins.
Distr: KOR, UN; pantropical.
KOR: Niamir 55 (fr) 10/1980; **UN:** Sherif A.4024 (fl fr)
9/1951.

Fimbristylis schweinfurthiana Boeckeler
F.P.S. 3: 362 (1956); F.W.T.A. 3(2): 323 (1972).
Slender perennial or annual herb. Seasonal swamps,
depressions in rock outcrops.
Distr: BAG; Mali to C.A.R.
BAG: Schweinfurth 1824 (fr) 7/1869.

Fimbristylis squarrosa Vahl
F.P.S. 3: 360 (1956); F.T.E.A. Cyper.: 63 (2010).
Small annual herb. Seasonal wetlands, sometimes on
bare mud or sand.
Distr: BAG; widespread in the tropics & subtropics.
BAG: Schweinfurth 1638 (fl) 4/1869.

Fuirena ciliaris (L.) Roxb.
F.P.S. 3: 363 (1956); F.E.E. 6: 395 (1997); F.T.E.A. Cyper.:
17 (2010).
Robust annual herb. Seasonal swamp margins, wet
grassland.
Distr: UN; palaeotropical.
UN: J.M. Lock 81/8 (fl) 2/1981.

Fuirena leptostachya Oliv.
F.P.S. 3: 363 (1956); F.E.E. 6: 395 (1997); Biol. Skr. 51:
540 (2005); F.T.E.A. Cyper.: 14 (2010).
Slender annual herb. Seasonally wet grassland & wooded
grassland, swamp margins.
Distr: BAG, EQU; Mali to Ethiopia, S to Namibia & South
Africa.
BAG: Schweinfurth 2504 (fl) 10/1869.

Fuirena pubescens (Poir.) Kunth
F.P.S. 3: 362 (1956); F.J.M.: 163 (1976); F.E.E. 6: 395
(1997); F.T.E.A. Cyper.: 11 (2010).
Rhizomatous perennial herb. Wet grassland, swamp &
stream margins.
Distr: DAR, UN; Nigeria to Ethiopia, S to South Africa;
Madagascar, Mediterranean to India.
DAR: Jackson 3353 (fl) 12/1954; **UN:** J.M. Lock 82/47
(fl) 8/1982.
Note: the varieties of this species are difficult to
distinguish (see note in F.T.E.A.) and are not upheld here.

Fuirena stricta Steud. subsp. *stricta*
F.P.S. 3: 362 (1956); F.E.E. 6: 394 (1997); Biol. Skr. 51:
540 (2005); F.T.E.A. Cyper.: 9 (2010).
Tufted perennial herb with short rhizome. Wet grassland,

swamp margins, often in shallow water.
Distr: EQU; Senegal to Ethiopia, S to South Africa;
Madagascar, Comoros.
EQU: Friis & Vollesen 1313 (fl) 3/1982.

Fuirena umbellata Rottb.

F.P.S. 3: 363 (1956); F.E.E. 6: 395 (1997); Biol. Skr. 51:
540 (2005); F.T.E.A. Cyper.: 20 (2010).
Rhizomatous perennial herb. Wet grassland, swamps,
river margins.
Distr: BAG, UN, EQU; widespread in the tropics.
EQU: Friis & Vollesen 1243 (fl) 3/1982.

Isolepis costata Hochst. ex A.Rich.

F.E.E. 6: 423 (1997); Biol. Skr. 51: 540 (2005); F.T.E.A.
Cyper.: 123 (2010).
Tufted annual or short-lived perennial herb. Swamps in
montane forest, streambanks.
Distr: EQU; Ethiopia to Namibia & South Africa;
Madagascar.
EQU: Friis & Vollesen 1012 (fr) 2/1982.

Isolepis fluitans (L.) R.Br.

F.E.E. 6: 424 (1997); Biol. Skr. 51: 541 (2005); F.T.E.A.
Cyper.: 121 (2010).
Mat-forming perennial herb. Floating in shallow water or
terrestrial in boggy areas.
Distr: EQU; palaeotropical & subtropical.
EQU: Friis & Vollesen 1016 (fr) 2/1982.

Kyllinga alba Nees subsp. *alba*

F.J.M.: 163 (1976); F.T.E.A. Cyper.: 331 (2010).
Syn: *Cyperus alatus* (Nees) F.Muell. subsp. *albus* (Nees)
Lye – F.E.E. 6: 477 (1997).
Syn: *Cyperus cristatus* (Kunth) Mattf. & Kük. var.
nigritanus (C.B.Clarke) Kük. – F.P.S. 3: 357 (1956); Bot.
Exp. Sud.: 44 (1970).
Tussocky perennial herb with short rhizome. Grassland,
woodland & bushland on sandy soils.
Distr: RS, DAR; Ivory Coast to Ethiopia, S to Namibia &
South Africa; Seychelles.
RS: Carter 1833 (fl) 11/1987.

Kyllinga brevifolia Rottb. var. *brevifolia*

F.T.E.A. Cyper.: 313 (2010).
Syn: *Cyperus erectus* sensu Andrews, non (Schumach.)
Mattf. & Kük. – F.P.S. 3: 357 (1956).
Rhizomatous perennial herb. Seasonally wet grassland,
swamp margins, forest margins.
Distr: BAG, EQU; pantropical.
EQU: Sillitoe 74 (fl) 1919.

Kyllinga bulbosa P.Beauv.

F.T.E.A. Cyper.: 336 (2010).
Syn: *Cyperus purpureoglandulosus* Mattf. & Kük. –
F.P.S. 3: 354 (1956).
Syn: *Cyperus richardii* Steud. – F.E.E. 6: 470 (1997).
Syn: *Cyperus triceps* (Rottb.) Endl. var. *obtusiflorus*
(Boeckeler) Kük. – F.P.S. 3: 355 (1956).
Stoloniferous or rhizomatous perennial herb. Damp
grassland, lawns, roadsides.
Distr: ?ES, EQU; Senegal to Ethiopia, S to Mozambique;
India to China.

EQU: Simpson 7585 (fl) 11/1930.
Note: Andrews records *Cyperus triceps* var. *obtusiflorus*
from both Equatoria and Kassala: Gallabat; we have
not seen material from the latter locality, which was
presumably based on a Schweinfurth collection.

Kyllinga chlorotropis Steud.

F.J.M.: 163 (1976); F.T.E.A. Cyper.: 345 (2010).
Syn: *Cyperus chlorotropis* (Steud.) Mattf. & Kük. –
F.E.E. 6: 472 (1997).
Tufted perennial herb. Grassland, shallow soils over
rocks.
Distr: DAR; Ethiopia, ?E Africa; Arabia.
DAR: Wickens 2402 (fl) 9/1964.
Note: there is some uncertainty over the status of this
species in East Africa (see F.T.E.A.).

Kyllinga crassipes Boeckeler

F.T.E.A. Cyper.: 342 (2010).
Syn: *Cyperus bulbipes* Mattf. & Kük. – F.P.S. 3: 355
(1956).
Shortly rhizomatous perennial herb. Wet grassland,
woodland clearings, lawns.
Distr: ?EQU; Uganda to Zimbabwe & Mozambique.
Note: Andrews records this species from Equatoria but
we have not seen the material on which this was based
and it may be a misidentification.

Kyllinga erecta Schumach. var. *africana* (Kük.) S.S.Hooper

Kew Bull. 26: 580 (1972).
Syn: *Cyperus obtusatus* Steud. var. *africanus* Kük. –
F.P.S. 3: 355 (1956).
Rhizomatous perennial herb. Swamps, lake margins, wet
grassland.
Distr: DAR, ?EQU; Mali to D.R. Congo.
DAR: Wickens 1878 (fl) 7/1964.
Note: Andrews recorded this taxon from Equatoria but
we have seen no material to verify this. *Myers* 6940 from
Yambio is determined as sp. nr. *erecta*.

Kyllinga melanosperma Nees

F.T.E.A. Cyper.: 323 (2010).
Syn: *Cyperus melanospermus* (Nees) Suringar – F.P.S.
3: 355 (1956).
Rhizomatous perennial herb. Wet grassland, roadside
ditches.
Distr: ?EQU; Nigeria to D.R. Congo, S to South Africa;
Madagascar, India to Malesia.
Note: Andrews recorded this species from Equatoria
but we have seen no material with which to verify this
record.

Kyllinga odorata Vahl var. *cylindrica* (Nees) Kük.

F.J.M.: 164 (1964); F.T.E.A. Cyper.: 341 (2010).
Syn: *Cyperus sesquiflorus* (Torr.) Mattf. & Kük. subsp.
cylindricus (Nees) Koyama – F.E.E. 6: 476 (1997).
Tufted perennial herb. Grassland including seasonally wet
or disturbed areas.
Distr: DAR, EQU; palaeotropical.
DAR: Wickens 2129 (fl) 8/1964; **EQU:** Andrews 1999
(fl) 6/1939.

Kyllinga odorata Vahl var. **major** (C.B.Clarke) Chiov.

F.T.E.A. Cyper.: 340 (2010).
Syn: Cyperus sesquiflorus (Torr.) Mattf. & Kük. – F.P.S. 3: 357 (1956).
Syn: Cyperus sesquiflorus (Torr.) Mattf. & Kük. subsp. **appendiculatus** (K.Schum.) Lye – F.E.E. 6: 475 (1997); Biol. Skr. 51: 538 (2005).
Tufted perennial herb. Forest margins, woodland & wooded grassland.
Distr: EQU; Togo, Cameroon to Somalia, S to Tanzania.
EQU: Friis & Vollesen 386 (fl) 11/1980.

Kyllinga polyphylla Willd. ex Kunth var. **polyphylla**

F.T.E.A. Cyper.: 318 (2010).
Syn: Cyperus aromaticus (Ridl.) Mattf. & Kük. – F.P.S. 3: 355 (1956).
Syn: Kyllinga erecta Schumach. var. **polyphylla** (Willd. ex Kunth) S.S.Hooper – F.J.M.: 163 (1976).
Rhizomatous perennial herb. Moist grassland, swamp & river margins.
Distr: DAR, EQU; Nigeria to Somalia, S to South Africa; Mauritius.
DAR: Wickens 2223 (fl) 3/1964; **EQU:** Sillitoe 85 (fl) 1919.

Kyllinga polyphylla Willd. ex Kunth var. **elatior** (Kunth) Kük.

F.T.E.A. Cyper.: 319 (2010).
Syn: Cyperus pinguis (C.B.Clarke) Mattf. & Kük. – F.E.E. 6: 475 (1997); Biol. Skr. 51: 536 (2005).
Perennial herb. Damp grassland, forest margins, disturbed damp ground.
Distr: EQU; Cameroon to Ethiopia, S to South Africa; Madagascar.
EQU: Friis & Vollesen 385 (fl) 11/1980.

Kyllinga pulchella Kunth

F.T.E.A. Cyper.: 316 (2010).
Syn: Cyperus bracheilema (Steud.) Mattf. & Kük. – Biol. Skr. 51: 528 (2005).
Rhizomatous or stoloniferous perennial herb. Wet grassland, thin soils over rock.
Distr: EQU; Eritrea, Ethiopia, Kenya, Tanzania, South Africa, Lesotho.
EQU: Kielland-Lund 918 6/1984 (n.v.).

Kyllinga pumila Michx.

F.J.M.: 164 (1976); F.T.E.A. Cyper.: 325 (2010).
Syn: Cyperus densicaespitosus Michx. – F.E.E. 6: 476 (1997).
Syn: Cyperus tenuifolius sensu Andrews pro parte, non (Steud.) Dandy s.s. – F.P.S. 3: 357 (1956).
Annual herb. Streamsides, ditches, sandy riverbeds & hollows.
Distr: DAR, UN; pantropical.
DAR: Wickens 1521 (fl) 5/1964; **UN:** J.M. Lock 81/238 (fl) 8/1981.

Kyllinga squamulata Vahl

F.T.E.A. Cyper.: 326 (2010).
Syn: Cyperus metzii (Hochst. ex Steud.) Mattf. & Kük. – F.P.S. 3: 358 (1956); F.E.E. 6: 478 (1997).

Small tussocky annual herb. Streambanks, lake shores, disturbed sandy ground.
Distr: KOR, BAG, UN; palaeotropical.
KOR: Wickens 737 (fl) 10/1961; **UN:** J.M. Lock 81/164 (fl) 8/1981.

Kyllinga tenuifolia Steud. var. **tenuifolia**

F.T.E.A. Cyper.: 329 (2010).
Syn: Cyperus triceps (Rottb.) Endl. – F.P.S. 3: 354 (1956); F.E.E. 6: 471 (1997); Biol. Skr. 51: 538 (2005).
Tufted perennial herb. Seasonal swamps, streamsides.
Distr: BAG, UN; palaeotropical.
UN: J.M. Lock 81/270 (fl) 9/1981.

Kyllinga tenuifolia Steud. var. **ciliata** (Boeck.) Beentje

F.T.E.A. Cyper.: 330 (2010).
Syn: Cyperus controversus (Steud.) Mattf. & Kük. var. **subexalatus** (C.B.Clarke) Kük. – F.P.S. 3: 357 (1956).
Syn: Cyperus welwitschii (Ridl.) Lye – F.E.E. 6: 471 (1997).
Syn: Kyllinga welwitschii Ridl. – F. Darfur Nord-Occ. & J. Gourgeil: 142 (1969).
Tufted perennial herb. Seasonally wet grassland, drainage lines.
Distr: DAR, KOR, CS; Mauritania to Somalia, S to Namibia & Zimbabwe.
CS: Andrews 201 (fl) 8/1931.

Lipocarpha albiceps Ridl.

F.T.E.A. Cyper.: 354 (2010).
Shortly rhizomatous perennial herb. Wet grassland.
Distr: EQU; Senegal to Uganda, S to Angola & Zimbabwe.
EQU: Myers 7547 (fl) 8/1937.

Lipocarpha chinensis (Osbeck) J.Kern

F.E.E. 6: 489 (1997); F.T.E.A. Cyper.: 354 (2010).
Syn: Lipocarpha senegalensis (Lam.) T.Durand & H.Durand – F.P.S. 3: 363 (1956).
Tufted perennial herb. Swamps, lake & stream margins, ditches, wet depressions.
Distr: DAR, BAG, EQU; palaeotropical.
DAR: Wickens 1897 (fl) 7/1965; **EQU:** Andrews 1078 (fl) 3/1937.

Lipocarpha hemisphaerica (Roth) Goetgh.

F.E.E. 6: 490 (1997); Biol. Skr. 51: 541 (2005); F.T.E.A. Cyper.: 348 (2010).
Syn: Scirpus isolepis (Nees) Boeckeler – F.P.S. 3: 366 (1956).
Tufted annual herb. Wet grassland, shallow soil over rock, stream & swamp margins.
Distr: ES; Senegal to Ethiopia, S to South Africa; India, Thailand.
ES: Schweinfurth 2044 (fl) 9/1865.

Lipocarpha kernii (Raymond) Goetgh.

F.T.E.A. Cyper.: 352 (2010).
Syn: Scirpus squarrosus sensu Andrews, non L. – F.P.S. 3: 366 (1956).
Tufted annual herb. Swamps, woodland.
Distr: ES, BAG; Senegal to Ethiopia, S to South Africa; India.
ES: Schweinfurth 3003 (fl) 10/1865; **BAG:** Schweinfurth 2572 (fl) 10/1869.

Lipocarpha nana (A.Rich.) Cherm.

F.E.E. 6: 489 (1997); Biol. Skr. 51: 541 (2005); F.T.E.A.
Cyper.: 351 (2010).
Tufted annual herb. Wet grassland, shallow soils over
rock, moist depressions.
Distr: EQU; Guinea to Ethiopia, S to South Africa;
Madagascar.
EQU: Friis & Vollesen 416 (fr) 11/1980.

Lipocarpha prieuriana Steud.

F.P.S. 3: 364 (1956); F.T.E.A. Cyper.: 353 (2010).
Tufted annual herb. Wet sand.
Distr: ?EQU; Senegal to Ethiopia, S to Zimbabwe.
Note: the Sudan folder at Kew for this species is empty
and it is possible that Andrews' record from Equatoria is
incorrect, although this species is likely to occur in South
Sudan.

Pycreus acuticarinatus (Kük.) Cherm.

F.W.T.A. 3(2): 302 (1972).
Tufted perennial herb. Wet grassland.
Distr: EQU; Senegal to Cameroon.
EQU: Sillitoe 76 (fr) 1919.

Pycreus aethiops (Welw. ex Ridl.) C.B.Clarke

F.T.E.A. Cyper.: 301 (2010).
Syn: *Cyperus aethiops* Welw. ex Ridl. – F.P.S. 3: 350
(1956); F.E.E. 6: 482 (1997); Biol. Skr. 51: 527 (2005).
Tufted perennial herb. Swampy grassland.
Distr: BAG, EQU; Ivory Coast to Ethiopia, S to Botswana
& South Africa.
BAG: Schweinfurth 1462 (fr) 4/1869.

Pycreus capillifolius (A.Rich.) C.B.Clarke

F.J.M.: 164 (1976); F.T.E.A. Cyper.: 284 (2010).
Syn: *Cyperus capillifolius* A.Rich. – F.P.S. 3: 349 (1956);
F.E.E. 6: 482 (1997); Biol. Skr. 51: 529 (2005).
Tufted annual herb. Wet grassland.
Distr: BAG, EQU; Senegal to Ethiopia, S to Angola &
Zambia; Madagascar, Brazil.
EQU: Friis & Vollesen 569 (fr) 11/1980.

Pycreus diloloensis Kük. ex Cherm.

Rev. Zool. Bot. Africaines 24: 296 (1934).
Syn: *Cyperus fibrillosus* sensu Friis & Vollesen, non
Kük. – Biol. Skr. 51: 533 (2005).
Perennial herb. Burnt *Loudetia* grassland.
Distr: EQU; Cameroon to D.R. Congo, S to Malawi.
EQU: Friis & Vollesen 1204 (fr) 3/1982.
Note: the determination of the two South Sudan
specimens requires confirmation.

Pycreus elegantulus (Steud.) C.B.Clarke

F.J.M.: 164 (1976); F.T.E.A. Cyper.: 303 (2010).
Syn: *Cyperus elegantulus* Steud. – F.P.S. 3: 350 (1956);
F.E.E. 6: 480 (1997); Biol. Skr. 51: 533 (2005).
Tufted perennial herb, often with slender stolons.
Swamps, wet grassland, river margins, forest margins.
Distr: DAR, CS, EQU; Nigeria to Eritrea, S to South
Africa; Arabia, tropical America.
DAR: Wickens 2594 (fl) 9/1964; **EQU:** Friis &
Vollesen 1032 (fr) 2/1982.

Pycreus flavescens (L.) P.Beauv. ex Rchb. subsp. *flavescens*

F.J.M.: 164 (1976); F.T.E.A. Cyper.: 292 (2010).
Syn: *Cyperus flavescens* L. subsp. *flavescens* – F.P.S.
3: 351 (1956); F.E.E. 6: 487 (1997); Biol. Skr. 51: 534
(2005).
Tufted annual herb. Swamps, seepage over rocks,
riversides, lake margins.
Distr: DAR, EQU; pantropical, also in Europe.
DAR: Wickens 1896 (fr) 7/1964; **EQU:** Sillitoe 62 (fl)
1919.

Pycreus lanceolatus (Poir.) C.B.Clarke

F.J.M.: 164 (1976); F.T.E.A. Cyper.: 291 (2010).
Syn: *Cyperus lanceolatus* Poir. – F.P.S. 3: 348 (1956);
F.E.E. 6: 482 (1997); Biol. Skr. 51: 535 (2005).
Tufted perennial herb. Wet grassland, swamps, river
margins.
Distr: DAR; tropical Africa; Madagascar, tropical America.
DAR: Wickens 2174 (fr) 8/1964.

Pycreus macrostachyos (Lam.) J.Raynal

F.J.M.: 165 (1976); F.T.E.A. Cyper.: 288 (2010).
Syn: *Cyperus macrostachyos* Lam. subsp. *tremulus*
(Poir.) Lye – Biol. Skr. 51: 535 (2005).
Syn: *Cyperus tremulus* Poir. – F.P.S. 3: 350 (1956).
Syn: *Pycreus macrostachyos* (Lam.) J.Raynal var. *tenuis*
(Boeckeler) Wickens – F.J.M.: 165 (1976).
Robust annual herb. Seasonal lakes & ponds, river margins.
Distr: DAR, KOR, BAG, EQU; tropical Africa; Madagascar,
tropical America.
DAR: Wickens 2076 (fr) 8/1964; **EQU:** Myers 8599 (fl)
2/1938.

Pycreus mundtii Nees var. *mundtii*

F.J.M.: 165 (1956); F.T.E.A. Cyper.: 284 (2010).
Syn: *Cyperus mundtii* (Nees) Kunth – Bot. Exp. Sud.: 45
(1970); F.E.E. 6: 479 (1997).
Stoloniferous perennial herb. Swamps, lake margins, wet
grassland, forest margins, sometimes floating.
Distr: DAR, BAG, UN; Senegal to Ethiopia, S to South
Africa; Madagascar, Spain, Egypt, India.
DAR: Jackson 3296 (fl) 12/1954; **UN:** J.M. Lock 82/15
(fr) 2/1982.

Pycreus nigricans (Steud.) C.B.Clarke

F.T.E.A. Cyper.: 305 (2010).
Syn: *Cyperus nigricans* Steud. – F.E.E. 6: 481 (1997);
Biol. Skr. 51: 535 (2005).
Robust tussocky perennial herb. Swamps and bogs.
Distr: EQU; Ethiopia to Malawi; Madagascar.
EQU: Jackson 1527 (fr) 6/1950.

Pycreus nitidus (Lam.) J.Raynal

F.T.E.A. Cyper.: 307 (2010).
Syn: *Cyperus lanceus* Thunb. – F.P.S. 3: 351 (1956).
Syn: *Cyperus nitidus* Lam. – F.E.E. 6: 481 (1997).
Robust stoloniferous perennial herb. Swamps, wet
grassland, lake margins.
Distr: BAG, EQU; Senegal to Ethiopia, S to South Africa;
Madagascar, India.
BAG: Simpson 7661 (fr) 3/1930.

Pycreus nuerensis (Boeckeler) S.S.Hooper

F.T.E.A. Cyper.: 302 (2010).
Syn: *Cyperus nuerensis* Boeckeler – F.P.S. 3: 351 (1956).
Tufted perennial herb. Swamps, wet grassland, ditches.
Distr: EQU; Nigeria to Kenya, Tanzania, Zimbabwe.
EQU: Myers 6927 (fr) 7/1937.

Pycreus pelophilus (Ridl.) C.B.Clarke

F.T.E.A. Cyper.: 280 (2010).
Syn: *Cyperus pelophilus* Ridl. – F.P.S. 3: 349 (1956).
Slender annual herb. Swamps, wet depressions, pool margins.
Distr: UN; Somalia, Uganda to South Africa; Madagascar.
UN: J.M. Lock 81/197 (fr) 8/1981.

Pycreus polystachyos (Rottb.) P.Beauv. var. polystachyos

F.T.E.A. Cyper.: 289 (2010).
Syn: *Cyperus polystachyos* Rottb. – F.P.S. 3: 349 (1956); F.E.E. 6: 482 (1997).
Annual or short-lived perennial herb. Swamps, wet grassland, lake margins.
Distr: BAG, UN; palaeotropical.
UN: J.M. Lock 82/49 (fr) 8/1982.
Note: var. *laxiflorus* (Benth.) C.B.Clarke may also occur in Sudan.

Pycreus pumilus (L.) Nees

F.J.M.: 165 (1976); F.T.E.A. Cyper.: 283 (2010).
Syn: *Cyperus pumilus* L. – F.E.E. 6: 484 (1997).
Syn: *Cyperus pumilus* L. var. *muticus* (Boeckeler) C.B.Clarke – F.P.S. 3: 349 (1956).
Tufted annual herb. Pool margins, swamps, wet grassland, ditches, often on shallow soil over rocks.
Distr: DAR, KOR, UN; palaeotropical.
KOR: Jackson 2408 (fl) 10/1952; **UN:** J.M. Lock 81/264 (fl) 9/1981.
Note: (1) F.T.E.A. gives Domnin as the author of this name; this is believed to be in error; (2) we have been unable to trace the varietal name given by Andrews.

Pycreus unioloides (R.Br.) Urb.

F.J.M.: 165 (1976); F.T.E.A. Cyper.: 305 (2010).
Syn: *Cyperus unioloides* R.Br. – F.P.S. 3: 350 (1956); F.E.E. 6: 481 (1997); Biol. Skr. 51: 539 (2005).
Short-lived tufted perennial herb. Swamps, wet grassland, riverbanks, ditches.
Distr: DAR, EQU; pantropical.
DAR: Wickens 2257 (fr) 8/1964; **EQU:** Myers 6932 (fl) 7/1939.

[Pycreus sp. ?nov.]

Note: Wickens lists *Pycreus* sp. nov. aff. *P. subtrigonus* C.B.Clarke from Jebel Marra based on *Wickens* 2858; we have not located that specimen.

Rhynchospora corymbosa (L.) Britton

F.P.S. 3: 364 (1956); F.E.E. 6: 493 (1997); Biol. Skr. 51: 542 (2005); F.T.E.A. Cyper.: 360 (2010).
Tufted perennial herb with woody rhizome. Swamps, lake shores, riverbanks.
Distr: BAG, EQU; pantropical & warm temperate regions.
EQU: Friis & Vollesen 1244 (fr) 3/1982.

Schoenoplectiella articulata (L.) Lye

F.T.E.A. Cyper.: 30 (2010).
Syn: *Schoenoplectus articulatus* (L.) Palla – F.E.E. 6: 400 (1997).
Syn: *Scirpus articulatus* L. – F.P.S. 3: 367 (1956); F.J.M.: 165 (1976).
Tufted annual or short-lived perennial herb. Margins of pools & watercourses.
Distr: DAR, UN; palaeotropical.
DAR: Jackson 3268 (fr) 11/1954; **UN:** J.M. Lock 82/60 (fr) 8/1982.

Schoenoplectiella juncea (Willd.) Lye

F.T.E.A. Cyper.: 34 (2010).
Tufted annual herb. Swamps, shallow pools, ditches.
Distr: BAG, UN; Senegal to Uganda, Somalia, S to Tanzania.
UN: J.M. Lock 81/9 (fr) 2/1981.

Schoenoplectiella oxyjulos (S.S.Hooper) Lye

Lidia 6: 26 (2003).
Tufted annual herb. Wet depressions on shallow soil over rock.
Distr: EQU; Guinea, Sierra Leone, Nigeria, Cameroon, C.A.R.
EQU: Andrews 1605A (fr) 6/1939.
Note: a widespread but scarce species.

Schoenoplectiella senegalensis (Steud.) Lye

F.T.E.A. Cyper.: 31 (2010).
Syn: *Schoenoplectus senegalensis* (Steud.) Palla – F.E.E. 6: 400 (1997); Biol. Skr. 51: 542 (2005).
Syn: *Scirpus praelongatus* sensu Andrews, non Poir. – F.P.S. 3: 367 (1956).
Annual herb. Pool margins including ephemeral pools on rock outcrops, swamps.
Distr: KOR, BAG, EQU; Senegal to Somalia, S to Namibia & South Africa; tropical Asia & Australia.
KOR: Kotschy 56 (fl) 9/1839; **EQU:** Kielland-Lund 569 1/1984 (n.v.).

Schoenoplectiella supina (L.) Lye

Lidia 6: 27 (2003).
Syn: *Scirpus supinus* L. – F.P.S. 3: 366 (1956).
Annual herb. Wet grassland.
Distr: ?DAR; Mali to Nigeria; Madagascar, N Africa, Europe to Himalayas, Brazil to Argentina.
Note: Andrews recorded this species from Darfur; we have seen no material with which to verify this record.

Schoenoplectus corymbosus (Roth ex Roem. & Schult.) J.Raynal

F.E.E. 6: 399 (1997); F.T.E.A. Cyper.: 25 (2010).
Syn: *Schoenoplectus corymbosus* (Roth ex Roem. & Schult.) J.Raynal var. *brachyceras* (Hochst. ex A.Rich.) Lye – F.E.E. 6: 399 (1997); Biol. Skr. 51: 542 (2005).
Syn: *Scirpus brachyceras* Hochst. ex A.Rich. – F.P.S. 3: 368 (1956); F.J.M.: 165 (1976).
Syn: *Scirpus inclinatus* (Delile) Asch. & Schweinf. – Bot. Exp. Sud.: 46 (1970).
Tufted perennial herb with short rhizome. Standing water such as shallow lakes & streamsides.
Distr: DAR, EQU; palaeotropical & subtropical.

DAR: <u>Wickens 2674</u> (fr) 9/1964; **EQU:** <u>Myers 11162</u> (fl) 4/1939.

Schoenoplectus subulatus (Vahl) Lye

F.E.E. 6: 398 (1997).
Syn: *Scirpus litoralis* sensu Ahti et al., non Schrad. – Ann. Bot. Fennici 10: 156 (1973).
Stoloniferous perennial herb. Lake margins, temporary pools including saline areas.
Distr: NS; palaeotropical and warm temperate regions.
NS: <u>Ahti 16571</u> (fr) 9/1962.
Note: in F.T.E.A. Cyper.: 29 (2010) the NE African material is treated as *S. scirpoides*, stating that all specimens have at least some ciliae near the apex of the glumes whilst in *S. subulatus* the glumes are glabrous. This seems a rather weak distinction and, in any case, the glumes of the single Sudanese specimen are glabrous.

Schoenoplectus cf. tabernaemontani (C.C.Gmel.) Palla

Perennial herb. In canals.
Distr: CS; species widespread in temperate & tropical regions but in Africa only in North Africa.
CS: <u>Andrews 276</u> (fr) 12/1937.
Note: the identity of the single Sudan specimen requires confirmation.

Schoenoxiphium sparteum (Wahlenb.) C.B.Clarke

F.E.E. 6: 501 (1997); F.T.E.A. Cyper.: 418 (2010).
Tufted perennial herb. Upland grassland, forest margins.
Distr: EQU; Ethiopia to South Africa.
EQU: <u>Jackson 1525</u> (fr) 6/1950.

Scleria bulbifera Hochst. ex A.Rich.

F.P.S. 3: 369 (1956); F.E.E. 6: 495 (1997); Biol. Skr. 51: 543 (2005); F.T.E.A. Cyper.: 386 (2010).
Stoloniferous or rhizomatous perennial herb. Grassland & open woodland, sometimes seasonally wet.
Distr: BAG, EQU; Senegal to Ethiopia, S to South Africa; Madagascar, Arabia.
EQU: <u>Andrews 1016</u> (fr) 5/1939.

Scleria distans Poir. var. distans

F.E.E. 6: 495 (1997); Biol. Skr. 51: 543 (2005); F.T.E.A. Cyper.: 383 (2010).
Syn: *Scleria hirtella* sensu Andrews, non Sw. – F.P.S. 3: 369 (1956).
Rhizomatous perennial herb. Swamps, wet or dry grassland & woodland.
Distr: EQU; Nigeria to Ethiopia, S to South Africa; Madagascar, Mauritius, tropical America.
EQU: <u>Andrews 1893</u> (fr) 6/1939.

Scleria foliosa Hochst. ex A.Rich.

F.P.S. 3: 371 (1956); F.E.E. 6: 498 (1997); Biol. Skr. 51: 543 (2005); F.T.E.A. Cyper.: 400 (2010).
Syn: *Scleria complanata* Boeckeler – F.P.S. 3: 371 (1956).
Robust annual herb. Wet flushes, depressions in grassland.
Distr: KOR, BAG, EQU; Senegal to Ethiopia, S to South Africa; Madagascar, India.
KOR: <u>Niamir 63</u> (fl) 10/1980; **EQU:** <u>Friis & Vollesen 629</u> (fr) 12/1980.

Note: *S. complanata* is here considered to be a form of *S. foliosa*; the type specimen at P was determined as such by E.A. Robinson in 1963.

Scleria globonux C.B.Clarke

F.P.S. 3: 371 (1956); F.T.E.A. Cyper.: 402 (2010).
Annual herb. Swamp margins.
Distr: BAG; Senegal to Uganda, Zambia.
BAG: <u>Schweinfurth 2500</u> (fr) 10/1869.

Scleria gracillima Boeckeler

F.P.S. 3: 371 (1956); F.T.E.A. Cyper.: 404 (2010).
Slender annual herb. Swampy grassland.
Distr: BAG; Senegal to Chad & C.A.R., Tanzania, Zambia; Brazil.
BAG: <u>Schweinfurth 189</u> (fr) 10/1869.

Scleria hispidior (C.B.Clarke) Nelmes

F.E.E. 6: 496 (1997); Biol. Skr. 51: 543 (2005); F.T.E.A. Cyper.: 398 (2010).
Slender annual herb. Shallow pools over rock, wet grassland.
Distr: EQU; Ethiopia, Uganda, Kenya.
EQU: <u>K. Lock K13</u> 8/1966.
Note: the single specimen from our region is from the Sudan-Uganda border on Mt Lonyili.

Scleria lagoensis Boeckeler

F.E.E. 6: 499 (1997); F.T.E.A. Cyper.: 407 (2010).
Syn: *Scleria canaliculato-triquetra* Boeckeler – F.P.S. 3: 372 (1956).
Tufted perennial herb with short rhizome. Moist grassland & woodland.
Distr: BAG, EQU; pantropical.
EQU: <u>Myers 7744</u> (fr) 9/1937.

Scleria melanomphala Kunth

F.P.S. 3: 372 (1956); F.E.E. 6: 499 (1997); F.T.E.A. Cyper.: 411 (2010).
Robust tussock-forming perennial herb with thick rhizome. Swamps, wet grassland.
Distr: EQU; Guinea to Ethiopia, S to South Africa; Madagascar, South America.
EQU: <u>Myers 6929</u> (fr) 7/1937.

Scleria melanotricha Hochst. & A.Rich.

F.P.S. 3: 369 (1956); F.E.E. 6: 496 (1997); Biol. Skr. 51: 543 (2005); F.T.E.A. Cyper.: 392 (2010).
Slender annual herb. Wet grassland, shallow soil over rock.
Distr: BAG, EQU; Guinea to Ethiopia, Tanzania, Zambia.
EQU: <u>Friis & Vollesen 555</u> (fr) 11/1980.

Scleria pergracilis (Nees) Kunth

F.P.S. 3: 368 (1956); F.E.E. 6: 497 (1997); F.T.E.A. Cyper.: 395 (2010).
Tufted annual herb. Wet grassland, wet flushes over rock.
Distr: BAG; Senegal to Nigeria, Ethiopia to Angola & South Africa; India, Sri Lanka, New Guinea.
BAG: <u>Schweinfurth 2472</u> (fr) 10/1869.

Scleria racemosa Poir.

F.P.S. 3: 369 (1956); F.E.E. 6: 499 (1997); Biol. Skr. 51: 544 (2005); F.T.E.A. Cyper.: 413 (2010).
Rhizomatous perennial herb. Lake shores, swamps including within forest; river margins.

Distr: BAG, EQU; Ethiopia to Zimbabwe & Mozambique; Madagascar, Comoros.
EQU: Myers 7750 (fr) 9/1937.

Scleria schimperiana Boeckeler

F.P.S. 3: 371 (1956); F.E.E. 6: 497 (1997); F.T.E.A. Cyper.: 399 (2010).
Syn: *Scleria schimperiana* Boeckeler var. *hypoxis* (Boeckeler) C.B.Clarke – F.P.S. 3: 371 (1956).
Robust annual herb. Swamps, depressions in grassland & open woodland.
Distr: ES; Ethiopia to Zimbabwe.
ES: Schweinfurth 2054 (fr) 9/1865.
Note: the single specimen seen is from the Ethiopia border region.

Scleria tessellata Willd. var. tessellata

Biol. Skr. 51: 544 (2005); F.T.E.A. Cyper.: 403 (2010).
Syn: *Scleria glandiformis* Boeckeler – F.P.S. 3: 371 (1956).
Tufted annual herb. Seasonally wet grassland.
Distr: BAG; Senegal to Nigeria, Tanzania, Zambia; Madagascar, India to SE Asia.
BAG: Harrison 1083 (fr) 10/1950.
Note: F.T.E.A. lists also var. *sphaerocarpa* E.A.Rob. for Sudan but we have seen no material.

Scleria woodii C.B.Clarke var. ornata (Cherm.) J.Schultze-Motel

F.T.E.A. Cyper.: 391 (2010).
Rhizomatous perennial herb. Grassland including seasonally wet areas, woodland, rocky hillsides.
Distr: EQU; Ghana to Uganda, S to Angola & Zambia.
EQU: Sillitoe 78 (fr) 1919.

Poaceae (Gramineae)

H. Pickering & I. Darbyshire

Acrachne racemosa (B.Heyne ex Roth) Ohwi

F.T.E.A. Gramin. 2: 258 (1974); F.E.E. 7: 137 (1995).
Syn: *Eleusine verticillata* Roxb. – F.P.S. 3: 445 (1956).
Tufted annual. Dry bushland, stony ground.
Distr: RS, DAR, EQU; palaeotropical.
RS: Schweinfurth 505 (fl) 7/1868; EQU: Padwa 262 6/1953.

Acritochaete volkensii Pilg.

F.P.S. 3: 385 (1956); F.T.E.A. Gramin. 3: 658 (1982); F.E.E. 7: 245 (1995); Biol. Skr. 51: 544 (2005).
Annual. Montane forest & bamboo thicket.
Distr: EQU; Nigeria, Bioko, Cameroon, Ethiopia to D.R. Congo & Tanzania.
EQU: Friis & Vollesen 505 (fl) 11/1980.

Acroceras amplectens Stapf

F.P.S. 3: 386 (1956); F.T.E.A. Gramin. 3: 565 (1982); F.E.E. 7: 210, in notes (1995).
Annual. Wet grassland.
Distr: UN, EQU; Mauritania to Tanzania.
EQU: Myers 7553 (fl) 8/1937.

Acroceras gabunense (Hack.) Clayton

F.T.E.A. Gramin. 3: 567 (1982); Biol. Skr. 51: 544 (2005).

Perennial. Forest.
Distr: EQU; Guinea to Uganda, S to Angola.
EQU: Friis & Vollesen 473 (fl) 11/1980.

Acroceras zizanioides (Kunth) Dandy

F.P.S. 3: 385 (1956); F.T.E.A. Gramin. 3: 565 (1982); F.E.E. 7: 210 (1995); Biol. Skr. 51: 545 (2005).
Perennial. Damp places in forest.
Distr: EQU; Senegal to Ethiopia, S to Mozambique; India to Vietnam, New Guinea, neotropics.
EQU: Friis & Vollesen 652 (fl) 12/1980.

Aeluropus lagopoides (L.) Trin. ex Thwaites var. lagopoides

F.P.S. 3: 386 (1956); F.E.E. 7: 93 (1995).
Tufted rhizomatous perennial. Sandy seashores, saline flats.
Distr: NS, RS; Eritrea, Ethiopia, Somalia; Mediterranean, Arabia to Sri Lanka.
RS: Wickens 3075 (fl) 5/1969.

Agrostis kilimandscharica Mez

F.T.E.A. Gramin. 1: 110 (1970); F.E.E. 7: 49 (1995); Biol. Skr. 51: 545 (2005).
Syn: *Agrostis sororia* C.E.Hubb. – F.P.S. 3: 386 (1956).
Perennial, sometimes tufted. Montane forest, grassland & bamboo thicket.
Distr: EQU; Congo-Brazzaville to Ethiopia, S to Tanzania.
EQU: Thomas 1856 (fl) 12/1935.

Agrostis lachnantha Nees

F.P.S. 3: 387 (1956); F.T.E.A. Gramin. 1: 106 (1970); F.J.M.: 165 (1976); F.E.E. 7: 47 (1995).
Tufted annual or short-lived perennial. Streamsides, damp grassland.
Distr: DAR; Eritrea to Namibia & South Africa; Yemen.
DAR: Wickens 1250 (fl) 2/1964.

Agrostis producta Pilg.

F.P.S. 3: 386 (1956); F.T.E.A. Gramin. 1: 108 (1970); F.E.E. 7: 50 (1995); Biol. Skr. 51: 545 (2005).
Tufted perennial. Montane grassland & swamps.
Distr: EQU; Uganda, Kenya, Tanzania, Malawi, Zimbabwe.
EQU: Friis & Vollesen 845 (fl) 12/1980.

Aira caryophyllea L.

F.T.E.A. Gramin. 1: 84 (1970); F.J.M.: 165 (1976); F.E.E. 7: 37 (1995); Biol. Skr. 51: 545 (2005).
Syn: *Aira caryophyllea* L. var. *latigluma* (Steud.) C.E. Hubb. – F.P.S. 3: 387 (1956); Bot. Exp. Sud.: 52 (1970).
Annual. Montane grassland.
Distr: RS, DAR; Nigeria, Cameroon, Ethiopia to Malawi, South Africa; Madagascar, Europe, Central Asia, Central America.
DAR: Jackson 3337 (fl) 12/1964.

Alloteropsis cimicina (L.) Stapf

F.P.S. 3: 388 (1956); F.J.M.: 166 (1976); F.T.E.A. Gramin. 3: 617 (1982); F.E.E. 7: 216 (1995); Biol. Skr. 51: 546 (2005).
Tufted annual. Open bushland & grassland, abandoned cultivation.
Distr: DAR, KOR, EQU; palaeotropical.
DAR: Wickens 2309 9/1964; EQU: Myers 14118 9/1941.

Alloteropsis semialata (R.Br.) Hitchc. subsp. **semialata**

F.P.S. 3: 387 (1956); F.T.E.A. Gramin. 3: 616 (1982); F.E.E. 7: 215 (1995); Biol. Skr. 51: 546 (2005).
Tufted perennial. Grassland, bushland, seasonal swamps.
Distr: ES, EQU; palaeotropical.
ES: Schweinfurth 1079 6/1865; **EQU:** Myers 10935 4/1939.

Andropogon amethystinus Steud.

F.T.E.A. Gramin. 3: 772 (1982); F.E.E. 7: 323 (1995); Biol. Skr. 51: 546 (2005).
Syn: Andropogon longipes Hack. – F.J.M.: 166 (1976).
Perennial. Montane forest margins & grassland.
Distr: DAR, ?EQU; Togo to Cameroon, Eritrea to Zambia & Malawi, South Africa; Yemen, India, Myanmar.
DAR: Wickens 1705 5/1964.
Note: Jackson recorded this species from the Imatong Mts but no collections have been seen (see Friis & Vollesen 2005: 546).

Andropogon distachyos L.

F.P.S. 3: 388 (1956); F.J.M.: 166 (1976); F.T.E.A. Gramin. 3: 770 (1982); F.E.E. 7: 322 (1995).
Tufted perennial. Montane grassland.
Distr: DAR; Cameroon to Eritrea, S to South Africa; Mediteranean, Arabia, Thailand.
DAR: Wickens 2488 9/1964.

Andropogon fastigiatus Sw.

F.J.M.: 166 (1976); F.T.E.A. Gramin. 3: 777 (1982); F.E.E. 7: 325 (1995).
Syn: Diectomis fastigiata (Sw.) Kunth – F.P.S. 3: 431 (1956).
Tufted annual. Bushland, wooded grassland.
Distr: DAR, KOR; pantropical.
DAR: Wickens 2792 10/1964.

Andropogon gayanus Kunth var. **gayanus**

F.P.S. 3: 389 (1956); F.W.T.A. 3(2): 488 (1972); Biol. Skr. 51: 546 (2005).
Tufted perennial. Bushland & woodland, swamps.
Distr: CS, BAG, UN, EQU; Cape Verde Is., Senegal to D.R. Congo.
CS: Halwazy s.n. (fl) 9/1957; **BAG:** Jackson 3989 10/1958.

Andropogon gayanus Kunth var. **bisquamulatus** (Hochst.) Hack.

F.P.S. 3: 391 pro parte (1956); F.W.T.A. 3(2): 489 (1972).
Tufted perennial. Wooded grassland.
Distr: KOR; Mauritania to C.A.R.
KOR: Kotschy 143 (fl) 10/1839.
Note: Andrews records this variety from Central & Southern Sudan but we have only seen material from Kordofan.

Andropogon gayanus Kunth var. **polycladus** (Hack.) Clayton

F.T.E.A. Gramin. 3: 777 (1982); F.E.E. 7: 325 (1995).
Syn: Andropogon gayanus Kunth var. **squamulatus** (Hochst.) Stapf – F.P.S. 3: 391 (1956); F.J.M.: 166 (1976).
Tufted perennial. Bushland & wooded grassland.
Distr: DAR, KOR, ES, BAG, UN, EQU; Mauritania to Ethiopia, S to Tanzania, Namibia, Botswana, South Africa.

DAR: Jackson 3309 (fl) 12/1954; **EQU:** Myers 9468 (fl) 9/1938.

Andropogon gayanus Kunth var. **tridentatus** Hack.

F.W.T.A. 3(2): 488 (1972); F.J.M.: 166 (1976).
Tufted perennial. Wooded grassland.
Distr: DAR, KOR; Senegal & Mauritania to Cameroon.
DAR: Harrison 84 9/1947.

Andropogon lima (Hack.) Stapf

F.T.E.A. Gramin. 3: 771 (1982); F.E.E. 7: 322 (1995); Biol. Skr. 51: 547 (2005).
Tufted perennial. Montane grassland.
Distr: EQU; Cameroon, Ethiopia to Malawi.
EQU: Myers 11600 7/1939.

Andropogon mannii Hook.f.

F.T.E.A. Gramin. 3: 774 (1982); F.E.E. 7: 324 (1995); Biol. Skr. 51: 547 (2005).
Tufted perennial. Montane grassland.
Distr: EQU; Sierra Leone, Ivory Coast, Bioko, Cameroon, Gabon, Ethiopia to South Africa.
EQU: Myers 11650 7/1939.

Andropogon schirensis Hochst. ex A.Rich.

F.P.S. 3: 389 (1956); F.J.M.: 166 (1976); F.T.E.A. Gramin. 3: 779 (1982); F.E.E. 7: 327 (1995); Biol. Skr. 51: 547 (2005).
Tufted perennial. Grassland & wooded grassland.
Distr: DAR, ES, BAG, EQU; Senegal to Ethiopia, S to Namibia & South Africa.
DAR: Blair 183 10/1963; **BAG:** Myers 7701 9/1937.

Andropogon tenuiberbis Hack.

F.P.S. 3: 388 (1956); F.T.E.A. Gramin. 3: 776 (1982).
Tufted perennial. Swamps, lake margins.
Distr: BAG; Senegal to Tanzania, S to Angola.
BAG: Schweinfurth 2600 11/1869.

Anthephora laevis Stapf & C.E.Hubb.

F.E.E. 7: 281 (1995).
Perennial. Dry, stony ground.
Distr: RS; Eritrea; Dead Sea.
IUCN: RD
RS: Khalid 61 1/1952.

Anthephora nigritana Stapf & C.E.Hubb.

F.T.E.A. Gramin. 3: 662 (1982); F.E.E. 7: 281 (1995).
Syn: Anthephora lynesii Stapf & C.E.Hubb. – F.P.S. 3: 391 (1956); F.J.M.: 167 (1976).
Syn: Anthephora hochstetteri sensu Wickens, non Nees – F.J.M.: 166 (1976).
Syn: Anthephora sp. aff. **nigritana** Stapf & C.E.Hubb. – F.J.M.: 167 (1976).
Tufted perennial. Rocky hillslopes, dry sandy ground.
Distr: DAR, KOR; Niger to Somalia & Kenya.
DAR: Wickens 2080 4/1964.

Anthephora pubescens Nees

F.T.E.A. Gramin. 3: 664 (1982); F.E.E. 7: 281 (1995).
Syn: Anthephora hochstetteri Nees – F.P.S. 3: 392 (1956); Bot. Exp. Sud.: 52 (1970).
Tufted perennial. Bushland on rocky hillslopes.
Distr: NS, RS, ?CS; Ethiopia, Somalia, Kenya, Angola,

Botswana, Namibia, South Africa.
NS: <u>Kotschy 367</u> 1837.

Aristida adoensis Hochst. ex A.Rich.
F.T.E.A. Gramin. 1: 144 (1970); F.J.M.: 167 (1976); F.E.E.
7: 80 (1995); Biol. Skr. 51: 548 (2005).
Tufted perennial. Grassland & bushland, overgrazed areas.
Distr: DAR, EQU; Eritrea to D.R. Congo & Tanzania.
DAR: <u>Jackson 3373</u> 12/1954; **EQU:** <u>Myers 11012</u>
4/1939.

Aristida adscensionis L.
F.P.S. 3: 395 (1956); F.T.E.A. Gramin. 1: 148 (1970);
F.J.M.: 167 (1976); F.E.E. 7: 78 (1995); Biol. Skr. 51: 548
(2005).
Syn: *Aristida coerulescens* Desf. – F.J.M.: 167 (1976).
Syn: *Aristida submucronata* Schumach. & Thonn. –
F.P.S. 3: 395 (1956); Bot. Exp. Sud.: 52 (1970).
Tufted annual. Open sandy & stony ground including
disturbed areas.
Distr: NS, RS, DAR, KOR, CS, ES, EQU; pantropical &
subtropical.
RS: <u>Carter 2012</u> 11/1987; **EQU:** <u>Myers 14092</u> 4/1941.

Aristida congesta Roem. & Schult.
F.T.E.A. Gramin. 1: 156 (1970); F.J.M.: 167 (1976); F.E.E.
7: 84 (1995).
Tufted perennial. Dry sandy or stony ground, bushland.
Distr: DAR; Chad to Eritrea, S to Namibia & South Africa;
Madagascar, N Africa, Arabia.
DAR: <u>Wickens 1245</u> 2/1964.

Aristida cumingiana Trin. & Rupr.
F.P.S. 3: 393 (1956); F.T.E.A. Gramin. 1: 146 (1970);
F.J.M.: 167 (1976); F.E.E. 7: 80 (1995).
Annual, sometimes tufted. Damp ground.
Distr: DAR, BAG; scattered from Senegal to Ethiopia, S
to Zimbabwe; India to New Guinea & Australia.
DAR: <u>Blair 15</u> 10/1963; **BAG:** <u>Schweinfurth 181</u>
11/1869.

Aristida funiculata Trin. & Rupr.
F.P.S. 3: 396 (1956); F.T.E.A. Gramin. 1: 152 (1970);
F.J.M.: 167 (1976); F.E.E. 7: 81 (1995).
Tufted annual. Dry bushland, open sandy or stony
ground.
Distr: RS, DAR, KOR, CS; Mauritania & Senegal to
Somalia & Kenya; Arabia to Myanmar.
RS: <u>Carter 2011</u> 11/1987.

Aristida hordeacea Kunth
F.P.S. 3: 395 (1956); F.T.E.A. Gramin. 1: 150 (1970);
F.J.M.: 167 (1976); F.E.E. 7: 81 (1995); Biol. Skr. 51: 548
(2005).
Annual. Bushland, dry wooded grassland, wadis.
Distr: RS, DAR, KOR, CS, UN, EQU; Mauritania to Eritrea,
S to Namibia & South Africa.
DAR: <u>Wickens 2370</u> 9/1964; **EQU:** <u>Myers 7910</u> 11/1937.

Aristida kenyensis Henrard
F.T.E.A. Gramin. 1: 150 (1970); F.E.E. 7: 78 (1995).
Annual. Bushland, dry stony ground, disturbed areas.
Distr: DAR; Eritrea, Ethiopia, Uganda, Kenya, Tanzania.
DAR: <u>Miehe 696</u> 10/1982.

Aristida mutabilis Trin. & Rupr.
F.P.S. 3: 397 (1956); F.T.E.A. Gramin. 1: 157 (1970); F.E.E.
7: 84 (1995); F.J.U. 1: 280 (1997).
Syn: *Aristida cassanellii* A.Terracc. – F.P.S. 3: 396
(1956).
Tufted annual. Dry bushland & semi-desert.
Distr: NS, RS, DAR, KOR, CS, ES; Mauritania to Somalia,
S to Tanzania; N Africa, Arabia, Pakistan, India.
RS: <u>Carter 1809</u> 11/1987.

Aristida recta Franch.
F.P.S. 3: 395 (1956); F.T.E.A. Gramin. 1: 145 (1970).
Tufted perennial. Moist hollows.
Distr: ?EQU; Guinea to Uganda, S to Angola & South
Africa.
Note: Andrews recorded this species from Equatoria; we
have seen no material with which to verify this record.

Aristida rhiniochloa Hochst.
F.P.S. 3: 393 (1956); F.T.E.A. Gramin. 1: 147 (1970); F.E.E.
7: 77 (1995).
Tufted annual. Dry bushland.
Distr: DAR, KOR, CS; Mauritania to Eritrea & Ethiopia,
Tanzania to Namibia & South Africa.
CS: <u>Jackson 1624</u> 1/1951.

Aristida sieberiana Trin.
F.T.E.A. Gramin. 1: 155 (1970).
Syn: *Aristida pallida* Steud. – F.P.S. 3: 396 (1956); Bot.
Exp. Sud.: 52 (1970).
Tufted perennial. Dry bushland.
Distr: DAR, KOR, CS; Mauritania & Senegal to Somalia &
Kenya; NW Africa, Palestine, Yemen.
KOR: <u>Harrison 998</u> 9/1947.

Aristida stipoides Lam.
F.P.S. 3: 396 (1956); F.T.E.A. Gramin. 1: 153 (1970).
Tufted annual. Abandoned cultivations, disturbed ground.
Distr: NS, DAR, KOR, CS; Mauritania to Ethiopia, S to
Namibia & Botswana.
CS: <u>Jackson 2359</u> 10/1952.
Note: Andrews records this species from Central &
Southern Sudan but this is considered to be in error; we
have seen no material from the South.

Arthraxon hispidus (Thunb.) Makino
Blumea 27: 280 (1981).
Syn: *Arthraxon micans* (Nees) Hochst. – F.T.E.A.
Gramin. 3: 742 (1982); F.E.E. 7: 312 (1995); Biol. Skr. 51:
548 (2005).
Syn: *Arthraxon quartinianus* (A.Rich.) Nash – F.P.S. 3:
399 (1956); F.J.M.: 167 (1976).
Mat-forming annual. Forest, woodland, damp rocks.
Distr: RS, DAR, EQU; scattered through the tropics &
warm temperate regions.
DAR: <u>Wickens 2744</u> 9/1964; **EQU:** <u>Friis & Vollesen 6</u>
11/1980.

Arthraxon lancifolius (Trin.) Hochst.
F.P.S. 3: 399 (1956); F.J.M.: 167 (1976); F.T.E.A. Gramin.
3: 742 (1982); F.E.E. 7: 312 (1995).
Mat-forming annual. Steep rocky banks & rock crevices.
Distr: DAR, ?KOR, CS; palaeotropical.
DAR: <u>Wickens 2320</u> 5/1964.

Arthraxon prionodes (Steud.) Dandy

F.P.S. 3: 399 (1956); F.T.E.A. Gramin. 3: 741 (1982); F.E.E. 7: 310 (1995).
Mat-forming perennial. Rocky hillslopes.
Distr: RS, ES; Eritrea to Tanzania; tropical Asia.
RS: Jackson 2967 4/1953.

Arundinella pumila (Hochst. ex A.Rich.) Steud.

F.J.M.: 167 (1976); F.E.E. 7: 285 (1995).
Annual. Moist rocks & shaded banks.
Distr: DAR; Guinea to Cameroon, Ethiopia; Oman, India to Indonesia.
DAR: Wickens 2506 9/1964.

Bothriochloa bladhii (Retz.) S.T.Blake

F.T.E.A. Gramin. 3: 719 (1982); F.E.E. 7: 306 (1995).
Syn: *Bothriochloa insculpta* (Hochst. ex A.Rich.) A.Camus var. *vegetior* (Hack.) C.E.Hubb. – F.P.S. 3: 403 (1956).
Tufted perennial. Streamsides, swamp margins.
Distr: DAR, ES; pantropical & warm temperate.
DAR: Wickens 2243 8/1964.

Bothriochloa insculpta (Hochst. ex A.Rich.) A.Camus

F.P.S. 3: 402 (1956); F.J.M.: 168 (1976); F.T.E.A. Gramin. 3: 720 (1982); F.E.E. 7: 306 (1995); Biol. Skr. 51: 584 (2005).
Syn: *Bothriochloa pertusa* (L.) A.Camus – F.P.S. 3: 402 (1956); Bot. Exp. Sud.: 52 (1970).
Tufted perennial. Grassland, disturbed ground.
Distr: NS, RS, DAR, BAG, EQU; Sierra Leone to Somalia, S to South Africa; Madagascar, Arabia, scattered elsewhere in the tropics.
RS: Andrews 3594 3/1949; **EQU:** Myers 9921 10/1938.

Bothriochloa radicans (Lehm.) A.Camus

F.P.S. 3: 402 (1956); F.T.E.A. Gramin. 3: 721 (1982); F.E.E. 7: 307 (1995).
Cushion-forming perennial. Grassland & wooded grassland, bushland.
Distr: EQU; Eritrea & Somalia, S to Namibia & South Africa; Arabia.
EQU: Jackson 759 5/1949.

Brachiaria arrecta (Hack. ex T.Durand & Schinz) Stent

F.T.E.A. Gramin. 3: 585 (1982); F.E.E. 7: 222 (1995).
Perennial. Riverbanks, swamps.
Distr: BAG, UN, EQU; Ethiopia to Namibia & South Africa; Madagascar, neotropics.
BAG: Simpson 7366 7/1929.

Brachiaria brizantha (Hochst. ex A.Rich.) Stapf

F.P.S. 3: 407 (1956); F.J.M.: 168 (1976); F.T.E.A. Gramin. 3: 587 (1982); F.E.E. 7: 222 (1995); Biol. Skr. 51: 549 (2005).
Tufted perennial. Woodland & wooded grassland.
Distr: DAR, ES, EQU; Gambia to Somalia, S to Namibia & South Africa; introduced elsewhere in the tropics.
DAR: Wickens 2184 8/1964; **EQU:** Friis & Vollesen 767 12/1980.

Brachiaria comata (Hochst. ex A.Rich.) Stapf

F.T.E.A. Gramin. 3: 593 (1982); F.E.E. 7: 227 (1995); Biol. Skr. 51: 549 (2005).

Syn: *Brachiaria kotschyana* (Steud.) Stapf – F.P.S. 3: 411 (1956); F.J.M.: 168 (1976).
Syn: *Brachiaria secernenda* (Hochst. ex Mez) Henrard – F.P.S. 3: 411 (1956); F.J.M.: 168 (1976).
Tufted annual. Bushland, disturbed ground.
Distr: RS, DAR, KOR, ES, BAG, UN, EQU; Senegal, Burkina Faso to Eritrea, S to Tanzania; Yemen.
DAR: Wickens 3151 8/1964; **EQU:** Sillitoe 7 1919.

Brachiaria deflexa (Schumach.) C.E.Hubb. ex Robyns

F.P.S. 3: 410 (1956); F.J.M.: 168 (1976); F.T.E.A. Gramin. 3: 598 (1982); F.E.E. 7: 228 (1995); Biol. Skr. 51: 550 (2005).
Tufted annual. Bushland & woodland, often in disturbed areas.
Distr: NS, RS, DAR, CS, UN, EQU; Mauritania to Ethiopia, S to Namibia & South Africa; Madagascar, Arabia, India.
CS: Farag 159 7/2010; **EQU:** Myers 9148 7/1938.

Brachiaria dictyoneura (Fig. & De Not.) Stapf

F.P.S. 3: 406 (1956); F.T.E.A. Gramin. 3: 582 (1982); F.E.E. 7: 221 (1995).
Tufted perennial. Wooded grassland, bushland.
Distr: KOR, CS; Ethiopia to Namibia & South Africa.
KOR: Figari s.n. 4/1844.

Brachiaria eruciformis (Sm.) Griseb.

F.P.S. 3: 406 (1956); F.T.E.A. Gramin. 3: 590 (1982); F.E.E. 7: 224 (1995).
Tufted annual. Damp grassland.
Distr: NS, DAR, CS, UN; pantropical & warm temperate.
CS: Andrews 130 1/1936; **UN:** Simpson 7215 6/1929.

Brachiaria humidicola (Rendle) Schweick.

F.J.M.: 168 (1976); F.T.E.A. Gramin. 3: 583 (1982); F.E.E. 7: 221 (1995).
Stoloniferous perennial. Wet grassland.
Distr: DAR; Nigeria, Ethiopia to Namibia & South Africa.
DAR: Blair 339 8/1964.

Brachiaria jubata (Fig. & De Not.) Stapf

F.P.S. 3: 406 (1956); F.J.M.: 168 (1976); F.T.E.A. Gramin. 3: 580 (1982); F.E.E. 7: 221 (1995); Biol. Skr. 51: 550 (2005).
Tufted perennial. Damp grassland & bushland.
Distr: DAR, ?KOR, CS, ?ES, BAG, UN, EQU; Mauritania to Ethiopia, S to Angola & Zimbabwe; Madagascar.
DAR: Blair 71 6/1963; **EQU:** Myers 11099 4/1939.

Brachiaria lata (Schumach.) C.E.Hubb.

F.P.S. 3: 408 (1956); F.J.M.: 168 (1976); F.T.E.A. Gramin. 3: 600, in notes (1982); F.E.E. 7: 229 (1995).
Annual. Open grassland.
Distr: RS, DAR, KOR, CS, ES, BAG, UN, EQU; Mauritania & Senegal to Ethiopia; Arabia, India.
DAR: Blair 326 7/1964; **BAG:** Myers 7412 7/1937.

Brachiaria leersioides (Hochst.) Stapf

F.P.S. 3: 411 (1956); F.T.E.A. Gramin. 3: 591 (1982); F.E.E. 7: 225 (1995); Biol. Skr. 51: 550 (2005).
Annual. Bushland, grassland, disturbed ground.
Distr: RS, BAG, EQU; Chad to Ethiopia, D.R. Congo, Tanzania; Arabia.

RS: <u>Schweinfurth 685</u> 9/1868; **BAG:** <u>Simpson 7271</u> 7/1929.

Brachiaria mutica (Forssk.) Stapf

F.P.S. 3: 407 (1965); F.T.E.A. Gramin. 3: 584 (1982); F.E.E. 7: 222, in notes (1995).
Syn: *Panicum infestum* sensu Wickens, non Andersson – F.J.M.: 180 (1976).
Perennial. Pasture.
Distr: DAR, CS, BAG, UN; pantropical, widely cultivated for forage and frequently naturalised.
CS: <u>Farag 162</u> 7/2010; **UN:** <u>J.M. Lock 80/9</u> 2/1980.

Brachiaria orthostachys (Mez) Clayton

F.E.E. 7: 229, in notes (1995).
Annual. Bushland.
Distr: DAR, KOR; Mauritania to Chad.
KOR: <u>Harrison 931</u> 9/1950.

Brachiaria ovalis Stapf

F.T.E.A. Gramin. 3: 599 (1982); F.E.E. 7: 227 (1995); Biol. Skr. 51: 551 (2005).
Syn: *Brachiaria glauca* Stapf – F.P.S. 3: 410 (1956).
Tufted annual. Dry bushland & semi-desert grassland.
Distr: RS, EQU; Chad, Eritrea, Ethiopia, Somalia, Kenya; Arabia, Pakistan.
RS: <u>Carter 1962</u> 11/1987; **EQU:** <u>Myers 14019</u> 9/1941.

Brachiaria ramosa (L.) Stapf

F.P.S. 3: 409 (1956); F.J.M.: 168 (1976); F.T.E.A. Gramin. 3: 599 (1982); F.E.E. 7: 228 (1995).
Tufted annual. Rocky hillslopes, sandy plains, disturbed ground.
Distr: RS, DAR, KOR, CS, UN; palaeotropical.
CS: <u>Andrews 70</u> 6/1936; **UN:** <u>Jackson 4008</u> 9/1959.

Brachiaria reptans (L.) C.A.Gardner & C.E.Hubb.

F.P.S. 3: 404 (1956); F.T.E.A. Gramin. 3: 591 (1982); F.E.E. 7: 225 (1995).
Tufted annual. Disturbed ground.
Distr: CS, UN; tropical Asia; introduced elsewhere in the tropics.
CS: <u>Ouren 29137</u> 7/1961; **UN:** <u>Harrison 20</u> 6/1947.

Brachiaria scalaris Pilg.

F.T.E.A. Gramin. 3: 594 (1982); F.E.E. 7: 227 (1995); Biol. Skr. 51: 551 (2005).
Annual. Disturbed ground.
Distr: EQU; Eritrea to Zimbabwe; Arabia.
EQU: <u>Friis & Vollesen 39</u> 11/1980.

Brachiaria semiundulata (Hochst. ex A.Rich.) Stapf

F.J.M.: 169 (1976); F.T.E.A. Gramin. 3: 592 (1982); F.E.E. 7: 227 (1995).
Tufted annual. Bushland & grassland, disturbed ground.
Distr: DAR; Ethiopia to Tanzania; India, Sri Lanka, China.
DAR: <u>Wickens 2158</u> 8/1964.

Brachiaria serrifolia (Hochst.) Stapf

F.P.S. 3: 410 (1956); F.J.M.: 169 (1976); F.T.E.A. Gramin. 3: 596 (1982); F.E.E. 7: 225 (1995); Biol. Skr. 51: 551 (2005).
Annual. Grassland.

Distr: DAR, KOR, EQU; Mali to Ethiopia, S to Tanzania, Zimbabwe.
DAR: <u>Blair 131</u> 8/1963; **EQU:** <u>Peers K028</u> 8/1953.

Brachiaria stigmatisata (Mez) Stapf

F.P.S. 3: 407 (1956); F.W.T.A. 3(2): 442 (1972).
Annual, sometimes tufted. Disturbed ground.
Distr: BAG, EQU; Senegal to Chad.
EQU: <u>Myers 7255</u> 7/1937.

Brachiaria villosa (Lam.) A.Camus

F.T.E.A. Gramin. 3: 593 (1982).
Annual, sometimes tufted. Disturbed ground.
Distr: DAR, KOR; Mauritania to Uganda & D.R. Congo; India to Japan & New Guinea.
DAR: <u>Blair 125</u> 8/1963.

Brachiaria xantholeuca (Hack.) Stapf

F.P.S. 3: 409 (1956); F.J.M.: 169 (1976); F.T.E.A. Gramin. 3: 597 (1982); F.E.E. 7: 229 (1995).
Syn: *Brachiaria pubifolia* Stapf – Bot. Exp. Sud.: 53 (1970).
Tufted annual. Bushland, disturbed ground.
Distr: DAR, KOR, CS, BAG, UN, EQU; Mauritania to Eritrea, S to Namibia & South Africa.
DAR: <u>Wickens 2312</u> 9/1964; **EQU:** <u>Myers 7552</u> 8/1937.

Brachypodium distachyon (L.) P.Beauv.

F.E.E. 7: 55 (1995).
Syn: *Trachynia distachya* (L.) Link – F.P.S. 3: 551 (1956).
Tufted annual. Dry scrub.
Distr: RS; Eritrea, Ethiopia; Mediterranean, Arabia to India, introduced elsewhere.
RS: <u>Andrews 3506</u> 3/1949.

Brachypodium flexum Nees

F.T.E.A. Gramin. 1: 71 (1970); F.E.E. 7: 57 (1995); Biol. Skr. 51: 551 (2005).
Syn: *Brachypodium flexum* Nees var. *abyssinicum* Hochst. – F.P.S. 3: 413 (1956).
Perennial, sometimes mat-forming. Montane forest & bushland, often in forest margins & clearings or bamboo thicket.
Distr: DAR, EQU; Sierra Leone, Nigeria, Bioko, Cameroon, Eritrea to South Africa; Madagascar, Yemen.
DAR: <u>de Wilde et al. 5556</u> 1/1965; **EQU:** <u>Friis & Vollesen 115</u> 11/1980.

Brachypodium sylvaticum (Huds.) P.Beauv.

F.T.E.A. Gramin. 1: 73, in notes (1970); F.J.M.: 169 (1976); F.E.E. 7: 58 (1995).
Tufted perennial. Shady places.
Distr: DAR; Eritrea; Europe, temperate Asia & mountains of tropical Asia.
DAR: <u>Blair 281</u> 5/1964.

Bromus diandrus Roth

F.T.E.A. Gramin. 1: 67 (1970).
Annual. Weed of disturbed ground & cultivation.
Distr: NS; Mediterranean region; widely naturalised as a weed elsewhere.
NS: <u>Boulos s.n.</u> 2/1964.
Note: the record of *B. rigidus* Roth in Ahti et al. (1973) may refer to this species.

Bromus leptoclados Nees

F.T.E.A. Gramin. 1: 68 (1970); F.J.M.: 169 (1976); F.E.E. 7: 54 (1995); Biol. Skr. 51: 552 (2005).
Syn: *Bromus runssoroensis* K.Schum. – F.P.S. 3: 414 (1956).
Tufted perennial. Montane grassland, bushland & forest clearings.
Distr: DAR, EQU; Bioko, Cameroon, Eritrea & Somalia to South Africa.
DAR: Wickens 2477 9/1964; **EQU:** Friis & Vollesen 135 11/1980.

Bromus pectinatus Thunb.

F.T.E.A. Gramin. 1: 68 (1970); F.J.M.: 169 (1976); F.E.E. 7: 54 (1995).
Syn: *Bromus adoensis* Hochst. ex Steud. – F.P.S. 3: 413 (1956); Bot. Exp. Sud.: 53 (1970).
Tufted annual. Grassland.
Distr: RS, DAR; Eritrea, Ethiopia, Uganda, Kenya; Madagascar, N Africa, Arabia to China.
DAR: Wickens 2605 9/1964.

Calamagrostis epigejos (L.) Roth var. capensis Stapf

F.T.E.A. Gramin. 1: 102 (1970); F.J.M.: 169 (1976); F.E.E. 7: 51 (1995).
Tufted perennial. Montane grassland & forest clearings.
Distr: DAR; Rwanda, Ethiopia, Uganda, Kenya, Tanzania, South Africa.
DAR: Wickens 1717 5/1964.

Capillipedium parviflorum (R.Br.) Stapf

F.J.M.: 169 (1976); F.T.E.A. Gramin. 3: 718 (1982); F.E.E. 7: 305 (1995).
Tufted perennial. Grassland.
Distr: DAR; Eritrea to South Africa; Oman, tropical Asia to Australia.
DAR: Blair 196 10/1963.

Castellia tuberculosa (Moris) Bor

F.E.E. 7: 17 (1995).
Annual. Shade of *Juniperus* forest.
Distr: RS; Eritrea, Djibouti, Somalia; Mediterranean, Arabia to India.
RS: Jackson 2861 4/1953.

Cenchrus biflorus Roxb.

F.P.S. 3: 415 (1956); F.J.M.: 170 (1976); F.T.E.A. Gramin. 3: 695 (1982); F.E.E. 7: 278 (1995).
Syn: *Cenchrus leptacanthus* A.Camus – F. Darfur Nord-Occ. & J. Gourgeil: 136 (1969).
Tufted annual. Weed of disturbed ground.
Distr: NS, RS, DAR, KOR, CS, ES, BAG, UN, EQU; Mauritania to Somalia, S to Namibia & South Africa; N Africa, Arabia, India.
KOR: Harrison 963 9/1950; **BAG:** Myers 7200 7/1937.

Cenchrus ciliaris L.

F.P.S. 3: 414 (1956); F.J.M.: 170 (1976); F.T.E.A. Gramin. 3: 691 (1982); F.E.E. 7: 276 (1995); Biol. Skr. 51: 552 (2005).
Tufted or mat-forming perennial. Wooded grassland, woodland & bushland.
Distr: NS, RS, DAR, KOR, CS, ES, UN, EQU; Mauritania to Somalia, S to Namibia & South Africa; N Africa to India, introduced elsewhere.
DAR: Blair 136 8/1963; **EQU:** Jackson 2218 6/1952.

Cenchrus pennisetiformis Steud.

F.P.S. 3: 414 (1956); F.T.E.A. Gramin. 3: 692 (1982); F.E.E. 7: 278 (1995).
Annual or short lived perennial. Dry grassland & semi-desert.
Distr: RS, ?KOR, CS, ES; Eritrea, Ethiopia, Djibouti, Somalia, Kenya; Egypt, Arabia to India & Myanmar.
CS: Andrews 193 2/1936.

Cenchrus prieurii (Kunth) Maire

F.P.S. 3: 414 (1956); F.E.E. 7: 276 (1995).
Tufted annual. Semi-desert, open sandy ground.
Distr: RS, DAR, KOR, ES, BAG; Mauritania to Ethiopia; Yemen, Pakistan, India, Myanmar.
KOR: Harrison 462 9/1950; **BAG:** Jackson 2355 10/1952.

Cenchrus setigerus Vahl

F.P.S. 3: 415 (1956); F.T.E.A. Gramin. 3: 694 (1982); F.E.E. 7: 278 (1995); Biol. Skr. 51: 552 (2005).
Perennial. Semi-desert grassland & bushland.
Distr: RS, CS, EQU; Ethiopia, Somalia, Kenya, Tanzania; Arabia, India.
RS: Andrews 48 6/1935; **EQU:** Wilson s.n. 4/1960.
Note: the South Sudan specimen cited is from Mt Lotuke on the Ugandan border.

Centropodia forskalii (Vahl) Cope subsp. forskalii

F.E.E. 7: 73 (1995); F.J.U. 1: 282 (1997).
Syn: *Asthenatherum forskalii* (Vahl) Nevski – F.P.S. 3: 400 (1956).
Tufted perennial or annual. Sandy & stony desert.
Distr: NS, DAR, KOR; N Africa, Arabia to Central Asia.
DAR: Shaw 34 12/1932.

Chasmopodium caudatum (Hack.) Stapf

F.P.S. 3: 415 (1956); F.W.T.A. 3(2): 509 (1972).
Annual. Wooded grassland.
Distr: BAG; Senegal to Chad, S to Angola.
BAG: Schweinfurth 2357 9/1869.

Chloris barbata Sw.

F.P.S. 3: 420 (1956); F.T.E.A. Gramin. 2: 345 (1974); F.E.E. 7: 168 (1995).
Tufted stoloniferous perennial. Disturbed areas.
Distr: ?DAR; Ivory Coast to Eritrea, S to Tanzania; Madagascar, NW Africa, Arabia to E Asia, introduced elsewhere.
Note: Andrews records this species from Darfur; we have seen no material with which to verify this record.

Chloris gayana Kunth

F.P.S. 3: 418 (1956); F.J.M.: 170 (1976); F.E.E. 7: 169 (1995); Biol. Skr. 51: 553 (2005).
Stoloniferous perennial. Woodland, bushland grassland.
Distr: DAR, CS, BAG, UN, EQU; Senegal to Somalia, S to Namibia & South Africa; introduced elsewhere in the tropics.
DAR: Wickens 1060 1/1964; **EQU:** Myers 13453 8/1940.

Chloris lamproparia Stapf

F.P.S. 3: 416 (1956); F.T.E.A. Gramin. 2: 340 (1974); F.J.M.: 170 (1976); F.E.E. 7: 168 (1995); Biol. Skr. 51: 553 (2005). Tufted annual. Bushland, woodland & wooded grassland.
Distr: DAR, KOR, UN, EQU; Mauritania to Ethiopia, S to Tanzania.
DAR: Wickens 2304 8/1964; **UN:** Sherif A.3945 8/1951.

Chloris pilosa Schumach. & Thonn.

F.P.S. 3: 418 (1956); F.T.E.A. Gramin. 2: 345 (1974); F.J.M.: 170 (1976); F.E.E. 7: 169 (1995); Biol. Skr. 51: 553 (2005). Annual. Bushland & disturbed ground.
Distr: DAR, UN, EQU; Mauritania to Eritrea, S to Angola & Malawi.
DAR: Blair 133 8/1963; **UN:** Sherif A.2862 10/1951.

Chloris pycnothrix Trin.

F.P.S. 3: 418 (1956); F.T.E.A. Gramin. 2: 340 (1974); F.J.M.: 170 (1976); F.E.E. 7: 169 (1995); Biol. Skr. 51: 553 (2005). Tufted annual. Woodland, bushland & grassland, disturbed ground.
Distr: NS, DAR, CS, EQU; Senegal to Somalia, S to South Africa; Arabia, tropical America.
DAR: Blair 18 5/1963; **EQU:** Jackson 4260 6/1961.

Chloris robusta Stapf

F.P.S. 3: 416 (1956); F.T.E.A. Gramin. 2: 342 (1974); F.J.M.: 170 (1976); Biol. Skr. 51: 554 (2005). Perennial. Sandy riverbeds in woodland & wooded grassland.
Distr: DAR, EQU; Senegal to Uganda.
DAR: Blair 206 1/1964; **EQU:** Myers 8496 2/1938.

Chloris roxburghiana Schult.

F.T.E.A. Gramin. 2: 339 (1974); F.E.E. 7: 168 (1995); Biol. Skr. 51: 554 (2005).
Tufted perennial. Woodland, bushland, disturbed ground.
Distr: EQU; Eritrea & Somalia, S to Angola & South Africa; Madagascar, Yemen, India, Myanmar.
EQU: Myers 11236 2/1938.

Chloris virgata Sw.

F.P.S. 3: 418 (1956); F.T.E.A. Gramin. 2: 343 (1974); F.J.M.: 170 (1976); F.E.E. 7: 168 (1995); Biol. Skr. 51: 554 (2005). Tufted annual. Wooded grassland, bushland, disturbed ground.
Distr: NS, RS, DAR, KOR, CS, ES, UN, EQU; pantropical & warm temperate.
DAR: Wickens 1786 6/1964; **EQU:** Myers 14015 9/1941.

Chrysopogon nigritanus (Benth.) Veldkamp

Austrobaileya 5: 526 (1999).
Syn: *Vetiveria nigritana* (Benth.) Stapf – F.P.S. 3: 557 (1956); F.T.E.A. Gramin. 3: 739 (1982).
Tufted perennial. Riverbanks, floodplains.
Distr: KOR, UN, EQU; Mauritania to Kenya, S to Namibia, Botswana & Mozambique; scattered in tropical Asia.
KOR: Andrews 200 11/1933; **EQU:** Myers 7094 7/1937.

Chrysopogon plumulosus Hochst.

F.T.E.A. Gramin. 3: 737 (1982); F.E.E. 7: 303 (1995); Biol. Skr. 51: 554 (2005).
Syn: *Chrysopogon aucheri* (Boiss.) Stapf var. *quinqueplumis* (A.Rich.) Stapf – F.P.S. 3: 420 (1956).

Perennial. Grassland, dry bushland, semi-desert.
Distr: RS, DAR, KOR, EQU; Niger to Somalia, S to Tanzania; Arabia.
KOR: Harrison 957 9/1950; **EQU:** Myers 14982 9/1941.

Cleistachne sorghoides Benth.

F.T.E.A. Gramin. 3: 734 (1982); F.E.E. 7: 303 (1995). Annual with prop roots. Abandoned cultivations, riverbanks.
Distr: ?Sudan (unplaced); D.R. Congo to Ethiopia, S to South Africa; Oman, India.
Note: recorded in both F.E.E. and F.T.E.A. as occuring in 'Sudan' but we have seen no material.

Coelachyrum brevifolium Hochst. & Nees

F.P.S. 3: 420 (1956); F.E.E. 7: 133 (1995).
Tufted annual. Semi-desert, sandy coastal plains.
Distr: RS, CS; Mauritania to Somalia; Egypt, Arabia.
RS: Farag 34 1/2010.

Coelorachis afraurita (Stapf) Stapf

F.P.S. 3: 420 (1956); F.T.E.A. Gramin. 3: 842 (1982); F.E.E. 7: 361 (1995).
Tufted perennial. Swampy grassland.
Distr: EQU; Senegal to Ethiopia, S to Angola, Zimbabwe & Mozambique.
EQU: Myers 6933 7/1937.

Crinipes longifolius C.E.Hubb.

F.P.S. 3: 421 (1956); F.T.E.A. Gramin. 1: 130 (1970); F.E.E. 7: 68 (1995); Biol. Skr. 51: 554 (2005).
Tufted perennial. Grassland, often in moist rock crevices.
Distr: EQU; Ethiopia, Uganda.
IUCN: RD
EQU: Jackson 1081 1/1950.

Crypsis vaginiflora (Forssk.) Opiz

F.E.E. 7: 142 (1995) F.J.U. 1: 283 (1997).
Syn: *Crypsis schoenoides* sensu auctt., non (L.) Lam. – F.P.S. 3: 421 (1956); F.T.E.A. Gramin. 2: 353 (1974).
Syn: *Heleochloa schoenoides* sensu Drar, non (L.) Host – Bot. Exp. Sud.: 56 (1970).
Mat forming annual. Seasonally flooded muds by rivers & lakes.
Distr: NS, CS; Senegal, Mauritania, Eritrea, Ethiopia, Somalia, Tanzania, Mozambique; Madagascar, N Africa to India, introduced in U.S.A.
NS: Léonard 5027 12/1968.

Ctenium elegans Kunth

F.P.S. 3: 421 (1956); F.W.T.A. 3(2): 398 (1972).
Tufted annual. Dry sandy soils.
Distr: DAR, KOR, BAG; Mauritania to Chad, S to Congo-Brazzaville; Saudi Arabia.
KOR: Wickens 646 10/1962; **BAG:** Chevalier 10069 10/1903.

Ctenium newtonii Hack.

F.T.E.A. Gramin. 2: 324 (1974); F.J.M.: 170 (1976).
Syn: *Ctenium schweinfurthii* Pilg. – F.P.S. 3: 421 (1956). Tufted perennial. Bushland on shallow soils.
Distr: DAR, BAG, EQU; Senegal to Uganda, S to Angola & Zambia.
DAR: Wickens 2784 10/1964; **BAG:** Myers 7331 7/1937.

Ctenium somalense (Chiov.) Chiov.

F.T.E.A. Gramin. 2: 324 (1974); F.J.M.: 171 (1976); F.E.E. 7: 163 (1995).
Tufted perennial. Bushland & grassland.
Distr: DAR; Ethiopia & Somalia, S to Angola & Zimbabwe; Madagascar.
DAR: Jackson 3328 12/1954.

Cymbopogon caesius (Nees ex Hook. & Arn.) Stapf

F.P.S. 3: 423 (1956); F.J.M.: 171 (1976); F.T.E.A. Gramin. 3: 761 (1982); F.E.E. 7: 328 (1995); Biol. Skr. 51: 555 (2005).
Syn: Cymbopogon excavatus (Hochst.) Stapf ex Burtt Davy – F.P.S. 3: 423 (1956); F.J.M.: 171 (1976).
Syn: Cymbopogon afronardus sensu Wickens, non Stapf – F.J.M.: 171 (1976).
Tufted perennial. Bushland, woodland & grassland.
Distr: RS, DAR, CS, UN, EQU; Ethiopia & Somalia, S to Namibia & South Africa; Madagascar, Yemen, Pakistan to Vietnam.
DAR: Wickens 1932 7/1964; **EQU:** Davies 20 1952.

Cymbopogon commutatus (Steud.) Stapf

F.P.S. 3: 422 (1956); F.J.M.: 171 (1976); F.T.E.A. Gramin. 3: 766 (1982); F.E.E. 7: 330 (1995).
Tufted perennial. Bushland, semi-desert grassland.
Distr: DAR; Mauritania, Senegal, Chad to Somalia, S to Tanzania; Arabia to India.
DAR: Wickens 2581 9/1964.

Cymbopogon giganteus Chiov.

F.P.S. 3: 425 (1956); F.J.M.: 171 (1976); F.T.E.A. Gramin. 3: 763 (1982); F.E.E. 7: 329 (1995); Biol. Skr. 51: 555 (2005).
Tufted perennial. Bushland & wooded grassland.
Distr: DAR, KOR, BAG, UN, EQU; Mauritania to Somalia, S to South Africa.
DAR: Robertson 8 11/1957; **EQU:** Friis & Vollesen 674 12/1980.

Cymbopogon nardus (L.) Rendle

F.T.E.A. Gramin. 3: 764 (1982); F.E.E. 7: 328 (1995); Biol. Skr. 51: 555 (2005).
Tufted perennial. Upland grassland & bushland.
Distr: EQU; Nigeria, Uganda & Kenya, S to Angola & South Africa; Madagascar, Egypt, India to SE Asia, scattered in the neotropics.
EQU: Jackson 1088 1/1950.

Cymbopogon nervatus (Hochst.) Chiov.

F.P.S. 3: 423 (1956); F.J.M.: 171 (1976); F.E.E. 7: 328 (1995).
Annual, sometimes tufted. Wooded grassland, open plains.
Distr: KOR, CS; Eritrea (see note).
KOR: Wickens 841 4/1962.
Note: in the e-Monocot database, this species is given a wider distribution, from C.A.R. to Somalia and Tanzania, Arabia and SE Asia.

Cymbopogon schoenanthus (L.) Spreng. subsp. proximus (Hochst. ex A.Rich.) Maire & Weiller

F.J.M.: 171 (1976); F.T.E.A. Gramin. 3: 765 (1982); F.E.E. 7: 329 (1995).

Syn: Cymbopogon proximus (Hochst. ex A.Rich.) Chiov. – F.P.S. 3: 422 (1956).
Syn: Cymbopogon proximus (Hochst. ex A.Rich.) Chiov. var. **sennarensis** (Hochst.) Drar – Bot. Exp. Sud.: 53 (1970).
Syn: Cymbopogon sennarensis (Hochst.) Chiov. – F.P.S. 3: 422 (1956); F.J.M.: 171 (1976).
Tufted perennial. Semi-desert bushland & open ground.
Distr: RS, DAR, KOR, CS, EQU; Mauritania to Eritrea, S to Kenya; Egypt.
KOR: Wickens 718 10/1962; **EQU:** Padwa 266 6/1953.

Cynodon aethiopicus Clayton & Harlan

F.T.E.A. Gramin. 2: 319 (1974); F.E.E. 7: 176 (1995).
Stoloniferous, mat-forming perennial. Streamsides, disturbed ground including abandoned cultivations.
Distr: KOR, CS; Ethiopia to South Africa.
KOR: Wickens 794 11/1962.

Cynodon dactylon (L.) Pers.

F.P.S. 3: 425 (1956); F.T.E.A. Gramin. 2: 318 (1974); F.J.M.: 171 (1976); F.E.E. 7: 175 (1995); Biol. Skr. 51: 556 (2005).
Rhizomatous mat-forming perennial. Weed of disturbed ground.
Distr: NS, RS, DAR, KOR, CS, BAG, UN, EQU; pantropical & warm temperate.
CS: Farag 174 7/2010; **EQU:** Jackson 755 5/1949.

Cynodon nlemfuensis Vanderyst var. nlemfuensis

F.T.E.A. Gramin. 2: 319 (1974); F.E.E. 7: 175 (1995).
Syn: Cynodon nlemfuensis Vanderyst var. **robustus** Clayton & Harlan – F.T.E.A. Gramin. 2: 319 (1974); F.E.E. 7: 176 (1995); Biol. Skr. 51: 556 (2005).
Stoloniferous mat-forming perennial. Bushland & grassland, seasonally flooded areas, disturbed ground.
Distr: EQU; Ethiopia & Somalia, S to Zimbabwe; introduced elsewhere in the tropics.
EQU: Peers 23 8/1953.

Cynodon transvaalensis Burtt Davy

F.T.E.A. Gramin. 2: 317 (1974); F.E.E. 7: 175 (1995).
Rhizomatous and stoloniferous mat-forming perennial. Weed of disturbed ground.
Distr: EQU; native to South Africa, cultivated elsewhere as a lawn grass and often naturalised.
EQU: Cartwright 6 1921.

Cyrtococcum chaetophoron (Roem. & Schult.) Dandy

F.P.S. 3: 427 (1956); F.T.E.A. Gramin. 3: 500 (1982); Biol. Skr. 51: 556 (2005).
Perennial. Forest.
Distr: EQU; Senegal to Uganda, S to Angola & Tanzania.
EQU: Myers 6514 5/1937.

Dactyloctenium aegyptium (L.) Willd.

F.P.S. 3: 427 (1956); F.T.E.A. Gramin. 2: 252 (1974); F.J.M.: 171 (1976); F.E.E. 7: 135 (1995); Biol. Skr. 51: 556 (2005).
Stoloniferous annual. Weed of grassland, woodland & disturbed ground.
Distr: NS, RS, DAR, KOR, CS, ES, BAG, UN, EQU; palaeotropical; introduced in the neotropics.
CS: Farag 156 7/2010; **UN:** Sherif A.3990 8/1951.

Dactyloctenium aristatum Link

F.P.S. 3: 427 (1956); F.T.E.A. Gramin. 2: 254 (1974); F.E.E. 7: 135 (1995).
Tufted annual. Coastal sands.
Distr: RS; Eritrea, Somalia, Kenya; Arabia to Pakistan & India.
RS: <u>Carter 1906</u> 11/1987.

Dactyloctenium germinatum Hack.

F.T.E.A. Gramin. 2: 255 (1974); F.E.E. 7: 137 (1995).
Syn: *Dactyloctenium bogdanii* S.M.Phillips – F.T.E.A. Gramin. 2: 256 (1974).
Stoloniferous mat-forming perennial. Open grassland, foot of bare hillslopes.
Distr: EQU; Kenya to South Africa.
EQU: <u>Myers 14097</u> 9/1941.

Dactyloctenium scindicum Boiss.

F.P.S. 3: 427 (1956); F.T.E.A. Gramin. 2: 255 (1974); F.E.E. 7: 135 (1995).
Stoloniferous mat-forming perennial. Damp hollows in dry grassland & bushland.
Distr: RS; Eritrea, Ethiopia, Somalia, Kenya; Egypt, Arabia to India.
RS: <u>Schweinfurth 312</u> 9/1868.

Danthoniopsis barbata (Nees) C.E.Hubb.

F.P.S. 3: 429 (1956); F.E.E. 7: 285 (1995).
Tufted perennial. Rocky hillslopes, coastal sands, semi-desert plains.
Distr: RS; Eritrea, Ethiopia, Somalia; Egypt, Arabia.
RS: <u>Schweinfurth 584</u> 10/1868.

Desmostachya bipinnata (L.) Stapf

F.P.S. 3: 429 (1956); F.J.M.: 172 (1976); F.E.E. 7: 131 (1995).
Tufted perennial, tussock-forming. Margins of watercourses in dry areas, saltmarshes.
Distr: NS, RS, DAR, KOR, ES; Mauritania to Somalia; N Africa to Indo-China.
RS: <u>Andrews 59</u> 6/1935.

Dichanthium annulatum (Forssk.) Stapf var. *annulatum*

F.P.S. 3: 429 (1956); Bot. Exp. Sud.: 54 (1970); F.T.E.A. Gramin. 3: 725 (1982); F.E.E. 7: 308 (1995).
Syn: *Dichanthium papillosum* (Hochst. ex A.Rich.) Stapf – F.J.M.: 172 (1976).
Tufted perennial. Open, disturbed ground.
Distr: RS, KOR, CS, ES; palaeotropical; introduced in southern Africa & the neotropics.
CS: <u>de Wilde et al. 258</u> 2/1965.

Dichanthium annulatum (Forssk.) Stapf var. *papillosum* (A.Rich.) de Wet & Harlan

F.T.E.A. Gramin. 3: 725 (1982); F.E.E. 7: 309 (1995).
Perennial. Wet depressions.
Distr: DAR, KOR, EQU; Ethiopia & Somalia, S to Namibia & South Africa.
KOR: <u>Wickens 89</u> 7/1962; **EQU:** <u>Davies E.6</u> 4/1952.

Dichanthium foveolatum (Delile) Roberty

F.T.E.A. Gramin. 3: 723 (1982); F.E.E. 7: 308 (1995).
Syn: *Eremopogon foveolatus* (Delile) Stapf – F.P.S. 3: 458 (1956).

Tufted perennial. Dry grassland & open bushland.
Distr: RS; Mauritania to Somalia, S to Tanzania; N Africa, Arabia to Myanmar.
RS: <u>Jackson 2772</u> 4/1953.

Digitaria abyssinica (Hochst. ex A.Rich.) Stapf

F.J.M.: 172 (1976); F.T.E.A. Gramin. 3: 641 (1982); F.E.E. 7: 252 (1995); Biol. Skr. 51: 557 (2005).
Syn: *Digitaria vestita* Fig. & De Not. – F.P.S. 3: 435 (1956).
Rhizomatous mat-forming perennial. Upland grassland, bushland & forest clearings, disturbed areas.
Distr: DAR, ?CS, EQU; Nigeria to Somalia, S to South Africa; Madagascar, Arabia, Sri Lanka.
DAR: <u>Wickens 1889</u> 7/1964; **EQU:** <u>Myers 6491</u> 4/1937.

Digitaria acuminatissima Stapf

F.T.E.A. Gramin. 3: 650 (1982).
Syn: *Digitaria ciliaris* sensu Wickens pro parte, non (Retz.) Koeler – F.J.M.: 172 (1976).
Annual. Riversides, damp rocks.
Distr: DAR, UN, EQU; scattered from Mauritania to Somalia, S to Namibia, Botswana & Mozambique.
DAR: <u>Blair 124</u> 8/1963; **UN:** <u>Harrison 35</u> 6/1947.

Digitaria ciliaris (Retz.) Koeler

F.J.M.: 172 pro parte (1976); F.T.E.A. Gramin. 3: 653 (1982); F.E.E. 7: 256 (1995); Biol. Skr. 51: 557 (2005).
Syn: *Digitaria adscendens* (Kunth) Henrard – F.P.S. 3: 438 (1956).
Syn: *Digitaria adscendens* (Kunth) Henrard subsp. *chrysoblephara* (Fig. & De Not.) Henrard – F.P.S. 3: 439 (1956).
Syn: *Digitaria adscendens* (Kunth) Henrard subsp. *nubica* (Stapf) Henrard – F.P.S. 3: 439 (1956).
Annual. Woodland, bushland & grassland in disturbed areas.
Distr: NS, RS, DAR, KOR, CS, BAG, UN, EQU; pantropical & warm temperate.
RS: <u>Harrison 1274</u> 10/1951; **UN:** <u>Evans-Pritchard 71</u> 9/1936.

Digitaria debilis (Desf.) Willd.

F.P.S. 3: 433 (1956); F.T.E.A. Gramin. 3: 637 (1982); F.E.E. 7: 252 (1995).
Annual. Along water courses, damp ground.
Distr: ?KOR, CS, BAG, UN; throughout Africa; Madagascar, Mediterranean.
CS: <u>Broun 801</u> 1/1906; **UN:** <u>J.M. Lock 81/168</u> 8/1981.

Digitaria diagonalis (Nees) Stapf

F.J.M.: 172 (1976); F.T.E.A. 3: 624 (1982); F.E.E. 7: 247 (1995); Biol. Skr. 51: 557 (2005).
Syn: *Digitaria diagonalis* (Nees) Stapf var. *hirsuta* (De Wild. & T.Durand) Troupin – F.T.E.A. Gramin. 3: 626 (1982); F.E.E. 7: 247 (1995); Biol. Skr. 51: 557 (2005).
Syn: *Digitaria diagonalis* (Nees) Stapf var. *uniglumis* (A.Rich.) Pilg. – F.T.E.A. Gramin. 3: 626 (1982); F.E.E. 7: 247 (1995); Biol. Skr. 51: 557 (2005).
Syn: *Digitaria uniglumis* (A.Rich.) Stapf – F.P.S. 3: 433 (1956).
Tufted perennial. Open grassy areas in both waterlogged soils & rocky hillslopes.

Distr: DAR, EQU; Senegal to Eritrea, S to Namibia & South Africa; Yemen.
DAR: Blair 35 6/1963; **EQU:** Friis & Vollesen 1314 10/1982.

Digitaria gayana (Kunth) A.Chev.

F.P.S. 3: 435 (1956); F.J.M.: 172 (1976); F.T.E.A. Gramin. 3: 627 (1982); F.E.E. 7: 248 (1995); Biol. Skr. 51: 558 (2005).
Tufted annual. Bushland, woodland, disturbed areas & abandoned cultivations.
Distr: DAR, KOR, BAG, UN; Mauritania to Kenya, S to Namibia, Botswana & Zimbabwe.
KOR: Harrison 990 8/1950; **BAG:** Myers 7471 8/1937.

Digitaria gazensis Rendle

F.J.M.: 172 (1976); F.T.E.A. Gramin. 3: 643 (1982); F.E.E. 7: 253 (1995).
Tufted perennial. Upland grassland & forest clearings, wooded grassland.
Distr: DAR; Ethiopia to Namibia & Mozambique; Madagascar.
DAR: Wickens 1871 7/1964.

Digitaria intecta Stapf

F.P.S. 3: 433 (1956); F.T.E.A. Gramin. 3: 623 (1982); F.E.E. 7: 247 (1995).
Annual. Bushland, grassland on black clay soils.
Distr: ES; Eritrea, Ethiopia, Uganda, Zambia.
ES: Schweinfurth 1156 1865.

Digitaria leptorhachis (Pilg.) Stapf

F.T.E.A. Gramin. 3: 644 (1982).
Annual or short-lived perennial. Disturbed areas.
Distr: DAR; Senegal to Chad, S to Zambia & Tanzania.
DAR: Blair 10 5/1963.

Digitaria longiflora (Retz.) Pers.

F.P.S. 3: 435 (1956); F.J.M.: 172 (1976); F.T.E.A. Gramin. 3: 634 (1982); F.E.E. 7: 249 (1995).
Syn: Digitaria argyrotricha sensu Andrews, non (Andersson) Chiov. – F.P.S. 3: 434 (1956).
Stoloniferous annual or short-lived perennial. Open bushland, disturbed areas.
Distr: DAR, KOR, UN, EQU; palaeotropical; introduced in the neotropics.
DAR: Wickens 1976 4/1964; **EQU:** Myers 6908 6/1937.

Digitaria macroblephara (Hack.) Paoli

F.P.S. 3: 438 (1956); F.T.E.A. Gramin. 3: 646 (1982); F.E.E. 7: 254 (1995).
Tufted perennial. Open bushland.
Distr: EQU; Ethiopia, Somalia, Uganda, Kenya, Tanzania.
EQU: Myers 13993 9/1941.

Digitaria nodosa Parl.

F.P.S. 3: 438 (1956); F.T.E.A. Gramin. 3: 647 (1982).
Tufted perennial. Bushland, often on black clay soils.
Distr: ?EQU; Cape Verde Is., Mauritania, Eritrea & Somalia to Tanzania; NW Africa, Arabia to India.
Note: Andrews records this species from "Equatoria: Alakapoy near Kapoeta" but we have not located the specimen on which this record is based.

Digitaria nuda Schumach.

F.T.E.A. Gramin. 3: 654 (1982); F.E.E. 7: 256 (1995); Biol. Skr. 51: 558 (2005).
Syn: Digitaria horizontalis sensu auctt., non Willd. – F.P.S. 3: 436 (1956); F.J.M.: 172 (1976).
Annual. Open disturbed ground.
Distr: DAR, EQU; Guinea-Bissau, Ghana, Cameroon to Ethiopia, S to South Africa; Madagascar, Comoros, Mauritius, SE Asia, neotropics.
DAR: Wickens 1329 1964; **EQU:** Myers 6341 3/1937.

Digitaria pennata (Hochst.) T.Cooke

F.P.S. 3: 433 (1956); F.T.E.A. Gramin. 3: 636 (1982); F.E.E. 7: 251 (1995).
Tufted perennial. Dry bushland.
Distr: ?RS; Eritrea & Somalia, S to Tanzania; Arabia, Pakistan, India.
Note: Andrews records this species from Red Sea District; we have seen no material with which to verify this record.

Digitaria sanguinalis (L.) Scop.

F.T.E.A. Gramin. 3: 650 (1982).
Syn: Digitaria sanguinalis (L.) Scop. subsp. aegyptiaca (Willd.) Henrard – F.P.S. 3: 438 (1956); Bot. Exp. Sud.: 54 (1970).
Annual. Weed of disturbed areas.
Distr: RS, DAR, CS, EQU; warm temperate & subtropical regions throughout the world.
DAR: Wickens 2463 9/1964; **EQU:** Myers 7544 8/1937.

Digitaria ternata (A.Rich.) Stapf

F.P.S. 3: 434 (1956); F.J.M.: 173 (1976); F.T.E.A. Gramin. 3: 630 (1982); F.E.E. 7: 284 (1995); Biol. Skr. 51: 558 (2005).
Tufted annual. Weed of disturbed areas & abandoned cultivations.
Distr: DAR, ?CS, UN, EQU; palaeotropical.
DAR: Wickens 2595 9/1964; **EQU:** Beshir 4 8/1951.
Note: Drar (1970) recorded D. tricostulata (Hack.) Henrard from Erkowit; this is now treated as a synonym of D. thouaresiana (Flüggé) A.Camus which occurs from Cameroon to Kenya and southwards and so is unlikely to occur in Sudan. D. thouaresiana is easily confused with D. ternata, which may therefore be the true identity of Drar's plant.

Digitaria velutina (Forssk.) P.Beauv.

F.P.S. 3: 435 (1956); F.J.M.: 173 (1976); F.T.E.A. Gramin. 3: 652 (1982); F.E.E. 7: 254 (1995); Biol. Skr. 51: 558 (2005).
Annual. Weed of disturbed ground.
Distr: RS, DAR, KOR, CS, BAG, EQU; Eritrea & Somalia, S to Namibia & South Africa; Madagascar, Egypt, Arabia.
RS: Harrison 1273 10/1951; **EQU:** Friis & Vollesen 1030 2/1982.

Digitaria xanthotricha (Hack.) Stapf

F.P.S. 3: 434 (1956).
Annual. Open wooded grassland, roadsides.
Distr: BAG, UN; C.A.R.
IUCN: RD
BAG: Myers 13967 (fr) 8/1941.

Diheteropogon amplectens (Nees) Clayton var. catangensis (Chiov.) Clayton

F.J.M.: 173 (1976); F.T.E.A. Gramin. 3: 784 (1982).
Shortly rhizomatous perennial. Bushland.
Distr: DAR; Mauritania to Kenya, S to South Africa; Madagascar.
DAR: Blair 182 10/1963.
Note: Wickens records both var. *amplectens* and var. *catangensis* from Jebel Marra, but it seems that only the latter truly occurs there.

Dinebra retroflexa (Vahl) Panz. var. retroflexa

F.P.S. 3: 439 (1956); F.T.E.A. Gramin. 2: 273 (1974); F.E.E. 3: 105 (1989).
Tufted annual. Seasonally wet grassland & wooded grassland.
Distr: NS, RS, DAR, KOR, CS, ES, UN, EQU; Mauritania to Ethiopia, S to Tanzania; Egypt to E Asia & Australia.
RS: Farag 68 2/2010; **UN:** Harrison 12 6/1947.

Drake-brockmania somalensis Stapf

F.P.S. 3: 441 (1956); F.T.E.A. Gramin. 2: 184 (1974); F.E.E. 7: 109 (1995).
Mat-forming annual. Bare seasonally flooded areas.
Distr: RS, DAR; Ethiopia, Djibouti, Somalia, Kenya, Tanzania; Arabia.
DAR: Massey s.n. 12/1912.

Echinochloa callopus (Pilg.) Clayton

F.T.E.A. Gramin. 3: 555 (1982).
Syn: *Brachiaria callopus* (Pilg.) Stapf – F.P.S. 3: 408 (1956).
Tufted annual. Swampy ground.
Distr: BAG, EQU; Senegal to D.R. Congo & Tanzania.
BAG: Harrison 1097 11/1950.

Echinochloa colona (L.) Link

F.P.S. 3: 441 (1956); F.J.M.: 173 (1976); F.T.E.A. Gramin. 3: 557 (1982); F.E.E. 7: 212 (1995).
Tufted annual. Bare damp & swampy ground.
Distr: NS, RS, DAR, KOR, CS, ES, BAG, UN, EQU; pantropical & warm temperate.
CS: Farag 80 2/2010; **BAG:** Myers 7418 7/1937.

Echinochloa crus-galli (L.) P.Beauv.

F.J.M.: 173 (1976); F.T.E.A. Gramin. 3: 557 (1982).
Annual. Weed of moist sandy ground.
Distr: DAR; widespread in temperate & tropical regions but scattered in tropical Africa.
DAR: Wabuyele 2032 11/1964.

Echinochloa crus-pavonis (Kunth) Schult.

F.P.S. 3: 441 (1956); F.T.E.A. Gramin. 3: 556 (1982); F.E.E. 7: 213 (1995).
Perennial. Wet grassland, swamps, streamsides.
Distr: EQU; widespread in the tropics.
EQU: Sillitoe 3 1919.

Echinochloa haploclada (Stapf) Stapf

F.T.E.A. Gramin. 3: 560 (1982); F.E.E. 7: 212 (1995); Biol. Skr. 51: 559 (2005).
Rhizomatous perennial. Wet grassland, streambanks & dry streambeds.
Distr: UN, EQU; Ethiopia & Somalia, S to Angola & South Africa.

EQU: Harrison 465 12/1948.

Echinochloa obtusiflora Stapf

F.P.S. 3: 441 (1956); F.W.T.A. 3(2): 439 (1972).
Tufted annual. Marshy grassland, shallow pools.
Distr: KOR; Senegal, Nigeria, Niger, Cameroon, Chad.
KOR: Andrews 3139 10/1946.

Echinochloa pyramidalis (Lam.) Hitchc. & Chase

F.P.S. 3: 443 (1956); Bot. Exp. Sud.: 54 (1970); F.J.M.: 173 (1976); F.T.E.A. Gramin. 3: 561 (1982); F.E.E. 7: 213 (1995).
Syn: *Echinochloa frumentacea* sensu Wickens, non Link – F.J.M.: 173 (1976).
Rhizomatous perennial. Shallow pools, swamps, riversides.
Distr: DAR, KOR, BAG, UN, EQU; Mauritania to Somalia, S to Namibia & South Africa; Madagascar, Egypt, Arabia.
DAR: Wickens 1223 8/1964; **UN:** J.M. Lock 80/13 2/1980.

Echinochloa rotundiflora Clayton

F.E.E. 7: 212 (1995).
Syn: *Brachiaria obtusiflora* (Hochst.) Stapf – F.P.S. 3: 408 (1956).
Annual. Marshy grassland, shallow watercourses.
Distr: RS, KOR, CS, ES, UN, EQU; Nigeria, Cameroon, Eritrea, Ethiopia.
ES: Beshir 132 4/1951; **UN:** Sherif A.4010 9/1951.

Echinochloa stagnina (Retz.) P.Beauv.

F.P.S. 3: 443 (1956); Bot. Exp. Sud.: 54 (1970); F.E.E. 7: 215 (1995).
Floating rhizomatous perennial or annual. Marshes.
Distr: DAR, KOR, CS, BAG, UN, EQU; Mauritania to Somalia, S to Namibia & South Africa; Madagascar, India to New Guinea.
DAR: Harrison 209 11/1947; **EQU:** Simpson 7282 7/1929.

Ehrharta erecta Lam.

F.T.E.A. Gramin. 1: 38 (1970); F.E.E. 7: 12 (1995); Biol. Skr. 51: 559 (2005).
Syn: *Ehrharta abyssinica* Hochst. – F.P.S. 3: 445 (1956).
Syn: *Ehrharta erecta* Lam. var. *abyssinica* (Hochst.) Pilg. – Biol. Skr. 51: 559 (2005).
Tufted perennial. Upland forest clearings & bushland.
Distr: EQU; Eritrea & Somalia to South Africa; Arabia, India to Australia.
EQU: Friis & Vollesen 856 12/1980.

Eleusine africana Kenn.-O'Byrne

F.E.E. 7: 139 (1995).
Syn: *Eleusine indica* L. subsp. *africana* (Kenn.-O'Byrne) S.M.Phillips – F.T.E.A. Gramin. 2: 263 (1974); F.J.M.: 173 (1976); Biol. Skr. 51: 560 (2005).
Syn: *Eleusine indica* sensu auctt., non L. – F.P.S. 3: 445 pro parte (1956); F. Darfur Nord-Occ. & J. Gourgeil: 138 (1969); Bot. Exp. Sud.: 54 ?pro parte (1970).
Tufted annual. Weed of disturbed & cultivated ground.
Distr: DAR, KOR, CS, ES, UN, EQU; Senegal to Somalia, S to Namibia & South Africa; Madagascar, Egypt, Arabia.
KOR: Wickens 824 11/1962; **UN:** Sherif A.3980 8/1951.

Eleusine coracana (L.) Gaertn.

F.T.E.A. Gramin. 2: 260 (1974); F.E.E. 7: 139 (1995); Biol. Skr. 51: 559 (2005).
Tufted annual. Cultivated & occasionally naturalised.
Distr: DAR, UN, EQU; cultivated throughout tropical Africa & elsewhere in the tropics.
DAR: Wickens 1741 6/1964; **EQU:** Simpson 7351 7/1929.
Note: the cultivated 'finger millet'.

Eleusine indica (L.) Gaertn.

F.P.S. 3: 445 pro parte (1956); F.T.E.A. Gramin. 2: 262 (1974); F.E.E. 7: 141 (1995); Biol. Skr. 51: 560 (2005).
Tufted annual. Weed of disturbed ground & cultivation.
Distr: UN, EQU; pantropical & subtropical.
UN: J.M. Lock 81/191 8/1981.

Eleusine multiflora Hochst. ex A.Rich.

F.T.E.A. Gramin. 2: 261 (1974); F.E.E. 7: 139 (1995).
Tufted annual. Bushland & grassland, disturbed ground.
Distr: RS; Eritrea, Ethiopia, Kenya, Tanzania, South Africa; Arabia.
RS: Carter 1783 11/1987.

Elionurus hirtifolius Hack.

F.P.S. 3: 447 (1956); F.W.T.A. 3(2): 504 (1972); F.J.M.: 173 (1976).
Tufted perennial. Grassland amongst rocks, woodland, plantations.
Distr: DAR, KOR, ?EQU; Guinea Bissau to Chad, S to Congo-Brazzaville & C.A.R.
DAR: Blair 220 1/1964.
Note: Andrews records this species from Equatoria but we have seen no material from South Sudan.

Elionurus muticus (Spreng.) Kuntze

F.T.E.A. Gramin. 3: 837 (1982); F.E.E. 7: 358 (1995); Biol. Skr. 51: 560 (2005).
Tufted perennial. Upland bushland & grassland.
Distr: EQU; Guinea to Somalia, S to Namibia & South Africa; Yemen, neotropics.
EQU: Wilson 834 2/1960.
Note: the only specimens from our region are from Mt Lonyili on the Uganda-South Sudan border and may have been collected from the Ugandan side.

Elionurus royleanus Nees ex A.Rich.

F.P.S. 3: 447 (1956); F.T.E.A. Gramin. 3: 835 (1982); F.E.E. 7: 359 (1995).
Annual. Semi-desert, dry open ground.
Distr: RS, KOR, CS; Cape Verde Is., Mauritania, Senegal, Eritrea, Ethiopia, Djibouti, Somalia, Kenya; Morocco, Egypt, Arabia to India.
CS: Ferguson 356 4/1948.

Elytrophorus spicatus (Willd.) A.Camus

F.P.S. 3: 447 (1956); F.T.E.A. Gramin. 1: 135 (1970); F.E.E. 7: 66 (1995).
Annual. Grassland.
Distr: DAR, KOR, BAG, UN; Mauritania to Eritrea, S to Namibia & Botswana; India to Australia.
KOR: Jackson 2487 1/1953; **BAG:** Harrison 1050 10/1950.

Enneapogon cenchroides (Licht. ex Roem. & Schult.) C.E.Hubb.

F.P.S. 3: 448 (1956); F.T.E.A. Gramin. 1: 169 (1970); F.E.E. 7: 89 (1995).
Tufted annual or short-lived perennial. Grassland.
Distr: DAR, KOR, CS; Niger to Somalia, S to Namibia & South Africa; Madagascar, Arabia to India, Ascension Is.
CS: Farag 160 7/2010.

Enneapogon desvauxii P.Beauv.

F.T.E.A. Gramin. 1: 167 pro parte (1970); F.E.E. 7: 88 (1995).
Syn: Enneapogon brachystachyus (Jaub. & Spach) Stapf – F.P.S. 3: 447 (1956).
Tufted annual or perennial. Bushland, semi-desert.
Distr: NS, RS, DAR, CS; widespread in the tropics & temperate regions.
CS: Jackson 3225 11/1954.

Enneapogon lophotrichus Chiov. ex H.Scholz & P.Konig

F.E.E. 7: 91 (1995).
Syn: Enneapogon desvauxii sensu Clayton pro parte, non P.Beauv. – F.T.E.A. Gramin. 1: 167 (1970).
Syn: Enneapogon elegans sensu auctt., non (Nees) Stapf – F.P.S. 3: 448 (1956); F. Darfur Nord-Occ. & J. Gourgeil: 138 (1969).
Tufted perennial. Semi-desert, dry open bushland.
Distr: RS, DAR, KOR, CS, ES; Chad, Eritrea, Ethiopia, Djibouti, Somalia, Kenya; Arabia.
CS: Farag 165 7/2010.
Note: this is treated as a synonym of E. desvauxii in the e-Monocot database.

Enneapogon persicus Boiss.

F.E.E. 7: 89 (1995).
Syn: Enneapogon glumosus (Hochst.) Maire & Weiller – F. Darfur Nord-Occ. & J. Gourgeil: 138 (1969).
Syn: Enneapogon schimperianus (Hochst. ex A.Rich.) Renvoize – F.T.E.A. Gramin. 1: 169 (1970).
Tufted perennial. Semi-desert, dry bushland & grassland.
Distr: DAR, KOR, CS; Mali to Somalia, S to Tanzania; Spain, Algeria, Egypt, Arabia to Burma.
KOR: Harrison 1408 9/1950.

Enteropogon macrostachyus (Hochst. ex A.Rich.) Munro ex Benth.

F.P.S. 3: 448 (1956); F.T.E.A. Gramin. 2: 332 (1974); F.J.M.: 173 (1976); F.E.E. 7: 172 (1995); Biol. Skr. 51: 560 (2005).
Tufted perennial. Bushland, disturbed areas.
Distr: DAR, EQU; Mauritania to Somalia, S to Namibia & South Africa; Arabia.
DAR: Blair 347 9/1964; **EQU:** Beshir 18 8/1951.

Enteropogon prieurii (Kunth) Clayton

F.E.E. 7: 171 (1995).
Syn: Chloris prieurii Kunth – F.P.S. 3: 419 (1956); F.T.E.A. Gramin. 2: 342 (1974); F.J.M.: 170 (1976).
Annual. Grassland.
Distr: NS, RS, DAR, KOR, CS; Cape Verde Is., Mauritania to Eritrea, Tanzania, Namibia; Macaronesia, Arabia, India.
DAR: Blair 580 8/1963.

Enteropogon rupestris (J.A.Schmidt) A.Chev.

F.T.E.A. Gramin. 2: 332 (1974); F.E.E. 7: 172 (1995).
Tufted perennial. Grassland.
Distr: DAR; Cape Verde Is., Mauritania to Somalia, S to Namibia & Botswana; Morocco.
DAR: Pfund 580 9/1875.

Eragrostis aegyptiaca (Willd.) Delile

F.P.S. 3: 454 (1956); F.W.T.A. 3(2): 391 (1972).
Tufted annual. Riverbanks.
Distr: NS, RS, CS, ES; Mauritania to Chad; E Europe to C Asia, N Africa, Arabia to Iran.
RS: Jackson 4086 12/1960.

Eragrostis amabilis (L.) Wight & Arn.

Syn: *Eragrostis tenella* (L.) Roem. & Schult. – F.T.E.A. Gramin. 2: 206 (1974); F.J.M.: 175 (1976).
Tufted annual. Disturbed areas & cultivation.
Distr: DAR; pantropical & subtropical.
DAR: Wickens 2040 7/1964.

Eragrostis arenicola C.E.Hubb.

F.P.S. 3: 452 (1956); F.T.E.A. Gramin. 2: 207 (1974); F.J.M.: 174 (1976); Biol. Skr. 51: 561 (2005).
Tufted annual. Open sandy areas, disturbed ground.
Distr: DAR, EQU; Togo to Tanzania, S to Namibia & South Africa.
DAR: Blair 81 6/1963; **EQU:** Jackson 1138 2/1950.

Eragrostis aspera (Jacq.) Nees

F.P.S. 3: 452 (1956); F.T.E.A. Gramin. 2: 209 (1974); F.J.M.: 174 (1976); F.E.E. 7: 114 (1995); Biol. Skr. 51: 561 (2005).
Tufted annual. Disturbed areas & abandoned cultivations.
Distr: RS, DAR, KOR, BAG, UN, EQU; Mauritania to Somalia, S to Namibia & South Africa; Madagascar, Arabia, India to Malesia.
KOR: Wickens 762 11/1962; **BAG:** Harrison 1085 10/1950.

Eragrostis atrovirens (Desf.) Trin. ex Steud.

F.P.S. 3: 456 (1956); F.T.E.A. Gramin. 2: 217 (1974); F.J.M.: 174 (1976).
Tufted perennial. Swamps, wet grassland.
Distr: DAR, BAG, EQU; palaeotropical; introduced elsewhere.
DAR: Blair 148 8/1963; **BAG:** Myers 7283 7/1937.

Eragrostis barrelieri Daveau

F.P.S. 3: 454 (1956); F.E.E. 7: 121 (1995).
Tufted annual. Disturbed & cultivated ground, often in damp areas.
Distr: NS, RS; Mauritania to Somalia; Mediterranean, N Africa, Arabia to Pakistan, introduced elsewhere.
NS: Pettet 78 9/1962.

Eragrostis chapelieri (Kunth) Nees

F.P.S. 3: 454 (1956); F.T.E.A. Gramin. 2: 225 (1974); F.E.E. 7: 116 (1995).
Tufted perennial. Woodland & bushland margins & clearings.
Distr: EQU; Ethiopia to Namibia & South Africa.
EQU: Myers 8021 12/1937.

Eragrostis cilianensis (All.) Janch.

F.P.S. 3: 456 (1956); F.T.E.A. Gramin. 2: 232 (1974); F.J.M.: 174 (1976); F.E.E. 7: 119 (1995); Biol. Skr. 51: 561 (2005).
Tufted annual. Disturbed areas, overgrazed grassland & bushland.
Distr: NS, RS, DAR, KOR, CS, ES, BAG, UN, EQU; palaeotropical & warm temperate; introduced in the New World.
RS: Farag 79 2/2010; **EQU:** Myers 6894 6/1937.

Eragrostis ciliaris (L.) R.Br.

F.P.S. 3: 450 (1956); F.T.E.A. Gramin. 2: 204 (1974); F.J.M.: 174 (1976); F.E.E. 7: 113 (1995); Biol. Skr. 51: 561 (2005).
Syn: *Eragrostis ciliaris* (L.) R.Br. var. **brachystachya** Boiss. – F.P.S. 3: 451 (1956).
Tufted annual. Disturbed areas, overgrazed grassland.
Distr: NS, RS, DAR, KOR, CS, BAG, UN, EQU; pantropical.
RS: Farag 129 3/2010; **EQU:** Jackson 727 5/1949.

Eragrostis cylindriflora Hochst.

F.T.E.A. Gramin. 2: 239 (1974); F.J.M.: 174 (1976); F.E.E. 7: 125 (1995); Biol. Skr. 51: 562 (2005).
Tufted annual. Disturbed areas, overgrazed grassland.
Distr: DAR; Ghana, Niger, Eritrea to Namibia & South Africa; Madagascar, Yemen.
DAR: Blair 137 8/1963.

Eragrostis elegantissima Chiov.

F.E.E. 7: 118 (1995).
Tufted annual. Dry rocky areas.
Distr: KOR; Mali to Eritrea.
KOR: Wickens 448 9/1962.
Note: a widespread but scarce and rarely collected species.

Eragrostis gangetica (Roxb.) Steud.

F.P.S. 3: 454 (1956); F.T.E.A. Gramin. 2: 217 (1974); F.J.M.: 174 (1976); F.E.E. 7: 127 (1995).
Tufted annual. Disturbed areas.
Distr: DAR, BAG, UN; palaeotropical.
DAR: Blair 290 5/1964; **BAG:** Harrison 1077 10/1950.

Eragrostis hispida K.Schum.

F.P.S. 3: 452 (1956); F.T.E.A. Gramin. 2: 201 (1974); F.E.E. 7: 113 (1995); Biol. Skr. 51: 562 (2005).
Tufted perennial. Damp grassland, seepages over rock outcrops.
Distr: EQU; Ethiopia to Angola, Zimbabwe & Mozambique.
EQU: Friis & Vollesen 321 11/1980.

Eragrostis japonica (Thunb.) Trin.

F.E.E. 7: 114 (1995).
Syn: *Eragrostis namaquensis* Nees – F.P.S. 3: 452 (1956); F.T.E.A. Gramin. 2: 208 (1974); F.J.M.: 174 (1976).
Syn: *Eragrostis namaquensis* Nees var. **diplachnoides** (Steud.) Clayton – F.T.E.A. Gramin. 2: 209 (1974); F.J.M.: 174 (1976).
Syn: *Eragrostis diplachnoides* Steud. – F.P.S. 3: 452 (1956); Bot. Exp. Sud.: 55 (1970).

Tufted annual or short-lived perennial. Seasonally wet hollows, floodplains, streamsides.
Distr: NS, RS, DAR, KOR, CS, ES, BAG, UN, EQU; palaeotropical & subtropical; introduced in the neotropics.
DAR: Harrison 210 11/1947; **EQU:** Myers 8022 12/1937.

Eragrostis lepida (A.Rich.) Hochst. ex Steud.
F.E.E. 7: 113 (1995).
Syn: *Eragrostis tenella* sensu auctt., non (L.) P.Beauv. ex Roem. & Schult. – F.P.S. 3: 452 (1956); Bot. Exp. Sud.: 55 (1970); F.T.E.A. Gramin. 2: 206 pro parte (1974).
Annual. Open sandy ground.
Distr: RS; Eritrea, Somalia, Kenya; Arabia.
RS: Khalid 40 12/1951.

Eragrostis macilenta (A.Rich.) Steud.
F.P.S. 3: 454 (1956); F.T.E.A. Gramin. 2: 235 (1974); F.E.E. 7: 118 (1995); Biol. Skr. 51: 562 (2005).
Tufted annual. Bushland, disturbed ground in shade.
Distr: EQU; Ivory Coast to Eritrea, S to Angola, Zimbabwe & Mozambique; Madagascar, Yemen, India.
EQU: Friis & Vollesen 1033 2/1982.

Eragrostis minor Host
F.T.E.A. Gramin. 2: 234 (1974); F.E.E. 7: 121 (1995).
Tufted annual. Weed of disturbed ground.
Distr: RS; Ethiopia, subtropical & warm temperate regions of the Old World, introduced elsewhere.
RS: Harrison 1313 10/1951.

Eragrostis olivacea K.Schum.
F.T.E.A. Gramin. 2: 201 (1974); F.E.E. 7: 114 (1995); Biol. Skr. 51: 562 (2005).
Tufted perennial. Upland grassland & bushland, often in rocky areas.
Distr: EQU; Ethiopia to Zambia & Mozambique.
EQU: Jackson 1452 4/1950.

Eragrostis paniciformis (A.Braun) Steud.
F.P.S. 3: 458 (1956); F.T.E.A. Gramin. 2: 219 (1974); F.E.E. 7: 127 (1995).
Tufted perennial. Wet grassland, streamsides, wet areas in bushland.
Distr: EQU; Eritrea to Zambia.
EQU: Andrews 1080 5/1939.

Eragrostis papposa (Roem. & Schult.) Steud.
F.P.S. 3: 456 (1956); F.T.E.A. Gramin. 2: 239 (1974); F.E.E. 7: 122 (1995).
Tufted or cushion-forming short-lived perennial. Dry bushland & grassland.
Distr: RS; Chad to Somalia, S to Tanzania; Iberia, N Africa, Arabia to India & Myanmar.
RS: Jackson 2838 4/1953.

Eragrostis patula (Kunth) Steud.
Syn: *Eragrostis tenuifolia* (A.Rich.) Hochst. ex Steud. – F.P.S. 3: 456 (1956); Bot. Exp. Sud.: 55 (1970); F.T.E.A. Gramin. 2: 238 (1974); F.E.E. 7: 122 (1995); Biol. Skr. 51: 563 (2005).
Tufted perennial. Weed of disturbed areas.
Distr: EQU; pantropical & subtropical.
EQU: Andrews 1219 5/1939.

Eragrostis pilosa (L.) P.Beauv.
F.P.S. 3: 454 (1956); F.T.E.A. Gramin. 2: 214 (1974); F.J.M.: 175 (1976); F.E.E. 7: 125 (1995).
Tufted annual. Weed of disturbed ground & cultivation.
Distr: RS, DAR, KOR, CS, ES, BAG, UN, EQU; palaeotropical & warm temperate; introduced in the New World.
KOR: Harrison 902 8/1950; **UN:** Harrison 1053 10/1950.

Eragrostis porosa Nees
F. Darfur Nord-Occ. & J. Gourgeil: 139 (1969); F.T.E.A. Gramin. 2: 240 (1974); F.E.E. 7: 124 (1995).
Tufted annual. Rocky semi-desert & bushland.
Distr: DAR; Chad to Ethiopia & Kenya, Angola to Mozambique, S to South Africa.
Note: the record for this species is from Quézel (1969); we have not seen the specimen on which it is based.

Eragrostis racemosa (Thunb.) Steud.
F.P.S. 3: 456 (1956); F.T.E.A. Gramin. 2: 230 (1974); F.E.E. 7: 117 (1995); Biol. Skr. 51: 562 (2005).
Tufted perennial. A wide variety of habitats, often in disturbed areas.
Distr: EQU; Ethiopia to Angola & South Africa; Madagascar.
EQU: Friis & Vollesen 1219 3/1982.

Eragrostis schweinfurthii Chiov.
Biol. Skr. 51: 563 (2005).
Tufted annual or short-lived perennial. Upland forest & grassland, often in rocky areas.
Distr: EQU; Eritrea to D.R. Congo, Tanzania & Malawi; Yemen, Sri Lanka.
EQU: Friis & Vollesen 38 11/1980.

Eragrostis superba Peyr.
F.P.S. 3: 450 (1956); F.T.E.A. Gramin. 2: 211 (1974); F.E.E. 7: 112 (1995); Biol. Skr. 51: 563 (2005).
Tufted perennial. Bushland & wooded grassland.
Distr: BAG, EQU; Gabon to Somalia, S to Angola & South Africa; introduced elsewhere.
EQU: Myers 7918 11/1937.

Eragrostis tef (Zucc.) Trotter
F.E.E. 7: 125 (1995).
Tufted annual. Cultivated.
Distr: CS; extensively cultivated in Ethiopia as a staple cereal crop; widely cultivated elsewhere for fodder.
CS: Andrews K.28 1933.
Note: in Sudan this species has been cultivated at Gezira for fodder trials.

Eragrostis tremula Hochst. ex Steud.
F.P.S. 3: 456 (1956); F.T.E.A. Gramin. 2: 236 (1974); F.J.M.: 175 (1976); F.E.E. 7: 118 (1995); Biol. Skr. 51: 563 (2005).
Tufted annual. Weed of disturbed areas & cultivation.
Distr: DAR, KOR, CS, BAG, UN, EQU; Mauritania to Eritrea, S to Angola & South Africa; Arabia, Pakistan to Vietnam.
KOR: Jackson 4333 9/1961; **EQU:** Myers 6778 5/1937.

Eragrostis turgida (Schumach.) De Wild.
F.P.S. 3: 458 (1956); F.T.E.A. Gramin. 2: 223 (1974); F.J.M.: 175 (1976).
Tufted annual. Grassland.

Distr: DAR, BAG, UN, EQU; Mauritania to Uganda & D.R. Congo; Arabia.
DAR: Blair 172 9/1963; **BAG:** Myers 7287 7/1937.

Eriochloa fatmensis (Hochst. & Steud.) Clayton

F.J.M.: 175 (1976); F.T.E.A. Gramin. 3: 571 (1982); F.E.E. 7: 218 (1995); Biol. Skr. 51: 564 (2005).
Syn: *Eriochloa nubica* (Steud.) Hack. & Stapf ex Thell. – F.P.S. 3: 460 (1956); Bot. Exp. Sud.: 56 (1970).
Tufted annual. Wet grassland & damp depressions.
Distr: RS, DAR, CS, BAG, UN, EQU; Mauritania to Somalia, S to South Africa; Madagascar, Arabia, India to Myanmar.
DAR: Wickens 2291 8/1964; **UN:** Sherif A.3943 8/1951.

Eriochloa meyeriana (Nees) Pilg. subsp. *meyeriana*

F.T.E.A. Gramin. 3: 569 (1982); F.E.E. 7: 216 (1995).
Syn: *Panicum meyerianum* Nees – F.P.S. 3: 497 (1956).
Perennial. Riverbanks, swamps.
Distr: NS, KOR, CS, BAG, UN, EQU; Ghana, C.A.R. to Somalia, S to Angola & South Africa; Madagascar, Arabia.
CS: Andrews 216 3/1936; **BAG:** Harrison 1062 10/1950.

Euclasta condylotricha (Hochst. ex Steud.) Stapf

F.P.S. 3: 460 (1956); F.J.M.: 175 (1976); F.T.E.A. Gramin. 3: 722 (1982); F.E.E. 7: 307 (1995).
Annual. Wooded grassland & bushland.
Distr: DAR, CS, ES; Senegal to Ethiopia, S to Angola & Mozambique; Madagascar, India, neotropics.
DAR: Wickens 2557 9/1964.
Note: Andrews records this species from Central & Southern Sudan but we have seen no material from the South.

Exotheca abyssinica (Hochst. ex A.Rich.) Andersson

F.P.S. 3: 460 (1956); F.T.E.A. Gramin. 3: 821 (1982); F.E.E. 7: 353 (1995); Biol. Skr. 51: 564 (2005).
Tufted perennial. Upland grassland.
Distr: EQU; Eritrea to Zambia & Mozambique; Vietnam.
EQU: Myers 11580 7/1939.

Festuca abyssinica Hochst. ex A.Rich.

F.T.E.A. Gramin. 1: 60 (1970); F.J.M.: 175 (1976); F.E.E. 7: 24 (1995).
Tufted perennial. Montane grassland.
Distr: DAR; Cameroon, Ethiopia to Angola, Zimbabwe & Mozambique.
DAR: Blair 371 9/1964.

Festuca africana (Hack.) Clayton

Biol. Skr. 51: 565 (2005).
Syn: *Pseudobromus silvaticus* K.Schum. – F.P.S. 3: 521 (1956); F.T.E.A. Gramin. 1: 54 (1970).
Tufted perennial. Montane forest & bamboo thicket.
Distr: EQU; Uganda to South Africa.
EQU: Jackson 1102 1/1950.

Festuca chodatiana (St.-Yves) E.B.Alexeev

F.E.E. 7: 27 (1995); Biol. Skr. 51: 565 (2005).
Syn: *Festuca camusiana* St.-Yves subsp. *chodatiana* St.-Yves – F.T.E.A. Gramin. 1: 57 (1970).

Tufted perennial. Montane grassland & bushland.
Distr: EQU; Cameroon to Ethiopia, S to Tanzania.
EQU: Thomas 1717 12/1935.

Festuca elgonensis E.B.Alexeev

Bot. Zhurn. (Moscow & Leningrad) 72: 1266 (1987); Biol. Skr. 51: 565 (2005).
Tufted perennial. Montane grassland & bushland.
Distr: EQU; Uganda.
IUCN: RD
EQU: Jackson 939 (fr) 11/1949.
Note: known only from the Imatong Mts and Mt Elgon.

Festuca simensis Hochst. ex A.Rich.

F.P.S. 3: 461 (1956); F.T.E.A. Gramin. 1: 57 (1970); F.E.E. 7: 27 (1995); Biol. Skr. 51: 565 (2005).
Tufted rhizomatous perennial. Montane forest, bamboo thicket & grassland, streambanks.
Distr: EQU; Bioko, Cameroon, Congo-Brazzaville, D.R. Congo, Rwanda, Ethiopia, Uganda, Kenya.
EQU: Johnston s.n. 2/1936.

Festuca sudanensis E.B.Alexeev

Bot. Zhurn. (Moscow & Leningrad) 72: 1264 (1987); Biol. Skr. 51: 565 (2005).
Syn: *Festuca rigidula* sensu Andrews, non Steud. – F.P.S. 3: 461 (1956).
Tufted perennial. Montane grassland.
Distr: EQU; South Sudan endemic.
IUCN: RD
EQU: Myers 13497 (fr) 9/1940.

Gastridium phleoides (Nees & Meyen) C.E.Hubb.

F.T.E.A. Gramin. 1: 100 (1970); F.J.M.: 175 (1976); F.E.E. 7: 43 (1995).
Syn: *Gastridium ventricosum* sensu auctt., non (Gouan) Schinz & Thell. – F.P.S. 3: 461 (1956); Bot. Exp. Sud.: 56 (1970).
Annual, sometimes tufted. Rocky hillslopes, disturbed areas.
Distr: RS, DAR; Eritrea, Ethiopia, Kenya; Mediterranean, Arabia, introduced elsewhere.
RS: Jackson 2730 3/1953.

Hackelochloa granularis (L.) Kuntze

F.P.S. 3: 461 (1956); F.J.M.: 175 (1976); F.T.E.A. Gramin. 3: 849 (1982); F.E.E. 7: 363 (1995).
Annual. Weed of disturbed areas.
Distr: DAR, KOR, ES, BAG, UN, EQU; pantropical.
DAR: Wickens 2235 8/1964; **EQU:** Myers 7071 7/1937.

[*Hainardia*]

Note: Drar (1970) records *Lepturus cylindricus* (Willd.) Trin. (= *Hainardia cylindrica* (Willd.) Greuter) from Jebel Marra; this is a Mediterranean species extending to Iran and seems unlikely to occur in Sudan. We have not seen the specimen on which the record was based.

Halopyrum mucronatum (L.) Stapf

F.T.E.A. Gramin. 2: 183 (1974); F.E.E. 7: 94 (1995).
Tufted stoloniferous perennial. Coastal sand dunes.
Distr: RS; Eritrea to Mozambique; Egypt, Arabia to India & Sri Lanka.

RS: Tackholm et al. 640 1962.
Note: the single collection seen is from the Hala'ib Triangle; it is, however, very likely to also occur along the Red Sea coast of Sudan proper.

Harpachne schimperi Hochst. ex A.Rich.

F.P.S. 3: 463 (1956); F.T.E.A. Gramin. 2: 270 (1974); F.J.M.: 176 (1976); F.E.E. 7: 129 (1995); Biol. Skr. 51: 566 (2005).
Tufted perennial. Dry grassland & bushland, disturbed areas.
Distr: RS, DAR, EQU; Eritrea & Somalia, S to Zambia & Zimbabwe; Arabia.
DAR: Wickens 2616 9/1964; **EQU:** Kielland-Lund 373 12/1983 (n.v.).

Helictotrichon elongatum (Hochst. ex A.Rich.) C.E.Hubb.

F.P.S. 3: 463 (1956); F.T.E.A. Gramin. 1: 89 (1970); F.J.M.: 176 (1976); F.E.E. 7: 31 (1995); Biol. Skr. 51: 566 (2005).
Tufted perennial. Montane forest, bushland & grassland.
Distr: DAR, EQU; Nigeria, Cameroon, Eritrea to Zimbabwe; Madagascar, Yemen.
DAR: Blair 363 9/1964; **EQU:** Friis & Vollesen 4 11/1980.

Helictotrichon umbrosum (Hochst. ex Steud.) C.E.Hubb.

F.T.E.A. Gramin. 1: 88 (1970); F.E.E. 7: 32 (1995); Biol. Skr. 51: 566 (2005).
Syn: *Helictotrichon thomasii* C.E.Hubb. – F.P.S. 3: 463 (1956).
Rhizomatous perennial. Montane grassland, bushland & bamboo thicket.
Distr: EQU; Ethiopia, Uganda, Kenya, Tanzania.
EQU: Thomas 1862 9/1935.

Hemarthria altissima (Poir.) Stapf & C.E.Hubb.

F.T.E.A. Gramin. 3: 853 (1982); F.E.E. 7: 363 (1995).
Syn: *Hemarthria natans* sensu Wickens, non Stapf – F.J.M.: 176 (1976).
Stoloniferous perennial. Shallow water, muddy edges of streams & ponds.
Distr: DAR; scattered throughout the tropics & subtropics.
DAR: Wickens 2965 5/1965.

Heteropogon contortus (L.) P.Beauv. ex Roem. & Schult.

F.P.S. 3: 463 (1956); F.J.M.: 176 (1976); F.T.E.A. Gramin. 3: 827 (1982); F.E.E. 7: 356 (1995); Biol. Skr. 51: 566 (2005).
Tufted perennial. Bushland & wooded grassland.
Distr: RS, DAR, KOR, CS, BAG, UN, EQU; pantropical & warm temperate.
DAR: Blair 175 10/1963; **UN:** Sherif A.3965 8/1951.

Heteropogon melanocarpus (Elliott) Benth.

F.J.M.: 176 (1976); F.T.E.A. Gramin. 3: 827 (1982); F.E.E. 7: 356 (1995).
Annual. Bushland, wooded grassland, disturbed areas.
Distr: DAR, KOR; widespread in the tropics.
DAR: Wickens 2755 10/1964.

Hyparrhenia anamesa Clayton

F.J.M.: 176 (1976); F.T.E.A. Gramin. 3: 800 (1982); F.E.E. 7: 341 (1995).

Tufted perennial. Grassland, dry hillslopes.
Distr: DAR; Ethiopia to Angola & South Africa.
DAR: Wickens 2991 6/1965.

Hyparrhenia anthistirioides (Hochst. ex A.Rich.) Andersson ex Stapf

F.P.S. 3: 470 (1956); F.J.M.: 176 (1976); F.T.E.A. Gramin. 3: 804 (1982); F.E.E. 7: 341 (1995); Biol. Skr. 51: 567 (2005).
Syn: *Hyparrhenia pseudocymbaria* (Steud.) Stapf – F.P.S. 3: 470 (1956); Bot. Exp. Sud.: 56 (1970).
Annual. Open disturbed areas, open bushland.
Distr: DAR, KOR, CS, ES; Eritrea & Somalia, S to Zambia & Malawi.
DAR: Wickens 1059 1/1964.

Hyparrhenia bracteata (Humb. & Bonpl. ex Willd.) Stapf

F.T.E.A. Gramin. 3: 814 (1982); Biol. Skr. 51: 567 (2005).
Tufted perennial. Wet grassland.
Distr: EQU; Mali to Kenya, S to South Africa; neotropics.
EQU: Friis & Vollesen 201 11/1980.

Hyparrhenia coleotricha (Steud.) Andersson ex Clayton

F.T.E.A. Gramin. 3: 812 (1982); F.E.E. 7: 348 (1995).
Annual. Disturbed areas, dry bushland.
Distr: EQU; Eritrea, Ethiopia, Tanzania; Yemen.
EQU: Myers 7732 9/1937.

Hyparrhenia collina (Pilg.) Stapf

F.J.M.: 176 (1976); F.T.E.A. Gramin. 3: 811 (1982); F.E.E. 7: 347 (1995).
Tufted rhizomatous perennial. Upland grassland, bushland & woodland.
Distr: DAR; Nigeria, Cameroon, Ethiopia to South Africa.
DAR: Jackson 2559 1/1953.

Hyparrhenia confinis (Hochst. ex A.Rich.) Andersson ex Stapf var. nudiglumis (Hack.) Clayton

F.P.S. 3: 473 (species only) (1956); F.J.M.: 176 (1976); F.E.E. 7: 348 (1995).
Annual. Heavy clay soils.
Distr: DAR, KOR, CS, UN; Ethiopia.
IUCN: RD
KOR: Andrews 34 11/1933; **UN:** Sherif A.4003 9/1951.

Hyparrhenia confinis (Hochst. ex A.Rich.) Andersson ex Stapf var. pellita (Hack.) Stapf

F.P.S. 3: 474 (1956); F.E.E. 7: 348 (1995).
Annual. Heavy clay soils.
Distr: CS, ES; Sudan/Ethiopia border region.
IUCN: RD
CS: Lea 190 9/1953.

Hyparrhenia cymbaria (L.) Stapf

F.P.S. 3: 471 (1956); F.J.M.: 177 (1976); F.T.E.A. Gramin. 3: 804 (1982); F.E.E. 7: 343 (1995); Biol. Skr. 51: 567 (2005).
Tufted rhizomatous perennial. Tall grassland.
Distr: DAR, UN, EQU; Nigeria to Eritrea, S to Angola & South Africa; Madgascar, Comoros.
DAR: Pettet 39 12/1962; **EQU:** Myers 7734 9/1937.

Hyparrhenia dichroa (Steud.) Stapf

F.J.M.: 177 (1976); F.T.E.A. Gramin. 3: 796 (1982); F.E.E. 7: 337 (1995).
Tufted perennial. Bushland & wooded grassland.
Distr: DAR, ES; Ethiopia & Djibouti, S to Namibia & South Africa.
DAR: <u>Wickens 2771</u> 10/1964.

Hyparrhenia diplandra (Hack.) Stapf

F.P.S. 3: 474 (1956); F.T.E.A. Gramin. 3: 818 (1982); F.E.E. 7: 349 (1995); Biol. Skr. 51: 567 (2005).
Tufted perennial. Wooded grassland & bushland, often in moist areas.
Distr: EQU; Mauritania to Ethiopia, S to Angola, Zimbabwe & Mozambique; Madagascar, Brazil, SE Asia.
EQU: <u>Friis & Vollesen 670</u> 12/1980.

Hyparrhenia dregeana (Nees) Stapf ex Stent

F.J.M.: 177 (1976); F.T.E.A. Gramin. 3: 809 (1982); F.E.E. 7: 346 (1995); Biol. Skr. 51: 568 (2005).
Tufted perennial. Upland grassland, often in damp areas.
Distr: DAR, EQU; Eritrea to Namibia & South Africa;. Yemen.
DAR: <u>Blair 36</u> 6/1963; **EQU:** <u>Myers 13498</u> 9/1940.

Hyparrhenia exarmata (Stapf) Stapf

F.P.S. 3: 468 (1956); F.T.E.A. Gramin. 3: 794 (1982).
Tufted perennial or annual. Margins of swamps, wooded grassland.
Distr: EQU; Mali to Kenya.
EQU: <u>Myers 7049</u> 7/1938.

Hyparrhenia familiaris (Steud.) Stapf

F.T.E.A. Gramin. 3: 802 (1982).
Tufted perennial. Open bushland.
Distr: EQU; Guinea to Uganda, S to Angola & Tanzania; Vietnam.
EQU: <u>Simpson 7552</u> 11/1930.

Hyparrhenia figariana (Chiov.) Clayton

F.T.E.A. Gramin. 3: 802 (1982); Biol. Skr. 51: 568 (2005).
Syn: *Hyparrhenia barteri* (Hack.) Stapf var. *calvascens* (Hack.) Stapf – F.P.S. 3: 469 (1956).
Annual. Bushland & wooded grassland.
Distr: BAG, UN, EQU; Nigeria to Uganda, S to D.R. Congo & Tanzania.
EQU: <u>Myers 6909</u> 6/1937.

Hyparrhenia filipendula (Hochst.) Stapf

F.P.S. 3: 469 (1956); F.J.M.: 177 (1976); F.T.E.A. Gramin. 3: 803 (1982); F.E.E. 7: 341 (1995); Biol. Skr. 51: 568 (2005).
Tufted rhizomatous perennial. Open disturbed areas.
Distr: DAR, BAG, UN, EQU; Guinea to Ethiopia, S to Namibia & South Africa; Madagascar, SE Asia.
DAR: <u>Wickens 2804</u> 11/1964; **EQU:** <u>Beshir 442</u> 9/1951.

Hyparrhenia finitima (Hochst.) Andersson ex Stapf

F.P.S. 3: 466 (1956); F.T.E.A. Gramin. 3: 797 (1982); F.E.E. 7: 340 (1995).
Tufted perennial. Bushland & wooded grassland, amongst rocks or in disturbed areas.
Distr: ?ES; Sierra Leone, Ethiopia to Angola, Namibia & South Africa.

Note: Andrews records this species from Gallabat, presumably based on a Schweinfurth collection, but we have seen no material with which to verify this record.

Hyparrhenia gazensis (Rendle) Stapf

F.T.E.A. Gramin. 3: 797 (1982); Biol. Skr. 51: 568 (2005).
Tufted perennial. Disturbed areas.
Distr: EQU; D.R. Congo to Kenya, S to South Africa.
EQU: <u>Shigeta 97</u> 1979 (n.v.).
Note: the specimen cited was not traced by Friis & Vollesen.

Hyparrhenia griffithii Bor

F.J.M.: 177 (1976); F.T.E.A. Gramin. 3: 799 (1982).
Tufted perennial. Grassland.
Distr: DAR; Kenya, Tanzania, Zambia; Madagascar, India, SE Asia.
DAR: <u>Blair 59</u> 6/1963.

Hyparrhenia hirta (L.) Stapf

F.P.S. 3: 468 (1956); F.J.M.: 177 (1976); F.T.E.A. Gramin. 3: 798 (1982); F.E.E. 7: 340 (1995).
Tufted perennial. Bushland & grassland, often on hillslopes.
Distr: RS, DAR, CS; Mauritania to Somalia, S to South Africa; Mediterranean, Arabia to Pakistan, introduced in Australia & the neotropics.
RS: <u>Andrews 3554</u> 3/1949.

Hyparrhenia madaropoda Clayton

F.J.M.: 177 (1976); F.T.E.A. Gramin. 3: 812 (1982); Biol. Skr. 51: 568 (2005).
Annual. Bushland.
Distr: CS, UN, EQU; D.R. Congo to Kenya, S to Zambia & Mozambique.
CS: <u>Wilson-Jones 170</u> 11/1952; **UN:** <u>Harrison 1041</u> 10/1950.

Hyparrhenia multiplex (Hochst. ex A.Rich.) Andersson ex Stapf

F.J.M.: 177 (1976); F.E.E. 7: 352 (1995).
Tufted annual. Hillslopes.
Distr: DAR; Ethiopia.
IUCN: RD
DAR: <u>Blair 378</u> 9/1964.

Hyparrhenia niariensis (Franch.) Clayton var. *macrarrhena* (Hack.) Clayton

F.T.E.A. Gramin. 3: 813 (species only) (1982).
Syn: *Hyparrhenia macrarrhena* (Hack.) Stapf – F.P.S. 3: 474 (1956).
Annual. Wooded grassland.
Distr: BAG; Cameroon to Uganda, S to Angola & Zambia.
BAG: <u>Harrison 1412</u> 10/1950.

Hyparrhenia nyassae (Rendle) Stapf

F.J.M.: 178 (1976); F.T.E.A. Gramin. 3: 793 (1982); F.E.E. 7: 338 (1995).
Tufted perennial. Bushland, wooded grassland, swamp margins.
Distr: DAR; Ivory Coast to Ethiopia, S to Angola & South Africa; Madagascar, Thailand, Vietnam.
DAR: <u>Blair 174</u> 10/1963.

Hyparrhenia papillipes (Hochst. ex A.Rich.) Andersson ex Stapf

F.P.S. 3: 473 (1956); F.T.E.A. Gramin. 3: 808 (1982); F.E.E. 7: 345 (1995).

Tufted perennial. Bushland & upland grassland, often on rocky hillslopes.

Distr: DAR, ?EQU; Ethiopia, Uganda, Kenya, Tanzania; Madagascar, Yemen.

DAR: Miehe 658 9/1982; **EQU:** Myers 13472 9/1940.

Note: the Myers collection cited is on two sheets; one is filed at Kew as *H. dregeana*, the second as *H. papillipes*, yet the two plants look very similar. It is possible that both refer to *H. dregeana*. There are no other specimens of *H. papillipes* from South Sudan.

Hyparrhenia pilgeriana C.E.Hubb.

F.J.M.: 178 (1976); F.T.E.A. Gramin. 3: 807 (1982); F.E.E. 7: 345 (1995); Biol. Skr. 51: 569 (2005).

Tufted rhizomatous perennial. Upland grassland & bushland, forest margins.

Distr: DAR, EQU; Ethiopia to South Africa.

DAR: Jackson 2628; **EQU:** Friis & Vollesen 429 11/1980.

Hyparrhenia poecilotricha (Hack.) Stapf

F.J.M.: 178 (1976); F.T.E.A. Gramin. 3: 796 (1982); F.E.E. 7: 338 (1995); Biol. Skr. 51: 559 (2005).

Tufted perennial. Bushland, woodland & grassland.

Distr: DAR, EQU; Guinea to Ethiopia, S to Namibia & South Africa.

DAR: Blair 185 10/1963; **EQU:** Jackson 362 10/1948.

Hyparrhenia quarrei Robyns

F.T.E.A. Gramin. 3: 799 (1982); F.E.E. 7: 340 (1995).

Tufted perennial. Bushland, wooded grassland, forest margins.

Distr: DAR; scattered from Nigeria to Eritrea, S to Namibia & South Africa; Yemen.

DAR: de Wilde et al. 5494 1/1965.

Hyparrhenia rudis Stapf

F.J.M.: 178 (1976); F.T.E.A. Gramin. 3: 811 (1982); F.E.E. 7: 347 (1995).

Tufted perennial. Upland grassland & bushland.

Distr: DAR; Burkina Faso to Ethiopia, S to Angola & South Africa; Madagascar.

DAR: Wickens 1457 4/1964.

Hyparrhenia rufa (Nees) Stapf var. *rufa*

F.P.S. 3: 468 (1956); F.J.M.: 178 (1976); F.T.E.A. Gramin. 3: 794 (1982); F.E.E. 7: 337 (1995); Biol. Skr. 51: 569 (2005).

Syn: *Hyparrhenia altissima* Stapf – F.P.S. 3: 468 (1956).

Tufted perennial. Wooded grassland & bushland, often abundant.

Distr: DAR, KOR, BAG, UN, EQU; Mauritania to Eritrea, S to Namibia & South Africa: Madagascar, introduced elsewhere in the tropics.

DAR: Wickens 2720 9/1964; **EQU:** Friis & Vollesen 174 11/1980.

Hyparrhenia schimperi (Hochst. ex A.Rich.) Andersson ex Stapf

F.J.M.: 178 (1976); F.T.E.A. Gramin. 3: 808 (1982); F.E.E. 7: 345 (1995); Biol. Skr. 51: 569 (2005).

Tufted perennial. Open bushland & wooded grassland including moist areas.

Distr: DAR; Ethiopia to Angola & South Africa; Madagascar.

DAR: Pettet s.n. 12/1962.

Note: this species is also likely to occur in South Sudan, having been recorded from the Ugandan side of the Imatong Mts; see Friis & Vollesen (2005: 569).

Hyparrhenia subplumosa Stapf

F.J.M.: 178 (1976); F.T.E.A. Gramin. 3: 818 (1982).

Tufted perennial. Wet grassland.

Distr: DAR; Guinea to Chad, S to Angola & Zimbabwe.

DAR: Jackson 3381 12/1954.

Hyparrhenia tamba (Hochst. ex Steud.) Andersson ex Stapf

F.J.M.: 178 (1976); F.T.E.A. Gramin. 3: 810 (1982); F.E.E. 7: 347 (1995); Biol. Skr. 51: 570 (2005).

Tufted perennial. Upland grassland.

Distr: DAR, EQU; D.R. Congo, Eritrea, Ethiopia, Kenya, Zimbabwe, South Africa.

DAR: Blair 47 6/1963; **EQU:** Johnston 1493 2/1936.

Hyparrhenia umbrosa (Hochst.) Andersson ex Clayton

F.P.S. 3: 473 (1956); F.T.E.A. Gramin. 3: 810 (1982); Biol. Skr. 51: 570 (2005).

Tufted perennial. Upland grassland, disturbed areas.

Distr: ?DAR, EQU; Nigeria, Cameroon, Uganda, Kenya, Tanzania, South Africa.

EQU: Thomas 1806 12/1935.

Note: a widespread but uncommon species. Andrews also records it from Darfur but we have seen no material with which to verify this and it may be an error.

Hyparrhenia variabilis Stapf

F.P.S. 3: 471 (1956); F.J.M.: 178 (1976); F.T.E.A. Gramin. 3: 805 (1982); F.E.E. 7: 344 (1995); Biol. Skr. 51: 570 (2005).

Tufted perennial. Upland bushland & woodland.

Distr: DAR, UN, EQU; Ethiopia to Angola & South Africa; Madgascar, Comoros, Yemen, Java.

DAR: Wickens 1567 5/1964; **EQU:** Jackson 4216 6/1961.

Hyparrhenia welwitschii (Rendle) Stapf

F.T.E.A. Gramin. 3: 813 (1982).

Syn: *Hyparrhenia gracilescens* Stapf – F.P.S. 3: 474 (1956).

Annual. Wooded grassland.

Distr: BAG, EQU; Guinea to Uganda, S to Angola, Zimbabwe & Mozambique; Comoros.

BAG: Jackson 3985 10/1958.

Hyperthelia cornucopiae (Hack.) Clayton

Kew Bull. 20: 446 (1966).

Syn: *Hyparrhenia cornucopiae* (Hack.) Stapf – F.P.S. 3: 476 (1956).

Annual. Wooded grassland.

Distr: BAG, UN, EQU; Chad, C.A.R.

IUCN: RD

EQU: Myers 13531 10/1940.

Hyperthelia dissoluta (Nees ex Steud.) Clayton

F.T.E.A. Gramin. 3: 766 (1982); F.E.E. 7: 333 (1995); Biol. Skr. 51: 570 (2005).
Syn: *Hyparrhenia dissoluta* (Nees ex Steud.) C.E.Hubb. – F.P.S. 3: 469 (1956).
Tufted perennial. Bushland & wooded grassland, often in disturbed areas.
Distr: BAG, UN, EQU; Senegal to Ethiopia, S to South Africa; Madagascar.
UN: Myers 12051 10/1939.

Hyperthelia edulis (C.E.Hubb.) Clayton

Kew Bull. 20: 447 (1966).
Syn: *Hyparrhenia edulis* C.E.Hubb. – F.P.S. 3: 470 (1956).
Annual. Grassland on rocky hillslopes.
Distr: EQU; South Sudan endemic.
IUCN: RD
EQU: Myers 7154 (fr) 7/1937.

Hyperthelia macrolepis (Hack.) Clayton

Kew Bull. 20: 445 (1966).
Syn: *Hyparrhenia macrolepis* (Hack.) Stapf – F.P.S. 3: 470 (1956).
Annual. Habitat not known.
Distr: BAG; ?South Sudan endemic.
IUCN: RD
BAG: Schweinfurth 2361 (fr) 9/1869.
Note: this is possibly just an annual variant of *H. dissoluta*.

Imperata cylindrica (L.) Raeusch.

F.J.M.: 178 (1976); F.T.E.A. Gramin. 3: 700 (1982); F.E.E. 7: 292 (1995); Biol. Skr. 51: 571 (2005).
Syn: *Imperata cylindrica* (L.) Raeusch. var. *africana* (Andersson) C.E.Hubb. – F.P.S. 3: 476 (1956); Bot. Exp. Sud.: 56 (1970); F.J.M.: 178 (1976).
Rhizomatous perennial. Woodland & grassland, dominating frequently burnt or disturbed sites.
Distr: NS, DAR, BAG, UN, EQU; Senegal to Ethiopia, S to South Africa; Madagascar, Mediterranean to Afghanistan, introduced elsewhere.
NS: Pettet 100 10/1962; **EQU:** Myers 6338 3/1937.

Isachne mauritiana Kunth

F.T.E.A. Gramin. 2: 434 (1974); F.E.E. 7: 283 (1995); Biol. Skr. 51: 571 (2005).
Perennial. Upland forest.
Distr: EQU; Ghana to Kenya, S to Zimbabwe & Mozambique, ?Ethiopia.
EQU: Friis & Vollesen 243 11/1980.

Ischaemum afrum (J.F.Gmel.) Dandy

F.P.S. 3: 476 (1956); F.T.E.A. Gramin. 3: 747 (1982); F.E.E. 7: 314 (1995); Biol. Skr. 51: 571 (2005).
Tufted rhizomatous perennial. Grassland & bushland on black clay soils.
Distr: KOR, CS, ES, UN, EQU; Burkina Faso to Somalia, S to Namibia & South Africa; India.
CS: Andrews 141 1/1936; **EQU:** Myers 13445 8/1940.

Koeleria capensis (Steud.) Nees

F.T.E.A. Gramin. 1: 79 (1970); F.E.E. 7: 38 (1995); Biol. Skr. 51: 571 (2005).

Syn: *Koeleria cristata* (L.) Pers. var. *brevifolia* (Nees) C.E.Hubb. – F.P.S. 3: 477 (1956).
Tufted perennial. Montane grassland & bare ground.
Distr: EQU; Cameroon, Ethiopia to South Africa; Yemen.
EQU: Friis & Vollesen 1220 3/1982.

Koordersiochloa longiarista (A.Rich.) Veldkamp

Reinwardtia 13: 301 (2012).
Syn: *Streblochaete longiarista* (A.Rich.) Pilg. – F.T.E.A. Gramin. 1: 74 (1970); F.E.E. 7: 29 (1995); Biol. Skr. 51: 594 (2005).
Tufted perennial. Upland forest including margins & associated bushland.
Distr: EQU; Nigeria, Cameroon, Ethiopia to South Africa; Réunion, Indonesia, Philippines.
EQU: Jackson 905 11/1949.

Lasiurus scindicus Henrard

F.E.E. 7: 359 (1995).
Syn: *Lasiurus hirsutus* (Forssk.) Boiss. – F.P.S. 3: 479 (1956).
Tufted rhizomatous perennial. Open desert plains.
Distr: NS, RS, KOR, ES; Mauritania to Somalia; N Africa, Arabia to India.
RS: Jackson 3906 8/1958.

Leersia angustifolia Prodoehl

F.P.S. 3: 482 (1956); Biol. Skr. 51: 572 (2005).
Tufted annual. Swamps.
Distr: DAR, BAG, UN; C.A.R., D.R. Congo.
DAR: Wickens 3149 8/1972; **BAG:** Harrison 911 10/1950.

Leersia drepanothrix Stapf

F.P.S. 3: 482 (1956); F.T.E.A. Gramin. 1: 25 (1970).
Tufted annual or perennial, sometimes rhizomatous. Swamps.
Distr: UN, EQU; Guinea to Uganda.
UN: Harrison 1039 10/1950.

Leersia friesii Melderis

F.T.E.A. Gramin. 1: 27 (1970).
Rhizomatous perennial. Swamps, lake margins.
Distr: BAG; D.R. Congo, Uganda, Tanzania, Angola, Zambia, Botswana.
BAG: Harrison 1406 3/1949.

Leersia hexandra Sw.

F.P.S. 3: 479 (1956); F.T.E.A. Gramin. 1: 25 (1970); F.J.M.: 179 (1976); F.E.E. 7: 9 (1995); Biol. Skr. 51: 572 (2005).
Rhizomatous perennial. Swamps & open water.
Distr: DAR, UN, EQU; pantropical & subtropical.
DAR: Blair 341 8/1964; **UN:** Harrison 1027 10/1950.

Leptaspis zeylanica Nees ex Steud.

F.E.E. 7: 8 (1995); Biol. Skr. 51: 572 (2005).
Syn: *Leptaspis cochleata* Thwaites – F.P.S. 3: 482 (1956); F.T.E.A. Gramin. 1: 21 (1970).
Rhizomatous perennial. Forest.
Distr: EQU; Guinea to Ethiopia, S to Zimbabwe & Mozambique; Madagascar, Sri Lanka, Malesia, Solomon Is.
EQU: Myers 10623 8/1939.

Leptocarydion vulpiastrum (De Not.) Stapf

F.P.S. 3: 482 (1956); F.T.E.A. Gramin. 2: 281 (1974); F.E.E. 7: 104 (1995); Biol. Skr. 51: 572 (2005).
Tufted annual. Wooded grassland & bushland.
Distr: ?KOR, EQU; Eritrea to Namibia & South Africa; Madagascar, Yemen.
EQU: Harrison 517 12/1948.
Note: Andrews records this species from Kordofan; this may be based on Figari's type from 'Upper Nubia', for which the exact collecting locality is unknown.

Leptochloa caerulescens Steud.

F.P.S. 3: 484 (1956); F.E.E. 7: 102 (1995).
Semi-aquatic stoloniferous annual. Shallow water, riverbanks.
Distr: UN, EQU; Senegal to Ethiopia, S to Angola & Zambia; Madagascar.
EQU: Myers 6938 7/1937.

Leptochloa fusca (L.) Kunth

F.E.E. 7: 101 (1995).
Syn: Diplachne fusca (L.) P.Beauv. ex Roem. & Schult. – F.P.S. 3: 439 (1956); F.T.E.A. Gramin. 2: 281 (1974).
(Semi-) Aquatic rhizomatous perennial. Shallow water & marshy ground.
Distr: KOR, CS; palaeotropical & subtropical.
CS: Jackson 3089 3/1954.

Leptochloa obtusiflora Hochst.

F.T.E.A. Gramin. 2: 278 (1974); F.E.E. 7: 102 (1995).
Perennial. Bushland & grassland, disturbed ground.
Distr: RS, DAR; Ethiopia & Somalia, S to Angola & Tanzania; Arabia, India.
RS: Jackson 3977 4/1959.

Leptochloa panicea (Retz.) Ohwi

F.T.E.A. Gramin. 2: 279 (1974).
Syn: Leptochloa chinensis sensu Andrews, non (L.) Nees – F.P.S. 3: 484 (1956).
Tufted annual. Bushland, wooded grassland, disturbed areas.
Distr: KOR, UN, EQU; C.A.R., Somalia to Namibia & South Africa; tropical Asia & America.
KOR: Niamir 49 8/1980; **UN:** Sherif A.3931 8/1951.

Leptothrium senegalense (Kunth) Clayton

F.T.E.A. Gramin. 2: 402 (1974); F.E.E. 7: 181 (1995).
Syn: Latipes senegalensis Kunth – F.P.S. 3: 479 (1956).
Tufted short-lived perennial. Dry bushland, semi-desert grassland.
Distr: RS, DAR, KOR; Mauritania to Somalia, S to Tanzania; Egypt, Arabia to Pakistan.
RS: Farag 103 3/2010.

Lintonia nutans Stapf

F.P.S. 3: 484 (1956); F.T.E.A. Gramin. 2: 302 (1974); F.E.E. 7: 159 (1995); Biol. Skr. 51: 572 (2005).
Tufted stoloniferous perennial. Grassland & bushland, often on black clay soils.
Distr: EQU; Ethiopia & Somalia, S to South Africa.
EQU: Myers 14012 9/1941.

Loudetia annua (Stapf) C.E.Hubb.

F.P.S. 3: 486 (1956); F.T.E.A. Gramin. 2: 421 (1974);
F.J.M.: 179 (1976).
Tufted annual. Shallow soils over rock.
Distr: DAR, KOR, BAG, UN, EQU; Senegal to Uganda.
DAR: Wickens 2564 9/1964; **EQU:** Myers 7333 7/1937.

Loudetia arundinacea (Hochst. ex A.Rich.) Steud.

F.P.S. 3: 486 (1956); F.T.E.A. Gramin. 2: 417 (1974); F.E.E. 7: 288 (1995); Biol. Skr. 51: 573 (2005).
Syn: Loudetia arundinacea (Hochst. ex A.Rich.) Steud. var. **hensii** (De Wild.) C.E.Hubb. ex Pichi-Serm. – F.P.S. 3: 486 (1956).
Tufted perennial. Grassland & bushland on thin soils over rock.
Distr: UN, EQU; Senegal to Ethiopia, S to Angola & Mozambique.
EQU: Jackson 488 11/1949.

Loudetia coarctata (A.Camus) C.E.Hubb.

F.P.S. 3: 486 (1956); F.W.T.A. 3(2): 417 (1972).
Tufted perennial. Swamps, wet grassland.
Distr: BAG, EQU; Guinea, Nigeria to C.A.R., S to Zambia.
EQU: Jackson 2296 6/1952.

Loudetia esculenta C.E.Hubb.

Kew Bull. 4: 352 (1949); F.P.S. 3: 485 (1956).
Tufted annual. Grassland on rocky hillslopes.
Distr: EQU; South Sudan endemic.
IUCN: RD
EQU: Andrews 26/85 9/1947.
Note: in the African Plants Database this is treated in *Tristachya*, as *T. esculenta* (C.E.Hubb.) Conert.

Loudetia flavida (Stapf) C.E.Hubb.

F.T.E.A. Gramin. 2: 416 (1974); F.E.E. 7: 286 (1995); Biol. Skr. 51: 573 (2005).
Tufted perennial. Grassland, bushland & woodland, often on sandy or rocky soils.
Distr: EQU; Burkina Faso & Ghana to Ethiopia, S to South Africa; Oman.
EQU: Myers 13447 8/1940.

Loudetia hordeiformis (Stapf) C.E.Hubb.

F.W.T.A. 3(2): 417 (1972).
Tufted annual. Disturbed sandy soils.
Distr: DAR; Senegal to Chad.
DAR: Harrison 109 9/1947.

Loudetia phragmitoides (Peter) C.E.Hubb.

F.P.S. 3: 485 (1956); F.T.E.A. Gramin. 2: 415 (1974); F.E.E. 7: 286 (1995).
Tufted perennial. Swamps.
Distr: BAG, UN, EQU; Senegal to Ethiopia, S to Angola & Mozambique.
EQU: Harrison 610 12/1948.

Loudetia simplex (Nees) C.E.Hubb.

F.P.S. 3: 486 (1956); F.T.E.A. Gramin. 2: 418 (1974); F.J.M.: 179 (1976); F.E.E. 7: 288 (1995); Biol. Skr. 51: 573 (2005).
Tufted perennial. Grassland & bushland, often dominant on shallow soils.
Distr: DAR, EQU; Senegal to Ethiopia, S to Angola & South Africa; Madagascar.
DAR: Wickens 2137 8/1964; **EQU:** Beshir 10 8/1951.

Loudetia togoensis (Pilg.) C.E.Hubb.

F.P.S. 3: 487 (1956); F.W.T.A. 3(2): 416 (1972); F.J.M.: 179 (1976).
Tufted annual. Grassland on shallow soils.
Distr: DAR, KOR, CS; Mauritania & Senegal to Chad & C.A.R.
KOR: Wickens 396 9/1962.

Louisiella fluitans C.E.Hubb. & J.Léonard

Bull. Jard. Bot. État Bruxelles 22: 317 (1952); F.P.S. 3: 487 (1956).
Aquatic perennial. Floating in water.
Distr: BAG, UN; Cameroon, C.A.R., D.R. Congo.
BAG: Myers 13889 7/1941.

[*Loxodera*]

Note: *Loxodera ledermannii* (Pilg.) Launert is likely to occur in South Sudan, having been recorded on the Ugandan side of the Imatong Mts; see Friis & Vollesen (2005: 574).

Melanocenchris abyssinica (R.Br. ex Fresen.) Hochst.

F.P.S. 3: 487 (1956); F.E.E. 7: 176 (1995).
Tufted annual. Dry grassland.
Distr: RS, CS; Chad to Somalia; Egypt, Arabia to India.
CS: Farag 193 7/2010.

Melinis longiseta (A.Rich.) Zizka subsp. *longiseta*

F.E.E. 7: 187 (1995).
Syn: *Rhynchelytrum longisetum* (A.Rich.) Stapf & C.E.Hubb. – F.J.M.: 183 (1976); F.T.E.A. Gramin. 3: 512 (1982).
Syn: *Rhynchelytrum minutiflorum* (Rendle) Stapf & C.E.Hubb. var. *melinioides* (Stent) Stapf & C.E.Hubb. – F.J.M.: 183 (1976).
Syn: *Melinis ambigua* sensu Wickens, non Hack. – F.J.M.: 179 (1976).
Tufted perennial. Wooded grassland.
Distr: DAR; Ethiopia, Angola to Tanzania, S to Mozambique.
DAR: Blair 188 10/1963.
Note: we have not seen all the specimens cited by Wickens under *M. ambigua* but those that we have seen are referable here.

Melinis macrochaeta Stapf & C.E.Hubb.

F.P.S. 3: 487 (1956); F.T.E.A. Gramin. 3: 508 (1982); F.E.E. 7: 189 (1995); Biol. Skr. 51: 574 (2005).
Tufted annual or short-lived perennial. Grassland, streamsides, disturbed areas.
Distr: EQU; Ivory Coast, Benin to Ethiopia, S to Angola & South Africa.
EQU: Friis & Vollesen 780 12/1980.

Melinis minutiflora P.Beauv.

F.P.S. 3: 488 (1956); F.T.E.A. Gramin. 3: 506 (1982); F.E.E. 7: 189 (1995); Biol. Skr. 51: 574 (2005).
Syn: *Melinis tenuinervis* Stapf – F.P.S. 3: 488 (1956); F.J.M.: 179 (1976).
Tufted perennial. Grassland.
Distr: DAR, EQU; Cape Verde Is., Senegal to Somalia, S to Angola & South Africa; Madagascar, introduced elsewhere in the tropics.

DAR: Pettet 41 12/1962; **EQU:** Friis & Vollesen 282 11/1980.

Melinis repens (Willd.) Zizka subsp. *grandiflora* (Hochst.) Zizka

F.E.E. 7: 186 (1995); Biol. Skr. 51: 575 (2005).
Syn: *Rhynchelytrum grandiflorum* Hochst – F.P.S. 3: 522 (1956); F.J.M.: 183 (1976); F.T.E.A. Gramin. 3: 512 (1982).
Syn: *Rhynchelytrum villosum* (Parl.) Chiov. – F.P.S. 3: 522 (1956).
Tufted annual. Grassland & open woodland.
Distr: RS, DAR, KOR, ES, EQU; Cape Verde Is., Ghana, Niger to Somalia, S to Namibia & South Africa; Madagascar, Arabia.
DAR: Blair 198 10/1963; **EQU:** Myers 14006 9/1941.

Melinis repens (Willd.) Zizka subsp. *repens*

F.E.E. 7: 186 (1995); Biol. Skr. 51: 574 (2005).
Syn: *Rhynchelytrum repens* (Willd.) C.E.Hubb. – F.P.S. 3: 522 (1956); Bot. Exp. Sud.: 58 (1970); F.J.M.: 183 (1976); F.T.E.A. Gramin. 3: 515 (1982).
Tufted short-lived perennial. Weed of disturbed areas.
Distr: RS, DAR, EQU; Cape Verde Is., Mauritania to Somalia, S to Namibia & South Africa; Yemen.
DAR: Wickens 1911 9/1964; **EQU:** Myers 7725 9/1937.

Microchloa indica (L.f.) P.Beauv.

F.P.S. 3: 488 (1956); F.T.E.A. Gramin. 2: 314 (1974); F.J.M.: 179 (1976); F.E.E. 7: 172 (1995).
Tufted annual. Bare ground in bushland & grassland.
Distr: DAR, KOR, BAG; pantropical.
DAR: Wickens 2358 9/1964; **BAG:** Schweinfurth 2359 9/1869.
Note: *Microchloa caffra* Nees may well also occur in South Sudan, having been recorded from the Ugandan side of the Imatong Mts (Friis & Vollesen 2005: 575).

Microchloa kunthii Desv.

F.T.E.A. Gramin. 2: 314 (1974); F.J.M.: 179 (1976); F.E.E. 7: 174 (1995); Biol. Skr. 51: 575 (2005).
Syn: *Microchloa abyssinica* Desv. – F. Darfur Nord-Occ. & J. Gourgeil: 139 (1969).
Tufted perennial. Bare ground & amongst rocks in bushland & grassland.
Distr: DAR, EQU; pantropical.
DAR: Wickens 2586 9/1964; **EQU:** Friis & Vollesen 261 11/1980.

Monocymbium ceresiiforme (Nees) Stapf

F.P.S. 3: 491 (1956); F.T.E.A. Gramin. 3: 825 (1982); F.E.E. 7: 331 (1995).
Perennial, sometimes tufted. Grassy hillslopes.
Distr: BAG, EQU; Guinea to Ethiopia, S to Namibia & South Africa.
BAG: Schweinfurth 2191 7/1869.

Ochthochloa compressa (Forssk.) Hilu

F.E.E. 7: 108 (1995).
Syn: *Eleusine compressa* (Forssk.) Asch. & Schweinf. ex C.Chr. – F.P.S. 3: 445 (1956).
Stoloniferous perennial. Sandy or stony desert.
Distr: RS, ES; Niger to Somalia; N Africa, Arabia to India.
RS: Jackson 3910 8/1958.

Olyra latifolia L.

F.P.S. 3: 491 (1956); F.T.E.A. Gramin. 1: 17 (1970); F.E.E.
7: 6 (1995); Biol. Skr. 51: 576 (2005).
Tufted perennial. Forest.
Distr: EQU; Senegal to Ethiopia, S to Angola, Zimbabwe
& Mozambique; Madagascar, neotropics.
EQU: Jackson 395 10/1958.

Ophiuros papillosus Hochst.

F.P.S. 3: 491 (1956); F.E.E. 7: 365 (1995).
Annual. Weed of Sorghum fields, open plains.
Distr: RS, KOR, CS, ES, UN; Eritrea, Ethiopia; introduced
in India.
KOR: Jackson 2372 10/1952; **UN:** Sherif A.4009 9/1951.
Note: a very local species but within its range it can be
abundant and even a troublesome weed.

Oplismenus burmannii (Retz.) P.Beauv.

F.P.S. 3: 493 (1956); F.T.E.A. Gramin. 3: 542 (1982); F.E.E.
7: 192 (1995).
Annual. Moist shady places.
Distr: EQU; palaeotropical; introduced in the neotropics.
EQU: Wyld 267 8/1937.

Oplismenus compositus (L.) P.Beauv.

F.P.S. 3: 493 (1956); F.T.E.A. Gramin. 3: 542 (1982); F.E.E.
7: 192 (1995); Biol. Skr. 51: 576 (2005).
Perennial. Forest.
Distr: EQU; Eritrea to Zimbabwe & Mozambique;
widespread in tropical & subtropical Asia & America.
EQU: Thomas 1688 12/1935.

Oplismenus hirtellus (L.) P.Beauv. subsp. *hirtellus*

F.P.S. 3: 493 (1956); F.J.M.: 180 (1976); F.T.E.A. Gramin.
3: 542 (1982); F.E.E. 7: 192 (1995); Biol. Skr. 51: 576
(2005).
Stoloniferous perennial. Forest.
Distr: DAR, EQU; pantropical.
DAR: Blair 205 1/1964; **EQU:** Jackson 4287 7/1961.

Oropetium minimum (Hochst.) Pilg

F.T.E.A. Gramin. 2: 307 (1974); F.E.E. 7: 100 (1995).
Syn: *Chaetostichium majusculum* C.E.Hubb. – F. Darfur
Nord-Occ. & J. Gourgeil: 136 (1969).
Tufted perennial. Dry grassland, open bushland, rocky
areas.
Distr: DAR, CS; Chad to Somalia, S to Tanzania.
DAR: Quézel & Bourreil s.n. 9/1967.

Oropetium thomaeum (L.f.) Trin.

F. Darfur Nord-Occ. & J. Gourgeil: 140 (1969); F.T.E.A.
Gramin. 2: 306 (1974); F.E.E. 7: 98 (1995).
Tiny tufted perennial. Dry open bushland & amongst rocks.
Distr: DAR; Chad to Somalia, S to Tanzania; Arabia,
Pakistan to Vietnam.
Note: this record is based on Quézel (1969) who
recorded it as common in barren rocky areas of NW
Darfur. We have not seen the material on which this
record is based.

Oryza barthii A.Chev.

F.P.S. 3: 493 (1956); F.T.E.A. Gramin. 1: 30 (1970); F.E.E.
7: 10 (1995); Biol. Skr. 51: 576 (2005).
Syn: *Oryza breviligulata* A.Chev. & Roehr. – F.P.S. 3:
494 (1956).
Tufted annual. Swamps, weed of rice fields.
Distr: DAR, KOR, EQU; Mauritania to Ethiopia, S to
Angola, Botswana & Zimbabwe.
KOR: Jackson 4028 9/1959; **EQU:** Myers 7554 8/1937.

Oryza brachyantha A.Chev. & Roehr.

F.P.S. 3: 494 (1956); F.W.T.A. 3(2): 365 (1972).
Tufted annual. Shallow pools.
Distr: BAG, EQU; Mauritania to Chad, S to Zambia.
BAG: Myers 13562 10/1940.

Oryza longistaminata A.Chev. & Roehr.

F.T.E.A. Gramin. 1: 30 (1970); F.J.M.: 180 (1976); F.E.E.
7: 10 (1995).
Rhizomatous perennial. Swamps, wet grassland.
Distr: DAR, CS, BAG, UN; Senegal to Somalia, S to
Namibia & South Africa; Madagascar.
DAR: Blair 202 12/1963; **BAG:** Myers 13560 10/1940.

Oryza punctata Kotschy ex Steud.

F.P.S. 3: 494 (1956); F.T.E.A. Gramin. 1: 31 (1970).
Syn: *Oryza schweinfurthiana* Prodoehl – F.P.S. 3: 494
(1956).
Tufted annual or perennial. Open swampy areas.
Distr: KOR, UN, EQU; Ivory Coast to Kenya, Tanzania,
Zimbabwe, South Africa; Madagascar, Thailand.
KOR: Kotschy 136 10/1839; **EQU:** Jackson 336 9/1948.

Oxytenanthera abyssinica (A.Rich.) Munro

F.P.S. 3: 494 (1956); Bot. Exp. Sud.: 57 (1970); F.T.E.A.
Gramin. 1: 11 (1970); F.J.M.: 180 (1976); T.S.S.: 455
(1990); F.E.E. 7: 6 (1995); Biol. Skr. 51: 577 (2005).
Clump-forming bamboo. Riverbanks in wooded
grassland.
Distr: DAR, KOR, CS, ES, BAG, EQU; Senegal to Eritrea, S
to Angola & South Africa.
DAR: Wickens 2925 4/1966; **EQU:** Chipp 51 2/1929.

Panicum anabaptistum Steud.

F.P.S. 3: 501 (1956); F.W.T.A. 3(2): 432 (1972).
Rhizomatous perennial. Wet grassland.
Distr: DAR, EQU; Mauritania & Senegal to Chad & C.A.R.
DAR: Wickens 3148 8/1972; **EQU:** Myers 7254 7/1937.

Panicum atrosanguineum Hochst. ex A.Rich.

F.P.S. 3: 502 (1956); F.T.E.A. Gramin. 3: 488 (1982); F.E.E.
7: 204 (1995); Biol. Skr. 51: 577 (2005).
Syn: *Panicum sociale* Stapf – F.P.S. 3: 502 (1956).
Tufted annual. Disturbed bushland & abandoned
cultivations.
Distr: NS, ES, EQU; Eritrea & Somalia, S to Botswana;
Arabia; Pakistan to Vietnam.
ES: Schweinfurth 1585 10/1865; **EQU:** Myers 7023
7/1937.

Panicum bambusiculme Friis & Vollesen

Kew Bull. 37: 475 (1982); Biol. Skr. 51: 577 (2005).
Rhizomatous perennial. Lowland forest.
Distr: EQU; South Sudan endemic.
IUCN: RD
EQU: Friis & Vollesen 760 (fr) 12/1980.

Panicum brevifolium L.

F.P.S. 3: 507 (1956); F.T.E.A. Gramin. 3: 496 (1982); Biol. Skr. 51: 577 (2005).
Annual. Forest.
Distr: EQU; Senegal to Kenya, S to South Africa; Madagascar, Mascarenes, India to SE Asia.
EQU: Friis & Vollesen 542 11/1980.

Panicum callosum Hochst. ex A.Rich.

F.P.S. 3: 501 (1956); F.J.M.: 180 (1976); F.E.E. 7: 203 (1995).
Annual. Shallow soils amongst rocks.
Distr: DAR, KOR, ES; Cameroon, Ethiopia.
DAR: Wickens 2571 9/1964.
Note: a widely distributed but very scarce species.

Panicum calvum Stapf

F.T.E.A. Gramin. 3: 493 (1982); F.E.E. 7: 209 (1995); Biol. Skr. 51: 577 (2005).
Stoloniferous perennial. Upland forest.
Distr: EQU; Guinea to Ethiopia, S to South Africa.
EQU: Friis & Vollesen 26 11/1980.

Panicum carneovaginatum Renvoize

F.T.E.A. Gramin. 3: 489 (1982).
Tufted perennial. Wet grassland.
Distr: EQU; Uganda, Tanzania, Malawi, Zambia, Mozambique.
EQU: Myers 7548 8/1937.

Panicum chionachne Mez

F.T.E.A. Gramin. 3: 495 (1982); F.E.E. 7: 207 (1995); Biol. Skr. 51: 578 (2005).
Annual or short-lived perennial. Forest.
Distr: EQU; D.R. Congo to Kenya, S to Zambia & Malawi.
EQU: Friis & Vollesen 307 11/1980.

Panicum coloratum L.

F.P.S. 3: 504 (1956); F.T.E.A. Gramin. 3: 485 (1982); F.E.E. 7: 201 (1995); Biol. Skr. 51: 578 (2005).
Syn: *Panicum coloratum* L. var. *minus* Stapf ex Chiov. – F.P.S. 3: 505 (1956); F.T.E.A. Gramin. 3: 486 (1982); Biol. Skr. 51: 578 (2005).
Tufted perennial. Dry grassland & bushland.
Distr: RS, DAR, ES, UN, EQU; Burkina Faso to Somalia, S to Namibia & South Africa; introduced elsewhere in the tropics.
DAR: Wickens 2198 8/1964; **UN:** Harrison 1055 10/1950.

Panicum comorense Mez

F.P.S. 3: 497 (1956); F.T.E.A. Gramin. 3: 492 (1982); F.E.E. 7: 206 (1995); Biol. Skr. 51: 578 (2005).
Annual. Forest.
Distr: EQU; Guinea to Ethiopia, S to South Africa; Madagascar, Comoros.
EQU: Friis & Vollesen 551 11/1980.

Panicum delicatulum Fig. & De Not.

F.T.E.A. Gramin. 3: 491 (1982); F.E.E. 7: 206 (1995).
Annual. Open damp ground, forest margins.
Distr: CS; D.R. Congo to Ethiopia, S to Zambia & Malawi.
CS: Figari s.n. (n.v.).

Note: known in Sudan only from the type from 'Blue Nile'; both F.T.E.A. and F.E.E. note that the location of this specimen is unknown.

Panicum deustum Thunb.

F.P.S. 3: 497 (1956); F.T.E.A. Gramin. 3: 468 (1982); F.E.E. 7: 202 (1995); Biol. Skr. 51: 578 (2005).
Tufted rhizomatous perennial. Forest & bushland.
Distr: UN, EQU; Eritrea & Somalia, S to South Africa.
EQU: Thomas 1561 12/1935.

Panicum dregeanum Nees

F.P.S. 3: 501 (1956); F.T.E.A. Gramin. 3: 478 (1982); F.E.E. 7: 199 (1995).
Tufted perennial. Wet grassland.
Distr: BAG, EQU; Senegal to Ethiopia, S to Angola & South Africa; Madagascar.
EQU: Myers 7095 7/1937.

Panicum fluviicola Steud.

F.J.M.: 180 (1976); F.T.E.A. Gramin. 3: 478 (1982).
Syn: *Panicum aphanoneurum* Stapf – F.P.S. 3: 501 (1956).
Syn: *Panicum dregeanum* sensu Wickens, non Nees – F.J.M.: 180 (1976).
Tufted perennial. Wet grassland, riverbanks.
Distr: DAR, BAG; Mauritania to Kenya, S to Namibia, Botswana & Mozambique.
DAR: Blair 228 2/1964; **BAG:** Myers 7284 7/1937.

Panicum griffonii Franch.

F.T.E.A. Gramin. 3: 479 (1982); Biol. Skr. 51: 579 (2005).
Annual. Woodland with rocky outcrops.
Distr: EQU; Senegal to Uganda, S to Angola & Tanzania.
EQU: Friis & Vollesen 585 11/1980.

Panicum haplocaulos Pilg.

F.P.S. 3: 502 (1956); F.T.E.A. Gramin. 3: 487 (1982); F.E.E. 7: 204 (1995).
Annual. Swamps.
Distr: BAG; Senegal to Ethiopia, Tanzania, Malawi, Zambia.
BAG: Schweinfurth 2003 7/1869.

Panicum hygrocharis Steud.

F.P.S. 3: 505 (1956); F.E.E. 7: 202 (1995).
Syn: *Panicum repens* sensu Clayton & Renvoize pro parte, non L. – F.T.E.A. Gramin. 3: 481 (1982).
Aquatic rhizomatous perennial. Shallow water, marshes.
Distr: UN, EQU; Eritrea & Somalia, S to Namibia & South Africa.
UN: Simpson 7624 4/1930.
Note: (1) Andrews records this species from Northern & Central Sudan but this appears to be in error; we have only seen material from the South; (2) *P. hymeniochilum* Nees and *P. infestum* Andersson are both also likely to occur in South Sudan, having been recorded from the Ugandan side of the Imatong Mts (Friis & Vollesen 2005: 579).

Panicum issongense Pilg.

F.T.E.A. Gramin. 3: 495 (1982); Biol. Skr. 51: 579 (2005).
Perennial. Forest.
Distr: EQU; Uganda, Kenya, Tanzania.
EQU: Friis & Vollesen 729 12/1980.

Panicum laetum Kunth

F.J.M.: 180 (1976); F.T.E.A. Gramin. 3: 488 (1982); F.E.E.
7: 204 (1995).
Annual. Damp grassland.
Distr: DAR, KOR, CS, ES, UN; Mauritania & Senegal to
Eritrea, Angola, Tanzania; Madagascar.
CS: Andrews 127 1/1936; **UN:** Harrison 1004 10/1950.

Panicum maximum Jacq.

F.P.S. 3: 497 (1956); F.J.M.: 180 (1976); F.T.E.A. Gramin.
3: 471 (1982); F.E.E. 7: 198 (1995); Biol. Skr. 51: 579
(2005).
Tufted perennial. Woodland & bushland, disturbed areas.
Distr: RS, DAR, BAG, UN, EQU; Senegal to Somalia, S to
Namibia & South Africa; Madagascar, Arabia, introduced
elsewhere.
DAR: Wickens 2301 8/1964; **UN:** Harrison 245 1/1948.

Panicum monticola Hook.f.

F.T.E.A. Gramin. 3: 494 (1982); F.E.E. 7: 209 (1995); Biol.
Skr. 51: 580 (2005).
Perennial. Forest & associated bushland.
Distr: EQU; Nigeria to Ethiopia, S to South Africa.
EQU: Friis & Vollesen 236 11/1980.

Panicum nervatum (Franch.) Stapf

F.T.E.A. Gramin. 3: 475 (1982); F.E.E. 7: 206 (1995); Biol.
Skr. 51: 580 (2005).
Syn: Panicum fulgens Stapf – F.P.S. 3: 499 (1956).
Perennial. Wooded grassland, rocky hillslopes.
Distr: EQU; Guinea to Ethiopia, S to Angola &
Zimbabwe.
EQU: Myers 7778 9/1937.

Panicum pansum Rendle

F.T.E.A. Gramin. 3: 487 (1982); F.E.E. 7: 203 (1995).
Syn: Panicum kerstingii Mez – F.P.S. 3: 502 (1956).
Annual. Wooded grassland, disturbed areas.
Distr: EQU; Mauritania to Ethiopia, S to Angola and
Zambia.
EQU: Myers 7024 7/1937.

Panicum paucinode Stapf

F.W.T.A. 3(2): 433 (1972).
Tufted annual. Shallow soils on moist ground.
Distr: DAR, KOR; Guinea, Ghana, Nigeria, Cameroon,
Chad.
KOR: Wickens 653 10/1962.

Panicum porphyrrhizos Steud.

F.P.S. 3: 504 (1956); F.J.M.: 180 (1976); F.T.E.A. Gramin.
3: 484 (1982); F.E.E. 7: 201 (1995).
Annual or short-lived perennial. Wet grassland,
riverbanks.
Distr: DAR, KOR, CS, ES, UN, EQU; Senegal to Ethiopia,
S to Botswana & Mozambique.
KOR: Wickens 941 2/1963; **EQU:** Myers 7537 8/1937.

Panicum praealtum Afzel. ex Sw.

F.W.T.A. 3(2): 431 (1972).
Syn: Panicum carinifolium Stapf – F.P.S. 3: 499 (1956).
Perennial. Wooded grassland.
Distr: BAG; Senegal to Cameroon.
BAG: Schweinfurth 2288 8/1869.

Panicum pusillum Hook.f.

F.P.S. 3: 505 (1956); F.J.M.: 181 (1976); F.T.E.A. Gramin. 3:
490 (1982); F.E.E. 7: 204 (1995); Biol. Skr. 51: 580 (2005).
Annual. Upland bushland, grassland & open forest.
Distr: DAR, EQU; Sierra Leone, Nigeria to Ethiopia, S to
Malawi.
DAR: Wickens 2479 9/1964; **EQU:** Friis & Vollesen 25
11/1980.

Panicum repens L.

F.P.S. 3: 504 (1956); F.T.E.A. Gramin. 3: 481 pro parte
(1982); F.E.E. 7: 202 (1995).
Rhizomatous perennial. Marshes, shallow water.
Distr: DAR, KOR, CS, BAG, UN, EQU; Ethiopia,
palaeotropical & warm temperate; introduced in the
neotropics.
KOR: Wickens 705 10/1962; **UN:** Harrison 1028 10/1950.
Note: Andrews recorded this species as "an important
weed infesting the banks of the Gezira canals".

Panicum subalbidum Kunth

F.J.M.: 181 (1976); F.T.E.A. Gramin. 3: 484 (1982); F.E.E.
7: 201 (1995); Biol. Skr. 51: 580 (2005).
Syn: Panicum glabrescens Steud. – F.P.S. 3: 505 (1956).
Annual or short-lived perennial. Swamps, river & lake
margins.
Distr: EQU; Mauritania to Somalia, S to Namibia & South
Africa; Madagascar.
DAR: Wickens 2013 7/1964; **EQU:** Friis & Vollesen 318
11/1980.

Panicum tenellum Lam.

Biol. Skr. 51: 580 (2005).
Syn: Panicum lindleyanum Nees ex Steud. – F.W.T.A.
3(2): 431 (1972); F.T.E.A. Gramin. 3: 475, in notes (1982).
Annual. Shallow pools & wet flushes over rock.
Distr: EQU; Senegal to C.A.R., S to Zambia &
Mozambique.
EQU: Friis & Vollesen 584 11/1980.

Panicum trichocladum Hack. ex K.Schum.

F.P.S. 3: 499 (1956); F.T.E.A. Gramin. 3: 473 (1982); F.E.E.
7: 207 (1995); Biol. Skr. 51: 581 (2005).
Rhizomatous perennial. Forest.
Distr: EQU; Ethiopia to Zimbabwe & Mozambique.
EQU: Myers 8753 3/1938.

Panicum turgidum Forssk.

F.P.S. 3: 504 (1956); F.E.E. 7: 199 (1995).
Perennial. Sandy desert & semi-desert.
Distr: NS, RS, DAR, KOR, CS; Mauritania to Somalia; N
Africa, Arabia to India.
RS: Carter 1914 11/1987.

Panicum wiehei Renvoize

F.T.E.A. Gramin. 3: 499 (1982); Biol. Skr. 51: 581 (2005).
Annual. Upland forest margins & riverine woodland.
Distr: EQU; Uganda, Malawi, Zimbabwe, Mozambique.
EQU: Friis & Vollesen 8 11/1980.

Parahyparrhenia annua (Hack.) Clayton

F.W.T.A. 3(2): 498 (1972).
Syn: Andropogon annuus Hack. – F.P.S. 3: 391 (1956).
Tufted annual. Shallow pools over rock.

Distr: BAG; Senegal to Chad & C.A.R.; Algeria.
BAG: Schweinfurth III.183 9/1869.

Paspalidium dersertorum (A.Rich.) Stapf

F.P.S. 3: 507 (1956); F.T.E.A. Gramin. 3: 552 (1982); F.E.E. 7: 243 (1995).
Tufted perennial. Seasonally damp grassland, dry riverbeds.
Distr: RS, KOR, CS; Eritrea, Ethiopia, Somalia, Kenya; Arabia.
CS: Harrison 1211 10/1951.

Paspalidium geminatum (Forssk.) Stapf

F.P.S. 3: 507 (1956); F.T.E.A. Gramin. 3: 552 (1982); F.E.E. 7: 243 (1995).
Rhizomatous mat-forming perennial. Marshes, shallow water.
Distr: NS, BAG, UN; widespread in the tropics.
NS: Ahti 16593 10/1962; **UN:** Harrison 1032 10/1950.

Paspalum glumaceum Clayton

F.T.E.A. Gramin. 3: 612 (1982); F.E.E. 7: 233 (1995).
Tufted perennial. Forest margins, damp areas.
Distr: EQU; Ethiopia to Tanzania, Zimbabwe; Madagascar.
EQU: Myers 7777 9/1937.

Paspalum lamprocaryon K.Schum.

F.E.E. 7: 234 (1995).
Syn: *Paspalum auriculatum* sensu Andrews, non J.Presl – F.P.S. 3: 509 (1956).
Perennial. Swampy grassland, stream margins.
Distr: EQU; Senegal to Ethiopia, S to Zimbabwe.
EQU: Schweinfurth 3786 5/1870.

Paspalum scrobiculatum L.

F.J.M.: 181 (1976); F.T.E.A. Gramin. 3: 610 (1982); F.E.E. 7: 233 (1995); Biol. Skr. 51: 581 (2005).
Syn: *Paspalum commersonii* Lam. – F.P.S. 3: 509 (1956).
Syn: *Paspalum polystachyum* R.Br. – F.P.S. 3: 509 (1956).
Tufted or mat-forming perennial. Damp disturbed ground.
Distr: DAR, CS, BAG, UN, EQU; pantropical.
DAR: Wickens 2747 9/1964; **EQU:** Myers 6765 5/1937.

Paspalum vaginatum Sw.

F.T.E.A. Gramin. 3: 609 (1982); F.E.E. 7: 234 (1995).
Stoloniferous perennial. Damp saline sites, irrigation ditches.
Distr: NS; native to the neotropics, now widespread in the palaeotropics & subtropics.
NS: Ahti 16382 10/1962.
Note: this is usually a coastal species but can extend inland on saline soils; the record is taken from Ahti et al. (1973).

Pennisetum glaucum (L.) R.Br.

F.T.E.A. Gramin. 3: 672 (1982); F.E.E. 7: 263 (1995).
Syn: *Setaria glauca* (L.) P.Beauv. – Bot. Exp. Sud.: 58 (1970).
Annual. Cultivated, occasionally naturalised.
Distr: RS, DAR; widely cultivated – in Africa particularly in the Sahel zone.
DAR: Drar 2280 5/1938 (n.v.).

Note: said to be the cultivated derivative of *P. violaceum*; an important crop plant 'pearl millet' or 'bullrush millet'.

Pennisetum gracilescens Hochst.

F.P.S. 3: 513 (1956); F.J.M.: 181 (1976); F.E.E. 7: 267 (1995).
Tufted perennial. Damp rocks.
Distr: DAR; Eritrea, Ethiopia.
IUCN: RD
DAR: Blair 245 2/1964.

Pennisetum laxius (Clayton) Clayton

Kew Bull. 32: 580 (1978).
Annual. Rocky outcrops.
Distr: EQU; Ghana to Cameron, Annobon, São Tomé.
EQU: Myers 7971 12/1937.

Pennisetum macrourum Trin.

F.T.E.A. Gramin. 3: 689 (1982); F.E.E. 7: 271 (1995); Biol. Skr. 51: 581 (2005).
Syn: *Pennisetum giganteum* A.Rich. – F.P.S. 3: 512 (1956); F.J.M.: 181 (1976).
Syn: *Pennisetum stenorrhachis* Stapf & C.E.Hubb. – F.P.S. 3: 512 (1956); Bot. Exp. Sud.: 58 (1970); F.J.M.: 182 (1976).
Rhizomatous perennial. Swamps, riverbanks.
Distr: DAR, ES, EQU; Guinea to Somalia, S to Namibia & South Africa; Arabia.
DAR: Blair 260 2/1964; **EQU:** Myers 9437 9/1938.

Pennisetum mezianum Leeke

F.P.S. 3: 511 (1956); F.T.E.A. Gramin. 3: 686 (1982); F.E.E. 7: 269 (1995).
Bushy rhizomatous perennial. Open bushland & dry grassland including disturbed areas.
Distr: EQU; Ethiopia to Tanzania, Namibia, South Africa.
EQU: Myers 14073 9/1941.

Pennisetum nubicum (Hochst.) K.Schum. ex Engl.

F.E.E. 7: 275 (1995).
Syn: *Beckeropsis nubica* (Hochst.) Fig. & De Not. – F.P.S. 3: 401 (1956); Bot. Exp. Sud.: 52 (1970); F.J.M.: 168 (1976).
Tufted annual. Disturbed areas.
Distr: RS, DAR, KOR, CS, ES; Eritrea, Ethiopia; Arabia.
ES: Beshir 120 9/1951.

Pennisetum pedicellatum Trin. subsp. pedicellatum

F.P.S. 3: 517 (1956); F.J.M.: 181 (1976); F.T.E.A. Gramin. 3: 680 (1982); F.E.E. 7: 262 (1995).
Annual. Disturbed areas, bushland on rocky hillslopes.
Distr: DAR, KOR, CS, ES, UN, EQU; Mauritania & Senegal to Eritrea, S to Zambia; Madagascar, India to Vietnam.
DAR: Wickens 2716 9/1964; **UN:** Sherif A.4037 9/1951.

Pennisetum petiolare (Hochst.) Chiov.

F.E.E. 7: 275 (1995).
Syn: *Beckeropsis petiolaris* (Hochst.) Fig. & De Not. – F.P.S. 3: 401 (1956).
Tufted annual. Disturbed areas.
Distr: KOR, ES; Eritrea, Ethiopia.
ES: Schweinfurth 1004 10/1865.

Pennisetum pirottae Chiov.

F.E.E. 7: 275 (1995).
Syn: *Beckeropsis pirottae* (Chiov.) Stapf & C.E.Hubb. –
F.P.S. 3: 400 (1956).
Tufted perennial. Riverbanks.
Distr: CS, UN; Eritrea.
IUCN: RD
CS: <u>Jackson 1577</u> 12/1950; **UN:** <u>Ferguson 443</u> 1948.

Pennisetum polystachion (L.) Schult. subsp. *polystachion*

F.P.S. 3: 517 (1956); F.J.M.: 181 (1976); F.E.E. 7: 262
(1995); Biol. Skr. 51: 582 (2005).
Tufted annual or perennial. Disturbed areas.
Distr: DAR, KOR, BAG, UN, EQU; palaeotropical;
introduced in the neotropics.
DAR: <u>Wickens 2531</u> 9/1964; **EQU:** <u>Jackson 477</u> 11/1948.

Pennisetum purpureum Schumach.

F.P.S. 3: 514 (1956); Bot. Exp. Sud.: 57 (1970); F.T.E.A.
Gramin. 3: 677 (1982); F.E.E. 7: 263 (1995); Biol. Skr. 51:
582 (2005).
Stoloniferous perennial. Grassland, forest margins,
riverine fringes.
Distr: CS, EQU; Senegal to Ethiopia, S to Zimbabwe &
Mozambique; widely introduced elsewhere in the tropics
& subtropics.
CS: <u>Drar 239</u> 2/1938 (n.v.); **EQU:** <u>Harrison 489</u> 2/1949.

Pennisetum ramosum (Hochst.) Schweinf.

F.P.S. 3: 512 (1956); F.J.M.: 182 (1976); F.T.E.A. Gramin. 3:
684 (1982); F.E.E. 7: 267 (1995); Biol. Skr. 51: 582 (2005).
Tufted annual. Swamps, seasonally wet soils, often in
disturbed areas.
Distr: DAR, KOR, BAG, UN, EQU; Nigeria to Eritrea, S to
Tanzania.
KOR: <u>Wickens 855</u> 11/1962; **UN:** <u>Harrison 33</u> 6/1947.

Pennisetum schweinfurthii Pilg.

F.E.E. 7: 261 (1995).
Syn: *Pennisetum tetrastachyum* K.Schum. – F.P.S. 3:
517 (1956).
Annual or perennial. Clay soils.
Distr: KOR, ES, UN; Ethiopia.
IUCN: RD
KOR: <u>Andrews 56</u> 11/1933; **UN:** <u>Sherif A.4028</u> 9/1951.

Pennisetum setaceum (Forssk.) Chiov.

F.P.S. 3: 514 (1956); F.T.E.A. Gramin. 3: 675 (1982); F.E.E.
7: 265 (1995).
Syn: *Pennisetum erythraeum* Chiov. – F.P.S. 3: 513
(1956).
Tufted perennial. Rocky hillslopes.
Distr: RS; Chad to Somalia, S to Tanzania; N Africa,
Middle East, Arabia, Afghanistan, introduced elsewhere.
RS: <u>Jackson 3956</u> 4/1959.

Pennisetum sieberianum (Schltdl.) Stapf & C.E.Hubb.

Bot. Exp. Sud.: 58 (1970); F.T.E.A. Gramin. 3: 673 (1982);
F.E.E. 7: 264 (1995).
Syn: *Pennisetum niloticum* Stapf & C.E.Hubb. – Bot.
Exp. Sud.: 57 (1970).
Annual. Weed of cultivation of *P. glaucum*.

Distr: DAR, CS; Senegal to Eritrea & Ethiopia, Angola; N
Africa, Arabia.
DAR: <u>Drar 2265</u> 5/1938 (n.v.).
Note: this species is said to mimic the crop millet *P.
glaucum* and so survives as a weed within the crop, but
rarely persists beyond the period of cultivation. Drar
(1970) cites several collections, some of which he lists as
being cultivated for fodder.

Pennisetum thunbergii Kunth.

F.T.E.A. Gramin. 3: 687 (1982); F.E.E. 7: 269 (1995); Biol.
Skr. 51: 583 (2005).
Syn: *Pennisetum adoense* Steud. – F.P.S. 3: 513 (1956).
Syn: *Pennisetum glabrum* Steud. – F.P.S. 3: 512 (1956);
F.J.M.: 181 (1976).
Tufted rhizomatous perennial. Upland grassland, often in
disturbed areas.
Distr: DAR, EQU; Nigeria, Eritrea to Angola & South
Africa; Yemen, introduced in Sri Lanka.
DAR: <u>Jackson 3352</u> 12/1954; **EQU:** <u>Myers 14201</u>
10/1941.

Pennisetum trachyphyllum Pilg.

F.P.S. 3: 511 (1956); F.T.E.A. Gramin. 3: 682 (1982); F.E.E.
7: 272 (1995); Biol. Skr. 51: 583 (2005).
Perennial. Forest margins & clearings.
Distr: EQU; Cameroon, Ethiopia to D.R. Congo &
Tanzania.
EQU: <u>Myers 11779</u> 8/1939.

Pennisetum unisetum (Nees) Benth.

F.T.E.A. Gramin. 3: 681 (1982); F.E.E. 7: 273 (1995); Biol.
Skr. 51: 583 (2005).
Syn: *Beckeropsis uniseta* (Nees) K.Schum. – F.P.S. 3:
401 (1956); Bot. Exp. Sud.: 52 (1970); F.J.M.: 168 (1976).
Tufted perennial. Wooded grassland & bushland.
Distr: DAR, KOR, CS, UN, EQU; Senegal to Somalia, S to
Angola & South Africa; Arabia.
DAR: <u>Wickens 2785</u> 10/1964; **UN:** <u>Harrison 1034</u>
10/1950.

Pennisetum violaceum (Lam.) Rich.

F.J.M.: 182 (1976); F.E.E. 7: 264 (1995).
Syn: *Pennisetum darfuricum* Stapf & C.E.Hubb. – F.P.S.
3: 516 (1956).
Syn: *Pennisetum fallax* (Fig. & De Not.) Stapf &
C.E.Hubb. – F.P.S. 3: 514 (1956); F.J.M.: 181 (1976).
Syn: *Pennisetum mollissimum* Hochst. – F.P.S. 3: 516
(1956).
Syn: *Pennisetum ochrops* Stapf & C.E.Hubb. – F.P.S. 3:
516 (1956).
Annual. Disturbed areas.
Distr: DAR, KOR, CS; Mauritania & Senegal to Ethiopia;
Saharan Mts.
KOR: <u>Wickens 687</u> 10/1962.

Pentameris pictigluma (Steud.) Galley & H.P.Linder

Ann. Missouri Bot. Gard. 97: 334 (2010).
Syn: *Pentaschistis imatongensis* C.E.Hubb. – F.P.S. 3:
519 (1956).
Syn: *Pentaschistis pictigluma* (Steud.) Pilg. – F.J.M.:
182 (1976); F.E.E. 7: 70 (1995); Biol. Skr. 51: 584 (2005).
Tufted perennial. Montane grassland.

Distr: DAR, EQU; Cameroon, Ethiopia to Tanzania; Yemen.
DAR: <u>Jackson 3338</u> 12/1954; **EQU:** <u>Friis & Vollesen 835</u> 12/1980.

Perotis patens Gand.

F.P.S. 3: 519 (1956); F.T.E.A. Gramin. 2: 394 (1974); F.E.E. 7: 182 (1995); Biol. Skr. 51: 594 (2005).
Tufted annual or short-lived perennial. Disturbed areas on sandy soils.
Distr: BAG, UN, EQU; Guinea to Somalia, S to Namibia & South Africa; Madagascar.
EQU: <u>Myers 6343</u> 3/1937.

Phacelurus gabonensis (Steud.) Clayton

F.Z. 10(4): 160 (2002).
Syn: *Jardinea congoensis* (Hack.) Franch. – F.P.S. 3: 477 (1956).
Tufted perennial. Seasonally wet grassland, riverbanks.
Distr: EQU; Ghana to Chad, S to Angola & Zambia.
EQU: <u>Myers 7644</u> 9/1937.

Phaenanthoecium koestlinii (Hochst. ex A.Rich.) C.E.Hubb.

F.P.S. 3: 519 (1956); F.J.M.: 182 (1976); F.E.E. 7: 73 (1995).
Tufted perennial. Damp rock crevices & cliffs.
Distr: DAR; Eritrea, Ethiopia; Yemen.
DAR: <u>Lynes 44</u> 1/1922.

Phalaris minor Retz.

F.E.E. 7: 42 (1997).
Syn: *Phalaris minor* Retz. var. *gracilis* (Parl.) Pamp. – Bot. Exp. Sud.: 58 (1970).
Tufted annual. Dry sandy & stony ground.
Distr: RS; Eritrea; N Africa to India, introduced elsewhere.
RS: <u>Drar 299</u> 3/1938 (n.v.).

Phragmites australis (Cav.) Trin. ex Steud. subsp. *altissimus* (Benth.) Clayton

F.T.E.A. Gramin. 1: 118 (1970); F.E.E. 7: 64 (1995).
Rhizomatous perennial. Swamps, riverbanks, forming reedbeds.
Distr: NS; Mauritania & Senegal to Somalia & Kenya; N Africa, Arabia to Iran.
NS: <u>Kennedy-Shaw 9</u> 10/1932.

Phragmites karka (Retz.) Trin. ex Steud.

F.T.E.A. Gramin. 1: 118 (1970); F.J.M.: 182 (1976); F.E.E. 7: 65 (1995); Biol. Skr. 51: 584 (2005).
Robust perennial. Riverbanks, lake shores.
Distr: DAR, UN, EQU; palaeotropical.
DAR: <u>Blair 218</u> 1/1964; **UN:** <u>J.M. Lock 80/10</u> 2/1980.

Phragmites mauritianus Kunth

F.P.S. 3: 519 (1956); F.T.E.A. Gramin. 1: 120 (1970).
Rhizomatous perennial. Riverbanks & swamps, forming reedbeds.
Distr: ?DAR, CS, BAG, UN, EQU; C.A.R. to Ethiopia, S to Namibia & South Africa; Madagascar, Mascarenes.
CS: <u>Kotschy 414</u> 1837; **BAG:** <u>Myers 6299</u> 2/1937.
Note: Drar (1970: 58) records this species from Darfur; this may be based on a misidentification of *P. karka*.

Poa bulbosa L.

F. Libya: 68 (1989).
Syn: *Poa bulbosa* L. var. *vivipara* Koeler – F.J.M.: 182 (1976).
Perennial. Montane grassland.
Distr: DAR; Europe to Central Asia, S to N Africa & Arabia, South Africa; introduced in the Americas.
DAR: <u>Blair 355</u> 9/1964.

Poa leptoclada Hochst. ex A.Rich.

F.J.M.: 182 (1976); F.E.E. 7: 20 (1995).
Tufted perennial. Montane grassland.
Distr: DAR; Cameroon, Eritrea & Somalia, S to Zimbabwe, South Africa; Arabia.
DAR: <u>Blair 308</u> 5/1964.

Poa schimperiana Hochst. ex A.Rich.

F.T.E.A. Gramin. 1: 48 (1970); F.E.E. 7: 21 (1995); Biol. Skr. 51: 584 (2005).
Syn: *Poa leptoclada* sensu Andrews, non Hochst. ex A.Rich. – F.P.S. 3: 521 (1956).
Tufted perennial. Montane grassland, forest margins.
Distr: RS, EQU; Nigeria, Cameroon, Ethiopia to Malawi; Arabia.
RS: <u>Jackson 2860</u> 4/1953; **EQU:** <u>Friis & Vollesen 103</u> 11/1980.

Poecilostachys oplismenoides (Hack.) Clayton

F.E.E. 7: 194 (1995); Biol. Skr. 51: 585 (2005).
Syn: *Chloachne oplismenoides* (Hack.) Stapf ex Robyns – F.T.E.A. Gramin. 3: 545 (1982).
Perennial. Upland forest.
Distr: EQU; Nigeria, Cameroon, Ethiopia to Zimbabwe & Mozambique.
EQU: <u>Friis & Vollesen 121</u> 11/1980.

Pogonarthria squarrosa (Roem. & Schult.) Pilg.

F.T.E.A. Gramin. 2: 267 (1974); F.J.M.: 182 (1976); F.E.E. 7: 130 (1995).
Tufted perennial. Grassland, open woodland, disturbed open areas.
Distr: DAR; Ivory Coast to Eritrea, S to Namibia & South Africa; Madagascar.
DAR: <u>Jackson 3871</u> 11/1959.

Polypogon monspeliensis (L.) Desf.

F.P.S. 3: 521 (1956); F.T.E.A. Gramin. 1: 100 (1970); F.J.M.: 182 (1976); F.E.E. 7: 44 (1995).
Tufted annual. Riverbanks.
Distr: NS, DAR; Mauritania to Somalia, S to Tanzania; Europe to Japan; introduced in southern Africa, Australia & the Americas.
DAR: <u>Blair 216</u> 1/1964.

Pseudechinolaena polystachya (Kunth) Stapf

F.T.E.A. Gramin. 3: 547 (1982); F.E.E. 7: 194 (1995); Biol. Skr. 51: 585 (2005).
Mat-forming annual. Forest.
Distr: EQU; pantropical.
EQU: <u>Friis & Vollesen 91</u> 11/1980.

Rhytachne rottboellioides Desv. ex Ham.

F.P.S. 3: 523 (1956); F.T.E.A. Gramin. 3: 843 (1982).
Tufted perennial. Swamps, wet grassland.

Distr: BAG; Senegal to Kenya, S to Angola & South Africa; Madagascar, introduced in the neotropics.
BAG: Schweinfurth 1493 4/1869.

Rhytachne triaristata (Steud.) Stapf

F.P.S. 3: 523 (1956); F.W.T.A. 3(2): 511 (1972).
Tufted annual. Shallow soils over rock outcrops, disturbed ground.
Distr: BAG; Senegal to C.A.R., Zambia.
BAG: Schweinfurth 2485 10/1869.

Rostraria cristata (L.) Tzvelev

F.E.E. 7: 40 (1995).
Syn: *Koeleria phleoides* (Vill.) Pers. – F.P.S. 3: 477 (1956); Bot. Exp. Sud.: 56 (1970).
Annual, sometimes tufted. Weed of disturbed ground.
Distr: RS, DAR; Mauritania, Eritrea; Mediterranean, N Africa, Arabia to India, introduced elsewhere.
DAR: Wickens 2794 10/1964.

Rostraria pumila (Desf.) Tzvelev

F.Egypt 4: 164 (2005).
Syn: *Koeleria pumila* (Desf.) Domin – Bot. Exp. Sud.: 56 (1970).
Annual. Weed of cultivation, alluvial plains.
Distr: RS; Spain, N Africa to India.
RS: Drar 417 3/1938 (n.v.).

Rottboellia cochinchinensis (Lour.) Clayton

F.T.E.A. Gramin. 3: 853 (1982); F.E.E. 7: 365 (1995); Biol. Skr. 51: 585 (2005).
Syn: *Rottboellia exaltata* L.f. – F.P.S. 3: 523 (1956); F.J.M.: 183 (1976).
Annual with prop roots. Disturbed & cultivated ground.
Distr: DAR, ES, UN, EQU; palaeotropical; introduced in the neotropics.
DAR: Wickens 2491 9/1964; **UN:** Sherif A.2840 9/1951.

Saccharum spontaneum L. subsp. *spontaneum*

F.T.E.A. Gramin. 3: 704 (1982); F.E.E. 7: 293 (1995).
Syn: *Saccharum spontaneum* L. subsp. *aegyptiacum* sensu Wickens, non (Willd.) Hack. – F.J.M.: 183 (1976).
Rhizomatous perennial. Riverbanks, alluvial plains.
Distr: NS, DAR, CS; Burkina Faso to Ethiopia; Egypt, Arabia, tropical & warm temperate Asia.
DAR: Blair 217 1/1964.

Saccharum spontaneum L. subsp. aegyptiacum (Willd.) Hack.

F.P.S. 3: 525 (1956); F.T.E.A. Gramin. 3: 704 (1982); F.E.E. 7: 293 (1995).
Rhizomatous perennial. Riverbanks, alluvial plains.
Distr: CS, EQU; Burkina Faso & Ghana to Somalia, S to Malawi; N Africa, Middle East, Arabia.
CS: Broun 782 1/1906; **EQU:** Simpson 7297 7/1929.

Sacciolepis africana C.E.Hubb. & Snowden

F.P.S. 3: 526 (1956); F.T.E.A. Gramin. 3: 455 (1982); F.E.E. 7: 195 (1995).
Rhizomatous perennial. Swamps, shallow water.
Distr: UN, EQU; Senegal to Ethiopia, S to Namibia & South Africa; Madagascar.
UN: J.M. Lock 81/206 8/1981.

Sacciolepis chevalieri Stapf

F.P.S. 3: 525 (1956); F.T.E.A. Gramin. 3: 459 (1982).
Tufted perennial. Swamps.
Distr: ?BAG/EQU; Senegal to Kenya, S to Angola & South Africa; Madagascar.
Note: Andrews records this species from Equatoria but we have seen no specimens with which to verify this record. It is quite likely to occur in the wetlands of Bahr el Ghazal.

Sacciolepis ciliocincta (Pilg.) Stapf

F.P.S. 3: 525 (1956); F.W.T.A. 3(2): 425 (1972).
Annual. Seasonal wetlands over rock.
Distr: BAG, UN; Senegal to D.R. Congo.
BAG: Harrison 1095 11/1950.

Sacciolepis indica (L.) Chase

F.T.E.A. Gramin. 3: 458 (1982); F.E.E. 7: 195 (1995); Biol. Skr. 51: 586 (2005).
Syn: *Sacciolepis auriculata* Stapf – F.P.S. 3: 526 (1956).
Annual. Swamps, streamsides.
Distr: EQU; Guinea Bissau to Ethiopia, S to Angola & South Africa; Madagascar, SE Asia.
EQU: Myers 6936 7/1937.

Sacciolepis micrococca Mez

F.P.S. 3: 525 (1956); F.T.E.A. Gramin. 3: 458 (1982).
Tufted annual. Swamps.
Distr: BAG, UN; Senegal to Uganda, S to Angola & Zimbabwe; Madagascar.
BAG: Harrison 1081 10/1950.

Sacciolepis myosuroides (R.Br.) A.Camus

Syn: *Sacciolepis huillensis* (Rendle) Stapf – F.T.E.A. Gramin. 3: 458 (1982).
Syn: *Sacciolepis spiciformis* (Hochst. ex A.Rich.) Stapf – F.E.E. 7: 196 (1995).
Tufted annual. Grassland.
Distr: BAG; Cameroon to Ethiopia, S to Namibia & South Africa; Madagascar, India to SE Asia.
BAG: Schweinfurth 2591 1869.

Sacciolepis typhura (Stapf) Stapf

F.P.S. 3: 526 (1956); F.T.E.A. Gramin. 3: 460 (1982).
Rhizomatous perennial. Swamps.
Distr: BAG; Ivory Coast, Benin, Cameroon to Kenya, S to Namibia & South Africa.
BAG: Harrison 704 3/1949.

Schizachyrium brevifolium (Sw.) Nees ex Büse

F.P.S. 3: 527 (1956); F.J.M.: 183 (1976); F.T.E.A. Gramin. 3: 754 (1982); F.E.E. 7: 317 (1995); Biol. Skr. 51: 586 (2005).
Syn: *Schizachyrium brevifolium* (Sw.) Nees ex Büse var. *flaccidum* (A.Rich.) Stapf – F.P.S. 3: 527 (1956).
Annual. Grassland, woodland, forest clearings.
Distr: DAR, UN, EQU; pantropical.
DAR: Blair 176 10/1963; **EQU:** Thomas 1611 12/1935.

Schizachyrium exile (Hochst.) Pilg.

F.P.S. 3: 529 (1956); Bot. Exp. Sud.: 58 (1970); F.J.M.: 183 (1976); F.T.E.A. Gramin. 3: 756 (1982); F.E.E. 7: 319 (1995).
Tufted annual. Open bushland & woodland.
Distr: DAR, KOR, CS, ES, BAG; Mauritania to Eritrea, S to Namibia & South Africa; India to Thailand.

DAR: Wickens 2739 9/1964; BAG: Schweinfurth 2416 9/1869.

Schizachyrium platyphyllum (Franch.) Stapf
F.P.S. 3: 527 (1956); F.T.E.A. Gramin. 3: 755 (1982).
Annual or perennial. Marshes.
Distr: EQU; Senegal to Kenya, S to Mozambique; Madagascar.
EQU: Myers 7879 10/1937.

Schizachyrium sanguineum (Retz.) Alston
F.T.E.A. Gramin. 3: 756 (1982); F.E.E. 7: 319 (1995).
Tufted perennial. Open bushland & woodland.
Distr: EQU; pantropical.
EQU: Myers 14054 9/1941.

Schizachyrium schweinfurthii (Hack.) Stapf
F.P.S. 3: 529 (1956); F.W.T.A. 3(2): 481 (1972).
Tufted perennial. Moist shallow soils over rock.
Distr: BAG, UN; Mali & Ivory Coast to Cameroon.
BAG: Harrison 1096 11/1950.

Schizachyrium urceolatum (Hack.) Stapf
F.P.S. 3: 527 (1956); F.E.E. 6: 319 (1997).
Annual. Grassland.
Distr: ES; Senegal to Ethiopia.
ES: Schweinfurth 1031 10/1865.

Schmidtia kalahariensis Stent
F.Z. 10(1): 152 (1971).
Tufted annual. Grassland, open woodland.
Distr: DAR, KOR; Chad, Angola, Namibia, Botswana, South Africa.
KOR: Harrison 918 9/1950.

Schmidtia pappophoroides Steud. ex J.A.Schmidt
F.P.S. 3: 529 (1956); F.T.E.A. Gramin. 1: 165 (1970); F.E.E. 7: 91 (1995).
Rhizomatous & often stoloniferous perennial. Open ground in dry bushland.
Distr: DAR, KOR; Cape Verde Is., Mauritania & Senegal to Somalia, S to Namibia & South Africa; Egypt, Pakistan.
KOR: Wickens 223 8/1962.

Schoenefeldia gracilis Kunth
F.P.S. 3: 530 (1956); F.J.M.: 183 (1976); F.E.E. 7: 163 (1995).
Annual. Dry grassland.
Distr: RS, DAR, KOR, CS, ES; Senegal to Ethiopia; Arabia, India.
DAR: Blair 162 9/1963.

Schoenefeldia transiens (Pilg.) Chiov.
F.T.E.A. Gramin. 2: 309 (1974); F.E.E. 7: 163 (1995).
Tufted perennial. Dry grassland, wooded grassland & bushland.
Distr: EQU; Ethiopia & Somalia, S to South Africa.
EQU: Peers M.O.9 9/1953.

Sehima ischaemoides Forssk.
F.P.S. 3: 531 (1956); F.J.M.: 184 (1976); F.T.E.A. Gramin. 3: 750 (1982); F.E.E. 7: 316 (1995).
Tufted annual. Dry grassland & bushland, semi-desert.
Distr: RS, DAR, KOR, CS, ES; Mali to Somalia, S to

Namibia & South Africa; Arabia, Pakistan, India.
DAR: Wickens 2768 10/1964.

Sehima nervosum (Rottler) Stapf
F.P.S. 3: 531 (1956); F.T.E.A. Gramin. 3: 750 (1982); F.E.E. 7: 316 (1995); Biol. Skr. 51: 586 (2005).
Tufted perennial. Dry grassland & bushland, semi-desert.
Distr: EQU; Eritrea & Somalia, S to Mozambique; Arabia, Pakistan to SE Asia.
EQU: Beshir 23 8/1951.

Setaria acromelaena (Hochst.) T.Durand & Schinz
F.T.E.A. Gramin. 3: 527 (1982); F.E.E. 7: 238 (1995).
Tufted annual. Seasonally damp grassland including disturbed areas.
Distr: CS, ES, UN, EQU; Eritrea & Somalia, S to D.R. Congo & Tanzania.
ES: Harrison 1224 10/1951; EQU: Myers 14020 9/1941.

Setaria atrata Hack.
F.P.S. 3: 534 (1956); F.T.E.A. Gramin. 3: 524 (1982); F.E.E. 7: 236 (1995); Biol. Skr. 51: 586 (2005).
Tufted perennial. Swampy grassland.
Distr: EQU; Ethiopia to Malawi, Angola.
EQU: Myers 11696 1/1939.

Setaria barbata (Lam.) Kunth
F.P.S. 3: 537 (1956); F.J.M.: 184 (1976); F.T.E.A. Gramin. 3: 536 (1982); F.E.E. 7: 241 (1995); Biol. Skr. 51: 587 (2005).
Tufted annual. Disturbed grassland & bushland, often in damp areas.
Distr: DAR, ?KOR, UN, EQU; widespread in the tropics.
DAR: Wickens 2191 8/1964; EQU: Jackson 322 9/1948.

Setaria homonyma (Steud.) Chiov.
Bot. Exp. Sud.: 58 (1970); F.T.E.A. Gramin. 3: 536 (1982); F.E.E. 7: 241 (1995); Biol. Skr. 51: 587 (2005).
Syn: Setaria lancea Stapf ex Massey – F.P.S. 3: 538 (1956).
Tufted annual. Weed of disturbed areas & cultivation.
Distr: RS, UN; Cameroon to Ethiopia, S to Namibia & South Africa; India to Myanmar.
RS: Drar 446 3/1938 (n.v.); UN: Broun 1114 6/1906.
Note: the record for the Red Sea region is based on Drar (1970) who collected two specimens from Erkowit; it is odd that no other collectors have picked this species up at this well botanised location.

Setaria incrassata (Hochst.) Hack.
F.T.E.A. Gramin. 3: 525 (1982); F.E.E. 7: 237 (1995); Biol. Skr. 51: 586 (2005).
Syn: Setaria lynesii Stapf & C.E.Hubb. – F.P.S. 3: 532 (1956).
Tufted perennial. Swamps, wet grassland over clay, forest.
Distr: DAR, KOR, CS, ES, BAG, UN, EQU; Nigeria to Somalia, S to Namibia & South Africa.
DAR: Wickens 2015 7/1964; EQU: Myers 13894 7/1941.

Setaria kagerensis Mez
F.T.E.A. Gramin. 3: 538 (1982); F.E.E. 7: 242 (1995); Biol. Skr. 51: 587 (2005).
Short-lived, mat-forming perennial. Riverine forest, woodland & bushland.

Distr: EQU; Ethiopia to D.R. Congo & Tanzania.
EQU: Jackson 3887 6/1958.

Setaria longiseta P.Beauv.

F.P.S. 3: 536 (1956); F.T.E.A. Gramin. 3: 535 (1982); F.E.E.
7: 241 (1995); Biol. Skr. 51: 587 (2005).
Syn: Setaria lasiothyrsa Stapf ex Massey – F.P.S. 3: 536
(1956).
Tufted perennial. Bushland & woodland.
Distr: EQU; Guinea to Ethiopia, S to Angola, Zimbabwe
& Mozambique.
EQU: Sillitoe 21 1919.

Setaria megaphylla (Steud.) T.Durand & Schinz

F.P.S. 3: 536 (1956); F.T.E.A. Gramin. 3: 539 (1982); F.E.E.
7: 242 (1995); Biol. Skr. 51: 588 (2005).
Syn: Setaria chevalieri Stapf – F.P.S. 3: 537 (1956).
Syn: Setaria plicatilis (Hochst.) Hack. ex Engl. – F.J.M.:
184 (1976).
Tufted perennial. Forest margins & clearings.
Distr: DAR, UN, EQU; Senegal to Somalia, S to Angola &
South Africa; Madagascar, Arabia, India, Myanmar.
DAR: Blair 23 6/1963; **EQU:** Myers 7746 9/1937.

Setaria poiretiana (Schult.) Kunth

F.T.E.A. Gramin. 3: 540 (1982); F.E.E. 7: 243 (1995); Biol.
Skr. 51: 588 (2005).
Syn: Setaria caudula Stapf – F.P.S. 3: 537 (1956).
Tufted perennial. Forest margins & clearings.
Distr: EQU; Nigeria to Ethiopia, S to Zambia; India,
Myanmar, scattered in the neotropics.
EQU: Friis & Vollesen 65 11/1980.

Setaria pumila (Poir.) Roem. & Schult.

F.T.E.A. Gramin. 3: 530 (1982); F.E.E. 7: 238 (1995); Biol.
Skr. 51: 588 (2005).
Syn: Setaria pallide-fusca (Schumach.) Stapf &
C.E.Hubb. – F.P.S. 3: 535 (1956); Bot. Exp. Sud.: 58
(1970); F.J.M.: 184 (1976).
Tufted annual. Weed of disturbed areas & cultivation.
Distr: RS, DAR, KOR, CS, ES, BAG, UN, EQU;
palaeotropical & warm temperate; introduced in the
Americas.
DAR: Harrison 94 9/1947; **EQU:** Myers 7535 8/1937.

Setaria restioidea (Franch.) Stapf

F.P.S. 3: 534 (1956); F.T.E.A. Gramin. 3: 524 (1982).
Syn: Setaria schweinfurthii R.A.W.Herrm. – F.P.S. 3:
534 (1956).
Tufted perennial. Swamps.
Distr: UN, EQU; Chad to Gabon, D.R. Congo & Uganda.
EQU: Jackson 3819 9/1957.

Setaria sagittifolia (Hochst. ex A.Rich.) Walp.

F.T.E.A. Gramin. 3: 533 (1982); F.E.E. 7: 235 (1995); Biol.
Skr. 51: 589 (2005).
Tufted annual. Woodland & bushland.
Distr: EQU; Ethiopia, Somalia, S to South Africa; Arabia.
EQU: Peers 30 8/1953.

Setaria sphacelata (Schumach.) Stapf & C.E.Hubb. ex Moss

F.P.S. 3: 533 (1956); F.J.M.: 184 (1976); F.T.E.A. Gramin.
3: 527 (1982); F.E.E. 7: 238 (1995).

Syn: Setaria anceps Stapf ex Massey var. **sericea** Stapf
– F.P.S. 3: 533 (1956).
Syn: Setaria sphacelata (Schumach.) Stapf & C.E.Hubb.
ex Moss var. **sericea** (Stapf) Clayton – F.T.E.A. Gramin. 3:
529 (1982); F.E.E. 7: 238 (1995); Biol. Skr. 51: 589 (2005).
Syn: Setaria sphacelata (Schumach.) Stapf & C.E.Hubb.
ex Moss var. **aurea** (A.Braun) Clayton – F.T.E.A. Gramin. 3:
528 (1982); F.E.E. 7: 238 (1995); Biol. Skr. 51: 589 (2005).
Syn: Setaria sphacelata (Schumach.) Stapf & C.E.Hubb.
ex Moss var. **splendida** (Stapf) Clayton – F.T.E.A. Gramin.
3: 530 (1982); Biol. Skr. 51: 590 (2005).
Syn: Setaria splendida Stapf – F.P.S. 3: 533 (1956).
Syn: Setaria trinervia Stapf – F.P.S. 3: 533 (1956).
Tufted perennial. Wide variety of habitats from dry
bushland to swamps.
Distr: DAR, BAG, UN, EQU; Mauritania to Somalia, S
to Namibia & South Africa; Madagascar, introduced
elsewhere in the tropics.
DAR: Wickens 2070 8/1964; **EQU:** Myers 10866 4/1939.

Setaria verticillata (L.) P.Beauv.

F.P.S. 3: 535 (1956); F.J.M.: 184 (1976); F.T.E.A. Gramin. 3:
522 (1982); F.E.E. 7: 236 (1995); Biol. Skr. 51: 590 (2005).
Tufted annual. Weed of disturbed, often damp or shady
ground.
Distr: RS, DAR, KOR, CS, ES, BAG, UN, EQU;
palaeotropical & warm temperate; introduced in the
Americas.
KOR: Wickens 521 9/1962; **UN:** Harrison 1030 10/1950.

Setaria viridis (L.) P.Beauv.

F.P.S. 3: 535 pro parte (1956); F.Egypt 4: 301 (2005).
Tufted annual. Weed of disturbed areas & cultivation.
Distr: RS; warm temperate & subtropical regions,
introduced in the Americas.
RS: Newberry 332 1/1928.
Note: the identity of the single specimen seen, held at
BM, requires confirmation; it may refer to the hybrid S. ×
verticilliformis.

Setaria × verticilliformis Dumort.

F.E.E. 7: 236 (1995).
Syn: Setaria ambigua sensu Andrews, non Guss. –
F.P.S. 3: 535 ?pro parte (1956).
Annual. Dry grassland.
Distr: RS, DAR, KOR, CS; Ethiopia; Mediterranean,
Middle East, Arabia.
RS: Andrews 3604 3/1949.
Note: this is the hybrid between S. verticillata and S. viridis.

Snowdenia polystachya (Fresen.) Pilg.

F.P.S. 3: 538 (1956); F.J.M.: 184 (1976); F.T.E.A. Gramin.
3: 670 (1982); F.E.E. 7: 257 (1994).
Tufted annual. Weed of disturbed & open areas.
Distr: DAR, CS; Eritrea, Ethiopia; Arabia, introduced in
East Africa.
DAR: Wickens 2632 8/1964.

Sorghastrum incompletum (J.Presl) Nash

Ann. Missouri Bot. Gard. 76: 1171 (1989).
Syn: Sorghastrum bipennatum (Hack.) Pilg. – F.P.S. 3:
538 (1956); F.T.E.A. Gramin. 3: 731 (1982); F.E.E. 7: 301
(1995).
Tufted annual. Wooded grassland.

Distr: BAG; Senegal to Ethiopia, S to Zimbabwe & Mozambique; neotropics.
BAG: Schweinfurth 2486 10/1869.

Sorghum arundinaceum (Desv.) Stapf

F.T.E.A. Gramin. 3: 727 (1982); F.E.E. 7: 299 (1995); Biol. Skr. 51: 591 (2005).
Syn: *Sorghum aethiopicum* (Hack.) Rupr. ex Stapf – F.P.S. 3: 543 (1956); F.J.M.: 185 (1976).
Syn: *Sorghum lanceolatum* Stapf – F.P.S. 3: 540 (1956); F.E.E. 7: 299 (1995).
Syn: *Sorghum macrochaetum* Snowden – F.P.S. 3: 543 (1956).
Syn: *Sorghum usambarense* Snowden – F.J.M.: 185 (1976).
Syn: *Sorghum verticilliflorum* (Steud.) Stapf – F.P.S. 3: 540 (1956); F.J.M.: 185 (1976).
Annual or short-lived perennial. Wet grassland & woodland.
Distr: NS, DAR, KOR, CS, ES, BAG, UN, EQU; Mauritania to Somalia, S to Namibia & South Africa; Madagascar, Egypt, India, Venezuela.
DAR: Wickens 2709 9/1964; **BAG:** Myers 14129 9/1941.
Note: a very variable species with several rather distinct forms sometimes treated as separate taxa – see F.E.E. for further discussion.

Sorghum bicolor (L.) Moench

F.E.E. 7: 300 (1995).
Syn: *Sorghum cernuum* (Ard.) Host – Bot. Exp. Sud.: 59 (1970).
Syn: *Sorghum dochna* (Forssk.) Snowden – Bot. Exp. Sud.: 59 (1970).
Syn: *Sorghum durra* (Forssk.) Trab. – Bot. Exp. Sud.: 59 (1970).
Annual. Seasonally inundated floodplains.
Distr: NS, CS, EQU; originating from northern sub-Saharan Africa, now widely cultivated in tropical & warm temperate regions.
NS: Pettet 85 9/1962; **EQU:** Drar 947 4/1938 (n.v.).
Note: the commercial cereal grain sorghum.

Sorghum halepense (L.) Pers.

F.P.S. 3: 545 (1956); F.E.E. 7: 299, in notes (1995).
Rhizomatous perennial. Weed of disturbed areas.
Distr: ?CS; N Africa to Central & SE Asia; widely introduced elsewhere.
Note: Andrews recorded this as an introduced grass in Khartoum; we have not seen any specimens to confirm this record.

Sorghum purpureosericeum (Hochst. ex A.Rich.) Schweinf. & Asch.

F.P.S. 3: 543 (1956); F.T.E.A. Gramin. 3: 730 (1982); F.E.E. 7: 301 (1995); Biol. Skr. 51: 591 (2005).
Syn: *Sorghum dimidiatum* Stapf – F.P.S. 3: 545 (1956).
Annual. Alluvial grassland, riverbanks.
Distr: DAR, KOR, CS, ES, EQU; Mali to Somalia, S to Tanzania; Arabia, India.
CS: Andrews 3277 11/1948; **EQU:** Myers 9904 10/1938.

Sorghum virgatum (Hack.) Stapf

F.P.S. 3: 539 (1956); F.E.E. 7: 300 (1995).
Annual. Marshes, seasonally inundated grassland & bushland, irrigated areas.

Distr: NS, DAR, CS, ES; Mauritania & Senegal to Ethiopia; Egypt, Middle East, Arabia.
CS: Jackson 4190 5/1961.
Note: this species is included within the *S. arundinaceum* complex in F.E.E. and F.T.E.A.

Sorghum × drummondii (Nees ex Steud.) Millsp. & Chase

F.T.E.A. Gramin. 3: 726 (1982); F.E.E. 7: 299 (1995).
Syn: *Sorghum niloticum* (Stapf ex Piper) Snowden – F.P.S. 3: 540 (1956).
Syn: *Sorghum sudanense* (Piper) Stapf – F.P.S. 3: 540 (1956); F.J.M.: 185 (1976).
Annual. Riverbanks, seasonally inundated grassland.
Distr: NS, DAR, KOR, CS, ES, UN; scattered in northern tropical Africa; widely introduced elsewhere.
KOR: Andrews 3843 9/1949; **UN:** Harrison 40 6/1947.
Note: this is the hybrid between *S. bicolor* and *S. arundinaceum*; it is also grown commercially under the name *S. sudanense*: 'Sudan grass'.

Sporobolus angustifolius A.Rich.

F.T.E.A. Gramin. 2: 377 (1974); F.J.M.: 185 (1976); F.E.E. 7: 153 (1995); Biol. Skr. 51: 591 (2005).
Tufted perennial. Open areas in woodland & bushland on shallow, moist soils.
Distr: DAR, EQU; Eritrea & Somalia, S to Zambia & Malawi; Arabia.
DAR: Blair 39 6/1963; **EQU:** Myers 10937 4/1939.
Note: *S. africanus* (Poir.) Robyns & Tournay is also likely to occur in South Sudan, having been recorded from the Ugandan side of the Imatong Mts (Friis & Vollesen 2005: 591).

Sporobolus centrifugus (Trin.) Nees

F.P.S. 3: 548 (1956); F.T.E.A. Gramin. 2: 365 (1974); F.E.E. 7: 148 (1995); Biol. Skr. 51: 591 (2005).
Tufted perennial. Grassland & wooded grassland, often on thin soils over rock.
Distr: EQU; D.R. Congo & Kenya, S to South Africa; Madagascar.
EQU: Friis & Vollesen 1318 3/1982.

Sporobolus consimilis Fresen.

Bot. Exp. Sud.: 59 (1970); F.T.E.A. Gramin. 2: 371 (1974); F.E.E. 7: 151 (1995); Biol. Skr. 51: 592 (2005).
Syn: *Sporobolus robustus* sensu Andrews, non Kunth – F.P.S. 3: 548 (1956).
Tufted shortly rhizomatous perennial. Wet grassland, lake shores.
Distr: RS, DAR, CS, ES, EQU; Chad to Somalia, S to Namibia & South Africa; Arabia.
RS: Schweinfurth 222 9/1868; **EQU:** Shigeta 16 1979 (n.v.).
Note: Friis & Vollesen (2005) cite the single specimen from South Sudan but note that they were unable to trace it at EA.

Sporobolus cordofanus (Hochst. ex Steud.) Hérincq ex Coss.

F.T.E.A. Gramin. 3: 362 (1982); F.E.E. 7: 147 (1995); Biol. Skr. 51: 592 (2005).
Syn: *Sporobolus humifusus* (Kunth) Kunth var. *cordofanus* (Hochst. ex Steud.) Massey – F.P.S. 3: 547 (1965); Bot. Exp. Sud.: 59 (1970).

Tufted annual. Sandy grassland.
Distr: NS, RS, DAR, KOR, CS, ES, UN, EQU; Senegal,
Niger to Eritrea, S to Botswana & Mozambique.
CS: Jackson 4298 8/1961; **EQU:** Beshir 13 8/1951.

Sporobolus coromandelianus (Retz.) Kunth

F.T.E.A. Gramin. 2: 363 (1974); F.E.E. 7: 147 (1995).
Tufted annual. Bare ground.
Distr: RS, KOR, CS; Eritrea, Somalia, Kenya, Angola,
Namibia, Botswana, South Africa; Madagascar, Arabia,
Afghanistan to Thailand.
CS: Farag 181 7/2010.

Sporobolus festivus Hochst. ex A.Rich.

F.P.S. 3: 549 (1956); F.T.E.A. Gramin. 2: 384 (1974); F.J.M.:
185 (1976); F.E.E. 7: 155 (1995); Biol. Skr. 51: 592 (2005).
Tufted perennial. Grassland & wooded grassland, often
on rock outcrops.
Distr: DAR, KOR, CS, EQU; Mauritania to Somalia, S to
Namibia & South Africa; Madagascar, Arabia, India.
DAR: Wickens 3154 8/1972; **EQU:** Friis & Vollesen 262
11/1980.

Sporobolus fimbriatus (Nees ex Trin.) Nees

F.T.E.A. Gramin. 2: 377 (1974); F.E.E. 7: 151 (1995); Biol.
Skr. 51: 592 (2005).
Tufted perennial. Open bushland.
Distr: EQU; Ethiopia & Somalia, S to Namibia & South
Africa; introduced to India.
EQU: Peers KM20 9/1953.

Sporobolus helvolus (Trin.) T.Durand & Schinz

F.P.S. 3: 548 (1956); F.T.E.A. Gramin. 2: 371 (1974); F.E.E.
7: 151 (1995); Biol. Skr. 51: 593 (2005).
Tufted perennial. Bushland.
Distr: RS, DAR, KOR, CS, UN, EQU; Mauritania to
Somalia, S to Tanzania; Arabia, Pakistan to Myanmar.
KOR: Wickens 274 8/1962; **EQU:** Peers M06 9/1953.

Sporobolus ioclados (Trin.) Nees

F.T.E.A. Gramin. 2: 367 (1974); F.J.M.: 185 (1976); F.E.E.
7: 148 (1995); Biol. Skr. 51: 593 (2005).
Syn: *Sporobolus marginatus* Hochst. ex A.Rich. – F.P.S.
3: 547 (1956); Bot. Exp. Sud.: 59 (1970).
Tufted perennial. Bushland, often over hardpans or in
saline soils.
Distr: RS, DAR, CS, UN, EQU; Mauritania to Somalia, S to
Namibia & South Africa; Arabia to India & Sri Lanka.
DAR: Wickens 1989 7/1964; **UN:** Simpson 7234 7/1929.

Sporobolus macranthelus Chiov.

F.E.E. 7: 151 (1956); F.T.E.A. 2: 380 (1974).
Tufted perennial. Bushland, streambanks.
Distr: EQU; Somalia to Namibia.
EQU: Myers 14072 2/1941.

Sporobolus micranthus (Steud.) T.Durand & Schinz

F.E.E. 7: 145 (1995).
Syn: *Sporobolus paniculatus* sensu Wickens, non
(Trin.) T.Durand & Schinz – F.J.M.: 185 (1976).
Annual. Dry grassland.
Distr: DAR, KOR; Eritrea, Ethiopia, Somalia, S to Zimbabwe.
DAR: Blair 171 9/1963.

Sporobolus microprotus Stapf

F.T.E.A. Gramin. 2: 363 (1974); F.J.M.: 185 (1976); F.E.E.
7: 147 (1995); Biol. Skr. 51: 593 (2005).
Syn: *Sporobolus scabriflorus* Stapf ex Massey – F.P.S.
3: 547 (1956).
Tufted annual. Open bushland, disturbed areas.
Distr: RS, DAR, UN, EQU; Senegal to Ethiopia, Uganda
& Kenya.
DAR: Blair 145 8/1963; **EQU:** Myers 9367 8/1938.

Sporobolus minutus Link

F.E.E. 7: 147 (1995).
Tufted ephemeral. Saline flats.
Distr: RS; Eritrea, Somalia, Socotra, Saudi Arabia.
RS: Jackson 3928 4/1959.

Sporobolus panicoides A.Rich.

F.P.S. 3: 548 (1956); F.T.E.A. Gramin. 2: 359 (1974);
F.J.M.: 185 (1976); F.E.E. 7: 145 (1995); Biol. Skr. 51: 593
(2005).
Tufted annual. Bushland including seasonally moist sites.
Distr: DAR; Eritrea & Somalia, S to Namibia & South
Africa; Yemen.
DAR: Wickens 2203 8/1964.

Sporobolus pectinellus Mez

F.P.S. 3: 549 (1956); F.T.E.A. Gramin. 2: 385 (1974); F.E.E.
7: 156 (1995); Biol. Skr. 51: 593 (2005).
Tufted annual. Grassland on rock outcrops & stony soils.
Distr: EQU; Senegal to Ethiopia, S to D.R. Congo &
Tanzania.
EQU: Friis & Vollesen 567 11/1980.

Sporobolus pellucidus Hochst.

F.T.E.A. Gramin. 2: 374 (1974); F.J.M.: 186 (1976); F.E.E.
7: 150 (1995).
Tufted perennial. Open bushland & dry grassland.
Distr: RS, DAR; Burkian Faso to Somalia, S to Zambia,
Namibia; Arabia.
DAR: Blair 147 8/1963.

Sporobolus piliferus (Trin.) Kunth

F.T.E.A. Gramin. 2: 361 (1974); F.E.E. 7: 145 (1995); Biol.
Skr. 51: 593 (2005).
Tufted annual. Bushland on rock outcrops or in shallow
soils.
Distr: EQU; Guinea to Ethiopia, S to Angola, Zimbabwe
& Mozambique; Madagascar, India, SE Asia, scattered in
the neotropics.
EQU: K. Lock K15 8/1966.
Note: the collection cited is from Mt Lonyili on the South
Sudan-Uganda border and may have been collected from
the Ugandan side.

Sporobolus pyramidalis P.Beauv.

F.P.S. 3: 548 (1956); F.J.M.: 186 (1976); F.E.E. 7: 148
(1995); Biol. Skr. 51: 594 (2005).
Tufted perennial. Grassland & bushland, often in
disturbed areas.
Distr: DAR, UN, EQU; Senegal to Ethiopia, S to Namibia
& South Africa; Madagascar, Mauritius, Arabia,
introduced in the neotropics.
DAR: Blair 79 6/1963; **EQU:** Beshir 9 8/1951.

Sporobolus sanguineus Rendle

F.T.E.A. Gramin. 2: 364 (1974).
Syn: *Sporobolus schweinfurthii* Stapf – F.P.S. 3: 547 (1956).
Tufted perennial. Bushland & wooded grassland.
Distr: EQU; Guinea to Kenya, S to Angola & South Africa.
EQU: Schweinfurth 3971 7/1870.

Sporobolus spicatus (Vahl) Kunth

F.P.S. 3: 547 (1956); F.T.E.A. Gramin. 2: 369 (1974); F.E.E. 7: 151 (1995).
Mat-forming or tufted perennial. Grassland & bushland on saline soils.
Distr: RS, UN; Mauritania to Somalia, S to Namibia & South Africa; N Africa, Arabia, India.
RS: Jackson 2760 5/1953; **UN:** J.M. Lock 82/16 2/1982.

Sporobolus stapfianus Gand.

F.P.S. 3: 548 (1956); F.J.M.: 186 (1976); F.E.E. 7: 156 (1995); Biol. Skr. 51: 594 (2005).
Tufted perennial. Wooded grassland & bushland on rock outcrops.
Distr: DAR, UN, EQU; Nigeria to Ethiopia, S to Namibia & South Africa; Madagascar.
DAR: Wickens 2624 9/1964; **EQU:** Myers 7127 7/1937.

Stipa tigrensis Chiov.

F.E.E. 7: 15 (1995).
Tufted perennial. Rock crevices, montane grassland.
Distr: ?DAR; Eritrea, Uganda; Saudi Arabia, Yemen.
IUCN: RD
Note: this species is tentatively included here on the basis of the record in F.E.E. from Jebel Marra; however, we have seen no Sudanese material and it was not mentioned by Wickens in his Flora of Jebel Marra.

Stipagrostis acutiflora (Trin. & Rupr.) De Winter

F.J.U. 1: 299 (1997); F.Egypt 4: 229 (2005).
Syn: *Aristida acutiflora* Trin. & Rupr. – F.P.S. 3: 398 (1956).
Tufted perennial, sometimes stoloniferous. Desert sands.
Distr: NS, ?RS, KOR; Mauritania to Chad; N Africa, Arabia.
KOR: Wickens 575 10/1962.

Stipagrostis ciliata (Desf.) De Winter

F.Egypt 4: 225 (2005).
Syn: *Aristida ciliata* Desf. – F.P.S. 3: 397 (1956).
Tufted perennial. Dry sandy & gravelly areas.
Distr: RS; Mauritania to Chad; Macaronesia, N Africa, Arabia to Pakistan.
RS: Schweinfurth 109 9/1868.

Stipagrostis hirtigluma (Steud. ex Trin. & Rupr.) De Winter

F.T.E.A. Gramin. 1: 140 (1970); F.E.E. 7: 86 (1995).
Syn: *Aristida hirtigluma* Steud. ex Trin. & Rupr. – F.P.S. 3: 398 (1956).
Tufted annual or short-lived perennial. Dry bushland & semi-desert grassland.
Distr: RS, DAR, KOR, CS, EQU; Mauritania to Somalia, S to Tanzania, Namibia, Botswana; Arabia to India.
KOR: Harrison 955 9/1950; **EQU:** Peers 3 9/1953.

Stipagrostis plumosa (L.) Munro ex T.Anderson

F.J.U. 1: 301 (1997); F.Egypt 4: 231 (2005).
Syn: *Aristida plumosa* L. – F.P.S. 3: 397 (1956).
Tufted perennial. Dry sandy & stony areas.
Distr: NS; Mauritania & Senegal to Chad; N Africa, SW Asia, Arabia to India.
NS: Kotschy 360 4/1840.

Stipagrostis raddiana (Savi) De Winter

F.Egypt 4: 229 (2005).
Tufted perennial. Sandy & rocky desert.
Distr: ?RS; Egypt to Arabia & Afghanistan.
Note: recorded from the Gebel Elba region in the Hala'ib Triangle by Boulos (l.c.) – we have seen no material to confirm this.

Stipagrostis rigidifolia (H.Scholz) H.Scholz

F.J.U. 1: 302 (1997).
Tufted perennial. Dry sandy & stony areas.
Distr: NS; Algeria, Libya.
NS: Léonard 4943 12/1968.
Note: treated as a synonym of *S. plumosa* in F.Egypt 4: 231 (2005). The specimen cited has not been seen by us; it was determined by Léonard (1997).

Stipagrostis uniplumis (Licht. ex Roem. & Schult.) De Winter

F.T.E.A. Gramin. 1: 138 (1970); F.E.E. 7: 86 (1995).
Syn: *Aristida papposa* Trin. & Rupr. – F.P.S. 3: 398 (1956).
Tufted perennial. Dry bushland.
Distr: DAR, KOR, CS; Mauritania to Somalia, S to Namibia & South Africa; Morocco, Egypt, Arabia to Pakistan.
KOR: Wickens 206 8/1962.

Streptogyna crinita P.Beauv.

F.P.S. 3: 549 (1956); F.T.E.A. Gramin. 1: 23 (1970); F.E.E. 7: 8 (1995); Biol. Skr. 51: 595 (2005).
Rhizomatous perennial. Forest.
Distr: EQU; Senegal to Ethiopia, S to Angola & Tanzania; India, Sri Lanka.
EQU: Friis & Vollesen 484 11/1980.

Suddia sagittifolia Renoize

Kew Bull. 39: 455 (1984).
Rhizomatous perennial. Swamps.
Distr: UN; Uganda.
IUCN: VU
UN: J.M. Lock 82/21 5/1982.

Tetrapogon cenchriformis (A.Rich.) Clayton

F.T.E.A. Gramin. 2: 348 (1974); F.J.M.: 186 (1976); F.E.E. 7: 159 (1995); Biol. Skr. 51: 595 (2005).
Syn: *Tetrapogon spathaceus* (Hochst. ex Steud.) Hack. ex T.Durand & Schinz – F.P.S. 3: 549 (1956).
Tufted annual or short-lived perennial. Dry, open grassland & bushland.
Distr: RS, DAR, KOR, ES, UN, EQU; Cape Verde Is., Mauritania to Somalia, S to Tanzania; Arabia.
DAR: Blair 142 8/1963; **EQU:** Myers 14096 9/1941.

Tetrapogon tenellus (J.Koenig ex Roxb.) Chiov.

F.T.E.A. Gramin. 2: 348 (1974); F.E.E. 7: 161 (1995).
Tufted annual or short-lived perennial. Dry bushland

& grassland.
Distr: RS, EQU; Eritrea to Angola, Namibia & South
Africa; Arabia, Pakistan, India, Bangladesh.
RS: Khalid 42 12/1951; **EQU:** Myers 14021 9/1941.

Tetrapogon villosus Desf.
F.P.S. 3: 549 (1956); F.T.E.A. Gramin. 2: 350 (1974); F.E.E.
7: 161 (1995).
Tufted perennial. Rocky hillsides.
Distr: RS; Cape Verde Is., Mauritania to Somalia;
Macaronesia, N Africa, Arabia to India.
RS: Jackson 2839 4/1953.

Thelepogon elegans Roem. & Schult.
F.P.S. 3: 551 (1956); F.T.E.A. Gramin. 3: 744 (1982); F.E.E.
7: 313 (1995); Biol. Skr. 51: 595 (2005).
Annual with prop roots. Grassland on black clay soils.
Distr: ?CS, UN, EQU; Senegal to Eritrea, S to Namibia,
Zimbabwe & Mozambique; India to SE Asia.
EQU: Myers 10079 11/1938.
Note: Andrews records this species from Central Sudan
but we have only seen material from South Sudan.

Themeda triandra Forssk.
F.J.M.: 186 (1976); F.T.E.A. Gramin. 3: 829 (1982); F.E.E.
7: 353 (1995); Biol. Skr. 51: 595 (2005).
Syn: *Themeda triandra* Forssk. var. *hispida* Stapf –
F.P.S. 3: 551 (1956).
Tufted perennial. Fire-prone grassland & bushland.
Distr: DAR, EQU; palaeotropical.
DAR: Blair 105 7/1963; **EQU:** Beshir 8 8/1951.

Trachypogon spicatus (L.f.) Kuntze
F.T.E.A. Gramin. 3: 709 (1982); F.E.E. 7: 297 (1995); Biol.
Skr. 51: 596 (2005).
Tufted perennial. Grassland, woodland & bushland, often
on rocky hillslopes.
Distr: EQU; Mali & Ivory Coast to Ethiopia, S to Namibia
& South Africa; Madagascar, neotropics.
EQU: Friis & Vollesen 1261 3/1982.

Tragus berteronianus Schult.
F.P.S. 3: 553 (1956); F.T.E.A. Gramin. 2: 400 (1974); F.E.E.
7: 178 (1995); Biol. Skr. 51: 596 (2005).
Tufted annual. Disturbed & overgrazed areas in grassland
& bushland.
Distr: RS, DAR, KOR, CS, ES, UN, EQU; Mauritania to
Somalia, S to Namibia & South Africa; Egypt, Arabia to
Pakistan, China, neotropics.
RS: Farag 31 1/2010; **EQU:** Jackson 4205 6/1961.

Tragus racemosus (L.) All.
F.P.S. 3: 553 (1956); F.T.E.A. Gramin. 2: 400 (1974); F.E.E.
7: 178 (1995).
Syn: *Tragus paucispinus* Hack. – F.P.S. 3: 551 (1956).
Tufted annual. Weed of disturbed grassland & open areas.
Distr: NS, RS, DAR, KOR, CS, BAG; Mauritania to
Somalia & Kenya, Angola to South Africa; S Europe, N
Africa to Pakistan.
CS: Pettet 56 9/1962; **BAG:** Chevalier 10012 9/1903.

Tricholaena teneriffae (L.f.) Link
F.P.S. 3: 553 (1956); Bot. Exp. Sud.: 60 (1970); F.T.E.A.
Gramin. 3: 504 (1982); F.E.E. 7: 191 (1995).

Syn: *Tricholaena leucantha* (A .Rich.) Stapf &
C.E.Hubb. – F.P.S. 3: 553 (1956).
Tufted perennial. Dry rocky hillslopes.
Distr: RS, DAR; Cape Verde Is., Mauritania, Chad to
Somalia, S to Tanzania; Macaronesia, Italy, N Africa,
Arabia to India.
RS: Jackson 2817 4/1953.
Note: Andrews also records this species from "White
Nile" without a more precise locality; we have only seen
material from the Red Sea region.

Trichoneura mollis (Kunth) Ekman
F.T.E.A. Gramin. 2: 297 (1974); F.E.E. 7: 105 (1995).
Syn: *Trichoneura arenaria* (Steud.) Ekman – F.P.S. 3:
553 (1956).
Tufted annual. Dry bushland.
Distr: RS, DAR, KOR; Mauritania & Senegal to Somalia &
Kenya; Egypt, Arabia.
KOR: Wickens 668 11/1962.

Tripogon leptophyllus (A.Rich.) Cufod.
F.J.M.: 186 (1976); F.E.E. 7: 97 (1995).
Syn: *Tripogon abyssinicus* Nees ex Steud. – F. Darfur
Nord-Occ. & J. Gourgeil: 142 (1969).
Tufted perennial. Amongst rocks on grassy hillslopes.
Distr: DAR; Eritrea, Ethiopia; Yemen, Oman.
DAR: Wickens 2448 9/1964.

Tripogon major Hook.f.
F.T.E.A. Gramin. 2: 291 (1974); F.E.E. 7: 97 (1995); Biol.
Skr. 51: 596 (2005).
Syn: *Tripogon snowdenii* C.E.Hubb. – F.P.S. 3: 554
(1956).
Tufted perennial. Rock crevices in upland grassland.
Distr: EQU; Sierra Leone, Nigeria, Cameroon, Ethiopia
to Malawi.
EQU: Myers 11651 7/1939.

Tripogon minimus (A.Rich.) Hochst. ex Steud.
F.P.S. 3: 554 (1956); F.T.E.A. Gramin. 2: 289 (1974);
F.J.M.: 186 (1976); F.E.E. 7: 96 (1995).
Tufted perennial. Bushland & wooded grassland, often at
base of rocky outcrops.
Distr: DAR, KOR, ES; Cape Verde Is., Mauritania to
Ethiopia, S to Namibia & South Africa; Madagascar.
KOR: Wickens 355 9/1962.

Tripogon montanus Chiov.
F.T.E.A. Gramin. 2: 291 (1974); F.J.M.: 186 (1976); F.E.E.
7: 97 (1995).
Tufted perennial. Rock crevices in upland grassland.
Distr: DAR; Eritrea, Ethiopia, Uganda; Arabia.
DAR: Wickens 2475 9/1964.

Tripogon multiflorus Miré & H.Gillet
F. Darfur Nord-Occ. & J. Gourgeil: 142 (1969); F.T.E.A.
Gramin. 2: 292 (1974); F.E.E. 7: 98 (1995).
Tufted perennial. Rocky ground including seasonally wet
areas.
Distr: DAR; Niger (Aïr), Chad (Tibesti, Ennedi), Eritrea,
Kenya; Arabia.
Note: the record of this species is from Quézel (1969) who
records it from Umm Burru, Musbat and Jebel Gourgeil;
we have not seen the specimens on which this is based.

Triraphis pumilio R.Br.

F.P.S. 3: 554 (1956); F.Egypt 4: 247 (2005).
Annual. Dry sandy & stony areas.
Distr: RS, KOR; Mauritania to Chad, Angola, Namibia;
Morocco, Egypt, Arabia.
KOR: Harrison 968 9/1950.

Tristachya superba (De Not.) Schweinf. & Asch.

F.T.E.A. Gramin. 2: 422 (1974).
Syn: *Loudetia superba* De Not. – F.P.S. 3: 485 (1956).
Tufted rhizomatous perennial. Fire-prone wooded
grassland.
Distr: KOR, BAG, EQU; Senegal to Uganda, S to Namibia
& South Africa.
KOR: Jackson 2406 10/1952; **BAG:** Myers 7202 7/1937.

Triticum turgidum L. subsp. durum (Desf.) Husn.

Syn: *Triticum durum* Desf. – F.E.E. 7: 62 (1995).
Syn: *Triticum pyramidale* Percival – Bot. Exp. Sud.: 60
(1970).
Annual. Cultivated, rarely naturalised.
Distr: DAR; widely cultivated particularly in the
Mediterranean, Middle East & N Africa.
DAR: Drar 2121 5/1938 (n.v.).
Note: the cultivated 'durum wheat', included here since
Drar noted his collection as "not cultivated" suggesting it
was from a naturalised population.

Urelytrum giganteum Pilg.

F.T.E.A. Gramin. 3: 833 (1982).
Syn: *Urelytrum thyrsioides* Stapf – F.P.S. 3: 554 (1956).
Tufted perennial. Damp ground.
Distr: EQU; Nigeria to D.R. Congo & Uganda.
EQU: Myers 7715 9/1937.

Urochloa oligotricha (Fig. & De Not.) Henrard

F.T.E.A. Gramin. 3: 606 (1982); F.E.E. 7: 232 (1995).
Syn: *Urochloa bolbodes* (Hochst. ex Steud.) Stapf –
F.P.S. 3: 555 (1956).
Tufted perennial. Wooded grassland, disturbed areas,
weed of cultivation.
Distr: ?CS; Eritrea to Angola, Namibia & South Africa.
Note: Andrews records this species from Blue Nile
Province; we have seen no material with which to verify
this record.

Urochloa panicoides P.Beauv.

F.P.S. 3: 555 (1956); F.T.E.A. Gramin. 3: 602 (1982); F.E.E.
7: 230 (1995); Biol. Skr. 51: 596 (2005).
Tufted annual. Bushland, disturbed areas.
Distr: NS, BAG, UN, EQU; Chad to Somalia, S to Namibia
& South Africa; Egypt, Arabia to SE Asia.
NS: Petherick s.n. 8/1862; **BAG:** Harrison 1078 10/1950.

Urochloa trichopus (Hochst.) Stapf

F.P.S. 3: 555 (1956); F.J.M.: 186 (1976); F.T.E.A. Gramin.
3: 604 (1982); F.E.E. 7: 231 (1995).
Tufted annual. Wooded grassland & bushland, disturbed
areas.
Distr: RS, DAR, KOR, CS, ES, UN, EQU; Mauritania to
Somalia, S to Namibia & South Africa; Arabia, India,
Myanmar.
DAR: Wickens 2307 8/1964; **EQU:** Myers 7543 8/1937.

Urochondra setulosa (Trin.) C.E.Hubb.

F.P.S. 3: 555 (1956); F.E.E. 7: 157 (1995).
Tufted perennial. Coastal sand dunes.
Distr: RS; Eritrea, Somalia; Arabia, Pakistan, India.
RS: Massey s.n. 6/1914.

Vossia cuspidata (Roxb.) Griff.

F.P.S. 3: 557 (1956); Bot. Exp. Sud.: 60 (1970); F.T.E.A.
Gramin. 3: 832 (1982); F.E.E. 7: 361 (1995).
Mat-forming perennial. Rivers & swamps, often floating.
Distr: KOR, CS, BAG, UN, EQU; Mauritania to Somalia,
S to Namibia, Botswana & Mozambique; Egypt, India,
SE Asia.
KOR: Wickens 733 10/1962; **EQU:** Myers 7556 8/1937.
Note: Andrews recorded this species as "a troublesome
weed of the Gezira canals and one of the principal
constituents of the Sudd".

Vulpia bromoides (L.) S.F.Gray

F.P.S. 3: 557 (1956); F.T.E.A. Gramin. 1: 64 (1970); F.J.M.:
186 (1976); F.E.E. 7: 28 (1995).
Annual, sometimes tufted. Montane grassland,
streambanks, roadsides & rocky ground.
Distr: RS, DAR; Cameroon, Eritrea, Ethiopia, Somalia,
Uganda, Kenya; Europe, N Africa, Yemen, introduced
elsewhere.
DAR: Blair 352 9/1964.

Yushania alpina (K.Schum.) W.C.Lin

Bull. Taiwan For. Res. Inst. 248: 14 (1974).
Syn: *Arundinaria alpina* K.Schum. – F.P.S. 3: 400
(1956); F.T.E.A. Gramin. 1: 9 (1970); T.S.S.: 455 (1990);
F.E.E. 7: 3 (1995).
Syn: *Sinarundinaria alpina* (K.Schum.) C.S.Chao &
Renvoize – Biol. Skr. 51: 590 (2005).
Clump-forming bamboo. Montane forest, sometimes
forming large dense thickets.
Distr: EQU; Cameroon to Ethiopia, S to Zambia & Malawi.
EQU: Chipp 94 2/1929.

Zea mays L.

Bot. Exp. Sud.: 60 (1970); F.T.E.A. Gramin. 3: 857 (1982);
F.E.E. 7: 365 (1995).
Annual. Cultivated, occasionally a naturalised escape.
Distr: CS; originating in the Americas, now widespread
in the tropics & temperate regions.
CS: Drar 244 2/1938 (n.v.).
Note: the cultivated maize, widely grown in Sudan and
South Sudan. The specimen cited was not noted to be
cultivated by Drar (1970) and so may have been
naturalised.

CERATOPHYLLALES

Ceratophyllaceae

I. Darbyshire

Ceratophyllum demersum L. var. demersum

F.P.S. 1: 14 (1950); F.J.M.: 83 (1976); F.T.E.A. Ceratophyll.:
3 (1985); F.Egypt 1: 153 (1999); F.E.E. 2(1): 36 (2000).
Submerged aquatic perennial herb. Lakes, ponds, rivers
& streams.

Distr: DAR, KOR, CS, BAG, UN, EQU; cosmopolitan.
KOR: Simpson 7760 (fr) 4/1930; **UN:** J.M. Lock 80/30 (fr) 2/1980.

Ceratophyllum muricatum Cham. subsp. muricatum

F.T.E.A. Ceratophyll.: 1 (1985); F.Egypt 1: 153 (1999); F.E.E. 2(1): 37 (2000).
Submerged aquatic perennial herb. Lakes & pools.
Distr: UN; scattered across Africa; Italy, Russia, India, China, Pacific Is., U.S.A. to N South America.
UN: J.M. Lock 83/2 (fr) 1/1983.

RANUNCULALES

Papaveraceae

I. Darbyshire

Argemone mexicana L.

F.P.S. 1: 26 (1950); F.T.E.A. Papaver.: 2 (1962); F.Egypt 1: 161 (1999); F.E.E. 2(1): 66 (2000).
Annual or perennial herb. Waste ground, riverbanks.
Distr: NS, RS, CS, ES; widespread weed of the tropics & subtropics; native of West Indies & C America.
RS: Jackson 2714 (fl) 3/1953.

Corydalis cornuta Royle

Biol. Skr. 51: 75 (1998); Trop. Afr. Fl. Pl. 1: 116 (2003).
Syn: Corydalis mildbraedii Fedde – F.T.E.A. Fumar.: 1 (1962); F.E.E. 2(1): 66 (2000).
Perennial herb. Montane forest, bushland & grassland.
Distr: EQU; Ethiopia to D.R. Congo & Tanzania; Himalayas.
EQU: Myers 11569 7/1939.
Note: the specimen cited has not been traced but mis-identification seems unlikely.

Fumaria abyssinica Hammar

F.T.E.A. Fumar.: 3 (1962); Biol. Skr. 51: 75 (1998); F.E.E. 2(1): 69 (2000).
Annual herb. Montane forest, bushland & grassland, often growing amongst rocks.
Distr: EQU; Eritrea to Somalia, S to D.R. Congo & Tanzania; Arabia.
EQU: Kielland-Lund 915 6/1984 (n.v.).

[Fumaria officinalis L.]

Note: Andrews lists this species from White Nile District but notes that the plant is "possibly introduced, or wrongly identified". We have not seen the specimen on which this record is based and are unsure of its likely identity.

Menispermaceae

I. Darbyshire

Chasmanthera dependens Hochst.

F.P.S. 1: 15 (1950); F.T.E.A. Menisperm.: 17 (1956); F.J.M.: 84 (1976); Biol. Skr. 51: 71 (1998); F.E.E. 2(1): 47 (2000).
Herbaceous or woody climber. Forest, riverine forest & thicket, rocky wooded hillslopes, wooded grassland.

Distr: DAR, BAG, UN, EQU; Sierra Leone to Eritrea & Somalia, S to Zambia.
DAR: Wickens 1977 (fl) 7/1964; **UN:** Sherif A.3905 (fl) 7/1951.

Cissampelos mucronata A.Rich.

F.P.S. 1: 17 (1950); F.T.E.A. Menisperm.: 27 (1956); F.J.M.: 84 (1976); Biol. Skr. 51: 71 (1998); F.E.E. 2(1): 49 (2000).
Herbaceous or woody climber. Forest, woodland, wooded grassland, riverbanks, disturbed areas.
Distr: DAR, UN, EQU; Senegal to Eritrea, S to Namibia & South Africa.
DAR: Wickens 2017 (fl) 7/1964; **UN:** Myers 10472 (fl) 2/1939.
Note: Andrews also records C. macrosepala Diels, but this is almost certainly based upon Schweinfurth III: 79, collected from Diomvonoo which is in current-day D.R. Congo, not South Sudan.

Cissampelos rigidifolia (Engl.) Diels var. rigidifolia

F.P.S. 1: 17 (1950); F.T.E.A. Menisperm.: 30 (1956).
Woody climber. Woodland & wooded grassland.
Distr: EQU; C.A.R. to Uganda, S to Zambia.
EQU: Jackson 2243 (fl) 6/1952.
Note: a scarce taxon but not considered threatened.

Cissampelos sp. cf. pareira L.

Biol. Skr. 51: 71 (1998).
Creeping herb. Secondary Dombeya thicket.
Distr: EQU.
EQU: Myers 14165 10/1941.
Note: the single collection is sterile and insufficient for confident determination. C. pareira is widely distributed in the tropics but has not so far been confirmed from Sudan / South Sudan.

Cocculus hirsutus (L.) Diels

F.P.S. 1: 17 (1950); F.T.E.A. Menisperm.: 12 (1956); T.S.S.: 19 (1990); Biol. Skr. 51: 71 (1998); F.E.E. 2(1): 45 (2000).
Woody climber. Thickets, riverine woodland, termitaria.
Distr: KOR, CS, ES, UN, EQU; Eritrea to Namibia & South Africa; Arabia to China.
CS: Broun 713 2/1912; **UN:** J.M. Lock 81/36 (fl) 2/1981.
Note: Drar's (1970: 84) record of Cocculus pendulus from Equatoria may well be based upon a misidentification of this species.

Cocculus pendulus (J.R.Forst. & G.Forst.) Diels

F.P.S. 1: 17 (1950); F.T.E.A. Menisperm.: 10 (1956); T.S.S.: 20 (1990); F.J.U. 2: 185 (1999); F.E.E. 2(1): 45 (2000).
Woody climber. Dry bushland, riverine thickets.
Distr: NS, RS, DAR, KOR, CS, ES; Cape Verde Is., Mauritania & Senegal to Somalia & Kenya; N Africa to Arabia, India.
RS: Sahni & Kamil 722 (fl fr) 4/1967.

Dioscoreophyllum cumminsii (Stapf) Diels var. cumminsii

F.W.T.A. 1(1): 73 (1954); Troupin, Monogr. Menispermaceae Afr.: 134 (1962); T.S.S.: 20 (1990).
Syn: Dioscoreophyllum sp. sensu Andrews – F.P.S. 1: 17 (1950).
Herbaceous climber. Riverine forest.

Distr: EQU; Guinea to C.A.R. & D.R. Congo.
EQU: Myers 9190 (fl) 8/1938.
Note: treated as a synonym of *D. volkensii* Engl. in the African Plants Database, but it is usually considered a good species (e.g. see Tropin, l.c.).

Stephania abyssinica (Quart.-Dill. & A.Rich.) Walp. var. *abyssinica*
F.P.S. 1: 19 (1950); F.T.E.A. Menisperm.: 22 (1956); Biol. Skr. 51: 72 (1998); F.E.E. 2(1): 51 (2000).
Herbaceous climber. Montane forest margins, montane bushland.
Distr: EQU; Guinea, Nigeria, Bioko & Cameroon, Eritrea to Angola & South Africa; Arabia.
EQU: Friis & Vollesen 961 (fl) 2/1982.

Stephania cynantha Welw. ex Hiern
F.T.E.A. Menisperm.: 21 (1956); Biol. Skr. 51: 72 (1998); F.E.E. 2(1): 53 (2000).
Herbaceous or woody climber. Riverine forest & thicket.
Distr: EQU; Bioko, Cameroon, Ethiopia to Angola & Zambia.
EQU: Friis & Vollesen 1051 (fr) 2/1982.
Note: a widespread but scarce species.

Tiliacora funifera (Miers) Oliv.
F.T.E.A. Menisperm.: 9 (1956); Biol. Skr. 51: 72 (1998); F.E.E. 2(1): 43 (2000).
Woody climber. Forest.
Distr: EQU; Ivory Coast to Somalia, S to Angola & South Africa.
EQU: Friis & Vollesen 248 (fl) 11/1980.

Tinospora bakis (A.Rich.) Miers
F.P.S. 1: 19 (1950); F.T.E.A. Menisperm.: 18 (1956); F.E.E. 2(1): 48 (2000).
Woody climber. *Acacia-Commiphora* bushland & woodland.
Distr: KOR, CS; Mauritania & Senegal to Somalia, S to Tanzania.
KOR: Kotschy 244 (fr) 11/1839.

Tinospora caffra (Miers) Troupin
F.T.E.A. Menisperm.: 19 (1956); Biol. Skr. 51: 72 (1998); F.E.E. 2(1): 49 (2000).
Syn: *Desmonema mucronulatum* Engl. – F.P.S. 1: 17 (1950).
Woody climber. Woodland, termitaria.
Distr: UN, EQU; C.A.R. to Ethiopia, S to Angola & South Africa.
EQU: Myers 11049 (fl) 4/1939.

Ranunculaceae

I. Darbyshire

Anemone thomsonii Oliv. var. *thomsonii*
F.P.S. 1: 10 (1950); F.T.E.A. Ranuncul.: 12 (1952); Biol. Skr. 51: 68 (1998); F.E.E. 2(1): 23 (2000).
Rhizomatous perennial herb. Montane grassland & bushland, often amongst rocks.
Distr: EQU; Ethiopia to D.R. Congo & Tanzania.
EQU: Myers 11656 (fl fr) 7/1939.
Note: a localised species but not considered threatened.

Clematis hirsuta Guill. & Perr. var. *hirsuta*
F.T.E.A. Ranuncul.: 6 (1952); F.J.M.: 83 (1976); F.E.E. 2(1): 19 (2000).
Syn: *Clematis glaucescens* Fresen. – F.P.S. 1: 10 (1950).
Syn: *Clematis tibestica* Quézel – Bot. Exp. Sud.: 98 (1970).
Syn: *Clematis brachiata* sensu auctt., non Thunb. – Biol. Skr. 51: 68 (1998); T.S.S.: 19 (1990).
Woody climber. Forest margins, montane bushland & grassland.
Distr: RS, DAR, KOR, ES, BAG, EQU; Senegal to Eritrea, S to Angola, Zimbabwe & Mozambique; Arabia.
DAR: Robertson 7 (fl) 10/1957; **EQU:** Friis & Vollesen 427 (fl) 11/1980.
Note: sometimes treated as a synonym of the southern Africa *C. brachiata* Thunb.

Clematis hirsuta Guill. & Perr. var. *inciso-dentata* (A.Rich.) W.T.Wang
Acta Phytotax. Sin. 38: 325 (2000).
Syn: *Clematis inciso-dentata* A.Rich. – F.P.S. 1: 10 (1950).
Woody climber. Forest margins, montane bushland & grassland.
Distr: RS; Eritrea, Ethiopia.
RS: Andrews 3469 (fl).
Note: often not recognised as distinct from var. *hirsuta* (e.g. in F.E.E.) but Wang (2000, l.c.) separates it based on the lanate-sericeous leaf indumentum and glabrous internal surface of the sepals.

Clematis simensis Fresen.
F.P.S. 1: 10 (1950); F.T.E.A. Ranuncul.: 2 (1952); F.J.M.: 83 (1976); T.S.S.: 19 (1990); Biol. Skr. 51: 69 (1998); F.E.E. 2(1): 21 (2000).
Woody climber. Forest margins, montane bushland & grassland.
Distr: DAR, EQU; Nigeria to Somalia, S to Angola, Zimbabwe & Mozambique; Madagascar, Arabia.
DAR: Jackson 2612 (fr) 1/1953; **EQU:** Chipp 101 (fr) 2/1929.

Clematis villosa DC. subsp. *oliveri* (Hutch.) Brummitt
Kew Bull. 55: 104 (2000).
Syn: *Clematopsis oliveri* Hutch. – F.P.S. 1: 12 (1950).
Syn: *Clematopsis scabiosifolia* (DC.) Hutch. **Group D** – F.T.E.A. Ranuncul.: 8 (1952).
Syn: *Clematopsis scabiosifolia* sensu Drar, non (DC.) Hutch. – Bot. Exp. Sud.: 98 (1970).
Perennial herb. Wooded grassland with periodic burning.
Distr: EQU; Nigeria, Cameroon, Uganda, S to D.R. Congo & Tanzania.
EQU: Myers 8402 (fl) 1/1938.

Delphinium dasycaulon Fresen.
F.P.S. 1: 12 (1950); F.T.E.A. Ranuncul.: 32 (1952); F.E.E. 2(1): 29 (2000).
Perennial herb. Montane grassland, forest margins & clearings.
Distr: RS; Nigeria, Cameroon, Eritrea to Zambia & Mozambique.
RS: Robbie 6 (fl).

Delphinium leroyi Franch. ex Huth

F.P.S. 1: 12 (1950); F.T.E.A. Ranuncul.: 20 (1952); Biol. Skr. 51: 69 (1998).
Perennial herb. Montane grassland including damp areas, rocky ground, montane bushland, forest margins.
Distr: EQU; D.R. Congo, Rwanda, Uganda, Kenya, Tanzania, Malawi.
EQU: Myers 11742 (fl) 8/1939.
Note: a rather scarce species but not considered threatened.

Nigella sativa L.

F.P.S. 1: 12 (1950); F.J.M.: 83 (1976); F.E.E. 2(1): 32 (2000).
Annual herb. Cultivated and occasionally naturalised.
Distr: ?NS, RS, DAR; probably native of SW Asia, widely cultivated in Europe, N & E Africa to India.
RS: Farag 523 (fl) 11/2010.
Note: the record for Northern Sudan is based on Andrews. This species is cultivated for its seeds 'black cumin'.

Ranunculus multifidus Forssk.

F.P.S. 1: 12 (1950); F.T.E.A. Ranuncul.: 19 (1952); F.J.M.: 83 (1976); Biol. Skr. 51: 69 (1998); F.E.E. 2(1): 27 (2000).
Perennial herb. Streamsides, marshes, irrigation channels, forest margins & montane grassland.
Distr: DAR, EQU; Nigeria, Bioko, Cameroon, Eritrea to Namibia & South Africa; Arabia.
DAR: Wickens 1243 (fl fr) 2/1964; **EQU:** Thomas 1796 (fl fr) 12/1935.

Ranunculus oreophytus Delile var. oreophytus

F.P.S. 1: 13 (1950); F.T.E.A. Ranuncul.: 14 (1952); Biol. Skr. 51: 70 (1998); F.E.E. 2(1): 25 (2000).
Perennial herb. Montane grassland, marshes.
Distr: DAR, EQU; Ethiopia to D.R. Congo & Tanzania.
DAR: Miehe 788 3/1983; **EQU:** Thomas 1880 (fl) 12/1935.

Thalictrum rhynchocarpum Quart.-Dill. & A.Rich. ex A.Rich. subsp. rhynchocarpum

F.P.S. 1: 13 (1950); F.T.E.A. Ranuncul.: 11 (1952); Biol. Skr. 51: 70 (1998); F.E.E. 2(1): 21 (2000).
Perennial herb. Forest margins, montane bushland.
Distr: EQU; Bioko & Cameroon, Ethiopia to South Africa.
EQU: Friis & Vollesen 86 (fr) 11/1980.

PROTEALES

Proteaceae

H. Pickering

Faurea rochetiana (A.Rich.) Chiov. ex Pic. Serm.

F.T.E.A. Prot.: 3 (1993); Biol. Skr. 51: 113 (1998).
Syn: *Faurea speciosa* Welw. – F.P.S. 1: 153 (1950); T.S.S.: 51 (1990); F.E.E. 2(1): 437 (2000).
Shrub. Woodland & wooded grassland.
Distr: EQU; Togo to Nigeria, Eritrea to Angola & South Africa.
EQU: Friis & Vollesen 936 2/1982.

Faurea saligna Harv.

F.T.E.A. Prot.: 5 (1993).
Tree. Wooded grassland.
Distr: EQU; Uganda & Kenya, S to Angola & South Africa.
EQU: Andrews 1046 5/1939.

Protea gaguedi J.F.Gmel.

F.P.S. 1: 154 (1950); T.S.S.: 53 (1990); F.T.E.A. Prot.: 17 (1993); Biol. Skr. 51: 113 (1998); F.E.E. 2(1): 438 (2000).
Shrub. Wooded grassland.
Distr: EQU; Eritrea to Angola & South Africa.
EQU: Friis & Vollesen 976 2/1982.

Protea madiensis Oliv. subsp. madiensis

F.P.S. 1: 154 (1950); T.S.S.: 53 (1990); F.T.E.A. Prot.: 10 (1993); Biol. Skr. 51: 114 (1998); F.E.E. 2(1): 438 (2000).
Tree. Upland grassland.
Distr: ES, BAG, EQU; Nigeria to Ethiopia, S to Angola & Mozambique.
ES: Schweinfurth 2396 6/1865; **EQU:** Andrews 1970 6/1939.

GUNNERALES

Gunneraceae

I. Darbyshire

Gunnera perpensa L.

F.P.S. 1: 147 (1950); F.T.E.A. Halorag.: 3 (1973); Biol. Skr. 51: 109 (1998); F.E.E. 2(1): 422 (2000).
Rhizomatous perennial herb. Montane swamps, streamsides & moorland.
Distr: EQU; Ethiopia to South Africa; Madagascar.
EQU: Myers 11695A (fl) 7/1939.

DILLENIALES

Dilleniaceae

R. Vanderstricht & I. Darbyshire

Tetracera masuiana De Wild. & T.Durand

F.P.S. 1: 155 (1950); F.T.E.A. Dillen.: 3 (1968); T.S.S.: 54 (1990).
Shrub. Grassland & woodland.
Distr: BAG, EQU; Cameroon, Chad & C.A.R., S to Angola & Zambia.
EQU: Schweinfurth 3985 (fl) 6/1870.

Tetracera stuhlmanniana Gilg

F.P.S. 1: 154 (1950); F.T.E.A. Dillen.: 6 (1968); Biol. Skr. 51: 114 (1998); F.E.E. 2(1): 439 (2000).
Syn: *Tetracera alnifolia* sensu El Amin, non Willd. – T.S.S.: 53 (1990).
Syn: *Tetracera potatoria* sensu Andrews, non G.Don – F.P.S. 1: 154 (1950).
Woody climber or shrub. Lowland & mid-altitude forest.
Distr: BAG, EQU; Senegal to Ethiopia, S to Tanzania.
EQU: Myers 11919 (fl) 9/1939.

SAXIFRAGALES

Hamamelidaceae

I. Darbyshire

Trichocladus ellipticus Eckl. & Zeyh. subsp. *malosanus* (Baker) Verdc.

F.T.E.A. Hamamelid.: 2 (1971); F.E.E. 3: 252 (1989); T.S.S.: 249 (species only) (1990); Biol. Skr. 51: 237 (1998).
Syn: *Trichocladus malosanus* Baker – F.P.S. 2: 249 (1952).
Shrub or small tree. Upland & riverine forest.
Distr: EQU; Ethiopia to Angola & Mozambique.
EQU: Myers 10986 (fl) 4/1939.

Crassulaceae

H. Pickering & I. Darbyshire

Bryophyllum pinnatum (Lam.) Oken

F.P.S. 1: 75 (1950); F.T.E.A. Crassul.: 28 (1987); F.E.E. 3: 25 (1989).
Syn: *Kalanchoe pinnata* (Lam.) Pers. – Bot. Exp. Sud.: 41 (1970).
Succulent perennial herb. Naturalised in riverine & rocky habitats.
Distr: RS, ?EQU; ?native to Africa, now widely distributed in the tropics.
RS: Drar 325 3/1938 (n.v.).
Note: no South Sudanese material has been seen but Andrews lists this distinctive species from Equatoria. In the African Plants Database, the accepted name is given as *Kalanchoe pinnata*.

Cotyledon barbeyi Schweinf. ex Baker

F.T.E.A. Crassul.: 58 (1987); F.E.E. 3: 18 (1989).
Shrub. Bushland.
Distr: RS; Eritrea & Somalia, S to South Africa; Yemen.
RS: Jackson 2918 4/1953.
Note: Drar (1970: 40) lists *Cotyledon arabicum* Schweinf. ex Penzig. from Erkowit; this name is not recognised in the Plant List; the record presumably refers to *C. barbeyi*.

Crassula alata (Viv.) A.Berger subsp. *alata*

F.E.E. 3: 8, in notes (1989); F.Egypt 1: 239 (1999).
Annual herb. Open sandy soils & rocky hillslopes.
Distr: RS; E Mediterranean to India.
RS: Farag & Zubair 3 (fl) 2/2012.

Crassula alata (Viv.) A.Berger subsp. *pharnaceoides* (Fisch. & C.A.Mey.) Wickens & Bywater

F.T.E.A. Crassul.: 6 (1987); F.E.E. 3: 8 (1989).
Syn: *Crassula pharnaceoides* Fisch. & C.A.Mey. – F.J.M.: 87 (1976).
Annual herb. Moist rock crevices in grassland.
Distr: RS, DAR, CS; Cameroon, Ethiopia, Somalia, Uganda, Kenya.
DAR: Wickens 2535 9/1964.

Crassula alsinoides (Hook.f.) Engl.

F.P.S. 1: 76 (1950); F.T.E.A. Crassul.: 12 (1987); F.E.E. 3: 6 (1989); Biol. Skr. 51: 85 (1998).
Creeping perennial herb. Moist open areas.

Distr: EQU; Cameroon to Somalia, S to South Africa; Madagascar, Yemen.
EQU: Friis & Vollesen 45 11/1980.

Crassula schimperi Fisch. & C.A.Mey. subsp. *schimperi*

F.J.M.: 87 (1976); F.T.E.A. Crassul.: 7 (1987).
Syn: *Crassula pentandra* (Royle ex Edgew.) Schönland – F.P.S. 1: 76 (1950).
Perennial herb. On moist rocks.
Distr: RS, DAR; Cameroon to Ethiopia, S to Zambia; Arabia, India.
DAR: Wickens 2378 9/1964.

Crassula schimperi Fisch. & C.A.Mey. subsp. *phyturus* (Mildbr.) R.Fern.

F.T.E.A. Crassul.: 8 (1987); F.E.E. 3: 8 (1989); Biol. Skr. 51: 86 (1998).
Perennial herb. Upland rocky outcrops.
Distr: EQU; Ethiopia, Uganda, Kenya, Tanzania; Socotra, Yemen.
EQU: Friis & Vollesen 1100 3/1982.

Crassula vaginata Eckl. & Zeyh.

F.T.E.A. Crassul.: 13 (1987); F.E.E. 3: 8 (1989); Biol. Skr. 51: 86 (1998).
Syn: *Crassula alba* sensu Andrews, non Forssk. – F.P.S. 1: 76 (1950).
Perennial herb. Montane rocky outcrops.
Distr: EQU; Nigeria to Ethiopia, S to Angola & South Africa.
EQU: Myers 13508 9/1940.

Kalanchoe densiflora Rolfe var. *densiflora*

F.T.E.A. Crassul.: 39 (1987); F.E.E. 3: 23 (1989); Biol. Skr. 51: 86 (1998).
Syn: *Kalanchoe petitiana* sensu Andrews, non A.Rich. – F.P.S. 1: 76 (1950).
Perennial herb. Moist open grassland.
Distr: EQU; D.R. Congo to Ethiopia, S to Tanzania.
EQU: Friis & Vollesen 916 2/1982.

Kalanchoe glaucescens Britten

F.P.S. 1: 76 (1950); F.T.E.A. Crassul.: 39 (1987); F.E.E. 3: 22 (1989); Biol. Skr. 51: 87 (1998).
Perennial herb. Rocky open bushland.
Distr: RS; D.R. Congo to Somalia, S to Tanzania; Arabia.
RS: Jackson 2629A 4/1953.

Kalanchoe laciniata (L.) DC.

F.T.E.A. Crassul.: 52 (1987); F.E.E. 3: 25 (1989); Biol. Skr. 51: 87 (1998).
Perennial herb. Bushland.
Distr: EQU; Nigeria to Somalia, S to Angola & Mozambique; Arabia, India.
EQU: Jackson 1109 1/1950.

Kalanchoe lanceolata (Forssk.) Pers.

F.P.S. 1: 79 (1950); F.J.M.: 87 (1976); F.T.E.A. Crassul.: 49 (1987); F.E.E. 3: 23 (1989); Biol. Skr. 51: 87 (1998).
Syn: *Kalanchoe lanceolata* (Forssk.) Pers. var. *glandulosa* (Hochst. ex A.Rich.) Cufod. – F.J.M.: 88 (1976).
Syn: *Kalanchoe modesta* Kotschy & Peyr. – F.P.S. 1: 79 (1950).

Annual or perennial herb. Open woodland.
Distr: DAR, UN, EQU; Mali to Somalia, S to Namibia &
South Africa; Madagascar, Arabia, India.
DAR: Wickens 2828 11/1964; **EQU:** Friis & Vollesen 969
2/1982.

Kalanchoe marmorata Baker
F.T.E.A. Crassul.: 33 (1987); F.E.E. 3: 20 (1989).
Perennial herb. Rocky slopes.
Distr: EQU; D.R. Congo, Rwanda, Ethiopia, Somalia,
Tanzania; New Zealand.
EQU: Myers 14049 9/1941.

Kalanchoe prittwitzii Engl.
F.T.E.A. Crassul.: 38 (1987); F.E.E. 3: 22 (1989); Biol. Skr.
51: 87 (1998).
Perennial herb. Rocky forest margins.
Distr: EQU; Eritrea & Somalia, S to D.R. Congo &
Tanzania.
EQU: Myers 7972 12/1937.

Sedum ruwenzoriense Baker f.
F.T.E.A. Crassul.: 24 (1987); Biol. Skr. 51: 88 (1998).
Succulent perennial herb. Upland rock crevices.
Distr: EQU; D.R. Congo, Rwanda, Uganda, Kenya,
Ethiopia.
EQU: Myers 10899 4/1939.

Umbilicus botryoides Hochst. ex A.Rich.
F.P.S. 1: 79 (1950); F.J.M.: 87 pro parte (1976); F.T.E.A.
Crassul.: 60 (1987); F.E.E. 3: 16 (1989).
Succulent perennial herb. Rocky grassland.
Distr: RS, DAR; Cameroon to Somalia, S to D.R. Congo &
Tanzania; Egypt.
DAR: Wickens 2636 9/1964.

Umbilicus gaditanus Boiss.
Trav. Inst. Sci. Univ. Mohammed V, Sér. Bot. 37: 169
(2005).
Syn: *Umbilicus botryoides* sensu Wickens pro parte,
non Hochst. ex A.Rich. – F.J.M.: 88 (1976).
Succulent perennial herb. Rocky slopes.
Distr: DAR; Spain, Morocco, Algeria.
DAR: Macintosh 74 3/1930.

Umbilicus paniculiformis Wickens
Kew Bull. 33: 421 (1979).
Succulent perennial herb. Rocky slopes.
Distr: RS; Sudan endemic.
IUCN: RD
RS: Farag & Zubair 4 (fl) 2/2012.
Note: the African Plants Database records *U.
tropaeolifolius* Boiss. from Sudan but it is likely that this is
based upon a misidentification of *U. paniculiformis*.

Haloragaceae

I. Darbyshire

Laurembergia tetrandra (Schott) Kanitz subsp.
brachypoda (Welw. ex Hiern) A.Raynal var.
brachypoda
F.T.E.A. Halorag.: 6 (1973); F.J.M.: 95 (1976); F.E.E. 2(1):
425 (2000).

Syn: *Laurembergia engleri* Schindl. – F.P.S. 1: 147
(1950).
Perennial herb, sometimes mat-forming. Marshes.
Distr: DAR, BAG; Senegal to Ethiopia, S to Angola,
Zimbabwe & Mozambique.
DAR: Wickens 2681 (fl) 9/1964; **BAG:**
Schweinfurth 2582 (fl) 10/1869.
Note: in the African Plants Database, this taxon is treated
under *L. repens* (L.) P.J.Bergius subsp. *brachypoda* (Welw.
ex Hiern) Oberm., with the distribution including South
Africa and NW Africa.

Myriophyllum spicatum L.
F.T.E.A. Halorag.: 7 (1973); F.J.M.: 95 (1976); F.E.E. 2(1):
425 (2000).
Submerged aquatic herb. Lakes, ponds, ditches, streams.
Distr: DAR; scattered in tropical Africa; widely distributed
globally but absent from S America & Australia.
DAR: Wickens 1594 (fl fr) 5/1964.

VITALES

Vitaceae

I. Farag & I. Darbyshire

Ampelocissus abyssinica (Hochst. ex A.Rich.)
Planch.
F.P.S. 2: 304 (1952); F.E.E. 3: 401 (1989); F.T.E.A. Vit.: 3
(1993); Biol. Skr. 51: 269 (1998).
Syn: *Ampelocissus cavicaulis* (Baker) Planch. – F.P.S. 2:
304 (1952); Bot. Exp. Sud.: 111 (1970).
Syn: *Ampelocissus schimperiana* sensu Andrews,
non (Hochst. ex A.Rich.) Planch. – F.P.S. 2: 303 pro parte
(1952).
Herbaceous climber. Forest & woodland.
Distr: DAR, ES, EQU; Nigeria to Eritrea, S to Angola &
D.R. Congo.
DAR: Wickens 2933 (fr) 4/1965; **EQU:** Andrews 1263
(fr) 5/1939.

Ampelocissus africana (Lour.) Merr. var.
africana
F.J.M.: 121 (1976); F.T.E.A. Vit.: 5 (1993); Biol. Skr. 51:
270 (1998).
Syn: *Ampelocissus asarifolia* (Baker) Planch. – F.P.S. 2:
303 (1952).
Syn: *Ampelocissus grantii* (Baker) Planch. – Bot. Exp.
Sud.: 111 (1970).
Herbaceous or woody climber. Woodland, wooded
grassland, riverine forest.
Distr: DAR, ES, BAG, UN, EQU; Senegal to Kenya, S to
Botswana & South Africa.
DAR: Wickens 3147 (fr) 8/1974; **BAG:** Andrews 770 (fr)
4/1939.

Ampelocissus bombycina (Baker) Planch.
F.P.S. 2: 304 (1952); F.E.E. 3: 399 (1989); Biol. Skr. 51:
270 (1998).
Syn: *Ampelocissus cinnamochroa* Planch. – F.P.S. 2:
303 (1952).
Herbaceous climber. Bushland & woodland.
Distr: ES, BAG, EQU; Guinea to Eritrea, S to D.R. Congo.

ES: Beshir 5 (fr) 7/1951; **EQU:** Friis & Vollesen 1251 (fr) 3/1982.

Ampelocissus multistriata (Baker) Planch.

F.P.S. 2: 304 (1952); F.T.E.A. Vit.: 3 (1993).
Herbaceous or woody climber. Bushland, wooded grassland, riverine forest.
Distr: BAG; Senegal to Tanzania, S to Zambia & Mozambique.
BAG: Schweinfurth 1674 (fr) 4/1869.

Ampelocissus sarcocephala (Schweinf. ex Oliv.) Planch.

F.P.S. 2: 304 (1952); F.T.E.A. Vit.: 8 (1993); Biol. Skr. 51: 270 (1998).
Herbaceous climber. Thicket over ironstone.
Distr: BAG; Chad, C.A.R. & D.R. Congo.
IUCN: RD
BAG: Jackson 3076 (fr) 7/1953.
Note: Friis & Vollesen collected a specimen close to this species from the Imatong Mts, listing it under *A. sp. cf. A. sarcocephala*; they noted that the material is insufficient for exact determination.

Ampelocissus schimperiana (Hochst. ex A.Rich.) Planch.

F.P.S. 2: 303 pro parte (1952); F.E.E. 3: 400 (1989).
Herbaceous climber. Riverine forest and woodland.
Distr: ES, EQU; Ethiopia.
IUCN: RD
ES: Schweinfurth 1268 7/1865; **EQU:** Wyld 58 4/1937.

Cayratia debilis (Baker) Suess.

F.T.E.A. Vit.: 139 (1993); Biol. Skr. 51: 270 (1998).
Herbaceous climber. Forest.
Distr: EQU; Guinea to Uganda, S to Angola & D.R. Congo.
EQU: Jackson 881 10/1949.

Cayratia gracilis (Guill. & Perr.) Suess.

F.J.M.: 121 (1976); F.E.E. 3: 403 (1989); F.T.E.A. Vit.: 138 (1993); Biol. Skr. 51: 271 (1998).
Syn: *Cissus gracilis* Guill. & Perr. – F.P.S. 2: 312 (1952).
Herbaceous climber. Woodland & grassland, forest margins, riverine forest.
Distr: DAR, ES, EQU; Senegal to Eritrea, S to Angola & South Africa; Yemen.
DAR: Wickens 2187 (fr) 8/1964; **EQU:** Friis & Vollesen 88 (fr) 11/1980.

Cayratia ibuensis (Hook.f.) Suess.

F.E.E. 3: 404 (1989); F.T.E.A. Vit.: 136 (1993).
Syn: *Cissus ibuensis* Hook.f. – F.P.S. 2: 311 (1952); Bot. Exp. Sud.: 112 (1970).
Herbaceous climber. Riverine forest, forest margins, secondary bushland, grassland.
Distr: CS, ES, BAG, UN, EQU; Togo to Ethiopia, S to Angola & Mozambique; Egypt.
CS: Jackson 4141 (fr) 4/1961; **UN:** Denny 31 (fr) 4/1978.

Cissus aralioides (Baker) Planch.

F.P.S. 2: 310 (1952); F.T.E.A. Vit.: 45 (1993).
Large herbaceous or woody climber. Forest & thicket.
Distr: BAG; Senegal to Kenya, S to Angola & Mozambique.
BAG: Schweinfurth 2405 (fr) 4/1869.

Note: the Sudanese material is probably referable to the W African subsp. *aralioides* – see note in F.T.E.A.

Cissus cornifolia (Baker) Planch.

F.P.S. 2: 308 (1952); F.J.M.: 121 (1976); F.E.E. 3: 405 (1989); F.T.E.A. Vit.: 36 (1993); Biol. Skr. 51: 271 (1998).
Erect shrub. Grassland & woodland.
Distr: DAR, KOR, BAG, EQU; Ivory Coast to Ethiopia, S to Botswana & South Africa.
DAR: Wickens 3431 (fr) 4/1971; **EQU:** Friis & Vollesen 988 (fr) 2/1982.

Cissus diffusiflora (Baker) Planch.

F.T.E.A. Vit.: 24 (1993).
Syn: *Cissus afzelii* (Baker) Gilg & M.Brandt – F.P.S. 2: 308 (1952).
Herbaceous climber. Forest & forest margins.
Distr: EQU; Senegal to Uganda, S to Angola & D.R. Congo.
EQU: Myers 9195 (fr) 8/1938.

Cissus integrifolia (Baker) Planch.

F.T.E.A. Vit.: 15 (1993); Biol. Skr. 51: 271 (1998).
Herbaceous or woody climber. Grassland, woodland & forest.
Distr: EQU; D.R. Congo to Kenya, S to South Africa.
EQU: Fukui 84-73 1/1985.

Cissus palmatifida (Baker) Planch.

F.P.S. 2: 309 (1952); F.W.T.A. 1(2): 676 (1958).
Woody climber, at first erect. Wooded grassland, riverine forest.
Distr: BAG; Senegal to C.A.R.
BAG: Schweinfurth 2001 (fr) 7/1869.

Cissus petiolata Hook.f.

F.P.S. 2: 306 (1952); F.E.E. 3: 406 (1989); F.T.E.A. Vit.: 16 (1993); Biol. Skr. 51: 271 (1998).
Syn: *Cissus bignonioides* Gilg & M.Brandt – F.P.S. 2: 306 (1952).
Herbaceous or woody climber. Bushland, woodland & forest margins.
Distr: BAG, EQU; Guinea to Eritrea, S to Angola & Mozambique.
EQU: Friis & Vollesen 677 (fr) 12/1980.

Cissus populnea Guill. & Perr.

F.P.S. 2: 306 (1952); F.E.E. 3: 407 (1989); F.T.E.A. Vit.: 18 (1993); Biol. Skr. 51: 272 (1998).
Woody climber. Grassland & woodland on rocky outcrops.
Distr: DAR, KOR, ES, BAG, EQU; Senegal to Ethiopia, S to Tanzania.
KOR: Wickens 92 (fr) 7/1962; **EQU:** Myers 9147 (fr) 12/1938.

Cissus quadrangularis L.

F.P.S. 2: 308 (1952); F.J.M.: 121 (1976); F.E.E. 3: 407 (1989); T.S.S.: 295 (1990); F.T.E.A. Vit.: 40 (1993); Biol. Skr. 51: 272 (1998).
Succulent climber. Grassland, bushland & woodland.
Distr: RS, DAR, KOR, CS, ES, EQU; palaeotropical.
ES: Wickens 3062 (fr) 5/1969; **EQU:** Kielland-Lund 147 12/1983 (n.v.).

Cissus rotundifolia (Forssk.) Vahl

F.E.E. 3: 408 (1989); F.T.E.A. Vit.: 21 (1993); Biol. Skr. 51: 272 (1998).
Woody climber. Bushland, rocky hillslopes.
Distr: EQU; D.R. Congo to Eritrea & Somalia, S to South Africa; Yemen.
EQU: Jackson 4250 (fr) 6/1961.

Cissus rubiginosa (Welw. ex Baker) Planch.

F.P.S. 2: 306 (1952); F.T.E.A. Vit.: 26 (1993).
Herbaceous or woody climber. Woodland & grassland, forest margins.
Distr: EQU; Guinea to Uganda, S to Angola & Mozambique.
EQU: Sillitoe 471 (fr) 1919.

Cissus rufescens Guill. & Perr.

F.P.S. 2: 308 (1952); F.W.T.A. 1(2): 678 (1958).
Herbaceous climber. Wooded grassland, forest margins.
Distr: EQU; Senegal to D.R. Congo.
EQU: Andrews 1616 (fr) 6/1939.

Cyphostemma adenanthum (Fresen.) Desc.

F.E.E. 3: 410 (1989); F.T.E.A. Vit.: 83 (1993).
Syn: *Cissus adenantha* Fresen. – F.P.S. 2: 311 (1952).
Erect perennial herb. Woodland & wooded grassland, rocky hillslopes.
Distr: ES; Eritrea, Ethiopia.
IUCN: RD
ES: Schweinfurth 1256 (fl fr) 6/1865.

Cyphostemma adenocaule (Steud. ex A.Rich.) Desc. ex Wild & R.B.Drumm. subsp. adenocaule

F.J.M.: 121 (1976); F.E.E. 3: 412 (1989); F.T.E.A. Vit.: 89 (1993); Biol. Skr. 51: 272 (1998).
Syn: *Cissus adenocaulis* Steud. ex A.Rich. – F.P.S. 2: 312 (1952); Bot. Exp. Sud.: 111 (1970).
Herbaceous climber. Forest, bushland, woodland & grassland.
Distr: DAR, ES, BAG, EQU; Senegal to Eritrea, S to Angola & Mozambique.
DAR: Wickens 1908 (fl) 7/1964; **EQU:** Friis & Vollesen 1001 (fl) 2/1982.

Cyphostemma alnifolium (Schweinf. ex Planch.) Desc.

Trop. Afr. Fl. Pl. 5: 318 (2010).
Syn: *Cissus alnifolia* Schweinf. ex Planch. – F.P.S. 2: 308 (1952).
Erect perennial herb. Wooded grassland.
Distr: EQU; Cameroon, C.A.R.
IUCN: RD
EQU: Schweinfurth 3766 (fr) 5/1870.

Cyphostemma bambuseti (Gilg & M.Brandt) Desc. ex Wild & R.B.Drumm.

F.T.E.A. Vit.: 119 (1993); Biol. Skr. 51: 273 (1998).
Herbaceous climber. Forest, secondary bushland.
Distr: EQU; D.R. Congo, Rwanda, Uganda, Kenya.
EQU: Friis & Vollesen 272 (fl) 11/1980.
Note: a very local species, but not considered threatened.

Cyphostemma crinitum (Planch.) Desc.

F.J.M.: 122 (1976); Trop. Afr. Fl. Pl. 5: 324 (2010).
Syn: *Cissus crinita* Planch. – F.P.S. 2: 311 (1952).
Erect perennial herb. Woodland & wooded grassland.
Distr: DAR, BAG, EQU; ?Cameroon, C.A.R., D.R. Congo.
IUCN: RD
DAR: Wickens 1863 (fl fr) 7/1964; **EQU:** Myers 13844 (fl) 6/1941.

Cyphostemma crotalarioides (Planch.) Desc. ex Wild & R.B.Drumm.

F.J.M.: 122 (1976); F.T.E.A. Vit.: 81 (1993); Biol. Skr. 51: 273 (1998).
Syn: *Cissus crotalarioides* Planch. – F.P.S. 2: 309 (1952); Bot. Exp. Sud.: 111 (1970).
Erect or straggling perennial herb. Woodland & wooded grassland, rock outcrops.
Distr: DAR, BAG, EQU; Mali & Ivory Coast to Tanzania, S to Zimbabwe & Mozambique.
DAR: Wickens 1817 (fr) 7/1964; **BAG:** Schweinfurth 1554 (fr) 4/1869.

Cyphostemma cymosum (Schumach. & Thonn.) Desc. subsp. cymosum

F.J.M.: 122 (1976); F.T.E.A. Vit.: 95 (species only) (1993).
Syn: *Cissus cymosa* Schumach. & Thonn. – Bot. Exp. Sud.: 112 (1970).
Herbaceous climber. Woodland, bushland & wooded grassland.
Distr: DAR; Senegal to C.A.R.
DAR: Wickens 1746 (fl) 6/1964.

Cyphostemma cyphopetalum (Fresen.) Desc. ex Wild & R.B.Drumm. subsp. cyphopetalum

F.E.E. 3: 417 (1989); F.T.E.A. Vit.: 121 (1993); Biol. Skr. 51: 273 (1998).
Syn: *Cissus cyphopetala* Fresen. – F.P.S. 2: 311 (1952); Bot. Exp. Sud.: 112 (1970).
Herbaceous climber. Grassland, bushland, woodland & forest.
Distr: RS, EQU; Cameroon to Somalia, S to Zambia.
RS: Carter 1842 (fl) 11/1987; **EQU:** Friis & Vollesen 779 (fr) 12/1980.

Cyphostemma dembianense (Chiov.) Vollesen

F.E.E. 3: 415 (1989); Biol. Skr. 51: 275 (1998).
Syn: *Cissus adenocephala* Gilg & M.Brandt – F.P.S. 2: 311 (1952); Bot. Exp. Sud.: 111 (1970).
Herbaceous climber. Forest margins & secondary regrowth, scrubby grassland.
Distr: RS, ES, EQU; Ethiopia.
IUCN: RD
RS: Aylmer 654 (fr) 3/1936; **EQU:** Drar 1467b 4/1938 (n.v.).

Cyphostemma junceum (Webb) Wild & R.B.Drumm. subsp. junceum

F.E.E. 3: 410 (1989); F.T.E.A. Vit.: 62 (1993); Biol. Skr. 51: 274 (1998).
Syn: *Cissus juncea* Webb – F.P.S. 2: 310 (1952).
Erect perennial herb. Grassland, bushland & woodland, swamp margins.

Distr: CS, ES, EQU; Nigeria to Ethiopia, S to D.R. Congo & Kenya.
ES: Schweinfurth 1260 (fr) 6/1865; **EQU:** Sillitoe 470 (fr) 1919.

Cyphostemma kilimandscharicum (Gilg) Desc. ex Wild & R.B.Drumm. var. kilimandscharicum

F.E.E. 3: 416 (1989); F.T.E.A. Vit.: 108 (1993); Biol. Skr. 51: 274 (1998).
Herbaceous climber. Forest, forest margins & secondary scrub.
Distr: EQU; Ethiopia to Zimbabwe.
EQU: Myers 11760 (fl fr) 8/1939.

Cyphostemma rhodesiae (Gilg & M.Brandt) Desc. ex Wild & R.B.Drumm.

F.E.E. 3: 411 (1989); F.T.E.A. Vit.: 81 (1993).
Syn: Cissus triumfettioides Gilg & M.Brandt – F.P.S. 2: 309 (1952).
Erect perennial herb. Grassland, woodland & thicket.
Distr: BAG; Ethiopia to Zimbabwe & Mozambique.
BAG: Schweinfurth 1562 (fl) 4/1869.

Cyphostemma serpens (A.Rich.) Desc. subsp. serpens

F.T.E.A. Vit.: 86 (1993); Biol. Skr. 51: 274 (1998).
Syn: Cissus schweinfurthii Planch. – F.P.S. 2: 310 (1952).
Syn: Cissus serpens A.Rich. – F.P.S. 2: 310 (1952); Bot. Exp. Sud.: 112 (1970).
Syn: Cyphostemma sesquipedale (Gilg) Desc. – F.J.M.: 122 (1976).
Herbaceous climber or trailer. Grassland & woodland, often on black clay soils.
Distr: DAR, KOR, BAG, UN, EQU; Eritrea to D.R. Congo & Tanzania.
DAR: Wickens 1831 (fr) 7/1964; **EQU:** Jackson 2283 (fr) 6/1952.

Cyphostemma ternatum (Forssk.) Desc.

F.E.E. 3: 416 (1989); F.T.E.A. Vit.: 93 (1993).
Syn: Cissus ternata (Forssk.) J.G.Gmel. – F.P.S. 2: 310 (1952); Bot. Exp. Sud.: 112 (1970).
Trailing or climbing succulent herb. Bushland, grassland, riverine forest margins.
Distr: RS, EQU; Eritrea, Ethiopia, Somalia, Kenya; Yemen.
RS: Carter 1891 (fl) 11/1987; **EQU:** Drar 1394 4/1938 (n.v.).
Note: a rather local species but not considered threatened. Drar (1970) cites three specimens from Equatoria, including the one cited here, but all are sterile and their identification requires confirmation.

Cyphostemma ukerewense (Gilg) Desc.

F.T.E.A. Vit.: 100 (1993); Biol. Skr. 51: 274 (1998).
Herbaceous climber or trailer. Wet grassland, swamp margins, forest margins.
Distr: EQU; Cameroon to Congo-Brazzaville, Uganda to Zambia & Mozambique.
EQU: Friis & Vollesen 49 (fr) 11/1980.
Note: Verdcourt in F.T.E.A. suggested that the single specimen seen from the Imatong Mts may represent a distinct variety on account of its smaller bracts.

Leea guineensis G.Don

F.P.S. 2: 312 (1952); T.S.S.: 295 (1990); F.T.E.A. Leeaceae (in Vitaceae): 141 (1993); Biol. Skr. 51: 275 (1998).
Shrub. Forest & swamp forest, often in clearings or margins.
Distr: EQU; palaeotropical.
EQU: Friis & Vollesen 178 (fr) 11/1980.

Rhoicissus revoilii Planch.

F.J.M.: 122 (1976); F.E.E. 3: 401 (1989); F.T.E.A. Vit.: 11 (1993).
Scandent shrub or woody climber. Woodland, wooded grassland, dry forest margins.
Distr: DAR; Ghana, Ethiopia & Somalia, S to South Africa; Comoros, Arabia.
DAR: Wickens 1310 (fr) 2/1964.

Rhoicissus tridentata (L.f.) Wild & R.B.Drumm.

F.J.M.: 122 (1976); F.E.E. 3: 401 (1989); F.T.E.A. Vit.: 9 (1993).
Woody climber or shrub. Grassland, thicket, forest margins.
Distr: DAR; Nigeria to Eritrea, S to South Africa; Yemen.
DAR: Kassas 255 (fl) 12/1965.

ZYGOPHYLLALES

Zygophyllaceae

I. Darbyshire

Balanites aegyptiaca (L.) Delile var. aegyptiaca

F.P.S. 2: 318 (1952); F.J.M.: 93 (1976); F.E.E. 3: 433 (1989); T.S.S.: 299 (1990); Biol. Skr. 51: 102 (1998); F.T.E.A. Balanit.: 6 (2003).
Small tree. Dry wooded grassland & bushland, wadis.
Distr: NS, RS, DAR, KOR, CS, ?ES, UN, EQU; Mauritania & Senegal to Somalia, S to Zimbabwe; Egypt, Israel, Arabia.
NS: Pettet 92 (fl) 9/1962; **UN:** J.M. Lock 79/10 (fl) 3/1979.

Balanites rotundifolia (Tiegh.) Blatt. var. rotundifolia

F.E.E. 3: 434 (1989); Biol. Skr. 51: 102 (1998); F.T.E.A. Balanit.: 10 (2001).
Shrub or small tree. Dry wooded grassland & bushland.
Distr: EQU; Eritrea, Ethiopia, Djibouti, Somalia, Uganda, Kenya; Yemen.
EQU: Kielland-Lund 444 12/1983 (n.v.).

Fagonia arabica L.

F.J.U. 3: 250 (1999); F.Egypt 2: 214 (2000).
Syn: Fagonia arabica L. var. viscidissima Maire – F.Egypt 2: 14 (2000).
Syn: Fagonia thebaica Boiss. – F.J.U. 3: 249 (1999); F.Egypt 2: 15 (2000).
Shrub. Dry sandy plains, wadis.
Distr: NS; N Africa & Sahara to Palestine & Arabia.
NS: Léonard 5036 (fr) 1/1969.

Fagonia bruguieri DC.

F.Egypt. 2: 15 (2000).
Biennial or perennial shrublet. Dry sandy areas & waste ground.

Distr: NS; N Africa to Pakistan.
NS: Ahti 16469 10/1962.

Fagonia indica Burm.f.

F.T.E.A. Zygophyll.: 12 (1985); F.J.U. 3: 245 (1999); F.E.E. 2(1): 358 (2000).
Syn: *Fagonia cretica* sensu Andrews, non L. – F.P.S. 2: 124 (1952).
Shrub. Dry open rocky & sandy plains, wadis.
Distr: NS, RS, DAR, KOR, CS, ES; Cape Verde Is., Mauritania to Somalia & Kenya; N Africa, Arabia to India.
CS: de Wilde et al. 5803 (fr) 2/1965.

Fagonia latifolia Delile subsp. isotricha (Murb.) Ozenda & Quézel

Fl. Tunisie ed. 2: 327 (2010).
Syn: *Fagonia isotricha* Murb. var. *isotricha* – F.E.E. 2(1): 358 (2000); F.Egypt 2: 19 (2000).
Shrub. Dry sandy & stony ground.
Distr: ?NS; N Africa & Sahara from Cape Verde Is., Mauritania & Morocco to Egypt.
Note: no Sudanese material has been seen but it is listed as occurring in Sudan by Boulos in F.Egypt.

Fagonia paulayana Wagner & Vierh.

F.T.E.A. Zygophyll.: 12 (1985); F.E.E. 2(1): 358 (2000).
Syn: *Fagonia indica* L. var. *schweinfurthii* Hadidi – F.Egypt 2: 17 (2000).
Syn: *Fagonia schweinfurthii* (Hadidi) Hadidi – F.E.E. 2(1): 358 (2000).
Shrub or perennial herb. Open dry sandy plains & rocky areas, sparse bushland.
Distr: RS; Eritrea to Somalia, S to Kenya; Egypt, Arabia & Pakistan.
RS: Andrews 3557 (fl fr).
Note: Beier (Syst. Biodiv. 3: 249 (2005)) treats *F. schweinfurthii* as a trifoliolate variant of *F. paulayana*. Both forms occur in Sudan.

Seetzenia lanata (Willd.) Bullock

F.Egypt 2: 22 (2000).
Syn: *Seetzenia orientalis* Decne. – F.P.S. 2: 124 (1952).
Prostrate annual or perennial herb. Dry open sandy plains.
Distr: RS, DAR; Somalia, South Africa; Libya, Egypt, Arabia to India.
RS: Carter 1939 (fl) 11/1987.

Tetraena alba (L.f.) Beier & Thulin

Pl. Syst. Evol. 240: 35 (2003).
Syn: *Zygophyllum album* L.f. – F.P.S. 2: 127 (1952); F.E.E. 2(1): 357 (2000).
Shrub. Saline coastal plains, mudflats.
Distr: RS; Eritrea, Somalia; N Africa & Arabia.
RS: Johnston 15 (fl fr) 5/1885.

Tetraena coccinea (L.) Beier & Thulin

Pl. Syst. Evol. 240: 35 (2003).
Syn: *Zygophyllum coccineum* L. – F.P.S. 2: 127 (1952); F.E.E. 2(1): 355 (2000).
Shrub. Sandy plains, salt pans.
Distr: RS; Eritrea; Egypt, Arabia.
RS: Jackson 2788 (fl fr) 4/1953.

Tetraena decumbens (Delile) Beier & Thulin

Pl. Syst. Evol. 240: 35 (2003).
Syn: *Zygophyllum decumbens* Delile – F.P.S. 2: 127 (1952); F.Egypt 2: 24 (2000).
Shrub. Dry open saline plains, wadis, coastal thicket.
Distr: RS; Somalia; Egypt, Arabia, Socotra.
RS: Jackson 2773 (fr) 4/1953.

Tetraena simplex (L.) Beier & Thulin

Pl. Syst. Evol. 240: 36 (2003).
Syn: *Zygophyllum simplex* L. – F.P.S. 2: 126 (1952); Bot. Exp. Sud.: 113 (1970); F.T.E.A. Zygophyll.: 10 (1985); F.E.E. 2(1): 355 (2000).
Annual or short-lived perennial herb or subshrub. Sandy plains, wadis, bushland & open woodland.
Distr: NS, RS, KOR, CS; Cape Verde Is., Mauritania to Somalia, S to Kenya, Angola to South Africa; N Africa & Arabia to India.
RS: Drummond & Hemsley 960 (fl) 1/1953.

[Tribulus]

Note: species delimitation in *Tribulus* is problematic and this account should be considered provisional; a critical revision in Sudan and the surrounding region is desirable.

Tribulus bimucronatus Viv.

F.T.E.A. Zygophyll.: 3 (1985); F.J.U. 3: 253 (1999); F.E.E. 2(1): 360 (2000).
Annual herb. Sandy plains & wadi beds.
Distr: NS, CS; Chad, Eritrea, Ethiopia, Somalia, Kenya; Libya, Egypt & Arabia.
NS: Shaw 13 (fl fr) 10/1932.

Tribulus cistoides L.

F.T.E.A. Zygophyll.: 4 (1985); F.E.E. 2(1): 361 (2000).
Perennial herb. Open bushland & dry wooded grassland.
Distr: EQU; Eritrea & Somalia, S to Mozambique; Madagascar, Mascarenes.
EQU: Carr 802 (fl fr) 6/1970.
Note: the cited specimen is also close to *T. zeyheri* Sond. subsp. *macranthus* (Hassk.) Hadidi and it is doubtful as to whether these two are separate taxa; see also the note to *T. terrestris* L. in F.E.E. 2(1): 361 (2000).

Tribulus macropterus Boiss. var. macropterus

F.P.S. 2: 126 (1952); F.J.U. 3: 255 (1999); F.Egypt 2: 27 (2000).
Annual herb. Open woodland, semi-desert, wadis.
Distr: NS, RS; N Africa & Sahara, Israel to India.
NS: Ahti 16739 (fl fr) 10/1962.

Tribulus megistopterus Kralik subsp. pterocarpus (Ehrenb. ex C.Müll.) H.Hosni

F.Egypt 2: 28 (2000).
Syn: *Tribulus pterocarpus* Ehrenb. ex C.Müll. – F.P.S. 2: 126 (1952).
Annual herb. Sandy plains.
Distr: RS; Egypt, Arabia.
RS: Schweinfurth 649 (fl fr) 10/1868.
Note: a localised taxon but not considered threatened.

Tribulus mollis Ehrenb. ex Schweinf.

F.P.S. 2: 126 pro parte (1952); F.Egypt 2: 30 (2000).
Annual herb. Sandy plains.

Distr: RS, CS; N Africa & Sahara, Arabia to Pakistan.
RS: Schweinfurth 646 (fl fr) 10/1868.

Tribulus pentandrus Forssk.

F.P.S. 2: 125 (1952); F.T.E.A. Zygophyll.: 3 (1985); F.E.E.
2(1): 360 (2000).
Syn: Tribulus longipetalus Viv. – F.P.S. 2: 126 (1952);
Bot. Exp. Sud.: 113 (1970).
Syn: Tribulus mollis sensu Andrews pro parte, non
Ehrenb. ex Schweinf. – F.P.S. 2: 126 (1952).
Annual or short-lived perennial herb. Sandy plains,
roadsides.
Distr: NS, RS, KOR, CS; Eritrea, Somalia, Kenya; N Africa
& Sahara, Palestine, Arabia to India.
RS: Carter 1960 (fl fr) 11/1987.
Note: *Tribulus parvispinus* Presl. is also likely to occur in
Sudan but we have not seen any.

Tribulus terrestris L.

F.P.S. 2: 125 (1952); F.T.E.A. Zygophyll.: 7 (1985); Biol.
Skr. 51: 102 (1998); F.E.E. 2(1): 360 (2000).
Annual herb. Disturbed ground, weed of cultivation.
Distr: NS, RS, DAR, KOR, CS, UN, EQU; throughout the
tropical & warm temperate regions.
NS: Ahti 15756 (fl fr) 9/1962; **UN:** J.M. Lock 81/48 (fl fr)
5/1981.

FABALES

Leguminosae: Papilionoideae
(Fabaceae: Faboideae)

H. Pickering & I. Darbyshire

Abrus canescens Welw. ex Baker

F.P.S. 2: 169 (1952); F.T.E.A. Leg.-Pap.: 117 (1971); F.E.E.
3: 108 (1989); Biol. Skr. 51: 198 (1998).
Woody climber. Grassland, often by water.
Distr: EQU; Gambia to Ethiopia, S to Angola & Zambia.
EQU: Friis & Vollesen 1253 3/1982.

Abrus melanospermus Hassk. subsp.
tenuiflorus (Benth.) D.K.Harder

Novon 10: 124 (2000).
Syn: Abrus pulchellus Wall. ex Thwaites var.
tenuiflorus (Benth.) Verdc. – T.S.S.: 221 (1990).
Woody climber. Forest.
Distr: BAG, EQU; Senegal to Uganda, S to Angola &
Mozambique; Brazil & Surinam.
BAG: Schweinfurth 2345 9/1869.

Abrus precatorius L. subsp. africanus Verdc.

F.P.S. 2: 169 (1952); F.T.E.A. Leg.-Pap.: 114 (1971); F.E.E.
3: 107 (1989); T.S.S.: 221 (species only) (1990); Biol. Skr.
51: 198 (1998).
Woody climber. Wooded grassland.
Distr: DAR, BAG, UN, EQU; Cape Verde Is., Senegal
to Eritrea, S to Namibia & South Africa; Madagascar,
Mascarenes, introduced elsewhere.
DAR: Wickens 3118 2/1971; **EQU:** Myers 7967 12/1937.
Note: the single collection seen from southern Darfur is
identified as *A. precatorius* sensu lato.

Abrus schimperi Hochst. ex Baker subsp. schimperi

F.P.S. 2: 169 (1952); F.T.E.A. Leg.-Pap.: 115 (1971); F.E.E. 3:
107 (1989); T.S.S.: 223 (1990); Biol. Skr. 51: 198 (1998).
Shrub. Bushland & woodland.
Distr: KOR, CS, EQU; C.A.R., Eritrea, Ethiopia, Uganda.
KOR: Turner 235 7/1938; **EQU:** Thomas 1569 12/1935.

Adenocarpus mannii (Hook.f.) Hook.f.

F.P.S. 2: 169 (1952); F.T.E.A. Leg.-Pap.: 1009 (1971); F.E.E.
3: 222 (1989); Biol. Skr. 51: 199 (1998).
Shrub. Montane grassland, bushland & forest clearings.
Distr: EQU; Nigeria, Cameroon, Ethiopia to Angola &
Zambia.
EQU: Friis & Vollesen 830 12/1980.

Adenodolichos paniculatus (Hua) Hutch. & Dalziel

F.P.S. 2: 170 (1952); F.T.E.A. Leg.-Pap.: 703 (1971).
Subshrub. Wooded grassland.
Distr: EQU; Guinea to Uganda.
EQU: Andrews 1094 5/1939.

Aeschynomene abyssinica (A.Rich.) Vatke

F.P.S. 2: 173 (1952); F.T.E.A. Leg.-Pap.: 396 (1971); F.E.E.
3: 144 (1989); Biol. Skr. 51: 199 (1998).
Perennial herb or shrub. Woodland & grassland.
Distr: EQU; Nigeria to Ethiopia, S to Zimbabwe.
EQU: Friis & Vollesen 254 11/1980.

Aeschynomene afraspera J.Léonard

F.W.T.A. 1(2): 579 (1958).
Syn: Aeschynomene aspera sensu Andrews, non L. –
F.P.S. 2: 171 (1952).
Annual or perennial herb or shrub. Swamps, riverbanks,
floodplains.
Distr: BAG, UN, EQU; Senegal to D.R. Congo, S to
Mozambique.
BAG: Myers 13905 7/1941.

Aeschynomene cristata Vatke var. cristata

F.P.S. 2: 171 (1952); F.T.E.A. Leg.-Pap.: 378 (1971).
Shrub or perennial herb. Swamps, riverbanks.
Distr: UN; Uganda to Botswana.
UN: J.M. Lock 82/14 2/1982.

Aeschynomene cristata Vatke var. pubescens J.Léonard

F.T.E.A. Leg.-Pap.: 379 (1971); F.E.E. 3: 144 (1989).
Shrub or perennial herb. Swamps, riverbanks.
Distr: BAG, UN, EQU; Benin to Uganda, S to Angola &
Mozambique.
BAG: Aylmer 27/22 2/1926.

Aeschynomene elaphroxylon (Guill. & Perr.) Taub.

F.P.S. 2: 173 (1952); Bot. Exp. Sud.: 66 (1970); F.T.E.A.
Leg.-Pap.: 375 (1971); F.E.E. 3: 144 (1989).
Shrub. Swamps, riverbanks.
Distr: BAG, UN, EQU; Senegal to Ethiopia, S to Angola &
Mozambique; Madagascar.
UN: Harrison 292 1/1948.

161

Aeschynomene indica L.

F.P.S. 2: 171 (1952); F.T.E.A. Leg.-Pap.: 373 (1971); F.J.M.: 110 (1976); F.E.E. 3: 144 (1989).
Shrubby annual or perennial herb. Swamps, floodplain grassland & woodland.
Distr: DAR, KOR, BAG, UN, EQU; palaeotropical & N America.
DAR: Wickens 2292 8/1964; **UN:** J.M. Lock 81/110 8/1981.

Aeschynomene nilotica Taub.

F.P.S. 2: 173 (1952); F.T.E.A. Leg.-Pap.: 378 (1971).
Perennial herb or subshrub. Swamps, floodplain grassland.
Distr: UN, EQU; Mali to Tanzania, S to Namibia & Zimbabwe.
EQU: Jackson 2095 9/1951.

Aeschynomene pfundii Taub.

F.P.S. 2: 173 (1952); F.T.E.A. Leg.-Pap.: 380 (1971); F.E.E. 3: 144 (1989).
Subshrub. Swampy grassland.
Distr: KOR, UN; Mali to Ethiopia, S to Zambia & Mozambique.
KOR: Wickens 933 1/1963; **UN:** Simpson 7449 1/1930.

Aeschynomene schimperi Hochst. ex A.Rich.

F.P.S. 2: 171 (1952); F.T.E.A. Leg.-Pap.: 376 (1971); F.J.M.: 110 (1976); F.E.E. 3: 144 (1989).
Perennial herb. Swampy grassland.
Distr: DAR, ES, EQU; Senegal to Mali & Ivory Coast, D.R. Congo to Ethiopia, S to Zimbabwe & Mozambique; Madagascar.
DAR: Wickens 2221 8/1964; **EQU:** Myers 7706 9/1937.

Aeschynomene sensitiva Sw.

F.P.S. 2: 171 (1952); F.T.E.A. Leg.-Pap.: 373 (1971).
Perennial herb or subshrub. Swampy grassland.
Distr: EQU; Senegal to Uganda, S to Malawi & Mozambique; Madagascar, Mascarenes, West Indies, tropical America.
EQU: Schweinfurth 3177 2/1870.

Aeschynomene uniflora E.Mey. var. uniflora

F.P.S. 2: 170 (1952); F.T.E.A. Leg.-Pap.: 372 (1971); F.J.M.: 110 (1976).
Perennial herb. Swampy grassland.
Distr: DAR, BAG, UN, EQU; Senegal to Kenya, S to Angola & South Africa; Madagascar, Mascarenes.
DAR: Wickens 2746 9/1964; **UN:** J.M. Lock 81/150 8/1981.

Alhagi maurorum Medik. subsp. graecorum (Boiss.) Awmack & Lock

F.P.S. 2: 173 (species only) (1952); Kew Bull. 57: 441 (2002).
Perennial herb. Disturbed ground, edges of cultivation.
Distr: NS; SE Europe, Algeria, Libya, Egypt to Iran.
NS: Ahti 16815 9/1962.

Alysicarpus glumaceus (Vahl) DC. subsp. glumaceus

F.P.S. 2: 175 pro parte (1952); F.T.E.A. Leg.-Pap.: 497 (1971); F.J.M.: 110 (1976); F.E.E. 3 155 (1989).

Annual herb. Grassland, abandoned cultivations.
Distr: DAR, ES, BAG, UN; Senegal to Eritrea, S to Angola & South Africa; Arabia.
DAR: Wickens 2727 9/1964; **UN:** Simpson 7387 7/1929.

Alysicarpus monilifer (L.) DC.

F.P.S. 2: 174 (1952); F.E.E. 3: 154 (1989).
Annual herb. Sandy grassland.
Distr: NS, DAR, KOR, CS; Niger, Ethiopia, Somalia; Pakistan, India; introduced in Madagascar.
KOR: Wickens 549 9/1962.

Alysicarpus ovalifolius (Schumach. & Thonn.) J.Léonard

F.T.E.A. Leg.-Pap.: 493 (1971); F.E.E. 3: 154 (1989); Biol. Skr. 51: 199 (1998).
Annual herb. Grassland, woodland, weed of cultivation.
Distr: DAR, KOR, UN, EQU; Cape Verde Is., Mauritania to Ethiopia, S to Angola & Mozambique; Madagascar, Comoros, SE Asia.
DAR: Aglen 1 8/1944; **UN:** J.M. Lock 81/149 8/1981.

Alysicarpus rugosus (Willd.) DC. subsp. rugosus

F.T.E.A. Leg.-Pap.: 495 (1971); F.J.M.: 111 (1976); F.E.E. 3: 155 (1989); Biol. Skr. 51: 200 (1998).
Syn: *Alysicarpus glumaceus* sensu Andrews pro parte, non (Vahl) DC. – F.P.S. 2: 175 (1952).
Annual or perennial herb. Wooded grassland including seasonally wet areas.
Distr: DAR, KOR, CS, ES, UN, EQU; Mali to Ethiopia, S to Zimbabwe & Mozambique; Madagascar, Arabia.
DAR: Wickens 1882 7/1964; **UN:** J.M. Lock 81/170 8/1981.

Alysicarpus vaginalis (L.) DC. var. vaginalis

F.P.S. 2: 174 (1952); F.T.E.A. Leg.-Pap.: 493 (1971).
Perennial herb. Grassland, woodland, disturbed ground.
Distr: BAG, UN, EQU; palaeotropical, introduced in America.
BAG: Simpson 7132 6/1929.

Alysicarpus zeyheri Harv.

F.P.S. 2: 175 (1952); F.T.E.A. Leg.-Pap.: 494 (1971); F.E.E. 3: 154 (1989).
Perennial herb. Grassland including swampy areas.
Distr: UN, EQU; Sierra Leone to Ethiopia, S to Angola & South Africa.
EQU: Sillitoe 179 1919.

Amphicarpaea africana (Hook.f.) Harms

F.T.E.A. Leg.-Pap.: 511 (1971); F.E.E. 3: 155 (1989); Biol. Skr. 51: 200 (1998).
Perennial herb. Forest & forest clearings.
Distr: EQU; Cameroon to Ethiopia, S to Zambia & Malawi.
EQU: Friis & Vollesen 326 11/1980.

Amphimas pterocarpoides Harms

F.P.S. 2: 110 (1952); F.W.T.A. 1(2): 448 (1958); T.S.S.: 223 (1990).
Tree. Riverbanks, riverine forest & woodland.
Distr: EQU; Guinea to D.R. Congo.
EQU: Andrews 1539 5/1939.

Andira inermis (W.Wright) DC. subsp. rooseveltii (De Wild.) J.B.Gillett ex Polhill

F.P.S. 2: 175 (1952); F.T.E.A. Leg.-Pap.: 63 (1971); T.S.S.: 223 (species only) (1990); Biol. Skr. 51: 200 (1998).
Tree. Woodland, riverine forest.
Distr: BAG, UN, EQU; Ivory Coast to Uganda.
EQU: Friis & Vollesen 1042 2/1982.

Antopetitia abyssinica A.Rich.

F.T.E.A. Leg.-Pap.: 1049 (1971); F.E.E. 3: 227 (1989); Biol. Skr. 51: 201 (1998).
Annual herb. Montane grassland.
Distr: EQU; Nigeria, Cameroon, Eritrea to Zimbabwe & Mozambique.
EQU: Friis & Vollesen 105 11/1980.

Arachis hypogaea L.

F.P.S. 2: 175 (1952); F.T.E.A. Leg.-Pap.: 442 (1971); F.E.E. 3: 147 (1989).
Annual herb. Cultivated, occasionally a casual escapee.
Distr: BAG, EQU; pantropical.
BAG: Schweinfurth 2425 9/1869.
Note: the cultivated 'peanut' or 'groundnut'.

Argyrolobium arabicum (Decne.) Jaub. & Spach

F.J.M.: 111 (1976); F.E.E. 3: 222 (1989).
Syn: *Argyrolobium abyssinicum* Jaub. & Spach – F.P.S. 2: 175 (1952); Bot. Exp. Sud.: 67 (1970).
Annual or short-lived perennial herb. Rocky hillslopes, weed of cultivation.
Distr: RS, DAR; Mauritania to Eritrea & Ethiopia; Algeria, Egypt, Arabia, Iran.
RS: Carter 1974 11/1987.

Argyrolobium fischeri Taub.

F.T.E.A. Leg.-Pap.: 1002 (1971); F.E.E. 3: 220 (1989); Biol. Skr. 51: 201 (1998).
Perennial herb or subshrub. Montane grassland, bushland & forest clearings.
Distr: EQU; Ethiopia to Zambia & Mozambique.
EQU: Friis & Vollesen 305 11/1980.

Argyrolobium rupestre (E.Mey.) Walp. subsp. aberdaricum (Harms) Polhill

F.T.E.A. Leg.-Pap.: 1006 (1971); F.E.E. 3: 220 (1989).
Perennial herb. Montane grassland.
Distr: EQU; D.R. Congo to Ethiopia, S to Malawi & Zambia.
EQU: Friis & Vollesen 836 12/1980.

Argyrolobium rupestre (E.Mey.) Walp. subsp. kilimandscharicum (Taub.) Polhill

F.T.E.A. Leg.-Pap.: 1005 (1971); Biol. Skr. 51: 201 (1998).
Perennial herb. Montane grassland & forest margins.
Distr: EQU; Uganda, Kenya, Tanzania.
EQU: Jackson 1321.

Astragalus atropilosulus (Hochst.) Bunge subsp. abyssinicus (Steud. ex A.Rich.) J.B.Gillett var. abyssinicus

F.J.M.: 111 (1976); F.E.E. 3: 231 (1989).
Syn: *Astragalus abyssinicus* Steud. ex A.Rich. – F.P.S. 2: 176 pro parte (1952); Bot. Exp. Sud.: 67 (1970).
Perennial herb. Montane grassland.

Distr: RS, DAR; Ethiopia to South Africa; Arabia.
DAR: Wickens 2524 9/1964.

Astragalus atropilosulus (Hochst.) Bunge subsp. burkeanus (Harv.) J.B.Gillett var. burkeanus

F.T.E.A. Leg.-Pap.: 1056 (1971); F.E.E. 3: 231 (1989); Biol. Skr. 51: 202 (1998).
Syn: *Astragalus abyssinicus* sensu Andrews pro parte, non Steud. ex A.Rich. – F.P.S. 2: 176 (1952).
Perennial herb. Grassland, bushland, forest margins.
Distr: EQU; Ethiopia & Somalia, S to South Africa.
EQU: Friis & Vollesen 303 11/1980.

Astragalus eremophilus Boiss.

F.P.S. 2: 176 (1952); F.E.E. 3: 231 (1989).
Annual herb. Open sandy or stony ground.
Distr: RS; Eritrea; Egypt, Arabia to Pakistan.
RS: Andrews 3634.

Astragalus pelecinus (L.) Barneby subsp. leiocarpus (A.Rich.) Podlech

Syn: *Biserrula pelecinus* L. subsp. *leiocarpa* (A.Rich.) J.B.Gillett – F.P.S. 2: 177 (species only) (1952); F.T.E.A. Leg.-Pap.: 1059 (1971); F.J.M.: 111 (1976); F.E.E. 3: 231 (1989).
Annual herb. Grassland, scrub.
Distr: DAR; Eritrea, Ethiopia, Tanzania.
DAR: Wickens 1484 4/1964.

Astragalus vogelii (Webb) Bornm.

F.P.S. 2: 176 (1952); F.J.U. 3: 221 (1999).
Syn: *Astragalus fatmensis* Hochst. ex Chiov. – F.E.E. 3: 231 (1989).
Annual herb. Dry woodland.
Distr: NS, RS, KOR, CS; Eritrea, Ethiopia; N Africa, Arabia to India.
NS: Léonard 5002 12/1968.

Baphia abyssinica Brummitt

F.E.E. 3: 104 (1989); T.S.S.: 225 (1990).
Tree. Forest.
Distr: UN; Ethiopia.
IUCN: VU
UN: Myers 10438 11/1939.
Note: Andrews also records *B. wollastonii* Baker f. from the Imatong Mts but this record is considered likely to be incorrect – see Brummitt in Kew Bull. 22: 535 (1968).

Cajanus cajan (L.) Millsp.

F.P.S. 2: 164 (1952); F.T.E.A. Leg.-Pap.: 709 (1971); F.E.E. 3: 181 (1989); Biol. Skr. 51: 202 (1998).
Shrub. Cultivated, naturalised in abandoned cultivations.
Distr: KOR, CS, EQU; native to eastern India, now widely cultivated in the tropics.
KOR: Pfund 256 5/1875; **EQU:** Schweinfurth 4005 7/1870.
Note: the 'pigeon pea', widely cultivated as a food source.

Calpurnia aurea (Aiton) Benth. subsp. aurea

F.T.E.A. Leg.-Pap.: 47 (1971); F.E.E. 3: 104 (1989); Biol. Skr. 51: 202 (1998).
Syn: *Calpurnia subdecandra* (L'Her.) Schweick. – F.P.S. 2: 177 (1952).

Shrub or small tree. Bushland, forest margins.
Distr: EQU; C.A.R. to Eritrea, S to Angola & South Africa.
EQU: Myers 11149 4/1939.

Canavalia africana Dunn

F.T.E.A. Leg.-Pap.: 574 (1971); F.E.E. 3: 165 (1989).
Climbing perennial herb. Grassland, bushland.
Distr: EQU; Senegal to Somalia, S to Angola & South
Africa; Arabia, Socotra, India.
EQU: Sahni & Kamil 226 4/1966.

Canavalia ensiformis (L.) DC.

F.P.S. 2: 177 (1952); Bot. Exp. Sud.: 68 (1970); F.T.E.A.
Leg.-Pap.: 572 (1971); F.E.E. 3: 165 (1989).
Annual or perennial herbaceous climber. Cultivated and
sometimes naturalised.
Distr: RS, CS, ES, EQU; native to South America, now
widely cultivated in the tropics.
ES: Schweinfurth 1902 1865; **EQU:** Drar 1242 4/1938
(n.v.).
Note: cultivated, mainly for animal feed, known as 'jack-
bean'.

Canavalia regalis Piper & Dunn

Legumes of Africa: 389 (1989).
Herbaceous climber. Cultivated.
Distr: KOR; perhaps native to tropical Africa but now
known only from cultivation.
KOR: Broun 1375 1910.

Cicer arietinum L.

F.T.E.A. Leg.-Pap.: 1065 (1971); F.J.M.: 111 (1976); F.E.E.
3: 246 (1989).
Annual herb. Cultivated, potentially becoming
naturalised.
Distr: NS, DAR; probably native to the Mediterranean,
now widely cultivated.
DAR: Wickens 1146 2/1964.
Note: the commonly cultivated 'chickpea', this species is
widely grown in Sudan.

Cicer cuneatum A.Rich.

F.E.E. 3: 248 (1989).
Annual herb. Grassland, weed of cultivation.
Distr: RS; Eritrea, Ethiopia.
RS: Drar 231 1933 (n.v.).
Note: an apparently wild chickpea from NE Africa, the
record for Sudan is from Drar (1936) and is from Gebel
Elba in the disputed Hala'ib Triangle.

Clitoria ternatea L.

F.P.S. 2: 177 (1956); F.T.E.A. Leg.-Pap.: 515 (1971); F.E.E.
3: 156 (1989).
Climbing perennial herb. Grassland, bushland,
abandoned cultivations, rocky hillslopes.
Distr: RS, KOR, CS, ES, BAG, UN, EQU; native to tropical
Asia, now pantropical.
CS: Lea 12 10/1952; **BAG:** Myers 6297 2/1937.

Clitoriopsis mollis R.Wilczek

Bull. Jard. Bot. État Bruxelles 24: 412 (1954); Trop. Afr. Fl.
Pl. 3: 218 (2008).
Subshrub. Wooded grassland.
Distr: EQU; D.R. Congo.

IUCN: RD
EQU: Wyld 822 7/1920.

Cordyla richardii Milne-Redh.

F.P.S. 2: 177 (1952); F.T.E.A. Leg.-Caes.: 223 (1967);
T.S.S.: 225 (1990); Biol. Skr. 51: 180 (1998).
Tree. Woodland on rocky hillslopes.
Distr: BAG, UN, EQU; Uganda.
IUCN: VU
EQU: Friis & Vollesen 1186 3/1982.

Craibia brownii Dunn

F.T.E.A. Leg.-Pap.: 152 (1971); Biol. Skr. 51: 203 (1998).
Tree. Forest.
Distr: EQU; D.R. Congo, Uganda, Kenya, Tanzania.
EQU: Friis & Vollesen 743 12/1980.

Craibia grandiflora (Micheli) Baker f.

F.P.S. 2: 178 (1952); F.T.E.A. Leg.-Pap.: 153 (1971); T.S.S.:
227 (1990).
Shrub or tree. Riverine forest.
Distr: EQU; C.A.R., D.R. Congo, Tanzania, Zambia.
EQU: Andrews 1390 5/1939.

Craibia laurentii (De Wild.) De Wild.

F.T.E.A. Leg.-Pap.: 149 (1971); F.E.E. 3: 109 (1989); T.S.S.:
227 (1990); Biol. Skr. 51: 203 (1998).
Syn: *Craibia utilis* M.B.Moss – F.P.S. 2: 178 (1952).
Tree. Forest & bushland.
Distr: EQU; Cameroon, D.R. Congo, Ethiopia, Uganda,
Kenya.
EQU: Myers 9579 10/1938.

Crotalaria aculeata De Wild. subsp. *claessensii* (De Wild.) Polhill

Biol. Skr. 51: 203 (1998).
Syn: *Crotalaria aculeata* De Wild. var. *claessensii* (De
Wild.) R.Wilczek – F.T.E.A. Leg.-Pap.: 970 (1971).
Syn: *Crotalaria claessensii* De Wild. – F.P.S. 2: 184
(1952).
Syn: *Crotalaria spinosa* sensu Andrews pro parte, non
Hochst. ex Benth. – F.P.S. 2: 184 (1952).
Perennial. Bushland, abandoned cultivations.
Distr: KOR, BAG, UN, EQU; D.R. Congo, Uganda, Kenya.
KOR: Kotschy 552 1837; **UN:** Sherif A.3979 8/1951.

Crotalaria aegyptiaca Benth.

F.P.S. 2: 182 (1952); F.Egypt 1: 254 (1999).
Subshrub. Wadis, open plains.
Distr: ?RS; Somalia; Egypt, Palestine, Arabia.
Note: Andrews records this species from Northern Sudan;
it is believed to occur in the NE Red Sea Hills but no
material has been seen by us.

Crotalaria alexandri Baker f.

F.T.E.A. Leg.-Pap.: 992 (1971); F.E.E. 3: 217 (1989); Biol.
Skr. 51: 203 (1998).
Annual herb. Woodland & bushland.
Distr: EQU; Ethiopia to Zimbabwe & Mozambique.
EQU: Friis & Vollesen 259 11/1980.

Crotalaria atrorubens Hochst. ex Benth.

F.P.S. 2: 185 (1952); F.J.M.: 111 (1976); Trop. Afr. Fl. Pl.
3: 226 (2008).

Syn: *Argyrolobium schimperianum* sensu Andrews, non Hochst. ex A.Rich. – F.P.S. 2: 176 (1952).
Annual herb. Sandy grassland & wooded grassland.
Distr: DAR, KOR; Senegal to Chad.
DAR: Wickens 2565 9/1964.

Crotalaria barkae Schweinf. subsp. *barkae*

F.T.E.A. Leg.-Pap.: 866 (1971); F.J.M.: 111 (1976); F.E.E. 3: 210 (1989).
Syn: *Crotalaria taubertii* Baker f. – F.P.S. 2: 190 (1952).
Annual or short-lived perennial herb. Bushland & grassland, disturbed ground.
Distr: DAR, KOR, UN, EQU; Senegal to Eritrea, S to Namibia & South Africa.
DAR: Wickens 2562 9/1964; **EQU:** Andrews 108 11/1933.

Crotalaria bongensis Baker f.

F.P.S. 2: 182 (1952); F.T.E.A. Leg.-Pap.: 945 (1971); F.E.E. 3: 213 (1989).
Annual herb. Grassland.
Distr: BAG, EQU; Guinea to Ethiopia, S to Angola & Zambia.
BAG: Schweinfurth 2135 7/1869.

Crotalaria brevidens Benth. var. *brevidens*

F.P.S. 2: 186 (1952); F.T.E.A. Leg.-Pap.: 909 (1971); F.E.E. 3: 207 (1989).
Syn: *Crotalaria intermedia* Kotschy var. *abyssinica* Taub. ex Engl. – F.P.S. 2: 186 (1952).
Annual or short-lived perennial herb. Wooded grassland.
Distr: KOR, BAG, UN, EQU; Ethiopia.
KOR: Andrews 51 11/1933; **EQU:** Myers 7562 8/1937.

Crotalaria brevidens Benth. var. *intermedia* (Kotschy) Polhill

F.P.S. 2: 186 (1952); F.T.E.A. Leg.-Pap.: 909 (1971); F.J.M.: 111 (1976); F.E.E. 3: 207 (1989); Biol. Skr. 51: 204 (1998).
Syn: *Crotalaria intermedia* Kotschy – F.P.S. 2: 186 (1952).
Annual or short-lived perennial herb. Grassland, bushland & woodland.
Distr: DAR, BAG, UN, EQU; Nigeria to Ethiopia, S to Tanzania.
DAR: Kassas 903; **BAG:** Schweinfurth 2320 8/1869.

Crotalaria calycina Schrank

F.P.S. 2: 182 (1952); F.T.E.A. Leg.-Pap.: 908 (1971); F.E.E. 3: 214 (1989).
Annual herb. Bushland, woodland & grassland.
Distr: BAG; Senegal to Ethiopia, S to Angola & Mozambique; tropical Asia, Australia.
BAG: Schweinfurth 2325 4/1869.

Crotalaria cephalotes Steud. ex A.Rich.

F.P.S. 2: 184 (1952); F.T.E.A. Leg.-Pap.: 983 (1971); F.E.E. 3: 217 (1989); Biol. Skr. 51: 204 (1998).
Annual herb. Grassland & woodland, disturbed areas.
Distr: EQU; Senegal to Ethiopia, S to Angola, Zimbabwe & Mozambique.
EQU: Myers 7804 9/1937.

Crotalaria chrysochlora Baker f. ex Harms

F.P.S. 2: 188 (1952); F.T.E.A. Leg.-Pap.: 902 (1971); Biol. Skr. 51: 204 (1998).
Perennial herb. Montane grassland.

Distr: EQU; Cameroon, D.R. Congo to Kenya, S to Angola & Zambia.
EQU: Friis & Vollesen 1263 3/1982.

Crotalaria comosa Baker

F.J.M.: 112 (1976); F.E.E. 3: 210 (1989); Biol. Skr. 51: 204 (1998).
Syn: *Crotalaria petitiana* sensu Andrews pro parte, non (A.Rich.) Walp. – F.P.S. 2: 190 (1952).
Annual or short-lived perennial herb. Woodland & wooded grassland.
Distr: DAR, BAG, UN, EQU; Senegal to Eritrea, S to Angola & Mozambique.
DAR: Wickens 2170 8/1964; **EQU:** Friis & Vollesen 667 12/1980.

Crotalaria congesta Polhill

Crotalaria Afr. & Madagascar: 315 (1982); Biol. Skr. 51: 205 (1998).
Annual herb. Forest, woodland & grassland.
Distr: EQU; D.R. Congo, Uganda, Tanzania, Malawi.
EQU: Friis & Vollesen 368 11/1980.
Note: a rather scarce species but not considered threatened.

Crotalaria cylindrica A.Rich. subsp. *cylindrica*

F.T.E.A. Leg.-Pap.: 920 (1971); F.E.E. 3: 209 (1989); Biol. Skr. 51: 205 (1998).
Perennial herb. Bushland & grassland, often in disturbed areas.
Distr: EQU; Eritrea, Ethiopia, Kenya.
EQU: Friis & Vollesen 257 11/1980.

Crotalaria deserticola Taub. ex Baker f. subsp. *deserticola* var. *deserticola*

F.T.E.A. Leg.-Pap.: 960 (1971); F.E.E. 3: 216 (1989); Biol. Skr. 51: 205 (1998).
Syn: *Crotalaria kikangaensis* De Wild. – F.P.S. 2: 190 (1952).
Annual or short-lived perennial herb. Grassland, bushland, disturbed areas.
Distr: EQU; Ethiopia to Zimbabwe & Mozambique.
EQU: Friis & Vollesen 777 12/1980.

Crotalaria emarginella Vatke

F.T.E.A. Leg.-Pap.: 956 (1971); F.E.E. 3: 214 (1989).
Syn: *Crotalaria laxa* Franch. – F.P.S. 2: 190 (1952).
Perennial herb or subshrub. Bushland on dry stony ground.
Distr: RS; Ethiopia, Somalia, Kenya; Yemen.
RS: Andrews 3486.

Crotalaria glauca Willd.

F.P.S. 2: 182 (1952); F.T.E.A. Leg.-Pap.: 885 (1971); F.E.E. 3: 202 (1989); Biol. Skr. 51: 205 (1998).
Annual or short-lived perennial herb. Grassland, bushland & woodland.
Distr: EQU; Senegal to Ethiopia, S to Angola, Zimbabwe & Mozambique.
EQU: Friis & Vollesen 250 11/1980.

Crotalaria goreensis Guill. & Perr.

F.P.S. 2: 191 (1952); F.T.E.A. Leg.-Pap.: 875 (1971); F.J.M.: 112 (1976); F.E.E. 3: 201 (1989); Biol. Skr. 51: 205 (1998).

Annual herb. Grassland & woodland.
Distr: DAR, KOR, CS, EQU; Senegal to Ethiopia, S to Angola, Zimbabwe & Mozambique.
DAR: Wickens 1427 4/1964; **EQU:** Friis & Vollesen 666 12/1980.

Crotalaria hyssopifolia Klotzsch
F.T.E.A. Leg.-Pap.: 991 (1971); F.E.E. 3: 218 (1989); Biol. Skr. 51: 206 (1998).
Annual or short-lived perennial herb. Woodland & grassland.
Distr: DAR, ES, EQU; Senegal to Ethiopia, S to Mozambique.
ES: Beshir 149 9/1951; **EQU:** Friis & Vollesen 665 12/1980.

Crotalaria impressa Nees ex Walp.
F.P.S. 2: 188, pro parte (1952); F.E.E. 3: 208 (1989).
Annual or short-lived perennial herb. Woodland & bushland.
Distr: RS; Eritrea, Ethiopia, Djibouti, Somalia; Egypt.
RS: Aylmer 248 3/1932.

Crotalaria incana L. subsp. *purpurascens* (Lam.) Milne-Redh.
F.T.E.A. Leg.-Pap.: 870 (1971); F.E.E. 3: 201 (1989); Biol. Skr. 51: 206 (1998).
Annual or short-lived perennial herb. Forest margins, montane grassland & bushland, disturbed areas.
Distr: EQU; Cameroon, Ethiopia to Mozambique; Madagascar, Arabia.
EQU: Friis & Vollesen 410 11/1980.

Crotalaria intonsa Polhill
F.E.E. 3: 209 (1989); Biol. Skr. 51: 206 (1998).
Perennial herb. Montane forest, often in clearings.
Distr: EQU; Ethiopia.
IUCN: RD
EQU: Friis & Vollesen 15 11/1980.

Crotalaria juncea L.
Bot. Exp. Sud.: 69 (1970); F.T.E.A. Leg.-Pap.: 950 (1971); F.E.E. 3: 218 (1989).
Annual herb. Cultivated, sometimes naturalised in disturbed grassland & bushland.
Distr: EQU; native to India, now pantropical as a crop plant.
EQU: Drar 1436 4/1938 (n.v.).
Note: cultivated for its fibre and as a green manure.

Crotalaria laburnifolia L. subsp. *laburnifolia*
F.P.S. 2: 185 (1952); F.T.E.A. Leg.-Pap.: 856 (1971); F.J.M.: 112 (1976); F.E.E. 3: 200 (1989); Biol. Skr. 51: 206 (1998).
Perennial herb. Woodland, bushland & grassland, disturbed areas.
Distr: DAR, KOR, BAG, EQU; Chad to Somalia, S to Zimbabwe & Mozambique; Madagascar, Mascarenes, tropical Asia to Australia.
DAR: Wickens 1039 1/1964; **EQU:** Myers 11090 4/1939.

Crotalaria lachnocarpoides Engl.
F.P.S. 2: 185 (1952); F.T.E.A. Leg.-Pap.: 879 (1971); F.E.E. 3: 202 (1989); Biol. Skr. 51: 207 (1998).
Perennial herb or subshrub. Montane grassland & bushland.

Distr: EQU; Ethiopia to Zimbabwe & Mozambique.
EQU: Friis & Vollesen 251 11/1980.

Crotalaria lachnophora A.Rich.
F.P.S. 2: 191 (1952); F.T.E.A. Leg.-Pap.: 877 (1971); F.E.E. 3: 202 (1989); Biol. Skr. 51: 207 (1998).
Perennial herb or shrub. Grassland & wooded grassland.
Distr: EQU; Senegal to Eritrea, S to Angola, Zimbabwe & Mozambique.
EQU: Jackson 3829 9/1957.

Crotalaria lachnosema Stapf
F.P.S. 2: 185 (1952); F.W.T.A. 1(2): 550 (1958); F.J.M.: 112 (1976).
Annual herb. Grassland, forest margins.
Distr: DAR, BAG; Guinea to C.A.R., S to Angola.
DAR: Wickens 1430 4/1964; **BAG:** Hoyle 489 1/1939.

Crotalaria leprieurii Guill. & Perr.
F.T.E.A. Leg.-Pap.: 944 (1971); Biol. Skr. 51: 207 (1998).
Syn: Crotalaria vogelii Benth. – F.P.S. 2: 182 (1952).
Annual herb. Bushland & woodland, often on rocky ground.
Distr: BAG, EQU; Senegal to Uganda, S to Angola & D.R. Congo.
BAG: Schweinfurth 2377 9/1869.

Crotalaria longithyrsa Baker f. var. *latifolia* R.Wilczek
Trop. Afr. Fl. Pl. 3: 264 (2008).
Annual or perennial herb. Grassland, disturbed areas.
Distr: EQU; D.R. Congo.
EQU: Myers 7148 7/1937.

Crotalaria microcarpa Hochst. ex Benth.
F.P.S. 2: 190 (1952); F.T.E.A. Leg.-Pap.: 943 (1971); F.E.E. 3: 213 (1989).
Syn: Crotalaria microcarpa Hochst. ex Benth. var. *dawei* Baker f. – F.P.S. 2: 191 (1952).
Syn: Crotalaria microcarpa Hochst. ex Benth. var. *sudanica* Baker f. – F.P.S. 2: 191 (1952).
Annual or short-lived perennial herb. Bushland, wooded grassland.
Distr: DAR, KOR, ES, BAG, UN; Mali to Eritrea, S to Angola, Zimbabwe & Mozambique.
KOR: Jackson 3256 11/1954; **UN:** Simpson 7247 7/1929.

Crotalaria microphylla Vahl
F.P.S. 2: 190 (1952); F.E.E. 3: 213 (1989).
Annual herb. Semi-desert, open sandy plains.
Distr: RS, ?CS; Cape Verde Is., Mauritania to Somalia; Egypt, Arabia.
RS: Carter 1946 11/1987.
Note: Andrews records this species from Northern and Central Sudan but we have only seen material from Red Sea District.

Crotalaria naragutensis Hutch.
F.W.T.A. 1(2): 551 (1958); F.J.M.: 112 (1976).
Syn: Crotalaria lynesii Baker f. & Martin – F.P.S. 2: 188 (1952); Bot. Exp. Sud.: 70 (1970).
Perennial herb. Wooded grassland.
Distr: DAR, ?EQU; Mali to Chad.

DAR: Blair 263 2/1964.

Note: Drar (1970) records this species from Equatoria based on two sterile specimens, but this seems unlikely and is probably a misidentification.

Crotalaria natalitia Meissn. var. *rutshuruensis* De Wild.

F.P.S. 2: 191 (species only) (1952); F.T.E.A. Leg.-Pap.: 872 (1971); F.E.E. 3: 201 (1989); Biol. Skr. 51: 207 (1998).
Perennial herb or subshrub. Woodland & bushland.
Distr: EQU; Ethiopia to Angola & Zimbabwe; Arabia.
EQU: Friis & Vollesen 265 11/1980.

Crotalaria ochroleuca G.Don

F.T.E.A. Leg.-Pap.: 908 (1971); F.J.M.: 112 (1976); F.E.E. 3: 207 (1989); Biol. Skr. 51: 208 (1998).
Syn: **Crotalaria cannabina** Schweinf. ex Baker f. – F.P.S. 2: 188 (1952).
Annual or short-lived perennial herb. Seasonally wet grassland & woodland.
Distr: DAR, UN, EQU; Senegal to Ethiopia, S to Namibia, Botswana & Mozambique; introduced in Madagascar.
DAR: Robertson 37 10/1957; **EQU:** Friis & Vollesen 537 11/1980.

Crotalaria onobrychis A.Rich.

F.P.S. 2: 188 (1952); F.E.E. 3: 208 (1989); Biol. Skr. 51: 208 (1998).
Syn: **Crotalaria impressa** sensu Andrews pro parte, non Nees ex Walp. – F.P.S. 2: 188 (1952).
Annual herb. Montane grassland & bushland.
Distr: EQU; Ethiopia, D.R. Congo, Malawi.
EQU: Andrews 1974 6/1939.

Crotalaria ononoides Benth.

F.P.S. 2: 185 (1952); F.T.E.A. Leg.-Pap.: 892 (1971); F.E.E. 3: 202 (1989); Biol. Skr. 51: 208 (1998).
Annual herb. Grassland.
Distr: BAG, EQU; Guinea to Ethiopia, S to Angola & Zambia; Madagascar.
EQU: Myers 7072 7/1937.

Crotalaria pallida Aiton var. *obovata* (G.Don) Polhill

F.T.E.A. Leg.-Pap.: 906 (1971); F.E.E. 3: 207 (1989).
Annual or short-lived perennial herb. Floodplain grassland, riverbanks.
Distr: BAG; pantropical, probably native to the eastern U.S.A.
BAG: Simpson 7532 11/1930.

Crotalaria pallida Aiton var. *pallida*

F.T.E.A. Leg.-Pap.: 905 (1971); F.J.M.: 112 (1976); F.E.E. 3: 205 (1989); Biol. Skr. 51: 208 (1998).
Syn: **Crotalaria mucronata** Desv. – F.P.S. 2: 187 (1952); Bot. Exp. Sud.: 70 (1970).
Annual or short-lived perennial herb. Floodplain grassland, riverbanks, disturbed areas.
Distr: DAR, ES, BAG, EQU; pantropical, often naturalised from cultivation.
DAR: Wickens 1940 7/1964; **BAG:** Schweinfurth 2605 3/1869.

Crotalaria petitiana (A.Rich.) Walp.

F.P.S. 2: 190, pro parte (1952); F.T.E.A. Leg.-Pap.: 944 (1971); F.E.E. 3: 210 (1989).
Annual or short-lived perennial herb. Grassland & bushland including seasonally wet areas.
Distr: BAG, EQU; D.R. Congo, Ethiopia, Uganda, Kenya, Tanzania.
EQU: Myers 7798 9/1937.

Crotalaria podocarpa DC.

F.P.S. 2: 191 (1952); F.T.E.A. Leg.-Pap.: 876 (1971); F.J.M.: 112 (1976); F.E.E. 3: 201 (1989).
Annual herb. Woodland & bushland.
Distr: DAR, KOR; Senegal to Eritrea, S to Namibia & South Africa.
DAR: Wickens 2336 9/1964.
Note: Andrews records this species from Central & Southern Sudan but we have seen no material from South Sudan, and it was not recorded from there by Polhill in his revision of African *Crotalaria*.

Crotalaria polysperma Kotschy

F.T.E.A. Leg.-Pap.: 864 (1971); F.E.E. 3: 200 (1989).
Syn: Crotalaria grantii Baker – F.P.S. 2: 186 (1952).
Annual herb. Woodland, bushland & grassland.
Distr: BAG, UN; Eritrea to Zimbabwe & Mozambique.
UN: J.M. Lock 81/44 2/1981.

Crotalaria pseudotenuirama Torre

F.T.E.A. Leg.-Pap.: 989 (1971); F.E.E. 3: 218 (1989).
Annual herb. Seasonally wet grassland, marshes.
Distr: BAG; Senegal to Ethiopia, S to Angola & Zambia.
BAG: Schweinfurth 2604 11/1869.

Crotalaria pycnostachya Benth. subsp. *pycnostachya*

F.P.S. 2: 189 (1952); F.T.E.A. Leg.-Pap.: 903 (1971); F.E.E. 3: 205 (1989); Biol. Skr. 51: 208 (1998).
Annual herb. Grassland & bushland, often in disturbed areas.
Distr: ?DAR, CS, EQU; Ethiopia, Somalia, Uganda, Kenya, Tanzania; Arabia.
CS: Andrews 391 3/1938; **EQU:** Peers 2 8/1953.
Note: the record for Darfur is from Quézel (1969: 115) and requires confirmation.

Crotalaria quartiniana A.Rich.

F.T.E.A. Leg.-Pap.: 863 (1971); F.E.E. 3: 200 (1989); Biol. Skr. 51: 209 (1998).
Annual herb. Montane bushland at forest margins.
Distr: EQU; Nigeria, Cameroon, Eritrea to Angola & D.R. Congo; Arabia.
EQU: Friis & Vollesen 764 12/1980.

Crotalaria recta Steud. ex A.Rich.

F.P.S. 2: 188 (1952); F.T.E.A. Leg.-Pap.: 957 (1971); F.E.E. 3: 216 (1989).
Perennial herb. Grassland, bushland.
Distr: DAR, ?EQU; Nigeria to Ethiopia, S to Angola & South Africa.
DAR: Wickens 1386 4/1964.
Note: Andrews records this species from Jebel Marra and from Equatoria – we have only seen material from the

former. Polhill, in his revision of African *Crotalaria*, places a dot for this species just over the border in D.R. Congo and this may be based on a Schweinfurth collection that Andrews mistakenly placed in South Sudan.

Crotalaria retusa L. var. *retusa*

F.P.S. 2: 182 (1952); F.T.E.A. Leg.-Pap.: 958 (1971); F.E.E. 3: 216 (1989); Biol. Skr. 51: 209 (1998).
Annual or short-lived perennial herb. Grassland, woodland, disturbed ground, often naturalised from cultivation.
Distr: EQU; pantropical.
EQU: Kielland-Lund 255 12/1983 (n.v.).
Note: cultivated for its fibres.

Crotalaria saltiana Andrews

F.P.S. 2: 184 (1952); F.T.E.A. Leg.-Pap.: 900 (1971); F.E.E. 3: 204 (1989).
Short-lived perennial herb. Bushland.
Distr: NS, CS, KOR; Ethiopia, Somalia, Kenya; Arabia.
CS: Schweinfurth 859 11/1868.

Crotalaria senegalensis (Pers.) Bacle ex DC.

F.P.S. 2: 186 (1952); F.T.E.A. Leg.-Pap.: 967 (1971); F.E.E. 3: 217 (1989); Biol. Skr. 51: 209 (1998).
Syn: *Crotalaria senegalensis* (Pers.) Bacle ex DC. var. **carinata** (Steud. ex A.Rich.) Baker f. – F.P.S. 2: 187 (1952).
Syn: *Crotalaria maxillaris* sensu Andrews, non Klotzsch. – F.P.S. 2: 187 (1952).
Annual or short-lived perennial herb. Woodland & wooded grassland, often in disturbed areas.
Distr: NS, RS, DAR, KOR, ES, BAG, UN, EQU; Cape Verde Is., Senegal to Eritrea, S to Angola, Botswana & Mozambique; Egypt, Yemen.
KOR: Jackson 4325 9/1961; **UN:** Simpson 7434 1/1930.

Crotalaria shirensis (Baker f.) Milne-Redh.

F.T.E.A. Leg.-Pap.: 946 (1971); Biol. Skr. 51: 209 (1998).
Annual herb. Wooded grassland.
Distr: EQU; Cameroon to Kenya, S to Angola, Zimbabwe & Mozambique.
EQU: Friis & Vollesen 287 11/1980.

Crotalaria spectabilis Roth

F.T.E.A. Leg.-Pap.: 959 (1971).
Annual herb. Cultivated, sometimes naturalised as a weed.
Distr: EQU; pantropical, native to Asia.
EQU: Kosper 146 11/1982.

Crotalaria sphaerocarpa Perr. ex DC. subsp. *sphaerocarpa*

F.P.S. 2: 184 (1952); F.T.E.A. Leg.-Pap.: 942 (1971).
Syn: *Crotalaria sphaerocarpa* Perr. ex DC. var. **angustifolia** Hochst. ex Kuntze – F.P.S. 2: 185 (1952).
Syn: *Crotalaria sphaerocarpa* Perr. ex DC. var. **grandiflora** Schweinf. ex Baker f. – F.P.S. 2: 185 (1952).
Annual herb. Bushland, wooded grassland, sometimes in damp sites.
Distr: DAR, KOR; Senegal to C.A.R., Angola to Tanzania, S to Namibia & South Africa.
KOR: Harrison 930 9/1950.
Note: Andrews records this species from Central & Southern Sudan but we have only seen material from

Kordofan, and no South Sudan records are noted by Polhill in his revision of African *Crotalaria*. The record for Darfur is from Quézel (1969: 116).

Crotalaria spinosa Hochst. ex Benth.

F.P.S. 2: 184 pro parte (1952); F.T.E.A. Leg.-Pap.: 970 (1971); F.J.M.: 113 (1976); F.E.E. 3: 217 (1989).
Annual or short-lived perennial herb. Bushland & grassland, often in seasonally damp sites, disturbed areas.
Distr: DAR; Senegal, Eritrea to Angola & Zimbabwe; Yemen, Socotra.
DAR: Wickens 3001 6/1965.

Crotalaria stenorhampha Harms

F.P.S. 2: 184 (1952); F.E.E. 3: 218 (1989).
Perennial herb. Woodland & wooded grassland.
Distr: EQU; Cameroon to Ethiopia & Uganda.
EQU: Myers 8472 2/1938.

Crotalaria steudneri Schweinf.

F.T.E.A. Leg.-Pap.: 936 (1971); F.J.M.: 113 (1976); F.E.E. 3: 212 (1989).
Annual herb. Bushland, grassland, disturbed areas.
Distr: DAR; Eritrea, Tanzania to Namibia & South Africa.
DAR: Wickens 2095 8/1964.

Crotalaria thebaica (Delile) DC.

F.P.S. 2: 182 (1952); F. Egypt 1: 253 (1999); F.J.U. 3: 222 (1999).
Subshrub. Sandy plains, wadis.
Distr: NS, RS, DAR, KOR, CS, ES; Chad; Libya, Egypt, Saudi Arabia.
KOR: Wickens 162 7/1962.

Crotalaria vallicola Baker f.

F.P.S. 2: 188 (1952); F.T.E.A. Leg.-Pap.: 919 (1971).
Annual or short-lived perennial herb. Grassland & bushland, often in rocky areas, disturbed sites.
Distr: ?EQU; D.R. Congo, Rwanda, Uganda, Kenya, Tanzania.
Note: Andrews recorded this species from Equatoria but we have not seen the material on which this was based.

Crotalaria vatkeana Engl.

F.T.E.A. Leg.-Pap.: 924 (1971); F.E.E. 3: 209 (1989); Biol. Skr. 51: 210 (1998).
Annual herb. Montane bushland, forest margins.
Distr: EQU; Ethiopia, Kenya, Tanzania.
EQU: Thomas 1845 12/1935.

Crotalaria verdcourtii Polhill

F.T.E.A. Leg.-Pap.: 869 (1971); Biol. Skr. 51: 210 (1998).
Annual herb. Bushland & grassland.
Distr: EQU; Ethiopia, Kenya.
EQU: Peers 8 9/1953.

Cullen plicatum (Delile) C.H.Stirt.

Bothalia 13: 317 (1981); Legumes of Africa: 454 (1989).
Syn: *Psoralea plicata* Delile – F.P.S. 2: 225 (1952); F.E.E. 3: 223 (1989); T.S.S.: 241 (1990).
Perennial herb. Sandy wadi beds & plains.
Distr: NS, KOR, CS, ES, ?SOUTHSUD; Mauritania to Somalia; N Africa, Arabia, Iran, Pakistan.

CS: <u>Aylmer 385</u> 5/1964.
Note: El Amin in T.S.S. records this species from Central & South Sudan; we have seen no material from the South.

Cyamopsis senegalensis Guill. & Perr.
F.P.S. 2: 191 (1952); F.T.E.A. Leg.-Pap.: 328 (1971); F.E.E. 3: 139 (1989).
Annual herb. Disturbed grassland.
Distr: RS, KOR, ES; Senegal to Eritrea, Tanzania, Namibia, Botswana; Arabia.
ES: <u>Beshir 48</u> 8/1951.

Dalbergia boehmii Taub. subsp. *boehmii*
F.T.E.A. Leg.-Pap.: 106 (1971); T.S.S.: 229 (1990).
Syn: *Dalbergia elata* Harms – F.P.S. 2: 192 (1952).
Shrub or tree. Bushland, woodland, riverbanks.
Distr: KOR; scattered from Senegal to Kenya, S to Angola, Zimbabwe & Mozambique.
KOR: <u>Broun 1295</u> 12/1907.

Dalbergia lactea Vatke
F.P.S. 2: 192 (1952); F.T.E.A. Leg.-Pap.: 111 (1971); F.E.E. 3: 107 (1989); T.S.S.: 229 (1990); Biol. Skr. 51: 210 (1998).
Scrambling shrub or small tree. Forest.
Distr: EQU; Ivory Coast to Gabon, Ethiopia to Zambia & Mozambique.
EQU: <u>Thomas 1693</u> 12/1935.

Dalbergia melanoxylon Guill. & Perr.
F.P.S. 2: 192 (1952); Bot. Exp. Sud.: 70 (1970); F.T.E.A. Leg.-Pap.: 100 (1971); F.J.M.: 113 (1976); F.E.E. 3: 105 (1989); T.S.S.: 229 (1990); Biol. Skr. 51: 210 (1998).
Tree. Woodland & wooded grassland.
Distr: DAR, KOR, CS, ES, BAG, UN, EQU; Senegal to Eritrea, S to Angola & South Africa; India.
IUCN: NT
DAR: <u>Wickens 2930</u> 4/1965; **UN:** <u>Harrison 276</u> 4/1948.

Dalbergia sissoo Roxb. ex DC.
Bot. Exp. Sud.: 70 (1970); F.E.E. 3: 107 (1989) ; T.S.S.: 231 (1990).
Tree. Cultivated ornamental, occasionally naturalised.
Distr: CS, EQU; native to SE Asia, cultivated elsewhere.
CS: <u>de Wilde et al. 5833</u> 3/1965; **EQU:** <u>Myers 6365</u> 3/1937.

Desmodium adscendens (Sw.) DC. var. *robustum* B.G.Schub.
F.P.S. 2: 194 (species only) (1952); F.T.E.A. Leg.-Pap.: 462 (1971); F.E.E. 3: 150 (1989); Biol. Skr. 51: 211 (1998).
Perennial herb. Forest.
Distr: EQU; Senegal to Ethiopia, S to Angola & Mozambique.
EQU: <u>Myers 7801</u> 9/1937.

Desmodium barbatum (L.) Benth. var. *dimorphum* (Welw. ex Baker) B.G.Schub.
F.T.E.A. Leg.-Pap.: 478 (1971).
Syn: *Desmodium dimorphum* Welw. ex Baker – F.P.S. 2: 195 (1952).
Perennial herb. Grassland & woodland.
Distr: CS, BAG; Nigeria, Cameroon, D.R. Congo to Tanzania, S to Angola & South Africa; Madagascar.

CS: <u>Kotschy 556</u> 1837; **BAG:** <u>Schweinfurth 4274</u> 10/1870.

Desmodium dichotomum (Willd.) DC.
F.P.S. 2: 195 (1952); F.T.E.A. Leg.-Pap.: 471 (1971); F.E.E. 3: 151 (1989).
Perennial herb or subshrub. Grassland & woodland.
Distr: KOR, CS, ES, EQU; Eritrea, Ethiopia, Uganda; India to China, Java, Celebes.
KOR: <u>Wickens 703</u> 10/1962; **EQU:** <u>Peers 7</u> 9/1953.

Desmodium gangeticum (L.) DC.
F.P.S. 2: 193 (1952); F.T.E.A. Leg.-Pap.: 467 (1971); F.E.E. 3: 151 (1989); Biol. Skr. 51: 211 (1998).
Perennial herb. Woodland, grassland & dry forest.
Distr: DAR, ES, EQU; palaeotropical.
DAR: <u>Wickens 1596</u> 4/1964; **EQU:** <u>Friis & Vollesen 1019</u> 2/1982.

Desmodium hirtum Guill. & Perr. var. *delicatulum* (A.Rich.) Harms ex Baker f.
F.E.E. 3: 150 (1989).
Syn: *Desmodium delicatulum* A.Rich. – F.P.S. 2: 194 (1952).
Perennial herb or subshrub. Grassland.
Distr: UN; Senegal to Ethiopia, S to Zimbabwe & Mozambique (but not in E Africa); Madagascar.
UN: <u>Sherif A.2856</u> 10/1951.

Desmodium ospriostreblum Chiov.
F.P.S. 2: 195 (1952); F.T.E.A. Leg.-Pap.: 475 (1971); F.E.E. 3: 151 (1989).
Annual herb. Woodland & bushland.
Distr: CS, ES; Cape Verde Is., Senegal to Eritrea, S to Zimbabwe; Arabia; possibly of American origin.
ES: <u>Schweinfurth 1725</u> 10/1865.

Desmodium ramosissimum G.Don
F.P.S. 2: 194 (1952); F.T.E.A. Leg.-Pap.: 464 (1971); F.E.E. 3: 150 (1989).
Perennial herb or shrub. Grassland & woodland.
Distr: EQU; Senegal to Ethiopia, S to Angola & Mozambique; Madagascar, Mascarenes.
EQU: <u>Myers 6532</u> 5/1937.

Desmodium repandum (Vahl) DC.
F.P.S. 2: 194 (1952); F.T.E.A. Leg.-Pap.: 465 (1971); F.J.M.: 113 (1976); F.E.E. 3: 150 (1989); Biol. Skr. 51: 211 (1998).
Perennial herb or subshrub. Forest & forest-grassland transition.
Distr: DAR, EQU; palaeotropical.
DAR: <u>Jackson 3384</u> 12/1954; **EQU:** <u>Myers 9754</u> 10/1938.

Desmodium salicifolium (Poir.) DC. var. *salicifolium*
F.P.S. 2: 195 (1952); F.T.E.A. Leg.-Pap.: 469 (1971); F.J.M.: 113 (1976); F.E.E. 3: 151 (1989); Biol. Skr. 51: 212 (1998).
Perennial herb or subshrub. Forest margins & associated woodland & grassland, swampy ground.
Distr: DAR, BAG, EQU; Senegal to Ethiopia, S to Angola & South Africa; Madagascar, Mascarenes.
DAR: <u>Wickens 1065</u> 1/1964; **EQU:** <u>Friis & Vollesen 809</u> 12/1980.

Desmodium schweinfurthii Schindl.

F.P.S. 2: 194 (1952); F.W.T.A. 1(2): 585 (1958).
Annual herb. Laterite pans.
Distr: BAG, EQU; Mali, Nigeria, Chad, C.A.R., D.R. Congo.
BAG: Harrison 1094 11/1950.

Desmodium setigerum (E.Mey.) Benth. ex Harv.

F.T.E.A. Leg.-Pap.: 460 (1971); Biol. Skr. 51: 212 (1998).
Perennial herb. Grassland.
Distr: EQU; Guinea to Kenya, S to Angola & South Africa.
EQU: Friis & Vollesen 637 12/1980.

Desmodium triflorum (L.) DC.

F.P.S. 2: 193 (1952); F.T.E.A. Leg.-Pap.: 459 (1971).
Annual or perennial herb. Grassland, weed of cultivation.
Distr: CS, BAG, EQU; pantropical.
CS: de Wilde et al. 5708 1/1965; **EQU:** Myers 6764 5/1937.

Desmodium velutinum (Willd.) DC.

F.T.E.A. Leg.-Pap.: 466 (1971); F.E.E. 3: 151 (1989); Biol. Skr. 51: 212 (1998).
Syn: *Desmodium lasiocarpum* (P.Beauv.) DC. – F.P.S. 2: 193 (1952).
Perennial herb or subshrub. Woodland, grassland, disturbed ground.
Distr: EQU; Senegal to Ethiopia, S to Angola & Mozambique; tropical Asia, introduced in Madagascar.
EQU: Friis & Vollesen 532 11/1980.

Dolichos compressus R.Wilczek

F.T.E.A. Leg.-Pap.: 690 (1971); Biol. Skr. 51: 213 (1998).
Perennial herb. Grassland, wooded grassland & bushland.
Distr: EQU; D.R. Congo, Uganda, Kenya.
EQU: Jackson 3207 5/1954.
Note: a rather local species, but perhaps not distinct from the more widespread *D. schweinfurthii*. *Kosper* 348 is det. at Kew as *D. gululu* De Wild. but it is surely conspecific with *Jackson* 3207.

Dolichos kilimandscharicus Taub. subsp. kilimandscharicus

F.T.E.A. Leg.-Pap.: 684 (1971); F.E.E. 3: 179 (1989); Biol. Skr. 51: 213 (1998).
Syn: *Dolichos malosanus* Baker – F.P.S. 2: 196 (1952).
Perennial herb. Grassland & woodland.
Distr: EQU; Ethiopia to Angola, Zimbabwe & Mozambique.
EQU: Friis & Vollesen 1178 3/1982.

Dolichos oliveri Schweinf.

F.P.S. 2: 196 (1952); F.T.E.A. Leg.-Pap.: 683 (1971); F.E.E. 3: 179 (1989).
Perennial herb or shrub. Grassland & woodland.
Distr: EQU; Eritrea to Zimbabwe.
EQU: Myers 14066 9/1941.

Dolichos schweinfurthii Harms

F.P.S. 2: 196 (1952); F.T.E.A. Leg.-Pap.: 689 (1971); F.E.E. 3: 179 (1989); Biol. Skr. 51: 213 (1998).
Perennial herb. Grassland.
Distr: EQU; Senegal to Ethiopia, S to Tanzania.
EQU: Friis & Vollesen 919 2/1982.

Dolichos sericeus E.Mey. subsp. sericeus

F.T.E.A. Leg.-Pap.: 680 (1971); F.E.E. 3: 179 (1989); Biol. Skr. 51: 213 (1998).
Perennial herb. Forest margins & associated grassland.
Distr: EQU; Cameroon, Eritrea to South Africa.
EQU: Friis & Vollesen 152 11/1980.

Dolichos sericeus E.Mey. subsp. formosus (A.Rich.) Verdc.

F.T.E.A. Leg.-Pap.: 680 (1971); F.E.E. 3: 179 (1989); Biol. Skr. 51: 214 (1998).
Syn: *Dolichos formosus* A.Rich. – F.P.S. 2: 196 (1952).
Perennial herb. Forest margins & associated woodland & grassland.
Distr: EQU; Ethiopia, Uganda, Kenya.
EQU: Myers 9649 10/1938.

Dolichos trilobus L. subsp. trilobus var. trilobus

F.T.E.A. Leg.-Pap.: 679 (1971); F.J.M.: 113 (1976); F.E.E. 3: 178 (1989).
Perennial climbing herb. Grassland, bushland & forest.
Distr: DAR; Ethiopia to Angola & South Africa; Arabia, tropical Asia.
DAR: Wickens 2337 9/1964.

Dolichos trilobus L. subsp. occidentalis Verdc.

F.T.E.A. Leg.-Pap.: 680 (1971); Biol. Skr. 51: 214 (1998).
Perennial climbing herb. Forest.
Distr: EQU; Ghana to Uganda.
EQU: Friis & Vollesen 507 11/1980.

Dorycnopsis abyssinicum (A.Rich.) Tikhom. & D.D.Sokoloff

Feddes Repert. 108: 342 (1997).
Syn: *Helminthocarpon abyssinicum* A.Rich. – F.P.S. 2: 203 (1952); Bot. Exp. Sud.: 72 (1970).
Syn: *Vermifrux abyssinica* (A.Rich.) J.B.Gillett – F.J.M.: 117 (1976).
Perennial herb. Montane grassland & bushland.
Distr: DAR; Eritrea, Ethiopia, Somalia; Yemen.
DAR: Wickens 2871 3/1965.

Eriosema chrysadenium Taub. var. chrysadenium

F.T.E.A. Leg.-Pap.: 801 (1971); Biol. Skr. 51: 214 (1998).
Perennial herb. Montane grassland.
Distr: EQU; Cameroon to Uganda, S to Angola & Zimbabwe.
EQU: Friis & Vollesen 1321 3/1982.

Eriosema cordifolium Hochst. ex A.Rich.

F.T.E.A. Leg.-Pap.: 805 (1971); F.E.E. 3: 193 (1989).
Perennial herb. Grassland & woodland.
Distr: EQU; D.R. Congo, Ethiopia, Uganda, Kenya.
EQU: Friis & Vollesen 433 11/1980.
Note: Friis & Vollesen named the cited specimen "*E.* sp. cf. *E. cordifolium* A.Rich. vel forma" (Biol. Skr. 51: 215). It looks close to Ethiopian material of *E. cordifolium* but also to *E. verdickii*.

Eriosema flemingioides Baker

F.P.S. 2: 198 (1952); F.T.E.A. Leg.-Pap.: 783 (1971); Biol. Skr. 51: 214 (1998).
Perennial herb or subshrub. Moist wooded grassland.

Distr: EQU; Nigeria to Kenya.
EQU: Friis & Vollesen 668 12/1980.

Eriosema glomeratum (Guill. & Perr.) Hook.f.

F.T.E.A. Leg.-Pap.: 770 (1971).
Perennial herb or subshrub. Grassland & wooded grassland.
Distr: UN; Senegal to Kenya, S to Angola & Tanzania.
UN: J.M. Lock 81/86 5/1981.

Eriosema griseum Baker

F.P.S. 2: 199 (1952); F.T.E.A. Leg.-Pap.: 770 (1971).
Subshrub. Grassland & wooded grassland.
Distr: BAG; Senegal to Uganda, S to Angola & D.R. Congo.
BAG: Schweinfurth 1388 4/1869.

Eriosema linifolium Baker f.

F.P.S. 2: 198 (1952); F.W.T.A. 1(2): 557 (1958).
Perennial herb. Wooded grassland.
Distr: EQU; Burkina Faso to C.A.R.
EQU: Schweinfurth 3888 6/1870.

Eriosema macrostipulum Baker f. var. macrostipulum

F.P.S. 2: 199 (1952); F.T.E.A. Leg.-Pap.: 789 (1971).
Perennial herb. Grassland & woodland.
Distr: BAG, EQU; Mali to Kenya, S to Zambia & Zimbabwe.
EQU: Myers 6659 5/1937.

Eriosema montanum Baker f. var. montanum

F.T.E.A. Leg.-Pap.: 779 (1971); F.E.E. 3: 190 (1989); Biol. Skr. 51: 215 (1998).
Perennial herb. Grassland & bushland.
Distr: EQU; Nigeria to Ethiopia, S to Angola & Zimbabwe.
EQU: Friis & Vollesen 13 11/1980.

Eriosema nutans Schinz

F.T.E.A. Leg.-Pap.: 995 (1971); F.J.M.: 113 (1976); F.E.E. 3: 190 (1989); Biol. Skr. 51: 215 (1998).
Syn: *Eriosema richardii* Benth. ex Baker f. – F.P.S. 2: 198 (1952).
Perennial herb. Grassland & wooded grassland, often in seasonal swamps.
Distr: DAR, EQU; Eritrea to South Africa.
DAR: Wickens 2610 9/1964; **EQU:** Andrews 1918 6/1939.

Eriosema pauciflorum Klotzsch var. pauciflorum

F.P.S. 2: 199 (1952); F.T.E.A. Leg.-Pap.: 792 (1971).
Perennial herb. Seasonally burnt grassland & wooded grassland.
Distr: EQU; Ghana to Kenya, S to Angola & South Africa.
EQU: Schweinfurth 2908 2/1870.

Eriosema psoraleoides (Lam.) G.Don

F.P.S. 2: 198 (1952); F.T.E.A. Leg.-Pap.: 772 (1971); F.J.M.: 113 (1976); F.E.E. 3: 190 (1989).
Perennial herb. Grassland, open bushland.
Distr: DAR, BAG, UN, EQU; Senegal to Ethiopia, S to Angola & South Africa; Madagascar.
DAR: Robertson 2 10/1957; **EQU:** Myers 6995 7/1937.

Eriosema pulcherrimum Taub.

F.P.S. 2: 197 (1952); F.W.T.A. 1(2): 559 (1958).
Perennial herb. Wooded grassland.
Distr: BAG, EQU; Ivory Coast to D.R. Congo.
EQU: Myers 6545 5/1937.

Eriosema rhodesicum R.E.Fr. var. rhodesicum

F.T.E.A. Leg.-Pap.: 795 (1971); F.E.E. 3: 193 (1989); Biol. Skr. 51: 215 (1998).
Perennial herb. Seasonally burnt grassland.
Distr: EQU; Nigeria to Ethiopia, S to Angola, Zimbabwe & Mozambique.
EQU: Friis & Vollesen 1073 3/1982.

Eriosema schweinfurthii Baker f.

F.P.S. 2: 198 (1952); F.T.E.A. Leg.-Pap.: 785 (1971).
Perennial herb. Wooded grassland, rocky outcrops.
Distr: BAG; C.A.R., D.R. Congo, Uganda.
IUCN: RD
BAG: Schweinfurth 1797 5/1869.

Eriosema sparsiflorum Baker f. var. sparsiflorum

F.P.S. 2: 199 (1952); F.T.E.A. Leg.-Pap.: 791 (1971); F.E.E. 3: 192 (1989).
Pyrophytic perennial herb. Seasonally burnt wooded grassland.
Distr: BAG, EQU; Nigeria, Ethiopia, Kenya.
BAG: Schweinfurth 1976 6/1869.

Eriosema verdickii De Wild.

F.T.E.A. Leg.-Pap.: 803 (1971); F.E.E. 3: 193 (1989); Biol. Skr. 51: 215 (1998).
Syn: *Eriosema schoutedenianum* Staner & De Craene – F.P.S. 2: 197 (1952).
Perennial herb. Wooded grassland.
Distr: EQU; Cameroon, Ethiopia to Zambia & Malawi.
EQU: Andrews 1902 6/1939.

Erythrina abyssinica Lam. ex DC. subsp. abyssinica

F.P.S. 2: 201 (1952); F.T.E.A. Leg.-Pap.: 555 (1971); F.E.E. 3: 161 (1989); T.S.S.: 231 (1990); Biol. Skr. 51: 215 (1998).
Syn: *Erythrina comosa* Hua – F.P.S. 2: 201 (1952); T.S.S.: 233 (1990).
Tree. Wooded grassland.
Distr: DAR, KOR, CS, BAG, UN, EQU; D.R. Congo to Ethiopia, S to Zimbabwe & Mozambique.
KOR: Wickens 54 7/1962; **EQU:** Myers 9033 5/1938.

Erythrina excelsa Baker

F.P.S. 2: 202 (1952); F.T.E.A. Leg.-Pap.: 559 (1971); T.S.S.: 233 (1990); Biol. Skr. 51: 216 (1998).
Syn: *Erythrina senegalensis* sensu auctt., non A.DC. – F.P.S. 2: 201 (1952); T.S.S.: 235 (1990).
Tree. Riverine & swamp forest.
Distr: EQU; Ivory Coast to Kenya, S to Zambia.
EQU: Myers 10607 3/1939.

Erythrina sigmoidea Hua

F.W.T.A. 1(2): 562 (1958); F.J.M.: 114 (1976); T.S.S.: 235 (1990).
Syn: *Erythrina eriotricha* Harms – F.P.S. 2: 201 (1952); Bot. Exp. Sud.: 72 (1970); T.S.S.: 233 (1990).

Syn: *Erythrina sudanica* Baker f. – F.P.S. 2: 201 (1952); Bot. Exp. Sud.: 72 (1970).
Shrub or small tree. Wooded grassland, open forest.
Distr: DAR; Senegal to C.A.R.
DAR: Wickens 1823 7/1964.

Flemingia grahamiana Wight & Arn.

F.T.E.A. Leg.-Pap.: 806 (1971); F.J.M.: 114 (1976); F.E.E. 3: 193 (1989).
Perennial herb or subshrub. Wooded grassland.
Distr: DAR; Cameroon to Eritrea, S to South Africa; Yemen, India.
DAR: Wickens 1435 4/1964.

Galactia striata (Jacq.) Urb. var. *villosa* (Wight & Arn.) Verdc.

F.Z. 3(5): 42 (2001).
Syn: *Galactia tenuiflora* (Klein ex Willd.) Wight & Arn. var. *villosa* (Wight & Arn.) Benth. – F.T.E.A. Leg.-Pap.: 579 (1971); F.E.E. 3: 165 (1989).
Syn: *Galactia tenuiflora* sensu Drar, non (Klein ex Willd.) Wight & Arn. – Bot. Exp. Sud.: 72 (1970).
Perennial climbing herb. Wooded grassland.
Distr: BAG, EQU; Ivory Coast to Kenya, S to Angola & South Africa; Mascarenes, India, China.
BAG: Schweinfurth 2101 7/1869.

Glycyrrhiza glabra L.

F.P.S. 2: 203 (1952); F. Libya: 84 (1989).
Perennial herb. Dry open scrub, often along drainage channels.
Distr: ?NS; Eurasia, N Africa, widely cultivated elsewhere.
Note: this is the cultivated 'liquorice'; Andrews recorded it from Northern Sudan, where its status is uncertain.

Hippocrepis constricta Kunze

F.Egypt 1: 303 (1999).
Annual herb. Sandy plains, rocky hillslopes.
Distr: RS; Cape Verde Is., Mauritania, Chad; Macaronesia, N Africa, Egypt to Pakistan.
RS: Farag & Zubair 11 (fl fr) 2/2012.

Humularia sudanica P.A.Duvign

Trop. Af. Fl. Pl. 4: 64 (2008).
Syn: *Geissaspis* sp. – F.P.S. 2: 202 (1952).
Shrub. Near river.
Distr: EQU; South Sudan endemic.
IUCN: RD
EQU: Myers 9369 8/1938.
Note: only known from the type specimen.

Indigastrum costatum (Guill. & Perr.) Schrire subsp. *costatum*

Bothalia 22: 168 (1992).
Syn: *Indigofera costata* Guill. & Perr. subsp. *costata* – F.P.S. 2: 218 (1952); F.T.E.A. Leg.-Pap.: 320, in notes (1971); F.J.M.: 114 (1976); F.E.E. 3: 137 (1989).
Annual herb. Grassland.
Distr: DAR; Senegal to Eritrea.
DAR: Wickens 2131 8/1964.

Indigastrum parviflorum (B.Heyne ex Wight & Arn.) Schrire

Bothalia 22: 168 (1992).

Syn: *Indigofera parviflora* B.Heyne ex Wight & Arn. – F.P.S. 2: 218 (1952); Bot. Exp. Sud.: 73 (1970); F.T.E.A. Leg.-Pap.: 321 (1971); F.J.M.: 114 (1976); F.E.E. 3: 137 (1989).
Annual herb. Grassland & wooded grassland.
Distr: NS, RS, DAR, CS, ES; Ethiopia to Namibia & South Africa; India, Australia.
ES: Beshir 94 9/1951.

[*Indigofera*]

Note: Andrews records two additional species of *Indigofera* which are surely erroneous since they are southern African taxa: *I. oxalidea* Welw. ex Baker from Central & Southern Sudan and *I. alternans* DC. from Jebel Marra in Darfur. However, we are unsure as to which species these records truly refer. Wickens made no reference to the latter record in his Flora of Jebel Marra.

Indigofera achyranthoides Taub.

F.P.S. 2: 209 (1952); Bot. Exp. Sud.: 72 (1970); Trop. Afr. Fl. Pl. 4: 68 (2008).
Perennial herb. Wooded grassland.
Distr: BAG, EQU; D.R. Congo.
IUCN: RD
BAG: Schweinfurth 1624 4/1869.

Indigofera ambelacensis Schweinf.

F.T.E.A. Leg.-Pap.: 247 (1971); F.E.E. 3: 124 (1989); Biol. Skr. 51: 216 (1998).
Annual herb. Rocky grassland, woodland, disturbed areas.
Distr: EQU; Eritrea to D.R. Congo & Tanzania; Arabia.
EQU: Peers 18 8/1953.

Indigofera andrewsiana J.B.Gillett

F.T.E.A. Leg.-Pap.: 237 (1971); Biol. Skr. 51: 216 (1998).
Annual herb. Grassland on black clay soils.
Distr: UN, EQU; Nigeria, Uganda.
IUCN: RD
UN: Simpson 7441 1/1930.

Indigofera argentea Burm.f.

F.P.S. 2: 214 (1952); F.E.E. 3: 128 (1989).
Syn: *Indigofera semitrijuga* sensu Andrews, non Forssk. – F.P.S. 2: 215 (1952).
Annual or perennial herb. Sandy plains.
Distr: NS, RS, KOR; Mauritania to Somalia; N Africa, Arabia to India.
NS: Beshir 985 9/1950.

Indigofera arrecta Hochst. ex A.Rich.

F.P.S. 2: 215 (1952); F.T.E.A. Leg.-Pap.: 307 (1971); F.J.M.: 114 (1976); F.E.E. 3: 135 (1989); Biol. Skr. 51: 217 (1998).
Perennial herb or shrub. Grassland, bushland & waste ground, also cultivated.
Distr: DAR, EQU; Senegal to Somalia, S to South Africa; Madagascar, Arabia, SE Asia; its wild distribution obscured by cultivation.
DAR: Wickens 2452 9/1964; **EQU:** Friis & Vollesen 283 11/1980.
Note: widely cultivated for indigo dye.

Indigofera articulata Gouan

F.P.S. 2: 216, pro parte (1952); F.E.E. 3: 134 (1989).
Shrub. Dry grassland, bushland.

Distr: RS, KOR; Eritrea, Ethiopia, Djibouti, Somalia; Algeria, Egypt, Arabia to Pakistan.
KOR: <u>Kotschy 28</u> 1842.

Indigofera aspera Perr. ex DC.

F.P.S. 2: 219 (1952); F.W.T.A. 1(2): 538 (1958).
Annual herb. Sandy ground.
Distr: DAR, KOR, BAG; Mauritania & Senegal to Chad.
KOR: <u>Wickens 127</u> 7/1962; **BAG:** <u>Andrews 3124</u> 10/1946.

Indigofera astragalina DC.

F.P.S. 2: 218 (1952); F.E.E. 3: 135 (1989).
Annual herb. Sandy ground, disturbed areas.
Distr: DAR, KOR; Cape Verde Is., Senegal to Ethiopia, S to Namibia & South Africa; Pakistan, India, Sri Lanka.
KOR: <u>Jackson 3252</u> 11/1954.

Indigofera atriceps Hook.f. subsp. *atriceps*

F.T.E.A. Leg.-Pap.: 282 (1971); F.E.E. 3: 130 (1989); Biol. Skr. 51: 217 (1998).
Syn: *Indigofera alboglandulosa* Engl. – F.P.S. 2: 217 (1952).
Perennial herb. Montane grassland.
Distr: EQU; Guinea to Liberia, Nigeria, Cameroon, Uganda & Kenya, S to Zimbabwe & Mozambique.
EQU: <u>Friis & Vollesen 1218</u> 3/1982.

Indigofera atriceps Hook.f. subsp. *kaessneri* (Baker f.) J.B.Gillett

F.T.E.A. Leg.-Pap.: 283 (1971); F.E.E. 3: 130 (1989); Biol. Skr. 51: 217 (1998).
Perennial herb. Montane grassland & bushland.
Distr: EQU; Ethiopia to Zambia & Tanzania; Arabia.
EQU: <u>MacDonald 61</u> 1/1939.

Indigofera barteri Hutch. & Dalziel

F.P.S. 2: 217 (1952).
Perennial herb. Rocky crevices.
Distr: EQU; Ghana and Nigeria.
EQU: <u>Wyld 246</u> 7/1937.

Indigofera biglandulosa J.B.Gillett

F.T.E.A. Leg.-Pap.: 257 (1971).
Syn: *Indigofera pilosa* sensu Andrews pro parte, non Poir. – F.P.S. 2: 211 (1952).
Annual herb. Sandy ground.
Distr: KOR; Tanzania, Zambia.
IUCN: RD
KOR: <u>Broun 1353</u> 12/1907.

Indigofera binderi Kotschy

F.P.S. 2: 214 (1952); F.T.E.A. Leg.-Pap.: 297 (1971); F.E.E. 3: 131 (1989).
Shrub. Woodland, bushland.
Distr: UN, EQU; C.A.R., Ethiopia, Uganda, Kenya.
EQU: <u>Jackson 2257</u> 6/1952.

Indigofera bongensis Kotschy & Peyr.

F.P.S. 2: 209 (1952); Trop. Afr. Fl. Pl. 4: 72 (2008).
Perennial herb. Grassland.
Distr: BAG, EQU; C.A.R., D.R. Congo.
EQU: <u>Myers 7001</u> 7/1937.
Note: a local species but probably not threatened.

Indigofera bracteolata DC.

F.P.S. 2: 210, pro parte (1952); Trop. Afr. Fl. Pl. 4: 72 (2008).
Annual herb. Sandy ground including seasonally flooded areas.
Distr: KOR, CS; Senegal to Chad.
KOR: <u>Andrews 104</u> 11/1933.

Indigofera brevifilamenta J.B.Gillett

F.T.E.A. Leg.-Pap.: 268 (1971).
Annual herb. Grassland.
Distr: EQU; Guinea Bissau, Nigeria, Cameroon, C.A.R., Tanzania, Malawi, Zambia.
EQU: <u>Myers 7147</u> 7/1937.
Note: the identification of the single Sudanese specimen is tentative. A widespread but very scattered and apparently scarce species.

Indigofera capitata Kotschy

F.P.S. 2: 208 (1952); F.T.E.A. Leg.-Pap.: 232 (1971).
Annual herb. Grassland, open ground.
Distr: ?CS, BAG, EQU; Gambia to Kenya, S to Angola & Zambia.
BAG: <u>Schweinfurth 2431</u> 10/1869.
Note: Andrews records this species from Central & Southern Sudan but we have only seen material from the South.

Indigofera coerulea Roxb. var. *coerulea*

F.E.E. 3: 134 (1989).
Perennial herb. Arid coastal plains.
Distr: RS; Eritrea, Somalia; Arabia to India & Sri Lanka.
RS: <u>Newberry 112</u> 1/1928.

Indigofera coerulea Roxb. var. *occidentalis* J.B.Gillett & Ali

F.T.E.A. Leg.-Pap.: 305 (1971); F.E.E. 3: 134 (1989).
Syn: *Indigofera articulata* sensu Andrews pro parte, non Gouan – F.P.S. 2: 216 (1952).
Perennial herb. Dry coastal plains, semi-desert, dry bushland.
Distr: RS, DAR, KOR, CS, ES; Niger to Somalia, S to Uganda & Kenya; Mauritius, Arabia to Pakistan.
RS: <u>Carter 1927</u> 11/1987.

Indigofera colutea (Burm.f.) Merr. var. *colutea*

F.T.E.A. Leg.-Pap.: 266 (1971); F.E.E. 3: 127 (1989); Biol. Skr. 51: 217 (1998).
Syn: *Indigofera viscosa* Lam. – F.P.S. 2: 211 (1952).
Annual herb. Grassland, bushland, weed of cultivation.
Distr: RS, DAR, KOR, CS, UN, EQU; Cape Verde Is., Senegal to Somalia, S to South Africa; Egypt, Yemen, India to Australia.
KOR: <u>Jackson 4360</u> 11/1961; **EQU:** <u>Sillitoe 186</u> 1919.

Indigofera congesta Welw. ex Baker

F.T.E.A. Leg.-Pap.: 234 (1971).
?Perennial herb. Grassland & wooded grassland, including seasonally flooded areas.
Distr: EQU; Guinea Bissau to Kenya, S to Angola, Zambia & Mozambique.
EQU: <u>Wyld 701</u> 1/1940.

Indigofera congolensis De Wild. & T.Durand var. *bongensis* (Baker f.) J.B.Gillett

F.T.E.A. Leg.-Pap.: 250 (species only) (1971); F.E.E. 3: 126 (species only) (1989).
Syn: *Indigofera sparsa* Baker var. *bongensis* Baker f. – F.P.S. 2: 210 (1952).
Annual herb. Sparse grassland.
Distr: BAG; ?South Sudan endemic.
IUCN: RD
BAG: Schweinfurth 2526 10/1869.
Note: Gillett in Kew Bull. Add. Ser. 1: 48 (1958) notes that this variety is somewhat intermediate between typical *I. congolensis* and *I. sparsa*. It may also occur in C.A.R.

Indigofera conjugata Baker var. *conjugata*

F.P.S. 2: 209 (1952); F.T.E.A. Leg.-Pap.: 319 (1971); F.E.E. 3: 137 (1989); Biol. Skr. 51: 218 (1998).
Perennial herb. Wooded grassland.
Distr: UN, EQU; Guinea to Ethiopia, S to Angola & Tanzania.
EQU: Friis & Vollesen 1020 2/1982.

Indigofera conjugata Baker var. *schweinfurthii* (Taub.) J.B.Gillett

Kew Bull. Add. Series 1: 122 (1958).
Perennial herb. Grassland.
Distr: BAG, EQU; South Sudan endemic.
IUCN: RD
EQU: Jackson 4277 6/1961.

Indigofera cordifolia B.Heyne ex Roth

F.P.S. 2: 208 (1952); F.E.E. 3: 123 (1989).
Annual or perennial herb. Dry grassland, wadis.
Distr: ?NS, KOR, CS, ES; Cape Verde Is., Mauritania to Eritrea; Egypt to India, Java, Timor, N Australia.
KOR: Wickens 629 10/1962.

Indigofera deightonii J.B.Gillett subsp. *deightonii*

F.W.T.A. 1(2): 542 (1958).
Annual herb. Grassland, including on rock outcrops and on floodplains, disturbed areas.
Distr: EQU; Sierra Leone to Nigeria.
EQU: Andrews 1487 5/1939.

Indigofera dendroides Jacq.

F.P.S. 2: 212 (1952); F.T.E.A. Leg.-Pap.: 242 (1971); F.E.E. 3: 124 (1989); Biol. Skr. 51: 218 (1998).
Annual herb. Seasonally wet grassland, wooded grassland.
Distr: EQU; Senegal to Ethiopia, S to Angola & Mozambique.
EQU: Myers 7726 9/1937.

Indigofera diphylla Vent.

F.P.S. 2: 209 (1952); F.W.T.A. 1(2): 542 (1958).
Perennial herb. Sandy ground.
Distr: DAR, KOR; Senegal to Chad.
KOR: Wickens 367 9/1962.

Indigofera emarginella Steud. ex A.Rich.

F.P.S. 2: 217 pro parte (1952); F.T.E.A. Leg.-Pap.: 300 (1971); F.E.E. 3: 131 (1989).

Shrub. Grassland, woodland.
Distr: BAG, EQU; Nigeria to Eritrea, S to Angola & Mozambique.
EQU: Jackson 3835 9/1957.

Indigofera fulvopilosa Benan

F.T.E.A. Leg.-Pap.: 257 (1971).
Syn: *Indigofera pilosa* Poir. var. *multiflora* Baker f. – F.P.S. 2: 211 (1952).
Annual or short-lived perennial herb. Grassland, forest margins, disturbed areas.
Distr: EQU; Sierra Leone to Uganda, S to Angola, Zimbabwe & Mozambique.
EQU: Wyld 473.

Indigofera garckeana Vatke

F.P.S. 2: 217 (1952); F.T.E.A. Leg.-Pap.: 296 (1971); F.E.E. 3: 130 (1989); Biol. Skr. 51: 219 (1998).
Shrub. Woodland, grassland.
Distr: EQU; Senegal, Chad, Eritrea to Zimbabwe & Mozambique.
EQU: Andrews 933 4/1939.

Indigofera hendecaphylla Jacq.

F.P.S. 2: 212, pro parte (1952); Kew Bull. 48: 731 (1993); Trop. Afr. Fl. Pl. 4: 84 (2008).
Syn: *Indigofera spicata* sensu auctt. pro parte, non Forssk. – F.E.E. 3: 138 (1989); Biol. Skr. 51: 220 (1998).
Perennial herb. Grassland, disturbed areas.
Distr: BAG, UN, EQU; palaeotropical, introduced in the neotropics.
EQU: Myers 6763 5/1937.

Indigofera hirsuta L. var. *hirsuta*

F.P.S. 2: 215 (1952); F.T.E.A. Leg.-Pap.: 310 (1971); F.E.E. 3: 135 (1989); Biol. Skr. 51: 219 (1998).
Annual or short-lived perennial herb. Cultivated & waste ground.
Distr: KOR, BAG, UN, EQU; palaeotropical, introduced in the neotropics.
KOR: Broun 1355 11/1907; **EQU:** Myers 8720 3/1938.

Indigofera hochstetteri Baker subsp. *hochstetteri*

F.P.S. 2: 219 (1952); F.T.E.A. Leg.-Pap.: 223 (1971); F.J.M.: 114 (1976); F.E.E. 3: 122 (1989).
Syn: *Indigofera arenaria* sensu Andrews pro parte, non A.Rich. – F.P.S. 2: 208 (1952).
Annual herb. Grassland.
Distr: NS, RS, DAR, KOR, CS, ES, UN; Mauritania to Somalia, S to Zambia; N Africa, Arabia to India.
DAR: Wickens 2105 8/1964; **UN:** Sherif A.2860 10/1951.

Indigofera knoblecheri Kotschy

F.P.S. 2: 209 (1952); Trop. Afr. Fl. Pl. 4: 86 (2008).
Perennial herb. Wooded grassland, riverbanks.
Distr: EQU; South Sudan endemic.
IUCN: RD
EQU: Myers 7097 7/1937.

Indigofera letestui Tisser.

Trop. Afr. Fl. Pl. 4: 88 (2008).
Syn: *Indigofera mittuensis* Baker f. – F.P.S. 2: 210 (1952).

Perennial herb. Grassland, burnt ground.
Distr: EQU; C.A.R.
IUCN: RD
EQU: Schweinfurth 2777 12/1869.

Indigofera linifolia (L.f.) Retz.

F.P.S. 2: 207 (1952); F.E.E. 3: 123 (1989).
Annual or perennial herb. Dry grassland, bushland.
Distr: RS, ES; Eritrea, Ethiopia; S Asia to China, Australia.
RS: Drummond & Hemsley 940 1/1953.

Indigofera linnaei Ali

Bot. Not. 111: 549 (1958); Ann. Bot. Fennici 10: 150 (1973).
Perennial herb. Sandy grassland.
Distr: NS, KOR, CS; Pakistan to Australia.
KOR: Wickens 40 6/1962.
Note: this predominantly Asian species is only known in Africa from Sudan.

Indigofera lotononoides Baker f.

F.P.S. 2: 208 (1952); F.Egypt 1: 309 (1999); Trop. Afr. Fl. Pl. 4: 88 (2008).
Shrub. Burnt bushland, sandy plains.
Distr: DAR, BAG; C.A.R.; Libya, Egypt.
IUCN: RD
DAR: de Wilde et al. 5521 1/1965; **BAG:** Schweinfurth 63 2/1871.

Indigofera microcarpa Desv.

F.P.S. 2: 217 (1952); F.T.E.A. Leg.-Pap.: 312 (1971); F.E.E. 3: 135 (1989).
Annual or perennial herb. Riverbanks, weed of cultivation.
Distr: RS, DAR; Senegal to Nigeria, Ethiopia to Angola & Zimbabwe; Madagascar, tropical America.
DAR: Wickens 3133 4/1971.

Indigofera mildbraediana J.B.Gillett

F.T.E.A. Leg.-Pap.: 237 (1971).
Perennial herb. Wooded grassland.
Distr: EQU; Nigeria, Gabon, Congo-Brazzaville, C.A.R., Angola, Tanzania.
EQU: Schweinfurth 2728 12/1869.
Note: a widespread but very scattered species.

Indigofera nigritana Hook.f.

F.W.T.A. 1(2): 539 (1958).
Perennial herb. Flooded grassland, riverbanks.
Distr: EQU; Mali to D.R. Congo.
EQU: Broun s.n. 3/1903.

Indigofera nummulariifolia (L.) Livera ex Alston

F.T.E.A. Leg.-Pap.: 217 (1971).
Syn: *Indigofera echinata* Willd. – F.P.S. 2: 219 (1952).
Annual herb. Weed of cultivated & waste ground.
Distr: KOR, UN; Senegal to Tanzania, S to Namibia & Zimbabwe; Madagascar, India, Sri Lanka, Indo-China.
KOR: Musselman 6279 8/1983; **UN:** J.M. Lock 81/152 8/1981.

Indigofera oblongifolia Forssk.

F.P.S. 2: 214 (1952); F.E.E. 3: 136 (1989); T.S.S.: 235 (1990).
Subshrub. Grassland, often on seasonally wet ground.

Distr: NS, RS, DAR, KOR, CS, ES; Senegal to Somalia, Angola; Egypt, Arabia to Java.
ES: Jackson 2678 3/1953.

Indigofera oubanguiensis Tisser.

Trop. Afr. Fl. Pl. 4: 98 (2008).
Subshrub. Laterite.
Distr: BAG; Cameroon, Gabon, C.A.R.
IUCN: RD
BAG: Hoyle 375 1/1939.

Indigofera pilosa Poir. var. pilosa

F.P.S. 2: 211, pro parte (1952); F.E.E. 3: 217 (1989).
Annual or short-lived perennial herb. Wooded grassland, waste ground.
Distr: KOR; Senegal to Eritrea.
KOR: Pfund 44 9/1875.

Indigofera polysphaera Baker

F.P.S. 2: 208 (1952); F.T.E.A. Leg.-Pap.: 231 (1971); F.E.E. 3: 219 (1989); Biol. Skr. 51: 219 (1998).
Annual herb. Wooded grassland, overgrazed areas.
Distr: EQU; Mali & Ivory Coast to Uganda, S to Angola, Zambia & Tanzania.
EQU: Myers 7741 9/1937.

Indigofera prieureana Guill. & Perr.

F.E.E. 3: 123 (1989).
Syn: *Indigofera stenophylla* Guill. & Perr. var. *latifolia* A.Rich. – F.P.S. 2: 218 (1952).
Annual herb. Wooded grassland.
Distr: KOR, ES; Senegal to Eritrea & Ethiopia.
KOR: Wickens 494 9/1962.

Indigofera pseudosubulata Baker f.

F.P.S. 2: 214 (1952); Trop. Afr. Fl. Pl. 4: 100 (2008).
Subshrub. Riverine habitats.
Distr: EQU; Senegal, C.A.R., D.R. Congo.
IUCN: RD
EQU: Schweinfurth 2703 12/1869.

Indigofera pulchra Willd.

F.P.S. 2: 210 (1952); F.T.E.A. Leg.-Pap.: 235 (1971); F.J.M.: 115 (1976).
Syn: *Indigofera bracteolata* sensu Andrews pro parte, non DC. – F.P.S. 2: 210 (1952).
Annual herb. Grassland, disturbed ground.
Distr: CS, BAG; Senegal to Uganda, S to Angola & Zambia.
CS: Kotschy 576; **BAG:** Schweinfurth 2455 10/1868.

Indigofera schimperi Jaub. & Spach

F.P.S. 2: 214 (1952); F.T.E.A. Leg.-Pap.: 313 (1971); F.E.E. 3: 136 (1989).
Perennial herb. Swampy grassland, bushland & woodland.
Distr: RS; Eritrea & Somalia, S to Namibia & South Africa.
RS: Bent s.n. 1896.

Indigofera secundiflora Poir. var. secundiflora

F.P.S. 2: 217 (1952); F.T.E.A. Leg.-Pap.: 269 (1971); Biol. Skr. 51: 219 (1998).
Annual herb. Grassland, woodland & disturbed ground.
Distr: KOR, BAG, EQU; Senegal to C.A.R.

KOR: Jackson 4374 9/1961; **BAG:** Schweinfurth 76 7/1869.

Note: Andrews recorded the otherwise West African *I. barteri* Hutch. & Dalziel from Equatoria based on *Wyld* 246, but Gillett in Kew Bull. Add. Ser. 1: 67 (1958) notes that, in the absence of fruits, this specimen could be a form of *I. secundiflora*.

Indigofera secundiflora Poir. var. rubripilosa De Wild.

F.T.E.A. Leg.-Pap.: 269 (1971); F.E.E. 3: 128 (1989); Biol. Skr. 51: 219 (1998).
Perennial herb. Grassland, woodland & disturbed ground.
Distr: KOR, ES, EQU; C.A.R. to Ethiopia, S to Zambia.
ES: Schweinfurth 1837 6/1865; **EQU:** Friis & Vollesen 625 12/1980.

Indigofera sessiliflora DC.

F.E.E. 3: 123 (1989); F.J.U. 3: 226 (1999).
Annual herb. Sandy plains.
Distr: NS, DAR, KOR, CS; Senegal to Eritrea; Libya, Egypt, Arabia, Pakistan, India.
KOR: Harrison 940 9/1950.

Indigofera simplicifolia Lam.

F.P.S. 2: 208 (1952); F.T.E.A. Leg.-Pap.: 236 (1971).
Annual herb. Short grassland, bushland.
Distr: BAG, EQU; Senegal to Uganda, S to Angola & Tanzania.
BAG: Schweinfurth 2471 10/1869.

Indigofera spicata Forssk. var. spicata

F.T.E.A. Leg.-Pap.: 317, pro parte (1971); F.J.M.: 115 (1976); F.E.E. 3: 136, pro parte (1989); Biol. Skr. 51: 220, pro parte (1998).
Syn: *Indigofera parvula* Delile – F.P.S. 2: 212 (1952).
Syn: *Indigofera hendecaphylla* sensu Andrews pro parte, non Jacq. – F.P.S. 2: 212 (1952).
Perennial herb. Wooded grassland, disturbed ground.
Distr: DAR, ES, UN, EQU; Senegal to Ethiopia, S to South Africa; Madagascar, Yemen, India, Australia.
DAR: Wickens 984 1/1964; **EQU:** Friis & Vollesen 999 2/1982.

Indigofera spiniflora Hochst. & Steud. ex Boiss.

F.E.E. 3: 132 (1989).
Subshrub. Dry grassland, bushland.
Distr: RS, KOR; Eritrea, Somalia; Egypt, Yemen, Socotra.
RS: Bally 6962 4/1949.

Indigofera spinosa Forssk.

F.P.S. 2: 210 (1952); F.T.E.A. Leg.-Pap.: 253 (1971); F.E.E. 3: 126 (1989).
Shrublet. Semi-desert, dry bushland.
Distr: RS, DAR, ES, BAG, EQU; Eritrea & Somalia, S to Tanzania; Egypt, Arabia.
RS: Carter 1816 11/1987; **EQU:** Carr 815 6/1970.

Indigofera stenophylla Guill. & Perr. var. stenophylla

F.P.S. 2: 218 (1952); F.T.E.A. Leg.-Pap.: 236 (1971); F.J.M.: 115 (1976); Biol. Skr. 51: 220 (1998).
Annual herb. Grassland.

Distr: DAR, KOR, BAG, UN, EQU; Senegal to Tanzania & D.R. Congo.
KOR: Jackson 4014 9/1959; **EQU:** Myers 7050 7/1937.

Indigofera strobilifera (Hochst.) Hochst. ex Baker subsp. strobilifera

F.P.S. 2: 210 (1952); F.T.E.A. Leg.-Pap.: 230 (1971).
Annual herb. Woodland & bushland on sandy soils.
Distr: KOR, UN; Mali to Chad, Somalia to Zimbabwe.
KOR: Jackson 3240 11/1954; **UN:** J.M. Lock 82/42 8/1982.

Indigofera suaveolens Jaub. & Spach

F.P.S. 2: 211 (1952); F.E.E. 3: 127 (1989).
Shrublet. Woodland, bushland & dry grassland.
Distr: DAR, KOR; Mali to Somalia.
KOR: Pfund 165 8/1875.
Note: Andrews records this species from Central & Southern Sudan but we have not seen any material from the South.

Indigofera suffruticosa Mill.

F.P.S. 2: 216 (1952); Trop. Afr. Fl. Pl. 4: 108 (2008).
Perennial herb or subshrub. Weed of disturbed areas.
Distr: ?CS; native to tropical America, cultivated elsewhere and sometimes naturalised
Note: cultivated for indigo dye. Andrews recorded it from Jebel Moya in Blue Nile Prov.; we have not seen the specimen on which this record is based.

Indigofera swaziensis Bolus var. swaziensis

F.T.E.A. Leg.-Pap.: 301 (1971).
Syn: *Indigofera emarginella* sensu Andrews pro parte, non Steud. ex A.Rich. – F.P.S. 2: 217 (1952).
Shrub. Upland forest margins & bushland.
Distr: EQU; Uganda & Kenya, S to South Africa.
EQU: Dale 323 2/1943.

Indigofera tinctoria L.

F.P.S. 2: 215 (1952); F.T.E.A. Leg.-Pap.: 308 (1971); F.J.M.: 115 (1976); F.E.E. 3: 135 (1989).
Annual or perennial herb. Bush margins, cultivated & disturbed ground.
Distr: NS, DAR, KOR, UN; palaeotropical.
KOR: Jackson 3102 2/1954; **UN:** Sherif 3981 8/1951.
Note: formerly commonly cultivated for indigo dye.

Indigofera trita L.f. var. maffei (Chiov.) Ali

F.T.E.A. Leg.-Pap.: 304 (1971); F.E.E. 3: 134 (1989).
Syn: *Indigofera subulata* sensu Andrews pro parte, non Vahl ex Poir. – F.P.S. 2: 215 (1952).
Perennial herb. Bushland.
Distr: CS, ES; Ethiopia to Mozambique; India.
CS: Andrews 121 12/1935.

Indigofera trita L.f. var. scabra (Roth) Ali

F.T.E.A. Leg.-Pap.: 304 (1971); F.E.E. 3: 134 (1989); Biol. Skr. 51: 220 (1998).
Perennial herb. Wooded grassland, bushland.
Distr: EQU; Nigeria to Somalia, S to South Africa; Madagascar, Arabia, India, Central America.
EQU: Friis & Vollesen 808 12/1980.

Indigofera trita L.f. subsp. *subulata* (Vahl ex Poir.) Ali var. *nubica* (J.B.Gillett) Boulos & Schrire

F.Egypt 1: 313 (1999).
Syn: *Indigofera subulata* sensu Andrews pro parte, non Vahl ex. Poir. – F.P.S. 2: 215 pro parte (1952).
Syn: *Indigofera quartiniana* sensu Andrews, non A.Rich. – F.P.S. 2: 214 (1952).
Subshrub. Rocky hillslopes, semi-desert.
Distr: RS; Somalia; Egypt.
RS: Schweinfurth 1812 5/1865.

Indigofera vicioides Jaub. & Spach

F.T.E.A. Leg.-Pap.: 277 (1971); F.J.M.: 115 (1976); F.E.E. 3: 130 (1989).
Annual herb. Stony grassland.
Distr: DAR, ?EQU; Congo-Brazzaville to Eritrea, S to Namibia & South Africa.
DAR: Wickens 2972 5/1965.
Note: (1) the single specimen seen is only tentatively identified; (2) Andrews records *I. rogersii* R.E.Fr. from Equatoria; this taxon is now treated as a variety of *I. vicioides*. We have seen no material of this species from South Sudan.

Indigofera volkensii Taub.

F.T.E.A. Leg.-Pap.: 315 (1971); F.E.E. 3: 136 (1989); Biol. Skr. 51: 220 (1998).
Syn: *Indigofera insularis* Chiov. – F.P.S. 2: 212 (1952); F.T.E.A. Leg.-Pap.: 315 (1971).
Perennial herb. Woodland, bushland.
Distr: RS, EQU; Eritrea & Somalia, S to Tanzania.
RS: Schweinfurth s.n. 1868; **EQU:** Peers KO17 8/1953.

Indigofera zenkeri Harms ex Baker f.

F.P.S. 2: 211 (1952); F.T.E.A. Leg.-Pap.: 265 (1971); F.E.E. 3: 128 (1989).
Perennial bushy herb. Woodland, grassland, disturbed areas.
Distr: KOR, ?EQU; Cameroon to Ethiopia, S to Zimbabwe & Mozambique.
KOR: Kotschy 49.
Note: Andrews records this species from Equatoria; we have not been able to trace the material on which this was based and it may be a misidentification.

Kotschya aeschynomenoides (Welw. ex Baker) Dewit & P.A.Duvign.

F.T.E.A. Leg.-Pap.: 421 (1971); Biol. Skr. 51: 221 (1998).
Syn: *Smithia volkensii* Taub. – F.P.S. 2: 233 (1952).
Shrub. Grassland, bushland.
Distr: EQU; D.R. Congo to Kenya, S to Angola & Zambia.
EQU: Friis & Vollesen 1142 3/1982.

Kotschya africana Endl. var. *africana*

F.T.E.A. Leg.-Pap.: 412 (1971); F.E.E. 3: 146 (1989).
Syn: *Smithia africana* (Endl.) Taub. – F.P.S. 2: 233 (1952).
Shrub. Grassland, woodland, forest margins.
Distr: CS; Ethiopia to D.R. Congo & Zambia.
CS: Kotschy 524 1837.

Kotschya schweinfurthii (Taub.) Dewit & P.A.Duvign.

F.W.T.A. 1(2): 581 (1958).
Syn: *Smithia schweinfurthii* Taub. – F.P.S. 2: 233 (1952).
Subshrub. Wooded grassland.
Distr: EQU; Ghana, Nigeria, Cameroon, C.A.R., D.R. Congo.
EQU: Schweinfurth 2673 11/1869.

Lablab purpureus (L.) Sweet subsp. *purpureus*

F.T.E.A. Leg.-Pap.: 696 (1971).
Syn: *Dolichos lablab* L. – Bot. Exp. Sud.: 71 (1970).
Syn: *Lablab niger* Medik. – F.P.S. 2: 219, pro parte (1952).
Perennial herb, often climbing. Cultivated, sometimes naturalised.
Distr: NS, RS, DAR, CS; widely cultivated throughout the tropics.
CS: Schweinfurth 949 1/1869.

Lablab purpureus (L.) Sweet subsp. *uncinatus* Verdc.

F.T.E.A. Leg.-Pap.: 696 (1971); F.J.M.: 115 (1976); F.E.E. 3: 179 (1989); Biol. Skr. 51: 221 (1998).
Syn: *Lablab niger* sensu Andrews pro parte, non Medik. – F.P.S. 2: 219 (1952).
Perennial herb, often climbing. Forest.
Distr: DAR, ES, BAG, EQU; Senegal to Ethiopia, S to South Africa.
DAR: Wickens 1417 4/1964; **EQU:** Friis & Vollesen 693 12/1980.

Lathyrus hygrophilus Taub.

F.P.S. 2: 219 (1952); F.T.E.A. Leg.-Pap.: 1077 (1971); Biol. Skr. 51: 221 (1998).
Annual climbing herb. Montane grassland & bushland, especially in swampy areas.
Distr: EQU; D.R. Congo to Kenya, S to Malawi.
EQU: Myers 11699 7/1939.

Lathyrus sativus L.

F.J.M.: 115 (1976); F.E.E. 3: 250 (1989).
Annual herb. Grassland on black clay soil, cultivated.
Distr: NS, DAR; Ethiopia; widely cultivated in the warm temperate regions.
DAR: Wickens 1339 3/1964.
Note: cultivated for its edible seeds, eaten by both humans and livestock.

Leobordea platycarpa (Viv.) B.E.van Wyk & Boatwr.

Taxon 60: 173 (2011).
Syn: *Lotononis platycarpa* (Viv.) Pic.Serm. – F.P.S. 2: 221 (1952); F.T.E.A. Leg.-Pap.: 813 (1971); F.J.M.: 115 (1976); F.E.E.: 3: 193 (1989).
Annual herb. Dry rocky areas, weed of cultivation.
Distr: NS, RS, DAR; Eritrea to Namibia & South Africa; N Africa, Arabia, India.
RS: Bent s.n. 1896.

Lotus arabicus L.

F.P.S. 2: 222 pro parte (1952); F.T.E.A. Leg.-Pap.: 1048 (1971); F.J.M.: 116 (1976); F.E.E. 3: 225 (1989).

Annual herb. Riverbanks.
Distr: NS, DAR, KOR, CS, ES; Senegal to Eritrea, S to Angola & South Africa; Arabia, Iran.
DAR: <u>Wickens 2896</u> 4/1965.

Lotus becquetii Boutique

F.T.E.A. Leg.-Pap.: 1044 (1971); Biol. Skr. 51: 222 (1998).
Perennial herb. Montane grassland & bushland.
Distr: EQU; D.R. Congo, Uganda, Kenya.
EQU: <u>Friis & Vollesen 375</u> 11/1980.

Lotus glinoides Delile

F.P.S. 2: 221 (1952); F.Egypt 1: 298 (1999).
Syn: *Lotus schimperi* Steud. ex Boiss. – F.E.E. 3: 225 (1989).
Annual herb. Sandy plains, wadis.
Distr: RS; Mauritania to Eritrea; N Africa, Palestine, Arabia, Pakistan.
RS: <u>Broun 1552</u> 1/1908.

Lotus hebecarpus J.B.Gillett

F.E.E. 3: 225 (1989).
Annual or short-lived perennial herb. Rocky or sandy ground.
Distr: RS; Eritrea, Djibouti.
IUCN: RD
RS: <u>Sahni & Kamil 649</u> 4/1967.

Lotus hebranicus Hochst. ex Brand

F.Egypt 1: 297 (1999).
Syn: *Lotus deserti* Täckh. & Boulos – Bot. Exp. Sud.: 74 (1970).
Syn: *Lotus arabicus* sensu Andrews pro parte, non L. – F.P.S. 2: 222 (1952).
Perennial shrublet. Desert wadis.
Distr: RS; Egypt.
IUCN: RD
RS: <u>Bent s.n.</u> 1896.
Note: Andrews' record of *L. garcinii* DC. from Central Sudan is in error according to Trop. Afr. Fl. Pl. 4: 134 (2008).

Lotus nubicus Hochst. ex Baker

F.Egypt 1: 298 (1999).
Syn: *Lotus montanus* sensu Andrews pro parte, non A.Rich. – F.P.S. 2: 221 (1952).
Annual herb. Desert plains.
Distr: RS, CS; Egypt.
IUCN: RD
CS: <u>Schweinfurth 900</u> 1/1869.

Lotus quinatus (Forssk.) J.B.Gillett var. *brachycarpus* (Hochst. & Steud. ex A.Rich.) J.B.Gillett

F.E.E. 3: 224 (1989).
Syn: *Lotus montanus* A.Rich. – F.P.S. 2: 221 pro parte (1952).
Perennial herb. Grassland, bushland, abandoned cultivations.
Distr: CS; Eritrea, Ethiopia.
CS: <u>de Wilde et al. 5818</u> 2/1965.

Lotus schoelleri Schweinf.

F.E.E. 3: 223 (1989).

Lotus torulosus (Chiov.) Fiori

F.E.E. 3: 225 (1989).
Syn: *Lotus corniculatus* L. var. *eremanthus* Chiov. – F.T.E.A. Leg.-Pap.: 1043 (1971); F.J.M.:116 (1976).
Syn: *Lotus corniculatus* sensu auctt., non L. – F.P.S. 2: 221 (1952); Bot. Exp. Sud.: 74 (1970).
Perennial herb. Montane grassland including damp areas.
Distr: DAR; Eritrea, Ethiopia, Kenya, Tanzania.
DAR: <u>Wickens 2883</u> 3/1965.

Macrotyloma axillare (E.Mey.) Verdc. var. *glabrum* (E.Mey.) Verdc.

F.T.E.A. Leg.-Pap.: 586 (1971); F.E.E. 3: 166 (1989); Biol. Skr. 51: 222 (1998).
Syn: *Dolichos biflorus* sensu Andrews, non L. – F.P.S. 2: 196 (1952).
Perennial climbing or trailing herb. Wooded grassland.
Distr: EQU; D.R. Congo to Somalia, S to South Africa; Madagascar, Mascarenes, India.
EQU: <u>Friis & Vollesen 269</u> 11/1980.

Macrotyloma biflorum (Schumach. & Thonn.) Hepper var. *biflorum*

F.E.E. 3: 166 (1989).
Perennial climbing or trailing herb. Wooded grassland, roadsides.
Distr: EQU; Senegal to Ethiopia, S to Angola & Zambia.
EQU: <u>Myers 9314</u> 8/1938.

Macrotyloma daltonii (Webb) Verdc.

F.E.E. 3: 166 (1989).
Annual or short-lived perennial, trailing or climbing herb. Wooded grassland.
Distr: KOR, CS; Cape Verde Is., Senegal, Niger to Eritrea, S to Namibia & Botswana.
KOR: <u>Wickens 809</u> 11/1962.

Macrotyloma schweinfurthii Verdc.

F.E.E. 3: 168 (1989).
Annual herb. Woodland, roadsides.
Distr: ES; Togo, Nigeria, Ethiopia.
ES: <u>Schweinfurth 1887</u> 10/1865.
Note: a widespread but apparently scarce species.

Macrotyloma stenophyllum (Harms) Verdc.

F.T.E.A. Leg.-Pap.: 588 (1971); F.E.E. 3: 168 (1989); Biol. Skr. 51: 223 (1998).
Annual or perennial herb. Wooded grassland.
Distr: EQU; Senegal to Ethiopia, S to Angola & Tanzania.
EQU: <u>Friis & Vollesen 455</u> 11/1980.

Medicago laciniata (L.) Mill.

Bot. Exp. Sud.: 74 (1970); F.T.E.A. Leg.-Pap.: 1037 (1971); F.E.E. 3: 245 (1989).
Annual herb. Grassland & woodland.
Distr: RS; Ethiopia & Somalia, S to Tanzania, South Africa; Macaronesia, N Africa to India.
RS: <u>Drar 489</u> 3/1938 (n.v.).

Note: we have not seen the cited specimen; the identity requires confirmation.

Medicago minima (L.) L.

F.P.S. 2: 222 (1952); F.E.E. 3: 245 (1989).
Annual herb. Open sandy & rocky ground including disturbed areas.
Distr: RS; Eritrea, Ethiopia, Somalia; Europe, N Africa, S Asia, Socotra.
RS: Farag & Zubair 14 (fr) 2/2012.

Medicago sativa L.

Bot. Exp. Sud.: 74 (1970); F.Egypt 1: 275 (1999).
Perennial herb. Cultivated, occasionally naturalised in waste ground.
Distr: NS, RS, CS; probably native to W Asia, now widely cultivated particularly in Europe & N Africa.
NS: Ahti 16998 9/1962.
Note: Ahti et al. (1973) recorded this species as "rarely cultivated, a few escapes" at Wadi Halfa. Drar (1970) records further, apparently naturalised, specimens from Khartoum and Sinkat.

Microcharis disjuncta (J.B.Gillett) Schrire var. disjuncta

Bothalia 22: 166 (1992).
Syn: *Indigofera arenaria* sensu Andrews pro parte, non A.Rich. – F.P.S. 2: 208 (1952).
Annual herb. Sandy grassland, rocky slopes.
Distr: NS, RS; Mauritania, Niger, Chad.
RS: Schweinfurth 662 10/1868.

Microcharis longicalyx (J.B.Gillett) Schrire

Bothalia 22: 166 (1992).
Syn: *Indigofera longicalyx* J.B.Gillett – F.J.M.: 114 (1976).
Annual herb. Sandy soils, fallow land.
Distr: DAR; Guinea, Sierra Leone, Nigeria, Cameroon, Chad, C.A.R.
DAR: Wickens 2273 8/1964.

Microcharis praetermissa (Baker f.) Schrire

Bothalia 22: 167 (1992).
Syn: *Indigofera praetermissa* Baker f. – Legumes of Africa: 320 (1989).
Annual herb. Wooded grassland.
Distr: EQU; C.A.R., D.R. Congo, Zambia.
EQU: Schweinfurth 2538 1869.

Microcharis tritoides (Baker) Schrire

Bothalia 22: 167 (1992).
Syn: *Indigofera tritoides* Baker – F.P.S. 2: 209 (1952); F.E.E. 3: 138 (1989).
Perennial herb. Bushland, semi-desert, rocky hillslopes.
Distr: RS; Ethiopia, Djibouti, Somalia; Egypt, Arabia.
RS: Schweinfurth 297 9/1868.

Microcharis welwitschii (Baker) Schrire

Bothalia 22: 167 (1992).
Syn: *Indigofera welwitschii* Baker – F.P.S. 2: 211 (1952); F.T.E.A. Leg.-Pap.: 326 (1971); F.E.E. 3: 138 (1989); Biol. Skr. 51: 220 (1998).
Annual herb. Wooded grassland.
Distr: EQU; C.A.R. to Ethiopia, S to Angola & Zimbabwe.
EQU: Friis & Vollesen 580 11/1980.

Mildbraediodendron excelsum Harms

F.P.S. 2: 222 (1952); F.T.E.A. Leg.-Caes.: 225 (1967); T.S.S.: 237 (1990); Biol. Skr. 51: 181 (1998).
Tree. Forest.
Distr: EQU; Ghana, Nigeria, Cameroon, D.R. Congo, Uganda.
EQU: Myers 6515 5/1937.

Millettia barteri (Benth.) Dunn

F.P.S. 2: 222 (1952); F.W.T.A. 1(2): 526 (1958).
Shrub or woody climber. Swamp forest & streambanks.
Distr: EQU; Senegal to D.R. Congo.
EQU: Andrews 1438 5/1939.
Note: El Amin in T.S.S.: 239 (1990) records *M. psilopetala* Harms from Yambio; this is believed to be in error and probably based on a misidentification of *M. barteri*.

Mucuna poggei Taub. var. poggei

F.P.S. 2: 223 (1952); F.T.E.A. Leg.-Pap.: 565 (1971).
Woody climber. Swamp forest & swampy grassland.
Distr: EQU; D.R. Congo to Kenya, S to Angola & Zambia.
EQU: Myers 9385 8/1938.

Mucuna poggei Taub. var. pesa (De Wild.) Verdc.

F.T.E.A. Leg.-Pap.: 566 (1971); Biol. Skr. 51: 223 (1998).
Woody climber. Swamp forest, bushland.
Distr: EQU; D.R. Congo to Kenya, S to Zimbabwe & Mozambique
EQU: Friis & Vollesen 181 11/1980.

Mucuna pruriens (L.) DC. var. pruriens

F.P.S. 2: 223 (1952); F.T.E.A. Leg.-Pap.: 567 (1971); F.E.E. 3: 164 (1989); Biol. Skr. 51: 223 (1998).
Annual or short-lived perennial climbing herb. Wooded grassland.
Distr: ?CS, BAG, EQU; pantropical.
BAG: Schweinfurth 2714 12/1869.
Note: Andrews records this species from Central & Southern Sudan but we have only seen the single specimen cited.

Mundulea sericea (Willd.) A.Chev.

F.P.S. 2: 223 (1952); F.T.E.A. Leg.-Pap.: 155 (1971); F.J.M.: 116 (1976); T.S.S.: 239 (1990).
Small tree. Woodland, bushland, forest margins.
Distr: DAR, KOR, CS, EQU; Guinea to Somalia, S to Namibia & South Africa; Madagascar, India to New Guinea.
DAR: Wickens 2816 11/1964; **EQU:** Andrews 851 4/1939.

Neonotonia wightii (Wight & Arn.) J.A.Lackey subsp. wightii var. longicauda (Schweinf.) J.A.Lackey

Biol. Skr. 51: 223 (1998); Trop. Afr. Fl. Pl. 4: 171 (2008).
Syn: *Glycine wightii* (Wight & Arn.) Verdc. subsp. *wightii* – Bot. Exp. Sud.: 72 (1970); F.T.E.A. Leg.-Pap.: 529 (1971); F.J.M.: 114 (1976); F.E.E. 3: 158 (1989).
Perennial climbing or trailing herb. Montane grassland & bushland.
Distr: RS, DAR, ES, EQU; Sierra Leone to Ethiopia, S to Angola & South Africa; Arabia.

DAR: Wickens 2603 9/1964; **EQU:** Friis & Vollesen 772 12/1980.
Note: Quézel's (1969: 116) record of *Glycine javanica* L. is presumably referable here.

Neonotonia wightii (Wight & Arn.) J.A.Lackey subsp. *pseudojavanica* (Taub.) J.A.Lackey var. *pseudojavanica*

Biol. Skr. 51: 223 (1998); Trop. Afr. Fl. Pl. 4: 171 (2008).
Syn: *Glycine wightii* (Wight & Arn.) Verdc. subsp. *pseudojavanica* (Taub.) Verdc. var. *pseudojavanica* – F.T.E.A. Leg.-Pap.: 530 (1971).
Syn: *Glycine javanica* sensu Andrews, non L. – F.P.S. 2: 203 (1952).
Perennial climbing or trailing herb. Montane grassland & bushland.
Distr: EQU; Sierra Leone to Uganda, Kenya & Tanzania.
EQU: Andrews 1743 6/1939.

Neorautanenia mitis (A.Rich.) Verdc.

F.T.E.A. Leg.-Pap.: 700 (1971); F.E.E. 3: 181 (1989); Biol. Skr. 51: 224 (1998).
Syn: *Neorautanenia pseudopachyrhiza* (Harms) Milne-Redh. – F.P.S. 2: 223 (1952); Bot. Exp. Sud.: 74 (1970).
Perennial herb, sometimes climbing. Grassland, bushland, open woodland.
Distr: KOR, ES, BAG, UN, EQU; Ivory Coast to Ethiopia, S to Angola & South Africa.
KOR: Broun 69 6/1906; **EQU:** Andrews 1304 5/1939.

Ophrestia hedysaroides (Willd.) Verdc.

F.T.E.A. Leg.-Pap.: 526 (1971).
Syn: *Glycine hedysaroides* Willd. – F.P.S. 2 203 (1952).
Perennial climbing herb. Bushland, grassland, old cultivations.
Distr: ?EQU; Ivory Coast to Tanzania, S to Angola.
Note: Andrews recorded this species from Equatoria but we have seen no material with which to verify this, and Verdcourt in F.T.E.A. did not include Sudan in the distribution of this species; the identification may be erroneous.

Ormocarpum pubescens (Hochst.) Cufod.

F.E.E. 3: 143 (1989).
Shrub. Woodland, bushland.
Distr: RS; Senegal to Cameroon, Eritrea, Ethiopia.
RS: Kennedy-Cooke 206 3/1936.
Note: El Amin's record of *O. trichocarpum* from Red Sea region in T.S.S.: 239 (1990) is presumably referable here.

Ormocarpum trichocarpum (Taub.) Harms

F.P.S. 2: 224 (1952); F.T.E.A. Leg.-Pap.: 359 (1971); F.E.E. 3: 143 (1989); T.S.S.: 239, pro parte (1990); Biol. Skr. 51: 224 (1998).
Shrub. Wooded grassland & bushland.
Distr: EQU; Ethiopia to South Africa.
EQU: Simpson 7322 7/1929.

Pericopsis laxiflora (Benth.) Meeuwen

T.S.S.: 241 (1990); Trop. Afr. Fl. Pl. 4: 182 (2008).
Syn: *Afrormosia laxiflora* (Benth.) Harms – F.P.S. 2: 173 (1952).
Tree. Wooded grassland, often in rocky areas.
Distr: ?DAR, BAG, UN; Senegal to C.A.R.

BAG: Myers 13811 4/1941.
Note: the record for Darfur is from El Amin in T.S.S.; we have seen no specimen with which to verify this.

Phaseolus lunatus L.

F.P.S. 2: 164 (1952); F.T.E.A. Leg.-Pap.: 615 (1971); F.E.E. 3: 171 (1989).
Perennial climbing herb or shrub. Grassland & bushland.
Distr: BAG, EQU; native to tropical America, now widely cultivated & naturalised in the tropics.
EQU: Myers 10199 7/1938.

Philenoptera laxiflora (Guill. & Perr.) Roberty

Trop. Afr. Fl. Pl. 4: 182 (2008).
Syn: *Lonchocarpus laxiflorus* Guill. & Perr. – F.P.S. 2: 221 (1952); Bot. Exp. Sud.: 73 (1970); F.T.E.A. Leg.-Pap.: 67 (1971); F.J.M.: 115 (1976); F.E.E. 3: 104 (1989); T.S.S.: 237 (1990); Biol. Skr. 51: 222 (1998).
Tree. Woodland, often in rocky areas.
Distr: RS, DAR, KOR, CS, ES, BAG, UN, EQU; Cape Verde Is., Senegal to Ethiopia & Uganda.
DAR: Wickens 1178 2/1964; **EQU:** Andrews 701 4/1939.

Pseudarthria confertiflora (A.Rich.) Baker

F.P.S. 2: 225 (1952); F.T.E.A. Leg.-Pap.: 484 (1971); F.E.E. 3: 152 (1989); Biol. Skr. 51: 225 (1998).
Perennial herb or subshrub. Grassland & wooded grassland.
Distr: EQU; Ivory Coast to Ethiopia, S to Angola & Tanzania.
EQU: Friis & Vollesen 255 11/1980.

Pseudarthria hookeri Wight & Arn. var. *hookeri*

F.P.S. 2: 225 (1952); F.T.E.A. Leg.-Pap.: 484 (1971); F.E.E. 3: 153 (1989); Biol. Skr. 51: 225 (1998).
Perennial herb or subshrub. Grassland, abandoned cultivations.
Distr: EQU; Cameroon to Ethiopia, S to Angola & South Africa.
EQU: Myers 9152 7/1938.

Pseudarthria hookeri Wight & Arn. var. *argyrophylla* Verdc.

F.T.E.A. Leg.-Pap.: 485 (1971).
Perennial herb or subshrub. Bushland, abandoned cultivations.
Distr: BAG; Senegal to Cameroon, Uganda.
BAG: Schweinfurth 1489 4/1869.

Pseudoeriosema andongense (Baker) Hauman

F.T.E.A. Leg.-Pap.: 521 (1971); Biol. Skr. 51: 225 (1998).
Syn: *Glycine holophylla* (Baker f.) Taub. – F.P.S. 2: 202 (1952).
Perennial herb or subshrub. Grassland & wooded grassland.
Distr: EQU; Benin to Uganda, S to Angola & Zambia.
EQU: Andrews 974 5/1939.

Pseudoeriosema borianii (Schweinf.) Hauman subsp. *borianii*

F.E.E. 3: 156 (1989).
Syn: *Glycine borianii* (Schweinf.) Baker – F.P.S. 2: 202 (1952).

Perennial herb. Grassland & wooded grassland including seasonally flooded areas.
Distr: CS, ES, UN; Ethiopia to Zambia.
ES: <u>Beshir 29</u> 8/1951; **UN:** <u>Sherif A.3961</u> 8/1951.

Psophocarpus obovalis Tisser.
Trop. Afr. Fl. Pl. 4: 192 (2008).
Prostrate perennial herb. Wooded grassland, roadsides.
Distr: EQU; C.A.R
IUCN: RD
EQU: <u>Myers 9292</u> 8/1938.

Psophocarpus palustris Desv.
F.W.T.A. 1(2): 572 (1958).
Perennial herb. Riverine forest, bushland.
Distr: EQU; Senegal to C.A.R.
EQU: <u>Schweinfurth 2852</u> 12/1869.

Psophocarpus scandens (Endl.) Verdc.
F.T.E.A. Leg.-Pap.: 603 (1971).
Syn: *Psophocarpus palmettorum* sensu Andrews, non Guill. & Perr. – F.P.S. 2: 225 (1952).
Perennial climbing herb. Swamp forest, riverbanks, damp areas in woodland.
Distr: BAG; Nigeria to Uganda, S to Angola & Mozambique; Madagascar, Mascarenes.
BAG: <u>Broun 1863</u> 3/1903.

Pterocarpus lucens Lepr. ex Guill. & Perr. subsp. *lucens*
F.P.S. 2: 226 (1952); F.T.E.A. Leg.-Pap.: 82 (1971); F.J.M.: 116 (1976); F.E.E. 3: 105 (1989); T.S.S.: 243 (1990); Biol. Skr. 51: 225 (1998).
Tree. Wooded grassland.
Distr: DAR, CS, ES, BAG, EQU; Senegal to Eritrea & Ethiopia.
DAR: <u>Wickens 1557</u> 5/1964; **EQU:** <u>Myers 8255</u> 1/1938.

Requienia obcordata (Lam. ex Poir.) DC.
F.P.S. 2: 226 (1952); Bot. Exp. Sud.: 75 (1970); Trop. Afr. Fl. Pl. 4: 200 (2008).
Perennial herb or subshrub. Sandy soils.
Distr: DAR, KOR; Senegal to Chad.
KOR: <u>Wickens 160</u> 7/1962.

Retama raetam (Forssk.) Webb
F.Egypt 1: 258 (1999).
Shrub. Wadis, sandy desert.
Distr: NS; N Africa, E Mediterranean.
Note: the inclusion of this species here is based on a record and photograph by Malterer (2013) in his report on the survey for the Merowe Dam Project – he noted this as the only species occuring in areas of shifting desert sands.

Rhynchosia densiflora (Roth) DC. subsp. *chrysadenia* (Taub.) Verdc.
F.T.E.A. Leg.-Pap.: 724 (1971); F.E.E. 3: 183 (1989).
Syn: *Rhynchosia schweinfurthii* Harms – F.P.S. 2: 227 (1952).
Perennial herb, often climbing. Grassland & thicket.
Distr: BAG; Ethiopia to Namibia & South Africa.
BAG: <u>Schweinfurth 2335</u> 9/1869.

Rhynchosia densiflora (Roth.) DC. subsp. *debilis* (G.Don) Verdc.
F.T.E.A. Leg.-Pap.: 723 (1971); F.E.E. 3: 183 (1989); Biol. Skr. 51: 226 (1998).
Perennial herb, often climbing. Forest margins & associated bushland.
Distr: EQU; Ghana to Uganda, S to Angola & Zambia.
EQU: <u>Friis & Vollesen 482</u> 11/1980.

Rhynchosia elegans A.Rich. var. *elegans*
F.T.E.A. Leg.-Pap.: 739 (1971); F.E.E. 3: 184 (1989); Biol. Skr. 51: 226 (1998).
Perennial climbing herb. Montane grassland, bushland & forest margins.
Distr: EQU; Eritrea & Somalia, S to Malawi.
EQU: <u>Friis & Vollesen 394</u> 11/1980.

Rhynchosia hirta (Andrews) Meikle & Verdc.
F.T.E.A. Leg.-Pap.: 720 (1971); F.E.E. 3: 189 (1989).
Syn: *Rhynchosia albiflora* (Sims) Alston – F.P.S. 2: 227 (1952).
Perennial climbing or trailing herb. Forest margins.
Distr: EQU; Nigeria to Ethiopia, S to South Africa; Mascarenes, India, Sri Lanka.
EQU: <u>Schweinfurth 3121</u> 5/1870.

Rhynchosia malacophylla (Spreng.) Bojer
F.T.E.A. Leg.-Pap.: 753 (1971); F.J.M.: 116 (1976); F.E.E. 3: 187 (1989).
Syn: *Rhynchosia sennaarensis* Hochst. ex Schweinf. – F.P.S. 2: 227 (1952); Bot. Exp. Sud.: 76 (1970).
Perennial climbing or trailing herb. Bushland.
Distr: RS, DAR, CS, KOR, UN, EQU; Eritrea to Tanzania; Egypt, Yemen.
CS: <u>Andrews A.9</u> 6/1935; **EQU:** <u>Hepper 18</u> 4/1984.

Rhynchosia minima (L.) DC. var. *minima*
F.P.S. 2: 228 (1952); F.T.E.A. Leg.-Pap.: 757 (1971); F.E.E. 3: 187 (1989).
Perennial climbing or trailing herb. Wooded grassland & bushland.
Distr: RS, KOR, UN, EQU; widespread in the tropics.
RS: <u>Broun 1221</u> 5/1907; **UN:** <u>J.M. Lock 81/134</u> 8/1981.

Rhynchosia minima (L.) DC. var. *memnonia* (Delile) T.Cooke
F.T.E.A. Leg.-Pap.: 758, in notes (1971); F.J.M.: 116 (1976); F.E.E. 3: 188 (1989).
Syn: *Rhynchosia memnonia* (Delile) DC. – F.P.S. 2: 228 (1952).
Perennial climbing or trailing herb. Wooded grassland & bushland.
Distr: NS, DAR, KOR, CS; Cape Verde Is., Senegal & Mauritania to Somalia; Egypt, Arabia.
DAR: <u>Wickens 1850</u> 7/1964.

Rhynchosia minima (L.) DC. var. *nuda* (DC.) Kuntze
F.T.E.A. Leg.-Pap.: 758 (1971); F.E.E. 3: 187 (1989); Biol. Skr. 51: 226 (1998).
Perennial climbing or trailing herb. Wooded grassland & bushland.

Distr: EQU; Cape Verde Is., Senegal to Ethiopia, S to Tanzania; Mauritius, India to Australia.
EQU: Friis & Vollesen 996 2/1982.

Rhynchosia minima (L.) DC. var. *prostrata* (Harv.) Meikle

F.T.E.A. Leg.-Pap.: 758 (1971); F.J.M.: 116 (1976); F.E.E. 3: 188 (1989); Biol. Skr. 51: 226 (1998).
Perennial climbing or trailing herb. Wooded grassland & bushland.
Distr: NS, DAR, ES, BAG, UN, EQU; Senegal to Somalia, S to South Africa; Yemen.
DAR: Wickens 1350 3/1964; **EQU:** Schweinfurth 2488 12/1869.

Rhynchosia nyasica Baker

F.P.S. 2: 230 (1952); F.T.E.A. Leg.-Pap.: 736 (1971); F.E.E. 3: 184 (1989); Biol. Skr. 51: 227 (1998).
Perennial herb or subshrub. Wooded grassland & bushland.
Distr: EQU; Sierra Leone to Ethiopia, S to Angola & Zimbabwe.
EQU: Friis & Vollesen 1113 3/1982.

Rhynchosia orthobotrya Harms

F.P.S. 2: 228 (1952); F.T.E.A. Leg.-Pap.: 735 (1971); F.E.E. 3: 183 (1989); Biol. Skr. 51: 227 (1998).
Perennial herb or subshrub. Wooded grassland & bushland.
Distr: EQU; Senegal, Nigeria to Eritrea, S to Tanzania.
EQU: Friis & Vollesen 994 2/1982.

Rhynchosia procurrens (Hiern) K.Schum. subsp. *latisepala* (Hauman) Verdc.

F.T.E.A. Leg.-Pap.: 725 (1971).
Perennial climbing or trailing herb. Bushland, grassland.
Distr: EQU; Mali to Kenya, ?Zambia.
EQU: Petherick s.n. 7/1863.

Rhynchosia pseudoviscosa Harms

F.T.E.A. Leg.-Pap.: 734 (1971).
Perennial climbing or trailing herb. Bushland & grassland, particularly in swampy places.
Distr: UN; Uganda to Mozambique.
UN: Broun 1729 12/1909.

Rhynchosia pulverulenta Stocks

F.T.E.A. Leg.-Pap.: 755 (1971); F.E.E. 3: 187 (1989).
Perennial herb. Bushland, woodland & grassland.
Distr: NS, RS, KOR; Eritrea, Ethiopia, Somalia, Kenya, Tanzania; Egypt, Arabia, Socotra, Pakistan, India.
KOR: Wickens 173 8/1962.

Rhynchosia resinosa (Hochst. ex A.Rich.) Baker

F.P.S. 2: 227 (1952); F.T.E.A. Leg.-Pap.: 727 (1971); F.J.M.: 116 (1976); F.E.E. 3: 183 (1989); Biol. Skr. 51: 227 (1998).
Perennial scrambling herb. Bushland & wooded grassland, disturbed areas.
Distr: DAR, EQU; Guinea, Nigeria to Eritrea, S to Namibia & South Africa.
DAR: Wickens 978 1/1964; **EQU:** Friis & Vollesen 139 3/1982.

Rhynchosia splendens Schweinf.

F.P.S. 2: 228 (1952); F.E.E. 3: 188 (1989).
Subshrub. Habitat unknown.
Distr: ES; Sudan/Ethiopia endemic.
IUCN: RD
ES: Schweinfurth 1870 1865.
Note: known only from the type collection which is from the Sudan/Ethiopia border region.

Rhynchosia sublobata (Schumach. & Thonn.) Meikle

F.P.S. 2: 228 (1952); F.T.E.A. Leg.-Pap.: 751 (1971); F.E.E. 3: 185 (1989); Biol. Skr. 51: 227 (1998).
Perennial climbing or trailing herb. Grassland & wooded grassland.
Distr: ?DAR, ?CS, BAG, UN, EQU; Senegal to Eritrea & Somalia, S to Namibia & South Africa; Madagascar, Comoros, Yemen.
EQU: Andrews 557 4/1939.
Note: Andrews records this species from Central & Southern Sudan but we have only seen material from the South. Quézel (1969: 117) records it from Jebel Gourgeil in Darfur.

Rhynchosia teramnoides Harms

F.P.S. 2: 230 (1952); Trop. Afr. Fl. Pl. 4: 214 (2008).
Perennial twining herb. Habitat not known.
Distr: BAG; Chad.
IUCN: RD
BAG: Schweinfurth 7642 11/1869.

Rhynchosia totta (Thunb.) DC. var. *fenchelii* Schinz

Trop. Afr. Fl. Pl. 4: 214 (2008).
Syn: *Rhynchosia lynesii* Baker f. & W.Martin – F.P.S. 2: 230 (1952).
Syn: *Rhynchosia totta* (Thunb.) DC. var. *venulosa* (Hiern) Verdc. – F.T.E.A. Leg.-Pap.: 748 (1971); F.J.M.: 116 (1976); F.E.E. 3: 185 (1989).
Perennial herb. Grassland.
Distr: DAR; Eritrea to South Africa.
DAR: de Wilde et al. 5400 1/1965.

Rhynchosia tricuspidata Baker f. subsp. *imatongensis* Verdc.

F.T.E.A. Leg.-Pap.: 731 (1971); Biol. Skr. 51: 227 (1998).
Shrub. Montane grassland, often on rocky outcrops.
Distr: EQU; Uganda.
IUCN: RD
EQU: Friis & Vollesen 978 2/1982.

Rhynchosia viscosa (Roth) DC. subsp. *violacea* (Hiern) Verdc.

F.P.S. 2: 228 (1952); F.T.E.A. Leg.-Pap.: 733 (1971).
Perennial climbing or trailing herb. Forest margins, swampy ground.
Distr: UN; Cameroon to Uganda, S to Angola & D.R. Congo.
UN: Sherif A.3989 8/1951.

Rothia hirsuta (Guill. & Perr.) Baker

F.P.S. 2: 230 (1952); F.T.E.A. Leg.-Pap.: 811 (1971); F.E.E. 3: 195 (1989).

Annual herb. Weed of disturbed & cultivated ground, short grassland.
Distr: DAR, KOR; Senegal to Eritrea, S to Namibia & South Africa.
KOR: <u>Kotschy 420</u> 1840.

Sesbania leptocarpa DC.

F.E.E. 3: 142 (1989).
Syn: Sesbania arabica sensu Andrews pro parte, non E.Phillips & Hutch. – F.P.S. 2: 231 (1952).
Annual herb. Seasonally flooded areas, pool margins.
Distr: KOR, CS, UN; Cape Verde Is., Senegal to Somalia, Zambia, Zimbabwe, Mozambique; Yemen.
CS: <u>Aylmer 362</u> 5/1934; **UN:** <u>Broun 1737</u> 1903.
Note: Drar (1970) recorded both *S. aculeata* Pers. (= *S. bispinosa* (Jacq.) W.Wight) and *S. speciosa* Taub. from the Sudd region but did not make specimens. Both these records are considered to be erroneous.

Sesbania macrantha Welw. ex Phill. var. macrantha

F.P.S. 2: 232 (1952); F.T.E.A. Leg.-Pap.: 341 (1971); Biol. Skr. 51: 228 (1998).
Woody herb or tree. Swampy grassland.
Distr: ?EQU; Cameroon to Kenya, S to Angola & South Africa.
Note: Andrews records this species from the Imatong Mts; this is probably based on *Maxwell Forbes* 113 which is actually from the Ugandan side (see Friis & Vollesen) but it may well also occur on the South Sudanese side.

Sesbania microphylla Harms

F.P.S. 2: 231 (1952); F.T.E.A. Leg.-Pap.: 348 (1971); T.S.S.: 243 (1990).
Annual herb. Swampy areas, riverbanks.
Distr: BAG, UN, EQU; Chad to Uganda, S to Namibia, Botswana & Zimbabwe.
EQU: <u>Myers 7673</u> 8/1937.
Note: Quézel (1969: 117) records this species from El Fasher in Darfur but this is considered likely to be a misidentification.

Sesbania pachycarpa DC. subsp. pachycarpa

F.P.S. 2: 231 pro parte (1952); Bot. Exp. Sud.: 76 (1970); F.T.E.A. Leg.-Pap.: 345 (1971); F.E.E. 3: 142 (1989); T.S.S.: 245 (1990).
Syn: Sesbania cannabina sensu Andrews, non (Retz.) Poir. – F.P.S. 2: 231 (1952).
Annual or biennial herb. Seasonal swamps, riverbanks, roadsides.
Distr: DAR, KOR, CS, ES, UN, EQU; Cape Verde Is., Senegal to Eritrea, S to Angola & Zambia.
KOR: <u>Jackson 4016</u> 9/1959; **UN:** <u>Freeman & Lucas s.n.</u>

Sesbania rostrata Bremeck. & Oberm.

F.T.E.A. Leg.-Pap.: 343 (1971); F.E.E. 3: 140 (1989); T.S.S.: 245 (1990); Biol. Skr. 51: 228 (1998).
Syn: Sesbania pachycarpa sensu Andrews, non DC. – F.P.S. 2: 231 pro parte (1952).
Annual or biennial herb. Pool margins, swamps, riverbanks.
Distr: KOR, UN; Senegal to Ethiopia, S to Botswana & Zimbabwe.
KOR: <u>Pfund 404</u> 10/1875; **UN:** <u>J.M. Lock 81/136</u> 8/1981.

Sesbania sericea (Willd.) Link

Bot. Exp. Sud.: 76 (1970); F.T.E.A. Leg.-Pap.: 350 (1971); F.E.E. 3: 142 (1989).
Syn: Sesbania pubescens DC. – F.P.S. 2: 232 (1952).
Annual or biennial herb. Marshes, riverbanks.
Distr: BAG, UN; Senegal to Somalia, S to Angola & Malawi; Arabia, Sri Lanka, tropical America.
UN: <u>Harrison 295</u> 1/1948.

Sesbania sesban (L.) Merr. subsp. sesban var. sesban

F.P.S. 2: 232 (1952); F.T.E.A. Leg.-Pap.: 339, in notes (1971); T.S.S.: 245 (1990).
Perennial herb. Swamps, riverbanks, also cultivated.
Distr: NS, KOR, CS; Egypt, India.
CS: <u>Kotschy 347</u> 3/1840.
Note: this variety is probably native to Asia and has been cultivated in Sudan and Egypt.

Sesbania sesban (L.) Merr. subsp. sesban var. bicolor (Wight & Arn.) F.W.Andrews

F.P.S. 2: 232 (1952); Bot. Exp. Sud.: 76 (1970); T.S.S.: 247 (1990); Trop. Af. Fl. Pl. 4: 224 (2008).
Short-lived shrub or tree. Swamps, riverbanks.
Distr: KOR, CS; wider distribution unclear.
CS: <u>Schweinfurth 796</u> 11/1868.
Note: close to var. *sesban*.

Sesbania sesban (L.) Merr. subsp. sesban var. nubica Chiov.

F.T.E.A. Leg.-Pap.: 339 (1971); F.J.M.: 117 (1976); T.S.S.: 247 (1990); Biol. Skr. 51: 228 (1998).
Short-lived shrub or tree. Swamps, riverbanks.
Distr: DAR, KOR, CS, ES, UN, EQU; Chad to Somalia, S to South Africa.
DAR: <u>Wickens 1024</u> 1/1964; **EQU:** <u>Myers 7916</u> 11/1937.

Sesbania sesban (L.) Merr. subsp. puncata (DC.) J.B.Gillett

T.S.S.: 247 (1990) as var. *punctata*; Trop. Afr. Fl. Pl. 4: 224 (2008).
Syn: Sesbania punctata DC. – F.P.S. 2: 232 (1952).
Short-lived shrub or tree. Swamps, riverbanks.
Distr: EQU; Cape Verde Is., Senegal to C.A.R.
EQU: <u>Myers 10330</u> 1/1939.

Sesbania sudanica J.B.Gillett

Trop. Af. Fl. Pl. 4: 226 (2008).
Syn: Sesbania arabica Hochst. & Steud. ex E.Phillips & Hutch. – F.P.S. 2: 231 pro parte (1952).
Syn: Sesbania hepperi J.B.Gillett – F.E.E. 3: 142 (1989).
Short-lived subwoody herb. Damp sites in grassland & woodland, swampy ground.
Distr: CS, ES, UN; Liberia & Mali to Ethiopia.
ES: <u>Beshir 135</u> 9/1951; **UN:** <u>Simpson 7443</u> 1/1930.

Sesbania tetraptera Hochst. ex Baker subsp. tetraptera

F.P.S. 2: 232 (1952); F.T.E.A. Leg.-Pap.: 351 (1971); F.E.E. 3: 142 (1989); T.S.S.: 247 (1990); Biol. Skr. 51: 228 (1998).
Annual or biennial herb. Seasonally flooded areas.
Distr: KOR, CS, ES; Niger to Ethiopia, S to Malawi.
ES: <u>Jackson 3399</u> 1/1955.

Smithia elliotii Baker f. var. *elliotii*

F.P.S. 2: 233 (1952); F.T.E.A. Leg.-Pap.: 408 (1971); F.E.E. 3: 146 (1989); Biol. Skr. 51: 229 (1998).
Perennial herb. Swampy ground, damp montane grassland.
Distr: EQU; Cameroon, Ethiopia to Zambia & Mozambique; Madagascar.
EQU: Friis & Vollesen 317 11/1980.

Sphenostylis schweinfurthii Harms subsp. *schweinfurthii*

F.P.S. 2: 233 (1952); F.W.T.A. 1(2): 565 (1958).
Subshrub. Grassland.
Distr: BAG, EQU; Guinea to C.A.R.
EQU: Myers 13926 8/1941.

Sphenostylis stenocarpa (Hochst. ex A.Rich.) Harms

F.P.S. 2: 233 (1952); F.T.E.A. Leg.-Pap.: 672 (1971); F.E.E. 3: 178 (1989).
Perennial climbing or trailing herb. Wooded grassland & bushland.
Distr: ES, ?EQU; Guinea to Ethiopia, S to Angola & Zimbabwe.
ES: Schweinfurth 1899 8/1865.
Note: Andrews records this species from Gallabat and from Equatoria but we have only seen material from the former; the latter may be based on a misidentification of *S. schweinfurthii*.

Stylosanthes fruticosa (Retz.) Alston

F.T.E.A. Leg.-Pap.: 437 (1971); F.J.M.: 117 (1976); F.E.E. 3: 147 (1989); Biol. Skr. 51: 229 (1998).
Syn: *Stylosanthes flavicans* Baker – F.P.S. 2: 234 (1952).
Perennial herb. Bushland, grassland.
Distr: DAR, KOR, EQU; Cape Verde Is., Mauritania to Somalia, S to Namibia & South Africa; Madgascar, Arabia, India, Sri Lanka.
KOR: Wickens 414 9/1962; **EQU:** Kielland-Lund 110 11/1983 (n.v.).

Taverniera aegyptiaca Boiss.

F.P.S. 2: 234 (1952); F.E.E. 3: 228 (1989).
Subshrub. Saline sandy areas.
Distr: RS; Eritrea; Egypt, Saudi Arabia.
RS: Bent s.n. 1896.

Taverniera lappacea (Forssk.) DC.

F.P.S. 2: 234 (1952); F.E.E. 3: 228 (1989).
Shrub. Coastal plains, semi-desert.
Distr: RS; Eritrea, Somalia; Egypt, Arabia, Pakistan.
RS: Carter 2026 11/1987.

Tephrosia bracteolata Guill. & Perr. var. *bracteolata*

F.P.S. 2: 240 (1952); F.T.E.A. Leg.-Pap.: 199 (1971); F.J.M.: 117 (1976); F.E.E. 3: 116 (1989); Biol. Skr. 51: 229 (1998).
Annual herb. Seasonally swampy grassland & woodland.
Distr: DAR, KOR, CS, ES, EQU; Cape Verde Is., Senegal to Ethiopia, S to Tanzania.
ES: Beshir 42 8/1951; **EQU:** Myers 7545 8/1937.

Tephrosia cordatistipula J.B.Gillett

F.T.E.A. Leg.-Pap.: 209 (1971).
?Annual herb. Granite outcrops.

Distr: EQU; Uganda.
IUCN: RD
EQU: Jackson 3848 9/1957.

Tephrosia elegans Schumach.

F.P.S. 2: 236 (1952); F.T.E.A. Leg.-Pap.: 169 (1971).
Annual or short-lived perennial herb. Grassland.
Distr: BAG; Senegal to Uganda, S to Angola & Tanzania.
BAG: Schweinfurth 2584 10/1869.

Tephrosia emeroides A.Rich.

F.P.S. 2: 177 (1952); F.T.E.A. Leg.-Pap.: 177 (1971); F.E.E. 3: 113 (1989).
Perennial herb. Upland woodland, bushland & grassland.
Distr: ?CS; Chad, Ethiopia, Somalia, Uganda, Kenya.
Note: Andrews records this species from Central Sudan but we have seen no material with which to verify this.

Tephrosia fulvinervis Hochst. ex A.Rich.

F.P.S. 2: 239 (1952); F.E.E. 3: 113 (1989).
Annual herb. Grassland & woodland.
Distr: ?KOR; Eritrea, Ethiopia.
IUCN: RD
Note: Andrews recorded this species from Kordofan but we have seen no material with which to verify this; in F.E.E. it is treated as an Eritrean/Ethiopian endemic.

Tephrosia gracilipes Guill. & Perr.

F.P.S. 2: 238 (1952); F.E.E. 3: 114 (1989).
Annual herb. Wooded grassland on laterite.
Distr: KOR, ES; Senegal to Niger, Eritrea, Ethiopia; Arabia.
ES: Schweinfurth 1885 9/1865.

Tephrosia humilis Guill. & Perr.

F.E.E. 3: 118 (1989).
Syn: *Tephrosia platycarpa* sensu Andrews, non Guill. & Perr. – F.P.S. 2: 238 (1952).
Annual herb. Grassland.
Distr: BAG; Senegal to Ethiopia.
BAG: Schweinfurth 2289 6/1869.

Tephrosia interrupta Hochst. & Steud. ex Engl. subsp. *interrupta*

F.T.E.A. Leg.-Pap.: 206 (1971); F.E.E. 3: 118 (1989); Biol. Skr. 51: 229 (1998).
Syn: *Tephrosia dichrocarpa* sensu Andrews, non Steud. ex A.Rich. – F.P.S. 2: 239 (1952).
Perennial herb or shrub. Montane grassland & bushland, often on rocky outcrops.
Distr: EQU; Ethiopia, Uganda, Kenya, Tanzania.
EQU: Friis & Vollesen 813 12/1980.

Tephrosia interrupta Hochst. & Steud. ex Engl. subsp. *mildbraedii* (Harms) J.B.Gillett

F.T.E.A. Leg.-Pap.: 206 (1971); Biol. Skr. 51: 230 (1998).
Syn: *Tephrosia atroviolacea* Baker f. – F.P.S. 2: 240 (1952).
Perennial herb or shrub. Montane grassland, bushland & forest margins.
Distr: EQU; D.R. Congo, Uganda, Tanzania, Malawi.
EQU: Chipp 96 2/1929.

Tephrosia lebrunii Cronquist

F.T.E.A. Leg.-Pap.: 209 (1971); Biol. Skr. 51: 230 (1998).
Perennial herb. Montane grassland.

Distr: EQU; D.R. Congo, Uganda.
IUCN: RD
EQU: <u>Andrews 1914</u> 6/1939.

Tephrosia linearis (Willd.) Pers.

F.P.S. 2: 238 (1952); F.T.E.A. Leg.-Pap.: 179 (1971); F.E.E. 3: 113 (1989); Biol. Skr. 51: 230 (1998).
Annual herb. Grassland, rocky outcrops, seasonal swampy areas.
Distr: KOR, BAG, UN, EQU; Senegal to Ethiopia, S to South Africa; Madagascar.
KOR: <u>Jackson 2420</u> 10/1952; **EQU:** <u>Sillitoe 18</u> 1919.

Tephrosia lupinifolia DC.

F.P.S. 2: 236 (1952); F.T.E.A. Leg.-Pap.: 167 (1971); F.E.E. 3: 167 (1989).
Annual or perennial herb. Open sandy areas, grassland.
Distr: DAR, KOR; Senegal to Chad, S to Namibia & South Africa.
DAR: <u>Harrison 165</u> 9/1947.

Tephrosia nana Kotschy ex Schweinf.

F.P.S. 2: 238 (1952); F.T.E.A. Leg.-Pap.: 208 (1971); F.E.E. 3: 118 (1989).
Syn: *Tephrosia barbigera* Welw. ex Baker – F.P.S. 2: 240 (1952).
Annual or short-lived perennial herb. Grassland, swampy areas, disturbed ground.
Distr: KOR, CS, ES, UN, EQU; Senegal to Ethiopia, S to Angola & Mozambique.
KOR: <u>Wickens 780</u> 11/1962; **EQU:** <u>Myers 9167</u> 8/1938.

Tephrosia nubica (Boiss.) Baker subsp. nubica

F.P.S. 2: 241 (1952); F.T.E.A. Leg.-Pap.: 192 (1971); F.E.E. 3: 116 (1989).
Perennial herb. Dry bushland, damp sandy areas.
Distr: RS, DAR, KOR; Mauritania to Eritrea, S to Uganda & Kenya; Egypt.
DAR: <u>Aglen 16</u> 8/1944.

Tephrosia pentaphylla (Roxb.) G.Don

F.P.S. 2: 238 (1952); F.T.E.A. Leg.-Pap.: 169 (1971); F.E.E. 3: 112 (1989).
Annual or short-lived perennial herb. Grassland & woodland on black clays or stony hillslopes.
Distr: KOR, ES; Eritrea & Somalia to Mozambique; Arabia, India.
ES: <u>Jackson 2470</u> 11/1952.
Note: Andrews' record of *T. pseudolongipes* Baker f. (= *T. longipes* Meisn.) is clearly incorrect since this is a southern African species but we are unsure as to which species this record truly refers.

Tephrosia pumila (Lam.) Pers. var. pumila

F.T.E.A. Leg.-Pap.: 184 (1971); F.E.E. 3: 115 (1989); Biol. Skr. 51: 230 (1998).
Syn: *Tephrosia uniflora* sensu Andrews pro parte, non Pers. – F.P.S. 2: 236 (1952).
Annual or short-lived perennial herb. Grassland, bushland, weed of cultivation.
Distr: RS, KOR, UN, EQU; Ghana, Eritrea to Angola & South Africa; Madagascar.
RS: <u>Yeates 22</u> 3/1988; **EQU:** <u>Friis & Vollesen 560</u> 11/1980.

Tephrosia purpurea (L.) Pers. subsp. apollinea (Delile) Hosni & El Karemy

F.Egypt 1: 318 (1999).
Syn: *Tephrosia apollinea* (Delile) Link – F.P.S. 2: 239 (1952); F.E.E. 3: 116 (1989).
Annual or short-lived perennial herb. Wadis, riverbanks.
Distr: NS, RS, KOR, CS, ES; Eritrea, Somalia; Egypt, Arabia, Socotra.
RS: <u>Carter 2002</u> 11/1987.

Tephrosia purpurea (L.) Pers. subsp. leptostachya (DC.) Brummitt var. leptostachya

F.P.S. 2: 239 (species only) (1952); F.T.E.A. Leg.-Pap.: 186 (1971); F.E.E. 3: 115 (1989).
Annual or short-lived perennial herb. Grassland, disturbed ground.
Distr: DAR, KOR, CS; Senegal to Nigeria, Eritrea to Angola & South Africa; N Africa, Arabia.
KOR: <u>Wickens 586</u> 10/1962.

Tephrosia purpurea (L.) Pers. subsp. leptostachya (DC.) Brummitt var. pubescens Baker

F.T.E.A. Leg.-Pap.: 188 (1971); F.J.M.: 117 (1976); F.E.E. 3: 115 (1989).
Annual or short-lived perennial herb. Grassland, disturbed ground.
Distr: DAR, KOR, UN; Mali to Eritrea, S to Namibia & South Africa.
DAR: <u>Wickens 2265</u> 8/1964; **UN:** <u>Simpson 7088</u> 6/1929.

Tephrosia subtriflora Hochst. ex Baker

F.P.S. 2: 236 (1952); F.T.E.A. Leg.-Pap.: 170 (1971); F.E.E. 3: 112 (1989).
Annual or short-lived perennial herb. Woodland & bushland.
Distr: RS, DAR, KOR, CS; Cape Verde Is., Niger to Eritrea & Somalia, S to Tanzania, Angola; Madagascar, Arabia, Pakistan to Burma.
CS: <u>Andrews 75</u> 6/1935.

Tephrosia uniflora Pers. subsp. uniflora

F.P.S. 2: 236 (1952); F.T.E.A. Leg.-Pap.: 171 (1971); F.E.E. 3: 112 (1989).
Syn: *Tephrosia vicioides* A.Rich. – F.P.S. 2: 238 (1952).
Annual or short-lived perennial herb. Dry grassland & bushland, often in rocky areas, or on disturbed ground.
Distr: RS, DAR, KOR, CS, ES; Cape Verde Is., Mauritania to Somalia, S to Namibia & South Africa; N Africa.
KOR: <u>Jackson 4032</u> 9/1959.

Tephrosia villosa (L.) Pers. subsp. ehrenbergiana (Schweinf.) Brummitt

F.T.E.A. Leg.-Pap.: 190 (1971); F.E.E. 3: 116 (1989).
Syn: *Tephrosia ehrenbergiana* Schweinf. – F.P.S. 2: 240 (1952).
Annual or short-lived perennial herb. Open dry bushland & woodland.
Distr: ?CS, EQU; Eritrea to Namibia & South Africa; Madagascar, Egypt.
EQU: <u>Broun 1664</u> 5/1909.
Note: Andrews records this species from Central & Southern Sudan but we have only seen the single specimen from Equatoria.

Tephrosia vogelii Hook.f.

F.P.S. 2: 240 (1952); F.T.E.A. Leg.-Pap.: 210 (1971); F.E.E. 3: 119 (1989); Biol. Skr. 51: 231 (1998).
Perennial herb or shrub. Grassland & bushland, often in disturbed areas, also cultivated.
Distr: EQU; Senegal to Ethiopia, S to Angola & South Africa; Madagascar, India to Indonesia.
EQU: Myers 6578 5/1937.

Teramnus labialis (L.f.) Spreng. subsp. arabicus Verdc.

F.P.S. 2: 241 (species only) (1952); F.T.E.A. Leg.-Pap.: 537 (1971); F.E.E. 3: 160 (1989).
Perennial herb. Grassland, woodland & bushland.
Distr: KOR, CS, BAG, UN; Mali, Príncipe, Eritrea to Zimbabwe & Mozambique; Madagascar, Comoros, Mascarenes, Arabia, Caribbean, Guyana.
CS: Andrews 101 11/1935; **UN:** J.M. Lock 81/153 8/1981.

Teramnus micans (Welw. ex Baker) Baker f. var. fagifolius Hauman

F.T.E.A. Leg.-Pap.: 540 (1971); Biol. Skr. 51: 231 (species only) (1998).
Perennial climbing herb. Grassland & wooded grassland.
Distr: EQU; D.R. Congo, Uganda.
EQU: Myers 7803 9/1937.

Teramnus repens (Taub.) Baker f. subsp. repens

F.T.E.A. Leg.-Pap.: 538 (1971).
Perennial herb. Grassland.
Distr: UN; Uganda & Kenya, S to Angola & Zimbabwe.
UN: J.M. Lock 81/160 8/1981.

Teramnus repens (Taub.) Baker f. subsp. gracilis (Chiov.) Verdc.

F.T.E.A. Leg.-Pap.: 538 (1971); F.E.E. 3: 160 (1989).
Perennial herb. Bushland.
Distr: UN; Eritrea, Somalia, Kenya, Tanzania; Socotra, Arabia, ?India & Pakistan.
UN: Broun 1446 11/1909.

Teramnus uncinatus (L.) Sw. subsp. axilliflorus (Kotschy) Verdc.

F.E.E. 3: 160 (1989); Biol. Skr. 51: 231 (1998).
Syn: Teramnus axilliflorus (Kotschy) Baker f. – F.P.S. 2: 241 (1952).
Perennial climbing or trailing herb. Grassland & bushland including seasonally flooded areas.
Distr: UN; Senegal to Uganda, S to Angola & Zambia.
UN: Binder 1 1840.

Trifolium arvense L.

F.P.S. 2: 242 (1952); F.E.E. 3: 242 (1989).
Annual herb. Montane grassland.
Distr: RS; Eritrea, Ethiopia; Eurasia, N Africa.
RS: Jackson 2880 4/1953.

Trifolium campestre Schreb.

F.P.S. 2: 242 (1952); F.E.E. 3: 242 (1989).
Annual herb. Grassland, rocky hillslopes.
Distr: RS; Eritrea, Ethiopia, South Africa; Europe, N Africa, SW Asia.
RS: Jackson 2728 3/1953.

Trifolium elgonense J.B.Gillett

F.T.E.A. Leg.-Pap.: 1035 (1971); F.E.E. 3: 241 (1989); Biol. Skr. 51: 231 (1998).
Annual herb. Montane grassland, bushland & forest clearings.
Distr: EQU; Ethiopia, Uganda, Kenya.
EQU: Friis & Vollesen 848 12/1980.

Trifolium fragiferum L.

F.P.S. 2: 242 (1952); F.E.E. 3: 242 (1989).
Perennial herb. Streamsides, moist ground.
Distr: ?RS; Eritrea, Ethiopia; Europe, N Africa, W Asia, Arabia.
Note: Andrews records this species from Erkowit in the Red Sea Hills but we have seen no material with which to verify this; it is possible that it is based on a misidentification of T. tomentosum.

Trifolium multinerve (Hochst.) A.Rich.

F.P.S. 2: 242 (1952); F.T.E.A. Leg.-Pap.: 1034 (1971); F.E.E. 3: 241 (1989); Biol. Skr. 51: 231 (1998).
Annual herb. Montane grassland, bushland.
Distr: EQU; D.R. Congo, Eritrea, Ethiopia, Uganda, Kenya.
EQU: Friis & Vollesen 379 11/1980.

Trifolium polystachyum Fresen. var. polystachyum

F.P.S. 2: 242 (1952); F.T.E.A. Leg.-Pap.: 1024 (1971); F.E.E. 3: 236 (1989); Biol. Skr. 51: 232 (1998).
Annual or short-lived perennial herb. Swampy grassland, forest margins, montane bushland.
Distr: EQU; Eritrea, Ethiopia, Kenya.
EQU: Myers 13487 9/1940.

Trifolium rueppellianum Fresen.

F.P.S. 2: 243 (1952); F.T.E.A. Leg.-Pap.: 1030 (1971); F.E.E. 3: 238 (1989); Biol. Skr. 51: 232 (1998).
Annual herb. Montane grassland, bushland & forest clearings.
Distr: EQU; Bioko, Cameroon, Ethiopia to D.R. Congo & Tanzania.
EQU: Thomas 1655 12/1935.

Trifolium simense Fresen.

F.P.S. 2: 242 (1952); F.T.E.A. Leg.-Pap.: 1022 (1971); F.E.E. 3: 236 (1989); Biol. Skr. 51: 232 (1998).
Perennial herb. Montane grassland.
Distr: EQU; Bioko, Cameroon, Gabon, Eritrea to Zambia & Malawi.
EQU: Myers 11693 7/1939.

Trifolium tomentosum L.

F.P.S. 2: 242 (1952); F.Egypt 1: 281 (1999).
Annual herb. Sandy plains.
Distr: RS; Mediterranean, N Africa, Turkey, Iraq.
RS: Aylmer 195 3/1932.

Trifolium usambarense Taub.

F.T.E.A. Leg.-Pap.: 1024 (1971); F.E.E. 3: 236 (1989); Biol. Skr. 51: 232 (1998).
Annual or short-lived perennial herb. Forest clearings, montane grassland, often along streams.
Distr: EQU; Cameroon, Ethiopia to Zambia & Malawi.
EQU: Friis & Vollesen 92 11/1980.

Trigonella foenum-graecum L.

Bot. Exp. Sud.: 77 (1970); F.E.E. 3: 246 (1989).
Annual herb. Cultivated and occasionally naturalised.
Distr: DAR, KOR; widely cultivated; native to S Europe
and Asia.
KOR: Kotschy 646 1842.

Trigonella glabra Thunb.

Trop. Afr. Fl. Pl. 4: 266 (2008).
Syn: Trigonella hamosa L. – F.P.S. 2: 243 (1952); Bot.
Exp. Sud.: 77 (1970).
Annual herb. Sandy soils, weed of disturbed & cultivated
ground.
Distr: NS, CS; Namibia, South Africa; E Mediterranean
to India.
CS: Aylmer 273 1/1933.

Trigonella laciniata L.

F.P.S. 2: 243 (1952); F.Egypt 1: 265 (1999).
Annual herb. Sandy soils, weed of disturbed & cultivated
ground.
Distr: NS, CS; N Africa, Middle East, Saudi Arabia.
CS: Pettet 135 2/1963.

Trigonella occulta Ser.

F.P.S. 2: 243 (1952); F.Egypt 1: 264 (1999).
Annual herb. Sandy soils.
Distr: NS, CS; Egypt.
IUCN: RD
CS: Jackson 4099 12/1960.
Note the identity of the Sudanese material is
questionable; it may be a form of *T. laciniata* (see Trop.
Afr. Fl. Pl. 4: 268).

Uraria picta (Jacq.) DC.

F.P.S. 2: 243 (1952); F.T.E.A. Leg.-Pap.: 479 (1971); F.E.E.
3: 152 (1989).
Perennial herb or subshrub. Grassland & wooded
grassland.
Distr: BAG, UN, EQU; Senegal to Ethiopia, S to Angola,
Zimbabwe & Mozambique; Asia, Australia.
EQU: Myers 6660 5/1937.

Vatovaea pseudolablab (Harms) J.B.Gillett

F.T.E.A. Leg.-Pap.: 609 (1971); F.E.E. 3: 169 (1989).
Woody climber. Dry grassland.
Distr: EQU; Ethiopia, Somalia, Uganda, Kenya, Tanzania;
Arabia.
EQU: Padwa 245 6/1953.

Vicia faba L.

F.T.E.A. Leg.-Pap.: 1067 (1971); F.E.E. 3: 249 (1989).
Annual herb. Cultivated & occasionally naturalised on
riverbanks.
Distr: NS; originating from the Mediterranean or SW
Asia, now very widely cultivated.
NS: Ahti 15704 10/1962.
Note: the cultivated 'broad bean'; Ahti et al (1973)
noted it as "certainly cultivated in the area [of the Nile
in northern Sudan] but in the study season only clearly
naturalised plants seen along the muddy waterline of the
Nile, especially on the islands".

Vicia hirsuta (L.) Gray

F.T.E.A. Leg.-Pap.: 1072 (1971); F.E.E. 3: 249 (1989); Biol.
Skr. 51: 233 (1998).
Annual climbing or trailing herb. Montane grassland,
bushland & forest margins.
Distr: EQU; Eritrea to Tanzania; Europe, N Africa, Asia.
EQU: Friis & Vollesen 846 12/1980.

Vicia sativa L. var. angustifolia L.

F.T.E.A. Leg.-Pap.: 1069 (1971); F.E.E. 3: 249 (1989); Biol.
Skr. 51: 233 (1998).
Annual herb. Montane grassland, bushland & forest
margins.
Distr: EQU; Eritrea to Zimbabwe; Europe, N Africa &
Arabia to India.
EQU: Friis & Vollesen 860 12/1980.

Vigna comosa Baker subsp. comosa

F.T.E.A. Leg.-Pap.: 630 (1971); Biol. Skr. 51: 233 (1989).
Perennial climbing or trailing herb. Montane grassland.
Distr: EQU; Sierra Leone to Kenya, S to Angola &
Mozambique.
EQU: Friis & Vollesen 268 11/1980.

Vigna frutescens A.Rich. subsp. frutescens var. frutescens

F.T.E.A. Leg.-Pap.: 648 (1971); F.E.E. 3: 176 (1989); Biol.
Skr. 51: 234 (1998).
Syn: Vigna fragrans Baker f. – F.P.S. 2: 246 pro parte
(1952).
Syn: Vigna sudanica Baker f. – F.P.S. 2: 248 (1952).
Perennial climbing or trailing herb. Grassland & bushland.
Distr: EQU; Nigeria to Ethiopia, S to Angola & South
Africa.
EQU: Friis & Vollesen 1308 3/1982.

Vigna frutescens A.Rich. subsp. kotschyi (Schweinf.) Verdc.

F.T.E.A. Leg.-Pap.: 647 (1971); F.J.M.: 117 (1976); F.E.E.
3: 176 (1989); Biol. Skr. 51: 234 (1998).
Syn: Vigna kotschyi Schweinf. – F.P.S. 2: 247 (1952).
Syn: Vigna fragrans sensu Andrews pro parte, non
Baker f. – F.P.S. 2: 246 (1952).
Perennial climbing or trailing herb. Grassland.
Distr: DAR, UN, EQU; Ethiopia.
DAR: Wickens 1824 7/1964; **EQU:** Friis & Vollesen 1205
3/1982.

Vigna gracilis (Guill & Perr.) Hook.f.

F.W.T.A. 1(2): 569 (1958); Trop. Afr. Fl. Pl. 4: 274 (2008).
Syn: Vigna occidentalis Baker – F.P.S. 2: 246 (1952).
Annual twining herb. Wooded grassland.
Distr: EQU; Senegal to C.A.R.
EQU: Dandy 542 2/1934.

Vigna heterophylla A.Rich.

Biol. Skr. 51: 234 (1998); Trop. Afr. Fl. Pl. 4: 274 (2008).
Syn: Vigna ambacensis Welw. ex Baker – F.P.S. 2:
245 (1952); F.T.E.A. Leg.-Pap.: 632 (1971); F.E.E. 3: 173
(1989); Biol. Skr. 51: 233 (1998).
Annual or perennial climbing herb. Montane grassland
& bushland.

Distr: KOR, CS, BAG, UN, EQU; Senegal to Ethiopia, S to Angola & Malawi; Egypt, Arabia.
KOR: <u>Broun 1362</u> 12/1907; **EQU:** <u>Friis & Vollesen 561</u> 11/1980.

Vigna kirkii (Baker) J.B.Gillett
F.T.E.A. Leg.-Pap.: 637 (1971).
Perennial climbing herb. Swampy grassland.
Distr: UN, EQU; Senegal, D.R. Congo, Kenya, Tanzania, Zambia, Mozambique.
EQU: <u>Andrews 1010</u> 5/1939.

Vigna luteola (Jacq.) Benth.
F.T.E.A. Leg.-Pap.: 660 (1971); F.E.E. 3: 172 (1989); Biol. Skr. 51: 234 (1998).
Syn: *Vigna nilotica* (Delile) Hook.f. – F.P.S. 2: 245 (1952); Bot. Exp. Sud.: 77 (1970).
Perennial herb. Swampy grassland, lake shores.
Distr: ?DAR, BAG, UN; pantropical.
BAG: <u>Simpson 7118</u> 6/1929.
Note: Drar (1970) records three sterile specimens from Jebel Marra. These are probably misidentifications.

Vigna membranacea A.Rich. subsp. *membranacea*
F.P.S. 2: 247 (1952); F.T.E.A. Leg.-Pap.: 639 (1971); F.E.E. 3: 174 (1989); Biol. Skr. 51: 235 (1998).
Annual or perennial climbing herb. Grassland, bushland.
Distr: RS, EQU; Eritrea, Ethiopia, Uganda, Kenya, Tanzania.
RS: <u>Aylmer 209</u> 3/1932; **EQU:** <u>Friis & Vollesen 155</u> 11/1980.

Vigna multinervis Hutch. & Dalziel
F.T.E.A. Leg.-Pap.: 637 (1971); F.E.E. 3: 173 (1989).
Perennial climbing herb. Swampy grassland.
Distr: BAG; Ivory Coast to Ethiopia, S to Angola & Zambia.
BAG: <u>Schweinfurth 2578</u> 10/1869.

Vigna oblongifolia A.Rich. var. *oblongifolia*
F.T.E.A. Leg.-Pap.: 629 (1971); F.J.M.: 117 (1976); F.E.E. 3: 172 (1989).
Syn: *Vigna lancifolia* A.Rich. – F.P.S. 2: 245 (1952).
Annual herb. Damp grassland.
Distr: DAR, KOR, UN, EQU; Nigeria, Eritrea to South Africa.
DAR: <u>Wickens 2260</u> 8/1964; **UN:** <u>J.M. Lock 81/154</u> 8/1981.

Vigna racemosa (G.Don) Hutch. & Dalziel
F.T.E.A. Leg.-Pap.: 633 (1971); Biol. Skr. 51: 235 (1998).
Perennial climbing herb. Grassland & woodland.
Distr: EQU; Senegal to Uganda, S to Angola & Zambia.
EQU: <u>Friis & Vollesen 620</u> 12/1980.

Vigna radiata (L.) R.Wilczek var. *radiata*
F.T.E.A. Leg.-Pap.: 655 (1971); F.E.E. 3: 177 (1989).
Syn: *Phaseolus trinervius* Wight & Arn. – F.P.S. 2: 225 (1952).
Annual herb. Naturalised in grassland & bushland, abandoned cultivations.
Distr: CS, UN, EQU; cultivated throughout the tropics; native to India.
CS: <u>Lea 15</u> 10/1952; **EQU:** <u>Myers 13904</u> 7/1941.

Vigna reticulata Hook.f.
F.P.S. 2: 247 (1952); F.T.E.A. Leg.-Pap.: 650 (1971); F.E.E. 3: 176 (1989); Biol. Skr. 51: 235 (1998).
Annual or perennial trailing or climbing herb. Grassland & bushland, often in swampy areas.
Distr: BAG, UN, EQU; Senegal to Kenya, S to Angola, Zimbabwe & Mozambique.
UN: <u>Broun 931</u> 5/1906.

Vigna schimperi Baker
F.P.S. 2: 246 (1952); F.T.E.A. Leg.-Pap.: 628 (1971); F.E.E. 3: 172 (1989); Biol. Skr. 51: 235 (1998).
Perennial herb. Montane grassland, bushland & forest margins.
Distr: ?RS, EQU; D.R. Congo, Ethiopia, Uganda, Kenya, Tanzania.
EQU: <u>Friis & Vollesen 260</u> 11/1980.
Note: Andrews records this species from Erkowit and the Imatong Mts, but we have only seen specimens from the latter.

Vigna subterranea (L.) Verdc. var. *subterranea*
F.E.E. 3: 173 (1989).
Syn: *Voandzeia subterranea* (L.) DC. – F.P.S. 2: 248 (1952); F.T.E.A. Leg.-Pap.: 668 (1971).
Annual herb. Cultivated.
Distr: BAG; cultivated throughout the tropics; native to W Africa.
BAG: <u>Schweinfurth 2424</u> 9/1869.
Note: Andrews records this species from Central & Southern Sudan but we have only seen the single collection cited. Widely cultivated for its edible seed.

Vigna unguiculata (L.) Walp. subsp. *unguiculata*
F.P.S. 2: 246 pro parte (1952); F.T.E.A. Leg.-Pap.: 642 (1971); F.E.E. 3: 174 (1989).
Syn: *Vigna unguiculata* (L.) Walp. subsp. *cylindrica* (L.) Verdc. – Biol. Skr. 51: 236 (1998).
Annual or perennial herb. Cultivated.
Distr: BAG, UN, EQU; widely cultivated.
UN: <u>J.M. Lock 81/120</u> 8/1981.
Note: Andrews (1952) and Drar (1970) record *V. coerulea* Baker from Equatoria and Central Sudan respectively; this is a synonym of *V. unguiculata* subsp. *tenuis* (E.Mey.) Mithen which is a southern African subspecies; the true identity of the Sudanese plant must surely fall within one of the other subspecies recorded here. Drar's (1970: 75) record of *Phaseolus trilobus* Ait. is possibly also referable here. Quézel (1969: 116) records *V. unguiculata* from Darfur but it is unclear as to which subspecies this refers.

Vigna unguiculata (L.) Walp. subsp. *dekindtiana* (Harms) Verdc.
F.P.S. 2: 246 (species only, pro parte) (1952); F.T.E.A. Leg.-Pap.: 644 (1971); F.E.E. 3: 174 (1989).
Annual or perennial trailing or climbing herb. Grassland, bushland & forest margins, also cultivated.
Distr: ES, UN, EQU; pantropical, native to tropical Africa.
ES: <u>Schweinfurth 1895</u> 9/1865; **EQU:** <u>Friis & Vollesen 624</u> 12/1980.

Vigna unguiculata (L.) Walp. subsp. pubescens (R.Wilczek) Pasquet

Trop. Afr. Fl. Pl. 4: 280 (2008).
Syn: Vigna pubescens R.Wilczek – F.T.E.A. Leg.-Pap.: 646 (1971).
Perennial herb. Grassland & bushland.
Distr: EQU; D.R. Congo, Uganda, Kenya, Tanzania, Mozambique.
EQU: Myers 13901 7/1941.

Vigna vexillata (L.) A.Rich.

F.P.S. 2: 246 (1952); F.T.E.A. Leg.-Pap.: 652 (1971); F.J.M.: 118 (1976); F.E.E. 3: 176 (1989); Biol. Skr. 51: 236 (1998).
Syn: Vigna angustifolia (Schumach. & Thonn.) Hook.f. – F.P.S. 2: 247 (1952).
Perennial climbing or trailing herb. Grassland, bushland, forest margins, disturbed areas.
Distr: DAR, KOR, CS, ES, BAG, UN, EQU; pantropical.
DAR: Wickens 2036 7/1964; **EQU:** Friis & Vollesen 295 11/1980.

Wajira grahamiana (Wight & Arn.) Thulin & Lavin

Trop. Afr. Fl. Pl. 4: 283 (2008).
Syn: Vigna macrorhyncha (Harms) Milne-Redh. – F.P.S. 2: 245 (1952); F.T.E.A. Leg.-Pap.: 658 (1971); F.J.M.: 117 (1976); F.E.E. 3: 177 (1989); Biol. Skr. 51: 235 (1998).
Perennial climbing herb. Grassland, bushland & woodland.
Distr: DAR, BAG, EQU; Nigeria to Somalia, S to Zimbabwe; Arabia, India, Sri Lanka.
DAR: Wickens 2865 3/1965; **BAG:** Schweinfurth 2391 9/1869.

Zornia durumuensis De Wild.

F.P.S. 2: 248 (1952); F.W.T.A. 1(2): 576 (1958).
Perennial herb. Grassland, disturbed ground.
Distr: EQU; Mali to D.R. Congo.
EQU: Myers 8464 2/1938.

Zornia glochidiata Rchb. ex DC.

F.P.S. 2: 248 (1952); F.T.E.A. Leg.-Pap.: 444 (1971); F.J.M.: 118 (1976); F.E.E. 3: 148 (1989); Biol. Skr. 51: 236 (1998).
Annual herb. Grassland & woodland.
Distr: DAR, KOR, UN, EQU; Cape Verde Is., Senegal to Eritrea, S to Namibia & South Africa; Madagascar.
DAR: Wickens 2540 9/1964; **EQU:** Myers 6775 5/1937.

Zornia pratensis Milne-Redh. subsp. pratensis var. pratensis

F.T.E.A. Leg.-Pap.: 445 (1971); F.E.E. 3: 148 (1989); Biol. Skr. 51: 237 (1998).
Perennial herb. Wooded grassland & woodland.
Distr: UN, EQU; Ethiopia to Angola & Zimbabwe.
UN: Jackson 4117 2/1961.

Leguminosae: Caesalpinioideae (Fabaceae: Caesalpinioideae)

H. Pickering & I. Darbyshire

Afzelia africana Sm. ex Pers.

F.P.S. 2: 125 (1952); F.T.E.A. Leg.-Caes.: 125 (1967); T.S.S.: 191 (1990); Biol. Skr. 51: 179 (1998).
Tree. Wooded grassland.
Distr: BAG, EQU; Senegal to Uganda.
IUCN: VU
EQU: Friis & Vollesen 1086 3/1982.
Note: Drar (1970) records an *Afzelia* sp. based on sterile material from Sinkat in the Red Sea Hills – we have not seen the specimen, which will surely prove not to be an *Afzelia*.

Bauhinia rufescens Lam.

F.P.S. 2: 111 (1952); F.W.T.A. 1(2): 444 (1958); Bot. Exp. Sud.: 68 (1970); F.J.M.: 105 (1976); T.S.S.: 191 (1990).
Small tree. Wooded grassland.
Distr: DAR, KOR, SOUTHSUD; Mauritania & Senegal to Chad & C.A.R.; Algeria.
DAR: Wickens 1380 4/1964.
Note: El Amin in T.S.S. records this species from "S. Kordofan, Darfur and S. Sudan" but does not say from which region in the South and we have not seen any material with which to verify this.

Bauhinia tomentosa L.

F.T.E.A. Leg.-Caes.: 209 (1967); F.E.E. 3: 68 (1989); T.S.S.: 191 (1990).
Shrub or small tree. Wooded grassland.
Distr: ES, ?EQU; Ethiopia to Angola & South Africa.
ES: Sahni & Kamil 635 4/1967.
Note: the record for Equatoria is from El Amin in T.S.S. who recorded it from Juba. We have not seen any material from South Sudan.

Burkea africana Hook.

F.P.S. 2: 113 (1952); F.T.E.A. Leg.-Caes.: 21 (1967); T.S.S.: 193 (1990).
Tree. Wooded grassland.
Distr: DAR, KOR, BAG, UN, EQU; Senegal to Uganda, S to Namibia & South Africa.
DAR: Wickens 3637 6/1977; **EQU:** von Ramm 148 3/1956.

Caesalpinia decapetala (Roth) Alston

F.T.E.A. Leg.-Caes.: 36 (1967); F.E.E. 3: 53 (1989).
Scrambling shrub. Secondary forest.
Distr: EQU; Uganda to Angola & South Africa.
EQU: Myers 9455 9/1938.

Cassia arereh Delile

F.P.S. 2: 116 (1952); F.E.E. 3: 56 (1989); T.S.S.: 197 (1990).
Tree. Wooded grassland.
Distr: NS, KOR, CS, BAG, EQU; Nigeria to Ethiopia.
CS: Aylmer 12 3/1922; **BAG:** Broun 260 (fr) 3/1902.
Note: *C. fistula* L. has been introduced in Sudan as a decorative street tree but has not naturalised.

Cassia mannii Oliv.

F.P.S. 2: 116 (1952); F.T.E.A. Leg.-Caes.: 58 (1967); T.S.S.: 197 (1990); Biol. Skr. 51: 179 (1998).
Tree. Forest.
Distr: EQU; Nigeria to D.R. Congo & Uganda.
EQU: Myers 13126 2/1940.

Cassia sieberiana DC.

F.P.S. 2: 116 (1952); F.T.E.A. Leg.-Caes.: 61 (1967); F.J.M.: 106 (1976); T.S.S.: 203 (1990).

Tree. Wooded grassland.
Distr: NS, DAR, KOR, CS, BAG, UN, EQU; Senegal to Uganda.
KOR: Wickens 3080 6/1969; **BAG:** Myers 6295 2/1937.

Chamaecrista absus (L.) H.S.Irwin & Barneby

F.J.M.: 105 (1976); F.E.E. 3: 64 (1989); Biol. Skr. 51: 179 (1998).
Syn: Cassia absus L. – F.P.S. 2: 118 (1952); F.T.E.A. Leg.-Caes.: 81 (1967); T.S.S.: 195 (1990).
Annual or perennial herb. Grassland.
Distr: DAR, KOR, CS, ES, BAG, UN, EQU; palaeotropical.
KOR: Wickens 382 9/1961; **EQU:** Andrews 1530 5/1939.

Chamaecrista kirkii (Oliv.) Standl. var. guineensis (Steyaert) Lock

F.E.E. 3: 65 (1989); Biol. Skr. 51: 179 (1998).
Syn: Cassia kirkii Oliv. var. **guineensis** Steyaert – F.T.E.A. Leg.-Caes.: 92 (1967).
Annual herb. Wooded grassland.
Distr: EQU; Gabon to Ethiopia, S to Zambia.
EQU: Sillitoe 433 1919.

Chamaecrista mimosoides (L.) Greene

F.J.M.: 105 (1976); F.E.E. 3: 66 (1989); Biol. Skr. 51: 180 (1998).
Syn: Cassia mimosoides L. – F.P.S. 2: 121 (1952); F.T.E.A. Leg.-Caes.: 100 (1967); T.S.S.: 199 (1990).
Annual herb. Grassland.
Distr: DAR, KOR, CS, ES, BAG, UN, EQU; palaeotropical.
DAR: Harrison 161 9/1947; **EQU:** von Ramm 246 5/1956.

Chamaecrista nigricans (Vahl) Greene

F.J.M.: 106 (1976); F.E.E. 3: 65 (1989); Biol. Skr. 51: 180 (1998).
Syn: Cassia nigricans Vahl – F.P.S. 2: 121 (1952); F.T.E.A. Leg.-Caes.: 81 (1967); T.S.S.: 199 (1990).
Annual herb. Bushland.
Distr: DAR, KOR, CS, ES, BAG, UN, EQU; Senegal to Eritrea, S to Angola & Zambia; Arabia, India.
KOR: Harrison 986 9/1950; **EQU:** Sillitoe 175 1919.

Daniellia oliveri (Rolfe) Hutch. & Dalziel

F.P.S. 2: 121 (1952); F.T.E.A. Leg.-Caes.: 132 (1967); T.S.S.: 207 (1990).
Tree. Wooded grassland & forest margins.
Distr: BAG, EQU; Senegal to Uganda.
BAG: Aylmer 27/12 12/1926.

Delonix elata (L.) Gamble

F.P.S. 2: 121 (1952); F.T.E.A. Leg.-Caes.: 23 (1967); T.S.S.: 209 (1990).
Tree. Thickets & bushland.
Distr: RS, ES; Eritrea to Tanzania; Egypt, Arabia to India.
RS: Jackson 3933 4/1959.
Note: *D. regia* (Bojer ex Hook.) Raf. from Madagascar has been introduced in Sudan and South Sudan as a decorative street tree but has not naturalised.

Detarium microcarpum Guill. & Perr.

F.W.T.A. 1(2): 457 (1958).
Syn: Detarium senegalense sensu Andrews pro parte, non J.F.Gmel. – F.P.S. 2: 123 (1952).
Tree. Dry forest, wooded grassland.

Distr: EQU; Senegal to C.A.R.
EQU: Andrews 1007 5/1939.

Detarium senegalense J.F.Gmel.

F.P.S. 2: 123, pro parte (1952); F.W.T.A. 1(2): 457 (1958); Bot. Exp. Sud.: 71 (1970); T.S.S.: 209 (1990).
Tree. Forest.
Distr: ?KOR, BAG, EQU; Senegal to D.R. Congo.
EQU: Myers 11381 5/1939.
Note: Drar (1970) records either this species or *D. microcarpum* (which he treats in synonymy) from the Nuba Mts in Kordofan, but no specimen was collected.

Dialium excelsum Steyaert

F.T.E.A. Leg.-Caes.: 104 (1967); Biol. Skr. 51: 180 (1998).
Syn: Dialium sp. – F.P.S. 2: 123 (1952).
Large tree. Forest.
Distr: ?EQU; Uganda, D.R. Congo.
IUCN: RD
Note: the occurrence of this species in South Sudan requires confirmation; Jackson recorded it from Lotti Forest and Andrews recorded a *Dialium* sp. from Azza Forest – El Amin in T.S.S.: 211 records this as *Dialium* nr. *excelsum*.

Erythrophleum africanum (Welw. ex Benth.) Harms

F.P.S. 2: 125 (1952); F.T.E.A. Leg.-Caes.: 20 (1967); T.S.S.: 211 (1990).
Tree. Woodland.
Distr: BAG, EQU; Senegal to Tanzania, S to Namibia & Zimbabwe.
EQU: Myers 7228 7/1937.

Erythrophleum suaveolens (Guill. & Perr.) Brenan

F.T.E.A. Leg.-Caes.: 18 (1967); T.S.S.: 211 (1990); Biol. Skr. 51: 181 (1998).
Syn: Erythrophleum guineense G.Don – F.P.S. 2: 123 (1952).
Tree. Lowland & riverine forest.
Distr: BAG, UN, EQU; Senegal to Uganda, S to Mozambique.
EQU: Myers 10553 2/1939.

Haematoxylum campechianum L.

F.T.E.A. Leg.-Caes.: 16 (1967); Bot. Exp. Sud.: 72 (1970); T.S.S.: 213 (1990).
Small tree. Cultivated, occasionally naturalised.
Distr: RS, BAG/UN; native to tropical America & the Caribbean, cultivated in tropical Africa.
RS: Sahni & Kamil 676 4/1967.
Note: Drar (1970) records this species as an escape in the Sudd region.

Isoberlinia angolensis (Welw. ex Benth.) Hoyle & Brenan

F.P.S. 2: 125 (1952); F.T.E.A. Leg.-Caes.: 139 (1967); T.S.S.: 213 (1990).
Tree. Woodland.
Distr: EQU; Cameroon, D.R. Congo, Tanzania, Angola, Malawi, Zambia.
EQU: Andrews 1195 5/1939.

Isoberlinia doka Craib & Stapf

F.P.S. 2: 125 (1952); F.T.E.A. Leg.-Caes.: 142 (1967);
T.S.S.: 215 (1990).
Tree. Woodland.
Distr: ?DAR, BAG, EQU; Guinea to Uganda.
EQU: Myers 10227 7/1938.
Note: the record for Darfur is from El Amin in T.S.S.; we
have seen no material with which to verify this.

Isoberlinia tomentosa (Harms) Craib & Stapf

F.P.S. 2: 125 (1952); F.T.E.A. Leg.-Caes.: 142 (1967);
T.S.S.: 215 (1990).
Tree. Woodland.
Distr: BAG, EQU; Guinea to C.A.R., S to Zambia &
Tanzania.
EQU: Myers 8486 2/1938.

Parkinsonia aculeata L.

F.P.S. 2: 127 (1952); F.T.E.A. Leg.-Caes.: 43 (1967); Bot.
Exp. Sud.: 74 (1970); F.E.E. 3: 56 (1989); T.S.S.: 215 (1990).
Small tree. Cultivated & naturalised.
Distr: NS, RS, DAR, KOR, CS; native to tropical America;
cultivated in tropical Africa.
KOR: Wickens 721 10/1962.

Peltophorum pterocarpum (DC.) K.Heyne

F.T.E.A. Leg.-Caes.: 17 (1967); T.S.S.: 217 (1990).
Tree. Cultivated.
Distr: UN; native to tropical Asia & Australia; cultivated
elsewhere in the tropics.
UN: Smith 6 5/1933.

Piliostigma reticulatum (DC.) Hochst.

F.P.S. 2: 127 (1952); F.W.T.A. 1(2): 444 (1958); F.J.M.: 106
(1976); T.S.S.: 217 (1990).
Syn: *Bauhinia reticulata* DC. – F. Darfur Nord-Occ. & J.
Gourgeil: 113 (1969).
Tree. Grassland.
Distr: DAR, KOR; Senegal to Chad & C.A.R.
DAR: Wickens 1615 5/1964.

Piliostigma thonningii (Schumach.) Milne-Redh.

F.P.S. 2: 127 (1952); F.T.E.A. Leg.-Caes.: 206 (1967);
F.J.M.: 106 (1976); F.E.E. 3: 68 (1989); T.S.S.: 217 (1990);
Biol. Skr. 51: 181 (1998).
Small tree. Wooded grassland.
Distr: DAR, KOR, CS, ES, BAG, UN, EQU; Senegal to
Ethiopia, S to South Africa.
CS: Lea 113 11/1952; **UN:** Sherif 3985 8/1951.

Pterolobium stellatum (Forssk.) Brenan

F.T.E.A. Leg.-Caes.: 42 (1967); F.E.E. 3: 54 (1989); T.S.S.:
219 (1990); Biol. Skr. 51: 182 (1998).
Syn: *Pterolobium exosum* (J.F.Gmel.) Baker f. – F.P.S. 2:
118 (1952).
Scrambling shrub. Dry forest.
Distr: RS, EQU; Eritrea to South Africa; Arabia.
RS: Kennedy-Cooke 258 3/1938; **EQU:** Friis &
Vollesen 399 11/1980.

Senna alexandrina Mill.

F.E.E. 3: 63 (1989).
Syn: *Cassia senna* L. – F.P.S. 2: 118 (1952); F.T.E.A.
Leg.-Caes.: 65 (1967); Bot. Exp. Sud.: 69 (1970); T.S.S.:
201 (1990).

Shrub. Dry grassland & bushland.
Distr: NS, RS, KOR, CS, ES; Mauritania to Somalia &
Kenya; N Africa, Arabia, India.
RS: Sahni & Kamil 710 4/1967.

Senna auriculata (L.) Roxb.

Legumes of Africa: 36 (1989).
Syn: *Cassia auriculata* L. – F.T.E.A. Leg.-Caes.: 76
(1967); T.S.S.: 205 (1990).
Shrub or small tree. Wooded grassland.
Distr: BAG; native to tropical Asia, occasionally cultivated
& naturalised elsewhere in the tropics.
BAG: Myers 7703 9/1937.
Note: used for tannin production from its bark.

Senna bicapsularis (L.) Roxb.

F.E.E. 3: 60 (1989); Biol. Skr. 51: 182 (1998).
Syn: *Cassia bicapsularis* L. – F.T.E.A. Leg.-Caes.: 71
(1967); Bot. Exp. Sud.: 68 (1970); T.S.S.: 205 (1990).
Shrub. Wooded grassland, bushland.
Distr: BAG, EQU; native to West Indies & South America;
cultivated & naturalised elsewhere.
EQU: Myers 6400 3/1937.

Senna didymobotrya (Fresen.) Irwin & Barneby

F.E.E. 3: 61 (1989); Biol. Skr. 51: 182 (1998).
Syn: *Cassia didymobotrya* Fresen. – F.T.E.A. Leg.-Caes.:
66 (1967); T.S.S.: 197 (1990).
Shrub. Montane wooded grassland.
Distr: EQU; Ethiopia to Angola & South Africa.
EQU: Myers 9456 9/1938.

Senna holosericea (Fresen.) Greuter

F.E.E. 3: 63 (1989).
Syn: *Cassia holosericea* Fresen. – F.P.S. 2: 117 (1952);
F.T.E.A. Leg.-Caes.: 66 (1967).
Perennial herb. Semi-desert.
Distr: NS, RS; Ethiopia, Somalia; Egypt, Arabia to
Pakistan.
RS: Drummond & Hemsley 983 1/1953.

Senna italica Mill. subsp. *italica*

F.E.E. 3: 63 (1989).
Syn: *Cassia italica* (Mill.) Spreng. subsp. *italica* – F.P.S.
2: 117 (1952); F.T.E.A. Leg.-Caes.: 65 (1967); Bot. Exp.
Sud.: 68 (1970); F.J.M.: 105 (1976); T.S.S.: 197 (1990).
Perennial herb. Wooded grassland.
Distr: NS, RS, DAR, KOR, CS; palaeotropical.
RS: Carter 1885 11/1987.

Senna italica Mill. subsp. *micrantha* (Brenan)
Lock

F.E.E. 3: 63 (1989).
Syn: *Cassia italica* (Mill.) Spreng. subsp. *micrantha*
Brenan – F.T.E.A. Leg.-Caes.: 65 (1967); Bot. Exp. Sud.:
68 (1970).
Herb. Wooded grassland.
Distr: NS, RS, CS, UN, EQU; Senegal to Somalia, S to
South Africa; India, Pakistan.
RS: Carter 1886 11/1987; **EQU:** Carr 738 8/1969.

Senna obtusifolia (L.) H.S.Irwin & Barneby

F.E.E. 3: 60 (1989); Biol. Skr. 51: 182 (1998).
Syn: *Cassia obtusifolia* L. – F.T.E.A. Leg.-Caes.: 77

(1967); Bot. Exp. Sud.: 68 (1970); T.S.S.: 199 (1990).
Syn: *Cassia tora* sensu Andrews pro parte, non L. –
F.P.S. 2: 116 (1952).
Perennial herb. Grassland.
Distr: DAR, KOR, CS, ES, BAG, UN, EQU; pantropical.
DAR: <u>Wickens 1745</u> 6/1964; **EQU:** <u>Myers 8636</u> 2/1938.

Senna occidentalis (L.) Link
F.E.E. 3: 60 (1989); Biol. Skr. 51: 183 (1998).
Syn: *Cassia occidentalis* L. – F.P.S. 2: 116 (1952);
F.T.E.A. Leg.-Caes.: 78 (1967); Bot. Exp. Sud.: 69 (1970);
F.J.M.: 106 (1976); T.S.S.: 201 (1990).
Perennial herb. Disturbed ground.
Distr: NS, DAR, KOR, CS, ES, BAG, UN, EQU; pantropical.
DAR: <u>Wickens 2045</u> 7/1964; **UN:** <u>Sherif 3982</u> 8/1951.

Senna petersiana (Bolle) Lock
F.E.E. 3: 59 (1989); Biol. Skr. 51: 183 (1998).
Syn: *Cassia petersiana* Bolle – F.P.S. 2: 117 (1952);
F.T.E.A. Leg.-Caes.: 72 (1967); T.S.S.: 201 (1990).
Shrub. Bushland.
Distr: BAG, UN, EQU; Cameroon to Ethiopia, S to South
Africa; Madagascar.
EQU: <u>Myers 11015</u> 4/1939.

Senna polyphylla (Jacq.) H.S. Irwin & Barneby
Legumes of Africa: 39 (1989).
Syn: *Cassia polyphylla* Jacq. – F.T.E.A. Leg.-Caes.: 50
(1967).
Shrub. Cultivated.
Distr: CS; native to West Indies; introduced in drier areas
of Africa.
CS: <u>de Wilde et al. 5784</u> 11/1965.

Senna siamea (Lam.) H.S.Irwin & Barneby
F.E.E. 3: 60 (1989).
Syn: *Cassia siamea* Lam. – F.T.E.A. Leg.-Caes.: 50
(1967); T.S.S.: 207 (1990).
Tree. Cultivated, possibly becoming naturalised.
Distr: DAR; native to SE Asia, cultivated in tropical Africa.
DAR: <u>Wickens 1752</u> 6/1964.
Note: widely cultivated elsewhere in the Sudan region.

Senna singueana (Delile) Lock
F.E.E. 3: 59 (1989); Biol. Skr. 51: 183 (1998).
Syn: *Cassia goratensis* Fresen. – Bot. Exp. Sud.: 68
(1970).
Syn: *Cassia singueana* Delile – F.P.S. 2: 117 (1952);
F.T.E.A. Leg.-Caes.: 73 (1967); T.S.S.: 203 (1990).
Shrub. Woodland.
Distr: DAR, KOR, BAG, EQU; Mali to Ethiopia, S to
Namibia & Zimbabwe.
KOR: <u>Kotschy 563</u> 1837; **EQU:** <u>Myers 10982</u> 4/1939.

Senna tora (L.) Roxb.
Legumes of Africa: 40 (1989).
Syn: *Cassia tora* L. – F.P.S. 2: 116, pro parte (1952);
F.T.E.A. Leg.-Caes.: 77, in notes (1967); Bot. Exp. Sud.:
69 (1970).
Shrub. Grassland.
Distr: DAR, KOR, CS, EQU; native to India; occasionally
cultivated elsewhere.
KOR: <u>Simpson 7084</u> 6/1929; **EQU:** <u>Drar 975</u> 4/1938
(n.v.).

Tamarindus indica L.
F.P.S. 2: 129 (1952); F.T.E.A. Leg.-Caes.: 153 (1967); Bot.
Exp. Sud.: 76 (1970); F.J.M.: 106 (1976); F.E.E. 3: 66
(1989); T.S.S.: 219 (1990); Biol. Skr. 51: 183 (1998).
Tree. Wooded grassland.
Distr: DAR, KOR, CS, ES, BAG, UN, EQU; palaeotropical.
DAR: <u>Wickens 2942</u> 4/1965; **UN:** <u>Simpson 7233</u> (fr)
7/1929.

Tylosema fassoglensis (Schweinf.) Torre & Hillc.
F.T.E.A. Leg.-Caes.: 213 (1967); F.E.E. 3: 70 (1989); T.S.S.:
221 (1990); Biol. Skr. 51: 184 (1998).
Syn: *Bauhinia fassoglensis* Schweinf. – F.P.S. 2: 110
(1952); Bot. Exp. Sud.: 68 (1970).
Climbing shrub. Wooded grassland.
Distr: CS, ES, UN, EQU; Ethiopia to Angola & South
Africa.
ES: <u>Kennedy-Cooke 39</u> 6/1931; **EQU:** <u>Myers 6376</u>
3/1937.

Leguminosae: Mimosoideae (Fabaceae: Mimosoideae)

H. Pickering & I. Darbyshire

[Acacia]
Note: under the recently proposed alternative classification
of *Acacia* and allies, of the species recorded in Sudan and
South Sudan only the introduced and invasive Australian
species *A. mearnsii* would remain within the genus. All
the native species recorded here would fall under either
Senegalia or *Vachellia*. However, we prefer to maintain
Acacia in its broad sense here, particularly since the true
type species of the genus is African (*A. scorpioides* (L.)
W.F. Wright = *A. nilotica* (L.) Delile).

Acacia abyssinica Hochst. ex Benth. subsp. calophylla Brenan
F.T.E.A. Leg.-Mim.: 112 (1959); F.E.E. 3: 89 (1989); T.S.S.:
148 (species only) (1990); Biol. Skr. 51: 184 (1998).
Syn: *Acacia xiphocarpa* sensu Andrews, non Hochst. ex
Benth. – F.P.S. 2: 148 (1952).
Tree. Wooded grassland.
Distr: EQU; D.R. Congo & Uganda to Mozambique.
EQU: <u>Myers 11147</u> 4/1939.

Acacia amythethophylla Steud. ex A.Rich.
F.E.E. 3: 86 (1989); Biol. Skr. 51: 185 (1998).
Syn: *Acacia macrothyrsa* Harms – F.P.S. 2: 148 (1952);
F.T.E.A. Leg.-Mim.: 101 (1959).
Tree. Woodland.
Distr: EQU; Mali & Ghana to Ethiopia, S to Angola &
Mozambique.
EQU: <u>Myers 6912</u> 6/1937.

Acacia asak (Forssk.) Willd.
F.E.E. 3: 80 (1989); T.S.S.: 149 (1990).
Syn: *Acacia glaucophylla* Steud. ex A.Rich. – F.P.S. 2:
133 (1952).
Shrub or tree. Bushland.
Distr: RS, KOR; Eritrea, Ethiopia; Egypt, Arabia.
RS: <u>Schweinfurth II.89</u> 9/1868.

Acacia ataxacantha DC.

F.P.S. 2: 140 (1952); F.T.E.A. Leg.-Mim.: 82 (1959); F.J.M.: 107 (1976); T.S.S.: 151 (1990); Biol. Skr. 51: 185 (1998).
Shrub, sometimes climbing. Riverine forest & woodland.
Distr: DAR, ?EQU; Senegal to Kenya, S to Namibia & South Africa.
DAR: Wickens 2742 9/1964.
Note: the record for EQU is derived from Jackson who recorded this species from the Didinga Mts; see Friis & Vollesen (1998: 185).

Acacia brevispica Harms subsp. brevispica

F.T.E.A. Leg.-Mim.: 96 (1959); F.E.E. 3: 78 (1989); T.S.S.: 151 (1990); Biol. Skr. 51: 185 (1998).
Shrub, sometimes climbing. Bushland, rocky hillslopes.
Distr: BAG, EQU; Ethiopia & Somalia, S to Tanzania.
EQU: Myers 10869 4/1939.

Acacia dolichocephala Harms

F.T.E.A. Leg.-Mim.: 79 (1959); F.E.E. 3: 85 (1989); T.S.S.: 152 (1990); Biol. Skr. 51: 186 (1998).
Tree. Riverine woodland, rocky outcrops.
Distr: EQU; Ethiopia, Uganda, Kenya, Tanzania.
EQU: Myers 11219 4/1939.

Acacia drepanolobium Harms ex Y.Sjöstedt

F.P.S. 2: 143 (1952); F.T.E.A. Leg.-Mim.: 121 (1959); F.E.E. 3: 88 (1989); T.S.S.: 152 (1990); Biol. Skr. 51: 186 (1998).
Shrub. Grassland.
Distr: CS, ES, UN; Ethiopia to Tanzania.
ES: Kennedy-Cooke 72 8/1931; **UN:** J.M. Lock 79/19 4/1979.

Acacia ehrenbergiana Hayne

F.P.S. 2: 144 (1952); F.E.E. 3: 86 (1989); T.S.S.: 153 (1990); F.J.U. 2: 187 (1999).
Syn: *Acacia flava* (Forssk.) Schweinf. – F. Darfur Nord-Occ. & J. Gourgeil: 114 (1969).
Shrub. Rocky and sandy areas near water courses.
Distr: NS, RS, DAR, KOR, ES, EQU; Mauritania to Ethiopia; N Africa, Arabia.
RS: Schweinfurth 666 10/1868; **EQU:** Padwa 226 5/1953.

Acacia elatior Brenan subsp. turkanae Brenan

F.T.E.A. Leg.-Mim.: 114 (1959); T.S.S.: 153 (1990).
Tree. Sandy riverbanks.
Distr: EQU; Uganda, Kenya.
EQU: Myers 10421 2/1939.

Acacia etbaica Schweinf. subsp. etbaica

F.P.S. 2: 148 (1952); F.E.E. 3: 86 (1989); Biol. Skr. 51: 186 (1998).
Tree. Bushland.
Distr: NS, RS, ES, ?EQU; Eritrea, Somalia.
RS: Sahni & Kamil 674 4/1967.
Note: the record by El Amin from near Kapoeta, Equatoria (cited by Friis & Vollesen 1998) is likely to be incorrect since this taxon is otherwise restricted to the north.

[Acacia farnesiana (L.) Willd.]

Note: this South American species of *Acacia* has been introduced in Sudan as an ornamental and a shelterbelt species. Vogt (1995: 54) indicates that it may have

naturalised in Sudan but is not very common. It was not mentioned by El Amin (1990).

Acacia gerrardii Benth. var. gerrardii

F.T.E.A. Leg.-Mim.: 119 (1959); F.J.M.: 107 (1976); F.E.E. 3: 88 (1989); T.S.S.: 154 (1990); Biol. Skr. 51: 186 (1998).
Syn: *Acacia hebecladoides* Harms – F.P.S. 2: 144 (1952).
Tree or shrub. Woodland & wooded grassland.
Distr: DAR, KOR, CS, UN, EQU; Nigeria to Ethiopia, S to South Africa.
DAR: Wickens 2754 10/1964; **EQU:** Myers 11229 9/1939.

Acacia hecatophylla Steud. ex A.Rich.

F.P.S. 2: 137 (1952); F.T.E.A. Leg.-Mim.: 87 (1959); F.E.E. 3: 84 (1989); T.S.S.: 155 (1990); Biol. Skr. 51: 187 (1998).
Tree. Woodland.
Distr: CS, EQU; D.R.Congo, Ethiopia, Uganda.
CS: Aylmer 27/16 3/1922; **EQU:** Myers 8395 1/1938.

Acacia hockii De Wild.

F.T.E.A. Leg.-Mim.: 104 (1959); F.E.E. 3: 86 (1989); T.S.S.: 155 (1990); Biol. Skr. 51: 187 (1998).
Syn: *Acacia seyal* Delile var. *multijuga* Schweinf. ex Baker f. – F.P.S. 2: 144 (1952).
Shrub or tree. Woodland.
Distr: ?RS, ?CS, EQU; Guinea to Eritrea, S to Zimbabwe; Arabia.
EQU: Myers 6906 6/1937.
Note: El Amin in T.S.S. records this species from the Red Sea Hills and Central and South Sudan but we have only seen material from Equatoria.

Acacia horrida (L.) Willd. subsp. benadirensis (Chiov.) Hillc. & Brenan

F.T.E.A. Leg.-Mim.: 81 (1959); F.E.E. 3: 84 (1989); T.S.S.: 155 (1990); Biol. Skr. 51: 187 (1998).
Shrub. Bushland.
Distr: EQU; Ethiopia, Somalia, Uganda, Kenya.
EQU: Myers 14087 9/1941.

Acacia kamerunensis Gand.

F.T.E.A. Leg.-Mim.: 98 (1959); Biol. Skr. 51: 188 (1998).
Syn: *Acacia pennata* sensu Andrews, non (L.) Willd. – F.P.S. 2: 150 (1952).
Woody climber or shrub. Forest.
Distr: EQU; Sierra Leone to D.R. Congo & Uganda.
EQU: Jackson 748 4/1949.

Acacia kirkii Oliv. subsp. mildbraedii (Harms) Brenan

F.T.E.A. Leg.-Mim.: 107 (1959); T.S.S.: 157 (1990).
Syn: *Acacia mildbraedii* Harms – F.P.S. 2: 147 (1952).
Tree. Riverine & swamp forest.
Distr: EQU; D.R. Congo to Tanzania.
EQU: Aylmer 27/15 10/1926.

Acacia laeta R.Br. ex Benth.

F.P.S. 2: 137 (1952); F.T.E.A. Leg.-Mim.: 83 (1959); F.E.E. 3: 81 (1989); T.S.S.: 157 (1990).
Tree or shrub. Bushland.
Distr: NS, RS, KOR, CS, ES; Mali to Somalia, S to Tanzania; N Africa, Israel, Arabia.
KOR: Jackson 4368 9/1961.

Acacia macrostachya Rchb. ex DC.

F.P.S. 2: 140 (1952); F.W.T.A. 1(2): 501 (1958); T.S.S.: 159 (1990).
Tree or shrub, sometimes climbing. Bushland.
Distr: BAG, UN, EQU; Senegal to Chad.
BAG: Myers 7335 7/1937.

Acacia mearnsii De Wild.

F.T.E.A. Leg.-Mim.: 95 (1959); F.E.E. 3: 91 (1989).
Tree. Cultivated, naturalised along river banks.
Distr: DAR, EQU; native to Australia, introduced & naturalised elsewhere in the palaeotropics.
DAR: Francis 74 12/1957; **EQU:** Howard I.M.89 3/1976.

Acacia mellifera (Vahl) Benth. subsp. mellifera

F.P.S. 2: 136 (1952); F.T.E.A. Leg.-Mim.: 84 (1959); F.J.M.: 107 (1976); F.E.E. 3: 81 (1989); T.S.S.: 159 (1990); Biol. Skr. 51: 188 (1998).
Tree or shrub. Bushland.
Distr: RS, DAR, KOR, CS, ES, EQU; Ethiopia to Tanzania, Angola, Namibia; N Africa, Arabia.
DAR: Harrison 120 10/1947; **EQU:** Padwa 241 6/1953.

Acacia nilotica (L.) Willd. ex Delile subsp. nilotica

F.P.S. 2: 147 (1952); F.J.M.: 107 (1976); F.E.E. 3: 87 (1989); T.S.S.: 160 (1990).
Syn: *Acacia arabica* (Lam.) Willd. – F.P.S. 2: 147 (1952); Bot. Exp. Sud.: 64 (1970).
Syn: *Acacia scorpioides* (L.) W.Wight var. *nilotica* (L.) A.Chev. – F. Darfur Nord-Occ. & J. Gourgeil: 114 (1969).
Tree. Woodland.
Distr: NS, RS, DAR, KOR, CS; Senegal to Eritrea; N Africa.
CS: Schweinfurth 927 1/1869.

Acacia nilotica (L.) Willd. ex Delile subsp. adstringens (Schumach. & Thonn.) Roberty

F.W.T.A. 1(2): 500 (1958); T.S.S.: 160 (1990).
Syn: *Acacia arabica* (Lam.) Willd. var. *adansoniana* Dubard – Bot. Exp. Sud.: 65 (1970).
Syn: *Acacia nilotica* (L.) Willd. ex Delile subsp. *adansonii* (Guill. & Perr.) Brenan – F.J.M.: 107 (1976).
Tree. Woodland.
Distr: DAR, KOR, BAG, UN; Senegal to Chad, cultivated in Somalia.
DAR: Wickens 963 1/1964; **BAG:** Myers 7498 8/1937.

Acacia nilotica (L.) Willd. ex Delile subsp. indica (Benth.) Brenan

F.E.E. 3: 87 (1989).
Tree. Cultivated & occasionally naturalised, in Sudan on riverbanks.
Distr: NS; native to India, cultivated elsewhere.
NS: Bristow 10 3/1984.

Acacia nilotica (L.) Willd. ex Delile subsp. subalata (Vatke) Brenan

F.T.E.A. Leg.-Mim.: 110 (1959); F.E.E. 3: 87 (1989); T.S.S.: 160 (1990); Biol. Skr. 51: 188 (1998).
Syn: *Acacia arabica* (Lam.) Willd. var. *adansoniana* sensu Andrews, non Dubard – F.P.S. 2: 147 (1952).
Tree. Bushland & wooded grassland.
Distr: EQU; Ethiopia, Somalia, Uganda, Kenya, Tanzania.
EQU: Myers 13451 8/1940.

Acacia nilotica (L.) Willd. ex Delile subsp. tomentosa (Benth.) Brenan

F.W.T.A. 1(2): 500 (1958); T.S.S.: 160 (1990).
Tree. Wooded grassland.
Distr: KOR, CS, ES; Senegal to Djibouti; Egypt.
KOR: Wickens 890 11/1963

Acacia oerfota (Forssk.) Schweinf.

F.J.M.: 107 (1976); F.E.E. 3: 90 (1989); Biol. Skr. 51: 188 (1998).
Syn: *Acacia nubica* Benth. – F.P.S. 2: 148 (1952); F.T.E.A. Leg.-Mim.: 129 (1959); T.S.S.: 160 (1990).
Shrub. Bushland.
Distr: RS, DAR, KOR, CS, ES, EQU; Chad to Somalia, S to Tanzania; Egypt, Arabia.
KOR: Wickens 3053 5/1969; **EQU:** Padwa 261 6/1953.

Acacia paolii Chiov.

F.T.E.A. Leg.-Mim.: 130 (1959); F.E.E. 3: 90 (1989); T.S.S.: 161 (1990); Biol. Skr. 51: 189 (1998).
Shrub. Dry bushland.
Distr: EQU; Ethiopia, Somalia, Kenya.
EQU: Myers 13999 9/1941.

Acacia pentagona (Schumach. & Thonn.) Hook.f.

F.T.E.A. Leg.-Mim.: 100 (1959); F.E.E. 3: 78 (1989); T.S.S.: 161 (1990).
Woody climber. Lowland forest, swamp and riverine forest.
Distr: EQU; Sierra Leone to Ethiopia, S to Angola & Mozambique.
EQU: Wyld 845 3/1941.

Acacia persiciflora Pax

F.T.E.A. Leg.-Mim.: 86 (1959); F.E.E. 3: 81 (1989); T.S.S.: 162 (1990); Biol. Skr. 51: 189 (1998).
Syn: *Acacia eggelingii* Baker f. – F.P.S. 2: 137 (1952).
Tree. Bushland & wooded grassland.
Distr: EQU; D.R. Congo, Ethiopia, Uganda, Kenya.
EQU: Myers 10977 4/1939.

Acacia polyacantha Willd. subsp. campylacantha (Hochst. ex A.Rich.) Brenan

F.T.E.A. Leg.-Mim.: 88 (1959); F.J.M.: 107 (1976); F.E.E. 3: 81 (1989); T.S.S.: 162 (1990); Biol. Skr. 51: 189 (1998).
Syn: *Acacia campylacantha* Hochst. ex A.Rich. – F.P.S. 2: 137 (1952); Bot. Exp. Sud.: 65 (1970).
Tree. Woodland & wooded grassland, often riverine.
Distr: DAR, KOR, CS, ES, UN, EQU; Senegal to Eritrea, S to Angola & South Africa.
KOR: Wickens 815 11/1962; **UN:** Simpson 7074 6/1929.

Acacia reficiens Wawra subsp. misera (Vatke) Brenan

F.T.E.A. Leg.-Mim.: 116 (1959); F.E.E. 3: 88 (1989); T.S.S.: 162 (species only) (1990); Biol. Skr. 51: 190 (1998).
Shrub. Dry scrub.
Distr: EQU; Ethiopia, Somalia, Uganda, Kenya.
EQU: Myers 13437 8/1940.

Acacia schweinfurthii Brenan & Exell var. schweinfurthii

F.T.E.A. Leg.-Mim.: 99 (1959); T.S.S.: 163 (1990); Biol. Skr. 51: 190 (1998).

Shrub or liana. Riverine forest, woodland, bushland.
Distr: BAG, UN, EQU; Uganda to South Africa.
UN: J.M. Lock 81/223 8/1981.

Acacia senegal (L.) Willd. var. senegal

F.P.S. 2: 135 (1952); F.T.E.A. Leg.-Mim.: 93 (1959); F.J.M.:
108 (1976); F.E.E. 3: 78 (1989); T.S.S.: 163 (1990); Biol.
Skr. 51: 190 (1998).
Tree. Wooded grassland, bushland.
Distr: DAR, KOR, CS, ES, UN, EQU; Senegal to Eritrea, S
to Angola & South Africa; N Africa, Arabia.
DAR: Francis 1 11/1957; **EQU:** Simpson 7292 7/1929.

Acacia seyal Delile var. seyal

F.P.S. 2: 144 (1952); F.T.E.A. Leg.-Mim.: 103 (1959);
F.J.M.: 108 (1976); F.E.E. 3: 68 (1989); T.S.S.: 163 (1990);
Biol. Skr. 51: 190 (1998).
Tree. Wooded grassland.
Distr: NS, RS, DAR, KOR, UN, EQU; Senegal to Ethiopia,
S to Tanzania; Egypt, Arabia.
DAR: Wickens 1118 2/1964; **EQU:** Friis & Vollesen 1164
3/1982.

Acacia seyal Delile var. fistula (Schweinf.) Oliv.

F.T.E.A. Leg.-Mim.: 103 (1959); F.E.E. 3: 86 (1989); T.S.S.:
165 (1990); Biol. Skr. 51: 190 (1998).
Syn: *Acacia fistula* Schweinf. – F.P.S. 2: 142 (1952); Bot.
Exp. Sud.: 65 (1970).
Tree. Wooded grassland, bushland.
Distr: KOR, UN, EQU; Ethiopia & Somalia, S to Zambia &
Mozambique.
KOR: Wickens 699 10/1962; **UN:** Schweinfurth 1084
1/1869.

Acacia sieberiana DC. var. sieberiana

F.T.E.A. Leg.-Mim.: 127 (1959); F.P.S. 2: 140 (1952); Bot.
Exp. Sud.: 66 (1970); F.E.E. 3: 89 (1989); T.S.S.: 165
(1990); Biol. Skr. 51: 191 (1998).
Tree. Woodland.
Distr: DAR, KOR, CS, BAG, UN, EQU; Senegal to
Ethiopia, S to Mozambique.
DAR: Harrison 121 11/1947; **EQU:** Myers 8593 2/1938.

Acacia sieberiana DC. var. villosa A.Chev.

F.W.T.A. 1(2): 499 (1958); F.J.M.: 108 (1976); T.S.S.: 167
(1990).
Tree. Woodland.
Distr: DAR; Senegal to C.A.R.
DAR: Wickens 1568 4/1964.

Acacia sieberiana DC. var. woodii (Burtt-Davy) Keay & Brenan

F.T.E.A. Leg.-Mim.: 128 (1959); F.J.M.: 108 (1976); F.E.E.
3: 90 (1989); Biol. Skr. 51: 191 (1998).
Syn: *Acacia sieberiana* DC. var. *vermoesenii* (De
Wild.) Keay & Brenan – T.S.S.: 167 (1990).
Tree. Woodland & bushland.
Distr: DAR, EQU; Senegal to Ethiopia, S to South Africa.
DAR: Wickens 2835 2/1965; **EQU:** Kosper 308 4/1983.
Note: El Amin in T.S.S. records this variety (under
vermoesenii) as more widespread than recorded here but
this may be based on misidentification.

Acacia tortilis (Forssk.) Hayne subsp. tortilis

F.E.E. 3: 87 (1989); T.S.S.: 167 (1990); F.J.U. 2: 190 (1999).
Tree. Dry bushland.
Distr: NS, RS, KOR, CS; Eritrea, Somalia; Egypt, Middle
East.
KOR: Wickens 891 1/1963.

Acacia tortilis (Forssk.) Hayne subsp. raddiana (Savi) Brenan

F.T.E.A. Leg.-Mim.: 117 (1952); T.S.S.: 167 (1990); Biol.
Skr. 51: 191 (1998).
Syn: *Acacia raddiana* Savi – F.P.S. 2: 140 (1952).
Tree. Bushland.
Distr: NS, RS, DAR, CS, ES, ?EQU; Senegal to Somalia &
Kenya; Algeria, Egypt, Israel, Arabia.
CS: Aylmer 84 6/1931.
Note: the record of this subspecies from the Imatong Mts is
almost certainly incorrect (see Friis & Vollesen 1998: 191).

Acacia tortilis (Forssk.) Hayne subsp. spirocarpa (Hochst. ex A.Rich.) Brenan

F.T.E.A. Leg.-Mim.: 117 (1959); F.J.M.: 108 (1976); F.E.E.
3: 87 (1989); T.S.S.: 167 (1990); Biol. Skr. 51: 192 (1998).
Tree. Wooded grassland, bushland.
Distr: NS, RS, DAR, KOR, EQU; Ethiopia & Somalia, S to
Angola, Zimbabwe & Mozambique.
RS: Aylmer 633 3/1936; **EQU:** Myers 13433 8/1940.

Acacia venosa Hochst. ex Benth.

F.P.S. 2: 133 (1952); F.E.E. 3: 81 (1989).
Small tree. Woodland, bushland.
Distr: ?ES; Eritrea, Ethiopia.
IUCN: RD
Note: Andrews records this species from Eastern Sudan
but we have been unable to locate the material on which
this record is based. It is usually considered endemic to
Ethiopia and Eritrea.

Albizia adianthifolia (Schumach.) W.Wight

F.T.E.A. Leg.-Mim.: 160 (1959).
Syn: *Albizia intermedia* De Wild. & T.Durand – T.S.S.:
173 (1990).
Tree. Forest, woodland & wooded grassland.
Distr: EQU; Senegal to Kenya, S to Angola & South Africa.
EQU: Gadokia 476 3/1948.

Albizia amara (Roxb.) Boivin subsp. sericocephala (Benth.) Brenan

F.T.E.A. Leg.-Mim.: 152 (1959); Bot. Exp. Sud.: 67
(species only) (1970); F.J.M.: 109 (1976); F.E.E. 3: 94
(1989); T.S.S.: 169 (1990); Biol. Skr. 51: 192 (1998).
Syn: *Albizia sericocephala* Benth. – F.P.S. 2: 154 (1952);
Bot. Exp. Sud.: 67 (1970).
Tree. Wooded grassland.
Distr: NS, DAR, KOR, CS, ES, BAG, UN, EQU; Ethiopia to
South Africa.
DAR: Harrison 118 11/1947; **EQU:** Friis & Vollesen 1191
3/1982.

Albizia anthelmintica Brongn.

F.P.S. 2: 151 (1952); F.T.E.A. Leg.-Mim.: 148 (1959);
F.J.M.: 109 (1976); F.E.E. 3: 94 (1989); T.S.S.: 169 (1990);
Biol. Skr. 51: 192 (1998).
Tree or shrub. Woodland & wooded grassland.

Distr: DAR, KOR, CS, ES, BAG, UN, EQU; Eritrea to Namibia & South Africa.
DAR: <u>Wickens 962</u> 1/1964; <u>Broun 1132</u> 4/1907.

Albizia aylmeri Hutch.

F.P.S. 2: 154 (1952); F.J.M.: 109 (1976); T.S.S.: 171 (1990).
Tree. Woodland, rocky hillslopes.
Distr: DAR, KOR, CS; Sudan endemic.
IUCN: RD
DAR: <u>Wickens 3003</u> 6/1965.

Albizia coriaria Welw. ex Oliv.

F.P.S. 2: 151 (1952); F.T.E.A. Leg.-Mim.: 143 (1959); F.E.E. 3: 93 (1989); T.S.S.: 171 (1990); Biol. Skr. 51: 193 (1998).
Tree. Wooded grassland, riverine thicket.
Distr: EQU; Ivory Coast to Ethiopia, S to Angola & Zambia.
EQU: <u>Myers 10186</u> 12/1938.

Albizia glaberrima (Schumach. & Thonn.) Benth. var. glaberrima

F.T.E.A. Leg.-Mim.: 156 (1959); T.S.S.: 171 (1990); Biol. Skr. 51: 193 (1998).
Syn: *Albizia warneckei* Harms. – F.P.S. 2: 154 (1952).
Tree or shrub. Forest.
Distr: KOR, EQU; Guinea to Kenya, S to Mozambique.
KOR: <u>Simpson 7734</u> 4/1930; **EQU:** <u>Myers 10042</u> 11/1938.

Albizia grandibracteata Taub.

F.P.S. 2: 154 (1952); F.T.E.A. Leg.-Mim.: 161 (1959); F.E.E. 3: 96 (1989); T.S.S.: 173 (1990); Biol. Skr. 51: 193 (1998).
Tree. Forest, often riverine.
Distr: EQU; Ethiopia to D.R. Congo & Tanzania.
EQU: <u>Friis & Vollesen 1273</u> 3/1982.

Albizia grandibracteata x gummifera

F.T.E.A. Leg.-Mim.: 162 (1959); Biol. Skr. 51: 194 (1998).
Tree. Forest.
Distr: EQU; D.R. Congo, Uganda, Kenya.
EQU: <u>Andrews 1911</u> 6/1939.

Albizia gummifera (J.F.Gmel.) C.A.Sm. var. gummifera

F.P.S. 2: 155 (1952); F.T.E.A. Leg.-Mim.: 157 (1959); F.E.E. 3: 96 (1989); T.S.S.: 173 (1990); Biol. Skr. 51: 194 (1998).
Tree. Forest, including secondary growth.
Distr: EQU; Nigeria, Cameroon, Ethiopia to Zimbabwe; Madagascar.
EQU: <u>Myers 10966</u> 4/1939.

Albizia lebbeck (L.) Benth.

F.T.E.A. Leg.-Mim.: 147 (1959); Bot. Exp. Sud.: 67 (1970); F.E.E. 3: 94 (1989); T.S.S.: 175 (1990).
Tree. Cultivated & naturalised.
Distr: NS, UN; native to tropical Asia, now widespread in tropical Africa.
NS: <u>Drar 2543</u> 5/1938 (n.v.); **UN:** <u>Speke & Grant s.n.</u> 4/1863.

Albizia malacophylla (A.Rich.) Walp. var. ugandensis Baker f.

F.P.S. 2: 151 (1952); F.T.E.A. Leg.-Mim.: 145 (1959); F.J.M.: 109 (1976); F.E.E. 3: 93 (1989); T.S.S.: 175 (species only) (1990).
Tree. Wooded grassland.

Distr: DAR, CS, ?BAG; Mali to Ethiopia & Uganda.
DAR: <u>Wickens 2935</u> 4/1965.
Note: the record for Bahr el Ghazal is from El Amin in T.S.S.; we have seen no material from South Sudan.

Albizia schimperiana Oliv. var. schimperiana

F.P.S. 2: 150 (1952); F.T.E.A. Leg.-Mim.: 154 (1959); F.E.E. 3: 96 (1989); T.S.S.: 177 (1990); Biol. Skr. 51: 194 (1998).
Syn: *Albizia maranguensis* Taub. – F.P.S. 2: 151 (1952).
Tree. Forest including disturbed areas.
Distr: EQU; Ethiopia to Zimbabwe & Mozambique.
EQU: <u>Friis & Vollesen 1272</u> 3/1982.

Albizia schimperiana Oliv. var. tephrocalyx Brenan

F.T.E.A. Leg.-Mim.: 155 (1959); Biol. Skr. 51: 195 (1998).
Tree. Riverine forest & forest margins.
Distr: EQU; Uganda.
IUCN: RD
EQU: <u>Jackson 1359</u> 3/1950.

Albizia zygia (DC.) J.F.Macbr.

F.P.S. 2: 154 (1952); F.T.E.A. Leg.-Mim.: 161 (1959); T.S.S.: 177 (1990); Biol. Skr. 51: 195 (1998).
Tree. Forest & woodland.
Distr: DAR, KOR, BAG, UN, EQU; Senegal to Kenya, S to D.R. Congo.
DAR: <u>Wickens 1570</u> 5/1964; **EQU:** <u>Friis & Vollesen 1235</u> 3/1982.

Amblygonocarpus andongensis (Welw. ex Oliv.) Exell & Torre

F.T.E.A. Leg.-Mim.: 34 (1959); T.S.S.: 177 (1990).
Syn: *Amblygonocarpus schweinfurthii* Harms – F.P.S. 2: 155 (1952).
Tree. Wooded grassland.
Distr: DAR, BAG, EQU; Burkina Faso to Uganda, S to Angola, Zimbabwe & Mozambique.
DAR: <u>Wickens 3639</u> 6/1977; **BAG:** <u>Turner 87</u> 8/1931.

Cathormion altissimum (Hook.f.) Hutch. & Dandy

F.P.S. 2: 155 (1952); F.T.E.A. Leg.-Mim.: 166 (1959); T.S.S.: 179 (1990).
Shrub or tree. Swamp forest.
Distr: EQU; Sierra Leone to Uganda, S to Angola & Zambia.
EQU: <u>Myers 7996</u> 12/1937.

Cathormion eriorhachis (Harms) Dandy

F.P.S. 2: 155 (1952); Trop. Afr. Fl. Pl. 3: 62 (2008).
Tree. Wooded grassland & woodland.
Distr: ?BAG; Cameroon, C.A.R.
Note: recorded by Andrews from "Equatoria: road between Raga and Said Bundas" which is in Bahr el Ghazal; we have not seen the specimen on which this record is based.

Dichrostachys cinerea (L.) Wight & Arn.

F.T.E.A. Leg.-Mim.: 36 (1959); Bot. Exp. Sud.: 71 (1970); F.J.M.: 109 (1976); F.E.E. 3: 74 (1989); T.S.S.: 179 (1990); Biol. Skr. 51: 195 (1998).
Syn: *Dichrostachys glomerata* (Forssk.) Chiov. – F.P.S. 2: 156 (1952).

Shrub or small tree. Woodland & bushland, often dominating in overgrazed areas.
Distr: DAR, KOR, CS, ES, BAG, UN, EQU; palaeotropical.
DAR: <u>Wickens 2808</u> 11/1964; **EQU:** <u>Friis & Vollesen 164</u> 11/1980.
Note: a number of subspecies and varieties have previously been recognised in this widespread and variable species, several of which occur in Sudan and South Sudan, but they are difficult to distinguish and are not upheld here.

Entada abyssinica Steud. ex A.Rich.

F.P.S. 2: 158 (1952); F.T.E.A. Leg.-Mim.: 13 (1959); F.J.M.: 109 (1976); F.E.E. 3: 72 (1989); T.S.S.: 181 (1990); Biol. Skr. 51: 196 (1998).
Tree. Woodland & wooded grassland.
Distr: DAR, EQU; Guinea to Eritrea, S to Angola & Mozambique.
DAR: <u>Wickens 3616</u> 6/1977; **EQU:** <u>Myers 10987</u> 4/1939.

Entada abyssinica x africana

F.J.M.: 110 (1976).
Tree. Riverine forest.
Distr: DAR.
DAR: <u>Wickens 1190</u> 2/1964.

Entada africana Guill. & Perr.

F.T.E.A. Leg.-Mim.: 12 (1959); F.J.M.: 110 (1976); F.E.E. 3: 72 (1989); T.S.S.: 181 (1990); Biol. Skr. 51: 196 (1998).
Syn: Entada sudanica Schweinf. – F.P.S. 2: 156 (1952).
Shrub or tree. Woodland.
Distr: DAR, KOR, CS, ES, UN, EQU; Senegal to Ethiopia, S to D.R. Congo & Uganda.
DAR: <u>Wickens 2924</u> 4/1965; **UN:** <u>Myers 10430</u> 2/1939.

Entada gigas Fawc. & Rendle

F.T.E.A. Leg.-Mim.: 11 (1959); T.S.S.: 181 (1990).
Syn: Entada phaseoloides sensu Andrews, non (L.) Merrill – F.P.S. 2: 158 (1952).
Woody climber. Riverine forest.
Distr: EQU; Guinea to Ivory Coast, Cameroon to Uganda, S to Angola & D.R. Congo; Caribbean, central America, Columbia.
EQU: <u>Myers 13263</u> 5/1940.

Entada rheedei Spreng.

Biol. Skr. 51: 197 (1998); Trop. Afr. Fl. Pl. 3: 66 (2008).
Syn: Entada phaseoloides sensu Brenan, non (L.) Merrill – F.T.E.A. Leg.-Mim.: 12 (1959).
Woody climber. Lowland forest.
Distr: EQU; Guinea to Kenya, S to South Africa; tropical Asia.
EQU: <u>Friis & Vollesen 801</u> 12/1980.

Entada wahlbergii Harv.

F.T.E.A. Leg.-Mim.: 18 (1959); T.S.S.: 181 (1990); Biol. Skr. 51: 197 (1998).
Syn: Entada flexuosa Hutch. & Dalziel – F.P.S. 2: 157 (1952).
Woody climber. Wooded grassland.
Distr: BAG, EQU; Guinea to Tanzania, S to Zambia & South Africa.
BAG: <u>Schweinfurth 1868</u> 7/1869.

Faidherbia albida (Delile) A.Chev.

Trop. Afr. Fl. Pl. 3: 68 (2008).
Syn: Acacia albida Delile – F.P.S. 2: 133 (1952); F.T.E.A. Leg.-Mim.: 78 (1959); Bot. Exp. Sud.: 64 (1970); F.E.E. 3: 84 (1989); T.S.S.: 149 (1990); Biol. Skr. 51: 185 (1998).
Tree. Wooded grassland.
Distr: NS, RS, DAR, KOR, CS, UN; Mauritania to Somalia, S to Namibia & South Africa; N Africa.
DAR: <u>Wickens 1537</u> 5/1964; **UN:** <u>Speke & Grant 771</u> 3/1863.

Leucaena leucocephala (Lam.) de Wit

F.E.E. 3: 75 (1989); T.S.S.: 183 (1990).
Syn: Leucaena glauca Benth. – F.P.S. 2: 129 (1952); F.T.E.A. Leg.-Mim.: 48 (1959); Bot. Exp. Sud.: 73 (1970).
Shrub or tree. Naturalised in secondary bushland.
Distr: KOR; native to Central America, now cultivated throughout the tropics.
KOR: <u>Kotschy s.n.</u> 1842.

Mimosa pigra L.

F.P.S. 2: 158 (1952); F.T.E.A. Leg.-Mim.: 43 (1959); F.J.M.: 110 (1976); F.E.E. 3: 74 (1989); T.S.S.: 183 (1990); Biol. Skr. 51: 197 (1998).
Shrub. Swampy grassland, riverine scrub.
Distr: NS, DAR, CS, ES, BAG, UN, EQU; native to the neotropics, now invasive throughout the tropics.
DAR: <u>Wickens 2944</u> 4/1965; **UN:** <u>Harrison 308</u> 3/1948.

Mimosa pudica L.

F.T.E.A. Leg.-Mim.: 46 (1959); T.S.S.: 183 (1990).
Annual or perennial herb. Weed of irrigated cultivation, roadsides.
Distr: EQU; native to the neotropics, now a pantropical weed.
EQU: <u>Jackson 2989</u> 6/1953.

Neptunia olearacea Lour.

F.P.S. 2: 158 (1952); F.T.E.A. Leg.-Mim.: 40 (1959); F.E.E. 3: 71 (1989).
Aquatic perennial herb. Freshwater pools.
Distr: KOR, UN; pantropical.
KOR: <u>Pfund 29</u> 7/1875; **UN:** <u>Simpson 7028</u> 6/1929.

Parkia biglobosa (Jacq.) R.Br. ex G.Don

F.P.S. 2: 159 (1952); F.W.T.A. 1(2): 487 (1958); T.S.S.: 185 (1990).
Syn: Parkia clappertoniana Keay – T.S.S.: 185 (1990).
Syn: Parkia oliveri J.F.Macbr. – F.P.S. 2: 159 (1952); T.S.S.: 187 (1990).
Tree. Wooded grassland.
Distr: BAG, UN; Senegal to Chad.
BAG: <u>Andrews 728</u> 4/1939.
Note: Andrews' and El Amin's records of P. bicolor A.Chev. from Equatoria are surely in error.

Parkia filicoidea Welw. ex Oliv.

F.P.S. 2: 158 (1952); F.T.E.A. Leg.-Mim.: 7 (1959); T.S.S.: 187 (1990); Biol. Skr. 51: 197 (1998).
Tree. Riverine forest.
Distr: EQU; Ivory Coast to Somalia, S to Angola and Mozambique.
EQU: <u>Andrews 1191</u> 5/1939.

Piptadeniastrum africanum (Hook.f.) Brenan

F.T.E.A. Leg.-Mim.: 21 (1959); T.S.S.: 187 (1990).
Syn: *Piptadenia africana* Hook.f. – F.P.S. 2: 159 (1952).
Large tree. Forest.
Distr: EQU; Senegal to Uganda, S to Angola & D.R.
Congo.
EQU: Sillitoe 173 1919.

Pithecellobium dulce (Roxb.) Benth.

Bot. Exp. Sud.: 75 (1970); F.J.M.: 110 (1976); F.T.E.A.
Leg.-Mim.: 86 (1959); T.S.S.: 187 (1990).
Tree. Naturalised in wooded grassland.
Distr: DAR, CS; native to Mexico, now widely cultivated
in the tropics.
DAR: Wickens 1115 1964.

Prosopis africana (Guill. & Perr.) Taub.

F.P.S. 2: 163 (1952); F.T.E.A. Leg.-Mim.: 36 (1959); F.J.M.:
110 (1976); T.S.S.: 189 (1990).
Tree. Wooded grassland.
Distr: DAR, KOR, BAG, UN, EQU; Senegal to Uganda.
DAR: Wickens 2926 4/1965; **BAG:** Schweinfurth 1691
3/1869.

Prosopis alba Griseb.

Legumes of Africa: 96 (1989).
Tree. Cultivated, ?naturalised on riverbanks.
Distr: NS, RS; native to South America, introduced
elsewhere.
NS: Ahti 16804 9/1962.

Prosopis glandulosa Tor.

Common Trees & Shrubs Dryland Sudan: 75 (1995).
Syn: *Prosopis chilensis* sensu El Amin pro parte, non
(Molina) Stuntz – T.S.S.: 189 (1990).
Small tree. Widely planted, naturalised in bushland &
irrigated areas.
Distr: ?NS, RS, CS, ES; native to America, now widely
naturalised.
RS: Drummond & Hemsley 984 1/1953.
Note: cultivated and now becoming a troublesome
invasive in much of Sudan. The separation of *P. glandulosa*
and *P. juliflora* remains somewhat confused such that
their distributions in Sudan are not well documented.

Prosopis juliflora (Sw.) DC.

F.E.E. 3: 72 (1989); Common Trees & Shrubs Dryland
Sudan: 77 (1995).
Syn: *Prosopis chilensis* sensu El Amin pro parte, non
(Molina) Stuntz – T.S.S.: 189 (1990).
Tree or shrub. Widely planted, naturalised in e.g. wadi
margins & on saline soils.
Distr: ?NS, DAR, KOR, CS, ?ES; native to America,
naturalised in Africa from Eritrea to South Africa.
CS: de Wilde et al. 5801 2/1965.
Note: commonly planted and now becoming a
troublesome invasive in much of Sudan. Of the several
species of *Prosopis* cultivated in the past, *P. juliflora*
appears to be the most widespread and well established,
mainly in the Sahel zone.

Tetrapleura tetraptera (Schumach. & Thonn.)
Taub.

F.P.S. 2: 163 (1952); F.T.E.A. Leg.-Mim.: 32 (1959); T.S.S.:
189 (1990); Biol. Skr. 51: 198 (1998).
Tree. Forest.
Distr: DAR, EQU; Senegal to Uganda, S to Angola &
Tanzania.
DAR: Haekstra 195 9/1972; **EQU:** Andrews 1396 5/1939.

Polygalaceae

I. Darbyshire

Carpolobia goetzei Gürke

Biol. Skr. 51: 83 (1998); F.T.E.A. Polygal.: 3 (2007).
Syn: *Carpolobia conradsiana* Engl. – T.S.S.: 43 (1990).
Syn: *Carpolobia sp.* – F.P.S. 1: 69 (1950).
Shrub or small tree. Forest.
Distr: EQU; Uganda & Kenya, S to Zambia &
Mozambique; Madagascar.
EQU: Jackson 1275 (fl fr) 3/1950.

Polygala abyssinica R.Br. ex Fresen.

F.P.S. 1: 71 (1950); F.J.M.: 87 (1976); Biol. Skr. 51: 83
(1998); F.E.E. 2(1): 185 (2000); F.T.E.A. Polygal.: 57
(2007).
Perennial herb. Montane grassland, wooded grassland,
open stony ground.
Distr: RS, DAR, CS, EQU; Eritrea to Somalia, S to
Tanzania; Arabia, India.
DAR: Jackson 2588 (fl fr) 1/1953; **EQU:** Thomas 1814 (fl
fr) 12/1935.

Polygala acicularis Oliv.

F.P.S. 1: 70 (1950); F.T.E.A. Polygal.: 49 (2007).
Shrublet, rarely an annual herb. Wet grassland, wooded
grassland.
Distr: BAG, EQU; Nigeria to Uganda, S to Angola & D.R.
Congo.
EQU: Jackson 2233 (fl fr) 6/1952.

Polygala albida Schinz subsp. *stanleyana*
(Chodat) Paiva

Biol. Skr. 51: 84 (1998); F.E.E. 2(1): 184 (2000); F.T.E.A.
Polygal.: 37 (2007).
Syn: *Polygala stanleyana* Chodat – F.P.S. 1: 72 (1950).
Annual herb. Grassland, wooded grassland, weed of
cultivation.
Distr: EQU; Guinea to Ethiopia, S to Angola &
Mozambique.
EQU: Myers 7808 (fl) 9/1937.

Polygala arenaria Willd.

F.P.S. 1: 72 (1950); F.J.M.: 87 (1976); F.E.E. 2(1): 184
(2000); F.T.E.A. Polygal.: 33 (2007).
Annual herb. Open woodland, grassland, disturbed
areas.
Distr: DAR, BAG, UN, EQU; Senegal to Ethiopia, S to
Angola & Mozambique.
DAR: Wickens 2729 (fl) 9/1964; **EQU:** Andrews 75A (fl)
11/1933.

Polygala erioptera DC.

F.P.S. 1: 73 (1950); Biol. Skr. 51: 84 (1998); F.E.E. 2(1):
182 (2000); F.T.E.A. Polygal.: 25 (2007).
Annual or perennial herb or shrublet. Woodland to semi-
desert, disturbed areas.

Distr: NS, RS, DAR, KOR, CS, ES, EQU; Cape Verde Is., Senegal to Somalia, S to Namibia & South Africa; N Africa to India.
KOR: Harrison 976 (fl fr) 9/1950; **EQU:** Kielland-Lund 545 1/1984 (n.v.).
Note: the African Plants Database records also *P. ehlersii* Gürke from Sudan but this is an error, since *P. ehlersii* is known only from Kenya and Tanzania.

Polygala irregularis Boiss.
F.P.S. 1: 73 (1950); F.E.E. 2(1): 187 (2000); F.T.E.A. Polygal.: 52 (2007).
Shrublet or perennial herb. Open sandy & rocky areas.
Distr: RS, DAR, KOR, CS; Senegal to Eritrea, S to Kenya; Egypt, Arabia.
KOR: Wickens 124 (fl fr) 8/1962.

Polygala macrostigma Chodat
F.T.E.A. Polygal.: 40 (2007).
Annual herb. Woodland & wooded grassland.
Distr: EQU; Uganda & Kenya, S to Angola & Mozambique.
EQU: Sillitoe 172 (fl) 1919.

Polygala muratii Jacq.-Fél.
F.E.E. 2(1): 187 (2000); F.T.E.A. Polygal.: 51 (2007).
Syn: *Polygala liniflora* sensu Andrews, non Bojer ex Chodat – F.P.S. 1: 70 (1950).
Perennial or annual herb. Open woodland, grassland, often over rocks.
Distr: DAR; Chad, Ethiopia, Uganda, Kenya, Tanzania.
DAR: Lynes 308 (fl fr) 2/1922.

Polygala myriantha Chodat
F.E.E. 2(1): 188 (2000); F.T.E.A. Polygal.: 24 (2007).
Annual herb. Montane grassland.
Distr: DAR; Nigeria to Ethiopia, S to Angola & Zimbabwe.
DAR: de Wilde et al. 5392 (fl fr) 1/1965.

Polygala obtusissima Chodat
F.E.E. 2(1): 180 (2000); F.T.E.A. Polygal.: 28 (2007).
Syn: *Polygala senensis* sensu Andrews, non Klotzsch – F.P.S. 1: 72 (1950).
Shrublet. Dry bushland.
Distr: RS; Ethiopia, Somalia; Arabia.
RS: Ahti 16497 10/1962 (n.v.).
Note: the specimen cited has not been seen; it was recorded by Ahti et al. in Ann. Bot. Fennici 10: 144 (1973) under *P. senensis*; the taxonomy here follows F.E.E.

Polygala persicariifolia DC.
F.P.S. 1: 72 (1950); F.J.M.: 87 (1976); Biol. Skr. 51: 84 (1998); F.E.E. 2(1): 184 (2000); F.T.E.A. Polygal.: 38 (2007).
Annual herb. Forest margins, woodland, secondary bushland, weed of cultivation.
Distr: DAR, ES, EQU; Guinea to Eritrea, S to Zambia & Mozambique; India.
DAR: Wickens 2518 (fl fr) 9/1964; **EQU:** Friis & Vollesen 765 (fl fr) 12/1980.

Polygala petitiana A.Rich. subsp. *petitiana* var. *petitiana*
F.J.M.: 87 (species only) (1976); F.E.E. 2(1): 188 (2000); F.T.E.A. Polygal.: 47 (2007).

Syn: *Polygala nilotica* Chodat – F.P.S. 1: 71 (see note) (1950).
Annual herb. Grassland & woodland.
Distr: DAR, ?BAG (see note); Mali to Ethiopia, S to Angola & Mozambique.
DAR: Wickens 2501 (fl fr) 9/1964.
Note: Andrews records *P. nilotica* based upon *Schweinfurth* 2130 from Bongo, near the Bahr el Ghazal River; Paiva in Fontqueria 50: 311 (1998) notes that the type has not been located but that the description and illustration suggest that *P. nilotica* is a synonym of *P. petitiana* s.s. – this would be the only record of this taxon for South Sudan.

Polygala petitiana A.Rich. subsp. *parviflora* (Exell) Paiva
Biol. Skr. 51: 84 (1998); F.T.E.A. Polygal.: 47 (2007).
Syn: *Polygala petitiana* sensu Andrews, non A.Rich. s.s. – F.P.S. 1: 70 (1950).
Annual herb. Grassland, woodland & scrub.
Distr: EQU; Uganda to Zambia, Zimbabwe & Mozambique.
EQU: Friis & Vollesen 194 (fl fr) 11/1980.

Polygala sadebeckiana Gürke
F.E.E. 2(1): 183 (2000); F.T.E.A. Polygal.: 26 (2007).
Syn: *Polygala kassasii* Chrtek – Fontqueria 50: 176 (1998).
Perennial herb. Wooded grassland, forest margins, grassland on clays.
Distr: ES; Ethiopia to Malawi & Mozambique.
ES: Kassas 319 (fl fr) 12/1967.
Note: the Sudanese *P. kassasii* was maintained as a distinct species by Paiva in his revision of the genus (l.c.) based on its longer seed caruncle appendage. However, Gilbert in F.E.E. treats it as a synonym of the variable *P. sadebeckiana*.

Polygala schweinfurthii Chodat
F.P.S. 1: 72 (1950); F.T.E.A. Polygal.: 38 (2007).
Annual herb. Woodland, grassland, disturbed areas.
Distr: BAG, EQU; Cameroon to Uganda, S to D.R. Congo & Burundi.
EQU: Myers 7763 (fl) 9/1937.

Polygala septentrionalis Troupin
Fontqueria 50: 200 (1998).
Annual herb. Wooded grassland over ironstone.
Distr: EQU; Cameroon, C.A.R., D.R. Congo.
IUCN: RD
EQU: Myers 9371 (fl) 8/1938.

Polygala sphenoptera Fresen.
Biol. Skr. 51: 85 (1998); F.E.E. 2(1): 180 (2000); F.T.E.A. Polygal.: 31 (2007).
Annual or often perennial herb. Open bushland & woodland, open rocky ground.
Distr: EQU; Eritrea & Somalia, S to Angola, Botswana & South Africa.
EQU: Myers 11107 (fl) 4/1939.

Securidaca longipedunculata Fresen. var. *longipedunculata*
F.P.S. 1: 75 (1950); F.J.M.: 87 (1976); T.S.S.: 45 (1990); Biol. Skr. 51: 85 (1998); F.E.E. 2(1): 177 (2000); F.T.E.A. Polygal.: 6 (2007).

Shrub or small tree. Woodland, wooded grassland.
Distr: DAR, KOR, CS, ES, BAG, EQU; Senegal to Ethiopia, S to Namibia & South Africa.
DAR: Kamil 1069 (fl) 5/1968; **EQU:** Myers 8485 (fl) 2/1938.

Securidaca welwitschii Oliv.
Biol. Skr. 51: 85 (1998); F.T.E.A. Polygal.: 4 (2007).
Woody climber or scandent shrub. Forest.
Distr: EQU; Guinea to Uganda, S to Angola & Zambia.
EQU: Friis & Vollesen 923 (fr) 2/1982.

ROSALES

Rosaceae

I. Darbyshire

Alchemilla abyssinica Fresen.
F.E.E. 3: 40 (1989); Biol. Skr. 51: 175 (1998).
Stoloniferous perennial herb. Damp areas in montane forest & ericaceous bushland.
Distr: EQU; Ethiopia, Uganda, Kenya.
EQU: Friis & Vollesen 122 (fl) 11/1980.

Alchemilla argyrophylla Oliv. subsp. argyrophylla
F.T.E.A. Ros.: 7 (1960); T.S.S.: 143 (1990); Biol. Skr. 51: 175 (1998).
Subshrub. Montane bushland, dry rocky ground, montane grassland.
Distr: EQU; Uganda, Kenya.
EQU: Howard I.M.55 (fl) 2/1976.

Alchemilla cryptantha Steud. ex A.Rich.
F.P.S. 2: 101 (1952); F.T.E.A. Ros.: 15 (1960); F.E.E. 3: 39 (1989); Biol. Skr. 51: 175 (1998).
Short-lived perennial herb. Montane grassland, forest clearings, streamsides.
Distr: EQU; Cameroon to Eritrea, S to South Africa; Madagascar, Arabia.
EQU: Andrews 2011 (fr) 6/1939.

Alchemilla ellenbeckii Engl. subsp. ellenbeckii
F.T.E.A. Ros.: 12 (1960); F.E.E. 3: 38 (1989); Biol. Skr. 51: 176 (1998).
Scrambling perennial herb. Montane swamp, streamsides, moist ground in ericaceous bushland.
Distr: EQU; D.R. Congo, Ethiopia, Uganda, Kenya, Tanzania.
EQU: Thomas 1803 12/1935.

Hagenia abyssinica (Bruce) J.F.Gmel.
F.P.S. 2: 101 (1952); F.T.E.A. Ros.: 43 (1960); F.E.E. 3: 43 (1989); T.S.S.: 143 (1990); Biol. Skr. 51: 176 (1998).
Tree. Montane forest & upper forest margin where often dominant.
Distr: EQU; Eritrea to Zambia & Malawi.
EQU: Thomas 1888 (fl) 12/1935.

Potentilla supina L.
F.P.S. 2: 106 (1952); F.Egypt 1: 246 (1999).
Annual herb. Alluvial flats, riverbanks.

Distr: NS, CS; central & S Europe, N Africa to Tibet.
CS: de Wilde et al. 5794 (fl fr) 2/1965.

Prunus africana (Hook.f.) Kalkman
F.E.E. 3: 41 (1989); Biol. Skr. 51: 176 (1998).
Syn: *Pygeum africanum* Hook.f. – F.P.S. 2: 106 (1952); T.S.S.: 145 (1990).
Tree, often large. Montane & riverine forest.
Distr: EQU; Cameroon to Ethiopia, S to Angola & South Africa.
IUCN: VU
EQU: Thomas 1682 (fl) 12/1935.

Rosa abyssinica Lindl.
F.E.E. 3: 35 (1989).
Shrub, often scandent. Montane forest & margins, secondary scrub, roadsides, rock outcrops.
Distr: RS; Eritrea, Ethiopia, Somalia; Yemen.
RS: Jackson 3971 4/1959.
Note: the single Sudanese collection is sterile; fertile material is desirable to confirm the identity.

Rubus apetalus Poir.
F.T.E.A. Ros.: 39 (1960); F.E.E. 3: 34 (1989); Biol. Skr. 51: 177 (1998).
Scrambling shrub. Forest margins, secondary scrub, streamsides.
Distr: EQU; Nigeria to Eritrea, S to Malawi; Madagascar, Mascarenes.
EQU: Friis & Vollesen 1216 (fr) 3/1982.

Rubus niveus Thunb.
F.T.E.A. Ros.: 40 (1960); F.J.M.: 105 (1976); F.E.E. 3: 34 (1989).
Shrub. Naturalised on riverbank.
Distr: DAR; native to India & Malaya, introduced in E Africa.
DAR: Wickens 1479 (fl fr) 4/1964.

Rubus pinnatus Willd.
F.T.E.A. Ros.: 37 (1960); Biol. Skr. 51: 177 (1998).
Syn: *Rubus pinnatus* Willd. var. *afrotropicus* (Engl.) Gust. – Biol. Skr. 51: 177 (1998).
Shrub, often scrambling. Forest margins, secondary scrub.
Distr: EQU; Guinea to Uganda, S to South Africa; St Helena, Ascension Is.
EQU: Friis & Vollesen 150 (fr) 11/1980.

Rubus steudneri Schweinf. var. steudneri
F.P.S. 2: 106 (1952); F.T.E.A. Ros.: 26 (1960); F.E.E. 3: 34 (1989); T.S.S.: 145 (1990); Biol. Skr. 51: 177 (1998).
Syn: *Rubus steudneri* Schweinf. var. *aberensis* Engl. ex Gust. – F.P.S. 2: 106 (1952); Biol. Skr. 51: 177 (1998).
Scandent shrub. Montane forest & forest margins, secondary scrub.
Distr: EQU; D.R. Congo, Ethiopia, Uganda, Kenya, Tanzania.
EQU: Friis & Vollesen 397 (fr) 11/1980.

Rhamnaceae

I. Darbyshire

Berchemia discolor (Klotzsch) Hemsl.

F.P.S. 2: 298 (1952); F.T.E.A. Rhamn.: 32 (1972); F.E.E. 3: 393 (1989); T.S.S.: 289 (1990); Biol. Skr. 51: 266 (1998). Small tree. Wooded grassland, bushland, riverine forest, often on rock outcrops.
Distr: EQU; Eritrea to Namibia & South Africa; Madagascar, Arabia.
EQU: Jackson 676 4/1949 (n.v.).
Note: the specimen cited has not been seen by us; the identification follows Friis & Vollesen.

Gouania longispicata Engl.

F.P.S. 2: 299 (1952); F.T.E.A. Rhamn.: 12 (1972); F.E.E. 3: 386 (1989); T.S.S.: 289 (1990); Biol. Skr. 51: 267 (1998). Scandent shrub or woody climber. Forest, often on margins & in disturbed areas.
Distr: UN, EQU; Nigeria to Ethiopia, S to Zimbabwe & Mozambique.
EQU: Thomas 1669 (fr) 12/1935.

Gouania sp. A

Woody climber. Habitat not recorded.
Distr: UN.
UN: Sherif A.3843 (fr) 9/1951.
Note: the specimen at K is identified as *G. leptostachya* DC. vel. sp. aff.; *G. leptostachya* is an Indian species and has longer and stouter peduncles than the Sudan specimen.

Lasiodiscus mildbraedii Engl.

F.P.S. 2: 299 (1952); F.T.E.A. Rhamn.: 5 (1972); T.S.S.: 289 (1990); Biol. Skr. 51: 267 (1998).
Small tree or shrub. Forest understorey.
Distr: EQU; Ivory Coast to Uganda, S to Tanzania & Mozambique.
EQU: Friis & Vollesen 788 (fr) 12/1980.

Maesopsis eminii Engl. subsp. *eminii*

F.P.S. 2: 299 (1952); F.T.E.A. Rhamn.: 38 (1972); T.S.S.: 290 (1990); Biol. Skr. 51: 267 (1998).
Tree. Forest, sometimes as a pioneer.
Distr: BAG, EQU; Uganda to Angola & Tanzania.
EQU: Sommerlatte 80 5/1984 (n.v.).
Note: the specimen cited has not been seen by us; the identification follows Friis & Vollesen.

Rhamnus prinoides L'Hér.

F.P.S. 2: 299 (1952); F.T.E.A. Rhamn.: 18 (1972); F.E.E. 3: 390 (1989); T.S.S.: 290 (1990); Biol. Skr. 51: 268 (1998). Shrub or small tree. Montane forest, often along margins, secondary bushland.
Distr: EQU; Cameroon to Eritrea, S to Angola & South Africa; Arabia.
EQU: Andrews 1881 (fr) 6/1939.

Scutia myrtina (Burm.f.) Kurz

F.T.E.A. Rhamn.: 21 (1972); F.E.E. 3: 390 (1989); T.S.S.: 290 (1990); Biol. Skr. 51: 268 (1998).

Shrub or small tree. Forest, evergreen bushland, woodland.
Distr: EQU; Ethiopia to South Africa; Madagascar, Mascarenes, India to Vietnam.
EQU: Jackson 654 4/1949 (n.v.).
Note: the specimen cited has not been seen by us; the identification follows Friis & Vollesen.

Ziziphus abyssinica Hochst. ex A.Rich.

F.P.S. 2: 300 (1952); F.T.E.A. Rhamn.: 27 (1972); F.J.M.: 121 (1976); F.E.E. 3: 396 (1989); T.S.S.: 291 (1990); Biol. Skr. 51: 268 (1998).
Shrub or small tree. Woodland & wooded grassland, margins of riverine forest, rock outcrops.
Distr: DAR, KOR, CS, ES, BAG, UN, EQU; Senegal to Eritrea, S to Angola & Mozambique.
KOR: Broun 1339 (fr) 12/1907; **EQU:** Andrews 1813 (fl) 6/1939.

Ziziphus mauritiana Lam.

F.P.S. 2: 300 (1952); F.T.E.A. Rhamn.: 29 (1972); F.J.M.: 121 (1976); F.E.E. 3: 396 (1989); T.S.S.: 291 (1990); Biol. Skr. 51: 269 (1998).
Shrub or tree. *Acacia* woodland, termite mounds in wooded grassland, riverbanks.
Distr: NS, ?DAR, KOR, CS, ES, BAG, UN, EQU; Senegal to Ethiopia, S to Tanzania, widely cultivated elsewhere, possibly native in parts of Asia.
CS: Lea 37 (fl) 10/1952; **EQU:** Simpson 7323 (fr) 7/1929.
Note: the record from Darfur is unconfirmed (see Wickens, l.c.).

Ziziphus mucronata Willd. subsp. *mucronata*

F.P.S. 2: 300 (1952); F.T.E.A. Rhamn.: 25 (1972); F.E.E. 3: 393 (1989); T.S.S.: 293 (1990).
Shrub or tree. Bushland & woodland, dry riverine forest.
Distr: DAR, KOR, CS, ES, BAG, UN, EQU; Senegal to Somalia, S to South Africa; Madagascar, Arabia.
DAR: Kamil 1036 5/1968; **EQU:** Myers 14074 9/1941.

Ziziphus pubescens Oliv.

F.P.S. 2: 302 (1952); F.T.E.A. Rhamn.: 24 (1972); F.E.E. 3: 393 (1989); T.S.S.: 293 (1990); Biol. Skr. 51: 269 (1998). Tree. Dry forest including riverine fringes, woodland.
Distr: BAG, UN, EQU; Ethiopia to Angola & Zimbabwe.
UN: J.M. Lock 81/78 (fl) 5/1981.

Ziziphus spina-christi (L.) Desf.

F.P.S. 2: 300 (1952); F.T.E.A. Rhamn.: 30 (1972); F.J.M.: 121 (1976); F.E.E. 3: 395 (1989); T.S.S.: 293 (1990).
Tree or shrub. Woodland & bushland, rocky hillslopes.
Distr: NS, RS, DAR, CS, ES; native to Near & Middle East & N Africa, probably introduced in E Africa where well naturalised.
DAR: Wickens 2899 (fl) 4/1965.
Note: El Amin in T.S.S. records both var. *spina-christi* and var. *microphylla* A.Rich. from Sudan whilst Wickens in F.J.M. only records var. *spina-christi*. The latter is considered to be the native variety in Africa, but the varieties are not maintained in F.E.E.

Ulmaceae

I. Darbyshire

Chaetachme aristata E.Mey ex Planch.

F.P.S. 2: 254 (1952); F.T.E.A. Ulm.: 12 (1966); T.S.S.: 255 (1990); Biol. Skr. 51: 240 (1998).
Shrub or small tree. Forest & forest margins, shaded rocks.
Distr: EQU; Sierra Leone to Kenya, S to Angola & Tanzania; Madagascar.
EQU: Myers 11039 (fl) 4/1939.

Holoptelea grandis (Hutch.) Mildbr.

F.P.S. 2: 255 (1952); F.T.E.A. Ulm.: 3 (1966); T.S.S.: 255 (1990).
Large tree. Lowland forest including riverine fringes.
Distr: EQU; Ivory Coast to Uganda.
EQU: Turner 154 (fr) 2/1936.

Cannabaceae

I. Darbyshire

Cannabis sativa L.

F.P.S. 2: 280 (1952); F.T.E.A. Cannab.: 1 (1975); F.E.E. 3: 327 (1989); Biol. Skr. 51: 254 (1998).
Annual herb. Cultivated & occasionally naturalised as a weed.
Distr: EQU; originating from Asia, now widely cultivated.
EQU: Jackson 834 7/1949.

Celtis adolfi-friderici Engl.

F.P.S. 2: 253 (1952); F.T.E.A. Ulm.: 9 (1966); T.S.S.: 251 (1990).
Large tree. Forest.
Distr: EQU; Ivory Coast to Uganda.
EQU: Andrews 1395 5/1939.

Celtis africana Burm.f.

F.T.E.A. Ulm.: 4 (1966); F.E.E. 3: 266 (1989); T.S.S.: 251 (1990); Biol. Skr. 51: 238 (1998).
Syn: Celtis kraussiana Bernh. – F.P.S. 2: 253 (1952).
Tree. Forest including riverine fringes, evergreen bushland, often a pioneer.
Distr: EQU; Ghana to Ethiopia, S to Angola & South Africa; Yemen.
EQU: Myers 11033 (fr) 4/1939.

Celtis gomphophylla Baker

F.T.E.A. Ulm.: 6 (1966); F.E.E. 3: 267 (1989); Biol. Skr. 51: 239 (1998).
Small tree. Forest.
Distr: EQU; Ivory Coast to Ethiopia, S to Angola & South Africa; Madagascar, Comoros.
EQU: Howard U.T.T.17 (fr) 11/1981.

Celtis mildbraedii Engl.

F.T.E.A. Ulm.: 7 (1966); T.S.S.: 253 (1990); Biol. Skr. 51: 239 (1998).
Syn: Celtis soyauxii Engl. – F.P.S. 2: 251 (1952).
Tree. Lowland forest.
Distr: ?EQU; Guinea to Kenya, S to Angola & South Africa.
Note: no material has been seen from South Sudan but this species has been recorded from, for example,

Talanga in the Imatong Mts (see Friis & Vollesen 1998) and these records are likely to be accurate.

Celtis philippensis Blanco

F.T.E.A. Ulm.: 9 (1966); F.E.E. 3: 267 (1989); Biol. Skr. 51: 239 (1998).
Syn: Celtis scotellioides A.Chev. – F.P.S. 2: 253 (1952).
Syn: Celtis wightii Planch. – T.S.S.: 253 (1990).
Tree or shrub. Forest, sometimes in secondary growth.
Distr: EQU; Senegal to Ethiopia, S to Angola & Mozambique; tropical Asia & N Australia.
EQU: Myers 11368 (fr) 5/1939.

Celtis toka (Forssk.) Hepper & J.R.I.Wood

F.T.E.A. Ulm.: 9 (1966); F.E.E. 3: 267 (1989); Biol. Skr. 51: 239 (1998).
Syn: Celtis integrifolia Lam. – F.P.S. 2: 251 (1952); Bot. Exp. Sud.: 108 (1970); F.J.M.: 118 (1976); T.S.S.: 253 (1990).
Tree. Riverine forest.
Distr: DAR, KOR, CS, BAG, UN, EQU; Senegal to Ethiopia & Uganda; Yemen.
DAR: Wickens 2897 (fr) 4/1965; **EQU:** Myers 8509 (fr) 2/1938.

Celtis zenkeri Engl.

F.P.S. 2: 251 (1952); F.T.E.A. Ulm.: 8 (1966); F.E.E. 3: 267 (1989); T.S.S.: 255 (1990); Biol. Skr. 51: 240 (1998).
Tree. Forest.
Distr: EQU; Guinea to Ethiopia, S to Angola & Tanzania.
EQU: Andrews 1752 (fr) 6/1939.

Trema orientalis (L.) Blume

F.J.M.: 118 (1976); F.E.E. 3: 268 (1989); T.S.S.: 257 (1990); Biol. Skr. 51: 240 (1998).
Syn: Trema guineensis (Schum. & Thonn.) Ficalho – F.P.S. 2: 256 (1952).
Small tree or shrub. Pioneer of forest margins & clearings, persisting in secondary bushland.
Distr: DAR, EQU; Senegal to Ethiopia, S to South Africa; Madagascar, Mascarenes, Arabia, tropical Asia.
DAR: Wickens 3691 6/1977; **EQU:** Myers 7965 (fl fr) 12/1937.

Moraceae

I. Darbyshire

Antiaris toxicaria Lesch. subsp. welwitschii (Engl.) C.C.Berg var. africana (Engl.) A.Chev.

F.T.E.A. Mor.: 13 (1989).
Syn: Antiaris africana Engl. – F.P.S. 2: 257 (1952); T.S.S.: 257 (1990).
Large tree. Wooded grassland, amongst rocks.
Distr: BAG, EQU; Senegal to Uganda.
EQU: Myers 10360 (fr) 1/1939.

Dorstenia annua Friis & Vollesen

Kew Bull. 37: 473 (1982); Biol. Skr. 51: 241 (1998).
Annual herb. Upland forest.
Distr: EQU; South Sudan endemic.
IUCN: RD
EQU: Friis & Vollesen 29 (fl fr) 11/1980.
Note: known only from the type collection cited here.

Dorstenia barnimiana Schweinf.

F.P.S. 2: 259 (1952); F.T.E.A. Mor.: 41 (1989); F.E.E. 3: 278 (1989); Biol. Skr. 51: 242 (1998).
Syn: *Dorstenia barnimiana* Schweinf. var. *ophioglossoides* (Hochst. ex Bureau) Engl. – F.P.S. 2: 260 (1952); F.T.E.A. Mor.: 42 (1989).
Syn: *Dorstenia palmata* Engl. – F.P.S. 2: 260 (1952).
Tuberous perennial herb. Grassland & wooded grassland, often on shallow soil over rock.
Distr: BAG; Cameroon to Somalia, S to Zambia; Yemen.
BAG: Schweinfurth 1830 (fr) 5/1869.

Dorstenia benguellensis Welw.

F.T.E.A. Mor.: 38 (1989).
Tuberous perennial herb. Woodland, wooded grassland, amongst rocks.
Distr: EQU; Cameroon to Kenya, S to Angola, Zimbabwe & Mozambique.
EQU: Jackson 3204 (fl).

Dorstenia brownii Rendle

F.T.E.A. Mor.: 30 (1989); F.E.E. 3: 275 (1989); Biol. Skr. 51: 242 (1998).
Rhizomatous perennial herb. Wet forest, often amongst rocks.
Distr: EQU; D.R. Congo, Ethiopia, Uganda, Kenya, Tanzania.
EQU: Friis & Vollesen 119 (fl fr) 11/1980.

Dorstenia cuspidata Hochst. ex A.Rich. var. cuspidata

F.T.E.A. Mor.: 40 (1989); F.E.E. 3: 277 (1989).
Syn: *Dorstenia walleri* Hemsl. – F.P.S. 2: 259 (1952); F.J.M.: 118 (1976).
Tuberous perennial herb. Woodland, often amongst rocks.
Distr: DAR, KOR, BAG, EQU; Senegal to Ethiopia, S to Zimbabwe & Mozambique.
KOR: Wickens 654 (fr) 10/1962; BAG: Jackson 3089a (fr) 7/1953.

Dorstenia foetida (Forssk.) Schweinf. var. obovata (A.Rich.) Schweinf. & Engl.

F.P.S. 2: 277 (1952); F.T.E.A. Mor.: 41 (1989); F.E.E. 3: 277 (1989).
Perennial herb with swollen stem base. Rock outcrops in bushland.
Distr: ?RS; Ethiopia; Yemen.
Note: no material has been seen from Sudan but both Andrews and F.E.E. list this taxon from the Red Sea Hills.

Dorstenia psilurus Welw. var. psilurus

F.P.S. 2: 259 (1952); F.T.E.A. Mor.: 31 (1989); Biol. Skr. 51: 242 (1998).
Syn: *Dorstenia bicornis* Schweinf. – F.P.S. 2: 258 (1952).
Rhizomatous perennial herb. Wet forest.
Distr: EQU; Cameroon to Uganda, S to Angola, Zimbabwe & Mozambique.
EQU: Jackson 3023 (fl) 6/1953.

Dorstenia tropaeolifolia (Schweinf.) Bureau

F.E.E. 3: 278 (1989).
Syn: *Dorstenia barnimiana* Schweinf. var. *tropaeolifolia* (Schweinf.) Rendle – F.P.S. 2: 260 (1952); F.T.E.A. Mor.: 42 (1989).

Tuberous perennial herb. Grassland & wooded grassland, often on shallow soil over rock.
Distr: ?ES, EQU; Cameroon to Somalia & Kenya.
EQU: Myers 6552 (fl fr) 5/1937.
Note: Andrews recorded this species from Gallabat, but the specimen on which this record is based (*Schweinfurth 364*) was actually collected from the Gendua River which is in Ethiopia. Of course, this species is also likely to occur on the Sudan side.

Ficus abutilifolia (Miq.) Miq.

F.P.S. 2: 272 (1952); F.T.E.A. Mor.: 67 (1989); F.E.E. 3: 296 (1989); T.S.S.: 261 (1990); Biol. Skr. 51: 243 (1998).
Tree. Rocky slopes.
Distr: KOR, CS, EQU; Guinea to Ethiopia, Zambia & Mozambique to South Africa.
KOR: Simpson 7746 (fl) 4/1930; EQU: Myers 9878 (fr) 10/1938.

Ficus asperifolia Miq.

F.T.E.A. Mor.: 53 (1989); Biol. Skr. 51: 243 (1998).
Syn: *Ficus urceolaris* Welw. ex Hiern – F.P.S. 2: 268 (1952); T.S.S.: 273 (1990).
Scandent or straggling shrub. Forest, often along margins, streamsides.
Distr: EQU; Senegal to Uganda, S to Angola & Zambia.
EQU: Thomas 1591 (fr) 12/1935.

Ficus capreifolia Delile

F.P.S. 2: 268 (1952); F.T.E.A. Mor.: 53 (1989); F.E.E. 3: 281 (1989); T.S.S.: 263 (1990); Biol. Skr. 51: 243 (1998).
Syn: *Ficus antithetophylla* Steud. ex Miq. – T.S.S.: 263 (1990).
Shrub or small tree. Riverbanks, riverine scrub.
Distr: NS, CS, ES, BAG, UN, EQU; Senegal to Somalia, S to Angola & South Africa.
ES: Schweinfurth 550 (fl) 6/1865; EQU: Thomas 1597 (fr) 12/1935.

Ficus cordata Thunb. subsp. salicifolia (Vahl) C.C.Berg

F.T.E.A. Mor.: 63 (1989); Biol. Skr. 51: 244 (1998).
Syn: *Ficus salicifolia* Vahl – F.P.S. 2: 265 (1952); Bot. Exp. Sud.: 86 (1970); F.E.E. 3: 290 (1989); T.S.S.: 271 (1990).
Tree or shrub. Rocky hillslopes, riverine forest & scrub, woodland.
Distr: RS, DAR, KOR, CS, ES; Ethiopia to Zambia & South Africa; Egypt, Arabia.
DAR: Wickens 1296 (fr) 2/1964.

Ficus cyathistipula Warb.

F.T.E.A. Mor.: 83 (1989); T.S.S.: 265 (1990); Biol. Skr. 51: 244 (1998).
Tree, hemi-epiphytic. Forest, often riverine or on rocks.
Distr: EQU; Ivory Coast to Uganda, S to Angola, Zambia & Tanzania.
EQU: Friis & Vollesen 552 (fr) 11/1980.

Ficus dicranostyla Mildbr.

F.P.S. 2: 265 (1952); F.T.E.A. Mor.: 59 (1989); F.E.E. 3: 289 (1989); T.S.S.: 265 (1990); Biol. Skr. 51: 244 (1998).
Tree or shrub. Wooded grassland, woodland.
Distr: BAG, EQU; Senegal to Ethiopia, S to D.R. Congo & Zambia.

EQU: <u>Myers 9287</u> (fr) 8/1938.

Ficus exasperata Vahl
F.P.S. 2: 268 (1952); F.T.E.A. Mor.: 52 (1989); F.E.E. 3: 284 (1989); T.S.S.: 265 (1990); Biol. Skr. 51: 244 (1998).
Tree or shrub. Forest, often along margins, secondary scrub.
Distr: ?CS, EQU; Senegal to Djibouti, S to Angola & Mozambique; Yemen, India & Sri Lanka.
EQU: <u>Turner 191</u> (fr).
Note: the record for Central Sudan is from El Amin in T.S.S.; no specimen has been seen to verify this.

Ficus glumosa Delile
F.P.S. 2: 270 (1952); F.J.M.: 118 (1976); F.T.E.A. Mor.: 65 (1989); F.E.E. 3: 293 (1989); T.S.S.: 267 (1990); Biol. Skr. 51: 245 (1998).
Syn: *Ficus glumosa* Delile var. *glaberrima* Martelli – F.P.S. 2: 270 (1952).
Tree or shrub. Rock outcrops & screes, open wooded grassland.
Distr: RS, DAR, KOR, CS, ES, BAG, EQU; Senegal to Ethiopia, S to Namibia & South Africa; Yemen.
DAR: <u>Wickens 1650</u> (fl) 5/1964; **EQU:** <u>Friis & Vollesen 442</u> (fr) 11/1980.

Ficus ingens (Miq.) Miq.
F.P.S. 3: 268 (1952); F.J.M.: 119 (1976); F.T.E.A. Mor.: 60 (1989); F.E.E. 3: 291 (1989); T.S.S.: 267 (1990); Biol. Skr. 51: 245 (1998).
Syn: *Ficus ingens* (Miq.) Miq. var. *tomentosa* Hutch. – F.J.M.: 119 (1976); T.S.S.: 267 (1990).
Syn: *Ficus ingentoides* Hutch. – F.P.S. 2: 268 (1952); Bot. Exp. Sud.: 85 (1970).
Tree or shrub. Wooded grassland, rock outcrops, riverine forest margins.
Distr: DAR, KOR, CS, ES, BAG, EQU; Senegal to Somalia, S to South Africa; Yemen & Saudi Arabia.
KOR: <u>Broun 1365</u> 11/1907; **EQU:** <u>Friis & Vollesen 443</u> (fr) 11/1980.

Ficus lutea Vahl
F.T.E.A. Mor.: 69 (1989); F.E.E. 3: 294 (1989); Biol. Skr. 51: 245 (1998).
Tree, hemi-epiphytic. Forest, persisting in secondary bush.
Distr: EQU; Cape Verde Is., Senegal to Ethiopia, S to Angola & South Africa; Madagascar, Seychelles.
EQU: <u>Jackson 1725</u>.

Ficus mucuso Welw. ex Ficalho
F.T.E.A. Mor.: 56 (1989); Biol. Skr. 51: 246 (1998).
Tree. Forest, sometimes in secondary regrowth.
Distr: ?EQU; Guinea Bissau to Uganda, S to Angola & Tanzania.
Note: no South Sudanese material has been seen but Jackson reported this species from the Imatong Mts; this record needs confirming.

Ficus ovata Vahl
F.P.S. 2: 270 (1952); F.T.E.A. Mor.: 81 (1989); F.E.E. 3: 292 (1989); T.S.S.: 269 (1990); Biol. Skr. 51: 246 (1998).
Tree, shrub or woody climber; often hemi-epiphytic. Forest, often riverine, woodland & wooded grassland.
Distr: ?DAR, BAG, EQU; Senegal to Ethiopia, S to Angola & Mozambique.

DAR: <u>Wickens 3689</u> 6/1977; **EQU:** <u>Turner 11</u> 6/1930.
Note: the cited specimen from Darfur is only tentatively identified as *F.* sp. cf. *ovata*.

Ficus palmata Forssk.
F.P.S. 2: 263 (1952); Bot. Exp. Sud.: 85 (1970); F.J.M.: 119 (1976); F.E.E. 3: 281 (1989); T.S.S.: 269 (1990).
Shrub or small tree. Riverine forest, streambanks, upland scrub & grassland.
Distr: ?RS, DAR, ?BAG, ?EQU; Eritrea, Ethiopia, Somalia; Egypt to India.
DAR: <u>Wickens 2444</u> (fr) 9/1964; **EQU:** <u>Drar 1378</u> 4/1938 (n.v.).
Note: the records from the Red Sea region and Equatoria are based on sterile material collected by Drar (1970). The record for Bahr el Ghazal is from El Amin in T.S.S.

Ficus platyphylla Delile
F.P.S. 2: 272 (1952); F.J.M.: 119 (1976); F.T.E.A. Mor.: 64 (1989); F.E.E. 3: 296 (1989); T.S.S.: 269 (1990); Biol. Skr. 51: 246 (1998).
Tree. Riverine forest & scrub, rocky areas in woodland or grassland.
Distr: NS, DAR, KOR, CS, BAG, UN, EQU; Senegal to Somalia & Uganda.
DAR: <u>Wickens 3127</u> (fr) 4/1971; **BAG:** <u>Schweinfurth 1288</u> 3/1869.

Ficus polita Vahl subsp. *polita*
F.P.S. 2: 269 (1952); F.T.E.A. Mor.: 80 (1989); T.S.S.: 269 (1990).
Tree, often hemi-epiphytic. Forest.
Distr: BAG, EQU; Senegal to Uganda, S to Angola & South Africa; Madagascar.
EQU: <u>Schweinfurth 3134</u> (fr) 2/1870.

Ficus populifolia Vahl
F.P.S. 2: 270 (1952); F.T.E.A. Mor.: 68 (1989); F.E.E. 3: 295 (1989); T.S.S.: 271 (1990); Biol. Skr. 51: 246 (1998).
Tree or shrub. Rocky slopes, bushland & wooded grassland.
Distr: DAR, KOR, CS, EQU; Ghana to Somalia, S to Tanzania; Yemen.
KOR: <u>Wickens 763</u> 11/1962; **EQU:** <u>Myers 7903</u> (fr) 11/1937.

Ficus saussureana DC.
F.T.E.A. Mor.: 69 (1989); Biol. Skr. 51: 246 (1998).
Syn: *Ficus eriobotryoides* Kunth & Bouché – T.S.S.: 265 (1990).
Tree, hemi-epiphytic. Forest including margins.
Distr: EQU; Guinea to Uganda & Tanzania.
EQU: <u>Andrews 1181</u> 5/1939.

Ficus sur Forssk.
F.J.M.: 119 (1976); F.T.E.A. Mor.: 56 (1989); F.E.E. 3: 287 (1989); T.S.S.: 271 (1990); Biol. Skr. 51: 247 (1998).
Syn: *Ficus capensis* Thunb. – F.P.S. 2: 265 (1952); Bot. Exp. Sud.: 85 (1970); F.J.M.: 118 (1976); T.S.S.: 263 (1990).
Tree, often large. Forest including riverine fringes, often in clearings, wooded grassland.
Distr: DAR, BAG, EQU; Cape Verde Is., Senegal to Eritrea, S to South Africa; Yemen.
DAR: <u>de Wilde et al. 5448</u> (fr) 1/1965; **EQU:** <u>Friis & Vollesen 297</u> (fl) 11/1980.

Ficus sycomorus L.

F.P.S. 2: 263 (1952); F.J.M.: 119 (1976); F.T.E.A. Mor.: 54 (1989); F.E.E. 3: 285 (1989); T.S.S.: 271 (1990); Biol. Skr. 51: 247 (1998).
Syn: *Ficus gnaphalocarpa* (Miq.) Steud. ex A.Rich. – F.P.S. 2: 265 (1952); Bot. Exp. Sud.: 85 (1970); T.S.S.: 267 (1990).
Syn: *Ficus sycomorus* L. subsp. *gnaphalocarpa* (Miq.) C.C.Berg – F.E.E. 3: 285 (1989).
Large tree. Riverine & upland forest margins, lakesides, rocky slopes, wooded grassland with localised moisture.
Distr: NS, RS, DAR, KOR, CS, ES, BAG, UN, EQU; Cape Verde Is. to Eritrea, S to Namibia & South Africa; Madagascar, Comoros, Egypt, Arabia.
ES: Jackson 4148 (fr) 4/1961; **EQU:** Friis & Vollesen 1145 (fr) 3/1982.

Ficus thonningii Blume

F.P.S. 2: 270 (1952); F.J.M.: 119 (1976); F.T.E.A. Mor.: 73 (1989); F.E.E. 3: 298 (1989); T.S.S.: 273 (1990); Biol. Skr. 51: 248 (1998).
Syn: *Ficus dekdekena* (Miq.) A.Rich. – F.P.S. 2: 272 (1952); Bot. Exp. Sud.: 85 (1970); T.S.S.: 265 (1990).
Syn: *Ficus hochstetteri* (Miq.) A.Rich. – F.J.M.: 119 (1976).
Syn: *Ficus iteophylla* Miq. – F.P.S. 2: 272 (1952); F.J.M.: 119 (1976).
Tree or shrub, sometimes hemi-epiphytic. Forest including riverine strips, rock outcrops, wooded grassland.
Distr: RS, DAR, KOR, CS, BAG, EQU; Cape Verde Is., Senegal to Ethiopia, S to Angola & South Africa.
DAR: Wickens 1321 3/1964; **EQU:** Friis & Vollesen 1210 (fr) 3/1982.

Ficus trichopoda Baker

F.T.E.A. Mor.: 68 (1989).
Syn: *Ficus congensis* Engl. – F.P.S. 2: 272 (1952) ; T.S.S.: 263 (1990).
Tree or shrub. Swampy forest & margins, riverine forest.
Distr: DAR, KOR, CS, BAG, EQU; Senegal to Uganda, S to South Africa.
DAR: Miehe 253 9/1982; **EQU:** Myers 13591 (fr) 11/1940.

Ficus vallis-choudae Delile

F.P.S. 2: 265 pro parte (1952); F.T.E.A. Mor.: 58 (1989); F.E.E. 3: 289 (1989); T.S.S.: 273 (1990); Biol. Skr. 51: 248 (1998).
Tree. Riverine forest & scrub.
Distr: EQU; Guinea to Ethiopia, S to Zimbabwe & Mozambique.
EQU: Friis & Vollesen 1237 (fr) 3/1982.
Note: Andrews records this species from Central & Southern Sudan; it is likely that some of the material on which he based this distribution was misidentified.

Ficus variifolia Warb.

F.T.E.A. Mor.: 59 (1989); T.S.S.: 275 (1990); Biol. Skr. 51: 248 (1998).
Syn: *Ficus sciarophylla* Warb. – F.P.S. 2: 269 (1952).
Large tree. Secondary forest regrowth.
Distr: BAG, EQU; Sierra Leone to Uganda, S to Angola & Tanzania.
EQU: Andrews 1759 6/1939.

Ficus vasta Forssk.

F.P.S. 2: 271 (1952); F.T.E.A. Mor.: 64 (1989); F.E.E. 3: 296 (1989); T.S.S.: 275 (1990); Biol. Skr. 51: 249 (1998).
Large tree. Forest & forest margins, often riverine, rocky outcrops, secondary bushland.
Distr: NS, RS, ES, EQU; Ethiopia, Somalia, Uganda, Kenya; Arabia.
RS: Aylmer 243 3/1932; **EQU:** Myers 10912 (fr) 4/1939.

Milicia excelsa (Welw.) C.C.Berg

F.T.E.A. Mor.: 4 (1989); F.E.E. 3: 272 (1989); Biol. Skr. 51: 249 (1998).
Syn: *Chlorophora excelsa* (Welw.) Benth. – F.P.S. 2: 257 (1952); T.S.S.: 259 (1990).
Large canopy tree. Lowland & riverine forest including disturbed areas.
Distr: BAG, UN, EQU; Guinea Bissau to Ethiopia, S to Angola, Zimbabwe & Mozambique.
IUCN: NT
BAG: Turner 117 (fl) 2/1935.
Note: Drar (1970: 84) records this species from Jebel Marra, Darfur, based on two sterile specimens which we have not seen. This seems an unlikely record and we consider this likely to be a misidentification.

Morus alba L.

Bot. Exp. Sud.: 86 (1970); F.T.E.A. Mor.: 2 (1989); F.E.E. 3: 272 (1989); T.S.S.: 275 (1990).
Syn: *Morus acidosa* Griff. – F.J.M.: 119 (1976); T.S.S.: 275 (1990).
Small tree. Naturalised hedgerow escape in swamp with cultivation.
Distr: DAR, EQU; native to warm temperate Asia, widely introduced and sometimes naturalised elsewhere.
DAR: Wickens 1052 (fl) 1/1964; **EQU:** Drar 1787 4/1938 (n.v.).
Note: Drar (1970: 86) also records *M. nigra* L. from a sterile specimen collected from Shambe, Upper Nile. This is the cultivated 'black mulberry'.

Morus australis Poir.

F.T.E.A. Mor.: 2 (1989).
Small tree. Naturalised in streamside forest near cultivation.
Distr: DAR; native to SE Asia, widely cultivated in the tropics.
DAR: Miehe 406.

Morus mesozygia Stapf

F.P.S. 2: 273 (1952); F.T.E.A. Mor.: 2 (1989); F.E.E. 3: 272 (1989); T.S.S.: 275 (1990); Biol. Skr. 51: 249 (1998).
Syn: *Morus lactea* (Sim) Mildbr. – T.S.S.: 275 (1990).
Tree. Forest including riverine fringes.
Distr: UN, EQU; Senegal to Ethiopia, S to Angola & South Africa.
EQU: Myers 11418 6/1939.

Treculia africana Decne. subsp. *africana* var. *africana*

F.P.S. 2: 273 (1952); F.T.E.A. Mor.: 10 (1989); T.S.S.: 277 (1990).
Large tree. Forest, often riverine.
Distr: EQU; Senegal to Uganda, S to Mozambique.
EQU: Andrews 1192 (fl) 5/1939.

Trilepisium madagascariense DC.

F.T.E.A. Mor.: 17 (1989); F.E.E. 3: 299 (1989); Biol. Skr.
51: 250 (1998).
Syn: Bosqueia angolensis Ficalho – T.S.S.: 257 (1990).
Tree. Wet forest.
Distr: EQU; Guinea to Ethiopia, S to Angola & South
Africa; Madagascar, Seychelles.
EQU: Jackson 1798 3/1951 (n.v.).
Note: the specimen cited has not been seen by us; the
identification follows Friis & Vollesen.

Urticaceae

I. Darbyshire

Boehmeria macrophylla Hornem.

F.T.E.A. Urtic.: 44 (1989); F.E.E. 3: 314 (1989); Biol. Skr.
51: 250 (1998).
Syn: Boehmeria platyphylla D.Don – F.P.S. 2: 275
(1952); T.S.S.: 279 (1990).
Perennial herb or shrub. Forest & forest margins, wooded
grassland.
Distr: BAG, EQU; Guinea to Ethiopia, S to Angola &
Mozambique; Madagascar, tropical Asia to SW China.
EQU: Jackson 3006 (fl) 6/1953.

Droguetia iners (Forssk.) Schweinf. subsp. *iners*

F.P.S. 2: 276 (1952); F.T.E.A. Urtic.: 56 (1989); F.E.E. 3:
322 (1989); Biol. Skr. 51: 251 (1998).
Perennial herb. Montane forest & evergreen bushland.
Distr: EQU; Bioko, Cameroon, Ethiopia to Angola &
South Africa; Yemen.
EQU: Jackson 1037 1/1950.

Elatostema monticola Hook.f.

F.T.E.A. Urtic.: 41 (1989); F.E.E. 3: 314 (1989); Biol. Skr.
51: 251 (1998).
Syn: Elatostema orientale Engl. – F.P.S. 2: 276 (1952).
Annual or perennial herb. Montane forest, often
amongst rocks.
Distr: EQU; Nigeria, Bioko, Cameroon, Ethiopia to
Zimbabwe.
EQU: Friis & Vollesen 244 (fl) 11/1980.

Forsskaolea tenacissima L.

F.P.S. 2: 277 (1952); F.E.E. 3: 322 (1989); F.J.U. 3: 243
(1999).
Perennial or annual herb. Dry rocky slopes, sandy wadi
beds, sand dunes.
Distr: NS, RS, CS; Eritrea, Ethiopia, Djibouti, Somalia; S
Spain, N Africa, Arabia to India.
CS: Jackson 3226 (fl) 11/1954.

Forsskaolea viridis Ehrenb. ex Webb

F.P.S. 2: 278 (1952); F.T.E.A. Urtic.: 54 (1989); F.E.E. 3:
322 (1989).
Annual herb. Rock crevices, rocky slopes, sandy wadi
beds.
Distr: RS, ?CS; Cape Verde Is., Eritrea, Somalia to
Tanzania, Angola, Namibia; Egypt, Arabia & Socotra.
RS: Jackson 2809 (fl fr) 4/1953.

Girardinia bullosa (Steud.) Wedd.

F.T.E.A. Urtic.: 12 (1989); F.E.E. 3: 307 (1989); Biol. Skr.
51: 251 (1998).
Annual herb. Forest margins & clearings, montane
grassland, abandoned cultivation.
Distr: EQU; D.R. Congo, Rwanda, Burundi, Ethiopia,
Kenya, Tanzania,.
EQU: Friis & Vollesen s.n. 1980.
Note: the record from the Imatong Mts is based on a
sight record and photograph (Friis & Vollesen 1998).

Girardinia diversifolia (Link) Friis

F.T.E.A. Urtic.: 13 (1989); F.E.E. 3: 308 (1989); Biol. Skr.
51: 251 (1998).
Syn: Girardinia condensata (Steud.) Wedd. – F.P.S. 2:
278 (1952).
Syn: Girardinia heterophylla Decne. – F.J.M.: 120
(1976).
Annual herb. Forest, particularly in clearings, shaded
rocks.
Distr: DAR, EQU; Guinea to Ethiopia, S to South Africa;
Madagascar, Yemen, India to China & Indonesia.
DAR: Pettet 163 (fr) 12/1962; **EQU:** Thomas 1721 (fr)
12/1935.

Laportea aestuans (L.) Chew

F.T.E.A. Urtic.: 23 (1989); F.E.E. 3: 309 (1989); Biol. Skr.
51: 252 (1998).
Syn: Fleurya aestuans (L.) Gaudich. – F.P.S. 2: 277
(1952).
Annual herb. Secondary bushland & farmland, often
weedy.
Distr: KOR, BAG, EQU; pantropical.
KOR: Simpson 7783 4/1930; **EQU:** Friis & Vollesen 487
11/1980.
Note: the single specimen seen from Kordofan is very
poor; more ample material is required to confirm the
presence of this species in Sudan.

Laportea interrupta (L.) Chew

F.T.E.A. Urtic.: 17 (1989); F.E.E. 3: 308 (1989).
Syn: Fleurya interrupta (L.) Gaudich. – F.P.S. 2: 277
(1952).
Annual herb. Lowland forest & forest margins, swamps,
weed of cultivation.
Distr: EQU; scattered throughout the tropics.
EQU: Wyld 460 (fl) 4/1938.

Laportea ovalifolia (Schumach.) Chew

F.T.E.A. Urtic.: 18 (1989).
Syn: Fleurya ovalifolia (Schumach.) Dandy – F.P.S. 2:
277 (1952).
Stoloniferous perennial herb. Forest, sometimes along
streams.
Distr: EQU; Sierre Leone to Kenya, S to Angola &
Zimbabwe.
EQU: Andrews 1405 (fl) 5/1939.

Myrianthus arboreus P.Beauv.

F.P.S. 2: 273 (1952); F.T.E.A. Mor. & Cecrop: 87 (1989);
T.S.S.: 277 (1990); Biol. Skr. 51: 250 (1998).
Tree, more rarely a shrub. Wet forest, usually as a pioneer.
Distr: EQU; Guinea to Ethiopia, S to Angola & Tanzania.
EQU: Andrews 1706 6/1939.

Parietaria alsinefolia Delile

F.Egypt 1: 18 (1999).
Annual herb. Rock crevices.
Distr: ?NS; N Africa to C Asia.
Note: the inclusion of this species here is based on a record by Malterer (2013) in his report on the survey for the Merowe Dam Project.

Parietaria debilis G.Forst.

F.P.S. 2: 278 (1952); F.J.M.: 120 (1976); F.T.E.A. Urtic.: 52 (1989); F.E.E. 3: 320 (1989).
Annual herb. Dry montane forest & bushland, often on rocks.
Distr: RS, DAR; throughout montane tropical and warm temperate regions.
RS: Jackson 2805 (fl) 4/1953.

Pilea angolensis (Hiern) Rendle subsp. *angolensis*

F.T.E.A. Urtic.: 29 (1989); F.E.E. 3: 310 (1989); Biol. Skr. 51: 252 (1998).
Annual herb. Shade under rocks.
Distr: EQU; Guinea to Ethiopia, S to Angola & Tanzania.
EQU: Friis & Vollesen 554 (fl) 11/1980.

Pilea johnstonii Oliv. subsp. *johnstonii*

F.P.S. 2: 279 (1952); F.T.E.A. Urtic.: 34 (1989); F.E.E. 3: 312 (1989); Biol. Skr. 51: 252 (1998).
Rhizomatous perennial herb. Montane forest, splash zone of waterfalls.
Distr: EQU; Ethiopia to Malawi.
EQU: Friis & Vollesen 402 (fl) 11/1980.

Pilea rivularis Wedd.

F.T.E.A. Urtic.: 29 (1989); F.E.E. 3: 312 (1989); Biol. Skr. 51: 252 (1998).
Syn: *Pilea ceratomera* Wedd. – F.P.S. 2: 279 (1952).
Rhizomatous perennial herb. Forest including rocky stream margins.
Distr: EQU; Nigeria to Ethiopia, S to South Africa; Madagascar, Comoros.
EQU: Thomas 1697 (fl) 12/1935.

Pilea tetraphylla (Steud.) Blume

F.P.S. 2: 279 (1952); F.J.M.: 120 (1976); F.T.E.A. Urtic.: 27 (1989); F.E.E. 3: 310 (1989); Biol. Skr. 51: 253 (1998).
Annual herb. Forest & forest margins, sometimes on rocks, weed in gardens.
Distr: RS, DAR, EQU; Bioko & Cameroon to Ethiopia, S to Angola & Malawi; Madagascar.
DAR: Wickens 2125 (fl) 8/1964; **EQU:** Friis & Vollesen 47 (fl) 11/1980.

Pouzolzia guineensis Benth.

F.T.E.A. Urtic.: 47 (1989); F.E.E. 3: 317 (1989); Biol. Skr. 51: 253 (1998).
Annual or short-lived perennial herb. Clearings & margins of lowland forest & woodland.
Distr: EQU; Senegal to Ethiopia, S to Angola & Tanzania.
EQU: Friis & Vollesen 456 (fr) 11/1980.

Pouzolzia mixta Solms

F.P.S. 2: 279 (1952); F.T.E.A. Urtic.: 48 (1989); F.E.E. 3: 317 (1989); T.S.S.: 279 (1990).

Shrub. Rocky slopes with wooded grassland, edges of riverine forest.
Distr: CS; Ethiopia to Namibia & South Africa; Yemen.
CS: Cienkowsky s.n. (n.v.).
Note: the single collection from Sudan is the type specimen, from close to the Ethiopian border at Jebel Fazughli (Fazokel).

Pouzolzia parasitica (Forssk.) Schweinf.

F.T.E.A. Urtic.: 51 (1989); F.E.E. 3: 317 (1989); Biol. Skr. 51: 253 (1998).
Rhizomatous perennial herb. Forest margins & clearings, evergreen bushland & moist woodland.
Distr: EQU; Sierra Leone to Ethiopia, S to South Africa; Arabia.
EQU: Friis & Vollesen 234 11/1980 (n.v.).
Note: the specimen cited has not been seen by us; the identification follows Friis & Vollesen.

Urera hypselodendron (Hochst. ex A.Rich.) Wedd.

F.T.E.A. Urtic.: 9 (1989); F.E.E. 3: 304 (1989); T.S.S.: 279 (1990); Biol. Skr. 51: 253 (1998).
Woody climber. Montane forest & forest margins.
Distr: EQU; Ethiopia to Malawi.
EQU: Andrews 1745 (fr) 6/1939.

Urera trinervis (Hochst.) Friis & Immelman

F.T.E.A. Urtic.: 6 (1989); F.E.E. 3: 304 (1989); Biol. Skr. 51: 254 (1998).
Woody climber. Forest & forest clearings.
Distr: EQU; Ghana to Ethiopia, S to South Africa; Madagascar.
EQU: Friis & Vollesen 1058 (fl) 3/1982.

Urtica urens L.

F.T.E.A. Urtic.: 5 (1989); F.E.E. 3: 303 (1989).
Annual herb. Weed of cultivation & villages.
Distr: EQU; almost cosmopolitan.
EQU: Andrews 856 4/1939.

FAGALES

Myricaceae

I. Darbyshire

Morella salicifolia (Hochst. ex A.Rich.) Verdc. & Polhill

F.T.E.A. Myric.: 5 (2000).
Syn: *Myrica salicifolia* Hochst. ex A.Rich. – F.E.E. 3: 261 (1989); T.S.S.: 247 (1990).
Syn: *Myrica humilis* sensu Friis & Vollesen, non Cham. & Schltdl. – Biol. Skr. 51: 238 (1998).
Shrub or tree. Montane bushland, woodland & open forest, lightly wooded grassland.
Distr: EQU; Bioko, Cameroon, Eritrea to Malawi; Arabia.
EQU: Friis & Vollesen 1211 (fl) 3/1982.
Note: a very variable species with several subspecies sometimes recognised (see F.T.E.A.).

Casuarinaceae

I. Darbyshire

[*Casuarina equisetifolia* L.]

Note: this species, native to Australia, is widely grown in Sudan and South Sudan as a shade tree and as a shelterbelt species, being particularly successful in coastal regions and in sandy areas, but is not considered to have naturalised.

CUCURBITALES

Cucurbitaceae

I. Darbyshire

Citrullus colocynthis (L.) Schrad.

F.T.E.A. Cucurbit.: 46 (1967); F.E.E. 2(2): 48 (1995).
Syn: *Colocynthis vulgaris* Schrad. – F.P.S. 1: 166 (1950).
Perennial trailing herb. Semi-desert bushland & grassland.
Distr: NS, RS, DAR, KOR, CS, ES; dry palaeotropical regions.
RS: Carter 1934 (fr) 11/1987.

Citrullus lanatus (Thunb.) Matsum. & Nakai

F.T.E.A. Cucurbit.: 46 (1967); F.J.M.: 97 (1976); F.E.E. 2(2): 48 (1995).
Syn: *Citrullus vulgaris* Schrad. ex Eckl. & Zeyh. – Bot. Exp. Sud.: 42 (1970).
Syn: *Colocynthis citrullus* (L.) Kuntze – F.P.S. 1: 168 (1950).
Annual climbing or trailing herb. Weed of cultivated & disturbed areas; also cultivated.
Distr: NS, DAR, KOR, CS, ES; widespread in the palaeotropics, its native range clouded by widespread cultivation.
NS: Ahti 16818 (fl) 9/1962.
Note: the cultivated 'water melon'.

Coccinia adoensis (Hochst. ex A.Rich.) Cogn.

F.T.E.A. Cucurbit.: 65 (1967); F.J.M.: 97 (1976); F.E.E. 2(2): 52 (1995).
Syn: *Coccinia djurensis* Gilg – F.P.S. 1: 166 (1950).
Syn: *Coccinia diversifolia* sensu Andrews, non (Naud.) Cogn. – F.P.S. 1: 166 (1950).
Climbing or trailing perennial herb. Woodland & grassland, riverbanks.
Distr: DAR, BAG, UN, EQU; Ghana to Ethiopia, S to Namibia & South Africa.
DAR: Wickens 3607 (fl) 6/1977; **UN:** Sherif A.3892 6/1951.

Coccinia barteri (Hook.f.) Keay

F.T.E.A. Cucurbit.: 60 (1967).
Climbing or trailing perennial herb. Forest & forest margins.
Distr: ?EQU; Senegal to Uganda, S to Angola & Mozambique.
Note: no material has been seen for Sudan; the record is based on Jeffrey in F.T.E.A. who listed the distribution as "...eastwards to the Sudan Republic"; if confirmed, this species will probably be restricted to Equatoria.

Coccinia grandis (L.) Voigt

F.P.S. 1: 165 (1950); F.T.E.A. Cucurbit.: 68 (1967); F.J.M.: 97 (1976); F.E.E. 2(2): 54 (1995).
Climbing or trailing perennial herb. Woodland, bushland & wooded grassland, riverbanks, also as a weed.
Distr: RS, DAR, KOR, CS, ES, BAG, UN, EQU; palaeotropical.
DAR: Wickens 2016 7/1964; **UN:** J.M. Lock 81/43 (fl fr) 2/1981.

Coccinia schliebenii Harms

F.T.E.A. Cucurbit.: 63 (1967); F.E.E. 2(2): 52 (1995); Biol. Skr. 51: 118 (1998).
Robust climbing perennial herb. Forest including riverine strips & clearings, evergreen woodland & bushland.
Distr: EQU; Ethiopia, Tanzania, Mozambique.
EQU: Myers 10918 (fl) 4/1939.

Coccinia subsessiliflora Cogn.

Bull. Jard. Bot. Brux. 4: 225 (1914); Biol. Skr. 51: 118 (1998).
Syn: *Coccinia sp. D* – F.T.E.A. Cucurbit.: 70 (1967).
Climbing perennial herb. Forest.
Distr: EQU; C.A.R., D.R. Congo, Burundi, Uganda.
IUCN: RD
EQU: Jackson 3026 (fl) 6/1953.

Corallocarpus boehmii (Cogn.) C.Jeffrey

F.T.E.A. Cucurbit.: 144 (1967).
Climbing or trailing perennial herb. Open woodland, bushland, grassland.
Distr: BAG; Mauritania to Somalia, S to Zimbabwe.
BAG: Schweinfurth 2441 (fr) 11/1869.

Corallocarpus epigaeus (Rottler) C.B.Clarke

F.T.E.A. Cucurbit.: 141 (1967); F.E.E. 2(2): 25 (1995).
Syn: *Corallocarpus corallinus* (Naudin) Cogn. – F.P.S. 1: 168 (1950).
Climbing perennial herb. Woodland & bushland.
Distr: KOR; Nigeria, Eritrea & Somalia, S to Rwanda & Tanzania; Oman, Pakistan, India.
KOR: Kotschy 162 (fr) 10/1839.

Corallocarpus schimperi (Naudin) Hook.f.

F.T.E.A. Cucurbit.: 144 (1967); F.E.E. 2(2): 25 (1995).
Syn: *Corallocarpus erostris* (Schweinf.) Hook.f. – F.P.S. 1: 168 (1950).
Syn: *Corallocarpus velutinus* (Dalzell & A.Gibson) Hook.f. – F.P.S. 1: 170 (1950).
Climbing perennial herb. *Acacia-Commiphora* woodland.
Distr: RS, CS; Eritrea, Ethiopia, Somalia, Kenya; Egypt, Arabia to India.
RS: Schweinfurth 61 (fr) 4/1865.
Note: the African Plants Database records this species as also occurring in Equatoria but we have seen no material from there.

Ctenolepis cerasiformis (Stocks) Hook.f.

F.P.S. 1: 170 (1950); F.T.E.A. Cucurbit.: 92 (1967); F.E.E. 2(2): 38 (1995).
Slender climbing herb. Bushland & thicket.
Distr: RS, DAR, KOR, CS, BAG, UN; Mauritania to Ethiopia, S to South Africa; Arabia to India.

KOR: Pfund 66 9/1875; **UN:** J.M. Lock 80/4 (fr) 2/1980.

[*Cucumeropsis*]

Note: the record of *Cucumeropsis edulis* (Hook.f.) Cogn.
(= *C. mannii* Naudin) in F.P.S. 1: 170 (1950) is believed to
be based on *Schweinfurth* 3462 from Munza which is in
modern-day D.R. Congo, not South Sudan.

Cucumis dipsaceus Ehrenb. ex Spach

F.P.S. 1: 172 (1950); F.T.E.A. Cucurbit.: 105 (1967); F.E.E.
2(2): 37 (1995).
Annual trailing herb. Woodland, wooded-grassland &
bushland, cultivation.
Distr: CS; Eritrea & Somalia, S to Tanzania; Arabia.
CS: Kassas 630 (fr) 8/1954.
Note: Andrews records this species as widespread in
Sudan but we have only seen a single, possibly cultivated
specimen from Khartoum.

Cucumis ficifolius A.Rich.

F.P.S. 1: 174 (1950); F.T.E.A. Cucurbit.: 100 (1967); F.J.M.:
97 (1976); F.E.E. 2(2): 34 (1995).
Syn: *Cucumis pustulatus* Naudin ex Hook.f. – F.P.S. 1:
172 (1950); F.E.E. 2(2): 34 (1995); Biol. Skr. 51: 119 (1998).
Syn: *Cucumis figarei* sensu Drar, non Delile ex Naudin –
Bot. Exp. Sud.: 43 (1970).
Perennial trailing or climbing herb. Woodland & wooded
grassland; weed of cultivation & waste ground.
Distr: RS, DAR, KOR, CS, BAG, UN, EQU; Nigeria to
Somalia, S to Tanzania; Arabia.
RS: Schweinfurth 392 9/1868; **EQU:** J.M. Lock 80/1 (fl
fr) 2/1980.
Note: Andrews records also *C. hirsutus* Sond.; this is
believed to be based on *Schweinfurth* 3304 from Assika
which is in modern-day D.R. Congo, not South Sudan.

Cucumis kirkbridei Ghebretinsae & Thulin

Novon 17: 177 (2007).
Syn: *Cucumella engleri* sensu auctt., non (Gilg)
C.Jeffrey s.s. – F.T.E.A. Cucurbit.: 113 pro parte (1967);
Biol. Skr. 51: 119 (1998).
Perennial trailing or climbing herb. Woodland & wooded
grassland including recently burnt areas.
Distr: EQU; C.A.R. to Ethiopia, S to Malawi.
EQU: Myers 6686 (fl) 5/1937.

Cucumis maderaspatanus L.

Blumea 52: 167 (2007).
Syn: *Melothria maderaspatana* (L.) Cogn. – F.P.S. 1:
178 (1950).
Syn: *Mukia maderaspatana* (L.) M.Roem. – F.T.E.A.
Cucurbit.: 115 (1967); Bot. Exp. Sud.: 44 (1970); F.E.E.
2(2): 29 (1995); Biol. Skr. 51: 121 (1998).
Perennial climbing or trailing herb. Woodland &
bushland, forest margins, weed of cultivation.
Distr: NS, KOR, CS, ES, BAG, UN, EQU; Senegal to
Eritrea, S to Angola & South Africa; Madagascar,
Mascarenes, tropical Asia & Australia.
CS: Lea 30 (fl fr) 10/1952; **EQU:** Myers 7645 9/1937.

Cucumis melo L.

F.P.S. 1: 172 (1950); F.T.E.A. Cucurbit.: 106 (1967); F.E.E.
2(2): 33 (1995).

Trailing or climbing annual herb. Open woodland,
riverbanks, weed of cultivation & waste ground.
Distr: NS, RS, DAR, KOR, CS, ES, BAG, UN, EQU;
palaeotropical, widely cultivated elsewhere.
CS: Andrews 261 (fr) 9/1937; **UN:** J.M. Lock 81/208 8/1981.
Note: the wild plants in Sudan are referable to subsp.
agrestis (Naudin) Pangalo, but subsp. *melo*, the cultivated
melon, is also recorded from at least RS, DAR, KOR & ES,
presumably as a casual following escape from cultivation.

Cucumis metuliferus E.Mey. ex Naudin

F.P.S. 1: 172 (1950); F.T.E.A. Cucurbit.: 98 (1967); F.E.E.
2(2): 34 (1995).
Annual climbing or trailing herb. Woodland, bushland &
grassland.
Distr: BAG, UN, EQU; Nigeria to Ethiopia, S to Namibia &
South Africa; Yemen.
UN: Sherif A.2842 (fl) 9/1951.

Cucumis oreosyce H.Schaef.

Blumea 52: 171 (2007).
Syn: *Oreosyce africana* Hook.f. – F.T.E.A. Cucurbit.: 110
(1967); F.E.E. 2(2): 30 (1995); Biol. Skr. 51: 122 (1998).
Annual climbing herb. Montane forest margins.
Distr: EQU; Bioko, Cameroon, Ethiopia to Angola &
Mozambique; Madagascar.
EQU: Friis & Vollesen 390 (fl) 11/1980.

Cucumis prophetarum L. subsp. *prophetarum*

F.P.S. 1: 175 (1950); F.T.E.A. Cucurbit.: 103 (1967); F.E.E.
2(2): 36 (1995).
Perennial trailing herb. Dry bushland & semi-desert.
Distr: RS, KOR, CS; Senegal to Somalia & Kenya; Arabia
& Syria to India.
CS: Kassas 131 (fr) 2/1954.

Cucumis prophetarum L. subsp. *dissectus* (Naudin) C.Jeffrey

F.T.E.A. Cucurbit.: 103 (1967); F.E.E. 2(2): 39 (1995).
Perennial trailing herb. *Acacia-Commiphora* bushland &
grassland.
Distr: RS; Senegal to Somalia, S to Tanzania; Arabia.
RS: Schweinfurth II.143 (fl fr) 9/1868.

Cucurbita moschata Duchesne ex Lam.

F.T.E.A. Cucurbit.: 2 (1967); Bot. Exp. Sud.: 43 (1970);
F.E.E. 2(2): 58 (1995).
Annual herb. Cultivated, occasionally naturalised.
Distr: DAR, CS; native to the neotropics, widely
cultivated throughout the world.
CS: Drar 163 2/1938 (n.v.).
Note: inclusion of this species here is based on Drar (1970);
his collections are likely to be based on naturalised plants
since he did not note them as cultivated.

Diplocyclos palmatus (L.) C.Jeffrey

F.T.E.A. Cucurbit.: 73 (1967); F.E.E. 2(2): 56 (1995).
Syn: *Bryonopsis laciniosa* sensu Andrews, non (L.)
Naud. – F.P.S. 1: 165 (1950).
Perennial climbing herb. Woodland, forest including
swampy areas, wet grassland.
Distr: ?RS, EQU; palaeotropical.
EQU: Broun 1450 (fr) 11/1908.

Note: Andrews recorded *Bryonopsis laciniosa* from "NE Sudan, 21 degrees N", presumably based on a Bent specimen. However, we have only seen the specimen from Equatoria.

Eureiandra formosa Hook.f.

Kew Bull. 15: 353 (1961).
Syn: *Eureiandra schweinfurthii* Cogn. – F.P.S. 1: 175 (1950).
Climbing perennial herb. Forest & forest margins.
Distr: BAG, EQU; Cameroon to D.R. Congo, S to Angola.
EQU: Andrews 1299 (fl) 5/1939.

Kedrostis foetidissima (Jacq.) Cogn.

F.P.S. 1: 175 (1950); F.T.E.A. Cucurbit.: 137 (1967); F.E.E. 2(2): 23 (1995); Biol. Skr. 51: 119 (1998).
Climbing or trailing perennial herb. Evergreen bushland, wooded grassland, forest margins.
Distr: UN; Senegal to Somalia, S to Angola & South Africa; Arabia, India.
UN: J.M. Lock 81/119 (fl fr) 8/1981.
Note: Andrews records this species from Central & Southern Sudan but we have only seen material from Upper Nile.

Kedrostis gijef (J.F.Gmel.) C.Jeffrey

F.T.E.A. Cucurbit.: 136 (1967); F.E.E. 2(2): 23 (1995).
Syn: *Corallocarpus gijef* (J.F.Gmel.) Hook.f. – F.P.S. 1: 168 (1950).
Woody climber. *Acacia-Commiphora* bushland.
Distr: RS, KOR, ES; Ethiopia & Somalia to Tanzania; Arabia.
RS: Schweinfurth 671 (fr) 9/1868.

Kedrostis leloja (Forssk. ex J.F.Gmel.) C.Jeffrey

F.T.E.A. Cucurbit.: 134 (1967); F.E.E. 2(2): 23 (1995).
Syn: *Kedrostis hirtella* (Naudin) Cogn. – F.J.M.: 97 (1976).
Climbing perennial herb. Dry bushland, rock crevices.
Syn: *Momordica multiflora* sensu Andrews, non Hook.f. – F.P.S. 1: 181 (1950).
Distr: DAR, KOR, CS, UN; Senegal to Somalia, S to South Africa; Yemen.
DAR: Wickens 1843 (fl) 7/1964; **UN:** J.M. Lock 81/143 8/1981.

Lagenaria abyssinica (Hook.f.) C.Jeffrey

F.T.E.A. Cucurbit.: 50 (1967); F.E.E. 2(2): 49 (1995); Biol. Skr. 51: 119 (1998).
Climbing or trailing perennial herb. Forest & bushland, often riverine.
Distr: EQU; D.R. Congo to Ethiopia, S to Tanzania.
EQU: Andrews 1870 (fl) 6/1939.

Lagenaria breviflora (Benth.) Roberty

F.T.E.A. Cucurbit.: 49 (1967).
Syn: *Adenopus breviflorus* Benth. – F.P.S. 1: 164 (1950).
Climbing perennial herb. Wet forest, swampy areas.
Distr: EQU; Senegal to Uganda, S to Angola & Mozambique.
EQU: Schweinfurth 3179 (fr) 2/1870.
Note: Andrews records this species from Central & Southern Sudan but we have only seen two collections from Equatoria.

Lagenaria siceraria (Molina) Standl.

F.P.S. 1: 175 (1950); F.T.E.A. Cucurbit.: 51 (1967); F.J.M.: 97 (1976); F.E.E. 2(2): 50 (1995).
Climbing or trailing perennial herb. Grassland & bushland, also cultivated.
Distr: NS, RS, DAR, KOR; pantropical, probably native to Africa & Asia.
DAR: Lynes 555A (fl) 9/1921.

Luffa cylindrica (L.) M.Roem.

F.T.E.A. Cucurbit.: 76 (1967); F.J.M.: 97 (1976); F.E.E. 2(2): 56 (1995); Biol. Skr. 51: 120 (1998).
Syn: *Luffa aegyptiaca* Mill. – F.P.S. 1: 175 (1950).
Climbing or trailing annual herb. Cultivated, naturalised in a variety of habitats.
Distr: DAR, CS, BAG, UN, EQU; widespread in the tropics, probably native to the Old World.
DAR: Wickens 1750 (fl) 6/1964; **EQU:** Myers 7914 11/1937.

Momordica balsamina L.

F.P.S. 1: 181 (1950); F.T.E.A. Cucurbit.: 32 (1967); F.J.M.: 97 (1976); F.E.E. 2(2): 42 (1995).
Climbing or trailing annual herb. Dry bushland, riverbanks, dry riverbeds, weed of cultivation.
Distr: RS, DAR, KOR, CS; Mauritania to Somalia, S to Namibia & South Africa; widespread in the palaeotropics.
DAR: Wickens 1377 (fl fr) 4/1964.
Note: *M. boivinii* Baill. is also likely to occur in South Sudan, having been recorded from just over the border in the Ugandan Imatong Mts (Friis & Vollesen 1998).

Momordica charantia L.

F.P.S. 1: 181 (1950); F.T.E.A. Cucurbit.: 31 (1967); F.J.M.: 97 (1976); F.E.E. 2(2): 42 (1995); Biol. Skr. 51: 120 (1998).
Climbing annual herb. Forest & forest margins, secondary bushland, often weedy.
Distr: DAR, UN, EQU; Senegal to Ethiopia, S to Angola & South Africa; widespread in the tropics but native to the palaeotropics.
DAR: Wickens 1762 6/1964; **EQU:** Simpson 7367 (fl fr) 7/1929.

Momordica cissoides Benth.

F.P.S. 1: 179 (1950); F.T.E.A. Cucurbit.: 26 (1967); Biol. Skr. 51: 120 (1998).
Climbing or trailing herb. Forest, often riverine.
Distr: EQU; Guinea to Uganda & Kenya, S to Angola & Tanzania.
EQU: Friis & Vollesen 726 (fl) 12/1980.

Momordica cymbalaria Fenzl ex Hook.f.

F.T.E.A. Cucurbit.: 34 (1967); F.E.E. 2(2): 44 (1995).
Syn: *Momordica tuberosa* (Roxb.) Cogn. – F.P.S. 1: 183 (1950).
Trailing perennial herb. Bushland & grassland.
Distr: RS, CS; Eritrea to Tanzania; Pakistan, India.
CS: Bashir s.n. (fl fr) 5/1954.

Momordica foetida Schumach.

F.P.S. 1: 181 (1950); F.T.E.A. Cucurbit.: 29 (1967); F.E.E. 2(2): 42 (1995); Biol. Skr. 51: 121 (1998).
Syn: *Momordica schimperiana* Naudin – F.P.S. 1: 181 (1950).

Trailing or climbing perennial herb. Forest clearings & margins, secondary bushland, disturbed areas.
Distr: CS, BAG, UN, EQU; Sierra Leone to Eritrea, S to Angola & South Africa.
CS: <u>Jackson 3111</u> 3/1954; **UN:** <u>Simpson 7171</u> (fr) 6/1929.

Momordica pterocarpa A.Rich.

F.P.S. 1: 179 (1950); F.T.E.A. Cucurbit.: 23 (1967); F.E.E. 2(2): 41 (1995); Biol. Skr. 51: 121 (1998).
Syn: *Momordica runssorica* Gilg – F.P.S. 1: 179 (1950).
Trailing or climbing perennial herb. Montane grassland, forest margins, bushland.
Distr: RS, EQU; D.R. Congo to Eritrea, S to Malawi & Mozambique.
RS: <u>Carter 1817</u> 11/1987; **EQU:** <u>Friis & Vollesen 1213</u> (fl) 3/1982.

Peponium cienkowskii (Schweinf.) Engl.

F.P.S. 1: 183 (1950); F.T.E.A. Cucurbit.: 83 (1967); F.E.E. 2(2): 47 (1995); Biol. Skr. 51: 122 (1998).
Trailing or climbing perennial herb. Montane rock outcrops & grassland.
Distr: EQU; Ethiopia, Uganda, Kenya.
IUCN: RD
EQU: <u>Friis & Vollesen 975</u> (fl fr) 2/1982.

Peponium vogelii (Hook.f.) Engl.

F.E.E. 2(2): 46 (1995); Biol. Skr. 51: 122 (1998).
Climbing or trailing perennial herb. Forest & forest margins, evergreen bushland.
Distr: EQU; Ghana to Ethiopia, S to Angola, Zimbabwe & Mozambique; Seychelles.
EQU: <u>Friis & Vollesen 1063</u> (fr) 3/1982.

Raphidiocystis phyllocalyx C.Jeffrey & Keraudren

Fl. Cameroun 6: 102 (1967).
Climbing ?perennial herb. Forest.
Distr: EQU; Cameroon to D.R. Congo.
EQU: <u>Andrews 1151</u> (fl fr) 5/1939.

Trochomeria macrocarpa (Sond.) Hook.f. subsp. *macrocarpa*

F.T.E.A. Cucurbit.: 87 (1967); F.E.E. 2(2): 37 (1995); Biol. Skr. 51: 122 (1998).
Syn: *Trochomeria djurensis* Schweinf. & Gilg – F.P.S. 1: 183 (1950).
Climbing or trailing perennial herb. Woodland on rocky slopes; wooded grassland.
Distr: BAG, EQU; Senegal to Eritrea, S to South Africa.
EQU: <u>Friis & Vollesen 1112</u> (fl) 3/1982.

Zehneria anomala C.Jeffrey

F.T.E.A. Cucurbit.: 125 (1967); F.E.E. 2(2): 27 (1995).
Woody climber. *Acacia-Commiphora* bushland.
Distr: RS; Niger, Chad, Ethiopia, Kenya; Egypt, Arabia.
RS: <u>Bent s.n.</u> 1896.

Zehneria minutiflora (Cogn.) C.Jeffrey

F.T.E.A. Cucurbit.: 126 (1967); F.J.M.: 98 (1976); F.E.E. 2(2): 27 (1995); Biol. Skr. 51: 123 (1998).
Syn: *Zehneria peneyana* sensu Wickens, non (Naudin) Schweinf. & Asch. – F.J.M.: 98 (1976).

Climbing or trailing herb. Swamps, river margins, forest margins.
Distr: DAR, UN, EQU; Cameroon to Ethiopia, S to Zimbabwe & Mozambique.
DAR: <u>Wickens 974</u> (fl) 1/1964; **UN:** <u>J.M. Lock 81/39</u> 2/1981.

Zehneria peneyana (Naudin) Schweinf. & Asch.

F.T.E.A. Cucurbit.: 127 (1967).
Syn: *Melothria peneyana* (Naudin) Cogn. – F.P.S. 1: 177 (1950).
Syn: *Melothria deltoidea* sensu Andrews, non (Schumach. & Thonn.) Benth. – F.P.S. 1: 177 (1950).
Trailing or climbing perennial herb. Swamps, river margins, seasonally flooded grassland.
Distr: ?KOR, BAG, UN; Kenya, Tanzania; Madagascar.
UN: <u>J.M. Lock 82/57</u> (fl fr) 8/1982.
Note: Andrews recorded this species from Kordofan but we have seen no material from Sudan.

Zehneria scabra (L.f.) Sond. subsp. *scabra*

F.T.E.A. Cucurbit.: 122 (1967); F.E.E. 2(2): 27 (1995); Biol. Skr. 51: 123 (1998).
Syn: *Melothria cordata* (Thunb.) Cogn. – F.P.S. 1: 178 (1950).
Syn: *Melothria longepedunculata* Hochst. ex Cogn. – F.P.S. 1: 178 (1950).
Syn: *Melothria scrobiculata* (A.Rich.) Cogn. – F.P.S. 1: 178 (1950).
Climbing or trailing perennial herb. Forest margins, woodland, secondary vegetation including abandoned cultivation.
Distr: ?KOR, UN, EQU; Ivory Coast to Somalia, S to South Africa; Arabia, India, Java, Philippines.
EQU: <u>Friis & Vollesen 902</u> (fr) 2/1982.
Note: Andrews recorded this species from Kordofan, under his *Melothria scrobiculata* and *M. longepedunculata*; it is possible that they were misapplied but we have not seen any Sudanese material annotated with these names.

Zehneria thwaitesii (Schweinf.) C.Jeffrey

F.T.E.A. Cucurbit.: 128 (1967); Biol. Skr. 51: 123 (1998); F.E.E. 2(1): 451 (2000).
Syn: *Melothria tridactyla* Hook.f. – F.P.S. 1: 177 (1950).
Climbing or trailing herb. Grassland, swamps, river margins, forest clearings.
Distr: BAG, EQU; Uganda & Kenya, S to Angola & Mozambique; Madagascar, India, Sri Lanka.
BAG: <u>Schweinfurth 2724</u> (fl fr) 12/1869.

Begoniaceae

I. Darbyshire

Begonia eminii Warb.

F.P.S. 1: 183 (1950); Biol. Skr. 51: 124 (1998); F.T.E.A. Begon.: 8 (2006).
Succulent perennial herb, sometimes epiphytic, sometimes a climber. Forest.
Distr: EQU; Nigeria to Kenya, S to Angola & Tanzania.
EQU: <u>Andrews 1781</u> (fl fr) 6/1939.

CELASTRALES

Celastraceae

I. Farag & I. Darbyshire

Campylostemon angolense Welw. ex Oliv.

F.T.E.A. Celastr.: 73 (1994).
Woody climber. Riverine & secondary forest.
Distr: EQU; Guinea to Uganda, S to Angola & Zambia.
EQU: Andrews 1394 (fl) 4/1939.

Catha edulis (Vahl) Forssk. ex Endl.

F.E.E. 3: 340 (1989); F.T.E.A. Celastr.: 25 (1994); Biol. Skr. 51: 255 (1998).
Tree. Dry forest & margins, woodland on rocky slopes, often naturalised from cultivation.
Distr: CS, EQU; Eritrea to Angola & South Africa; Arabia.
CS: Boulos s.n. 5/1970; **EQU:** Kielland-Lund 366 12/1983 (n.v.).
Note: the specimen from Central Sudan is from a plant cultivated in the Botanic Garden in Khartoum.

Elaeodendron buchananii (Loes.) Loes.

F.P.S. 2: 280 (1952); F.E.E. 3: 344 (1989); F.T.E.A. Celastr.: 33 (1994); Biol. Skr. 51: 255 (1998).
Syn: *Cassine buchananii* Loes. – T.S.S.: 280 (1990).
Shrub or tree. Mid-altitude & montane forest & bushland.
Distr: BAG, EQU; Ghana to Ethiopia, S to Angola & Zambia.
BAG: Turner 102 (fl) 2/1935.

Gymnosporia buchananii Loes.

Trop. Afr. Fl. Pl. 5: 104 (2010).
Syn: *Maytenus buchananii* (Loes.) R.Wilczek – F.J.M.: 120 (1976); F.E.E. 3: 335 (1989); F.T.E.A. Celastr.: 11 (1994); Biol. Skr. 51: 256 (1998).
Shrub. Forest margins & bushland.
Distr: DAR, EQU; Ivory Coast to Ethiopia, S to Angola & Mozambique.
DAR: Miehe 482 (fr) 8/1982; **EQU:** Friis & Vollesen 338 (fl) 11/1980.

Gymnosporia buxifolia (L.) Szyszyl.

Trop. Afr. Fl. Pl. 5: 105 (2010).
Syn: *Maytenus cymosa* (Sol.) Exell – F.P.S. 2: 281 (1952).
Syn: *Maytenus heterophylla* sensu auctt., non (Eckl. & Zeyh.) N.Robson s.s. – F.E.E. 3: 336 (1989); F.T.E.A. Celastr.: 14 (1994); Biol. Skr. 51: 257 (1998).
Shrub. Riverine forest & margins, scrub & grassland.
Distr: BAG, EQU; Ethiopia & Somalia, S to Namibia & South Africa.
EQU: Myers 8754 (fl) 3/1938.

Gymnosporia gracilipes (Welw. ex Oliv.) Loes. subsp. *arguta* (Loes.) Jordaan

Trop. Afr. Fl. Pl. 5: 105 (2010).
Syn: *Maytenus gracilipes* (Welw. ex Oliv.) Exell subsp. *arguta* (Loes.) Sebsebe – F.E.E. 3: 332 (1989); F.T.E.A. Celastr.: 5 (1994); Biol. Skr. 51: 256 (1998).
Syn: *Maytenus ovata* (Wall. ex Wight & Arn.) Loes. var. *arguta* (Loes.) Blakelock – Bot. Exp. Sud.: 29 (1970); T.S.S.: 281 (1990).

Syn: *Maytenus gracilipes* sensu Andrews, non (Welw. ex Oliv.) Exell s.s. – F.P.S. 2: 281 (1952).
Shrub. Mid-altitude & montane forest, forest margins & bushland.
Distr: EQU; Ethiopia to D.R. Congo & Tanzania.
EQU: Friis & Vollesen 52 (fr) 11/1980.

Gymnosporia obscura (A.Rich.) Loes.

Trop. Afr. Fl. Pl. 5: 108 (2010).
Syn: *Maytenus obscura* (A.Rich.) Cufod. – F.E.E. 3: 333 (1989); F.T.E.A. Celastr.: 11 (1994); Biol. Skr. 51: 257 (1998).
Shrub or tree. Montane forest margins & bushland.
Distr: EQU; Ethiopia to Burundi & Tanzania.
EQU: Jackson 172 1/1947.

Gymnosporia senegalensis (Lam.) Loes.

Trop. Afr. Fl. Pl. 5: 110 (2010).
Syn: *Maytenus senegalensis* (Lam.) Exell – F.P.S. 2: 281 (1952); F.J.M.: 120 (1976); F.E.E. 3: 336 (1989); T.S.S.: 281 (1990); F.T.E.A. Celastr.: 18 (1994); Biol. Skr. 51: 257 (1998).
Shrub or small tree. Woodland & bushland, riverbanks.
Distr: RS, DAR, KOR, CS, ES, BAG, UN, EQU; palaeotropical.
RS: Carter 1844 (fr) 11/1987; **UN:** J.M. Lock 80/48 (fr) 3/1980.

Helictonema velutinum (Afzel.) Pierre ex N.Hallé

F.T.E.A. Celastr.: 46 (1994).
Syn: *Hippocratea velutina* Afzel. – F.P.S. 2: 284 (1952); T.S.S.: 284 (1990).
Woody climber. Forest margins, thickets.
Distr: EQU; Guinea to Uganda, S to D.R. Congo.
EQU: Turner 176 (fr) 11/1936.

Loeseneriella africana (Willd.) N.Hallé var. *richardiana* (Cambess.) N.Hallé

F.T.E.A. Celastr.: 69 (1994); Biol. Skr. 51: 255 (species only) (1998).
Syn: *Hippocratea africana* (Willd.) Loes. var. *richardiana* (Cambess.) N.Robson – F.E.E. 3: 342 (1989).
Syn: *Hippocratea richardiana* Cambess. – F.P.S. 2: 285 (1952).
Syn: *Hippocratea africana* sensu El Amin, non (Willd.) Loes. s.s. – T.S.S.: 283 (1990).
Woody climber. Forest & bushland.
Distr: DAR, KOR, BAG, EQU; Mauritania & Senegal to Ethiopia, S to South Africa.
DAR: Wickens 3114 (fr) 4/1971; **EQU:** Friis & Vollesen 357 (fl) 11/1980.

Maytenus undata (Thunb.) Blakelock

F.E.E. 3: 338 (1989); T.S.S.: 283 (1990); F.T.E.A. Celastr.: 19 (1994); Biol. Skr. 51: 258 (1998).
Syn: *Maytenus lancifolia* (Thonn.) Loes. – F.P.S. 2: 283 (1952).
Syn: *Maytenus luteola* (Delile) Andrews – F.P.S. 2: 283 (1952); T.S.S.: 281 (1990).
Shrub or tree. Forest & bushland.
Distr: RS, EQU; Guinea to Eritrea, S to South Africa; Madagascar, Comoros, Arabia.
RS: Andrews 3587 (fr); **EQU:** Friis & Vollesen 71 (fl) 11/1980.

Mystroxylon aethiopicum (Thunb.) Loes.

F.P.S. 2: 284 (1952); F.E.E. 3: 338 (1989); F.T.E.A. Celastr.: 21 (1994).

Syn: *Cassine aethiopica* Thunb. – T.S.S.: 280 (1990).

Shrub or tree. Forest, woodland & scrub.

Distr: BAG, EQU; Cameroon, Ethiopia to Angola & South Africa.

BAG: <u>Schweinfurth 1653</u> (fl) 4/1869.

Pristimera graciliflora (Welw. ex Oliv.) N.Hallé subsp. *graciliflora*

F.T.E.A. Celastr.: 61 (1994); Biol. Skr. 51: 259 (1998).

Woody climber. Forest & forest margins.

Distr: EQU; Cameroon to Uganda, S to Angola & Tanzania.

EQU: <u>Friis & Vollesen 933</u> (fl) 2/1982.

Reissantia indica (Willd.) N.Hallé var. *loeseneriana* (Hutch. & M.B.Moss) N.Hallé

F.T.E.A. Celastr.: 52 (1994); Biol. Skr. 51: 259 (1998).

Woody climber. Forest, riverine forest, woodland.

Distr: EQU; Senegal to Ethiopia, S to Angola & D.R. Congo.

EQU: <u>Kielland-Lund 743B</u> 5/1984 (n.v.).

Note: the specimen cited has not been seen; according to Friis & Vollesen (1998) it shares the same number as a specimen of *Coptosperma graveolens* (Rubiaceae), hence we have given it the suffix 'B' here.

Salacia cerasifera Welw. ex Oliv.

F.T.E.A. Celastr.: 41 (1994); Biol. Skr. 51: 259 (1998).

Woody climber or shrub. Forest.

Distr: EQU; Guinea to Kenya, S to Angola & Tanzania.

EQU: <u>Friis & Vollesen 1057</u> (fr) 3/1982.

Salacia congolensis De Wild. & T.Durand

F.E.E. 3: 346 (1989).

Woody climber or shrub. Riverine forest.

Distr: EQU; C.A.R., D.R. Congo, Ethiopia & Angola.

EQU: <u>Schweinfurth 3921</u> 6/1870.

Salacia ducis-wuertembergiae Hochst.

F.P.S. 2: 285 (1952); T.S.S.: 284 (1990); Trop. Afr. Fl. Pl. 5: 132 (2010).

Shrub. By watercourses.

Distr: ?CS; ?Sudan endemic.

IUCN: RD

Note: no extant material has been seen, this species was treated as imperfectly known in Trop. Afr. Fl. Pl.

Salacia pyriformis (Sabine) Steud.

F.P.S. 2: 286 (1952); T.S.S.: 284 (1990); F.T.E.A. Celastr.: 40 (1994).

Shrub or woody climber. Forest & forest margins.

Distr: EQU; Guinea to Tanzania.

EQU: <u>Hoyle 508</u> 1/1939.

Simirestis brianii N.Hallé

F.T.E.A. Celastr.: 62 (1994); Biol. Skr. 51: 260 (1998).

Woody climber. Forest.

Distr: EQU; D.R. Congo, Uganda, Kenya.

EQU: <u>Thomas 1696</u> (fl fr) 12/1935.

OXALIDALES

Connaraceae

I. Darbyshire & H. Pickering

Agelaea pentagyna (Lam.) Baill.

Agric. Univ. Wageningen Papers 89(6): 144 (1989); Biol. Skr. 51: 293 (1998).

Syn: *Agelaea heterophylla* Gilg – F.T.E.A. Connar.: 11 (1956).

Syn: *Agelaea ugandensis* Schellenb. – F.P.S. 2: 353 (1952); F.T.E.A. Connar.: 12 (1956); T.S.S.: 344 (1990).

Woody climber. Forest.

Distr: EQU; Senegal to Kenya, S to Angola & Mozambique; Madagascar.

EQU: <u>Sillitoe</u> 376 (fl) 1919.

Note: *A. rubiginosa* Gilg has been collected just over the border in D.R. Congo (*Schweinfurth* 3099) and is likely to occur in Equatoria.

Cnestis ferruginea DC.

F.P.S. 2: 353 (1952); Agric. Univ. Wageningen Papers 89(6): 196 (1989); T.S.S.: 344 (1990).

Woody climber or shrub. Forest, bushland & disturbed areas.

Distr: EQU; Gambia to C.A.R., S to Angola & D.R. Congo.

EQU: <u>Schweinfurth 2948</u> (fl fr) 2/1870.

Cnestis mildbraedii Gilg

Agric. Univ. Wageningen Papers 89(6): 212 (1989); Biol. Skr. 51: 293 (1998).

Syn: *Cnestis ugandensis* Schellenb. – F.T.E.A. Connar.: 2 (1956); T.S.S.: 344 (1990).

Shrub or small tree. Forest.

Distr: EQU; D.R. Congo, Uganda.

IUCN: RD

EQU: <u>Thomas 1759</u> (fr) 12/1935.

Cnestis urens Gilg

Agric. Univ. Wageningen Papers 89 (6): 231 (1989); T.S.S.: 345 (1990).

Woody climber. Forest & secondary regrowth.

Distr: EQU; Gabon, Congo-Brazzaville, D.R. Congo & C.A.R.

EQU: <u>Wyld 672</u> (fr) 1/1940.

Connarus longistipitatus Gilg

F.T.E.A. Connar.: 24 (1956); Biol. Skr. 51: 294 (1998).

Tree, shrub or woody climber. Forest & secondary regrowth.

Distr: ?EQU; Nigeria to Kenya, S to Angola & Mozambique.

Note: no specimen has been seen, its inclusion here is based upon site records by Jackson from Lotti and Talanga, Imatong Mts (see Friis & Vollesen 1998).

Rourea minor (Gaertn.) Alston

Agric. Univ. Wageningen Papers 89(6): 337 (1989).

Syn: *Santaloides gudjuanum* (Gilg) Schellenb. – F.P.S. 2: 355 (1952).

Syn: *Santaloides splendidum* (Gilg) Engl. – F.T.E.A. Connar.: 13 (1956) ; T.S.S.: 345 (1990).

Small tree, shrub or woody climber. Forest & wooded grassland.

Distr: BAG, EQU; Senegal to Kenya, S to Angola & Mozambique; Madagascar.
EQU: Sillitoe 227 (fr) 1919.

Rourea thomsonii (Baker) Jongkind

Agric. Univ. Wageningen Papers 89(6): 359 (1989); Biol. Skr. 51: 294 (1998).
Syn: *Jaundea monticola* (Gilg) Schellenb. – F.P.S. 2: 354 (1952).
Syn: *Jaundea pinnata* (P.Beauv.) Schellenb. – F.P.S. 2: 354 (1952); T.S.S.: 345 (1990).
Woody climber, shrub or small tree. Forest.
Distr: EQU; Guinea Bissau to Kenya, S to Angola & Zambia.
EQU: Thomas 1745 (fl) 12/1935.

Oxalidaceae

I. Darbyshire

Biophytum abyssinicum Steud. ex A.Rich.

F.T.E.A. Oxalid.: 12 (1971); F.J.M.: 93 (1976); Biol. Skr. 51: 104 (1998); F.E.E. 2(1): 381 (2000).
Annual herb. Wooded grassland, rocky slopes, weed of cultivation.
Distr: DAR, EQU; Nigeria to Eritrea, S to Namibia, Zimbabwe & Mozambique.
DAR: Wickens 2276 (fl fr) 8/1964; **EQU:** Friis & Vollesen 702 (fl fr) 12/1980.

Biophytum umbraculum Welw.

Biol. Skr. 51: 104 (1998); F.E.E. 2(1): 380 (2000).
Syn: *Biophytum petersianum* Klotzsch – F.P.S. 1: 133 (1950); F.J.M.: 93 (1976).
Annual herb. Woodland, bushland & grassland, weed of disturbed ground.
Distr: DAR, BAG, EQU; palaeotropical.
DAR: Wickens 2227 8/1964; **EQU:** Friis & Vollesen 523 (fl) 11/1980.
Note: *Petherick* s.n. from Neangara (EQU), originally named '*B. sensitivum*', is mounted on a sheet containing five plants and three labels and it is unclear which label refers to which specimen. There are three plants of *B. crassipes* Engl., at least two of which are associated with a Speke & Grant label from Unyamwezi, Tanzania. There are also two plants of *B. umbraculum* at least one of which is associated with a second Speke & Grant label, from Unyoro, Uganda. An earlier, unidentified researcher, has drawn an arrow from the Petherick label to the third plant of *B. crassipes*, but I do not think this is correct since (1) *B. crassipes* is unlikely to ever have been confused with *B. sensitivum* and (2) it is otherwise recorded only from considerably further south. It is much more likely that the Petherick collection is of one of the two pieces of *B. umbraculum* (a species which has previously been confused with *B. sensitivum*), or that only two collections have been mounted and that the mounting of three labels is in error.

Oxalis anthelmintica A.Rich.

F.P.S. 1: 133 (1950); F.T.E.A. Oxalid.: 7 (1971); Biol. Skr. 51: 104 (1998); F.E.E. 2(1): 385 (2000).
Bulbous perennial herb. Montane grassland, stony slopes.
Distr: RS, EQU; Eritrea to D.R. Congo, Zimbabwe & Mozambique; Egypt.

RS: Crowfoot s.n. (fl) 9/1918; **EQU:** Myers 11050 (fl) 4/1939.

Oxalis corniculata L.

F.P.S. 1: 133 (1950); F.T.E.A. Oxalid.: 3 (1971); F.J.M.: 94 (1976); Biol. Skr. 51: 104 (1998); F.E.E. 2(1): 382 (2000).
Syn: *Oxalis radicosa* A.Rich. – F.J.M.: 94 (1976); Biol. Skr. 51: 105 (1998); F.E.E. 2(1): 384 (2000).
Creeping perennial herb. Forest margins, wooded grassland, rock crevices, weed of cultivation & disturbed ground.
Distr: NS, RS, DAR, EQU; widespread in the tropics & warm-temperate regions.
DAR: Wickens 2859 (fl fr) 3/1965; **EQU:** Myers 11089 (fl) 4/1939.

Oxalis latifolia Kunth

F.T.E.A. Oxalid.: 10 (1971); F.J.M.: 94 (1976); F.E.E. 2(1): 386 (2000).
Bulbous perennial herb. Weed of cultivation.
Distr: DAR; native to central America, cultivated as an ornamental and sometimes naturalised elsewhere.
DAR: Wickens 1482 (fl) 4/1964.

Oxalis obliquifolia Steud. ex A.Rich.

F.P.S. 1: 136 (1950); F.T.E.A. Oxalid.: 6 (1971); Biol. Skr. 51: 105 (1998); F.E.E. 2(1): 385 (2000).
Bulbous perennial herb. Grassland, bushland & woodland, weed of cultivation.
Distr: EQU; Eritrea to Angola & South Africa.
EQU: Andrews 1973 (fl) 6/1939.

MALPIGHIALES

Rhizophoraceae

I. Darbyshire & I. Farag

Bruguiera gymnorhiza (L.) Savigny

F.P.S. 1: 211 (1950); F.T.E.A. Rhizophor.: 6 (1956); T.S.S.: 97 (1990); Field Guide Mangrove Trees Afr. & Madagascar: 52 (2007).
Mangrove shrub or tree. Mangroves.
Distr: ?RS; Indian Ocean coast to the Pacific.
Note: no Sudanese specimen has been seen by us; Andrews records it from the "muddy shores of the Red Sea south of Suakin" but this record requires confirmation; it is possibly based on a misidentification of *Rhizophora mucronata*.

Cassipourea malosana (Baker) Alston

F.T.E.A. Rhizophor.: 10 (1956); T.S.S.: 97 (1990); F.E.E. 2(2): 134 (1995); Biol. Skr. 51: 135 (1998).
Syn: *Cassipourea elliottii* (Engl.) Alston – F.P.S. 1: 211 (1950).
Tree. Montane forest.
Distr: EQU; Eritrea & Somalia, S to South Africa.
EQU: Friis & Vollesen 359 (fl) 11/1980.

Cassipourea ruwenzoriensis (Engl.) Alston

F.T.E.A. Rhizophor.: 14 (1956); T.S.S.: 97 (1990); Biol. Skr. 51: 135 (1998).
Tree or shrub. Forest.

Distr: EQU; C.A.R. to Kenya, S to Congo-Brazzaville, D.R. Congo & Tanzania.
EQU: Jackson 1375 (fl) 4/1950.

Rhizophora mucronata Lam.

F.P.S. 1: 211 (1950); F.T.E.A. Rhizophor.: 2 (1956); T.S.S.: 98 (1990); F.E.E. 2(2): 133 (1995); Field Guide Mangrove Trees Afr. & Madagascar: 64 (2007).
Mangrove tree. Mangroves.
Distr: RS; Indian Ocean coast, E to Polynesia.
RS: Kassas 930 12/1966.

Erythroxylaceae

I. Darbyshire

Erythroxylum fischeri Engl.

F.P.S. 2: 44 (1952); F.T.E.A. Erythroxyl.: 5 (1984); T.S.S.: 120 (1990); F.E.E. 2(2): 264 (1995); Biol. Skr. 51: 154 (1998).
Shrub or small tree. Forest, dense woodland.
Distr: BAG, EQU; Ethiopia to D.R. Congo & Tanzania.
EQU: Andrews 537 (fl) 4/1939.

Peraceae

H. Pickering & I. Darbyshire

Clutia abyssinica Jaub. & Spach

F.P.S. 2: 60 (1952); F.T.E.A. Euphorb.: 333 (1987); T.S.S.: 125 (1990); F.E.E. 2(2): 286 (1995); Biol. Skr. 51: 161 (1998).
Perennial herb or shrub. Wooded grassland, forest margins & secondary bushland.
Distr: EQU; D.R. Congo to Somalia, S to South Africa.
EQU: Friis & Vollesen 128 (fr) 11/1980.

Clutia lanceolata Forssk. subsp. lanceolata

F.E.E. 2(2): 286 (1995).
Syn: *Clutia richardiana* Müll.-Arg. – T.S.S.: 126 (1990).
Perennial herb or shrub. Margins of *Juniperus* forest & bushland.
Distr: ?RS; Eritrea, Ethiopia; Arabia.
Note: the record is from El Amin in T.S.S. and requires confirmation; we have seen no specimens from Sudan.

Euphorbiaceae

H. Pickering & I. Darbyshire

Acalypha acrogyna Pax

F.T.E.A. Euphorb.: 195 (1987); F.E.E. 2(2): 300 (1995); Biol. Skr. 51: 154 (1998).
Shrub. Forest & clearings.
Distr: EQU; D.R. Congo to Ethiopia, S to Zimbabwe.
EQU: Friis & Vollesen 604 (fl) 12/1980.

Acalypha bipartita Müll.Arg.

F.P.S. 2: 53 (1952); F.T.E.A. Euphorb.: 205 (1987); T.S.S.: 121 (1990); Biol. Skr. 51: 154 (1998).
Perennial herb or shrub. Forest & associated bushland.
Distr: EQU; D.R. Congo, Uganda, Kenya, Tanzania.
EQU: Friis & Vollesen 512 11/1980.

Acalypha brachiata C.Krauss

Trop. Afr. Fl. Pl. 2: 17 (2006).

Syn: *Acalypha senensis* Klotzsch – F.P.S. 2: 50 (1952); T.S.S.: 122 (1990).
Syn: *Acalypha villicaulis* Hochst. ex A.Rich. – F.P.S. 2: 50 (1952); Bot. Exp. Sud.: 48 (1970); F.J.M.: 103 (1976); F.T.E.A. Euphorb.: 192 (1987); F.E.E. 2(2): 301 (1995); Biol. Skr. 51: 157 (1998).
Perennial herb or subshrub. Woodland & grassland.
Distr: DAR, KOR, BAG, UN, EQU; Senegal, Mali, Chad to Eritrea, S to Namibia & South Africa.
DAR: Wickens 1949 7/1964; **EQU:** Andrews 555 (fl) 4/1939.

Acalypha ciliata Forssk.

F.P.S. 2: 51 (1952); F.J.M.: 103 (1976); F.T.E.A. Euphorb.: 197 pro parte (1987); F.E.E. 2(2): 303 (1995); Biol. Skr. 51: 155 (1998).
Annual herb. Woodland & riverine forest.
Distr: DAR, KOR, CS, UN, EQU; Senegal to Eritrea, S to Namibia & South Africa; Arabia, Pakistan, India & Sri Lanka.
DAR: Wickens 2286 8/1964; **UN:** Sherif A.3936 (fl) 8/1951.

Acalypha crenata Hochst. ex A.Rich.

F.P.S. 2: 52 (1952); F.T.E.A. Euphorb.: 200 (1987); F.E.E. 2(2): 303 (1995).
Perennial herb. Disturbed ground.
Distr: DAR, KOR, CS, EQU; palaeotropical.
CS: Lea 24 (fr) 10/1952; **EQU:** Carr 820 1970.

Acalypha fimbriata Schumach. & Thonn.

Trop. Afr. Fl. Pl. 2: 20 (2006).
Syn: *Acalypha ciliata* sensu Radcliffe-Smith pro parte, non Forssk. – F.T.E.A. Euphorb.: 197 (1987).
Annual or perennial herb. Woodland, thicket, disturbed areas.
Distr: EQU; Cape Verde Is., Senegal to Uganda, S to Namibia & South Africa.
EQU: Jackson 4244 (fr) 6/1961.

Acalypha fruticosa Forssk. var. fruticosa

F.P.S. 2: 50 (1952); F.T.E.A. Euphorb.: 206 (1987); T.S.S.: 121 (1990); F.E.E. 2(2): 301 (1995); Biol. Skr. 51: 155 (1998).
Shrub or small tree. Bushland, riverine thicket.
Distr: RS, BAG, UN, EQU; Eritrea to Mozambique, Namibia; Arabia, India, Sri Lanka, Myanmar.
RS: Jackson 2827 (fl) 4/1953; **EQU:** Friis & Vollesen 1192 (fl) 3/1982.

Acalypha fruticosa Forssk. var. eglandulosa Radcl.-Sm.

F.T.E.A. Euphorb.: 209 (1987); F.E.E. 2(2): 301 (1995); Biol. Skr. 51: 155 (1998).
Shrub or small tree. Bushland.
Distr: EQU; general distribution as for var. *fruticosa*.
EQU: Myers 11213 8/1940.

Acalypha hochstetteriana Müll.Arg.

F.P.S. 2: 51 (1952); Trop. Afr. Fl. Pl. 2: 20 (2006).
Annual herb. Swampy ground.
Distr: KOR, CS; Sudan endemic.
IUCN: RD
CS: Broun 998 8/1907.

Acalypha indica L.

F.P.S. 2: 52 (1952); F.T.E.A. Euphorb.: 199 (1987); F.E.E. 2(2): 303 (1995); Biol. Skr. 51: 156 (1998).
Annual or perennial herb. Scrub, rocky hillslopes, disturbed areas.
Distr: ?DAR, UN, EQU; palaeotropical, introduced in the neotropics.
UN: Broun 1740 (fl) 12/1909.
Note: the record for Darfur is from Quézel (1969: 111); we have not seen the specimen on which it is based.

Acalypha lanceolata Willd. var. glandulosa (Müll.Arg.) Radcl.-Sm.

F.T.E.A. Euphorb.: 202 (1987); F.E.E. 2(2): 304 (1995).
Syn: *Acalypha glomerata* Hutch. – F.P.S. 2: 52 (1952).
Annual herb. Rocky ground, open grassland, disturbed areas.
Distr: ES, BAG, UN, EQU; palaeotropical.
ES: Beshir 16 8/1951; **BAG:** Simpson 7249 (fl) 7/1929.

Acalypha neptunica Müll.Arg. var. neptunica

F.P.S. 2: 52 (1952); F.T.E.A. Euphorb.: 210 (1987); T.S.S.: 121 (1990); Biol. Skr. 51: 156 (1998).
Shrub. Forest & associated bushland.
Distr: EQU; Ghana to Kenya & Tanzania.
EQU: Andrews 1726 (fl) 6/1939.

Acalypha neptunica Müll.Arg. var. pubescens (Pax) Hutch.

F.T.E.A. Euphorb.: 211 (1987).
Shrub. Forest & associated bushland.
Distr: EQU; Ghana to Kenya, S to Malawi.
EQU: Andrews 1746 (fl) 6/1939.

Acalypha ornata Hochst. ex A.Rich.

F.P.S. 2: 50 (1952); F.T.E.A. Euphorb.: 190 (1987); T.S.S.: 121 (1990); F.E.E. 2(2): 300 (1995); Biol. Skr. 51: 156 (1998).
Syn: *Acalypha grantii* Baker & Hutch. – F.P.S. 2: 49 (1952).
Shrub. Forest margins & woodland.
Distr: BAG, UN, EQU; Nigeria to Eritrea, S to Angola & South Africa.
UN: Sherif A.3910 7/1951.

Acalypha psilostachya Hochst. ex A.Rich. var. psilostachya

F.P.S. 2: 51 (1952); F.T.E.A. Euphorb.: 204 (1987); F.E.E. 2(2): 302 (1995); Biol. Skr. 51: 157 (1998).
Perennial herb or subshrub. Forest margins, disturbed ground.
Distr: EQU; Ethiopia to Angola & Mozambique.
EQU: Friis & Vollesen 347 (fl) 11/1980.

Acalypha racemosa Baill.

F.T.E.A. Euphorb.: 187 (1987); T.S.S.: 121 (1990); F.E.E. 2(2): 300 (1995); Biol. Skr. 51: 157 (1998).
Syn: *Acalypha paniculata* Miq. – F.P.S. 2: 49 (1952).
Perennial herb or subshrub. Forest.
Distr: BAG, EQU; Ivory Coast to Ethiopia, S to Angola & Mozambique; Arabia, India, Sri Lanka, Indonesia.
EQU: Andrews 1695 (fl) 6/1939.

Acalypha segetalis Müll.Arg.

F.J.M.: 103 (1976); F.T.E.A. Euphorb.: 200 (1987); F.E.E. 2(2): 304 (1995).

Perennial herb. Grassland, floodplains, disturbed ground.
Distr: DAR; Sierra Leone to Ethiopia, S to Namibia & South Africa.
DAR: Wickens 2178 8/1964.

Acalypha supera Forssk.

Trop. Afr. Fl. Pl. 2: 24 (2006).
Syn: *Acalypha brachystachya* Hornem. – F.P.S. 2: 51 (1952); F.T.E.A. Euphorb.: 203 (1987); F.E.E. 2(2): 303 (1995).
Annual herb. Woodland, forest margins, disturbed ground.
Distr: UN; Cameroon to Eritrea, S to Angola & Mozambique; tropical Asia.
UN: Simpson 7040 6/1929.

Acalypha volkensii Pax

F.P.S. 2: 54 (1952); F.T.E.A. Euphorb.: 205 (1987); F.E.E. 2(2): 302 (1995); Biol. Skr. 51: 158 (1998).
Perennial herb or shrub. Montane forest & associated bushland.
Distr: EQU; Ethiopia, Uganda, Kenya, Tanzania.
EQU: Andrews 2008 (fl) 6/1939.

Alchornea cordifolia (Schumach. & Thonn.) Müll.Arg.

F.P.S. 2: 54 (1952); F.T.E.A. Euphorb.: 252 (1987); T.S.S.: 122 (1990).
Climbing shrub or small tree. Forest margins & secondary bushland.
Distr: EQU; Senegal to Kenya, S to Angola & Tanzania.
EQU: Sillitoe 375 (fr) 1919.

Alchornea floribunda Müll.Arg.

F.P.S. 2: 54 (1952); F.T.E.A. Euphorb.: 253 (1987); T.S.S.: 122 (1990).
Shrub or small tree. Forest clearings.
Distr: EQU; Senegal to D.R. Congo & Uganda.
EQU: Wyld 547 6/1939.

Alchornea laxiflora (Benth.) Pax & K.Hoffm.

F.P.S. 2: 54 (1952); F.T.E.A. Euphorb.: 257 (1987); T.S.S.: 122 (1990); F.E.E. 2(2): 291 (1995).
Shrub or tree. Forest, bushland, riverine thickets, woodland.
Distr: EQU; Nigeria to Ethiopia, S to South Africa.
EQU: Andrews 1268 5/1939.

Alchornea yambuyaensis De Wild.

F.P.S. 2: 55 (1952); F.T.E.A. Euphorb.: 259 (1987).
Shrub. Forest, riparian woodland.
Distr: ?EQU; D.R. Congo, Tanzania, Angola, Zambia & Mozambique.
Note: Andrews recorded this species from Equatoria but we have been unable to trace the material on which this record was based.

Argomuellera macrophylla Pax

F.P.S. 2: 56 (1952); F.T.E.A. Euphorb.: 225 (1987); T.S.S.: 123 (1990); F.E.E. 2(2): 290 (1995); Biol. Skr. 51: 158 (1998).
Shrub. Upland forest.
Distr: EQU; Sierra Leone to Ethiopia, S to Angola & Mozambique.
EQU: Chipp 36 (fl) 2/1929.

Astraea lobata (L.) Klotzsch

Syn: *Croton lobatus* L. – F.P.S. 2: 60 (1952); F.J.M.: 104 (1976); F.E.E. 2(2): 326 (1995).
Annual herb, often woody at the base. Weed of cultivation.
Distr: NS, DAR, KOR; Mauritania & Senegal to Eritrea, S to Cameroon & C.A.R.; Arabia, S America.
KOR: Wickens 373 (fr) 9/1962.

Caperonia fistulosa Beille var. *fistulosa*

F.T.E.A. Euphorb.: 166 (1987).
Syn: *Caperonia palustris* sensu Andrews, non (L.) A.St.-Hil. – F.P.S. 2: 57 (1952).
Annual herb. Riverbanks, seasonally wet ground.
Distr: UN; Mali, Chad to Somalia, S to Botswana.
UN: J.M. Lock 82/55 8/1982.

Caperonia serrata (Turcz.) C.Presl

F.P.S. 2: 57 (1952); F.T.E.A. Euphorb.: 164 (1987); F.E.E. 2(2): 287 (1995); Biol. Skr. 51: 160 (1998).
Annual herb. Swampy & seasonally wet ground.
Distr: KOR, CS, ES, BAG, UN; Senegal to Ethiopia, S to D.R. Congo & Kenya.
CS: Lea 173 8/1953; **UN:** Sherif A.4014 9/1951.

Cephalocroton cordofanus Hochst.

F.P.S. 2: 58 (1952); T.S.S.: 125 (1990); F.E.E. 2(2): 291 (1995).
Shrub. Riverbanks, seasonally wet ground.
Distr: KOR, CS, ES, UN; Ethiopia, Somalia, Kenya, Tanzania; Madagascar.
CS: Lea 149 (fl) 3/1953; **UN:** Simpson 7004 6/1929.

Chrozophora brocchiana (Vis.) Schweinf.

F.P.S. 2: 58 (1952); F.Egypt 2: 43 (2000).
Subshrub. Dry pools, sandy soils.
Distr: NS, DAR, KOR; Cape Verde Is., Mauritania to Chad; Morocco, Algeria, Egypt, Arabia.
DAR: Jackson 2489 1/1953.

Chrozophora oblongifolia (Delile) A.Juss. ex Spreng.

F.P.S. 2: 58 (1952); F.E.E. 2(2): 288 (1995).
Perennial herb. Wadis, semi-desert.
Distr: RS, KOR, ES; Eritrea, Ethiopia, Somalia; Egypt, Socotra to India.
ES: Jackson 2647 (fr) 3/1953.

Chrozophora plicata (Vahl) A.Juss. ex Spreng.

F.P.S. 2: 58 (1952); F.T.E.A. Euphorb.: 161 (1987); F.E.E. 2(2): 288 (1995); Biol. Skr. 51: 160 (1998).
Annual herb or short-lived subshrub. Seaasonally flooded ground, riverbanks.
Distr: NS, CS, ES, BAG, UN, EQU; Mauritania & Senegal to Eritrea, S to Zimbabwe & Mozambique; Egypt, Syria, Palestine, Arabia.
ES: Jackson 4160 (fl) 4/1961; **EQU:** Myers 10358 1/1939.

Croton gratissimus Burch.

Kew Bull. 45: 556 (1990); Trop. Afr. Fl. Pl. 2: 46 (2006).
Syn: *Croton zambesicus* Müll.Arg. – F.P.S. 2: 61 (1952); Bot. Exp. Sud.: 48 (1970); F.T.E.A. Euphorb.: 138 (1987); T.S.S.: 127 (1990); F.E.E. 2(2): 325 (1995); Biol. Skr. 51: 162 (1998).

Shrub or tree. Forest margins, woodland & thicket.
Distr: DAR, KOR, UN, EQU; Gambia to Ethiopia, Uganda & Kenya, Angola to Mozambique & South Africa.
KOR: Wickens 895 1/1963; **EQU:** Myers 14034 (fl) 9/1941.

Croton leuconeurus Pax

F.P.S. 2: 61 (1952); F.T.E.A. Euphorb.: 156 (1987); T.S.S.: 126 (1990).
Tree. Riverine forest & thicket.
Distr: EQU; Cameroon, D.R. Congo to Tanzania, Zambia, Zimbabwe.
EQU: Schweinfurth 2831 (fr) 12/1869.

Croton macrostachyus Hochst. ex Delile

F.P.S. 2: 60 (1952); F.T.E.A. Euphorb.: 149 (1987); T.S.S.: 126 (1990); F.E.E. 2(2): 326 (1995); Biol. Skr. 51: 161 (1998).
Tree. Forest margins.
Distr: EQU; Guinea to Eritrea, S to Angola & Mozambique; Madagascar, Arabia.
EQU: Myers 11359 (fr) 5/1939.

Croton polytrichus Pax

F.P.S. 2: 60 (1952); F.T.E.A. Euphorb.: 148 (1987); T.S.S.: 127 (1990).
Shrub or tree. Dry forest & thicket.
Distr: BAG, EQU; Kenya, Tanzania, Zambia.
BAG: Schweinfurth 1845 (fl fr) 5/1869.

Croton sylvaticus Hochst.

F.T.E.A. Euphorb.: 155 (1987); F.E.E. 2(2): 326 (1995); Biol. Skr. 51: 161 (1998).
Syn: *Croton oxypetalus* Müll.Arg. – T.S.S.: 127 (1990).
Shrub or tree. Lowland forest & forest margins.
Distr: ?KOR, EQU; Guinea to Ethiopia, S to Angola & South Africa.
EQU: Jackson 811 5/1949.
Note: the record for Kordofan is from El Amin in T.S.S.; we have seen no material with which to verify this.

Dalechampia parvifolia Lam.

F.E.E. 2(2): 312 (1995); Biol. Skr. 51: 162 (1998).
Syn: *Dalechampia scandens* L. var. *cordofana* (Hochst. ex Webb) Müll.Arg. – F.P.S. 2: 61 (1952); F.T.E.A. Euphorb.: 288 (1987).
Herbaceous climber. Bushland, often in rocky areas.
Distr: DAR, KOR, CS, ES, UN; Cape Verde Is., Senegal to Somalia, S to Angola & Mozambique; Arabia to India.
ES: Jackson 3397 (fl) 1/1955; **UN:** Simpson 7007 6/1929.

Erythrococca atrovirens (Pax) Prain

F.P.S. 2: 63 (1952); F.T.E.A. Euphorb.: 277 (1987); T.S.S.: 128 (1990).
Shrub or small tree. Forest margins & associated bushland.
Distr: UN, EQU; Cameroon to Kenya, S to Zambia.
EQU: Schweinfurth 3056 2/1870.

Erythrococca bongensis Pax

F.P.S. 2: 63 (1952); F.T.E.A. Euphorb.: 267 (1987); T.S.S.: 129 (1990); F.E.E. 2(2): 298 (1995).
Shrub or small tree. Riverine forest & associated bushland.

Distr: BAG, EQU; Ethiopia to D.R. Congo & Tanzania.
BAG: Schweinfurth 2296 8/1869.

Erythrococca trichogyne (Müll.Arg.) Prain

F.T.E.A. Euphorb.: 274 (1987); F.E.E. 2(2): 298 (1995);
Biol. Skr. 51: 163 (1998).
Shrub. Forest & associated bushland.
Distr: EQU; Ethiopia to Angola, Zimbabwe &
Mozambique.
EQU: Friis & Vollesen 110 11/1980.

Euphorbia abyssinica J.F.Gmel.

F.P.S. 2: 66 (1952); T.S.S.: 129 (1990); F.E.E. 2(2): 334
(1995).
Tree. Rocky hillslopes.
Distr: RS; Eritrea, Ethiopia, Somalia.
RS: Schweinfurth 226 4/1868.
Note: El Amin in T.S.S. also records this species from Bahr
el Jebel and Darfur but these are considered likely to be
misidentifications.

Euphorbia acalyphoides Hochst. ex Boiss.
subsp. **acalyphoides**

F.P.S. 2: 68 (1952); F.T.E.A. Euphorb.: 447 (1988); F.E.E.
2(2): 351 (1995).
Annual herb. Open bushland, disturbed ground.
Distr: DAR, CS; Chad to Eritrea & Somalia, S to Tanzania,
Angola; Arabia.
CS: Andrews 96 (fr) 11/1935.

Euphorbia agowensis Hochst. ex Boiss.

F.P.S. 2: 76 (1952); F.T.E.A. Euphorb.: 449 (1988); F.E.E.
2(2): 354 (1995).
Annual or short-lived perennial herb. Wooded & bushed
grassland.
Distr: ?CS; Eritrea, Ethiopia, Somalia, Uganda, Kenya,
Tanzania; Yemen, India.
Note: Andrews records this species from Fazoghli in Fung
District, but we have been unable to trace the specimen
on which this is based.

Euphorbia ampliphylla Pax

F.E.E. 2(2): 334 (1995); Biol. Skr. 51: 163 (1998).
Succulent tree. Montane forest on rocky hillslopes.
Distr: EQU; Eritrea to Zambia & Malawi.
EQU: Thomas 1889 12/1935.

Euphorbia arabica Hochst. & Steud. ex
T.Anderson

F.P.S. 2: 71 (1952); F.T.E.A. Euphorb.: 428 (1988); F.E.E.
2(2): 376 (1995).
Annual herb. Arid bare ground, open dry bushland.
Distr: RS; Eritrea, Ethiopia, Djibouti, Somalia, Kenya;
Egypt, Arabia.
RS: Carter 1918 11/1987.
Note: Quézel (1969: 112) records this species from
Darfur but this is likely to be based on a misidentification.

Euphorbia arguta Banks & Sol.

F.P.S. 2: 75 (1952); F.Egypt 2: 59 (2000).
Annual herb. Woodland, weed of cultivation.
Distr: NS, CS; Libya, Egypt, E Mediterranean, SW Asia.
CS: Eldin 2 2/1962.

Euphorbia balsamifera Aiton subsp. **adenensis**
(Deflers) P.R.O.Bally

F.Som. 1: 314 (1993).
Succulent shrub. Open rocky hillslopes.
Distr: RS; Somalia; Arabia.
RS: Carter 1881 11/1987.

Euphorbia bongensis Kotschy & Peyr. ex Boiss.

F.T.E.A. Euphorb.: 464 (1988); Biol. Skr. 51: 163 (1998).
Perennial herb. Grassland & wooded grassland, often
after burning.
Distr: BAG, EQU; Rwanda, Uganda, Kenya, Tanzania,
Zambia.
EQU: Friis & Vollesen 989 (fl) 2/1982.

Euphorbia breviarticulata Pax var.
breviarticulata

F.T.E.A. Euphorb.: 492 (1988); F.E.E. 2(2): 337 (1995);
Biol. Skr. 51: 164 (1998).
Succulent shrub. Dry bushland.
Distr: EQU; Ethiopia, Somalia, Uganda, Kenya, Tanzania.
EQU: Jackson 690 4/1949.

Euphorbia brunellii Chiov.

F.T.E.A. Euphorb.: 531 (1988); F.E.E. 2(2): 349 (1995).
Geophytic perennial herb. Bushland & grassland.
Distr: BAG; Ethiopia, Uganda, Kenya.
BAG: Schweinfurth 1929 (fr) 5/1869.

Euphorbia cactus Ehrenb. ex Boiss.

F.E.E. 2(2): 337 (1995).
Succulent shrub. Saline sandy coastal plains.
Distr: RS; Eritrea, Ethiopia; Arabia.
RS: Khattab 6357 5/1928.

Euphorbia candelabrum Trémaux ex Kotschy
var. **candelabrum**

F.P.S. 2: 66 (1952); F.J.M.: 104 (1976); F.T.E.A. Euphorb.:
485 (1988); T.S.S.: 130 (1990); F.E.E. 2(2): 336 (1995);
Biol. Skr. 51: 164 (1998).
Succulent tree. Dry woodland & bushland, rocky ground.
Distr: RS, DAR, KOR, UN, EQU; Eritrea & Somalia, S to
Zambia & Malawi.
DAR: Wickens 2815 (fl) 11/1964; **EQU:**
Schweinfurth 2824 12/1869.

Euphorbia collenetteae Al-Zahrani & El-Karemy

Edinburgh J. Bot. 64: 131 (2007).
Syn: *Euphorbia fractiflexa* sensu El Amin, non S.Carter
& J.R.I.Wood. – T.S.S.: 129 (1990).
Succulent shrub. Cracks in rocks of dry coastal plains.
Distr: RS; Eritrea; Saudi Arabia (Farasan Archipelago).
IUCN: RD
RS: Khattab 6457 5/1928.

Euphorbia convolvuloides Hochst. ex Benth.

F.P.S. 2: 72 (1952); F.E.E. 2(2): 374, in notes (1995); Trop.
Afr. Fl. Pl. 2: 89 (2006).
Annual herb. Open sandy ground.
Distr: KOR; Mauritania & Senegal to Chad.
KOR: Jackson 4031 (fl) 9/1959.

Euphorbia crotonoides Boiss. subsp. *crotonoides*

F.P.S. 2: 73 (1952); F.T.E.A. Euphorb.: 446 (1988); F.E.E. 2(2): 352 (1995).
Annual herb. Disturbed ground, open bushland, grassland.
Distr: KOR; Ethiopia to Namibia & South Africa.
KOR: Wickens 444 (fr) 9/1962.

Euphorbia cuneata Vahl subsp. *cuneata*

F.P.S. 2: 68 (1952); F.T.E.A. Euphorb.: 466 (1988); T.S.S.: 130 (1990); F.E.E. 2(2): 359 (1995).
Shrub. Rocky, open ground, sandy coastal plains, widely cultivated for hedging in E Africa.
Distr: RS; Eritrea, Somalia to Mozambique (mainly cultivated in the southern part of its range); Egypt, Arabia.
RS: Jackson 3934 (fl) 9/1959.

Euphorbia cyparissioides Pax

F.P.S. 2: 75 (1952); F.T.E.A. Euphorb.: 439 (1988); F.E.E. 2(2): 367 (1995); Biol. Skr. 51: 164 (1998).
Perennial herb. Grassland, especially after burning.
Distr: EQU; Ethiopia to Angola & Mozambique.
EQU: Myers 11140 (fr) 4/1939.

Euphorbia depauperata Hochst. ex A.Rich. var. *depauperata*

F.P.S. 2: 73 (1952); F.T.E.A. Euphorb.: 439 (1988); F.E.E. 2(2): 366 (1995); Biol. Skr. 51: 164 (1998).
Perennial herb. Grassland.
Distr: EQU; Sierra Leone to Ivory Coast, Nigeria, Cameroon, Eritrea to Zimbabwe & Mozambique.
EQU: Myers 11582 7/1939.

Euphorbia depauperata Hochst. ex A.Rich. var. *laevicarpa* Friis & Vollesen

F.T.E.A. Euphorb.: 440 (1988); F.E.E. 2(2): 367, in notes (1995); Biol. Skr. 51: 165 (1998).
Perennial herb. Montane grassland.
Distr: EQU; Uganda.
IUCN: RD
EQU: Friis & Vollesen 1259 (fl) 3/1982.

Euphorbia dracunculoides Lam. subsp. *dracunculoides*

F.P.S. 2: 74 (1952); F.Egypt 2: 63 (2000).
Annual herb. Open sandy ground including disturbed areas.
Distr: RS; Chad, ?Eritrea; Madagascar, Mauritius, Spain, N Africa, Arabia, India to China.
RS: Schweinfurth 863 5/1865.

Euphorbia forsskalii J.Gay

F.E.E. 2(2): 375 (1995); F.Egypt 2: 53 (2000).
Syn: *Euphorbia aegyptiaca* Boiss. – F.P.S. 2: 73 (1952); Bot. Exp. Sud.: 49 (1970).
Annual herb. Riverbanks, disturbed ground.
Distr: NS, RS, KOR, CS, ES; Cape Verde Is., Mauritania & Senegal to Eritrea; Macaronesia, N Africa to Arabia.
CS: Pettet 10 (fl) 8/1962.

Euphorbia granulata Forssk.

F.P.S. 2: 72 (1952); F.T.E.A. Euphorb.: 424 (1988); F.E.E. 2(2): 377 (1995); F.J.U. 2: 175 (1999).

Syn: *Euphorbia granulata* Forssk. var. *glabrata* (J.Gay) Boiss. – F.P.S. 2: 72 (1952); Bot. Exp. Sud.: 49 (1970); F.T.E.A. Euphorb. 2: 424 (1988); F.E.E. 2(2): 377 (1995).
Annual herb. Open sandy ground.
Distr: NS, RS, DAR, KOR; Cape Verde Is., Mauritania to Somalia, S to Kenya; Macaronesia, N Africa, Arabia to India.
RS: Carter 1786 11/1987.

Euphorbia heterophylla L.

F.P.S. 2: 76 (1952); F.J.M.: 104 (1976); F.T.E.A. Euphorb.: 431 (1988); F.E.E. 2(2): 373 (1995); Biol. Skr. 51: 165 (1998).
Syn: *Euphorbia prunifolia* Jacq. – Bot. Exp. Sud.: 49 (1970).
Annual herb. Weed of cultivation & waste ground.
Distr: NS, DAR, KOR, CS, EQU; pantropical, native to central America.
DAR: Wickens 2708 (fr) 9/1964; **EQU:** von Ramm 156 4/1956.

Euphorbia hirta L.

F.P.S. 2: 71 (1952); F.J.M.: 104 (1976); F.T.E.A. Euphorb.: 415 (1988); F.E.E. 2(2): 374 (1995); Biol. Skr. 51: 165 (1998).
Annual herb. Weed of cultivation & waste ground.
Distr: NS, RS, DAR, KOR, CS, ES, BAG, UN, EQU; pantropical, native to central America.
CS: Lea 184 (fl) 9/1953; **EQU:** Myers 6710 5/1937.

Euphorbia inaequilatera Sond.

F.P.S. 2: 72 (1952); F.T.E.A. Euphorb.: 426 (1988); F.E.E. 2(2): 376 (1995).
Annual herb. Dry bushland, disturbed ground.
Distr: ?RS; Mauritania to Somalia, S to Namibia & South Africa; N Africa, Arabia.
Note: Andrews recorded this species from the Red Sea District at Suakin and from "21 degrees N" (the latter presumably based on a Bent collection), but we have been unable to locate the material on which these records are based.

Euphorbia indica Lam.

F.P.S. 2: 71 (1952); F.T.E.A. Euphorb.: 417 (1988); F.E.E. 2(2): 374 (1995); Biol. Skr. 51: 165 (1998).
Annual herb. Wet grassland, disturbed areas.
Distr: NS, KOR, CS, ES, BAG, UN, EQU; pantropical, native to India.
ES: Jackson 3403 (fl) 1/1955; **UN:** Sherif A.3913 7/1951.

Euphorbia macrophylla Pax

F.P.S. 2: 76 (1952); F.W.T.A. 1(2): 421 (1958).
Perennial herb. Wooded grassland.
Distr: BAG; Senegal to C.A.R.
BAG: Schweinfurth 2006 (fr) 7/1869.

Euphorbia magnicapsula S.Carter var. *lacertosa* S.Carter

F.T.E.A. Euphorb.: 490 (1988); Biol. Skr. 51: 166 (1998).
Succulent tree. Amongst rocks in bushland.
Distr: EQU; Uganda, Kenya, Tanzania.
IUCN: RD
EQU: Friis & Vollesen 1156 3/1982.

Euphorbia monacantha Pax

F.P.S. 2: 78 (1952); F.E.E. 2(2): 346 (1995).
Succulent perennial herb. Dry woodland.
Distr: ?RS; Ethiopia.
RS: Kassas 463 12/1966.
Note: this is treated in F.E.E. as an Ethiopian endemic and the identification of the collection cited, held at KHU, must be in doubt. Plowes (1989) suggests that the record of this species from Sudan may be based on a misidentification of *E. triaculeata*.

Euphorbia nubica N.E.Br.

F.P.S. 2: 68 (1952); F.J.M.: 104 (1976); F.T.E.A. Euphorb.: 473 (1988); F.E.E. 2(2): 364 (1995).
Syn: *Euphorbia consobrina* N.E.Br. – F.P.S. 2: 69 (1952).
Syn: *Euphorbia schimperi* var. *nubica* sensu El Amin, non Bally – T.S.S.: 130 (1990).
Scrambling shrub. Dry open woodland & bushland.
Distr: RS, DAR, KOR, CS; Eritrea, Ethiopia, Djibouti, Somalia, Uganda, Kenya.
DAR: Wickens 1491 (fr) 4/1964.

Euphorbia peplus L.

F.P.S. 2: 72 (1952); F.T.E.A. Euphorb.: 432 (1988); F.E.E. 2(2): 370 (1995).
Annual herb. Weed of cultivation & shady disturbed ground.
Distr: NS, RS; widespread weed of temperate & subtropical zones, native to Eurasia.
RS: Andrews 3537.

Euphorbia polyacantha Boiss.

F.E.E. 2(2): 342 (1995).
Syn: *Euphorbia infausta* N.E.Br. – F.P.S. 2: 78 (1952); Bot. Exp. Sud.: 49 (1970).
Syn: *Euphorbia thi* Schweinf. – F.P.S. 2: 78 (1952); Bot. Exp. Sud.: 49 (1970) ; T.S.S.: 131 (1990).
Syn: *Euphorbia thi* Schweinf. var. *subinarticulata* (Schweinf.) N.E.Br. – F.P.S. 2: 78 (1952).
Succulent shrub. Rocky hillslopes, open bushland.
Distr: RS, ?DAR; Eritrea, Ethiopia; Egypt.
RS: Schweinfurth 339 (fl) 9/1868.
Note: Drar (1970) records *E. infausta* from Jebel Marra based on a sterile specimen; this was probably misidentified.

Euphorbia polycnemoides Hochst. ex Boiss.

F.P.S. 2: 71 (1952); F.J.M.: 104 (1976); F.T.E.A. Euphorb.: 428 (1988); F.E.E. 2(2): 376 (1995).
Annual or short-lived perennial herb. Wooded grassland, disturbed ground.
Distr: DAR, KOR, CS, ES; Mauritania & Senegal to Ethiopia, S to Zambia & Malawi.
ES: Beshir 224 8/1951.

Euphorbia prostrata Aiton

F.T.E.A. Euphorb.: 421 (1988); F.E.E. 2(2): 375 (1995).
Annual herb. Disturbed ground, sandy soils.
Distr: DAR, CS, ES; pantropical, native to the Caribbean.
CS: Jackson 4184 5/1961.

Euphorbia rivae Pax

F.T.E.A. Euphorb.: 430 (1988); F.E.E. 2(2): 378 (1995).
Perennial herb. Wet grassland.

Distr: EQU; Ethiopia, Kenya.
IUCN: RD
EQU: Myers 8468 2/1938.

Euphorbia schimperiana Scheele var. schimperiana

F.P.S. 2: 75 (1952); F.T.E.A. Euphorb.: 433 (1988); F.E.E. 2(2): 370 (1995); Biol. Skr. 51: 166 (1998).
Annual or short-lived perennial herb. Forest margins, grassland, weed of cultivation.
Distr: EQU; Cameroon to Somalia, S to Zimbabwe; Arabia.
EQU: Friis & Vollesen 337 11/1980.
Note: var. *velutina* N.E.Br. is also likely to occur in South Sudan since it has been recorded on the Ugandan side of the Imatong Mts (Friis & Vollesen 1998).

Euphorbia schimperiana Scheele var. pubescens (N.E.Br.) S.Carter

F.T.E.A. Euphorb. 2: 435 (1988).
Annual or short-lived perennial herb. Forest margins, grassland.
Distr: CS; Nigeria, Cameroon, Uganda & Kenya, S to Zambia & Malawi.
CS: Broun 798 1/1906.

Euphorbia scordifolia Jacq.

F.P.S. 2: 73 (1952); F.E.E. 2(2): 375 (1995).
Annual or short-lived perennial herb. Sandy plains, disturbed sandy ground.
Distr: NS, RS, DAR, KOR, CS; Cape Verde Is., Mauritania & Senegal to Somalia; Macaronesia, N Africa, Arabia.
KOR: Wickens 189 8/1962.

Euphorbia teke Schweinf. ex Pax

F.P.S. 2: 67 (1952); F.T.E.A. Euphorb.: 481 (1988); T.S.S.: 131 (1990); Biol. Skr. 51: 166 (1998).
Succulent tree. Swampy forest.
Distr: EQU; Cameroon to Uganda, S to Angola & Tanzania.
EQU: Jackson 745 (fl) 4/1949.

Euphorbia tirucalli L.

F.P.S. 2: 67 (1952); F.T.E.A. Euphorb. 2: 471 (1988); T.S.S.: 131 (1990); F.E.E. 2(2): 364 (1995).
Succulent shrub or tree. Grassland, thickets, naturalised around habitation where used as hedging.
Distr: DAR, KOR, CS, ES; pantropical, native to tropical Africa.
KOR: Broun 1298 12/1907.

Euphorbia triaculeata Forssk.

F.P.S. 2: 78 (1952); F.E.E. 2(2): 345 (1995).
Succulent shrub. Open rocky hillslopes.
Distr: RS; Eritrea, Djibouti; Arabia.
RS: Carter 1894 11/1987.

Euphorbia umbellata (Pax) Bruyns

Euphorbia World 3(1): 5 (2007).
Syn: *Synadenium grantii* Hook.f. – F.P.S. 2: 98 (1952); F.T.E.A. Euphorb.: 535 (1988); T.S.S.: 141 (1990); Biol. Skr. 51: 174 (1998).
Shrub or tree. Dry woodland, rocky hillslopes.
Distr: EQU; D.R. Congo, Uganda, Kenya, Tanzania.

EQU: Andel s.n. (n.v.).
Note: the specimen cited was listed by Friis & Vollesen but has not been traced.

Euphorbia venenifica Tremaux ex Kotschy
F.P.S. 2: 67 (as *E. venefica*) (1952); F.T.E.A. Euphorb.: 494 (1988); T.S.S.: 132 (1990); F.E.E. 2(2): 334 (1995); Biol. Skr. 51: 167 (1998).
Succulent shrub. Open rocky hillslopes.
Distr: KOR, CS, BAG, EQU; Ethiopia, Uganda.
IUCN: RD
KOR: Wickens 3090 6/1969; **BAG:** Schweinfurth 1375 4/1869.

Euphorbia wellbyi N.E.Br. var. glabra S.Carter
F.T.E.A. Euphorb.: 437 (1988); F.E.E. 2(2): 369 (1995); Biol. Skr. 51: 167 (1998).
Annual or short-lived perennial herb. Montane grassland & moorland.
Distr: EQU; Ethiopia, Uganda, Kenya, Tanzania.
EQU: Kielland-Lund 926 6/1984 (n.v.).

Jatropha aceroides (Pax & K.Hoffm.) Hutch.
F.P.S. 2: 83 (1952); T.S.S.: 133 (1990); F.E.E. 2(2): 321 (1995).
Shrub. Sandy plains, wadis, rocky hillslopes.
Distr: RS; Eritrea.
IUCN: RD
RS: Carter 1875 (fr) 11/1987.

Jatropha aethiopica Müll.Arg.
F.P.S. 2: 85 (1952); F.E.E. 2(2): 322 (1995).
Perennial herb. Woodland.
Distr: CS; Ethiopia.
IUCN: RD
CS: Kotschy 398 (fl fr).
Note: the type specimen, which is the only Sudanese specimen we have seen, is a mixed collection with a *Cyphostemma* species.

Jatropha afrotuberosa Radcl.-Sm. & Govaerts
F.T.E.A. Euphorb.: 353 (1988).
Syn: *Jatropha tuberosa* Pax – F.P.S. 2: 83 (1952).
Perennial herb. Wooded grassland.
Distr: BAG; Uganda.
IUCN: RD
BAG: Schweinfurth 1850 (fr) 5/1869.

Jatropha curcas L.
F.P.S. 2: 45 (1952); F.J.M.: 104 (1976); F.T.E.A. Euphorb.: 356 (1987); T.S.S.: 133 (1990); F.E.E. 2(2): 324 (1995); Biol. Skr. 51: 168 (1998).
Shrub. Naturalised in abandoned cultivation & secondary woodland.
Distr: DAR, KOR, CS, ES, BAG, UN, EQU; native to the neotropics, now widely cultivated in the tropics.
KOR: Wickens 38 (fl) 6/1962; **EQU:** Friis & Vollesen 1090 3/1982.

Jatropha gallabatensis Schweinf.
F.P.S. 2: 84 (1952); T.S.S.: 133 (1990); F.E.E. 2(2): 322 (1995).
Shrub. Dry wooded grassland.
Distr: ES; Ethiopia.

IUCN: RD
ES: Schweinfurth 932 7/1865.

Jatropha glauca Vahl
F.P.S. 2: 83 (1952); T.S.S.: 135 (1990); F.E.E. 2(2): 322 (1995).
Subshrub. Semi-desert, open bushland.
Distr: RS, KOR, CS; Eritrea, Ethiopia, Djibouti, Somalia; Egypt, Arabia.
RS: Carter 1792 (fr) 11/1987.

Jatropha melanosperma Pax
F.P.S. 2: 83 (1952); Trop. Afr. Fl. Pl. 2: 160 (2006).
Perennial herb. Habitat unknown.
Distr: BAG; South Sudan endemic.
IUCN: RD
BAG: Schweinfurth 1952 (fr) 6/1869.

Jatropha neriifolia Müll.Arg.
Trop. Afr. Fl. Pl. 2: 162 (2006).
Perennial herb. Woodland.
Distr: DAR; Nigeria, Cameroon, C.A.R.
IUCN: RD
DAR: Wickens 3630 (fl) 6/1977.

Jatropha pelargoniifolia Courbon var. pelargoniifolia
F.T.E.A. Euphorb.: 358 (1987); F.E.E. 2(2): 321 (1995).
Syn: *Jatropha villosa* (Forssk.) Müll.Arg. – F.P.S. 2: 83 (1952); T.S.S.: 135 (1990).
Shrub. Dry bushland.
Distr: RS; Eritrea, Ethiopia, Somalia, Kenya; Arabia.
RS: Schweinfurth 934 4/1865.

Jatropha schweinfurthii Pax subsp. schweinfurthii
F.P.S. 2: 85 (1952); F.T.E.A. Euphorb.: 358 (1987).
Perennial herb. Wooded grassland, bushland.
Distr: BAG, EQU; Uganda.
IUCN: RD
BAG: Schweinfurth 1930 (fr) 1/1869.

Macaranga capensis (Baill.) Benth. ex Sim var. kilimandscharica (Pax) Friis & M.G.Gilbert
Trop. Afr. Fl. Pl. 2: 170 (2006).
Syn: *Macaranga kilimandscharica* Pax – F.P.S. 2: 85 (1952); F.T.E.A. Euphorb.: 245 (1987); T.S.S.: 135 (1990); F.E.E. 2(2): 295 (1995); Biol. Skr. 51: 168 (1998).
Tree. Forest & forest margins.
Distr: EQU; Ethiopia to South Africa.
EQU: Friis & Vollesen 232 (fr) 11/1980.

Macaranga schweinfurthii Pax
F.P.S. 2: 85 (1952); F.T.E.A. Euphorb.: 241 (1987); T.S.S.: 135 (1990); Biol. Skr. 51: 169 (1998).
Shrub. Swamp forest.
Distr: EQU; Guinea to Kenya, S to Angola & Zambia.
EQU: Andrews 1292 5/1939.

Mallotus oppositifolius (Geiseler) Müll.Arg.
F.P.S. 2: 85 (1952); F.T.E.A. Euphorb.: 236 (1987); T.S.S.: 136 (1990); F.E.E. 2(2): 304 (1995); Biol. Skr. 51: 169 (1998).
Shrub or small tree. Secondary forest & thicket.

Distr: BAG, EQU; Senegal to Ethiopia, S to Angola &
Mozambique; Madagascar.
EQU: <u>Friis & Vollesen 1060</u> (fl) 3/1982.

Manihot esculenta Crantz

F.P.S. 2: 86 (1952); F.T.E.A. Euphorb.: 367 (1987); Bot. Exp.
Sud.: 50 (1970); T.S.S.: 136 (1990); F.E.E. 2(2): 315 (1995).
Woody herb or subshrub. Cultivated, occasionally
becoming semi-naturalised.
Distr: EQU; native to S America, now widely cultivated in
the tropics, being the staple food in many regions.
EQU: <u>Drar 1322</u> 4/1938 (n.v.).
Note: El Amin in T.S.S. also records *M. glaziovii* Müll.-Arg.
as cultivated in Kordofan and Equatoria.

Manniophyton fulvum Müll.Arg.

F.W.T.A. 1(1): 400 (1954); T.S.S.: 137 (1990).
Woody climber. Forest.
Distr: EQU; Sierra Leone to D.R.Congo, S to Angola.
EQU: <u>Wyld 80</u> 4/1937.

Micrococca mercurialis (L.) Benth.

F.P.S. 2: 86 (1952); F.J.M.: 105 (1976); F.T.E.A. Euphorb.:
261 (1987); F.E.E. 2(2): 298 (1995).
Annual herb. Open woodland & scrub, disturbed ground.
Distr: DAR, UN, EQU; palaeotropical.
DAR: <u>Wickens 2061</u> (fl) 8/1964; **UN:** <u>Simpson 7054</u>
6/1929.

Neoboutonia melleri (Müll.Arg.) Prain

F.P.S. 2: 88 (1952); F.T.E.A. Euphorb.: 234 (1987); T.S.S.:
138 (1990); Biol. Skr. 51: 170 (1998).
Syn: *Neoboutonia canescens* Pax – F.P.S. 2: 88 (1952).
Syn: *Neoboutonia macrocalyx* sensu auctt., non Pax –
F.P.S. 2: 88 (1952); T.S.S.: 137 (1990).
Tree. Forest, often in riverine or swamp forest.
Distr: EQU; Nigeria to Kenya, S to Angola & Mozambique.
EQU: <u>Andrews 1198</u> (fl) 5/1939.

Pycnocoma chevalieri Beille

F.P.S. 2: 95 (1952); F.T.E.A. Euphorb.: 229 (1987); T.S.S.:
139 (1990).
Shrub. Riverine forest.
Distr: ?EQU; C.A.R., D.R. Congo, Uganda.
Note: Andrews recorded this species from near the
source of the R. Yubu in Equatoria; we have not seen
the specimen on which this record is based but it is also
recorded in South Sudan in F.T.E.A.

Ricinodendron heudelotii (Baill.) Pierre ex Heckel subsp. *africanum* (Müll.Arg.) J.Léonard

F.P.S. 2: 95 (1952); F.T.E.A. Euphorb.: 326 (1987); T.S.S.:
140 (species only) (1990); Biol. Skr. 51: 172 (1998).
Tree. Forest & secondary regrowth.
Distr: EQU; Senegal to Kenya, S to Angola &
Mozambique.
EQU: <u>Myers 11367</u> 5/1939.

Ricinus communis L.

F.P.S. 2: 96 (1952); F.J.M.: 105 (1976); F.T.E.A. Euphorb.:
322 (1987); T.S.S.: 140 (1990); F.E.E. 2(2): 293 (1995);
Biol. Skr. 51: 173 (1998).
Annual or perennial herb or shrub. Waste ground &
open, disturbed habitats.

Distr: NS, RS, DAR, KOR, CS, BAG, EQU; widely planted
& naturalised in the tropics & warm temperate regions,
possibly native to NE Africa.
KOR: <u>Kotschy 243</u> 11/1839; **EQU:** <u>Shigeta 35</u> 1/1979
(n.v.).

Shirakiopsis elliptica (Hochst.) Esser

Kew Bull. 56: 1018 (2001); Trop. Afr. Fl. Pl. 2: 226 (2006).
Syn: *Sapium ellipticum* (Hochst.) Pax – F.P.S. 2: 96
(1952); Bot. Exp. Sud.: 51 (1970); F.T.E.A. Euphorb.: 390
(1987); T.S.S.: 140 (1990); F.E.E. 2(2): 328 (1995); Biol.
Skr. 51: 173 (1998).
Shrub or tree. Forest & forest margins.
Distr: DAR, EQU; Guinea to Ethiopia, S to Angola &
South Africa.
DAR: <u>Miehe 466</u> 8/1982; **EQU:** <u>Myers 8028</u> (fl) 12/1937.

Suregada procera (Prain) Croizat

F.T.E.A. Euphorb.: 376 (1987); F.E.E. 2(2): 317 (1995);
Biol. Skr. 51: 173 (1998).
Shrub or tree. Forest & forest margins.
Distr: EQU; Ethiopia to South Africa.
EQU: <u>Friis & Vollesen 246</u> (fr) 11/1980.

Tragia benthamii Baker

F.P.S. 2: 99 (1952); F.T.E.A. Euphorb.: 303 (1987); F.E.E.
2(2): 307 (1995).
Herbaceous climber. Forest margins, montane scrub.
Distr: EQU; Ivory Coast to Uganda, S to Angola &
Mozambique.
EQU: <u>Schweinfurth 2984</u> (fl) 2/1870.

Tragia bongolana Prain

F.P.S. 2: 99 (1952); Trop. Afr. Fl. Pl. 2: 236 (2006).
Herbaceous climber. Habitat unknown.
Distr: BAG; South Sudan endemic.
IUCN: RD
BAG: <u>Schweinfurth 2729</u> (fl) 12/1869.

Tragia brevipes Pax

F.T.E.A. Euphorb.: 302 (1987); F.E.E. 2(2): 307 (1995);
Biol. Skr. 51: 174 (1998).
Straggling perennial herb or shrub. Forest margins,
wooded grassland.
Distr: EQU; Eritrea to Zimbabwe.
EQU: <u>Friis & Vollesen 694</u> (fl) 12/1980.
Note: *Tragia cinerea* (Pax) M.G.Gilbert & Radl.-Sm. is
recorded in F.E.E. 2(2): 308 (1995) as occurring in Sudan
but we have seen no material with which to verify this
record.

Tragia mitis Hochst. ex A.Rich.

F.P.S. 2: 99 (1952); F.E.E. 2(2): 310 (1995).
Herbaceous twiner. Bushland.
Distr: ES; Eritrea, Ethiopia.
IUCN: RD
ES: <u>Schweinfurth 873</u> (fl) 7/1865.

Tragia plukenetii Radcl.-Sm.

F.T.E.A. Euphorb.: 296 (1987); F.E.E. 2(2): 306 (1995).
Syn: *Tragia cannabina* L.f. – F.P.S. 2: 98 (1952); Bot.
Exp. Sud.: 51 (1970).
Syn: *Tragia cannabina* L.f. var. *brouniana* (Prain) Prain
– F.P.S. 2: 98 (1952).

Syn: *Tragia cannabina* L.f. var. *intermedia* (Müll.Arg.) Prain – F.P.S. 2: 98 (1952).
Syn: *Tragia gallabatensis* Prain – F.P.S. 2: 98 (1952). Annual herb. Floodplains, seasonally wet ground, disturbed areas.
Distr: KOR, ES, UN; Cameroon to Eritrea, S to Zimbabwe; India, Sri Lanka.
ES: Beshir 28 (fl) 8/1951; **UN:** Drar 914 4/1938 (n.v.).
Note: there is considerable confusion around the names included within synonymy here, with some authors considering *T. cannabina* var. *brouniana* to be a distinct species or a variety of *T. hildebrandtii* Müll.-Arg, but the specimens appear to us to match *T. plukenetii* sensu lato.

Tragia schweinfurthii Baker

F.P.S. 2: 100 (1952); Trop. Afr. Fl. Pl. 2: 242 (2006). Perennial herb. Habitat unknown.
Distr: BAG; South Sudan endemic.
IUCN: RD
BAG: Schweinfurth III.153 (fl fr) 1/1871.

Tragia tenuifolia Benth.

F.T.E.A. Euphorb.: 295 (1987); Biol. Skr. 51: 174 (1998). Perennial herb, sometimes climbing. Forest & forest margins.
Distr: EQU; Sierra Leone to Uganda, S to Zimbabwe.
EQU: Andrews 1403 (fl) 5/1939.

Tragia volubilis L.

F.T.E.A. Euphorb.: 294 (1987); Biol. Skr. 51: 175 (1998). Perennial herb, usually scandent. Forest margins.
Distr: EQU; Sierra Leone to Uganda, S to Angola & Zimbabwe; perhaps introduced from the neotropics where widespread.
EQU: Friis & Vollesen 722 12/1980.

Tragiella natalensis (Sond.) Pax & K.Hoffm.

F.P.S. 2: 100 (1952); F.T.E.A. Euphorb.: 318 (1987); Biol. Skr. 51: 175 (1998). Herbaceous climber. Forest, forest margins & secondary regrowth.
Distr: EQU; Uganda & Kenya, S to South Africa.
EQU: Myers 11030 (fl) 4/1939.

Ochnaceae

I. Darbyshire

Campylospermum densiflorum (De Wild. & T.Durand) Farron

Bot. Helv. 95: 69 (1985); Biol. Skr. 51: 124 (1998).
Syn: *Gomphia densiflora* (De Wild. & T.Durand) Verdc. – F.T.E.A. Ochn.: 50 (2005).
Syn: *Ouratea densiflora* De Wild. & T.Durand – F.P.S. 1: 188 (1950); T.S.S.: 69 (1990).
Syn: *Campylospermum calanthum* sensu Friis & Vollesen, non (Gilg) Farron – Biol. Skr. 51: 124 (1998).
Syn: *Ouratea calantha* sensu auctt., non Gilg – F.P.S. 1: 188 (1950); T.S.S.: 69 (1990).
Shrub or small tree. Forest.
Distr: BAG, EQU; Nigeria to Kenya, S to Zambia.
EQU: Turner 136 (fl fr) 2/1936.

Campylospermum likimiense (De Wild.) I.Darbysh. & Kordofani

Kew Bull. 66: 606 (2012).
Syn: *Gomphia likimiensis* (De Wild.) Verdc. – F.T.E.A. Ochn.: 46 (2005).
Syn: *Gomphia sp.* – F.E.E. 2(2): 69 (1995).
Syn: *Campylospermum bukobense* sensu Friis & Vollesen, non (Gilg) Farron s.s. – Biol. Skr. 51: 124 (1998).
Syn: *Ouratea bukobensis* sensu auctt., non Gilg – F.P.S. 1: 187 (1950); T.S.S.: 67 (1990).
Syn: *Ouratea flava* sensu auctt., non (Schumach. & Thonn.) Stapf – F.P.S. 1: 188 (1950); T.S.S.: 69 (1990).
Shrub or small tree. Forest & forest margins.
Distr: EQU; D.R. Congo, Ethiopia, Uganda, Kenya, Tanzania.
EQU: Andrews 1211 (fl) 5/1939.
Note: Andrews lists *Ouratea flava* (Schumach. & Thonn.) Hutch. & Dalziel ex Stapf (= *Campylospermum flavum* (Schumach. & Thonn.) Farron) from Azza Forest in Equatoria. We have not seen this material but *C. flavum* is a West African species, not recorded from Sudan by Farron who revised the genus. Based upon the similiarity in the conspicuously serrate leaf margin of the two species, we think the South Sudanese plants named *O. flava* belong here.

Campylospermum reticulatum (P.Beauv.) Farron

Bot. Helv. 95: 71 (1985).
Syn: *Gomphia reticulata* P.Beauv. – F.T.E.A. Ochn.: 45 (2005).
Syn: *Ouratea reticulata* (P.Beauv.) Engl. – F.P.S. 1: 187 (1950); T.S.S.: 69 (1990).
Shrub or small tree. Forest.
Distr: ?EQU; Senegal to Kenya & Tanzania.
Note: no specimen has been seen; the record is based on Andrews and may be a misidentification, although this widespread species is quite likely to occur in Equatoria.

Lophira lanceolata Tiegh. ex Keay

T.S.S.: 65 (1990); Biol. Skr. 51: 125 (1998); F.T.E.A. Ochn.: 53 (2005).
Syn: *Lophira alata* sensu auctt., non Banks ex C.F.Gaertn. – F.P.S. 1: 185 (1950); Bot. Exp. Sud.: 89 (1970); T.S.S.: 65 (1990).
Tree. Fire-prone wooded grassland.
Distr: BAG, EQU; Senegal to Uganda.
BAG: Schweinfurth 2847 1/1870.

Ochna afzelii R.Br. ex Oliv. subsp. *afzelii*

F.P.S. 1: 185 (1950); T.S.S.: 66 (1990); F.T.E.A. Ochn.: 34 (2005).
Syn: *Ochna mossambicensis* sensu Andrews, non Klotzsch – F.P.S. 1: 187 (1950).
Shrub or tree. Grassland & wooded grassland, often over rocks.
Distr: BAG, EQU; Guinea to Cameroon, Uganda to Angola & Zambia.
BAG: Myers 13838 (fl fr) 6/1941.

Ochna bracteosa Robyns & Lawalrée

F.E.E. 2(2): 67 (1995); Biol. Skr. 51: 125 (1998); F.T.E.A. Ochn.: 17 (2005).
Shrub. Evergreen forest.
Distr: EQU; Cameroon, D.R. Congo, Ethiopia, Uganda.

EQU: Andrews 1765 (fr) 6/1939.

Ochna holstii Engl.
F.P.S. 1: 187 (1950); T.S.S.: 66 (1990); F.E.E. 2(2): 66 (1995);
Biol. Skr. 51: 125 (1998); F.T.E.A. Ochn.: 23 (2005).
Tree (or shrub). Montane & submontane forest, forest-grassland transition.
Distr: EQU; Ethiopia to South Africa.
EQU: Friis & Vollesen 1076 (fl fr) 3/1982.

Ochna leptoclada Oliv.
T.S.S.: 66 (1990); F.T.E.A. Ochn.: 28 (2005).
Shrub or suffrutex. Fire-prone woodland.
Distr: BAG; Uganda to Zambia & Mozambique.
BAG: Turner 281 (fr).
Note: El Amin in T.S.S.: 66 records this species from
Darfur but we consider this likely to be in error and so
have omitted it here.

Ochna leucophloeos A.Rich. subsp. *leucophloeos*
F.P.S. 1: 187 (1950); T.S.S.: 61 (1990); F.E.E. 2(2): 67
(1995); F.T.E.A. Ochn.: 21 (2005).
Syn: *Ochna ardisioides* Webb – F.P.S. 1: 187 (1950).
Tree. Woodland on rocky slopes.
Distr: ?NS, CS, ES; Eritrea, Ethiopia.
IUCN: RD
CS: Sahni & Kamil s.n. (fl) 4/1966.

Ochna micrantha Schweinf. & Gilg
F.P.S. 1: 185 (1950).
Suffrutex. Wooded grassland over ironstone.
Distr: BAG; South Sudan endemic.
IUCN: RD
BAG: Myers 7226 (fr) 7/1937.
Note: the specimen cited was det. "? *O. micrantha*"
by N.K. Robson at Kew, probably due to the lack of
reference material, the types having apparently been
destroyed at Berlin.

Ochna schweinfurthiana F.Hoffm.
F.P.S. 1: 185 (1950); T.S.S.: 67 (1990); F.E.E. 2(2): 67
(1995); F.T.E.A. Ochn.: 33 (2005).
Shrub or tree. Woodland & wooded grassland, often on
rocky hillslopes.
Distr: DAR, EQU; Mali to Ethiopia, S to Angola,
Zimbabwe & Mozambique.
DAR: Wickens 3125 (fl fr) 4/1971; **EQU:**
Schweinfurth 2761 (fr) 1/1870.

Rhabdophyllum arnoldianum (De Wild. & T.Durand) Tiegh.
Adansonia, sér. 3, 30: 125 (2008).
Syn: *Ouratea arnoldiana* De Wild. & T.Durand – T.S.S:
67 (1990).
Tree. Forest.
Distr: EQU; Nigeria to C.A.R. & D.R. Congo.
EQU: Wyld 675 1/1940.

Sauvagesia erecta L.
F.P.S. 1: 188 (1950); F.T.E.A. Ochn.: 57 (2005).
Annual or perennial herb. Wet grassland, marsh edges,
irrigation channels.

Distr: BAG, EQU; Senegal to Kenya, S to Angola &
Mozambique; Madagascar, W Indies, S America.
EQU: Schweinfurth 2805 (fl) 1/1870.

Phyllanthaceae

H. Pickering & I. Darbyshire

Andrachne aspera Spreng. var. *glandulosa* Hochst. ex A.Rich.
F.P.S. 2: 55 (1952); F.J.M.: 103 (1976); F.T.E.A. Euphorb.:
9 (1987); F.E.E. 2(2): 271 (1995).
Perennial herb. Rocky open ground.
Distr: NS, RS, DAR; Cameroon, Eritrea, Ethiopia, Kenya;
Morocco, Egypt, SW Asia to India.
DAR: Wickens 2867 (fl) 3/1965.

Antidesma membranaceum Müll.Arg.
F.P.S. 2: 55 (1952); F.T.E.A. Euphorb.: 574 (1988); T.S.S.:
123 (1990); Biol. Skr. 51: 158 (1998).
Tree or shrub. Forest margins, wooded grassland.
Distr: EQU; Senegal to Kenya, S to South Africa.
EQU: Jackson 3893 (fr) 7/1958.

Antidesma venosum E.Mey. ex Tul.
F.P.S. 2: 55 (1952); F.T.E.A. Euphorb.: 573 (1988); T.S.S.:
123 (1990); F.E.E. 2(2): 270 (1995).
Tree or shrub. Riverine forest & forest margins.
Distr: BAG, EQU; Gambia to Ethiopia, S to Namibia &
South Africa.
EQU: Myers 3878 (fl) 1/1938.

Bridelia atroviridis Müll.Arg.
F.P.S. 2: 56 (1952); F.T.E.A. Euphorb.: 125 (1987); T.S.S.:
124 (1990); F.E.E. 2(2): 268 (1995).
Shrub. Forest, particularly margins, secondary scrub.
Distr: ?KOR, EQU; Sierra Leone to Ethiopia, S to Angola
& Zimbabwe.
EQU: Friis & Vollesen 707 12/1980.
Note: the record for Kordofan is from El Amin in T.S.S.;
we have seen no material with which to verify this.

Bridelia brideliifolia (Pax) Fedde
F.P.S. 2: 57 (1952); F.T.E.A. Euphorb.: 126 (1987); T.S.S.:
124 (1990); Biol. Skr. 51: 159 (1998).
Tree. Upland forest & associated bushland.
Distr: EQU; D.R. Congo, Rwanda, Burundi, Uganda,
Tanzania, Malawi.
EQU: Friis & Vollesen 242 (fr) 11/1980.

Bridelia cathartica G.Bertol.
F.T.E.A. Euphorb.: 123 (1987); F.E.E. 2(2): 268 (1995).
Tree. Woodland.
Distr: CS; Ethiopia & Somalia, S to Namibia & South
Africa.
CS: Jackson 4104 1/1961.

Bridelia micrantha (Hochst.) Baill.
F.P.S. 2: 56 (1952); F.T.E.A. Euphorb.: 127 (1987); T.S.S.:
124 (1990); F.E.E. 2(2): 269 (1995).
Shrub or tree. Open woodland, forest margins &
associated bushland.
Distr: DAR, BAG, EQU; Senegal to Ethiopia, S to Angola
& South Africa; Réunion.

EQU: Andrews 1613 6/1939.

Note: the record for Darfur is from El Amin in T.S.S.; we have seen no material with which to verify this but it is likely to be correct in view of the fact that its hybrid with *B. ndellensis* is recorded from Darfur (see below).

Bridelia micrantha × *ndellensis*

F.J.M.: 104 (1976); F.T.E.A. Euphorb.: 130, in notes (1987). Tree. Streambanks.

Distr: DAR.

DAR: Kamil 1160 5/1968.

Note: Wickens' record of this hybrid is verified by Radcliffe-Smith in F.T.E.A.

Bridelia ndellensis Beille

F.J.M.: 104 (1976); F.T.E.A. Euphorb.: 129 (1987); T.S.S.: 124 (1990); F.E.E. 2(2): 267 (1995); Biol. Skr. 51: 159 (1998).

Syn: *Bridelia aubrevillei* sensu Andrews, non Pellegr. – F.P.S. 2: 56 (1952).

Small tree. Open bushland, secondary forest.

Distr: DAR, BAG, EQU; Benin to D.R. Congo & Uganda.

DAR: Wickens 1947 7/1964; **EQU:** Myers 8759 3/1938.

Bridelia scleroneura Müll.Arg. subsp. *scleroneura*

F.J.M.: 104 (1976); F.T.E.A. Euphorb.: 122 (1987); T.S.S.: 125 (1990); F.E.E. 2(2): 267 (1995); Biol. Skr. 51: 160 (1998).

Syn: *Bridelia scleroneuroides* Pax – F.P.S. 2: 56 (1952). Tree or shrub. Open woodland.

Distr: DAR, KOR, BAG, UN, EQU; Ghana to Eritrea, S to Angola & D.R. Congo; Yemen.

KOR: Wickens 820 (fr) 11/1961; **EQU:** Myers 6958 7/1937.

Flueggea virosa (Roxb. ex Willd.) Voigt subsp. *virosa*

F.T.E.A. Euphorb.: 70 (1987); F.E.E. 2(2): 272 (1995); Biol. Skr. 51: 167 (1998).

Syn: *Securinega virosa* (Roxb. ex Willd.) Baill. – F.P.S. 2: 97 (1952); Bot. Exp. Sud.: 51 (1970); F.J.M.: 105 (1976); T.S.S.: 141 (1990).

Tree or shrub. Woodland, grassland, disturbed ground.

Distr: DAR, KOR, CS, ES, BAG, UN, EQU; Senegal to Eritrea, S to Namibia & South Africa; Madagascar, Arabia, Asia.

DAR: Wickens 1903 7/1964; **EQU:** Friis & Vollesen 1159 3/1982.

Hymenocardia acida Tul. var. *acida*

F.P.S. 2: 80 (1952); F.J.M.: 104 (1976); F.T.E.A. Euphorb.: 579 (1988); T.S.S.: 132 (1990); F.E.E. 2(2): 285 (1995); Biol. Skr. 51: 168 (1998).

Shrub or tree. Woodland, wooded grassland.

Distr: DAR, KOR, BAG, EQU; Senegal to Ethiopia, S to Angola & South Africa.

KOR: Wickens 483 9/1962; **EQU:** Myers 6427 (fr) 4/1837.

Hymenocardia ulmoides Oliv.

F.P.S. 2: 80 (1952); F.T.E.A. Euphorb.: 577 (1988); T.S.S.: 132 (1990); Biol. Skr. 51: 168 (1998).

Tree. Dry forest & associated bushland & wooded grassland.

Distr: ?EQU; Cameroon to Uganda, S to Angola & South Africa.

Note: Andrews recorded this species from Equatoria, with El Amin in T.S.S. noting it from Katire. No material has been seen from South Sudan, although it is likely to occur there.

Margaritaria discoidea (Baill.) G.L.Webster var. *fagifolia* (Pax) Radcl.-Sm.

F.T.E.A. Euphorb.: 66 (1987); T.S.S.: 137 (species only) (1990); F.E.E. 2(2): 284 (1995); Biol. Skr. 51: 170 (1998).

Syn: *Phyllanthus discoideus* (Baill.) Müll.Arg. – F.P.S. 2: 90 (1952).

Shrub or tree. Forest & secondary bushland.

Distr: EQU; Guinea Bissau to Ethiopia, S to Angola & South Africa.

EQU: Sillitoe 345 (fr) 1919.

Phyllanthus amarus Schumach. & Thonn.

F.P.S. 2: 93 (1952); F.T.E.A. Euphorb.: 58 (1987); F.E.E. 2(2): 282 (1995); Biol. Skr. 51: 170 (1998).

Annual herb. Wooded grassland, disturbed ground.

Distr: BAG; pantropical, originating in tropical America.

BAG: Schweinfurth 2063 7/1869.

Phyllanthus boehmii Pax var. *boehmii*

F.T.E.A. Euphorb.: 54 (1987); F.E.E. 2(2): 282 (1995); Biol. Skr. 51: 170 (1998).

Annual or perennial herb. Montane grassland.

Distr: EQU; Ethiopia to Malawi.

EQU: Friis & Vollesen 371 11/1980.

Phyllanthus chevalieri Beille

F.T.E.A. Euphorb.: 50 (1987).

Annual herb. Seasonally wet grassland.

Distr: UN; Chad, Kenya, Tanzania.

IUCN: RD

UN: J.M. Lock 81/56 5/1981.

Phyllanthus fischeri Pax

F.T.E.A. Euphorb.: 42 (1987); F.E.E. 2(2): 280 (1995); Biol. Skr. 51: 170 (1998).

Annual or perennial herb. Forest, bushland.

Distr: EQU; Ethiopia, Uganda, Kenya, Tanzania.

EQU: Friis & Vollesen 819 12/1980.

Phyllanthus fraternus G.L.Webster

F.T.E.A. Euphorb.: 49 (1987); F.E.E. 2(2): 281 (1995).

Syn: *Phyllanthus niruri* sensu auctt., non L. – F.P.S. 2: 93 (1952); Bot. Exp. Sud.: 50 (1970).

Annual herb. Seasonally wet bushland & woodland.

Distr: NS, RS, KOR, CS; Cape Verde Is., Gambia to Somalia, S to Namibia & South Africa; Arabia, W Asia to India.

CS: Lea 25 10/1952.

Phyllanthus inflatus Hutch.

F.P.S. 2: 89 (1952); F.T.E.A. Euphorb.: 25 (1987); T.S.S.: 138 (1990); Biol. Skr. 51: 171 (1998).

Shrub. Riverine forest.

Distr: EQU; Uganda & Kenya, S to Zambia & Zimbabwe.

EQU: Sillitoe 321 (fr) 1919.

Phyllanthus leucanthus Pax

F.T.E.A. Euphorb.: 48 (1987); F.E.E. 2(2): 281 (1995); Biol.
Skr. 51: 171 (1998).
Annual or perennial herb. Rocky areas in grassland &
woodland.
Distr: EQU; Eritrea & Somalia, S to Zambia & Zimbabwe.
EQU: Friis & Vollesen 663 12/1980.

Phyllanthus limmuensis Cufod.

F.E.E. 2(2): 278 (1995); Biol. Skr. 51: 171 (1998).
Scandent shrub. Forest margins.
Distr: EQU; Ethiopia.
IUCN: RD
EQU: Friis & Vollesen 708 12/1980.

Phyllanthus maderaspatensis L.

F.P.S. 2: 93 (1952); F.T.E.A. Euphorb.: 18 (1987); F.E.E.
2(2): 276 (1995); Biol. Skr. 51: 171 (1998).
Perennial herb. Woodland, bushland & grassland, weed
of disturbed ground.
Distr: RS, DAR, KOR, CS, ES, UN, EQU; palaeotropical &
subtropical.
RS: Carter 1824 11/1987; **EQU:** Sillitoe 460 1919.
Note: Andrews also records *P. prostratus* Welw. ex Müll.
Arg. from Khartoum but this is surely a misidentification
since this is a southern African species; we have not been
able to locate the material on which this record is based.

Phyllanthus muellerianus (Kuntze) Exell

F.P.S. 2: 89 (1952); F.T.E.A. Euphorb.: 24 (1987); T.S.S.:
138 (1990); Biol. Skr. 51: 171 (1998).
Shrub, often scandent. Woodland & riverine forest.
Distr: EQU; Guinea to Kenya, S to Angola, Zambia &
Mozambique.
EQU: von Ramm 110 (fl fr) 5/1956.

Phyllanthus nummariifolius Poir.

F.P.S. 2: 90 (1952); F.T.E.A. Euphorb.: 28 (1987); Biol. Skr.
51: 172 (1998).
Syn: *Phyllanthus capillaris* Schumach. & Thonn. – F.P.S.
2: 90 (1952).
Perennial herb or shrub. Woodland, forest margins,
disturbed ground.
Distr: EQU; Sierra Leone to Ethiopia, S to Angola &
South Africa; Madagascar, Mascarenes, Seychelles & India.
EQU: Friis & Vollesen 173 11/1980.

Phyllanthus odontadenius Müll.Arg.

F.P.S. 2: 95 (1952); F.T.E.A. Euphorb.: 47 (1987).
Annual or perennial herb. Forest margins, grassland,
bushland, waste ground.
Distr: ?UN, EQU; Sierra Leone to Kenya, S to Angola &
Zimbabwe.
EQU: Wyld 531 5/1939.
Note: Andrews recorded this species from Jongol's Post
in Upper Nile; we have not seen the specimen on which
this record was based. On the African Plants Database
this name is treated as a synonym of the W African *P.
gagnioevae* Brunel & J.P.Roux.

Phyllanthus ovalifolius Forssk.

F.T.E.A. Euphorb.: 32 (1987); F.E.E. 2(2): 276 (1995); Biol.
Skr. 51: 172 (1998).
Syn: *Phyllanthus guineensis* Pax – T.S.S.: 138 (1990).

Shrub, sometimes scandent. Forest margins, woodland,
thickets.
Distr: EQU; Nigeria, Ethiopia to Angola, Zimbabwe &
Mozambique.
EQU: Friis & Vollesen 296 (fr) 11/1980.

Phyllanthus pentandrus Schumach. & Thonn.

F.P.S. 2: 90 (1952); F.T.E.A. Euphorb.: 31 (1987); F.E.E.
2(2): 277 (1995).
Annual or short-lived perennial herb. Grassland,
woodland, often in sandy areas.
Distr: DAR, KOR, BAG, UN; Senegal to Ethiopia, S to
Namibia & South Africa.
KOR: Jackson 4349 9/1961; **UN:** J.M. Lock 81/106
8/1981.

Phyllanthus pseudoniruri Müll.Arg.

F.P.S. 2: 93 (1952); F.T.E.A. Euphorb.: 48 (1987); F.E.E.
2(2): 280 (1995).
Annual herb. Grassland & disturbed ground, often in
seasonally wet areas.
Distr: EQU; Cameroon, Ethiopia & Somalia, S to Zimbabwe.
EQU: Myers 6769 5/1937.

Phyllanthus reticulatus Poir. var. reticulatus

F.P.S. 2: 89 (1952); F.J.M.: 105 (1976); F.T.E.A. Euphorb.:
34 (1987); T.S.S.: 139 (1990); F.E.E. 2(2): 277 (1995).
Shrub or small tree. Riverbanks, floodplain grassland &
woodland.
Distr: NS, RS, DAR, KOR, CS, ES, BAG, UN, EQU;
palaeotropical.
DAR: Wickens 3111 4/1971; **UN:** Simpson 7654 3/1930.

Phyllanthus rotundifolius L.

F.P.S. 2: 95 (1952); F.T.E.A. Euphorb.: 51 (1987); F.E.E.
2(2): 280 (1995).
Annual herb. Open bushland, rocky hillslopes, wadis,
desert scrub.
Distr: RS, EQU; Cape Verde Is., Senegal & Mauritania to
Somalia, S to Tanzania; Egypt, Arabia, Socotra, Pakistan,
India, Sri Lanka.
RS: Carter 1828 11/1987; **EQU:** Padwa 277 6/1953.

Phyllanthus sepialis Müll.Arg.

F.T.E.A. Euphorb.: 32 (1987); F.E.E. 2(2): 278 (1995); Biol.
Skr. 51: 172 (1998).
Shrub. Woodland, bushland, forest margins.
Distr: EQU; Ethiopia, Uganda, Kenya, Tanzania.
EQU: Dale 318 2/1943.

Phyllanthus trichotepalus Brenan

F.T.E.A. Euphorb.: 46 (1987).
Annual herb. Forest & forest margins, disturbed ground.
Distr: EQU; Rwanda, Burundi, Uganda.
IUCN: RD
EQU: Andrews 1214 5/1939.

Spondianthus preussii Engl. subsp. glaber (Engl.) J.Léonard & Nkounkou

Bull. Jard. Bot. Nat. Belg. 59: 143 (1989); T.S.S.: 141
(species only) (1990).
Syn: *Spondianthus glaber* Engl. – F.P.S. 2: 97 (1952).
Syn: *Spondianthus preussii* Engl. var. *glaber* (Engl.)
Engl. – F.T.E.A. Euphorb.: 105 (1987).

Tree. Swamp & riverine forest.
Distr: EQU; Guinea to Uganda, S to Angola & D.R. Congo.
EQU: Sillitoe 301 (fr) 1919.

Uapaca sansibarica Pax

F.P.S. 2: 100 (1952); T.S.S.: 141 (1990); F.Z. 9(4): 98 (1996).
Tree. Riverine forest & woodland, wooded grassland.
Distr: EQU; Congo-Brazzaville to Burundi, S to Zimbabwe & Mozambique.
EQU: Myers 8477 2/1938.

Elatinaceae

R. Vanderstricht & I. Darbyshire

Bergia ammannioides Roxb. ex Roth

F.P.S. 1: 85 (1950); F.T.E.A. Elatin.: 3 (1968).
Syn: *Bergia capensis* sensu Andrews, non L. – F.P.S. 1: 83 (1950).
Annual herb. Swampy areas, shallow pools, riverbanks & damp grassland.
Distr: KOR, CS, ES, UN; Mauritania & Senegal to Somalia, S to Namibia & South Africa; Egypt, Arabia to India, China to Australia.
CS: Lea 221 (fl) 10/1954; **UN:** Simpson 7433 (fl fr) 1/1930.
Note: Andrews recorded *B. capensis* from Kordofan; this record is believed to be based on *Kotschy* 65 which is more likely a depauperate form of *B. ammannioides*.

Bergia suffruticosa (Delile) Fenzl

F.P.S. 1: 83 (1950); F.T.E.A. Elatin.: 4 (1968); F.E.E. 2(1): 195 (2000).
Subshrub. Open, seasonally wet depressions.
Distr: NS, KOR, CS, ES, EQU; Mauritania & Senegal to Nigeria; Somalia, Kenya; Egypt, Arabia, Pakistan & India.
CS: de Wilde et al. 5857 (fl) 3/1965; **EQU:** Carr 551 (fl) 8/1968.

Malpighiaceae

I. Darbyshire

Acridocarpus ugandensis Sprague

F.P.S. 2: 43 (1952); F.T.E.A. Malpigh.: 9 (1968); F.E.E. 2(2): 258 (1995); Biol. Skr. 51: 154 (1998).
Woody climber. Riverine scrub; bushland on rocky hillslopes.
Distr: EQU; Ethiopia, ?Tanzania.
IUCN: RD
EQU: Myers 13349 (fl) 6/1940.
Note: *A. scheffleri* Engl. from NE Tanzania is treated as a synonym of *A. ugandensis* in the African Plants Database but that species is known from wet forest and the two are likely to be distinct from one another.

Flabellaria paniculata Cav.

F.P.S. 2: 44 (1952); F.T.E.A. Malpigh.: 23 (1968); F.E.E. 2(2): 261 (1995).
Woody climber. Riverine scrub & forest margins.
Distr: EQU; Senegal to Ethiopia, S to Angola & Tanzania.
EQU: Myers 11293 (fr) 5/1939.

Dichapetalaceae

I. Darbyshire

Dichapetalum angolense Chodat

Meded. Land. Wag. 73(13): 55 (1973); F.T.E.A. Dichapetal.: 4 (1988).
Woody climber or shrub. Forest.
Distr: EQU; Liberia to Uganda, S to Angola & D.R. Congo.
EQU: Wyld 353 11/1937.

Dichapetalum heudelotii (Planch. ex Oliv.) Baill. var. *heudelotii*

F.W.T.A. 1(2): 438 (1958); Meded. Land. Wag. 79(16): 27 (1979); T.S.S.: 146 (1990).
Syn: *Dichapetalum schweinfurthii* Engl. – F.P.S. 2: 107 (1952).
Woody climber, shrub or treelet. Forest.
Distr: EQU; Guinea Bissau to Uganda, S to Angola & Zambia.
EQU: Sillitoe 386 (fr) 1919.

Dichapetalum madagascariense Poir. var. *madagascariense*

Agric. Univ. Wag. Papers 86(3): 39 (1986); F.T.E.A. Dichapetal.: 11 (1988); Biol. Skr. 51: 178 (1998).
Woody climber, shrub or small tree. Forest.
Distr: EQU; Senegal to Kenya, S to Angola & Mozambique; Madagascar, Comoros.
EQU: Andrews 1773 (fl) 6/1939.

Dichapetalum staudtii Hiern

F.W.T.A. 1(2): 438 (1958); Meded. Land. Wag. 82(8): 28 (1982).
Woody climber, shrub or small tree. Forest including swamp forest.
Distr: EQU; Ivory Coast, Nigeria to D.R. Congo & Angola.
EQU: Andrews 1349 (fl fr) 5/1939.

Dichapetalum ugandense M.B.Moss

Meded. Land. Wag. 82(8): 65 (1982); F.T.E.A. Dichapetal.: 15 (1988); Biol. Skr. 51: 178 (1998).
Woody climber, shrub or treelet. Forest.
Distr: EQU; D.R. Congo, Uganda, Tanzania.
EQU: Jackson 1347 (fl) 4/1950.
Note: a local species but not considered threatened.

Tapura fischeri Engl.

F.P.S. 2: 107 (1952); F.T.E.A. Dichapetal.: 16 (1988); F.E.E. 3: 47 (1989); T.S.S.: 146 (1990); Biol. Skr. 51: 178 (1998).
Tree or shrub. Forest, bushland, termitaria.
Distr: UN, EQU; Ivory Coast to Ethiopia, S to South Africa.
UN: J.M. Lock 81/66 (fl) 5/1981.

Chrysobalanaceae

I. Darbyshire

Maranthes polyandra (Benth.) Prance

Sp. Pl.: Fl. World, Chyrsobalan. 2: 65 (2003).
Syn: *Parinari polyandra* Benth. – F.P.S. 2: 105 (1952); F.T.E.A. Ros.: 53 (excl. subsp. *floribunda*) (1960); Bot. Exp. Sud.: 29 (1970) ; T.S.S.: 144 (1990).

Tree. Woodland & wooded grassland.
Distr: BAG, EQU; Mali & Ivory Coast to C.A.R.
EQU: <u>Myers 7107</u> (fl) 7/1937.

Parinari curatellifolia Planch. ex Benth.

F.P.S. 2: 104 (1952); F.T.E.A. Ros.: 50 (1960); T.S.S.: 144 (1990).
Syn: *Parinari curatellifolia* Planch. ex Benth. subsp.
mobola (Oliv.) R.A.Graham – F.T.E.A. Ros.: 51 (1960).
Tree. Woodland, often on rocky slopes.
Distr: DAR, BAG, EQU; Senegal to Kenya, S to Namibia & South Africa.
DAR: <u>Wickens 3126</u> (fl) 4/1971; **BAG:** <u>Turner 16</u> (fl) 4/1931.

Parinari excelsa Sabine

F.P.S. 2: 106 (1952); F.T.E.A. Ros.: 49 (1960); T.S.S.: 144 (1990).
Syn: *Parinari excelsa* Sabine subsp. ***holstii*** (Engl.)
R.A.Graham – F.T.E.A. Ros.: 50 (1960); Biol. Skr. 51: 178 (1998).
Syn: *Parinari tenuifolia* A.Chev. – F.P.S. 2: 105 (1952).
Large tree. Forest.
Distr: EQU; Senegal to Uganda, S to Angola & Mozambique.
EQU: <u>Myers 11808</u> 8/1939.

Putranjivaceae

H. Pickering

Drypetes bipindensis (Pax) Hutch.

F.T.E.A. Euphorb. 1: 96 (1987).
Shrub or small tree. Forest, often in rocky gorges.
Distr: EQU; Cameroon, D.R. Congo, Uganda.
EQU: <u>Andrews 1294</u> 5/1939.
Note: Andrews lists *D. mildbraedii* (Pax) Hutch. from Equatoria, almost certainly based on *Schweinfurth* 3266 from [Turu] Yuru River which is in present day D.R. Congo. It is, however, quite likely that this species also occurs in South Sudan. This record was repeated in T.S.S.: 128.

Drypetes gerrardii Hutch. var. ***tomentosa*** Radcl.-Sm.

F.T.E.A. Euphorb. 1: 98 (1987); T.S.S.: 128 (species only) (1990); Biol. Skr. 51: 162 (1998).
Shrub or small tree. Forest.
Distr: EQU; Rwanda, Uganda, Kenya, Tanzania, Zambia, South Africa.
EQU: <u>Myers 11812</u> 8/1939.

Drypetes natalensis (Harv.) Hutch.

F.T.E.A. Euphorb. 1: 92 (1987); T.S.S.: 128 (1990).
Shrub or small tree. Dry forest, riverine forest.
Distr: EQU; Kenya, Tanzania, Malawi, Mozambique, Zimbabwe, South Africa.
EQU: <u>Andrews 1589</u> 5/1939.

Drypetes ugandensis (Rendle) Hutch.

F.T.E.A. Euphorb. 1: 93 (1987); T.S.S.: 128 (1990); Biol. Skr. 51: 162 (1998).
Small tree. Forest.
Distr: EQU; Cameroon, D.R. Congo, Uganda.
EQU: <u>Jackson 1383</u> 4/1950.

Note: a rather scarce species; the African Plants Database does not include Cameroon in its distribution.

Passifloraceae

I. Darbyshire

Adenia cissampeloides (Planch. ex Hook.) Harms

F.T.E.A. Passiflor.: 49 (1975); Biol. Skr. 51: 117 (1998).
Herbaceous climber, or sometimes a woody climber.
Forest.
Distr: EQU; Guinea to Kenya, S to Angola, D.R. Congo & Tanzania.
EQU: <u>Jackson 4242</u> (fl) 6/1961.

Adenia lanceolata Engl. subsp. ***lanceolata***

F.P.S. 1: 163 (1950); F.T.E.A. Passiflor.: 43 (1975).
Perennial herb, suberect or climbing. Dry woodland, bushland & wooded grassland.
Distr: BAG; Uganda, Tanzania.
BAG: <u>Andrews 425</u> (fl) 4/1939.

Adenia lobata (Jacq.) Engl. subsp. ***rumicifolia*** (Engl. & Harms) Lye

Lidia 4: 92 (1998).
Syn: *Adenia rumicifolia* Engl. & Harms – F.J.M.: 96 (1976); F.E.E. 2(2): 10 (1995).
Syn: *Adenia schweinfurthii* sensu Andrews pro parte, non Engl. – F.P.S. 1: 162 (1950).
Woody climber. Forest margins, secondary thicket.
Distr: DAR, ?EQU; Senegal to Ethiopia, S to Angola & Mozambique.
DAR: <u>de Wilde et al. 5427</u> (n.v.); **EQU:** <u>Andrews 965</u> 5/1939.
Note: the specimen from Darfur has not been seen; the identification follows Wickens. The single specimen seen from South Sudan is sterile and was determined as *A.* cf. *rumicifolia* var. *rumicifolia* by W.J.J.O. de Wilde in 1968 at K.

Adenia lobata (Jacq.) Engl. subsp. *schweinfurthii* (Engl.) Lye

Lidia 4(3): 92 (1998).
Syn: *Adenia schweinfurthii* Engl. – F.P.S. 1: 162 pro parte (1950); F.E.E. 2(2): 9 (1995); Biol. Skr. 51: 117 (1998).
Woody climber. Forest & forest margins.
Distr: EQU; Cameroon to Ethiopia, S to D.R. Congo & Tanzania.
EQU: <u>Sillitoe 385</u> (fl) 1919.

Adenia venenata Forssk.

F.P.S. 1: 161 (1950); F.T.E.A. Passiflor.: 30 (1975); F.J.M.: 96 (1976); T.S.S.: 63 (1990); F.E.E. 2(2): 7 (1995); Biol. Skr. 51: 118 (1998).
Woody climber. Dry woodland, bushland & grassland.
Distr: RS, DAR, KOR, BAG, EQU; Nigeria to Somalia, S to Tanzania; Yemen.
DAR: <u>Wickens 1368</u> (fl fr) 4/1964; **EQU:** <u>Friis & Vollesen 1193</u> (fl) 3/1982.

Basananthe hanningtoniana (Mast.) W.J.de Wilde

F.T.E.A. Passiflor.: 59 (1975); F.E.E. 2(2): 12 (1995); Biol.

Skr. 51: 118 (1998).
Annual or perennial herb, usually climbing. Forest margins & clearings, bushland.
Distr: EQU; Ethiopia to Zimbabwe & Mozambique.
EQU: Sillitoe 235 (fr) 1919.

Passiflora edulis Sims

F.T.E.A. Passiflor.: 15 (1975); F.E.E. 2(2): 15 (1995); Biol. Skr. 51: 118 (1998).
Perennial climbing herb. Naturalised from cultivation in, for example, village margins.
Distr: DAR, EQU; native to S America, naturalised elsewhere in the tropics.
DAR: Kamil 1186 (fl) 5/1968; **EQU:** Jackson 3160 5/1954.

Streptopetalum serratum Hochst.

F.T.E.A. Turner.: 17 (1954); F.E.E. 2(1): 71 (2000).
Annual herb. Dry woodland & bushland on sandy soil.
Distr: RS; Eritrea to Namibia & South Africa.
RS: Bally 6969 (fl fr) 4/1949.

Tricliceras bivinianum (Tul.) R.Fern.

F.E.E. 2(1): 72 (2000).
Syn: Wormskioldia biviniana Tul. – F.T.E.A. Turner.: 10 (1954).
Annual herb. Woodland, dry thicket, weed of cultivation.
Distr: UN; Ethiopia, Tanzania.
UN: Simpson 7386 (fl fr) 7/1929.
Note: a very localised species, but not considered threatened as it tends to be weedy.

Tricliceras lobatum (Urb.) R.Fern.

F.T.E.A. Turner.: 10 (1954); Biol. Skr. 51: 75 (1998).
Syn: Wormskioldia lobata Urb. – F.P.S. 1: 29 (1950); F.T.E.A. Turner.: 10 (1954).
Annual herb. Grassland, woodland, bushland & rocky outcrops, on sandy soils.
Distr: BAG, EQU; Gabon, D.R. Congo to Uganda & Kenya, S to Namibia, Zimbabwe & Mozambique.
EQU: Andrews 514 (fl) 4/1939.

Tricliceras pilosum (Willd.) R.Fern.

F.E.E. 2(1): 72 (2000).
Syn: Wormskioldia pilosa (Willd.) Schweinf. ex Urb. – F.P.S. 1: 30 (1950); F.T.E.A. Turner.: 11 (1954); F.J.M.: 84 (1976).
Annual herb. Grassland & woodland on sandy or rocky soils.
Distr: DAR, KOR, ES, BAG, UN, EQU; Senegal to Ethiopia, S to D.R. Congo & Tanzania.
DAR: Wickens 2011 (fl fr) 7/1964; **UN:** Sherif A.3906 (fl fr) 7/1951.

Salicaceae

I. Darbyshire

Casearia barteri Mast.

Bull. Jard. Bot. Nat. Belg. 41: 406 (1971); F.J.M.: 96 (1976).
Tree or shrub. Riverine forest.
Distr: DAR; Sierre Leone to D.R. Congo.
DAR: Wickens 986 (fl) 1/1964.

Casearia runssorica Gilg

F.T.E.A. Flacourt.: 48 (1975); T.S.S.: 57 (1990); Biol. Skr. 51: 115 (1998).
Syn: Casearia engleri sensu Andrews, non Gilg – F.P.S. 1: 158 (1950).
Tree. Forest, including swamp forest.
Distr: BAG, EQU; D.R. Congo, Rwanda, Burundi, Tanzania.
IUCN: RD
Note: no material has been seen but there is a sight record from Lotti Forest by Jackson (see Friis & Vollesen 1998) and El Amin in T.S.S. records it from Bahr el Ghazal. The status of this taxon is uncertain; in the African Plants Database it is treated as a synonym of *C. barteri* Mast.

Dovyalis abyssinica (A.Rich.) Warb.

F.T.E.A. Flacourt.: 61 (1975); Biol. Skr. 51: 116 (1998); F.E.E. 2(1): 448 (2000).
Shrub or tree. Dry montane forest & bushland, riverine scrub.
Distr: EQU; Ethiopia & Somalia, S to Malawi; Socotra.
EQU: Friis & Vollesen 1050 (fl) 2/1982.

Dovyalis macrocalyx (Oliv.) Warb.

F.P.S. 1: 157 (1950); F.T.E.A. Flacourt.: 64 (1975); T.S.S.: 58 (1990); Biol. Skr. 51: 116 (1998).
Shrub or small tree. Montane forest & bushland.
Distr: EQU; C.A.R. to Kenya, S to Angola & Mozambique.
EQU: Friis & Vollesen 1080 3/1982.

Flacourtia indica (Burm.f.) Merr.

F.T.E.A. Flacourt.: 57 (1975); Biol. Skr. 51: 116 (1998); F.E.E. 2(1): 446 (2000).
Shrub or tree. Dry forest & forest margins, riverine forest, montane bushland.
Distr: EQU; Senegal to Ethiopia, S to Botswana & South Africa; Madagascar, Mascarenes, Seychelles, India, Indonesia.
EQU: Myers 11739 (fr) 7/1939.

Homalium abdessammadii Asch. & Schweinf.

F.P.S. 1: 158 (1950); F.T.E.A. Flacourt.: 42 (1975); T.S.S.: 58 (1990).
Tree, more rarely a shrub. Riverine forest & forest margins.
Distr: EQU; Cameroon to Kenya, S to Namibia (Caprivi) & Mozambique.
EQU: Myers 13797 (fl) 4/1941.

Oncoba spinosa Forssk.

F.P.S. 1: 158 (1950); F.T.E.A. Flacourt.: 16 (1975); T.S.S.: 55 (1990); Biol. Skr. 51: 117 (1998); F.E.E. 2(1): 443 (2000).
Shrub or tree. Woodland, riverine forest & margins, rocky hillslopes.
Distr: CS, ES, BAG, EQU; Senegal to Somalia, S to Botswana & South Africa; Arabia.
CS: Lane 8 (fl) 5/1966; **EQU:** Myers 8487 (fl) 2/1938.

Salix mucronata Thunb.

F.Egypt 1: 13 (1999).
Syn: Salix murielii Skan – F.P.S. 2: 250 (1952); Bot. Exp. Sud.: 101 (1970) ; T.S.S.: 249 (1990).

Syn: *Salix schweinfurthii* Skan – F.P.S. 2: 250 (1952); Bot. Exp. Sud.: 101 (1970).
Syn: *Salix subserrata* Willd. – F.P.S. 2: 250 (1952); Bot. Exp. Sud.: 102 (1970); F.J.M.: 118 (1976); F.T.E.A. Salic.: 1 (1985); F.E.E. 3: 258 (1989); T.S.S.: 251 (1990).
Shrub or tree. Riverbanks, streamsides, swamps.
Distr: NS, DAR, KOR, CS; Eritrea to Namibia & South Africa; Egypt, Arabia.
DAR: <u>Wickens 2665</u> (fl) 9/1964.
Note: in South Africa this species is divided into several subspecies (see, for example, Jordaan in Bothalia 35: 15 (2005)); in this case, Sudanese material would fall under subsp. *subserrata* (Willd.) R.H.Archer & Jordaan.

Scolopia theifolia Gilg

F.T.E.A. Flacourt.: 33 (1975); Biol. Skr. 51: 117 (1998); F.E.E. 2(1): 446 (2000).
Tree. Dry montane forest & bushland.
Distr: EQU; Ethiopia, Uganda to Tanzania.
EQU: <u>Myers 10960</u> (fl) 4/1939.

Violaceae

I. Darbyshire

Hybanthus enneaspermus (L.) F.Muell. var. *enneaspermus*

F.P.S. 1: 62 (1950); F.T.E.A. Viol.: 32 (1986); F.E.E. 2(1): 168 (2000).
Annual or perennial herb. Dry woodland, bushland & grassland, sometimes weedy.
Distr: CS, UN; palaeotropical.
CS: <u>Lea 27</u> (fl fr) 10/1952; **UN:** <u>Simpson 7053</u> (fl fr) 6/1929.

Rinorea brachypetala (Turcz.) Kuntze

F.T.E.A. Viol.: 7 (1986); Biol. Skr. 51: 82 (1998).
Syn: *Rinorea poggei* Engl. – F.P.S. 1: 63 (1950); T.S.S.: 43 (1990).
Shrub or small tree. Forest.
Distr: EQU; Sierre Leone to Kenya, S to Angola & Zambia.
EQU: <u>Friis & Vollesen 712</u> 12/1980.

Rinorea dentata (P.Beauv.) Kuntze

F.T.E.A. Viol.: 12 (1986); T.S.S.: 42 (1990).
Shrub or small tree. Forest including swamp forest & margins.
Distr: ?EQU; Liberia to Uganda, S to Angola & Tanzania.
Note: El Amin in T.S.S records this species as "common in tall grass savanna in Equatoria (Torit, Talanga forests)" but we have seen no material with which to verify this record. The habitat description seems incorrect, but *R. dentata* would not be unexpected in South Sudan. Friis & Vollesen (1998) do not mention the Talanga record in their checklist of the plants of the Imatong Mts.

Rinorea ilicifolia (Welw. ex Oliv.) Kuntze var. *ilicifolia*

F.P.S. 1: 63 (1950); F.T.E.A. Viol.: 4 (1986); T.S.S.: 42 (1990); Biol. Skr. 51: 82 (1998); F.E.E. 2(1): 163 (2000).
Shrub or small tree. Forest.
Distr: UN, EQU; Guinea to Ethiopia, S to Angola & South Africa.
UN: <u>Myers 10484</u> (fl) 2/1939.

Rinorea oblongifolia (C.H.Wright) Marquand ex Chipp

F.P.S. 1: 65 (1950); F.T.E.A. Viol.: 9 (1986); T.S.S.: 43 (1990); Biol. Skr. 51: 82 (1998).
Shrub or tree. Forest.
Distr: EQU; Guinea to Uganda & D.R. Congo.
EQU: <u>Andrews 1755</u> (fl) 6/1939.

Viola abyssinica Steud. ex Oliv.

F.T.E.A. Viol.: 25 (1986); Biol. Skr. 51: 82 (1998); F.E.E. 2(1): 165 (2000).
Perennial herb. Montane forest, bushland & grassland.
Distr: EQU; Nigeria, Bioko & Cameroon, Ethiopia to South Africa; Madagascar.
EQU: <u>Friis & Vollesen 366</u> (fl) 11/1980.

Viola cinerea Boiss. var. *stocksii* (Boiss.) W.Becker

F.Som. 1: 74 (1993).
Syn: *Viola etbaica* Schweinf. – F.P.S. 1: 65 (1950).
Annual or short-lived perennial herb. Stony ground, rock crevices.
Distr: RS; Somalia; Arabia to India.
RS: <u>Bent s.n.</u> (fl) 1896.

Viola eminii (Engl.) R.E.Fr.

F.P.S. 1: 65 (1950); F.T.E.A. Viol.: 27 (1986); Biol. Skr. 51: 83 (1998).
Perennial herb. Montane forest margins, bushland & grassland.
Distr: EQU; D.R. Congo, Rwanda, Burundi, Uganda, Kenya, Tanzania.
EQU: <u>Thomas 1857</u> (fl fr) 12/1935.

Achariaceae

I. Darbyshire

Caloncoba crepiniana (De Wild. & T.Durand) Gilg

F.T.E.A. Flacourt.: 24 (1975); T.S.S.: 57 (1990); Biol. Skr. 51: 115 (1998).
Syn: *Caloncoba schweinfurthii* Gilg – F.P.S. 1: 156 (1950).
Tree or shrub. Forest & forest margins.
Distr: EQU; C.A.R., D.R. Congo, Uganda.
EQU: <u>Cartwright 10</u> (fl) 3/1916.

Lindackeria bukobensis Gilg

F.T.E.A. Flacourt.: 25 (1975).
Shrub or tree. Forest & forest margins.
Distr: EQU; Uganda & Somalia, S to Zambia & Malawi.
EQU: <u>Andrews 1200</u> (fr) 5/1939.

Lindackeria schweinfurthii Gilg

F.P.S. 1: 157 (1950); F.T.E.A. Flacourt.: 27 (1975); T.S.S.: 59 (1990); Biol. Skr. 51: 116 (1998).
Shrub or small tree. Forest, sometimes riverine.
Distr: EQU; Cameroon to Uganda, S to D.R. Congo.
EQU: <u>Myers 9191</u> (fl) 8/1938.

Rawsonia lucida Harv. & Sond.

F.T.E.A. Flacourt.: 4 (1975); T.S.S.: 59 (1990); Biol. Skr. 51: 117 (1998).
Shrub or tree. Forest, sometimes riverine.
Distr: EQU; D.R. Congo to Somalia, S to Angola & South Africa.
EQU: Myers 10961 (fl fr) 4/1939.

Scottellia orientalis Gilg

Blumea 20: 277 (1973).
Tree. Riverine & swamp forest.
Distr: EQU; Nigeria, C.A.R., D.R. Congo.
EQU: Andrews 1631 6/1939.

Irvingiaceae

I. Darbyshire

Irvingia smithii Hook.f.

F.P.S. 2: 320 (1952); T.S.S.: 303 (1990); Sp. Pl., Fl. World 1, Irving.: 16 (1999).
Large tree. Riverine forest.
Distr: BAG, EQU; Nigeria to C.A.R. & D.R. Congo, S to Angola.
BAG: Aylmer 27/47 (fr) 10/1926.

Irvingia wombolu Vermoesen

Biol. Skr. 51: 279 (1998); Sp. Pl., Fl. World 1, Irving.: 15 (1999).
Syn: *Irvingia barteri* sensu Andrews, non Hook.f. – F.P.S. 2: 320 (1952).
Syn: *Irvingia gabonensis* sensu auctt., non (Aubry-LeComte ex O'Rorke) Baill. – F.T.E.A. Ixonanth.: 7 pro parte (1984); T.S.S.: 303 (1990); Biol. Skr. 51: 279 (1998).
Large tree. Forest.
Distr: EQU; Senegal to Uganda, S to Angola.
EQU: Myers 10567 (fr) 2/1939.

Klainedoxa gabonensis Pierre ex Engl.

F.P.S. 2: 321 (1952); F.T.E.A. Ixonanth.: 4 (1984); T.S.S.: 305 (1990); Biol. Skr. 51: 279 (1998); Sp. Pl., Fl. World 1, Irving.: 5 (1999).
Large tree. Forest.
Distr: EQU; Guinea Bissau to Uganda, S to D.R. Congo and Zambia.
EQU: Andrews 1177 (fr) 5/1939.

Linaceae

I. Darbyshire

Hugonia platysepala Welw. ex Oliv.

F.P.S. 1: 123 (1950); F.T.E.A. Lin.: 2 (1966); T.S.S.: 47 (1990); Biol. Skr. 51: 101 (1998).
Shrub, woody climber or small tree. Forest.
Distr: EQU; Sierra Leone to Uganda & D.R. Congo.
EQU: Jackson 1500 5/1950.

Linum strictum L.

F.P.S. 1: 123 (1950); F.E.E. 2(1): 352 (2000).
Syn: *Linum strictum* L. subsp. *corymbulosum* (Rchb.) Rouy – F.J.M.: 93 (1976).
Annual herb. Montane grassland & bushland; weed of irrigated fields & fallow land.

Distr: RS, DAR; Eritrea, Ethiopia; S Europe, Macaronesia, N Africa to NW India.
DAR: Wickens 2606 (fl) 9/1964.
Note: material from Sudan is quite variable but is considered to represent a single taxon here (see note in F.E.E. on the variation within this species).

Ixonanthaceae

I. Darbyshire

Phyllocosmus africanus (Hook.f.) Klotzsch

Fl. Cameroun 14: 62 (1972).
Tree. Forest, often in riverine or swamp forest.
Distr: EQU; Guinea Bissau to C.A.R. & D.R. Congo.
EQU: Andrews 1541 (fl).

Clusiaceae

H. Gibreel & I. Darbyshire

Garcinia buchananii Baker

F.P.S. 1: 214 (1950); F.T.E.A. Guttif.: 24 (1978); F.E.E. 2(2): 142 (1995); Biol. Skr. 51: 136 (1998).
Syn: *Garcinia huillensis* sensu El Amin, non Welw. ex Oliv. – T.S.S.: 102 (1990).
Tree or shrub. Forest & dense bushland, often riverine.
Distr: EQU; Ethiopia to Zimbabwe & Mozambique.
EQU: Andrews 1132 (fl fr) 5/1939.

Garcinia ovalifolia Oliv.

F.P.S. 1: 214 (1950); F.T.E.A. Guttif.: 22 (1978); T.S.S.: 102 (1990); F.E.E. 2(2): 142 (1995).
Shrub or tree. Forest, often riverine.
Distr: EQU; Senegal to Ethiopia, S to Angola & D.R. Congo.
EQU: Hoyle 487 1/1939.

Podostemaceae

I. Farag & I. Darbyshire

Ledermanniella ramosissima Hauman ex C.Cusset

Adansonia, Ser. 4, 6(3): 258 (1984); F.T.E.A. Podostem.: 4, in notes (2005).
Rheophytic perennial herb. Habitat not recorded, but probably on rocks in fast-flowing streams.
Distr: EQU; South Sudan endemic.
IUCN: RD
EQU: Thomas 1660 (fr) 12/1935.

Sphaerothylax abyssinica (Wedd.) Warm.

Biol. Skr. 51: 88 (1998); F.E.E. 2(1): 194 (2000); F.T.E.A. Podostem.: 7 (2005).
Syn: *Inversodicraea sp.* sensu Andrews – F.P.S. 1: 82 (1950).
Rheophytic perennial herb. On rocks in fast-flowing streams.
Distr: EQU; Ethiopia to Zimbabwe; Madagascar.
EQU: Friis & Vollesen 420 (fr) 11/1980.

Tristicha trifaria (Bory ex Willd.) Spreng.

F.P.S. 1: 83 (1950); F.J.M.: 88 (1976); Biol. Skr. 51: 88 (1998); F.E.E. 2(1): 191 (2000); F.T.E.A. Podostem.: 2 (2005).
Rheophytic perennial herb. On rocks in streams & rivers, waterfalls.
Distr: DAR, KOR, BAG, EQU; Sierra Leone to Ethiopia, S to South Africa; Madagascar, Australia, tropical America.
DAR: <u>Wickens 2994</u> 6/1965; **EQU:** <u>Friis & Vollesen 421</u> 11/1980.

Hypericaceae

H. Gibreel & I. Darbyshire

Harungana madagascariensis Lam. ex Poir.

F.P.S. 1: 212 (1950); F.T.E.A. Hyperic.: 19 (1953); T.S.S.: 99 (1990); Biol. Skr. 51: 136 (1998).
Tree or shrub. Forest margins & disturbed areas.
Distr: BAG, EQU; Senegal to Kenya, S to Angola & South Africa; Madagascar, Mascarenes.
EQU: <u>Friis & Vollesen 298</u> (fl) 11/1980.

Hypericum annulatum Moris subsp. **intermedium** (Steud. ex A.Rich.) N.Robson

F.T.E.A. Hyperic.: 6 (1953); F.E.E. 2(2): 138 (1995).
Perennial herb. Dry montane grassland & rocky slopes.
Distr: RS; Eritrea, Ethiopia; Saudi Arabia.
RS: <u>Jackson 2867</u> (fl) 4/1953.

Hypericum lalandii Choisy

F.P.S. 1: 212 (1950); F.T.E.A. Hyperic.: 7 (1953); F.E.E. 2(1): 452 (2000).
Perennial herb. Marshy grassland.
Distr: EQU; Nigeria, Ethiopia to Angola & South Africa; Madagascar.
EQU: <u>Andrews 432</u> (fr) 11/1939.

Hypericum peplidifolium A.Rich.

F.P.S. 1: 212 (1950); F.T.E.A. Hyperic.: 9 (1953); F.E.E. 2(2): 140 (1995); Biol. Skr. 51: 136 (1998).
Syn: *Hypericum riparium* A.Chev. – F.W.T.A. 1(1): 287 (1954).
Shrub. Damp grassland, streamsides, forest clearings.
Distr: EQU; Nigeria, Bioko, Cameroon, Eritrea to Angola & Mozambique.
EQU: <u>Friis & Vollesen 367</u> (fl) 11/1981.

Hypericum perforatum L.

F.P.S. 1: 212 (1950); F.J.M.: 99 (1976).
Perennial herb. Grassland, streamsides, weed of irrigated land.
Distr: DAR; South Africa; Europe, N Africa, Middle East.
DAR: <u>Wickens 1463</u> (fl) 4/1964.

Hypericum quartinianum A.Rich.

F.T.E.A. Hyperic.: 3 (1953); F.E.E. 2(2): 136 (1995); Biol. Skr. 51: 137 (1998).
Shrub or tree. Rocky grassland & bushland.
Distr: EQU; Ethiopia to Zambia & Mozambique; Arabia.
EQU: <u>Friis & Vollesen 335</u> (fl) 11/1980.

Hypericum revolutum Vahl subsp. **revolutum**

T.S.S.: 99 (1990); F.E.E. 2(2): 136 (1995); Biol. Skr. 51: 137 (1998).

Syn: *Hypericum lanceolatum* Lam. – F.T.E.A. Hyperic.: 4 (1953).
Syn: *Hypericum leucoptychodes* Steud. ex A.Rich. – F.P.S. 1: 213 (1950).
Shrub or small tree. Open montane forest, montane bushland.
Distr: EQU; Nigeria, Bioko, Cameroon, Eritrea to South Africa; Madagascar, Mascarenes, Arabia.
EQU: <u>Myers 10893</u> (fl) 4/1939.

Hypericum roeperianum G.W.Schimp. ex A.Rich.

F.T.E.A. Hyperic.: 3 (1953); T.S.S.: 101 (1990); F.E.E. 2(2): 136 (1995); Biol. Skr. 51: 137 (1998).
Shrub or small tree. Montane forest.
Distr: EQU; Guinea, Nigeria, Cameroon, Ethiopia to Angola & South Africa; Madagascar, Comoros, Macarenes.
EQU: <u>Jackson 256</u> 2/1948.

Psorospermum febrifugum Spach

F.T.E.A. Hyperic.: 16 (1953); T.S.S.: 101 (1990); F.E.E. 2(2): 140 (1995); Biol. Skr. 51: 138 (1998).
Syn: *Psorospermum campestre* Engl. – F.P.S. 1: 213 (1950).
Syn: *Psorospermum febrifugum* Spach var. **ferrugineum** (Hook.f.) Keay & Milne-Redh. – F.T.E.A. Hyperic.: 18 (1953).
Syn: *Psorospermum salicifolium* Engl. – F.P.S. 1: 214 (1950).
Shrub or tree. Woodland.
Distr: DAR, BAG, EQU; Sierra Leone to Ethiopia, S to Angola & South Africa.
DAR: <u>Miehe 270</u> (fl) 6/1982; **EQU:** <u>Friis & Vollesen 1065</u> (fl) 3/1982.
Note: Andrews records also *P. tenuifolium* Hook.f. from "central and southern Sudan, often common near swamps". The record from South Sudan is believed to be based on *Schweinfurth* 3480, collected from Munza which is in present-day D.R. Congo. We have not seen the specimen(s) from central Sudan, or those cited by Drar from Darfur (Bot. Exp. Sud.: 61, 1970), but they are almost certainly based on misidentifications, perhaps of *P. febrifugum*.

Psorospermum suffruticosum Engl.

F.P.S. 1: 213 (1950); Trop. Afr. Fl. Pl. 1: 624 (2003).
Subshrub. Wooded grassland.
Distr: ?BAG; Cameroon (see note).
IUCN: DD
Note: Andrews recorded this little-known species from "Said Bundas District"; we have not seen the material on which this record is based. In Trop. Afr. Fl. Pl. this species is treated as of uncertain status, possibly close to *P. corymbiferum* Hochr.

[*Vismia*]

Note: Andrews records *Psorospermum guineense* (L.) Hochr. (= *Vismia guineensis* (L.) Choisy) from Equatoria; this is likely to be a misidentification but we have been unable to locate the material on which the record was based. The most likely *Vismia* species in South Sudan is *V. laurentii* De Wild. which extends to NE D.R. Congo.

GERANIALES

Geraniaceae

I. Darbyshire

Erodium cf. *glaucophyllum* (L.) L'Hér.

F.Egypt 2: 6 (2000).
Syn: *Erodium malacoides* sensu Wickens pro parte, non (L.) L'Hér. – F.J.M.: 93 (1976).
Perennial herb. Upland fallow.
Distr: DAR; (for *E. glaucophyllum*) N Africa, Arabia, Palestine to Iran.
DAR: Wickens 1696 (fl) 5/1964.
Note: the single specimen was identified by G. Guittonneau in 1989 as *E. ? glaucophyllum*; more material is needed from Jebel Marra.

Erodium malacoides (L.) L'Hér.

F.P.S. 1: 128 pro parte (1950); F.Som. 1: 188 (1993).
Annual or biennial herb. Rocky slopes.
Distr: RS; Somalia; Mediterranean to Pakistan.
RS: Jackson 2734 (fl fr) 3/1953.
Note: of the Sudanese material named as this species, only the specimen cited has glands on the mericarps (see key to species in *Flora of Somalia*).

Erodium neuradifolium Delile ex Godr.

F.Egypt 2: 6 (2000).
Syn: *Erodium sutrilobum* (Lange) Jord. – Bot. Exp. Sud.: 51 (1970).
Syn: *Erodium malacoides* sensu Andrews pro parte, non (L.) L'Hér. – F.P.S. 1: 128 (1950).
Annual or biennial herb. Open rocky slopes & plains.
Distr: RS; Somalia; Canary Is., Mediterranean to Afghanistan.
RS: Jackson 2903 (fr) 4/1953.
Note: it appears that most of the material from the Red Sea region previously identified as *E. malacoides* is actually this species, but the two are closely allied and difficult to separate.

Erodium oreophilum Quézel

Fl. Sahara (ed. 2): 557 (1983); Acta Bot. Gallica 140: 303 (1993).
Syn: *Erodium malacoides* sensu auctt., non (L.) L'Hér. – F.P.S. 1: 128 pro parte (1950); F.J.M.: 93 pro parte (1976).
Annual herb. Montane grassland, open rocky ground, seasonally moist soils.
Distr: DAR; Chad (Tibesti).
IUCN: RD
DAR: Jackson 3363 (fl fr) 12/1954.
Note: the African Plants Database records *E. chium* (L.) Willd. from Sudan (Red Sea region) but this is believed to be in error, probably a misplacement on the map of the single record from Eritrea.

Geranium aculeolatum Oliv.

F.P.S. 1: 128 (1950); F.T.E.A. Geran.: 2 (1971); Biol. Skr. 51: 365 (1998); F.E.E. 2(1): 365 (2000).
Perennial herb. Montane forest margins & secondary growth.
Distr: EQU; Ethiopia to D.R. Congo & Mozambique.
EQU: Friis & Vollesen 1027 (fl fr) 2/1982.

Geranium arabicum Forssk. subsp. *arabicum*

F.T.E.A. Geran.: 9 (1971); Biol. Skr. 51: 103 (1998); F.E.E. 2(1): 365 (2000).
Syn: *Geranium simense* Hochst. ex A.Rich. – F.P.S. 1: 128 (1950).
Perennial herb. Montane forest & forest-grassland transition.
Distr: EQU; Nigeria, Cameroon, Eritrea to Zimbabwe; Madagascar, Egypt, Arabia.
EQU: Myers 11581 (fl) 7/1939.

Geranium biuncinatum Kokwaro

F.E.E. 2(1): 369 (2000).
Annual herb. Bushland on rocky slopes.
Distr: RS; Eritrea, Somalia; Egypt, Yemen.
IUCN: RD
RS: Andrews 3528 (fl fr).

Geranium elamellatum Kokwaro

F.T.E.A. Geran.: 6 (1971); F.E.E. 2(1): 367 (2000).
Annual herb. Montane forest margins, streamsides.
Distr: RS; D.R. Congo, Eritrea, Kenya, Tanzania.
RS: Jackson 2961 (fl fr) 4/1953.
Note: we here follow F.E.E. in keeping this species distinct from *G. purpureum* Vill.

Geranium favosum Hochst.

F.P.S. 1: 129 pro parte (1950); F.T.E.A. Geran.: 6 (1971); F.E.E. 2(1): 367 (2000).
Syn: *Geranium ocellatum* sensu Wickens pro parte, non Cambess. – F.J.M.: 93 (1976).
Annual herb. Streamsides & streambeds, riverine forest.
Distr: RS, DAR; Eritrea, Ethiopia, Djibouti; Egypt, Saudi Arabia.
DAR: Wickens 3690 (fl fr) 6/1977.

Geranium mascatense Boiss.

F.E.E. 2(1): 367 (2000).
Annual herb. Montane slopes, forest margins.
Distr: RS; Eritrea, Somalia; Arabia, Iran.
RS: Aylmer 221 (fl fr) 3/1932.
Note: we here follow F.E.E. in treating *G. favosum*, *G. mascatense* and *G. ocellatum* as distinct species. The latter two are treated as a single species in the African Plants Database, but the mericarps of *G. mascatense* appear closer to *G. favosum* (see key in F.E.E.) from which it is difficult to separate, at least in Sudan.

Geranium ocellatum Cambess. var. *sublaeve* (Oliv.) Milne-Redh.

F.T.E.A. Geran.: 6 (1971); F.J.M.: 93 pro parte (species only) (1976); Biol. Skr. 51: 103 (1998); F.E.E. 2(1): 367 (2000).
Syn: *Geranium favosum* sensu Andrews pro parte, non Hochst. – F.P.S. 1: 129 (1950).
Annual herb. Montane forest, bushland & grassland.
Distr: DAR, EQU; Cameroon, Eritrea to Somalia, S to Zimbabwe; Yemen.
DAR: Wickens 1682 (fl fr) 5/1964; **EQU:** Friis & Vollesen 156 (fl fr) 11/1980.

Geranium trilophum Boiss.

F.P.S. 1: 129 (1950); F.E.E. 2(1): 368 (2000).
Annual herb. Bushland, scrub on hillslopes, grassland.

Distr: RS; Eritrea, Ethiopia, Somalia; Egypt, Arabia & Iran.
RS: Kennedy-Cooke 9 (fl fr) 3/1938.

Monsonia heliotropioides (Cav.) Boiss.

F.P.S. 1: 131 (1950); F.Egypt 2: 11 (2000).
Perennial herb. Deserts, rocky ground.
Distr: RS; N Africa, Palestine, Arabia to India.
RS: Bent s.n. (fl) 1896.

Monsonia nivea (Decne.) Decne. ex Webb

F.P.S. 1: 131 (1950); F.J.U. 2: 178 (1999); F.Egypt 2: 11 (2000).
Perennial herb. Open sandy plains, wadis.
Distr: NS, RS; N Africa, Palestine, Arabia to Pakistan.
RS: Carter 1947 (fl fr) 11/1987.

Monsonia senegalensis Guill. & Perr.

F.P.S. 1: 131 (1950); F.T.E.A. Geran.: 10 (1971); F.J.M.: 93 (1976); F.E.E. 2(1): 370 (2000).
Annual herb. Rocky hillslopes, open bushland & grassland.
Distr: DAR, KOR, CS, ES; Senegal & Mauritania to Somalia, S to Namibia & South Africa; Egypt.
KOR: Jackson 4013 (fl fr) 9/1959.

Pelargonium multibracteatum Hochst. ex A.Rich.

F.P.S. 1: 131 (1950); F.T.E.A. Geran.: 21 (1971); F.E.E. 2(1): 373 (2000).
Perennial herb. Rocky hillslopes, clefts in rocks.
Distr: RS; Eritrea & Somalia, S to Tanzania; Arabia.
RS: Aylmer 638 (fl fr) 3/1936.

Melianthaceae

I. Darbyshire

Bersama abyssinica Fresen.

F.P.S. 2: 344 (1952); F.T.E.A. Melianth.: 2 (1958); F.E.E. 3: 511 (1989); T.S.S.: 333 (1990); Biol. Skr. 51: 288 (1998).
Tree or shrub. Montane forest & forest margins, secondary bushland.
Distr: EQU; Sierra Leone to Eritrea, S to Zambia & Zimbabwe; Arabia.
EQU: Andrews 1128 (fl) 5/1939.
Note: this very variable species has in the past been subdivided into a number of subspecies and varieties (see, for example, F.T.E.A. Melianthaceae) but is treated as a single entity here.

MYRTALES

Combretaceae

I. Farag & I. Darbyshire

Anogeissus leiocarpa (DC.) Guill. & Perr.

F.J.M.: 98 (1976); T.S.S.: 79 (1990); F.E.E. 2(2): 130 (1995).
Syn: *Anogeissus schimperi* Hochst. ex Hutch. & Dalziel – F.P.S. 1: 197 (1950); Bot. Exp. Sud.: 30 (1970).
Tree. Woodland, wooded grassland & bushland, sometimes dominant.
Distr: DAR, KOR, CS, ES, BAG, UN, EQU; Mauritania to Eritrea, S to D.R. Congo.
DAR: Wickens 1925 (fr) 7/1964; **EQU:** Andrews 1509 (fr) 5/1939.

Combretum aculeatum Vent.

F.P.S. 1: 199 (1950); F.T.E.A. Combret.: 55 (1973); F.J.M.: 98 (1976); T.S.S.: 81 (1990); F.E.E. 2(2): 120 (1995); Biol. Skr. 51: 130 (1998).
Shrub or woody climber. Woodland, bushland, riverine forest.
Distr: DAR, KOR, CS, ES, BAG, UN, EQU; Mauritania to Somalia, S to Tanzania; Morocco.
ES: Beshir 55 (fl) 8/1951; **EQU:** Friis & Vollesen 857 (fr) 12/1980.

Combretum adenogonium Steud. ex A.Rich.

F.P.S. 1: 204 (1950); T.S.S.: 81 (1990); F.E.E. 2(2): 117 (1995); Biol. Skr. 51: 130 (1998).
Syn: *Combretum fragrans* F.Hoffm. – Bot. Exp. Sud.: 31 (1970); T.S.S.: 83 (1990).
Syn: *Combretum ghasalense* Engl. & Diels – F.P.S. 1: 206 (1950).
Syn: *Combretum multispicatum* Engl. & Diels – F.P.S. 1: 204 (1950).
Syn: *Combretum undulatum* Engl. & Diels – F.P.S. 1: 204 (1950).
Tree. Woodland & wooded grassland.
Distr: NS, RS, DAR, KOR, CS, ES, BAG, UN, EQU; Senegal to Eritrea, S to Botswana & Mozambique.
ES: Kennedy-Cooke 29 (fl) 6/1931; **EQU:** Myers 8256 (fl) 1/1938.

Combretum capituliflorum Fenzl ex Schweinf.

F.P.S. 1: 203 (1950); F.T.E.A. Combret.: 46 (1973); T.S.S.: 81 (1990); F.E.E. 2(2): 120 (1995); Biol. Skr. 51: 130 (1998).
Syn: *Combretum undulato-marginatum* De Wild. & Exell – F.P.S. 1: 203 (1950).
Woody climber. Riverine forest, riverbanks.
Distr: ?CS, BAG, EQU; C.A.R., D.R. Congo, Ethiopia, Uganda, Kenya.
EQU: Friis & Vollesen 1045 (fl) 2/1982.
Note: a local species but not considered threatened. The record for Central Sudan is from El Amin in T.S.S.; no specimen has been seen to verify it.

Combretum collinum Fresen. subsp. *collinum*

F.T.E.A. Combret.: 24 (1973); F.E.E. 2(2): 116 (1995).
Syn: *Combretum bongense* Engl. – F.P.S. 1: 201 (1950).
Tree. Woodland & wooded grassland.
Distr: KOR, CS, BAG; Eritrea, Ethiopia.
KOR: Wickens 930 (fr) 1/1963; **BAG:** Schweinfurth 2773 (fl) 12/1869.

Combretum collinum Fresen. subsp. binderianum (Kotschy) Okafor

F.T.E.A. Combret.: 26 (1973); F.J.M.: 98 (1976); T.S.S.: 82 (1990); F.E.E. 2(2): 116 (1995); Biol. Skr. 51: 131 (1998).
Syn: *Combretum binderianum* Kotschy – F.P.S. 1: 199 (1950).
Tree. Woodland & wooded grassland.
Distr: DAR, CS, ES, BAG, UN, EQU; Ivory Coast to Ethiopia, S to Tanzania.

ES: Francis 84 (fr) 12/1957; **BAG:** Schweinfurth 1374 (fr) 4/1869.

Combretum collinum Fresen. subsp. elgonense (Exell) Okafor

Bot. Exp. Sud.: 30 (1970); F.T.E.A. Combret.: 25 (1973); T.S.S.: 83 (1990); F.E.E. 2(2): 116 (1995); Biol. Skr. 51: 131 (1998).
Syn: *Combretum kabadense* Exell – F.P.S. 1: 199 (1950).
Syn: *Combretum laboniense* M.B.Moss – F.P.S. 1: 202 (1950).
Tree. Woodland & wooded grassland.
Distr: ?CS, EQU; Ethiopia to D.R. Congo & Zambia.
CS: Drar 721 3/1938 (n.v.); **EQU:** Friis & Vollesen 1236 (fl) 3/1982.
Note: the single record from Central Sudan is based on sterile material and requires confirmation.

Combretum collinum Fresen. subsp. hypopilinum (Diels) Okafor

F.T.E.A. Combret.: 25 (1973); F.J.M.: 99 (1976); T.S.S.: 83 (1990); F.E.E. 2(2): 117 (1995).
Syn: *Combretum verticillatum* Engl. & Diels – F.P.S. 1: 203 (1950).
Tree. Woodland & wooded grassland.
Distr: DAR, KOR, CS, EQU; Guinea to Ethiopia & Uganda.
DAR: Wickens 3214 (fl) 6/1977; **EQU:** Myers 6371 (fl) 3/1937.

Combretum comosum G.Don

Trop. Afr. Fl. Pl. 1: 574 (2003).
Syn: *Combretum rhodanthum* Engl. & Diels – F.T.E.A. Combret.: 59 (1973); T.S.S.: 87 (1990).
Woody climber or scandent shrub. Forest, secondary regrowth.
Distr: EQU; Senegal to Uganda.
EQU: Schweinfurth 2998 (fr) 2/1870.

Combretum glutinosum Perr. ex DC.

F.P.S. 1: 202 (1950); F.W.T.A. 1(1): 271 (1954); F.J.M.: 99 (1976); T.S.S.: 77 (1990).
Syn: *Combretum cordofanum* Engl. & Diels – F.P.S. 1: 202 (1950).
Shrub or tree. Woodland & wooded grassland.
Distr: DAR, KOR, CS, BAG; Mauritania & Sierra Leone to C.A.R.
DAR: Wickens 3107 (fr) 4/1971; **BAG:** Andrews 624 (fr) 4/1939.

Combretum hartmannianum Schweinf.

F.P.S. 1: 204 (1950); T.S.S.: 85 (1990); F.E.E. 2(2): 117 (1995).
Tree. Woodland & wooded grassland.
Distr: DAR, KOR, CS, ES, ?BAG, ?UN, ?EQU; Eritrea & Ethiopia.
IUCN: VU
KOR: Wickens 3082 (fr) 6/1968.
Note: the records for South Sudan are derived from El Amin in T.S.S. and require confirmation.

Combretum hereroense Schinz subsp. *grotei* (Exell) Wickens

F.T.E.A. Combret.: 41 (1973).
Small tree. Wooded grassland, bushland.

Distr: EQU; Uganda, Kenya, Tanzania.
EQU: Myers 14033 (fr) 9/1941.

Combretum molle R.Br. ex G.Don

F.P.S. 1: 201 (1950); F.T.E.A. Combret.: 33 (1973); F.J.M.: 99 (1976); T.S.S.: 86 (1990); F.E.E. 2(2): 118 (1995); Biol. Skr. 51: 131 (1998).
Syn: *Combretum gueinzii* Sond. – F.P.S. 1: 202 (1950).
Syn: *Combretum lepidotum* A.Rich. – F.P.S. 1: 201 (1950).
Tree. Woodland & wooded grassland where often dominant, dry forest.
Distr: RS, DAR, KOR, CS, ES, BAG, EQU; Senegal to Somalia, S to Angola & South Africa; Madagascar, Yemen.
DAR: Wickens 2873 (fr) 3/1965; **EQU:** Myers 7899 (fr) 11/1937.

Combretum mucronatum Schumach. & Thonn.

F.T.E.A. Combret.: 49 (1973).
Syn: *Combretum smeathmannii* G.Don – F.P.S. 1: 204 (1950).
Scrambling shrub or woody climber. Riverine forest, secondary forest.
Distr: EQU; Senegal to Uganda.
EQU: Hoyle 744 12/1939.

Combretum nigricans Lepr. ex Guill. & Perr.

F.E.E. 2(2): 117 (1995).
Syn: *Combretum elliotii* Engl. & Diels – F.P.S. 1: 201 (1950).
Syn: *Combretum nigricans* Lepr. ex Guill. & Perr. var. *elliotii* (Engl. & Diels) Aubrév. – T.S.S.: 86 (1990).
Tree. Woodland.
Distr: DAR, BAG, EQU; Senegal to Ethiopia.
DAR: Wickens 3123 (fl) 4/1971; **EQU:** Myers 7560 (fr) 7/1937.
Note: the varieties recognised within this species in West Africa are not easily separated in the eastern part of its range.

Combretum paniculatum Vent.

F.P.S. 1: 203 (1950); F.T.E.A. Combret.: 53 (1973); F.J.M.: 99 (1976); T.S.S.: 86 (1990); F.E.E. 2(2): 120 (1995); Biol. Skr. 51: 132 (1998).
Woody climber or shrub. Forest, particularly along margins and in clearings, woodland.
Distr: DAR, EQU; Senegal to Ethiopia, S to Angola & Mozambique.
DAR: Wickens 2900 (fr) 4/1965; **EQU:** Myers 8508 (fr) 2/1938.

Combretum platypterum (Welw.) Hutch. & Dalziel

F.P.S. 1: 204 (1950); F.W.T.A. 1(1): 274 (1954); T.S.S.: 87 (1990).
Woody climber or scandent shrub. Forest, forest margins & secondary growth.
Distr: EQU; Guinea to Angola & D.R. Congo.
EQU: Andrews 1483 (fr) 5/1939.

Combretum racemosum P.Beauv.

F.P.S. 1: 203 (1950); F.T.E.A. Combret.: 55 (1973); T.S.S.: 87 (1990); Biol. Skr. 51: 133 (1998).

Woody climber. Riverine forest & forest margins, secondary bushland.
Distr: EQU; Sierra Leone to Uganda, S to Angola & D.R. Congo.
EQU: Myers 8206 (fl) 1/1938.

Combretum rochetianum A.Rich. ex A.Juss.

F.E.E. 2(2): 117 (1995).
Syn: Combretum gallabatense Schweinf. – F.P.S. 1: 202 (1950); T.S.S.: 85 (1990);.
Tree. Woodland & bushland.
Distr: CS, ES, UN, EQU; Eritrea, Ethiopia.
IUCN: VU
ES: Kennedy-Cooke 35 (fl) 6/1931; **EQU:** Simpson 7594 (fr) 11/1930.

Combretum schweinfurthii Engl. & Diels

F.P.S. 1: 203 (1950); F.T.E.A. Combret.: 30 (1973).
Shrub or small tree. Wooded grassland.
Distr: BAG, EQU; D.R. Congo, Uganda.
IUCN: RD
EQU: Schweinfurth 2886 (fr) 6/1870.

Combretum sericeum G.Don

F.P.S. 1: 199 (1950); F.W.T.A. 1(1): 270 (1954).
Syn: Combretum parvulum Engl. & Diels – Engl. Monogr. Afrik. Pflanzenfam. 3: 67 (1899).
Suffruticose shrub. Wooded grassland.
Distr: BAG; Senegal to C.A.R.
BAG: Schweinfurth III.64 (fl fr) 1/1871.
Note: *C. parvulum* (type from South Sudan) is maintained by the African Plants Database, albeit with uncertain status; we consider it to be a synonym of *C. sericeum*.

Combretum umbricola Engl.

F.T.E.A. Combret.: 20 (1973).
Syn: Combretum sp. cf. **C. tanaense** sensu Friis & Vollesen – Biol. Skr. 51: 133 (1998).
Woody climber. Forest including riverine fringes.
Distr: EQU; D.R. Congo, Uganda, Tanzania, Mozambique.
EQU: Friis & Vollesen 724 (fr) 12/1980.
Note: a scarce species with a scattered distribution; habitat loss, particularly in the coastal parts of its range, may threaten this species.

[Conocarpus lancifolius Engl. ex Engl. & Diels]

Note: this species, native to Somalia and Yemen, is fairly widely planted in Sudan as a shelterbelt species. It is not considered to have naturalised.

Guiera senegalensis J.F.Gmel.

F.P.S. 1: 206 (1950); F.W.T.A. 1(1): 275 (1954); Bot. Exp. Sud.: 31 (1970); F.J.M.: 99 (1976); T.S.S.: 89 (1990).
Shrub. Open sandy areas, secondary woodland.
Distr: RS, DAR, KOR, CS, BAG, UN; Mauritania & Senegal to Chad.
DAR: Wickens 398 (fl fr) 9/1962; **UN:** Simpson 7718 (fl, fr) 4/1930.
Note: the record from Red Sea region is based on a sterile collection by Drar (1970) and may be incorrect. The record from Upper Nile is from between Tonga and Amira, which will fall close to the Sudan/South Sudan border.

Terminalia arjuna (Roxb.) Wight & Arn.

Checkl. Fl. Pl. Sub-Saharan Africa: 224 (2006).
Tree. Planted as an ornamental.
Distr: ES; native to India, widely planted in W Africa from Senegal to C.A.R.
ES: Sahni & Kamil 640 4/1967.

Terminalia brownii Fresen.

F.P.S. 1: 210 (1950); F.T.E.A. Combret.: 90 (1973); F.J.M.: 99 (1976); T.S.S.: 91 (1990); F.E.E. 2(2): 127 (1995); Biol. Skr. 51: 133 (1998).
Tree. Woodland, wooded grassland, bushland & dry riverine forest.
Distr: DAR, KOR, CS, ES, EQU; Nigeria to Somalia, S to Tanzania; Arabia.
DAR: Wickens 1820 (fl) 7/1964; **EQU:** Friis & Vollesen 697 (fr) 12/1980.

Terminalia laxiflora Engl. & Diels

F.P.S. 1: 208 (1950); F.T.E.A. Combret.: 87 (1973); F.J.M.: 99 (1976); T.S.S.: 93 (1990); F.E.E. 2(2): 128 (1995); Biol. Skr. 51: 134 (1998).
Syn: Terminalia schweinfurthii Engl. & Diels – F.P.S. 1: 210 (1950).
Tree. Woodland & wooded grassland.
Distr: DAR, KOR, CS, ES, BAG, UN, EQU; Senegal to Ethiopia, S to D.R. Congo & Uganda.
DAR: Kamil 1038 (fl) 5/1968; **EQU:** Friis & Vollesen 166 (fr) 11/1980.

Terminalia macroptera Guill. & Perr.

F.P.S. 1: 208 (1950); F.T.E.A. Combret.: 87 (1973); T.S.S.: 93 (1990); F.E.E. 2(2): 127 (1995).
Tree or shrub. Woodland & wooded grassland.
Distr: DAR, CS, BAG, EQU; Senegal to Ethiopia, S to D.R. Congo & Uganda.
DAR: Wickens 3124 (fr) 4/1971; **EQU:** Myers 11522 (fr) 4/1941.

Terminalia mollis M.A.Lawson

F.P.S. 1: 210 (1950); F.T.E.A. Combret.: 88 (1973); T.S.S.: 95 (1990); Biol. Skr. 51: 134 (1998).
Syn: Terminalia splendida sensu Andrews, non Engl. & Diels – F.P.S. 1: 208 (1950).
Small tree. Wooded grassland.
Distr: BAG, EQU; Senegal to Kenya, S to Angola & Zambia.
BAG: Myers 6294 (fl) 2/1937.

Terminalia schimperiana Hochst.

F.E.E. 2(2): 128 (1995); Biol. Skr. 51: 134 (1998).
Syn: Terminalia glaucescens Planch. ex Benth. – F.P.S. 1: 210 (1950); T.S.S.: 93 (1990).
Syn: Terminalia salicifolia Schweinf. – F.P.S. 1: 208 (1950).
Syn: Terminalia avicennioides sensu Andrews, non Guill. & Perr. – F.P.S. 1: 210 (1950).
Tree. Woodland & wooded grassland.
Distr: DAR, ES, BAG, EQU; Guinea Bissau to Eritrea, S to Tanzania.
DAR: Wickens 3641 (fr) 6/1977; **EQU:** Myers 10949 (fl) 4/1939.

Terminalia spinosa Engl.

F.P.S. 1: 208 (1950); F.T.E.A. Combret.: 82 (1973); T.S.S.: 95 (1990); F.E.E. 2(2): 125 (1995); Biol. Skr. 51: 135 (1998).
Tree. Dry woodland & bushland with *Acacia*.
Distr: UN, EQU; Ethiopia, Somalia, Uganda, Kenya, Tanzania.
UN: Simpson 7603 (fl) 11/1930.

Lythraceae

I. Darbyshire

Ammannia auriculata Willd.

F.J.M.: 94 (1976); F.T.E.A. Lythr.: 37 (1994); Biol. Skr. 51: 107 (1998); F.E.E. 2(1): 403 (2000).
Syn: *Ammannia senegalensis* sensu Andrews pro parte, non Lam. – F.P.S. 1: 138 (1950).
Annual herb. Muddy margins of ponds & streams, wet grassland, weed of irrigated cultivation.
Distr: NS, DAR, KOR, CS, ES, UN, EQU; widespread in the tropics & subtropics.
DAR: Kamil 1166 (fr) 5/1968; **UN:** Simpson 7662 (fl fr) 3/1930.
Note: the form with condensed inflorescences, sometimes separated as var. *bojeriana* Koehne (e.g. in F.E.E.), is recorded from Sudan and South Sudan but in view of the large variability in this species it seems inappropriate to recognise varieties. The African Plants Database records *A. senegalensis* Lam. from Sudan but we have seen no genuine records of this species, though it is likely to occur.

Ammannia baccifera L. subsp. *baccifera*

F.P.S. 1: 138 (1950); F.J.M.: 94 (1976); F.T.E.A. Lythr.: 39 (1994); Biol. Skr. 51: 107 (1998); F.E.E. 2(1): 405 (2000).
Syn: *Ammannia apiculata* Koehne – F.P.S. 1: 139 (1950); F.T.E.A. Lythr.: 47, in notes (1994).
Syn: *Ammannia attenuata* Hochst. ex A.Rich. – F.P.S. 1: 139 (1950).
Annual herb. Muddy margins of pools & streams, seasonally wet grassland & woodland.
Distr: DAR, KOR, CS, ES, BAG, EQU; palaeotropical & subtropical; introduced elsewhere.
CS: Lea 225 10/1954; **BAG:** Schweinfurth II.106 (fl) 11/1869.
Note: *A. apiculata* is not treated in the African Plants Database, but it is listed as a synonym of *A. baccifera* in The Plant List.

Ammannia baccifera L. subsp. *aegyptiaca* (Willd.) Koehne

F.E.E. 2(1): 405 (2000).
Syn: *Ammannia aegyptiaca* Willd. – F.P.S. 1: 139 (1950); Bot. Exp. Sud.: 80 (1970).
Annual herb. Margins of pools & streams, wet flushes on rocks.
Distr: RS, KOR, CS; D.R. Congo, Eritrea, Ethiopia, Zambia, Zimbabwe, Angola; S Europe, Egypt, widespread in Asia.
RS: Jackson 2782 (fr) 4/1953.
Note: the specimens from South Sudan listed by Drar (1970) under *A. aegyptiaca* are possibly of subsp. *baccifera*.

Ammannia coccinea Rottb.

Annual herb. Wet grassland on riverbanks.
Distr: CS; Americas, Caribbean; introduced elsewhere.
CS: de Wilde et al. 5661 (fl fr) 2/1965.
Note: this is a New World species, scattered as an introduction in the Old World. The African Plants Database lists *A. coccinea* as a synonym of *A. latifolia* L. but The Plant List maintains it as distinct; the Sudan collection was identified as *A. coccinea* by B. Verdcourt at Kew.

Ammannia prieuriana Guill. & Perr.

F.J.M.: 94 (1976); F.T.E.A. Lythr.: 40 (1994); F.E.E. 2(1): 404 (2000).
Annual herb. Muddy margins of ponds & streams, boggy ground.
Distr: DAR, CS, EQU; Mauritania & Senegal to Ethiopia, S to Angola & South Africa.
DAR: Kassas 861 (n.v.); **EQU:** Sillitoe 218 (fl fr) 1919.

Ammannia urceolata Hiern

F.P.S. 1: 139 (1950); F.T.E.A. Lythr.: 46 (1994).
Annual herb. Swamp margins, wet grassland.
Distr: KOR; Niger, Somalia, Kenya.
KOR: Kotschy 173 p.p. 10/1839.
Note: this is a scarce and scattered species.

Lawsonia inermis L.

F.P.S. 1: 139 (1950); T.S.S.: 49 (1990); F.T.E.A. Lythr.: 10 (1994); F.E.E. 2(1): 396 (2000).
Tangled shrub or small tree. Seasonally flooded water courses, riverine thicket, sometimes naturalised from planting.
Distr: NS, RS, KOR, CS, ES, ?UN; widespread in the palaeotropics & subtropics.
NS: Pettersson 16439 (fl) 10/1962.
Note: widely cultivated for the dye henna; Sudanese records appear to be of either cultivated or naturalised plants. The occurrence of this species in South Sudan is based on *Muriel* L/87 from the Blue Nile, with the label stating "observed also from Sobat River..."; this record requires confirmation.

Lythrum rotundifolium Hochst. ex A.Rich.

F.P.S. 1: 141 (1950); F.T.E.A. Lythr.: 12 (1994); Biol. Skr. 51: 107 (1998); F.E.E. 2(1): 399 (2000).
Perennial herb. Swamps, streams, wet montane grassland & bushland.
Distr: EQU; Ethiopia to D.R. Congo & Tanzania.
EQU: Thomas 1875 (fl) 12/1935.

Nesaea cordata Hiern

F.P.S. 1: 141 (1950); F.T.E.A. Lythr.: 24 (1994); Biol. Skr. 51: 107 (1998).
Annual herb. Wet flushes, seasonally wet wooded grassland.
Distr: BAG, EQU; Mali to Uganda, S to South Africa.
EQU: Friis & Vollesen 586 (fl) 11/1980.
Note: Verdcourt in F.T.E.A. notes the similiarity to *N. erecta*.

Nesaea dodecandra (DC.) Koehne

F.J.M.: 94 (1976); F.T.E.A. Lythr.: 26 (1994); F.E.E. 2(1): 401 (2000).
Perennial herb. Seasonally wet soils in *Acacia* woodland.
Distr: DAR, ES; Senegal, Ethiopia, Uganda.

DAR: <u>Wickens 2022</u> (fl) 7/1964.
Note: widespread but scattered and scarce.

Nesaea erecta Guill. & Perr. fa. *erecta*
F.P.S. 1: 141 (1950); F.T.E.A. Lythr.: 23 (1994); F.E.E. 2(1): 402 (2000).
Annual herb. Seasonally wet grassland, pond & streamsides.
Distr: UN; Mauritania to Eritrea, S to Angola & Mozambique; Madagascar.
UN: <u>Broun 1437</u> (fl) 11/1908.

Nesaea icosandra Kotschy & Peyr.
F.P.S. 1: 141 (1950).
Perennial herb. Seasonally wet grassland.
Distr: BAG; Chad, C.A.R., Ethiopia.
IUCN: RD
BAG: <u>Broun s.n.</u> (fl) 2/1902.

Nesaea sp. A
Syn: *Ammannia* sp. aff. *senegalensis* sensu Wickens, non Lam. – F.J.M.: 94 (1976).
?Annual herb. Moist ground by stream.
Distr: DAR, UN; not known elsewhere.
DAR: <u>Wickens 2967</u> (fl) 5/1965; **UN:** <u>J.M. Lock 83/5</u> (fl) 2/1983.
Note: *Lock* 83/5 appears to match *Wickens* 2967 which was labelled by Verdcourt as "probably a *Nesaea* but not matched".

Nesaea sp. aff. *heptamera* Hiern
Subshrubby perennial herb. Seasonally wet soils in *Acacia* woodland, seasonally wet grassland.
Distr: CS, ES, UN; not known elsewhere.
CS: <u>Lea 175</u> (fl) 8/1953; **UN:** <u>Simpson 7218</u> (fl fr) 6/1929.
Note: known from four collections, of which *Simpson* 7218 is labelled as *N. heptamera*, but Verdcourt only records that species from Tanzania southwards; this taxon requires further investigation.

Rotala densiflora (Roth) Koehne
Boissiera 29: 82 (1979).
Annual herb. Habitat not known in Sudan.
Distr: BAG; Himalayas, India, Sri Lanka, Sabah, Australia; rarely introduced elsewhere.
BAG: <u>Schweinfurth 2768</u> (fl fr) 12/1869.
Note: the single Sudan specimen is not mentioned in C.D.K. Cook's revision (l.c.); if truly *R. densiflora*, it must surely be introduced.

Rotala mexicana Cham. & Schltdl.
F.P.S. 1: 142 (1950); F.T.E.A. Lythr.: 50 (1994); F.E.E. 2(1): 406 (2000).
Annual herb. Wet grassland.
Distr: BAG; widespread in the tropics & subtropics.
BAG: <u>Schweinfurth 2434</u> 10/1869.

Rotala repens (Hochst.) Koehne
F.T.E.A. Lythr.: 51 (1994); F.E.E. 2(1): 406 (2000).
Aquatic perennial herb. Rheophyte of rocks in fast-flowing streams & waterfall spray-zones.
Distr: BAG; D.R. Congo, Ethiopia, Uganda, Kenya.
BAG: <u>Buxton s.n.</u> (fl) 12/1949.
Note: a scarce species, but probably not globally threatened.

Rotala serpiculoides Welw. ex Hiern
F.P.S. 1: 142 (1950); F.T.E.A. Lythr.: 54 (1994).
Annual herb. Seasonally wet grassland & woodland, muddy hollows.
Distr: CS, BAG; C.A.R. to Uganda, S to Angola & Zimbabwe.
CS: <u>Pettet 111</u> (fl) 1/1963; **BAG:** <u>Schweinfurth 2575</u> (fl) 10/1869.

Rotala stagnina Hiern
F.P.S. 1: 143 (1950); F.E.E. 2(1): 407 (2000).
Annual herb. Pool margins, streamsides.
Distr: BAG; Senegal to Ethiopia.
BAG: <u>Schweinfurth 2498</u> (fl) 10/1869.

Rotala tenella (Guill & Perr.) Hiern
F.P.S. 1: 141 (1950); F.J.M.: 94 (1976); F.T.E.A. Lythr.: 51 (1994); F.E.E. 2(1): 407 (2000).
Aquatic annual or ?perennial herb. Edges of muddy pools, marshes.
Distr: DAR; Senegal to Ethiopia, S to Namibia & South Africa; Madagascar.
DAR: <u>Wickens 1590</u> 5/1964.

Rotala welwitschii Exell
Bol. Soc. Brot., Ser. 2, 30: 70 (1956).
Annual herb. Seasonally wet woodland.
Distr: UN; Senegal to C.A.R., Gabon, Angola.
UN: <u>Simpson 7699</u> (fl fr) 3/1930.

Trapa natans L. var. *bispinosa* (Roxb.) Makino
F.T.E.A. Trap.: 3 (1953).
Syn: *Trapa bispinosa* Roxb. – F.P.S. 1: 147 (1950).
Floating aquatic annual herb. Lakes, ponds, papyrus swamps.
Distr: BAG, UN; Guinea Bissau, Tanzania to Angola & South Africa; India to China & Japan.
UN: <u>J.M. Lock 82/17</u> 2/1982.

Woodfordia uniflora (A.Rich.) Koehne
F.P.S. 1: 143 (1950); F.J.M.: 94 (1976); T.S.S.: 49 (1990); F.T.E.A. Lythr.: 4 (1994); F.E.E. 2(1): 396 (2000).
Slender shrub or small tree. Rocky streamsides, rocky slopes, woodland.
Distr: DAR, KOR, CS; Nigeria to Eritrea, S to Uganda & Kenya.
DAR: <u>Wickens 1169</u> (fl) 2/1964.
Note: widespread but uncommon.

Onagraceae

I. Darbyshire

Epilobium hirsutum L.
F.P.S. 1: 143 (1950); F.T.E.A. Onagr.: 2 (1953); F.J.M.: 95 (1976); F.E.E. 2(1): 413 (2000).
Rhizomatous perennial herb. Marshy areas & wet grassland by streams.
Distr: DAR; Eritrea to Namibia & South Africa; widespread in Europe, N Africa & Asia.
DAR: <u>Jackson 3292</u> (fl fr) 12/1954.

Epilobium salignum Hausskn.

F.T.E.A. Onagr.: 5 (1953); Biol. Skr. 51: 108 (1998); F.E.E. 2(1): 414 (2000).
Rhizomatous perennial herb. Montane swamps & wet grassland.
Distr: EQU; Nigeria, Cameroon, Ethiopia to Angola & South Africa; Madagascar.
EQU: Friis & Vollesen 315 (fl) 11/1980.

Epilobium stereophyllum Fresen.

F.T.E.A. Onagr.: 4 (1953); Biol. Skr. 51: 108 (1998); F.E.E. 2(1): 413 (2000).
Stoloniferous perennial herb. Montane swamps & wet grassland.
Distr: EQU; Ethiopia to D.R. Congo & Tanzania.
EQU: Myers 11594 (fl fr) 7/1939.

Ludwigia abyssinica A.Rich.

Biol. Skr. 51: 108 (1998); F.E.E. 2(1): 417 (2000).
Syn: *Jussiaea abyssinica* (A.Rich.) Dandy & Brenan – F.P.S. 1: 145 (1950); F.T.E.A. Onagr.: 18 (1953).
Perennial herb. Swamps, streamsides.
Distr: EQU; Senegal to Ethiopia, S to Angola & South Africa; Madagascar.
EQU: Sillitoe 239 (fl fr) 1919.

Ludwigia adscendens (L.) H.Hara subsp. diffusa (Forssk.) P.H.Raven

Kew Bull. 15: 476 (1962).
Syn: *Jussiaea repens* L. var. *diffusa* (Forssk.) H.Hara – F.T.E.A. Onagr.: 19 (1953); Bot. Exp. Sud.: 90 (1970).
Syn: *Ludwigia stolonifera* (Guill. & Perr.) P.H.Raven – F.J.M.: 95 (1976); F.E.E. 2(1): 419 (2000).
Creeping or often floating perennial herb. Swamps, pools, riversides.
Distr: DAR, CS, BAG, UN, EQU; Mauritania to Eritrea, S to Namibia & South Africa; Madagascar, Mauritius, N Africa to Middle East.
DAR: Kamil 1168 (fl) 5/1968; **EQU:** Myers 8038 (fl) 1/1938.

Ludwigia erecta (L.) H.Hara

Biol. Skr. 51: 109 (1998); F.E.E. 2(1): 416 (2000).
Syn: *Jussiaea erecta* L. – F.P.S. 1: 145 (1950); F.T.E.A. Onagr.: 12 (1953); Bot. Exp. Sud.: 90 (1970).
Annual herb. Riversides, marshes.
Distr: NS, CS, BAG, EQU; Mauritania to Somalia, S to Angola & Mozambique; Madagascar, Comoros, Seychelles, tropical America.
CS: Schweinfurth 846 (fl fr) 11/1868; **EQU:** Simpson 7284 (fl fr) 7/1929.

Ludwigia hyssopifolia (G.Don) Exell

Reinwardtia 6: 385 (1963).
Annual or short-lived perennial herb. Marshes, riverbanks, muddy flats.
Distr: NS, KOR; Cape Verde Is., Senegal to D.R. Congo; India to Australia & Pacific Islands.
NS: Briston 17 (fl fr) 3/1984.

Ludwigia leptocarpa (Nutt.) H.Hara

Reinwardtia 6: 375 (1963); F.E.E. 2(1): 417 (2000).
Syn: *Jussiaea leptocarpa* Nutt. – F.P.S. 1: 144 (1950); F.T.E.A. Onagr.: 16 (1953); Bot. Exp. Sud.: 90 (1970).

Perennial herb. Marshes, riversides, rock pools.
Distr: KOR, ES, BAG, UN, EQU; Senegal to Ethiopia, S to Angola & South Africa; Madagascar, tropical America.
KOR: Simpson 7759 (fl fr) 4/1930; **EQU:** Myers 6666 (fl) 5/1937.
Note: Andrews records also *L. jussiaeoides* Desr. from Equatoria but we have seen no Sudanese material and it is possibly a misidentification.

Ludwigia octovalvis (Jacq.) P.H.Raven

Reinwardtia 6: 356 (1963); F.E.E. 2(1): 416 (2000).
Syn: *Jussiaea suffruticosa* L. – F.P.S. 1: 145 (1950); F.T.E.A. Onagr.: 14 (1953).
Syn: *Jussiaea suffruticosa* L. var. *linearis* (Willd.) Oliv. ex Kuntze – Bot. Exp. Sud.: 91 (1970).
Syn: *Ludwigia octovalvis* (Jacq.) P.H.Raven subsp. *brevisepala* (Brenan) P.H.Raven – F.J.M.: 95 (1976); Biol. Skr. 51: 109 (1998).
Perennial herb or subshrub. Swamps, seasonally wet grassland, streamsides, ditches.
Distr: DAR, KOR, BAG, UN, EQU; pantropical.
DAR: Wickens 2745 (fl) 9/1964; **EQU:** Andrews 808 (fl fr) 4/1939.

Ludwigia perennis L.

Reinwardtia 6: 367 (1963); F.E.E. 2(1): 416 (2000).
Syn: *Jussiaea perennis* (L.) Brenan – F.T.E.A. Onagr.: 13 (1953).
Syn: *Ludwigia parviflora* Roxb. – F.P.S. 1: 145 (1950).
Annual or short-lived perennial herb. Swamps, lake margins.
Distr: KOR, BAG, ?UN; palaeotropical.
KOR: Andrews 162 (fl fr) 11/1933; **BAG:** Schweinfurth 2535 (fl fr) 10/1869.

Ludwigia senegalensis (DC.) Troch.

Reinwardtia 6: 371 (1963); F.Z. 4: 339 (1978).
Creeping perennial herb. Wetlands, sometimes submerged.
Distr: BAG; Senegal to D.R. Congo, S to Angola & Zimbabwe.
BAG: Schweinfurth 2690 (fr) 11/1869.

Ludwigia stenorraphe (Brenan) H.Hara subsp. stenorraphe

Reinwardtia 6: 351 (1963).
Syn: *Jussiaea stenorraphe* Brenan var. *stenorraphe* – F.T.E.A. Onagr.: 10 (1953); Bot. Exp. Sud.: 91 (1970).
Perennial herb or subshrub. Swamps, riversides, flooded grassland.
Distr: ?DAR, BAG, EQU; Guinea to Kenya, S to Angola & Zambia.
DAR: Drar 2415 5/1938 (n.v.); **EQU:** Sandison 29 (fl fr) 12/1935.
Note: the record from Darfur is based on Drar (1970).

Ludwigia stenorraphe (Brenan) H.Hara subsp. macrosepala (Brenan) P.H.Raven

Reinwardtia 6: 353 (1963).
Syn: *Jussiaea stenorraphe* Brenan var. *macrosepala* Brenan – F.T.E.A. Onagr.: 11 (1953).
Perennial herb or subshrub. Swamps, riversides, flooded grassland.
Distr: EQU; Uganda to Malawi.

EQU: Sillitoe 426 (fl) 1919.
Note: the identity of the single South Sudanese specimen is unconfirmed; Brenan in F.T.E.A. records it as "Anglo-Egyptian Sudan (probable, but not certain)"; Raven in his account of the genus (Reinwardtia 6: 327-427 (1963)) does not record this subspecies from Sudan.

Myrtaceae

I. Darbyshire

[*Callistemon lanceolatus* (Sm.) Sweet]
Note: a native of Australia, El Amin (1990: 71) notes this species to be widely grown on deep, well-drained silty soils in the Sudan region. It is not known to naturalise.

[*Eucalyptus*]
Note: *Eucalyptus* species are widely grown in forestry plantations and shelter beds in Sudan and South Sudan but are not known to naturalise; El Amin in T.S.S. (1990) provides a useful account of the six most widely grown species.

[*Psidium guajava* L.]
Note: 'guava' is widely grown in the Sudan region for its edible fruits, but it has not become naturalised.

Syzygium guineense (Willd.) DC. subsp. *guineense*
F.P.S. 1: 190 (species only) (1950); F.J.M.: 98 (1976); F.E.E. 2(2): 77 (1995); Biol. Skr. 51: 126 (1998); F.T.E.A. Myrt.: 78 (2001).
Tree. Riverine forest.
Distr: DAR, KOR, CS, BAG, EQU; Senegal to Somalia, S to South Africa; Arabia.
DAR: Kamil 1033 (fl fr) 5/1968; **EQU:** Friis & Vollesen 1169 (fl fr) 3/1982.
Note: not all the material of *S. guineense* from Sudan and South Sudan is easily identifiable to subspecies; for example, much of the material from the riverine forests of Jebel Marra has markedly longer petioles than typical subsp. *guineense*.

Syzygium guineense (Willd.) DC. subsp. *afromontanum* F.White
T.S.S.: 75 (1990), pro parte; F.E.E. 2(2): 78 (1995); Biol. Skr. 51: 127 (1998); F.T.E.A. Myrt.: 79 (2001).
Large tree. Montane forest, sometimes co-dominant.
Distr: EQU; Ethiopia & Uganda, S to Angola & Zimbabwe.
EQU: Howard I.M.64 (fr) 2/1976.
Note: El Amin in T.S.S. mistakenly places all *S. guineense* from Sudan and South Sudan within this subspecies.

Syzygium guineense (Willd.) DC. subsp. *macrocarpum* (Engl.) F.White
F.E.E. 2(2): 78 (1995); F.T.E.A. Myrt.: 79 (2001).
Tree or shrub. Fire-prone woodland & wooded grassland.
Distr: BAG, EQU; Senegal to C.A.R. & D.R. Congo.
BAG: Simpson 7635 (fl) 3/1930.

Melastomataceae

I. Darbyshire

Antherotoma debilis (Sond.) Jacq.-Fél.
Bull. Mus. Natl. Hist. Nat. B, Adansonia sér. 4, 16: 270 (1995).
Syn: *Dissotis debilis* (Sond.) Triana – F.P.S. 1: 192 (1950); F.T.E.A. Melastomat.: 35 (1975); F.E.E. 2(1): 452 (2000).
Annual or perennial herb. Wet grassland.
Distr: BAG; Uganda & Kenya, S to Namibia & South Africa.
BAG: Schweinfurth 4289 (fl fr) 11/1869.
Note: the single Sudan specimen seen is of the annual var. *postpluvialis* (Gilg) A.Fern. & R.Fern. (see F.T.E.A.: 36) but the varieties are not maintained by all authors.

Antherotoma naudinii Hook.f.
F.P.S. 1: 191 (1950); F.T.E.A. Melastomat.: 9 (1975); F.E.E. 2(2): 111 (1995); Biol. Skr. 51: 127 (1998).
Annual herb. Wet grassland & wooded grassland.
Distr: BAG, EQU; Guinea to Ethiopia, S to Angola & South Africa; Madagascar.
EQU: Friis & Vollesen 208 (fl fr) 11/1980.

Antherotoma phaeotricha (Hochst.) Jacq.-Fél.
Bull. Mus. Natl. Hist. Nat. B, Adansonia sér. 4, 16: 270 (1995).
Syn: *Dissotis phaeotricha* (Hochst.) Hook.f. – F.P.S. 1: 192 (1950); F.T.E.A. Melastomat.: 36 (1975).
Perennial herb. Wet grassland.
Distr: BAG; Senegal to C.A.R., S to Angola & South Africa.
BAG: Schweinfurth 4041 (fl fr) 7/1870.

Antherotoma senegambiensis (Guill & Perr.) Jacq.-Fél.
Bull. Mus. Natl. Hist. Nat., B, Adansonia sér. 4, 16: 270 (1995).
Syn: *Dissotis senegambiensis* (Guill & Perr.) Triana – F.T.E.A. Melastomat.: 48 (1975); T.S.S.: 77 (1990); F.E.E. 2(2): 109 (1995); Biol. Skr. 51: 129 (1998).
Syn: *Osbeckia abyssinica* Gilg – F.P.S. 1: 194 (1950).
Syn: *Osbeckia saxicola* Gilg – F.P.S. 1: 194 (1950).
Syn: *Osbeckia senegambiensis* Guill. & Perr. – F.P.S. 1: 194 (1950).
Perennial herb or subshrub. Forest margins, montane grassland & scrub, wet grassland.
Distr: BAG, EQU; Senegal to Ethiopia, S to Mozambique.
EQU: Thomas 1623 (fl fr) 12/1935.
Note: two distinct forms are recorded in our region and the form previously separated as *Osbekia saxicola* may require reinstatement as a good taxon.

Dissotis brazzae Cogn.
F.T.E.A. Melastomat.: 41 (1975); F.E.E. 2(2): 109 (1995); Biol. Skr. 51: 128 (1998).
Perennial herb. Grassland, bushland, forest margins.
Distr: EQU; Sierra Leone to Ethiopia, S to Angola & Zambia.
EQU: Friis & Vollesen 200 (fl) 11/1980.

Note: Cufodontis (Enum. Pl. Aeth.: 630 (1959)) lists *D. princeps* (Kunth) Triana from Sudan but we have seen no material and this record was not confirmed by e.g. Wickens in F.T.E.A.

Dissotis perkinsiae Gilg

F.T.E.A. Melastomat.: 56 (1975); T.S.S.: 77 (1990); Biol. Skr. 51: 128 (1998).
Syn: *Dissotis scabra* Gilg – F.P.S. 1: 192 (1950).
Syn: *Dissotis schweinfurthii* Gilg – F.P.S. 1: 192 (1950).
Perennial herb or shrub. Grassland & wooded grassland, abandoned cultivation, bamboo forest.
Distr: BAG, EQU; Ghana to Uganda.
EQU: Friis & Vollesen 198 (fl) 11/1980.
Note: an uncommon species but not considered to be threatened.

Dissotis speciosa Taub.

F.T.E.A. Melastomat.: 57 (1975).
Syn: *Dissotis macrocarpa* Gilg – F.P.S. 1: 194 (1950).
Perennial herb or shrub. Wet grassland, swamps.
Distr: ?EQU; D.R. Congo to Kenya, S to Zambia.
Note: no South Sudanese material has been seen; the record is based on Andrews who recorded it from the Aloma Plateau.

Dissotis tubulosa (Sm.) Triana

Trans. Linn. Soc. 28: 58 (1871).
Syn: *Derosiphia tubulosa* (Sm.) Raf. – F.J.M.: 98 (1976).
Annual herb. Rocky slopes, open ground, roadsides.
Distr: DAR; Senegal to C.A.R.
DAR: Wickens 2756 (fl fr) 10/1964.
Note: this species is sometimes treated in the genus *Osbekia* (e.g. see African Plants Database) but we prefer to keep it in *Dissotis* here.

Heterotis cinerascens (Hutch.) Jacq.-Fél.

Adansonia sér. 2, 20: 419 (1981).
Syn: *Dissotis cinerascens* Hutch. – F.W.T.A. 1(1): 257 (1954).
Annual or perennial herb. Swamp margins, wet grassland.
Distr: EQU; Nigeria, Cameroon, C.A.R.
IUCN: RD
EQU: Jackson 2307 (fl fr) 6/1952.
Note: a fairly widespread but scarce species.

Heterotis decumbens (P.Beauv.) Jacq.-Fél.

Adansonia, sér. 2, 20: 418 (1981).
Syn: *Dissotis decumbens* (P.Beauv.) Triana – F.P.S. 1: 194 (1950); F.T.E.A. Melastomat.: 38 (1975); F.E.E. 2(2): 111 pro parte (1995); Biol. Skr. 51: 128 (1998).
Perennial herb. Wet grassland, forest margins.
Distr: EQU; Nigeria to Ethiopia, S to Angola, D.R. Congo & Tanzania.
EQU: Friis & Vollesen 638 (fl fr) 12/1980.

Heterotis rotundifolia (Sm.) Jacq.-Fél.

Adansonia, sér. 2, 20: 417 (1981).
Syn: *Dissotis rotundifolia* (Sm.) Triana – F.T.E.A. Melastomat.: 39 (1975); Biol. Skr. 51: 128 (1998).
Syn: *Dissotis decumbens* sensu Gilbert pro parte, non (P.Beauv.) Triana – F.E.E. 2(2): 111 (1995).

Perennial herb. Riverine forest & scrub, wet grassland.
Distr: EQU; Sierra Leone to Kenya, S to Angola & Mozambique.
EQU: Friis & Vollesen 1298 (fl fr) 3/1982.

Melastomastrum capitatum (Vahl) A.Fern. & R.Fern.

F.T.E.A. Melastomat.: 20 (1975); F.E.E. 2(2): 111 (1995); Biol. Skr. 51: 129 (1998).
Syn: *Dissotis erecta* (Guill. & Perr.) Dandy – F.P.S. 1: 192 (1950).
Syn: *Dissotis petiolata* Hook.f. – F.P.S. 1: 192 (1950).
Perennial herb or shrub. Forest margins & clearings, wet grassland.
Distr: BAG, EQU; Senegal to Ethiopia, S to Angola & Zambia.
EQU: Myers 9462 (fl) 9/1938.

Tristemma leiocalyx Cogn.

F.T.E.A. Melastomat.: 16 (1975).
Syn: *Tristemma roseum* Gilg – F.P.S. 1: 194 (1950).
Shrub. Riverine & swamp forest, including clearings.
Distr: EQU; Cameroon to Uganda, S to Congo-Brazzaville & D.R. Congo.
EQU: Andrews 1598 (fr) 5/1939.
Note: Andrews lists also *T. littorale* Benth. (F.P.S. 1: 195, repeated by El Amin in T.S.S.) but this is surely in error since this is a West African species.

Tristemma mauritianum J.F.Gmel.

F.T.E.A. Melastomat.: 17 (1975); F.E.E. 2(2): 112 (1995).
Syn: *Tristemma incompletum* R.Br. – F.P.S. 1: 195 (1950); T.S.S.: 77 (1990).
Syn: *Tristemma grandifolium* (Cogn.) Gilg – F.P.S. 1: 195 (1950); Bot. Exp. Sud.: 83 (1970).
Perennial herb or shrub. Clearings in forest, riverbanks, often weedy.
Distr: BAG, EQU; Ivory Coast to Ethiopia, S to Angola & Mozambique; Madagascar & Mascarenes.
EQU: Myers 9193 (fl) 8/1938.

Warneckea jasminoides (Gilg) Jacq.-Fél.

Biol. Skr. 51: 129 (1998).
Syn: *Lijndenia jasminoides* (Gilg) Borhidi – Opera Bot. 121: 151 (1993).
Syn: *Memecylon jasminoides* Gilg – F.T.E.A. Melastomat.: 83 (1975).
Shrub or small tree. Forest.
Distr: EQU; Cameroon to Uganda, S to Angola & Tanzania.
EQU: Shigeta 136 (fr) 1/1979 (n.v.).

Penaeaceae

I. Darbyshire

Olinia rochetiana Juss.

F.T.E.A. Olin.: 2 (1975); T.S.S.: 367 (1990); Biol. Skr. 51: 108 (1998); F.E.E. 2(1): 409 (2000).
Small tree or shrub. Montane forest, especially margins.
Distr: EQU; Ethiopia to Angola & South Africa.
EQU: Friis & Vollesen 1138 (fr) 3/1982.

SAPINDALES

Nitrariaceae

I. Darbyshire

Nitraria retusa (Forrsk.) Asch.

F.Egypt 2: 32 (2000).
Shrub. Saline soils including edges of saltmarshes.
Distr: RS; N Africa to Arabia & Pakistan.
RS: <u>Carter 1957</u> fl 11/1987.

Burseraceae

H. Pickering

Boswellia papyrifera (Delile) Hochst.

F.P.S. 2: 321 (1952); F.J.M.: 122 (1976); F.E.E. 3: 443
(1989); T.S.S.: 305 (1990); F.T.E.A. Burser.: 5 (1991); Biol.
Skr. 51: 280 (1998).
Tree. Dry woodland & wooded grassland.
Distr: DAR, KOR, CS, BAG, UN, EQU; Nigeria to Ethiopia
& Uganda.
DAR: <u>Wickens 3647</u> 6/1977; **EQU:** <u>Friis & Vollesen 1154</u>
(fl) 3/1982.

Canarium schweinfurthii Engl.

F.P.S. 2: 323 (1952); T.S.S.: 305 (1990); F.T.E.A. Burser.: 3
(1991); Biol. Skr. 51: 280 (1998).
Large tree. Forest, often along edges in clearings & along
lake shores.
Distr: EQU; Senegal to Ethiopia, S to Zambia.
EQU: <u>Cartwright 1161</u> (fl) 3/1919.

Commiphora africana Engl. var. *africana*

F.P.S. 2: 323 (1952); F.J.M.: 122 (1976); F.E.E. 3: 454
(1989); T.S.S.: 307 (1990); F.T.E.A. Burser.: 46 (1991);
Biol. Skr. 51: 280 (1998).
Shrub or small tree. Bushland & wooded grassland.
Distr: RS, DAR, KOR, CS, EQU; Mauritania to Somalia, S
to Namibia & South Africa.
KOR: <u>Wickens 628</u> 10/1962; **EQU:** <u>Friis & Vollesen 1144</u>
3/1982.

Commiphora confusa Vollesen

F.E.E. 3: 462 (1989); F.T.E.A. Burser.: 49 (1991).
Shrub or tree. Bushland.
Distr: EQU; Ethiopia, Kenya, Tanzania.
EQU: <u>Padwa 250</u> 6/1953.

Commiphora gileadensis (L.) C.Chr.

F.E.E. 3: 475 (1989); T.S.S.: 309 (1990).
Syn: *Commiphora opobalsamum* (L.) Engl. – F.P.S. 2:
324 (1952).
Shrub or tree. Semi-desert.
Distr: NS, RS; Ethiopia, Somalia; Egypt.
RS: <u>Kassas 947</u> 12/1954.

Commiphora kataf (Forssk.) Engl.

F.T.E.A. Burser.: 81 (1991).
Syn: *Commiphora erythraea* (Ehrenb.) Engl. – F.P.S. 2:
325 (1952); F.E.E. 3: 450 (1989); T.S.S.: 307 (1990).
Small tree. Hillsides with open bushland.
Distr: RS; Ethiopia & Somalia, S to Tanzania.
RS: <u>Schweinfurth S.2303</u> (fr) 5/1864.

Commiphora pedunculata Engl.

F.P.S. 2: 323 (1952); F.J.M.: 122 (1976); F.E.E. 3: 457
(1989); T.S.S.: 309 (1990); F.T.E.A. Burser.: 73 (1991).
Small tree. Woodland & wooded grassland.
Distr: DAR, KOR, CS, BAG, UN, EQU; Mali to Ethiopia.
DAR: <u>Wickens 3627</u> (fl) 6/1977; **BAG:** <u>Myers 14146</u>
10/1941.

Commiphora quadricincta Schweinf.

F.P.S. 2: 325 (1952); F.E.E. 3: 471 (1989); T.S.S.: 309 (1990).
Syn: *Commiphora habessinica* sensu auctt., non
(O.Berg.) Engl. – F.P.S. 2: 323 (1952); T.S.S.: 307 (1990)
as *C. abyssinica*.
Shrub or tree. Bushland, wooded grassland, rocky
hillslopes.
Distr: RS, DAR, KOR, ES; Nigeria, Chad, Eritrea, Ethiopia;
Egypt, Arabia.
KOR: <u>Kassas 42</u> (fr) 2/1954.
Note: Gillett in F.T.E.A. noted that Andrews' record of *C.
habessinica* (= *C. kua* (R.Br. ex Royle) Vollesen) was based
on misidentified material of *C. quadricincta*. *C. kua* is,
however, likely to occur in South Sudan, having been
recorded from the Ugandan side of the Imatong Mts
(Friis & Vollesen 1998: 281).

Commiphora samharensis Schweinf.

F.E.E. 3: 452 (1989); F.T.E.A. Burser.: 40 (1991).
Spiny tree. Bushland.
Distr: RS; Ethiopia, Somalia, Uganda, Kenya, Tanzania.
RS: <u>Bally 6973</u> 4/1949.

Commiphora schimperi (Berg.) Engl.

F.J.M.: 122 (1976); F.E.E. 3: 454 (1989); F.T.E.A. Burser.:
43 (1991).
Shrub or tree. Stony bushland.
Distr: DAR; Eritrea & Somalia, S to South Africa.
DAR: <u>Wickens 1593</u> 1964.

Anacardiaceae

M. Kordofani, R. Vanderstricht & I. Darbyshire

Anacardium occidentale L.

F.J.M.: 123 (1976); F.T.E.A. Anacard.: 2 (1986); T.S.S.: 335
(1990).
Shrub or tree. Cultivated & naturalised.
Distr: DAR; native of tropical America, cultivated
throughout the tropics & sometimes naturalised.
DAR: <u>Kamil 1150</u> (fr) 5/1968.

Haematostaphis barteri Hook.f.

F.P.S. 2: 345 (1952); F.W.T.A. 1(2): 733 (1958); T.S.S.: 335
(1990).
Small tree. Wooded grassland, rocky hillslopes.
Distr: EQU; Ghana to Cameroon, S to Gabon & Cabinda.
EQU: <u>Myers 6329</u> (fl) 2/1937.

Lannea barteri (Oliv.) Engl.

F.T.E.A. Anacard.: 21 (1986); F.E.E. 3: 518 (1989); T.S.S.:
336 (1990); Biol. Skr. 51: 289 (1998).
Syn: *Lannea kerstingii* Engl. & K.Krause – F.P.S. 2: 348
(1952); F.J.M.: 124 (1976).
Tree. Woodland, wooded grassland, riverine forest
margins.

Distr: RS, DAR, KOR, CS, BAG, EQU; Guinea to Ethiopia, S to D.R. Congo & Uganda.
DAR: <u>Wickens 1523</u> (fl) 5/1964; **EQU:** <u>Myers 8377</u> (fl) 1/1938.

Lannea fruticosa (Hochst. ex A.Rich.) Engl.

F.P.S. 2: 346 (1952); F.J.M.: 123 (1976); F.T.E.A. Anacard.: 23 (1986); F.E.E. 3: 518 (1989); T.S.S.: 336 (1990); Biol. Skr. 51: 289 (1998).
Tree or shrub. Woodland.
Distr: RS, DAR, KOR, CS, ES, EQU; Nigeria to Eritrea, S to D.R. Congo & Uganda; Yemen.
DAR: <u>Francis 42</u> (fr) 4/1957; **EQU:** <u>Myers 7907</u> (fl) 11/1937.

Lannea fulva (Engl.) Engl.

F.T.E.A. Anacard.: 15 (1986); T.S.S.: 336 (1990); Biol. Skr. 51: 289 (1998).
Shrub or tree. Woodland & wooded grassland, often on rocky hillslopes.
Distr: EQU; D.R. Congo, Rwanda, Burundi, Uganda, Kenya, Tanzania.
EQU: <u>Friis & Vollesen 1152</u> 3/1982.

Lannea humilis (Oliv.) Engl.

F.P.S. 2: 348 (1952); F.J.M.: 124 (1976); F.T.E.A. Anacard.: 18 (1986); F.E.E. 3: 516 (1989); T.S.S.: 337 (1990); Biol. Skr. 51: 290 (1998).
Shrub or small tree. Woodland.
Distr: DAR, KOR, CS, ES, UN, EQU; Senegal to Ethiopia, S to Zimbabwe.
DAR: <u>Miehe 153</u> (fr) 6/1982; **EQU:** <u>Jackson 1289</u> (fl) 3/1950.

Lannea schimperi (Hochst. ex A.Rich.) Engl.

F.P.S. 2: 346 (1952); F.J.M.: 124 (1976); F.T.E.A. Anacard.: 19 (1986); F.E.E. 3: 516 (1989); T.S.S.: 337 (1990); Biol. Skr. 51: 290 (1998).
Small to medium tree. Woodland.
Distr: RS, DAR, KOR, BAG, EQU; Nigeria to Eritrea, S to Zambia & Mozambique.
KOR: <u>Simpson 7781</u> (fr) 4/1930; **EQU:** <u>Andrews 559</u> (fr) 4/1939.

Lannea schweinfurthii (Engl.) Engl. var. schweinfurthii

F.P.S. 2: 348 (1952); F.T.E.A. Anacard.: 25 (1986); F.E.E. 3: 518 (1989); T.S.S.: 337 (1990); Biol. Skr. 51: 290 (1998).
Shrub or tree. Woodland & wooded grassland.
Distr: ?DAR, ?KOR, BAG, UN, EQU; Ethiopia & Somalia, S to Namibia & South Africa.
BAG: <u>Myers 7259</u> (fl) 7/1937.
Note: Andrews lists this species also from Darfur and El Amin records it from the Nuba Mts in Kordofan; we have not seen any material with which to verify these records.

Mangifera indica L.

F.T.E.A. Anacard.: 3 (1986); F.E.E. 3: 519 (1989).
Tree. Cultivated, occasionally naturalised.
Distr: EQU; native of India, cultivated throughout the tropics.
Note: widely cultivated in the wetter areas, with occasional semi-naturalised plants.

Ozoroa insignis Delile subsp. insignis

F.J.M.: 124 (1976); F.T.E.A. Anacard.: 7 (1986); F.E.E. 3: 521 (1989); T.S.S.: 337 (1990).
Syn: *Heeria insignis* (Delile) Kuntze – F.P.S. 2: 345 (1952); Bot. Exp. Sud.: 15 (1970).
Tree or shrub. Woodland, often on rocky hillslopes.
Distr: RS, DAR, KOR, CS, BAG, UN, EQU; Senegal to Somalia; Yemen.
KOR: <u>Musselman 6251</u> (fl) 8/1983; **UN:** <u>Myers 10432</u> (fl) 2/1939.

Ozoroa insignis Delile subsp. reticulata (Baker f.) J.B.Gillett

F.T.E.A. Anacard.: 5 (1986); Biol. Skr. 51: 291 (1998).
Syn: *Heeria reticulata* (Baker f.) Engl. – F.P.S. 2: 346 (1952).
Tree or shrub. Woodland & wooded grassland, often on rocky hillslopes.
Distr: EQU; ?Ethiopia, Kenya to Angola & South Africa.
EQU: <u>Friis & Vollesen 1114</u> (fl fr) 3/1982.

Ozoroa pulcherrima (Schweinf.) R.Fern. & A.Fern.

F.T.E.A. Anacard.: 8 (1986); F.E.E. 3: 521 (1989).
Syn: *Heeria pulcherrima* (Schweinf.) Kuntze – F.P.S. 2: 346 (1952); Bot. Exp. Sud.: 15 (1970).
Suffrutex. Woodland & wooded grassland.
Distr: RS, KOR, ES, BAG, EQU; Senegal to Ethiopia, S to Uganda.
ES: <u>Schweinfurth 1279</u> (fl) 1865; **EQU:** <u>Andrews 447</u> 4/1939.
Note: the records for Red Sea region and Kordofan are based on Drar (1970) and are unconfirmed.

Pistacia falcata Mart.

F.P.S. 2: 348 (1952); F.E.E. 3: 532 (1989); T.S.S.: 338 (1990).
Tree. Woodland & bushland on dry rocky slopes.
Distr: RS, ES; Eritrea, Ethiopia, Somalia; Arabia.
RS: <u>Kennedy-Cooke 257</u> (fl) 3/1938.

Pistacia cf. khinjuk Stocks

F.Egypt 2: 75 (2000).
Small tree. Rocky hillslopes.
Distr: RS; (for species) Egypt, Palestine, Arabia.
RS: <u>Schweinfurth 332</u> 9/1868.
Note: the single Sudan specimen is sterile; it was identified as *P. ?khinjuk* by J. Linczevski at Kew.

Pseudospondias microcarpa Engl.

F.P.S. 2: 348 (1952); F.T.E.A. Anacard.: 53 (1986); T.S.S.: 338 (1990); Biol. Skr. 51: 291 (1998).
Tree. Riverine & swamp forest.
Distr: UN, EQU; Senegal to Kenya, S to Angola & Zambia.
EQU: <u>Myers 6393</u> (fr) 3/1937.

[Rhus]

Note: Moffett in Bothalia 37: 165-176 (2007) transferred all tropical African *Rhus* to the genus *Searsia*. However, the African Plants Database currently maintains *Rhus* over *Searsia*, stating that the taxonomy of *Rhus* is complex and so further research is required before *Searsia* can be recognised.

Rhus flexicaulis Baker

F.P.S. 2: 351 (1952); T.S.S.: 339 (1990); F.Egypt 2: 75 (2000).
Shrub. Open bushland, rocky hillslopes.
Distr: RS; Egypt, Arabia.
RS: Jackson 3962 (fr) 4/1959.
Note: J.R.I. Wood (Handbook Yemen Flora) noted that the African plants given this name are probably a different species to that from Arabia.

Rhus glutinosa Hochst. ex A.Rich. subsp. abyssinica (Hochst. ex Oliv.) M.G.Gilbert

F.E.E. 3: 528 (1989); T.S.S.: 339 (species only) (1990).
Syn: Rhus abyssinica Hochst. ex Oliv. – F.P.S. 2: 349 (1952); Bot. Exp. Sud.: 16 (1970); T.S.S.: 339 (1990).
Shrub or tree. Bushland.
Distr: RS, ?DAR; Eritrea, Ethiopia; Egypt, Arabia.
RS: Sahni & Kamil 666 (fr) 8/1967.
Note: the record from Darfur is based on a sterile specimen by Drar (1970) which we have not seen; it may have been misidentified. El Amin in T.S.S. also records this taxon from Equatoria but this is likely to be in error.

Rhus longipes Engl. var. longipes

F.T.E.A. Anacard.: 37 (1986); F.E.E. 3: 528 (1989).
Shrub or small tree. Open woodland, riverine forest.
Distr: EQU; Ethiopia to Angola & Mozambique.
EQU: Kosper 27 (fl) 5/1982.

Rhus natalensis Bernh. ex C.Krauss

F.P.S. 2: 319 (1952); F.T.E.A. Anacard.: 28 (1986); F.E.E. 3: 523 (1989); T.S.S.: 339 (1990); Biol. Skr. 51: 292 (1998).
Shrub, sometimes scandent, or small tree. Bushland & woodland.
Distr: KOR, CS, BAG, EQU; Guinea to Somalia, S to South Africa; Arabia.
CS: Lea 108 11/1952; **EQU:** Myers 8025 (fl) 12/1937.

Rhus pyroides Burch. var. pyroides

F.T.E.A. Anacard.: 32 (1986).
Syn: Rhus vulgaris Meikle – F.P.S. 2: 349 (1952); F.E.E. 3: 529 (1989); T.S.S.: 341 (1990).
Shrub or small tree. Forest margins, montane bushland & woodland.
Distr: DAR, BAG, EQU; Cameroon to Ethiopia, S to Namibia & South Africa.
DAR: Wickens 1308 (fl fr) 2/1964; **EQU:** Andrews 895 (fl fr) 4/1939.

Rhus retinorrhaea Steud. ex Oliv.

F.P.S. 2: 351 (1952); F.E.E. 3: 529 (1989); T.S.S.: 341 (1990).
Shrub or small tree. Bushland on dry rocky slopes.
Distr: RS; Eritrea, Ethiopia, Somalia; Arabia.
RS: Kennedy-Cooke 262 3/1938.

Rhus ruspolii Engl.

F.T.E.A. Anacard.: 29 (1986); F.E.E. 3: 528 (1989); Biol. Skr. 51: 292 (1998).
Shrub or small tree. Montane bushland & forest margins.
Distr: EQU; D.R. Congo, Ethiopia, Uganda, Kenya.
EQU: Friis & Vollesen 1134 (fl) 3/1982.

Rhus tripartita (Ucria) Grande

F.Egypt 2: 73 (2000).
Syn: Rhus oxyacantha Schousb. ex Cav. – F.P.S. 2: 349 (1952); Bot. Exp. Sud.: 16 (1970); T.S.S.: 341 (1990).
Shrub. Wadis, dry rocky hillslopes.
Distr: RS, DAR; Sicily, Sahara & N Africa to W Asia.
RS: Bent s.n. (fl fr) 1896.
Note: the record from Darfur is based on a sterile specimen by Drar (1970) which we have not seen.

[Schinus molle L.]

Note: this species is native to S America but is quite widely planted in Sudan and South Sudan as an ornamental and for timber and firewood. It is not known to have naturalised. El Amin (1990) also records S. terebinthifolius Raddi as planted in South Sudan.

Sclerocarya birrea (A.Rich.) Hochst. subsp. birrea

F.P.S. 2: 351 (1952); F.J.M.: 124 (1976); F.T.E.A. Anacard.: (1986); T.S.S.: 343 (1990); F.E.E. 3: 519 (1989); Biol. Skr. 51: 292 (1998).
Tree. Woodland & wooded grassland, often on rocky hillslopes.
Distr: DAR, KOR, CS, ES, BAG, UN, EQU; Senegal to Eritrea, S to Tanzania.
DAR: Wickens 1294 2/1964; **EQU:** Andrews 2038 6/1939.
Note: Wickens 394 from Jebel Kaldu, Kordofan, is tentatively identified as subsp. caffra (Sond.) Kokwaro.

Sorindeia grandifolia Engl.

Adansonia 25: 103 (2003).
Syn: Sorindeia schweinfurthii Engl. – F.P.S. 2: 351 (1952); T.S.S.: 343 (1990).
Tree. Forest.
Distr: BAG; Guinea to C.A.R. & D.R. Congo, São Tomé.
BAG: Schweinfurth III:100 (fl) 1/1871.
Note: the single Sudanese specimen was detroyed in the Berlin fire during WWII.

Spondias mombin L.

F.T.E.A. Anacard.: 1 (1986); T.S.S.: 343 (1990) as S. mombiro.
Tree. Forest patches in wooded grassland, also planted.
Distr: CS, EQU; native to the neotropics, now widely introduced and naturalised in the tropics.
EQU: Turner 58 (st) 1931.
Note: El Amin in T.S.S. noted that this species is probably introduced, though his records indicate that it is naturalised. The only specimen we have seen (cited here) is from a garden in Amadi and noted as "probably introduced".

Sapindaceae

R. Vanderstricht & I. Darbyshire

Allophylus abyssinicus (Hochst.) Radlk.

F.E.E. 3: 499 (1989); T.S.S.: 325 (1990); F.T.E.A. Sapind.: 78 (1998); Biol. Skr. 51: 285 (1998).
Tree. Montane forest & forest margins.
Distr: EQU; Eritrea to Zimbabwe.
EQU: Friis & Vollesen 973 (fr) 2/1982.

Allophylus africanus P.Beauv. var. *africanus*

F.P.S. 2: 335 (1952); F.E.E. 3: 499 (1989); T.S.S.: 325 (1990); F.T.E.A. Sapind.: 81 (1998).
Shrub or tree. Forest, forest margins, wooded grassland, thicket.
Distr: DAR, CS, ES, BAG, EQU; Gambia to Ethiopia, S to Tanzania.
DAR: Wickens 3105 4/1971; **EQU:** Myers 7100 (fl) 7/1937.

Allophylus ferrugineus Taub. var. *ferrugineus*

F.T.E.A. Sapind.: 86 (1998).
Syn: *Allophylus macrobotrys* Gilg – F.E.E. 3: 499 (1989); T.S.S.: 325 (1990); Biol. Skr. 51: 285 (1998).
Syn: *Allophylus welwitschii* Gilg – F.P.S. 2: 335 (1952); T.S.S.: 326 (1990).
Shrub or tree. Forest, often by streams.
Distr: EQU; Cameroon to Ethiopia, S to Angola.
EQU: Andrews 1713 (fr) 6/1939.

Allophylus rubifolius (Hochst.) Engl. var. *rubifolius*

F.P.S. 2: 335 (1952); F.E.E. 3: 502 (1989); T.S.S.: 326 (1990); F.T.E.A. Sapind.: 88 (1998); Biol. Skr. 51: 285 (1998).
Shrub or small to medium sized tree. Woodland, bushland, grassland & riverine forest.
Distr: KOR, CS, ES, BAG, UN, EQU; Eritrea to South Africa; tropical Arabia.
KOR: Broun 46 4/1906; **EQU:** Andrews 1804 (fl) 6/1939.

Aporrhiza paniculata Radlk.

F.P.S. 2: 336 (1952); T.S.S.: 327 (1990); F.T.E.A. Sapind.: 20 (1998).
Syn: *Aporrhiza nitida* Gilg –T.S.S.: 327 (1990).
Tree. Riverine forest, sometimes growing in the water.
Distr: EQU; Nigeria to Kenya, S to Mozambique.
EQU: Schweinfurth 3041 (fr) 2/1870.

Blighia unijugata Baker

F.P.S. 2: 336 (1952); F.E.E. 3: 508 (1989); T.S.S.: 327 (1990); F.T.E.A. Sapind.: 25 (1998); Biol. Skr. 51: 285 (1998).
Tree. Riverine forest & forest margins, also woodland & bushland.
Distr: EQU; Sierra Leone to Eritrea, S to South Africa.
EQU: Friis & Vollesen 1109 3/1982.

Blighia welwitschii (Hiern) Radlk.

F.T.E.A. Sapind.: 26 (1998); Biol. Skr. 51: 286 (1998).
Tree. Forest.
Distr: ?EQU; Sierra Leone to Uganda, S to Angola & D.R. Congo.
Note: no Sudanese material has been seen; its inclusion here is based on a site record by Jackson from Lotti, Imatong Mts.

Cardiospermum corindum L.

F.P.S. 2: 339 (1952); F.E.E. 3: 497 (1989); F.T.E.A. Sapind.: 101 (1998).
Herbaceous to somewhat woody climber. Bushland & woodland, sometimes riverine.
Distr: RS, ES; Eritrea to South Africa.
RS: Bent s.n. (fr) 1896.

Cardiospermum grandiflorum Sw.

F.P.S. 2: 338 (1952); F.T.E.A. Sapind.: 98 (1998); Biol. Skr. 51: 286 (1998).
Herbaceous or somewhat woody climber. Riverine forest.
Distr: EQU; Guinea to Uganda, S to Angola & D.R. Congo; tropical America, introduced into Australia, Sri Lanka and S Europe.
EQU: Friis & Vollesen 750 (fr) 12/1980.

Cardiospermum halicacabum L. var. *halicacabum*

F.P.S. 2: 339 (1952); F.J.M.: 123 (1976); F.E.E. 3: 498 (1989); F.T.E.A. Sapind.: 100 (1998); Biol. Skr. 51: 286 (1998).
Herbaceous climber. Riverine forest, riverbanks, wooded grassland, secondary scrub, overgrazed woodland & bushland.
Distr: NS, RS, DAR, KOR, UN, EQU; pantropical.
NS: Pettet 66 9/1962; **UN:** Simpson 7081 6/1929.

Cardiospermum halicacabum L. var. *microcarpum* (Kunth) Blume

F.E.E. 3: 498 (1989); F.T.E.A. Sapind.: 101 (1998).
Syn: *Cardiospermum microcarpum* Kunth – F.P.S. 2: 339 (1952).
Herbaceous climber. Habitat as for var. *halicacabum*.
Distr: UN; pantropical.
UN: J.M. Lock 81/224 (fl) 8/1981.

Deinbollia cf. *fulvo-tomentella* Baker f.

F.P.S. 2: 339 (1952); F.T.E.A. Sapind.: 68 (1998).
Tree. Forest, mountain gullies.
Distr: EQU; (for species) Cameroon to C.A.R., S to Tanzania & Angola.
EQU: Andrews 1504 5/1939.
Note: probably this species but the two specimens seen are sterile. Andrews and El Amin list also *D. grandifolia* Hook.f. but in F.W.T.A. 1(2): 715 (1958) the authors note that this is based on a misidentification of *Hoyle* 826 (BM).

Dodonaea viscosa (L.) Jacq.

F.P.S. 2: 339 (1952); T.S.S.: 329 (1990); F.T.E.A. Sapind.: 8 (1998).
Syn: *Dodonaea angustifolia* L.f. – F.E.E. 3: 491 (1989).
Small tree or shrub. Secondary bushland, grassland & forest margins.
Distr: NS, RS, ES; pantropical & subtropical.
RS: Aylmer 128 (fr) 2/1932.
Note: the varieties sometimes recognised in this species are not upheld here.

Eriocoelum kerstingii Gilg ex Engl.

F.P.S. 2: 340 (1952); T.S.S.: 329 (1990); F.T.E.A. Sapind.: 28 (1998).
Tree. Riverine forest.
Distr: EQU; Guinea Bissau to Uganda, S to Gabon & D.R. Congo.
EQU: Kosper 179b (fr) 2/1983.

Glenniea africana (Radlk.) Leenh.

F.T.E.A. Sapind.: 52 (1998); Biol. Skr. 51: 286 (1998).
Syn: *Melanodiscus oblongus* Radlk. ex Taub. – T.S.S.: 331 (1990).

Syn: *Melanodiscus* **sp.** sensu Andrews – F.P.S. 2: 342 (1952).
Shrub or tree. Forest & bushland.
Distr: EQU; Cameroon to Kenya, S to Mozambique.
EQU: Jackson 1231 1960.

Haplocoelum foliolosum Hiern
F.P.S. 2: 341 (1952); F.E.E. 3: 507 (1989); T.S.S.: 330 (1990); F.T.E.A. Sapind.: 44 (1998).
Shrub or tree. Dry forest, riverine forest, woodland, bushland.
Distr: EQU; Cameroon to Ethiopia, S to South Africa.
EQU: Dale 277 2/1943.

Lecaniodiscus cupanioides Planch.
F.P.S. 2: 341 (1952); T.S.S.: 330 (1990); F.T.E.A. Sapind.: 40 (1998).
Understory shrub or small tree. Lowland forest, periodically inundated forest & riverine fringes.
Distr: EQU; Senegal to Uganda, S to Angola and D.R. Congo.
EQU: Myers 11419 (fr) 6/1939.

Lepisanthes senegalensis (Juss. ex Poir.) Leenh.
F.E.E. 3: 504 (1989); F.T.E.A. Sapind.: 50 (1998); Biol. Skr. 51: 287 (1998).
Syn: *Aphania senegalensis* (Juss. ex Poir.) Radlk. – F.P.S. 2: 336 (1952); T.S.S.: 326 (1990).
Tree. Riverine & swamp forest.
Distr: BAG, EQU; Ethiopia & Somalia, S to Mozambique; India & Malesia.
EQU: Hoyle 539 (fr) 1/1939.

Lychnodiscus cerospermus Radlk.
F.T.E.A. Sapind.: 22 (1998); Biol. Skr. 51: 287 (1998).
Tree. Forest.
Distr: EQU; D.R. Congo, Uganda.
IUCN: RD
EQU: Jackson 1226 (fl) 3/1950.
Note: assessed as Near Threatened by Kalema & Beentje, Cons. Checkl. Trees of Uganda: 145 (2012) since it has experienced widespread decline through habitat loss.

Majidea fosteri (Sprague) Radlk.
F.P.S. 2: 341 (1952); T.S.S.: 330 (1990); F.T.E.A. Sapind.: 18 (1998); Biol. Skr. 51: 287 (1998).
Tree. Forest.
Distr: EQU; Ivory Coast to Uganda, S to Tanzania.
EQU: Myers 10551 (fl) 2/1939.

Pappea capensis Eckl. & Zeyh.
F.P.S. 2: 342 (1952); F.E.E. 3: 508 (1989); T.S.S.: 331 (1990); F.T.E.A. Sapind.: 35 (1998).
Syn: *Pappea* **sp.** sensu Andrews – F.P.S. 2: 342 (1952).
Tree. Woodland.
Distr: EQU; Eritrea to South Africa.
EQU: Myers 14064 9/1941.

Paullinia pinnata L.
F.P.S. 2: 343 (1952); F.J.M.: 123 (1976); F.E.E. 3: 493 (1989); T.S.S.: 331 (1990); F.T.E.A. Sapind.: 102 (1998); Biol. Skr. 51: 288 (1998).
Woody climber. Forest margins, riverine forest, moist thicket & scrub.

Distr: DAR, BAG, EQU; Senegal to Ethiopia, S to South Africa; Madagascar, neotropics.
DAR: Wickens 1183 2/1964; **EQU:** Andrews 947 (fl) 4/1939.

[*Placodiscus*]
Note: Andrews records a *Placodiscus* sp. from Azza Forest in Equatoria. El Amin apparently saw the specimen on which this was based and called it *P.* sp. near *pseudostipularis* Radlk. but noted that the material was not sufficient to confirm the identity. We have been unable to trace this specimen.

Zanha golungensis Hiern
F.P.S. 2: 343 (1952); F.E.E. 3: 493 (1989); T.S.S.: 333 (1990); F.T.E.A. Sapind.: 14 (1998); Biol. Skr. 51: 288 (1998).
Tree. Forest.
Distr: BAG, EQU; Senegal to Ethiopia, S to Mozambique.
EQU: Friis & Vollesen 1108 3/1982.

Rutaceae

R. Vanderstricht & I. Darbyshire

Aeglopsis eggelingii M.Taylor
F.P.S. 2: 313 (1952); F.T.E.A. Rut.: 7 (1982); T.S.S.: 296 (1990).
Shrub or small tree. Forest margins & riverine forest.
Distr: EQU; D.R. Congo, Uganda.
EQU: Andrews 1361 5/1939.
Note: a very localised species but not considered threatened.

Citropsis articulata (Spreng.) Swingle & M.Kellerm.
F.T.E.A. Rut.: 33 (1982).
Syn: *Citropsis schweinfurthii* (Engl.) Swingle & M.Kellerm. – F.P.S. 2: 314 (1952).
Shrub or small tree. Woodland, bushland & thicket.
Distr: EQU; Sierra Leone to Uganda, D.R. Congo & Tanzania.
EQU: Schweinfurth 3656 6/1870.

[*Citrus*]
Note: several *Citrus* species are widely grown for their edible fruits in irrigation schemes and gardens in Sudan, particularly in the drier northern regions. They are not known to naturalise.

Clausena anisata (Willd.) Hook.f. ex Benth. var. *anisata*
F.P.S. 2: 314 (1952); F.T.E.A. Rut.: 49 (1982); F.E.E. 3: 429 (1989); T.S.S.: 296 (1990); Biol. Skr. 51: 275 (1998).
Shrub or tree. Montane forest margins.
Distr: EQU; Guinea to Ethiopia, S to South Africa.
EQU: Friis & Vollesen 441 (fr) 11/1980.

Fagaropsis angolensis (Engl.) Dale
F.P.S. 2: 315 (1952); F.T.E.A. Rut.: 45 (1982); F.E.E. 3: 422 (1989); T.S.S.: 297 (1990); Biol. Skr. 51: 276 (1998).
Tree. Forest.
Distr: EQU; Ethiopia to Angola & Mozambique.
EQU: Friis & Vollesen 215 11/1980.

Haplophyllum tuberculatum (Forssk.) A.Juss.
F.Egypt 2: 67 (2000).
Syn: ***Ruta tuberculata*** Forssk. – F.P.S. 2: 315 (1952).
Shrub. Arid ground, coastal plains.
Distr: NS, RS, CS, ES; Somalia; Egypt, Arabia to
Afghanistan.
ES: Jackson 2941 (fl) 4/1953.

Harrisonia abyssinica Oliv.
F.P.S. 2: 320 (1952); F.E.E. 3: 437 (1989); T.S.S.: 303 (1990);
Biol. Skr. 51: 279 (1998); F.T.E.A. Simaroub.: 1 (2000).
Small tree or shrub. Dry evergreen forest, thickets &
riverine vegetation.
Distr: ?CS, BAG, UN, EQU; Cameroon to Ethiopia, S to
Mozambique.
UN: J.M. Lock 81/79 (fl) 5/1981.
Note: the record for Central Sudan is from El Amin in
T.S.S.; we have seen no material with which to verify this.

Toddalia asiatica (L.) Lam.
F.P.S. 2: 317 (1952); F.T.E.A. Rut.: 3 (1982); F.E.E. 3: 424
(1989); T.S.S.: 299 (1990); Biol. Skr. 51: 276 (1998).
Climbing shrub. Forest margins, bushland & wooded
grassland.
Distr: EQU; Ethiopia to South Africa; Madagascar,
Mascarenes & India.
EQU: Howard I.M.51 (fl) 2/1976.

Vepris glomerata F.Hoffm. var. ***glabra*** Kokwaro
F.T.E.A. Rut.: 23 (1982); F.E.E. 3: 426 (1989); Biol. Skr. 51:
276 (1998).
Shrub or small tree. Woodland & bushland.
Distr: EQU; Ethiopia, Uganda, Kenya, Tanzania.
EQU: Jackson 4230 6/1961.

Vepris grandifolia (Engl.) Mziray
Biol. Skr. 51: 276 (1998); Trop. Afr. Fl. Pl. 5: 386 (2010).
Syn: ***Teclea grandifolia*** Engl. – F.T.E.A. Rut.: 27 (1982).
Shrub or tree. Forest.
Distr: ?EQU; Cameroon to Kenya, S to Angola & Zambia.
Note: no Sudanese material has been seen; its inclusion
here is based on the site record by Jackson from Lotti and
Talanga in the Imatong Mts (see Friis & Vollesen 1998).

Vepris nobilis (Delile) Mziray
Biol. Skr. 51: 277 (1998); Trop. Afr. Fl. Pl. 5 388 (2010).
Syn: ***Teclea nobilis*** Delile – F.P.S. 2: 317 (1952); Bot.
Exp. Sud.: 101 (1970); F.J.M.: 122 (1976); F.T.E.A. Rut.:
26 (1982); F.E.E. 3: 427 (1989); T.S.S.: 297 (1990).
Shrub or tree. Upland forest & adjacent woodland.
Distr: RS, DAR, KOR, BAG, EQU; Ethiopia to Zimbabwe.
DAR: Wickens 1317 2/1964; **EQU:** Friis & Vollesen 118
(fr) 11/1980.

Zanthoxylum chalybeum Engl. var. ***molle***
Kokwaro
F.T.E.A. Rut.: 38 (1982); F.E.E. 3: 420 (1989); Biol. Skr. 51:
277 (1998).
Shrub or tree. Dry bushland & wooded grassland.
Distr: EQU; Uganda, Tanzania.
EQU: Jackson 4231 6/1961.

Zanthoxylum gilletii (De Wild.) P.G.Waterman
F.T.E.A. Rut.: 38 (1982); Biol. Skr. 51: 278 (1998).

Syn: ***Fagara macrophylla*** Engl. – T.S.S.: 297 (1990).
Tree. Lowland & mid-altitude forest.
Distr: EQU; Sierra Leone to Kenya, S to Zimbabwe.
EQU: Myers 11307 (fl) 5/1939.

Zanthoxylum leprieurii Guill. & Perr.
F.T.E.A. Rut.: 39 (1982); F.E.E. 3: 420 (1989); Biol. Skr. 51:
278 (1998).
Syn: ***Fagara angolensis*** Engl. – F.P.S. 2: 315 (1952).
Syn: ***Fagara leprieurii*** (Guill. & Perr.) Engl. – T.S.S.: 296
(1990).
Tree. Forest.
Distr: EQU; Senegal to Ethiopia, S to South Africa.
EQU: Andrews 1350 (fr) 5/1939.

Simaroubaceae
R. Vanderstricht

[*Ailanthus excelsa* Roxb.]
Note: a native of India, widely planted in drier parts of
central Sudan as a timber, fodder and shelterbelt tree,
particularly along rivers.

Brucea antidysenterica J.F.Mill.
F.P.S. 2: 319 (1952); F.E.E. 3: 440 (1989); Biol. Skr. 51:
278 (1998); F.T.E.A. Simaroub.: 7 (2000).
Syn: ***Brucea ferruginea*** L'Hér. – T.S.S.: 301 (1990).
Shrub or tree. Upland forest, evergreen bushland.
Distr: EQU; Guinea to Eritrea, S to Angola & Zambia;
Arabia.
EQU: Myers 10892 (fl) 4/1939.

Hannoa schweinfurthii Oliv.
F.P.S. 2: 320 (1952); T.S.S.: 301 (1990).
Syn: ***Quassia schweinfurthii*** (Oliv.) Noot. – Blumea 11:
520 (1962); Trop. Afr. Fl. Pl. 5: 410 (2010).
Shrub or small tree. Wooded grassland.
Distr: BAG, EQU; D.R. Congo.
IUCN: RD
EQU: Schweinfurth 2843 (fr) 2/1870.

Meliaceae
R. Vanderstricht

Azadirachta indica A.Juss.
F.E.E. 3: 485 (1989); T.S.S.: 311 (1990); F.T.E.A. Mel.: 27
(1991).
Tree. Introduced as a plantation & ornamental tree,
naturalised.
Distr: EQU; native to India and Myanmar, introduced
widely across drier parts of Africa.
EQU: Kosper 61 (fr) 7/1982.
Note: El Amin in T.S.S. records this as commonly grown
in Sudan and so it may be more widely naturalised than
documented here.

Ekebergia capensis Sparrm.
F.E.E. 3: 489 (1989); T.S.S.: 311 (1990); F.T.E.A. Mel.: 38
(1991); Biol. Skr. 51: 281 (1998).
Syn: ***Ekebergia rueppelliana*** (Fresen.) A.Rich. – F.P.S.
2: 326 (1952).
Syn: ***Ekebergia senegalensis*** A.Juss. – F.P.S. 2: 327
(1952); T.S.S.: 313 (1990).

Medium-sized tree. Upland & riverine forest, often along margins.
Distr: BAG, EQU; Senegal to Eritrea, S to South Africa.
EQU: Friis & Vollesen 904 (fl) 2/1982.

Entandrophragma angolense (Welw.) C.DC.
F.P.S. 2: 327 (1952); T.S.S.: 313 (1990); F.T.E.A. Mel.: 56 (1991); Biol. Skr. 51: 282 (1998).
Large tree. Lowland & mid-altitude forest.
Distr: EQU; Guinea to Uganda, S to Angola & Kenya.
IUCN: VU
EQU: Myers 11771 8/1939.

Khaya anthotheca C.DC.
T.S.S.: 313 (1990); F.T.E.A. Mel.: 47 (1991).
Large tree. Lowland & riverine forest.
Distr: EQU; Sierra Leone to Uganda & Tanzania, S to Mozambique.
EQU: Andrews 1196 5/1939.

Khaya grandifoliola C.DC.
F.P.S. 2: 329 (1952); T.S.S.: 315 (1990); F.T.E.A. Mel.: 49 (1991); Biol. Skr. 51: 282 (1998).
Medium-sized to large tree. Lowland & mid-altitude forest.
Distr: BAG, EQU; Guinea Bissau to Uganda, S to Angola & D.R. Congo.
IUCN: VU
EQU: Turner 184 (fl) 2/1936.

Khaya senegalensis A.Juss.
F.P.S. 2: 329 (1952); F.J.M.: 123 (1976); T.S.S.: 315 (1990); F.T.E.A. Mel.: 49 (1991).
Large tree. Wooded grassland, often in rocky places, riverine woodland.
Distr: DAR, BAG, EQU; Senegal to Uganda; cultivated in Ethiopia.
IUCN: VU
DAR: Wickens 1584 (fl) 5/1964; **BAG:** Young 11 3/1914.

Lepidotrichilia volkensii (Gürke) J.-F.Leroy
F.E.E. 3: 485 (1989); F.T.E.A. Mel.: 37 (1991); Biol. Skr. 51: 282 (1998).
Syn: *Trichilia volkensii* Gürke – F.P.S. 2: 333 (1952); T.S.S.: 321 (1990).
Shrub, small or medium-sized tree. Upland forest.
Distr: EQU; Ethiopia to Malawi.
EQU: Friis & Vollesen 67 (fl) 11/1980.

Melia azedarach L.
F.E.E. 3: 485 (1989); T.S.S.: 317 (1990); F.T.E.A. Mel.: 22 (1991).
Medium-sized tree. Introduced as an ornamental tree, naturalised.
Distr: DAR; native from India to Australia; widely planted elsewhere in the tropics and subtropics.
DAR: Wickens 1327 (fl fr) 3/1964.

Pseudocedrela kotschyi (Schweinf.) Harms
F.P.S. 2: 331 (1952); F.J.M.: 123 (1976); F.E.E. 3: 479 (1989); T.S.S.: 317 (1990); F.T.E.A. Mel.: 56 (1991); Biol. Skr. 51: 283 (1998).
Tree. Woodland & wooded grassland.
Distr: DAR, KOR, CS, ES, BAG, UN, EQU; Senegal to Ethiopia & Uganda.

DAR: Wickens 1641 (fl fr) 5/1964; **EQU:** Andrews 1085 5/1939.

Trichilia dregeana Sond.
F.E.E. 3: 483 (1989); T.S.S.: 319 (1990); F.T.E.A. Mel.: 34 (1991); Biol. Skr. 51: 283 (1998).
Large tree. Forest.
Distr: EQU; Guinea to Ethiopia, S to South Africa.
EQU: Andrews 1379 5/1939.

Trichilia emetica Vahl subsp. emetica
F.P.S. 2: 331 (1952); F.J.M.: 123 (1976); T.S.S.: 319 (1990); F.T.E.A. Mel.: 34 (1991); Biol. Skr. 51: 283 (1998).
Syn: *Trichilia roka* (Forssk.) Chiov. – Bot. Exp. Sud.: 84 (1970).
Tree. Riverine forest & woodland, more rarely on rocky outcrops or in wooded grassland.
Distr: DAR, KOR, CS, BAG, UN, EQU; Eritrea to Somalia, S to South Africa; Madagascar, Yemen.
DAR: Wickens 2888 3/1965; **EQU:** Myers 8634 (fl fr) 2/1938.

Trichilia emetica Vahl subsp. suberosa J.J.DeWilde
F.J.M.: 123 (1976); T.S.S.: 319 (1990); F.T.E.A. Mel.: 34 (1991).
Tree. Riverine woodland, more rarely on rocky outcrops or in wooded grassland.
Distr: DAR; Senegal to Uganda.
DAR: Chevalier 556 (fl) 3/1899.

Trichilia prieureana A.Juss.
F.P.S. 2: 331 (1952); F.E.E. 3: 483 (1989); T.S.S.: 319 (1990); F.T.E.A. Mel.: 30 (1991); Biol. Skr. 51: 284 (1998).
Tree. Lowland & riverine forest.
Distr: BAG, EQU; Senegal to Ethiopia, S to Zambia.
EQU: Andrews 1404 5/1939.

Trichilia retusa Oliv.
F.P.S. 2: 332 (1952); F.E.E. 3: 483 (1989); T.S.S.: 321 (1990).
Tree. Riverine forest.
Distr: BAG, EQU; Nigeria to Ethiopia, S to D.R. Congo.
BAG: Simpson 76422 (fl) 3/1930.

Turraea floribunda Hochst.
T.S.S.: 321 (1990); F.T.E.A. Mel.: 21 (1991); Biol. Skr. 51: 284 (1998).
Shrub or small tree. Lowland & mid-altitude forest.
Distr: EQU; D.R. Congo to Kenya, S to South Africa.
EQU: Friis & Vollesen 736 12/1980.

Turraea holstii Gürke
F.P.S. 2: 333 (1952); F.E.E. 3: 481 (1989); T.S.S.: 323 (1990); F.T.E.A. Mel.: 13 (1991); Biol. Skr. 51: 284 (1998).
Shrub or small tree. Upland forest.
Distr: EQU; Ethiopia & Somalia, S to Malawi; Arabia.
EQU: Andrews 1864 (fl) 6/1939.

Turraea nilotica Kotschy & Peyr.
F.P.S. 2: 333 (1952); F.E.E. 3: 481 (1989); T.S.S.: 323 (1990); F.T.E.A. Mel.: 10 (1991).
Shrub or small tree. Woodland, bushland & wooded grassland.
Distr: KOR, BAG, UN, EQU; Ethiopia to South Africa.

KOR: Broun 1396 1/1908; **UN:** J.M. Lock 81/74 (fl) 5/1981.

Turraea vogelii Hook.f. ex Benth.

F.P.S. 2: 333 (1952); T.S.S.: 323 (1990); F.T.E.A. Mel.: 14 (1991).
Woody climber. Evergreen forest & forest margins.
Distr: EQU; Ghana to Uganda, S to Angola & D.R. Congo.
EQU: Myers 11457 (fl) 6/1939.

MALVALES

Neuradaceae

I. Darbyshire

Neurada procumbens L.

F.P.S. 2: 103 (1952); F.E.E. 3: 45 (1989).
Annual herb. Dry plains.
Distr: NS, RS, DAR; Mali, Eritrea; Cyprus, N Africa to Pakistan.
RS: Bent s.n. (fr) 1896.

Malvaceae: Bombacoideae (Bombacaceae)

R. Vanderstricht

Adansonia digitata L.

F.P.S. 2: 10 (1952); F.J.M.: 101 (1976); F.T.E.A. Bombac.: 4 (1989); T.S.S.: 113 (1990); F.E.E. 2(2): 186 (1995).
Large tree. Dry woodland, wooded grassland, semi-desert.
Distr: DAR, KOR, CS, UN, BAG; Mauritania to Somalia, S to Namibia & South Africa; Madagascar.
KOR: Wickens 3093 (fl fr) 6/1969.
Note: no South Sudanese material has been seen; the records are from El Amin in T.S.S.

Ceiba pentandra (L.) Gaertn.

F.P.S. 2: 10 (1952); F.J.M.: 101 (1976); F.T.E.A. Bombac.: 7 (1989); T.S.S.: 113 (1990); F.E.E. 2(2): 186 (1995); Biol. Skr.: 147 (1998).
Large spiny tree. Lowland forest, groundwater forest, also planted.
Distr: DAR, CS, ES, EQU; widespread in tropical Africa, America & SE Asia; also widely cultivated.
DAR: Wickens 1116 (fl) 2/1964; **EQU:** Friis & Vollesen 647 12/1980.

Malvaceae: Malvoideae (Malvaceae)

H. Pickering & I. Darbyshire

Abelmoschus esculentus (L.) Moench

F.E.E. 2(2): 212 (1995); F.T.E.A. Malv.: 76 (2009).
Syn: *Hibiscus esculentus* L. – F.P.S. 2: 31 (1952); Bot. Exp. Sud.: 82 (1970).
Annual herb. Grassland including seasonally flooded areas, sometimes weedy, also cultivated.
Distr: CS, ES, BAG, UN; widely cultivated & naturalised in the tropics, probably of Asian origin.
ES: Beshir 148 9/1951; **BAG:** Simpson 7114 6/1929.
Note: the cultivated 'okra'. It appears likely that

Abelmoschus species will be transferred back into *Hibiscus* in the future; see note in F.T.E.A. Malv.: 1 (2009).

Abelmoschus ficulneus (L.) Wight & Arn.

F.E.E. 2(2): 212 (1995); Biol. Skr. 51: 147 (1998); F.T.E.A. Malv.: 76 (2009).
Syn: *Hibiscus ficulneus* L. – F.P.S. 2: 31 (1952).
Annual herb, sometimes subshrubby. Grassland including seasonally waterlogged areas, riverbanks, sometimes weedy.
Distr: CS, UN, EQU; Nigeria to Ethiopia, S to Zambia; Madagascar, India, Sri Lanka, Indonesia to Australia.
CS: Andrews s.n. 11/1947; **EQU:** Myers 14161 10/1941.

Abutilon angulatum (Guill & Perr.) Mast. var. angulatum

F.P.S. 2: 12 (1952); F.J.M.: 101 (1976); F.E.E. 2(2): 242 (1995); F.T.E.A. Malv.: 121 (2009).
Shrubby perennial herb or shrub. Wooded grassland & bushland.
Distr: DAR, CS, SOUTHSUD; Senegal to Eritrea, S to Namibia & South Africa.
CS: Jackson 3108 3/1954; **SOUTH SUD:** Ruttledge 2a (fl fr)
Note: the single collection seen from South Sudan has no locality details save for 'Southern Sudan'.

Abutilon bidentatum (Hochst.) A.Rich.

F.P.S. 2: 15 (1952); Bot. Exp. Sud.: 81 (1970); F.E.E. 2(2): 246 (1995); F.T.E.A. Malv.: 132 (2009).
Shrubby perennial herb. Riverine forest & woodland, wooded grassland.
Distr: NS, RS, DAR; Eritrea & Somalia to Tanzania; Egypt, Arabia to India.
RS: Schweinfurth 1608 4/1864.

Abutilon erythraeum Mattei

F.E.E. 2(2): 246 (1995).
Annual herb. Bushland, sometimes coastal.
Distr: RS, CS; Eritrea.
IUCN: RD
CS: Figari s.n. (n.v.)
Note: a poorly known species, only known in Sudan from two original syntypes which we have not seen. It was, however, maintained as distinct by Vollesen in F.E.E.

Abutilon eufigarii Chiov.

Boll. Soc. Bot. Ital.: 23 (1917).
Herb. Habitat unknown.
Distr: CS; ?Sudan endemic (see note).
CS: Figari s.n. 1867 (n.v.).
Note: the status of this taxon is uncertain; it is known only from the type which we have not seen. It may well prove to be conspecific with one of the other Sudanese species of *Abutilon*.

Abutilon fruticosum Guill. & Perr. var. fruticosum

F.P.S. 2: 15 (1952); F.E.E. 2(2): 244 (1995); F.T.E.A. Malv.: 128 (2009).
Perennial herb. Wooded grassland & bushland.
Distr: RS, DAR, KOR; Mauritania & Senegal to Somalia, S to Namibia & South Africa; Egypt, Arabia, Pakistan, India.
KOR: Wickens 276 (fl) 8/1962.

Abutilon guineense (Schumach.) Baker f. & Exell

F.T.E.A. Malv.: 139 (2009).
Annual or perennial herb or subshrub. Wooded grassland.
Distr: UN; Ghana, Togo, Somalia to Angola & South Africa; Madagascar.
UN: J.M. Lock 81/165 8/1981.

Abutilon hirtum (Lam.) Sweet

F.P.S. 2: 15 (1952); F.E.E. 2(2): 248 (1995); F.T.E.A. Malv.: 122 (2009).
Shrubby perennial herb or shrub. Bushland & grassland, sometimes weedy.
Distr: ES, UN; pantropical.
ES: Schweinfurth 1418 1865; **UN:** J.M. Lock 81/144 8/1981.

Abutilon longicuspe Hochst. ex A.Rich. var. longicuspe

F.P.S. 2: 12 (1952); F.E.E. 2(2): 241 (1995); Biol. Skr. 51: 147 (1998); F.T.E.A. Malv.: 118 (2009).
Shrubby perennial herb. Forest margins & secondary thicket.
Distr: RS, EQU; Eritrea to Zimbabwe & Mozambique; Arabia.
RS: Aylmer 205 3/1932; **EQU:** Myers 10990 4/1939.

Abutilon mauritianum (Jacq.) Medik. subsp. mauritianum

F.P.S. 2: 15 (1952); F.E.E. 2(2): 246 (1995); Biol. Skr. 51: 148 (1998); F.T.E.A. Malv.: 133 (2009).
Shrub. Grassland, woodland, forest margins.
Distr: BAG, EQU; Senegal to Ethiopia, S to South Africa; Mauritius.
EQU: Myers 10916 (fl) 4/1939.

Abutilon pannosum (G.Forst.) Schltdl. var. pannosum

F.P.S. 2: 14 (1952); F.E.E. 2(2): 246 (1995); F.T.E.A. Malv.: 123 (2009).
Syn: *Abutilon glaucum* (Cav.) Webb – F. Darfur Nord-Occ. & J. Gourgeil: 110 (1969).
Perennial herb or shrub. Seasonally wet grassland, bushland & depressions, riverbanks, sometimes in saline areas.
Distr: NS, RS, DAR, KOR, CS, ES; Cape Verde Is., Mauritania & Senegal to Ethiopia; N Africa, Arabia to India & Sri Lanka.
NS: Ahti 16694 (fr) 9/1962.
Note: Verdcourt in F.T.E.A. also recorded his var. *scabrum* from the Red Sea Hills. Whilst some very hairy specimens are recorded from RS (e.g. *Wickens* 3072 from Sinkat), this form appears to merge into typical *pannosum* in Sudan. The stem hairs are not so rough as in plants of var. *scabrum* from coastal E Africa, where the distinction is more convincing.

Abutilon pannosum (G.Forst.) Schltdl. var. figarianum (Webb) Verdc.

F.T.E.A. Malv.: 124 (2009).
Syn: *Abutilon figarianum* Webb – F.P.S. 2: 15 (1952); F.E.E. 2(2): 246 (1995); Biol. Skr. 51: 147 (1998).
Annual or perennial herb. Grassland & bushland, often in seasonally wet areas.

Distr: NS, RS, KOR, CS, ES, UN; Ethiopia, Somalia, Uganda, Kenya, Tanzania; Egypt, Arabia, Pakistan.
RS: Jackson 2780 4/1953; **UN:** Simpson 7094 6/1929.

Abutilon ramosum (Cav.) Guill. & Perr.

F.P.S. 2: 15 (1952); F.E.E. 2(2): 242 (1995); F.T.E.A. Malv.: 143 (2009).
Perennial herb. Riverine forest & woodland, riverbeds.
Distr: KOR, UN; Senegal to Somalia, S to Namibia & South Africa; Arabia, Pakistan, India.
KOR: Kotschy 278 11/1839; **UN:** J.M. Lock 81/163 8/1981.

Cienfuegosia digitata Cav.

F.P.S. 2: 17 (1952); F.W.T.A. 1(2): 343 (1958).
Shrubby perennial herb. Dry bushland & scrub, rocky hillslopes.
Distr: ?DAR, ?CS; Mauritania & Senegal to Chad, Angola, Zambia, Zimbabwe, South Africa.
Note: Andrews recorded this species from Central Sudan and Quézel (1969: 110) recorded it from Darfur; we have not seen the material on which these records are based.

Gossypium anomalum Wawra ex Wawra & Peyr. subsp. senarense (Fenzl ex Wawra & Peyr.) Vollesen

F.P.S. 2: 17 (species only) (1952); T.S.S.: 115 (species only) (1990); F.E.E. 2(2): 219 (1995).
Perennial herb or shrub. Seasonally wet bushland & grassland.
Distr: DAR, KOR, CS; Cape Verde Is., Mauritania to Eritrea.
KOR: MacMichael 1536 2/1909.

Gossypium arboreum L.

F.P.S. 2: 17 (1952); F.E.E. 2(2): 222 (1995); F.T.E.A. Malv.: 97 (2009).
Shrub. Cultivated, ocasionally naturalised.
Distr: RS, KOR, CS, BAG, UN, EQU; origin unclear, formerly widely cultivated but now less so.
KOR: Broun 1343 (fr) 12/1907; **UN:** Simpson 7157 6/1929.
Note: much of the material from Sudan has been previously treated as *G. soudanense* Watt and is sometimes considered native, but it is likely to be an introduced variant of *G. arboreum*.

Gossypium barbadense L.

F.P.S. 2: 17 (1952); F.E.E. 2(2): 222 (1995); F.T.E.A. Malv.: 98 (2009).
Shrub. Cultivated, occasionally a casual weed.
Distr: BAG, EQU; originally from S America, now widely cultivated in the tropics & subtropics.
BAG: Broun 67A 4/1907.

Gossypium herbaceum L.

F.P.S. 2: 17 (1952); F.E.E. 2(2): 222 (1995).
Shrub. Cultivated, occasionally a casual weed.
Distr: DAR, KOR; widely cultivated in the palaeotropics in the past, now more rarely so.
KOR: Pfund 474 3/1875.

Gossypium hirsutum L.

F.E.E. 2(2): 222 (1995); F.T.E.A. Malv.: 97 (2009).
Shrub. Cultivated, occasionally a casual weed.

Distr: NS, KOR, BAG, UN, EQU; originally from central America, Caribbean & Pacific Is., now widely cultivated in the tropics & subtropics.
KOR: Pfund 496 1875; **UN:** Myers 10477 11/1939.

Gossypium longicalyx Hutch. & B.J.S.Lee
F.T.E.A. Malv.: 94 (2009).
Scandent shrub. Bushland & woodland in seasonally wet areas.
Distr: UN; Uganda, Tanzania.
IUCN: RD
UN: Broun 43 1910.
Note: a scarce species which may prove to be threatened.

Gossypium somalense (Gürke) J.B.Hutch.
F.P.S. 2: 17 (1952); T.S.S.: 117 (1990); F.E.E. 2(2): 220 (1995); F.T.E.A. Malv.: 94 (2009).
Shrub. Bushland, rocky hillslopes.
Distr: DAR, CS; Niger to Somalia, Uganda & Kenya.
CS: Unknown collector s.n. 5/1949.

Hibiscus acetosella Welw. ex Hiern
F.E.E. 2(2): 198 (1995); Biol. Skr. 51: 148 (1998); F.T.E.A. Malv.: 32 (2009).
Annual or perennial herb or subshrub. Weed of cultivation & disturbed ground.
Distr: EQU; native to Angola, Zambia and perhaps elsewhere in C Africa, now widely cultivated & naturalised.
EQU: Myers 11897 1938.

Hibiscus aponeurus Sprague & Hutch.
F.E.E. 2(2): 209 (1995); Biol. Skr. 51: 148 (1998); F.T.E.A. Malv.: 52 (2009).
Shrubby perennial herb. Grassland, secondary bushland, forest margins.
Distr: EQU; Ethiopia to D.R. Congo & Mozambique; Arabia.
EQU: Myers 11214 4/1939.

Hibiscus articulatus Hochst. ex A.Rich.
F.P.S. 2: 26 (1952); F.J.M.: 102 (1976); F.E.E. 2(2): 204 (1995); Biol. Skr. 51: 148 (1998); F.T.E.A. Malv.: 71 (2009).
Perennial herb. Seasonally wet grassland & woodland.
Distr: DAR, CS, ES, BAG, UN, EQU; Burkina Faso to Ethiopia, S to Angola & South Africa.
DAR: Wickens 2030 7/1964; **EQU:** Andrews 860 4/1939.

Hibiscus calyphyllus Cav.
F.P.S. 2: 27 (1952); F.E.E. 2(2): 192 (1995); Biol. Skr. 51: 149 (1998); F.T.E.A. Malv.: 61 (2009).
Syn: *Hibiscus owariensis* P.Beauv. – F.P.S. 2: 27 (1952).
Shrub. Woodland, forest margins.
Distr: CS, EQU; Ghana to Ethiopia, S to Botswana & South Africa; Arabia, Madagascar, Mascarenes.
CS: Hamid 11 11/1972; **EQU:** Friis & Vollesen 715 12/1980.

Hibiscus cannabinus L.
F.P.S. 2: 28 (1952); F.J.M.: 102 (1976); F.E.E. 2(2): 198 (1995); Biol. Skr. 51: 149 (1998); F.T.E.A. Malv.: 41 (2009).
Syn: *Hibiscus asper* Hook.f. – F.P.S. 2: 28 (1952); T.S.S.: 117 (1990).

Annual herb. Wooded grassland & bushland.
Distr: DAR, KOR, CS, ES, BAG, UN, EQU; palaeotropical.
DAR: Wickens 1749 6/1964; **EQU:** Myers 10194 7/1938.

Hibiscus corymbosus Hochst. ex A.Rich.
F.P.S. 2: 24 (1952); F.E.E. 2(2): 204 (1995); Biol. Skr. 51: 149 (1998); F.T.E.A. Malv.: 71 (2009).
Perennial herb. Woodland, bushland & grassland.
Distr: EQU; Eritrea to D.R. Congo & Tanzania; Arabia.
EQU: Andrews 1305 5/1939.

Hibiscus crassinervius Hochst. ex A.Rich.
F.P.S. 2: 28 (1952); F.E.E. 2(2): 209 (1995); Biol. Skr. 51: 149 (1998); F.T.E.A. Malv.: 52 (2009).
Perennial herb. Wooded grassland & bushland.
Distr: RS, EQU; Eritrea, Ethiopia, Somalia, Uganda, Kenya.
RS: Jackson 2850 4/1953; **EQU:** Myers 10938 4/1939.

Hibiscus dictyocarpus Webb
F.P.S. 2: 22 (1952).
Syn: *Fioria dictyocarpa* (Webb) Mattei – F.E.E. 2(2): 214 (1995).
Syn: *Roifia dictyocarpa* (Webb) Verdc. – F.T.E.A. Malv.: 79 (2009).
Perennial herb or subshrub. Grassland, wooded grassland & bushland.
Distr: KOR, CS; Ethiopia, Somalia, Kenya; Saudi Arabia.
KOR: Kotschy 124 10/1869.

Hibiscus diversifolius Jacq. subsp. *diversifolius*
F.P.S. 2: 24 (1952); F.E.E. 2(2): 196 (1995); F.T.E.A. Malv.: 45 (2009).
Shrub or perennial herb. Swampy grassland, riverbanks.
Distr: BAG, UN; pantropical.
UN: J.M. Lock 83/13 2/1983.

Hibiscus dongolensis Delile
F.P.S. 2: 26 (1952); F.E.E. 2(2): 192 (1995); F.T.E.A. Malv.: 66 (2009).
Shrub. Wooded grassland & bushland.
Distr: KOR, CS, UN, EQU; Eritrea & Somalia, S to Namibia & South Africa; Arabia.
CS: Aylmer 65 12/1929; **UN:** Sherif A.3978 8/1951.

Hibiscus eriospermus Hochst. ex Cufod.
F.E.E. 2(2): 209 (1995).
Syn: *Hibiscus aponeurus* sensu auctt., non Sprague & Hutch. – F.P.S. 2: 31 (1952); Bot. Exp. Sud.: 82 (1970).
Shrubby perennial herb. Bushland on rocky hillslopes.
Distr: RS; Eritrea, Ethiopia.
IUCN: RD
RS: Maffey 13 1928.

Hibiscus furcatus Willd.
F.P.S. 2: 26 (1952); Trop. Afr. Fl. Pl. 1: 726 (2003).
Perennial herb. Habitat in Sudan unknown.
Distr: EQU; native to India & Thailand.
EQU: Dandy 485 2/1934.
Note: Lebrun & Stork in Trop. Afr. Fl. Pl. note that Andrews' record of this Asian species requires further study.

Hibiscus lobatus (Murray) Kuntze
F.P.S. 2: 31 (1952); F.J.M.: 102 (1976); F.E.E. 2(2): 205 (1995); F.T.E.A. Malv.: 74 (2009).

Annual herb. Riverine woodland & forest, wooded grassland, sometimes weedy.
Distr: DAR, CS; palaeotropical, rather scattered in Africa.
DAR: Wickens 2180 8/1964.

Hibiscus macranthus Hochst. ex A.Rich.

F.E.E. 2(2): 194 (1995); Biol. Skr. 51: 149 (1998); F.T.E.A. Malv.: 65 (2009).
Syn: *Hibiscus ludwigii* sensu Andrews, non Eckl. & Zeyh. – F.P.S. 2: 27 (1952).
Shrub. Wooded grassland, bushland, forest margins.
Distr: EQU; Cameroon to Ethiopia, S to Zimbabwe; Arabia.
EQU: Kosper 196 1/1982.

Hibiscus micranthus L.f.

F.P.S. 2: 28 (1952); F.J.M.: 102 (1976); F.E.E. 2(2): 210 (1995); Biol. Skr. 51: 150 (1998); F.T.E.A. Malv.: 47 (2009).
Perennial herb or subshrub. Grassland & bushland, sometimes weedy.
Distr: RS, DAR, KOR, CS, ES, EQU; Mauritania & Senegal to Somalia, S to Namibia & South Africa; Madagascar, Comoros, Arabia, India.
RS: Carter 1841 11/1987; **EQU:** Friis & Vollesen 1198 3/1982.

Hibiscus mongallaensis Baker f.

F.P.S. 2: 27 (1952); Trop. Afr. Fl. Pl. 1: 730 (2003).
Erect herb. Gravelly riverbed in gorge.
Distr: EQU; South Sudan endemic.
IUCN: RD
EQU: Imp. Forestry Inst. 55238 (fr).
Note: known only from the type, collected from Naramum in the Ilemi Triangle; in the protologue it was noted as being close to *H. dongolensis*.

Hibiscus muhamedis Webb

Fragm. Fl. Aethiop.: 46 (1854).
Herb. Habitat unknown.
Distr: KOR; ?Sudan endemic.
KOR: Figari s.n. (n.v.).
Note: the status of this taxon is uncertain.

Hibiscus noldeae Baker f.

F.E.E. 2(2): 200 (1995); F.T.E.A. Malv.: 39 (2009).
Annual or perennial herb or shrub. Woodland, riverbanks.
Distr: DAR; Sierra Leone to Ethiopia, S to Angola & Zambia.
DAR: Wickens 2576 9/1964.

Hibiscus obtusilobus Garcke

F.P.S. 2: 22 (1952); F.E.E. 2(2): 204 (1995); F.T.E.A. Malv.: 70 (2009).
Annual or perennial herb or subshrub. Seasonally flooded grassland & woodland.
Distr: KOR, CS; Mauritania, Niger, Ethiopia, Kenya; Pakistan, India.
CS: Aylmer 399 7/1934.

Hibiscus ovalifolius (Forssk.) Vahl

F.E.E. 2(2): 192 (1995); F.T.E.A. Malv.: 63 (2009).
Perennial herb or shrub. Woodland, bushland & grassland.

Distr: CS, UN; Senegal, Eritrea & Somalia to South Africa; Yemen.
CS: Univ. Khartoum Ingassana Hills Exp. 440 10/1973;
UN: Sherif A.2861 10/1951.
Note: in the African Plants Database, this taxon is treated as a synonym of *H. micranthus*.

Hibiscus palmatus Forssk.

F.E.E. 2(2): 205 (1995); F.T.E.A. Malv.: 56 (2009).
Syn: *Hibiscus aristivalvis* Garcke – F.P.S. 2: 24 (1952).
Annual or perennial herb. Wooded grassland & bushland.
Distr: RS, EQU; Eritrea to Namibia & South Africa; Arabia, India.
RS: Andrews 3636; **EQU:** Broun 1445 (fl fr) 11/1908.

Hibiscus panduriformis Burm.f.

F.P.S. 2: 28 (1952); F.E.E. 2(2): 202 (1995); F.T.E.A. Malv.: 67 (2009).
Perennial herb or shrub. Grassland & woodland, often in seasonally flooded areas.
Distr: KOR, ES, BAG, UN, EQU; palaeotropical.
KOR: Andrews 181 11/1933; **EQU:** Myers 9903 10/1938.

Hibiscus physaloides Guill. & Perr.

F.P.S. 2: 24 (1952); F.E.E. 2(2): 204 (1995); F.T.E.A. Malv.: 70 (2009).
Annual herb. Dry woodland & bushland.
Distr: BAG, EQU; Cape Verde Is., Senegal to Eritrea, S to Angola & South Africa; Madagascar, Comoros, Seychelles.
BAG: Broun 1682 6/1909.

Hibiscus rhabdotospermus Garcke

F.P.S. 2: 24 (1952); F.E.E. 2(2): 202 (1995); F.T.E.A. Malv.: 69 (2009).
Annual herb. Rocky hillslopes, riverbanks, woodland.
Distr: KOR, CS, ES; scattered from Chad to Eritrea, S to Namibia & Zimbabwe.
CS: Lea 53 10/1952.

Hibiscus rostellatus Guill. & Perr.

F.P.S. 2: 26 (1952); F.E.E. 2(2): 198 (1995); F.T.E.A. Malv.: 39 (2009).
Shrub, sometimes scandent. Forest margins, riverine forest, swamp margins.
Distr: EQU; Senegal to Ethiopia, S to Angola & Mozambique.
EQU: Myers 7824 10/1937.

Hibiscus sabdariffa L.

F.P.S. 2: 28 (1952); F.E.E. 2(2): 198 (1995); Biol. Skr. 51: 150 (1998); F.T.E.A. Malv.: 32 (2009).
Annual herb. Cultivated, sometimes a weed of disturbed ground.
Distr: DAR, KOR, UN, EQU; probably native to India, now widely cultivated in the tropics.
KOR: Kotschy 422 1842; **UN:** Sherif A.2871 10/1951.

Hibiscus schweinfurthii Gürke

F.P.S. 2: 27 (1952); Trop. Afr. Fl. Pl. 1: 736 (2003).
Shrub. Wooded grassland, seasonally waterlogged clays.
Distr: BAG, EQU; South Sudan endemic.
IUCN: RD

BAG: Myers 7702 (fl) 9/1937.

Hibiscus sidiformis Baill.

F.J.M.: 102 (1976); F.E.E. 2(2): 205 (1995); F.T.E.A. Malv.: 74 (2009).

Syn: Hibiscus ternifoliolus F.W.Andrews – F.P.S. 2: 31 (1952).

Annual herb. Woodland, bushland & grassland, often in rocky areas.

Distr: DAR, KOR; Mauritania & Senegal to Eritrea, S to Namibia & South Africa; Madagascar.

DAR: Wickens 2338 9/1964.

Note: Andrews records this species from Central & Southern Sudan but we have seen no South Sudanese specimens.

Hibiscus sudanensis Hochr.

Annuaire Conserv. Jard. Bot. Genève 10: 18 (1906); F.P.S. 2: 26 (1952).

Shrub or woody climber. Swamp & forest margins, humid woodland.

Distr: ?EQU; C.A.R., D.R. Congo.

IUCN: RD

Note: Andrews recorded this species from Equatoria but we have seen no South Sudanese material; despite its name, it was described from C.A.R.

Hibiscus surattensis L.

F.P.S. 2: 26 (1952); F.E.E. 2(2): 198 (1995); Biol. Skr. 51: 150 (1998); F.T.E.A. Malv.: 38 (2009).

Annual herb. Wooded grassland, bushland, forest margins, often in disturbed areas.

Distr: EQU; palaeotropical.

EQU: Sillitoe 382 1919.

Hibiscus trionum L.

F.P.S. 2: 22 (1952); F.E.E. 2(2): 200 (1995); F.T.E.A. Malv.: 60 (2009).

Spreading annual herb. Swampy grassland, weed of disturbed ground.

Distr: ?NS, KOR, CS; palaeotropical & subtropical.

CS: Pettet 141 2/1963.

Hibiscus vitifolius L.

F.P.S. 2: 22 (1952); F.E.E. 2(2): 205 (1995); F.T.E.A. Malv.: 56 (2009).

Shrub. Forest margins & clearings, upland bushland & wooded grassland.

Distr: RS, UN, EQU; palaeotropical, introduced in the neotropics.

RS: Carter 1848 11/1987; **UN:** Sherif A.2852 10/1951.

Note: material from Red Sea region is of the hairy subsp. *vulgaris* Brenan & Exell, though this subspecies is not maintained in the African Plants Database.

Kosteletzkya adoensis (Hochst. ex A.Rich.) Mast.

F.P.S. 2: 32 (1952); F.E.E. 2(2): 215 (1995); Biol. Skr. 51: 150 (1998); F.T.E.A. Malv.: 85 (2009).

Perennial herb or subshrub. Upland grassland, bushland & forest margins, sometimes weedy.

Distr: ?CS, EQU; Sierra Leone to Ethiopia, S to Zimbabwe; Madagascar, Arabia.

EQU: Friis & Vollesen 3 11/1980.

Note: Andrews recorded this species from Central Sudan but we have seen no Sudanese material.

Kosteletzkya buettneri Gürke

F.P.S. 2: 33 (1952); F.T.E.A. Malv.: 82 (2009).

Perennial herb or subshrub. Swampy grassland, riverbanks.

Distr: EQU; Senegal to Uganda, S to Namibia & Botswana.

EQU: Myers 6662 5/1937.

Kosteletzkya grantii (Mast.) Garcke

F.P.S. 2: 33 (1952); Biol. Skr. 51: 150 (1998); F.T.E.A. Malv.: 87 (2009).

Perennial herb or subshrub. Grassland, woodland, forest margins, disturbed ground.

Distr: BAG, EQU; Guinea Bissau to Uganda, S to Angola & Tanzania.

EQU: Myers 13600 11/1940.

Malachra radiata (L.) L.

F.P.S. 2: 33 (1952); F.W.T.A. 2(1): 342 (1958).

Perennial herb. Swamps & wet ground.

Distr: BAG; Mali & Ivory Coast to D.R. Congo.

BAG: Schweinfurth III.10 7/1870.

Malva parviflora L. var. parviflora

F.P.S. 2: 33 (1952); F.E.E. 2(2): 237 (1995); F.T.E.A. Malv.: 109 (2009).

Annual herb. Grassland, disturbed ground.

Distr: NS, RS, DAR; native to Eurasia, now a cosmopolitan weed.

RS: Andrews 2718.

Malva verticillata L.

F.P.S. 2: 33 (1952); F.J.M.: 102 (1976); F.E.E. 2(2): 237 (1995); F.T.E.A. Malv.: 107 (2009).

Annual or biennial herb. Bushland, forest margins, disturbed ground.

Distr: DAR; widespread in the palaeotropics & warm temperate regions of the Old World.

DAR: Wickens 2657 9/1964.

Malvastrum coromandelianum (L.) Garcke

F.P.S. 2: 34 (1952); F.T.E.A. Malv.: 110 (2009).

Annual or perennial herb. Weed of cultivation & disturbed areas.

Distr: ?CS; native to the Americas, now widely naturalised elsewhere in the tropics & subtropics.

Note: Andrews recorded this species from Central Sudan; whilst we have seen no material to confirm this, it is quite likely to occur as a weed in Sudan. More doubtful is his record of *M. americanum* (L.) Torr. which S.R. Hill, in his revision of the genus (Rhodora 84: 1-83, 159-264 & 317-409 (1982)), does not record at all from continental Africa.

Pavonia arabica Hochst. & Steud. ex Boiss.

F.P.S. 2: 36 (1952); F.E.E. 2(2): 232 (1995); F.T.E.A. Malv.: 26 (2009).

Perennial herb. Bushland, woodland, rocky hillslopes.

Distr: RS, DAR; Eritrea, Ethiopia, Somalia, Uganda, Kenya, Tanzania; Egypt, Arabia, Socotra, Pakistan, India.

DAR: Lynes 309 2/1922.

Pavonia arenaria (Murray) Roth var. *microphylla* (Ulbr.) Verdc.

F.T.E.A. Malv.: 25 (2009).
Syn: *Pavonia zeylanica* sensu auctt., non Cav. – F.E.E. 2(2): 233 (1995).
Annual or perennial herb. Woodland, bushland on rocky hillslopes.
Distr: EQU; Senegal to Somalia, S to Tanzania.
EQU: Carr 841 7/1970.

Pavonia burchellii (DC.) R.A.Dyer

F.P.S. 2: 36 (1952); F.E.E. 2(2): 225 (1995); Biol. Skr. 51: 151 (1998); F.T.E.A. Malv.: 18 (2009).
Syn: *Pavonia patens* sensu auctt., non (Andrews) Chiov. – F.P.S. 2: 34 pro parte (1952); Bot. Exp. Sud.: 82 (1970); F.J.M.: 102 (1976).
Shrubby perennial herb. Wooded grassland, forest margins, rocky hillslopes.
Distr: RS, DAR, EQU; Cameroon to Somalia, S to Namibia & South Africa; Egypt, Arabia.
RS: Aylmer 637 3/1936; **EQU:** Myers 11215 4/1939.

Pavonia flavoferruginea (Forssk.) Hepper & J.R.I.Wood var. *flavoferruginea*

F.T.E.A. Malv.: 16 (2009).
Syn: *Pavonia glechomifolia* (A.Rich.) Garcke – F.E.E. 2(2): 226 (1995).
Syn: *Pavonia patens* sensu Andrews pro parte, non (Andrews) Chiov. – F.P.S. 2: 34 (1952).
Perennial herb or subshrub. Wooded grassland, bushland.
Distr: RS, DAR, KOR; Eritrea, Ethiopia, Somalia, Uganda, Kenya, Tanzania; Arabia, Pakistan, India.
RS: Jackson 2953 4/1953.

Pavonia kotschyi Hochst. ex Webb

F.P.S. 2: 36 (1952); F.E.E. 2(2): 234 (1995); F.T.E.A. Malv.: 30 (2009).
Perennial herb or subshrub. Dry bushland & woodland.
Distr: RS, DAR, KOR; Mauritania & Senegal to Somalia & Kenya; Egypt, Arabia, Pakistan, India.
DAR: Aglen 10 8/1944.

Pavonia propinqua Garcke

F.E.E. 2(2): 230 (1995); F.T.E.A. Malv.: 20 (2009).
Shrub. Bushland & wooded grassland, rocky hillslopes.
Distr: EQU; Ethiopia, Somalia, Uganda, Kenya, Tanzania; Pakistan, India.
EQU: Myers 14081 (fl) 9/1941.

Pavonia schimperiana Hochst. ex A.Rich.

F.E.E. 2(2): 228 (1995); Biol. Skr. 51: 151 (1998); F.T.E.A. Malv.: 13 (2009).
Syn: *Pavonia urens* Cav. var. *glabrescens* (Ulbr.) Brenan – F.P.S. 2: 36 (1952).
Shrubby perennial herb. Wet grassland, swamp margins, forest margins.
Distr: EQU; Guinea to Ethiopia, S to D.R. Congo & Tanzania.
EQU: Myers 9753 10/1938.

Pavonia senegalensis (Cav.) Leistner

F.T.E.A. Malv.: 15 (2009).
Syn: *Pavonia hirsuta* Guill. & Perr. – F.P.S. 2: 36 (1952); F.J.M.: 102 (1976).

Perennial herb or shrub. Woodland, grassland, dry riverbeds, sometimes in damp areas.
Distr: DAR, KOR; Senegal to Chad, D.R. Congo & Tanzania, S to Namibia & South Africa.
KOR: Wickens 594 10/1962.

Pavonia triloba Guill. & Perr.

F.P.S. 2: 39 (1952); F.E.E. 2(2): 232 (1995); F.T.E.A. Malv.: 26 (2009).
Syn: *Pavonia zeylanica* sensu Andrews, non Cav. – F.P.S. 2: 37 (1952).
Annual or biennial herb. Woodland, bushland & grassland, including seasonally wet areas.
Distr: NS, RS, DAR, KOR; Mauritania & Senegal to Eritrea, S to Kenya; Egypt, Arabia, Pakistan, India.
KOR: Wickens 262 8/1962.

Pavonia urens Cav. var. *urens*

F.P.S. 2: 36 (1952); F.E.E. 2(2): 228 (1995); Biol. Skr. 51: 151 (1998); F.T.E.A. Malv.: 11 (2009).
Subshrub. Forest margins & secondary bushland.
Distr: EQU; Guinea to Ethiopia, S to South Africa; Madagascar.
EQU: Friis & Vollesen 812 12/1980.

Senra incana Cav.

F.P.S. 2: 39 (1952); F.E.E. 2(2): 216 (1995); F.T.E.A. Malv.: 89 (2009).
Annual or perennial herb. Wooded grassland, bushland.
Distr: RS, KOR; Eritrea, Ethiopia, Djibouti, Somalia, Kenya; Arabia, Socotra, Pakistan, India.
KOR: Kotschy 125 10/1839.

Sida acuta Burm.f.

F.P.S. 2: 41 (1952); F.E.E. 2(2): 252 (1995); F.T.E.A. Malv.: 155 (2009).
Shrubby perennial or annual herb. Grassland, bushland, forest margins.
Distr: EQU; pantropical.
EQU: von Ramm 112 3/1956.

Sida alba L.

F.P.S. 2: 41 (1952); F.J.M.: 103 (1976); F.E.E. 2(2): 251 (1995); Biol. Skr. 51: 152 (1998); F.T.E.A. Malv.: 154 (2009).
Perennial herb. Grassland, woodland, weed of disturbed ground.
Distr: NS, RS, DAR, KOR, CS, ES, BAG, UN, EQU; pantropical.
NS: Pettet 73 9/1962; **UN:** Sherif A.4015 9/1951.

Sida cordifolia L. subsp. *cordifolia*

F.E.E. 2(2): 251 (species only) (1995); F.T.E.A. Malv.: 149 (2009).
Annual or perennial herb or subshrub. Bushland, weed of disturbed ground.
Distr: KOR; scattered in tropical Africa where probably introduced; widespread elsewhere in the palaeotropics & subtropics.
KOR: Wickens 599 (fl fr) 10/1962.

Sida cordifolia L. subsp. *maculata* (Cav.) Marais

F.P.S. 2: 41 (species only) (1952); Biol. Skr. 51: 152 (species only) (1998); F.T.E.A. Malv.: 150 (2009).

Annual or perennial herb or subshrub. Grassland, bushland, woodland, weed of disturbed ground.
Distr: KOR, BAG, EQU; widespread in the tropics (see note).
KOR: Jackson 4327 9/1961; **BAG:** Schweinfurth 1372 4/1869.
Note: Verdcourt believes this to be the native subspecies in tropical Africa where it is common. Putative hybridisation with subsp. *cordifolia* is not infrequent. For example in South Sudan, *Muriel* S/89 from Gondokoro appears somewhat intermediate between the two; however, true subsp. *cordifolia* has not so far been recorded from South Sudan.

Sida javensis Cav.
F.E.E. 2(2): 251 (1995); Biol. Skr. 51: 152 (1998); F.T.E.A. Malv.: 146 (2009).
Syn: *Sida veronicifolia* sensu Andrews, non Lam. – F.P.S. 2: 41 (1952).
Perennial herb. Grassland, woodland, forest margins, weed of disturbed ground.
Distr: BAG, EQU; palaeotropical.
EQU: Friis & Vollesen 811 12/1980.

Sida ovata Forssk.
F.P.S. 2: 41 (1952); F.J.M.: 103 (1976); F.E.E. 2(2): 254 (1995); Biol. Skr. 51: 152 (1998); F.T.E.A. Malv.: 162 (2009).
Perennial herb or subshrub. Grassland, bushland, often in rocky areas, weed of disturbed ground.
Distr: RS, DAR, KOR, ES, UN, EQU; Mauritania & Senegal to Somalia, S to South Africa; Egypt, Arabia to India.
DAR: Wickens 2254 4/1964; **UN:** Simpson 7245 7/1929.

Sida rhombifolia L. var. maderensis (Lowe) Lowe
F.T.E.A. Malv.: 159 (2009).
Perennial herb. Disturbed ground, thickets.
Distr: CS; Eritrea, Ethiopia, Kenya, Tanzania, South Africa; widespread elsewhere in the tropics particularly in Asia.
CS: Figari s.n. (n.v.).

Sida rhombifolia L. var. petherickii Verdc.
F.P.S. 2: 41 (species only, pro parte) (1952); F.J.M.: 103 (species only, pro parte) (1976); F.T.E.A. Malv.: 160 (2009).
Syn: *Sida sp. 10* – F.E.E. 2(2): 252 (1995).
Perennial herb. Woodland, disturbed ground.
Distr: DAR, KOR, ES, BAG; W Africa (not clearly delimited) to Ethiopia, Uganda & Tanzania.
DAR: Wickens 2266 8/1964; **BAG:** Macintosh 13.K (fl fr) 1932.

Sida rhombifolia L. var. riparia Burtt Davy
F.P.S. 2: 41 (species only, pro parte) (1952); Biol. Skr. 51: 153 (species only) (1998); F.T.E.A. Malv.: 158 (2009).
Perennial herb. Grassland, woodland, forest margins.
Distr: EQU; widespread in tropical & southern Africa (not fully delimited).
EQU: Friis & Vollesen 714 12/1980.

Sida rhombifolia L. var. serratifolia (R.Wilczek & Steyaert) Verdc.
F.J.M.: 103 (species only, pro parte) (1976); F.T.E.A. Malv.: 158 (2009).

Syn: *Sida serratifolia* R.Wilczek & Steyaert – F.E.E. 2(2): 254 (1995).
Perennial herb. Grassland & open woodland.
Distr: DAR; Ethiopia to South Africa.
DAR: Wickens 1576 (fl fr) 5/1964.

Sida urens L.
F.P.S. 2: 41 (1952); F.J.M.: 103 (1976); F.E.E. 2(2): 249 (1995); F.T.E.A. Malv.: 148 (2009).
Perennial herb or subshrub. Wooded grassland, forest margins, weed of disturbed ground.
Distr: DAR, BAG, EQU; Cape Verde Is., Senegal to Eritrea, S to Angola & Zimbabwe; Arabia, Madagascar, tropical America, W Indies.
DAR: Wickens 1342 3/1964; **BAG:** Schweinfurth 2494 10/1869.

Sidastrum paniculatum (L.) Fryxell
F.T.E.A. Malv.: 164 (2009).
Syn: *Sida paniculata* L. – F.P.S. 2: 42 (1952); Biol. Skr. 51: 152 (1998).
Perennial herb. Wooded grassland, forest margins, disturbed ground.
Distr: EQU; C.A.R., D.R. Congo, Uganda, Cabinda; C & S America, W Indies.
EQU: Friis & Vollesen 536 11/1980.

Thespesia garckeana F.Hoffm. var. garckeana
F.P.S. 2: 42 (1952); F.T.E.A. Malv.: 104 (2009).
Syn: *Azanza garckeana* (F.Hoffm.) Exell & Hillc. – Bot. Exp. Sud.: 81 (1970); F.J.M.: 102 (1976); T.S.S.: 115 (1990).
Small tree or shrub. Wooded grassland.
Distr: DAR, KOR, BAG, UN, EQU; Nigeria, Kenya to South Africa.
KOR: Wickens 830 11/1962; **EQU:** Drar 1701b 4/1938 (n.v.).

Thespesia populnea (L.) Corrêa var. populnea
F.T.E.A. Malv.: 102 (2009).
Shrub or small tree. Saline sandy areas.
Distr: RS; pantropical.
RS: Sahni & Kamil 723 4/1967.

Urena lobata L. var. lobata
F.P.S. 2: 43 (1952); F.E.E. 2(2): 224 (1995); Biol. Skr. 51: 153 (1998); F.T.E.A. Malv.: 4 (2009).
Annual or perennial herb or subshrub. Wide variety of habitats, often in disturbed areas.
Distr: DAR, CS, BAG, UN, EQU; pantropical.
CS: Jackson 1648 1/1951; **EQU:** Andrews 721 4/1939.

Wissadula rostrata (Schumach. & Thonn.) Hook.f.
F.J.M.: 103 (1976); F.E.E. 2(2): 239 (1995); Biol. Skr. 51: 153 (1998); F.T.E.A. Malv.: 112 (2009).
Syn: *Wissadula amplissima* (L.) R.E.Fr. var. *rostrata* (Schumach. & Thonn.) R.E.Fr. – F.P.S. 2: 43 (1952).
Perennial herb or subshrub. Grassland, bushland & woodland, often riverine.
Distr: DAR, ES, UN, EQU; Cape Verde Is., Senegal to Eritrea, S to Angola & South Africa; Yemen.
DAR: Wickens 2830 2/1965; **EQU:** Myers 10080 11/1938.

Malvaceae: Grewioideae (Sparrmanniaceae)

H. Pickering & I. Darbyshire

Clappertonia ficifolia (Willd.) Decne.
F.P.S. 1: 215 (1950); T.S.S.: 102 (1990); F.T.E.A. Til.: 97 (2001).
Shrub. Swampy grassland, forest margins.
Distr: BAG, EQU; Senegal to Uganda, S to Mozambique.
EQU: Myers 7048 7/1937.

Corchorus aestuans L.
F.P.S. 1: 217 (1950); F.E.E. 2(2): 156 (1995); F.T.E.A. Til.: 108 (2001).
Annual herb. Grassland.
Distr: KOR, UN; pantropical.
KOR: Kotschy 397 11/1839; **UN:** J.M. Lock 81/159 8/1981.

Corchorus depressus (L.) Stocks
F.P.S. 1: 217 (1950); F.E.E. 2(2): 156 (1995); F.J.U. 3: 242 (1999).
Annual or perennial herb. Open bushland.
Distr: NS, RS, KOR, CS; Cape Verde Is., Mauritania to Somalia; N Africa (Sahara), Arabia, Afghanistan, India.
RS: Carter 1945 11/1987.

Corchorus fascicularis Lam.
F.P.S. 1: 217 (1950); F.E.E. 2(2): 155 (1995); F.T.E.A. Til.: 107 (2001).
Annual herb. Seasonally wet grassland.
Distr: KOR, CS, ES, BAG; tropical Africa; Arabia, India, Australia.
CS: Lea 134 12/1952; **BAG:** Schweinfurth 2469 10/1869.

Corchorus olitorius L.
F.P.S. 1: 217 (1950); F.J.M.: 100 (1976); F.E.E. 2(2): 155 (1995); F.T.E.A. Til.: 110 (2001).
Annual herb. Swampy grassland, weed of cultivation.
Distr: NS, RS, DAR, KOR, CS, ES, BAG, UN, EQU; pantropical.
DAR: Wickens 1795 6/1964; **EQU:** von Ramm 259 5/1956.

Corchorus pseudocapsularis Schweinf.
F.E.E. 2(2): 156 (1995); F.T.E.A. Til.: 109 (2001).
Syn: *Corchorus hochstetteri* Milne-Redh. – F.P.S. 1: 216 (1950).
Annual herb. Grassland, woodland.
Distr: KOR, ES; Ethiopia to Zambia.
ES: Beshir 129 9/1951.

Corchorus tridens L.
F.P.S. 1: 217 (1950); F.E.E. 2(2): 156 (1995); Biol. Skr. 51: 138 (1998); F.T.E.A. Til.: 113 (2001).
Annual herb. Woodland, bushland, damp ground.
Distr: NS, RS, DAR, KOR, CS, BAG; Senegal to Eritrea, S to Namibia & South Africa; Egypt, Arabia, tropical Asia.
RS: Schweinfurth 663 8/1868; **BAG:** Schweinfurth 2317 8/1869.

Corchorus trilocularis L.
F.P.S. 1: 217 (1950); F.J.M.: 100 (1976); F.E.E. 2(2): 155

(1995); Biol. Skr. 51: 138 (1998); F.T.E.A. Til.: 111 (2001).
Annual herb. Grassland, woodland, damp ground, often weedy.
Distr: NS, RS, DAR, CS, ES, BAG, UN, EQU; pantropical.
ES: Beshir 26 8/1951; **BAG:** Simpson 7112 6/1929.

Glyphaea brevis (Spreng.) Monach.
T.S.S.: 103 (1990); Biol. Skr. 51: 139 (1998); F.T.E.A. Til.: 99 (2001).
Syn: *Glyphaea lateriflora* (G.Don) Hutch. & Dalziel – F.P.S. 1: 218 (1950).
Shrub. Secondary bushland, forest regrowth.
Distr: EQU; Guinea Bissau to Uganda, S to Angola & Tanzania.
EQU: Andrews 1352 5/1939.

Grewia bicolor Juss.
F.P.S. 1: 222 (1950); F.J.M.: 100 (1976); T.S.S.: 105 (1990); F.E.E. 2(2): 146 (1995); Biol. Skr. 51: 139 (1998); F.T.E.A. Til.: 42 (2001).
Small tree. Woodland, bushland.
Distr: RS, DAR, KOR, CS, EQU; Senegal to Somalia, S to Zimbabwe; Arabia, India.
KOR: Wickens 426 9/1962; **EQU:** Friis & Vollesen 1188 3/1982.

Grewia erythraea Schweinf.
F.P.S. 1: 223 (1950); T.S.S.: 105 (1990); F.E.E. 2(2): 152 (1995); F.T.E.A. Til.: 18 (2001).
Shrub. *Acacia-Commiphora* bushland.
Distr: RS, KOR; Eritrea, Ethiopia, Somalia, Kenya; Egypt, Arabia, Afghanistan.
RS: Jackson 4352 9/1961.

Grewia ferruginea A.Rich.
F.P.S. 1: 223 (1950); T.S.S.: 105 (1990); F.E.E. 2(2): 150 (1995); F.T.E.A. Til.: 22 (2001).
Shrub or small tree. Riverine forest, woodland.
Distr: RS, KOR, CS; Eritrea, Ethiopia, Kenya.
RS: Broun 1168 5/1905.
Note: we have only seen material from Red Sea region; the other regional records are from El Amin in T.S.S.

Grewia flavescens Juss.
F.P.S. 1: 223 (1950); F.J.M.: 100 (1976); T.S.S.: 106 (1990); F.E.E. 2(2): 149 (1995); F.T.E.A. Til.: 33 (2001).
Shrub. Woodland.
Distr: RS, DAR, KOR, CS, ES, UN, EQU; Mauritania & Senegal to Eritrea, S to Namibia & South Africa; Arabia, India.
DAR: Francis 77 12/1957; **UN:** Sherif A.3904 7/1951.

Grewia mollis Juss.
F.P.S. 1: 220 (1950); F.J.M.: 100 (1976); T.S.S.: 106 (1990); F.E.E. 2(2): 146 (1995); Biol. Skr. 51: 139 (1998); F.T.E.A. Til.: 44 (2001).
Shrub or tree. Woodland & wooded grassland.
Distr: NS, RS, DAR, KOR, CS, ES, BAG, UN, EQU; Senegal to Somalia, S to Zambia; Arabia.
DAR: Wickens 1556 5/1964; **BAG:** Myers 6290 2/1937.
Note: El Amin in T.S.S.: 105 records *G. carpinifolia* Juss. from Red Sea region; this is clearly in error but the correct identity is unclear. *G. carpinifolia* has previously been misapplied to specimens of *G. mollis*.

Grewia oligoneura Sprague

F.P.S. 1: 223 (1950); F.W.T.A. 1(2): 303 (1958).
Shrub or tree. Swamp forest.
Distr: EQU; Ivory Coast to C.A.R. & D.R. Congo.
EQU: Andrews 1548 5/1939.

Grewia seretii De Wild.

F.T.E.A. Til.: 59 (2001).
Syn: *Grewia floribunda* sensu Andrews, non Mast. –
F.P.S. 1: 223 (1950).
Shrub tree or woody climber. Forest edges, riverine forest.
Distr: EQU; Cameroon, Gabon, Congo-Brazzaville, D.R. Congo, Uganda.
EQU: Myers 11373 5/1939.

Grewia stolzii Ulbr.

Biol. Skr. 51: 140 (1998); F.T.E.A. Til.: 22 (2001).
Small tree or shrub, often scandent. Riverine forest, dense bushland.
Distr: EQU; Uganda & Kenya, S to Zimbabwe & Mozambique.
EQU: Friis & Vollesen 343 11/1980.

Grewia tembensis Fresen.

F.P.S. 1: 224 (1950); F.E.E. 2(2): 150 (1995).
Shrub. Woodland on rocky ground.
Distr: ?RS; Eritrea, Ethiopia, Somalia; Egypt, Arabia.
Note: the material on which Andrews based his record from the Red Sea Hills has not been traced; this species is, however, likely to occur there. Quézel's (1969: 109) record of *G. populifolia* Vahl (= *G. tembensis*) as frequent in Darfur is considered to be in error.

Grewia tenax (Forssk.) Fiori

F.P.S. 1: 222 (1950); F.J.M.: 100 (1976); T.S.S.: 106 (1990); F.E.E. 2(2): 152 (1995); Biol. Skr. 51: 140 (1998); F.T.E.A. Til.: 14 (2001).
Shrub. Dry wooded grassland & bushland.
Distr: RS, DAR, KOR, CS, ES, BAG, UN, EQU; Mauritania & Senegal to Somalia, S to South Africa; N Africa, Arabia to India & Sri Lanka.
KOR: Wickens 86 7/1962; **EQU:** Myers 8612 2/1938.

Grewia trichocarpa Hochst. ex A.Rich.

F.E.E. 2(2): 148 (1995); Biol. Skr. 51: 140 (1998); F.T.E.A. Til.: 45 (2001).
Shrub or small tree. Riverine forest, wooded grassland.
Distr: RS, EQU; Eritrea & Somalia, S to Tanzania; Arabia.
RS: Aylmer 646 3/1936; **EQU:** Myers 11216 4/1939.

Grewia velutina (Forssk.) Lam.

F.E.E. 2(2): 148 (1995); F.T.E.A. Til.: 44 (2001).
Shrub or small tree. Rocky hillsides, open woodland.
Distr: RS; Eritrea, Ethiopia, Somalia, Uganda, Kenya; Yemen.
RS: Robbie 14.

Grewia villosa Willd.

F.P.S. 1: 220 (1950); F.J.M.: 100 (1976); T.S.S.: 107 (1990); F.E.E. 2(2): 153 (1995); Biol. Skr. 51: 141 (1998); F.T.E.A. Til.: 30 (2001).
Shrub. Dry rocky hillsides, woodland.
Distr: RS, DAR, KOR, CS, ES, BAG, UN, EQU; Senegal

& Mauritania to Somalia, S to Namibia & South Africa; Arabia, India.
KOR: Wickens 239 8/1962; **EQU:** Friis & Vollesen 1200 3/1982.

Grewia sp. cf. *similis* K.Schum.

Biol. Skr. 51: 141 (1998).
Habit & habitat not known.
Distr: EQU; (for *G. similis*) Ethiopia to D.R. Congo & Tanzania.
EQU: Shigeta 183 1979 (n.v.).
Note: the single specimen from the Imatong Mts was not traced by Friis & Vollesen (1998) or by us and so the identification remains uncertain.

Sparrmannia ricinocarpa (Eckl. & Zeyh.) Kuntze var. *ricinocarpa*

F.P.S. 1: 224 (1950); F.E.E. 2(2): 158 (1995); Biol. Skr. 51: 141 (1998); F.T.E.A. Til.: 93 (2001).
Syn: *Urena ricinocarpa* Eckl. & Zeyh. – T.S.S.: 117 (1990).
Shrub or perennial herb. Forest clearings, riverine thicket.
Distr: EQU; Cameroon, Eritrea to Angola & South Africa; Madagascar, Réunion.
EQU: Myers 11578 7/1939.

Triumfetta annua L. fa. *annua*

F.P.S. 1: 227 (1950); F.J.M.: 100 (1976); F.E.E. 2(2): 160 (1995); Biol. Skr. 51: 144 (1998); F.T.E.A. Til.: 79 (2001).
Annual herb. Forest margins, woodland, weed of cultivation.
Distr: RS, DAR, EQU; Eritrea to Angola, Namibia & South Africa; Madagascar, tropical Asia to China.
DAR: Wickens 2556 9/1964; **EQU:** Friis & Vollesen 301 11/1980.

Triumfetta brachyceras K.Schum.

F.P.S. 1: 227 (1950); F.E.E. 2(2): 161 (1995); Biol. Skr. 51: 142 (1998); F.T.E.A. Til.: 89 (2001).
Shrub. Forest margins.
Distr: EQU; D.R. Congo to Ethiopia, S to Tanzania.
EQU: Myers 8718 3/1938.

Triumfetta cordifolia A.Rich. var. *tomentosa* Sprague

F.P.S. 1: 227 (1950); F.E.E. 2(2): 161 (species only) (1995); Biol. Skr. 51: 142 pro parte (1998); F.T.E.A. Til.: 87 (2001).
Shrub. Forest margins, woodland, disturbed ground.
Distr: EQU; Nigeria to Uganda, S to Zimbabwe.
EQU: Kosper 113 (fl) 10/1982.

Triumfetta flavescens Hochst. ex A.Rich.

F.P.S. 1: 227 (1950); F.E.E. 2(2): 164 (1995); F.T.E.A. Til.: 81 (2001).
Subshrub. Dry bushland, woodland.
Distr: NS, RS, ?KOR, CS, ?UN, ?EQU; Burkino Faso to Eritrea, S to Tanzania; Egypt, Arabia.
RS: Carter 1986 11/1987.
Note: Friis & Vollesen (1998) record this species from the Boma Plateau in Upper Nile but we have not seen the material on which this record is based. Andrews also recorded it from Kordofan and the Imatong Mts but it is possible that these records were based on misidentifications.

Triumfetta lepidota K.Schum.

F.P.S. 1: 226 (1950); F.W.T.A. 1(2): 309 (1958).
Shrub. Sandy grassland.
Distr: EQU; Burkina Faso to C.A.R.
EQU: Schweinfurth 4012 6/1870.

Triumfetta pentandra A.Rich.

F.P.S. 1: 227 (1950); F.J.M.: 100 (1976); F.E.E. 2(2): 162 (1995); F.T.E.A. Til.: 85 (2001).
Annual herb. Bushland, wooded grassland, disturbed areas.
Distr: DAR, KOR, UN, EQU; Cape Verde Is., Senegal to Eritrea, S to Namibia & South Africa; Arabia, India, Taiwan.
DAR: Wickens 2356 9/1964; **UN:** Sherif A.4002 9/1951.

Triumfetta pilosa Roth

F.E.E. 2(2): 162 (1995); Biol. Skr. 51: 142 (1998); F.T.E.A. Til.: 91 (2001).
Shrub. Montane woodland, forest margins, riverbanks.
Distr: EQU; pantropical.
EQU: Kielland-Lund 249 12/1983 (n.v.).

Triumfetta rhomboidea Jacq.

F.P.S. 1: 227 (1950); F.J.M.: 101 (1976); F.E.E. 2(2): 162 (1995); Biol. Skr. 51: 142 (1998); F.T.E.A. Til.: 83 (2001).
Perennial herb. Forest margins, secondary scrub, weed of cultivation.
Distr: DAR, BAG, UN, EQU; pantropical.
DAR: Wickens 1418 4/1964; **EQU:** Simpson 7587 11/1930.

Triumfetta setulosa Mast.

F.E.E. 2(2): 160 (1995); Biol. Skr. 51: 143 (1998); F.T.E.A. Til.: 76 (2001).
Syn: *Triumfetta buettneriacea* K.Schum. – F.P.S. 1: 226 (1950).
Syn: *Triumfetta micrantha* K.Schum. – F.P.S. 1: 226 (1950).
Annual herb or subshrub. Wooded grassland, disturbed ground.
Distr: BAG, EQU; Senegal to Ethiopia, S to Angola & Zimbabwe.
EQU: Friis & Vollesen 528 11/1980.

Triumfetta tomentosa Bojer

F.P.S. 1: 226 (1950); F.E.E. 2(2): 161 (1995); F.T.E.A. Til.: 90 (2001).
Syn: *Triumfetta cordifolia* var. *tomentosa* sensu Friis & Vollesen pro parte, non Sprague – Biol. Skr. 51: 142 (1998).
Shrub. Forest & forest margins, bushland, disturbed ground.
Distr: EQU; pantropical.
EQU: Friis & Vollesen 810 12/1980.

Triumfetta trichocarpa Hochst. ex A.Rich.

F.J.M.: 101 (1976); F.E.E. 2(2): 160 (1995); F.T.E.A. Til.: 80 (2001).
Annual herb. Wooded grassland, weed of cultivation.
Distr: DAR; C.A.R. to Ethiopia, S to Zambia & Mozambique.
DAR: Wickens 2327 9/1964.

Malvaceae: Brownlowioideae (Brownlowiaceae)

H. Pickering

Christiana africana DC.

F.P.S. 1: 215 (1950); T.S.S.: 102 (1990); F.T.E.A. Til.: 4 (2001).
Tree. Forest.
Distr: BAG, EQU; Senegal to Kenya, S to Cabinda & Tanzania; Madagascar, Comoros, S America.
EQU: Myers 11371 5/1939.

Malvaceae: Dombeyoideae (Pentapetaceae)

H. Pickering & I. Darbyshire

Dombeya buettneri K.Schum.

F.E.E. 2(2): 168 (1995); Biol. Skr. 51: 144 (1998); F.T.E.A. Stercul.: 63 (2007).
Syn: *Dombeya bagshawei* Baker f. – F.P.S. 2: 4 (1952); T.S.S.: 108 (1990).
Shrub or small tree. Wooded grassland, secondary forest, rocky hillslopes.
Distr: EQU; Guinea to Ethiopia, S to Zambia.
EQU: Wyld 351 11/1937.

Dombeya burgessiae Gerrard ex Harv.

T.S.S.: 108 (1990); Biol. Skr. 51: 144 (1998); F.T.E.A. Stercul.: 64 (2007).
Syn: *Dombeya mastersii* Hook.f. – F.P.S. 2: 4 (1952).
Small tree or shrub. Forest margins, bushland.
Distr: EQU; Uganda to South Africa.
EQU: Jackson 234 11/1947.

Dombeya quinqueseta (Delile) Exell

F.P.S. 2: 2 (1952); F.J.M.: 101 (1976); T.S.S.: 109 (1990); F.E.E. 2(2): 170 (1995); Biol. Skr. 51: 144 (1998); F.T.E.A. Stercul.: 68 (2007).
Syn: *Dombeya multiflora* (Endl.) Planch. – F.P.S. 2: 4 (1952).
Syn: *Dombeya mukole* sensu auctt., non Sprague – F.P.S. 2: 3 (1952); T.S.S.: 109 (1990).
Small tree or shrub. Wooded grassland.
Distr: DAR, KOR, CS, BAG, EQU; Senegal to Eritrea, S to Kenya.
DAR: Kamil 1082 5/1968; **EQU:** Myers 10931 4/1939.

Dombeya torrida (J.F.Gmel.) Bamps subsp. torrida

F.E.E. 2(2): 168 (1995); Biol. Skr. 51: 145 (1998); F.T.E.A. Stercul.: 68 (2007).
Syn: *Dombeya elliotii* K.Schum. & Engl. – F.P.S. 2: 4 (1952); T.S.S.: 109 (1990).
Syn: *Dombeya goetzenii* K.Schum. – F.P.S. 2: 5 (1952); T.S.S.: 109 (1990).
Tree. Montane forest & bushland.
Distr: EQU; Eritrea to D.R. Congo & Tanzania; Arabia.
EQU: Myers 1157 4/1939.

Melhania albiflora (Hiern) Exell & Mendonça ex Hill & Salisb.

Trop. Afr. Fl. Pl. 1: 694 (2003).
Syn: *Melhania steudneri* Schweinf. – F.P.S. 2: 6 (1952); F.E.E. 2(2): 175 (1995).
Shrub. Dry grassland, sandy ground.
Distr: RS; Eritrea, Ethiopia, Somalia, Angola; Egypt, Arabia, Socotra, India, Australia.
RS: Jackson 2835 4/1953.
Note: we here follow the African Plants Database in treating *M. incana* B.Heyne ex Wight & Arn (including *M. steudneri*) as a synonym of the SW African *M. albiflora*.

Melhania denhamii R.Br.

F.P.S. 2: 6 (1952); F.E.E. 2(2): 176 (1995).
Perennial herb or shrub. Bushland.
Distr: RS, DAR, KOR; Mauritania & Senegal to Somalia; Algeria, Egypt, Arabia, India.
KOR: Jackson 4354 4/1961.

Melhania ovata (Cav.) Spreng.

F. Darfur Nord-Occ. & J. Gourgeil:109 (1969); F.E.E. 2(2): 173 (1995); F.T.E.A. Stercul.: 80 (2007).
Perennial herb or subshrub. Wooded grassland & bushland.
Distr: ?DAR; Cape Verde Is., Mauritania, Senegal, Eritrea to Somalia, S to Tanzania; Arabia, India, Australia.
Note: the record of this species is based on Quézel (1969) who recorded it from several areas in NW Darfur; we have not seen the material on which this is based. In F.T.E.A. it is also noted as possibly occurring in South Sudan.

Melhania phillipsiae Baker f.

F.E.E. 2(2): 176 (1995).
Syn: *Melhania grandibracteata* (K.Schum.) K.Schum. –F. Darfur Nord-Occ. & J. Gourgeil:109 (1969).
Subshrub. Rocky hillslopes, bushland.
Distr: RS, DAR; Niger to Somalia, S to Kenya; Arabia.
RS: Shabetai F.1440.
Note: the collection cited is from Gebel Elba in the Hala'ib Triangle but the African Plants Database also records this species from elsewhere in Red Sea region.

Melhania velutina Forssk.

F.E.E. 2(2): 174 (1995); Biol. Skr. 51: 145 (1998); F.T.E.A. Stercul.: 75 (2007).
Syn: *Melhania ferruginea* A.Rich. – F.P.S. 2: 5 (1952).
Perennial herb. Wooded grassland, bushland.
Distr: EQU; Eritrea & Somalia, S to Angola & Tanzania; Arabia.
EQU: Russell 51951 8/1921.

Malvaceae: Byttnerioideae (Byttneriaceae)

H. Pickering & I. Darbyshire

Byttneria catalpifolia Jacq. subsp. *africana* (Mast.) Exell & Mendonça

F.E.E. 2(2): 166 (1995); T.S.S.: 107 (1990); Biol. Skr. 51: 143 (1998); F.T.E.A. Stercul.: 96 (2007).
Syn: *Byttneria africana* Mast. – F.P.S. 2: 2 (1952).
Climbing shrub. Forest.

Distr: EQU; Ghana to Ethiopia, S to Angola & D.R. Congo.
EQU: Friis & Vollesen 543 11/1980.

Hermannia modesta (Ehrenb.) Mast.

F.P.S. 2: 5 (1952); F.E.E. 2(2): 181 (1995); F.T.E.A. Stercul.: 108 (2007).
Syn: *Hermannia arabica* (Ehrenb.) Planch. – F. Darfur Nord-Occ. & J. Gourgeil: 109 (1969).
Annual or short-lived perennial herb. Semi-desert scrub, dry woodland.
Distr: NS, RS, DAR; Chad, Eritrea, Somalia, Kenya, Angola to Zimbabwe; Arabia.
NS: Kotschy 71 9/1839.

Hermannia quartiniana A.Rich.

F.E.E. 2(2): 181 (1995).
Syn: *Hermannia abyssinica* (Hochst. ex Harv.) K.Schum. – F. Darfur Nord-Occ. & J. Gourgeil: 109 (1969).
Perennial herb. Rocky hillslopes, disturbed ground.
Distr: DAR; Chad, Eritrea, Ethiopia, Angola & Zambia to South Africa.
DAR: Miehe 688 (fl fr) 9/1982.
Note: the single Sudanese specimen seen is rather depauperate and lacks stem bases so that the habit cannot be confirmed, but it looks to match this species. Quézel also records this species from Jebel Gourgeil.

Hermannia tigreensis Hochst. ex A.Rich.

F.P.S. 2: 5 (1952); F.J.M.: 101 (1976); F.E.E. 2(2): 181 (1995); F.T.E.A. Stercul.: 108 (2007).
Annual herb or short-lived perennial. Open woodland.
Distr: DAR, KOR, CS, UN; Senegal to Eritrea, S to Namibia & Zimbabwe.
DAR: Wickens 2183 8/1964; UN: J.M. Lock 82/43 8/1982.

Leptonychia chrysocarpa K.Schum.

F.P.S. 2: 5 (1952); T.S.S.: 111 (1990); Trop. Afr. Fl. Pl. 1: 692 (2003).
Shrub. Riverine forest, secondary forest.
Distr: BAG, EQU; D.R. Congo.
IUCN: RD
EQU: Schweinfurth 3089 (fr) 2/1870.
Note: the single collection seen is from the Sudan – D.R. Congo border.

Leptonychia cf. *mildbraedii* Engl.

F.T.E.A. Stercul.: 100 (*L. mildbraedii*) (2007).
Shrub or small tree. Riverbanks.
Distr: EQU; (for species) D.R. Congo, Uganda, possibly also C.A.R. & Tanzania.
EQU: Andrews 1547 (fr) 5/1939.
Note: known from two collections made by Andrews in Equatoria, one sterile the other with immature fruits. The genus is in need of revision in Africa.

Melochia corchorifolia L.

F.P.S. 2: 6 (1952); F.E.E. 2(2): 178 (1995); Biol. Skr. 51: 146 (1998); F.T.E.A. Stercul.: 123 (2007).
Annual or perennial herb. Roadsides, waste ground, rocky & sandy ground.
Distr: KOR, BAG, UN, EQU; palaeotropical, introduced in the neotropics.
KOR: Pfund 284 7/1875; UN: Sherif A.2857 10/1951.

Melochia melissifolia Benth. var. *mollis* K.Schum.

Biol. Skr. 51: 146 (species only) (1998); F.T.E.A. Stercul.: 125 (2007).
Syn: *Melochia mollis* (K.Schum.) Hutch. & Dalziel – F.P.S. 2: 6 (1952).
Shrub. Swamps.
Distr: EQU; Senegal to Uganda, S to Angola & Botswana; Madagascar, Mascarenes.
EQU: Andrews 1580 5/1939.

Waltheria indica L.

F.P.S. 2: 8 (1952); F.J.M.: 101 (1976); F.E.E. 2(2): 178 (1995); Biol. Skr. 51: 146 (1998); F.T.E.A. Stercul.: 127 (2007).
Perennial herb or subshrub. Weed in grassland, bushland & woodland.
Distr: DAR, KOR, UN, EQU; pantropical.
KOR: Wickens 706 10/1962; **EQU:** Andrews 553 4/1939.

Malvaceae: Sterculioideae (Sterculiaceae)

H. Pickering

Cola gigantea A.Chev.

T.S.S.: 107 (1990); Biol. Skr. 51: 143 (1998); F.T.E.A. Stercul.: 24 (2007).
Syn: *Cola cordifolia* sensu Andrews, non (Cav.) R.Br. – F.P.S. 2: 2 (1952).
Tree. Forest.
Distr: BAG, EQU; Senegal to D.R. Congo & Uganda.
EQU: Myers 6350 3/1937.

Sterculia africana (Lour.) Fiori

F.P.S. 2: 8 (1952); T.S.S.: 111 (1990); F.E.E. 2(2): 184 (1995); F.T.E.A. Stercul.: 18 (2007).
Tree. Dry bushland & woodland, often on rocky hillslopes.
Distr: RS; Eritrea & Somalia, S to Namibia, Botswana & Mozambique; Egypt.
RS: Kennedy-Cooke 162 11/1934.

Sterculia cinerea A.Rich.

F.E.E. 2(2): 184 (1995).
Syn: *Sterculia setigera* sensu auctt., non Delile – F.P.S. 2: 7 (1952), pro parte; T.S.S.: 111 (1990), pro parte.
Tree. Wooded grassland.
Distr: KOR, CS, ES; Ethiopia.
IUCN: RD
KOR: Wickens 59 (fl) 7/1962.

Sterculia setigera Delile

F.P.S. 2: 7 pro parte (1952); F.J.M.: 101 (1976); T.S.S.: 111 (1990) pro parte; F.E.E. 2(2): 183 (1995); Biol. Skr. 51: 146 (1998); F.T.E.A. Stercul.: 14 (2007).
Tree. Wooded grassland.
Distr: DAR, BAG, EQU; Senegal to Eritrea, S to Tanzania, Angola.
DAR: Wickens 3615 6/1977; **EQU:** Friis & Vollesen 1187 3/1982.

Sterculia stenocarpa H.Winkl.

F.E.E. 2(2): 184 (1995); F.T.E.A. Stercul.: 16 (2007).
Tree. Bushland on rocky hillslopes.

Distr: EQU; Ethiopia, Somalia, Uganda, Kenya, Tanzania.
EQU: Padwa 239 5/1953.

Thymelaeaceae

I. Darbyshire

Craterosiphon aff. *beniense* Domke

F.T.E.A. Thymel.: 8 (*C. beniense*) (1978).
Shrub or woody climber. Forest & forest margins.
Distr: EQU; (for *C. beniense*) Cameroon, D.R. Congo, Uganda.
EQU: Sillitoe 442 (fl) 1919.
Note: the determination is courtesy of Zachary Rogers (MO) who notes that the specimen is rather poor; more material is required for confirmation. *C. beniense* is a rather scarce species but would not be unexpected in South Sudan.

Gnidia apiculata (Oliv.) Gilg

F.P.S. 1: 150 (1950); F.T.E.A. Thymel.: 22 (1978).
Subshrub. Grassland, dry hillslopes.
Distr: EQU; Cameroon & Gabon to Kenya & Tanzania.
EQU: Schweinfurth 3901 (fr) 6/1870.

Gnidia chrysantha Gilg

F.P.S. 1: 148 (1950); F.T.E.A. Thymel.: 26 (1978); F.E.E. 2(1): 431 (2000).
Suffrutex. Grassland including seasonally wet areas, woodland.
Distr: CS, BAG; Guinea, Nigeria, Cameroon, Ethiopia to Angola & Zimbabwe.
CS: Cienkowsky s.n. (n.v.); **BAG:** Schweinfurth III:113 1/1871 (n.v.).
Note: we have seen no extant material from Sudan or South Sudan – the records are based on type material from the Berlin herbarium, destroyed in WWII.

Gnidia glauca (Fresen.) Gilg

F.P.S. 1: 150 (1950); F.T.E.A. Thymel.: 32 (1978); Biol. Skr. 51: 110 (1998); F.E.E. 2(1): 433 (2000).
Syn: *Lasiosiphon glaucus* Fresen. – T.S.S.: 51 (1990).
Shrub or small tree. Montane grassland, woodland & scrub, upper forest margins.
Distr: EQU; Cameroon, Ethiopia to D.R. Congo & Zambia.
EQU: Howard I.M.84 (fl) 3/1976.

Gnidia involucrata Steud. ex A.Rich.

F.T.E.A. Thymel. 25 (1978); Biol. Skr. 51: 110 (1998); F.E.E. 2(1): 430 (2000).
Syn: *Gnidia macrorrhiza* Gilg – F.P.S. 1: 148 (1950).
Syn: *Gnidia schweinfurthii* Gilg – F.P.S. 1: 148 (1950).
Perennial herb or subshrub. Montane grassland & woodland with periodic burning.
Distr: BAG, EQU; Nigeria to Eritrea, S to Angola, Zimbabwe & Mozambique.
EQU: Myers 11139 (fl) 4/1939.

Gnidia kraussiana Meisn.

F.P.S. 1: 150 (1950); F.T.E.A. Thymel.: 29 (1978); Biol. Skr. 51: 110 (1998).
Syn: *Lasiosiphon kraussianus* (Meisn.) Burtt Davy – F.J.M.: 95 (1976).
Perennial herb. Grassland & woodland with periodic

burning, rock crevices.
Distr: DAR, KOR, BAG; Senegal to Kenya, S to Namibia & South Africa.
DAR: Wickens 2664 (fl) 9/1964; **BAG:** Turner 257 (fl).

Gnidia lamprantha Gilg
F.P.S. 1: 150 (1950); F.T.E.A. Thymel.: 31 (1978); Biol. Skr. 51: 111 (1998); F.E.E. 2(1): 435 (2000).
Shrub or small tree. Montane wooded grassland & bushland.
Distr: EQU; Ethiopia, Uganda, Kenya & Tanzania.
EQU: Thomas 1751 (fl) 12/1935.
Note: a very local species but probably not threatened globally.

Gnidia subcordata Meisn.
F.T.E.A. Thymel.: 21 (1978); Biol. Skr. 51: 111 (1998).
Syn: *Englerodaphne leiosiphon* Gilg – F.P.S. 1: 148 (1950).
Shrub. Wooded grassland.
Distr: EQU; Kenya, Tanzania, South Africa.
EQU: Myers 11048 (fl) 4/1939.
Note: a scarce species with a disjunct distribution.

Peddiea fischeri Engl.
F.P.S. 1: 150 (1950); F.T.E.A. Thymel.: 12 (1978); T.S.S.: 51 (1990); Biol. Skr. 51: 111 (1998).
Shrub or small tree. Forest.
Distr: EQU; Guinea, Cameroon to Kenya, S to Angola & Zambia.
EQU: Friis & Vollesen 328 (fl) 11/1980.

Bixaceae

I. Darbyshire

Bixa orellana L.
F.T.E.A. Bix.: 1 (1975); T.S.S.: 55 (1990); F.E.E. 2(1): 441 (2000).
Tree or shrub. Cultivated as an ornamental, sometimes escaping & persisting.
Distr: EQU; native to tropical America, widely introduced elsewhere in the tropics.
EQU: Andrews 1584 (fl fr) 5/1939.

Cochlospermum tinctorium Perr. ex A.Rich.
F.P.S. 1: 155 (1950); F.T.E.A. Cochlosperm.: 1 (1975); F.J.M.: 96 (1976).
Suffruticose perennial herb. Fire-prone grassland & wooded grassland.
Distr: DAR, ?KOR, BAG, EQU; Senegal to Uganda.
DAR: Wickens 3628 (fl) 6/1977; **EQU:** Andrews 448 (fl) 4/1939.
Note: the record for Kordofan is from Andrews.

Cistaceae

I. Darbyshire

Helianthemum lippii (L.) Dum.Cours.
F.J.U. 2: 151 (1999); F.Egypt 2: 122 (2000).
Subshrub. Wadis.
Distr: NS; Sahara & N Africa to Pakistan.
NS: Léonard 4928 (fl fr) 11/1968 (n.v.).
Note: the specimen cited has not been seen by us; the identification follows Léonard.

Dipterocarpaceae

I. Darbyshire

Monotes kerstingii Gilg
F.P.S. 1: 189 (1950); F.W.T.A. 1(1): 235 (1954); T.S.S.: 71 (1990); Biol. Skr. 51: 126 (1998).
Shrub or tree. Woodland & wooded grassland.
Distr: BAG, EQU; Guinea & Mali to C.A.R.
EQU: Myers 9052 (fl) 5/1938.

BRASSICALES

Moringaceae

I. Darbyshire

Moringa oleifera Lam.
F.P.S. 1: 54 (1950); F.T.E.A. Moring.: 3 (1986); F.E.E. 2(1): 157 (2000).
Small tree or shrub. Cultivated land, woodland & bushland – naturalised.
Distr: RS, KOR, CS, BAG, EQU; widely cultivated and naturalised in the tropics, originating from India.
CS: Lea 119 (fl) 11/1952; **EQU:** Myers 6364 (fl) 3/1937.

Moringa peregrina (Forssk.) Fiori
F.P.S. 1: 54 (1950); F.T.E.A. Moring.: 7 (1986); F.Egypt 1: 238 (1999); F.E.E. 2(1): 157 (2000).
Tree or shrub. Open woodland on rocky slopes, wadis.
Distr: RS, DAR, KOR, CS; Eritrea, Ethiopia, Somalia; Egypt to Palestine & Arabia.
RS: Jackson 3915 (fl) 4/1959.

Caricaceae

I. Darbyshire

[*Carica papaya* L.]
Note: this species, native to Mexico, is widely grown in Sudan and South Sudan for its edible fruits 'papaya' and may be encountered in areas of abandoned cultivation but is not considered to have naturalised.

Salvadoraceae

I. Darbyshire

Dobera glabra (Forssk.) Poir.
F.P.S. 2: 287 (1952); F.T.E.A. Salvador.: 4 (1968); F.E.E. 3: 353 (1989); T.S.S.: 285 (1990).
Shrub or tree. *Acacia* woodland & scrub, often on rocky hillslopes.
Distr: NS, RS, DAR, KOR, CS, ES, EQU; Eritrea, Ethiopia, Somalia, Uganda, Kenya; Arabia, India.
KOR: Wickens 946 (fl fr) 2/1963; **EQU:** Myers 10428 (fl) 2/1939.

Salvadora persica L. var. *persica*
F.P.S. 2: 287 (1952); F.T.E.A. Salvador.: 7 (1968); F.E.E. 3: 354 (1989); T.S.S.: 287 (1990).
Shrub or small tree. Dense thornbush, fringing woodland, open dry sandy plains, coastal scrub, sometimes dominant.

Distr: NS, RS, DAR, KOR, CS, ES, BAG, UN, EQU; Mauritania to Somalia, S to Namibia & Zimbabwe; N Africa, Arabia to India & Sri Lanka.
KOR: <u>Wickens 3055</u> (fr) 5/1969; **BAG:** <u>Simpson 7490</u> (fl) 2/1930.

Resedaceae

I. Darbyshire

Caylusea abyssinica (Fresen.) Fisch. & C.A.Mey.

F.P.S. 1: 66 (1950); F.T.E.A. Resed.: 3 (1958); Biol. Skr. 51: 83 (1998); F.E.E. 2(1): 170 (2000).
Annual or biennial herb. Rocky ground, dry grassland, weed of cultivation.
Distr: ?RS; Eritrea, Ethiopia, Uganda, Kenya, Tanzania.
RS: <u>Bent s.n.</u> 1896 (n.v.).
Note: the Bent specimen, on which the record in Andrews is based, has not been found and no other Sudan material has been seen. This species is also likely to occur in Equatoria, having been recorded from the Ugandan side of the Imatong Mts.

Caylusea hexagyna (Forssk.) M.L.Green

F.P.S. 1: 66 (1950); F.E.E. 2(1): 172 (2000).
Syn: *Caylusea canescens* A.St.-Hil. – Bot. Exp. Sud.: 98 (1970).
Annual or perennial herb. Desert, open sandy areas, weed of cultivation.
Distr: RS; Eritrea; Crete, N Africa to Palestine & Iraq.
RS: <u>Jackson 2748</u> (fl fr) 3/1953.

Ochradenus baccatus Delile

F.P.S. 1: 67 (1950); F.J.U. 3: 231 (1999); F.E.E. 2(1): 172 (2000).
Shrub. Dry stony or sandy ground, open woodland.
Distr: NS, RS, DAR; Chad, Eritrea, Ethiopia, Somalia; N Africa, Arabia to Pakistan.
NS: <u>Léonard 4905</u> (fl) 11/1968.

Oligomeris linifolia (Vahl ex Hornem.) J.F.Macbr.

F.P.S. 1: 67 (1950); F.Egypt 1: 237 (1999).
Annual herb. Deserts, saline soils.
Distr: ?NS, ?CS; Chad, Somalia, Namibia, Botswana; Macaronesia, N Africa to India, N & C America.
Note: Andrews lists this species from "Northern and Central Sudan" but we have not seen any Sudanese material – *Bromfield* 59, collected on 10/1/1851 from "tropical Nubia, bank of Nile between Deyr and Ibraheen", is believed to be from southern Egypt.

Reseda amblycarpa Fresen.

F.E.E. 2(1): 175 (2000).
Perennial herb or subshrub. Wadis, rocky slopes, dry open bushland.
Distr: RS; Eritrea, Djibouti, Somalia; Yemen.
RS: <u>Jackson 3959</u> (fl fr) 4/1959.

Reseda pruinosa Delile

F.P.S. 1: 69 (1950); F.Egypt 1: 236 (1999).
Annual herb. Deserts, rocky slopes.
Distr: RS; Sahara, Egypt, ?Arabia.
RS: <u>Hemming 88/1</u> (fl fr) 3/1988.

Capparaceae

H. Pickering & I. Darbyshire

Boscia angustifolia A.Rich. var. *angustifolia*

F.P.S. 1: 33 (1950); F.T.E.A. Capparid.: 55 (1964); T.S.S.: 21 (1990); F.E.E. 2(1): 114 (2000).
Small tree. Wooded grassland.
Distr: RS, DAR, KOR, CS, ES, EQU; Senegal to Somalia, S to Mozambique; Egypt, Arabia.
KOR: <u>Wickens 759</u> 11/1962; **EQU:** <u>Padwa 263</u> 6/1953.

Boscia coriacea Pax

F.T.E.A. Capparid.: 56 (1964); T.S.S.: 23 (1990); F.E.E. 2(1): 117 (2000).
Shrub. Bushland, semi-desert.
Distr: ES, UN, EQU; Ethiopia, Djibouti, Somalia, Uganda, Kenya, Tanzania.
ES: <u>Kennedy-Cooke 172</u> 9/1934; **EQU:** <u>Padwa 252</u> 6/1953.

Boscia salicifolia Oliv.

F.P.S. 1: 32 (1950); F.T.E.A. Capparid.: 52 (1964); F.J.M.: 84 (1976); T.S.S.: 23 (1990); Biol. Skr. 51: 76 (1998); F.E.E. 2(1): 116 (2000).
Small tree. Wooded grassland.
Distr: DAR, KOR, CS, UN, EQU; Senegal to Eritrea, S to Zimbabwe and Mozambique.
KOR: <u>Wickens 24</u> 6/1962; **UN:** <u>Myers 10448</u> 2/1939.

Boscia senegalensis (Pers.) Lam. ex Poir.

F.P.S. 1: 31 (1950); F.J.M.: 84 (1976); T.S.S.: 23 (1990); F.E.E. 2(1): 116 (2000).
Syn: *Boscia firma* Radlk. – F.P.S. 1: 33 (1950).
Shrub or small tree. Semi-desert scrub.
Distr: NS, RS, DAR, KOR, CS, ES, BAG, UN, EQU; Senegal to Ethiopia; N Africa.
KOR: <u>Wickens 3054</u> 5/1969; **BAG:** <u>Simpson 7625</u> 3/1930.

Cadaba farinosa Forssk. subsp. *farinosa*

F.P.S. 1: 35 (1950); F.T.E.A. Capparid.: 74 (1964); F.J.M.: 84 (1976); T.S.S.: 25 (1990); Biol. Skr. 51: 76 (1998); F.E.E. 2(1): 88 (2000).
Shrub. Grassland, bushland, semi-desert.
Distr: NS, RS, DAR, KOR, CS, ES, BAG, UN, EQU; Senegal to Somalia, S to Angola & Tanzania; Egypt, Arabia, Iran, Pakistan.
ES: <u>Beshir 54</u> 5/1951; **EQU:** <u>Friis & Vollesen 1163</u> 3/1982.

Cadaba farinosa Forssk. subsp. *adenotricha* (Gilg & Gilg-Ben.) R.A.Graham

F.T.E.A. Capparid.: 76 (1964).
Shrub. Grassland, bushland.
Distr: BAG; D.R. Congo, Rwanda, Uganda, Kenya, Tanzania.
BAG: <u>Myers 13914</u> 7/1941.

Cadaba gillettii R.A.Graham

F.T.E.A. Capparid.: 81 (1964); T.S.S.: 25 (1990); F.E.E. 2(1): 89 (2000).
Scrambling shrub. Dry woodland, riverine scrub, dry rocky terrain.
Distr: EQU; Ethiopia, Kenya.

IUCN: RD
EQU: Padwa 256 (fl) 6/1953.

Cadaba glandulosa Forssk.

F.P.S. 1: 34 (1950); F.T.E.A. Capparid.: 74 (1964); F.J.M.:
85 (1976); T.S.S.: 27 (1990); F.E.E. 2(1): 91 (2000).
Shrub. Bushland, grassland.
Distr: NS, RS, DAR, KOR, CS; Mauritania to Somalia, S to
Tanzania; Egypt, Arabia.
KOR: Wickens 865 11/1962.

Cadaba kassasii Chrtek

Novit. Bot. (Inst. Horto. Bot. Univ. Carol.): 3 (1971).
Shrub. Habitat not known.
Distr: CS; Sudan endemic.
IUCN: RD
CS: Kassas E.750 (fl fr) 12/1967 (n.v.).
Note: known only from the type collection which we
have not seen; this species requires further investigation.
Cadaba sp. 13 (= Baldrati 2146) of F.E.E., recorded from
Eritrea and N Ethiopia, is noted to match the description
of C. kassasii in leaf and fruit characters.

Cadaba longifolia DC.

F.P.S. 1: 37 (1950); F.T.E.A. Capparid.: 78 (1964); T.S.S.:
27 (1990); F.E.E. 2(1): 89 (2000).
Scrambling shrub or small tree. Dry bushland, rocky
hillslopes, dry riverbeds.
Distr: RS, ES; Eritrea, Ethiopia, Djibouti, Somalia, Kenya;
Arabia, Socotra.
RS: Jackson 3913 (fl) 4/1959.

Cadaba mirabilis Gilg

F.T.E.A. Capparid.: 73 (1964); T.S.S.: 27 (1990); F.E.E.
2(1): 91 (2000).
Shrub. Bushland, semi-desert, dry rocky hillslopes.
Distr: EQU; Ethiopia, Somalia, Kenya; Arabia.
EQU: Padwa 223 (fl fr) 5/1953.

Cadaba rotundifolia Forssk.

F.P.S. 1: 36 (1950); F.T.E.A. Capparid.: 75 (1964); T.S.S.:
29 (1990); F.E.E. 2(1): 88 (2000).
Shrub. Open woodland, bushland, semi-desert.
Distr: NS, RS, DAR, KOR, CS, ES; Eritrea, Ethiopia,
Djibouti, Somalia, Kenya; Egypt, Arabia.
ES: Sahni & Kamil 609 4/1967.

Capparis cartilagenea Decne.

F.P.S. 1: 39 (1950); F.T.E.A. Capparid.: 59 (1964); F.J.M.:
85 (1976); T.S.S.: 31 (1990); F.E.E. 2(1): 94 (2000).
Syn: Capparis galeata Fresen. – F.P.S. 1: 39 (1950).
Syn: Capparis spinosa sensu auctt., non L. – F.P.S. 1: 39
(1950); T.S.S.: 33 (1990).
Scrambling shrub. Rocky outcrops, bushland.
Distr: NS, RS, DAR; Chad to Somalia, S to Tanzania;
Libya, Egypt, Arabia to India.
RS: Schweinfurth II.108 9/1871.

Capparis decidua (Forssk.) Edgew.

F.P.S. 1: 39 (1950); Bot. Exp. Sud.: 25 (1970); F.J.M.: 85
(1976); T.S.S.: 31 (1990); F.E.E. 2(1): 94 (2000).
Large shrub or small tree. Wooded grassland, semi-desert.
Distr: NS, RS, DAR, KOR, CS, ES, BAG, EQU; Mauritania
to Somalia; N Africa, Arabia, Iran, Pakistan, India.

KOR: Wickens 787 11/1962; **EQU:** Drar 1142 4/1938
(n.v.).
Note: the record for South Sudan is based on Drar (1970).

Capparis erythrocarpos Isert var. **erythrocarpos**

F.P.S. 1: 42 (1950); F.T.E.A. Capparid.: 60 (1964); T.S.S.:
31 (1990), pro parte; Biol. Skr. 51: 76 (1998); F.E.E. 2(1):
95 (2000).
Scrambling shrub. Wooded grassland, bushland.
Distr: BAG, EQU; Guinea to Ethiopia, S to Angola &
Zambia.
EQU: Myers 8923 4/1938.
Note: El Amin in T.S.S. records this taxon as much more
widespread in Sudan but this is considered to be in error.

Capparis fascicularis DC. var. **fascicularis**

F.T.E.A. Capparid.: 65 (1964); F.J.M.: 85 (1976); T.S.S.: 31
(1990); Biol. Skr. 51: 77 (1998); F.E.E. 2(1): 96 (2000).
Syn: Capparis rothii Oliv. – F.P.S. 1: 41 (1950); Bot. Exp.
Sud.: 26 (1970).
Scrambling shrub. Wooded grassland, bushland.
Distr: DAR, KOR, CS, BAG, UN, EQU; Senegal to
Somalia, S to Uganda & Kenya.
DAR: Wickens 1320 3/1964; **EQU:** Myers 8594 2/1938.

Capparis fascicularis DC. var. **elaeagnoides** (Gilg) DeWolf

F.T.E.A. Capparid.: 66 (1964); Biol. Skr. 51: 77 (1998).
Scrambling shrub. Wooded grassland, bushland, dry
forest.
Distr: EQU; D.R. Congo to Kenya, S to Botswana &
Zimbabwe.
EQU: Fukui 84-70 7/1985.

Capparis micrantha A.Rich.

F.P.S. 1: 42 (1950); F.J.M.: 85 (1976); T.S.S.: 32 (1990);
Biol. Skr. 51: 77 (1998); F.E.E. 2(1): 97 (2000).
Shrub. Bushland, wooded grassland.
Distr: DAR, KOR, BAG, UN, EQU; Ethiopia.
DAR: Kamil 877 5/1968; **EQU:** Friis & Vollesen 1039
2/1982.
Note: a local species but not considered to be threatened.

Capparis sepiaria L. var. **fisheri** (Pax) DeWolf

F.T.E.A. Capparid.: 62 (1964); F.J.M.: 85 (1976); T.S.S.: 32
(1990); Biol. Skr. 51: 77 (1998); F.E.E. 2(1): 97 (2000).
Syn: Capparis djurica Gilg & Gilg-Ben. – F.P.S. 1: 42
(1950); Bot. Exp. Sud.: 25 (1970).
Syn: Capparis corymbosa sensu Andrews, non Lam. –
F.P.S. 1: 42 (1950); Bot. Exp. Sud.: 25 (1970).
Scrambling shrub. Bushland.
Distr: DAR, KOR, CS, BAG; Senegal to Somalia, S to
Angola & South Africa; Madagascar, India and Malaysia.
DAR: Wickens 2905 4/1965; **BAG:** Schweinfurth 1454
4/1869.

Capparis tomentosa Lam.

F.P.S. 1: 41 (1950); F.T.E.A. Capparid.: 62 (1964); T.S.S.:
33 (1990); Biol. Skr. 51: 78 (1998); F.E.E. 2(1): 95 (2000).
Scrambling shrub or woody climber. Bushland, wooded
grassland.
Distr: NS, RS, DAR, KOR, CS, BAG, UN, EQU; Senegal
to Somalia, S to Namibia & South Africa; Arabia,
Mascarenes.

KOR: Wickens 944 2/1963; **EQU:** Friis & Vollesen 1158 3/1982.

Crateva adansonii DC. subsp. adansonii

F.P.S. 1: 47 (1950); F.T.E.A. Capparid.: 20 (1964); F.J.M.: 85 (1976); T.S.S.: 33 (1990); Biol. Skr. 51: 79 (1998); F.E.E. 2(1): 99 (2000).
Large shrub or tree. Riverine forest, wooded grassland.
Distr: NS, RS, DAR, KOR, CS, ES, BAG, UN, EQU; Senegal to Eritrea, S to Zambia; N Africa.
DAR: Wickens 2950 5/1965; **EQU:** Myers 8497 2/1938.

Dipterygium glaucum Decne.

F.P.S. 1: 48 (1950); F.E.E. 2(1): 118 (2000).
Subshrub. Sandy & silty plains, wadis.
Distr: NS, RS; Eritrea, Djibouti, Somalia; Egypt, Arabia to Pakistan.
RS: Andrews 238 (fl) 11/1936.

Maerua aethiopica (Fenzl) Oliv.

F.P.S. 1: 51 (1950); F.T.E.A. Capparid.: 34 (1964); F.E.E. 2(1): 110 (2000).
Shrub. Tall grassland, woodland & thicket.
Distr: CS, BAG; Cameroon to Ethiopia, S to Zambia & Mozambique.
CS: Kotschy 549 (fl fr) 1837; **BAG:** Schweinfurth 1965 6/1869.

Maerua angolensis DC. subsp. angolensis

F.P.S. 1: 52 (1950); F.T.E.A. Capparid.: 29 (1964); F.J.M.: 86 (1976); Biol. Skr. 51: 79 (1998); F.E.E. 2(1): 108 (2000).
Shrub or small tree. Bushland, wooded grassland.
Distr: RS, DAR, KOR, BAG, UN, EQU; Senegal to Somalia, S to Angola & South Africa.
DAR: Wickens 3008 6/1965; **BAG:** Turner 127 1/1936.

Maerua crassifolia Forssk.

F.P.S. 1: 50 (1950); F.T.E.A. Capparid.: 40 (1964); F.E.E. 2(1): 109 (2000).
Shrub or tree. Woodland & bushland near rivers.
Distr: NS, RS, DAR, KOR, ES, EQU; Mauritania to Somalia, S to Tanzania; N Africa, Arabia, Iran, Pakistan.
RS: Sahni & Kamil 696 4/1967; **EQU:** Myers 14104 9/1941.

Maerua duchesnei (De Wild.) F.White

F.T.E.A. Capparid.: 29 (1964); Biol. Skr. 51: 79 (1998).
Syn: Capparis duchesnei De Wild. – F.P.S. 1: 42 (1950).
Scrambling shrub. Lowland forest.
Distr: EQU; Sierra Leone to D.R. Congo & Tanzania.
EQU: Myers 13130 2/1940.

Maerua oblongifolia (Forssk.) A.Rich.

F.P.S. 1: 52 (1950); F.T.E.A. Capparid.: 37 (1964); F.J.M.: 86 (1976); Biol. Skr. 51: 79 (1998); F.E.E. 2(1): 112 (2000).
Syn: Maerua dolichobotrys Gilg & Gilg-Ben. – F.P.S. 1: 52 (1950).
Syn: Maerua virgata Gilg – F.P.S. 1: 52 (1950); Bot. Exp. Sud.: 27 (1970).
Scrambling shrub. Bushland, semi-desert.
Distr: NS, RS, DAR, CS, UN, EQU; Senegal to Somalia, S to Uganda & Kenya; Egypt, Arabia.
CS: Lea 129 2/1952; **EQU:** Friis & Vollesen 1041 2/1982.

Maerua parvifolia Pax

F.T.E.A. Capparid.: 39 (1964).
Syn: Maerua harmsiana Gilg – F.P.S. 1: 50 (1950).
Perennial herb or shrub. Dry woodland & bushland on rocky ground.
Distr: BAG; Uganda & Kenya, S to Namibia & South Africa.
BAG: Hoyle 615 10/1939.

Maerua pseudopetalosa (Gilg & Gilg-Ben.) DeWolf

F.T.E.A. Capparid.: 43 (1964); F.J.M.: 86 (1976); Biol. Skr. 51: 80 (1998); F.E.E. 2(1): 104 (2000).
Syn: Courbonia virgata Brongn. – F.P.S. 1: 46 (1950); Bot. Exp. Sud.: 26 (1970).
Subshrub. Tall grassland.
Distr: RS, DAR, KOR, BAG, UN, EQU; Senegal to Eritrea, S to Uganda & Kenya.
DAR: Wickens 971 1/1964; **EQU:** Friis & Vollesen 1038 2/1982.

Maerua subcordata (Gilg) DeWolf

F.T.E.A. Capparid.: 41 (1964); F.E.E. 2(1): 103 (2000).
Shrub. Dry wooded grassland, rocky terrain.
Distr: EQU; Ethiopia, Somalia, Uganda, Kenya, Tanzania.
EQU: Padwa 246 (fl) 6/1953.

Maerua triphylla A.Rich. var. triphylla

F.P.S. 1: 50 (1950); F.T.E.A. Capparid.: 43 (1964); Biol. Skr. 51: 80 (1998); F.E.E. 2(1): 105 (2000).
Syn: Maerua jasminifolia Gilg & Gilg-Ben. – F.P.S. 1: 50 (1950).
Scandent shrub or small tree. Bushland, grassland.
Distr: BAG, UN, EQU; Eritrea & Somalia, S to D.R. Congo & Tanzania.
EQU: Friis & Vollesen 1147 3/1982.

Maerua triphylla A.Rich. var. calophylla (Gilg) DeWolf

F.T.E.A. Capparid.: 48 (1964); Biol. Skr. 51: 80 (1998); F.E.E. 2(1): 106 (2000).
Shrub or small tree. Bushland, riverine forest margins, often on rocky ground.
Distr: EQU; Ethiopia & Somalia, S to Tanzania; Yemen.
EQU: Jackson 1061 1/1950.

Ritchiea albersii Gilg

F.P.S. 1: 52 (1950); F.T.E.A. Capparid.: 21 (1964); Biol. Skr. 51: 81 (1998); F.E.E. 2(1): 101 (2000).
Syn: Ritchiea pentaphylla sensu Andrews, non Gilg & Gilg-Ben. – F.P.S. 1: 52 (1950).
Shrub or tree. Montane forest.
Distr: EQU; Nigeria, Cameroon, Ethiopia to Zimbabwe.
EQU: Jackson 1363 (fr) 3/1950.

Cleomaceae

H. Pickering & I. Darbyshire

Cleome amblyocarpa Barratte & Murb.

F.Egypt 1: 180 (1999); F.J.U. 2: 141 (1999).
Syn: Cleome arabica sensu Andrews, non L. – F.P.S. 1: 44 (1950).
Annual or perennial herb. Rocky ground.

Distr: NS, RS, DAR; Mauritania to Chad; N Africa, Arabia, Iraq, Iran, Pakistan.
NS: Léonard 4934 11/1968.

Cleome angustifolia Forssk.
F.E.E. 2(1): 79 (2000).
Syn: *Cleome diandra* Burch. – F.P.S. 1: 44 (1950); F.T.E.A. Capparid.: 5 (1964).
Annual herb. *Acacia* woodland.
Distr: ?DAR, EQU; Niger to Ethiopia, S to Namibia & South Africa; Yemen.
EQU: Myers 14028 9/1941.
Notes: (1) the record for Darfur is based on Quézel (1969: 101) for which we have not seen the specimen. Andrews also records this species from Central Sudan but the specimen on which his record appears to be based (*Myers* 14028) is from near Moru Yakipi in Equatoria. (2) the South Sudan material may be referable to subsp. *petersiana* (Klotzsch) Kers but the single specimen seen is too poor to be certain.

Cleome brachycarpa Vahl ex DC.
F.P.S. 1: 44 (1950); F.T.E.A. Capparid.: 7 (1964); F.E.E. 2(1): 81 (2000).
Annual or perennial herb. Bushland, semi-desert.
Distr: NS, RS, KOR; Cape Verde Is., Mauritania to Somalia, S to Kenya; N Africa, Arabia to India.
KOR: Wickens 172 8/1962.

Cleome chrysantha Decne.
F.P.S. 1: 44 (1950); F.E.E. 2(1): 77 (2000).
Annual or short-lived perennial herb. Rocky outcrops, weed of disturbed ground.
Distr: NS, RS; Chad, Ethiopia; Libya, Egypt, Arabia, Iran.
RS: Carter 1950 11/1987.

Cleome coeruleo-rosea Gilg & Gilg-Ben.
F.P.S. 1: 46 (1950).
Syn: *Cleome foliosa* sensu Andrews, non Hook.f. – F.P.S. 1: 46 (1950).
Annual herb. Dry grassland, seasonally wet depressions.
Distr: UN; Cameroon, Chad.
UN: Simpson 7034 6/1929.
Note: a very local species.

Cleome droserifolia (Forssk.) Delile
F.P.S. 1: 43 (1950); F.Egypt 1: 177 (1999).
Compact subshrub. Desert wadis.
Distr: NS, RS; Djibouti; Libya, Egypt, Arabia.
RS: Carter 1959 11/1987.

Cleome gynandra L.
Biol. Skr. 51: 78 (1998); F.E.E. 2(1): 78 (2000).
Syn: *Gynandropsis gynandra* (L.) Briq. – F.P.S. 1: 49 (1950); F.T.E.A. Capparid.: 18 (1964); Bot. Exp. Sud.: 27 (1970); F.J.M.: 85 (1976).
Annual herb. Dry river beds, rocky & disturbed ground.
Distr: NS, DAR, CS, UN, EQU; pantropical & subtropical.
NS: Bristow 31 4/1984; **UN:** J.M. Lock 81/63 5/1981.

Cleome hanburyana Penz.
F.P.S. 1: 46 (1950); F.T.E.A. Capparid.:14 (1964); F.E.E. 2(1): 84 (2000).
Annual or perennial herb. Semi-desert, bushland.

Distr: RS; Eritrea, Ethiopia, Somalia, Kenya; Egypt & Arabia.
RS: Carter 2008 11/1987.

Cleome monophylla L.
F.P.S. 1: 44 (1950); F.T.E.A. Capparid.: 5 (1964); F.J.M.: 85 (1976); Biol. Skr. 51: 78 (1998); F.E.E. 2(1): 78 (2000).
Annual herb. Woodland, bushland & grassland.
Distr: DAR, KOR, ES, BAG, UN, EQU; Senegal to Somalia, S to Namibia & South Africa; Madagascar, Arabia, India & Sri Lanka.
DAR: Wickens 1957 7/1964; **EQU:** Myers 6776 5/1937.

Cleome niamniamensis Schweinf.
Bot. Jahrb. Syst. 33: 203 (1902); F.P.S. 1: 46 (1950).
Herb. Rocky outcrops.
Distr: EQU; South Sudan endemic.
IUCN: RD
EQU: Schweinfurth 3932 6/1870.
Note: only known from the type. It may be a depauperate specimen of *C. polyanthera*, also recorded from Equatoria.

Cleome paradoxa R.Br. ex DC.
F.P.S. 1: 44 (1950); F.E.E. 2(1): 81 (2000).
Annual herb. Dry rocky areas.
Distr: NS, RS, KOR; Mauritania to Eritrea & Djibouti; Libya, Egypt, Arabia.
RS: Schweinfurth 334 9/1868.
Note: Drar (1970) records this species from Yambio to Karika, Equatoria, but this was based on a sterile specimen and it is likely to have been misidentified.

Cleome parvipetala R.A.Graham
F.T.E.A. Capparid.: 15 (1964); F.E.E. 2(1): 86 (2000).
Annual or short-lived perennial herb. Bushland, semi-desert.
Distr: EQU; Eritrea, Ethiopia, Somalia, Kenya.
EQU: Myers 14002 9/1941.

Cleome polyanthera Schweinf. & Gilg
F.P.S. 1: 46 (1950); F.W.T.A. 1(1): 87 (1954).
Annual herb. Dry grassland.
Distr: BAG, UN, EQU; Nigeria to D.R. Congo.
EQU: Myers 6572 5/1937.

Cleome ramosissima Webb ex Parl.
F.E.E. 2(1): 81 (2000).
Syn: *Cleome schweinfurthii* Gilg – F.P.S. 1: 44 (1950); Bot. Exp. Sud.: 26 (1970).
Annual or perennial herb. Rocky, dry ground.
Distr: RS; Eritrea, Ethiopia, Somalia; Arabia.
RS: Andrews 2716.

Cleome rutidosperma DC.
F.T.E.A. Capparid.: 11 (1964).
Syn: *Cleome ciliata* Schumach. & Thonn. – F.P.S. 1: 46 (1950); Bot. Exp. Sud.: 26 (1970).
Annual herb. Disturbed ground.
Distr: DAR, EQU; Cape Verde Is., Senegal to Uganda, S to Angola & Tanzania.
DAR: Drar 2482 5/1938 (n.v.); **EQU:** Myers 8602 2/1938.
Note: the identification of the Sudanese specimen cited follows Drar (1970) who listed it under the synonym *C. ciliata*.

Cleome scaposa DC.

F.P.S. 1: 43 (1950); F.T.E.A. Capparid.: 4 (1964); F.E.E. 2(1): 78 (2000).
Annual herb. Semi-desert, bushland.
Distr: RS, KOR, CS, ES; Cape Verde Is., Mauritania to Somalia, S to Kenya; Libya, Egypt, Arabia, Pakistan, India.
CS: Jackson 3227 11/1954.

Cleome tenella L.

F.P.S. 1: 44 (1950); F.T.E.A. Capparid.: 6 (1964); Bot. Exp. Sud.: 26 (1970); F.E.E. 2(1): 6 (2000).
Annual herb. Semi-desert & bushland.
Distr: DAR, KOR, CS, EQU; Mauritania to Somalia & Kenya; Masdagascar, Arabia, India.
KOR: Jackson 4322 9/1961; **EQU:** Drar 1628 4/1938 (n.v.).
Note: the record for South Sudan is from Drar (1970).

Cleome usambarica Pax

F.T.E.A. Capparid.: 13 (1964); F.E.E. 2(1): 84 (2000).
Annual herb. Rocky ground, bushland, forest margins, swamp margins.
Distr: EQU; Ethiopia, Kenya, Tanzania.
EQU: Dale S.315 2/1943.

Cleome viscosa L.

F.P.S. 1: 46 (1950); F.T.E.A. Capparid.: 10 (1964); F.E.E. 2(1): 82 (2000).
Annual herb. Wooded grassland.
Distr: DAR, KOR, CS, ES; Cape Verde Is., Mauritania to Eritrea, S to Tanzania; Madagascar, Arabia, tropical Asia, Australia.
KOR: Wickens 673 10/1962.

Brassicaceae (Cruciferae)

H. Pickering & I. Darbyshire

Anastatica hierochuntica L.

F.P.S. 1: 56 (1950); F.E.E. 2(1): 138 (2000).
Annual herb. Dry sandy areas.
Distr: RS; Eritrea; N Africa to Pakistan.
RS: Andrews 229 11/1936.

Arabidopsis thaliana (L.) Heynh.

F.T.E.A. Crucifer.: 45 (1982); F.E.E. 2(1): 144 (2000).
Annual or short-lived perennial herb. Montane bushland & moorland.
Distr: DAR; Eritrea to D.R. Congo & Tanzania; Europe, N Africa, Asia.
DAR: Wickens 2382 9/1964.

Biscutella didyma L. var. elbensis (Chrtek) El Naggar

F.Egypt 1: 229 (1999).
Annual herb. Rocky wadis.
Distr: RS; Sudan endemic (see note).
IUCN: RD
RS: Drar 205 1933 (n.v.).
Note: the cited specimen is listed by Drar (1936). This variety is endemic to Gebel Elba in the disputed Hala'ib Triangle.

Brassica nigra (L.) W.D.J.Koch

F.P.S. 1: 56 (1950); F.E.E. 2(1): 123 (2000).

Annual herb. Weed of cultivation.
Distr: NS, CS, ?KOR; Ethiopia; Europe, Macaronesia, N Africa, Asia.
CS: de Wilde et al. 5741 2/1965.
Note: the record for Kordofan is from Andrews.

Capsella bursa-pastoris (L.) Medik.

F.J.M.: 86 (1976); F.T.E.A. Crucifer.: 29 (1982); F.E.E. 2(1): 135 (2000).
Annual herb. Roadsides, weed of cultivation & waste ground.
Distr: DAR; cosmopolitan.
DAR: Wickens 2683 9/1964.

Cardamine africana L.

F.T.E.A. Crucifer.: 38 (1982); Biol. Skr. 51: 81 (1998); F.E.E. 2(1): 142 (2000).
Perennial herb. Forest & forest-grassland transition.
Distr: EQU; mountains of tropical Africa, S Asia & tropical America.
EQU: Jackson 3136 5/1954.

Diceratella elliptica (R.Br. ex DC.) Jonsell

F.T.E.A. Crucifer.: 60 (1982); F.E.E. 2(1): 150 (2000).
Syn: *Matthiola elliptica* R.Br. ex DC. – F.P.S. 1: 60 (1950).
Small shrub. *Acacia-Commiphora* bushland.
Distr: RS; Eritrea, Ethiopia, Somalia, Kenya.
RS: Carter 1990 11/1987.

Diplotaxis erucoides (L.) DC.

F.P.S. 1: 56 (1950); Bot. Exp. Sud.: 41 (1970); F.Egypt 1: 210 (1999).
Annual herb. Waste ground, open sandy soils.
Distr: ?RS, ?CS; Europe, N Africa to Arabia & Iran.
CS: Drar 14 2/1938 (n.v.).
Note: Andrews recorded this species from "Red Sea District: Erkowit, Karora Hills" but we have not seen the material on which this was based; the Drar specimen cited is sterile and the identification is tentative.

Enarthrocarpus lyratus (Forssk.) DC.

F.P.S. 1: 57 (1950); F.Egypt 1: 219 (1999).
Annual herb. Weed of cultivation & waste ground.
Distr: ?NS; E Mediterranean, Egypt, Arabia.
Note: no Sudanese material has been seen; the record is from Andrews.

Eremobium aegyptiacum (Spreng.) Asch. & Schweinf. ex Boiss.

F.Egypt 1: 196 (1999).
Syn: *Malcolmia aegyptiaca* Spreng. – F.P.S. 1: 59 (1950).
Annual or short-lived perennial herb. Wadis, desert plains.
Distr: ?NS, ?KOR; Egypt, Palestine, Arabia to Pakistan.
Note: no Sudanese specimens have been seen but Andrews records this species from "Libyan and Nubian Deserts. Kordofan".

Eruca vesicaria (L.) Cav. subsp. sativa (Mill.) Thell.

Syn: *Eruca sativa* Mill. – Bot. Exp. Sud.: 41 (1970); F.E.E. 2(1): 129 (2000).
Annual herb. Cutlivated, naturalised weed of cultivation & waste ground.

Distr: NS, RS, EQU; widely cultivated in temperate and tropical regions.
NS: <u>Ahti 16686</u> 1962; **EQU:** <u>Myers 6761</u> 5/1937.

Erucastrum arabicum Fisch. & C.A.Mey.

F.T.E.A. Crucifer.: 9 (1982); Biol. Skr. 51: 81 (1998); F.E.E. 2(1): 125 (2000).
Syn: *Brassica arabica* (Fisch. & C.A.Mey.) Fiori – Bot. Exp. Sud.: 41 (1970).
Annual herb. Bushland, weed of cultivation & waste ground.
Distr: RS, DAR, EQU; Eritrea to D.R. Congo & Tanzania, Namibia to Zimbabwe; Arabia.
RS: <u>Jackson 2802</u> 4/1953; **EQU:** <u>Shigeta 199</u> 2/1979 (n.v.).

Farsetia longisiliqua Decne.

F.P.S. 1: 57 pro parte (1950); F.E.E. 2(1): 140 (2000).
Syn: *Farsetia longistyla* sensu auctt., non Baker f. – F.P.S. 1: 58 (1950); Bot. Exp. Sud.: 41 (1970).
Shrub. Dry rocky ground.
Distr: RS; Eritrea, Ethiopia, Somalia; Egypt, Arabia.
RS: <u>Carter 1806</u> 11/1987.

Farsetia stenoptera Hochst. subsp. *stenoptera*

F.P.S. 1: 57 (1950); F.J.M.: 86 (1976); F.T.E.A. Crucifer.: 35 (1982); F.E.E. 2(1): 140 (2000).
Syn: *Farsetia longisiliqua* sensu auctt., non Decne. – F.P.S. 1: 57 pro parte (1950); F.J.M.: 86 (1976).
Annual or perennial herb. Open bushland & grassland, disturbed ground.
Distr: DAR, KOR, CS; Mauritania to Somalia, S to Tanzania.
DAR: <u>Wickens 1999</u> 7/1964.

Farsetia stylosa R.Br.

Bull. Jard. Nat. Belg. 67:165 (1999); F.E.E. 2(1): 140 (2000).
Syn: *Farsetia ramosissima* Hochst. ex E.Fourn. – F.P.S. 1: 58 (1950); Bot. Exp. Sud.: 41 (1970).
Syn: *Farsetia aegyptia* sensu Andrews, non Turra – F.P.S. 1: 57 (1950).
Perennial herb. Dry bushland, sandy & rocky ground.
Distr: NS, RS, DAR, KOR, CS; Eritrea, Ethiopia, Somalia; N Africa, Arabia, Pakistan, India.
KOR: <u>Harrison 941</u> 4/1950.

Lepidium africanum (Burm.f.) DC.

F.T.E.A. Crucifer.: 21 (1982); F.E.E. 2(1): 133 (2000).
Annual or perennial herb. Weed of cultivation, open areas & waste ground.
Distr: RS; Eritrea to Namibia & South Africa.
RS: <u>Jackson 2883</u> 4/1953.

Lepidium coronopus (L.) Al-Shehbaz

Novon 14: 156 (2004).
Syn: *Coronopus squamatus* (Forssk.) Asch. – Ann. Bot. Fennici 10: 144 (1973).
Annual or biennial herb. Moist ground.
Distr: ?NS; Europe, Mediterranean, N Africa to W Asia, introduced elsewhere.
Note: inclusion of this species here is based on a record from "by the 2nd cataract" by Rikli & Rübel, cited in Ahti et al. (1973: 144).

Lepidium niloticum (Delile) Sieber ex Steud.

Novon 12: 5–11 (2002).
Syn: *Coronopus niloticus* (Delile) Spreng. – F.P.S. 2: 56 (1952); Bot. Exp. Sud.: 41 (1970); F.Egypt 1: 227 (1999).
Annual herb. Riverbanks, weed of cultivation.
Distr: NS, KOR, CS; Egypt, ?Arabia.
CS: <u>Jackson 4096</u> 7/1960.

Lepidium sativum L.

F.P.S. 1: 59 (1950); F.J.M.: 86 (1976); F.T.E.A. Crucifer.: 18 (1982); F.E.E. 2(1): 131 (2000).
Annual herb. Weed of cultivation.
Distr: DAR, KOR; ?native to NE Africa & the Mediterranean, widely cultivated & naturalised elsewhere.
KOR: <u>Pfund 703</u> 11/1875.

Morettia philaeana (Delile) DC.

F.P.S. 1: 60 (1950); F.J.U. 2: 167 (1999); F.Egypt 1: 199 (1999).
Perennial herb. Wadis, stony plains.
Distr: NS, RS, DAR, CS; Mauritania to Chad; N Africa, Arabia.
CS: <u>Jackson 3218</u> 11/1954.
Note: Andrews also records *M. canescens* Boiss. from "Northern and Central Sudan" but we have seen no material and it is possible that this was based on a misidentification of *M. philaeana*.

Notoceras bicorne (Aiton) Amo

F.Egypt 1: 202 (1999).
Annual herb. Sandy & stony deserts.
Distr: RS; Macaronesia, Mediterranean, N Africa to India.
RS: <u>Farag & Zubair 63</u> (fl fr) 2/2012.

Raphanus sativus L.

F.J.M.: 87 (1976); F.E.E. 2(1): 127 (2000).
Annual herb. Cultivated & naturalised as a weed.
Distr: DAR, CS; cultivated throughout the temperate regions & widely naturalised.
DAR: <u>Wickens 1144</u> 2/1964.

Rorippa micrantha (Roth) Jonsell

F.T.E.A. Crucifer.: 55 (1982); F.E.E. 2(1): 146 (2000).
Syn: *Nasturtium brachypus* Webb – F.P.S. 1: 60 (1950).
Syn: *Rorippa indica* sensu Andrews pro parte, non (L.) Hiern – F.P.S. 1: 60 (1950).
Annual herb. Seasonally wet ground by rivers.
Distr: CS, ES; Congo-Brazzaville, Ethiopia to Angola & Mozambique; Madagascar, Egypt, India, Sri Lanka.
CS: <u>Pettet 102</u> 1/1962.

Rorippa microphylla (Boenn. ex Rchb.) Hyl.

F.T.E.A. Crucifer.: 57 (1982); F.E.E. 2(1): 146 (2000).
Syn: *Nasturtium microphyllum* Boenn. ex Rchb. – F.J.M.: 86 (1976).
Perennial herb. Streambanks, ditches, shallow water.
Distr: DAR; native to Europe & W Asia, perhaps also in Ethiopia; widely cultivated and introduced.
DAR: <u>Steele 30</u> (n.v.).
Note: the specimen cited has not been seen by us; the identification follows Wickens in F.J.M.

Rorippa nasturtium-aquaticum (L.) Hayek

F.E.E. 2(1): 146 (2000).
Syn: *Nasturtium officinale* R.Br. – F.P.S. 1: 60 (1950);
Bot. Exp. Sud.: 42 (1970).
Perennial herb. Streams, ditches.
Distr: DAR; Ethiopia; Eurasia, widely cultivated &
naturalised elsewhere.
DAR: Dandy 149 1/1934.

Rorippa palustris (L.) Besser

F.E.E. 2(1): 146 (2000).
Syn: *Rorippa indica* sensu Andrews pro parte, non (L.)
Hiern – F.P.S. 1: 60 (1950).
Annual or short-lived perennial herb. Stream & riverbeds,
moist soils.
Distr: CS; Ethiopia; Eurasia & N America, widely introduced.
CS: Jackson 1567 12/1950.

Schouwia purpurea (Forssk.) Schweinf.

F.P.S. 1: 61 (1950); F.E.E. 2(1): 131 (2000).
Syn: *Schouwia purpurea* (Forssk.) Schweinf. subsp.
schimperi (Jaub. & Spach) Maire – F.J.U. 2: 169 (1999).
Syn: *Schouwia schimperi* Jaub. & Spach – F.P.S. 1: 62
(1950).
Syn: *Schouwia thebaica* Webb – Bot. Exp. Sud.: 42
(1970).
Syn: *Eruca sativa* sensu Andrews, non Mill – F.P.S. 1:
57 (1950).
Annual herb. Desert plains, wadis.
Distr: RS, CS, ES; Mauritania to Somalia; N Africa, Arabia.
CS: Aylmer 507 2/1936.

Sinapis arvensis L.

F.Egypt 1: 213 (1999).
Annual herb. Sandy soils, weed of cultivation & waste
ground.
Distr: CS; Europe, N Africa, Arabia to Afghanistan;
introduced elsewhere.
CS: Jones 44 3/1949.

Sisymbrium erysimoides Desf.

F.P.S. 1: 61 (1950); F.T.E.A. Crucifer.: 64 (1982); F.E.E.
2(1): 150 (2000).
Annual herb. Rocky & grassy hillslopes, margins of dry
forest.
Distr: RS; Eritrea & Somalia, S to Tanzania; Macaronesia,
N Africa to Arabia & Pakistan.
RS: Carter 1981 11/1987.

Sisymbrium irio L.

F.E.E. 2(1): 150 (2000).
Syn: *Sisymbrium pinnatifidum* Forssk. – F.P.S. 1: 62
(1950).
Annual herb. Rocky hillslopes.
Distr: ?RS; Eritrea, Somalia; Europe, N Africa to C Asia.
Note: no Sudanese specimens have been seen by us;
Andrews recorded this species from Red Sea District but
this may be based on a misidentification of *S. erysimoides*.

Zilla spinosa (L.) Prantl subsp. spinosa

F.P.S. 1: 62 (1950); F.Egypt 1: 218 (1999).
Spiny shrub. Sandy & stony plains.
Distr: RS; N Africa, SW Asia.
RS: Bent s.n. 1896.

SANTALALES

Balanophoraceae

H. Pickering

Thonningia sanguinea Vahl

F.P.S. 2: 298 (1952); F.E.E. 3: 384 (1989); F.T.E.A.
Balanophor.: 4 (1993); Biol. Skr. 51: 266 (1998).
Herbaceous root parasite. Forest floor.
Distr: EQU; Senegal to Ethiopia, S to Angola & Zambia.
EQU: Myers 6623 5/1937.

Olacaceae

I. Darbyshire

Note: Olacaceae is here treated in the broad sense,
including Strombosiaceae and Ximeniaceae.

Olax gambecola Baill.

F.T.E.A. Olac.: 7 (1968); Biol. Skr. 51: 261 (1998).
Shrub. Forest, especially by streams.
Distr: EQU; Guinea to Uganda, S to Angola & Zambia.
EQU: Friis & Vollesen 470 (fr) 11/1980.

Strombosia grandifolia Hook.f.

F.W.T.A. 1(2): 648 (1958); Biol. Skr. 51: 261 (1998).
Tree. Forest.
Distr: EQU; Benin to D.R. Congo.
EQU: Jackson 1024 1/1950 (n.v.).
Note: we have not seen the cited material; the
identification follows Friis & Vollesen (1998).

Strombosia scheffleri Engl.

F.T.E.A. Olac.: 11 (1968); T.S.S.: 287 (1990); Biol. Skr. 51:
261 (1998).
Tree. Forest.
Distr: EQU; Nigeria to Kenya, S to Angola, Zimbabwe &
Mozambique.
EQU: Friis & Vollesen 746 (fl) 12/1980.

Ximenia americana L. var. americana

F.P.S. 2: 290 (1952); F.T.E.A. Olac.: 3 (1968); F.J.M.: 120
(1976); F.E.E. 3: 356 (1989); T.S.S.: 287 (1990); Biol. Skr.
51: 262 (1998).
Small tree or shrub. Dry woodland, bushland & wooded
grassland.
Distr: RS, DAR, KOR, CS, ES, BAG, UN, EQU; widespread
in the tropics.
DAR: Wickens 1289 (fl) 2/1964; **BAG:** Turner 130 (fl)
1/1936.

Ximenia caffra Sond.

F.T.E.A. Olac.: 5 (1968); F.E.E. 3: 356 (1989).
Syn: *Ximenia americana* L. var. *caffra* (Sond.) Engl. –
Biol. Skr. 51: 262 (1998).
Small tree or shrub. Dry woodland, bushland & wooded
grassland.
Distr: EQU; Ethiopia to Namibia & South Africa (?also W
Africa); Madagascar.
EQU: Kielland-Lund 876 6/1984 (n.v.).
Note: sometimes treated as a variety of *X. americana*.
We have not seen the cited material; the identification
follows Friis & Vollesen (1998).

Opiliaceae

I. Darbyshire

Opilia amentacea Roxb.

F.T.E.A. Opil.: 1 (1968); F.E.E. 3: 358 (1989); Biol. Skr. 51: 262 (1998).
Syn: *Opilia celtidifolia* (Guill. & Perr.) Endl. ex Walp.
– F.P.S. 2: 290 (1952); Bot. Exp. Sud.: 91 (1970) ; T.S.S.: 288 (1990).
Woody climber or scrambling shrub. Woodland & thicket, often riverine.
Distr: DAR, KOR, CS, BAG, UN, EQU; palaeotropical.
DAR: Kamil 895 (fr) 5/1968; **EQU:** Friis & Vollesen 1044 (fr) 2/1982.

Santalaceae

H. Pickering

Osyris lanceolata Hochst. & Steud.

F.E.E. 3: 382 (1989); F.T.E.A. Santal.: 25 (2005).
Syn: *Osyris quadripartita* Salzm. ex Decne. – Biol. Skr. 51: 265 (1998).
Syn: *Osyris compressa* sensu auctt., non (P.J.Bergius) A.DC. – F.P.S. 2: 295 (1952); T.S.S.: 288 (1990).
Small tree or shrub. Upland & riverine forest & associated bushland.
Distr: KOR, EQU; Eritrea & Somalia, S to Namibia & South Africa; Iberia, Macaronesia, NW Africa, Arabia, India to China.
KOR: Wickens 858 11/1962; **EQU:** Myers 11017 4/1939.

Thesium schweinfurthii Engl.

F.P.S. 2: 295 (1952); F.E.E. 3: 380 (1989); Biol. Skr. 51: 265 (1998); F.T.E.A. Santal.: 18 (2005).
Perennial herb. Montane woodland & grassland.
Distr: BAG, EQU; Nigeria, Ethiopia to D.R. Congo, Zambia & Malawi.
EQU: Friis & Vollesen 1264 3/1982.

Thesium stuhlmannii Engl.

F.E.E. 3: 380 (1989); Biol. Skr. 51: 266 (1998); F.T.E.A. Santal.: 13 (2005).
Perennial herb. Montane grassland, rocky ground.
Distr: EQU; Rwanda, Ethiopia, Uganda, Kenya, Tanzania.
EQU: Prowse 390 6/1953.

Thesium viride A.W.Hill

F.P.S. 2: 296 (1952); F.W.T.A. 1(2): 666 (1958).
Perennial herb. Wooded grassland, bushland.
Distr: ?BAG; Sierra Leone to Cameroon.
Note: no South Sudanese material has been seen; the record is based on Andrews who lists it from "between Jebel Manda and Jebel Yukanga near River Boro".

Thesium sp. cf. *leucanthum* Gilg

Biol. Skr. 51: 266 (1998).
Shrub. *Loudetia* grassland with scattered trees.
Distr: EQU.
EQU: Friis & Vollesen 376 11/1980.
Note: *T. leucanthum* is known from Nigeria and Angola only. The specimen from the Imatong Mts requires further study.

Viscum tuberculatum A.Rich.

F.E.E. 3: 375 (1989); Biol. Skr. 51: 265 (1998); Mistletoes of Africa: 290 (1998); F.T.E.A. Visc.: 11 (1999).
Parasitic shrub. Montane woodland & forest margins.
Distr: EQU; Eritrea & Somalia, S to Angola, Namibia & South Africa.
EQU: Friis & Vollesen 1037 2/1982.
Note: *V. triflorum* DC. is also likely to occur in South Sudan, having been recorded from the Ugandan side of the Imatong Mts (Friis & Vollesen 1998).

Loranthaceae

H. Pickering & I. Darbyshire

Agelanthus djurensis (Engl.) Polhill & Wiens

Mistletoes of Africa: 157 (1998); F.T.E.A. Loranth.: 52 (1999).
Syn: *Loranthus djurensis* Engl. – F.P.S. 2: 293 (1952).
Parasitic shrub. Forest.
Distr: BAG, EQU; Cameroon to Uganda, S to Angola.
EQU: Myers 7101 7/1937.

Agelanthus dodoneifolius (DC.) Polhill & Wiens

Mistletoes of Africa: 162 (1998); F.T.E.A. Loranth.: 55 (1999).
Syn: *Loranthus dodoneifolius* DC. – F.P.S. 2: 293 (1952).
Syn: *Tapinanthus* sp. sensu Wickens – F.J.M.: 120 (1976).
Parasitic shrub. Wooded grassland.
Distr: DAR, KOR, EQU; Senegal to Uganda.
KOR: Wickens 502 9/1962; **EQU:** Jackson 4274 6/1961.

Agelanthus heteromorphus (A.Rich.) Polhill & Wiens

Mistletoes of Africa: 176 (1998).
Syn: *Loranthus heteromorphus* A.Rich. – F.P.S. 2: 295 (1952).
Syn: *Tapinanthus heteromorphus* (A.Rich.) Danser – F.E.E. 3: 373 (1989).
Parasitic shrub. Woodland & wooded grassland, often on *Combretum* or *Terminalia*.
Distr: ?CS; Togo to Eritrea & Ethiopia.
Note: no Sudanese material has been seen; Andrews recorded this species from Fazoghli in Central Sudan but this record requires confirmation. However, this species is likely to occur in Sudan in view of its wider distribution.

Agelanthus oehleri (Engl.) Polhill & Wiens

Mistletoes of Africa: 166 (1998); Biol. Skr. 51: 263 (1998); F.T.E.A. Loranth.: 61 (1999).
Parasitic shrublet. Bushland.
Distr: EQU; Kenya, Tanzania.
EQU: Friis & Vollesen 1077 3/1982.

Agelanthus platyphyllus (Hochst. ex A.Rich.) Balle

Mistletoes of Africa: 147 (1998); F.T.E.A. Loranth.: 44 (1999).
Syn: *Loranthus platyphyllus* Hochst. ex A.Rich. – F.P.S. 2: 293 (1952).
Syn: *Schimperina platyphylla* (Hochst. ex A.Rich.)

Tiegh. – F.E.E. 3: 372 (1989).
Parasitic shrub. Woodland & wooded grassland on *Combretum* or *Terminalia*.
Distr: ES, BAG, UN, EQU; Ethiopia, Uganda, Kenya.
ES: <u>Kennedy-Cooke 56</u> 6/1931; **EQU:** <u>Myers 3680</u> 3/1937.

Agelanthus cf. *scassellatii* (Chiov.) Polhill & Wiens

Parasitic shrub. Growing on *Ficus palmata* & *Faidherbia albida*.
Distr: DAR.
DAR: <u>Wickens 2872</u> 3/1965.
Note: both *Lynes* 11 and *Wickens* 2872 from Jebel Marra were previously identified as *Loranthus celtidifolius* Engl. which is a synoynym of *A. scassellatii*, a species recorded from Somalia to Tanzania. R. Polhill has written on the folder of these specimens that they are either a glandular form of *A. schweinfurthii* or an extension of the range of *A. scassellatii*; these specimens are not discussed or listed in Mistletoes of Africa but *A. schweinfurthii* is recorded from Jebel Marra.

Agelanthus schweinfurthii (Engl.) Polhill & Wiens

Mistletoes of Africa: 182 (1998); F.T.E.A. Loranth.: 71 (1999).
Syn: *Loranthus schweinfurthii* Engl. – F.P.S. 2: 293 (1952).
Parasitic shrub. Riverine forest, wooded grassland & bushland.
Distr: KOR, BAG, EQU; C.A.R., D.R. Congo, Uganda, Tanzania.
KOR: <u>Musselman 6275</u> 12/1983; **EQU:** <u>Myers 7109</u> 7/1937.

Erianthemum dregei (Eckl. & Zeyh.) Tiegh.

F.E.E. 3: 366 (1989); Mistletoes of Africa: 237 (1998); F.T.E.A. Loranth.: 95 (1999).
Syn: *Loranthus dregei* Eckl. & Zeyh. – Bot. Exp. Sud.: 80 (1970).
Syn: *Loranthus dregei* Eckl. & Zeyh. var. *kerenicus* Sprague – F.P.S. 2: 293 (1952).
Parasitic shrub. Forest margins.
Distr: CS, ES; Eritrea to Angola & South Africa.
ES: <u>Schweinfurth 2193</u> 7/1865.

Globimetula braunii (Engl.) Danser

Mistletoes of Africa: 211 (1998); Biol. Skr. 51: 263 (1998); F.T.E.A. Loranth.: 78 (1999).
Parasitic shrub. Forest & plantations.
Distr: EQU; Ivory Coast to Kenya & D.R. Congo.
EQU: <u>Friis & Vollesen 183</u> 11/1980.

Oliverella hildebrandtii (Engl.) Tiegh.

F.E.E. 3: 371 (1989); Mistletoes of Africa: 115 (1998); F.T.E.A. Loranth.: 23 (1999).
Syn: *Loranthus hildebrandtii* Engl. – F.P.S. 2: 293 (1952).
Small parasitic shrub. Bushland.
Distr: EQU; Ethiopia, Uganda, Kenya, Tanzania, Mozambique.
EQU: <u>Schweinfurth 2858</u> (fl fr) 12/1869.

Note: Polhill & Wiens (Mistletoes of Africa) note that the presence of this species in South Sudan requires reconfirmation.

Oncocalyx fischeri (Engl.) M.G.Gilbert

F.E.E. 3: 369 (1989); Mistletoes of Africa: 106 (1998); Biol. Skr. 51: 263 (1998); F.T.E.A. Loranth.: 16 (1999).
Parasitic shrub. Dry woodland & dry forest.
Distr: EQU; Ethiopia, Somalia, Uganda, Kenya, Tanzania.
EQU: <u>Myers 11145</u> 4/1939.

Phragmanthera polycrypta (Didr.) Balle subsp. *subglabriflora* Balle ex Polhill & Wiens

Mistletoes of Africa: 265 (1998); Biol. Skr. 51: 263 (1998); F.T.E.A. Loranth.: 111 (1999).
Parasitic shrub. Forest.
Distr: EQU; Gabon, D.R. Congo, Uganda, Tanzania.
EQU: <u>Friis & Vollesen 1255</u> 3/1982.

Phragmanthera usuiensis (Oliv.) M.G.Gilbert

F.E.E. 3: 368 (1989); Mistletoes of Africa: 259 (1998); Biol. Skr. 51: 264 (1998); F.T.E.A. Loranth.: 264 (1999).
Syn: *Loranthus usuiensis* Oliv. – F.P.S. 2: 293 (1952); Bot. Exp. Sud.: 80 (1970).
Parasitic shrub. Forest.
Distr: EQU; Cameroon to Ethiopia, S to Angola & Mozambique.
EQU: <u>Andrews 1938</u> 6/1939.

Plicosepalus acaciae (Zucc.) Wiens & Polhill

F.E.E. 3: 363 (1989); Mistletoes of Africa: 92 (1998).
Syn: *Loranthus acaciae* Zucc. – F.P.S. 2: 293 (1952); Bot. Exp. Sud.: 80 (1970).
Parasitic shrub. Wooded grassland, often on *Acacia*.
Distr: NS, RS, DAR, KOR, CS, UN; Eritrea, Ethiopia; Egypt, Arabia.
KOR: <u>Wickens 845</u> 11/1962.

Plicosepalus curviflorus (Benth. ex Oliv.) Tiegh.

F.E.E. 3: 364 (1989); Mistletoes of Africa: 88 (1998); Biol. Skr. 51: 265 (1998); F.T.E.A. Loranth.: 9 (1999).
Syn: *Loranthus curviflorus* Benth. ex Oliv. – F.P.S. 2: 293 (1952); Bot. Exp. Sud.: 80 (1970).
Parasitic herb. Deciduous woodland, usually on *Acacia*.
Distr: RS, ?DAR, EQU; Eritrea & Somalia, S to Angola & Tanzania; Egypt, Middle East, Arabia.
RS: <u>Bent s.n.</u> 1896; **EQU:** <u>Friis & Vollesen 182</u> 11/1980.
Note: the record from Darfur is based on a sight record by Drar (1970).

Tapinanthus constrictiflorus (Engl.) Danser

Mistletoes of Africa: 189 (1998); Biol. Skr. 51: 264 (1998); F.T.E.A. Loranth.: 73 (1999).
Parasitic shrub. Forest.
Distr: EQU; Gabon to Kenya, S to Angola & Tanzania.
EQU: <u>Andrews 1880</u> 6/1939.

Tapinanthus globiferus (A.Rich.) Tiegh.

F.J.M.: 120 (1976); F.E.E. 3: 373 (1989); Mistletoes of Africa: 201 (1998); Biol. Skr. 51: 264 (1998).
Syn: *Loranthus globiferus* A.Rich. – F.P.S. 2: 293 (1952); Bot. Exp. Sud.: 80 (1970).

Syn: *Loranthus verrucosus* Engl. – F.P.S. 2: 293 (1952); Bot. Exp. Sud.: 80 (1970).
Parasitic shrub. Forest margins, woodland.
Distr: RS, DAR, KOR, CS, ES, BAG, UN, EQU; Mauritania & Senegal to Eritrea, Ethiopia & Djibouti; Yemen.
DAR: <u>Wickens 1075</u> 1/1964; **EQU:** <u>Friis & Vollesen 1153</u> 3/1982.

CARYOPHYLLALES

Frankeniaceae

I. Darbyshire

Frankenia pulverulenta L.

F.P.S. 1: 159 (1950); F.E.E. 2(2): 2, in notes (1995); F.Egypt 2: 130 (2000).
Annual herb. Saline sandy soils, salt pans, edges of salt marshes.
Distr: RS, ?CS; Macaronesia, Mediterranean, Egypt, Arabia to central & N Asia.
RS: <u>Jackson 3929</u> (fl) 4/1959.
Note: Andrews records this species from Khartoum, where it must be introduced; it is native along the Red Sea coast and lowlands.

Tamaricaceae

M. Kordofani, I. Farag & I. Darbyshire

[*Tamarix*]

Note: the taxonomy of the genus *Tamarix* is confused, with much disagreement between different authors. We have followed Baum's (1978) revision (The Genus *Tamarix*) but note that this is rejected by some more recent authors (see, for example, Mesfin Tadesse in Fl. Somalia 1: 209 (1993)). The *T. nilotica-arborea-mannifera* complex is particularly problematic, with the three being treated as conspecific by Boulos in Fl. Egypt 2: 127 (2000). Clearly, a full critical revision is required.

Tamarix amplexicaulis Ehrenb.

Baum, Genus *Tamarix*: 153 (1978); F.Egypt 2: 129 (2000).
Shrub. Saline soils in deserts, riverbanks.
Distr: ?NS; Morocco & Mauritania to Palestine & Arabia.
Note: no Sudanese specimens have been seen; the record for Sudan is based on Boulos in F.Egypt (l.c.).

Tamarix aphylla (L.) H.Karst.

F.T.E.A. Tamaric.: 1 (1966); Baum, Genus *Tamarix*: 81 (1978); T.S.S.: 61 (1990); F.E.E. 2(2): 3 (1995).
Syn: *Tamarix orientalis* Forssk. – F.P.S. 1: 161 (1950).
Tree or shrub. Wadis, sandy or saline riverbanks, dry woodland.
Distr: NS, RS, CS, ES; Senegal, Eritrea, Ethiopia, Somalia, Kenya; N Africa to India.
NS: <u>Pettet 96</u> (fl) 9/1962.

Tamarix arborea (Sieber ex Ehrenb.) Bunge

Baum, Genus *Tamarix*: 53 (1978); F.E.E. 2(2): 5 (1995); F.J.U. 3: 241 (1999).
Tree or shrub. Saline flats, riverbanks.
Distr: NS, RS; Ethiopia; Tunisia to Egypt, Socotra.
NS: <u>Léonard 4838</u> (fl fr) 11/1968.

Tamarix mannifera (Ehrenb.) Bunge

F.P.S. 1: 161 (1950); Baum, Genus *Tamarix*: 70 (1978); T.S.S.: 62 (1990).
Small tree or shrub. Saline flats & deserts, wadis.
Distr: NS, RS, CS; Egypt to Jordan.
NS: <u>Ahti 16313</u> (fl) 9/1962.

Tamarix nilotica (Ehrenb.) Bunge

F.P.S. 1: 161 (1950); F.T.E.A. Tamaric.: 3 (1966); Baum, Genus *Tamarix*: 72 (1978); T.S.S.: 62 (1990); F.E.E. 2(2): 4 (1995).
Tree or shrub. Riverine woodland.
Distr: NS, RS, CS, ES; Eritrea & Somalia, S to Tanzania; Egypt to Arabia.
CS: <u>Schweinfurth 722</u> (fl fr) 10/1868.

Tamarix passerinoides Desv.

Baum, Genus *Tamarix*: 166 (1978); F.Egypt 2: 129 (2000).
Shrub. Saline soils in desert.
Distr: ?NS; Algeria to Arabia, Afghanistan & Russia.
Note: no Sudanese specimens have been seen; the record for Sudan is based on Boulos in F.Egypt (l.c.).

Plumbaginaceae

M. Kordofani, I. Farag & I. Darbyshire

Ceratostigma abyssinicum (Hochst.) Schweinf. & Asch.

F.T.E.A. Plumbag.: 5 (1976); F.E.E. 5: 33 (2006).
Small shrub. Rocky hillslopes, open dry bushland.
Distr: RS; Eritrea, Ethiopia, Somalia, Kenya.
RS: <u>Jackson 3981</u> (fl fr) 4/1959.

Limonium axillare (Forssk.) Kuntze

F.P.S. 3: 66 (1956); T.S.S.: 420 (1990); F.E.E. 5: 29 (2006).
Small shrub. Salt marshes, coastal sands.
Distr: RS; Eritrea; Red Sea coast of Egypt & Arabia.
RS: <u>Sahni & Kamil 712</u> (fl fr) 4/1967.

Plumbago zeylanica L.

F.P.S. 3: 68 (1956); F.J.M.: 139 (1976); F.T.E.A. Plumbag.: 6 (1976); T.S.S.: 420 (1990); Biol. Skr. 51: 404 (2005); F.E.E. 5: 31 (2006).
Perennial herb. Disturbed ground, bushland, woodland & wooded grassland.
Distr: RS, DAR, KOR, UN, EQU; native of tropical Asia, widely introduced in the palaeotropics.
DAR: <u>Wickens 1894</u> (fl fr) 7/1964; **UN:** <u>J.M. Lock 81/34</u> (fl fr) 2/1981.

Polygonaceae

I. Darbyshire

Calligonum polygonoides L. subsp. *comosum* (L'Hér.) Soskov

F.Egypt 1: 24 (1999).
Shrub. Sandy desert.
Distr: RS; N Africa to Arabia & Pakistan.
RS: <u>Drar 245</u> 1932 (n.v.).
Note: the record of this species is from Drar (1936) and is from Gebel Elba. Its presence there is confirmed by Boulos (1999).

Emex spinosa (L.) Campd.

F.T.E.A. Polygon.: 3 (1958); F.E.E. 2(1): 336 (2000).
Annual herb. Waste ground, around habitation.
Distr: RS; Eritrea, Ethiopia, Somalia, Kenya, South Africa;
Macaronesia & N Africa.
RS: Jackson 2699 (fl fr) 3/1953.

Fagopyrum snowdenii (Hutch. & Dandy) S.P.Hong

Biol. Skr. 51: 93 (1998); F.E.E. 2(1): 337 (2000).
Syn: *Harpagocarpus snowdenii* Hutch. & Dandy –
F.T.E.A. Polygon.: 4 (1958).
Annual climbing herb. Montane forest, often along
streams.
Distr: EQU; Cameroon, D.R. Congo, Uganda, Kenya,
Tanzania.
EQU: Friis & Vollesen 403 (fl fr) 11/1980.
Note: a local species but not considered to be threatened.

Oxygonum sinuatum (Hochst. & Steud. ex Meisn.) Dammer

F.T.E.A. Polygon.: 34 (1958); F.J.M.: 90 (1976); F.E.E. 2(1):
340 (2000).
Syn: *Oxygonum atriplicifolium* (Meisn.) Martelli var.
sinuatum (Hochst. & Steud. ex Meisn.) Baker – F.P.S. 1:
101 (1950); Bot. Exp. Sud.: 94 (1970).
Syn: *Oxygonum atriplicifolium* sensu Andrews, non
(Meisn.) Martelli – F.P.S. 1: 101 (1950).
Annual herb. Sandy soils in grassland & bushland, waste
ground, weed of cultivation.
Distr: RS, DAR, KOR, ES, UN, EQU; Niger to Eritrea, S to
Namibia & South Africa; Egypt, Arabia, India.
DAR: Wickens 2700 (fl fr) 9/1964; **UN:** J.M. Lock 81/51
(fl fr) 5/1981.

Persicaria barbata (L.) H.Hara

F.Z. 9(3): 14 (2006).
Syn: *Polygonum barbatum* L. – F.P.S. 1: 103 (1950);
Bot. Exp. Sud.: 94 (1970).
Syn: *Polygonum setulosum* sensu Wickens, non *P.*
setosulum A.Rich. – F.J.M.: 90 (1976).
Perennial herb. Damp ground, riversides.
Distr: DAR, CS, EQU; native to China, widespread in
Asia, scattered in tropical Africa & Madagascar.
DAR: Wickens 1199 (fl fr) 2/1964; **EQU:** Drar 1765
4/1938 (n.v.).
Note: the record for South Sudan is from Drar (1970).

Persicaria decipiens (R.Br.) K.L.Wilson

Biol. Skr. 51: 93 (1998); F.E.E. 2(1): 345 (2000).
Syn: *Polygonum salicifolium* Brouss. ex Willd. – F.P.S.
1: 102 (1950); F.T.E.A. Polygon.: 17 (1958); Bot. Exp.
Sud.: 94 (1970); F.J.M.: 90 (1976).
Perennial herb. Riverbanks, pools, swamps.
Distr: DAR, ES, UN, EQU; widespread in the tropics &
subtropics.
DAR: Jackson 4126 (fl) 12/1954; **EQU:** Andrews 1202 (fl
fr) 5/1939.

Persicaria glabra (Willd.) M.Gómez

F.E.E. 2(1): 345 (2000).
Syn: *Polygonum glabrum* Willd. – F.P.S. 1: 103 pro
parte (1950); Bot. Exp. Sud.: 94 (1970).
Perennial herb. Streams, riverbanks.

Distr: NS, KOR, CS, ?ES; Ethiopia; Socotra, widespread in
Asia, Pacific Is. & tropical America.
CS: Bromfield s.n. (fl fr) 3/1851.
Note: *Jackson* 4126 from Rahad River, Eastern Sudan
is identified as either a form of *P. glabra* or a hybrid
between *P. glabra* and *P. decipiens*.

Persicaria limbata (Meisn.) H.Hara

F.E.E. 2(1): 345 (2000).
Syn: *Polygonum limbatum* Meisn. – F.P.S. 1: 103
(1950); F.T.E.A. Polygon.: 20 (1958).
Perennial herb. Riverbanks, streamsides.
Distr: KOR, BAG; Senegal to Ethiopia, S to Namibia;
Egypt, India.
KOR: Pfund 668 (fl fr) 7/1875; **BAG:** Schweinfurth 1361
(fl) 4/1869.

Persicaria madagascariensis (Meisn.) S.Ortiz & Paiva

F.Z. 9(3): 12 (2006).
Syn: *Persicaria attenuata* (R.Br.) Soják subsp. *africana*
K.L.Wilson – F.E.E. 2(1): 347 (2000).
Syn: *Polygonum pulchrum* sensu Graham, non Blume
– F.T.E.A. Polygon.: 20 (1958).
Syn: *Polygonum tomentosum* sensu auctt., non Willd.
– F.P.S. 1: 103 (1950); Bot. Exp. Sud.: 94 (1970).
Rhizomatous perennial herb. Riversides, pools, swamps,
wet grassland.
Distr: NS, DAR, ES, BAG, UN, EQU; tropical & southern
Africa; Madagascar, tropical Asia.
ES: Jackson 4127 (fl) 4/1961; **UN:** Harrison 289 (fl) 1/1948.

Persicaria nepalensis (Meisn.) H.Gross

F.E.E. 2(1): 343 (2000).
Syn: *Polygonum nepalense* Meisn. – F.T.E.A. Polygon.:
12 (1958); Bot. Exp. Sud.: 94 (1970); F.J.M.: 90 (1976).
Annual herb. Weed of cultivation, waste ground, forest
margins.
Distr: RS, DAR; palaeotropical.
RS: Jackson 2966 (fl) 4/1953.

Persicaria senegalensis (Meisn.) Soják fa. *senegalensis*

F.E.E. 2(1): 345 (species only) (2000); F.Z. 9(3): 10 (2006).
Syn: *Polygonum senegalense* Meisn. fa. *senegalense*
– F.T.E.A. Polygon.: 18 (1958).
Perennial herb. Riverbanks, pools, swamps, drainage
channels.
Distr: NS, CS, BAG, UN, EQU; Senegal to Eritrea, S to
Namibia & South Africa; Madagascar, N Africa.
NS: Pettersson 16789 (fl) 10/1962; **EQU:** Myers 6663 (fl)
5/1937.
Note: Hedberg in F.E.E. does not separate the two
forms recognised here, but they are usually readily
distinguishable.

Persicaria senegalensis (Meisn.) Soják fa. *albotomentosa* (R.A.Graham) K.L.Wilson

F.Z. 9(3): 11 (2006).
Syn: *Polygonum senegalense* Meisn. fa.
albotomentosum R.A.Graham – F.T.E.A. Polygon.: 19
(1958); Bot. Exp. Sud.: 94 (1970).
Syn: *Polygonum lanigerum* sensu Andrews, non R.Br.
– F.P.S. 1: 103 (1950).

Perennial herb. Habitat as fa. *senegalensis*.
Distr: UN, EQU; global distribution as fa. *senegalensis*.
EQU: Sillitoe 284 (fl fr) 1919.

Persicaria setosula (A.Rich.) K.L.Wilson
Biol. Skr. 51: 93 (1998); F.E.E. 2(1): 347 (2000).
Syn: *Polygonum setosulum* A.Rich. – F.T.E.A. Polygon.: 22 (1958).
Syn: *Polygonum acuminatum* sensu Andrews, non Kunth – F.P.S. 1: 103 (1950).
Syn: *Polygonum senegalense* sensu Andrews, non Meisn. – F.P.S. 1: 103 (1950).
Perennial herb. Swamps, wet grassland.
Distr: EQU; Sierra Leone to Eritrea, S to Angola & South Africa.
EQU: Myers 11559 (fl fr) 7/1939.

Polygonum plebeium R.Br.
F.P.S. 1: 102 (1950); F.T.E.A. Polygon.: 13 (1958); Bot. Exp. Sud.: 94 (as *plebejum*) (1970); F.J.M.: 90 (1976); F.E.E. 2(1): 341 (2000).
Syn: *Polygonum aviculare* sensu Wickens, non L. – F.J.M.: 90 (1976).
Mat-forming annual herb. *Acacia* woodland, waste ground, weed of cultivation.
Distr: ?NS, DAR, CS; palaeotropical & subtropical.
DAR: Wickens 1141 (fl) 2/1964.
Note: the record for Northern Sudan is from Andrews.

Rumex abyssinicus Jacq.
F.P.S. 1: 104 (1950); F.T.E.A. Polygon.: 7 (1958); F.J.M.: 90 (1976); Biol. Skr. 51: 94 (1998); F.E.E. 2(1): 339 (2000).
Perennial herb. Forest margins, weed of cultivation & fallow.
Distr: DAR, EQU; Nigeria to Eritrea, S to Angola & Mozambique; Madagascar.
DAR: Lynes 157 (fr) 11/1921; **EQU:** Myers 11616 (fl fr) 7/1939.

Rumex crispus L.
F.T.E.A. Polygon.: 10 (1958); Bot. Exp. Sud.: 94 (1970).
Perennial herb. Weed of disturbed ground & damp sites.
Distr: DAR; native to Europe & W Asia, widely introduced elsewhere.
DAR: Drar 2154 5/1938 (n.v.).

Rumex nepalensis Spreng.
F.P.S. 1: 105 (1950); Biol. Skr. 51: 94 (1998); F.E.E. 2(1): 338 (2000).
Syn: *Rumex bequaertii* De Wild. – F.T.E.A. Polygon.: 8 (1958); F.J.M.: 90 (1976).
Perennial herb. Forest margins, streamsides, disturbed areas, weed of cultivation.
Distr: DAR, EQU; Cameroon to Eritrea, S to South Africa; Madagascar, SW Europe, W Asia to China.
DAR: Wickens 1440 (fr) 10/1964; **EQU:** Friis & Vollesen 93 (fr) 11/1980.

Rumex nervosus Vahl
F.P.S. 1: 105 (1950); F.E.E. 2(1): 339 (2000).
Shrub. Wadi beds, open rocky ground, waste places.
Distr: RS, ES; Eritrea, Ethiopia; Arabia.
RS: Jackson 3964 (fl) 4/1959.

Rumex simpliciflorus Murb.
F.Egypt 1: 34 (1999).
Annual herb. Desert wadis.
Distr: RS; Cape Verde Is., Mauritania, Chad; Macaronesia, N Africa.
RS: Drar 251 1933 (n.v.).
Note: the record of this species is derived from Drar (1936) and is from Gebel Elba. Its presence there is confirmed by Boulos (1999).

Rumex vesicarius L.
F.P.S. 1: 104 (1950); F.E.E. 2(1): 338 (2000).
Shrubby herb. Rocky & sandy ground, wadi beds.
Distr: RS, DAR; Eritrea, Somalia; N Africa, SE Europe, SW Asia, Afghanistan & India.
RS: Pettet 164 (fr) 12/1962.

Droseraceae
R. Vanderstricht & I. Darbyshire

Aldrovanda vesiculosa L.
F.P.S. 1: 82 (1950); F.Z. 4: 62 (1978).
Submerged aquatic herb. Shallow freshwater.
Distr: BAG; Ghana, Togo, Chad to Ethiopia, S to Botswana; scattered from Central Europe to Japan & to Australia.
IUCN: EN
BAG: Simpson 7632 3/1930.
Note: a very widespread but scattered and scarce species.

Drosera indica L.
F.P.S. 1: 82 (1950); F.T.E.A. Droser.: 2 (1959); Biol. Skr. 51: 88 (1998); F.E.E. 2(1): 189 (2000).
Annual herb. Marshy areas, wet rocks.
Distr: UN, EQU; Senegal to Ethiopia, S to Mozambique; Madagascar, tropical Asia, Australia.
UN: J.M. Lock 81/265 (fl) 9/1981.

Caryophyllaceae
I. Farag & I. Darbyshire

Arenaria leptoclados (Rchb.) Guss.
F.P.S. 1: 86 (1950); F.J.M.: 88 (1976); F.E.E. 2(1): 213 (2000).
Annual herb. Arable weed, margins of montane forest & thicket.
Distr: RS, DAR; Eritrea & Somalia, S to Tanzania; C & S Europe, Canary Is., N Africa to Afghanistan.
RS: Andrews 3507 (fr).

Arenaria serpyllifolia L.
F.Egypt 1: 72 (1999); F.E.E. 2(1): 214 (2000).
Annual herb. Arable weed, margins of bushland.
Distr: RS; Ethiopia; Eurasia, N Africa, introduced elsewhere.
RS: Drar 152 1932 (n.v.).
Note: the record for this species is from Drar (1936) and the cited specimen is from Gebel Elba in the disputed Hala'ib Triangle; Boulos (1999) confirms its presence there.

Cerastium afromontanum T.C.E.Fr. & Weim.
F.T.E.A. Caryophyll.: 21 (1956); Biol. Skr. 51: 89 (1998); F.E.E. 2(1): 216 (2000).
Perennial herb. Open montane bushland, grassland & forest margins.

Distr: EQU; Ethiopia, Uganda, Kenya, Tanzania.
EQU: Friis & Vollesen 850 (fl fr) 12/1980.

Cerastium fontanum Baumg. subsp. *triviale* (Link) Jalas

F.J.M.: 88 (1976).
Syn: *Cerastium vulgatum* sensu Andrews, non L. – F.P.S. 1: 86 (1950).
Syn: *Cerastium holosteoides* Fr. – Bot. Exp. Sudan: 28 (1970).
Perennial herb. Damp montane grassland.
Distr: DAR; South Africa; widespread in temperate regions of the world.
DAR: Lynes 84 (fl) 4/1921.

Cerastium octandrum Hochst. ex A.Rich. var. *octandrum*

F.P.S. 1: 86 (1950); F.T.E.A. Caryophyll.: 21 (1956); F.J.M.: 88 (1976); Biol. Skr. 51: 89 (1998); F.E.E. 2(1): 216 (2000).
Annual herb. Weed of cultivation, disturbed open ground often near water.
Distr: RS, DAR, EQU; Cameroon, Eritrea to D.R. Congo & Tanzania.
DAR: Wickens 2685 (fl fr) 9/1964; **EQU:** Myers 11653 (fl fr) 7/1939.

Cometes abyssinica R.Br. ex Wall.

F.P.S. 1: 105 (1950); F.J.M.: 91 (1976); F.E.E. 2(1): 199 (2000).
Subshrub. Rocky slopes, dry grassland, weed of cultivation & fallow.
Distr: RS, DAR; Eritrea, Ethiopia, Somalia; Egypt, Arabia.
DAR: Jackson 3313 (fl fr) 12/1954.

Drymaria cordata (L.) Willd. ex Roem. & Schult.

F.P.S. 1: 86 (1950); F.T.E.A. Caryophyll.: 9 (1956); F.J.M.: 88 (1976); Biol. Skr. 51: 89 (1998); F.E.E. 2(1): 209 (2000).
Perennial herb. Montane bushland, forest margins, riverbanks, weed of cultivation.
Distr: DAR, EQU; pantropical.
DAR: Wickens 997 (fl) 1/1964; **EQU:** Andrews 2027 (fr) 6/1939.

Gymnocarpos sclerocephalus (Decne.) Ahlgren & Thulin

Edinburgh J. Bot. 59: 233 (2002).
Syn: *Sclerocephalus arabicus* Decne. – F.E.E. 2(1): 199 (2000).
Annual herb. Dry riverbeds on sand or silt.
Distr: RS; Cape Verde Is., Eritrea, Somalia; Macaronesia, N Africa & Arabia to Iran.
RS: Bent s.n. 1896.

Herniaria hirsuta L.

F.E.E. 2(1): 199 (2000).
Prostrate annual herb. Open stony or sandy ground.
Distr: RS; Eritrea, Ethiopia; Europe, Mediterranean region, Middle East, Arabia to India.
RS: Farag & Zubair 37 (fl fr) 2/2012.
Note: the specimen cited is the first record for Sudan.

Minuartia filifolia (Forssk.) Mattf.

F.P.S. 1: 86 (1950); F.T.E.A. Caryophyll.: 17 (1956); F.J.M.: 89 (1976); Biol. Skr. 51: 90 (1998); F.E.E. 2(1): 211 (2000).
Mat-forming perennial herb. Montane grassland, often in damp places, rock crevices.
Distr: DAR, EQU; Eritrea, Ethiopia, Tanzania; Arabia.
DAR: Wickens 1675 (fr) 5/1964; **EQU:** Jackson 932 11/1949.

Polycarpaea corymbosa (L.) Lam.

F.P.S. 1: 87 pro parte (1950); F.T.E.A. Caryophyll.: 8 (1956); F.J.M.: 89 (1976); Biol. Skr. 51: 90 (1998); F.E.E. 2(1): 205 (2000).
Annual or perennial herb. Open woodland, grassland, weed of cultivation.
Distr: DAR, KOR, ES, BAG, EQU; pantropical.
KOR: Wickens 570 (fl) 10/1962; **EQU:** Sillitoe 211 (fr) 1919.
Note: Andrews records also *P. stellata* (Willd.) DC. which is otherwise a scarce West African species and may have been misidentified by Andrews – no Sudanese material has been seen.

Polycarpaea eriantha Hochst. ex A.Rich. var. *eriantha*

F.P.S. 1: 87 (1950); F.T.E.A. Caryophyll.: 7 (1956); F.J.M.: 89 (1976); Biol. Skr. 51: 90 (1998); F.E.E. 2(1): 205 (2000).
Annual herb. Sandy soils in grassland & bushland, sometimes in seasonally wet areas.
Distr: DAR, KOR, BAG, UN, EQU; Mauritania & Senegal to Eritrea, S to South Africa.
DAR: Wickens 2333 (fl fr) 9/1964; **EQU:** Simpson 7565 (fl fr) 2/1930.

Polycarpaea linearifolia (DC.) DC.

F.T.E.A. Caryophyll.: 7 (1956); F.E.E. 2(1): 205 (2000).
Syn: *Polycarpaea corymbosa* sensu Andrews pro parte, non (L.) Lam. – F.P.S. 1: 87 (1950).
Annual herb. Grassland & wooded grassland.
Distr: DAR, KOR, BAG, UN; Senegal to Ethiopia, Tanzania.
KOR: Jackson 3255 (fl) 2/1954; **BAG:** Broun 1425 (fl) 11/1908.
Note: closely allied to, and perhaps conspecific with, *P. corymbosa* – see note in F.E.E.

Polycarpaea repens (Forssk.) Asch. & Schweinf.

F.P.S. 1: 88 (1950); F.Egypt 1: 81 (1999).
Perennial herb. Dry sandy soils.
Distr: RS; Mauritania to Chad; N Africa to Iraq.
RS: Drummond & Hemsley 946 (fl fr) 1/1953.

Polycarpaea robbairea (Kuntze) Greuter & Burdet subsp. *robbairea*

F.J.U. 2: 147 (1999); F.E.E. 2(1): 206 (2000).
Syn: *Robbairea delileana* Milne-Redh. – F.P.S. 1: 89 (1950).
Syn: *Spergularia rubra* sensu Andrews, non (L.) J.Presl. & C.Presl. – F.P.S. 1: 90 (1950).
Annual or perennial herb. Dry sandy & stony plains.
Distr: NS, RS, KOR; Mauritania to Somalia; N Africa to Arabia & Iraq.
RS: Drummond & Hemsley 963 (fl fr) 1/1953.

Polycarpaea spicata Wight ex Arn.

F.P.S. 1: 88 (1950); F.E.E. 2(1): 207 (2000).
Perennial herb. Sandy soils & rocks, coastal.
Distr: RS; Eritrea, Somalia, Kenya; Egypt to Australia.
RS: <u>Jackson 2766</u> (fl) 4/1953.

Polycarpaea tenuifolia (Willd.) DC.

F.W.T.A. 1(1): 132 (1954).
Perennial herb. Rocky open ground.
Distr: BAG; Sierra Leone & Mali to Benin.
BAG: <u>Schweinfurth 2491</u> (fl) 10/1869.
Note: the single specimen from South Sudan is disjunct
from the remainder of this species' range.

Polycarpon prostratum (Forssk.) Asch. & Schweinf.

F.P.S. 1: 89 (1950); F.T.E.A. Caryophyll.: 5 (1956); F.E.E.
2(1): 204 (2000).
Annual herb. Damp open ground, sandy riverbanks.
Distr: KOR, CS, BAG; pantropical.
CS: <u>Kotschy 467</u> (fl fr) 3/1839; **BAG:** <u>Schweinfurth 1646</u>
(fl fr) 4/1869.

Polycarpon tetraphyllum (L.) L. subsp. *tetraphyllum*

F.P.S. 1: 89 (1950); F.E.E. 2(1): 203 (2000).
Annual herb. Open stony & sandy ground.
Distr: RS; Eritrea, Ethiopia, Somalia, South Africa;
widespread elsewhere.
RS: <u>Jackson 2745</u> (fl) 3/1953.

Sagina abyssinica Hochst. ex A.Rich.

F.P.S. 1: 89 (1950); F.T.E.A. Caryophyll.: 5 (1956); Biol. Skr.
51: 90 (1998); F.E.E. 2(1): 212 (2000).
Perennial herb. Montane grassland, often in moist places.
Distr: EQU; Cameroon, Ethiopia to D.R. Congo & Tanzania.
EQU: <u>Thomas 1809</u> 12/1935.

Silene burchellii Otth

F.T.E.A. Caryophyll.: 33 (1956); F.J.M.: 89 (1976); Biol.
Skr. 51: 90 (1998); F.E.E. 2(1): 223 (2000).
Syn: *Silene chirensis* A.Rich. – F.P.S. 1: 90 (1950).
Annual or perennial herb. Montane grassland.
Distr: RS, DAR, EQU; Eritrea & Somalia to South Africa;
Egypt & Arabia.
DAR: <u>Wickens 2658</u> (fl) 9/1964; **EQU:** <u>Friis & Vollesen 834</u> (fl) 12/1980.

Silene lynesii Norman

F.P.S. 1: 90 (1950); F.J.M.: 89 (1976).
Perennial herb. Grassland, sandy wadi beds.
Distr: DAR; Chad, Algeria.
IUCN: RD
DAR: <u>Wickens 2588</u> (fl) 9/1964.

Silene macrosolen Steud. ex A.Rich.

F.P.S. 1: 89 (1950); F.T.E.A. Caryophyll.: 34 (1956); F.J.M.:
89 (1976); F.E.E. 2(1): 223 (2000).
Perennial herb. Rocky hillslopes.
Distr: DAR; Eritrea to Tanzania.
DAR: <u>Lynes 1064</u> (fl fr) 12/1921.

Silene villosa Forssk.

F.Egypt 1: 62 (1999).

Annual herb. Sandy deserts.
Distr: RS; N Africa to Arabia & Iran.
RS: <u>Drar 260</u> 1933 (n.v.).
Note: the record for this species is from Drar (1936) and
the cited specimen is from Gebel Elba in the disputed
Hala'ib Triangle; Boulos (1999) confirms its presence there.

Spergula fallax (Lowe) E.H.L.Krause

F.P.S. 1: 90 (1950); F.Som. 1: 102 (1993).
Annual herb. Rocky hillslopes & wadis.
Distr: RS; Somalia; N Africa to India.
RS: <u>Yeates 8</u> (fl fr) 2/1988.

Spergularia rubra (L.) J.Presl & C.Presl

Bot. Exp. Sud.: 28 (1970); F.E.E. 2(1): 207 (2000).
Annual herb. Disturbed areas.
Distr: RS; Eritrea, Ethiopia, South Africa; Europe, N
Africa, Middle East, introduced elsewhere.
RS: <u>Drar 530</u> 3/1938 (n.v.).
Note: the identity of the cited specimen needs
confirming, but this species is likely to occur in Sudan.

Sphaerocoma hookeri T.Anderson subsp. *intermedia* J.B.Gillett

F.P.S. 1: 90 (species only) (1950); F.Egypt 1: 89 (1999);
F.E.E. 2(1): 198 (species only) (2000).
Subshrub. Wadis, stony & sandy ground.
Distr: RS; Somalia; Egypt, Arabia.
RS: <u>Palmer 142</u> (fl fr) 1/1933.
Note: this subspecies is close to *S. aucheri* Boiss. from W
Asia, and *Bent* s.n. from Sudan has been re-named as
that species. However, this specimen is clearly conspecific
with the other Sudanese material.

Stellaria mannii Hook.f.

F.T.E.A. Caryophyll.: 24 (1956); Biol. Skr. 51: 91 (1998);
F.E.E. 2(1): 218 (2000).
Trailing or scandent herb. Forest margins & clearings.
Distr: EQU; Bioko, Cameroon, Ethiopia to Zimbabwe &
Mozambique; Madagascar, Comoros.
EQU: <u>Friis & Vollesen 615</u> (fl) 12/1980.

Stellaria media (L.) Vill.

F.T.E.A. Caryophyll.: 24 (1956); Bot. Exp. Sud.: 28 (1970);
F.E.E. 2(1): 218 (2000).
Annual herb. Weed of disturbed areas & cultivation.
Distr: RS; almost cosmopolitan.
RS: <u>Drar 598</u> 3/1938 (n.v.).
Note: the record of this species is from Drar (1970).

Amaranthaceae

H. Gibreel & I. Darbyshire

Achyranthes aspera L. var. *aspera*

F.P.S. 1: 113 (species only) (1950); F.J.M.: 91 (species only)
(1976); F.T.E.A. Amaranth.: 101 (1985); Biol. Skr. 51: 96
(1998).
Perennial herb. In a wide variety of habitats; commonly a
weed of cultivation & disturbed ground.
Distr: RS, ?DAR, KOR, CS, BAG, UN, EQU; pantropical.
RS: <u>Andrews 3488</u> (fl); **EQU:** <u>Simpson 7267</u> (fl) 7/1929.
Note: not all the Sudanese material can be readily placed
to variety.

Achyranthes aspera L. var. *pubescens* (Moq.) C.C.Towns.

F.T.E.A. Amaranth.: 102 (1985); F.E.E. 2(1): 329 (2000).
Perennial herb. Habitat as for var. *aspera*.
Distr: UN, EQU; pantropical.
UN: Evans-Pritchard 1 (fl) 11/1936.

Achyranthes aspera L. var. *sicula* L.

F.T.E.A. Amaranth.: 104 (1985); Biol. Skr. 51: 96 (1998);
F.E.E. 2(1): 330 (2000).
Perennial herb. Habitat as for var. *aspera*.
Distr: RS, DAR, KOR, ES, BAG, UN, EQU; pantropical &
warm temperate regions.
DAR: Lynes 62b (fl fr) 12/1921; **EQU:** Thomas 1637 (fl)
12/1935.

Aerva javanica (Burm.f.) Juss. ex Schult. var. *javanica*

F.P.S. 1: 113 (1950); F.J.M.: 91 (1976); F.T.E.A. Amaranth.:
83 (1985); F.E.E. 2(1): 328 (2000).
Perennial herb or shrub. Sandy & stony plains, rocky
hillslopes, dry scrub & grassland.
Distr: NS, RS, DAR, KOR, CS, ES; palaeotropical &
subtropical.
CS: Pettet 2 (fl) 8/1962.

Aerva javanica (Burm.f.) Juss. ex Schult. var. *bovei* Webb

F.P.S. 1: 113 (1950); F.Egypt 1: 137 (1999).
Syn: *Aerva ruspolii* sensu Andrews, non Lopr. – F.P.S. 1:
113 (1950).
Perennial herb or subshrub. Dry sandy & stony ground,
wadis.
Distr: NS, RS, KOR; localised within the range of var.
javanica.
NS: Shaw 11 (fl) 10/1932.

Aerva lanata (L.) Juss. ex Schult.

F.P.S. 1: 115 (1950); F.T.E.A. Amaranth.: 85 (1985); F.E.E.
2(1): 328 (2000).
Syn: *Aerva lanata* (L.) Juss. ex Schult. var. *oblongata*
Asch. – F.P.S. 1: 115 (1950); F.T.E.A. Amaranth.: 86 (1985).
Perennial herb. Dry rocky hillslopes, weed of cultivation &
disturbed ground, also bushland & woodland.
Distr: RS, EQU; palaeotropical & subtropical.
RS: Jackson 2732 (fl) 3/1953; **EQU:** Myers 6779 (fl)
5/1937.

Alternanthera nodiflora R.Br.

F.P.S. 1: 115 (1950); F.J.M.: 91 (1976); F.T.E.A. Amaranth.:
125 (1985); F.E.E. 2(1): 334 (2000).
Annual or perennial herb. Damp ground, margins of
pools, ditches.
Distr: DAR, KOR, CS, ES, BAG, UN; Senegal to Eritrea, S
to Namibia & South Africa; Egypt, Australia, naturalised
in E Asia.
CS: Andrews 177 (fl fr) 2/1936; **BAG:** Myers 6293 (fr)
2/1937.

Alternanthera pungens Kunth

F.T.E.A. Amaranth.: 122 (1985); Biol. Skr. 51: 97 (1998);
F.E.E. 2(1): 333 (2000).

Syn: *Alternanthera repens* (L.) Link – F.P.S. 1: 116
(1950); Bot. Exp. Sud.: 13 (1970).
Perennial herb. Open disturbed ground, weed of gardens.
Distr: CS, ES, BAG, UN, EQU; native of America, now a
pantropical weed.
ES: Carter 1778 (fr) 11/1987; **UN:** Sherif A.3896 (fl)
7/1951.

Alternanthera sessilis (L.) R.Br. ex DC.

F.P.S. 1: 115 (1950); F.T.E.A. Amaranth.: 126 (1985); F.E.E.
2(1): 334 (2000).
Annual or perennial herb. Damp ground, streamsides &
margins of pools, roadsides.
Distr: NS, DAR, CS, BAG, UN, EQU; pantropical &
subtropical.
CS: Pettet 105 (fr) 1/1963; **EQU:** Andrews 1372 (fl fr)
5/1939.

Amaranthus blitum L.

Syn: *Amaranthus ascendens* Loisel. – Bot. Exp. Sud.:
14 (1970).
Syn: *Amaranthus lividus* L. – F.P.S. 1: 117 (1950);
F.T.E.A. Amaranth.: 34 (1985); F.E.E. 2(1): 308 (2000).
Annual herb. Cultivated, sometimes becoming weedy in
disturbed ground.
Distr: CS, EQU; widespread in the tropics & warm
temperate regions.
CS: Drar 22 2/1938 (n.v.); **EQU:** Drar 964 4/1938 (n.v.).
Note: we have seen no material from the Sudan region;
the records here are from Drar (1970) under the synonym
A. ascendens; Andrews recorded *A. lividus* from Blue Nile
Prov. which supports Drar's records from Central Sudan.

Amaranthus caudatus L.

F.P.S. 1: 116 pro parte (1950); F.T.E.A. Amaranth.: 23
(1985); F.E.E. 2(1): 304 (2000).
Annual herb. Weed of disturbed ground, riverbanks, also
cultivated as an ornamental.
Distr: DAR; widespread in the tropics & warm temperate
regions.
DAR: Lynes 596 (fr) 8/1921.
Note: Andrews records this species from Central &
Southern Sudan but we have only seen the cited specimen
from Darfur; it appeas that Andrews may have widely
misapplied this name (see also *A. hybridus* subsp. *cruentus*).

Amaranthus dubius Mart. ex Thell.

F.T.E.A. Amaranth.: 26 (1985); Biol. Skr. 51: 97 (1998);
F.E.E. 2(1): 306 (2000).
Annual herb. Weed of disturbed ground, riverbanks.
Distr: EQU; probably native of the neotropics, now
pantropical.
EQU: Myers 6733 (fr) 5/1937.

Amaranthus graecizans L. subsp. *graecizans*

F.P.S. 1: 117 (1950); F.T.E.A. Amaranth.: 30 (1985); F.E.E.
2(1): 307 (2000).
Annual herb. Weed of cultivation, waste ground,
seasonally wet open ground.
Distr: NS, RS, KOR, CS, ES, UN; palaeotropical & warm
temperate regions.
ES: Carter 1802 (fr) 11/1987; **UN:** J.M. Lock 81/54 (fl)
5/1981.

Amaranthus graecizans L. subsp. *sylvestris* (Vill.) Brenan

F.T.E.A. Amaranth.: 31 (1985); F.E.E. 2(1): 307 (2000).
Syn: *Amaranthus silvestris* Vill. – Bot. Exp. Sud.: 14 (1970).
Annual herb. Weed of cultivation & waste ground.
Distr: DAR, CS, UN; palaeotropical & warm temperate regions.
DAR: Drar 2106 (fr) 5/1938 (n.v.); **UN:** Simpson 7229 (fl fr) 6/1929.
Note: the records for Sudan are from Drar (1970). Some material from the Sudan region looks intermediate between this and subsp. *graecizans*.

Amaranthus hybridus L. subsp. *hybridus*

F.P.S. 1: 116 (1950); F.J.M.: 92 (1976); F.T.E.A. Amaranth.: 25 (1985); Biol. Skr. 51: 97 (1998); F.E.E. 2(1): 305 (2000).
Annual herb. Weed of cultivation & disturbed ground.
Distr: DAR, KOR, ES, UN, EQU; native to the New World, now found throughout the tropics & subtropics.
DAR: Wickens 2554 (fl fr) 9/1964; **UN:** Andrews 1693 (fr) 6/1939.

Amaranthus hybridus L. subsp. *cruentus* (L.) Thell.

F.T.E.A. Amaranth.: 26 (1985); Biol. Skr. 51: 97 (1998); F.E.E. 2(1): 306 (2000).
Syn: *Amaranthus hybridus* L. subsp. *incurvatus* sensu Wickens, non *A. incurvatus* Gren. & Godr. – F.J.M.: 92 (1976).
Syn: *Amaranthus paniculatus* L. – Bot. Exp. Sud.: 14 (1970).
Syn: *Amaranthus caudatus* sensu Andrews pro parte, non L. – F.P.S. 1: 116 (1950).
Annual herb. Cultivated, naturalised on waste ground.
Distr: DAR, KOR, CS, EQU; believed to be native to central America, now widespread in the tropics & subtropics.
DAR: Wickens 1879 (fr) 7/1964; **EQU:** Myers 11916 (fl fr) 10/1938.

Amaranthus sparganiocephalus Thell.

F.T.E.A. Amaranth.: 32 (1985); F.E.E. 2(1): 307 (2000).
Annual herb. Dry open areas including dry riverbeds.
Distr: KOR; Eritrea to Tanzania; Arabia, Socotra.
KOR: Wickens 393 (fr) 10/1962.

Amaranthus spinosus L.

F.P.S. 1: 116 (1950); F.J.M.: 92 (1976); F.T.E.A. Amaranth.: 26 (1985); Biol. Skr. 51: 97 (1998); F.E.E. 2(1): 306 (2000).
Annual herb. Weed of cultivation & waste ground, riverbanks.
Distr: NS, RS, DAR, ?CS, UN, EQU; native to the New World, now found throughout the tropics & subtropics.
DAR: Wickens 1140 (fr) 2/1964; **UN:** Sherif A.2853 (fr) 10/1951.
Note: Andrews records this species from Blue Nile Prov. but we have seen no material from Central Sudan.

Amaranthus tricolor L.

F.P.S. 1: 116 (1950); F.T.E.A. Amaranth.: 28 (1985); F.E.E. 2(1): 306 (2000).
Annual herb. Cultivated, sometimes becoming a weed.

Distr: CS, ES; native to tropical Asia, now widespread in the palaeotropics.
CS: Schweinfurth 833 (fr) 11/1868.

Amaranthus viridis L.

F.P.S. 1: 117 (1950); F.J.M.: 92 (1976); F.T.E.A. Amaranth.: 35 (1985); F.E.E. 2(1): 309 (2000).
Erect herb. Weed of cultivation & disturbed ground.
Distr: NS, ?RS, DAR, CS; widespread in the tropics & warm temperate regions.
DAR: Wickens 1792 (fr) 1/1964.
Note: Andrews records this species from Red Sea region but we have seen no material from there.

Anabasis ehrenbergii Boiss.

F.E.E. 2(1): 298 (2000).
Perennial herb. Sandy seashores.
Distr: RS; Eritrea, Somalia; Red Sea coast of Arabia.
RS: Schweinfurth 744 (fl) 5/1864.

Anabasis setifera Moq.

F.Egypt 1: 125 (1999).
Subshrub. Dry bushland on rocky slopes.
Distr: RS; Egypt, Arabia to Pakistan.
RS: Carter 2005 11/1987.

Arthrocnemum macrostachyum (Moric.) K.Koch

F.E.E. 2(1): 289 (2000).
Syn: *Arthrocnemum glaucum* (Delile) Ung.-Sternb. – F.P.S. 1: 108 (1950).
Syn: *Salicornia fruticosa* sensu Andrews, non (L.) L. – F.P.S. 1: 111 (1950).
Fleshy shrub. Foreshores, coastal saline pools, mangroves, salt pans.
Distr: RS; Senegal, Eritrea, Somalia; Mediterranean & N African coast to Arabia & Jordan valley.
RS: Jackson 2768 (fl) 4/1953.

Atriplex farinosa Forssk. subsp. *farinosa*

F.P.S. 1: 108 (1950); F.T.E.A. Chenopod.: 15 (1954); F.E.E. 2(1): 288 (2000).
Perennial herb or shrub. Sandy seashores.
Distr: RS; Eritrea, Somalia; Egypt.
RS: Broun 1545 (fl) 1/1909.

Beta vulgaris L.

Bot. Exp. Sud.: 29 (1970); F.E.E. 2(1): 278 (2000).
Annual or perennial herb. Cultivated & sometimes naturalised as a weed.
Distr: RS; Europe, N Africa, Asia.
RS: Drar 268 3/1938 (n.v.).
Note: widely grown and probably more widely semi-naturalised in Sudan.

Celosia argentea L.

F.P.S. 1: 117 (1950); F.J.M.: 92 (1976); F.T.E.A. Amaranth.: 19 (1985); Biol. Skr. 51: 98 (1998); F.E.E. 2(1): 303 (2000).
Annual herb. Weed of cultivation, disturbed ground & open scrub.
Distr: NS, RS, DAR, KOR, CS, ES, UN, EQU; almost pantropical, perhaps originating in Africa.
KOR: Wickens 697 (fl fr) 10/1962; **EQU:** Myers 13411 (fl fr) 8/1940.

Celosia isertii C.C.Towns.

F.T.E.A. Amaranth.: 15 (1985).
Syn: *Celosia laxa* sensu Andrews, non Schumach. &
Thonn. – F.P.S. 1: 117 (1950).
Syn: *Celosia trigyna* sensu Friis & Vollesen, non L. –
Biol. Skr. 51: 98 (?pro parte) (1998).
Perennial herb. Forest & forest margins.
Distr: EQU; Senegal to Uganda, S to Angola & Zambia.
EQU: Andrews 1161 (fr) 5/1939.

Celosia polystachia (Forssk.) C.C.Towns.

F.T.E.A. Amaranth.: 11 (1985); F.E.E. 2(1): 301 (2000).
Syn: *Celosia populifolia* Moq. – F.P.S. 1: 118 (1950).
Perennial herb. Rocky hillslopes, riverine scrub & forest,
bushland.
Distr: DAR, KOR, CS, EQU; Chad to Somalia, S to Kenya;
Yemen.
CS: Schweinfurth 1048 (fl fr) 1/1869; **EQU:** Carr 381 (fl
fr) 7/1968.

Celosia schweinfurthiana Schinz

F.P.S. 1: 118 (1950); F.T.E.A. Amaranth.: 9 (1985); Biol.
Skr. 51: 98 (1998); F.E.E. 2(1): 300 (2000).
Perennial herb. Forest, often riverine, also a weed of
cultivation.
Distr: ?CS, EQU; Ethiopia to Angola & Mozambique.
EQU: Jackson 3025 (fr) 6/1953.
Note: Drar (1970: 14) records this species from Ghabat
Station, Sennar which is likely to be a misidentification.

Celosia trigyna L.

F.P.S. 1: 117 (1950); F.J.M.: 92 (1976); F.T.E.A. Amaranth.:
12 (1985); Biol. Skr. 51: 98 (1998); F.E.E. 2(1): 302 (2000).
Annual herb. Weed of disturbed ground, riverbanks,
open woodland.
Distr: RS, DAR, KOR, CS, ES, BAG, UN, EQU; Senegal
to Eritrea, S to Namibia & South Africa; Madagascar,
Madeira, Arabia.
KOR: Wickens 1410 (fl fr) 4/1964; **EQU:** Myers 13950
(fl) 8/1941.

Centrostachys aquatica (R.Br.) Wall.

F.P.S. 1: 118 (1950); F.J.M.: 91 (1976); F.T.E.A. Amaranth.:
106 (1985); F.E.E. 2(1): 330 (2000).
Aquatic perennial herb. Margins of rivers & pools,
swamps.
Distr: DAR, KOR, BAG, UN; Nigeria to Ethiopia, S to
Zimbabwe & Mozambique; India to Java.
DAR: Wickens 1362 (fr) 3/1964; **UN:** J.M. Lock 81/193
(fl fr) 8/1981.

Chenopodium murale L.

F.P.S. 1: 109 (1950); F.T.E.A. Chenopod.: 7 (1954); F.J.M.:
91 (1976); F.E.E. 2(1): 283 (2000).
Annual or perennial herb. Weed of disturbed sites.
Distr: NS, RS, DAR, KOR, CS, ES; cosmopolitan.
DAR: Wickens 1162 (fr) 2/1964.

Chenopodium opulifolium Schrad. ex
W.D.J.Koch & Ziz

F.T.E.A. Chenopod.: 11 (1954); F.J.M.: 91 (1976); Biol.
Skr. 51: 95 (1998); F.E.E. 2(1): 280 (2000).
Syn: *Chenopodium mucronatum* Thunb. var.
subintegrum Aellen – F.P.S. 1: 109 (1950).

Annual or short-lived perennial herb. Weed of disturbed
habitats.
Distr: DAR, EQU; Cameroon to Eritrea, S to Angola
& Zimbabwe; S Europe, N Africa, Arabia to India &
Mongolia; introduced in N America.
DAR: Kassas 896 (n.v.); **EQU:** Myers 11104 (fl) 4/1939.

Chenopodium procerum Hochst. ex Moq.

F.T.E.A. Chenopod.: 11 (1954); Biol. Skr. 51: 96 (1998);
F.E.E. 2(1): 285 (2000).
Annual herb. Weed of cultivation, montane grassland.
Distr: EQU; Ethiopia to Malawi; Yemen.
EQU: Myers 11103 (fl) 4/1939.

Chenopodium pumilio R.Br.

F.T.E.A. Chenopod.: 13 (1954); Biol. Skr. 51: 96 (1998).
Annual herb. Disturbed areas in forest.
Distr: EQU; native of Australia & New Zealand, widely
introduced through the tropics.
EQU: Friis & Vollesen 409 (fl) 11/1980.

Chenopodium schraderianum Roem. & Schult.

F.P.S. 1: 109 (1950); F.T.E.A. Chenopod.: 12 (1954);
F.J.M.: 91 (1976); F.E.E. 2(1): 285 (2000).
Annual herb. Montane grassland, disturbed sites.
Distr: DAR, EQU; Eritrea to Angola & South Africa;
introduced in Europe.
DAR: Wickens 2456 (fr) 9/1964; **EQU:** Thomas 1614 (fl
fr) 12/1935.

Cornulaca ehrenbergii Asch.

F.P.S. 1: 110 (1950); F.E.E. 2(1): 295 (2000).
Subshrub. Sandy seashores.
Distr: RS; Eritrea, Somalia; Egypt, Red Sea coast of
Arabia.
RS: Schweinfurth 723 (fl).

Cornulaca monacantha Delile

F.P.S. 1: 109 (1950); F.J.U. 2: 150 (1999); F.Egypt 1: 127
(1999).
Shrub. Wadis, sandy & stony plains.
Distr: NS; Sahara, N Africa to Pakistan.
NS: Boulos 3412 (n.v.).
Note: the specimen cited has not been seen by us; it was
recorded under *C. monacantha* by Léonard in F.J.U. (l.c.).

Cyathula achyranthoides (Kunth) Moq.

F.T.E.A. Amaranth.: 60 (1985); Biol. Skr. 51: 98 (1998);
F.E.E. 2(1): 317 (2000).
Annual herb. Forest.
Distr: EQU; Sierra Leone to Ethiopia, S to D.R. Congo &
Tanzania; Madagascar, tropical America.
EQU: Friis & Vollesen 755 (fl fr) 12/1980.

Cyathula cylindrica Moq.

F.P.S. 1: 119 (1950); F.T.E.A. Amaranth.: 64 (1985); Biol.
Skr. 51: 99 (1998); F.E.E. 2(1): 317 (2000).
Perennial herb. Wide variety of disturbed habitats.
Distr: EQU; Bioko, Cameroon, Eritrea to South Africa.
EQU: Johnston 1503 (fl fr) 2/1936.

Cyathula orthacantha (Hochst. ex Asch.) Schinz

F.T.E.A. Amaranth.: 66 (1985); Biol. Skr. 51: 99 (1998);
F.E.E. 2(1): 319 (2000).

Annual herb. Grassland & woodland, stony ground, often near rivers.
Distr: UN, EQU; Eritrea to Namibia & Botswana.
UN: Sherif A.2850 (fl fr) 9/1951.

Cyathula prostrata (L.) Blume var. *prostrata*

F.P.S. 1: 119 (1950); F.T.E.A. Amaranth.: 59 (1985); Biol. Skr. 51: 99 (1998); F.E.E. 2(1): 316 (2000).
Annual or short-lived perennial herb. Forest clearings, disturbed ground.
Distr: EQU; pantropical.
EQU: Friis & Vollesen 445 (fl fr) 11/1980.

Cyathula prostrata (L.) Blume var. *pedicellata* (C.B.Clarke) Cavaco

F.T.E.A. Amaranth.: 59 (1985); Biol. Skr. 51: 99 (1998); F.E.E. 2(1): 317 (2000).
Annual or short-lived perennial herb. Forest, often in disturbed areas.
Distr: EQU; Sierra Leone to Ethiopia, S to D.R. Congo & Mozambique.
EQU: Friis & Vollesen 596 (fl fr) 12/1980.

Cyathula uncinulata (Schrad.) Schinz

F.T.E.A. Amaranth.: 65 (1985); Biol. Skr. 51: 100 (1998); F.E.E. 2(1): 319 (2000).
Perennial herb, sometimes climbing. A wide variety of habitats including forest margins & disturbed bushland.
Distr: EQU; Cameroon, Angola, Eritrea to South Africa; Madagascar.
EQU: Myers 14200 (fl) 10/1941.

Digera muricata (L.) Mart. subsp. *trinervis* C.C.Towns. var. *trinervis*

F.T.E.A. Amaranth.: 37 (1985); Biol. Skr. 51: 100 (1998); F.E.E. 2(1): 311 (2000).
Syn: *Digera alternifolia* sensu auctt., non (L.) Asch. – F.P.S. 1: 119 (1950); Bot. Exp. Sud.: 14 (1970).
Annual herb. Weed of disturbed ground, open woodland.
Distr: NS, RS, DAR, KOR, CS, ES, BAG, UN, EQU; Eritrea, Ethiopia, Somalia, Uganda, Kenya, Tanzania; Socotra.
ES: Carter 1795 (fl) 11/1987; **BAG:** Simpson 7102 (fl fr) 6/1929.

Gomphrena celosioides Mart.

F.T.E.A. Amaranth. (1985); Biol. Skr. 51: 100 (1998); F.E.E. 2(1): 334 (2000).
Perennial herb. Weed of disturbed ground, open sandy & rocky areas.
Distr: CS, EQU; native to S America, now pantropical.
CS: de Wilde et al. 5835 (fl) 3/1965; **EQU:** Kielland-Lund 47 11/1983 (n.v.).

Halopeplis perfoliata (Forssk.) Bunge ex Asch. & Schweinf.

F.P.S. 1: 110 (1950); F.E.E. 2(1): 289 (2000).
Fleshy subshrub. Seashores, saltmarshes, salt pans.
Distr: RS; Eritrea; Egypt, Arabia to Pakistan.
RS: Carter 1888 (fl) 11/1987.

Kochia indica Wight

Acta Bot. Hung. 50: 184 (2008).
Syn: *Bassia indica* (Wight) A.J.Scott – F.Egypt 1: 107 (1999).
Annual herb. Waste ground, open sandy areas.

Distr: RS; Libya, Egypt, Palestine to India.
RS: Yeates 14 (fl) 3/1988.

Nothosaerva brachiata (L.) Wight & Arn.

F.P.S. 1: 119 (1950); F.J.M.: 92 (1976); F.T.E.A. Amaranth.: 88 (1985); F.E.E. 2(1): 326 (2000).
Annual herb. Damp bare soil, ditches.
Distr: DAR, KOR, CS, UN; Senegal to Somalia, S to Angola & Zimbabwe; India, Sri Lanka, Myanmar.
DAR: Wickens 1151 (fl fr) 2/1964; **UN:** Sherif A.2883 (fl) 10/1951.

Pandiaka angustifolia (Vahl) Hepper

F.J.M.: 92 (1976); F.T.E.A. Amaranth.: 113 (1985); F.E.E. 2(1): 332 (2000).
Syn: *Pandiaka heudelotii* (Moq.) Hiern – F.P.S. 1: 120 (1950).
Annual herb. Woodland & wooded grassland on sandy soils.
Distr: DAR, KOR, BAG, UN; Mauritania & Senegal to Ethiopia, S to Angola & Malawi.
KOR: Wickens 412 (fr) 9/1962; **BAG:** Schweinfurth 2309 (fr) 7/1869.

Pandiaka elegantissima (Schinz) Dandy

F.P.S. 1: 119 (1950); Kew Bull. 34: 428 (1979).
Syn: *Pandiaka oblanceolata* (Schinz) C.B.Clarke – F.P.S. 1: 120 (1950).
Perennial herb. Wooded grassland, pathsides.
Distr: BAG; South Sudan endemic.
IUCN: RD
BAG: Myers 7307 (fr) 7/1937.
Note: Cavaco in Mem. Mus. Nat. Hist. Nat., Sér. B, 13: 129 (1962) described var. *macrantha* (under the synonym *P. cylindrica* Hook.f. ex Baker & C.B.Clarke) from C.A.R.; the relationship of this variety to the Sudanese plants needs further investigation.

Pandiaka involucrata (Moq.) B.D.Jacks.

F.T.E.A. Amaranth.: 115 (1985).
Annual herb. Weed of cultivation, rocky ground.
Distr: KOR; Senegal to Tanzania, S to Zambia & Zimbabwe.
KOR: Wickens 352 (fl fr) 9/1962.

Pandiaka welwitschii (Schinz) Hiern

F.T.E.A. Amaranth.: 117 (1985).
Syn: *Pandiaka schweinfurthii* (Schinz) C.B.Clarke – F.P.S. 1: 120 (1950).
Perennial herb. Woodland & grassland.
Distr: BAG; Cameroon, C.A.R., D.R. Congo, Rwanda, Tanzania, Zambia, Angola.
BAG: Schweinfurth 2185 (fl fr) 7/1869.

Psilotrichum elliotii Baker & C.B.Clarke

F.T.E.A. Amaranth.: 97 (1985); Biol. Skr. 51: 100 (1998); F.E.E. 2(1): 324 (2000).
Perennial herb. Dry grassland, bushland & woodland.
Distr: EQU; Ethiopia & Somalia, S to D.R. Congo & Tanzania; India, Sri Lanka.
EQU: Friis & Vollesen 1233 (fl fr) 3/1982.

Psilotrichum gnaphalobryum (Hochst.) Schinz

F.P.S. 1: 120 (1950); F.T.E.A. Amaranth.: 91 (1985); F.E.E. 2(1): 322 (2000).

Perennial herb. Rocky hillslopes, dry grassland & wooded grassland.
Distr: RS; Eritrea, Ethiopia, Somalia, Kenya; Egypt, Arabia.
RS: <u>Jackson 2801</u> (fl fr) 4/1953.

Psilotrichum schimperi Engl.

F.P.S. 1: 120 (1950); F.T.E.A. Amaranth.: 100 (1985); F.E.E. 2(1): 324 (2000).
Annual herb. Damp ground, swamps.
Distr: ?CS; Rwanda, Ethiopia, Uganda, Kenya, Tanzania, Zambia.
Note: the inclusion of this species is based on Andrews who recorded it from Central Sudan; more recent treatments of this species do not include Sudan within the distribution.

Pupalia grandiflora Peter

F.T.E.A. Amaranth.: 74 (1985); Biol. Skr. 51: 101 (1998); F.E.E. 2(1): 320 (2000).
Perennial herb. Forest & forest margins, often riverine, woodland.
Distr: EQU; Ethiopia to D.R. Congo & Tanzania; Arabia.
EQU: <u>Myers 11034</u> (fr) 4/1939.

Pupalia lappacea (L.) A.Juss. var. *velutina* (Moq.) Hook.f.

F.P.S. 1: 122 (species only) (1950); F.T.E.A. Amaranth.: 75 (1985); Biol. Skr. 51: 101 (1998); F.E.E. 2(1): 320 (2000).
Annual or perennial herb. Weed of disturbed ground, open woodland, rocky slopes.
Distr: RS, DAR, KOR, ES, BAG, UN, EQU; palaeotropical.
ES: <u>Beshir 49</u> (fl) 8/1951; **UN:** <u>J.M. Lock 81/183</u> (fr) 8/1981.

Salsola imbricata Forssk.

F.E.E. 2(1): 296 (2000).
Syn: *Salsola baryosma* (Schult. ex Roem. & Schult.) Dandy – F.P.S. 1: 111 (1950).
Syn: *Kochia cana* sensu Andrews, non Bunge ex Boiss. – F.P.S. 1: 111 (1950).
Sprawling shrub. Silty saline soils.
Distr: NS, RS, KOR; Eritrea, Somalia; Arabia, Iraq, Iran & Pakistan.
RS: <u>Carter 1923</u> (fr) 11/1987.

Salsola spinescens Moq.

F.E.E. 2(1): 296 (2000).
Spiny shrub. Seasonally damp sites in open sand & on rocky slopes.
Distr: NS, RS; Eritrea, Somalia; Egypt, Arabia.
RS: <u>Andrews 231</u> (fr) 11/1936.

Salsola vermiculata L.

F.P.S. 1: 111 (1950); Fl. Sahara: 232 (1977).
Shrub. Deserts, riverbanks.
Distr: NS; N Africa & Mediterranean.
NS: <u>Ahti 16713</u> 10/1962 (n.v.).
Notes: (1) the specimen cited has not been seen by us; it was cited by Ahti et al. (1973). (2) Boulos (F.Egypt 1: 120) records also *S. cyclophylla* Baker from both Sudan and Ethiopia; however, this species is not included in F.E.E. Clearly, more work is required on this genus in NE Africa.

Sericostachys scandens Gilg & Lopr.

F.T.E.A. Amaranth.: 42 (1985); Biol. Skr. 51: 101 (1998); F.E.E. 2(1): 312 (2000).
Woody climber. Forest.
Distr: EQU; Ivory Coast to Ethiopia, S to Angola & Malawi.
EQU: <u>Howard I.M.10</u> (fl fr) 1/1976.

Sevada schimperi Moq.

F.E.E. 2(1): 292 (2000).
Syn: *Suaeda volkensii* sensu Andrews pro parte, non C.B.Clarke – F.P.S. 1: 112 (1950).
Syn: *Suaeda schimperi* (Moq.) Martelli – F.P.S. 1: 112 (1950).
Subshrub. Open saline sands & silts, salt pans.
Distr: RS; Eritrea, Djibouti, Somalia; Egypt, Arabia.
RS: <u>Jackson 3924</u> (fl) 4/1959.

Suaeda aegyptiaca (Hasselq.) Zohary

F.E.E. 2(1): 291 (2000).
Annual or perennial herb or subshrub. Saltmarshes, wadis.
Distr: RS; Eritrea, Somalia; Libya to Arabia & Iran.
RS: <u>Jackson 3982</u> 4/1959.

Suaeda monoica Forssk. ex J.F.Gmel.

F.P.S. 1: 112 (1950); T.S.S.: 45 (1990); F.E.E. 2(1): 290 (2000).
Syn: *Suaeda fruticosa* sensu Andrews, non Forssk. ex J.F.Gmel. – F.P.S. 1: 111 (1950).
Fleshy shrub. Saltmarshes, saline soils in arid areas.
Distr: RS, CS; Eritrea & Somalia to Mozambique; Egypt to Syria, E to India & Sri Lanka.
RS: <u>Sahni & Kamil 713</u> (fr) 2/1966.

Suaeda vermiculata Forssk. ex J.F.Gmel.

F.P.S. 1: 112 (1950); F.E.E. 2(1): 290 (2000).
Syn: *Suaeda volkensii* C.B.Clarke – F.P.S. 1: 112, pro parte (1950); T.S.S.: 47 (1990).
Fleshy shrub. Coastal bushland on saline soils.
Distr: RS; Mauritania & Senegal to Niger, Eritrea, Somalia; Canary Is., N Africa, Palestine & Arabia to India.
RS: <u>Jackson 3923</u> (fr) 4/1959.

Volkensinia prostrata (Volkens ex Gilg) Schinz

F.T.E.A. Amaranth.: 81 (1985); F.E.E. 2(1): 322 (2000).
Perennial herb or subshrub. Sub-desert, *Acacia-Commiphora* bushland, open sandy & gravelly areas.
Distr: ?EQU; Ethiopia, Kenya, Tanzania.
Note: no specimen has been seen from South Sudan but it is listed as occurring in 'SE Sudan' in F.T.E.A.

Limeaceae

M. Kordofani, H. Gibreel, I. Farag & I. Darbyshire

Limeum diffusum (J.Gay) Schinz

F.P.S. 1: 93 (1950); F.W.T.A. 1(1): 134 (1954).
Annual herb. Wadi beds, sandy plains.
Distr: ?DAR; Mauritania to Chad.
Note: no specimens have been seen but Quézel (1969: 103) records this species from El Fasher whilst Andrews lists it from "Central Sudan".

Limeum obovatum Vicary

F.J.U. 2: 194 (1999); F.E.E. 2(1): 230 (2000).
Syn: *Limeum indicum* Stocks ex T.Anderson – F.P.S. 1: 94 (1950); Bot. Exp. Sud.: 13 (1970).
Annual or short-lived perennial herb. Wadi beds, sandy plains.
Distr: NS, KOR, CS; Mauritania to Eritrea; Libya, Egypt, Arabia to Pakistan.
CS: Kassas 575 8/1954.

Limeum pterocarpum (J.Gay) Heimerl var. pterocarpum

F.P.S. 1: 94 (1950); F.W.T.A. 1(1): 134 (1954).
Annual herb. Sandy plains, fallow land & disturbed areas.
Distr: DAR, KOR; Mauritania to Chad, Namibia to Zimbabwe, S to South Africa.
KOR: Wickens 165 (fl fr) 8/1962.

Limeum viscosum (J.Gay) Fenzl subsp. viscosum var. viscosum

F.T.E.A. Aizo.: 6 (1961); F.E.E. 2(1): 230 (2000).
Syn: *Limeum kotschyii* (Moq.) Schellenb. – F.P.S. 1: 94 (1950).
Annaual or short-lived perennial herb. Weed of cultivation.
Distr: DAR, KOR, UN; Senegal to Ethiopia, S to Namibia & South Africa.
KOR: Wickens 525 (fr) 9/1962; UN: Simpson 7363 7/1929.

Lophiocarpaceae

M. Kordofani, H. Gibreel, I. Farag & I. Darbyshire

Corbichonia decumbens (Forssk.) Exell

F.P.S. 1: 91 (1950); F.T.E.A. Aizo.: 9 (1961); F.E.E. 2(1): 233 (2000).
Annual or short-lived perennial herb. Sandy & stony ground, sometimes seasonally wet.
Distr: RS, UN, EQU; Mauritania to Somalia, S to Namibia & South Africa; N Africa, tropical Asia.
RS: Schweinfurth 465 (fl fr) 9/1868; UN: Sherif A.3950 (fr) 8/1951.

Gisekiaceae

M. Kordofani, H. Gibreel & I. Darbyshire

Gisekia pharnaceoides L. var. pharnaceoides

F.P.S. 1: 91 (1950); F.T.E.A. Aizo.: 5 (1961); Biol. Skr. 51: 91 (1998); F.E.E. 2(1): 238 (2000).
Syn: *Gisekia rubella* Moq. – F.P.S. 1: 92 (1950).
Annual or short-lived perennial herb. Open woodland & bushland on sandy soils, waste ground, weed of cultivation.
Distr: NS, RS, DAR, KOR, ES, UN, EQU; palaeotropical.
KOR: Wickens 585 (fr) 10/1962; UN: J.M. Lock 81/50 (fr) 5/1981.

Aizoaceae

M. Kordofani, H. Gibreel & I. Darbyshire

Aizoon canariense L.

F.P.S. 1: 95 (1950); F.T.E.A. Aizo.: 29 (1961); F.E.E. 2(1): 241 (2000).

Perennial herb. Semi-desert.
Distr: NS, RS, CS; Chad, Eritrea, Somalia, Kenya, Namibia & Zimbabwe to South Africa; Macaronesia, N Africa to Pakistan.
RS: Jackson 2783 (fl fr) 4/1953.

Sesuvium hydaspicum (Edgew.) Gonç.

F.E.E. 2(1): 246 (2000).
Syn: *Trianthema polysperma* Hochst. ex Oliv. – F.P.S. 1: 95 (1950).
Annual or short-lived perennial herb. Open sandy ground.
Distr: RS, KOR, CS; Cape Verde Is., Senegal to Eritrea, S to South Africa; India.
CS: Pettet 34 (fl fr) 9/1962.

Trianthema portulacastrum L.

F.T.E.A. Aizo.: 23 (1961); F.E.E. 2(1): 245 (2000).
Annual succulent herb. Weed of cultivation, waste ground, sandy plains.
Distr: RS, CS, ES, BAG, UN, EQU; pantropical.
ES: Carter 1779 (fl fr) 11/1987; UN: J.M. Lock 81/62 (fl) 5/1981.

Trianthema salsoloides Fenzl ex Oliv.

F.P.S. 1: 96 (1950); F.T.E.A. Aizo.: 26 (1961).
Annual herb. Open areas, waste ground.
Distr: KOR; Kenya, Tanzania.
KOR: Wickens 608 (fr) 10/1962.
Note: Gilbert et al. in F.E.E. record this species from Ethiopia, but according to Hartmann et al. (Pl. Ecol. Evol. 144) it is restricted to Sudan, Kenya & Tanzania.

Trianthema sedifolia Vis.

F.P.S. 1: 96 (1950); Pl. Ecol. Evol. 144: 206 (2011).
Syn: *Trianthema triquetra* sensu Gilbert et al., non Willd. – F.E.E. 2(1): 244 pro parte (2000).
Annual herb. Open, disturbed stony ground.
Distr: NS, RS, KOR, CS, ES, EQU; Eritrea & Somalia, S to Tanzania; N Africa, Arabia.
CS: Schweinfurth 904 (fl fr) 1/1869; EQU: Peers B.M.1 (fl fr) 8/1953.

Trianthema sheilae A.G.Mill. & J.A.Nyberg

Pl. Ecol. & Evol. 144: 207 (2011).
Syn: *Trianthema crystallina* sensu auctt., non (Forssk.) Vahl – F.P.S. 1: 96 (1950); F.E.E. 2(1): 244 pro parte (2000).
Annual or perennial herb. Sandy & rocky plains, often near the sea.
Distr: NS, RS; Eritrea; Egypt, Arabia.
RS: Murray N.D.S.3705 (fl fr) 6/1925.

Zaleya pentandra (L.) C.Jeffrey

F.T.E.A. Aizo.: 28 (1961); F.E.E. 2(1): 246 (2000).
Syn: *Trianthema pentandra* L. – F.P.S. 1: 95 (1950).
Annual or perennial herb. Weed of cultivation, waste ground & open habitats.
Distr: NS, RS, DAR, KOR, CS, ES, BAG, UN, EQU; Senegal to Eritrea, S to South Africa; Madagascar, Egypt, Arabia.
ES: Jackson 4161 (fl fr) 4/1961; UN: J.M. Lock 81/61 (fl fr) 5/1981.

Phytolaccaceae

M. Kordofani, H. Gibreel & I. Darbyshire

Hilleria latifolia (Lam.) H.Walter

F.P.S. 1: 107 (1950); F.T.E.A. Phytolacc.: 6 (1971); Biol. Skr. 51: 95 (1998); F.E.E. 2(1): 276 (2000).
Shrubby perennial herb. Forest including disturbed patches.
Distr: EQU; Guinea to Kenya, S to Angola & Mozambique; Madagascar, Mascarenes, tropical America.
EQU: Myers 9659 (fl) 10/1938.

Phytolacca dioica L.

F.E.E. 2(1): 274 (2000).
Tree. Cultivated ornamental, sometimes naturalised.
Distr: ES; native to the subtropical Americas; widely planted in S Europe & N Africa.
ES: Sahni & Kamil 637 (fl) 4/1967.

Phytolacca dodecandra L'Hér.

F.T.E.A. Phytolacc.: 2 (1971); T.S.S.: 45 (1990); Biol. Skr. 51: 95 (1998); F.E.E. 2(1): 274 (2000).
Shrub. Disturbed ground, forest margins, bushland.
Distr: EQU; Guinea to Eritrea, S to South Africa; Madagascar.
EQU: Jackson 602 1/1949.

Nyctaginaceae

H. Gibreel, M. Kordofani & I. Darbyshire

[Boerhavia]

Note: the taxonomy of this genus is difficult and was almost certainly confused by Andrews.

Boerhavia coccinea Mill.

F.J.M.: 96 (1976); F.T.E.A. Nyctag.: 5 (1996); F.E.E. 2(1): 268 (2000).
Syn: *Boerhavia repens* L. var. *viscosa* Choisy – F.P.S. 1: 152 (1950); Bot. Exp. Sud.: 88 (1970).
Annual or perennial herb. Open, disturbed ground; weed of cultivation.
Distr: NS, RS, DAR, KOR, CS, ES, BAG, UN; pantropical.
KOR: Pfund 675 (fl fr) 11/1875; **UN:** Simpson 7294 (fl) 7/1929.

Boerhavia diffusa L.

F.J.M.: 96 (1976); F.T.E.A. Nyctag.: 3 (1996); Biol. Skr. 51: 111 (1998); F.E.E. 2(1): 268 (2000).
Syn: *Boerhavia repens* L. var. *diffusa* (L.) Hook.f. – F.P.S. 1: 151 (1950).
Annual or short-lived perennial herb. Weed of cultivation, disturbed grassland & woodland.
Distr: NS, RS, DAR, KOR, CS, UN, EQU; pantropical.
DAR: Wickens 1753 (fl) 6/1964; **EQU:** Myers 10209 (fl) 12/1938.

Boerhavia elegans Choisy

F.P.S. 1: 151 (1950); Bot. Exp. Sud.: 88 (1970); F.T.E.A. Nyctag.: 7 (1996); F.E.E. 2(1): 269 (2000).
Shrublet or perennial herb. Semi-desert, wadis, dry bushland.
Distr: ?RS, EQU; Niger to Somalia & Kenya; Libya to Arabia & Pakistan.

EQU: Drar 956 4/1938 (n.v.).
Note: no material from the Sudan region has been seen; the record is based on Andrews who recorded this distinctive species from Red Sea District. The South Sudan record is based on Drar (1970).

Boerhavia erecta L.

F.T.E.A. Nyctag.: 2 (1996); Biol. Skr. 51: 112 (1998); F.E.E. 2(1): 298 (2000).
Annual or short-lived perennial herb. Weed of cultivation, waste ground & disturbed bushland.
Distr: KOR, CS, EQU; pantropical.
KOR: Wickens 416 (fl) 9/1962; **EQU:** von Ramm 162 (fl fr) 4/1956.

Boerhavia repens L. subsp. repens

F.P.S. 1: 151 pro parte (1950); F.T.E.A. Nyctag.: 6 (1996); F.E.E. 2(1): 268 (2000).
Annual or perennial herb. Weed of cultivation & disturbed areas.
Distr: NS, RS, DAR, KOR, CS, ES; pantropical.
CS: Pettet 57 (fl fr) 9/1962.

Boerhavia repens L. subsp. diandra (L.) Maire & Weiller

F.Egypt 1: 37 (1999).
Syn: *Boerhavia diandra* L. – Bot. Exp. Sud.: 88 (1970).
Syn: *Boerhavia vulvariifolia* Poir. – F. Darfur Nord-Occ. & J. Gourgeil: 107 (1969).
Annual or short-lived perennial herb. Weed of cultivation & waste ground.
Distr: NS, RS, DAR, CS, ES; N Africa, distribution elsewhere unclear.
RS: Kassas 2 (fl) 12/1966.

[Bougainvillea spectabilis Willd.]

Note: this species is very widely grown in Sudan and South Sudan as an ornamental but is not known to naturalise.

Commicarpus grandiflorus (A.Rich.) Standl.

F.T.E.A. Nyctag.: 9 (1996); Biol. Skr. 51: 112 (1998); F.E.E. 2(1): 271 (2000).
Syn: *Commicarpus pentandrus* (Burch.) Heimerl – F.J.M.: 96 (1976).
Syn: *Commicarpus africanus* sensu auctt., non (Lour.) Dandy – F.P.S. 1: 152 pro parte (1950); Bot. Exp. Sud.: 88 (1970); F.J.M.: 96 (1976).
Perennial herb. Disturbed grassland & woodland, roadsides.
Distr: NS, RS, DAR, ES; Eritrea & Somalia, S to Malawi; S Sahara, Egypt, Arabia, India.
DAR: Jackson 3306 (fl fr) 12/1954.

Commicarpus helenae (Schult.) Meikle

F.T.E.A. Nyctag.: 14 (1996); F.E.E. 2(1): 273 (2000).
Syn: *Commicarpus verticillatus* sensu Andrews, non (Poir.) Standl. – F.P.S. 1: 152 (1950).
Perennial herb, sometimes scandent. Dry woodland & bushland.
Distr: NS, RS, DAR, KOR, CS; Senegal to Somalia, S to Botswana; Macaronesia, Egypt & Arabia to India.
RS: Carter 1869 (fl fr) 11/1987.

Commicarpus montanus Miré, H.Gillet & Quézel

F. Darfur Nord-Occ. & J. Gourgeil: 107 (1969).
Perennial herb. Rocky slopes.
Distr: DAR; Chad (Tibesti); Libya.
IUCN: RD
Note: Quézel noted this species as widespread and common on Jebel Gourgeil; it is otherwise restricted to the eastern Saharan mountains. We have not seen Quézel's specimens.

Commicarpus pedunculosus (A.Rich.) Cufod.

F.T.E.A. Nyctag.: 8 (1996); F.E.E. 2(1): 270 (2000).
Perennial herb. Open ground, often on black cotton soils.
Distr: BAG; Eritrea & Somalia, S to Rwanda & Tanzania.
BAG: Myers 12052 (fl) 10/1939.
Note: the single specimen seen is only tentatively identified as this species, the inflorescences being immature.

Commicarpus plumbagineus (Cav.) Standl.

F.T.E.A. Nyctag.: 12 (1996); Biol. Skr. 51: 112 (1998); F.E.E. 2(1): 271 (2000).
Syn: *Commicarpus africanus* sensu Andrews pro parte, non (Lour.) Dandy – F.P.S. 1: 152 (1950).
Perennial herb, sometimes scandent. Woodland, bushland & wooded grassland, often in disturbed areas.
Distr: RS, DAR, CS, EQU; Senegal to Somalia, S to South Africa; S Spain, N Africa to Middle East & Arabia.
RS: Schweinfurth 272 (fl) 4/1868; **EQU:** Myers 7019 7/1937.

Mirabilis jalapa L.

F.P.S. 1: 152 (1950); F.T.E.A. Nyctag.: 15 (1996); F.E.E. 2(1): 265 (2000).
Perennial herb. Disturbed areas along roads & near habitation.
Distr: DAR, ES, EQU; pantropical, originally native of Peru.
DAR: Miehe 273 (fl) 7/1982; **EQU:** Hamdi 17 (fl) 1933.

Pisonia aculeata L.

F.T.E.A. Nyctag.: 17 (1996); Biol. Skr. 51: 112 (1998); F.E.E. 2(1): 265 (2000).
Woody climber. Forest & forest margins, often along water courses.
Distr: EQU; widespread in the tropics.
EQU: Friis & Vollesen s.n. 1980.
Note: the entry for South Sudan is based on a sight record by Friis & Vollesen (1998).

Molluginaceae

M. Kordofani, H. Gibreel, I. Farag & I. Darbyshire

Glinus lotoides L. var. lotoides

F.P.S. 1: 92 (1950); F.T.E.A. Aizo.: 15 (1961); F.J.M.: 89 (1976); F.E.E. 2(1): 234 (2000).
Annual or short-lived perennial herb. Open seasonally wet areas, usually in disturbed sites.
Distr: NS, RS, DAR, CS, BAG, UN; widespread in the tropics & subtropics.
DAR: Wickens 1343 (fl fr) 3/1964; **UN:** J.M. Lock 80/26 (fl fr) 2/1980.

Glinus lotoides L. var. virens Fenzl

F.E.E. 2(1): 234 (2000).
Syn: *Glinus lotoides* × *oppositifolius* sensu Wickens – F.J.M.: 89 (1976).
Annual or short-lived perennial herb. Open seasonally wet areas, usually in disturbed sites.
Distr: NS, DAR, EQU; mainly E & S Africa, but distribution not fully documented.
DAR: Wickens 1781 (fl fr) 6/1964; **EQU:** Schweinfurth 3733 (fl fr) 5/1870.

Glinus oppositifolius (L.) Aug.DC.

F.P.S. 1: 93 (1950); F.T.E.A. Aizo.: 13 (1961); F.E.E. 2(1): 234 (2000).
Annual or short-lived perennial herb. Wooded grassland, open sandy areas, weed of cultivation.
Distr: UN; pantropical.
UN: Douglas 7790 (fl fr) 4/1930.

Mollugo cerviana (L.) Ser. ex DC. var. cerviana

F.T.E.A. Aizo.: 17 (1961); F.J.U. 2: 196 (1999); F.E.E. 2(1): 236 (2000).
Annual herb. Sandy soils in bushland.
Distr: NS; Eritrea, Ethiopia; N Africa.
NS: Léonard 4835 11/1968.

Mollugo cerviana (L.) Ser. ex DC. var. spathulifolia Fenzl

F.P.S. 1: 94 (species only) (1950); F.T.E.A. Aizo.: 17 (1961); F.E.E. 2(1): 237 (2000).
Annual herb. Open sandy areas, waste ground, weed of cultivation.
Distr: NS, RS, DAR, KOR, ES, BAG, EQU; palaeotropical.
ES: Andrews 268 (fl fr) 12/1936; **EQU:** Sillitoe 459 (fl fr) 1919.

Mollugo nudicaulis Lam.

F.P.S. 1: 94 (1950); F.T.E.A. Aizo.: 17 (1961); Biol. Skr. 51: 91 (1998); F.E.E. 2(1): 236 (2000).
Annual or short-lived perennial herb. Rocky slopes, sandy areas, waste ground.
Distr: DAR, KOR, CS, ES, BAG, EQU; pantropical.
KOR: Wickens 409 (fl fr) 9/1962; **BAG:** Myers 7313 (fl fr) 7/1937.

Basellaceae

I. Darbyshire

Basella alba L.

F.T.E.A. Basell.: 2 (1968); F.E.E. 2(1): 348 (2000).
Syn: *Basella rubra* L. – F.P.S. 1: 123 (1950).
Annual or short-lived perennial climbing herb. Forest margins, thickets, swampy ground, often by streams.
Distr: EQU; widespread in the tropics, native to tropical Africa.
EQU: Schweinfurth 3040 (fl) 2/1870.
Note: cultivated as a spinach.

Talinaceae

H. Pickering

Talinum caffrum (Thunb.) Eckl. & Zeyh.

F.P.S. 1: 100 (1950); Biol. Skr. 51: 92 (1998); F.E.E. 2(1): 251 (2000); F.T.E.A. Portulac.: 30 (2002).
Perennial herb. Grassland, wooded grassland & bushland.
Distr: DAR; Eritrea to Angola & South Africa.
DAR: <u>Macintosh 128</u> 5/1930.

Talinum portulacifolium (Forssk.) Asch. ex Schweinf.

F.P.S. 1: 100 (1950); Biol. Skr. 51: 92 (1998); F.E.E. 2(1): 250 (2000); F.T.E.A. Portulac.: 29 (2002).
Perennial herb. Grassland, wooded grassland & bushland.
Distr: DAR, KOR, CS, UN, EQU; Nigeria to Somalia, S to Angola & South Africa; Arabia.
KOR: <u>Pfund 337</u> 5/1870; **EQU:** <u>Andrews 539</u> 4/1939.

Talinum triangulare (Jacq.) Willd.

F.W.T.A. 1(1): 136 (1954).
Succulent annual. Cultivated & sometimes naturalised in disturbed areas.
Distr: RS, EQU; native to tropical America, widely naturalised in W Africa.
RS: <u>Crossland S.G.H.1544</u> (fr) 3/1908; **EQU:** <u>Broun 1643</u> (fl fr) 5/1909.

Portulacaceae

H. Pickering & I. Darbyshire

Portulaca erythraeae Schweinf.

F.P.S. 1: 98 (1950); F.E.E. 2(1): 254 (2000).
Succulent ?perennial herb. Habitat unknown.
Distr: ?RS; Eritrea.
IUCN: RD
Note: no material has been seen from Sudan, and in F.E.E. it was said to be known only from the type from Eritrea, but Andrews recorded this species from Erkowit. It is closely allied to *P. oleracea*.

Portulaca foliosa Ker Gawl.

F.P.S. 1: 98 (1950); F.E.E. 2(1): 254 (2000); F.T.E.A. Portulac.: 15 (2002).
Succulent annual or perennial herb. Seasonally dry riverbeds, open sandy or rocky ground.
Distr: ?DAR, BAG, UN, EQU; Senegal to Ethiopia, S to Angola & South Africa.
UN: <u>Simpson 7607</u> 2/1930.
Note: the record for Darfur is from Quézel (1969: 104); we have not seen the specimen on which this is based.

Portulaca kermesina N.E.Br. var. *kermesina*

F.E.E. 2(1): 255 (2000); F.T.E.A. Portulac.: 24 (2002).
Succulent annual herb. Dry sandy or gravelly soils.
Distr: EQU; Eritrea to Somalia, S to Namibia & South Africa.
EQU: <u>Andrews 969</u> 5/1939.

Portulaca oleracea L.

F.P.S. 1: 97 (1950); Biol. Skr. 51: 92 (1998); F.E.E. 2(1): 253 (2000); F.T.E.A. Portulac.: 12 (2002).

Succulent annual herb. Weed of cultivation & disturbed ground, short grassland.
Distr: NS, RS, DAR, KOR, CS, ES, BAG, EQU; pantropical.
DAR: <u>Wickens 1397</u> 4/1964; **EQU:** <u>Kielland-Lund 735</u> 5/1984 (n.v.).

Portulaca quadrifida L.

F.P.S. 1: 98 (1950); Biol. Skr. 51: 92 (1998); F.E.E. 2(1): 257 (2000); F.T.E.A. Portulac.: 8 (2002).
Mat-forming annual or short-lived perennial herb. Weed of cultivation & disturbed ground, open sandy & rocky areas.
Distr: RS, DAR, KOR, CS, ES, UN, EQU; widespread in the tropics & warm temperate regions.
RS: <u>Aylmer 581</u> 3/1936; **EQU:** <u>Myers 11894</u> 1938.

Cactaceae

I. Darbyshire

Opuntia cochenillifera (L.) Mill.

F.T.E.A. Cact.: 3 (1968); F.J.M.: 98 (1976).
Succulent shrub. Rock crevices; naturalised.
Distr: DAR; native of Jamaica & tropical America; widely cultivated and naturalised.
DAR: <u>Wickens 1421</u> 4/1964.
Note: El Amin in T.S.S.: 59 also records *O. vulgaris* Mill. (= *O. ficus-indica* (L.) Mill.) as "cultivated and semi-wild" but does not say where it has naturalised.

Rhipsalis baccifera (J.Miller) Stearn

F.T.E.A. Cact.: 5 (1968); T.S.S.: 61 (1990); F.E.E. 2(1): 260 (2000).
Syn: *Rhipsalis cassutha* Gaertn. – F.P.S. 1: 5 (1968).
Epiphytic shrub. Forest.
Distr: EQU; Sierra Leone to Ethiopia, S to Angola & South Africa; Madagascar, Mascarenes, Sri Lanka, tropical America.
EQU: <u>Andrews 1165</u> 5/1939.
Note: Barthlott in Bradleya 13: 64 (1995) refers the African material to subsp. *mauritiana* (DC.) Barthlott.

CORNALES

Cornaceae

I. Darbyshire

Alangium chinense (Lour.) Harms

F.P.S. 2: 355 (1952); F.T.E.A. Alang.: 3 (1958); F.E.E. 3: 535 (1989); T.S.S.: 346 (1990); Biol. Skr. 51: 295 (1998).
Tree. Forest, especially clearings & margins.
Distr: EQU; Bioko, Cameroon, Ethiopia to Angola & Mozambique; tropical Asia.
EQU: <u>Myers 9602</u> (fr) 10/1938.

Cornus volkensii Harms

Engl. Pflanzenw. Ost-Afr. C: 301 (1895).
Syn: *Afrocrania volkensii* (Harms) Hutch. – F.T.E.A. Corn.: 1 (1958); T.S.S.: 427 (1990); Biol. Skr. 51: 294 (1998).
Tree. Montane forest, often beside streams.
Distr: EQU; D.R. Congo to Kenya, S to Zimbabwe.
EQU: <u>Howard I.M.40</u> (fr) 2/1976.

ERICALES

Balsaminaceae

R. Vandersticht & I. Darbyshire

Impatiens ethiopica Grey-Wilson

F.P.S. 1: 136 (1950); Biol. Skr. 51: 105 (1998); F.E.E. 2(1): 390 (2000).
Perennial herb. Forest, often along streams or swampy ground.
Distr: EQU; Ethiopia.
EQU: Friis & Vollesen 153 (fl) 11/1980.
Note: although having a narrow distribution, this species is locally common in Ethiopia.

Impatiens hochstetteri Warb. subsp. hochstetteri

F.P.S. 1: 137 (1950); F.T.E.A. Balsamin.: 45 (1982); Biol. Skr. 51: 105 (1998); F.E.E. 2(1): 390 (2000).
Annual or perennial herb. Forest, often by streams.
Distr: RS, EQU; Ethiopia to South Africa.
RS: Jackson 2962 (fl) 4/1953; **EQU:** Friis & Vollesen 34 (fl) 11/1980.

Impatiens irvingii Hook.f.

F.P.S. 1: 136 (1950); F.T.E.A. Balsamin.: 26 (1982).
Perennial herb. Forest, often by streams.
Distr: EQU; Guinea to Uganda, S to Angola & Zambia.
EQU: Myers 7775 (fl) 9/1937.

Impatiens meruensis Gilg subsp. septentrionalis Grey-Wilson

F.T.E.A. Balsamin.: 33 (1982); Biol. Skr. 51: 106 (1998).
Perennial herb. Forest, swampy ground.
Distr: EQU; Kenya.
IUCN: RD
EQU: Johnston 1526 (fl) 2/1936.

Impatiens niamniamensis Gilg

F.P.S. 1: 136 (1950); F.T.E.A. Balsamin.: 72 (1982).
Perennial herb. Forest including swampy ground.
Distr: EQU; Bioko & Cameroon to Kenya, S to Angola & Tanzania.
EQU: Sillitoe 429 (fl) 1919.

Impatiens tinctoria A.Rich. subsp. *tinctoria*

F.P.S. 1: 136 (1950); F.T.E.A. Balsamin.: 48 (1982); Biol. Skr. 51: 106 (1998); F.E.E. 2(1): 391 (2000).
Syn: *Impatiens elegantissima* sensu Andrews, non Gilg – F.P.S. 1: 136 (1950).
Perennial herb. Forest, often by streams.
Distr: UN, EQU; D.R. Congo, Eritrea, Ethiopia, Uganda.
UN: J.M. Lock 81/279 (fl) 6/1981.

Sapotaceae

H. Pickering & I. Darbyshire

Chrysophyllum albidum G.Don

F.P.S. 2: 374 (1952); F.T.E.A. Sapot.: 9 (1968); T.S.S.: 354 (1990); Biol. Skr. 51: 304 (1998).
Canopy tree. Lowland forest.
Distr: EQU; Sierra Leone to Kenya & Tanzania.
EQU: Friis & Vollesen 504 11/1980.

Chrysophyllum gorungosanum Engl.

F.T.E.A. Sapot.: 12 (1968); T.S.S.: 354 (1990); Biol. Skr. 51: 305 (1998).
Syn: *Chrysophyllum fulvum* S.Moore – F.P.S. 2: 374 (1952).
Tree. Forest.
Distr: ?EQU; Cameroon, Uganda & Kenya, S to Angola & Zimbabwe.
Note: recorded from the Imatong Mts by both Andrews and Jackson (Sudan J. Ecol. 44: 341-374 (1956)) but we have seen no material; Friis & Vollesen speculate that this may have been a misidentification of one of the two other *Chrysophyllum* species known from South Sudan.

Chrysophyllum muerense Engl.

F.T.E.A. Sapot.: 15 (1968); Biol. Skr. 51: 305 (1998).
Tree. Lowland forest.
Distr: EQU; D.R. Congo, Uganda.
EQU: Friis & Vollesen 498 11/1980.
Note: a very local and scarce species, but listed as Least Concern by Kalema & Beentje, Conservation Checklist of the Trees of Uganda: 207 (2012).

Englerophytum natalense (Sond.) T.D.Penn.

Genera Sapotaceae: 252 (1991).
Syn: *Bequaertiodendron natalense* (Sond.) Heine & J.H.Hemsl. – F.T.E.A. Sapot.: 19 (1968).
Syn: *Chrysophyllum natalense* Sond. – Biol. Skr. 51: 305 (1998).
Tree. Forest.
Distr: ?EQU; Uganda & Kenya, S to South Africa.
Note: the inclusion of this species here is based on Jackson (1956) who recorded it from the Imatong Mts. However, Friis & Vollesen note that this could have been a misidentification of *E. oblanceolatum* (S.Moore) T.D.Penn. or even *Chrysophyllum muerense*.

Manilkara butugi Chiov.

F.T.E.A. Sapot.: 67 (1968); T.S.S.: 355 (1990); Biol. Skr. 51: 305 (1998); F.E.E. 4(1): 57 (2003).
Tree. Montane forest.
Distr: EQU; Ethiopia, Uganda, Kenya.
EQU: Friis & Vollesen 76 11/1980.

Manilkara obovata (Sabine & G.Don) J.H.Hemsl.

F.T.E.A. Sapot.: 68 (1968); Genera Sapotaceae: 134 (1991).
Syn: *Manilkara multinervis* (Baker) Dubard – F.T.E.A. Sapot.: 70 (1968); T.S.S.: 355 (1990).
Syn: *Manilkara multinervis* (Baker) Dubard subsp. *schweinfurthii* (Engl.) J.H.Hemsl. – Bot. Exp. Sud.: 103 (1970); Biol. Skr. 51: 306 (1998).
Syn: *Manilkara schweinfurthii* (Engl.) Dubard – F.P.S. 2: 374 (1952).
Tree. Riverine & swamp forest, woodland.
Distr: BAG, UN, EQU; Sierra Leone to Uganda, S to Angola & Zambia.
EQU: Andrews 1588 5/1939.

Mimusops bagshawei S.Moore

F.T.E.A. Sapot.: 57 (1968); T.S.S.: 356 (1990); Biol. Skr. 51: 306 (1998).
Syn: *Mimusops ugandensis* Stapf – F.P.S. 2: 375 (1952).

Tree. Montane forest.
Distr: EQU; Rwanda, Uganda, Kenya, Tanzania.
EQU: Myers 13475 9/1940.

Mimusops kummel Bruce ex A.DC.

F.T.E.A. Sapot.: 54 (1968); T.S.S.: 356 (1990); Biol. Skr.
51: 306 (1998); F.E.E. 4(1): 55 (2003).
Syn: *Mimusops djurensis* Engl. – F.P.S. 2: 375 (1952).
Syn: *Mimusops fragrans* (Baker) Engl. – F.P.S. 2: 375
(1952); Bot. Exp. Sud.: 103 (1970).
Tree. Riverine forest.
Distr: BAG, UN, EQU; Guinea to Ethiopia, S to Malawi.
EQU: Friis & Vollesen 675 12/1980.

Pouteria adolfi-friedericii (Engl.) A.Meeuse

F.T.E.A. Sapot.: 31 (1968); Biol. Skr. 51: 307 (1998); F.E.E.
4(1): 61 (2003).
Syn: *Aningeria adolfi-friedericii* (Engl.) Robyns &
G.C.C.Gilbert – F.T.E.A. Sapot.: 28 (1968).
Tall tree. Montane forest.
Distr: EQU; D.R. Congo to Ethiopia, S to Zambia &
Zimbabwe.
EQU: Friis & Vollesen 1172 3/1982.

Pouteria alnifolia (Baker) Roberty

F.E.E. 4(1): 61 (2003).
Syn: *Malacantha alnifolia* (Baker) Pierre – F.T.E.A.
Sapot.: 24 (1968); T.S.S.: 354 (1990).
Syn: *Malacantha sp.* – F.P.S. 2: 374 (1952).
Tree. Lowland forest.
Distr: EQU; Senegal to Ethiopia, S to Mozambique.
EQU: Myers 11370 5/1939.

Pouteria altissima (A.Chev.) Baehni

Biol. Skr. 51: 307 (1998); F.E.E. 4(1): 62 (2003).
Syn: *Aningeria altissima* (A.Chev.) Aubrév. & Pellegr. –
F.P.S. 2: 371 (1952); F.T.E.A. Sapot.: 27 (1968); Bot. Exp.
Sud.: 103 (1970); T.S.S.: 353 (1990).
Tree. Lowland forest.
Distr: EQU; Guinea to Ethiopia, S to Zambia.
EQU: Friis & Vollesen 648 12/1980.

Synsepalum brevipes (Baker) T.D.Penn.

Biol. Skr. 51: 308 (1998).
Syn: *Pachystela brevipes* (Baker) Baill. ex Engl. – F.P.S. 2:
375 (1952); F.T.E.A. Sapot.: 36 (1968); T.S.S.: 356 (1990).
Tree. Lowland forest.
Distr: EQU; Guinea to Uganda, S to Angola &
Mozambique.
EQU: Friis & Vollesen 745 12/1980.

Synsepalum cerasiferum (Welw.) T.D.Penn.

Genera Sapotaceae: 248 (1991).
Syn: *Afrosersalisia cerasifera* (Welw.) Aubrév. ex Heine
– F.T.E.A. Sapot.: 42 (1968).
Tree. Lowland forest.
Distr: EQU; Guinea to Ivory Coast, Cameroon to
Uganda, S to Angola & Zambia.
EQU: Andrews 1470 5/1939.

Vitellaria paradoxa C.F.Gaertn.

Biol. Skr. 51: 308 (1998); F.E.E. 4(1): 57 (2003).
Syn: *Butyrospermum niloticum* Kotschy – F.P.S. 2: 373
(1952); Bot. Exp. Sud.: 103 (1970).

Syn: *Butyrospermum paradoxum* (C.F. Gaertn.)
Hepper – F.T.E.A. Sapot.: 49 (1968); T.S.S.: 353 (1990).
Tree. Wooded grassland.
Distr: DAR, BAG, UN, EQU; Senegal to Ethiopia & Uganda.
IUCN: VU
DAR: Kamil 1045 5/1968; **EQU:** Friis & Vollesen 1047
2/1982.
Note: the 'shea butter tree', over-exploited for its timber
which is widely used for charcoal. Sudanese plants are
sometimes treated as subsp. *nilotica* (Kotschy) A.N.Henry,
Chithra & N.C.Nair.

Ebenaceae

R. Vanderstricht & I. Darbyshire

Diospyros abyssinica (Hiern) F.White subsp. *abyssinica*

F.P.S. 2: 370 (1952); T.S.S.: 349 (1990); F.T.E.A. Eben.: 22
(1996); F.E.E. 4(1): 50 (2003).
Syn: *Maba abyssinica* Hiern – F.P.S. 2: 370 (1952).
Small tree. Forest & montane bushland.
Distr: ?KOR, ?CS, BAG, EQU; Mali to Ethiopia, S to
Zimbabwe.
EQU: Andrews 1770 (fr) 6/1939.
Note: El Amin records this species from "Blue Nile, White
Nile, Kordofan and Bahr el Ghazal". We have only seen
material from South Sudan.

Diospyros ferrea (Willd.) Bakh.

F.T.E.A. Eben.: 12 (1996).
Syn: *Diospyros heudelotii* sensu El Amin, non Hiern –
T.S.S.: 349 (1990).
Syn: *Maba lancea* sensu Andrews, non Hiern – F.P.S. 2:
370 (1952).
Shrub or small tree. Riverine forest.
Distr: BAG; Senegal to Kenya, S to Angola &
Mozambique; Madagascar, India to Australia & Hawaii.
BAG: Hoyle 458 (fr) 1/1939.

Diospyros mespiliformis Hochst. ex A.DC.

F.P.S. 2: 367 (1952); F.J.M.: 126 (1976); T.S.S.: 351
(1990); F.T.E.A. Eben.: 40 (1996); F.E.E. 4(1): 50 (2003).
Tree. Riverine forest & woodland.
Distr: RS, DAR, KOR, CS, ES, BAG, UN, EQU; Senegal to
Ethiopia, S to South Africa; Yemen.
DAR: Wickens 2938 4/1965; **UN:** J.M. Lock 80/43 (fl)
2/1980.
Note: El Amin in T.S.S.: 351 (1990) also records *D.
monbuttensis* Gürke from Equatoria. This is believed to
be based on the type specimen, *Schweinfurth* 3598,
collected from Welle which is in modern-day D.R. Congo.

Diospyros scabra (Chiov.) Cufod.

F.T.E.A. Eben.: 18 (1996); Biol. Skr. 51: 303 (1998); F.E.E.
4(1): 51 (2003).
Tree or shrub. Rocky hillslopes, seasonally dry
watercourses.
Distr: EQU; Ethiopia, Uganda, Kenya.
EQU: Padwa 233 (fr) 5/1953.

Euclea divinorum Hiern

F.P.S. 2: 370 (1952); T.S.S.: 351 (1990); F.T.E.A. Eben.: 47
(1996); F.E.E. 4(1): 52 (2003).

Shrub or small tree. Open montane forest & bushland.
Distr: EQU; Ethiopia to South Africa; Socotra.
EQU: Myers 14198 10/1941.

Euclea racemosa Murray subsp. *schimperi* (A.DC.) F.White

F.P.S. 2: 370 (1952); F.T.E.A. Eben.: 46 (1996); F.E.E. 4(1): 51 (2003).
Syn: *Euclea schimperi* A.DC. – F.P.S. 2: 370 (1952); Bot. Exp. Sud.: 47 (1970); T.S.S.: 353 (1990).
Tree. Dry forest margins, bushland & grassland.
Distr: RS, EQU; Eritrea to South Africa; Comoros, Egypt, Arabia.
RS: Broun 1135 (fl) 5/1907; **EQU:** Myers 10980 4/1939.

Primulaceae

I. Darbyshire

Ardisiandra sibthorpioides Hook.f.

F.T.E.A. Primul.: 2 (1958); Biol. Skr. 51: 404 (2005); F.E.E. 5: 26 (2006).
Syn: *Ardisiandra engleri* Weim. var. *microphylla* Weim. – F.P.S. 3: 65 (1956).
Prostrate perennial herb. Forest & montane bushland, streamside etc.
Distr: EQU; Bioko, Cameroon, Ethiopia to Malawi.
EQU: Thomas 1712 (fl) 12/1935.

[*Embelia*]

Note: the genus *Embelia* is in need of revision in Africa; naming of material is problematic at present.

Embelia schimperi Vatke

F.T.E.A. Myrsin.:13 (1984); T.S.S.: 357 (1990); Biol. Skr. 51: 309 (1998); F.E.E. 4(1): 69 (2003).
Woody climber or shrub. Montane forest & bushland.
Distr: DAR, EQU; Cameroon, Ethiopia to Angola & Zimbabwe.
DAR: Miehe 445 8/1982; **EQU:** Friis & Vollesen 142 (fr) 11/1980.
Note: the single collection seen from Jebel Marra is sterile but looks consistent with this species.

Embelia sp. A

Shrub. Riverine forest.
Distr: DAR.
DAR: Miehe 340a (fl) 8/1982.
Note: the single specimen seen was determined by N. Robson as *E.* cf. *rowlandii* Gilg; *E. rowlandii* is a West African species. The fertile shoot on *Miehe* 340a looks quite different to the sterile shoots on this and on 340 which have leaves with a more acute apex, a more toothed margin and drying green not blackish; two species may be involved.

Embelia sp. B

Shrub. Habitat not recorded.
Distr: EQU.
EQU: Andrews 1602 (fl) 5/1939.
Note: determined by A. Taton in 1976 as *E.* nr. *welwitschii* (Hiern) K.Schum.

Embelia sp. C

Syn: *Embelia* sp. sensu Wickens – F.J.M.: 126 (1976).
Scrambling shrub. Riverine forest.
Distr: DAR.
DAR: Wickens 1916 (fl) 7/1964.
Note: Wickens noted that this collection is near *E. welwitschii* (Hiern) K.Schum. but it is almost certainly different to the specimen identified as *E.* nr. *welwitschii* from Equatoria (*Andrews* 1602 – sp. B above).

Embelia sp. D

Habit and habitat not recorded.
Distr: EQU.
EQU: von Ramm 211 (fl) 4/1956.
Note: A. Taton identified this specimen as belonging near *E. nilotica* Oliv.

Lysimachia adoensis (Kunze) Klatt

Syn: *Asterolinon adoense* Kunze – F.T.E.A. Primul.: 9 (1958); F.J.M.: 139 (1976); F.E.E. 5: 20 (2006).
Annual herb. Open forest, sunny banks, weed of cultivation & fallow land.
Distr: RS, DAR; Eritrea, Uganda, Kenya, Tanzania, Angola.
DAR: Wickens 2698 (fl fr) 9/1964.

Lysimachia arvensis (L.) U.Manns & Anderb.

Willdenowia 39: 51 (2009).
Syn: *Anagallis arvensis* L. – F.P.S. 3: 65 (1956); F.T.E.A. Primul.: 10 (1958); Bot. Exp. Sud.: 97 (1970); F.E.E. 5: 21 (2006).
Annual or biennial herb. Weed of cultivation & fallow land, open sandy soils.
Distr: RS, CS; native to Europe, now widely naturalised in the temperate regions and high altitude tropics.
RS: Jackson 2737 (fl fr) 3/1953.

Lysimachia barbata (P.Taylor) U.Manns & Anderb.

Willdenowia 39: 51 (2009).
Syn: *Anagallis pumila* Sw. var. *barbata* P.Taylor – F.T.E.A. Primul.: 17 (1958); Biol. Skr. 51: 404 (2005); F.E.E. 5: 24, in notes (2006).
Annual herb. Wet flushes over rocks.
Distr: BAG, EQU; Mali to Kenya, S to Angola, Zimbabwe & Mozambique.
EQU: Friis & Vollesen 623 (fl fr) 12/1980.

Lysimachia djalonis (A.Chev.) U.Manns & Anderb.

Willdenowia 39: 51 (2009).
Syn: *Anagallis djalonis* A.Chev. – F.T.E.A. Primul.: 17 (1958); Biol. Skr. 51: 403 (2005).
Annual herb. Forest margins & clearings, secondary bushland.
Distr: EQU; Guinea to Kenya, S to Angola, Zambia & Malawi.
EQU: Friis & Vollesen 526 (fr) 11/1980.

Lysimachia ruhmeriana Vatke

F.T.E.A. Primul.: 5 (1958); Biol. Skr. 51: 404 (2005); F.E.E. 5: 18 (2006).
Syn: *Lysimachia africana* Engl. – F.P.S. 3: 66 (1956).
Perennial herb. Marshes, wet montane grassland & scrub.
Distr: EQU; Cameroon to Eritrea, S to South Africa;

Madagascar, Arabia.
EQU: <u>Myers 11601</u> (fl fr) 7/1939.

Lysimachia serpens (Hochst. ex A.DC.) U.Manns & Anderb. subsp. *serpens*
Willdenowia 39: 53 (2009).
Syn: *Anagallis serpens* Hochst. ex A.DC. – F.T.E.A. Primul.: 13 (1958); Biol. Skr. 51: 404 (2005); F.E.E. 5: 21 (2006).
Prostrate perennial herb. Wet montane grassland, marshes, streamsides.
Distr: EQU; Ethiopia, Zimbabwe; Arabia.
EQU: <u>Myers 11701</u> (fl) 7/1939.
Note: a localised species but not considered threatened.

Maesa lanceolata Forssk.
F.P.S. 2: 377 (1952); F.J.M.: 126 (1976); T.S.S.: 357 (1990); Biol. Skr. 51: 309 (1998); F.E.E. 4(1): 64 (2003).
Tree or shrub. Montane forest, particularly along margins & clearings, montane bushland & grassland.
Distr: DAR, EQU; Guinea to Ethiopia, S to South Africa; Madagascar, Arabia.
DAR: <u>Wickens 1072</u> (fr) 1/1964; **EQU:** <u>Myers 11719</u> (fl) 7/1939.

Maesa welwitschii Gilg
F.T.E.A. Myrsin.: 5 (1984); Biol. Skr. 51: 310 (1998).
Syn: *Maesa schweinfurthii* Mez – F.P.S. 2: 377 (1952).
Woody climber, shrub or small tree. Forest & secondary bushland.
Distr: EQU; Cameroon to Uganda, S to Angola & D.R. Congo.
EQU: <u>Friis & Vollesen 517</u> (fr) 11/1980.

Myrsine africana L.
F.T.E.A. Myrsin.: 6 (1984); T.S.S.: 357 (1990); Biol. Skr. 51: 310 (1998); F.E.E. 4(1): 66 (2003).
Shrub. Montane bushland & forest margins.
Distr: RS, EQU; Eritrea & Somalia, S to Angola & South Africa.
RS: <u>Jackson 3968</u> (fl fr) 4/1959; **EQU:** <u>Jackson 1327</u> 3/1950.

Rapanea melanophloeos (L.) Mez
F.T.E.A. Myrsin.: 8 (1984); Biol. Skr. 51: 310 (1998).
Syn: *Myrsine melanophloeos* (L.) R.Br. – F.E.E. 4(1): 67 (2003).
Syn: *Rapanea neurophylla* (Gilg) Mez – F.P.S. 2: 377 (1952).
Syn: *Rapanea pulchra* Gilg & Schellenb. – T.S.S.: 358 (1990).
Tree or shrub. Montane forest, often at upper margins, montane bushland.
Distr: EQU; Nigeria, Bioko & Cameroon, Ethiopia to Angola & South Africa; Madagascar, Comoros.
EQU: <u>Friis & Vollesen 1137</u> (fr) 3/1982.

Samolus valerandi L.
F.P.S. 3: 66 (1956); F.T.E.A. Primul.: 19 (1958); F.E.E. 5: 28 (2006).
Annual herb. Marshes, open wet mud.
Distr: RS; almost cosmopolitan but scattered.
RS: <u>Aylmer 580</u> (fl fr) 3/1936.

Ericaceae
I. Darbyshire

Agarista salicifolia (Comm. ex Lam.) G.Don
F.E.E. 4(1): 48 (2003); F.T.E.A. Eric.: 2 (2006).
Syn: *Agauria salicifolia* (Comm. ex Lam.) Hook.f. ex Oliv. – T.S.S.: 348 (1990); Biol. Skr. 51: 302 (1998).
Shrub or small tree. Montane bushland & forest margins.
Distr: EQU; Cameroon to Ethiopia, S to Zambia; Madagascar, Mascarenes.
EQU: <u>Friis & Vollesen 352</u> 11/1980.

Erica arborea L.
F.P.S. 2: 367 (1952); T.S.S.: 349 (1990); Biol. Skr. 51: 302 (1998); F.E.E. 4(1): 46 (2003); F.T.E.A. Eric.: 10 (2006).
Shrub or small tree. Scrub on montane rocky slopes, upper forest margins.
Distr: EQU; Chad, Eritrea & Somalia, S to D.R. Congo & Tanzania; Canary Is., N Africa & S Europe to Black Sea & Arabia.
EQU: <u>Thomas 1874</u> (fl fr) 12/1935.

Erica silvatica (Engl.) Beentje
F.T.E.A. Eric.: 19 (2006).
Syn: *Blaeria breviflora* Engl. – F.P.S. 2: 367 (1952).
Syn: *Blaeria spicata* Hochst. ex A.Rich. – F.P.S. 2: 366 (1952); Biol. Skr. 51: 302 (1998).
Syn: *Erica tenuipilosa* (Engl. ex Alm & T.C.E.Fr.) Cheek subsp. *spicata* (Hochst. ex A.Rich.) Cheek – F.E.E. 4(1): 47 (2003).
Shrub. Montane bushland, grassland & forest margins, sometimes dominant.
Distr: DAR, EQU; Guinea, Ivory Coast, Bioko, Cameroon, Ethiopia to Malawi & Zimbabwe.
DAR: <u>de Wilde et al. 5558</u> (fl) 1/1965; **EQU:** <u>Tothill 13520</u> (fl) 9/1940.

UNPLACED EUDICOTS

Icacinaceae
I. Darbyshire

Apodytes dimidiata E.Mey. ex Arn. subsp. *acutifolia* (Hochst. ex A.Rich.) Cufod.
T.S.S.: 284 (species only) (1990); Biol. Skr. 51: 260 (1998).
Syn: *Apodytes dimidiata* E.Mey. ex Arn. var. *acutifolia* (Hochst. ex A.Rich.) Boutique – F.T.E.A. Icac.: 4 (1968); F.E.E. 3: 348 (1989).
Tree. Montane forest.
Distr: EQU; Eritrea to D.R. Congo & Malawi; India, Sri Lanka.
EQU: <u>Howard I.M.26</u> (fl) 2/1976.
Note: subsp. *acutifolia* is treated as a variety in F.E.E. and F.T.E.A. but since it is allopatric from typical *dimidiata*, subspecies status seems more appropriate.

Icacina oliviformis (Poir.) J.Raynal
Adansonia, sér. 2 15: 194 (1975).
Syn: *Icacina senegalensis* Juss. – F.P.S. 2: 286 (1952); T.S.S.: 285 (1990).

Suffrutex or scandent shrub. Wooded grassland, forest, abandoned cultivations.
Distr: BAG; Senegal to C.A.R. & D.R. Congo.
BAG: Turner 266 (fr).

Pyrenacantha sylvestris S.Moore

F.T.E.A. Icac.: 14 (1968); F.E.E. 3: 351 (1989); Biol. Skr. 51: 260 (1998).
Woody climber. Forest.
Distr: EQU; Gabon to Ethiopia, S to Angola & D.R. Congo.
EQU: Friis & Vollesen 503 (fl) 11/1980.

Rhaphiostylis beninensis (Hook.f. ex Planch.) Planch. ex Benth.

F.T.E.A. Icac.: 9 (1968); F.E.E. 3: 350 (1989); Biol. Skr. 51: 261 (1998).
Woody climber or scandent shrub. Forest, evergreen bushland.
Distr: EQU; Senegal to Ethiopia, S to Angola, Zimbabwe & Mozambique.
EQU: Friis & Vollesen 932 (fl) 2/1982.

GENTIANALES

Rubiaceae

H. Pickering & I. Darbyshire

Afrocanthium lactescens (Hiern) Lantz

Bot. J. Linn. Soc. 146: 278 (2004).
Syn: *Canthium lactescens* Hiern – F.P.S. 2: 430 (1952); T.S.S.: 386 (1990); F.T.E.A. Rub. 3: 871 (1991); Biol. Skr. 51: 327 (1998); F.E.E. 4(1): 270 (2003).
Shrub or small tree. Upland bushland & woodland, often on rocky hillslopes.
Distr: BAG, EQU; Ethiopia to Angola & Zimbabwe.
BAG: Turner 228 (fl) 5/1936.

Aidia genipiflora (DC.) Dandy

F.P.S. 2: 424 (1952); F.W.T.A. 2: 114 (1963).
Small tree or shrub.
Distr: ?BAG; Guinea to C.A.R.
Note: recorded by Andrews from "Equatoria: Dar Fertit" which is in Bahr el Ghazal; we have not seen the specimen on which this record was based.

Anthospermum pachyrrhizum Hiern

F.P.S. 2: 425 (1952); F.J.M.: 128 (1976); F.E.E. 4(1): 233 (2003).
Dwarf shrub. Rocky ground.
Distr: DAR; Eritrea, Ethiopia; Yemen.
IUCN: RD
DAR: de Wilde et al. 5532 1/1965.

Anthospermum usambarense K.Schum.

F.P.S. 2: 425 (1952); F.T.E.A. Rub. 1: 331 (1976); Biol. Skr. 51: 326 (1998).
Shrub. Montane bushland, grassland & forest margins.
Distr: EQU; D.R. Congo, Uganda & Kenya, S to Malawi & Zimbabwe.
EQU: Jackson 1516 5/1950.

Argocoffeopsis rupestris (Hiern) Robbr. subsp. thonneri (Lebrun) Robbr.

Bull. Jard. Bot. Nat. Belg. 51: 369 (1981).
Woody climber. Forest.
Distr: EQU; C.A.R., D.R. Congo.
EQU: von Ramm 212 (fl) 2/1956.

Belonophora coffeoides Hook.f. subsp. hypoglauca (Welw. ex Hiern) S.E.Dawson & Cheek

Kew Bull. 55: 77 (2000).
Syn: *Belonophora glomerata* M.B.Moss – F.P.S. 2: 425 (1952); T.S.S.: 384 (1990).
Syn: *Belonophora hypoglauca* (Welw. ex Hiern) A.Chev. – F.T.E.A. Rub. 2: 728 (1988); Biol. Skr. 51: 326 (1998).
Shrub or small tree. Lowland forest.
Distr: BAG, EQU; Sierre Leone to Uganda, S to Angola & Zambia.
EQU: Chipp 46 (fl) 2/1929.

[Bertiera]

Note: Andrews and El Amin list *Bertiera aethiopica* Hiern from Equatoria but this is believed to be based on *Schweinfurth* 3274 which is from Yuroo in current-day D.R. Congo and no other South Sudanese records are known, though it is quite likely to extend into Equatoria.

Breonadia salicina (Vahl) Hepper & J.R.I.Wood

F.T.E.A. Rub. 2: 445 (1988); F.E.E. 4(1): 240 (2003).
Syn: *Adina microcephala* (Delile) Hiern – F.P.S. 2: 423 (1952); Bot. Exp. Sud.: 99 (1970); F.J.M.: 129 (1976); T.S.S.: 383 (1990).
Tree. Riverine forest & thicket.
Distr: DAR, KOR, CS, BAG; Mali to Ethiopia, S to Angola & South Africa; Madagascar, Yemen.
DAR: Wickens 1533 (fl) 5/1964; **BAG:** Schweinfurth III.238 (fl) 1/1871.

Calycosiphonia spathicalyx (K.Schum.) Robbr.

F.T.E.A. Rub. 2: 727 (1988); Biol. Skr. 51: 327 (1998).
Syn: *Coffea spathicalyx* K.Schum. – F.P.S. 2: 432 (1952).
Small tree or shrub. Forest.
Distr: EQU; Ivory Coast to Kenya, S to Angola & Tanzania.
EQU: Chipp 47 (fl) 2/1929.

Canthium oligocarpum Hiern subsp. oligocarpum

F.T.E.A. Rub. 3: 877 (1991); Biol. Skr. 51: 328 (1998); F.E.E. 4(1): 270 (2003).
Syn: *Canthium captum* sensu El Amin, non Bullock – T.S.S.: 385 (1990).
Shrub or small tree. Montane forest.
Distr: EQU; Ethiopia to D.R. Congo & Tanzania.
EQU: Myers 11143 (fl fr) 4/1939.
Note: El Amin in T.S.S. also records *C. glaucum* Hiern from Upper Nile which is surely based on a misidentification.

Catunaregam nilotica (Stapf) Tirveng.

F.T.E.A. Rub. 2: 497 (1988); Biol. Skr. 51: 328 (1998); F.E.E. 4(1): 252 (2003).
Syn: *Lachnosiphonium niloticum* (Stapf) Dandy – F.P.S. 2: 441 (1952); Bot. Exp. Sud.: 100 (1970).

Syn: *Xeromphis nilotica* (Stapf) Keay – F.J.M.: 131 (1976); T.S.S.: 414 (1990).
Shrub or small tree. Dry woodland.
Distr: DAR, KOR, CS, ES, BAG, UN, EQU; Guinea to Somalia, S to Tanzania.
DAR: Wickens 3104 (fl) 4/1971; **UN:** J.M. Lock 80/6 (fl) 2/1980.

Chassalia cristata (Hiern) Bremek.

F.T.E.A. Rub. 1: 122 (1976); Biol. Skr. 51: 328 (1998).
Syn: *Psychotria cristata* Hiern – F.P.S. 2: 460 (1952).
Woody climber or shrub. Forest.
Distr: EQU; Nigeria to Kenya, S to Angola & Tanzania.
EQU: Andrews 1353 (fl) 5/1939.
Note: El Amin in T.S.S. records *C. laxiflora* Benth. from Yambio; this was clearly a misidentification since *C. laxiflora* is a West African species.

Coffea arabica L.

F.T.E.A. Rub. 2: 712 (1988); F.E.E. 4(1): 266 (2003).
Shrub or small tree. Forest, sometimes naturalised from cultivation.
Distr: UN; Ethiopia, Kenya, now widely cultivated.
IUCN: RD
UN: Thomas 3771 (fr) 12/1941.

Coffea canephora Pierre ex A.Froehner

F.P.S. 2: 432 (1952); F.T.E.A. Rub. 2: 710 (1988); Biol. Skr. 51: 328 (1998); F.E.E. 4(1): 267 (2003).
Syn: *Coffea arabica* sensu auctt., non L. – F.P.S. 2: 432 (1952); Biol. Skr. 51: 328 (1998).
Shrub or tree. Lowland forest.
Distr: EQU; Senegal to Uganda, S to Angola & Tanzania, also widely cultivated.
EQU: Myers 6755 (fr) 5/1937.

Coffea eugenioides S.Moore

F.P.S. 2: 432 (1952); F.T.E.A. Rub. 2: 713 (1988).
Small tree. Forest.
Distr: EQU; D.R. Congo, Burundi, Rwanda, Uganda, Kenya, Tanzania.
EQU: Snowden 1699 5/1930.

Coffea liberica Hiern

F.T.E.A. Rub. 2: 706 (1988); Biol. Skr. 51: 329 (1998).
Syn: *Coffea excelsa* A.Chev. – F.P.S. 2: 432 (1952).
Syn: *Coffea liberica* Hiern var. *dewevrei* (De Wild. & T.Durand) Lebrun – F.T.E.A. Rub. 2: 709 (1988).
Tree. Lowland forest.
Distr: EQU; Guinea Bissau to D.R. Congo & Uganda, now widely cultivated.
EQU: Friis & Vollesen 497 (fr) 11/1980.

Coffea neoleroyi A.P.Davis

Phytotaxa 10: 43 (2010).
Syn: *Psilanthus leroyi* Bridson – F.T.E.A. Rub. 2: 727 (1988); F.E.E. 4(1): 268 (2003).
Treelet. Rock outcrops in combretaceous woodland & wooded grassland.
Distr: UN; Ethiopia, Uganda.
IUCN: RD
UN: Davis 6008 (fr) 4/2012.

Conostomium quadrangulare (Rendle) Cufod.

F.T.E.A. Rub. 1: 243 (1976); Biol. Skr. 51: 329 (1998); F.E.E. 4(1): 213 (2003).
Syn: *Oldenlandia dolichantha* Stapf – F.P.S. 2: 449 (1952).
Perennial herb. Dry bushland & wooded grassland.
Distr: EQU; Ethiopia, Uganda, Kenya.
EQU: Sillitoe 421 (fl fr) 1919.

Coptosperma graveolens (S.Moore) Degreef subsp. *graveolens* var. *graveolens*

Syst. Geogr. Pl. 71: 374 (2001).
Syn: *Tarenna graveolens* (S.Moore) Bremek. – F.T.E.A. Rub. 2: 598 (1988); T.S.S.: 411 (1990); Biol. Skr. 51: 348 (1998); F.E.E. 4(1): 262 (2003).
Shrub or small tree. Woodland & bushland, often on rocky outcrops.
Distr: BAG, EQU; Ethiopia & Djibouti, S to D.R. Congo & Tanzania.
EQU: Kielland-Lund 743 5/1984 (n.v.).

Cordylostigma virgatum (Willd.) Groeninckx & Dessein

Taxon 59: 1466 (2010).
Syn: *Kohautia virgata* (Willd.) Bremek. – F.T.E.A. Rub. 1: 234 (1976).
Syn: *Oldenlandia virgata* (Willd.) DC. – F.P.S. 2: 453 (1952).
Perennial herb. Grassland, woodland, weed of disturbed ground.
Distr: BAG, UN, EQU; Guinea to Kenya, S to Namibia & South Africa; Madagascar, Comoros.
EQU: Broun 1648 (fl) 5/1909.
Note: Andrews records this species from Central & Southern Sudan but we have only seen specimens from the South.

Craterispermum schweinfurthii Hiern

F.P.S. 2: 433 (1952); F.T.E.A. Rub. 1: 162 (1976); T.S.S.: 389 (1990); Biol. Skr. 51: 329 (1998); F.E.E. 4(1): 268 (2003).
Syn: *Craterispermum laurinum* sensu Andrews, non (Poir.) Benth. – F.P.S. 2: 433 (1952).
Shrub or tree. Upland forest.
Distr: EQU; Nigeria to Ethiopia, S to Angola, Zimbabwe & Mozambique.
EQU: Friis & Vollesen 329 (fl) 11/1980.

Cremaspora triflora (Thonn.) K.Schum. subsp. *triflora*

F.P.S. 2: 433 (1952); F.T.E.A. Rub. 1: 733 (1988); T.S.S.: 389 (1990); Biol. Skr. 51: 330 (1998).
Shrub, small tree or woody climber. Forest & moist woodland.
Distr: EQU; Cape Verde Is., Senegal to Kenya, S to Angola, Zimbabwe & Mozambique.
EQU: Myers 8707 (fl) 3/1938.

Crossopteryx febrifuga (Afzel. ex G.Don) Benth.

F.P.S. 2: 433 (1952); F.T.E.A. Rub. 2: 457 (1988); T.S.S.: 389 (1990); Biol. Skr. 51: 330 (1998); F.E.E. 4(1): 245 (2003).
Small tree. Woodland & wooded grassland.

Distr: CS, ES, BAG, UN, EQU; Senegal to Ethiopia, S to Namibia & South Africa.
CS: <u>Aylmer 26/11</u> (fr) 1/1926; **EQU:** <u>Myers 8727</u> (fl) 3/1938.

Diodella sarmentosa (Sw.) Bacigalupo & E.L.Cabral ex Borhidi

Rubiác. México: 186 (2006).
Syn: *Diodia sarmentosa* Sw. – F.T.E.A. Rub. 1: 336 (1976); F.E.E. 4(1): 223 (2003).
Syn: *Diodia scandens* sensu auctt., non Sw. – F.P.S. 2: 434 (1952); Bot. Exp. Sud.: 99 (as *Diocida scandens*) (1970).
Perennial herb. Forest, streamsides, plantations.
Distr: EQU; widespread in the tropics, mainly in Africa & America.
EQU: <u>Andrews 1162</u> (fr) 5/1939.

Dolichopentas decora (S.Moore) Karehed & B.Bremer var. *decora*

Taxon 56: 1075 (2007).
Syn: *Pentas decora* S.Moore var. *decora* – F.T.E.A. Rub. 1: 195 (1976); Biol. Skr. 51: 340 (1998).
Syn: *Pentas globifera* sensu Andrews, non Hutch. – F.P.S. 2: 458 (1952).
Perennial herb. Montane woodland & grassland.
Distr: EQU; D.R. Congo, Uganda & Kenya, S to Angola & Malawi.
EQU: <u>Myers 9771</u> (fl) 10/1938.

Euclinia longiflora Salisb.

F.T.E.A. Rub. 2: 495 (1988); T.S.S.: 390 (1990).
Syn: *Rothmannia macrantha* (Schult.) Robyns – F.P.S. 2: 462 (1952).
Shrub or small tree. Forest.
Distr: EQU; Guinea Bissau to D.R. Congo & Uganda.
EQU: <u>Sillitoe 399</u> 1919.

Fadogia ancylantha Schweinf.

F.T.E.A. Rub. 3: 799 (1991).
Syn: *Temnocalyx ancylanthus* (Schweinf.) Robyns – F.P.S. 2: 463 (1952); T.S.S.: 411 (1990).
Shrubby perennial herb. Grassland.
Distr: BAG, EQU; Nigeria to Uganda, S to Zambia & Zimbabwe.
EQU: <u>Sillitoe 225</u> (fl) 1919.

Fadogia cienkowskii Schweinf. var. *cienkowskii*

F.P.S. 2: 434 (1952); F.J.M.: 129 (1976); T.S.S.: 390 (1990); F.T.E.A. Rub. 3: 788 (1991); Biol. Skr. 51: 330 (1998); F.E.E. 4(1): 275 (2003).
Shrubby perennial herb. Grassland & wooded grassland.
Distr: DAR, CS, BAG, EQU; Mali to Ethiopia, S to Angola & South Africa.
DAR: <u>Wickens 2136</u> 8/1964; **BAG:** <u>Drar 1818</u> 4/1938.

Fadogia glaberrima Hiern

F.P.S. 2: 434 (1952); F.T.E.A. Rub. 3: 794 (1991).
Perennial herb or subshrub. Wooded grassland.
Distr: BAG, EQU; Cameroon to Uganda.
BAG: <u>Turner 209</u> (fl fr) 5/1936.
Note: a widespread but scarce species.

Fadogia leucophloea Schweinf. ex Hiern

F.T.A. 3: 153 (1877); F.P.S. 2: 434 (1952).

Small tree. Wooded grassland.
Distr: BAG; Chad, C.A.R.
IUCN: RD
BAG: <u>Schweinfurth 2628</u> (fl) 11/1869.

Feretia apodanthera Delile subsp. *apodanthera*

F.P.S. 2: 434 (1952); F.J.M.: 129 (1976); F.T.E.A. Rub. 2: 698 (1988); T.S.S.: 391 (1990); Biol. Skr. 51: 330 (1998); F.E.E. 4(1): 260 (2003).
Shrub. Riverine forest, woodland & bushland.
Distr: DAR, CS, BAG, UN, EQU; Mauritania & Senegal to Somalia.
DAR: <u>Wickens 1645</u> (fl) 5/1964; **EQU:** <u>Friis & Vollesen 1197</u> 3/1982.

Fleroya stipulosa (DC.) Y.F.Deng

Taxon 56: 247 (2007).
Syn: *Hallea stipulosa* (DC.) J.-F.Leroy – F.T.E.A. Rub. 2: 447 (1988).
Syn: *Mitragyna stipulosa* (DC.) Kuntze – F.P.S. 2: 444 (1952); T.S.S.: 397 (1990).
Tree. Swamp forest, often a pioneer or forming pure stands.
Distr: BAG, EQU; Senegal to Uganda, S to Angola & Zambia.
IUCN: VU
BAG: <u>Schweinfurth 1518</u> (fl) 4/1869.

Galiniera saxifraga (Hochst.) Bridson

F.T.E.A. Rub. 2: 696 (1988); Biol. Skr. 51: 331 (1998); F.E.E. 4(1): 258 (2003).
Syn: *Galiniera coffeoides* Delile – F.P.S. 2: 435 (1952); T.S.S.: 391 (1990).
Shrub or small tree. Montane forest & forest margins.
Distr: EQU; Ethiopia to Zambia & Malawi.
EQU: <u>Thomas 1799</u> (fr) 12/1935.

Galium acrophyum Hochst. ex Chiov.

F.E.E. 4(1): 237 (2003).
Perennial herb. Montane grassland & scrub.
Distr: ?RS; Ethiopia, Kenya, Tanzania.
Note: Andrews records this species from Red Sea District but we have seen no material with which to verify this record.

Galium chloroionanthum K.Schum.

F.P.S. 2: 436 (1952); F.T.E.A. Rub. 1: 388 (1976); Biol. Skr. 51: 331 (1998); F.E.E. 4(1): 236 (2003).
Scrambling or climbing perennial herb. Montane forest clearings.
Distr: EQU; Ethiopia to Zimbabwe; Madagascar.
EQU: <u>Friis & Vollesen 1316</u> 3/1982.

Galium setaceum Lam.

F.P.S. 2: 436 (1952); F.Egypt 2: 239 (2000).
Syn: *Galium decaisnei* Boiss. – F.P.S. 2: 436 (1952).
Syn: *Galium mollugo* sensu Andrews, non L. – F.P.S. 2: 436 (1952).
Annual herb. Dry hillslopes, wadis.
Distr: RS; Macaronesia, Mediterranean, N Africa to Pakistan & India.
RS: <u>Khattab 6327</u> (fr) 3/1928.

Galium simense Fresen.

F.T.E.A. Rub. 1: 391 (1976); Biol. Skr. 51: 331 (1998); F.E.E. 4(1): 237 (2003).
Climbing or scrambling perennial herb. Montane bushland & forest margins.
Distr: EQU; Bioko, Cameroon, Ethiopia to D.R. Congo & Tanzania.
EQU: <u>Myers 11560</u> (fr) 8/1939.

Galium spurium L. subsp. africanum Verdc.

F.T.E.A. Rub. 1: 390 (1976); F.E.E. 4(1): 237 (2003).
Syn: *Galium spurium* L. var. *echinospermum* (Wallr.) Desportes – F.J.M.: 129 (1976).
Syn: *Galium uniflorum* Quézel – F. Darfur Nord-Occ. & J. Gourgeil: 129 (1976).
Syn: *Galium aparine* sensu Andrews, non L. – F.P.S. 2: 435 (1952).
Scrambling annual herb. Disturbed areas in grassland, bushland & forest margins.
Distr: RS, DAR; Eritrea & Somalia, S to South Africa.
RS: <u>Jackson 2948</u> (fr) 4/1953.

Galium thunbergianum Eckl. & Zeyh. var. hirsutum (Sond.) Verdc.

F.J.M.: 129 (1976); F.T.E.A. Rub. 1: 388 (1976); F.E.E. 4(1): 236 (2003).
Syn: *Galium dasycarpum* Hochst. ex Schweinf. – F.P.S. 2: 435 (1952).
Climbing or scrambling perennial herb. Montane grassland, bushland & forest margins.
Distr: DAR; Bioko, Cameroon, Ethiopia to South Africa.
DAR: <u>Robertson 155</u> (fr) 12/1957.

Gardenia aqualla Stapf & Hutch.

F.P.S. 2: 437 (1952); F.T.E.A. Rub. 2: 505 (1988); T.S.S.: 393 (1990).
Shrub. Wooded grassland.
Distr: ?DAR, BAG, EQU; Mali to Uganda.
BAG: <u>Schweinfurth 1751</u> (fl fr) 6/1869.
Note: the record for Darfur is from El Amin in T.S.S.; we have seen no material with which to verify it.

Gardenia erubescens Stapf & Hutch.

F.P.S. 2: 437 (1952); F.T.E.A. Rub. 2: 505 (1988); T.S.S.: 393 (1990).
Shrub or small tree. Wooded grassland.
Distr: ?KOR, BAG, EQU; Senegal to Uganda.
EQU: <u>Myers 8415</u> (fl) 1/1938.
Note: the record for Kordofan is from El Amin in T.S.S.; we have seen no material with which to verify it.

Gardenia ternifolia Schumach. & Thonn. subsp. ternifolia

F.T.E.A. Rub. 2: 509 (1988); T.S.S.: 394 (species only) (1990); Biol. Skr. 51: 332 (1998); F.E.E. 4(1): 253 (2003).
Syn: *Gardenia lutea* sensu Andrews pro parte, non Fresen. – F.P.S. 2: 438 (1952).
Shrub or small tree. Woodland & bushland.
Distr: CS, EQU; Senegal to Ethiopia.
CS: <u>Lea 115</u> 11/1952; **EQU:** <u>Friis & Vollesen 1025</u> (fr) 2/1982.
Note: Drar (1970) and El Amin (1990) also recorded "*G. lutea*" from Kordofan – we are not sure which of the subspecies of *G. ternifolia* this will refer to since we have not seen the specimens on which his records are based.

Gardenia ternifolia Schumach. & Thonn. subsp. jovis-tonantis (Welw.) Verdc. var. jovis-tonantis

F.T.E.A. Rub. 2: 509 (1988); F.E.E. 4(1): 253 (2003).
Syn: *Gardenia jovis-tonantis* (Welw.) Hiern – F.P.S. 2: 436 (1952).
Syn: *Gardenia lutea* Fresen. – F.P.S. 2: 438, pro parte (1952); F.J.M.: 129 (1976); T.S.S.: 393 (1990), ?pro parte.
Shrub or small tree. Wooded grassland & woodland.
Distr: DAR, CS, UN, EQU; Nigeria, Cameroon, Ethiopia to Angola & Mozambique.
DAR: <u>Wickens 1592</u> 5/1964; **EQU:** <u>Myers 7960</u> (fl) 12/1937.

Gardenia ternifolia Schumach. & Thonn. subsp. jovis-tonantis (Welw.) Verdc. var. goetzei (Stapf & Hutch.) Verdc.

F.T.E.A. Rub. 2: 509 (1988).
Syn: *Gardenia triacantha* DC. – F.P.S. 2: 437 (1952); T.S.S.: 394 (1990).
Shrub or small tree. Woodland.
Distr: DAR, BAG, EQU; Senegal to Kenya, S to South Africa.
DAR: <u>Wickens 3103A</u> (fl fr) 4/1971; **BAG:** <u>Hoyle 648</u> (fl) 2/1929.

Gardenia tinneae Kotschy & Heuglin

F.P.S. 2: 430 (1952); F.T.E.A. Rub. 2: 510, in notes (1988).
Rhizomatous subshrub, almost stemless. Woodland.
Distr: BAG; Chad, C.A.R.
IUCN: RD
BAG: <u>Tinne & Heuglin 48</u> (fl fr) 11/1863.

Gardenia vogelii Hook.f. ex Planch.

F.P.S. 2: 437 (1952); F.T.E.A. Rub. 2: 505 (1988); T.S.S.: 394 (1990); Biol. Skr. 51: 332 (1998).
Shrub. Lowland forest.
Distr: BAG, EQU; Liberia to Uganda, S to Angola.
EQU: <u>Friis & Vollesen 783</u> (fr) 12/1980.

Geophila obvallata (Schumach.) Didr. subsp. involucrata (Hiern) Verdc.

F.T.E.A. Rub. 1: 113 (1976).
Creeping perennial herb. Forest.
Distr: EQU; D.R. Congo, Uganda.
IUCN: RD
EQU: <u>Schweinfurth 3670</u> 4/1870.

Geophila repens (L.) I.M.Johnst.

F.T.E.A. Rub. 1: 110 (1976); Biol. Skr. 51: 332 (1998); F.E.E. 4(1): 201 (2003).
Syn: *Geophila herbacea* (L.) K.Schum. – F.P.S. 2: 430 (1952).
Creeping perennial herb. Forest.
Distr: EQU; pantropical.
EQU: <u>Friis & Vollesen 465</u> 11/1980.

Heinsenia diervilleoides K.Schum. subsp. diervilleoides

F.T.E.A. Rub. 2: 730 (1988); T.S.S.: 394 (1990); Biol. Skr. 51: 332 (1998).
Shrub or small tree. Forest.
Distr: EQU; Uganda & Kenya, S to Zimbabwe & Mozambique
EQU: <u>Friis & Vollesen 922</u> (fl) 2/1982.

Hymenocoleus hirsutus (Benth.) Robbr.

F.T.E.A. Rub. 1: 115 (1976); Biol. Skr. 51: 332 (1998).
Creeping perennial herb. Forest.
Distr: EQU; Guinea to Uganda, S to Angola & Tanzania.
EQU: Friis & Vollesen 858 12/1980.

Hymenocoleus neurodictyon (K.Schum.) Robbr. var. orientalis (Verdc.) Robbr.

F.T.E.A. Rub. 1: 116 (1976); Biol. Skr. 51: 333 (1998).
Syn: Chassalia sp. sensu Andrews – F.P.S. 2: 431 (1952).
Perennial herb. Forest.
Distr: EQU; Ivory Coast to D.R. Congo & Tanzania.
EQU: Friis & Vollesen 464 (fr) 11/1980.

Hymenodictyon floribundum (Hochst. & Steud.) B.L.Rob.

F.P.S. 2: 441 (1952); F.T.E.A. Rub. 2: 452 (1988); T.S.S.: 395 (1990); Biol. Skr. 51: 333 (1998); F.E.E. 4(1): 245 (2003).
Shrub or small tree. Upland woodland, often on rocky outcrops.
Distr: BAG, EQU; Guinea to Ethiopia, S to Angola & Zimbabwe.
EQU: Howard I.M.77 (fl) 3/1976.

Hymenodictyon parvifolium Oliv. subsp. scabrum (Stapf) Verdc. var. scabrum

F.T.E.A. Rub. 2: 455 (1988); Biol. Skr. 51: 334 (1998).
Syn: Hymenodictyon scabrum Stapf – Bot. Exp. Sud.: 100 (1970); T.S.S.: 395 (1990).
Shrub or small tree. Woodland.
Distr: EQU; Uganda, Tanzania, Zambia.
EQU: Andrews 1680 (fr) 6/1939.

Ixora brachypoda DC.

F.W.T.A. 2: 144 (1963); F.T.E.A. Rub. 2: 611, in notes (1988).
Syn: Ixora radiata Hiern – F.P.S. 2: 441 (1952).
Shrub. Riverine & swamp forest.
Distr: EQU; Gambia to D.R. Congo, S to Angola & Zambia.
EQU: Turner 36 (fl) 5/1931.
Note: on the single Sudanese specimen seen, it is noted that this species is planted in the region.

Ixora mildbraedii K.Krause

F.T.E.A. Rub. 2: 615 (1988).
Small tree or scrambling shrub. Streamsides.
Distr: EQU; C.A.R., D.R. Congo, Uganda.
EQU: Hoyle 716 3/1939.
Note: El Amin in T.S.S.: 395 records *I. laxiflora* Sm. from Equatoria; this is clearly a misidentification, perhaps of *I. mildbraedii*.

Keetia gueinzii (Sond.) Bridson

F.T.E.A. Rub. 3: 911 (1991); Biol. Skr. 51: 334 (1998); F.E.E. 4(1): 272 (2003).
Syn: Canthium gueinzii Sond. – T.S.S.: 386 (1990).
Shrub or woody climber. Forest & woodland.
Distr: EQU; Cameroon to Ethiopia, S to Angola & South Africa.
EQU: Andrews 1296 (fl) 5/1939.

Notes: (1) Andrews recorded *Canthium venosissimum* Hutch. & Dalziel from Equatoria, but Bridson in Kew Bull. 41: 991 (1986) considers this to be based on misidentified material, perhaps of *K. gueinzii* or *K. venosum*. (2) El Amin recorded *Canthium multiflorum* (Schum. & Thonn.) Hiern (= *Keetia multiflora* (Schum. & Thonn.) Bridson) from Azza forest in Equatoria but Bridson (l.c.: 989) notes that this species has been much misunderstood in the past; we consider El Amin's record likely to be a misidentification.

Keetia mannii (Hiern) Bridson

Kew Bull. 41: 988 (1986).
Syn: Canthium afzelianum sensu auctt., non Hiern – F.P.S. 2: 431 (1952); T.S.S.: 385 (1990).
Climbing shrub. Forest, by streams.
Distr: BAG, EQU; Sierre Leone to C.A.R.
EQU: Myers 13291 (fr) 3/1938.

Keetia venosa (Oliv.) Bridson

F.T.E.A. Rub. 3: 914 (1991); Biol. Skr. 51: 334 (1998).
Syn: Canthium venosum (Oliv.) Hiern – F.P.S. 2: 431 (1952); Bot. Exp. Sud.: 99 (1970); T.S.S.: 387 (1990).
Shrub or woody climber. Forest margins & associated woodland.
Distr: ?DAR, BAG, EQU; Senegal to Kenya, S to Angola, Zimbabwe & Mozambique.
EQU: Myers 8026 (fr) 12/1937.
Note: the record for Darfur is based on El Amin in T.S.S. who records this species from Nyertete; it requires confirmation.

Keetia zanzibarica (Klotzsch) Bridson subsp. gentilii (De Wild.) Bridson

F.T.E.A. Rub. 3: 918 (1991); F.E.E. 4(1): 273 (2003); Biol. Skr. 51: 534 (2005).
Syn: Canthium zanzibaricum sensu auctt., non Klotzsch s.s. – F.P.S. 2: 430 (1952); T.S.S.: 388 (1990).
Shrub or woody climber. Dry forest & woodland.
Distr: EQU; C.A.R. to Ethiopia, S to D.R. Congo & Tanzania.
EQU: Thomas 1746 (fl) 12/1935.

Kohautia aspera (B.Heyne ex Roth) Bremek.

F.T.E.A. Rub. 1: 241 (1976); F.E.E. 4(1): 212 (2003).
Syn: Oldenlandia strumosa (A.Rich.) Hiern – F.P.S. 2: 452 (1952).
Annual herb. Grassland & open bushland, disturbed ground.
Distr: CS; Cape Verde Is., Senegal to Eritrea, S to Namibia & South Africa; SW Asia, Arabia, India, Australia.
CS: Andrews 94 (fr) 10/1935.

Kohautia caespitosa Schnizl. subsp. caespitosa

F.T.E.A. Rub. 1: 238 (1976); F.J.M.: 130 (1976); F.E.E. 4(1): 210 (2003).
Syn: Oldenlandia schimperi (C.Presl) T.Anderson – F.P.S. 2: 449 (1952).
Syn: Oldenlandia effusa sensu Andrews, non Oliv. – F.P.S. 2: 451 (1952).
Syn: Oldenlandia welwitschii sensu auctt., non Hiern – F.P.S. 2: 449 (1952); Bot. Exp. Sud.: 100 (1970).

Annual or perennial herb. Open bushland & woodland, arid plains.
Distr: RS, DAR, KOR, CS; Eritrea, Djibouti, Somalia, Kenya; Egypt, Arabia.
RS: <u>Jackson 2749</u> (fl) 3/1953.

Kohautia coccinea Royle

F.T.E.A. Rub. 1: 235 (1976); F.J.M.: 130 (1976); Biol. Skr. 51: 335 (1998); F.E.E. 4(1): 211 (2003).
Annual herb. Grassland, weed of disturbed ground.
Distr: DAR, EQU; Senegal to Ethiopia, S to Zimbabwe & Mozambique; Arabia & N India.
DAR: <u>Robertson 14</u> (fl) 10/1957; **EQU:** <u>Kielland-Lund 331</u> 12/1983 (n.v.).

Kohautia grandiflora DC.

F.T.E.A. Rub. 1: 236 (1976); F.J.M.: 130 (1976); F.E.E. 4(1): 209 (2003).
Syn: *Oldenlandia grandiflora* (DC.) Hiern – F.P.S. 2: 451 (1952).
Annual herb. Grassland & open woodland, often in seasonally flooded areas.
Distr: DAR, KOR, CS, ES, UN; Senegal to Ethiopia & Uganda.
DAR: <u>Wickens 2751</u> 10/1964; **UN:** <u>Simpson 7438</u> (fl fr) 1/1930.

Kohautia tenuis (Bowdich) Mabb.

F.E.E. 4(1): 209 (2003).
Syn: *Kohautia senegalensis* Cham. & Schltdl. – F.J.M.: 130 (1976).
Syn: *Oldenlandia noctiflora* (Hochst. ex A.Rich.) Hiern – F.P.S. 2: 450 (1952).
Syn: *Oldenlandia senegalensis* (Cham. & Schltdl.) Hiern – F.P.S. 2: 449 (1952).
Annual or perennial herb. Grassland & wooded grassland.
Distr: ?RS, DAR, KOR, CS, ES, UN; Cape Verde Is., Senegal to Somalia; Yemen.
DAR: <u>Wickens 3661</u> (fl) 6/1977; **UN:** <u>Sherif A.3967</u> (fl) 8/1951.

Leptactina platyphylla (Hiern) Wernham

F.P.S. 2: 442 (1952); F.T.E.A. Rub. 2: 689 (1988); T.S.S.: 396 (1990); Biol. Skr. 51: 335 (1998).
Shrub or small tree. Forest & woodland.
Distr: EQU; Cameroon to Kenya, S to Malawi & Mozambique.
EQU: <u>Jackson 4222</u> (fl) 6/1961.

Macrosphyra longistyla Hook.f.

F.P.S. 2: 443 (1952); F.T.E.A. Rub. 2: 491 (1988); T.S.S.: 396 (1990); F.E.E. 4(1): 252 (2003).
Shrub or woody climber. Rocky outcrops, forest margins.
Distr: EQU; Senegal to Ethiopia & Uganda.
EQU: <u>Myers 8760</u> (fl) 3/1938.

Meyna tetraphylla (Schweinf. ex Hiern) Robyns subsp. *tetraphylla*

F.P.S. 2: 443 (1952); T.S.S.: 396 (1990); F.T.E.A. Rub. 3: 859 (1991); Biol. Skr. 51: 335 (1998); F.E.E. 4(1): 277 (2003).
Spiny shrub, small tree or woody climber. Woodland, riverine scrub.

Distr: BAG, EQU; Ethiopia, Uganda, Kenya.
BAG: <u>Myers 13906</u> (fr) 7/1941.

Mitracarpus hirtus (L.) DC.

F.E.E. 4(1): 229 (2003).
Syn: *Mitracarpus scaber* Zucc. ex Schult. & Schult.f. – F.P.S. 2: 443 (1952); F.J.M.: 130 (1976).
Syn: *Mitracarpus villosus* (Sw.) DC. – F.T.E.A. Rub. 1: 375 (1976); Biol. Skr. 51: 336 (1998).
Annual herb. Weed of disturbed ground.
Distr: DAR, KOR, BAG, UN, EQU; pantropical.
KOR: <u>Pfund 725</u> (fl) 8/1875; **BAG:** <u>Schweinfurth 2371</u> 9/1869.

Mitragyna inermis (Willd.) Kuntze

F.P.S. 2: 443 (1952); F.W.T.A. 2: 161 (1963); F.J.M.: 130 (1976); T.S.S.: 397 (1990).
Shrub or small tree. Seasonally inundated areas in wooded grassland.
Distr: DAR, KOR, BAG, UN; Mauritania & Senegal to C.A.R. & D.R. Congo.
DAR: <u>Robertson 107</u> (fl) 12/1957; **BAG:** <u>Myers 7422</u> 7/1937.

Morelia senegalensis A.Rich. ex DC.

F.P.S. 2: 444 (1952); F.W.T.A. 2: 113 (1963); T.S.S.: 397 (1990).
Shrub or tree. Riverine woodland.
Distr: BAG, EQU; Senegal to D.R. Congo.
BAG: <u>Myers 11506</u> (fr) 7/1939.

Morinda lucida Benth.

F.P.S. 2: 444 (1952); F.T.E.A. Rub. 1: 146 (1976); T.S.S.: 399 (1990).
Tree. Grassland, bushland, open forest.
Distr: BAG, EQU; Senegal to Uganda, S to Angola & Tanzania.
EQU: <u>Myers 1661</u> (fr) 6/1939.

Morinda morindoides (Baker) Milne-Redh.

F.P.S. 2: 444 (1952); F.W.T.A. 2: 189 (1963); T.S.S.: 399 (1990).
Woody climber. Forest.
Distr: EQU; Senegal to C.A.R., S to Angola & D.R. Congo.
EQU: <u>von Ramm 231</u> 3/1956.

Multidentia crassa (Hiern) Bridson & Verdc. var. *crassa*

F.T.E.A. Rub. 3: 845 (1991); Biol. Skr. 51: 336 (1998).
Syn: *Canthium crassum* Hiern – F.P.S. 2: 430 (1952); T.S.S.: 385 (1990).
Shrub or small tree. Woodland.
Distr: EQU; Cameroon to Ethiopia, S to Angola, Zimbabwe & Mozambique.
EQU: <u>Myers 8726</u> (fr) 3/1938.

Multidentia dichrophylla (Mildbr.) Bridson

F.T.E.A. Rub. 3: 842 (1991); Biol. Skr. 51: 336 (1998).
Syn: *Vangueriopsis sillitoei* Bullock – F.P.S. 2: 466 (1952); T.S.S.: 414 (1990).
Small tree or climbing shrub. Forest.
Distr: EQU; Cameroon, D.R. Congo, Uganda.
EQU: <u>Sillitoe 377</u> (fr) 1919.

Mussaenda arcuata Lam. ex Poir.
F.P.S. 2: 446 (1952); F.J.M.: 130 (1976); F.T.E.A. Rub. 2: 461 (1988); Biol. Skr. 51: 336 (1998); F.E.E. 4(1): 246 (2003).
Syn: *Mussaenda arcuata* Lam. ex Poir. var. *pubescens* Wernham – F.P.S. 2: 446 (1952); T.S.S.: 399 (1990).
Woody climber or shrub. Forest & woodland.
Distr: DAR, BAG, EQU; Senegal to Ethiopia, S to Angola, Zimbabwe & Mozambique; Madagascar.
DAR: Kassas 245; **EQU:** Myers 773 (fl) 9/1937.

Mussaenda elegans Schumach. & Thonn.
F.P.S. 2: 446 (1952); F.T.E.A. Rub. 2: 462 (1988); T.S.S.: 400 (1990).
Shrub or woody climber. Forest & forest margins.
Distr: EQU; Guinea Bissau to Kenya, S to Angola & Tanzania.
EQU: Myers 1136 (fl fr) 5/1939.

Mussaenda erythrophylla Schumach. & Thonn.
F.P.S. 2: 446 (1952); F.T.E.A. Rub. 2: 463 (1988); T.S.S.: 400 (1990).
Shrub or woody climber. Forest, along rivers & in clearings.
Distr: EQU; Guinea to Kenya, S to Angola & Tanzania.
EQU: Sillitoe 273 (fl) 1919.

Oldenlandia affinis (Roem. & Schult.) DC. subsp. *fugax* (Vatke) Verdc.
F.T.E.A. Rub. 1: 291 (1976).
Perennial herb. Grassland, bushland, forest margins.
Distr: EQU; Liberia to Kenya, S to Angola & South Africa; Madagascar, Comoros.
EQU: Andrews 1092 (fl) 5/1939.

Oldenlandia capensis L.f. var. *capensis*
F.P.S. 2: 453 (1952); F.T.E.A. Rub. 1: 313 (1976); F.E.E. 4(1): 222 (2003).
Annual herb. Damp grassland, weed of cultivation.
Distr: CS, BAG, EQU; Senegal to Ethiopia, S to Namibia & South Africa; Madagascar, Balkans, Morocco, Egypt.
CS: Pettet 112 1/1963; **BAG:** Schweinfurth 2095 7/1869.

Oldenlandia capensis L.f. var. *pleiosepala* Bremek.
F.T.E.A. Rub. 1: 313 (1976); F.E.E. 4(1): 222 (2003).
Syn: *Oldenlandia hedyotoides* (Fisch. & C.A.Mey.) Boiss. – F.P.S. 2: 454 (1952).
Annual herb. Damp grassland, riverbanks, irrigated cultivations.
Distr: KOR, BAG, EQU; Guinea Bissau to Ethiopia, S to Angola, Botswana & Mozambique; Algeria, Egypt, Arabia to Iran.
KOR: Kotschy 41 1842; **EQU:** Schweinfurth 3734 5/1870.

Oldenlandia corymbosa L. var. *corymbosa*
F.P.S. 2: 453 (1952); F.T.E.A. Rub. 1: 308 (1976); F.J.M.: 130 (1976); Biol. Skr. 51: 337 (1998); F.E.E. 4(1): 220 (2003).
Annual herb. Grassland, woodland, weed of disturbed ground.
Distr: RS, DAR, CS, BAG, UN; pantropical & subtropical.
DAR: Wickens 2043 (fl) 7/1964; **BAG:** Schweinfurth 2344 9/1869.

Oldenlandia corymbosa L. var. *linearis* (DC.) Verdc.
F.T.E.A. Rub. 1: 309 (1976); F.E.E. 4(1): 221 (2003).
Annual herb. Habitat as for var. *corymbosa*.
Distr: KOR, BAG; Senegal, Ethiopia to Zimbabwe.
KOR: Simpson 2263 (fl) 5/1930; **BAG:** Schweinfurth 2396A 4/1869.

Oldenlandia echinulosa K.Schum. var. *pellucida* (Hiern) Verdc.
F.J.M.: 130 (species only) (1976); F.T.E.A. Rub. 3: 927 (1991).
Syn: *Oldenlandia pellucida* Hiern var. *pellucida* – F.T.E.A. Rub. 1: 286 (1976).
Herb. Moist sites on rocks & by rivers.
Distr: DAR; Sierra Leone, Nigeria, Cameroon, Tanzania to Angola & Zimbabwe.
DAR: Wickens 2515 9/1964.

Oldenlandia fastigiata Bremek. var. *fastigiata*
F.T.E.A. Rub. 1: 312 (1976); F.E.E. 4(1): 221 (2003).
Annual or perennial herb. Grassland & open woodland, often in disturbed or open areas.
Distr: UN; D.R. Congo to Somalia, S to Malawi & Mozambique.
UN: J.M. Lock 81/45 5/1981.

Oldenlandia goreensis (DC.) Summerh. var. *goreensis*
F.P.S. 2: 453 (1952); F.T.E.A. Rub. 1: 279 (1976); Biol. Skr. 51: 337 (1998); F.E.E. 4(1): 219 (2003).
Annual or short-lived perennial herb. Wet grassland, streamsides & swamps.
Distr: EQU; Senegal to Ethiopia, S to South Africa; Madagascar, Mascarenes.
EQU: Friis & Vollesen 314 (fl) 11/1980.
Note: Drar (1970) records this species from Erkowit in Red Sea region but this was based on a sterile specimen (not seen by us) and is almost certainly misidentified.

Oldenlandia herbacea (L.) Roxb. var. *herbacea*
F.P.S. 2: 451 (1952); F.T.E.A. Rub. 1: 305 (1976); F.J.M.: 130 (1976); Biol. Skr. 51: 337 (1998); F.E.E. 4(1): 220 (2003).
Annual herb. Grassland & woodland.
Distr: DAR, KOR, UN, EQU; palaeotropical.
DAR: Wickens 1694 5/1964; **EQU:** Myers 6773 5/1937.

Oldenlandia lancifolia (Schumach.) DC. var. *scabridula* Bremek.
F.P.S. 2: 453 (species only) (1952); Biol. Skr. 51: 338 (1998); F.E.E. 4(1): 219 (2003).
Perennial herb. Swamps, wet grassland, forest margins.
Distr: EQU; Sierre Leone to Ethiopia, S to South Africa.
EQU: Friis & Vollesen 370 (fr) 11/1980.
Note: Andrews records this species from Central & Southern Sudan but we have only seen material from Equatoria.

Otomeria elatior (A.Rich. ex DC.) Verdc.
F.T.E.A. Rub. 1: 214 (1976).
Syn: *Otomeria dilatata* Hiern – F.P.S. 2: 454 (1952).
Perennial herb. Swampy grassland.

Distr: EQU; Mali to Uganda, S to Angola, Zimbabwe & Mozambique.
EQU: <u>Myers 6681</u> (fl) 5/1937.

Otomeria madiensis Oliv.

F.P.S. 2: 454 (1952); F.T.E.A. Rub. 1: 215 (1976); F.E.E. 4(1): 206 (2003).
Perennial herb. Open woodland.
Distr: EQU; D.R. Congo, Uganda, Ethiopia.
EQU: <u>Schweinfurth 14</u> (fl) 6/1870.

Oxyanthus formosus Hook.f. ex Planch.

F.P.S. 2: 455 (1952); F.T.E.A. Rub. 2: 537 (1988); T.S.S.: 400 (1990).
Shrub or small tree. Forest.
Distr: EQU; Sierra Leone to Uganda, S to Angola.
EQU: <u>Sillitoe 233</u> (fl) 1919.

Oxyanthus lepidus S.Moore subsp. lepidus var. lepidus

F.T.E.A. Rub. 2: 531 (1988); T.S.S.: 401 (1990); F.E.E. 4(1): 257 (2003).
Syn: *Oxyanthus oxycarpus* S.Moore – F.P.S. 2: 455 (1952).
Shrub or small tree. Forest.
Distr: EQU; D.R. Congo, Burundi, Ethiopia, Uganda, Tanzania.
EQU: <u>Andrews 1652</u> (fr) 6/1939.

Oxyanthus speciosus DC. subsp. stenocarpus (K.Schum.) Bridson

F.T.E.A. Rub. 2: 529 (1988); T.S.S.: 401 (species only) (1990); Biol. Skr. 51: 338 (1998); F.E.E. 4(1): 257 (2003).
Shrub or small tree. Montane forest.
Distr: EQU; Ethiopia to South Africa.
EQU: <u>Friis & Vollesen 54</u> (fr) 11/1980.

Oxyanthus unilocularis Hiern

F.P.S. 2: 455 (1952); F.T.E.A. Rub. 2: 537 (1988); T.S.S.: 401 (1990); Biol. Skr. 51: 338 (1998).
Shrub or small tree. Forest.
Distr: BAG, EQU; Sierra Leone to Uganda, S to Angola.
EQU: <u>Myers 8318</u> (fl) 1/1938.

Pavetta abyssinica Fresen.

F.P.S. 2: 457 (1952); F.T.E.A. Rub. 2: 660 (1988); T.S.S.: 402 (1990); F.E.E. 4(1): 264 (2003).
Shrub or small tree. Montane forest & scrub.
Distr: RS, EQU; Eritrea to Tanzania.
RS: <u>Aylmer 249</u> 3/1932; **EQU:** <u>Jackson 1409</u> 4/1950.

Pavetta bilineata Bremek.

F.P.S. 2: 456 (1952); T.S.S.: 402 (1990); F.T.E.A. Rub. 2: 659, in notes (1988).
Tree. Forest, including streamsides.
Distr: EQU; South Sudan endemic.
IUCN: RD
EQU: <u>Andrews 1064</u> (fl) 5/1939.

Pavetta crassipes K.Schum.

F.P.S. 2: 457 (1952); F.T.E.A. Rub. 2: 670 (1988); T.S.S.: 402 (1990); Biol. Skr. 51: 339 (1998); F.E.E. 4(1): 265 (2003).
Shrub. Woodland.

Distr: DAR, KOR, EQU; Senegal to Ethiopia, S to Zambia & Mozambique.
KOR: <u>Simpson 7727</u> 4/1930; **EQU:** <u>Andrews 940</u> (fl) 4/1939.

Pavetta gardeniifolia A.Rich. var. gardeniifolia

F.T.E.A. Rub. 2: 677 (1988); T.S.S.: 403 (1990); Biol. Skr. 51: 339 (1998); F.E.E. 4(1): 265 (2003).
Syn: *Pavetta hochstetteri* Bremek. var. *glaberrima* Bremek. – F.P.S. 2: 458 (1952).
Shrub or small tree. Bushland, grassland & forest margins, often on rocky outcrops.
Distr: RS, DAR, KOR, EQU; Togo to Somalia, S to Namibia & South Africa.
DAR: <u>Wickens 1981</u> (fl) 7/1964; **EQU:** <u>Andrews 853</u> 4/1939.
Note: El Amin in T.S.S. records also var. *subtomentosa* K.Schum. from Sudan but this is considered likely to be an error.

Pavetta molundensis K.Krause

F.T.E.A. Rub. 2: 631 (1988); T.S.S.: 403 (1990); Biol. Skr. 51: 339 (1998).
Syn: *Pavetta insignis* Bremek. var. *glabra* Bremek. – F.P.S. 2: 456 (1952).
Tree or shrub. Forest.
Distr: BAG, EQU; Cameroon to Uganda, S to D.R. Congo & Tanzania.
EQU: <u>Friis & Vollesen 690</u> (fl fr) 12/1980.

Pavetta oliveriana Hiern var. oliveriana

F.P.S. 2: 457 (1952); F.T.E.A. Rub. 2: 644 (1988); T.S.S.: 404 (1990); Biol. Skr. 51: 340 (1998); F.E.E. 4(1): 263 (2003).
Shrub or small tree. Montane forest & grassland.
Distr: EQU; Eritrea to D.R. Congo & Tanzania.
EQU: <u>Friis & Vollesen 790</u> (fr) 12/1980.

Pavetta ruwenzoriensis S.Moore

F.P.S. 2: 456 (1952); F.T.E.A. Rub. 2: 659 (1988); T.S.S.: 404 (1990).
Shrub, sometimes scandent. Forest.
Distr: EQU; D.R. Congo, Rwanda, Uganda.
EQU: <u>Sillitoe 228</u> (fl) 1919.

Pavetta schweinfurthii Bremek.

F.P.S. 2: 436 (1952); T.S.S.: 404 (1990).
Shrub. Forest.
Distr: BAG; C.A.R.
IUCN: RD
BAG: <u>Schweinfurth 2114</u> (fr) 7/1869.

Pavetta subcana Hiern var. subcana

F.P.S. 2: 457 (1952); F.T.E.A. Rub. 2: 681 (1988); T.S.S.: 404 (1990).
Shrub. Wooded grassland, riverine forest.
Distr: ?KOR, BAG, UN; Nigeria to Kenya.
UN: <u>J.M. Lock 81/29</u> 2/1981.
Note: the record for Kordofan is from El Amin in T.S.S.; we have not seen any specimens to verify this.

Pavetta subcana Hiern var. longiflora (Vatke) Bridson

F.T.E.A. Rub. 2: 681 (1988); Biol. Skr. 51: 340 (1998); F.E.E. 4(1): 266 (2003).

Syn: *Pavetta albertina* S.Moore – F.P.S. 2: 457 (1952).
Shrub. Bushland & woodland, often along rivers.
Distr: DAR, KOR, CS, UN, EQU; C.A.R. to Eritrea, S to
D.R. Congo & Tanzania.
KOR: <u>Broun 1311</u> 12/1909; **EQU:** <u>Andrews 913</u> (fl)
4/1939.

Pavetta sp. nov. near *abyssinica* Fresen.
Biol. Skr. 51: 338 (1998).
Shrub. Montane forest & woodland.
Distr: EQU; ?South Sudan endemic.
IUCN: RD
EQU: <u>Friis & Vollesen 1212</u> (fl) 3/1982.

Pentanisia ouranogyne S.Moore
F.T.E.A. Rub. 1: 172 (1976); Biol. Skr. 51: 340 (1998);
F.E.E. 4(1): 282 (2003).
Perennial herb. Grassland, bushland & woodland, often
with periodic burning.
Distr: EQU; Ethiopia, Somalia, Uganda, Kenya, Tanzania.
EQU: <u>Kielland-Lund 947</u> 6/1984 (n.v.).

Pentanisia schweinfurthii Hiern
F.P.S. 2: 458 (1952); F.T.E.A. Rub. 1: 168 (1976).
Perennial pyrophytic herb. Burnt grassland & woodland.
Distr: EQU; Nigeria to Kenya, S to Angola & Zimbabwe.
EQU: <u>Myers 8467</u> (fl) 2/1938.

Pentas arvensis Hiern
F.P.S. 2: 458 (1952); F.T.E.A. Rub. 1: 208 (1976).
Perennial pyrophytic herb. Grassland subject to regular
burning.
Distr: EQU; Nigeria to Kenya.
EQU: <u>Schweinfurth 2775</u> (fl) 1/1870.

Pentas herbacea (Hiern) K.Schum.
F.P.S. 2: 459 (1952); Bull. Jard. Bot. Brux. 23: 312 (1953).
Annual or biennial herb. Woodland.
Distr: EQU; Guinea, C.A.R., D.R. Congo, Angola.
EQU: <u>Wyld 613</u> 10/1939.

Pentas lanceolata (Forssk.) Deflers subsp.
lanceolata
F.T.E.A. Rub. 1: 210 (1976); Biol. Skr. 51: 341 (1998);
F.E.E. 4(1): 205 (2003).
Perennial herb or subshrub. Montane grassland &
bushland.
Distr: RS, EQU; Eritrea to Tanzania; Arabia.
RS: <u>Jackson 2929</u> 4/1953; **EQU:** <u>Friis & Vollesen 195</u> (fl
fr) 11/1980.

Pentas lanceoata (Forssk.) Deflers subsp.
quartiniana (A.Rich.) Verdc.
F.T.E.A. Rub. 1: 210 (1996); Biol. Skr. 51: 341 (1998);
F.E.E. 4(1): 206 (2003).
Syn: *Pentas carnea* sensu Andrews, non Benth. – F.P.S.
2: 459 (1952).
Perennial herb or subshrub. Montane grassland.
Distr: EQU; Eritrea to D.R. Congo & Tanzania.
EQU: <u>Friis & Vollesen 94</u> (fl) 11/1980.
Note: our material would refer to var. *leucaster*
(K.Krause) Verdc. but varieties were not upheld in F.E.E.

Pentas purseglovei Verdc.
F.T.E.A. Rub. 1: 206 (1976); Biol. Skr. 51: 341 (1998).
Perennial herb. Montane grassland.
Distr: EQU; Uganda.
IUCN: RD
EQU: <u>Jackson 1505</u> (fl) 5/1950.

Pentodon pentandrus (Schumach. & Thonn.)
Vatke var. *pentandrus*
F.T.E.A. Rub. 1: 263 (1976); F.J.M.: 131 (1976); F.E.E. 4(1):
217 (2003).
Syn: *Oldenlandia macrophylla* DC. – F.P.S. 2: 454
(1952).
Annual or short-lived perennial herb. Swampy grassland.
Distr: DAR, BAG, UN, EQU; Cape Verde Is., Senegal to
Somalia, S to Namibia & Botswana; Madagascar, Arabia,
U.S.A. to Brazil.
DAR: <u>Wickens 1665</u> (fl) 5/1964; **EQU:** <u>Andrews 1407</u>
5/1939.

Phyllopentas schimperi (Hochst.) ined. subsp.
schimperi
Syn: *Pentas schimperi* (Hochst.) Wieringa subsp.
schimperi – Blumea 53: 567 (2008).
Syn: *Pentas schimperiana* (A.Rich.) Vatke subsp.
schimperiana – F.T.E.A. Rub. 1: 188 (1976); Biol. Skr. 51:
342 (1998); F.E.E. 4(1): 203 (2003).
Syn: *Phyllopentas schimperiana* (A.Rich.) Kårehed &
B.Bremer – Taxon 57: 668 (2008).
Shrub or woody perennial herb. Montane grassland.
Distr: EQU; Ethiopia to Zambia.
EQU: <u>Friis & Vollesen 202</u> (fl) 11/1980.
Note: the combination in *Phyllopentas* was made for *P.
schimperiana* but *schimperi* is an earlier name for this
speces and so has nomenclatural priority – the new
combination is yet to be formalised.

Plocama calycoptera (Decne.) M.Backlund &
Thulin
Taxon 56: 323 (2007).
Syn: *Gaillonia calycoptera* (Decne.) Jaub. & Spach –
F.P.S. 2: 435 (1952); T.S.S.: 391 (1990).
Syn: *Pterogaillonia calycoptera* (Decne.) Linchevskii –
F.Egypt 2: 233 (2000).
Subshrub. Wadis, stony deserts.
Distr: RS; Egypt, Palestine & Arabia to Pakistan.
RS: <u>Schweinfurth 1956</u> (fl fr) 4/1864.

Polysphaeria lanceolata Hiern subsp.
lanceolata var. *lanceolata*
F.T.E.A. Rub. 2: 575 (1988).
Syn: *Polysphaeria schweinfurthii* Hiern – F.P.S. 2: 459
(1952).
Shrub. Forest including riverine fringes.
Distr: EQU; Kenya to Zambia & Mozambique.
EQU: <u>Jackson 4280</u> (fr) 6/1961.
Note: (1) the status of *P. schweinfurthii* is doubtful –
Verdcourt in F.T.E.A. noted that it probably belongs here.
(2) El Amin in T.S.S. records *P. dischistocalyx* Brenan
from Equatoria; this is probably a misidentification of *P.
lanceolata* since *P. dischistiocalyx* is a species from further
south in tropical Africa.

Polysphaeria parvifolia Hiern

F.P.S. 2: 456 (1952); F.T.E.A. Rub. 2: 573 (1988); T.S.S.: 405 (1990); Biol. Skr. 51: 342 (1998); F.E.E. 4(1): 261 (2003).
Shrub or small tree. Woodland, dry forest, riverine scrub.
Distr: UN; Ethiopia, Somalia, Kenya, Tanzania.
UN: <u>Myers 10471</u> (fr) 2/1939.

Pseudomussaenda flava Verdc.

F.P.S. 2: 459 (1952); F.T.E.A. Rub. 2: 467 (1988); T.S.S.: 405 (1990); Biol. Skr. 51: 342 (1998); F.E.E. 4(1): 247 (2003).
Shrub. Wooded grassland & bushland, often in rocky outcrops.
Distr: CS, BAG, EQU; Nigeria to Ethiopia, Uganda & Kenya.
CS: <u>Aylmer 36</u> 12/1939; **EQU:** <u>Andrews 857</u> (fr) 4/1939.

Psychotria articulata (Hiern) E.M.A.Petit

F.W.T.A. 2: 202 (1963).
Syn: *Psychotria mahonii* C.H.Wright var. **puberula** (E.M.A.Petit) Verdc. – F.T.E.A. Rub. 1: 60 (1976); Biol. Skr. 51: 343 (1998).
Syn: *Psychotria mahonii* C.H.Wright var. **pubescens** (Robyns) Verdc. – F.T.E.A. Rub. 1: 61 (1976); Biol. Skr. 51: 343 (1998).
Syn: *Psychotria megistosticta* (S.Moore) E.M.A.Petit var. **imatongensis** E.M.A.Petit ined. – T.S.S.: 406 (1990).
Syn: *Psychotria robynsiana* E.M.A.Petit var. **pauciorinervata** E.M.A.Petit – T.S.S.: 407 (1990).
Syn: *Psychotria mahonii* sensu auctt., non C.H.Wright – F.T.E.A. Rub. 1: 58 (1976); Biol. Skr. 51: 343 (1998).
Shrub or tree. Upland forest & bushland.
Distr: EQU; Benin to Uganda, S to Zimbabwe.
EQU: <u>Friis & Vollesen 55</u> (fr) 11/1980.
Note: O. Lachenaud (pers. comm.) states that the true *P. mahonii* is a synonym of *P. capensis* (Eckl.) Vatke subsp. *riparia* (K.Schum. & K.Krause) Verdc. var. *puberula* (E.M.A.Petit) Verdc. and that the name has been misapplied in South Sudan. He considers the Sudanese plants to belong to *P. articulata* which is a highly polymorphic species.

Psychotria eminiana (Kuntze) E.M.A.Petit

F.T.E.A. Rub. 1: 41 (1976).
Syn: *Grumilea sulphurea* Hiern – F.P.S. 2: 439 (1952).
Shrub or small tree. Lowland forest.
Distr: EQU; Nigeria to C.A.R. & D.R. Congo, S to Angola & Mozambique.
EQU: <u>Andrews 935</u> (fl) 4/1939.
Note: El Amin in T.S.S.: 406 records *P. latistipula* from Equatoria; this is clearly a misidentification since this is a West African species; the true identity is unclear.

Psychotria fertitensis Schweinf. ex Verdc.

Kew Bull. 40: 647 (1985).
Suffruticose perennial herb. Wooded grassland.
Distr: BAG, EQU; Cameroon, C.A.R.
IUCN: RD
BAG: <u>Turner 283</u> (fr).
Note: in his forthcoming revision of the African *Psychotria*, O. Lachenaud will reduce this taxon to a subspecies of *P. moninensis* (Hiern) E.M.A.Petit.

Psychotria nubica Delile

Cent. Pl. Meroe: 66 (1826); F.P.S. 2: 460 (1952).
Shrub. Habitat unknown.
Distr: CS; ?Sudan endemic.
IUCN: RD
CS: <u>Cailliaud s.n.</u> (fl) 1827 (n.v.).
Note: this species is known only from the type and is of uncertain status; from the description it is in fact unlikely to be a *Psychotria* (O. Lachenaud, pers. comm.).

Psychotria orophila E.M.A.Petit

F.T.E.A. Rub. 1: 49 (1976); T.S.S.: 406 (1990); Biol. Skr. 51: 343 (1998); F.E.E. 4(1): 199 (2003).
Shrub or small tree. Montane forest.
Distr: EQU; D.R. Congo, Ethiopia, Uganda, Kenya, Tanzania.
EQU: <u>Friis & Vollesen 111</u> (fl) 11/1980.

Psychotria peduncularis (Salisb.) Steyerm. var. *ciliato-stipulata* Verdc.

F.T.E.A. Rub. 1: 74 (1976); Biol. Skr. 51: 344 (1998); F.E.E. 4(1): 199 (2003).
Shrub or subshrubby perennial herb. Forest.
Distr: EQU; Cameroon to Uganda & Tanzania.
EQU: <u>Friis & Vollesen 514</u> (fr) 11/1980.
Note: O. Lachenaud (pers. comm.) states that true *P. peduncularis* does not occur in Sudan and that var. *ciliato-stipulata* will become a synonym of '*P. ceratoloba*' for which the combination has not yet been formalised.

Psychotria peduncularis (Salisb.) Steyerm. var. *suaveolens* Lindau

F.T.E.A. Rub. 1: 73 (1976).
Syn: *Cephaelis peduncularis* Salisb. var. **suaveolens** (Schweinf. ex Hiern) Hepper – T.S.S.: 388 (1990).
Syn: *Cephaelis suaveolens* Schweinf. ex Hiern – F.P.S. 2: 431 (1952).
Shrub or subshrubby perennial herb. Riverine forest & forest margins.
Distr: BAG, EQU; Cameroon to Kenya & Tanzania.
EQU: <u>Andrews 1599</u> (fr) 5/1939.
Note: as per the note to var. *ciliato-stipulata* above, O. Lachenaud (pers. comm.) will transfer var. *suaveolens* to '*P. mildbraedii*' (combination not yet formalised) in his forthcoming revision. He also notes that W African plants previously treated under var. *suaveolens* are referable to a different species, *P. hypsophila* K.Schum. & K.Krause.

Psychotria psychotrioides (DC.) Roberty

F.W.T.A. 2: 202 (1963); T.S.S.: 407 (1990).
Syn: *Grumilea psychotrioides* DC. – F.P.S. 2: 439 (1952).
Shrub or small tree. Forest.
Distr: EQU; Senegal to C.A.R. & D.R. Congo.
EQU: <u>Andrews 1594</u> (fr) 5/1939.

Psychotria punctata Vatke

F.T.E.A. Rub. 1: 89 (1976).
Syn: *Psychotria kirkii* Hiern var. **mucronata** (Hiern) Verdc. – F.T.E.A. Rub. 1: 96 (1976); T.S.S.: 406 (species only) (1990); Biol. Skr. 51: 343 (1998).
Syn: *Psychotria mucronata* Hiern – F.P.S. 2: 460 (1952).
Syn: *Psychotria tarambassica* Bremek. – T.S.S.: 408 (1990).

Shrub or small tree. Forest.
Distr: BAG, EQU; D.R. Congo to Somalia, S to Mozambique; Comoros.
EQU: Friis & Vollesen 794 (fl) 12/1980.
Note: *Psychotria kirkii* is here treated as a synonym of *P. punctata* on the advice of O. Lachenaud who will take this approach in his revision of African *Psychotria*.

Psychotria schweinfurthii Hiern

F.P.S. 2: 460 (1952); F.T.E.A. Rub. 1: 64 (1976); T.S.S.: 407 (1990); Biol. Skr. 51: 344 (1998).
Syn: *Psychotria sodifera* De Wild. – T.S.S.: 407 (1990).
Shrub. Forest.
Distr: EQU; Ivory Coast to Uganda, S to Angola & D.R. Congo.
EQU: Andrews 1769 6/1939.

Psychotria succulenta (Schweinf. ex Hiern) E.M.A.Petit

F.T.E.A. Rub. 1: 57 (1976); T.S.S.: 408 (1990).
Syn: *Grumilea succulenta* Schweinf. ex Hiern – F.P.S. 2: 439 (1952).
Shrub. Forest & woodland.
Distr: EQU; Ghana to Uganda, S to Angola & Zambia.
EQU: Schweinfurth 2900 (fr) 2/1870.
Note: *P. vogeliana* Benth. is also likely to occur in South Sudan since it is recorded from close to the border in neighbouring D.R. Congo (O. Lachenaud, pers. comm.).

Psychotria sp. ?nov.

Biol. Skr. 51: 345 (1998).
Shrub. Forest.
Distr: EQU
EQU: Jackson 3022 6/1953 (n.v.).
Note: the cited specimen has not been seen by us; the determination follows Friis & Vollesen who noted that this is "a species with extremely long and narrow stipules".

Psydrax acutiflora (Hiern) Bridson

F.T.E.A. Rub. 3: 906 (1991).
Syn: *Canthium malacocarpum* sensu auctt., non (K.Schum. & K.Krause) Bullock – F.P.S. 2: 429 (1952); T.S.S.: 386 (1990).
Woody climber. Forest & forest margins.
Distr: BAG; Nigeria to Uganda, S to Cabinda & D.R. Congo.
BAG: Hoyle 494 1/1939.

Psydrax parviflora (Afzel.) Bridson subsp. parviflora

F.T.E.A. Rub. 3: 896 (1991); Biol. Skr. 51: 345 (1998); F.E.E. 4(1): 274 (2003).
Syn: *Canthium vulgare* (K.Schum.) Bullock – F.P.S. 2: 431 (1952); T.S.S.: 388 (1990).
Shrub or tree. Forest & associated bushland.
Distr: EQU; Senegal to Ethiopia & Kenya, S to Angola & Zambia.
EQU: Thomas 1567 (fl fr) 12/1935.

Psydrax parviflora (Afzel.) Bridson subsp. rubrocostata (Robyns) Bridson

F.T.E.A. Rub. 3: 898 (1991); Biol. Skr. 51: 345 (1998).
Syn: *Canthium rubrocostatum* Robyns – F.P.S. 2: 431

(1952); T.S.S.: 387 (1990).
Shrub or tree. Montane forest.
Distr: EQU; Uganda, Kenya, Tanzania, Malawi.
EQU: Myers 11720 7/1939.

Psydrax schimperiana (A.Rich) Bridson subsp. schimperiana

F.T.E.A. Rub. 2: 901 (1991); F.E.E. 4(1): 273 (2003).
Syn: *Canthium schimperianum* A.Rich. – T.S.S.: 387 (1990).
Syn: *Canthium euryoides* sensu Andrews, non Bullock ex Hutch. & Dalziel – F.P.S. 2: 429 (1952).
Shrub or tree. Forest, bushland.
Distr: BAG, EQU; Eritrea & Somalia, S to Zambia & Malawi; Yemen.
Note: the record here is based on El Amin who records this species from Yei River. Bridson in F.T.E.A. concluded that Andrews' record of *Canthium euryoides* was most likely to represent this species, although she had seen no Sudanese material.

Psydrax subcordata (DC.) Bridson var. subcordata

F.T.E.A. Rub. 3: 894 (1991).
Syn: *Canthium polycarpum* Schweinf. ex Hiern – F.P.S. 2: 430 (1952); T.S.S.: 387 (1990).
Tree. Forest.
Distr: EQU; Senegal to Uganda, S to Angola & Zambia.
EQU: Schweinfurth 3051 (fr) 2/1870.

Rothmannia longiflora Salisb.

F.T.E.A. Rub. 2: 515 (1988); T.S.S.: 408 (1990); Biol. Skr. 51: 345 (1998).
Shrub or small tree. Forest & woodland.
Distr: EQU; Guinea Bissau to Kenya, S to Angola & Malawi.
EQU: Snowden 1687 (fl fr) 4/1930.

Rothmannia urcelliformis (Hiern) Robyns

F.P.S. 2: 461 (1952); F.T.E.A. Rub. 2: 514 (1988); T.S.S.: 408 (1990); Biol. Skr. 51: 345 (1998); F.E.E. 4(1): 255 (2003).
Shrub or small tree. Forest.
Distr: BAG, EQU; Sierra Leone to Ethiopia, S to Angola, Zimbabwe & Mozambique.
EQU: Snowden 1688 (fr) 4/1930.

Rothmannia whitfieldii (Lindl.) Dandy

F.P.S. 2: 461 (1952); F.T.E.A. Rub. 2: 518 (1988); T.S.S.: 409 (1990); Biol. Skr. 51: 346 (1998).
Shrub or small tree. Forest.
Distr: BAG, UN, EQU; Senegal to Uganda, S to Angola & Zambia.
EQU: Myers 8008 (fl) 12/1937.

Rubia cordifolia L. subsp. conotricha (Gand.) Verdc.

F.T.E.A. Rub. 1: 381 (1976); Biol. Skr. 51: 346 (1998); F.E.E. 4(1): 234 (2003).
Climbing or scrambling perennial herb. Montane forest margins, associated bushland & grassland.
Distr: EQU; D.R. Congo to Somalia, S to South Africa.
EQU: Friis & Vollesen 1266 3/1982.

Rutidea olenotricha Hiern

F.P.S. 2: 462 (1952); F.W.T.A. 2: 146 (1963); T.S.S.: 409 (1990).
Climbing shrub. Forest.
Distr: EQU; Sierra Leone to D.R. Congo, S to Angola & Zambia.
EQU: Sillitoe 359 (fl) 1919.

Rutidea smithii Hiern subsp. smithii

F.P.S. 2: 462 (1952); F.T.E.A. Rub. 2: 608 (1988); T.S.S.: 409 (1990).
Woody climber or shrub. Riverine forest & forest margins.
Distr: EQU; Sierra Leone to Kenya, S to Angola & Zambia.
EQU: Hoyle 718 3/1939.

Rytigynia neglecta (Hiern) Robyns var. neglecta

F.T.E.A. Rub. 3: 813 (1991); Biol. Skr. 51: 346 (1998); F.E.E. 4(1): 278 (2003).
Shrub or small tree. Montane forest.
Distr: EQU; Ethiopia, Uganda, Kenya.
EQU: Andrews 1977 (fl fr) 6/1939.

Rytigynia pauciflora (Schweinf. ex Hiern) R.D.Good

F.P.S. 2: 462 (1952); T.S.S.: 409 (1990); F.T.E.A. Rub. 3: 810 (1991).
Shrub. Wooded grassland.
Distr: EQU; Cameroon to Uganda, Angola.
EQU: Myers 13774 (fr) 4/1941.

Rytigynia senegalensis Blume

F.P.S. 2: 462 (1952); F.W.T.A. 2: 186 (1963); T.S.S.: 408 (1990).
Scrambling shrub. Wooded grassland.
Distr: BAG, EQU; Senegal to C.A.R., S to Angola & Zambia.
EQU: Andrews 989 (fl) 5/1939.

Rytigynia umbellulata (Hiern) Robyns

F.T.E.A. Rub. 3: 808 (1991); Biol. Skr. 51: 347 (1998).
Syn: *Rytigynia perlucidula* Robyns – F.P.S. 2: 463 (1952); T.S.S.: 410 (1990).
Shrub or small tree. Forest.
Distr: EQU; Guinea Bissau to Uganda, S to Angola & Zambia.
EQU: Jackson 4223 (fr) 6/1961.

Sarcocephalus latifolius (Sm.) E.A.Bruce

F.T.E.A. Rub. 2: 439 (1988); Biol. Skr. 51: 347 (1998); F.E.E. 4(1): 241 (2003).
Syn: *Nauclea latifolia* Sm. – F.P.S. 2: 446 (1952); Bot. Exp. Sud.: 100 (1970); T.S.S.: 400 (1990).
Shrub or small tree. Woodland & forest.
Distr: DAR, KOR, CS, BAG, UN, EQU; Senegal to D.R. Congo, Uganda & Kenya.
KOR: Broun 1356 (fr) 12/1907; **EQU:** Myers 9120 (fr) 7/1938.

Sherardia arvensis L.

F.Z. 5(3): 399 (2003).
Annual herb. Grassland, open ground.
Distr: RS; largely cosmopolitan.
RS: Jackson 2938 4/1953.

Note: this species is scarce in tropical Africa; it is likely to have been introduced in Sudan where it is known only from the collection cited.

Sherbournia bignoniiflora (Welw.) Hua

F.T.E.A. Rub. 2: 521 (1988).
Syn: *Amaralia bignoniiflora* (Welw.) Hiern – T.S.S.: 383 (1990).
Woody climber or scrambling shrub. Forest.
Distr: EQU; Sierre Leone to Uganda, S to Uganda & Zambia.
EQU: Myers 6523 5/1937.

Spermacoce chaetocephala DC.

F.T.E.A. Rub. 1: 354 (1976); Biol. Skr. 51: 347 (1998); F.E.E. 4(1): 225 (2003).
Syn: *Borreria chaetocephala* (DC.) Hepper var. *chaetocephala* – F.J.M.: 129 (1976).
Syn: *Borreria compacta* (Hochst.) K.Schum. – F.P.S. 2: 428 (1952).
Syn: *Borreria hebecarpa* Hochst. ex A.Rich. – F.P.S. 2: 427 (1952).
Syn: *Borreria kotschyana* (Oliv.) K.Schum. – F.P.S. 2: 428 (1952).
Annual herb. Woodland, bushland, grassland, disturbed ground.
Distr: RS, DAR, KOR, BAG, UN, EQU; Senegal to Eritrea, S to Tanzania.
DAR: Wickens 2240 8/1964; **EQU:** Simpson 7300 7/1929.

Spermacoce filifolia (Schumach. & Thonn.) J.-P.Lebrun & Stork

F.T.E.A. Rub. 3: 927 (1991).
Syn: *Octodon filifolium* Schumach. & Thonn. – F.P.S. 2: 448 (1952).
Annual herb. Seasonally wet areas in grassland & woodland.
Distr: DAR, BAG, UN, EQU; Senegal to D.R. Congo & Tanzania, S to Zambia & Zimbabwe.
DAR: Haekstra 182 10/1971; **EQU:** Myers 7627 (fl) 8/1937.

Spermacoce hepperana Verdc.

F.T.E.A. Rub. 1: 358 (1976).
?Annual herb. Seasonally wet areas in grassland.
Distr: EQU; Senegal to Ghana, Tanzania.
EQU: Andrews 1429 5/1939.
Note: the single South Sudanese specimen seen is immature and so the identification is tentative.

Spermacoce minutiflora (K.Schum.) Verdc.

F.T.E.A. Rub. 1: 360 (1976); Biol. Skr. 51: 347 (1998).
Perennial herb. Upland grassland & open woodland.
Distr: EQU; Uganda, Kenya.
EQU: Wilson 896 4/1960.
Note: the single collection from our region is from Mt Lonyili on the Uganda-South Sudan border and may have been collected from the Ugandan side – indeed, it was cited from Uganda in F.T.E.A.

Spermacoce ocymoides Burm.f.

F.T.E.A. Rub. 1: 361 (1976); Biol. Skr. 51: 348 (1998); F.E.E. 4(1): 226 (2003).

Annual herb. Grassland & open forest.
Distr: EQU; Senegal to D.R. Congo, S to Angola & Zambia; tropical Asia, South America.
EQU: Friis & Vollesen 522 (fl) 11/1980.

Spermacoce phyteuma Schweinf. ex Hiern

F.T.E.A. Rub. 1: 372, in notes (1976).
Syn: *Borreria phyteuma* (Schweinf. ex Hiern) Dandy – F.P.S. 2: 426 (1952).
Annual herb. Grassland.
Distr: BAG, EQU; C.A.R.
IUCN: RD
BAG: Schweinfurth 2295 (fl) 8/1869.

Spermacoce princeae (K.Schum.) Verdc. var. princeae

F.T.E.A. Rub. 1: 362 (1976); Biol. Skr. 51: 348 (1998); F.E.E. 4(1): 227 (2003).
?Perennial herb. Upland grassland & woodland.
Distr: EQU; Bioko & Cameroon to Ethiopia, S to Zambia.
EQU: Friis & Vollesen 1201 3/1982.

Spermacoce pusilla Wall.

F.T.E.A. Rub. 1: 356 (1976); Biol. Skr. 51: 348 (1998); F.E.E. 4(1): 225 (2003).
Syn: *Borreria pusilla* (Wall.) DC. – F.P.S. 2: 427 (1952).
Annual herb. Grassland, woodland, rocky & disturbed ground.
Distr: BAG, EQU; Gambia to Ethiopia, S to South Africa; Madagascar, tropical Asia.
EQU: Friis & Vollesen 529 (fl) 11/1980.

Spermacoce radiata (DC.) Hiern

F.T.E.A. Rub. 1: 356 (1976).
Syn: *Borreria radiata* DC. – F.P.S. 2: 427 (1952).
Annual herb. Grassland.
Distr: DAR, KOR, BAG; Senegal to Kenya, Angola, Mozambique.
KOR: Wickens 526 9/1962; **BAG:** Schweinfurth 2580 11/1869.

Spermacoce ruelliae DC.

F.T.E.A. Rub. 1: 366 (1976).
Syn: *Borreria ruelliae* (DC.) Thoms – F.P.S. 2: 428 (1952).
Annual herb. Grassland, bare ground, rocky outcrops.
Distr: UN, EQU; Senegal to Uganda, S to Angola & D.R. Congo.
EQU: Jackson 3844 (fl) 9/1957.
Note: Quézel (1969: 122) records this species from Kutum in Darfur but we belive that this is based on a misidentification, most likely of *S. sphaerostigma*.

Spermacoce sphaerostigma (A.Rich.) Vatke

F.T.E.A. Rub. 1: 367 (1976); Biol. Skr. 51: 348 (1998); F.E.E. 4(1): 228 (2003).
Syn: *Hypodematium sphaerostigma* A.Rich. – F.P.S. 2: 441 (1952).
Syn: *Borreria senensis* sensu Andrews, non Klotzsch – F.P.S. 2: 426 (1952).
Annual herb. Grassland & woodland, often in disturbed areas.
Distr: DAR, KOR, CS, UN, EQU; Nigeria to Eritrea, S to Zambia & Mozambique.

KOR: Wickens 402 (fl) 9/1962; **UN:** Simpson 7056 6/1929.
Note: Andrews recorded *Borreria* (= *Spermacoce*) *dibrachiata* (Oliv.) K.Schum. from Jongol's Post, Upper Nile but this was surely based on misidentified material, since *S. dibrachiata* is otherwise known from much further south.

Spermacoce stachydea DC. var. stachydea

F.W.T.A. 2: 220 (1963); F.E.E. 4(1): 228, in notes (2003).
Syn: *Borreria leucadea* (Hochst.) K.Schum. – F.P.S. 2: 427 (1952).
?Annual herb. Grassland.
Distr: KOR, UN; Senegal to Cameroon.
KOR: Kotschy 259 11/1839; **UN:** J.M. Lock 82/41 8/1982.

Spermacoce tenuissima Hiern

F.T.A. 3: 234 (1877).
Syn: *Borreria tenuissima* (Hiern) K.Schum. – F.P.S. 2: 426 (1952).
Annual herb. Habitat unknown.
Distr: BAG, UN; C.A.R.
IUCN: RD
BAG: Schweinfurth 2385 (fr) 9/1869.

Spermacoce verticillata L.

Syn: *Borreria verticillata* (L.) G.Mey. – F.P.S. 2: 428 (1952); F.W.T.A. 2: 220 (1963).
Perennial herb. Weed of cultivation & disturbed ground.
Distr: ?CS; native to the neotropics; now widely naturalised in the palaeotropics; in Africa from Senegal to D.R. Congo & Angola.
Note: Andrews records this species from Central Sudan; we have not see the material on which this is based but this is a distinctive species and the record is considered likely to be correct.

Stipularia elliptica Schweinf. ex Hiern

F.P.S. 2: 463 (1952); Fl. Gabon 12: 158 (1966).
Syn: *Sabicea elliptica* (Schweinf. ex Hiern) Hepper – F.W.T.A. 2: 174 (1983); T.S.S.: 410 (1990).
Shrub. Wet places.
Distr: BAG, EQU; Nigeria to C.A.R. & D.R. Congo.
EQU: Myers 11477 (fl) 6/1939.
Note: the genus *Stipularia* is sometimes treated within *Sabicea*.

Tarenna nilotica Hiern

F.P.S. 2: 463 (1952); T.S.S.: 411 (1990); Opera Bot. Belg. 14: 102 (2006).
Shrub. Woodland.
Distr: BAG, EQU; Equatorial Guinea, Cameroon, C.A.R., D.R. Congo.
BAG: Turner 267 (fl) 4/1936.

Tarenna pavettoides (Harv.) Sim. subsp. gillmanii Bridson

F.T.E.A. Rub. 2: 590 (1988); Biol. Skr. 51: 349 (1998).
Syn: *Tarenna pavettoides* sensu Andrews, non (Harv.) Sim. s.s. – F.P.S. 2: 463 (1952); T.S.S.: 411 (1990).
Shrub or small tree. Montane forest & woodland.
Distr: EQU; D.R. Congo to Kenya, S to Zambia & Malawi.
EQU: Howard U.T.T.28 11/1981.

Thecorchus wauensis (Hiern) Bremek.

F.E.E. 4(1): 218 (2003).
Syn: *Oldenlandia wauensis* Hiern – F.P.S. 2: 454 (1952).
Annual herb. Riverbanks.
Distr: BAG; Senegal to Ethiopia.
BAG: Schweinfurth 1648 (fr) 4/1869.

Tricalysia niamniamensis Schweinf. ex Hiern var. djurensis (Schweinf. ex Hiern) Robbr.

Biol. Skr. 51: 349 (1998).
Syn: *Tricalysia djurensis* Schweinf. ex Hiern – F.P.S. 2: 464 (1952); T.S.S.: 412 (1990).
Shrub or small tree. Woodland & forest.
Distr: BAG, EQU; South Sudan endemic.
IUCN: RD
EQU: Friis & Vollesen 734 12/1980.

Tricalysia niamniamensis Schweinf. ex Hiern var. niamniamensis

F.P.S. 2: 464 (1952); F.T.E.A. Rub. 2: 552 (1988); T.S.S.: 412 (1990); Biol. Skr. 51: 349 (1998); F.E.E. 4(1): 258 (2003).
Shrub or small tree. Woodland & bushland, often on rocky outcrops.
Distr: BAG, EQU; D.R. Congo to Kenya, S to Zimbabwe.
BAG: Andrews 631 (fl) 4/1939.

Tricalysia okelensis Hiern var. okelensis

F.P.S. 2: 464 (1952); F.W.T.A. 2: 151 (1963); T.S.S.: 412 (1990).
Shrub or small tree. Woodland, riverine forest.
Distr: BAG; Sierra Leone to C.A.R. & D.R. Congo.
BAG: Schweinfurth 1737 (fr) 5/1869.

Uncaria africana G.Don subsp. africana

F.P.S. 2: 464 (1952); F.T.E.A. Rub. 2: 450 (1988); T.S.S.: 412 (1990).
Woody climber. Forest & woodland.
Distr: EQU; Guinea Bissau to C.A.R., S to D.R. Congo & Tanzania; Madagascar.
EQU: Dandy 674 3/1934.

Valantia hispida L.

F.P.S. 2: 466 (1952); F.E.E. 4(1): 239 (2003).
Annual herb. Rocky hillslopes, stony ground.
Distr: RS; ?Eritrea, ?Ethiopia, Somalia; Macaronesia, Mediterranean, N Africa, Arabia to Iran.
RS: Schweinfurth 1460 1865

Vangueria agrestis (Schweinf. ex Hiern) Lantz

Pl. Syst. Evol. 253: 179 (2005).
Syn: *Fadogia agrestis* Schweinf. ex Hiern – F.W.T.A. 2: 178 (1963); T.S.S.: 390 (1990).
Subshrub. Wooded grassland.
Distr: BAG; Guinea to C.A.R.
BAG: Schweinfurth 1312 (fr) 5/1869.

Vangueria apiculata K.Schum.

F.P.S. 2: 466 (1952); T.S.S.: 413 (1990); F.T.E.A. Rub. 3: 853 (1991); F.E.E. 4(1): 279 (2003); Biol. Skr. 51: 633 (2005).
Shrub or tree, sometimes scandent. Montane forest & woodland.
Distr: BAG, EQU; Ethiopia & Somalia, S to Zimbabwe & Mozambique.
EQU: Sillitoe 295 (fl) 1919.

Vangueria infausta Burchell subsp. rotundata (Robyns) Verdc. var. rotundata

F.T.E.A. Rub. 3: 852 (1991); F.E.E. 4(1): 279 (2003).
Syn: *Vangueria tomentosa* sensu auctt., non Hochst. – F.P.S. 2: 466 (1952); T.S.S.: 413 (1990).
Syn: *Vangueria* sp. aff. *linearisepala* sensu Wickens, non K.Schum. – F.J.M.: 131 (1976).
Shrub or small tree. Riverine forest & woodland, grassland.
Distr: ?RS, DAR, KOR, CS; Uganda & Kenya, S to Malawi & Mozambique.
DAR: Long 36 (fl fr) 6/1935.
Note: (1) Sudan specimens at Kew were determined by Verdcourt as *V. infausta* sensu lato, but he lists *V. tomentosa* sensu Andrews as a misapplied name under subsp. *rotundata* var. *rotundata* in F.T.E.A. (2) The record from Red Sea region is from El Amin in T.S.S. and requires confirmation.

Vangueria madagascariensis J.F.Gmel.

F.T.E.A. Rub. 3: 849 (1991); T.S.S.: 413 (1990); F.E.E. 4(1): 279 (2003); Biol. Skr. 51: 633 (2005).
Syn: *Vangueria venosa* Hochst. ex A.Rich. – F.P.S. 2: 466 (1952).
Shrub or small tree. Forest, woodland & bushland.
Distr: DAR, KOR, CS, ES, BAG, EQU; Ghana to Ethiopia, S to Angola & South Africa; Madagascar, cultivated elsewhere in the tropics.
DAR: Wickens 2817 (fr) 11/1964; **EQU:** Myers 13281 5/1940.

Vangueria volkensii K.Schum. var. volkensii

F.T.E.A. Rub. 3: 854 (1991); Biol. Skr. 51: 350 (1998); F.E.E. 4(1): 280 (2003).
Shrub or small tree. Forest, woodland & scrub.
Distr: EQU; Ethiopia to D.R. Congo & Tanzania.
EQU: Friis & Vollesen 1075 3/1982.

Vangueriella rhamnoides (Hiern) Verdc.

F.T.E.A. Rub. 3: 839 (1991).
Shrub, sometimes scandent. Forest.
Distr: EQU; D.R. Congo, Uganda, Angola.
EQU: Sillitoe 262B (fr) 1919.

Gentianaceae

T. Harris & I. Darbyshire

Anthocleista grandiflora Gilg

F.Z. 7(1): 329 (1983); Biol. Skr. 51: 311 (1998).
Syn: *Anthocleista pulcherrima* Gilg – F.P.S. 2: 380 (1952).
Syn: *Anthocleista zambesiaca* Baker – F.T.E.A. Logan.: 10 (1960); T.S.S.: 359 (1990).
Tree. Swamp forest.
Distr: EQU; Cameroon, D.R. Congo to Kenya, S to Angola & South Africa; Comoros.
EQU: Friis & Vollesen 349 (fl) 11/1980.

Anthocleista schweinfurthii Gilg

F.T.E.A. Logan.: 11 (1960); T.S.S.: 358 (1990); F.E.E. 4(1): 70 (2003).
Tree. Riverine forest.
Distr: EQU; Nigeria to Ethiopia, S to Cabinda, D.R. Congo & Tanzania.

EQU: Kosper 281 (fr) 3/1983.

Anthocleista vogelii Planch.

F.P.S. 2: 380 (1952); F.T.E.A. Logan.: 8 (1960); T.S.S.: 358 (1990).
Tree. Swamp forest including disturbed areas.
Distr: EQU; Sierra Leone to Uganda, S to Angola & Zambia.
EQU: Andrews 1568 (fl) 5/1939.

Canscora alata (Roth.) Wall.

Blumea 48: 6 (2003).
Syn: *Canscora decussata* (Roxb.) Roem. & Schult. –
F.P.S. 3: 64 (1956); F.T.E.A. Gentian.: 40 (2002); Biol. Skr. 51: 402 (2005); F.E.E. 5: 6 (2006).
Annual herb. Woodland & wooded grassland, rocky riverbanks, swamps.
Distr: BAG, EQU; palaeotropical.
EQU: Friis & Vollesen 531 (fl) 11/1980.

Canscora diffusa (Vahl) R.Br. ex Roem. & Schult.

F.P.S. 3: 64 (1956); F.J.M.: 139 (1976); F.T.E.A. Gentian.: 42 (2002); F.E.E. 5: 7 (2006).
Annual herb. Streambanks, woodland on rocky slopes.
Distr: DAR, KOR, BAG; palaeotropical.
DAR: Wickens 1526 (fl) 5/1964; **BAG:** Schweinfurth 2814 (fl) 1/1870.

Enicostema axillare (Lam.) A.Raynal subsp. *axillare*

F.T.E.A. Gentian.: 26 (2002); F.E.E. 5: 5 (2006).
Syn: *Enicostema verticillare* (Retz.) Baill. – F.P.S. 3: 64 (1956).
Perennial herb. Open rocky areas, grassland & bushland.
Distr: RS, KOR; Ethiopia to Namibia & South Africa; India.
RS: Farag 113 (fl fr) 3/2010.

Exacum oldenlandioides (S.Moore) Klack.

F.T.E.A. Gentian.: 4 (2002); Biol. Skr. 51: 402 (2005).
Annual herb. Moist sites in woodland & wooded grassland, riverbanks.
Distr: EQU; Senegal to Kenya, S to Angola & South Africa.
EQU: Friis & Vollesen 525 11/1980.

Faroa pusilla Baker

F.P.S. 3: 64 (1956); F.T.E.A. Gentian.: 34 (2002).
Annual herb. Shallow soil over rock.
Distr: BAG; Senegal to C.A.R., D.R. Congo, Tanzania, Zambia & Mozambique.
BAG: Schweinfurth 2513 (fl) 10/1869.

Neurotheca loeselioides (Spruce ex Prog.) Baill. subsp. *loeselioides*

F.T.E.A. Gentian.: 27 (2002); Biol. Skr. 51: 402 (2005).
Annual herb. Wet flushes over rock in woodland & grassland.
Distr: EQU; Senegal to Uganda, S to Angola; Madagascar, S America.
EQU: Friis & Vollesen 583 (fl fr) 11/1980.

Sebaea oligantha (Gilg) Schinz

F.T.E.A. Gentian.: 15 (2002); F.E.E. 5: 4 (2006).
Syn: *Belmontia sp.* – F.P.S. 3: 64 (1956).
Saprophytic annual herb. Forest.
Distr: EQU; Guinea to Ethiopia, S to Angola & Tanzania.
EQU: Sillitoe 390 (fl fr) 1919.
Note: a widespread but scarce species.

Sebaea pumila (Baker) Schinz

F.T.E.A. Gentian.: 16 (2002); Biol. Skr. 51: 402 (2005).
Annual herb. Damp sites in woodland & wooded grassland.
Distr: EQU; Nigeria, D.R. Congo to Tanzania, S to Zambia & Mozambique.
EQU: Friis & Vollesen 582 (fl fr) 11/1980.

Swertia abyssinica Hochst.

F.P.S. 3: 65 (1956); F.J.M.: 139 (1976); F.T.E.A. Gentian.: 48 (2002); F.E.E. 5: 12 (2006).
Annual herb. Montane grassland & forest-grassland transition, lakeshores, dry riverbeds.
Distr: DAR; Bioko, Cameroon, Eritrea to Zambia & Mozambique.
DAR: Miehe 649 (fl) 9/1982.

Swertia eminii Engl.

F.T.E.A. Gentian.: 55 (2002); Biol. Skr. 51: 403 (2005).
Annual herb. Woodland & bushland, wet grassland, swamp margins.
Distr: EQU; Cameroon, D.R. Congo to Kenya, S to Zimbabwe & Mozambique.
EQU: Myers 13416 (fl) 8/1940.

Swertia schimperi (Hochst.) Griseb.

F.T.E.A. Gentian.: 61 (2002); Biol. Skr. 51: 403 (2005); F.E.E. 5: 15 (2006).
Perennial herb. Montane grassland & ericaceous bushland.
Distr: EQU; Ethiopia, Uganda, Tanzania, Malawi.
EQU: Jackson 1543 (fl) 6/1950.

Gelsemiaceae

H. Pickering

Mostuea hirsuta (T.Anderson ex Benth. & Hook.f.) Baill. ex Baker

F.W.T.A. 2: 45 (1963).
Syn: *Coinochlamys schweinfurthii* Gilg. – F.P.S. 2: 380 (1952); T.S.S.: 359 (1990).
Shrub. Open forest, woodland.
Distr: BAG, EQU; Guinea Bissau to C.A.R., S to Angola & D.R. Congo.
EQU: Myers 6476 4/1937.

Loganiaceae

H. Pickering & I. Darbyshire

Strychnos henningsii Gilg

F.T.E.A. Logan.: 32 (1960); T.S.S.: 361 (1990); F.E.E. 4(1): 76 (2003).
Syn: *Strychnos holstii* Gilg – F.P.S. 2: 383, ?pro parte (1952).
Shrub or small tree. Woodland.

Distr: EQU; Ethiopia & Somalia, S to Angola & South
Africa; Madagascar.
EQU: <u>Dale 319</u> 2/1943.

Strychnos innocua Delile

F.P.S. 2: 381 (1952); F.T.E.A. Logan.: 25 pro parte (1960);
F.J.M.: 126 (1976); T.S.S.: 361 (1990); Biol. Skr. 51: 312
(1998); F.E.E. 4(1): 76 (2003).
Shrub or small tree. Woodland & wooded grassland.
Distr: DAR, KOR, CS, ES, BAG, EQU; Guinea to Eritrea, S
to Angola, Zimbabwe & Mozambique.
KOR: <u>Wickens 514</u> 9/1962; **EQU:** <u>Myers 6913</u> 6/1937.

Strychnos mitis S.Moore

F.T.E.A. Logan.: 17 (1960); T.S.S.: 362 (1990); Biol. Skr.
51: 312 (1998); F.E.E. 4(1): 76 (2003).
Tree. Forest.
Distr: EQU; Ethiopia to Angola & South Africa.
EQU: <u>Jackson 354</u> 1956.
Note: Friis & Vollesen (1998) stated that Andrews
misapplied the name *S. holstii*, the specimens being
referable to *S. mitis*. However, *Dale 319* from Lorienaton
(the locality cited by Andrews) which was previously
named *S. holstii* at Kew, is a specimen of *S. henningsii*, of
which *S. holstii* is a true synonym.

Strychnos spinosa Lam.

F.P.S. 2: 383 (1952); F.T.E.A. Logan.: 17 (1960); F.J.M.:
126 (1976); T.S.S.: 362 (1990); Biol. Skr. 51: 312 (1998);
F.E.E. 4(1): 78 (2003).
Syn: *Strychnos emarginata* Baker – F.P.S. 2: 383 (1952).
Syn: *Strychnos gracillima* Gilg – F.P.S. 2: 383 (1952).
Syn: *Strychnos spinosa* Lam. var. *pubescens* Baker –
F.P.S. 2: 383 (1952).
Shrub or small tree. Woodland & bushland.
Distr: DAR, BAG, EQU; Senegal to Ethiopia, S to Angola
& South Africa.
DAR: <u>Wickens 1672</u> 5/1964; **EQU:** <u>Andrews 1272</u>
5/1939.

Apocynaceae

D.J. Goyder & H. Pickering

Adenium obesum (Forssk.) Roem. & Schult.

F.J.M.: 127 (1976); T.S.S.: 369 (1990); Biol. Skr. 51: 316
(1998); F.T.E.A. Apoc. 1: 67 (2002); F.E.E. 4(1): 88 (2003).
Syn: *Adenium coetaneum* Stapf – F.P.S. 2: 390 (1952);
Bot. Exp. Sud.: 16 (1970).
Syn: *Adenium honghel* A.DC. – F.P.S. 2: 389 (1952);
Bot. Exp. Sud.: 17 (1970).
Syn: *Adenium speciosum* Fenzl – F.P.S. 2: 390 (1952).
Shrub. Semi-desert.
Distr: DAR, KOR, CS, UN, EQU; Senegal to Somalia,
Uganda, Kenya, Tanzania; Arabia.
DAR: <u>Wickens 1285</u> 2/1964; **EQU:** <u>Padwa 238</u> 5/1953.

Alafia multiflora (Stapf) Stapf

F.W.T.A. 2: 73 (1963); T.S.S.: 369 (1990).
Syn: *Holalafia multiflora* Stapf – F.P.S. 2: 392 (1952).
Woody climber. Rocky grassland.
Distr: EQU; Liberia to D.R. Congo.
EQU: <u>Andrews 1437</u> 4/1939.

Alstonia boonei De Wild.

F.P.S. 2: 390 (1952); T.S.S.: 369 (1990); Biol. Skr. 51: 316
(1998); F.T.E.A. Apoc. 1: 60 (2002); F.E.E. 4(1): 89 (2003).
Tree. Wet forest.
Distr: EQU; Senegal to Ethiopia, Uganda & Tanzania.
EQU: <u>Thomas 1761</u> 12/1935.

Ancylobotrys amoena Hua

F.W.T.A. 2: 60 (1963).
Syn: *Landolphia petersiana* (Klotsch) Dyer var.
schweinfurthiana (Hallier f.) Stapf – F.P.S. 2: 395 (1952).
Syn: *Landolphia petersiana* sensu auctt., non
(Klotzsch) Dyer – F.P.S. 2: 394 (1952); T.S.S.: 373 (1990).
Woody climber. Wet forest.
Distr: BAG, EQU; Nigeria to D.R. Congo.
BAG: <u>Turner 131</u> 1/1936.

Aspidoglossum angustissimum (K.Schum.) Bullock

Biol. Skr. 51: 319 (1998); F.T.E.A. Apoc. 2: 375 (2012).
Syn: *Schizoglossum angustissimum* K.Schum. – F.P.S.
2: 416 (1952).
Perennial herb. Marshy grassland.
Distr: EQU; Cameroon to Kenya, S to Zimbabwe.
EQU: <u>Sillitoe 449</u> 1919.

Aspidoglossum connatum (N.E.Br.) Bullock

F.T.E.A. Apoc. 2: 376 (2012).
Perennial herb. Marshy grassland.
Distr: EQU; Guinea Bissau, D.R. Congo to Kenya, S to
Zambia.
EQU: <u>Sillitoe 448</u> 1919.

Aspidoglossum interruptum (E.Mey.) Bullock

F.E.E. 4(1): 119 (2003); F.T.E.A. Apoc. 2: 378 (2012).
Syn: *Schizoglossum abyssinicum* K.Schum. – F.P.S. 2:
416 (1952).
Perennial herb. Grassland.
Distr: EQU; Guinea Bissau, Nigeria, S to Angola & South
Africa.
EQU: <u>Sillitoe 447</u> 1919.

Brachystelma lineare A.Rich.

F.P.S. 2: 402 (1952); Biol. Skr. 51: 319 (1998); F.E.E. 4(1):
169 (2003); F.T.E.A. Apoc. 2: 303 (2012).
Syn: *Brachystelma phyteumoides* K.Schum. – F.P.S. 2:
402 (1952).
Perennial herb. Wooded grassland.
Distr: EQU; Eritrea, Ethiopia, Kenya.
EQU: <u>Greenway & Hummel 7313</u> 4/1945.

Brachystelma sp. A

Syn: *Brachystelma* sp. ?nov. sensu Friis & Vollesen –
Biol. Skr. 51: 320 (1998).
Perennial herb. Recently burnt grassland with scattered
trees.
Distr: EQU.
EQU: <u>Friis & Vollesen 1202</u> 3/1982 (n.v.).
Note: the cited specimen, from Mt Konoro in the
Imatong Mts, has not been seen by us and its identity
remains uncertain; Friis & Vollesen recorded it as a
possible new species.

Buckollia tomentosa (E.A.Bruce) Venter & R.L.Verh.

Biol. Skr. 51: 320 (1998); F.E.E. 4(1): 103 (2003); F.T.E.A. Apoc. 2: 162 (2012).
Woody climber. Woodland, bushland & thicket.
Distr: ?EQU; Ethiopia, Uganda.
Note: this species is recorded from the Ugandan side of the Imatong Mts and may well occur in South Sudan; indeed, it is recorded as occurring in 'Sudan' in F.T.E.A. (l.c.) but we have seen no material to confirm this.

Calotropis procera (Aiton) W.T.Aiton

F.P.S. 2: 402 (1952); F.J.M.: 127 (1976); T.S.S.: 377 (1990); Biol. Skr. 51: 320 (1998); F.E.E. 4(1): 119 (2003); F.T.E.A. Apoc. 2: 368 (2012).
Shrub. Dry sandy areas.
Distr: NS, RS, DAR, KOR, CS, EQU; tropical Africa; Arabia to India.
RS: Schweinfurth 586 10/1868; **EQU:** Maxwell Forbes 92 1947.

Caralluma adscendens (Roxb.) Haw.

Syn. Pl. Succ.: 47 (1812).
Syn: *Caralluma vittata* sensu Quézel, non N.E.Br. – F. Darfur Nord-Occ. & J. Gourgeil: 120 (1969).
Syn: *Caralluma* sp. aff. *vittata* sensu Wickens – F.J.M.: 128 (1976).
Succulent herb. Semi-desert.
Distr: DAR, KOR; Mauritania to Sudan; Arabia, India
DAR: Wickens 2892 3/1965.
Note: we have taken a broad view of this species as indicated by annotations by Bruyns on Kew sheets and suggested by Müller & Albers (2002) in Albers & Meve (eds.), Illustrated Handbook of Succulent Plants: Asclepiadaceae. This concept includes *C. subulata* Forssk. ex Decne. from Arabia and *C. dalzielii* N.E.Br. from W & C Africa.

Caralluma darfurensis Plowes

Excelsa 21: 7 (2007).
Succulent herb. Dry wooded grassland.
Distr: DAR; Sudan endemic.
IUCN: RD
DAR: Plowes 7497 5/1987.

Caralluma dicapuae (Chiov.) Chiov.

F.E.E. 4(1): 173 (2003); F.T.E.A. Apoc. 2: 313 (2012).
Succulent herb. Dry bushland.
Distr: EQU; Eritrea, Ethiopia, Somalia, Kenya.
EQU: Padwa 227 5/1953.

Caralluma edulis (Edgew.) Meve & Liede

F.E.E. 4(1): 174 (2003).
Syn: *Caralluma longidens* N.E.Br. – F.P.S. 2: 403 (1952).
Syn: *Caralluma vittata* N.E.Br. – F.P.S. 2: 404 (1952); Bot. Exp. Sud.: 18 (1970).
Succulent herb. Open bushland.
Distr: RS; Ethiopia, Somalia; Saudi Arabia, Oman, Pakistan, India.
RS: Schweinfurth 441 10/1868.

Caralluma sinaica (Decne.) A.Berger

F.Egypt 2: 227 (2000).

Syn: *Caralluma maris-mortui* Zohary – Excelsa 14: 61 (1989).
Succulent herb. Dry rocky ground.
Distr: RS; Egypt (Sinai), Palestine, Saudi Arabia, Yemen.
Note: this record is from Plowes (1989: 61) who collected the species from the Red Sea Hills in November 1987; we have not seen the resultant specimen.

Caralluma sudanica Bruyns

Aloe 41(4): 79 (2004).
Succulent herb. Semi-desert.
Distr: KOR; Sudan endemic.
IUCN: RD
KOR: Wickens 27 6/1962.

Carissa spinarum L.

F.T.E.A. Apoc. 1: 12 (2002); F.E.E. 4(1): 90 (2003).
Syn: *Carissa edulis* (Forssk.) Vahl – F.P.S. 2: 390 (1952); Bot. Exp. Sud.: 17 (1970); F.J.M.: 127 (1976); T.S.S.: 371 (1990); Biol. Skr. 51: 317 (1998).
Syn: *Carissa edulis* (Forssk.) Vahl var. *tomentosa* (A.Rich.) Stapf – Bot. Exp. Sud.: 17 (1970).
Syn: *Acokanthera schimperi* sensu auctt., non (A.DC.) Schweinf. – F.P.S. 2: 388 (1952); T.S.S.: 367 (1990); Biol. Skr. 51: 316 (1998).
Spiny shrub. Wet forest.
Distr: RS, DAR, KOR, BAG, UN, EQU; Mali to Somalia, S to South Africa; Madagascar, Arabia, India, SE Asia, Australia.
RS: Sahni & Kamil 667 4/1967; **BAG:** Schweinfurth 1400 3/1869.

Ceropegia abyssinica Decne. var. *abyssinica*

Biol. Skr. 51: 320 (1998); F.E.E. 4(1): 160 (2003); F.T.E.A. Apoc. 2: 250 (2012).
Perennial herb. Wooded grassland.
Distr: EQU; C.A.R. to Ethiopia, S to Angola & Zimbabwe.
EQU: van Nordvijk s.n. 9/1981.

Ceropegia aristolochioides Decne. subsp. *aristolochioides*

F.E.E. 4(1): 164 (2003); F.T.E.A. Apoc. 2: 270 (2012).
Succulent climber. Wooded grassland.
Distr: UN; Senegal, Burkina Faso, Cameroon to Ethiopia, S to Tanzania.
UN: J.M. Lock 81/142 8/1981.

Ceropegia melanops H.Huber

Mem. Soc. Brot. 12: 158 (1957).
Slender twiner. Bushland.
Distr: BAG; Cameroon, C.A.R.
IUCN: RD
BAG: Myers 13966 8/1941.
Note: *C. meyeri-johannis* Engl. is also likely to occur in South Sudan, since it has been recorded from the Ugandan side of the Imatong Mts (Friis & Vollesen 1998: 320).

Ceropegia nilotica Kotschy

F.P.S. 2: 406 (1952); F.T.E.A. Apoc. 2: 283 (2012).
Herbaceous twiner. Bushland & forest margins.
Distr: EQU; Ghana to Kenya, S to Namibia & South Africa.
EQU: Knoblecher 35.

Ceropegia racemosa N.E.Br. var. *racemosa*

F.P.S. 2: 406 (1952); F.E.E. 4(1): 166 (2003); F.T.E.A. Apoc. 2: 261 (2012).
Slender twiner. Bushland.
Distr: ?DAR, BAG; Guinea to Somalia, S to Namibia & South Africa; Madagascar.
BAG: Schweinfurth 2105 7/1869.
Note: the record for Darfur is from Quézel (1969: 121); we have not seen the specimen on which it was based.

Ceropegia stenantha K.Schum.

F.P.S. 2: 406 (1952); F.T.E.A. Apoc. 2: 282 (2012).
Twiner. Wooded grassland.
Distr: BAG; D.R. Congo to Kenya, S to Namibia & South Africa.
BAG: Schweinfurth 2104 7/1869.

Conomitra linearis Fenzl

F.E.E. 4(1): 157 (2003); F.T.E.A. Apoc. 2: 212 (2012).
Syn: *Glossonema lineare* (Fenzl) Decne. – F.P.S. 2: 407 (1952).
Ephemeral herb. Sandy grassland.
Distr: KOR; Niger, Ethiopia, Somalia, Kenya.
IUCN: RD
KOR: Kotschy 78 9/1840.
Note: a rarely collected species, but probably under-recorded.

Cryptolepis oblongifolia (Meisn.) Schltr.

F.P.S. 2: 406 (1952); F.T.E.A. Apoc. 2: 130 (2012).
Syn: *Ectadiopsis oblongifolia* (Meisn.) Schltr. – Biol. Skr. 51: 322 (1998).
Subshrub. Wooded grassland.
Distr: EQU; Guinea Bissau to Kenya, S to South Africa.
EQU: Friis & Vollesen 1123 3/1982.

Cynanchum gerrardii (Harv.) Liede subsp. *gerrardii*

Biol. Skr. 51: 321 (1998); F.E.E. 4(1): 145 (2003); F.T.E.A. Apoc. 2: 476 (2012).
Succulent leafless climber. Dry bushland.
Distr: EQU; Eritrea & Somalia, S to South Africa; Madagascar, Arabia.
EQU: Kielland-Lund 145 12/1983 (n.v.).

Cynanchum polyanthum K.Schum.

F.P.S. 2: 407 (1952); F.T.E.A. Apoc. 2: 489 (2012).
Slender climber. Bushland, sandy open areas.
Distr: EQU; Cameroon, Gabon, D.R. Congo, Uganda, Angola.
EQU: Sillitoe 455 1919.

Cynanchum praecox S.Moore

Biol. Skr. 51: 321 (1998); F.T.E.A. Apoc. 2: 481 (2012).
Pyrophytic perennial herb. Wooded grassland.
Distr: EQU; Sierra Leone to Tanzania, S to Zimbabwe.
EQU: Friis & Vollesen 1082 3/1982.

Cynanchum viminale (L.) L.

F.T.E.A. Apoc. 2: 480 (2012).
Syn: *Sarcostemma viminale* (L.) W.T.Aiton – F.P.S. 2: 416 (1952); Bot. Exp. Sud.: 19 (1970); F.J.M.: 128 (1976); T.S.S.: 382 (1990); Biol. Skr. 51: 324 (1998); F.E.E. 4(1): 148 (2003).

Succulent shrub or twiner. Wooded grassland.
Distr: RS, DAR, BAG, EQU; palaeotropical.
DAR: Wickens 1493 4/1964; **EQU:** Myers 6635 5/1937.
Note: the subspecific status of the Sudanese material is unclear at present.

Desmidorchis retrospiciens Ehrenb.

Syn: *Caralluma acutangula* (Decne.) N.E.Br. – Biol. Skr. 51: 320 (1998); F.E.E. 4(1): 174 (2003).
Syn: *Caralluma retrospiciens* (Ehrenb.) N.E.Br. – F.P.S. 2: 404 (1952); Bot. Exp. Sud.: 18 (1970).
Syn: *Desmidorchis acutangula* Decne. – F.T.E.A. Apoc. 2: 320 (2012).
Succulent herb. Bushland.
Distr: RS, EQU; Mauritania to Somalia, Uganda, Kenya; Arabia.
RS: Schweinfurth 467 9/1868; **EQU:** Myers 13459 9/1940.

Desmidorchis penicillata (Deflers) Plowes

Haseltonia 3: 58 (1995).
Syn: *Caralluma penicillata* (Deflers) N.E.Br. – F.E.E. 4(1): 176 (2003).
Syn: *Caralluma penicillata* (Deflers) N.E.Br. var. *robusta* (N.E.Br.) A.C.White & B.Sloane – F.P.S. 2: 404 (1952).
Succulent herb. Sandy grassland.
Distr: RS; Ethiopia, Somalia; Arabia.
RS: Darvall 4 12/1937.

Desmidorchis speciosa (N.E.Br.) Plowes

F.T.E.A. Apoc. 2: 321 (2012).
Syn: *Caralluma speciosa* (N.E.Br.) N.E.Br. – F.E.E. 4(1): 176 (2003).
Succulent herb. Dry bushland.
Distr: RS, EQU; Ethiopia, Djibouti, Somalia, Uganda, Kenya, Tanzania.
RS: Ahti 16489 10/1962; **EQU:** Myers 13438 8/1940.

Duvalia sulcata N.E.Br. subsp. *sulcata*

F.P.S. 2: 407 (1952); F.E.E. 4(1): 190 (2003).
Succulent herb. Bushland.
Distr: RS; Ethiopia, Somalia; Saudi Arabia, Yemen.
RS: Carter 1880 11/1987.

Echidnopsis cereiformis Hook.f.

F.E.E. 4(1): 179 (2003); Pl. Syst. Evol. 265: 74 (2007).
Syn: *Echidnopsis nubica* N.E.Br. – F.P.S. 2: 407 (1952).
Succulent herb. Bushland.
Distr: RS; Eritrea, Ethiopia.
IUCN: RD
RS: Schweinfurth 228 9/1868.

Funtumia elastica (P.Preuss) Stapf

F.P.S. 2: 392 (1952); T.S.S.: 371 (1990); Biol. Skr. 51: 317 (1998); F.T.E.A. Apoc. 1: 88 (2002).
Tree. Wet forest.
Distr: EQU; Senegal to Uganda & Tanzania.
EQU: Turner 138 2/1936.

Glossonema boveanum (Decne.) Decne.

F.P.S. 2: 408 (1952); F.E.E. 4(1): 117 (2003).
Syn: *Glossonema nubicum* Decne. – F.P.S. 2: 407 (1952).
Perennial herb. Dry grassland.
Distr: RS, DAR, KOR; Mauritania to Somalia; N Africa, Arabia.
RS: Carter 1968 11/1987.

Gomphocarpus abyssinicus Decne.
F.E.E. 4(1): 124 (2003).
Syn: *Asclepias phillipsae* sensu auctt., non N.E.Br. –
F.P.S. 2: 402 (1952); F. Darfur Nord-Occ. & J. Gourgeil:
121 (1969).
Perennial herb. Open rocky ground.
Distr: DAR; Guinea, Cameroon, Eritrea, Ethiopia.
DAR: Jackson 4060 9/1960.

Gomphocarpus fruticosus (L.) W.T.Aiton
subsp. **flavidus** (N.E.Br.) Goyder
F.E.E. 4(1): 124 (2003); F.T.E.A. Apoc. 2: 426 (2012).
Syn: *Asclepias flavida* N.E.Br. – F.P.S. 2: 401 (1952);
Bot. Exp. Sud.: 18 (1970).
Syn: *Gomphocarpus fruticosus* sensu Wickens, non
(L.) W.T.Aiton s.s. – F.J.M.: 128 (1976).
Perennial herb. Disturbed areas.
Distr: RS, DAR; Ethiopia, Somalia, Uganda, Kenya.
RS: Andrews 3429.

Gomphocarpus semilunatus A.Rich.
F.E.E. 4(1): 123 (2003); F.T.E.A. Apoc. 2: 428 (2012).
Syn: *Asclepias semilunata* (A.Rich.) N.E.Br. – F.P.S. 2:
401 (1952).
Perennial herb. Seasonally flooded grasslands.
Distr: UN, EQU; Nigeria to Ethiopia, S to Angola &
Zambia.
UN: Simpson 7191 6/1929.

Holarrhena floribunda (G.Don) T.Durand &
Schinz
F.W.T.A. 2: 68 (1963).
Syn: *Holarrhena africana* A.DC. – T.S.S.: 371 (1990).
Syn: *Holarrhena wulfsbergii* Stapf – F.P.S. 2: 394
(1952); Bot. Exp. Sud.: 17 (1970).
Tree. Forest.
Distr: EQU; Senegal to D.R. Congo.
EQU: Myers 13605 12/1940.

Huernia macrocarpa (A.Rich.) Spreng.
F.P.S. 2: 408 (1952); F.E.E. 4(1): 191 (2003).
Succulent herb. Bushland.
Distr: RS; Ethiopia; Saudi Arabia.
RS: Schweinfurth 227 9/1868.

Huernia sudanensis Plowes
Haseltonia 19: 72 (2014).
Succulent herb. Sandy plains & bare rocky hills.
Distr: RS; Ethiopia.
IUCN: RD
RS: Plowes 7553 11/1987.
Note: this species falls within the *H. macrocarpa* complex
which some authors treat as a single more variable
species. The relationship between the two in Sudan
requires further study.

Kanahia laniflora (Forssk.) R.Br.
F.P.S. 2: 408 (1952); T.S.S.: 379 (1990); F.E.E. 4(1): 121
(2003); F.T.E.A. Apoc. 2: 370 (2012).
Perennial herb or shrub. River beds, rheophytic.
Distr: NS, CS, ES, UN, EQU; Ivory Coast to Somalia, S to
South Africa; Arabia.
ES: Schweinfurth 216 6/1865; **EQU:** Myers 8293 1/1938.

Landolphia buchananii (Hallier f.) Stapf
T.S.S.: 373 (1990); Biol. Skr. 51: 317 (1998); F.T.E.A.
Apoc. 1: 19 (2002); F.E.E. 4(1): 92 (2003).
Syn: *Landolphia ugandensis* (Stapf) Pichon – F.P.S. 2:
395 (1952).
Woody climber. Riverine forest.
Distr: EQU; Nigeria to Somalia, S to Mozambique.
EQU: Friis & Vollesen 925 2/1982.

Landolphia landolphioides (Hallier f.) A.Chev.
F.T.E.A. Apoc. 1: 21 (2002).
Woody climber. Wet forest.
Distr: EQU; Guinea to Uganda, S to Angola & D.R.
Congo.
EQU: Sillitoe 262 1919.

Landolphia owariensis P.Beauv.
T.S.S.: 373 (1990); F.T.E.A. Apoc. 1: 22 (2002).
Syn: *Landolphia owariensis* P.Beauv. var. *tomentella*
Stapf – F.P.S. 2: 395 (1952).
Woody climber. Forest.
Distr: BAG, EQU; Senegal to Uganda, S to Malawi.
EQU: Andrews 1635 6/1939.

Leptadenia arborea (Forssk.) Schweinf.
T.S.S.: 381 (1990); F.E.E. 4(1): 155 (2003).
Syn: *Leptadenia heterophylla* (Delile) Decne. – F.P.S.
2: 409 (1952).
Twining shrub. Dry river beds.
Distr: NS, RS, DAR, KOR, CS; Niger to Somalia; Algeria,
Egypt, Arabia.
CS: Andrews 65 6/1935.
Note: Drar (1970: 18) also records this species with
uncertainty from Darfur and Eastern Sudan, whilst
El Amin in T.S.S. records it from Equatoria which is
considered unlikely.

Leptadenia hastata (Pers.) Decne. subsp.
hastata
F.J.M.: 128 (1976); T.S.S.: 381 (1990); Biol. Skr. 51: 322
(1998); F.E.E. 4(1): 156 (2003); F.T.E.A. Apoc. 2: 210
(2012).
Syn: *Leptadenia lancifolia* (Schumach. & Thonn.)
Decne. – F.P.S. 2: 409 (1952).
Twining shrub. Dry river beds.
Distr: NS, RS, DAR, KOR, CS, BAG, UN, EQU; Mauritania
to Ethiopia, Uganda, Kenya.
DAR: Wickens 1287 4/1964; **EQU:** Friis & Vollesen 1043
2/1982.

Leptadenia pyrotechnica (Forssk.) Decne.
F.P.S. 2: 410 (1952); T.S.S.: 381 (1990); F.E.E. 4(1): 155
(2003).
Leafless shrub. Sandy desert.
Distr: NS, RS, DAR, KOR, CS, ES; Senegal to Ethiopia; N
Africa, Arabia, Pakistan, India.
KOR: Harrison 937 9/1950.

Margaretta rosea Oliv. subsp. *rosea*
F.P.S. 2: 411 (1952); F.T.E.A. Apoc. 2: 404 (2012).
Perennial herb. Grassland.
Distr: EQU; C.A.R., D.R. Congo, Rwanda, Burundi,
Uganda, Kenya, Tanzania.
EQU: Myers 6533 5/1937.

Marsdenia abyssinica (Hochst.) Schltr.

F.P.S. 2: 411 (1952); F.T.E.A. Apoc. 2: 204 (2012).
Syn: *Dregea abyssinica* (Hochst.) K.Schum. – F.J.M.: 128 (1976); Biol. Skr. 51: 321 (1998); F.E.E. 4(1): 151 (2003).
Syn: *Hoya africana* Decne. – T.S.S.: 379 (1990).
Woody climber. Riverine habitats.
Distr: DAR, KOR, ES, BAG, ?EQU; Senegal to Ethiopia, S to Zimbabwe; Arabia.
ES: Schweinfurth 239 1865; **BAG:** Schweinfurth 1696 5/1869.
Note: also likely to occur in Equatoria since it has been recorded on the Ugandan side of the Imatong Mts (Friis & Vollesen 1998: 321).

Marsdenia rubicunda (K.Schum.) N.E.Br.

F.P.S. 2: 411 (1952); F.T.E.A. Apoc. 2: 199 (2012).
Syn: *Dregea rubicunda* K.Schum. – F.J.M.: 128 (1976); T.S.S.: 379 (1990); F.E.E. 4(1): 151 (2003).
Woody climber. Woodland, bushland.
Distr: DAR, KOR, BAG, UN, EQU; Ethiopia, Somalia, Uganda, Kenya, Tanzania.
DAR: Wickens 1620 5/1964; **BAG:** Myers 13840 6/1941.

Marsdenia schimperi Decne.

F.T.E.A. Apoc. 2: 201 (2012).
Syn: *Dregea schimperi* (Decne.) Bullock – Biol. Skr. 51: 321 (1998); F.E.E. 4(1): 151 (2003).
Woody climber. Forest margins.
Distr: RS, EQU; Nigeria to Somalia, S to Angola and Tanzania; Arabia.
RS: Carter 1843 11/1987; **EQU:** Andrews 1952 6/1939.

Marsdenia sylvestris (Retz.) P.I.Forst.

F.T.E.A. Apoc. 2: 194 (2012).
Syn: *Gymnema sylvestris* (Retz.) R.Br. ex Schult. – F.P.S. 2: 408 (1952); T.S.S.: 379 (1990); F.E.E. 4(1): 153 (2003).
Woody climber. Sandy soil, near water.
Distr: BAG, UN, EQU; tropical Africa; Madagascar, Arabia, India.
UN: J.M. Lock 81/255 8/1981.

Mondia whitei (Hook.f.) Skeels

Biol. Skr. 51: 322 (1998); F.T.E.A. Apoc. 2: 145 (2012).
Woody climber. Forest.
Distr: EQU; Guinea Bissau to Kenya, S to South Africa.
EQU: Andrews 1739 6/1939.

Oncinotis pontyi Dubard

Biol. Skr. 51: 318 (1998); F.T.E.A. Apoc. 1: 107 (2002).
Woody climber. Wet forest.
Distr: EQU; Ivory Coast to Uganda.
EQU: Andrews 1732 6/1939.

Oncinotis tenuiloba Stapf

Biol. Skr. 51: 318 (1998); F.T.E.A. Apoc. 1: 108 (2002); F.E.E. 4(1): 92 (2003).
Climbing shrub. Swamp forest.
Distr: EQU; Nigeria to Ethiopia, S to South Africa.
EQU: Friis & Vollesen 355 11/1980.

Orbea decaisneana (Lem.) Bruyns

Aloe 37(4): 74 (2000).
Syn: *Angolluma sudanensis* Plowes – Excelsa 16: 109 (1994).

Orbea laticorona (M.G.Gilbert) Bruyns

Aloe 37(4): 75 (2000).
Syn: *Pachycymbium laticoronum* (M.G.Gilbert) M.G.Gilbert – F.E.E. 4(1): 186 (2003).
Succulent herb. Bushland.
Distr: NS, RS; Ethiopia.
IUCN: RD
RS: Plowes 7550 8/1988.

Orbea sprengeri (Schweinf.) Bruyns subsp. sprengeri

F.T.E.A. Apoc. 2: 345 (2012).
Syn: *Caralluma baldratii* sensu Andrews, non A.C.White & B.Sloane – F.P.S. 2: 406 (1952).
Succulent herb. Bushland.
Distr: RS; Eritrea, Ethiopia, Kenya.
IUCN: RD
RS: Robbie 17 5/1934.

Orthopichonia schweinfurthii (Stapf) H.Huber

F.W.T.A. 2: 58 (1963); T.S.S.: 374 (1990).
Syn: *Clitandra schweinfurthii* Stapf – F.P.S. 2: 392 (1952).
Woody climber. Forest.
Distr: BAG, EQU; Cameroon, Gabon, Congo-Brazzaville.
BAG: Schweinfurth III.68 2/1871.

Oxystelma bornouense R.Br.

F.P.S. 2: 412 (1952); F.E.E. 4(1): 149 (2003); F.T.E.A. Apoc. 2: 362 (2012).
Herbaceous climber. Riverbanks.
Distr: UN, BAG; Senegal to Ethiopia, Somalia & Kenya.
BAG: Myers 7302 7/1937.

Oxystelma esculentum (L.f.) Sm.

F.E.E. 4(1): 150 (2003); F.T.E.A. Apoc. 2: 362 (2012).
Syn: *Oxystelma esculentum* (L.f.) Sm. var. *alpini* (Decne.) N.E.Br. – F.P.S. 2: 412 (1952).
Herbaceous climber. Marshy plains, riverbanks.
Distr: NS, CS; Ethiopia, Tanzania; Egypt, tropical Asia, Australia.
NS: Pettet 63 9/1962.

Pachycarpus bisacculatus (Oliv.) Goyder

Biol. Skr. 51: 323 (1998); F.E.E. 4(1): 126 (2003); F.T.E.A Apoc. 2: 438 (2012).
Syn: *Asclepias lineolata* sensu Andrews, non (Decne.) Bullock – F.P.S. 2: 400 (1952).
Perennial herb. Seasonally flooded grassland.
Distr: UN, EQU; Senegal to Ethiopia, S to Angola & Mozambique.
EQU: Jackson 2290 5/1952.

Pachycarpus eximius (Schltr.) Bullock

Biol. Skr. 51: 323 (1998); F.T.E.A. Apoc. 2: 443 (2012).
Syn: *Schizoglossum eximium* (Schltr.) N.E.Br. – F.P.S. 2: 416 (1952).

Orbea sprengeri heading — Syn: Caralluma hesperidum

Perennial herb. Montane grassland.
Distr: EQU; D.R. Congo, Rwanda, Burundi, Uganda, Kenya, Tanzania.
EQU: Myers 11100 4/1939.

Pachycarpus grantii (Oliv.) Bullock subsp. *grantii*
F.T.E.A. Apoc. 2: 442 (2012).
Perennial herb. Grassland & wooded grassland.
Distr: EQU; D.R. Congo, Uganda, Kenya, Tanzania.
EQU: Myers 6451 4/1937.

Pachycarpus lineolatus (Decne.) Bullock
T.S.S.: 382 (1990); Biol. Skr. 51: 323 (1998); F.T.E.A. Apoc. 2: 437 (2012).
Syn: *Asclepias schweinfurthii* N.E.Br. – F.P.S. 2: 401 (1952).
Syn: *Pachycarpus schweinfurthii* (N.E.Br.) Bullock – T.S.S.: 382 (1990).
Perennial herb. Wooded grassland.
Distr: BAG, UN, EQU; Ivory Coast to Kenya, S to Malawi & Angola.
EQU: Andrews 958 5/1939.

Pachycarpus petherickianus (Oliv.) Goyder
F.E.E. 4(1): 128 (2003); F.T.E.A. Apoc. 2: 446 (2012).
Syn: *Schizoglossum petherickianum* Oliv. – F.P.S. 2: 416 (1952).
Perennial herb. Grassland.
Distr: BAG, EQU; Nigeria to Ethiopia, S to Burundi & Tanzania.
BAG: Hoyle 509 1/1939.

Parquetina calophylla (Baill.) Venter
F.T.E.A. Apoc. 2: 123 (2012).
Syn: *Omphalogonus nigritanus* N.E.Br. – F.P.S. 2: 411 (1952).
Woody climber. Wooded grassland.
Distr: EQU; Senegal to Kenya & Tanzania.
EQU: Sillitoe 452 1919.

Pentarrhinum insipidum E.Mey.
F.P.S. 2: 412 (1952); F.J.M.: 128 (1976); F.E.E. 4(1): 142 (2003); F.T.E.A. Apoc. 2: 469 (2012).
Herbaceous climber. Forest edges, wooded grassland.
Distr: DAR; Eritrea & Somalia, S to South Africa.
DAR: Wickens 2181 8/1964.

Pentatropis nivalis (J.F.Gmel.) D.V.Field & J.R.I.Wood
F.E.E. 4(1): 138 (2003); F.T.E.A. Apoc. 2: 494 (2012).
Syn: *Pentatropis spiralis* sensu auctt., non (Forssk.) Decne. – F.P.S. 2: 412 (1952); Bot. Exp. Sud.: 19 (1970).
Herbaceous climber. Dry sandy areas, riverine thicket.
Distr: RS, CS, UN, EQU; Senegal to Somalia, S to Tanzania; Madagascar, Arabia to India.
RS: Bally 6981 4/1949; **UN:** Simpson 7248 7/1929.

Pergularia daemia (Forssk.) Chiov. subsp. *daemia*
F.P.S. 2: 413 (1952); F.J.M.: 128 (1976); Biol. Skr. 51: 323 (1998); F.E.E. 4(1): 133 (2003); F.T.E.A. Apoc. 2: 364 (2012).
Herbaceous twiner. Dry bushland.
Distr: RS, DAR, KOR, CS, ES, UN, EQU; Senegal to

Eritrea, S to South Africa; Arabia, India.
KOR: Wickens 953 3/1963; **EQU:** Friis & Vollesen 658 12/1980.

Pergularia tomentosa L.
F.P.S. 2: 412 (1952); F.J.U. 2: 132 (1999); F.E.E. 4(1): 134 (2003).
Small shrub. Semi-desert.
Distr: NS, RS; Ethiopia; N Africa, Arabia, Pakistan.
NS: Léonard 4899 11/1968.

Periploca aphylla Decne.
F.P.S. 2: 413 (1952); F.Egypt 2: 215 (2000).
Leafless shrub. Semi-desert.
Distr: NS, RS; Egypt to Pakistan.
RS: Bent s.n. 1896.

Periploca linearifolia Quart.-Dill. & A.Rich.
F.P.S. 2: 413 (1952); Biol. Skr. 51: 324 (1998); F.E.E. 4(1): 103 (2003); F.T.E.A. Apoc. 2: 118 (2012).
Woody climber. Montane forest.
Distr: EQU; Ethiopia to Malawi & Zambia.
EQU: Friis & Vollesen 1098 3/1982.

Pleiocarpa pycnantha (K.Schum.) Stapf
F.T.E.A. Apoc. 1: 39 (2002).
Syn: *Pleiocarpa pycnantha* (K.Schum) Stapf var. *tubicina* (Stapf) Pichon – T.S.S.: 374 (1990).
Syn: *Pleiocarpa tubicina* Stapf – F.P.S. 2: 396 (1952).
Tree. Forest.
Distr: EQU; Senegal to Kenya, S to Angola & Mozambique.
EQU: Turner 160 2/1936.

Pleurostelma schimperi (Vatke) Liede
F.E.E. 4(1): 137 (2003).
Syn: *Podostelma schimperi* (Vatke) K.Schum. – F.P.S. 2: 413 (1952); Bot. Exp. Sud.: 19 (1970).
Perennial herb. Sandy hillsides.
Distr: RS, ES; Eritrea, Ethiopia, Somalia.
RS: Carter 1887 11/1987.

Raphionacme brownii Scott-Elliot
F.P.S. 2: 415 (1952); F.T.E.A. Apoc. 2: 148 (2012).
Perennial herb. Grassland.
Distr: EQU; Senegal to C.A.R. & Chad.
EQU: Myers 6546 4/1937.

Raphionacme splendens Schltr. subsp. *splendens*
F.T.E.A. Apoc. 2: 154 (2012).
Syn: *Raphionacme jurensis* N.E.Br. – F.P.S. 2: 413 (1952).
Perennial herb. Grassland & woodland.
Distr: BAG, UN; Cameroon to Ethiopia, S to Zimbabwe & Mozambique.
UN: J.M. Lock 80/50 3/1980.

Raphionacme splendens Schltr. subsp. *bingeri* (A.Chev.) Venter
S. African J. Bot. 75: 337 (2009).
Perennial herb. Woodland.
Distr: EQU; Senegal to Uganda.
EQU: Wyld 726 2/1940.

Rauvolfia caffra Sond.

Biol. Skr. 51: 318 (1998); F.T.E.A. Apoc. 1: 61 (2002).
Syn: *Rauvolfia oxyphylla* Stapf – F.P.S. 2: 396 (1952);
T.S.S.: 374 (1990).
Tree. Wet forest.
Distr: EQU; Ghana to Kenya, S to South Africa.
EQU: Friis & Vollesen 1121 3/1982.

Rauvolfia vomitoria Afzel.

F.P.S. 2: 396 (1952); T.S.S.: 374 (1990); Biol. Skr. 51: 318
(1998); F.T.E.A. Apoc. 1: 65 (2002).
Shrub. Wet forest.
Distr: BAG, EQU; Senegal to Uganda, S to Angola &
Tanzania.
EQU: Jackson 1190 3/1950.

Saba comorensis (Bojer ex A.DC.) Pichon

F.T.E.A. Apoc. 1: 34 (2002); F.E.E. 4(1): 93 (2003).
Syn: *Landolphia comorensis* (Bojer ex A.DC.) K.Schum.
– F.P.S. 2: 394 (1952).
Syn: *Landolphia comorensis* (Bojer ex A.DC.) K.Schum.
var. *florida* (Benth.) K.Schum. – F.P.S. 2: 394 (1952).
Syn: *Saba comorensis* (Bojer ex A.DC.) Pichon var.
florida (Benth.) Pichon – Biol. Skr. 51: 319 (1998).
Syn: *Saba florida* (Benth.) Bullock – Bot. Exp. Sud.: 17
(1970); F.J.M.: 127 (1976); T.S.S.: 375 (1990).
Syn: *Landolphia senegalensis* var. *glabriflora* sensu
Andrews, non Hua – F.P.S. 2: 395 (1952).
Syn: *Saba senegalensis* var. *glabriflora* sensu El Amin,
non (Hua) Pichon – T.S.S.: 375 (1990).
Woody climber. Forest.
Distr: DAR, KOR, CS, ES, BAG, UN, EQU; Senegal to
Ethiopia, S to Zimbabwe; Madagascar, Comoros.
DAR: Wickens 1551 5/1964; **EQU:** Myers 7704 9/1937.

Schizostephanus alatus Hochst. ex K.Schum.

Biol. Skr. 51: 325 (1998); F.E.E. 4(1): 136 (2003); F.T.E.A.
Apoc. 2: 466 (2012).
Succulent climbing shrub. Rocky ground.
Distr: EQU; Ethiopia & Somalia, S to South Africa.
EQU: Friis & Vollesen 1189 3/1982.

Secamone africana (Oliv.) Bullock

Biol. Skr. 51: 78 (1998); F.T.E.A. Apoc. 2: 175 (2012).
Woody climber. Forest margins.
Distr: EQU; Nigeria to Uganda, S to Angola & Tanzania.
EQU: Friis & Vollesen 781 12/1980.
Note: Andrews and El Amin list also *S. afzelii* (Schult.)
K.Schum. from Equatoria, but this is very likely to be
based on a misidentification since *S. afzelii* is a West
African species extending only as far as western D.R.
Congo.

Secamone parvifolia (Oliv.) Bullock

T.S.S.: 382 (1990); Biol. Skr. 51: 325 (1998); F.E.E. 4(1):
113 (2003); F.T.E.A. Apoc. 2: 171 (2012).
Syn: *Secamone schweinfurthii* K.Schum. – F.P.S. 2:
417 (1952).
Woody climber. Wooded grassland.
Distr: BAG; Ethiopia to South Africa.
BAG: Schweinfurth 2232 7/1869.

Secamone punctulata Decne.

F.E.E. 4(1): 113 (2003); F.T.E.A. Apoc. 2: 170 (2012).

Herbaceous climber or small shrub. Forest margins.
Distr: EQU; Ivory Coast, Ghana, Ethiopia & Somalia, S to
Namibia & D.R. Congo.
EQU: Andrews 1401 5/1939.

Solenostemma arghel (Delile) Hayne

F.P.S. 2: 418 (1952); F.Egypt 2: 218 (2000).
Perennial herb. Semi-desert.
Distr: NS, RS, KOR; Mali, Niger, Chad; N Africa, Palestine,
Arabia.
RS: Schweinfurth 670 9/1868.

Stathmostelma angustatum K.Schum. subsp. vomeriforme (S.Moore) Goyder

F.T.E.A. Apoc. 2: 419 (2012).
Perennial herb. Wooded grassland.
Distr: UN; Uganda, Tanzania.
IUCN: RD
UN: J.M. Lock 81/58 5/1981.

Stathmostelma pedunculatum (Decne.) K.Schum.

Bot. Exp. Sud. 19: (1970); F.J.M.: 128 (1976); F.E.E. 4(1):
130 (2003); F.T.E.A. Apoc. 2: 413 (2012).
Syn: *Asclepias pedunculata* (Decne.) Dandy – F.P.S. 2:
401 (1952).
Perennial herb. Grassland.
Distr: DAR, ES, EQU; Cameroon, D.R. Congo, Ethiopia,
Uganda, Kenya, Tanzania, Mozambique.
DAR: Wickens 2008 7/1964; **EQU:** Drar 1082 4/1938
(n.v.).
Note: the identity of the cited South Sudanese specimen
requires confirmation.

Stathmostelma rhacodes K.Schum.

F.E.E. 4(1): 130 (2003); F.T.E.A. Apoc. 2: 414 (2012).
Perennial herb. Seasonally waterlogged grassland.
Distr: UN; Ethiopia, Uganda, Kenya, Tanzania.
UN: J.M. Lock 81/57 5/1981.

Stathmostelma welwitschii Britten & Rendle var. bagshawei (S.Moore) Goyder

F.T.E.A. Apoc. 2: 417 (2012).
Perennial herb. Seasonally waterlogged grassland.
Distr: EQU; D.R. Congo, Uganda, Tanzania.
IUCN: RD
EQU: Myers 6555 5/1937.

Tabernaemontana pachysiphon Stapf

Biol. Skr. 51: 319 (1998); F.T.E.A. Apoc. 1:47 (2002).
Syn: *Conopharyngia holstii* (K.Schum.) Stapf – F.P.S. 2:
392 (1952).
Syn: *Tabernaemontana holstii* K.Schum. – T.S.S.: 375
(1990).
Small tree. Wet forest.
Distr: EQU; Ghana, Kenya to Angola and Malawi.
EQU: Andrews 1540 5/1939.

Tabernaemontana penduliflora K.Schum.

F.W.T.A. 2: 65 (1963).
Shrub. Forest.
Distr: EQU; Nigeria to D.R. Congo.
EQU: Schweinfurth 3271 5/1870.

Tacazzea apiculata Oliv.
F.P.S. 2: 418 (1952); T.S.S.: 377 (1990); F.E.E. 4(1): 106 (2003); F.T.E.A. Apoc. 2: 140 (2012).
Woody climber. Swamp forest & forest margins.
Distr: CS, BAG, UN, EQU; Senegal to Kenya, S to Angola & South Africa.
CS: Muriel 17 11/1900; **UN:** Sherif A.3902 7/1951.

Tacazzea conferta N.E.Br.
Biol. Skr. 51: 326 (1998); F.E.E. 4(1): 106 (2003); F.T.E.A. Apoc. 2: 142 (2012).
Woody climber. Forest margins.
Distr: EQU; Ivory Coast, C.A.R. to Ethiopia, S to Zambia & Malawi.
EQU: Friis & Vollesen 360 11/1980.

Tacazzea venosa Decne.
F.P.S. 2: 418 (1952); F.E.E. 4(1): 108 (2003).
Syn: *Tacazzea venosa* Decne. var. *martini* (Baill.) N.E.Br. – F.P.S. 2: 418 (1952); T.S.S.: 377 (1990).
Shrub. On sand or rock crevices on riverbanks.
Distr: ES; Ethiopia.
IUCN: RD
ES: Schweinfurth 194 (fl fr) 9/1865.
Note: the collection cited is from the Gallabat-Matemma region on the Sudan/Ethiopia border. Whilst the collection may well have been made from the Ethiopia side, Schweinfurth's notes on this part of his expedition are vague. In any case, this species is very likely to occur on the Sudan side of the border region.

Telosma africana (N.E.Br.) N.E.Br.
F.T.E.A. Apoc. 2: 208 (2012).
Woody climber. Scrub & forest margins.
Distr: UN; Sierra Leone to Ethiopia, S to Angola & South Africa; Madagascar.
UN: J.M. Lock 81/138 8/1981.

Tylophora caffra Meisn.
F.T.E.A. Apoc. 2: 499 (2012).
Syn: *Sphaerocodon caffrum* (Meisn.) Schltr. – Biol. Skr. 51: 325 (1998).
Perennial herb or climber. Wooded grassland.
Distr: EQU; Sierra Leone to Kenya, S to Namibia & South Africa.
EQU: Friis & Vollesen 1180 3/1982.

Tylophora sylvatica Decne.
F.E.E. 4(1): 136 (2003); F.T.E.A Apoc. 2: 508 (2012).
Herbaceous climber. Thickets.
Distr: EQU; Senegal to Cameroon, Ethiopia to Angola & Zambia.
EQU: Andrews 945 4/1939.

Voacanga africana Stapf
F.P.S. 2: 397 (1952); T.S.S.: 376 (1990); F.T.E.A. Apoc. 1: 42 (2002).
Tree. Wet forest.
Distr: BAG, UN, EQU; Senegal to Kenya, S to Angola & Mozambique.
EQU: Myers 8635 2/1938.

Voacanga chalotiana Pierre ex Stapf
Agric. Univ. Wageningen Papers 85(3): 31 (1985).
Tree. Forest.
Distr: UN; Congo-Brazzaville, D.R. Congo, Angola.
UN: J.M. Lock 80/7 2/1980.

Voacanga thouarsii Roem. & Schult.
T.S.S.: 376 (1990); F.T.E.A. Apoc. 1: 44 (2002).
Syn: *Voacanga obtusa* K.Schum. – F.P.S. 2: 397 (1952).
Tree. Riverine forest.
Distr: EQU; Senegal to Kenya, S to South Africa; Madagascar.
EQU: Andrews 1298 5/1939.

Xysmalobium heudelotianum Decne.
F.P.S. 2: 418 (1952); Biol. Skr. 51: 326 (1998); F.E.E. 4(1): 132 (2003).
Perennial herb. Wooded grassland.
Distr: ?EQU; Gambia to Kenya, S to Zimbabwe & Mozambique.
Note: we have seen no material from South Sudan; the entry here is based on Andrews who recorded it from Equatoria. It is also recorded from South Sudan in F.T.E.A. and it has been collected from the Ugandan side of the Imatong Mts (Friis & Vollesen 1998), so Andrews' record is likely to be genuine.

UNPLACED EUDICOTS

Vahliaceae

I. Darbyshire

Vahlia dichotoma (Murray) Kuntze
F.P.S. 1: 80 (1950); F.T.E.A. Vahl.: 2 (1975); F.J.M.: 88 (1976).
Annual herb. Wadis, weed of cultivation.
Distr: DAR, BAG, EQU; Mauritania & Senegal to Kenya, S to Mozambique.
DAR: Wickens 1790 (fl fr) 6/1964; **EQU:** Broun 1452 (fl fr) 11/1905.

Vahlia digyna (Retz.) Kuntze
F.P.S. 1: 79, pro parte (1950); F.T.E.A. Vahl.: 4 (1975); F.E.E. 3: 28 (1989).
Annual herb. Seasonal pools, riverbanks.
Distr: CS; Mauritania to Ethiopia, S to Botswana; Madagacar, Egypt, India & Pakistan.
CS: Lea 218 (fl fr) 10/1954.

Vahlia geminiflora (Delile) Bridson
F.T.E.A. Vahl.: 2 (1975); F.E.E. 3: 28 (1989).
Syn: *Vahlia digyna* sensu Andrews pro parte, non (Retz.) Kuntze – F.P.S. 1: 79 (1950).
Annual herb. Riverbanks, dry riverbeds, seasonal pools, weed of cultivation.
Distr: NS, KOR, CS, ES, UN; Mali, Nigeria, Eritrea, Ethiopia; Egypt, Iraq & Iran.
CS: de Wilde et al. 5771 (fl fr) 2/1965; **UN:** Harrison 318 (fl fr) 4/1948.

BORAGINALES

Boraginaceae

I. Darbyshire

Arnebia hispidissima (Lehm.) DC.

F.P.S. 3: 76 (1956); F.J.M.: 140 (1976); F.E.E. 5: 96 (2006).
Annual or perennial herb. Dry open habitats such as rocky outcrops.
Distr: NS, RS, DAR, KOR, CS; Eritrea, Ethiopia; N Africa, Arabia, Iran to Tibet.
DAR: Kamil 1089 (fl) 5/1968.

Brandella erythraea (Brand) R.R.Mill

F.E.E. 5: 94 (2006).
Annual or perennial herb. Rocky hillslopes, open woodland.
Distr: RS; Eritrea, Ethiopia; Egypt, Saudi Arabia.
IUCN: RD
RS: Farag & Zubair 1 (fl fr) 2/2012.

Coldenia procumbens L.

F.P.S. 3: 77 (1956); F.T.E.A. Borag.: 44 (1991); F.E.E. 5: 74 (2006).
Procumbent annual herb. Wet depressions, pond and pan margins, rice fields.
Distr: KOR, CS, BAG, UN, EQU; Mauritania to Ethiopia, S to Angola & Zimbabwe; Madagascar, tropical Asia & Australia.
CS: Kotschy 471 (fl fr) 3/1839; **UN:** Simpson 7179 (fr) 6/1929.

Cordia africana Lam.

F.T.E.A. Borag.: 31 (1991); Biol. Skr. 51: 409 (2005); F.E.E. 5: 68 (2006).
Syn: *Cordia abyssinica* R.Br. – F.P.S. 3: 77 (1956); F.J.M.: 140 (1976); T.S.S.: 421 (1990).
Tree, often small. Secondary & riverine forest & forest-wooded grassland transition.
Distr: DAR, KOR, CS, ES, BAG, UN, EQU; Guinea to Ethiopia, S to Angola & South Africa; Arabia.
KOR: Wickens 796 (fl) 11/1962; **EQU:** Myers 11018 (fl) 4/1939.

Cordia crenata Delile subsp. *crenata*

T.S.S.: 423 (1990); F.T.E.A. Borag.: 20 (1991); F.E.E. 5: 70 (2006).
Shrub or small tree. Dry riverine woodland & scrub, wadis, often associated with human settlement.
Distr: DAR, KOR, CS; Ethiopia, Somalia, Uganda; Egypt, Yemen, Oman, Iran, India.
CS: Broun 1502 3/1909.
Note: subsp. *meridionalis* Warfa is also likely in Equatoria as it occurs on the Ugandan side of the Imatong Mts (Friis & Vollesen 2005).

Cordia millenii Baker

F.P.S. 3: 77 (1956); T.S.S.: 423 (1990); F.T.E.A. Borag.: 13 (1991); Biol. Skr. 51: 409 (2005).
Tree. Forest.
Distr: EQU; Guinea to Kenya, S to Angola & Tanzania.
EQU: Myers 10613 (fl) 3/1939.

Cordia monoica Roxb.

F.T.E.A. Borag.: 15 (1991); Biol. Skr. 51: 410 (2005); F.E.E. 5: 68 (2006).
Syn: *Cordia ovalis* DC. & A.DC. – F.P.S. 3: 77 (1956); T.S.S.: 423 (1990).
Small tree or shrub. A wide range of habitats from forest to dry bushland & wooded grassland.
Distr: ?NS, RS, DAR, KOR, CS, ES, BAG, UN, EQU; Ethiopia to D.R. Congo, S to Namibia & South Africa; India & Sri Lanka.
CS: Muriel L/82 (fr) 12/1900; **UN:** J.M. Lock 81/73 (fl) 5/1981.
Note: the record for Northern Sudan is based on the distribution map by Warfa in Acta Univ. Upsal. 174 (1988).

Cordia quercifolia Klotzsch

F.T.E.A. Borag.: 22 (1991); Biol. Skr. 51: 410 (2005); F.E.E. 5: 67 (2006).
Shrub or small tree. Grassland with scattered trees, dry bushland.
Distr: DAR, KOR; Mauritania to Ethiopia, S to Zimbabwe & Mozambique; Arabia, India & Sri Lanka.
KOR: Wickens 638 (fr) 10/1962.
Note: also likely to occur in South Sudan since it is recorded from the Ugandan side of the Imatong Mts (Friis & Vollesen 2005).

Cordia sinensis Lam.

T.S.S.: 425 (1990); F.T.E.A. Borag.: 21 (1991); Biol. Skr. 51: 410 (2005); F.E.E. 5: 66 (2006).
Syn: *Cordia gharaf* (Forssk.) Ehrenb. ex Asch. – Bot. Exp. Sud.: 22 (1970).
Syn: *Cordia rothii* Roem. & Schult. – F.P.S. 3: 78 (1956); Bot. Exp. Sud.: 22 (1970).
Tree or shrub. Riverine forest & woodland, seasonally wet bushland.
Distr: RS, DAR, KOR, CS, ES, UN, EQU; Ghana to Somalia, Angola, Namibia; Egypt & Israel to India & Sri Lanka.
KOR: Wickens 790 (fr) 11/1962; **EQU:** Myers 8577 (fl) 2/1938.
Note: Verdcourt in F.T.E.A. Borag.: 15 (1991) suggests that *Cordia uncinulata* De Wild. probably also occurs in Sudan.

Cynoglossopsis latifolia (A.Rich.) Brand

F.P.S. 3: 78 (1956); F.E.E. 5: 84 (2006).
Annual herb. Dry bushland.
Distr: RS; Eritrea, Ethiopia.
IUCN: RD
RS: Crowfoot s.n. (fl fr) 2/1923.

Cynoglossum coeruleum A.DC. subsp. *johnstonii* (Baker) Verdc.

F.T.E.A. Borag.: 109 (1991); F.E.E. 5: 92 (2006).
Annual or perennial herb. Grassland including degraded areas.
Distr: ?DAR, ?EQU (see note); Cameroon, D.R. Congo, Ethiopia to Malawi, Angola.
DAR: Wickens 2427 (fl) 9/1964; **EQU:** Andrews 1345 (fl fr) 5/1939.
Note: the determination of the cited specimen from Equatoria is tentative; Sudan was not listed in the distribution of this complex taxon by Verdcourt in F.T.E.A.

(l.c.). Several flowering specimens at Kew from Jebel Marra (*Wickens* 1725, 2385, 2427), are determined as possibly *C. coeruleum* by Verdcourt but with fruits needed; these were all listed under *C. lanceolatum* by Wickens in F.J.M.: 141 (1976).

Cynoglossum lanceolatum Forssk.

F.P.S. 3: 78 (1956); F.J.M.: 141, pro parte (1976); F.T.E.A. Borag.: 106 (1991); Biol. Skr. 51: 411 (2005); F.E.E. 5: 93 (2006).
Annual or ?perennial herb. Grassland & bushland, often montane, also a weed of cultivation.
Distr: DAR, EQU; tropical & southern Africa; Arabia to Malaysia & China.
DAR: Kamil 1157 (fl fr) 5/1968; **EQU:** Myers 10947 (fl fr) 4/1939.

Cynoglossum sp. cf. aequinoctiale T.C.E.Fr.

F.T.E.A. Borag.: 113 (1991); Biol. Skr. 51: 411 (2005).
Herb. Open montane grassland.
Distr: EQU; ?South Sudan endemic (see note).
EQU: Myers 11141 (fl) 4/1939.
Note: the single specimen is in flower only; Verdcourt in F.T.E.A. (l.c.) states that it differs from *C. aequinoctiale* of Uganda and Kenya in the somewhat different sepals; fruits are needed to fully determine its identity.

Echiochilon persicum (Burm.f.) I.M.Johnst.

Bot. J. Linn. Soc. 130: 240 (1999).
Syn: *Echiochilon fruticosum* sensu Andrews, non Desf. – F.P.S. 3: 78 (1956).
Dwarf perennial herb. Dry bushland on sandy or stony ground.
Distr: RS; Djibouti, Somalia; Arabia, Iran, Pakistan.
RS: Schweinfurth 2107 (fl) 5/1864.
Note: *E. fruticosum* Desf. is recorded from SE-most Egypt and is likely to occur in NE Sudan (see Lönn in Bot. J. Linn. Soc. 130: 209).

Echium arenarium Guss.

F.P.S. 3: 81 (1956); Fl. Prat. Maroc 2: 398 (2007).
Annual or biennial herb. Open sandy areas.
Distr: ?CS; Mediterranean, N Africa.
Note: Andrews recorded this species from Wad Medani and Fung District; we have not seen the specimens on which these records were based and misidentification is quite possible.

Echium rauwolfii Delile

F.P.S. 3: 80 (1956); F.Egypt 2: 304 (2000).
Syn: *Echium longifolium* sensu Andrews, non Delile – F.P.S. 3: 81 (1956).
Annual herb. Sandy or stony plains, riverbanks, weed of cultivation, roadsides.
Distr: NS, RS, CS; Libya, Egypt, Middle East, Arabia.
CS: de Wilde et al. 5813 (fl) 2/1965.
Note: it is our opinion that all material named as *E. longifolium* and *E. rauwolfii* from Sudan that we have seen in fact belongs to a single taxon and that the latter is the correct name; certainly, the Sudan material lacks the short deflexed secondary stem indumentum listed as a key character for *E. longifolium* by El Hadidy & Boulos in F.Egypt.

Ehretia obtusifolia Hochst. ex A.DC.

F.T.E.A. Borag.: 35 (1991); F.E.E. 5: 73 (2006).
Syn: *Ehretia braunii* Vatke – F.P.S. 3: 81 (1956); T.S.S.: 425 (1990); Biol. Skr. 51: 411 (2005); F.E.E. 5: 73 (2006).
Shrub or small tree. Woodland & thicket, often in rocky areas, termite mounds.
Distr: RS, ?KOR, UN, EQU; Ethiopia & Somalia, S to Namibia & South Africa; Arabia, Socotra, Pakistan, India.
RS: Kennedy-Cooke 215 (st) 3/1936; **EQU:** Friis & Vollesen 1161 (fl) 3/1982.
Note: the record for Kordofan is from El Amin in T.S.S. and requires confirmation.

Ehretia sp. cf. amoena Klotzsch

F.T.E.A. Borag.: 34, in notes (1991); Biol. Skr. 51: 411 (2005).
Syn: *Ehretia cymosa* sensu El Amin, non Thonn. – T.S.S.: 425 (1990).
Syn: *Ehretia stuhlmannii* sensu auctt., non Gürke – F.P.S. 3: 81 (1956); T.S.S.: 425 (1990).
Shrub or tree. *Acacia* & *Combretum* woodland, often on termite mounds.
Distr: EQU; ?South Sudan endemic.
EQU: Friis & Vollesen 1195 (fl) 3/1982.
Note: the identity of the South Sudan specimens is uncertain; the true *E. amoena* is recorded from coastal E Africa and through SE Africa to South Africa.

Heliotropium aegyptiacum Lehm.

F.P.S. 3: 82 (1956); F.T.E.A. Borag.: 53 (1991); F.E.E. 5: 76 (2006).
Syn: *Heliotropium cinerascens* DC. & A.DC. – F.E.E. 5: 77 (2006).
Perennial herb. Dry open scrub, sometimes in overgrazed areas.
Distr: NS, RS, KOR, CS, ES; Eritrea, Ethiopia, Djibouti, Somalia, Kenya; Egypt, Arabia.
RS: Bally 7015 (fl fr) 4/1949.
Note: we follow Verdcourt in F.T.E.A. in treating *H. cinerascens* as a synonym. It is, however, recognised as a good species in F.E.E. with both recorded from Sudan; this requires further investigation.

Heliotropium arabinense Fresen.

F.P.S. 3: 87 (1956); F.Egypt 2: 276 (2000).
Perennial herb. Dry wadis, sandy soils.
Distr: RS, Ethiopia; Madagascar, Egypt to Afghanistan.
RS: Bally 6988 (fl fr) 4/1949.

Heliotropium bacciferum Forssk.

F.P.S. 3: 87 (1956); F.J.M.: 141 (1976); F.J.U. 2: 134 (1999); F.Egypt 2: 280 (2000).
Syn: *Heliotropium undulatum* Vahl – F. Darfur Nord-Occ. & J. Gourgeil: 126 (1969).
Perennial herb. Dry wadis, sandy soils, sometimes weedy.
Distr: NS, RS, DAR, KOR, CS, ES; Senegal to Somalia; N Africa, Arabia to Pakistan.
CS: Andrews 208 (fl fr) 3/1936.

Heliotropium indicum L.

F.P.S. 3: 85 (1956); F.T.E.A. Borag.: 69 (1991); F.E.E. 5: 83 (2006).
Annual or perennial herb. Riverbanks, weed of fallow land.

Distr: NS, CS, BAG, UN, EQU; largely pantropical, probably originally an American species.
CS: Andrews 175 (fl fr) 2/1936; **UN:** Simpson 7033 (fl fr) 6/1926.

Heliotropium lignosum Vatke

F.P.S. 3: 87 (1956).
Perennial herb. Dry hillslopes, sandy plains.
Distr: RS; Eritrea, Somalia; Arabia.
RS: Schweinfurth 2111 5/1864.
Note: Forther has labelled the Kew isotype of this taxon as "H. lignosum Vatke nom. illegit.". Thulin in Fl. Somalia 3: 50 (2006) treats plants previously named as H. lignosum under H. bacciferum, although he made no reference to the name lignosum. The two look, superficially at least, rather distinct but it may be that they fall under a single variable species.

Heliotropium longiflorum (A.DC.) Jaub. & Spach subsp. longiflorum var. stenophyllum Schwartz

F.P.S. 3: 87 (1956); F.T.E.A. Borag.: 68 (1991); F.E.E. 5: 79 (2006).
Annual or perennial herb or subshrub. Sandy and gravelly plains and wadis.
Distr: RS, ?EQU; Eritrea, Ethiopia, Somalia, Uganda, Kenya; Arabia.
RS: Jackson 2803 (fl fr) 4/1953.
Note: Drar (1970) also recorded this species from Karika in Equatoria; this record requires confirmation.

Heliotropium ovalifolium Forssk.

F.P.S. 3: 85 (1956); F.J.M.: 141 (1976); F.T.E.A. Borag.: 75 (1991); F.E.E. 5: 82 (2006).
Perennial herb. Riverbanks, seasonally wet soils in dry bushland, sometimes weedy.
Distr: NS, RS, DAR, KOR, CS, ES, UN, EQU; Senegal to Somalia, S to Angola & South Africa; Egypt, Arabia, Pakistan, India.
ES: Jackson 1933 (fl fr) 7/1951; **UN:** Simpson 7213 (fl fr) 6/1929.

Heliotropium pterocarpum (DC. & A.DC.) Bunge

F.P.S. 3: 85 (1956); F.T.E.A. Borag.: 51 (1991); F.E.E. 5: 80 (2006).
Perennial herb. Sandy plains, rocky areas with sparse vegetation.
Distr: RS; Eritrea, Somalia; Egypt, Arabia, Socotra.
RS: Naemi 8 (fl) 1/1949.

Heliotropium rariflorum Stocks subsp. rariflorum

F.P.S. 3: 88 (1956); F.T.E.A. Borag.: 70 (1991); F.E.E. 5: 80 (2006).
Perennial herb or subshrub. Acacia-Commiphora woodland, dry grassland including grazed areas.
Distr: NS, RS, DAR, CS, ES; Arabia, Iran, Pakistan; African distribution of this subspecies not well defined.
RS: Schweinfurth 696 (fl fr) 9/1868.
Note: both F.T.E.A. and F.E.E. record subsp. hereroense (Schinz) Verdc. as occurring in Sudan; however, the single specimen at Kew, cited above, is definitely subsp. rariflorum (and annotated as such by B. Verdcourt). The

remaining distribution listed for Sudan is from material housed at KHU and may be misidentified material.

Heliotropium steudneri Vatke

F.P.S. 3: 87 (1956); F.T.E.A. Borag.: 62 (1991); F.E.E. 5: 79 (2006).
Perennial herb or subshrub. Acacia-Commiphora woodland, dry grassland.
Distr: RS; Eritrea, Ethiopia, Somalia, Kenya, Tanzania.
RS: Schweinfurth 293 (fl) 9/1868.

Heliotropium strigosum Willd.

F.P.S. 3: 87 (1956); F.J.M.: 141 (1976); F.T.E.A. Borag.: 72 (1991); F.E.E. 5: 81 (2006).
Annual or perennial herb. Sandy washes, semi-desert, stony plains, open Acacia woodland.
Distr: RS, DAR, KOR, CS, ES, BAG, UN, EQU; Senegal to Ethiopia, S to Namibia & Mozambique; Egypt, Arabia, Afghanistan to China & Australia.
CS: Kassas 501 (fl fr) 8/1954; **EQU:** Andrews 548 (fl fr) 4/1939.

Heliotropium sudanicum Andrews

F.P.S. 3: 84 (1956); F.T.E.A. Borag.: 61, in notes (1991).
Annual herb. Weed of cultivation, irrigated fields & near streams.
Distr: CS; Sudan endemic.
IUCN: RD
CS: Wilson-Jones s.n. (fl fr) 10/1954.
Note: H. sudanicum appears endemic to the Gezira area. It is possibly a subspecies of H. pectinatum Vaupel which is recorded from Somalia to Zimbabwe.

Heliotropium supinum L.

F.P.S. 3: 85 (1956); F.J.M.: 141 (1976); F.T.E.A. Borag.: 76 (1991); F.E.E. 5: 82 (2006).
Annual herb. Edges of seasonally flooded areas, riverbanks, weed of cultivation.
Distr: NS, RS, DAR, KOR, CS, ES; Senegal to Ethiopia, S to South Africa; S Europe, N Africa, Arabia to India.
KOR: Wickens 939 (fl fr) 2/1963.

Heliotropium zeylanicum (Burm.f.) Lam.

F.P.S. 3: 82 (1956); F.T.E.A. Borag.: 56 (1991); Biol. Skr. 51: 411 (2005); F.E.E. 5: 76 (2006).
Perennial herb. Dry bushland, weed of fallow land and cultivation.
Distr: NS, RS, DAR, KOR, CS, ES, EQU; tropical & southern Africa; Arabia, Pakistan, India.
KOR: Wickens 105 (fl) 7/1962; **EQU:** Simpson 7550 (fl) 2/1930.

Lithospermum afromontanum Weim.

F.T.E.A. Borag.: 77 (1991); Biol. Skr. 51: 412 (2005); F.E.E. 5: 96 (2006).
Syn: Lithospermum officinale sensu Andrews, non L. – F.P.S. 3: 88 (1956).
Perennial herb or subshrub. Montane grassland & scrub.
Distr: EQU; Ethiopia to South Africa.
EQU: Thomas 1852 (fl) 12/1935.

Moltkiopsis ciliata (Forssk.) I.M.Johnst.

F.P.S. 3: 88 (1956); F.Egypt 2: 296 (2000).
Dwarf subshrub. Sandy plains, wadis.
Distr: ?NS/DAR; N Africa, Arabia, Iran.

Note: no material has been seen from Sudan but Andrews recorded this species from the Libyan Desert.

Myosotis abyssinica Boiss. & Reut.

F.J.M.: 141 (1976); F.T.E.A. Borag.: 87 (1991); F.E.E. 5: 100 (2006).
Annual herb. Forest-grassland and forest-bushland transition.
Distr: RS, DAR; Bioko, Cameroon, D.R. Congo, Ethiopia to Tanzania.
DAR: <u>Wickens 2400</u> (fl fr) 9/1964.

Trichodesma africanum (L.) Sm.

F.P.S. 3: 88 (1956); F.J.M.: 141 (1976); F.J.U. 2: 136 (1999); F.E.E. 5: 86 (2006).
Syn: *Trichodesma giganteum* Quézel – Bot. Exp. Sud.: 23 (1970).
Annual herb. Disturbed areas, open rocky ground.
Distr: NS, RS, DAR; Senegal to Ethiopia, S to South Africa; Libya, Arabia to India.
DAR: <u>Jackson 2542</u> (fl) 1/1953.
Note: if the varieties of this taxon are recognised, our plants would fall under var. *abyssinicum* Brand (see F.Egypt 2: 287), but this taxon is not maintained in F.E.E.

Trichodesma ambacense Welw. subsp. *hockii* (De Wild.) Brummitt

F.T.E.A. Borag.: 101 (1991); F.E.E. 5: 89 (2006).
Perennial herb. Regularly burnt woodland & grassland.
Distr: EQU; Nigeria to Uganda, S to Zimbabwe & Botswana.
EQU: <u>Myers 8762</u> (fl) 3/1938.

Trichodesma ehrenbergii Schweinf.

F.P.S. 3: 90 (1956); F.Egypt 2: 287 (2000).
Annual herb. Rocky hillslopes, wadis.
Distr: RS; Egypt, Arabia.
RS: <u>Kassas 719</u> (fl) 12/1966.

Trichodesma inaequale Edgew.

F.T.E.A. Borag.: 95 (1991); F.E.E. 5: 86 (2006).
Annual herb. *Acacia* & *Combretum-Terminalia* woodland, grassland.
Distr: KOR, CS, UN; Cameroon, Ethiopia, Tanzania; India.
CS: <u>Lea 229</u> (fl fr); **UN:** <u>Sherif A.3949</u> (fl fr) 8/1951.

Trichodesma physaloides (Fenzl) A.DC.

F.P.S. 3: 88 (1956); Bot. Exp. Sud.: 23 (1970); F.T.E.A. Borag.: 100 (1991); Biol. Skr. 51: 412 (2005); F.E.E. 5: 88 (2006).
Perennial herb. Woodland, bushland & grassland, particularly regularly burnt areas, disturbed ground.
Distr: ?DAR, CS, BAG, EQU; Ethiopia to South Africa.
CS: <u>Kotschy 577</u> (fl) 1837; **EQU:** <u>Chipp 34</u> (fl) 2/1929.
Note: the record for Darfur is from Drar (1970) and is based on sterile material.

Trichodesma zeylanicum (Burm.f.) R.Br.

F.P.S. 3: 90 (1956); F.T.E.A. Borag.: 92 (1991); Biol. Skr. 51: 412 (2005); F.E.E. 5: 85 (2006).
Annual or usually perennial herb or subshrub. Woodland & grassland, disturbed areas.
Distr: CS, BAG, UN, EQU; Ethiopia to Angola & South Africa; Madagascar, Mascarenes, India to Australia.
CS: <u>Ferguson 347</u> (fl fr) 1948; **EQU:** <u>Thomas 1570</u> (fl fr) 12/1935.

SOLANALES

Convolvulaceae

H. Pickering & I. Darbyshire

Astripomoea lachnosperma (Choisy) A.Meeuse

F.T.E.A. Convolv.: 77 (1963); F.J.M.: 142 (1976); F.E.E. 5: 201 (2006).
Syn: *Astrochleana lachnosperma* (Choisy) Hallier f. – F.P.S. 3: 103 (1956).
Annual herb. Woodland & bushland.
Distr: RS, DAR, KOR, EQU; Nigeria to Ethiopia, S to Namibia, Botswana & Mozambique.
KOR: <u>Wickens 346</u> 9/1962; **EQU:** <u>Andrews 22</u>.

Astripomoea malvacea (Klotzsch) A.Meeuse

F.T.E.A. Convolv.: 74 (1963); F.J.M.: 143 (1976); F.E.E. 5: 201 (2006).
Syn: *Astrochleana malvacea* (Klotzsch) Hallier f. – F.P.S. 3: 104 (1956).
Perennial herb or subshrub. Wooded grassland.
Distr: RS, DAR, KOR, CS, BAG, UN, EQU; Nigeria to Eritrea, S to South Africa.
KOR: <u>Jackson 2133</u> 4/1952; **EQU:** <u>Friis & Vollesen 778</u> 12/1980.
Note: several varieties are recognised in F.T.E.A., and the Sudanese material includes representatives of var. *malvacea*, var. *floccosa* (Vatke) Verdc. and var. *volkensii* (Dammer) Verdc. However, these varieties do not seem to be well defined in Sudan, with a good number of intermediate specimens seen.

Convolvulus arvensis L.

F.P.S. 3: 107 (1956); F.T.E.A. Convolv.: 41 (1963); F.E.E. 5: 184 (2006).
Perennial herb. Weed of disturbed areas.
Distr: NS, CS; widespread in temperate and subtropical regions, native to Europe.
NS: <u>Pettet 68</u> 9/1962.

Convolvulus auricomus (A.Rich.) Bhandari

Bull. Bot. Surv. India 6: 327 (1965).
Syn: *Convolvulus glomeratus* Choisy – F.P.S. 3: 106 (1956); Bot. Exp. Sud.: 38 (1970); F.E.E. 5: 181 (2006).
Perennial herb. Wooded grassland.
Distr: RS; Eritrea, Ethiopia, Djibouti, Somalia; N Africa, Arabia to India.
RS: <u>Schweinfurth 2167</u> 6/1864.
Note: the correct name for this species is usually given as *C. glomeratus* but the African Plants Database lists this as an illegitimate name.

Convolvulus fatmensis Kunze

F.P.S. 3: 108 (1956); Bot. Exp. Sud.: 38 (1970); F.Egypt 2: 252 (2000).
Annual or short-lived perennial herb. Open ground, weed of cultivation.
Distr: NS, DAR, KOR, CS, UN; N Africa, Middle East, Arabia, Iran.
KOR: <u>Pfund 34</u> (fl fr) 9/1875; **UN:** <u>Drar 827</u> 4/1938 (n.v.).

Convolvulus hamphilahensis A.Terracc.

F.E.E. 5: 181 (2006).

Annual herb. Open bushland, dry stony plains.
Distr: RS; Eritrea.
IUCN: RD
RS: Carter 1965 11/1987.
Note: closely allied to, and perhaps a form of, *C. rhyniospermus*.

Convolvulus hystrix Vahl
F.P.S. 3: 106 (1956); F.E.E. 5: 182 (2006).
Spiny shrub. Open sandy plains, wadis.
Distr: RS, Eritrea, Somalia; Egypt, Arabia.
RS: Bent s.n. 1896.

Convolvulus prostratus Forssk.
F.J.U. 2: 162 (1999); F.Egypt 2: 248 (2000).
Syn: *Convolvulus austro-aegyptiacus* Abdallah & Sa'ad – F.J.U. 2: 163 (1999).
Syn: *Convolvulus cancerianus* Abdallah & Sa'ad – F.J.U. 2: 161 (1999).
Syn: *Convolvulus deserti* Hochst. & Steud. ex Steud. – F.P.S. 3: 106 (1956).
Syn: *Convolvulus microphyllus* Sieber ex Spreng. – F.P.S. 3: 106 (1956).
Perennial herb. Open sandy plains.
Distr: NS, RS, ?KOR; Cape Verde Is., Mauritania, Somalia; N Africa, Arabia to Pakistan.
RS: Carter 2010 11/1987.

Convolvulus rhyniospermus Hochst. ex Choisy
F.P.S. 3: 106 (1956); F.T.E.A. Convolv.: 40 (1963); F.E.E. 5: 180 (2006).
Annual herb. Open sandy plains.
Distr: ?NS, KOR; Eritrea, Djibouti, Somalia, Kenya; Egypt, Socotra, Arabia, Pakistan.
KOR: Kotschy 235 (fr) 11/1839.

Convolvulus siculus L. subsp. *agrestis* (Hochst. ex Schweinf.) Verdc.
F.T.E.A. Convolv.: 41 (1963); F.E.E. 5: 184 (2006).
Syn: *Convolvulus agrestis* (Hochst. ex Schweinf.) Hallier f. – F.P.S. 3: 107 (1956); Bot. Exp. Sud.: 38 (1970).
Annual herb. Disturbed ground.
Distr: RS; Eritrea, Ethiopia, Kenya, Tanzania; Egypt.
RS: Bent s.n. 1896.

Cressa cretica L.
F.P.S. 3: 108 (1956); F.T.E.A. Convolv.: 33 (1963); F.E.E. 5: 179 (2006).
Perennial subshrubby herb. Sand dunes, saline sandy soils.
Distr: NS, RS, ?KOR, CS; widespread in the tropics.
RS: Andrews 221 11/1936.

Cuscuta australis R.Br.
F.T.E.A. Convolv.: 4 (1963); F.E.E. 5: 232 (2006).
Syn: *Cuscuta cordofana* (Engelm.) Yunck. – F.P.S. 3: 108 (1956).
Parasitic twiner. Swampy areas, riverbanks.
Distr: KOR; palaeotropical & warm temperate regions.
KOR: Figari s.n.

Cuscuta campestris Yunck.
F.T.E.A. Convolv.: 5 (1963); F.E.E. 5: 232 (2006).
Parasitic twiner. Weed of cultivation & plantations.
Distr: RS, DAR, CS; native to the Americas, now widely naturalised.

CS: Musselman 6270 11/1983.

Cuscuta chinensis Lam.
Notes Roy. Bot. Gard. Edinburgh 42: 37 (1984); F.Egypt 2: 265 (2000).
Parasitic twiner. Open areas, parasitising *Heliotropium*.
Distr: RS; Eritrea; Egypt, Arabia to Australia.
RS: Bally 6968 1/1983.

Cuscuta hyalina Roth
F.P.S. 3: 108 (1956); F.T.E.A. Convolv.: 8 (1963); F.E.E. 5: 233 (2006).
Syn: *Cuscuta hyalina* Roth var. *nubica* Yunck. – F.P.S. 3: 108 (1956).
Parasitic twiner. Open dry areas, usually parasitising *Tribulus* or *Trianthema*.
Distr: NS, RS, DAR, KOR, CS, ES; Ethiopia to South Africa; India.
KOR: Wickens 131 8/1962.

Cuscuta kilimanjari Oliv.
F.P.S. 3: 109 (1956); F.T.E.A. Convolv.: 6 (1963); Biol. Skr. 51: 416 (2005); F.E.E. 5: 233 (2006).
Parasitic twiner. Montane forest, parasitising e.g. Acanthaceae & Lamiaceae.
Distr: EQU; Ethiopia to South Africa.
EQU: Thomas 1647 12/1935.

Cuscuta pedicellata Ledeb.
Notes Roy. Bot. Gard. Edinburgh 42: 32 (1984); F.Egypt 2: 267 (2000).
Parasitic twiner. Weed of cultivation.
Distr: CS, BAG; N Africa to C Asia.
CS: Zubier s.n. (n.v.); **BAG:** Musselman 6236 1/1983.

Cuscuta planiflora Ten.
F.T.E.A. Convolv.: 9 (1963); F.J.M.: 143 (1976); F.E.E. 5: 233 (2006).
Syn: *Cuscuta brevistyla* A.Braun & A.Rich. – F.P.S. 3: 109 (1956).
Parasitic twiner. Woodland, parasitising e.g. Acanthaceae & Lamiaceae.
Distr: RS, DAR; palaeotropical & warm temperate regions.
RS: Jackson 2969 4/1953.

Dichondra repens J.R.Forst. & G.Forst.
F.P.S. 3: 109 (1956); F.T.E.A. Convolv.: 12 (1963); F.E.E. 5: 162 (2006).
Procumbent perennial herb. Moist grassland, forest margins.
Distr: RS, ?EQU; palaeotropical.
RS: Aylmer 636 3/1936.
Note: Andrews recorded this species from Equatoria but we have not traced the material on which this was based.

Evolvulus alsinoides (L.) L.
F.P.S. 3: 109 (1956); F.T.E.A. Convolv.: 18 (1963); F.J.M.: 143 (1976); Biol. Skr. 51: 417 (2005); F.E.E. 5: 166 (2006).
Annual or perennial herb. Woodland, grassland, weed of disturbed ground.
Distr: RS, DAR, KOR, BAG, UN, EQU; pantropical.
DAR: Wickens 2904 4/1965; **BAG:** Andrews 722 4/1939.

Evolvulus nummularius (L.) L.

F.P.S. 3: 109 (1956); F.T.E.A. Convolv.: 16 (1963); F.E.E. 5: 166 (2006).
Perennial herb. Short grassland & bare ground.
Distr: BAG, UN, EQU; Ivory Coast, Ethiopia to Angola & Zimbabwe; tropical America.
EQU: Myers 13287 5/1940.

Hewittia malabarica (L.) Suresh

Biol. Skr. 51: 417 (2005); F.E.E. 5: 187 (2006).
Syn: *Hewittia sublobata* (L.f.) Kuntze – F.P.S. 3: 109 (1956); F.T.E.A. Convolv.: 45 (1963); Bot. Exp. Sud.: 39 (1970).
Perennial herb. Open, disturbed ground, riverbanks, grassland.
Distr: UN, EQU; palaeotropical.
EQU: Myers 6785 5/1937.

Ipomoea alba L.

F.P.S. 3: 103 (1956); F.T.E.A. Convolv.: 130 (1963); F.E.E. 5: 222 (2006).
Annual or perennial herbaceous climber. Waste ground, also cultivated.
Distr: CS; native to the Americas, now widely naturalised from cultivation.
CS: Schweinfurth 288.

Ipomoea aquatica Forssk.

F.P.S. 3: 121 (1956); F.T.E.A. Convolv.: 120 (1963); F.J.M.: 143 (1976); F.E.E. 5: 219 (2006).
Annual or perennial herb. Rivers & swamps.
Distr: DAR, KOR, CS, ES, BAG, UN, EQU; pantropical.
KOR: Wickens 732 10/1962; **EQU:** Myers 7561 8/1937.

Ipomoea asarifolia (Desr.) Roem. & Schult.

F.W.T.A. 2: 348 (1963); F.J.M.: 143 (1976).
Syn: *Ipomoea repens* Lam. – F.P.S. 3: 120 (1956); Bot. Exp. Sud.: 40 (1970).
Annual herb. Riverbanks.
Distr: DAR, CS, BAG, UN, EQU; Senegal to C.A.R.; tropical America & Asia.
DAR: Harrison 301 10/1947; **EQU:** Myers 8688 2/1938.

Ipomoea blepharophylla Hallier f.

F.P.S. 3: 120 (1956); F.T.E.A. Convolv.: 96 (1963); Biol. Skr. 51: 417 (2005); F.E.E. 5: 208 (2006).
Perennial herb. Grassland & woodland.
Distr: ?CS, EQU; Senegal to Ethiopia, S to Angola & Mozambique.
EQU: Friis & Vollesen 993 2/1982.
Note: Andrews recorded this species from Fung District but we have been unable to trace the specimen on which this record is based.

Ipomoea cairica (L.) Sweet var. *cairica*

F.P.S. 3: 119 (1956); F.T.E.A. Convolv.: 125 (1963); Biol. Skr. 51: 417 (2005); F.E.E. 5: 220 (2006).
Perennial herb. Forest margins, woodland, grassland, often in disturbed areas.
Distr: NS, KOR, CS, BAG, UN, EQU; palaeotropical & warm temperate regions.
NS: Pettet 62 9/1962; **EQU:** Friis & Vollesen 1252 3/1982.

Ipomoea chrysochaetia Hallier f.

Fragm. Florist. Geobot. 37: 51 (1992).
Syn: *Ipomoea chaetocaulos* Hallier f. – F.P.S. 3: 112 (1956).
Syn: *Ipomoea pharbitiformis* Baker – F.T.E.A. Convolv.: 112 (1963).
Perennial climbing herb. Riverine forest, woodland.
Distr: BAG; Senegal to Tanzania, S to Angola & Zambia.
BAG: Schweinfurth 2607 11/1869.
Note: Lejoly & Lisowski (in the cited reference) recognise three varieties in this species. The type of *I. chaetocaulos* (*Schweinfurth* 2607) was destroyed in Berlin and we have seen no other Sudanese material, so its varietal placement is not clear.

Ipomoea convolvulifolia Hallier f.

Bot. Jahrb. Syst. 18: 126 (1894); F.P.S. 3: 114 (1956).
Perennial herb. Habitat unknown.
Distr: EQU; ?South Sudan endemic.
EQU: Schweinfurth 2926 2/1870 (n.v.).
Note: the status of this species, known only from the type (destroyed), is uncertain.

Ipomoea coptica (L.) Roth ex Roem. & Schult. var. *coptica*

F.P.S. 3: 115 (1956); F.T.E.A. Convolv.: 128 (1963); Biol. Skr. 51: 417 (2005); F.E.E. 5: 221 (2006).
Annual herb. Grassland & woodland, weed of waste ground & cultivation.
Distr: DAR, KOR, CS, UN, EQU; palaeotropical.
KOR: Wickens 563 9/1962; **UN:** J.M. Lock 81/117 8/1981.
Note: Andrews lists also var. *malvifolia* Hallier f. for which the type is from Kordofan (*Pfund* s.n.). We have not seen this specimen but the description in the protologue appears to match var. *coptica*.

Ipomoea cordofana Choisy

F.P.S. 3: 116 (1956); F.T.E.A. Convolv.: 99 (1963); F.E.E. 5: 210 (2006).
Perennial herb. Grassland & bare ground in seasonally wet areas.
Distr: NS, RS, KOR, CS, ES, UN, EQU; Eritrea, Ethiopia, Uganda, Kenya, Tanzania.
KOR: Wickens 271 8/1962; **EQU:** Myers 14084 9/1941.

Ipomoea coscinosperma Hochst. ex Choisy

F.P.S. 3: 113 (1956); F.T.E.A. Convolv.: 92 (1963); F.E.E. 5: 206 (2006).
Syn: *Ipomoea coscinosperma* Hochst. ex Choisy var. *glabra* Rendle – F.P.S. 3: 114 (1956).
Syn: *Ipomoea coscinosperma* Hochst. ex Choisy var. *hirsuta* A.Rich. – F.P.S. 3: 114 (1956).
Annual herb. Woodland, grassland, disturbed ground.
Distr: RS, DAR, KOR, CS, ES; Eritrea to Namibia & South Africa.
KOR: Jackson 4022 9/1959.

Ipomoea curtipes Rendle

F.T.A. 4(2): 140 (1905); F.P.S. 3: 116 (1956).
Annual herb. Habitat unknown.
Distr: BAG; South Sudan endemic.
IUCN: RD
BAG: Schweinfurth III.3 7/1869.

Ipomoea dichroa Choisy

F.E.E. 5: 213 (2006).
Syn: *Ipomoea aitoni* Lindl. – F.P.S. 3: 114 (1956).
Syn: *Ipomoea arachnosperma* Welw. – F.T.E.A.
Convolv.: 112 (1963); F.J.M.: 143 (1976).
Annual herb. Woodland, bushland & riverine thicket.
Distr: DAR, KOR, CS, ES, UN, EQU; Senegal to Ethiopia,
S to Namibia & South Africa; India.
DAR: Wickens 2575 9/1964; **EQU:** Myers 10040
11/1938.

Ipomoea eriocarpa R.Br.

F.P.S. 3: 113 (1956); F.T.E.A. Convolv.: 91 (1963); F.J.M.:
143 (1976); Biol. Skr. 51: 418 (2005); F.E.E. 5: 205
(2006).
Annual herbaceous climber. Woodland, grassland,
disturbed ground.
Distr: NS, RS, DAR, KOR, CS, ES, BAG, UN, EQU;
palaeotropical.
CS: Lea 91 11/1952; **EQU:** Friis & Vollesen 516 11/1980.

Ipomoea eurysepala Hallier f.

Bot. Jahrb. Syst. 18: 125 (1894); F.P.S. 3: 114 (1956).
Annual herb. Habitat unknown.
Distr: KOR; ?Sudan endemic.
IUCN: RD
KOR: Pfund s.n.
Note: the status of this species is uncertain.

Ipomoea fulvicaulis (Hochst. ex Choisy) Boiss. ex Hallier f. var. fulvicaulis

F.T.E.A. Convolv.: 97 (1963); Biol. Skr. 51: 418 (2005);
F.E.E. 5: 209 (2006).
Perennial herb. Grassland & woodland.
Distr: EQU; Ethiopia to Malawi.
EQU: Myers 10948 4/1939.

Ipomoea hederifolia L.

F.T.E.A. Convolv.: 132 (1963).
Annual twining herb. Naturalised in waste ground,
disturbed bushland.
Distr: EQU; native to tropical America, now widely
cultivated and naturalised.
EQU: von Ramm 160 (fl) 4/1956.

Ipomoea heterotricha Didr.

F.T.E.A. Convolv.: 107 (1963); Biol. Skr. 51: 417 (2005).
Syn: *Ipomoea amoenula* Dandy – F.P.S. 3: 112 (1956).
Annual herb. Woodland & grassland.
Distr: ?CS, EQU; Cape Verde Is., Senegal to Uganda, S to
Angola & Zambia.
EQU: Friis & Vollesen 632 12/1980.
Note: Andrews records this species from Central Sudan
but we have been unable to verify this record.

Ipomoea involucrata P.Beauv.

F.P.S. 3: 112 (1956); F.T.E.A. Convolv.: 104 (1963); F.J.M.:
143 (1976); Biol. Skr. 51: 419 (2005).
Annual or perennial herb. Woodland & grassland,
disturbed areas.
Distr: DAR, EQU; Senegal to Kenya, S to Angola & South
Africa.
DAR: Wickens 2500 9/1964; **EQU:** Friis & Vollesen 159
11/1980.

Ipomoea kotschyana Hochst. ex Choisy

F.P.S. 3: 114 (1956); F.T.E.A. Convolv.: 95 (1963); F.E.E. 5:
208 (2006).
Annual or biennial herb. Woodland, along water courses.
Distr: DAR, KOR; Mali, Eritrea & Somalia, S to
Mozambique; Socotra.
KOR: Wickens 518 9/1962.

Ipomoea marginata (Desr.) Verdc.

Kew Bull. 42: 658 (1987).
Syn: *Ipomoea hellebarda* Schweinf. ex Hallier f. – F.P.S.
3: 123 (1956).
Syn: *Ipomoea sepiaria* Roxb. – F.T.E.A. Convolv.: 117
(1963); Bot. Exp. Sud.: 40 (1970); F.E.E. 5: 218 (2006).
Twining perennial herb. Woodland, riverbanks.
Distr: ES, UN; palaeotropical.
ES: Schweinfurth 2176 1865; **UN:** Simpson 7119 6/1929.
Note: the single Sudanese specimen seen is from the
Ethiopian border region and may have been collected on
the Ethiopian side.

Ipomoea mauritiana Jacq.

F.T.E.A. Convolv.: 135 (1963); Biol. Skr. 51: 419 (2005).
Syn: *Ipomoea digitata* sensu Andrews, non L. – F.P.S.
3: 119 (1956).
Woody climber. Forest & secondary bushland.
Distr: BAG, EQU; pantropical.
EQU: Andrews 1314 5/1939.

Ipomoea mombassana Vatke

F.P.S. 3: 115 (1956); F.T.E.A. Convolv.: 99 (1963).
Annual twining herb. Woodland, bushland & grassland.
Distr: UN; Kenya, Tanzania.
UN: Simpson 7223 5/1929.

Ipomoea nil (L.) Roth

F.P.S. 3: 119 (1956); F.T.E.A. Convolv.: 113 (1963); F.J.M.:
144 (1976); F.E.E. 5: 215 (2006).
Annual twining herb. Naturalised in disturbed areas.
Distr: RS, DAR, CS, ES; native to N America, now widely
cultivated and naturalised in the tropics.
ES: Beshir 86 9/1951.

Ipomoea obscura (L.) Ker Gawl. var. obscura

F.T.E.A. Convolv.: 116 (1963); F.J.M.: 144 (1976); Biol.
Skr. 51: 419 (2005); F.E.E. 5: 216 (2006).
Syn: *Ipomoea acanthocarpa* (Choisy) Asch. &
Schweinf. – F.P.S. 3: 116 (1956).
Syn: *Ipomoea fragilis* Choisy – F.P.S. 3: 116 (1956).
Perennial twining herb. Woodland, bushland, grassland,
forest margins, weed of disturbed ground.
Distr: RS, DAR, KOR, ES, UN, EQU; palaeotropical.
DAR: Wickens 1014 1/1964; **UN:** J.M. Lock 81/30
2/1981.

Ipomoea ochracea (Lindl.) G.Don var. ochracea

F.T.E.A. Convolv.: 115 (1963); F.J.M.: 144 (1976); Biol.
Skr. 51: 419 (2005); F.E.E. 5: 215 (2006).
Syn: *Ipomoea kentrocarpa* Hochst. ex A.Rich. – F.P.S.
3: 115 (1956).
Perennial twining herb. Wooded grassland, secondary
bushland.
Distr: DAR, CS, EQU; Senegal to Eritrea, S to Angola and
Mozambique.

DAR: Wickens 1165 2/1964; **EQU:** Friis & Vollesen 776 12/1980.

Ipomoea pes-caprae (L.) R.Br.
F.T.E.A. Convolv.: 121 (1963); Bot. Exp. Sud.: 40 (1970); F.E.E. 5: 219 (2006).
Perennial herb. Seashores, lake shores, roadsides.
Distr: KOR; pantropical.
KOR: Drar 1983 5/1938 (n.v.).
Note: the record is based on Drar (1970). It is unusual to find this species so far inland but it is unlikely to have been misidentified – it was perhaps an escape from cultivation. The apparent absence of this species from the Red Sea coast is notable.

Ipomoea pes-tigridis L. var. pes-tigridis
F.P.S. 3: 113 (1956); F.T.E.A. Convolv.: 108 (1963).
Annual herb. Grassland & bushland.
Distr: KOR, CS; Somalia to South Africa; Mascarenes, tropical Asia.
KOR: Pfund 115 5/1898.

Ipomoea pileata Roxb.
F.P.S. 3: 112 (1956); F.T.E.A. Convolv.: 105 (1963).
Annual herb. Forest clearings, bushland, grassland, sometimes a weed of cultivation.
Distr: BAG; Kenya to Angola & Zimbabwe; Mascarenes, India to China & Malaysia.
BAG: Schweinfurth III.1 4/1869.

Ipomoea prismatosyphon Welw. var. prismatosyphon
F.T.E.A. Convolv.: 142 (1963); Biol. Skr. 51: 420 (2005).
Syn: *Ipomoea magnifica* Hallier f. – F.P.S. 3: 119 (1956).
Shrubby perennial herb. Woodland & grassland.
Distr: EQU; Nigeria to Uganda, S to Angola & Zambia.
EQU: Friis & Vollesen 1122 3/1982.

Ipomoea rubens Choisy
F.T.E.A. Convolv.: 134 (1963); F.E.E. 5: 224 (2006).
Syn: *Ipomoea riparia* G.Don – F.P.S. 3: 119 (1956); Bot. Exp. Sud.: 40 (1970).
Perennial climbing herb or subshrub. Swampy ground, riverbanks.
Distr: BAG, UN, EQU; palaeotropical, Guyana.
EQU: Harrison 306 1/1948.

Ipomoea shupangensis Baker
F.T.E.A. Convolv.: 132 (1963); F.E.E. 5: 224 (2006).
Prostrate woody herb. Secondary forest, woodland & wooded grassland.
Distr: EQU; Ethiopia to Angola, Zimbabwe & Mozambique.
EQU: Myers 8019 12/1937.

Ipomoea sinensis (Desr.) Choisy subsp. blepharosepala (Hochst. ex A.Rich.) Verdc. ex A.Meeuse
F.T.E.A. Convolv.: 101 (1963); F.J.M.: 144 (1976); Biol. Skr. 51: 420 (2005); F.E.E. 5: 210 (2006).
Syn: *Ipomoea blepharosepala* Hochst. ex A.Rich. – F.P.S. 3: 116 (1956); Bot. Exp. Sud.: 39 (1970).
Annual herb. Wooded grassland including seasonally flooded areas, weed of disturbed ground.

Distr: RS, DAR, KOR, CS, UN, EQU; Eritrea & Somalia, S to South Africa; Arabia.
DAR: Wickens 2305 8/1964; **UN:** Sherif A.4013 9/1951.
Note: some of the material from Sudan and South Sudan (including the Sherif specimen cited) approaches subsp. *sinensis* which is otherwise recorded from further south in Africa and in tropical Asia.

Ipomoea spathulata Hallier f.
F.T.E.A. Convolv.: 145 (1963); Biol. Skr. 51: 420 (2005); F.E.E. 5: 226 (2006).
Shrub, sometimes climbing. *Acacia-Commiphora* bushland.
Distr: EQU; Ethiopia, Uganda, Kenya.
EQU: Myers 13449 8/1940.

Ipomoea stenobasis Brenan
F.T.E.A. Convolv.: 121 (1963); Biol. Skr. 51: 420 (2005).
Perennial herb. Rocky woodland.
Distr: EQU; Nigeria to Uganda.
EQU: Friis & Vollesen 626 12/1980.

Ipomoea tenuipes Verdc.
F.T.E.A. Convolv.: 127 (1963).
Syn: *Ipomoea pulchella* sensu Andrews, non Roth – F.P.S. 3: 115 (1956).
Perennial herb. Bushland on seasonally damp ground.
Distr: KOR, ES, UN; Tanzania to Angola & South Africa; India.
ES: Andrews 58 6/1936; **UN:** Broun 1412 (fl fr) 12/1908.

Ipomoea tenuirostris Steud. ex Choisy subsp. tenuirostris
F.P.S. 3: 115 (1956); F.T.E.A. Convolv.: 101 (1963); Biol. Skr. 51: 421 (2005); F.E.E. 5: 211 (2006).
Perennial herb. Forest margins, bushland.
Distr: EQU; Cameroon to Eritrea, S to Zambia & Zimbabwe.
EQU: Friis & Vollesen 125 11/1980.

Ipomoea tuberculata Ker Gawl. var. tuberculata
F.P.S. 3: 118 (1956); F.T.E.A. Convolv.: 123 (1963); F.E.E. 5: 220 (2006).
Annual herb. Woodland & bushland.
Distr: ?RS, CS; Eritrea to South Africa; India, Sri Lanka.
CS: Broun 1008 1908.

Ipomoea turbinata Lag.
F.E.E. 5: 223 (2006).
Syn: *Ipomoea muricata* (L.) Jacq. – F.P.S. 3: 103 (1956).
Annual herb. Naturalised weed.
Distr: BAG; originating from America, now widely cultivated in the tropics.
BAG: Schweinfurth 2501 10/1869.

Ipomoea vagans Baker
F.P.S. 3: 113 (1956); F.W.T.A. 2: 349 (1963); F.J.M.: 144 (1976).
Annual herb. Open plains, rock crevices.
Distr: DAR, KOR; Senegal to Nigeria.
KOR: Wickens 397 9/1962.

Ipomoea verbascoidea Choisy
F.P.S. 3: 120 (1956); F.T.E.A. Convolv.: 140 (1963); F.J.M.: 144 (1976).
Syn: *Ipomoea fistulosa* sensu El Amin, non Mart. ex Choisy – T.S.S.: 427 (1990).

Shrub. Bushland.
Distr: DAR, KOR, BAG, EQU; Uganda to Namibia,
Botswana & Zimbabwe.
DAR: Wickens 1807 7/1964; **EQU:** Schweinfurth 4013
7/1870.

Ipomoea verticillata Forssk.

F.P.S. 3: 113 (1956); F.J.M.: 144 (1976); F.E.E. 5: 207
(2006).
Annual herb. Grassland including seasonally wet areas,
riverbanks, weed of cultivation.
Distr: RS, DAR, KOR; Eritrea, Ethiopia, Somalia; Arabia,
India.
DAR: Wickens 2285 8/1964.

Ipomoea welwitschii Vatke ex Hallier f.

F.T.E.A. Convolv.: 119 (1963).
Perennial herb. Woodland.
Distr: BAG, EQU; Tanzania to Namibia, Botswana &
Zimbabwe.
EQU: Andrews 421 4/1939.

Ipomoea wightii (Wall.) Choisy var. *wightii*

F.T.E.A. Convolv.: 110 (1963); Biol. Skr. 51: 421 (2005);
F.E.E. 5: 213 (2006).
Perennial herb. Forest clearings, bushland & grassland.
Distr: EQU; Uganda to South Africa; Madagascar,
tropical Asia.
EQU: Friis & Vollesen 291 11/1980.

Jacquemontia tamnifolia (L.) Griseb.

F.P.S. 3: 123 (1956); F.T.E.A. Convolv.: 35 (1963); Biol. Skr.
51: 421 (2005); F.E.E. 5: 188 (2006).
Annual climbing herb. Forest margins, woodland,
grassland, weed of disturbed ground.
Distr: RS, KOR, UN, EQU; Senegal to Ethiopia, S to
Namibia & South Africa; Madagascar, Mascarenes,
tropical America.
KOR: Wickens 633 10/1962; **UN:** Sherif A.2870
10/1951.

Lepistemon owariense (P.Beauv.) Hallier f.

F.P.S. 3: 123 (1956); F.T.E.A. Convolv.: 63 (1963).
Perennial herbaceous climber. Forest.
Distr: BAG, EQU; Guinea Bissau to Uganda, S to Angola
& Mozambique.
EQU: Myers 7995 12/1937.

Merremia aegyptia (L.) Urb.

F.P.S. 3: 125 (1956); F.E.E. 5: 192 (2006).
Syn: *Merremia pentaphylla* (L.) Hallier f. – F. Darfur
Nord-Occ. & J. Gourgeil: 126
Annual herb. Woodland & wooded grassland, rocky
hillslopes.
Distr: RS, DAR, KOR, CS, UN, EQU; widespread in the
tropics.
RS: Bally 7010 4/1949; **UN:** Sherif A.2865 10/1951.

Merremia dissecta (Jacq.) Hallier f.

F.P.S. 3: 124 (1956).
Perennial twining herb. Naturalised from cultivation.
Distr: NS, CS; native to tropical America, now widely
cultivated in the tropics.
CS: de Wilde et al. 5760 (fl) 2/1965.

Merremia emarginata (Burm.f.) Hallier f.

F.P.S. 3: 126 (1956); F.T.E.A. Convolv.: 55 (1963); F.E.E. 5:
192 (2006).
Perennial herb. Woodland.
Distr: CS, ES, BAG, UN; Cameroon, Burundi, Ethiopia,
Tanzania, Angola; tropical Asia.
ES: Carter 1781 11/1987; **UN:** Simpson 7806 11/1930.

Merremia gallabatensis Hallier f.

F.P.S. 3: 125 (1956); F.E.E. 5: 192 (2006).
Perennial herb. Woodland.
Distr: ES; Ethiopia.
IUCN: RD
ES: Schweinfurth 2182 1865.
Note: the collection cited is from the Sudan/Ethiopia
border and may have been collected on the Ethiopian
side.

Merremia hederacea (Burm.f.) Hallier f.

F.P.S. 3: 126 (1956); F.T.E.A. Convolv.: 54 (1963); F.E.E. 5:
191 (2006).
Perennial twining herb. Grassland, thickets.
Distr: ES, BAG, UN, EQU; palaeotropical.
ES: Jackson 4080 12/1960; **UN:** J.M. Lock 81/253 8/1981.

Merremia kentrocaulos (C.B.Clarke) Rendle

F.P.S. 3: 124 (1956); F.T.E.A. Convolv.: 59 (1963); Biol. Skr.
51: 421 (2005); F.E.E. 5: 196 (2006).
Perennial twining herb. Woodland & wooded grassland.
Distr: ES, UN, EQU; Senegal to Eritrea, S to Angola &
South Africa.
ES: Schweinfurth 2137 10/1865; **EQU:** Myers 11895
10/1938.

Merremia pinnata (Hochst. ex Choisy) Hallier f.

F.P.S. 3: 126 (1956); F.T.E.A. Convolv.: 55 (1963).
Annual herb. Dry grassland, rocky hillslopes.
Distr: DAR, KOR, CS, ?South Sudan; Senegal to Uganda,
S to Namibia & South Africa.
KOR: Wickens 432 7/1962.
Note: Andrews records this species from Central &
Southern Sudan but we have seen no material from the
South.

Merremia pterygocaulos (Choisy) Hallier f.

F.P.S. 3: 126 (1956); F.T.E.A. Convolv.: 57 (1963); F.J.M.:
144 (1976); Biol. Skr. 51: 422 (2005); F.E.E. 5: 196
(2006).
Perennial herb. Grassland, often by streams.
Distr: DAR, BAG, UN, EQU; Senegal to Ethiopia, S to
South Africa; Madagascar, Mascarenes.
DAR: Wickens 996 1/1964; **UN:** Jackson 1669 1/1951.

Merremia semisagitta (Peter) Dandy

F.P.S. 3: 125 (1956); F.T.E.A. Convolv.: 58 (1963); F.E.E. 5:
193 (2006).
Perennial herb. Open woodland.
Distr: RS; Ethiopia, Somalia, Uganda, Tanzania; Arabia.
RS: Bally 6985 4/1949.

Merremia tridentata (L.) Hallier f.

F.P.S. 3: 125 (1956); F.T.E.A. Convolv.: 51 (1963).
Syn: *Merremia tridentata* (L.) Hallier f. subsp. *alatipes*
(Dammer) Verdc. – Bot. Exp. Sud.: 40 (1970).

Syn: *Merremia tridentata* (L.) Hallier f. subsp.
angustifolia (Jacq.) Ooststr. – F.P.S. 3: 125 (1956);
F.T.E.A. Convolv.: 51 (1963); F.J.M.: 144 (1976).
Syn: *Xenostegia tridentata* (L.) D.F.Austin & Staples –
F.E.E. 5: 198 (2006).
Perennial herb. Woodland & wooded grassland.
Distr: DAR, KOR, BAG, UN, EQU; palaeotropical.
KOR: Jackson 3259 11/1954; **EQU:** Simpson 7368
7/1929.

Merremia tuberosa (L.) Rendle

F.T.E.A. Convolv.: 60 (1963); Bot. Exp. Sud.: 40 (1970).
Perennial herb or climber. Naturalised in disturbed areas
& bushland.
Distr: EQU; native to the neotropics, now widely
cultivated & naturalised.
EQU: Drar 1553 4/1938 (n.v.).
Note: the specimen cited is taken from Drar (1970); it
was presumably naturalised from cultivation.

Merremia xanthophylla (Hochst.) Hallier f.

F.P.S. 3: 125 (1956); F.E.E. 5: 196 (2006).
Trailing perennial herb. Stony grassland.
Distr: KOR, ?EQU; Ethiopia, Tanzania, Zambia, Zimbabwe.
KOR: Wickens 806 (fl) 11/1962.
Note: Andrews records this species from Equatoria but
we have seen no material from South Sudan.

Seddera arabica (Forssk.) Choisy

F.P.S. 3: 126 (1956); F.T.E.A. Convolv.: 27 (1963); F.E.E. 5:
174 (2006).
Perennial herb. Woodland & bushland.
Distr: RS; Eritrea, Ethiopia, Djibouti, Somalia, Uganda,
Kenya; Egypt, Arabia.
RS: Schweinfurth 257 9/1868.

Seddera bagshawei Rendle

F.P.S. 3: 127 (1956); F.T.E.A. Convolv.: 24 (1963); Biol. Skr.
51: 422 (2005); F.E.E. 5: 178 (2006).
Subshrub. Bushland, disturbed ground.
Distr: BAG, EQU; Ethiopia & Somalia, S to Tanzania.
BAG: Broun 1716 12/1909.

Seddera latifolia Hochst. & Steud.

F.P.S. 3: 126 (1956); F.T.E.A. Convolv.: 25 (1963); F.E.E. 5:
177 (2006).
Subshrub. Open woodland & bushland, often on stony
ground.
Distr: RS, DAR, CS; Mauritania, Niger, Eritrea to Somalia,
S to Tanzania; Arabia, Pakistan, India.
CS: Jackson 3230 11/1954.

Seddera virgata Hochst. & Steud.

F.P.S. 3: 127 (1956); F.E.E. 5: 178 (2006).
Shrub. Rocky hillslopes.
Distr: RS, CS; Eritrea; Arabia.
CS: Broun 439 9/1905.

Stictocardia beraviensis (Vatke) Hallier f.

F.P.S. 3: 127 (1956); F.T.E.A. Convolv.: 69 (1963); Biol. Skr.
51: 422 (2005); F.E.E. 5: 231 (2006).
Woody climber. Riverine forest, woodland & grassland.
Distr: EQU; Guinea Bissau to Ethiopia, S to Zambia;
Madagascar.

EQU: Friis & Vollesen 751 12/1980.

Turbina stenosiphon (Hallier f.) A.Meeuse var. *stenosiphon*

F.T.E.A. Convolv.: 152 (1963); T.S.S.: 427 (1990).
Syn: *Ipomoea stenosiphon* Hallier f. – F.P.S. 3: 120
(1956).
Woody climbing shrub. Dry bushland, often in rocky
areas.
Distr: EQU; Uganda & Kenya, S to South Africa.
EQU: Myers 9978 11/1938.

Solanaceae

I. Darbyshire

Capsicum frutescens L.

F.P.S. 3: 91 (1956); F.E.E. 5: 148 (2006); F.T.E.A. Solan.:
61 (2012).
Syn: *Capsicum abyssinicum* sensu Andrews, non
A.Rich. – F.P.S. 3: 91 (1956).
Shrub. Cultivated & occasionally naturalised in disturbed
areas.
Distr: KOR, CS, BAG, UN, EQU; native to tropical
America, now widely cultivated in the tropics.
KOR: Pfund 233 (fl fr) 6/1875; **EQU:** von Ramm 257 (fl
fr) 5/1956.
Note: *C. annuum* L. is also very likely to be naturalised in
parts of Sudan but we have seen no material – the record
of *C. abyssinicum* (a synonym of *C. annuum*) in F.P.S. is
incorrect and refers instead to *C. frutescens*.

Cestrum diurnum L.

F.T.E.A. Solan.: 14 (2012).
Shrub or small tree. Cultivated.
Distr: ES; native to S America, widely cultivated.
ES: Sahni & Kamil 636 (fl fr) 4/1967.

Datura innoxia Mill.

F.P.S. 3: 91 (1956); F.J.M.: 141 (1976); F.E.E. 5: 157
(2006); F.T.E.A. Solan.: 43 (2012).
Annual or short-lived perennial herb. Weed of disturbed
areas, riverbanks.
Distr: DAR, KOR, CS, ES; native to S America, now
widely cultivated & naturalised in the tropics.
DAR: Wickens 1780 (fl) 6/1964.

Datura metel L.

F.P.S. 3: 91 (1956); F.E.E. 5: 158 (2006); F.T.E.A. Solan.:
44 (2012).
Annual herb. Weed of disturbed ground, escape from
cultivation.
Distr: NS, RS, UN; native to S America, now widely
cultivated & sometimes naturalised.
RS: Ahti 16649 10/1962; **UN:** Aylmer 27/54 (fl) 12/1926.

Datura stramonium L.

F.P.S. 3: 91 (1956); F.J.M.: 141 (1976); F.E.E. 5: 158
(2006); F.T.E.A. Solan.: 40 (2012).
Annual herb. Weed of disturbed ground.
Distr: NS, RS, DAR, ES, EQU; native to Mexico, now
widespread in the tropics & warm temperate regions.
DAR: Wickens 2325 (fl) 9/1964; **EQU:** Andrews 1217
(fl) 5/1939.

Discopodium penninervium Hochst.

T.S.S.: 426 (1990); Biol. Skr. 51: 412 (2005); F.E.E. 5: 150 (2006); F.T.E.A. Solan.: 64 (2012).
Small tree or shrub. Montane forest, often along margins.
Distr: EQU; Bioko, Cameroon, D.R. Congo, Eritrea to Malawi & Mozambique.
EQU: Howard I.M.28 (fl fr) 1/1976.

Hyoscyamus muticus L.

F.P.S. 3: 93 (1956); F.Egypt 3: 50 (2002); F.E.E. 5: 160 (2006).
Perennial (or ?annual) herb or shrub. Wadis, desert plains, saline depressions.
Distr: NS, RS; N Africa, Arabia to India.
RS: Sahni & Kamil 725 (fl fr) 4/1967.

Lycium shawii Roem. & Schult.

F.E.E. 5: 110 (2006); F.T.E.A. Solan.: 50 (2012).
Syn: *Lycium persicum* Miers – F.P.S. 3: 94 (1956); Bot. Exp. Sud.: 105 (1970); T.S.S.: 426 (1990).
Shrub. Rocky slopes, open sandy plains, dry bushland & woodland.
Distr: RS, ES; Eritrea & Somalia, S to Namibia & South Africa; Mediterranean, Arabia to India.
RS: Carter 1856 (fl fr) 11/1987.

Nicotiana glauca Graham

F.E.E. 5: 106 (2006); F.T.E.A. Solan.: 25 (2012).
Shrub or small tree. Cultivated, sometimes naturalised on waste ground.
Distr: NS; native to S America, now widely naturalised in the tropics & warm temperate regions.
NS: Ahti 16785 10/1962.

Nicotiana rustica L.

F.P.S. 3: 94 (1956); F.E.E. 5: 106 (2006); F.T.E.A. Solan.: 24 (2012).
Annual herb. Cultivated, sometimes naturalised as a weed.
Distr: NS, DAR, KOR, UN; native to S America, widely cultivated in warmer parts of the world.
NS: Speke & Grant s.n. (fl) 4/1863; **UN:** Evans-Pritchard 36 (fl) 11/1936.
Note: previously widely cultivated but now usually replaced by *N. tabacum*.

Physalis angulata L.

F.J.M.: 141 (1976); F.E.E. 5: 152 (2006); F.T.E.A. Solan.: 75 (2012).
Annual herb. Weed of cultivation & disturbed ground.
Distr: NS, DAR, ES, UN; native to S America, now widely distributed in the tropics & warm temperate regions.
ES: Jackson 2454 (fl fr) 11/1952; **UN:** Sherif A.3897 (fl) 7/1951.

Physalis lagascae Roem. & Schult.

F.E.E. 5: 153 (2006); F.T.E.A. Solan.: 73 (2012).
Syn: *Physalis micrantha* sensu Wickens, non Link – F.J.M.: 142 (1976).
Syn: *Physalis minima* sensu Andrews, non L. – F.P.S. 3: 94 (1956).
Annual herb. Weed of cultivation & disturbed ground.
Distr: DAR, CS, ES, UN, EQU; native to C America, now widely distributed across the tropics & subtropics.

DAR: Wickens 1784 (fl fr) 6/1964; **EQU:** Myers 7779 (fl fr) 9/1937.

Physalis peruviana L.

F.J.M.: 142 (1976); Biol. Skr. 51: 413 (2005); F.E.E. 5: 153 (2006); F.T.E.A. Solan.: 71 (2012).
Perennial herb. Disturbed ground, weed of cultivation.
Distr: DAR, EQU; native to tropical America, now widely naturalised in the tropics.
DAR: Wickens 1409 (fl fr) 4/1964; **EQU:** Shigeta 219 2/1979 (n.v.).
Note: cultivated for its edible fruits.

Physalis philadelphica Lam.

F.T.E.A. Solan.: 74 (2012).
Syn: *Physalis angulata* sensu Andrews, non L. – F.P.S. 3: 94 (1956).
Annual herb. Weed of cultivation.
Distr: KOR; native to N America, occasionally naturalised in Africa.
KOR: Broun 849 (fl) 3/1906.

Schwenckia americana L.

F.P.S. 3: 94 (1956); F.T.E.A. Solan.: 35 (2012).
Annual or perennial herb. Weed of cultivation & disturbed ground, open woodland & forest margins.
Distr: BAG, EQU; native to S America, now widespread in tropical Africa.
EQU: Andrews 1519 (fl fr) 5/1939.

Solanum aculeastrum Dunal

Biol. Skr. 51: 413 (2005); F.E.E. 5: 138 (2006); F.T.E.A. Solan.: 198 (2012).
Tree or shrub. Forest margins, scrub, disturbed ground.
Distr: EQU; widely distributed in the tropics.
EQU: Friis & Vollesen 440 (fl fr) 11/1980.
Note: commonly cultivated as a hedge plant and sometimes naturalised.

Solanum aculeatissimum Jacq.

F.P.S. 3: 102 (1956); Biol. Skr. 51: 413 (2005); F.E.E. 5: 145 (2006); F.T.E.A. Solan.: 196 (2012).
Annual or perennial herb or shrub. Forest including margins & clearings, riverine thicket & bushland.
Distr: ?UN, EQU; Sierra Leone to Ethiopia, S to South Africa; India, Malay Peninsula, tropical America.
EQU: Myers 11101 (fl) 4/1939.
Note: Andrews records this species from Upper Nile but we have seen no material from that region.

Solanum adoense Hochst. ex A.Br.

F.E.E. 5: 131 (2006).
Shrub. Rocky hillslopes, disturbed ground.
Distr: RS; Eritrea, Ethiopia, ?Yemen.
RS: Jackson 2893 (fl fr) 4/1953.
Note: a very local species but not considered threatened since it favours disturbed or overgrazed areas.

Solanum aethiopicum L.

F.P.S. 3: 96 (1956); F.T.E.A. Solan.: 164 (2012).
Annual or perennial herb or shrub. Woodland & wooded grassland, cultivated.
Distr: BAG, EQU; Senegal to Somalia, S to Angola & Botswana; Madagascar, parts of Europe, Asia, S America.

EQU: <u>Myers 6587</u> (fr) 5/1937.
Note: a widely cultivated vegetable, 'garden eggs' or 'mock tomato'.

Solanum americanum Mill.

Biol. Skr. 51: 413 (2005); F.E.E. 5: 119 (2006); F.T.E.A. Solan.: 134 (2012).
Annual or short-lived perennial herb. Disturbed ground, weed of cultivation, open woodland, bushland & forest.
Distr: EQU; native to tropical America, widely naturalised elsewhere.
EQU: <u>Andrews 1415</u> (fl fr) 5/1939.
Note: possibly misidentified as *S. nigrum* by Andrews in F.P.S. since he lists *S. nodiflorum* Jacq. as a synonym of that species, this now being considered a synonym of *S. americanum*.

Solanum anguivi Lam.

Biol. Skr. 51: 414 (2005); F.E.E. 5: 129 pro parte (2006); F.T.E.A. Solan.: 167 (2012).
Syn: *Solanum indicum* L. subsp. *distichum* (Schumach. & Thonn.) Bitter – F.J.M.: 142 (1976).
Syn: *Solanum indicum* L. subsp. *distichum* (Schumach. & Thonn.) Bitter var. *monbuttorum* Bitter – Bot. Exp. Sud.: 105 (1970).
Syn: *Solanum anomalum* sensu Andrews, non Thonn. – F.P.S. 3: 102 (1956).
Shrub. Disturbed ground, woodland, bushland & grassland.
Distr: DAR, EQU; Senegal to Eritrea, S to Angola & South Africa; Madagascar.
DAR: <u>Kassas 338</u> (n.v.); **EQU:** <u>Friis & Vollesen 339</u> (fr) 11/1980.
Note: the specimen from Darfur has not been seen; the identification follows Wickens (1976).

Solanum campylacanthum Dunal

F.P.S. 3: 99 (1956); Biol. Skr. 51: 414 (2005); F.E.E. 5: 140 (2006); F.T.E.A. Solan.: 200 (2012).
Syn: *Solanum hybridum* sensu Andrews, non Jacq. – F.P.S. 3: 97 (1956).
Syn: *Solanum incanum* sensu Wickens pro parte, non L. – F.J.M.: 142 (1976).
Syn: *Solanum polyanthemum* sensu Andrews, non Hochst. ex A.Rich. – F.P.S. 3: 97 (1956).
Shrub or perennial herb. Woodland, forest margins & scrub, disturbed ground.
Distr: DAR, ES, UN, EQU; Eritrea & Somalia, S to Namibia & South Africa.
DAR: <u>Wickens 1835</u> (fl) 7/1964; **UN:** <u>J.M. Lock 81/118</u> (fl) 8/1981.

Solanum cerasiferum Dunal

F.P.S. 3: 99 (1956); F.J.M.: 142 (1976); F.E.E. 5: 142 (2006).
Syn: *Solanum cerasiferum* Dunal var. *cinereotomentosum* Dunal – F.P.S. 3: 99 (1956).
Syn: *Solanum xanthocarpum* sensu Andrews, non Schrad. & J.C.Wendl. – F.P.S. 3: 99 (1956).
Perennial herb or shrub. Disturbed ground, open woodland.
Distr: DAR, KOR, CS, ES, BAG, UN, EQU; Nigeria to Ethiopia.
DAR: <u>Lynes 505</u> (fl) 9/1921; **UN:** <u>Evans-Pritchard 28</u> (fl fr) 11/1936.

Solanum coagulans Forssk.

Biol. Skr. 51: 414 (2005); F.E.E. 5: 144 (2006); F.T.E.A. Solan.: 220 (2012).
Syn: *Solanum dubium* Fresen. – F.P.S. 3: 99 (1956); Bot. Exp. Sud.: 105 (1970).
Perennial herb. Wooded grassland, disturbed ground, weed of cultivation.
Distr: NS, RS, DAR, KOR, CS, ES, UN, EQU; Eritrea & Somalia, S to Uganda & Tanzania; Egypt, Arabia.
RS: <u>Carter 1793</u> (fl) 11/1987; **UN:** <u>Simpson 7222</u> (fl) 6/1929.

Solanum dasyphyllum Schum. & Thonn.

Biol. Skr. 51: 414 (2005); F.E.E. 5: 137 (2006); F.T.E.A. Solan.: 206 (2012).
Syn: *Solanum duplosinuatum* Klotzsch – F.P.S. 3: 99 (1956).
Perennial herb. Forest clearings, wooded grassland & grassland, distubed ground.
Distr: EQU; Gambia to Somalia, S to Angola & South Africa.
EQU: <u>Myers 6475</u> (fl) 4/1937.

Solanum forskalii Dunal

F.E.E. 5: 134 (2006); F.T.E.A. Solan.: 173 (2012).
Syn: *Solanum albicaule* Kotschy ex Dunal – F.P.S. 3: 97 (1956).
Subshrub, sometimes scandent. Rocky hillslopes.
Distr: RS, DAR, KOR; Senegal to Somalia & Kenya; Egypt, Arabia, India.
KOR: <u>Jackson 4011</u> (fl) 9/1959.

Solanum giganteum Jacq.

Biol. Skr. 51: 414 (2005); F.T.E.A. Solan.: 148 (2012).
Shrub or small tree. Montane forest & bushland.
Distr: EQU; Nigeria to Ethiopia, S to South Africa; India, Sri Lanka, Australia, S America.
EQU: <u>Friis & Vollesen 759</u> (fl fr) 12/1980.

Solanum hastifolium Hochst. ex Dunal

F.P.S. 3: 97 (1956); F.E.E. 5: 132 (2006); F.T.E.A. Solan.: 174 (2012).
Subshrub, sometimes scandent. Dry *Acacia* woodland, disturbed ground.
Distr: DAR, KOR, CS, UN, EQU; Eritrea & Somalia, S to Uganda & Tanzania.
CS: <u>Schweinfurth 1405</u> (fl fr) 10/1865; **EQU:** <u>Myers 14273</u> (fl) 11/1941.

Solanum incanum L.

F.P.S. 3: 98 (1956); F.J.M.: 142 pro parte (1976); Biol. Skr. 51: 415 (2005); F.E.E. 5: 140 (2006); F.T.E.A. Solan.: 208 (2012).
Syn: *Solanum unguiculatum* A.Rich. – F.P.S. 3: 102 (1956).
Perennial herb or shrub. Grassland, bushland & woodland, disturbed ground.
Distr: NS, RS, DAR, KOR, CS, ES, UN, EQU; Senegal to Somalia & Kenya; Egypt, Arabia, Middle East to India.
NS: <u>Pettet 61</u> (fl fr) 9/1962; **UN:** <u>Sherif A.2881</u> (fl) 10/1951.

Solanum macrocarpon L.

F.E.E. 5: 137 (2006); F.T.E.A. Solan.: 210 (2012).
Perennial herb or shrub. Cultivated, ?occasionally
naturalised.
Distr: EQU; cultivated throughout tropical Africa; Egypt,
Guatemala, Brazil.
EQU: Wyld 809 6/1948.
Note: this is the cultivated form of the wild *S.
dasyphyllum*; it is grown for its edible fruits and leaves.

Solanum memphiticum J.F.Gmel.

F.E.E. 5: 122 (2006); F.T.E.A. Solan.: 142 (2012).
Annual or perennial herb. Bushland, rocky hillslopes,
disturbed ground.
Distr: DAR; Cameroon to Eritrea & Somalia, S to Malawi
& South Africa; Egypt, Arabia.
DAR: de Wilde et al. 5504 (fl fr) 1/1965.

Solanum nigrum L. subsp. nigrum

F.P.S. 3: 96 pro parte (1956); F.J.M.: 142 (1976); Biol. Skr.
51: 415 (2005); F.E.E. 5: 117 (2006); F.T.E.A. Solan.: 124
(2012).
Annual or perennial herb. Disturbed ground, weed of
cultivation, open woodland.
Distr: NS, RS, DAR, KOR, CS, EQU; cosmopolitan.
CS: Jackson 4187 (fl fr) 5/1961; **EQU:** Friis &
Vollesen 1280 3/1982.
Note: Andrews records also *S. pruinosum* Dunal var.
pilosulum Dunal from Fung District, but this seems an
unlikely record for this W African species and it is likely to
be a misidentification of *S. nigrum* or one of its allies.

Solanum schimperianum Hochst. ex A.Rich.

F.P.S. 3: 96 (1956); F.E.E. 5: 126 (2006).
Syn: Solanum carense Dunal – F.P.S. 3: 97 (1956).
Shrub. Dry bushland, open rocky slopes, disturbed ground.
Distr: RS; Eritrea, Ethiopia, Somalia; Egypt, Yemen.
RS: Jackson 2700 (fl fr) 3/1953.

Solanum cf. sinaicum Boiss.

F.E.E. 5: 119 (2006).
Perennial herb. Rocky hillslopes, disturbed areas.
Distr: RS, DAR; (for species) Eritrea, Ethiopia; Middle
East, Arabia.
RS: Aylmer 517 (fl fr) 3/1936.
Note: this species forms a part of the difficult *S. nigrum*
complex and is closely allied to *S. villosum*. Several
collections at Kew have been provisionally determined
as *S.* cf. *sinaicum* but the status of this species in Sudan
requires confirmation.

Solanum terminale Forssk.

Biol. Skr. 51: 415 (2005); F.E.E. 5: 114 (2006); F.T.E.A.
Solan.: 107 (2012).
Woody climber. Forest & forest margins.
Distr: EQU; Bioko & Cameroon to Eritrea, S to South
Africa; Yemen.
EQU: Andrews 1430 (fr) 6/1939.

Solanum villosum Mill. subsp. miniatum (Bernh. ex Willd.) Edmonds

F.E.E. 5: 118 (2006); F.T.E.A. Solan.: 132 (2012).
Annual herb. Disturbed areas, bushland, woodland,
riverbanks.

Distr: RS, DAR; Nigeria, Eritrea & Somalia, S to Tanzania,
Angola; Eurasia, Macaronesia, Israel, Egypt.
RS: Jackson 2887 (fl fr) 4/1953.

Solanum welwitschii C.H.Wright

Biol. Skr. 51: 415 (2005); F.E.E. 5: 115 (2006); F.T.E.A.
Solan.: 113 (2012).
Syn: Solanum welwitschii C.H.Wright var. **strictum**
C.H.Wright – F.P.S. 3: 96 (1956).
Woody climber. Forest.
Distr: EQU; Guinea to Ethiopia, S to D.R. Congo &
Angola.
EQU: Jackson 3045 6/1953.

Withania somnifera (L.) Dunal

F.P.S. 3: 102 (1956); F.J.M.: 142 (1976); T.S.S.: 426
(1990); Biol. Skr. 51: 415 (2005); F.E.E. 5: 154 (2006);
F.T.E.A. Solan.: 227 (2012).
Annual or perennial herb or shrub. Grassland &
bushland, often in disturbed areas.
Distr: NS, RS, DAR, KOR, CS, ES, BAG, UN, EQU; Mali to
Somalia, S to South Africa; Macaronesia, Mediterranean,
Arabia to India & China.
DAR: Wickens 2822 (fl fr) 11/1964; **UN:** Muriel 5/97 (fr)
3/1901.

Sphenocleaceae

H. Pickering

Sphenoclea zeylanica Gaertn.

F.P.S. 3: 71 (1956); F.T.E.A. Sphenocl.: 2 (1968); F.E.E. 5:
37 (2006).
Annual herb. Marshes.
Distr: KOR, BAG, UN, EQU; pantropical, probably
introduced in the Americas.
KOR: Kotschy 189 (fr) 10/1839; **UN:** Sherif A.4001
9/1951.

Hydroleaceae

I. Darbyshire

Hydrolea floribunda Kotschy & Peyr.

F.P.S. 3: 74 (1956); F.T.E.A. Hydrophyll.: 4 (1989).
?Perennial herb. Seasonal swamps.
Distr: BAG, UN; Senegal to Cameroon, Uganda.
UN: Myers 13245 (fl fr) 4/1940.

Hydrolea macrosepala A.W.Benn.

F.P.S. 3: 74 (1956); F.W.T.A. 2: 317 (1963).
?Perennial herb. Swamps & seasonally wet grassland.
Distr: BAG; Senegal to Cameroon, Tanzania.
BAG: Schweinfurth 2540 (fl fr) 10/1869.

Hydrolea palustris (Aubl.) Raeusch.

Rhodora 90: 200 (1988).
Syn: Hydrolea glabra Schumach. & Thonn. – F.P.S. 3:
75 (1956).
?Perennial herb. Shallow pools & swamps, riverbanks.
Distr: EQU; Sierre Leone to D.R. Congo, Tanzania;
Madagascar.
EQU: Wyld 425 (fl fr) 4/1938.

LAMIALES

Oleaceae

I. Darbyshire

Chionanthus mildbraedii (Gilg & Schellenb.) Stearn

F.T.E.A. Olea.: 7 (1952); Biol. Skr. 51: 313 (1998); F.E.E. 4(1): 81 (2003).
Tree or shrub. Montane & riverine forest.
Distr: EQU; Cameroon, Ethiopia to Tanzania.
EQU: Friis & Vollesen 1258 (fr) 3/1982.

Chionanthus niloticus (Oliv.) Stearn

Bot. J. Linn. Soc. 80: 202 (1980).
Syn: *Linociera nilotica* Oliv. – F.T.E.A. Olea.: 14 (1952); F.P.S. 2: 385 (1952); T.S.S.: 365 (1990).
Shrub or tree. Riverine forest.
Distr: BAG, EQU; Ivory Coast to Uganda, S to Tanzania.
EQU: Myers 8383 (fl) 1/1938.

Jasminum abyssinicum Hochst. ex DC.

F.T.E.A. Olea.: 18 (1952); T.S.S.: 363 (1990); Biol. Skr. 51: 313 (1998); F.E.E. 4(1): 85 (2003).
Woody climber. Forest edges & secondary scrub, often near streams.
Distr: EQU; Ethiopia to South Africa.
EQU: Friis & Vollesen 962 (fl) 2/1982.

Jasminum dichotomum Vahl

F.T.E.A. Olea.: 23 (1952); F.P.S. 2: 385 (1952); F.J.M.: 127 (1976); T.S.S.: 363 (1990); Biol. Skr. 51: 313 (1998); F.E.E. 4(1): 84 (2003).
Scandent shrub. Forest, forest margins & secondary bushland.
Distr: DAR, BAG, EQU; Mali to Ethiopia, S to Zambia & Mozambique.
DAR: Wickens 1460 (fl) 4/1964; **BAG:** Schweinfurth 2665 (fr) 11/1869.

Jasminum fluminense Vell. subsp. gratissimum (Deflers) P.S.Green

F.T.E.A. Olea.: 19 (1952); F.E.E. 4(1): 86 (2003).
Woody climber. Rocky hillslopes & dry riverbeds.
Distr: RS; Eritrea, Ethiopia; Egypt, Yemen.
Note: no Sudanese material seen; P.S. Green (Kew Bull. 41: 416 (1986)) recorded it only from Gebel Elba in the Hala'ib Triangle, but Boulos (F.Egypt 2: 205 (2000)) records it from elsewhere in Sudan.

Jasminum grandiflorum L. subsp. floribundum (R.Br. ex Fresen.) P.S.Green

F.E.E. 4(1): 86 (2003).
Syn: *Jasminum floribundum* R.Br. ex Fresen. – F.P.S. 2: 385 (1952); Bot. Exp. Sud.: 90 (1970); T.S.S.: 365 (1990).
Syn: *Jasminum floribundum* R.Br. ex Fresen. var. *steudneri* (Schweinf. ex Baker) Gilg & Schellenb. – F.P.S. 2: 385 (1952); T.S.S.: 365 (1990).
Woody climber. Rocky wooded grassland, scrub.
Distr: RS; Eritrea, Ethiopia, Somalia, Uganda, Kenya; Arabia.
RS: Sahni & Kamil 679 (fr) 4/1967.

Jasminum pauciflorum Benth.

F.T.E.A. Olea.: 28 (1952); Biol. Skr. 51: 313 (1998).
Syn: *Jasminum dschuricum* Gilg – F.P.S. 2: 384 (1952).
Syn: *Jasminum schweinfurthii* Gilg – not of F.P.S. 2: 384 (1952).
Woody climber. Forest.
Distr: EQU; Sierra Leone to Uganda, S to Tanzania & Zambia.
EQU: Friis & Vollesen 1229 (fl) 3/1982.

Jasminum schimperi Vatke

Biol. Skr. 51: 314 (1998); F.E.E. 4(1): 84 (2003).
Syn: *Jasminum eminii* Gilg – T.S.S.: 363 (1990).
Woodly climber. Lowland & riverine forest, secondary bushland.
Distr: EQU; Ethiopia to D.R. Congo.
EQU: Friis & Vollesen 719 (fl fr) 12/1980.

Jasminum streptopus E.Mey.

Biol. Skr. 51: 314 (1998); F.E.E. 4(1): 84 (2003).
Syn: *Jasminum schweinfurthii* sensu auctt., non Gilg – F.P.S. 2: 384 (1952); T.S.S.: 365 (1990).
Woodly climber or scrambling shrub. Woodland, wooded grassland, often in rocky areas or on termite mounds.
Distr: BAG, EQU; Ethiopia to Angola & South Africa.
EQU: Andrews 827 (fl) 4/1939.

Olea capensis L. subsp. macrocarpa (C.H.Wright) I.Verd.

F.E.E. 4(1): 80 (2003).
Syn: *Olea capensis* L. subsp. *hochstetteri* (Baker) Friis & P.S.Green – Biol. Skr. 51: 315 (1998).
Syn: *Olea hochstetteri* Baker – F.T.E.A. Olea.: 10 (1952); F.P.S. 2: 386 (1952); T.S.S.: 366 (1990).
Tree, often large. *Juniperus-Podocarpus* forest, often co-dominant.
Distr: ?RS, EQU; Guinea to Somalia, S to South Africa; Madagascar, Comoros.
EQU: Friis & Vollesen 1106 (fr) 6/1982.
Note: the record for the Red Sea Hills is from El Amin in T.S.S.; we have seen no material with which to verify this.

Olea europaea L. subsp. cuspidata (Wall. ex G.Don) Cif.

Biol. Skr. 51: 314 (1998); F.E.E. 4(1): 79 (2003).
Syn: *Olea africana* Mill. – T.S.S.: 365 (1990).
Syn: *Olea chyrsophylla* Lam. – F.T.E.A. Olea.: 9 (1952); F.P.S. 2: 385 pro parte (1952); Bot. Exp. Sud.: 89 pro parte (1970).
Tree or rarely a shrub. *Juniperus-Podocarpus* forest.
Distr: RS, EQU; Eritrea to Angola & South Africa; Mascarenes, Egypt, Arabia to China.
RS: Sahni & Kamil 677 (fl) 4/1967; **EQU:** Jackson 698 4/1949.

Olea europaea L. subsp. laperrinei (Batt. & Trab.) Cif.

Kew Bull. 57: 97 (2002).
Syn: *Olea laperrinei* Batt. & Trab. – Bot. Exp. Sud.: 90 (1970); F.J.M.: 127 (1976); T.S.S.: 366 (1990).
Syn: *Olea chrysophylla* sensu Andrews pro parte, non Lam. – F.P.S. 2: 385 (1952).
Tree. Montane grassland, rocky outcrops.

Distr: DAR; Saharan Mts of Algeria & Niger.
IUCN: RD
DAR: Wickens 1461 (fl) 4/1964.
Note: a very local subspecies, potentially threatened by climate change and/or habitat degradation. Drar's (1970) records of *O. chrysophylla* from Darfur, based on sterile material, are probably referable here. El Amin's record of *O. laperrinei* from the Red Sea Hills is almost certainly in error.

Olea welwitschii (Knobl.) Gilg & Schellenb.

F.T.E.A. Olea.: 12 (1952); F.E.E. 4(1): 81 (2003); T.S.S.: 366 (1990).
Syn: *Olea capensis* L. subsp. *welwitschii* (Knobl.) Friis & P.S.Green – Biol. Skr. 51: 315 (1998).
Syn: *Steganthus welwitschii* (Knobl.) Knobl. – F.P.S. 2: 387 (1952).
Tree. Upland forest.
Distr: ?EQU; Cameroon to Ethiopia, S to Angola, Zambia & Mozambique.
Note: Andrews and El Amin list this species from the Dongotona Mts but we have seen no material with which to verify this. P.S. Green did not record this species from Sudan in his revision of *Olea* (Kew Bull. 57: 91–140 (2002)).

Schrebera arborea A.Chev.

F.T.E.A. Olea.: 2 (1952); T.S.S.: 367 (1990).
Syn: *Schrebera macrantha* Gilg & Schellenb. – F.P.S. 2: 387 (1952).
Syn: *Schrebera golungensis* sensu Friis & Vollesen, non Welw. – Biol. Skr. 51: 315 (1998).
Tree. Forest.
Distr: EQU; Senegal, Ivory Coast to Uganda, S to Angola.
EQU: Turner 174 (fl fr) 2/1936.

Gesneriaceae

I. Darbyshire

Epithema tenue C.B.Clarke

Biol. Skr. 51: 432 (2005); F.T.E.A. Gesner.: 2 (2006).
Annual herb. Rocks in forest.
Distr: EQU; Guinea to Uganda.
EQU: Friis & Vollesen 462 11/1980.

Streptocarpus elongatus Engl.

F.W.T.A. 2: 382 (1963); Biol. Skr. 51: 432 (2005).
Annual herb. Forest.
Distr: EQU; Guinea to Cameroon.
EQU: Friis & Vollesen 616 12/1980 (n.v.).
Note: the cited specimen has not been seen by us; the identification is based on Friis & Vollesen. If correct, the South Sudan population is highly disjunct from the remainder of this species' range.

Plantaginaceae

I. Darbyshire

Bacopa crenata (P.Beauv.) Hepper

F.T.E.A. Scroph.: 50 (2008).
Syn: *Bacopa calycina* (Benth.) Engl. ex De Wild. – F.P.S. 3: 132 (1956).
Annual or perennial herb. Wet grassland, swamps.

Distr: BAG; Senegal to Kenya, S to Angola & Mozambique; Madagascar.
BAG: Schweinfurth 2837 (fl fr) 1/1870.

Bacopa floribunda (R.Br.) Wettst.

F.P.S. 3: 132 (1956); F.T.E.A. Scroph.: 48 (2008).
Annual herb. Wet habitats, rice fields.
Distr: BAG, UN; Senegal to Uganda, S to Zimbabwe; Arabia, tropical Asia, Australia.
BAG: Schweinfurth 2588 (fl fr) 11/1869.

Bacopa punctata Engl.

Bot. Jahrb. Syst. 23: 499 (1897); F.P.S. 3: 132 (1956).
Annual herb. Wetlands.
Distr: BAG; apparently a South Sudan endemic.
IUCN: RD
BAG: Schweinfurth 2465 (fl fr) 10/1869.

Callitriche oreophila Schotsman

F.E.E. 2(1): 427 (2000); F.T.E.A. Callitrich.: 2 (2003).
Syn: *Callitriche stagnalis* sensu Friis & Vollesen, non Scop. – Biol. Skr. 51: 109 (1998).
Aquatic herb. Shallow stagnant water, damp exposed mud.
Distr: EQU; Cameroon, D.R. Congo to Ethiopia, S to Tanzania.
EQU: Friis & Vollesen 374 11/1980.

Dopatrium macranthum Oliv.

F.E.E. 5: 258 (2006); F.T.E.A. Scroph.: 53 (2008).
Syn: *Dopatrium schweinfurthii* Wettst. – F.P.S. 3: 136 (1956).
Syn: *Dopatrium tricolor* Wettst. – F.P.S. 3: 136 (1956).
Annual herb. Seasonal pools over rocks.
Distr: BAG, EQU; Guinea to Ethiopia, S to Zambia.
EQU: Myers 9163 (fl) 8/1938.

Kickxia aegyptiaca (L.) Nábělek subsp. aegyptiaca

F.J.U. 3: 237 (1999); F.Egypt 3: 65 (2002).
Perennial herb or subshrub. Wadis, desert plains.
Distr: NS; Libya, Egypt, Palestine.
NS: Léonard 4925 11/1968.

Kickxia aegyptiaca (L.) Nábělek subsp. virgata Wickens

F.J.M.: 145 (1976); Sutton, Rev. tribe Antirrhineae: 189 (1988).
Syn: *Linaria bentii* sensu Quézel, non Skan – F. Darfur Nord-Occ. & J. Gourgeil: 129 (1969).
Subshrub. Montane grassland, arable lands.
Distr: DAR; ?Sudan endemic
IUCN: RD
DAR: Robertson 129 12/1957.
Note: possibly endemic to Jebel Marra, but a similar specimen has been recorded from Chad (see Sutton, l.c.).

Limnophila heterophylla Benth.

Kew Bull. 24: 124 (1970).
Aquatic perennial herb. Shallow lakes.
Distr: CS; Pakistan to Borneo.
CS: Andrews WN117 (fl fr) 12/1938.
Note: presumably introduced in Sudan, only a single naturalised specimen seen.

Limnophila indica (L.) Druce

F.P.S. 3: 138 (1956); F.E.E. 5: 260 (2006); F.T.E.A. Scroph.: 41 (2008).
Perennial herb. Lakes, pools, margins of rivers & seasonal wetlands.
Distr: BAG, UN; Senegal to Ethiopia, S to South Africa; palaeotropical.
BAG: Simpson 7697 (fl fr) 3/1930.

Misopates marraicum D.A.Sutton

Sutton, Rev. Tribe Antirrhineae: 153 (1988).
Subshrub. Montane grassland, ericaceous scrub.
Distr: DAR; Sudan endemic.
IUCN: RD
DAR: Miehe 75 (fr) 5/1982.
Note: endemic to Jebel Marra, where Sutton (l.c.) says it replaces *M. orontium*, but it appears from material at Kew that both species occur on this massif.

Misopates orontium (L.) Rafin. subsp. *gibbosum* (Wall.) D.A.Sutton

F.P.S. 3: 140 (1956); F.J.M.: 146 (1976); F.E.E. 5: 239 (2006); F.T.E.A. Scroph.: 24 (2008).
Syn: *Antirrhinum orontium* sensu Drar, non L. s.s. – Bot. Exp. Sud.: 103 (1970).
Annual herb. Rocky grassland, arable land.
Distr: RS, DAR; Eritrea to Uganda & Kenya; Arabia.
RS: Aylmer 210 (fl fr) 3/1932.

Nanorrhinum hastatum (R.Br. ex Benth.) Ghebr.

F.E.E. 5: 243 (2006).
Syn: *Kickxia hastata* (R.Br. ex Benth.) Dandy – F.P.S. 3: 137 (1956).
Annual herb. Rocky slopes, cliff fissures, alluvial plains.
Distr: RS; Eritrea, Ethiopia, Djibouti, Somalia; ?Egypt, Arabia, Socotra.
RS: Jackson 2810 (fl) 4/1953.

Nanorrhinum heterophyllum (Schousb.) Ghebr.

F.E.E. 5: 242 (2006).
Syn: *Kickxia heterophylla* (Schousb.) Dandy – F.P.S. 3: 137 (1956).
Perennial herb or subshrub. Rocky slopes, cliff fissures.
Distr: RS; Macaronesia, N Africa to Somalia; Arabia.
RS: Jackson 2939 (fl) 4/1953.

Nanorrhinum macilentum (Decne.) Betsche

Nord. J. Bot. 20: 675 (2000).
Syn: *Kickxia bentii* (Skan) Dandy – F.P.S. 3: 138 (1956).
Syn: *Kickxia nubica* (Skan) Dandy – F.P.S. 3: 137 (1956).
Perennial herb or subshrub. Rocky slopes, cliff fissures, wadis.
Distr: RS; Egypt & Israel to Arabia.
RS: Bent s.n. (fl) 1896.

Nanorrhinum ramosissimum (Wall.) Betsche

F.E.E. 5: 242 (2006); F.T.E.A. Scroph.: 22 (2008).
Syn: *Kickxia diblophylla* Wickens – F.J.M.: 146 (1976).
Perennial herb or subshrub. Rocky slopes, cliff crevices, streamsides.
Distr: DAR; Ethiopia, Somalia, Kenya; Arabia.
DAR: Pettet 188 (fl fr) 12/1962.

Plantago afra L. var. *stricta* (Schousb.) Verdc.

F.T.E.A. Plantag.: 6 (1971); F.E.E. 5: 34 (2006).
Syn: *Plantago stricta* Schousb. – F.P.S. 3: 69 (1956).
Annual herb. Weed of cultivation, fallow land.
Distr: RS; Eritrea to Tanzania; Macaronesia, N Africa, Israel, Syria.
RS: Schweinfurth 1344 (fl fr) 2/1864.

Plantago amplexicaulis Cav.

F.Egypt 3: 117 (2002).
Annual herb. Sandy desert plains & wadis.
Distr: ?RS; Djibouti, Somalia; Spain, Macaronesia, N Africa to India.
RS: Drar 277 1932 (n.v.).
Note: the record of this species is from Drar (1936) and the cited specimen is from Gebel Elba in the disputed Hala'ib Triangle. Boulos in Fl. Egypt does not record this species from his 'GE' region, so the record must be in doubt.

Plantago ciliata Desf.

F.P.S. 3: 69 (1956); F.Egypt 3: 119 (2002).
Annual herb. Sandy wadis.
Distr: RS; Macaronesia, through N Africa & Arabia to Pakistan.
RS: Farag & Zubair 30 (fl) 2/2012.

Plantago lanceolata L.

F.P.S. 3: 69 (1956); F.T.E.A. Plantag.: 6 (1971); F.E.E. 5: 69 (2006).
Perennial herb. Weed of cultivated land, roadsides, waste ground.
Distr: RS; native of Europa & Asia, naturalised elsewhere.
RS: Jackson 2923 (fl) 4/1953.

Plantago palmata Hook.f.

F.T.E.A. Plantag.: 2 (1971); Biol. Skr. 51: 405 (2005); F.E.E. 5: 35 (2006).
Perennial herb. Clearings in montane forest.
Distr: EQU; Bioko, Cameroon, Ethiopia to Zimbabwe.
EQU: Friis & Vollesen 1011 (fl) 2/1982.

Schweinfurthia pterosperma (A.Rich.) A.Braun

F.P.S. 3: 141 (1956); F.E.E. 5: 238 (2006).
Annual or perennial herb. Rocky grassland.
Distr: RS; Eritrea, Somalia; Arabia.
RS: Bally 6989 (fl fr) 4/1949.

Scoparia dulcis L.

F.P.S. 3: 141 (1956); F.J.M.: 146 (1976); F.E.E. 5: 257 (2006); F.T.E.A. Scroph.: 98 (2008).
Perennial herb. Weed of cultivation & disturbed ground.
Distr: DAR, KOR, BAG, UN, EQU; pantropical.
DAR: Wickens 1789 (fl fr) 6/1964; **EQU:** Jackson 4279 (fl fr) 6/1961.

Stemodia serrata Benth.

F.P.S. 3: 143 (1956); F.E.E. 5: 260 (2006); F.T.E.A. Scroph.: 36 (2008).
Annual herb. Moist grassland, seasonally damp areas in woodland.
Distr: DAR, KOR, CS, UN; Senegal to Ethiopia, S to Mozambique; Madagascar, India.
DAR: Jackson 2511 (fl fr) 1/1953; **UN:** Broun 1722 (fl fr) 12/1909.

Veronica abyssinica Fresen.
F.P.S. 3: 189 (1956); Bot. Exp. Sud.: 104 (1970); Biol. Skr.
51: 430 (2005); F.E.E. 5: 281 (2006); F.T.E.A. Scroph.: 102
(2008).
Annual herb. Forest clearings & margins, montane
grassland.
Distr: ?CS, EQU; Nigeria & Cameroon, Ethiopia to D.R.
Congo & Malawi.
CS: Drar 24 2/1938 (n.v.); **EQU:** Thomas 1788 (fl fr)
12/1935.
Note: the identification of the cited Sudan specimen
follows Drar (1970). It was collected from the Blue Nile
in South Khartoum which seems an unlikely locality for
this species.

Veronica anagallis-aquatica L.
F.P.S. 3: 149 (1956); F.J.M.: 147 pro parte (1976); F.E.E. 5:
283 (2006); F.T.E.A. Scroph.: 101 (2008).
Annual or perennial herb. Streams & ponds.
Distr: NS, RS, DAR, CS; Eritrea to Zimbabwe; Arabia,
Europe, N America.
DAR: Wickens 2631 (fl fr) 9/1964.

Veronica beccabunga L.
F.E.E. 5: 284 (2006).
Syn: Veronica anagallis-aquatica sensu Wickens pro
parte, non L. – F.J.M.: 147 (1976).
Perennial herb. Montane wet grassland and streamsides,
irrigated land.
Distr: DAR; Ethiopia, Niger; widespread in temperate
regions.
DAR: Jackson 2619 (fl) 1/1953.

Scrophulariaceae

I. Darbyshire

Anticharis arabica Endl.
F.P.S. 3: 130 (1956); F.T.E.A. Scroph.: 10 (2008).
Annual herb. Desert.
Distr: RS; Egypt, Arabia, Socotra.
RS: Schweinfurth 702 (fl fr) 9/1868.
Note: Fischer in F.E.E. 5: 278 refers all material previously
identified as A. arabica from Eritrea & Ethiopia to A.
senegalensis.

Anticharis glandulosa Asch.
F.P.S. 3: 130 (1956); F.J.U. 3: 234 (1999); F.E.E. 5: 278 (2006).
Annual herb. Open Acacia woodland or wooded grassland.
Distr: NS, RS; Egypt to Somalia; Arabia, India.
RS: Carter 1951 (fl) 11/1987.

Anticharis senegalensis (Walp.) Bhandari
F.E.E. 5: 278 (2006); F.T.E.A. Scroph.: 10 (2008).
Syn: Anticharis linearis (Benth.) Hochst. ex Aschers. –
F.P.S. 3: 131 (1956).
Annual herb. Semi-desert, open grassland & bushland.
Distr: NS, RS, DAR, KOR, CS; Mali to Ethiopia, S to
Namibia & Botswana; Arabia, India.
CS: Andrews 35 (fl fr) 6/1935.

Aptosimum pumilum (Hochst.) Benth.
F.P.S. 3: 132 (1956); F.E.E. 5: 280 (2006); F.T.E.A. Scroph.:
8 (2008).

Annual or biennial herb. Semi-desert, rocky areas.
Distr: DAR, KOR; Mauritania, Niger, Ethiopia, Somalia,
Kenya; Yemen.
DAR: Lynes 341 (fl fr) 2/1921.

Buddleja polystachya Fresen.
F.T.E.A. Logan.: 36 (1960); T.S.S.: 359 (1990); F.E.E. 4(1):
73 (2003).
Shrub. Clearings in dry forest, woodland, wadis.
Distr: RS; Ethiopia, Kenya & Tanzania; Arabia.
RS: Jackson 3966 (fl).

Diclis ovata Benth.
Biol. Skr. 51: 425 (2005); F.E.E. 5: 256 (2006); F.T.E.A.
Scroph.: 21 (2008).
Annual herb. Montane grassland, woodland, forest
clearings.
Distr: CS, EQU; Cameroon to Ethiopia, S to Angola &
Mozambique; Madagascar, Mascarenes.
CS: Pettet 115 (fr) 1/1963; **EQU:** Friis & Vollesen 12 (fl
fr) 11/1980.

Hebenstretia angolensis Rolfe
Biol. Skr. 51: 456 (2005); F.E.E. 5: 254 (2006); F.T.E.A.
Scroph.: 197 (2008).
Syn: Hebenstretia dentata sensu Andrews, non L. –
F.P.S. 3: 192 (1956).
Perennial herb or subshrub. Montane grassland &
bushland.
Distr: EQU; Eritrea to Angola & Zimbabwe.
EQU: Thomas 1786 (fl) 12/1935.

Jamesbrittenia dissecta (Delile) Kuntze
F.P.S. 3: 137 (1956); F.Egypt 3: 76 (2002).
Annual herb. Riverbanks, weed of cultivation, waste
ground.
Distr: NS, CS; Egypt, India.
CS: de Wilde et al. 5790 (fl) 2/1965.

Rhabdotosperma brevipedicellata (Engl.) Hartl.
F.E.E. 5: 252 (2006).
Syn: Celsia brevipedicellata Engl. – F.P.S. 3: 134 (1956).
Syn: Verbascum brevipedicellatum (Engl.) Hub.-Mor. –
Biol. Skr. 51: 430 (2005); F.T.E.A. Scroph.: 16 (2008).
Perennial herb or subshrub. Montane grassland &
montane scrub.
Distr: EQU; Ethiopia to D.R. Congo, Rwanda & Burundi.
EQU: Johnston 1436 (fl fr) 2/1936.

Rhabdotosperma scrophulariifolia (Hochst. ex
A.Rich.) Hartl.
F.E.E. 5: 252 (2006).
Syn: Verbascum scrophulariifolium (Hochst. ex
A.Rich.) Hub.-Mor. – Biol. Skr. 51: 430 (2005); F.T.E.A.
Scroph.: 14 (2008).
Perennial herb. Montane grassland, montane scrub,
upper forest margins.
Distr: EQU; Cameroon, Ethiopia to Tanzania.
EQU: Myers 11619 (fl fr) 7/1939.

Scrophularia arguta Sol.
F.P.S. 3: 142 (1956); F.J.M.: 146 (1976); F.E.E. 5: 255 (2006).
Annual herb. Bare sandy soils, stone walls, weed of
cultivation.

Distr: RS, DAR; Eritrea; S Spain, N Africa to Arabia.
RS: Jackson 2970 (fl fr) 4/1953.

Verbascum nubicum Murb.

F.P.S. 3: 149 (1956).
Perennial herb. Rocky ground.
Distr: RS; Sudan endemic (but see note).
IUCN: RD
RS: Maffey 5 (fl) 1927.
Note: possibly endemic to the Red Sea Hills, but the
status of this taxon is in doubt; Wickens (F.J.M.: 147)
suggests that *V. nubicum* may be a synonym of *V. sinaiticum*.

Verbascum sinaiticum Benth.

F.P.S. 3: 147 (1956); F.J.M.: 147 (1976); F.E.E. 5: 246
(2006); F.T.E.A. Scroph.: 13 (2008).
Biennial herb. Montane & submontane grassland, waste
ground, fallow land.
Distr: DAR; Eritrea, Ethiopia, Somalia, Kenya; Sinai to
Afghanistan & S Russia.
DAR: Lynes 20A (fl fr) 2/1921.

Verbascum sudanicum (Murb.) Hepper

Kew Bull. 38: 598 (1984).
Syn: *Celsia sudanica* (Murb.) Wickens – F.J.M.: 145
(1976).
?Biennial herb. Open rocky areas, riverbanks, fallow land.
Distr: DAR; Sudan endemic.
IUCN: RD
DAR: Wickens 2998 (fl fr) 6/1965.
Note: Wickens listed a specimen probably belonging
to this species from Ethiopia (*Burger* 2084) but this
specimen is referred to *V. pedunculosum* (Benth.) Kuntze
in F.E.E. *V. sudanicum* is therefore considered endemic to
the Jebel Marra massif.

Stilbaceae

I. Darbyshire

Halleria lucida L.

T.S.S.: 428 (1990); Biol. Skr. 51: 425 (2005); F.E.E. 5: 255
(2006); F.T.E.A. Scroph.: 26 (2008).
Shrub or small tree. Montane forest & bushland.
Distr: EQU; Ethiopia to Angola & South Africa; Arabia.
EQU: Friis & Vollesen 350 (fl) 11/1980.

Nuxia congesta R.Br. ex Fresen.

F.T.E.A. Logan.: 44 (1960); T.S.S.: 359 (1990); Biol. Skr.
51: 311 (1998); F.E.E. 4(1): 74 (2003).
Syn: *Lachnopylis compacta* C.A.Sm. – F.P.S. 2: 381 (1952).
Syn: *Lachnopylis congesta* (R.Br. ex Fresen.) C.A.Sm. –
F.P.S. 2: 381 (1952).
Tree or shrub. Montane forest, particularly margins &
clearings.
Distr: RS, EQU; Guinea to Ethiopia, S to South Africa;
Arabia.
RS: Jackson 3970 (fl) 4/1959; **EQU:** Thomas 1778 (fl)
12/1935.

Nuxia oppositifolia (Hochst.) Benth.

F.T.E.A. Logan.: 43 (1960); T.S.S.: 361 (1990); F.E.E. 4(1):
74 (2003).

Syn: *Lachnopylis oppositifolia* Hochst. – F.P.S. 2: 381
(1952).
Shrub or small tree. Riverine woodland, rocky gullies.
Distr: RS; Ethiopia & Somalia, S to South Africa; Arabia.
RS: Kennedy-Cooke 246 (fl) 7/1936.

Linderniaceae

I. Darbyshire

Craterostigma plantagineum Hochst.

F.P.S. 3: 134 (1956); F.E.E. 5: 268 (2006); F.T.E.A. Scroph.:
61 (2008).
Perennial herb. Shallow soils over rock, in grassland &
open woodland.
Distr: RS, DAR, EQU; Niger, D.R. Congo to Ethiopia, S to
Namibia & South Africa; Arabia.
DAR: Wickens 1772 (fl) 6/1964; **EQU:** Myers 10868 (fl)
4/1939.

Craterostigma pumilum Hochst.

F.E.E. 5: 270 (2006); F.T.E.A. Scroph.: 60 (2008).
Perennial herb. Montane grassland, thin soil over rocks.
Distr: RS, DAR; Ethiopia & Somalia, S to Botswana;
Arabia.
RS: Aylmer 215 3/1932.

Crepidorhopalon debilis (Skan) Eb.Fisch.

Feddes Repert. 106: 10 (1995).
Syn: *Lindernia debilis* Skan – F.P.S. 3: 138 (1956); Biol.
Skr. 51: 426 (2005); F.T.E.A. Scroph.: 76 (2008).
Annual herb. Seasonally wet flushes over rocks.
Distr: BAG, EQU; Senegal, Ghana to Cameroon,
Uganda, Tanzania, South Africa.
EQU: Friis & Vollesen 556 (fl fr) 11/1980.

Crepidorhopalon schweinfurthii (Oliv.) Eb.Fisch.

Feddes Repert. 100: 443 (1989).
Syn: *Craterostigma schweinfurthii* (Oliv.) Engl. – F.P.S.
3: 134 (1956).
Syn: *Torenia schweinfurthii* Oliv. – F.T.E.A. Scroph.: 66
(2008).
Perennial herb. Wet flushes & swamps.
Distr: BAG, EQU; Mali to Uganda, S to Mozambique &
Angola.
EQU: Andrews 470 (fl fr) 4/1939.

Crepidorhopalon spicatus (Engl.) Eb.Fisch.

Feddes Repert. 100: 444 (1989).
Syn: *Torenia spicata* Engl. – F.P.S. 3: 147 (1956); F.T.E.A.
Scroph.: 63 (2008).
Annual herb. Seasonally wet flushes in grassland &
woodland.
Distr: BAG; Guinea to D.R. Congo, Tanzania to Angola
& South Africa.
BAG: Schweinfurth 4296 (fl) 11/1870.

Crepidorhopalon whytei (Skan) Eb.Fisch.

F.E.E. 5: 270 (2006).
Syn: *Lindernia whytei* Skan – Biol. Skr. 51: 428 (2005);
F.T.E.A. Scroph.: 71 (2008).
Annual or perennial herb. Montane seepage grassland.
Distr: EQU; Uganda to Angola & Mozambique.
EQU: Friis & Vollesen 316 (fl) 11/1980.

Lindernia abyssinica Engl.

Biol. Skr. 51: 425 (2005); F.E.E. 5: 273 (2006); F.T.E.A. Scroph.: 81 (2008).
Perennial herb. Montane grassland & woodland, often on rocks.
Distr: EQU; Nigeria, Cameroon, Ethiopia to Tanzania.
EQU: Jackson 1519 (fl fr) 5/1950.

Lindernia exilis Philcox

Biol. Skr. 51: 426 (2005); F.E.E. 5: 276 (2006).
Annual herb. Seasonally wet flushes amongst rocks.
Distr: EQU; Sierra Leone to Ethiopia, Burundi.
EQU: Friis & Vollesen 557 (fl fr) 11/1980.

Lindernia niamniamensis Eb.Fisch. & Hepper

Kew Bull. 46: 534 (1991); F.T.E.A. Scroph.: 87 (2008).
Perennial herb. Seasonally wet flushes amongst rocks.
Distr: EQU; Uganda.
IUCN: RD
EQU: Schweinfurth 3931 (fl fr) 6/1870.

Lindernia nummulariifolia (D.Don) Wettst.

F.J.M.: 146 (1976); Biol. Skr. 51: 426 (2005); F.E.E. 5: 272 (2006); F.T.E.A. Scroph.: 80 (2008).
Annual herb. Weed of cultivation, streamsides, wet grassland.
Distr: DAR, EQU; Sierra Leone to Ethiopia, S to Angola & Zimbabwe; Madagascar, India to Vietnam.
DAR: Wickens 2060 (fl) 8/1964; **EQU:** Friis & Vollesen 630 (fr) 12/1980.

Lindernia oliveriana Dandy

F.P.S. 3: 139 (1956); F.J.M.: 146 (1976); Biol. Skr. 51: 426 (2005); F.E.E. 5: 275 (2006); F.T.E.A. Scroph.: 74 (2008).
Annual herb. Swamps, seasonally wet pools on rocks.
Distr: DAR, BAG, EQU; Togo to Ethiopia, S to Angola & Zimbabwe.
DAR: Lynes 29a (fl) 2/1921; **EQU:** Myers 8399 (fl fr) 1/1938.

Lindernia parviflora (Roxb.) Haines

F.P.S. 3: 139 (1956); Biol. Skr. 51: 426 (2005); F.E.E. 5: 277 (2006); F.T.E.A. Scroph.: 83 (2008).
Annual herb. Small pools in grassland or forest.
Distr: BAG, UN, EQU; Senegal to Ethiopia, S to South Africa; Madagascar, India to Vietnam.
EQU: Friis & Vollesen 716 (fl fr) 12/1980.

Lindernia pulchella (Skan) Philcox

Biol. Skr. 51: 427 (2005); F.E.E. 5: 277 (2006); F.T.E.A. Scroph.: 86 (2008).
Perennial herb. Seasonally wet flushes amongst rocks.
Distr: EQU; Ethiopia to Angola & South Africa.
EQU: Friis & Vollesen 581 (fl fr) 11/1980.

Lindernia schweinfurthii (Engl.) Dandy

F.P.S. 3: 139 (1956); Biol. Skr. 51: 427 (2005); F.E.E. 5: 275 (2006); F.T.E.A. Scroph.: 88 (2008).
Syn: *Lindernia madiensis* Dandy – F.P.S. 3: 139 (1956).
Annual herb. Seasonally wet flushes amongst rocks.
Distr: EQU; Senegal to Ethiopia, S to Zambia.
EQU: Myers 9164 (fl fr) 8/1938.

Lindernia sudanica Eb.Fisch. & Hepper

Kew Bull. 46: 530 (1991); Biol. Skr. 51: 427 (2005).
Perennial herb. Montane wet grassland or wet flushes amongst rocks.
Distr: EQU; South Sudan endemic.
IUCN: RD
EQU: Myers 11744 (fl fr) 8/1939.
Note: endemic to the Imatong Mts.

Stemodiopsis buchananii Skan var. buchananii

F.T.E.A. Scroph.: 39 (2008).
Perennial herb. Rock crevices.
Distr: EQU; Somalia to Zambia & Mozambique.
EQU: Andrews 854 (fl fr) 4/1939.
Note: Andrews lists *Stemodiopsis* sp. from near the summit of Mt Konyi near Loka – we have not seen this specimen and it was not mentioned in the revision of the genus by Fischer (Bot. Jahrb. Syst. 119: 305-326 (1997)).

Stemodiopsis rivae Engl.

Biol. Skr. 51: 429 (2005); F.E.E. 5: 261 (2006); F.T.E.A. Scroph.: 39 (2008).
Perennial herb. Rock crevices.
Distr: EQU; Cameroon, Ethiopia to South Africa.
EQU: Friis & Vollesen 700 (fl fr) 12/1980.

Torenia thouarsii (Cham. & Schltdl.) Kuntze

F.P.S. 3: 147 (1956); F.T.E.A. Scroph.: 63 (2008).
Annual herb. Streamsides, wet hollows, roadside depressions.
Distr: BAG, EQU; throughout tropical Africa; Madagascar, tropical Asia and America.
EQU: Myers 6935 (fl fr) 7/1937.

Pedaliaceae

I. Darbyshire

Ceratotheca sesamoides Endl.

F.T.E.A. Pedal.: 14 (1953); F.P.S. 3: 160 (1956); F.J.M.: 148 (1976); Biol. Skr. 51: 434 (2005).
Annual herb. Weed of cultivation & waste ground, sand dunes.
Distr: DAR, KOR, BAG, EQU; Senegal to Uganda, S to Zimbabwe & Mozambique.
KOR: Wickens 45 (fl fr) 7/1962; **EQU:** Myers 10195 (fl fr) 12/1938.

Pedalium murex L.

F.T.E.A. Pedal.: 6 (1953); F.E.E. 5: 343 (2006).
Annual herb. Wadis, temporary watercourses, grassland, on saline soils.
Distr: RS; Ghana to Somalia, S to Mozambique; Madagascar, Socotra, India, Sri Lanka.
RS: Newberry 219 1/1928.

Pterodiscus ruspolii Engl.

F.T.E.A. Pedal.: 8 (1953); F.E.E. 5: 339 (2006).
Perennial herb. *Acacia-Commiphora* bushland, grassland, semi-desert.
Distr: EQU; Ethiopia, Somalia, Kenya.
EQU: Myers 14025 (fl fr) 9/1941.

Rogeria adenophylla J.Gay ex Delile

F.P.S. 3: 160 (1956); F.E.E. 5: 344 (2006).
Annual herb. Temporarily wet sites in arid areas.
Distr: RS, DAR, KOR, CS, ES; Cape Verde Is., Senegal to Eritrea, Angola & Namibia.
RS: Bally 7005 (fl fr) 4/1949.

Sesamum alatum Thonn.

F.T.E.A. Pedal.: 17 (1953); F.P.S. 3: 162 (1956); F.E.E. 5: 340 (2006).
Annual herb. Open bushland, weed of cultivation & waste ground.
Distr: RS, DAR, KOR, CS; Senegal to Eritrea & Kenya, Namibia to Zimbabwe.
KOR: Jackson 4338 (fl fr) 9/1961.

Sesamum angustifolium (Oliv.) Engl.

F.T.E.A. Pedal.: 19 (1953); F.P.S. 3: 163 (1956); F.J.M.: 148 (1976); F.E.E. 5: 341 (2006).
Annual or perennial herb. Bushland & woodland, waste ground, old cultivation.
Distr: DAR, KOR, CS, BAG, UN, EQU; Ethiopia to D.R. Congo & Mozambique.
KOR: Wickens 391 (fl) 9/1962; **EQU:** Myers 7092 (fl) 7/1937.

Sesamum latifolium Gillett

F.T.E.A. Pedal.: 21 (1953); F.E.E. 5: 342 (2006).
Syn: *Sesamum radiatum* sensu Andrews, non Schumach. & Thonn. – F.P.S. 3: 163 (1956).
Annual herb. Waste ground, old cultivation, open bushland & scrub.
Distr: CS, UN, EQU; Ethiopia, Uganda, Kenya.
CS: Kotschy 520 (fr) 1837; **EQU:** Myers 9922 (fl fr) 10/1938.

Sesamum orientale L.

F.E.E. 5: 340 (2006).
Syn: *Sesamum indicum* L. – F.T.E.A. Pedal.: 17 (1953); Bot. Exp. Sud.: 93 (1970); F.J.M.: 148 (1976).
Annual herb. Waste ground, cultivation.
Distr: NS, RS, DAR, KOR, CS, BAG; native of India & Africa (not Sudan), widely cultivated & naturalised elsewhere.
NS: Schweinfurth 748 (fl fr) 10/1868; **BAG:** Macintosh 100K (fl) 1932.

Lamiaceae (Labiatae)

I. Darbyshire

Achyrospermum axillare E.A.Bruce

Biol. Skr. 51: 462 (2005); F.T.E.A. Lam.: 39 (2009).
Shrub or perennial herb. Forest.
Distr: EQU; Uganda.
IUCN: RD
EQU: Friis & Vollesen 709 (fl) 12/1980.

Achyrospermum parviflorum S.Moore

Biol. Skr. 51: 462 (2005); F.E.E. 5: 532 (2006); F.T.E.A. Lam.: 39 (2009).
Perennial herb. Forest.
Distr: EQU; Ethiopia to D.R. Congo & Tanzania.
EQU: Friis & Vollesen 247 (fl) 11/1980.

Achyrospermum schimperi (Hochst. ex Briq.) Perkins

Biol. Skr. 51: 462 (2005); F.E.E. 5: 531 (2006); F.T.E.A. Lam.: 37 (2009).
Perennial herb. Forest.
Distr: EQU; Ethiopia to Burundi.
EQU: Shigeta 153 1/1979 (n.v.).

Aeollanthus ambustus Oliv.

F.T.E.A. Lam.: 365 (2009).
Syn: *Aeollanthus virgatus* Gürke – F.P.S. 3: 205 (1956).
Annual herb. Rock outcrops on shallow soil.
Distr: BAG; D.R. Congo, Uganda.
IUCN: RD
BAG: Schweinfurth 2225 (fl) 8/1869.

Aeollanthus densiflorus Ryding

Biol. Skr. 51: 462 (2005); F.E.E. 5: 586 (2006); F.T.E.A. Lam.: 376 (2009).
Subshrub. Rock outcrops in montane grassland.
Distr: EQU; Ethiopia, Uganda, Kenya, Tanzania.
EQU: Andrews 1844 (fl) 6/1939.

Aeollanthus myrianthus Baker subsp. *myrianthus*

Biol. Skr. 51: 463 (2005); F.E.E. 5: 585 (2006); F.T.E.A. Lam.: 369 (2009).
Subshrub. Rocky hillslopes.
Distr: EQU; Ethiopia to Zambia.
EQU: Myers 9503 (fl fr) 10/1938.

Aeollanthus repens Oliv.

F.P.S. 3: 205 (1956); Biol. Skr. 51: 463 (2005); F.T.E.A. Lam.: 366 (2009).
Perennial or ?annual herb. Rock outcrops in montane grassland.
Distr: EQU; D.R. Congo, Uganda, Tanzania.
EQU: Myers 11573 (fl) 8/1939.

Aeollanthus suaveolens Spreng.

Biol. Skr. 51: 463 (2005); F.T.E.A. Lam.: 367 (2009).
Syn: *Aeollanthus heliotropioides* Oliv. – F.P.S. 3: 205 (1956).
Annual or short-lived perennial herb. Rock outcrops.
Distr: EQU; Nigeria to Uganda, S to South Africa.
EQU: Thomas 1596 (fl fr) 12/1935.

Ajuga integrifolia Buch.-Ham. ex D.Don

F.P.S. 3: 205 (1956); F.E.E. 5: 528 (2006); F.T.E.A. Lam.: 15 (2009).
Perennial herb. Grassland or wooded grassland, ditches.
Distr: ?UN; Ethiopia to Tanzania; Arabia to E Asia.
Note: no material seen from South Sudan, but recorded by Andrews from Meshra el Zeraf, this species is unlikely to be mistaken.

Basilicum polystachyon (L.) Moench.

F.P.S. 3: 205 (1956); F.J.M.: 151 (1976); Biol. Skr. 51: 463 (2005); F.E.E. 5: 581 (2006); F.T.E.A. Lam.: 194 (2009).
Annual or short-lived perennial herb. Waste ground, seasonally flooded areas in woodland & bushland.
Distr: DAR, KOR, CS, ES, BAG, UN, EQU; palaeotropical.
DAR: Wickens 1634 (fl fr) 5/1964; **BAG:** Andrews 649 (fl fr) 4/1939.

Clerodendrum acerbianum (Vis.) Benth.

F.P.S. 3: 194 (1956); T.S.S.: 437 (1990); F.T.E.A. Verb.: 121 (1992); F.E.E. 5: 526 (2006).
Shrub, sometimes climbing. *Acacia* woodland, thicket, riverine scrub.
Distr: NS, CS, UN; Gambia, Guinea Bissau, Ethiopia & Somalia, S to Tanzania; Egypt.
CS: Andrews 80 (fl) 6/1935; **UN:** Simpson 7017 (fl fr) 6/1929.

Clerodendrum capitatum (Willd.) Schumach. & Thonn.

F.P.S. 3: 195 (1956); T.S.S.: 438 (1990); F.T.E.A. Verb.: 103 (1992); Biol. Skr. 51: 457 (2005).
Climbing or scrambling shrub. Forest, dense riverine thicket.
Distr: KOR, EQU; Gambia to Uganda, S to Zambia & Angola.
KOR: Wickens 776 (fl) 11/1962; **EQU:** Jackson 3804 (fl) 8/1957.

Clerodendrum formicarum Gürke

F.P.S. 3: 194 (1956); T.S.S.: 438 (1990); F.T.E.A. Verb.: 106 (1992).
Climbing shrub. Forest.
Distr: EQU; Guinea to Uganda, S to Angola & Zambia.
EQU: Sillitoe 267 (fl) 1919.

Clerodendrum johnstonii Oliv. subsp. johnstonii

F.T.E.A. Verb.: 118 (1992); Biol. Skr. 51: 457 (2005); F.E.E. 5: 523 (2006).
Shrub or small tree. Forest, forest margins & secondary thicket.
Distr: EQU; Ethiopia & Uganda, S to Zambia & Malawi.
EQU: Myers 11714 (fl) 8/1939.

Clerodendrum melanocrater Gürke

F.P.S. 3: 194 (1956); T.S.S.: 438 (1990); F.T.E.A. Verb.: 113 (1992).
Climbing shrub. Riverine & swamp forest, including margins.
Distr: EQU; Bioko, Cameroon, D.R. Congo, Rwanda, Uganda, Kenya, Tanzania.
EQU: Myers 9200 (fl) 8/1938.

Clerodendrum poggei Gürke

F.P.S. 3: 195 (1956); F.T.E.A. Verb.: 100 (1992).
Shrub or small tree. Forest.
Distr: ?EQU; Cameroon to Uganda & ?Ethiopia, S to Angola & Tanzania.
Note: Andrews records this species from Equatoria; we have seen no material with which to verify this, but Verdcourt in F.T.E.A. includes Sudan within its distribution and, since this is a distinctive species, the record is likely to be correct.

Clerodendrum rotundifolium Oliv.

F.T.E.A. Verb.: 99 (1992); Biol. Skr. 51: 458 (2005).
Shrub, sometimes climbing. Forest edges, wooded grassland & bushland, riverine thicket.
Distr: EQU; D.R. Congo, Uganda, S to Malawi & Mozambique.
EQU: Andrews 1028 (fl) 5/1939.

Clerodendrum schweinfurthii Gürke

F.P.S. 3: 194 (1956); T.S.S.: 439 (1990); F.T.E.A. Verb.: 112 (1992); Biol. Skr. 51: 458 (2005).
Woody climber. Riverine & swamp forest.
Distr: EQU; Sierra Leone to Uganda, S to Angola & Tanzania.
EQU: Myers 10580 (fl) 3/1939.

Clerodendrum triflorum Vis.

F.P.S. 3: 194 (1956); T.S.S.: 439 (1990).
Shrub, sometimes climbing. Termite mounds, wooded grassland.
Distr: CS, ES, UN; Sudan / South Sudan endemic.
IUCN: RD
CS: Lea 211 (fl) 8/1954; **UN:** J.M. Lock 81/228 (fl) 8/1981.
Note: *C. harnierianum* Schweinf. is listed as an accepted name in the World Checklist Series of Selected Plant Families (Govaerts 2003), but we follow Andrews in treating this as a synonym of *C. triflorum*.

Clerodendrum umbellatum Poir.

F.J.M.: 150 (1976); T.S.S.: 439 (1990); F.T.E.A. Verb.: 97 (1992); Biol. Skr. 51: 458 (2005); F.E.E. 5: 523 (2006).
Syn: *Clerodendrum cordifolium* (Hochst.) A.Rich. – F.P.S. 3: 195 (1956); Bot. Exp. Sud.: 110 (1970).
Subshrub or herbaceous climber. Wooded grassland, old cultivation, riverine thicket.
Distr: DAR, KOR, CS, BAG, EQU; Senegal to Ethiopia, S to Tanzania.
DAR: Lynes 568 (fl fr) 7/1921; **EQU:** Friis & Vollesen 997 (fl) 2/1982.

Clinopodium abyssinicum (Benth.) Kuntze var. abyssinicum

F.T.E.A. Lam.: 121 (2009).
Syn: *Micromeria abyssinica* Benth. – F.P.S. 3: 217 (1956).
Syn: *Satureja abyssinica* (Benth.) Briq. – Biol. Skr. 51: 471 (2005).
Annual or perennial herb or subshrub. Montane & mid-altitude grassland & bushland on stony ground.
Distr: RS, EQU; Ethiopia & Somalia, S to Tanzania; Arabia.
RS: Jackson 2895 (fl fr) 4/1953; **EQU:** Friis & Vollesen 289 (fl) 11/1980.

Clinopodium uhligii (Gürke) Ryding var. obtusifolium (Avetta) Ryding

F.T.E.A. Lam.: 123 (2009).
Syn: *Micromeria pseudosimensis* Brenan – F.E.E. 5: 555 (2006).
Syn: *Calamintha* sp. sensu Andrews – F.P.S. 3: 206 (1956).
Syn: *Satureja simensis* sensu Friis & Vollesen, non (Benth.) Briq. – Biol. Skr. 51: 472 (2005).
Perennial herb or subshrub. Montane forest-grassland transition, streamsides.
Distr: EQU; Bioko, Nigeria, Cameroon, Ethiopia to Mozambique.
EQU: Friis & Vollesen 132 (fl) 11/1980.

Endostemon tenuiflorus (Benth.) M.Ashby

F.E.E. 5: 579 (2006); F.T.E.A. Lam.: 205 (2009).
Annual or short-lived perennial herb. Dry rocky hillslopes & stony ground.

Distr: RS; Ethiopia to South Africa; Madagascar, Arabia & Socotra.
RS: Jackson 2826 (fl fr) 4/1953.
Note: Andrews (F.P.S. 3: 209) records also *E. gracilis* (Benth.) M.Ashby based upon *Andrews* 80 from the Nuba Mts, Kordofan. However, this specimen was mislabelled (the label data are for a grass species) and in view of the distribution of *E. gracilis*, it being a species of the coastal lowlands of E Africa, it is unlikely to occur in Sudan. This record was omitted from the revision of the genus by Paton et al. in Kew Bull. 49: 701 (1994).

Endostemon tereticaulis (Poir.) M.Ashby
F.P.S. 3: 209 (1956); F.E.E. 5: 579 (2006); F.T.E.A. Lam.: 206 (2009).
Annual or short-lived perennial herb. Dry open woodland & bushland.
Distr: DAR, ?KOR; Senegal to Somalia, S to South Africa; Arabia.
DAR: Aglen 51 (fl fr) 8/1944.

Fuerstia bartsioides (Baker) G.Taylor
J. Bot. 70: 272 (1932).
Syn: *Orthosiphon bartsioides* Baker – F.P.S. 3: 220 (1956).
Perennial herb. Habitat unknown.
Distr: EQU; South Sudan endemic.
IUCN: RD
EQU: Schweinfurth 2932 (fl) 2/1870.

[Gmelina arborea Roxb.]
Note: a native of tropical Asia, widely planted in both Kordofan and Equatoria for forestry (El Amin 1990: 440). It is not known to naturalise.

Haumaniastrum caeruleum (Oliv.) P.A.Duvign. & Plancke
F.J.M.: 151 (1976); F.E.E. 5: 569 (2006); F.T.E.A. Lam.: 213 (2009).
Syn: *Acrocephalus lilacinus* Oliv. – F.P.S. 3: 204 (1956).
Syn: *Acrocephalus schweinfurthii* Briq. – F.P.S. 3: 204 (1956).
Perennial herb. Rocky grassland, open wooded grassland.
Distr: DAR, BAG, EQU; Senegal to Ethiopia, S to Angola & Zimbabwe.
DAR: Pettet 191 (fl) 12/1962; **EQU:** Kosper 116 (fl) 10/1982.

Haumaniastrum villosum (Benth.) A.J.Paton
Biol. Skr. 51: 465 (2005); F.E.E. 5: 569 (2006); F.T.E.A. Lam.: 220 (2009).
Syn: *Acrocephalus galeopsifolius* Baker – F.P.S. 3: 204 (1956).
Syn: *Haumaniastrum galeopsifolium* (Baker) P.A.Duvign. & Plancke – F.J.M.: 151 (1976); F.T.E.A. Lam.: 222 (2009).
Annual or short-lived perennial herb. Moist grassland & woodland, forest edges, disturbed ground.
Distr: DAR, EQU; Guinea to Ethiopia, S to Angola & Mozambique; Madagascar.
DAR: Wickens 2515B (fl) 9/1964; **EQU:** Friis & Vollesen 292 (fl) 11/1980.

Hoslundia opposita Vahl
F.P.S. 3: 210 (1956); F.J.M.: 151 (1976); Biol. Skr. 51: 465 (2005); F.E.E. 5: 584 (2006); F.T.E.A. Lam.: 193 (2009).
Shrub or perennial herb. Open woodland, forest margins, grassland.
Distr: DAR, KOR, BAG, UN, EQU; Senegal to Eritrea, S to South Africa; Madagascar.
KOR: Musselman 6271 (fl fr) 8/1983; **BAG:** Simpson 7133 (fl fr) 6/1929.

Hyptis lanceolata Poir.
F.P.S. 3: 210 (1956); F.T.E.A. Lam.: 137 (2009).
Annual or short-lived perennial herb. Riverbanks, swamp margins, farmbush.
Distr: EQU; native to the Americas, naturalised in W Africa, extending to South Sudan & S to Malawi.
EQU: Sillitoe 490 (fl fr) 1919.

Hyptis pectinata (L.) Poit.
F.P.S. 3: 212 (1956); F.J.M.: 151 (1976); F.E.E. 5: 566 (2006); F.T.E.A. Lam.: 134 (2009).
Annual or short-lived perennial herb. Damp areas in grassland & thicket, riverbanks.
Distr: DAR, EQU; native to S America, widely naturalised in tropical Africa.
DAR: Wickens 2970 (fl fr) 5/1965; **EQU:** Jackson 3013 (fl fr) 6/1953.

Hyptis spicigera Lam.
F.P.S. 3: 210 (1956); Biol. Skr. 51: 465 (2005); F.E.E. 5: 567 (2006); F.T.E.A. Lam.: 133 (2009).
Annual herb. Wet ground by rivers & lakes, swamps, weed of cultivation.
Distr: KOR, CS, EQU; native to South America, widely naturalised in tropical Africa.
KOR: Andrews 61 (fl fr) 11/1933; **EQU:** Myers 8001 (fr) 12/1937.
Note: cultivated for its edible seeds.

Isodon ramosissimus (Hook.f.) Codd
Biol. Skr. 51: 465 (2005); F.E.E. 5: 565 (2006); F.T.E.A. Lam.: 139 (2009).
Syn: *Plectranthus ramosissimus* Hook.f. – F.P.S. 3: 223 (1956); Bot. Exp. Sud.: 63 (1970).
Perennial herb. Forest margins & clearings, secondary bushland.
Distr: EQU; Sierra Leone to Ethiopia, S to Angola & Zimbabwe.
EQU: Friis & Vollesen 176 (fl) 11/1980.

Isodon schimperi (Vatke) J.K.Morton
Biol. Skr. 51: 466 (2005); F.E.E. 5: 566 (2006); F.T.E.A. Lam.: 132 (2009).
Perennial herb. Forest margins & clearings.
Distr: EQU; Rwanda, Burundi, Ethiopia, Uganda.
EQU: Friis & Vollesen 85 (fl fr) 11/1980.

Lavandula antineae Maire subsp. marrana Upson & Jury
Kew Bull. 58: 895 (2003).
Syn: *Lavandula pubescens* sensu auctt., non Decne. – F.P.S. 3: 212 (1956); F.J.M.: 151 (1976); F. Darfur Nord-Occ. & J. Gourgeil: 132 (1969).
Perennial herb. Montane grassland on volcanic soils.

Distr: DAR; Chad.
IUCN: RD
DAR: Wickens 1224 (fl fr) 2/1964.
Note: locally abundant in suitable habitat on Jebel Marra.

Lavandula coronopifolia Poir.

F.P.S. 3: 212 (1956); F.E.E. 5: 564 (2006).
Syn: *Lavandula stricta* Delile – Bot. Exp. Sud.: 62 (1970).
Perennial herb or subshrub. Dry rocky areas, open bushland.
Distr: RS, DAR; Eritrea, Ethiopia; Madeira, N Africa, Arabia, Iran.
RS: Jackson 2690 (fl fr) 3/1953.

Lavandula saharica Upson & Jury

Kew Bull. 58: 896 (2003).
Syn: *Lavandula antineae* sensu Léonard, non Maire – F.J.U. 2: 181 (1999).
Perennial herb. Sandstone gorges in desert.
Distr: NS; Algeria, Libya.
IUCN: RD
NS: Léonard 4912 (fl) 11/1968.
Note: the locality in Northern Sudan, Djebel Uweinat, straddles the Sudan-Libya-Egypt border.

Leonotis nepetifolia (L.) R.Br. var. *nepetifolia*

F.P.S. 3: 212 (1956); Biol. Skr. 51: 466 (2005); F.E.E. 5: 549 (2006); F.T.E.A. Lam.: 105 (2009).
Annual or short-lived perennial herb. Disturbed grassland & bushland, waste ground.
Distr: UN; Sierra Leone to Kenya, S to Namibia & South Africa.
Notes: (1) in their revision of *Leonotis*, Iwarsson & Harvey (Kew Bull. 58: 635 (2003)) record a single Sudanese locality for this variety but do not cite the specimen; we have not seen this material. (2) we have not seen the specimens listed by Friis & Vollesen; they may refer to var. *africana*.

Leonotis nepetifolia (L.) R.Br. var. *africana* (P.Beauv.) J.K.Morton

F.J.M.: 151 (1976); F.E.E. 5: 549 (2006); F.T.E.A. Lam.: 105 (2009).
Syn: *Leonotis africana* (P.Beauv.) Briq. – F.P.S. 3: 212 (1956).
Annual or short-lived perennial herb. Weed of cultivation & waste ground.
Distr: NS, DAR, KOR, CS, ES, BAG, UN, EQU; Senegal to Ethiopia, S to Zambia & Botswana.
KOR: Wickens 335 (fl) 9/1962; **UN:** Sherif A.2854 (fl) 10/1951.

Leonotis ocymifolia (Burm.f.) Iwarsson var. *raineriana* (Vis.) Iwarsson

Biol. Skr. 51: 466 (2005); F.E.E. 5: 548 (2006); F.T.E.A. Lam.: 103 (2009).
Syn: *Leonotis raineriana* Vis. – F.P.S. 3: 213 (1956).
Shrub. Forest margins, montane bushland & grassland, disturbed ground.
Distr: CS, EQU; Eritrea, S to Angola & South Africa.
EQU: Thomas 1620 (fl) 12/1935.
Note: the record for Central Sudan is based upon the distribution map for this taxon in the revision of the genus by Iwarsson & Harvey (Kew Bull. 58: 626 (2003)).

Leucas calostachyus Oliv. var. *calostachyus*

F.P.S. 3: 215 (1956); Biol. Skr. 51: 466 (2005); F.E.E. 5: 545 (2006); F.T.E.A. Lam.: 69 (2009).
Shrub. Open wooded grassland, secondary bushland, forest margins.
Distr: EQU; Ethiopia to Tanzania.
EQU: Thomas 1894 (fl fr) 12/1935.

Leucas calostachyus Oliv. var. *schweinfurthii* (Gürke) Sebald

F.T.E.A. Lam.: 70 (2009).
Syn: *Leucas schweinfurthii* Gürke – F.P.S. 3: 215 (1956).
Shrub. Grassland.
Distr: EQU; D.R. Congo, Uganda.
EQU: Johnston 1400 (fl fr) 2/1936.
Note: a local variety, but not considered threatened.

Leucas deflexa Hook.f. var. *deflexa*

Biol. Skr. 51: 467 (2005); F.E.E. 5: 546 (2006); F.T.E.A. Lam.: 73 (2009).
Annual or short-lived perennial herb or subshrub. Forest margins.
Distr: EQU; Cameroon, D.R. Congo to Ethiopia, S to Angola & Zimbabwe.
EQU: Friis & Vollesen 212 (fl fr) 11/1980.

Leucas glabrata (Vahl) R.Br. ex Sm.

F.E.E. 5: 545 (2006); F.T.E.A. Lam.: 65 (2009).
Syn: *Leucas paucijuga* Baker – F.P.S. 3: 215 (1956); Bot. Exp. Sud.: 62 (1970).
Perennial herb or subshrub, rarely annual herb. Variously grassland to dry bushland or moist woodland.
Distr: RS; Eritrea & Somalia, S to Namibia & South Africa; Arabia.
RS: Aylmer 160 (fl) 3/1932.

Leucas inflata Benth.

F.P.S. 3: 213 (1956); F.E.E. 5: 542 (2006).
Subshrub. Dry bushland, rocky slopes.
Distr: RS; Eritrea, Ethiopia, Somalia; Egypt, Arabia.
RS: Carter 1857 (fl) 11/1987.

Leucas martinicensis (Jacq.) Ait.f.

F.P.S. 3: 215 (1956); F.J.M.: 152 (1976); F.E.E. 5: 547 (2006); F.T.E.A. Lam.: 94 (2009).
Annual herb. Weed of cultivation, disturbed grassland & woodland.
Distr: RS, DAR, KOR, CS, ES, BAG, UN, EQU; throughout tropical Africa; Madagascar, Yemen.
DAR: Wickens 1765 (fl fr) 9/1964; **BAG:** Schweinfurth 2406 (fl) 9/1869.

Leucas neuflizeana Courbon

F.P.S. 3: 215 (1956); F.E.E. 5: 542 (2006); F.T.E.A. Lam.: 58 (2009).
Annual or short-lived perennial herb. Grassland, dry bushland, rocky hillslopes.
Distr: RS; Eritrea to Namibia & South Africa; Egypt, Arabia, Socotra.
RS: Bent s.n. (fl fr) 1896.

Leucas nubica Benth.

F.P.S. 3: 213 (1956); F.E.E. 5: 545 (2006); F.T.E.A. Lam.: 64 (2009).

Annual herb. Dry grassland & bushland; weed of cultivation.
Distr: KOR; Ethiopia, Somalia, Kenya, Tanzania.
KOR: <u>Kotschy 111</u> (fl fr) 10/1839.

Leucas urticifolia (Vahl) R.Br. ex Sm. var. *urticifolia*

F.P.S. 3: 216 (1956); F.E.E. 5: 546 (2006); F.T.E.A. Lam.: 79 (2009).
Annual herb. Dry bushland; weed of cultivation.
Distr: RS, CS; Eritrea, Ethiopia, Somalia, Kenya; Arabia, Socotra, India.
RS: <u>Bent s.n.</u> (fl fr) 1896.

Mentha longifolia Briq. subsp. *longifolia*

F.J.M.: 152 (1976); F.E.E. 5: 551 (2006); F.T.E.A. Lam.: 115 (2009).
Syn: *Mentha longifolia* Briq. subsp. *schimperi* (Briq.) Briq. – F.P.S. 3: 217 (1956); Bot. Exp. Sud.: 62 (1970).
Perennial herb. Montane grassland, often in wet areas, streamsides.
Distr: DAR; Ethiopia to South Africa; Europe to N India & China.
DAR: <u>Robertson 112</u> (fl) 12/1957.

Micromeria imbricata (Forssk.) C.Chr. var. *imbricata*

F.T.E.A. Lam.: 117 (2009).
Syn: *Satureja imbricata* (Forssk.) Briq. – Biol. Skr. 51: 472 (2005); F.E.E. 5: 556 (2006).
Syn: *Satureja punctata* (Benth.) Briq. – F.J.M.: 152 (1976); F.E.E. 5: 557 (2006).
Syn: *Micromeria biflora* sensu auctt., non (Buch.-Ham. ex D.Don) Benth. – F.P.S. 3: 217 (1956); Bot. Exp. Sud.: 63 (1970); F. Darfur Nord-Occ. & J. Gourgeil: 131 (1969).
Perennial herb or subshrub. Dry grassland open woodland, often montane.
Distr: RS, DAR, EQU; Bioko, Nigeria, Cameroon, Eritrea to Somalia, S to Angola & South Africa; Egypt, Arabia, Socotra.
RS: <u>Jackson 2873</u> (fl) 4/1953; **EQU:** <u>Myers 11687</u> (fl fr) 7/1939.
Note: a very variable species; several names have been applied in the past on herbarium material from Sudan.

Nepeta azurea R.Br. ex Benth.

F.E.E. 5: 549 (2006); F.T.E.A. Lam.: 128 (2009).
Syn: *Nepeta ballotifolia* Hochst. ex A.Rich. – F.P.S. 3: 217 (1956); F.J.M.: 152 (1976).
Perennial herb. Montane grassland & open bushland, often in seasonally wet areas.
Distr: DAR; Eritrea to Tanzania; Arabia.
DAR: <u>Robertson 123</u> (fl) 12/1957.
Note: we here follow the circumscription of *N. azurea* in F.E.E. in which *N. ballotifolia* is treated in synonymy; Polhill in F.T.E.A. Lam. takes a rather narrower view of *N. azurea* and does not list Sudan in the distribution; presumably he considers *N. ballotifolia* to be a good species.

Nepeta sudanica F.W.Andrews

F.P.S. 3: 217 (1956).
Perennial herb or subshrub. Dry hillslopes.
Distr: RS; Sudan endemic.

IUCN: RD
RS: <u>Farag & Zubair 2</u> (fl) 2/2012.
Note: the distribution of this species extends into the Hala'ib Triangle.

Ocimum americanum L.

F.P.S. 3: 218 (1956); Biol. Skr. 51: 467 (2005); F.E.E. 5: 571 (2006); F.T.E.A. Lam.: 147 (2009).
Syn: *Ocimum dichotomum* Benth. – F.P.S. 3: 218 (1956); Bot. Exp. Sud.: 63 (1970).
Syn: *Ocimum hadiense* sensu Wickens pro parte, non Forssk. – F.J.M.: 152 (1976).
Annual or short-lived perennial herb. Disturbed & cultivated ground, grassland & bushland.
Distr: RS, DAR, KOR, CS, ES, BAG, UN, EQU; largely pantropical.
DAR: <u>Wickens 1148</u> (fl fr) 2/1964; **UN:** <u>Evans-Pritchard 33</u> (fl fr) 11/1936.

Ocimum basilicum L.

F.P.S. 3: 218 (1956); F.E.E. 5: 571 (2006); F.T.E.A. Lam.: 146 (2009).
Annual or short-lived perennial herb. Disturbed ground & grassland, often seasonally flooded.
Distr: NS, DAR, KOR, CS, ES; pantropical through cultivation and becoming naturalised, ?native to Ethiopia.
CS: <u>Lea 68</u> (fr) 10/1952.
Note: the varieties recognised in F.E.E. are not upheld here. *O. kilimandscharicum* Gürke is also cultivated in Khartoum but is not thought to have naturalised in Sudan.

Ocimum forskolei Benth.

F.E.E. 5: 571 (2006); F.T.E.A. Lam.: 149 (2009).
Syn: *Ocimum staminosum* Baker – F.P.S. 3: 220 (1956).
Syn: *Ocimum hadiense* sensu auctt., non Forssk. – F.P.S. 3: 218 (1956); F.J.M.: 152 pro parte (1976); F. Darfur Nord-Occ. & J. Gourgeil: 132 (1969).
Perennial herb. Rocky slopes, disturbed or grazed ground, bushland.
Distr: RS, DAR, KOR, CS; Eritrea to Kenya; Egypt, Arabia.
RS: <u>Carter 1991</u> (fl) 11/1987.
Note: Sebold (Stuttgarter Beitr. Naturk, ser. A 419: 1-74 (1988)) recorded *O. filamentosum* Forssk. (under *Becium*) from Sudan, citing the Helsinki specimen of *Ahti* 16435 from Red Sea region. However, the Kew specimen of this number has been redetermined as *Orthosiphon pallidus*, having been originally labelled *Becium* sp. nov. aff. *filamentosum* by Wickens and cited as such by Ahti et al. in Ann. Bot. Fennici 10: 155 (1973). The two species look rather similar and so the Helsinki specimen is believed to have been misidentified. Hence, whilst *O. filamentosum* is likely to occur in Sudan, there are no confirmed records to date.

Ocimum gratissimum L. subsp. *gratissimum* var. *gratissimum*

Biol. Skr. 51: 468 (2005); F.T.E.A. Lam.: 152 (2009).
Syn: *Ocimum suave* Willd. – F.P.S. 3: 218 (1956); F.J.M.: 152 (1976).
Perennial herb or shrub. Woodland, montane forest margins, disturbed areas.
Distr: DAR, EQU; tropical & subtropical Africa; India, naturalised in tropical America.

DAR: <u>Kassas 130</u> (n.v.); **EQU:** <u>Myers 9711</u> (fl fr) 10/1938. Note: we have not seen the material from Darfur, cited in F.J.M. under *O. suave*.

Ocimum obovatum E.Mey. ex Benth. subsp. *obovatum* var. *obovatum*

F.T.E.A. Lam.: 167 (2009).
Syn: *Becium affine* (Hochst. ex Benth.) Chiov. – F.P.S. 3: 206 (1956); Bot. Exp. Sud.: 62 (1970).
Syn: *Becium obovatum* (E.Mey. ex Benth.) N.E.Br. – F.P.S. 3: 206 (1956); F.J.M.: 151 (1976); Biol. Skr. 51: 464 (2005); F.E.E. 5: 577 (2006).
Syn: *Becium schweinfurthii* (Briq.) N.E.Br. ex Broun & Massey – F.P.S. 3: 206 (1956).
Syn: *Becium knyanum* sensu Andrews, non (Vatke) N.E.Br. – F.P.S. 3: 206 (1956).
Perennial herb or shrub. Fire-prone grassland & wooded grassland.
Distr: DAR, BAG, EQU; Guinea to Eritrea, S to South Africa.
DAR: <u>Wickens 1964</u> (fl) 7/1964; **EQU:** <u>Thomas 1652</u> (fl fr) 12/1935.
Note: the record of *Becium knyanum* by Andrews is believed to refer to *Schweinfurth* 1438 from 'Gir' which has been redetermined by A. Paton as *O. obovatum*.

Orthosiphon pallidus Royle ex Benth.

F.P.S. 3: 221 (1956); F.E.E. 5: 582 (2006); F.T.E.A. Lam.: 187 (2009).
Annual or perennial herb. Dry bushland, sometimes in seasonally waterlogged areas.
Distr: RS; Mali to Eritrea, S to Tanzania; Madagascar, Egypt, Arabia to India.
RS: <u>Ahti 16435</u> (fl fr) 10/1962.

Orthosiphon schimperi Benth.

F.P.S. 3: 221 (1956); F.E.E. 5: 582 (2006); F.T.E.A. Lam.: 182 (2009).
Syn: *Orthosiphon rubicundus* sensu Friis & Vollesen, non (D.Don) Benth. – Biol. Skr. 51: 468 (2005).
Perennial herb. Fire-prone grassland & wooded grassland, often in damp areas.
Distr: EQU; Ivory Coast to Ethiopia, S to South Africa.
EQU: <u>Myers 10921</u> (fl) 4/1939.

Orthosiphon thymiflorus (Roth) Sleesen

F.E.E. 5: 582 (2006); F.T.E.A. Lam.: 185 (2009).
Syn: *Orthosiphon australis* Vatke – F.P.S. 3: 221 (1956).
Syn: *Orthosiphon roseus* Briq. – F.P.S. 3: 221 (1956).
Perennial herb or subshrub. Wooded grassland & bushland, disturbed ground.
Distr: BAG, EQU; palaeotropical.
EQU: <u>Andrews 585</u> (fl fr) 4/1939.

Otostegia fruticosa (Forssk.) Schweinf. ex Penzig subsp. *fruticosa*

F.P.S. 3: 221 (1956); F.J.M.: 152 (1976); F.E.E. 5: 540 (2006).
Shrub. Rocky slopes, bushland.
Distr: RS, DAR; Cameroon, Eritrea, Ethiopia; Arabia.
DAR: <u>Kamil 1191</u> (fl) 5/1968.

Otostegia fruticosa (Forssk.) Schweinf. ex Penzig subsp. *schimperi* (Benth.) Sebald

F.E.E. 5: 540 (2006).
Shrub. Rocky slopes, dry bushland, wadis.
Distr: RS; Eritrea; Egypt, Arabia.
RS: <u>Robbie 43</u> (fl).

Otostegia tomentosa A.Rich. subsp. *ambigens* (Chiov.) Sebald

F.E.E. 5: 539 (2006).
Subshrub or shrub. Montane bushland.
Distr: CS; Ethiopia.
CS: <u>Cienkowsky 137</u> (fr) 4/1848.
Note: Andrews records *O. tomentosa* from the Red Sea Hills; this may be a misidentification – certainly, Sebald (Stuttgarter Beitr. Naturk. A 263: 1-84 (1973)) does not record this species from the Red Sea region.

Platostoma africanum P.Beauv.

F.P.S. 3: 222 (1956); Biol. Skr. 51: 468 (2005); F.E.E. 5: 567 (2006); F.T.E.A. Lam.: 228 (2009).
Annual herb. Forest & bushland, particularly along margins, often in seasonally wet areas.
Distr: BAG, EQU; Senegal to Ethiopia, S to Zambia & Malawi; India, Lesser Sunda Is.
EQU: <u>Andrews 1553</u> (fl fr) 5/1939.

Platostoma rotundifolium (Briq.) A.J.Paton

Biol. Skr. 51: 468 (2005); F.E.E. 5: 567 (2006); F.T.E.A. Lam.: 231 (2009).
Syn: *Geniosporum paludosum* Baker – F.P.S. 3: 210 (1956).
Perennial herb or shrub. Grassland including seasonally flooded areas, secondary scrub.
Distr: EQU; Sierra Leone to Ethiopia, S to South Africa.
EQU: <u>Myers 9751</u> (fl) 10/1938.

Plectranthus aegyptiacus (Forssk.) C.Chr.

F.T.E.A. Lam.: 322 (2009).
Syn: *Plectranthus tenuiflorus* (Vatke) Agnew – F.E.E. 5: 593 (2006).
Perennial herb or shrub. Dry woodland & thicket, amongst rocks.
Distr: RS; Eritrea & Somalia, S to Tanzania; Arabia.
RS: <u>Kassas 77</u> (fl) 4/1954.
Note: this record is based upon *Kassas* 77 at KHU, identified as *Coleus ghindanus* Baker which is a synonym of *P. aegyptiacus*.

Plectranthus alpinus (Vatke) Ryding

Biol. Skr. 51: 469 (2005); F.E.E. 5: 597 (2006); F.T.E.A. Lam.: 296 (2009).
Perennial herb. Moist forest, particularly margins & clearings.
Distr: EQU; Nigeria to Ethiopia, S to Malawi.
EQU: <u>Kielland-Lund 468</u> 12/1983 (n.v.).
Note: we have not seen the cited specimen; identification follows Friis & Vollesen (2005).

Plectranthus autranii (Briq.) A.J.Paton

F.T.E.A. Lam.: 326 (2009).
Syn: *Coleus darfurensis* R.D.Good – F.P.S. 3: 209 (1956); Bot. Exp. Sud.: 62 (1970).

Syn: *Solenostemon autranii* (Briq.) J.K.Morton – Biol. Skr. 51: 473 (2005); F.E.E. 5: 599 (2006).
Perennial herb or shrub. Forest, riverbanks, waterfall margins.
Distr: DAR, EQU; Ethiopia to Malawi, Mozambique & South Africa.
DAR: de Wilde et al. 5363 (fl fr) 1/1965; **EQU:** Johnston 1479 (fr) 2/1936.

Plectranthus barbatus Andr.
F.E.E. 5: 589 (2006); F.T.E.A. Lam.: 338 (2009).
Syn: *Coleus barbatus* (Andr.) Benth. – F.P.S. 3: 208 (1956); Bot. Exp. Sud.: 62 (1970).
Perennial herb or shrub. Dry grassland & woodland, abandoned cultivation.
Distr: RS; palaeotropical.
RS: Jackson 2710 (fr) 3/1953.
Note: Paton in F.T.E.A. records both var. *barbatus* and var. *grandis* from Sudan but the material at Kew is not labelled to variety.

Plectranthus bojeri (Benth.) Hedge
F.T.E.A. Lam.: 327 (2009).
Syn: *Coleus briquetii* Baker – F.P.S. 3: 209 (1956).
Syn: *Coleus latifolius* Hochst. ex Benth. – F.P.S. 3: 209 (1956).
Syn: *Solenostemon latifolius* (Hochst. ex Benth.) J.K.Morton – F.J.M.: 152 (1976); Biol. Skr. 51: 473 (2005); F.E.E. 5: 599 (2006).
Annual or short-lived perennial herb. Streamside in forest, woodland, grassland.
Distr: DAR, BAG, UN, EQU; tropical & southern Africa; Madagascar.
DAR: Wickens 3653 (fl) 6/1977; **EQU:** Myers 7756 (fl fr) 9/1937.

Plectranthus caninus Roth subsp. *flavovirens* (Gürke) A.J.Paton
F.J.M.: 152 (1976); Biol. Skr. 51: 469 (2005); F.E.E. 5: 588 (2006); F.T.E.A. Lam.: 345 (2009).
Annual herb. Rock outcrops, eroded grassland.
Distr: DAR, EQU; Ethiopia & Somalia, S to Namibia & Zimbabwe.
DAR: Wickens 2766 (fr) 10/1964; **EQU:** Kielland-Lund 30 11/1983 (n.v.).

Plectranthus defoliatus Hochst. ex Benth.
F.P.S. 3: 223 (1956); Biol. Skr. 51: 469 (2005); F.E.E. 5: 590 (2006); F.T.E.A. Lam.: 296 (2009).
Perennial herb or shrub. Margins and clearings in woodland & forest.
Distr: EQU; Ethiopia to Angola & Zambia.
EQU: Thomas 1629 (fl) 12/1935.

Plectranthus djalonensis (A.Chev.) A.J.Paton
F.T.E.A. Lam.: 286 (2009).
Syn: *Englerastrum schweinfurthii* Briq. – F.P.S. 3: 210 (1956); Biol. Skr. 51: 464 (2005); F.E.E. 5: 601 (2006).
Annual herb. Woodland & wooded grassland, often on damp soils.
Distr: EQU; Senegal to Ethiopia, S to Namibia & Mozambique.
EQU: Friis & Vollesen 530 (fl fr) 11/1980.

Plectranthus dupuisii (Briq.) A.J.Paton
F.T.E.A. Lam.: 329 (2009).
Syn: *Solenostemon porpeodon* (Baker) J.K.Morton – Biol. Skr. 51: 473 (2005); F.E.E. 5: 600 (2006).
Perennial herb or shrub. Wet rocks in forest & woodland including steamsides & torrents.
Distr: EQU; Ethiopia to Angola, Zimbabwe & South Africa.
EQU: Friis & Vollesen 154 (fl) 11/1980.

Plectranthus glandulosus Hook.f.
Biol. Skr. 51: 470 (2005); F.T.E.A. Lam.: 249 (2009).
Perennial herb. Forest & moist woodland.
Distr: EQU; Guinea-Bissau, Guinea, Mali, Nigeria to Ethiopia, Angola.
EQU: Friis & Vollesen 599 (fl) 12/1980.

Plectranthus gracillimus (T.C.E.Fr.) Hutch. & Dandy
Biol. Skr. 51: 470 (2005); F.T.E.A. Lam.: 290 (2009).
Annual herb. Rock crevices in *Terminalia* woodland.
Distr: EQU; Mali to D.R. Congo, S to Angola & Zimbabwe.
EQU: Friis & Vollesen 664 (fl fr) 12/1980.

Plectranthus grandicalyx (E.A.Bruce) J.K.Morton
Biol. Skr. 51: 470 (2005); F.E.E. 5: 588 (2006); F.T.E.A. Lam.: 32 (2009).
Syn: *Coleus grandicalyx* E.A.Bruce – F.P.S. 3: 208 (1956).
Perennial herb or shrub. Montane grassland, *Erica* thicket, open *Hagenia abyssinica* woodland.
Distr: EQU; Ethiopia, Uganda.
IUCN: RD
EQU: Friis & Vollesen 1007 (fl) 2/1982.

Plectranthus guerkei Briq.
F.T.E.A. Lam.: 317 (2009).
Syn: *Neohyptis paniculata* (Baker) J.K.Morton – Biol. Skr. 51: 467 (2005); F.E.E. 5: 601 (2006).
Perennial herb. Wet grassland & swamps.
Distr: EQU; Guinea Bissau to Ethiopia, S to Angola & Botswana.
EQU: Friis & Vollesen 1239 (fl) 3/1982.

Plectranthus igniarius (Schweinf.) Agnew
F.E.E. 5: 591 (2006); F.T.E.A. Lam.: 299 (2009).
Syn: *Coleus igniarius* Schweinf. – F.P.S. 3: 208 (1956).
Shrub. Dry rocky hillslopes.
Distr: RS; Eritrea, Ethiopia, Somalia, Kenya; Yemen.
RS: Broun 1203 (fl fr) 5/1907.

Plectranthus jebel-marrae Wickens & B.Mathew
F.J.M.: 152 (1976).
Syn: *Plectranthus glandulosus* sensu Quézel, non Hook.f. – F. Darfur Nord-Occ. & J. Gourgeil: 132 (1969).
Perennial herb. Rocky slopes, dry woodland.
Distr: DAR; Sudan endemic.
IUCN: RD
DAR: Wickens 2324 (fl) 9/1964.

Plectranthus lactiflorus (Vatke) Agnew
Biol. Skr. 51: 470 (2005); F.E.E. 5: 590 (2006); F.T.E.A. Lam.: 352 (2009).
Syn: *Coleus lactiflorus* Vatke – F.P.S. 3: 208 (1956).

(Sub)shrub. Rocky hillslopes, rocky grassland.
Distr: EQU; Ethiopia to Rwanda & Tanzania.
EQU: Thomas 1616 (fl) 12/1935.

Plectranthus lanuginosus (Benth.) Agnew

F.E.E. 5: 595 (2006); F.T.E.A. Lam.: 347 (2009).
Perennial herb. Rock outcrops in grassland & woodland.
Distr: RS; Eritrea & Somalia, S to D.R. Congo & Tanzania; Yemen.
RS: Jackson 2942 (fr) 4/1953.

Plectranthus laxiflorus Benth.

F.E.E. 5: 598 (2006); F.T.E.A. Lam.: 252 (2009).
Syn: *Plectranthus longipes* sensu Friis & Vollesen, non Baker – Biol. Skr. 51: 471 (2005).
Perennial herb, sometimes scandent. Forest margins & clearings, stream edges.
Distr: EQU; Ethiopia to South Africa.
EQU: Myers 10941 (fl) 4/1939.

Plectranthus monostachyus (P.Beauv.) B.J.Pollard subsp. perennis (J.K.Morton) B.J.Pollard

Kew Bull. 56: 981 (2001).
Syn: *Solenostemon monostachyus* (P.Beauv.) Briq. subsp. *perennis* J.K.Morton – F.J.M.: 153 (1976).
Perennial herb. Wet rocks, stream banks.
Distr: DAR; Ivory Coast, Ghana, Nigeria, Gabon, C.A.R.
DAR: Wickens 2128 (fl) 8/1964.
Note: the single specimen seen is depauperate; better material would be desirable to confirm the identity.

Plectranthus montanus Benth.

F.T.E.A. Lam.: 318 (2009).
Syn: *Plectranthus cylindraceus* Hochst. ex Benth. – F.P.S. 3: 223 (1956); F.E.E. 5: 594 (2006).
Perennial herb. Rock outcrops, dry bushland.
Distr: RS; Ethiopia & Somalia, S to Angola & Botswana; Arabia, Socotra.
RS: Robbie 10 (fl).
Note: Pollard in Kew Bull. 60: 146 (2005) records *P. occidentalis* B.J.Pollard from Sudan but this is based on *Schweinfurth* 3596 from Gadda which is in modern-day D.R. Congo, not South Sudan.

Plectranthus otostegioides (Schweinf. ex Gürke) Ryding

F.E.E. 5: 588 (2006); F.T.E.A. Lam.: 343 (2009).
Syn: *Capitanya otostegioides* Schweinf. ex Gürke – F.P.S. 3: 208 (1956).
Perennial herb or subshrub. Dry bushland in rocky areas.
Distr: RS; Eritrea & Somalia, S to Tanzania.
RS: Robbie 45 (fl).

Plectranthus tetradenifolius A.J.Paton

F.T.E.A. Lam.: 304 (2009).
Syn: *Plectranthus cyaneus* sensu Friis & Vollesen, non Gürke – Biol. Skr. 51: 469 (2005).
Syn: *Tetradenia riparia* sensu Friis & Vollesen pro parte, non (Hochst.) Codd – Biol. Skr. 51: 473 (2005).
(Sub)shrub. Rocky slopes.
Distr: EQU; Cameroon, Uganda, Kenya, Tanzania.
EQU: Friis & Vollesen 137 (fl fr) 11/1980.

Premna angolensis Gürke

T.S.S.: 442 (1990); F.T.E.A. Verb.: 70 (1992); Biol. Skr. 51: 460 (2005).
Syn: *Premna zenkeri* Gürke – F.P.S. 3: 198 (1956).
Tree or shrub. Forest margins, adjacent bushland & grassland.
Distr: EQU; Senegal to Gabon, São Tomé, Ethiopia to Tanzania & Angola.
EQU: Andrews 1320 (fl fr) 5/1939.

Premna resinosa (Hochst.) Schauer subsp. resinosa

F.P.S. 3: 198 (1956); T.S.S.: 442 (1990); F.T.E.A. Verb.: 72 (1992); F.E.E. 5: 521 (2006).
Shrub or small tree. *Acacia-Commiphora* bushland, dry grassland, rock outcrops.
Distr: RS, DAR, KOR, CS, ES, EQU; Ethiopia & Somalia, S to Tanzania; Arabia.
KOR: Jackson 4009 (fr) 9/1959; **EQU:** Carr 840 (fl) 7/1970.

Premna schimperi Engl.

F.P.S. 3: 198 (1956); T.S.S.: 442 (1990); F.T.E.A. Verb.: 76 (1992); Biol. Skr. 51: 460 (2005); F.E.E. 5: 519 (2006).
Shrub or tree. Degraded forest, secondary bushland & grassland.
Distr: EQU; Ethiopia, Somalia, Uganda, Kenya, Tanzania.
EQU: Myers 11716 (fl fr) 7/1939.

Pycnostachys batesii Baker

Biol. Skr. 51: 471 (2005); F.T.E.A. Lam.: 406 (2009).
Scandent herb or shrub. Lowland forest.
Distr: EQU; Cameroon, D.R. Congo, Uganda.
EQU: Friis & Vollesen 588 (fl fr) 12/1980.

Pycnostachys meyeri Gürke

F.P.S. 3: 224 (1956); Biol. Skr. 51: 471 (2005); F.E.E. 5: 602 (2006); F.T.E.A. Lam.: 412 (2009).
Shrub. Forest, particularly margins & clearing.
Distr: EQU; Guinea, Nigeria to Ethiopia, S to Tanzania.
EQU: Friis & Vollesen 144 (fl) 11/1980.

Pycnostachys niamniamensis Gürke

F.P.S. 3: 223 (1956); F.T.E.A. Lam.: 399 (2009).
Perennial herb. Swampy grassland.
Distr: EQU; Uganda, Kenya.
IUCN: RD
EQU: Andrews 1081 (fl fr) 5/1939.
Note: *Pycnostachys* sp.= *Mesfin & Kagnew* 2249 in F.E.E. 5: 602 (2006) is closely allied to and possibly a form of this species.

Pycnostachys schweinfurthii Briq.

F.P.S. 3: 223 (1956); F.T.E.A. Lam.: 391 (2009).
Annual herb. Wet grassland.
Distr: BAG; Ghana, Tanzania, ?D.R. Congo.
BAG: Schweinfurth 2770 (fl) 12/1869.

Rotheca alata (Gürke) Verdc.

Kew Bull. 55: 148 (2000).
Syn: *Clerodendrum alatum* Gürke – F.P.S. 3: 195 (1956); F.T.E.A. Verb.: 129 (1992); F.E.E. 5: 523 (2006).
Syn: *Clerodendrum myricoides* (Hochst.) R.Br. ex Vatke var. *floribundum* Baker – F.P.S. 3: 196 (1956).

Perennial herb or subshrub. Wooded grassland.
Distr: EQU; Ivory Coast, Benin, Nigeria to Ethiopia,
Tanzania.
EQU: Andrews 1459 (fl) 5/1939.

Rotheca cf. bukobensis (Gürke) Verdc.
Shrub. Sandy island.
Distr: UN; (for species) Uganda, D.R.Congo, Tanzania &
Zambia.
UN: Simpson 7159 (fl) 6/1929.
Note: the specimen cited was determined at Kew by E.
Persson as cf. *Clerodendrum bukobense* in 1990.

Rotheca myricoides (Hochst.) Steane & Mabb. subsp. myricoides
F.Som. 3: 316 (2006).
Syn: *Clerodendrum myricoides* (Hochst.) R.Br. ex Vatke
– Bot. Exp. Sud.: 110 (1970); T.S.S.: 438 (1990); F.T.E.A.
Verb.: 130 (1992); Biol. Skr. 51: 457 (2005); F.E.E. 5: 525
(2006).
Shrub or subshrub, sometimes climbing. Grassland,
wooded grassland, thicket.
Distr: RS, KOR, ES, BAG, UN, EQU; Uganda to Somalia, S
to Zimbabwe & Mozambique.
KOR: Turner 288 (fl) 7/1938; **EQU:** Jackson 3836 (fl)
9/1957.
Note: a very variable species which Verdcourt (F.T.E.A.
Verb.: 130) splits into several infraspecific taxa; both
var. *myricoides* and var. *discolor* (Klotzsch) Verdc.
occur in Sudan / South Sudan but there are also many
intermediates.

Salvia aegyptiaca L.
F.P.S. 3: 224 (1956); F.J.U. 2: 183 (1999); F.E.E. 5: 559
(2006).
Subshrub. Dry open areas.
Distr: NS, RS, DAR; Cape Verde Is., Eritrea, Somalia; N
Africa & Arabia to India.
RS: Carter 1835 (fl) 11/1987.

Salvia nilotica Jacq.
Biol. Skr. 51: 471 (2005); F.E.E. 5: 559 (2006); F.T.E.A.
Lam.: 110 (2009).
Perennial herb. Grassland & bushland, often on moist
soils, disturbed ground.
Distr: EQU; Ethiopia to Zimbabwe.
EQU: MacLeay 153 7/1947.

Scutellaria arabica Jaub. & Spach
F.E.E. 5: 530 (2006).
Perennial herb. *Juniperus* forest, wooded slopes.
Distr: RS; Eritrea, Ethiopia, Djibouti; Arabia.
RS: Jackson 2881 (fl) 4/1953.
Note: Andrews lists also *S. peregrina* L., a Mediterranean
species, recording it from Fung District. This may belong
here although the geography appears wrong.

Scutellaria schweinfurthii Briq. subsp. schweinfurthii
F.P.S. 3: 224 (1956); F.E.E. 3: 530 (1989); F.T.E.A. Lam.:
27 (2009).
Subshrub or perennial herb. Burnt grassland & wooded
grassland, rocky areas.
Distr: BAG, EQU; Ethiopia, Uganda.

IUCN: RD
EQU: Andrews 932 (fl fr) 4/1939.

Scutellaria schweinfurthii Briq. subsp. paucifolia (Baker) A.J.Paton
Biol. Skr. 51: 472 (2005); F.E.E. 5: 530 (2006); F.T.E.A.
Lam.: 27 (2009).
Syn: *Scutellaria paucifolia* Baker – F.P.S. 3: 224 (1956).
Perennial herb. Wooded grassland subject to burning.
Distr: BAG, EQU; Senegal to Ethiopia, S to Zimbabwe &
Mozambique.
EQU: Chipp 38 (fl fr) 2/1929.

Stachys schimperi Vatke
F.E.E. 5: 537 (2006).
Perennial herb. Rocky hillslopes, rock crevices.
Distr: RS; Ethiopia; Saudi Arabia.
IUCN: RD
RS: Jackson 2915 (fl fr) 4/1953.

Syncolostemon bracteosus (Benth.) D.F.Otieno
F.T.E.A. Lam.: 177 (2009).
Annual or perennial herb. Dry woodland or grassland.
Distr: KOR; Senegal to Sudan, S to Namibia & South
Africa.
KOR: Wickens 462 (fr) 9/1967.

[Tectona grandis L.f.]
Note: the important timber species 'teak', native of
Burma, is widely planted in South Sudan but is not
known to naturalise.

Tetradenia urticifolia (Baker) Phillipson
F.T.E.A. Lam.: 360 (2009).
Syn: *Iboza multiflora* sensu Andrews, non (Benth.)
E.A.Bruce – F.P.S. 3: 212 (1956).
Syn: *Iboza riparia* sensu El Amin, non (Hochst.) N.E.Br.
– T.S.S.: 445 (1990).
Syn: *Tetradenia riparia* sensu auctt., non (Hochst.)
Codd – Biol. Skr. 51: 474 (2005); F.E.E. 5: 565 (2006).
Shrub. Forest margins, woodland & bushland, often on
rock outcrops.
Distr: ?CS, EQU; Ethiopia to D.R. Congo & Tanzania.
EQU: Myers 7959 (fl) 12/1937.
Note: Andrews and El Amin record this species from Fung
District but we have only seen material from Equatoria.

Teucrium yemense Deflers
F.E.E. 5: 527 (2006).
Syn: *Teucrium sp.* – F.P.S. 3: 225 (1956); Bot. Exp. Sud.:
63 (1970).
Perennial herb or subshrub. Rocky hillslopes, open
bushland.
Distr: RS; Ethiopia, Djibouti, Somalia; Arabia.
RS: Jackson 2934 (fl) 4/1953.
Note: the Sudan plants are a rather distinct variant of this
species; see Ryding in Edinb. J. Bot. 55: 219 (1998).

Tinnea aethiopica Kotschy ex Hook.f. subsp. aethiopica
F.P.S. 3: 225 (1956); Bot. Exp. Sud.: 64 (1970); T.S.S.: 446
(1990); Biol. Skr. 51: 474 (2005); F.E.E. 5: 529 (2006);
F.T.E.A. Lam.: 19 (2009).

Shrub. Bushland, forest margins, secondary scrub.
Distr: ?RS, BAG, EQU; Mali to Ethiopia, S to Tanzania.
RS: Drar 585 3/1938 (n.v.); **EQU:** Myers 7150 (fl) 7/1937.
Note: the record from the Red Sea region is from Drar
(1970), based on a sterile specimen which we have not
seen; the identification requires confirmation.

Vitex doniana Sweet

F.P.S. 3: 200 (1956); F.J.M.: 151 (1976); T.S.S.: 443
(1990); F.T.E.A. Verb.: 62 (1992); Biol. Skr. 51: 461
(2005); F.E.E. 5: 521 (2006).
Tree. Wooded grassland, woodland, forest margins.
Distr: DAR, KOR, CS, BAG, EQU; Senegal to Ethiopia, S
to Angola & Mozambique; Comoros.
DAR: Kamil 1197 (fl) 5/1968; **EQU:** Myers 11036 (fl)
4/1939.

Vitex ferruginea Schumach. & Thonn.

F.T.E.A. Verb.: 66 (1992); Biol. Skr. 51: 461 (2005).
Syn: Vitex amboniensis Gürke – F.P.S. 3: 202 (1956);
T.S.S.: 443 (1990).
Tree. Forest, often riverine.
Distr: EQU; Guinea Bissau to Somalia, S to Angola &
South Africa.
EQU: Jackson 1257 (fl) 3/1950.
Note: Verdcourt in F.T.E.A. divides this species into two
subspecies, subsp. *amboniensis* (Gurke) Verdc. being
a species of bushland, woodland and dry forest in the
coastal E African lowlands.

Vitex fischeri Gürke

F.P.S. 3: 200 (1956); T.S.S.: 445 (1990); F.T.E.A. Verb.: 59
(1992); Biol. Skr. 51: 461 (2005).
Shrub or tree. Wooded grassland, riverine forest &
woodland, rock outcrops.
Distr: EQU; Uganda to Angola & Zambia.
EQU: Jackson 3884 (fr) 8/1958.

Vitex madiensis Oliv. subsp. *madiensis*

F.P.S. 3: 200 (1956); T.S.S.: 445 (1990); F.T.E.A. Verb.: 60
(1992); Biol. Skr. 51: 462 (2005).
Syn: Vitex simplicifolia Oliv. – F.P.S. 3: 200 (1956); Bot.
Exp. Sud.: 111 (1970) ; T.S.S.: 445 (1990).
Shrub or small tree. Bushland, woodland & wooded
grassland.
Distr: KOR, BAG, EQU; Gambia to Uganda, S to Angola,
Zambia & Mozambique.
KOR: Simpson 7713 (fr) 4/1930; **EQU:** Myers 10373 (fl)
2/1939.

Phrymaceae

I. Darbyshire

Mimulus gracilis R.Br.

F.P.S. 3: 140 (1956); F.J.M.: 146 (1976); F.E.E. 5: 267
(2006); F.T.E.A. Scroph.: 34 (2008).
Perennial herb. Montane streamsides, ponds, seasonally
wet open areas.
Distr: DAR; Nigeria, D.R. Congo, Ethiopia to South
Africa; Arabia, India, China, Australia.
DAR: Jackson 3379 (fl fr) 12/1954.

Orobanchaceae

I. Darbyshire

Alectra sessiliflora (Vahl) Kuntze

Biol. Skr. 51: 422 (2005); F.E.E. 5: 286 (2006); F.T.E.A.
Scroph.: 112 (2008).
Syn: Alectra communis Hemsl. – F.P.S. 3: 130 (1956).
Syn: Alectra sessiliflora (Vahl) Kuntze var.
senegalensis (Benth.) Hepper – F.J.M.: 145 (1976).
Syn: Alectra asperrima sensu Andrews, non Benth. –
F.P.S. 3: 130 (1956).
Annual herb. Grassland, woodland, forest margins,
swampy areas.
Distr: DAR, EQU; palaeotropical.
DAR: Wickens 1185 (fl fr) 2/1964; **EQU:** Thomas 162 (fl
fr) 12/1935.

Bartsia trixago L.

Biol. Skr. 51: 423 (2005); F.E.E. 5: 307 (2006); F.T.E.A.
Scroph.: 194 (2008).
Syn: Bellardia trixago (L.) All. – F.P.S. 3: 133 (1956);
F.J.M.: 145 (1976).
Annual herb. Montane grassland, rocky outcrops.
Distr: DAR, EQU; Ethiopia, Kenya, South Africa;
Mediterranean, South America.
DAR: Wickens 2793 (fl) 10/1964; **EQU:** Thomas 1839 (fl
fr) 12/1935.

Buchnera capitata Benth.

F.P.S. 3: 133 (1956); F.E.E. 5: 292 (2006); F.T.E.A. Scroph.:
126 (2008).
Annual herb. Wet grassland.
Distr: EQU; Ivory Coast to Ethiopia, S to Mozambique;
Madagascar.
EQU: Schweinfurth 3852 (fl fr) 5/1870.

Buchnera hispida D.Don

F.P.S. 3: 134 (1956); F.E.E. 5: 292 (2006); F.T.E.A. Scroph.:
134 (2008).
Annual herb. Grassland & open woodland.
Distr: DAR, KOR, CS, UN; tropical Africa; Madagascar,
Arabia, India & Nepal.
DAR: Wickens 2764 (fl fr) 10/1964; **UN:** Sherif A.3940
(fl) 8/1951.

Buchnera nuttii Skan

Biol. Skr. 51: 423 (2005); F.T.E.A. Scroph.: 125 (2008).
Perennial herb. Grassland, open woodland.
Distr: EQU; Kenya to Zambia & Malawi.
EQU: Andrews 1971 (fl) 6/1939.

Cistanche phelypaea (L.) Cout.

F.P.S. 3: 150 (1956); F.T.E.A. Orobanch.: 2, in notes
(1957); F.J.U. 3: 235 (1999); F.Som. 3: 295 (2006).
Parasitic fleshy herb. Semi-desert, dry bushland, often on
saline soils.
Distr: NS, RS, CS; S Europe, N Africa, Arabia.
CS: Musselman 6179 (fl fr) 1/1983.
Note: the separation of *C. phelypaea* and *C. tubulosa* is
problematic and although both names have been applied
to Sudanese material it may be that only one species is
involved; indeed, they are treated as conspecific in Fl.
Somalia and Musselman (1984) lists only *C. phelypaea* in
his review of parasitic plants in Sudan.

Cistanche tubulosa (Schenk) Hook.f.

F.T.E.A. Orobanch.: 2 (1957); F.E.E. 5: 309 (2006).
Parasitic fleshy herb. Semi-desert, saline flats including coastal areas.
Distr: RS, CS; Eritrea to Tanzania; N Africa, Arabia to India.
RS: Jackson 3922 (fl) 4/1959.

Cycnium adonense E.Mey. ex Benth. subsp. adonense

F.P.S. 3: 135 (1956); Biol. Skr. 51: 423 (2005); F.T.E.A. Scroph.: 168 (2008).
Perennial herb. Rock outcrops, wooded grassland.
Distr: BAG, EQU; Uganda to Angola & South Africa.
EQU: Johnston 12K (fl) 4/1932.
Note: Andrews (F.P.S. 3: 136) also lists *C. brachycalyx* Schweinf., which Hansen (Dansk. Bot. Arkiv. 32(3): 58 (1978)) treats as a doubtful species, the type having been destroyed.

Cycnium adonense E.Mey. ex Benth. subsp. camporum (Engl.) O.J.Hansen

Biol. Skr. 51: 423 (2005); F.E.E. 5: 303 (2006); F.T.E.A. Scroph.: 169 (2008).
Syn: *Cycnium camporum* Engl. – F.P.S. 3: 136 (1956).
Perennial herb. Rock outcrops, wooded grassland.
Distr: DAR, BAG, EQU; Guinea to Ethiopia, S to Angola & Mozambique.
DAR: Wickens 1935 (fl) 7/1964; **EQU:** von Ramm 227 (fl) 7/1956.

Cycnium erectum Rendle

Biol. Skr. 51: 424 (2005); F.E.E. 5: 301 (2006); F.T.E.A. Scroph.: 169 (2008).
Shrub. Montane forest margins, secondary scrub.
Distr: EQU; Ethiopia, Uganda, Kenya.
EQU: Jackson 3814 (fl) 8/1957.
Note: a local species but not considered to be threatened.

Cycnium recurvum (Oliv.) Engl.

Biol. Skr. 51: 424 (2005); F.T.E.A. Scroph.: 157 (2008).
Syn: *Ramphicarpa recurva* Oliv. – F.P.S. 3: 141 (1956).
Annual herb. Rock outcrops, grassland over shallow soils.
Distr: EQU; Uganda to Malawi.
EQU: Myers 9517 (fl fr) 10/1938.

Cycnium tenuisectum (Standl.) O.J.Hansen

Biol. Skr. 51: 424 (2005); F.E.E. 5: 303 (2006); F.T.E.A. Scroph.: 158 (2008).
Syn: *Ramphicarpa tenuisecta* Standl. – F.P.S. 3: 141 (1956).
Perennial herb. Montane grassland.
Distr: EQU; Ethiopia to Tanzania.
EQU: Thomas 1819 (fl fr) 12/1935.

Cycnium tubulosum (L.f.) Engl. subsp. tubulosum

F.E.E. 5: 301 (2006); F.T.E.A. Scroph.: 162 (2008).
Syn: *Ramphicarpa heuglinii* Hochst. ex Schweinf. – F.P.S. 3: 141 (1956).
Perennial herb. Swamps, wet grassland.
Distr: BAG, UN; Nigeria, Uganda to Angola & South Africa.
BAG: Schweinfurth 1241 (fl) 2/1869.

Note: *R. heuglinii* is only tentatively included in synonymy – see note in Dansk. Bot. Arkiv. 32(3): 33 (1978).

Cycnium tubulosum (L.f.) Engl. subsp. montanum (N.E.Br.) O.J.Hansen

F.E.E. 5: 301 (2006); F.T.E.A. Scroph.: 162 (2008).
Perennial herb. Dry or damp grassland or wooded grassland.
Distr: BAG, UN; Ethiopia to Zimbabwe & Mozambique.
BAG: Schweinfurth III.46 (fl) 6/1871.

Harveya helenae Buscal. & Muschl.

F.E.E. 5: 285 (2006).
Syn: *Harveya obtusifolia* sensu Hepper, non (Benth.) Vatke – F.T.E.A. Scroph.: 118 (2008).
Perennial parasitic herb. Grassland, open woodland.
Distr: EQU; Eritrea to Malawi; Yemen.
EQU: Lankester s.n. (fl) 11/1921.

Harveya sp. nov.

F.P.S. 3: 136 (1956); Biol. Skr. 51: 425 (2005).
Parasitic herb. Forest.
Distr: EQU; South Sudan endemic.
IUCN: RD
EQU: Myers 9755 (fl) 10/1938.
Note: allied to *H. liebuschiana* Skan from Tanzania.

Hedbergia abyssinica (Benth.) Molau

Biol. Skr. 51: 425 (2005); F.E.E. 5: 305 (2006); F.T.E.A. Scroph.: 195 (2008).
Perennial herb. Montane grassland.
Distr: EQU; Nigeria to Ethiopia, S to Zambia & Malawi.
EQU: Friis & Vollesen 320 (fl) 11/1980.

Lindenbergia indica (L.) Vatke

F.E.E. 5: 263 (2006); F.T.E.A. Scroph.: 34 (2008).
Syn: *Lindenbergia abyssinica* Hochst. ex Benth. – F.P.S. 3: 138 (1956).
Annual or perennial herb. Rocky slopes, wadis.
Distr: RS; Eritrea, Ethiopia, Somalia; Egypt, Arabia to Bangladesh.
RS: Robbie 23 (fl fr).

Micrargeria filiformis (Schum. & Thonn.) Hutch. & Dalziel

F.P.S. 3: 139 (1956); F.T.E.A. Scroph.: 187 (2008).
Annual herb. Wet grassland, swamps.
Distr: BAG; Uganda to Angola and Mozambique; Madagascar.
BAG: Schweinfurth 2730 (fl fr) 12/1869.

Orobanche cernua Loef.

F.T.E.A. Orobanch.: 6 (1957); F.E.E. 5: 311 (2006).
Syn: *Orobanche cernua* Loef. var. *desertorum* (Beck) Stapf – F.P.S. 3: 150 (1956); F.T.E.A. Orobanch.: 6 (1957); Bot. Exp. Sud.: 91 (1970).
Perennial herb. Parasite of cultivated crops.
Distr: NS, RS, DAR; Ethiopia & Somalia, S to Tanzania; S Europe, N Africa to China & Australia.
RS: Bent s.n. (fl fr) 1896.
Note: we follow C. Parker in F.E.E. in not recognising varietal taxa in Africa.

Orobanche minor Sm.

F.P.S. 3: 152 (1956); F.T.E.A. Orobanch.: 5 (1957); Biol.
Skr. 51: 428 (2005); F.E.E. 5: 312 (2006).
Parasitic fleshy herb. Parasite in cultivated land, grassland
& woodland.
Distr: RS, CS, EQU; tropical & southern Africa; Europe, N
Africa, Arabia, N America.
RS: Maffey 22 (fl fr) 1928; **EQU:** Thomas 1860 (fl)
12/1935.

Orobanche ramosa L.

F.P.S. 3: 150 (1956); F.T.E.A. Orobanch.: 3 (1957); F.J.M.:
147 (1976); F.E.E. 5: 311 (2006).
Parasitic annual herb. Parasite in cultivated fields,
roadsides, wooded grassland.
Distr: NS, RS, DAR, CS; Ethiopia; S Europe, NE Africa,
Arabia, sporadically introduced elsewhere.
CS: Musselman 198 (fl fr) 1/1983.
Note: *Mohamed Eff Ismail* 3491a from Erkowit is named
O. nana Reut., now considered a subspecies of *O.
ramosa*, but the specimen appears a close match for
other specimens of *O. ramosa* from the Red Sea Region
so is treated within this taxon in the broad sense here.

Parentucellia latifolia (L.) Caruel

F.J.M.: 146 (1976).
Syn: *Parentucellia latifolia* (L.) Caruel var. *flaviflora*
(Boiss.) Dandy – F.P.S. 3: 140 (1956).
Annual herb. Montane grassland.
Distr: DAR; S Europe & Mediterranean to Iran.
DAR: Dandy 108 1/1934.
Note: the single Sudan site (Jebel Marra) is highly disjunct
from the remainder of this species' range.

Rhamphicarpa elongata (Hochst.) O.J.Hansen

Bot. Tidsskr. 70: 117 (1975).
Annual herb. Seasonally wet flushes.
Distr: KOR; Sudan endemic.
IUCN: RD
KOR: Kotschy 77A (fl fr) 9/1839.

Rhamphicarpa fistulosa (Hochst.) Benth.

F.P.S. 3: 140 (1956); F.J.M.: 146 (1976); Biol. Skr. 51: 428
(2005); F.E.E. 5: 304 (2006); F.T.E.A. Scroph.: 154 (2008).
Annual herb. Wet flushes often over rock, lake & stream
margins.
Distr: DAR, KOR, BAG, UN, EQU; Senegal to Ethiopia, S
to South Africa; Madagascar, New Guinea, Australia.
DAR: Wickens 2956 (fl fr) 5/1965; **BAG:** Simpson 7410
(fl fr) 7/1929.

Sopubia parviflora Engl.

F.P.S. 3: 142 (1956); F.W.T.A. 2: 368 (1963).
Annual herb. Wet grassland over rock.
Distr: BAG, EQU; Guinea to D.R. Congo.
EQU: Jackson 3847 (fl fr) 9/1957.
Note: E. Fischer in F.E.E. 5: 290 (2006) records *S. mannii*
Skan. as occurring in Sudan but we have seen no
material to date.

Sopubia ramosa (Hochst.) Hochst.

F.P.S. 3: 142 (1956); F.J.M.: 146 (1976); Biol. Skr. 51: 428
(2005); F.E.E. 5: 288 (2006); F.T.E.A. Scroph.: 180 (2008).
Perennial herb. Montane grassland, rocky outcrops.

Distr: DAR, BAG, EQU; Cameroon, D.R. Congo to
Ethiopia, S to Malawi & Mozambique.
DAR: Wickens 2810 (fl fr) 11/1964; **EQU:** Thomas 1816
(fl) 12/1935.

Sopubia simplex (Hochst.) Hochst.

F.P.S. 3: 142 (1956); F.E.E. 5: 290 (2006); F.T.E.A. Scroph.:
178 (2008).
Perennial herb. Grassland & wooded grassland including
swampy areas.
Distr: EQU; Guinea to Ethiopia, S to South Africa.
EQU: Myers 6359 (fl fr) 3/1937.

Striga asiatica (L.) Kuntze

F.P.S. 3: 145 (1956); F.E.E. 5: 295 (2006); F.T.E.A. Scroph.:
148 (2008).
Annual herb. Grassland, open woodland, disturbed areas
and cultivation.
Distr: DAR, KOR, ES, BAG, EQU; palaeotropical.
KOR: Musselman 6260 (fl) 8/1983; **EQU:** Myers 6656
(fl) 5/1937.
Note: Mohamed et al. (Ann. Missouri Bot. Gard. 88:
60-103 (2001)) separate out *S. lutea* Lour. and *S. hirsuta*
Benth. from *S. asiatica*, with all three occuring in Sudan;
however, we follow F.E.E. and F.T.E.A. in treating these as
a single variable species.

Striga aspera (Willd.) Benth.

F.P.S. 3: 145 (1956); F.E.E. 5: 297 (2006); F.T.E.A. Scroph.:
142 (2008).
Annual herb. Grassland, parasitic on wild grass species.
Distr: KOR, CS, BAG, UN, EQU; Senegal to Ethiopia.
KOR: Musselman 6280 (fl) 8/1983; **BAG:** Myers 7329 (fl
fr) 7/1937.

Striga bilabiata (Thunb.) Kuntze subsp. *barteri* (Engl.) Hepper

F.T.E.A. Scroph.: 139 (2008).
Syn: *Striga barteri* Engl. – F.P.S. 3: 145 (1956).
Perennial herb. Seasonally wet grassland, parasitic on
grasses.
Distr: EQU; Senegal to Chad, Uganda to Tanzania.
EQU: Schweinfurth 2931 (fl) 2/1870.

Striga brachycalyx Skan

F.W.T.A. 2: 372 (1963); Kew Bull. 41: 211 (1986).
Annual herb. Grassland & wooded grassland, parasitic on
wild grasses.
Distr: BAG; Mali to Nigeria, D.R. Congo.
BAG: Aylmer 27/51 (n.v.).
Note: the cited specimen has not been seen by us;
identification follows Musselman & Hepper (1986, cited
above).

Striga forbesii Benth.

F.P.S. 3: 146 (1956); F.E.E. 5: 295 (2006); F.T.E.A. Scroph.:
146 (2008).
Annual herb. Parasite of seasonally damp grassland &
cultivated land.
Distr: KOR, ES, BAG, UN, EQU; Senegal to Somalia, S to
South Africa; Madagascar.
KOR: Bashi s.n. (fl) 1986; **UN:** J.M. Lock 82/31 (fl fr) 1981.

Striga gesnerioides (Willd.) Vatke ex Engl.

F.P.S. 3: 144 (1956); Biol. Skr. 51: 249 (1998); F.E.E. 5: 296 (2006); F.T.E.A. Scroph.: 144 (2008).
Annual or perennial herb. Parasite of grassland & cultivated land.
Distr: RS, DAR, KOR, CS, ES, UN, EQU; tropical & southern Africa; Arabia, India, Sri Lanka.
RS: Jackson 2722 (fl) 3/1953; **EQU:** Simpson 7361 (fl fr) 7/1929.

Striga hermonthica (Delile) Benth.

F.P.S. 3: 145 (1956); Biol. Skr. 51: 429 (2005); F.E.E. 5: 295 (2006); F.T.E.A. Scroph.: 142 (2008).
Syn: *Striga senegalensis* Benth. – F.P.S. 3: 145 (1956).
Annual herb. Parasitic on crops & wild grasses.
Distr: NS, DAR, KOR, CS, ES, BAG, UN, EQU; Senegal to Djibouti, S to Namibia & Tanzania; Madagascar, Egypt.
CS: Musselman 6139 (fl) 10/1982; **BAG:** Simpson 7406 (fl) 7/1929.
Note: a serious pest of grain crops.

Striga klingii (Engl.) Skan

F.W.T.A. 2: 371 (1963).
Annual herb. Parasite of grasses in wet grassland.
Distr: EQU; Senegal to C.A.R., Tanzania.
EQU: Myers 7867 (fl) 10/1937.

Striga linearifolia (Schumach. & Thonn.) Hepper

F.E.E. 5: 296 (2006); F.T.E.A. Scroph.: 141 (2008).
Syn: *Striga strictissima* Skan – F.P.S. 3: 145 (1956).
Perennial herb. Seasonally wet grassland, parasite on grasses.
Distr: BAG; Guinea to Ethiopia, S to Malawi.
BAG: Schweinfurth 1651 (fr) 4/1869.
Note: Mohamed et al. (Ann. Missouri Bot. Gard. 88 (2001)) treat this as a subsp. of *S. bilabiata* but we follow F.E.E. and F.T.E.A. here in maintaining it as a distinct species.

Striga macrantha (Benth.) Benth.

F.W.T.A. 2: 371 (1963).
Annual herb. Grassland & wooded grassland.
Distr: DAR; Senegal to C.A.R., Angola, Zambia.
DAR: Wickens 2807 (fl fr) 11/1964.

Striga passargei Engl.

F.T.E.A. Scroph.: 143 (2008).
Annual herb. Grassland, *Acacia* woodland, disturbed areas, parasite on grasses.
Distr: DAR, KOR, CS; Senegal to D.R. Congo, Tanzania, Zambia, Namibia; Arabia.
KOR: Musselman 6256 (fl fr) 8/1983.
Note: Andrews lists also *S. baumannii* Engl. but we have seen no material and it was not listed as occurring in Sudan by Musselman & Hepper (Kew Bull. 41: 205 (1986)) or in F.T.E.A.

Lentibulariaceae

I. Darbyshire

Utricularia andongensis Welw. ex Hiern

F.T.E.A. Lentib.: 6 (1973); Biol. Skr. 51: 430 (2005).
Terrestrial annual herb. Wet rocks, wet grassland.
Distr: EQU; Guinea to Uganda, S to Angola & Zambia.
EQU: Jackson 3855 (fl fr) 9/1957.

Utricularia arenaria A.DC.

F.T.E.A. Lentib.: 11 (1973); Biol. Skr. 51: 431 (2005); F.E.E. 5: 316 (2006).
Syn: *Utricularia exilis* Oliv. – F.P.S. 3: 153 (1956).
Syn: *Utricularia tribracteata* Hochst. ex A.Rich. – F.P.S. 3: 153, pro parte (1956).
Terrestrial annual herb. Wet rocks, marshy ground, wet grassland.
Distr: BAG, EQU; Senegal to Ethiopia, S to Namibia & South Africa; Madagascar, India.
EQU: Myers 7157 (fl fr) 7/1937.

Utricularia australis R.Br.

F.T.E.A. Lentib.: 20 (1973).
Aquatic perennial herb. Still or slow-flowing water.
Distr: BAG; palaeotropics & temperate regions.
BAG: Schweinfurth III.86 2/1869.

Utricularia cymbantha Oliv.

F.T.E.A. Lentib.: 22 (1973); Taylor, Genus Utricularia: 679 (1989).
Aquatic annual herb. Lakes & pools.
Distr: UNKNOWN; Ethiopia to South Africa; Madagascar.
Note: recorded from Sudan by P. Taylor (1989, l.c.) but without locality – we have not seen the specimen on which this record is based.

Utricularia firmula Welw. ex Oliv.

F.P.S. 3: 153 (1956); F.T.E.A. Lentib.: 9 (1973).
Terrestrial annual herb. Wet grassland & bogs.
Distr: BAG; Senegal to Kenya, S to Angola & South Africa; Madagascar.
BAG: Hope-Simpson 102 (fl fr) 1/1939.

Utricularia gibba L.

F.T.E.A. Lentib.: 21 (1973); F.E.E. 5: 319 (2006).
Syn: *Utricularia exoleta* R.Br. – F.P.S. 3: 154 (1956).
Syn: *Utricularia gibba* L. subsp. *exoleta* (R.Br.) P.Taylor – F.J.M.: 147 (1976).
Aquatic annual or perennial herb. Still or slow-flowing water & marginal mud.
Distr: DAR, ES, BAG, UN, EQU; pantropical & warm temperate regions.
DAR: Wickens 1488 (fl) 4/1964; **BAG:** Simpson 7681 (fl fr) 3/1930.

Utricularia inflexa Forssk.

F.T.E.A. Lentib.:16 (1973); F.E.E. 5: 318 (2006).
Syn: *Utricularia thoningii* Schumach. – F.P.S. 3: 154 (1956).
Aquatic ?perennial herb. Still or slow-flowing water.
Distr: KOR, CS, UN; Mauritania to Somalia, S to Namibia & South Africa; Madagascar, Egypt, India.
KOR: Kotschy 201 (fl) 10/1839; **UN:** J.M. Lock 81/195 (fl) 8/1981.

Utricularia livida E.Mey.

F.T.E.A. Lentib.: 11 (1973); Biol. Skr. 51: 431 (2005); F.E.E. 5: 316 (2006).
Syn: *Utricularia tribracteata* sensu Andrews pro parte, non Hochst. ex A.Rich. – F.P.S. 3: 153 (1956).
Terrestrial annual (or perennial) herb. Montane bogs, wet rocks.
Distr: EQU; Ethiopia & Somalia, S to Angola & South

Africa; Madagascar, Mexico.
EQU: Thomas 1882 (fl fr) 12/1935.

Utricularia pentadactyla P.Taylor

F.T.E.A. Lentib.: 9 (1973); Biol. Skr. 51: 431 (2005); F.E.E.
5: 316 (2006).
Terrestrial annual herb. Wet grassland, wet rocks.
Distr: EQU; Ethiopia to Angola & Zimbabwe.
EQU: Friis & Vollesen 418 (fl fr) 11/1980.

Utricularia raynalii P.Taylor

Taylor, Genus Utricularia: 641 (1989).
Aquatic annual herb. Lakes, pools, flooded grassland.
Distr: UNKNOWN; Senegal to C.A.R., Rwanda.
Note: recorded from Sudan by P. Taylor (1989, l.c.) but
without locality – we have not seen the specimen on
which this record is based. This is a widespread but rarely
collected species.

Utricularia reflexa Oliv.

F.P.S. 3: 155 (1956); F.T.E.A. Lentib.: 17 (1973); Biol. Skr.
51: 431 (2005); F.E.E. 5: 318 (2006).
Aquatic ?annual or perennial herb. Still or slow-flowing
water.
Distr: BAG, UN; Senegal to Uganda, S to Zimbabwe;
Madagascar.
UN: J.M. Lock 81/19 (fr) 2/1981.

Utricularia scandens Benj.

F.T.E.A. Lentib.: 4 (1973); F.E.E. 5: 316 (2006).
Syn: *Utricularia schweinfurthii* Baker ex Stapf – F.P.S.
3: 153 (1956).
Terrestrial annual herb. Wet grassland & bogs.
Distr: BAG, EQU; palaeotropical.
EQU: Myers 7156 (fl) 7/1937.

Utricularia stellaris L.f.

F.P.S. 3: 154 (1956); F.T.E.A. Lentib.: 16 (1973); F.J.M.:
147 (1976); Biol. Skr. 51: 432 (2005); F.E.E. 5: 318
(2006).
Aquatic ?perennial herb. Still or slow-flowing water.
Distr: DAR, KOR, CS, BAG, UN, EQU; palaeotropical.
DAR: Wickens 1359 (fl fr) 3/1964; **UN:** J.M. Lock 81/230
(fl fr) 8/1981.

Utricularia subulata L.

F.P.S. 3: 154 (1956); F.T.E.A. Lentib.: 14 (1973).
Terrestrial annual herb. Wet grassland, streamsides, wet
rocks.
Distr: BAG, EQU; tropical & southern Africa;
Madagascar, Portugal, Americas, Thailand & Borneo.
BAG: Schweinfurth 2559 (fl fr) 10/1869.

Acanthaceae

I. Darbyshire

Acanthopale pubescens (Engl.) C.B.Clarke

Biol. Skr. 51: 435 (2005); F.E.E. 5: 369 (2006); F.T.E.A.
Acanth.: 215 (2008).
Shrub or subshrub. Montane forest.
Distr: EQU; Ethiopia to Zimbabwe.
EQU: Friis & Vollesen 235 (fl) 11/1980.

Acanthus eminens C.B.Clarke

F.P.S. 3: 166 (1956); T.S.S.: 433 (1990); Biol. Skr. 51: 435
(2005); F.E.E. 5: 355 (2006); F.T.E.A. Acanth.: 94 (2008).
Shrub. Forest & forest margins.
Distr: EQU; Ethiopia, Uganda, Kenya.
EQU: Johnston 1427 (fl) 2/1936.

Acanthus polystachius Delile

F.T.E.A. Acanth.: 90 (2008).
Syn: *Acanthus arboreus* sensu Andrews, non Forssk. –
F.P.S. 3: 165 (1956).
Shrub. Forest margins.
Distr: CS; D.R. Congo to Ethiopia, S to Tanzania.
CS: Kotschy 489 (fl) 1837.
Note: El Amin's record (T.S.S.: 434 (1990) of *A. pubescens*
from Blue Nile Province is probably referable here.

Acanthus seretii De Wild.

Kew Bull. 62: 244 (2007).
Syn: *Acanthus pubescens* sensu Friis & Vollesen, non
(Oliv.) Engl. – Biol. Skr. 51: 436 (2005).
Shrub. Scrub at forest margins.
Distr: EQU; D.R. Congo, ?Uganda.
IUCN: RD
EQU: Thomas 1760 (fl) 12/1935.

Acanthus ueelensis De Wild.

F.T.E.A. Acanth.: 94 (2008).
Syn: *Acanthus montanus* sensu auctt., non (Nees)
T.Anderson – F.P.S. 3: 166 (1956); T.S.S.: 433 (1990).
Shrub. Forest, including riverine & montane forest.
Distr: EQU; D.R. Congo, Uganda S to Malawi.
EQU: Sillitoe 205 1919.

Anisotes pubinervis (T.Anderson) Heine

F.T.E.A. Acanth.: 654 (2010).
Syn: *Metarungia pubinervia* (T.Anderson) Baden –
Biol. Skr. 51: 450 (2005); F.E.E. 5: 488 (2006).
Shrub. Forest.
Distr: EQU; Nigeria, D.R. Congo to Ethiopia, S to South
Africa.
EQU: Friis & Vollesen 747 (fl) 12/1980.

Asystasia africana (S.Moore) C.B.Clarke

Biol. Skr. 51: 436 (2005); F.T.E.A. Acanth.: 462 (2010).
Subshrub. Forest.
Distr: EQU; Cameroon, Gabon, D.R. Congo, Uganda,
Tanzania, Angola.
IUCN: RD
EQU: Andrews 1763 (fl) 6/1939.

Asystasia gangetica (L.) T.Anderson subsp. micrantha (Nees) Ensermu

F.P.S. 3: 166 (1956); F.J.M.: 148 (1976); Biol. Skr. 51: 436
(2005); F.E.E. 5: 398 (2006); F.T.E.A. Acanth.: 459 (2010).
Perennial herb. Wooded grassland, woodland, forest
margins, disturbed areas.
Distr: DAR, CS, UN, EQU; Senegal to Somalia, S to South
Africa; Madagascar, Mascarenes, Arabia, introduced
elsewhere.
CS: Andrews 17 (fl fr) 6/1935; **EQU:** Myers 7748 (fl)
9/1937.

Asystasia mysorensis (Roth) T.Anderson

Biol. Skr. 51: 437 (2005); F.E.E. 5: 400 (2006); F.T.E.A. Acanth.: 466 (2010).
Annual herb. Disturbed areas in grassland & bushland.
Distr: EQU; Ethiopia to Namibia & South Africa; Yemen, India.
EQU: Kielland-Lund 556 1/1984 (n.v.).

Asystasia vogeliana Benth.

Biol. Skr. 51: 437 (2005); F.T.E.A. Acanth.: 448 (2010).
Perennial herb or subshrub. Forest & secondary growth.
Distr: EQU; Guinea to Uganda, S to D.R. Congo & Tanzania.
EQU: Friis & Vollesen 478 (fl) 11/1980.

Avicennia marina (Forssk.) Vierh.

F.P.S. 3: 193 (1956); T.S.S.: 435 (1990); F.E.E. 5: 515 (2006).
Mangrove shrub or tree. Mangroves.
Distr: RS; Egypt to South Africa; Arabia, coasts of Asia & W Pacific.
RS: Jackson 3926 4/1959.
Note: this is the common mangrove species in Sudan.

Barleria acanthoides Vahl

F.P.S. 3: 169 (1956); Biol. Skr. 51: 437 (2005); F.E.E. 5: 411 (2006); F.T.E.A. Acanth.: 349 (2010).
Syn: *Barleria candida* sensu Andrews, non Nees – F.P.S. 3: 167 (1956).
Spiny subshrub. Dry bushland, rocky semi-desert.
Distr: DAR, KOR, CS, EQU; Eritrea & Somalia, S to Tanzania; Egypt, Arabia, Pakistan, India.
KOR: Kotschy 58 (fl fr) 4/1837; **EQU:** Carr 808 (fr) 6/1970.

Barleria argentea Balf.f.

F.E.E. 5: 422 (2006); F.T.E.A. Acanth.: 406 (2010).
Perennial herb or subshrub. Dry bushland, semi-desert.
Distr: EQU; Ethiopia, Somalia, Uganda, Kenya, Tanzania; Arabia, Socotra.
EQU: Padwa 235 (fl) 5/1953.

Barleria brownii S.Moore

Biol. Skr. 51: 437 (2005); F.E.E. 5: 404 (2006); F.T.E.A. Acanth.: 358 (2010).
Scandent perennial herb or subshrub. Forest.
Distr: EQU; Ghana to Ethiopia, S to Angola & Tanzania.
EQU: Friis & Vollesen 731 (fl) 12/1980 (n.v.).

Barleria calophylla Lindau

F.P.S. 3: 169 (1956).
Suffrutex. Wooded grassland.
Distr: BAG, EQU; C.A.R.
IUCN: RD
EQU: Myers 9372 (fl) 8/1938.

Barleria eranthemoides C.B.Clarke var. *eranthemoides*

F.P.S. 3: 168 (1956); F.E.E. 5: 420 (2006); F.T.E.A. Acanth.: 436 (2010).
Spiny subshrub. Dry bushland & grassland, particularly degraded areas.
Distr: DAR, KOR; Nigeria to Somalia, S to Tanzania.
KOR: Wickens 837 (fl) 11/1962.

Barleria grandicalyx Lindau

F.P.S. 3: 169 (1956); F.E.E. 5: 408 (2006); F.T.E.A. Acanth.: 334 (2010).
Syn: *Barleria grandicalyx* Lindau var. *vix-dentata* C.B.Clarke – F.P.S. 3: 169 (1956).
Spiny suffrutex. Fire-prone grassland & woodland.
Distr: BAG, UN, EQU; C.A.R. to Ethiopia, S to D.R. Congo & Tanzania.
EQU: Johnston 14K (fl) 4/1932.

Barleria hochstetteri Nees

F.P.S. 3: 169 (1956); F.E.E. 5: 423 (2006); F.T.E.A. Acanth.: 412 (2010).
Subshrub. Dry bushland & grassland.
Distr: RS, DAR, KOR, CS; Eritrea, Ethiopia, Somalia, Kenya; Egypt, Arabia, India.
KOR: Jackson 4357 (fl fr) 9/1961.

Barleria lanceata (Forssk.) C.Chr.

F.E.E. 5: 408 (2006).
Syn: *Barleria triacantha* Nees – F.P.S. 3: 168 (1956).
Spiny subshrub. Dry woodland on rocky slopes, rock crevices.
Distr: RS; Eritrea, Djibouti, Somalia; Saudi Arabia, Yemen.
IUCN: RD
RS: Jackson 3939 (fr).

Barleria linearifolia Rendle

F.E.E. 5: 419, pro parte (2006); F.T.E.A. Acanth.: 433 (2010).
Spiny subshrub. Dry bushland & grassland.
Distr: EQU; Ethiopia, Somalia, Uganda, Kenya.
EQU: Padwa 243 (fl) 5/1953.

Barleria parviflora T.Anderson

F.P.S. 3: 169 (1956); F.E.E. 5: 423 (2006).
Perennial herb or subshrub. Dry bushland.
Distr: RS; Eritrea, Ethiopia, Somalia; Arabia.
RS: Schweinfurth 255 (fl) 9/1868.

Barleria steudneri C.B.Clarke

F.E.E. 5: 413 (2006); F.T.E.A. Acanth.: 351 (2010).
Spiny subshrub. Dry bushland & grassland.
Distr: KOR; Eritrea, Ethiopia, Uganda, Kenya.
KOR: Andrews 209 (fl) 11/1933.

Barleria submollis Lindau

Biol. Skr. 51: 437 (2005); F.E.E. 5: 407 (2006); F.T.E.A. Acanth.: 369 (2010).
Perennial herb or subshrub. Bushland & woodland on rocky slopes.
Distr: EQU; Ethiopia, Somalia, Uganda, Kenya, Tanzania.
EQU: Kielland-Lund 133 11/1983 (n.v.).

Barleria tetraglochin Milne-Redh.

F.P.S. 3: 167 (1956); F.T.E.A. Acanth.: 434, in notes (2010).
Syn: *Barleria linearifolia* sensu Ensermu pro parte, non Rendle – F.E.E. 5: 419 (2006).
Spiny suffrutex. Dry bushland on clay.
Distr: KOR, CS; Sudan endemic.
IUCN: RD
KOR: Dandy 322 (fl fr) 1/1934.
Note: Andrews lists also *B. prionitis* L. from the Nuba Mts (Jebel Daier). This is an Indian species; in Sudan the closest allies are *B. tetraglochin* and *B. trispinosa*.

Barleria trispinosa (Forssk.) Vahl subsp. trispinosa

F.P.S. 3: 168 (1956); F.E.E. 5: 415 (2006); F.T.E.A. Acanth.: 424 (2010).
Syn: *Barleria diacantha* Nees – F.P.S. 3: 168 (1956); Bot. Exp. Sud.: 11 (1970).
Spiny subshrub. Dry bushland & grassland, rocky semi-desert.
Distr: RS; Eritrea, Ethiopia; Arabia.
RS: Jackson 3973 (fl) 4/1959.

Barleria ventricosa Nees

F.P.S. 3: 170 (1956); Biol. Skr. 51: 437 (2005); F.E.E. 5: 404 (2006); F.T.E.A. Acanth.: 360 (2010).
Syn: *Barleria vix-dentata* C.B.Clarke – Biol. Skr. 51: 438 (2005).
Perennial herb, sometimes scandent. Montane forest & forest margins, transitional bushland.
Distr: KOR, BAG, EQU; Eritrea & Somalia, S to Zimbabwe & Mozambique.
KOR: Broun 1289 (fl fr) 12/1907; **EQU:** Andrews 1898 (fl) 6/1939.
Note: Sudanese material includes the form sometimes separated as *B. vix-dentata* C.B.Clarke as well as typical *B. ventricosa*.

Barleria sp. A

Suffrutex. Clay soils.
Distr: UN; South Sudan endemic.
IUCN: RD
UN: Sherif A.3958 (fl) 8/1951.
Note: this is a distinctive, apparently undescribed species of *Barleria* but the flowers are in bud only and more material is needed for formal description.

Blepharis edulis (Forssk.) Pers.

Biol. Skr. 51: 438 (2005); F.E.E. 5: 359 (2006); F.T.E.A. Acanth.: 102 (2008).
Syn: *Blepharis ciliaris* sensu Drar, non (L.) B.L.Burtt – Bot. Exp. Sud.: 11 (1970).
Syn: *Blepharis persica* sensu Andrews, non (Burm.f.) Kuntze – F.P.S. 3: 171 (1956).
Annual or perennial herb. *Acacia-Commiphora* woodland & bushland, semi-desert, *Combretum* woodland.
Distr: NS, RS, KOR, CS, EQU; Mauritania to Somalia, S to Tanzania; Egypt, Arabia, Iran.
RS: Jackson 2833 (fl) 4/1953.

Blepharis integrifolia (L.f.) Schinz var. integrifolia

Biol. Skr. 51: 438 (2005); F.E.E. 5: 361 (2006); F.T.E.A. Acanth.: 106 (2008).
Perennial herb. Grassland, dry woodland, disturbed areas.
Distr: EQU; Eritrea to Tanzania, Angola to Zimbabwe & South Africa; India & Sri Lanka.
EQU: Kielland-Lund 621 1/1984 (n.v.).

Blepharis involucrata Solms

F.T.E.A. Acanth.: 112 (2008).
Annual herb. *Acacia* woodland & bushland, on seasonally waterlogged soils or over hard pans.
Distr: KOR; Cameroon, Chad, Tanzania, Zambia, Mozambique & Zimbabwe.
KOR: Cienkowsky 360 9/1848.

Blepharis linariifolia Pers.

F.P.S. 3: 171 (1956); F.J.M.: 148 (1976); F.E.E. 5: 359 (2006).
Annual herb. Wooded grassland & bushland.
Distr: DAR, KOR, CS, ES, BAG; Mauritania & Senegal to Ethiopia.
DAR: Harrison 160 (fl) 9/1947.
Note: the record for South Sudan is based on the map in Vollesen, Blepharis: 101 (2000); we have not seen the specimen from Bahr el Ghazal.

Blepharis maderaspatensis (L.) Roth

F.P.S. 3: 171 (1956); F.J.M.: 148 (1976); Biol. Skr. 51: 438 (2005); F.E.E. 5: 363 (2006); F.T.E.A. Acanth.: 109 (2008).
Perennial herb. Woodland & wooded grassland, often in disturbed areas.
Distr: RS, DAR, KOR, CS, BAG, UN, EQU; Senegal to Somalia, S to South Africa; Madagascar, Comoros, Arabia to China & Vietnam.
DAR: Wickens 1086 (fl fr) 1/1964; **UN:** Jackson 4124 (fl) 2/1961.

Brachystephanus sudanicus (Friis & Vollesen) Champl.

Syst. Geogr. Pl. 79: 140 (2009).
Syn: *Oreacanthus sudanicus* Friis & Vollesen – Biol. Skr. 51: 452 (2005).
Perennial herb. Submontane forest.
Distr: EQU; South Sudan endemic.
IUCN: RD
EQU: Friis & Vollesen 448 (fl) 11/1980.

Brillantaisia cicatricosa Lindau

Biol. Skr. 51: 439 (2005); F.T.E.A. Acanth.: 159 (2008).
Perennial herb or shrub. Montane & submontane forest, forest margins.
Distr: EQU; D.R. Congo & Uganda, S to Mozambique & Zimbabwe.
EQU: Friis & Vollesen 58 (fl) 11/1980.

Brillantaisia lamium (Nees) Benth.

Biol. Skr. 51: 439 (2005); F.T.E.A. Acanth.: 156 (2008).
Syn: *Brillantaisia eminii* Lindau – F.P.S. 3: 172 (1956).
Perennial herb. Forest margins, swampy areas in forest, streamsides.
Distr: EQU; Guinea to Kenya, S to Angola & Tanzania.
EQU: Myers 7837 (fl) 10/1937.

Brillantaisia madagascariensis Lindau

Biol. Skr. 51: 439 (2005); F.E.E. 5: 372 (2006); F.T.E.A. Acanth.: 162 (2008).
Perennial herb. Forest, including margins & swampy areas.
Distr: UN, EQU; Guinea to Ethiopia, S to Zambia; Madagascar.
EQU: Friis & Vollesen 228 (fl fr) 11/1980.

Brillantaisia owariensis P.Beauv.

Biol. Skr. 51: 439 (2005); F.T.E.A. Acanth.: 157 (2008).
Syn: *Brillantaisia nyanzarum* Burkill – F.P.S. 3: 171 (1956).
Perennial herb or subshrub. Forest & forest margins.
Distr: EQU; Sierra Leone to Kenya, S to Angola & Tanzania.
EQU: Dandy 554 (fl) 2/1934.

Brillantaisia vogeliana (Nees) Benth.

Biol. Skr. 51: 440 (2005); F.T.E.A. Acanth.: 158 (2008).
Annual or short-lived perennial herb. Forest & forest margins.
Distr: EQU; Ghana to Kenya.
EQU: Friis & Vollesen 612 (fl fr) 12/1980.

Crabbea velutina S.Moore

Biol. Skr. 51: 440 (2005); F.E.E. 5: 426 (2006); F.T.E.A.
Acanth.: 297 (2010).
Perennial herb. Woodland & bushland on rocky slopes,
termite mounds.
Distr: EQU; Ethiopia & Somalia, S to Namibia & South
Africa.
EQU: Friis & Vollesen 1148 (fl) 3/1982.

Crossandra massaica Mildbr.

Biol. Skr. 51: 440 (2005); F.E.E. 5: 365 (2006); F.T.E.A.
Acanth.: 140 (2008).
Perennial herb or subshrub, sometimes scandent.
Bushland on rocky hillslopes.
Distr: EQU; Ghana, Ethiopia, Uganda, Kenya, Tanzania.
EQU: Jackson 1297 3/1950.

Crossandra nilotica Oliv.

Biol. Skr. 51: 440 (2005); F.E.E. 5: 364 (2006); F.T.E.A.
Acanth.: 139 (2008).
Perennial herb or subshrub. Dry forest including
degraded areas, riverine forest & scrub.
Distr: EQU; Ethiopia to Angola & Zambia.
EQU: Myers 14048 (fl) 9/1941.

Crossandra subacaulis C.B.Clarke

F.P.S. 3: 172 (1956); Biol. Skr. 51: 441 (2005); F.T.E.A.
Acanth.: 144 (2008).
Suffrutex. Fire-prone woodland & wooded grassland.
Distr: BAG, EQU; C.A.R., D.R. Congo, Uganda, Kenya,
Tanzania.
EQU: Johnston 10K (fl) 4/1932.

Dicliptera lanceolata (Lindau) I.Darbysh. & Kordofani

Kew Bull. 66: 605 (2012).
Syn: *Peristrophe lanceolata* (Lindau) Dandy – F.P.S. 3:
185 (1956).
Perennial herb. Open marshland.
Distr: BAG, EQU; South Sudan endemic.
IUCN: RD
BAG: Schweinfurth III.23 (fl fr) 2/1871.

Dicliptera latibracteata I.Darbysh.

F.T.E.A. Acanth.: 703 (2010).
Syn: *Dicliptera umbellata* sensu Friis & Vollesen pro
parte, non (Vahl) Juss. – Biol. Skr. 51: 442 (2005).
Perennial herb or subshrub. Montane forest & forest
margins.
Distr: EQU; Kenya.
IUCN: RD
EQU: Johnston 1461 (fl) 2/1936.

Dicliptera laxata C.B.Clarke

Biol. Skr. 51: 441 (2005); F.E.E. 5: 443 (2006); F.T.E.A.
Acanth.: 694 (2010).
Perennial herb or subshrub. Submontane & montane
forest, forest margins.

Distr: EQU; Nigeria, Bioko, Cameroon, Ethiopia to
Malawi.
EQU: Friis & Vollesen 117 (fl) 11/1980.

Dicliptera maculata Nees subsp. maculata

F.P.S. 3: 173 (1956); Biol. Skr. 51: 441 (2005); F.E.E. 5:
444 (2006); F.T.E.A. Acanth.: 701 (2010).
Syn: *Dicliptera umbellata* sensu auctt., non (Vahl) Juss.
– F.P.S. 3: 173 (1956); Biol. Skr. 51: 442 (2005).
Perennial herb, sometimes scandent. Montane &
submontane forest, forest margins.
Distr: EQU; Cameroon to Ethiopia, S to Angola &
Zambia.
EQU: Friis & Vollesen 684 (fl) 12/1980.

Dicliptera paniculata (Forssk.) I.Darbysh.

F.T.E.A. Acanth.: 716 (2010).
Syn: *Peristrophe bicalyculata* (Retz.) Nees – F.P.S. 3:
185 (1956); F.J.M.: 150 (1976).
Syn: *Peristrophe paniculata* (Forssk.) Brummitt – Biol.
Skr. 51: 452 (2005); F.E.E. 5: 447 (2006).
Annual herb. *Acacia-Commiphora* bushland, grassland,
roadsides, old cultivations.
Distr: RS, DAR, KOR, CS, EQU; Senegal to Somalia, S to
Namibia & Mozambique; Egypt, Arabia, India, Thailand.
DAR: Wickens 2796 (fl) 10/1964; **EQU:** Padwa 279 (fl)
6/1953.

Dicliptera pumila (Lindau) Dandy

F.P.S. 3: 173 (1956); Biol. Skr. 51: 441 (2005); F.E.E. 5:
445 (2006); F.T.E.A. Acanth.: 713 (2010).
Pyrophytic perennial herb. Fire-prone grassland & open
woodland.
Distr: EQU; D.R. Congo to Ethiopia, S to Angola &
Zimbabwe.
EQU: Myers 8466 (fl) 2/1938.

Dicliptera verticillata (Forssk.) C.Chr.

F.P.S. 3: 173 (1956); F.E.E. 5: 445 (2006); F.T.E.A. Acanth.:
697 (2010).
Annual herb. *Acacia-Commiphora* bushland & woodland,
old cultivations.
Distr: DAR, KOR, CS, UN; Cape Verde Is., Senegal to
Ethiopia, S to South Africa; Madagascar, Yemen, India.
CS: Andrews 14 (fl fr) 6/1935; **UN:** Evans-Pritchard 40
(fl) 11/1936.

Duosperma longicalyx (Deflers) Vollesen subsp. longicalyx

F.E.E. 5: 373 (2006); F.T.E.A. Acanth.: 246 (2008).
Perennial herb or shrub. *Acacia-Commiphora* bushland,
dry riverine scrub.
Distr: EQU; Ethiopia, Somalia, Uganda, Kenya.
EQU: Padwa 242 (fl) 6/1953.

Dyschoriste multicaulis (A.Rich.) Kuntze

F.E.E. 5: 376 (2006); F.T.E.A. Acanth.: 184 (2008).
Syn: *Dyschoriste clinopodioides* Mildbr. – Biol. Skr. 51:
442 (2005).
Perennial herb. Grassland, bushland, clearings in forest,
disturbed habitats.
Distr: EQU; Ethiopia to Burundi & Tanzania.
EQU: Friis & Vollesen 950 (fl) 2/1982.

Dyschoriste nagchana (Nees) Bennet

Biol. Skr. 51: 442 (2005); F.E.E. 5: 177 (2006); F.T.E.A.
Acanth.: 183 (2008).
Syn: *Dyschoriste perrottetii* (Nees) Kuntze – F.P.S. 3:
173 (1956); F.J.M.: 149 (1976).
Perennial herb. Grassland, swamps, riverine forest,
disturbed habitats.
Distr: DAR, CS, BAG, EQU; tropical Africa, India.
DAR: Wickens 1175 (fl) 2/1964; **EQU:** Friis &
Vollesen 676 (fl fr) 12/1980.

Dyschoriste radicans Nees

F.J.M.: 149 (1976); Biol. Skr. 51: 442 (2005); F.E.E. 5: 376
(2006); F.T.E.A. Acanth.: 186 (2008).
Perennial herb. Grassland, bushland & woodland
including disturbed areas.
Distr: RS, DAR; Ethiopia & Somalia, S to Burundi &
Tanzania; Yemen.
DAR: Wickens 2430 (fl fr) 9/1964.

Ecbolium gymnostachyum (Nees) Milne-Redh.

F.E.E. 5: 455 (2006).
Shrub. *Acacia* woodland & bushland.
Distr: RS; Eritrea, Ethiopia, Djibouti, Somalia; Arabia.
RS: Broun 1236A (fr) 5/1907.

Ecbolium viride (Forssk.) Alston

F.P.S. 3: 174 (1956); F.E.E. 5: 456 (2006); F.T.E.A. Acanth.:
665 (2010).
Shrub. *Acacia-Commiphora* bushland, dry rocky
outcrops, riverbeds.
Distr: RS; Eritrea, Ethiopia, Djibouti, Somalia, Kenya;
Arabia, India.
RS: Bally 7008 (fl fr) 4/1949.

Elytraria marginata Vahl

Biol. Skr. 51: 443 (2005); F.T.E.A. Acanth.: 14 (2008).
Annual or short-lived perennial herb. Forest.
Distr: EQU; Guinea to Uganda, S to Angola.
EQU: Schweinfurth III.225 (fl) 7/1870.

Eremomastax speciosa (Hochst.) Cufod.

Biol. Skr. 51: 443 (2005); F.E.E. 5: 378 (2006); F.T.E.A.
Acanth.: 200 (2008).
Syn: *Eremomastax polysperma* (Benth.) Dandy – F.P.S.
3: 174 (1956).
Perennial herb. Forest & forest margins, including
streamsides.
Distr: EQU; Guinea to Ethiopia, S to Tanzania;
Madagascar.
EQU: Thomas 1733 (fl) 12/1935.

Hygrophila abyssinica (Hochst. ex Nees) T.Anderson

F.J.M.: 149 (1976); F.E.E. 5: 381 (2006); F.T.E.A. Acanth.:
174 (2008).
Syn: *Hemigraphis abyssinica* (Nees) C.B.Clarke – F.P.S.
3: 174 (1956).
Syn: *Hemigraphis schweinfurthii* (S.Moore) C.B.Clarke
pro parte quoad *Schweinfurth* 2799 – F.P.S. 3: 174 (1956).
Annual or short-lived perennial herb. Seepage areas and
sandy soils in fire-prone woodland.
Distr: DAR, CS, UN, EQU; Ghana to Ethiopia, S to
Namibia & Botswana.

DAR: Wickens 1174 (fl) 2/1964; **UN:** Ferguson 380 (fr)
1948.

Hygrophila africana (T.Anderson) Heine

F.W.T.A. 2: 395 (1963).
Syn: *Hemigraphis schweinfurthii* (S.Moore) C.B.Clarke
pro parte quoad *Schweinfurth* 2708, 2764 – F.P.S. 3: 174
(1956).
Prostrate ?perennial herb. Seasonal swamps.
Distr: BAG; Mali, Burkina Faso, Ghana, Nigeria,
Cameroon, Tanzania.
BAG: Schweinfurth 2708 (fl) 12/1869.

Hygrophila caerulea (Hochst.) T.Anderson

F.P.S. 3: 176 (1956).
Prostrate herb. Dambo margins in wooded grassland.
Distr: KOR, UN; Sudan / South Sudan endemic.
IUCN: RD
KOR: Wickens 918 (fl fr) 1/1963; **UN:** J.M. Lock 80/49
(fl) 3/1980.

Hygrophila schulli (Buch.-Ham.) M.R.Almeida & S.M.Almeida

F.E.E. 5: 379 (2006); F.T.E.A. Acanth.: 176 (2008).
Syn: *Asteracantha longifolia* (L.) Nees – F.P.S. 3: 166
(1956).
Syn: *Hygrophila auriculata* (Schumach.) Heine – F.J.M.:
149 (1976); Biol. Skr. 51: 443 (2005).
Annual herb. Seasonally inundated & swampy grassland,
ditches.
Distr: DAR, KOR, CS, ES, BAG, UN, EQU; tropical Africa,
India.
DAR: Wickens 2735 (fl) 9/1964; **EQU:** Myers 7839 (fl)
10/1937.

Hygrophila uliginosa S.Moore

F.P.S. 3: 176 (1956); F.T.E.A. Acanth.: 166 (2008).
Syn: *Hygrophila acutisepala* Burkill – F.P.S. 3: 176 (1956).
Perennial herb. Seasonally inundated grassland, swamps,
riverbanks.
Distr: EQU; Nigeria to C.A.R., S to Angola & Zambia.
EQU: Petherick s.n. (fl).

Hypoestes aristata (Vahl) Roem. & Schult.

Biol. Skr. 51: 443 (2005); F.E.E. 5: 449 (2006); F.T.E.A.
Acanth.: 724 (2010).
Perennial herb or subshrub. Forest, forest margins &
secondary scrub.
Distr: EQU; Nigeria to Ethiopia, S to South Africa.
EQU: Friis & Vollesen 120 (fl) 11/1980.

Hypoestes cancellata Nees

F.P.S. 3: 176 (1956); F.T.E.A. Acanth.: 722 (2010).
Annual herb. Grassland & open bushland, roadsides.
Distr: BAG; Senegal to Ivory Coast, Nigeria to Uganda, S
to Angola & Tanzania.
BAG: Schweinfurth 2525 (fl fr) 10/1869.

Hypoestes forskaolii (Vahl) R.Br. subsp. forskaolii

F.J.M.: 149 (1976); Biol. Skr. 51: 444 (2005); F.E.E. 5: 449
(2006); F.T.E.A. Acanth.: 726 (2010).
Syn: *Hypoestes verticillaris* sensu Andrews, non (L.f.)
Roem. & Schult. – F.P.S. 3: 177 (1956).

Perennial herb. Forest margins, woodland, bushland & grassland, often in disturbed areas.
Distr: RS, DAR, KOR, ES, UN, EQU; tropical & southern Africa; Madagascar, Saharan highlands, Arabia.
DAR: <u>Wickens 1384</u> (fl fr) 4/1964; **EQU:** <u>Howard I.M.43</u> (fl) 2/1976.

Hypoestes strobilifera S.Moore var. *strobilifera*
F.P.S. 3: 177 (1956); F.T.E.A. Acanth.: 723 (2010).
Annual herb. Grassland, open bushland, roadsides.
Distr: BAG, EQU; D.R. Congo.
IUCN: RD
BAG: <u>Schweinfurth 2553</u> (fl) 10/1869.

Hypoestes triflora (Forssk.) Roem. & Schult.
F.P.S. 3: 177 (1956); Biol. Skr. 51: 444 (2005); F.E.E. 5: 450 (2006); F.T.E.A. Acanth.: 720 (2010).
Syn: *Hypoestes consanguinea* Lindau – Biol. Skr. 51: 444 (2005).
Annual or perennial herb. Forest & forest margins.
Distr: EQU; tropical & southern Africa; Arabia, India, China, Thailand.
EQU: <u>Friis & Vollesen 900</u> (fl) 2/1982.

Isoglossa membranacea C.B.Clarke subsp. *septentrionalis* I.Darbysh.
F.T.E.A. Acanth.: 623 (2010).
Syn: *Isoglossa punctata* sensu Friis & Vollesen, pro parte, non (Vahl) Brummitt & J.R.I.Wood – Biol. Skr. 51: 445 (2005).
Perennial herb or subshrub. Forest.
Distr: EQU; D.R. Congo, Uganda, Kenya.
IUCN: RD
EQU: <u>Friis & Vollesen 744</u> (fl) 12/1980.

Isoglossa punctata (Vahl) Brummitt & J.R.I.Wood
Biol. Skr. 51: 445, pro parte (2005); F.E.E. 5: 493 (2006); F.T.E.A. Acanth.: 620 (2010).
Perennial herb. Montane forest & margins.
Distr: EQU; D.R. Congo, Rwanda, Burundi, Ethiopia, Uganda, Tanzania; Yemen.
EQU: <u>Friis & Vollesen 31</u> (fl) 11/1980.

Isoglossa somalensis Lindau
Biol. Skr. 51: 445 (2005); F.E.E. 5: 494 (2006).
Syn: *Isoglossa ovata* sensu Andrews, non E.A.Bruce – F.P.S. 3: 177 (1956).
Perennial herb or shrub. Montane forest & forest margins.
Distr: EQU; Ethiopia.
EQU: <u>Friis & Vollesen 5</u> (fl) 11/1980.

Justicia afromontana Hedrén
Biol. Skr. 51: 445 (2005); F.T.E.A. Acanth.: 577 (2010).
Syn: *Justicia whytei* sensu Andrews pro parte, non S.Moore – F.P.S. 3: 180 (1956).
Perennial herb. Montane grassland, ericaceous bushland.
Distr: EQU; Uganda.
IUCN: RD
EQU: <u>Johnston 1510</u> (fl fr) 2/1936.
Note: Andrews' application of the name *J. whytei* is in part referable here; however, it is unclear to which taxon his Erkowit specimens of '*J. whytei*' refers.

Justicia anselliana (Nees) T.Anderson
Biol. Skr. 51: 446 (2005); F.T.E.A. Acanth.: 564 (2010).
Annual herb. Woodland, bushland & wooded grassland.
Distr: BAG, UN, EQU; Liberia to Kenya, S to Namibia & Zambia.
UN: <u>Simpson 7023</u> (fl) 6/1929.

Justicia betonica L.
F.P.S. 3: 178 (1956); Biol. Skr. 51: 446 (2005); F.E.E. 5: 468 (2006); F.T.E.A. Acanth.: 549 (2010).
Syn: *Justicia trinervia* Vahl – F.P.S. 3: 178 (1956).
Perennial herb. Woodland, often in disturbed areas.
Distr: ?KOR, BAG, UN, EQU; palaeotropical, introduced in the neotropics.
UN: <u>Jackson 4122</u> (fl) 2/1961.
Note: no material has been seen for Sudan; the record for Kordofan is based on Andrews.

Justicia bracteata (Hochst.) Zarb
F.T.E.A. Acanth.: 593 (2010).
Syn: *Monechma debile* sensu auctt., non (Forssk.) Nees – F.P.S. 3: 184, pro parte (1956); Bot. Exp. Sud.: 12 (1970); Biol. Skr. 51: 451 (2005); F.E.E. 5: 485, pro parte (2006).
Annual herb. Woodland, bushland & grassland, including disturbed areas.
Distr: RS, DAR, KOR, CS, ES, UN, EQU; Eritrea & Somalia, S to Namibia & South Africa; Arabia, India.
KOR: <u>Wickens 828</u> (fl fr) 11/1961; **UN:** <u>Sherif A.2844</u> (fr) 9/1951.

Justicia caerulea Forssk.
Biol. Skr. 51: 446 (2005); F.E.E. 5: 475 (2006); F.T.E.A. Acanth.: 537 (2010).
Perennial herb. *Acacia-Commiphora* woodland, wooded grassland.
Distr: EQU; Ethiopia & Somalia, S to Tanzania; Yemen.
EQU: <u>Padwa 232</u> (fl fr) 5/1953.

Justicia calyculata Deflers
F.E.E. 5: 482 (2006); F.T.E.A. Acanth.: 561 (2010).
Annual or short-lived perennial herb. Bushland & wooded grassland, disturbed areas.
Distr: EQU; Ethiopia, Somalia, Tanzania; Yemen.
EQU: <u>Padwa 275</u> (fl fr) 6/1953.

Justicia ciliata Jacq.
F.T.E.A. Acanth.: 592 (2010).
Syn: *Monechma ciliatum* (Jacq.) Milne-Redh. – F.P.S. 3: 184 (1956); F.J.M.: 150 (1976); Biol. Skr. 51: 450 (2005); F.E.E. 5: 485 (2006).
Annual herb. Woodland on black clay soils.
Distr: DAR, KOR, ES, BAG; Senegal to Ethiopia, S to Malawi & Zambia.
KOR: <u>Jackson 4019</u> (fr) 9/1959; **BAG:** <u>Myers 7279</u> (fl) 7/1937.

Justicia debilis (Forssk.) Vahl
F.T.E.A. Acanth.: 596 (2010).
Syn: *Monechma debile* (Forssk.) Nees – F.P.S. 3: 184, pro parte (1956); F.E.E. 5: 485, pro parte (2006).
Perennial herb. Dry woodland, bushland & grassland.
Distr: RS; Eritrea & Somalia to Tanzania; Arabia.
RS: <u>Sahni & Kamil 659</u> (fl fr) 4/1967.

Justicia depauperata T.Anderson

Syn: *Monechma depauperatum* (T.Anderson)
C.B.Clarke – F.W.T.A. 2: 429 (1963).
Perennial herb. Wet grassland, forest margins.
Distr: BAG; Senegal to C.A.R., S to Angola.
BAG: <u>Broun s.n.</u> (fl fr) 3/1902.
Note: the single specimen seen from South Sudan is a
small scrap; more ample material is needed to confirm
this record.

Justicia exigua S.Moore

Biol. Skr. 51: 446 (2005); F.E.E. 5: 482 (2006); F.T.E.A.
Acanth.: 562 (2010).
Annual herb. Woodland & bushland, often in disturbed
areas.
Distr: UN, EQU; Ethiopia to Namibia & South Africa.
EQU: <u>Kielland-Lund 201</u> (fl) 12/1983 (n.v.).

Justicia extensa T.Anderson

Biol. Skr. 51: 447 (2005); F.T.E.A. Acanth.: 511 (2010).
Perennial herb or shrub. Forest.
Distr: EQU; Guinea to Kenya, S to Angola.
EQU: <u>Friis & Vollesen 458</u> (fl) 11/1980.

Justicia flava (Vahl) Vahl

F.P.S. 3: 179 (1956); Biol. Skr. 51: 447 (2005); F.E.E. 5:
471 (2006); F.T.E.A. Acanth.: 532 (2010).
Syn: *Justicia palustris* (Hochst.) T.Anderson – F.P.S. 3:
179 (1956); F.E.E. 5: 474 (2006).
Perennial herb. Forest, forest margins, grassland, swampy
or disturbed ground.
Distr: RS, DAR, KOR, CS, ES, UN, EQU; Senegal to
Ethiopia, S to South Africa; Arabia.
RS: <u>Aylmer 238</u> (fl) 3/1932; **EQU:** <u>Andrews 1688</u> (fr)
6/1939.
Note: Ensermu in F.E.E. maintains *J. palustris* as distinct
but Vollesen in F.T.E.A. considers it to be a form of the
very variable *J. flava*.

Justicia grandis (T.Anderson) Vollesen

F.T.E.A. Acanth.: 547 (2010).
Syn: *Rungia grandis* T.Anderson – Biol. Skr. 51: 454
(2005); F.E.E. 5: 490 (2006).
Shrub. Forest.
Distr: EQU; Nigeria to Ethiopia, S to Angola & Uganda.
EQU: <u>Friis & Vollesen 1059</u> (fr) 12/1980.

Justicia heterocarpa T.Anderson subsp. heterocarpa

F.P.S. 3: 181 (1956); F.E.E. 5: 478 (2006); F.T.E.A. Acanth.:
581 (2010).
Annual herb. *Acacia-Commiphora* & *Combretum-
Terminalia* woodland, disturbed areas.
Distr: RS; Eritrea to Tanzania; Arabia, Pakistan.
RS: <u>Jackson 2715</u> (fl fr) 3/1953.

Justicia ladanoides Lam.

Biol. Skr. 51: 447 (2005); F.E.E. 5: 476 (2006); F.T.E.A.
Acanth.: 572 (2010).
Syn: *Justicia galeopsis* C.B.Clarke – F.P.S. 3: 180 (1956).
Syn: *Justicia kotschyi* (Hochst.) Dandy – F.P.S. 3: 180
(1956); F.J.M.: 149 (1976).
Syn: *Justicia schimperi* (Hochst.) Dandy – F.P.S. 3: 180
(1956).

Annual or short-lived perennial herb. Forest & bushland.
Distr: DAR, KOR, ES, BAG, UN, EQU; Senegal to
Ethiopia, S to D.R. Congo & Kenya.
KOR: <u>Musselman 6264</u> (fl) 8/1983; **UN:** <u>J.M.
Lock 81/148</u> (fl) 8/1981.

Justicia matammensis (Schweinf.) Oliv.

F.P.S. 3: 180 (1956); Biol. Skr. 51: 447 (2005); F.E.E. 5:
482 (2006); F.T.E.A. Acanth : 565 (2010).
Annual herb. *Acacia* & *Combretum-Terminalia* woodland,
wooded grassland.
Distr: CS, ES, BAG, UN, EQU; C.A.R. to Ethiopia, S to
South Africa
CS: <u>Lea 233</u> (fl fr); **EQU:** <u>Jackson 3841</u> (fl fr) 9/1959.

Justicia ndellensis Lindau

Bull. Soc. Bot. France 55 (Mem. 8b): 52 (1908).
Suffrutex. Crevices in large rock outcrops, dry rocky
woodland.
Distr: DAR, KOR; Senegal to C.A.R.
KOR: <u>Musselman 6273a</u> (fl) 8/1983.

Justicia nyassana Lindau

F.P.S. 3: 179 (1956); Biol. Skr. 51: 448 (2005); F.E.E. 5:
474 (2006); F.T.E.A. Acanth.: 539 (2010).
Perennial herb. Forest & forest margins.
Distr: EQU; C.A.R. to Ethiopia, S to Zimbabwe.
EQU: <u>Thomas 1672</u> (fl fr) 12/1935.

Justicia odora (Forssk.) Lam.

F.P.S. 3: 180 (1956); F.E.E. 5: 465 (2006); F.T.E.A. Acanth.:
565 (2010).
Perennial herb or shrub. *Acacia-Commiphora* woodland,
often on black cotton soils.
Distr: RS; Ethiopia & Somalia, S to Namibia & South
Africa; Arabia.
RS: <u>Robbie 3893</u> (fl) 1938.

Justicia ornatopila Ensermu

F.E.E. 5: 480 (2006); F.T.E.A. Acanth.: 559 (2010).
Subshrub. *Acacia-Commiphora* bushland, wooded
grassland.
Distr: EQU; Ethiopia & Somalia, S to Tanzania.
EQU: <u>Carr 835</u> 6/1970.

Justicia paxiana Lindau

F.T.E.A. Acanth.: 551 (2010).
Syn: *Rungia buettneri* Lindau – Biol. Skr. 51: 454
(2005).
Perennial herb or subshrub. Forest.
Distr: EQU; Guinea to Tanzania.
EQU: <u>Friis & Vollesen 756</u> (fl fr) 12/1980.

Justicia pinguior C.B.Clarke

Biol. Skr. 51: 448 (2005); F.T.E.A. Acanth.: 571 (2010).
Perennial herb. Submontane & montane forest, often in
clearings.
Distr: BAG, EQU; D.R. Congo, Uganda, Kenya.
BAG: <u>Dandy 609</u> (fl) 3/1934.
Note: a rather local species but not considered threatened.

Justicia scandens Vahl

Biol. Skr. 51: 448 (2005); F.E.E. 5: 461 (2006); F.T.E.A.
Acanth.: 513 (2010).

Syn: *Justicia glabra* Koen. ex Roxb. – F.P.S. 3: 179 (1956).
Perennial herb or subshrub, sometimes scandent. Dry forest, often in clearings.
Distr: EQU; Nigeria to Ethiopia, S to South Africa; Madagascar, India, SE Asia.
EQU: Jackson 1192 (fl) 3/1950.

Justicia tenella (Nees) T.Anderson
F.P.S. 3: 178 (1956); Biol. Skr. 51: 448 (2005); F.T.E.A. Acanth.: 556 (2010).
Annual herb. Disturbed areas in forest.
Distr: EQU; Senegal to Tanzania, S to Zimbabwe & Mozambique; Madagascar.
EQU: Friis & Vollesen 657 (fl fr) 12/1980.

Lankesteria elegans (P.Beauv.) T.Anderson
F.P.S. 3: 181 (1956); T.S.S.: 434 (1990); Biol. Skr. 51: 448 (2005); F.E.E. 5: 437 (2006); F.T.E.A. Acanth.: 273 (2008).
Perennial herb or shrub. Forest.
Distr: EQU; Sierra Leone to Ethiopia & Uganda.
EQU: Myers 10566 (fl) 2/1939.

Lepidagathis alopecuroides (Vahl) Griseb.
F.T.E.A. Acanth.: 304 (2010).
Syn: *Lepidagathis laguroidea* (Nees) T.Anderson – F.P.S. 3: 183 (1956).
Slender perennial herb. Moist forest, often in disturbed or trampled areas.
Distr: EQU; Senegal to D.R.Congo, S to Angola & Tanzania; neotropics.
EQU: Schweinfurth 2790 (fl) 12/1869.

Lepidagathis collina (Endl.) Milne-Redh.
F.P.S. 3: 182 (1956); F.T.E.A. Acanth.: 315 (2010).
Syn: *Lepidagathis hamiltoniana* Wall. subsp. *collina* (Endl.) J.K.Morton – Biol. Skr. 51: 449 (2005); F.E.E. 5: 427 (2006).
Syn: *Lepidagathis myrtifolia* S.Moore – F.P.S. 3: 182 (1956).
Syn: *Lepidagathis schweinfurthii* Lindau – F.P.S. 3: 182 (1956).
Syn: *Lepidagathis anobrya* sensu Quézel, non Nees – F. Darfur Nord-Occ. & J. Gourgeil: 131 (1969).
Syn: *Lepidagathis appendiculata* sensu Andrews, non Lindau – F.P.S. 3: 182 (1956).
Suffrutex. Fire-prone grassland, often on ironstone or rocky hillslopes.
Distr: DAR, KOR, BAG, UN, EQU; Senegal to Ethiopia & Kenya.
KOR: Jackson 2629 (fl fr) 1/1953; **EQU:** Dandy 453 (fl) 2/1934.

Lepidagathis diversa C.B.Clarke
F.P.S. 3: 183 (1956); Biol. Skr. 51: 449 (2005); F.E.E. 5: 427 (2006); F.T.E.A. Acanth.: 316 (2010).
Perennial herb. Grassland & wooded grassland.
Distr: EQU; Ghana to Ethiopia & Uganda.
EQU: Lankester s.n. (fl) 12/1920.

Lepidagathis glandulosa A.Rich.
Biol. Skr. 51: 449 (2005); F.E.E. 5: 429 (2006); F.T.E.A. Acanth.: 305 (2010).
Perennial herb. Forest margins, secondary bushland.

Distr: EQU; Cameroon, Ethiopia, D.R. Congo, Kenya, Tanzania, Zambia.
EQU: Friis & Vollesen 774 (fl) 12/1980.

Lepidagathis medusae S.Moore
F.P.S. 3: 183 (1956).
Suffrutex. Habitat not known.
Distr: BAG; South Sudan endemic.
IUCN: RD
BAG: Myers 11491 (fl) 6/1939.

Lepidagathis peniculifera S.Moore
F.P.S. 3: 183 (1956); F.T.E.A. Acanth.: 317 (2010).
Syn: *Lepidagathis perglabra* C.B.Clarke – F.P.S. 3: 182 (1956).
Suffrutex. *Loudetia* grassland over laterite, rocky hillslopes.
Distr: EQU; C.A.R., D.R. Congo, Uganda.
IUCN: RD
EQU: Kosper 97 (fl) 9/1982.

Lepidagathis scariosa Nees
F.P.S. 3: 182 (1956); F.T.E.A. Acanth.: 307 (2010).
Syn: *Lepidagathis aristata* (Vahl) Nees, nom. illegit. – F.E.E. 5: 428 (2006).
Perennial herb or subshrub. Dry bushland & bushy grassland.
Distr: DAR, KOR, CS, EQU; Mali, Benin to Somalia, S to Namibia & Zimbabwe; Arabia, India.
KOR: Jackson 3251 (fl) 11/1954; **EQU:** Mearns 2955 (fl) 2/1910.

Megalochlamys violacea (Vahl) Vollesen
F.E.E. 5: 457 (2006); F.T.E.A. Acanth.: 670 (2010).
Syn: *Ecbolium anisacanthus* (Schweinf.) C.B.Clarke – F.P.S. 3: 174 (1956).
Shrub. *Acacia-Commiphora* woodland, bushland & wooded grassland.
Distr: RS, EQU; Eritrea to Somalia, S to Kenya; Arabia.
RS: Schweinfurth 287 (fl fr) 9/1868; **EQU:** Carr 842 (fl fr).

Mellera lobulata S.Moore
Biol. Skr. 51: 449 (2005); F.E.E. 5: 382 (2006); F.T.E.A. Acanth.: 233 (2008).
Perennial herb or subshrub. Forest & forest clearings.
Distr: EQU; C.A.R. to Ethiopia, S to Zimbabwe & Mozambique.
EQU: Friis & Vollesen 956 (fl) 2/1982.

Mendoncia gilgiana (Lindau) Benoist
Biol. Skr. 51: 450 (2005); F.T.E.A. Acanth.: 78 (2008).
Woody climber. Forest.
Distr: EQU; Liberia to Kenya & Tanzania.
EQU: Friis & Vollesen 494 11/1980.

Mimulopsis solmsii Schweinf.
F.P.S. 3: 183 (1956); T.S.S.: 434 (1990); Biol. Skr. 51: 450 (2005); F.E.E. 5: 383 (2006); F.T.E.A. Acanth.: 222 (2008).
Mass-flowering perennial herb or subshrub. Forest & forest margins.
Distr: EQU; Guinea to Liberia, Nigeria to Ethiopia, S to Zimbabwe.
EQU: Friis & Vollesen 983 (fl) 2/1982.

Monothecium aristatum (Nees) T.Anderson
Biol. Skr. 51: 451 (2005); F.T.E.A. Acanth.: 494 (2010).
Perennial herb. Forest.
Distr: EQU; D.R. Congo, Uganda, Kenya, Tanzania, Angola; India.
EQU: Friis & Vollesen 480 (fl) 11/1980.

Monothecium glandulosum Hochst.
Biol. Skr. 51: 451 (2005); F.E.E. 5: 438 (2006); F.T.E.A. Acanth.: 493 (2010).
Perennial herb. Forest.
Distr: EQU; C.A.R. to Eritrea, S to D.R. Congo & Tanzania.
EQU: Friis & Vollesen 689 (fl) 12/1980.

Nelsonia canescens (Lam.) Spreng.
F.P.S. 3: 184 (1956); F.J.M.: 150 (1976); Biol. Skr. 51: 451 (2005); F.E.E. 5: 347 (2006); F.T.E.A. Acanth.: 18 (2008).
Perennial herb. Open scrub & disturbed habitats.
Distr: DAR, CS, BAG, UN, EQU; palaeotropical, introduced in the neotropics, the Pacific & Australia.
DAR: Wickens 1105 (fr) 1/1964; **UN:** J.M. Lock 80/12 (fl) 2/1980.

Nelsonia smithii Oerst.
Biol. Skr. 51: 451 (2005); F.E.E. 5: 347 (2006); F.T.E.A. Acanth.: 16 (2008).
Perennial herb. Forest, usually in swampy areas, riparian habitats.
Distr: EQU; Guinea to Ethiopia, S to Angola & Zambia.
EQU: Friis & Vollesen 606 (fl) 12/1980.

Neuracanthus niveus S.Moore
F.P.S. 3: 184 (1956); F.W.T.A. 2: 417 (1963).
Perennial herb. Grassland & wooded grassland.
Distr: BAG, EQU; Senegal, Guinea, Cameroon, Chad, C.A.R.
EQU: Andrews 972 (fl) 5/1939.
Note: a rather scarce and scattered species, but not considered threatened.

Phaulopsis barteri T.Anderson var. barteri
F.W.T.A. 2: 399 (1963); Symb. Bot. Ups. 31(2): 91 (1995).
Perennial herb. Dry forest & transitional woodland.
Distr: DAR; Senegal to Cameroon.
DAR: Kassas 368 (fr) 12/1965.
Note: the determination of the single Sudanese specimen is tentative; M. Manktelow noted that the leaves did not belong to the remainder of the material (fide Itambo Malombe, EA).

Phaulopsis barteri T. Anderson var. pauciglandula M.Manktelow
Symb. Bot. Ups. 31(2): 97 (1995).
Perennial herb. Grassland & wooded grassland.
Distr: BAG; Senegal to Cameroon.
BAG: Schweinfurth 2520 10/1869.
Note: the sheet at Kew of Schweinfurth 2520 is determined as P. savannicola; however, Manktelow cites the sheet at Stockholm of this number under P. barteri var. pauciglandula.

Phaulopsis ciliata (Willd.) Hepper
Symb. Bot. Ups. 31(2): 103 (1995).
Syn: *Phaulopsis falcisepala* C.B.Clarke – F.P.S. 3: 185 (1956).

Perennial herb. Forest, often in secondary growth & disturbed areas.
Distr: BAG; Senegal to C.A.R. & D.R. Congo.
BAG: Schweinfurth 1422 (fr) 4/1869.

Phaulopsis imbricata (Forssk.) Sweet subsp. imbricata
F.P.S. 3: 185 (1956); Biol. Skr. 51: 452 (2005); F.E.E. 5: 385 (2006); F.T.E.A. Acanth.: 260 (2008).
Perennial herb. Forest margins, secondary scrub, weed of disturbed areas.
Distr: RS, EQU; Ethiopia to South Africa; Madagascar, Yemen.
RS: Jackson 2891 (fl) 4/1953; **EQU:** Kosper 194 (fl) 1/1983.

Phaulopsis imbricata (Forssk.) Sweet subsp. poggei (Lindau) M.Manktelow
Biol. Skr. 51: 452 (2005); F.E.E. 5: 386 (2006); F.T.E.A. Acanth.: 261 (2008).
Perennial herb. Dry forest margins, riverine forest, secondary scrub, woodland & grassland.
Distr: EQU; Guinea to Ethiopia, S to Angola & Zambia.
EQU: Friis & Vollesen 769 (fl) 12/1980.

Phaulopsis savannicola M.Manktelow
Symb. Bot. Ups. 31(2): 86 (1995).
Perennial herb. Grassland & wooded grassland.
Distr: KOR, BAG, EQU; Mali, Nigeria to C.A.R. & D.R. Congo.
KOR: Jackson 2402 (fl) 10/1952; **EQU:** Kosper 151 (fl) 11/1982.

Phaulopsis talbotii S.Moore
Symb. Bot. Ups. 31(2): 162 (1995).
Annual herb. Dry forest, riverine forest, secondary scrub, weed of disturbed habitats.
Distr: EQU; Senegal to Chad; Jamaica & N South America where probably introduced.
EQU: Wyld 477 10/1938.

Pseuderanthemum ludovicianum (Büttner) Lindau
F.P.S. 3: 186 (1956); T.S.S.: 434 (1990); Biol. Skr. 51: 453 (2005); F.T.E.A. Acanth.: 480 (2010).
Perennial herb or shrub. Forest.
Distr: EQU; Liberia to D.R. Congo & Tanzania.
EQU: Myers 10217 (fl) 12/1938.

Pseuderanthemum subviscosum (C.B.Clarke) Stapf
F.T.E.A. Acanth.: 480 (2010).
Syn: *Pseuderanthemum tunicatum* sensu Friis & Vollesen, non (Afzel.) Milne-Redh. – Biol. Skr. 51: (1998).
Perennial herb or subshrub. Forest including riverine fringes.
Distr: EQU; Cameroon to Ethiopia, S to Angola & South Africa; Madagascar, Comoros.
EQU: Friis & Vollesen 602 (fl) 12/1980.

Ruellia bignoniiflora S.Moore
Biol. Skr. 51: 453 (2005); F.E.E. 5: 392 (2006); F.T.E.A. Acanth.: 210 (2008).
Perennial herb or shrub. Woodland & Acacia-Commiphora bushland.

Distr: EQU; Ethiopia to Angola & Zimbabwe.
EQU: Friis & Vollesen 1022 (fl fr) 2/1982.

Ruellia patula Jacq.

F.P.S. 3: 187 (1956); F.J.M.: 150 (1976); Biol. Skr. 51: 453 (2005); F.E.E. 5: 389 (2006); F.T.E.A. Acanth.: 207 (2008).
Syn: *Ruellia sudanica* (Schweinf.) Lindau – F.P.S. 3: 187 (1956).
Perennial herb or subshrub. Grassland, bushland & woodland including disturbed areas.
Distr: NS, RS, DAR, KOR, CS, ES, BAG, UN, EQU; Niger to Ethiopia, S to South Africa; Egypt, Arabia, India.
ES: Andrews 246 (fl fr) 12/1936; **EQU:** Padwa 257 (fl fr) 6/1953.

Ruellia praetermissa Lindau

F.P.S. 3: 187 (1956); F.T.E.A. Acanth.: 208 (2008).
Perennial herb. Grassland, bushland & woodland.
Distr: BAG, EQU; Senegal to Uganda, S to Zambia & Malawi.
EQU: von Ramm 256 (fl) 5/1956.
Note: Quézel (1969: 131) records this species from Darfur but this is considered to be in error.

Ruellia prostrata Poir.

Biol. Skr. 51: 453 (2005); F.E.E. 5: 391 (2006); F.T.E.A. Acanth.: 204 (2008).
Syn: *Ruellia genduana* (Schweinf.) C.B.Clarke – F.P.S. 3: 187 (1956).
Perennial herb. Grassland, bushland & woodland.
Distr: ES, BAG, EQU; Ethiopia & Somalia, S to Namibia & South Africa; Arabia, India, Sri Lanka, New Guinea, Pacific Is.
ES: Schweinfurth 131 (fl) 6/1865; **EQU:** Myers 9902 (fl fr) 10/1938.

Ruellia tuberosa L.

F.T.E.A. Acanth.: 204 (2008).
Perennial herb. Naturalised weed, grown as an ornamental.
Distr: CS, UN; native to the neotropics, naturalised elsewhere.
CS: DuBois D.B.0 (fl) 8/1951; **UN:** J.M. Lock 80/52 (fl) 2/1980.

Ruspolia decurrens (Nees) Milne-Redh.

F.P.S. 3: 187 (1956); F.J.M.: 150 (1976); Biol. Skr. 51: 454 (2005); F.E.E. 5: 434 (2006); F.T.E.A. Acanth.: 488 (2010).
Perennial herb or subshrub. Riverine woodland.
Distr: KOR, CS, BAG, UN, EQU; C.A.R. to Ethiopia, S to Zimbabwe & Mozambique.
KOR: Jackson 2384 (fl) 10/1952; **EQU:** Myers 10041 (fl fr) 11/1938.
Note: Andrews (1956) lists also *R. hypocrateriformis* (Vahl) Milne-Redh. but we have seen no material of this species from Sudan; the identification was most probably incorrect.

Satanocrater fellatensis Schweinf.

F.P.S. 3: 188 (1956); F.E.E. 5: 394 (2006).
Subshrub. Rocks in woodland.
Distr: ES; Guinea, Ethiopia.
IUCN: RD
ES: Schweinfurth 1306 (fl) 11/1865.
Note: the cited specimen is from the Sudan-Ethiopia border; no other Sudanese material is known.

Stenandrium guineense (Nees) Vollesen

Biol. Skr. 51: 454 (2005); F.T.E.A. Acanth.: 134 (2008).
Stoloniferous perennial herb. Forest.
Distr: EQU; Guinea to Uganda, S to Angola.
EQU: Friis & Vollesen 466 (fl) 11/1980.

Thunbergia alata Sims

F.P.S. 3: 189 (1956); Biol. Skr. 51: 455 (2005); F.E.E. 5: 353 (2006); F.T.E.A. Acanth.: 56 (2008).
Herbaceous climber. Bushland, forest margins, disturbed areas.
Distr: EQU; palaeotropical, introduced in the neotropics.
EQU: Andrews 1037 (fl) 5/1939.

Thunbergia annua Nees

F.P.S. 3: 189 (1956); F.E.E. 5: 355 (2006); F.T.E.A. Acanth.: 48 (2008).
Annual herb. Grassland & bushland on black clay soils.
Distr: KOR, CS, ES, UN; Niger, Ethiopia, Uganda, Kenya, Tanzania, Botswana.
CS: Andrews 259 (fl fr) 9/1937; **UN:** Broun 1711 (fr) 11/1909.

Thunbergia battiscombei Turrill

F.P.S. 3: 191 (1956); Biol. Skr. 51: 455 (2005); F.E.E. 5: 349 (2006); F.T.E.A. Acanth.: 58 (2008).
Suffrutex. Fire prone woodland & wooded grassland.
Distr: EQU; D.R. Congo, Ethiopia, Uganda, Kenya.
EQU: Dandy 472 (fl fr) 2/1934.

Thunbergia cycnium S.Moore

F.T.E.A. Acanth.: 74 (2008).
Suffrutex. Fire-prone woodland & wooded grassland.
Distr: BAG; C.A.R. to Uganda, S to Angola & Zambia.
BAG: Macintosh 26K (fl) 1932.

Thunbergia fasciculata Lindau

F.P.S. 3: 189 (1956); Biol. Skr. 51: 455 (2005); F.E.E. 5: 349 (2006); F.T.E.A. Acanth.: 43 (2008).
Herbaceous climber. Forest.
Distr: EQU; Togo to Ethiopia & Kenya.
EQU: Myers 9756 (fl) 10/1938.

Thunbergia hispida Solms

F.T.A. 5: 13 (1899); F.P.S. 3: 189 (1956).
Erect perennial herb. Habitat not known.
Distr: CS, ?UN; Sudan / South Sudan endemic.
IUCN: RD
CS: Cienkowsky s.n. (n.v.).
Note: no material of this poorly known species has been seen by the authors; Andrews cites two localities.

Thunbergia longifolia Lindau

Bot. Jahrb. Syst. 17: 91 (1893); F.P.S. 3: 191 (1956).
Erect perennial herb. Habitat not known.
Distr: EQU; South Sudan endemic.
IUCN: RD
EQU: Schweinfurth 3965 1870 (n.v.).
Note: no material has been seen of this poorly known species; the cited specimen is the type.

Thunbergia petersiana Lindau

F.T.E.A. Acanth.: 63 (2008).
Perennial herb or subshrub. Forest.

Distr: EQU; D.R. Congo to Kenya, S to Zimbabwe & Mozambique.
EQU: Jackson 3827 (fl) 9/1957.

Thunbergia reticulata Nees

F.E.E. 5: 353 (2006); F.T.E.A. Acanth.: 55 (2008).
Annual herb. Seepage grassland, bushland on rocky slopes.
Distr: RS; Eritrea, Ethiopia, Kenya, Zambia, S to Namibia & South Africa.
RS: Maffey 42a (fl).

Thunbergia schweinfurthii S.Moore

F.P.S. 3: 191 (1956).
?Perennial herb. Amongst rocks.
Distr: BAG, EQU; South Sudan endemic.
IUCN: RD
BAG: Turner 252 (fl).

Thunbergia vogeliana Benth.

T.S.S.: 435 (1990); Biol. Skr. 51: 456 (2005); F.T.E.A. Acanth.: 38 (2008).
Syn: *Thunbergia ikbaliana* sensu Andrews, non De Wild. – F.P.S. 3: 189 (1956).
Syn: *Thunbergia affinis* sensu Andrews, non S.Moore – F.P.S. 3: 188 (1956).
Shrub, sometimes scandent. Forest & forest margins.
Distr: BAG, EQU; Ghana to D.R. Congo, Uganda, Tanzania.
EQU: Friis & Vollesen 761 (fl fr) 12/1980.

Whitfieldia elongata (P.Beauv.) De Wild. & T.Durand

F.P.S. 3: 191 (1956); T.S.S.: 435 (1990); F.E.E. 5: 395 (2006); F.T.E.A. Acanth.: 268 (2008).
Shrub or scandent perennial herb. Forest & forest margins.
Distr: EQU; Nigeria to Ethiopia, S to Angola & Tanzania.
EQU: Thomas 1586 (fl) 12/1935.

Bignoniaceae

I. Darbyshire

Fernandoa adolfi-friderici (Gilg & Mildbr.) Heine

F.P.S. 3: 156 (1956); T.S.S.: 428 (1990).
Tree. Forest.
Distr: EQU; Cameroon, Gabon, C.A.R., D.R. Congo.
EQU: Turner 156 (fl) 2/1936.

Kigelia africana (Lam.) Benth. subsp. africana

F.J.M.: 148 (1976); T.S.S.: 429 (1990); Biol. Skr. 51: 432 (2005); F.E.E. 5: 325 (2006); F.T.E.A. Bignon.: 44 (2006).
Syn: *Kigelia aethiopum* (Fenzl) Dandy – F.P.S. 3: 156 (1956).
Syn: *Kigelia pinnata* DC. – Bot. Exp. Sud.: 20 (1970).
Shrub or tree. Wooded grassland, woodland, edges of riverine forest.
Distr: DAR, KOR, CS, ES, BAG, UN, EQU; Senegal to Ethiopia, S to Botswana & South Africa.
DAR: Wickens 1365 (fl) 4/1964; **EQU:** von Ramm 142 (fl) 3/1956.

Kigelia africana (Lam.) Benth. subsp. moosa (Sprague) Bidgood & Verdc.

F.T.E.A. Bignon.: 46 (2006).
Syn: *Kigelia moosa* Sprague – F.P.S. 3: 158 (1956); T.S.S.: 429 (1990); Biol. Skr. 51: 433 (2005).
Shrub or tree. Forest.
Distr: EQU; Sierra Leone to Kenya, S to Angola & Tanzania.
EQU: Jackson 1262 (fl) 3/1950.

Markhamia lutea (Benth.) K.Schum.

Biol. Skr. 51: 433 (2005); F.T.E.A. Bignon.: 34 (2006).
Syn: *Markhamia platycalyx* (Baker) Sprague – F.P.S. 3: 158 (1956); T.S.S.: 431 (1990).
Tree. Forest, wooded grassland.
Distr: EQU; Ghana to Kenya & Tanzania.
EQU: Myers 9757 (fl) 10/1937.

Spathodea campanulata P.Beauv. subsp. nilotica (Seem.) Bidgood

Biol. Skr. 51: 433 (2005); F.T.E.A. Bignon.: 31 (2006); F.E.E. 5: 327 (2006).
Syn: *Spathodea nilotica* Seem. – F.P.S. 3: 158 (1956); T.S.S.: 431 (1990).
Tree. Forest & forest margins, *Combretum* woodland.
Distr: EQU; Nigeria to Ethiopia, S to Tanzania.
EQU: Friis & Vollesen 177 (fl) 11/1980.

Stereospermum kunthianum Cham.

F.P.S. 3: 158 (1956); F.J.M.: 148 (1976); T.S.S.: 431 (1990); Biol. Skr. 51: 434 (2005); F.T.E.A. Bignon.: 37 (2006); F.E.E. 5: 323 (2006).
Shrub or tree. Wooded grassland & bushland.
Distr: DAR, KOR, CS, ES, BAG, UN, EQU; Senegal to Ethiopia, S to Zimbabwe & Mozambique; Arabia.
KOR: Burnett 12 (fl) 3/1933; **EQU:** Friis & Vollesen 1306 (fl) 3/1982.

[Tecoma stans (L.) Juss. ex Kunth]

Note: this species, native to tropical America, is commonly cultivated throughout Sudan and South Sudan but is not known to have naturalised.

Verbenaceae

I. Darbyshire

Chascanum laetum Fenzl ex Walp.

F.T.E.A. Verb.: 16 (1992); F.E.E. 5: 506 (2006).
Syn: *Svensonia laeta* (Fenzl ex Walp.) Moldenke – F.P.S. 3: 199 (1956).
Perennial herb. Open *Acacia* woodland, sandy soils.
Distr: RS, DAR, KOR, CS; Eritrea to Tanzania; Yemen.
KOR: Wickens 375 (fl fr) 9/1967.

Chascanum marrubiifolium Fenzl ex Walp.

F.P.S. 3: 193 (1956); F.T.E.A. Verb.: 12 (1992); F.E.E. 5: 505 (2006).
Syn: *Bouchea marrubiifolia* (Fenzl ex Walp.) Schauer – F. Darfur Nord-Occ. & J. Gourgeil: 131 (1969).
Perennial herb or subshrub. Dry thornbush, grassland, semi-desert scrub.
Distr: ?NS, RS, DAR, KOR, ES, EQU; Eritrea, Ethiopia, Kenya.
KOR: Jackson 2488 (fl fr) 1/1953; **EQU:** Myers 14108 (fl) 9/1941.

Duranta erecta L.

F.T.E.A. Verb.: 48 (1992); Biol. Skr. 51: 459 (2005); F.E.E. 5: 514 (2006).
Syn: *Duranta repens* L. – F.P.S. 3: 196 (1956); Bot. Exp. Sud.: 110 (1970); T.S.S.: 439 (1990).
Shrub or small tree, sometimes subscandent. Forest margins, riverine thicket.
Distr: EQU; native to the Americas, widely cultivated and naturalised in the palaeotropics & subtropics.
EQU: Myers 11016 (fl) 4/1939.
Note: El Amin notes that this species is widely grown in the Sudan region.

Lantana camara L.

F.J.M.: 150 (1976); T.S.S.: 440 (1990); F.T.E.A. Verb.: 39 (1992); F.E.E. 5: 507 (2006).
Syn: *Lantana antidotalis* Schumach. & Thonn. – F. Darfur Nord-Occ. & J. Gourgeil: 131 (1969).
Shrub. Roadsides, weed of abandoned cultivation, often highly invasive.
Distr: NS, DAR; native to the West Indies and/or tropical America, widely naturalised elsewhere.
DAR: Wickens 1335 (fl) 3/1964.
Note: cultivated elsewhere in Sudan and South Sudan and quite likely to be naturalised in some areas.

Lantana trifolia L.

F.P.S. 3: 196 (1956); T.S.S.: 440 (1990); F.T.E.A. Verb.: 46 (1992); Biol. Skr. 51: 459 (2005); F.E.E. 5: 507 (2006).
Shrub or perennial herb. Wooded grassland, forest margins, waste ground & margins of cultivation.
Distr: EQU; probably introduced from the Americas, now pantropical.
EQU: Myers 11011 (fl) 4/1939.

Lantana ukambensis (Vatke) Verdc.

F.T.E.A. Verb.: 43 (1992); F.E.E. 5: 507 (2006).
Syn: *Lantana rhodesiensis* Moldenke – F.P.S. 3: 196 (1956); F.J.M.: 150 (1976); T.S.S.: 440 (1990).
Subshrub or perennial herb. Wooded grassland, upland grassland, edges of gallery forest, often in disturbed areas.
Distr: RS, DAR, BAG, EQU; Guinea Bissau to Ethiopia, S to Zimbabwe & Mozambique.
DAR: Wickens 1898 (fl fr) 7/1964; **EQU:** Andrews 1003 (fl) 5/1939.

Lantana viburnoides (Forssk.) Vahl var. viburnoides

F.P.S. 3: 196 (1956); T.S.S.: 441 (1990); F.T.E.A. Verb.: 41 (1992); Biol. Skr. 51: 459 (2005); F.E.E. 5: 508 (2006).
Shrub. Dry thornbush & thicket, roadsides.
Distr: RS, DAR, BAG, EQU; Eritrea & Somalia, S to Angola & Mozambique.
RS: Bent s.n. (fl) 1896; **EQU:** Jackson 4237 (fl fr) 6/1961.

Lippia abyssinica (Otto & F.Dietr.) Cufod.

F.T.E.A. Verb.: 29 (1992); Biol. Skr. 51: 460 (2005); F.E.E. 5: 511 (2006).
Syn: *Lippia grandifolia* Hochst. ex Walp. – F.P.S. 3: 197 (1956); Bot. Exp. Sud.: 110 (1970); T.S.S.: 441 (1990) pro parte, as *L. grandiflora*.
Shrub or perennial herb. Wooded grassland, rough grassland, secondary bushland.
Distr: EQU; D.R. Congo to Ethiopia, S to Tanzania.

EQU: Friis & Vollesen 533 (fl) 11/1980.
Note: El Amin's record of '*L. grandiflora*' from Darfur is considered to be in error.

Lippia multiflora Moldenke

F.P.S. 3: 197 (1956); F.J.M.: 150 (1976); T.S.S.: 441 (1990); F.T.E.A. Verb.: 28 (1992).
Shrub or perennial herb. Wooded grassland, riverine thicket, weed of cultivation.
Distr: DAR, KOR, ES, BAG, EQU; Sierra Leone to D.R. Congo, S to Angola.
DAR: Robertson 6 (fl) 10/1957; **BAG:** Schweinfurth 2657 (fl) 11/1869.

Lippia radula Baker

F.T.A. 5: 279 (1900); F.P.S. 3: 197 (1956).
Subshrub. Habitat unknown.
Distr: BAG; South Sudan endemic.
IUCN: RD
BAG: Schweinfurth 2616 (fl fr) 11/1869.
Note: this species is listed as of uncertain status in the African Plants Database.

Lippia woodii Moldenke

F.T.E.A. Verb.: 33 (1992); Biol. Skr. 51: 460 (2005).
Subshrub or perennial herb. Grassland & wooded-grassland subject to seasonal burning.
Distr: EQU; Uganda to Zimbabwe & Mozambique.
EQU: Wilson 913 4/1960 (n.v.).
Note: the cited specimen has not been seen by us; identification follows Friis & Vollesen (2005).

Phyla nodiflora (L.) Greene

F.P.S. 3: 197 (1956); F.T.E.A. Verb.: 25 (1992); F.E.E. 5: 513 (2006).
Perennial herb. Muddy or sandy margins of pools, irrigated fields.
Distr: RS, CS; widespread in the tropics & subtropics.
RS: Jackson 2807 (fl) 4/1953.

Priva adhaerens (Forssk.) Chiov.

F.P.S. 3: 198 (1956); F.T.E.A. Verb.: 22 (1992); Fl. Somalia 3: 361 (2006).
Syn: *Priva cordifolia* (L.f.) Druce var. *abyssinica* (Jaub. & Spach) Moldenke – F.P.S. 3: 198 (1956).
Syn: *Priva tenax* Verdc. – F.E.E. 5: 503 (2006).
Perennial herb. Dry woodland.
Distr: RS, ?EQU; Eritrea & Somalia, S to Tanzania; Egypt, Yemen.
RS: Bent s.n. (fl fr) 1896; **EQU:** Andrews 1658 (fl) 6/1939.
Note: we here follow Thulin in Fl. Somalia in including *P. tenax* Verdc. within a broadly circumscribed *P. adhaerens* as the differences do not appear to hold up in Sudan. *Andrews* 1658 from Equatoria is tentatively included here – the material is in flower only and so too young for conclusive identification.

Priva flabelliformis (Moldenke) R.Fern.

F.T.E.A. Verb.: 21 (1992); F.E.E. 5: 503 (2006).
Syn: *Priva meyeri* sensu auctt., non Jaub. & Spach – F.P.S. 3: 199 (1956); Bot. Exp. Sud.: 110 (1970).
Perennial herb. Grassland & open bushland, weed of cultivation.
Distr: EQU; Ethiopia to South Africa; Madagascar.

EQU: <u>Drar & Mahdi 1750</u> 4/1938 (n.v.).
Note: the cited specimen was listed by Moldenke in
Phytologia 44: 108 (1979).

Stachytarpheta indica (L.) Vahl
F.T.E.A. Verb.: 20 (1992); F.E.E. 5: 506 (2006).
Syn: *Stachytarpheta angustifolia* (Mill.) Vahl – F.P.S. 3:
199 (1956).
Annual herb. Riverbanks, wet grassland, weed of
cultivation.
Distr: BAG, UN, EQU; tropical Africa; tropical America.
EQU: <u>Myers 7258</u> (fl fr) 7/1937.

Stachytarpheta urticifolia Sims
F.T.E.A. Verb.: 19 (1992).
Perennial herb or subshrub. Grassland, bushland, weed
of roadsides & cultivation.
Distr: EQU; largely pantropical, probably native to
tropical Asia; in Africa recorded from South Sudan to
South Africa.
EQU: <u>Jackson 2993</u> (fl fr) 6/1954.
Note: often mistakenly named *S. jamaicensis* (L.) Vahl,
which is a distinct species.

Verbena officinalis L. subsp. africana R.Fern. & Verdc.
F.P.S. 3: 199 (1956); Bot. Exp. Sud.: 111 (species only)
(1970); F.J.M.: 150 (1976); F.T.E.A. Verb.: 8 (1992); F.E.E.
5: 500 (2006).
Perennial herb. Disturbed grassland & bushland, weed of
cultivation.
Distr: DAR, EQU; Eritrea to South Africa.
DAR: <u>Wickens 3000</u> (fl fr) 6/1965; **EQU:** <u>Drar 1667</u>
4/1938 (n.v.).

Verbena supina L.
F.P.S. 3: 199 (1956); F.E.E. 5: 501 (2006).
Annual herb. Open seasonally flooded areas.
Distr: NS, KOR, CS; Mauritania & Senegal to Ethiopia; N
Africa, Spain, Middle East.
KOR: <u>Pfund 180</u> (fl fr) 9/1875.

Martyniaceae

M. Kordofani

Proboscidea parviflora Wooton & Standl.
Contrib. U.S. Nat. Herb.19: 602 (1915); J. Chem. &
Pharm. Res. 5: 381–386 (2013).
Annual herb with taproot. Disturbed area in semi-desert.
Distr: CS; native of U.S.A. & Mexico.
CS: <u>Nahla 88</u> (fl fr) 7/2007.
Note: this plant was found thoroughly naturalised in the
Feddasi area near Wad Medani.

AQUIFOLIALES

Cardiopteridaceae

I. Darbyshire

Leptaulus daphnoides Benth.
F.T.E.A. Icac.: 2 (1968); T.S.S.: 285 (1990); Biol. Skr. 51:
260 (1998).

Shrub or small tree. Forest.
Distr: EQU; Guinea Bissau to Uganda & Tanzania.
Note: no specimens are known for South Sudan but it is
recorded from Lotti Forest (Jackson) and the Imatong Mts
(El Amin) – see Friis & Vollesen (1998) for further details.

Aquifoliaceae

I. Darbyshire

Ilex mitis (L.) Radlk. var. mitis
F.T.E.A. Aquifol.: 1 (1968); F.E.E. 3: 329 (1989); T.S.S.:
279 (1990); Biol. Skr. 51: 254 (1998).
Tree or shrub. Montane forest.
Distr: EQU; Sierre Leone, Nigeria to Ethiopia, S to South
Africa; Madagascar.
EQU: <u>Friis & Vollesen 354</u> (fl) 11/1980.

ASTERALES

Campanulaceae

H. Pickering & I. Darbyshire

Campanula dimorphantha Schweinf.
F.P.S. 3: 70 (1956); F.E.E. 5: 57 (2006).
Annual herb. Moist ground.
Distr: NS, RS; Eritrea; Egypt, Afghanistan to China &
SE Asia.
RS: <u>Ehrenberg 182</u> 4/1871.

Campanula edulis Forssk.
F.P.S. 3: 70 (1956); F.J.M.: 140 (1976); F.T.E.A. Campanul.:
35 (1976); F.E.E. 5: 55 (2006).
Perennial herb. Montane forest glades, montane
grassland.
Distr: DAR; Chad to Somalia, S to Rwanda & Tanzania;
Arabia.
DAR: <u>Jackson 3326</u> 12/1954.

Campanula erinus L.
F.P.S. 3: 70 (1956); F.E.E. 5: 56 (2006).
Annual herb. Shady places.
Distr: RS; Eritrea, Ethiopia; Macaronesia, Mediterranean,
SW Asia, Arabia.
RS: <u>Aylmer 579</u> 3/1936.

Canarina abyssinica Engl.
F.T.E.A. Campanul.: 2 (1976); Biol. Skr. 51: 405 (2005);
F.E.E. 5: 49 (2006).
Perennial climbing herb. Upland wooded grassland,
forest clearings.
Distr: EQU; Ethiopia, Uganda, Kenya, Tanzania.
EQU: <u>Andrews 1989</u> 6/1939.
Note: Friis & Vollesen (2005) mistakenly cite *Andrews
1989* for *C. eminii* as well as *C. abyssinica*.

Canarina eminii Asch. & Schweinf.
F.P.S. 3: 70 (1956); F.T.E.A. Campanul.: 4 (1976); Biol. Skr.
51: 405 (2005); F.E.E. 5: 49 (2006).
Epiphytic or terrestrial perennial herb, sometimes
scandent. Montane forest.
Distr: EQU; D.R. Congo to Ethiopia, S to Malawi.
EQU: <u>MacLeay 29</u> 6/1947.

Lobelia dissecta M.B.Moss

F.P.S. 3: 74 (1956); F.T.E.A. Lobel.: 23 (1984); Biol. Skr. 51: 406 (2005).
Syn: Lobelia dissecta M.B.Moss subsp. **humidulorum** Friis & Vollesen – Biol. Skr. 51: 406 (2005).
Annual or perennial herb. Rocky ground, montane grassland & bushland.
Distr: EQU; Burundi, Uganda.
IUCN: RD
EQU: Myers 13517 9/1940.

Lobelia djurensis Engl. & Diels

F.P.S. 3: 74 (1956); F.W.T.A. 2: 313 (1963).
Annual herb. Grassland, muddy ground.
Distr: BAG; Senegal to C.A.R. & D.R. Congo.
BAG: Schweinfurth 2566 10/1869.

Lobelia erinus L.

F.T.E.A. Lobel.: 18 (1984); F.E.E. 5: 41 (2006).
Syn: Lobelia senegalensis A.DC. – F.P.S. 3: 74 (1956); Bot. Exp. Sud.: 24 (1970); F.J.M.: 140 (1976).
Annual or perennial herb. Montane grassland including boggy areas.
Distr: DAR; Senegal to Somalia, S to Namibia & South Africa.
DAR: Wickens 2983 6/1965.

Lobelia flaccida (C.Presl) A.DC. subsp. granvikii (T.C.E.Fr.) Thulin

F.T.E.A. Lobel.: 25 (1984); Biol. Skr. 51: 406 (2005).
Annual or short-lived perennial herb. Upland grassland, rocky areas, damp ground.
Distr: EQU; Uganda, Kenya.
EQU: Myers 11119 4/1939.

Lobelia giberroa Hemsl.

F.P.S. 3: 73 (1956); F.T.E.A. Lobel.: 9 (1984); Biol. Skr. 51: 407 (2005); F.E.E. 5: 40 (2006).
Large perennial herb. Montane forest & forest margins, streamsides.
Distr: EQU; D.R. Congo to Eritrea, S to Zambia & Malawi.
EQU: Thomas 1657 12/1935.
Note: a 'giant lobelia' species.

Lobelia inconspicua A.Rich.

F.T.E.A. Lobel.: 34 (1984); Biol. Skr. 51: 407 (2005); F.E.E. 5: 44 (2006).
Annual herb. Upland grassland & woodland.
Distr: EQU; Sierra Leone, Nigeria to Ethiopia, S to Malawi; India.
EQU: Friis & Vollesen 19 11/1980.

Lobelia molleri Henriq.

F.T.E.A. Lobel.: 31 (1984); Biol. Skr. 51: 407 (2005).
Annual or short-lived perennial herb. Upland forest & forest margins, streamsides.
Distr: BAG, EQU; Cameroon to Uganda, S to Zimbabwe.
EQU: Friis & Vollesen 28 11/1980.

Lobelia neumannii T.C.E.Fr.

F.T.E.A. Lobel.: 24 (1984); F.E.E. 5: 42 (2006).
Syn: Lobelia rubrimaris E.Wimm. – F.P.S. 3: 73 (1956).
Annual herb. Upland grassland, usually on bare or rocky ground.

Distr: RS; Cameroon, Ethiopia, Kenya.
RS: Maffey 15 1928 (n.v.).
Note: the single Sudanese collection, the type of *L. rubrimaris*, is currently missing at Kew. In the protologue, the collector was listed as Maffer but this should surely be Maffey.

Lobelia trullifolia Hemsl. subsp. trullifolia

Biol. Skr. 51: 407 (2005); F.E.E. 5: 42 (2006).
Annual herb. Upland grassland, forest margins, streamsides.
Distr: EQU; Ethiopia to Zambia & Mozambique.
EQU: Friis & Vollesen 160 11/1980.

Wahlenbergia campanuloides (Delile) Vatke

F.Egypt 3: 129 (2002).
Syn: Wahlenbergia cervicina A.DC. – Bot. Exp. Sud.: 24 (1970).
Annual herb. Margins of cultivation, damp bare ground.
Distr: BAG; Senegal, Chad, Zambia, Namibia, South Africa; Algeria, Egypt.
BAG: Drar 891 4/1938 (n.v.).
Note: this record is from Drar (1970).

Wahlenbergia erecta (Roth ex Roem. & Schult.) Tuyn

F.J.M.: 140 (1976); F.T.E.A. Campanul.: 23 (1976); F.E.E. 5: 53 (2006).
Syn: Cephalostigma erectum (Roth ex Roem. & Schult.) Vatke – F.P.S. 3: 71 (1956).
Annual herb. Wooded grassland.
Distr: DAR, ?EQU; Nigeria, Eritrea to Angola & Zimbabwe; India, Malaysia.
DAR: Miehe 680 9/1962.
Note: Andrews recorded this species from Equatoria, but we have seen no material from South Sudan.

Wahlenbergia flexuosa (Hook.f. & Thomson) Thulin

F.T.E.A. Campanul.: 24 (1976); Biol. Skr. 51: 407 (2005); F.E.E. 5: 53 (2006).
Annual herb. Wooded grassland.
Distr: EQU; Nigeria to Eritrea, S to Malawi; India.
EQU: Friis & Vollesen 701 12/1980.

Wahlenbergia hirsuta (Edgew.) Tuyn

F.T.E.A. Campanul.: 27 (1976); F.J.M.: 140 (1976); Biol. Skr. 51: 408 (2005); F.E.E. 5: 55 (2006).
Annual herb. Wooded grassland.
Distr: DAR, BAG, EQU; Senegal to Ethiopia, S to Zimbabwe; Madagascar, Comoros, Arabia, India, Nepal.
DAR: de Wilde et al. 5524b 2/1965 (n.v.); **EQU:** Friis & Vollesen 859 12/1980.
Note: the specimen from Dafur has not been seen and the identification requires confirmation.

Wahlenbergia lobelioides (L.f.) Link subsp. nutabunda (Guss.) Murb.

F.T.E.A. Campanul.: 11 (1976); F.E.E. 5: 51 (2006).
Syn: Wahlenbergia etbaica (Schweinf.) Vatke – F.P.S. 3: 72 (1956); Bot. Exp. Sud.: 24 (1970).
Annual herb. Dry grassland.
Distr: RS; Eritrea, Ethiopia, Kenya; Mediterranean, N Africa, Arabia.
RS: Jackson 2695 2/1953.

Wahlenbergia napiformis (A.DC.) Thulin

F.T.E.A. Campanul.: 18 (1976); Biol. Skr. 51: 408 (2005);
F.E.E. 5: 53 (2006).
Syn: *Lightfootia abyssinica* sensu Andrews, non
Hochst. ex A.Rich. – F.P.S. 3: 71 (1956).
Perennial herb. Woodland & grassland.
Distr: EQU; C.A.R. to Ethiopia, S to Namibia, Botswana
& Mozambique.
EQU: Myers 8380 1/1938.

Wahlenbergia perrottetii (A.DC.) Thulin

F.T.E.A. Campanul.: 31 (1976).
Syn: *Cephalostigma perrottetii* A.DC. – F.P.S. 3: 71
(1956).
Annual herb. Grassland, woodland, weed of disturbed
ground & cultivation.
Distr: EQU; Senegal to Uganda, S to Angola & Malawi;
Madagascar, Comoros, S America.
EQU: Wyld 474 10/1938.

Wahlenbergia silenoides Hochst. ex A.Rich.

F.P.S. 3: 72 (1956); F.T.E.A. Campanul.: 12 (1976); Biol.
Skr. 51: 408 (2005); F.E.E. 5: 51 (2006).
Perennial herb. Montane grassland.
Distr: EQU; Nigeria to Eritrea, S to Tanzania.
EQU: Friis & Vollesen 1290 3/1982.

Wahlenbergia virgata Engl.

F.P.S. 3: 72 (1956); F.T.E.A. Campanul.: 9 (1976); Biol. Skr.
51: 408 (2005); F.E.E. 5: 51 (2006).
Perennial herb. Upland grassland.
Distr: EQU; Ethiopia to South Africa.
EQU: Friis & Vollesen 974 2/1982.

Menyanthaceae

H. Pickering & I. Darbyshire

Nymphoides ezannoi Berhaut

Fl. Senegal 2: 427 (1967); Adansonia sér. 2, 14: 412 (1974).
Perennial herb. Shallow water, muddy pool margins.
Distr: DAR, KOR; southern fringes of the Sahara from
Senegal to Chad.
KOR: Wickens 935 3/1963.
Note: other species of *Nymphoides* may well occur
in Sudan and South Sudan, the most likely being *N.
brevipedicellata* (Vatke) A.Raynal and *N. indica* (L.)
Kuntze subsp. *occidentalis* A.Raynal.

Nymphoides forbesiana (Griseb.) Kuntze

F.T.E.A. Menyanth.: 4 (1996).
Syn: *Nymphoides nilotica* (Kotschy & Peyr.) J.Léonard –
F.P.S. 3: 65 (1956).
Perennial herb. Shallow pools & lake margins.
Distr: CS, EQU; Ivory Coast to Somalia, S to South Africa.
CS: Andrews s.n. (fl fr) 9/1939; **EQU:** Myers 7629 8/1937.

Asteraceae (Compositae)

H. Pickering & I. Darbyshire

Acanthospermum hispidum DC.

F.J.M.: 131 (1976); Biol. Skr. 51: 351 (1998); F.E.E. 4(2):
308 (2004); F.T.E.A. Comp. 3: 766 (2005).

Annual herb. Weed of cultivation & disturbed areas.
Distr: DAR, KOR, EQU; native to S America, now a
pantropical weed.
KOR: Jackson 4012 9/1959; **EQU:** Kielland-Lund 91
11/1983 (n.v.).

Acmella caulirhiza Delile

Biol. Skr. 51: 351 (1998); F.E.E. 4(2): 288 (2004); F.T.E.A.
Comp. 3: 730 (2005).
Syn: *Spilanthes caulirhiza* (Delile) DC. – F.P.S. 3: 54
(1956).
Annual or perennial herb. Streambanks, wet depressions,
forest margins.
Distr: ?CS, UN, EQU; Liberia to Eritrea, S to Angola &
South Africa; Madagascar.
EQU: Andrews 1091 5/1939.
Note: Andrews records this species from Central &
Southern Sudan but we have only seen specimens from
the South.

Adenostemma caffrum DC. var. asperum Brenan

F.P.S. 3: 8 (species only) (1956); F.J.M.: 131 (species only)
(1976); Biol. Skr. 51: 351 (species only) (1998); F.E.E. 4(2):
346 (species only) (2004); F.T.E.A. Comp. 3: 822 (2005).
Perennial herb. Swamps, wet disturbed ground.
Distr: DAR, EQU; Guinea to Ethiopia, S to Angola &
South Africa.
DAR: Wickens 1454 4/1964; **EQU:** Schweinfurth 3784
5/1870.

Adenostemma mauritianum DC.

F.P.S. 3: 8 (1956); Biol. Skr. 51: 351 (1998); F.E.E. 4(2):
345 (2004); F.T.E.A. Comp. 3: 824 (2005).
Perennial herb. Forest & associated bushland,
streambanks.
Distr: EQU; Sierra Leone to Ethiopia, S to South Africa;
Madagascar, Mascarenes.
EQU: Friis & Vollesen 408 11/1980.

Adenostemma viscosum J.R.Forst. & G.Forst.

F.T.E.A. Comp. 3: 820 (2005).
Syn: *Adenostemma perrottetii* DC. – F.P.S. 3: 8 (1956);
Biol. Skr. 51: 352 (1998); F.E.E. 4(2): 345 (2004).
Perennial herb. Swamps, riverbanks, forest margins.
Distr: EQU; Senegal to Eritrea, S to Angola & South
Africa; India, Sri Lanka, Indonesia, Australia, Pacific Is.
EQU: Friis & Vollesen 226 11/1980.

Ageratum conyzoides L.

F.P.S. 3: 8 (1956); Biol. Skr. 51: 352 (1998); F.E.E. 4(2):
347 (2004); F.T.E.A. Comp. 3: 827 (2005).
Annual herb. Weed of disturbed areas & cultivation.
Distr: NS, DAR, CS, UN, EQU; pantropical.
CS: Andrews 207 3/1936; **EQU:** Schweinfurth 2989
2/1870.

Ageratum houstonianum Mill. var. houstonianum

F.E.E. 4(2): 348 (2004); F.T.E.A. Comp. 3: 830 (2005).
Annual or short-lived perennial herb. Moist disturbed
areas, riverbanks.
Distr: RS; native to Mexico, now widespread in the

tropics; in Africa from Cameroon to Ethiopia, S to South Africa; Madagascar, Mascarenes.
RS: Andrews 3493.

Akeassia grangeoides J.-P.Lebrun & Stork
Candollea 48: 333 (1993).
Syn: *Microtrichia perrottetii* sensu auctt., non DC. – F.P.S. 3: 43 (1956); F.W.T.A. 2: 256 (1963).
Annual herb. Damp ground in wooded grassland.
Distr: BAG; Senegal to C.A.R.
BAG: Schweinfurth 1393 4/1869.

Ambrosia maritima L.
F.P.S. 3: 8 (1956); F.E.E. 4(2): 341 (2004); F.T.E.A. Comp. 3: 813 (2005).
Annual herb. Weed of lake shores & riverbanks.
Distr: NS, CS; a widespread weed in the palaeotropics and warm temperate regions.
CS: Jackson 1920 7/1951.

Anisopappus chinensis (L.) Hook. & Arn. subsp. africanus (Hook.f.) S.Ortíz & Paiva
F.E.E. 4(2): 119 (2004).
Syn: *Anisopappus africanus* (Hook.f.) Oliv. & Hiern – F.P.S. 3: 9 (1956); F.J.M.: 131 (1976); Biol. Skr. 51: 352 (1998).
Syn: *Anisopappus chinensis* (L.) Hook. & Arn. subsp. *buchwaldii* (O.Hoffm.) S.Ortíz, Paiva & Rodr.-Oubiña – F.T.E.A. Comp. 2: 344 (2002).
Perennial herb. Woodland & grassland.
Distr: DAR, EQU; Sierra Leone to Ethiopia, S to Angola, Zimbabwe & Mozambique.
DAR: de Wilde et al. 5399 1/1965; **EQU:** Friis & Vollesen 136 11/1980.

Aspilia africana (Pers.) C.D.Adams
Biol. Skr. 51: 352 (1998); F.T.E.A. Comp. 3: 750 (2005).
Syn: *Aspilia africana* (Pers.) C.D.Adams subsp. *magnifica* (Chiov.) Wild – F.E.E. 4(2): 296 (2004).
Syn: *Aspilia congoensis* S.Moore – Biol. Skr. 51: 352 (1998).
Syn: *Aspilia latifolia* Oliv. & Hiern – F.P.S. 3: 9 (1956).
Syn: *Wedelia africana* Pers. – F.P.S. 3: 62 (1956).
Perennial herb or subshrub. Grassland, woodland, forest margins.
Distr: BAG, EQU; Senegal to Ethiopia, S to Angola & Tanzania.
EQU: Jackson 3815 8/1957.

Aspilia ciliata (Schumach.) Wild
F.J.M.: 131 (1976); F.E.E. 4(2): 294 (2004); F.T.E.A. Comp. 3: 748 (2005).
Syn: *Aspilia helianthoides* (Schumach. & Thonn.) Oliv. & Hiern **subsp.** *prieuriana* (DC.) C.D.Adams – F. Darfur Nord-Occ. & J. Gourgeil: 123 (1969).
Syn: *Aspilia multiflora* Fenzl ex Oliv. & Hiern – F.P.S. 3: 9 (1956).
Syn: *Aspilia schimperi* (Sch. Bip. ex A.Rich.) Oliv. & Hiern – F.P.S. 3: 9 (1956).
Syn: *Blainvillea prieureana* DC. – F.P.S. 3: 13 (1956).
Annual herb. Moist sandy soils, disturbed areas.
Distr: DAR, KOR, CS, ES; Senegal to Ethiopia, S to Zambia.
KOR: Jackson 4010 9/1959.

Aspilia kotschyi (Sch.Bip.) Oliv. var. kotschyi
F.P.S. 3: 10 (1956); F.J.M.: 131 (1976); Biol. Skr. 51: 353 (1998); F.E.E. 4(2): 294 (2004); F.T.E.A. Comp. 3: 747 (2005).
Annual herb. Woodland, grassland, disturbed areas.
Distr: DAR, KOR, ES, BAG, UN, EQU; Senegal to Ethiopia, S to Angola, Zimbabwe & Mozambique; Arabia.
DAR: Wickens 1563 5/1964; **EQU:** Myers 7073 7/1937.

Aspilia kotschyi (Sch.Bip.) Oliv. var. alba Berhaut
F.E.E. 4(2): 294 (2004); F.T.E.A. Comp. 3: 747 (2005).
Annual herb. Woodland, grassland, disturbed areas.
Distr: CS, BAG, UN, EQU; Senegal to Ethiopia, S to Angola, Zimbabwe & Mozambique.
CS: Lea 228 1953; **EQU:** Jackson 3852 9/1957.

Aspilia mossambicensis (Oliv.) Wild
F.E.E. 4(2): 295 (2004); F.T.E.A. Comp. 3: 751 (2005).
Syn: *Wedelia abyssinica* Vatke – F.P.S. 3: 63 (1956) – see note.
Perennial herb or shrub. Woodland, grassland, forest margins, swamps, riverbanks, disturbed areas.
Distr: ?CS, EQU; Eritrea & Somalia, S to South Africa.
EQU: Peers KAM 34 9/1953.
Note: Andrews records *Wedelia abyssinica* Vatke from Fung District; this name appears on a specimen of *A. mossambicensis* from Mongalla but we have seen no material of *A. mossambicensis* from Central Sudan.

Aspilia pluriseta Schweinf.
Biol. Skr. 51: 353 (1998); F.T.E.A. Comp. 3: 749 (2005).
Perennial herb or subshrub. Grassland & woodland.
Distr: EQU; D.R. Congo & Uganda, S to South Africa.
EQU: Friis & Vollesen 820 12/1980.

Athrixia rosmarinifolia (Sch.Bip. ex Walp.) Oliv. & Hiern var. rosmarinifolia
F.P.S. 3: 10 (1956); T.S.S.: 415 (1990); Biol. Skr. 51: 353 (1998); F.T.E.A. Comp. 2: 456 (2002); F.E.E. 4(2): 159 (2004).
Perennial herb or subshrub. Montane grassland & bushland.
Distr: EQU; Ethiopia to Zimbabwe.
EQU: Friis & Vollesen 1095 3/1982.

Atractylis aristata Batt.
F.J.U. 2: 152 (1999); F.Egypt 3: 155 (2002).
Annual or short-lived perennial herb. Rock crevices, rocky hillslopes.
Distr: NS; Mauritania to Chad; Morocco, Algeria, Libya, Egypt, Arabia.
NS: Léonard 4873 11/1968.

Berkheya spekeana Oliv.
F.P.S. 3: 10 (1956); Biol. Skr. 51: 353 (1998); F.T.E.A. Comp. 1: 294 (2000); F.E.E. 4(2): 113 (2004).
Perennial herb. Upland woodland & wooded grassland.
Distr: EQU; Cameroon, Ethiopia to D.R. Congo & Tanzania.
EQU: Myers 11212 4/1939.

xs biternata (Lour.) Merr. & Sherff
F.P.S. 3: 12 (1956); F.E.E. 4(2): 320 (2004); F.T.E.A. Comp. 3: 804 (2005).

Syn: *Bidens biternata* (Lour.) Merr. & Sherff var. *glabrata* (Vatke) Sherff fa. *abyssinica* (Sch.Bip. ex Walp.) Sherff – F.J.M.: 132 (1976).
Syn: *Bidens bipinnata* sensu auctt., non L. – F.P.S. 3: 12 (1956); F.J.M.: 132 (1976).
Annual herb. Woodland, bushland, grassland, disturbed areas.
Distr: RS, DAR, KOR, ES, BAG, UN; palaeotropical.
DAR: Wickens 1399 6/1964; **BAG:** Schweinfurth 2240 8/1869.

Bidens borianiana (Sch.Bip. ex Schweinf. & Asch.) Cufod.

F.J.M.: 132 (1976); F.E.E. 4(2): 321 (2004).
Syn: *Coreopsis borianiana* Sch.Bip. ex Schweinf. & Asch. – F.P.S. 3: 19 (1956).
Annual herb. Woodland, wooded grassland, disturbed areas.
Distr: DAR, KOR, ES, UN; Senegal to Eritrea & Ethiopia.
DAR: Wickens 2570 9/1964; **UN:** Evans-Pritchard 29 11/1936.

Bidens buchneri (Klatt) Sherff

Biol. Skr. 51: 354 (1998); F.T.E.A. Comp. 3: 785 (2005).
Perennial herb or shrub. Grassland & wooded grassland.
Distr: EQU; D.R. Congo & Uganda, S to Angola & Tanzania.
EQU: Friis & Vollesen 669 12/1980.

Bidens camporum (Hutch.) Mesfin

F.E.E. 4(2): 323 (2004).
Syn: *Bidens chaetodonta* Sherff – F.J.M.: 132 (1976).
Perennial herb. Upland wooded grassland, forest margins, streambanks.
Distr: DAR; Nigeria, Cameroon, D.R. Congo, Eritrea, Ethiopia.
DAR: Wickens 2429 9/1964.
Note: Andrews records *Coreopsis schimperi* O.Hoffm. (= *Bidens carinata* Cufod. ex Mesfin) from Equatoria; this is believed to be in error since *B. carinata* only occurs in Eritrea and Ethiopia, but we are unsure as to the true identity of Andrews' plant.

Bidens chippii (M.B.Moss) Mesfin

Symb. Bot. Upsal. 24: 85 (1984); Biol. Skr. 51: 354 (1998).
Syn: *Bidens mossiae* Sherff – F.P.S. 3: 12 (1956).
Syn: *Coreopsis chippii* M.B.Moss – F.P.S. 3: 20 (1956); T.S.S.: 415 (1990).
Syn: *Coreopsis elgonensis* sensu Andrews, non Sherff – F.P.S. 3: 20 (1956).
Perennial herb or subshrub. Montane grassland.
Distr: EQU; South Sudan endemic.
IUCN: RD
EQU: Friis & Vollesen 1129 3/1982.

Bidens engleri O.E.Schulz

F.P.S. 3: 11 (1956).
Annual herb. Woodland.
Distr: BAG; Senegal, Guinea Bissau, C.A.R.
BAG: Schweinfurth 2596 11/1869.

Bidens isostigmatoides Sherff

Symb. Bot. Upsal. 24: 87 (1984); Biol. Skr. 51: 354 (1998).
Perennial herb. Montane grassland.

Distr: EQU; South Sudan endemic.
IUCN: RD
EQU: Johnston 1487 2/1936.
Note: *B. kilimandscharica* (O.Hoffm.) Sherff is also likely to occur in South Sudan, having been recorded from the Ugandan side of the Imatong Mts (Friis & Vollesen 1998: 355).

Bidens negriana (Sherff) Cufod.

Biol. Skr. 51: 355 (1998); F.E.E. 4(2): 327 (2004); F.T.E.A. Comp. 3: 799 (2005).
Syn: *Bidens natator* Friis & Vollesen – Kew Bull. 37: 468 (1982).
Annual herb. Upland grassland on rock outcrops or moist soils.
Distr: EQU; Ethiopia, Uganda.
EQU: Friis & Vollesen 322 11/1980.
Note: a local species but not considered to be threatened.

Bidens pilosa L.

F.P.S. 3: 11 (1956); F.J.M.: 132 (1976); Biol. Skr. 51: 355 (1998); F.E.E. 4(2): 329 (2004); F.T.E.A. Comp. 3: 806 (2005).
Annual herb. Weed of disturbed ground & cultivation, also in a variety of open habitats.
Distr: DAR, BAG, EQU; pantropical & warm temperate.
DAR: Wickens 1071 1/1964; **EQU:** Andrews 893 4/1939.

Bidens prestinaria (Sch. Bip. ex Walp.) Cufod.

F.J.M.: 132 (1976); Biol. Skr. 51: 355 (1998); F.E.E. 4(2): 330 (2004).
Syn: *Bidens chaetodonta* sensu Andrews, non Sherff – F.P.S. 3: 12 (1956) – see note.
Annual herb. Short grassland, hillslopes, riverbanks, disturbed areas.
Distr: DAR, KOR, ?CS, EQU; Eritrea, Ethiopia.
KOR: Wickens 862 11/1962; **EQU:** Friis & Vollesen 573 11/1980.
Note: it is likely that Andrews' record of *B. chaetodonta* from Fung District is referable here but we have not seen the material on which this record was based.

Bidens schimperi Sch.Bip. ex Walp.

F.P.S. 3: 11 (1956); F.J.M.: 132 (1976); Biol. Skr. 51: 356 (1998); F.E.E. 4(2): 331 (2004); F.T.E.A. Comp. 3: 803 (2005).
Annual herb. Grassland, woodland & bushland.
Distr: RS, DAR, EQU; Djibouti & Somalia, S to Angola & South Africa.
RS: Jackson 2713 3/1953; **EQU:** Jackson 528 11/1948.

Bidens setigera (Sch.Bip. ex Walp.) Sherff subsp. *setigera*

F.E.E. 4(2): 332 (2004).
Annual herb. Rocky hillslopes, riverbanks, disturbed areas.
Distr: RS; Eritrea, Ethiopia.
RS: Jackson 2925 4/1953.
Note: a range restricted species but widespread in Ethiopia.

Bidens somaliensis Sherff

Biol. Skr. 51: 356 (1998); F.E.E. 4(2): 333 (2004).
Syn: *Bidens imatongensis* Sherff – F.P.S. 3: 13 (1956).

Perennial herb. Upland grassland, bushland, woodland & forest margins.
Distr: EQU; Cameroon, Ethiopia.
EQU: Friis & Vollesen 175 11/1980.

Bidens ternata (Chiov.) Sherff var. *ternata*

Biol. Skr. 51: 356 (1998); F.E.E. 4(2): 334 (2004); F.T.E.A. Comp. 3: 792 (2005).
Perennial herb. Montane grassland, bushland & forest margins, disturbed areas.
Distr: EQU; Cameroon, D.R.Congo, Ethiopia, Uganda, Kenya, Tanzania.
EQU: Myers 7799 9/1937.

Bidens ternata (Chiov.) Sherff var. *vatkei* (Sherff) Mesfin

F.E.E. 4(2): 335 (2004).
Perennial herb. River valleys.
Distr: ?NS/RS; Ethiopia.
IUCN: RD
Note: Andrews records this taxon from "Northern Sudan: Nubia"; we have not seen the material on which this was based and Mesfin records this variety as an Ethiopian endemic but it may well occur in Sudan.

Bidens ugandensis (S.Moore) Sherff

Biol. Skr. 51: 356 (1998); F.T.E.A. Comp. 3: 790 (2005).
Syn: *Bidens schweinfurthii* Sherff – F.P.S. 3: 13 (1956).
Perennial herb. Upland grassland, bushland & woodland.
Distr: UN, EQU; Cameroon to Kenya, S to Burundi & Tanzania.
EQU: Friis & Vollesen 278 11/1980.

Blainvillea acmella (L.) Philipson

F.T.E.A. Comp. 3: 735 (2005).
Syn: *Blainvillea gayana* Cass. – Biol. Skr. 51: 356 (1998).
Syn: *Blainvillea rhomboidea* sensu Andrews, non Cass. – F.P.S. 3: 13 (1956).
Annual herb. Dry bushland, woodland & grassland including disturbed areas.
Distr: RS, DAR, KOR; palaeotropical.
KOR: Kotschy 191 10/1839.

Blumea axillaris (Lam.) DC.

F.T.E.A. Comp. 2: 319 (2002); F.E.E. 4(2): 126 (2004).
Syn: *Blumea mollis* (D.Don) Merrill – Bot. Exp. Sud.: 33 (1970).
Syn: *Blumea solidaginoides* (Poir.) DC. – F.J.M.: 132 (1976).
Syn: *Blumea lacera* sensu Andrews, non (Burm.f.) DC. – F.P.S. 3: 14 (1956).
Annual or short-lived perennial herb. Swampy grassland & riverbanks, roadsides.
Distr: DAR, KOR, BAG, UN, EQU; palaeotropical.
DAR: Kamil 1155 5/1968; **UN:** Simpson 7473 2/1930.

Blumea braunii (Vatke) J.-P.Lebrun & Stork

Énum. Pl. Fleurs Afr. Trop. 4: 261 (1997).
Syn: *Laggera braunii* Vatke – F.J.M.: 137 (1976); F.E.E. 4(2): 143 (2004).
Annual herb. Woodland & bamboo thicket.
Distr: DAR; Ethiopia.
DAR: Pettet 155 12/1962.
Note: a range-restricted species but frequent in Ethiopia.

Blumea oloptera DC.

Syn: *Laggera oblonga* Oliv. & Hiern – F.P.S. 3: 40 (1956).
Syn: *Laggera oloptera* (DC.) C.D.Adams – F.J.M.: 136 (1976).
Perennial herb. Grassland.
Distr: DAR, ?BAG; Senegal to D.R. Congo, S to Angola.
DAR: Wickens 1656 5/1963.
Note: Andrews records *Laggera oblonga* from Raga which is in Bahr el Ghazal, but we have only seen material from Darfur.

Bothriocline congesta (M.Taylor) Wech.

Biol. Skr. 51: 357 (1998); F.T.E.A. Comp. 1: 151 (2000).
Perennial herb. Montane grassland.
Distr: EQU; Uganda.
IUCN: RD
EQU: Friis & Vollesen 843 12/1980.

Bothriocline imatongensis (M.Taylor) C.Jeffrey

Biol. Skr. 51: 357 (1998).
Syn: *Erlangea imatongensis* M.Taylor – F.P.S. 3: 28 (1956).
Perennial herb. Montane grassland.
Distr: EQU; South Sudan endemic.
IUCN: RD
EQU: Myers 13510 9/1940.

Bothriocline monticola (M.Taylor) Wech.

Biol. Skr. 51: 357 (1998); F.T.E.A. Comp. 1: 151 (2000).
Shrub. Montane grassland & forest margins.
Distr: EQU; Uganda, Kenya.
IUCN: RD
EQU: Friis & Vollesen 1130 3/1982.

Carduus nyassanus (S.Moore) R.E.Fr. subsp. *nyassanus*

Biol. Skr. 51: 358 (species only) (1998); F.T.E.A. Comp. 1: 49 (2000); F.E.E. 4(2): 25 (2004).
Syn: *Carduus kikuyorum* sensu Andrews, non R.E.Fr. – F.P.S. 3: 14 (1956).
Perennial herb. Upland forest margins, wet grassland.
Distr: EQU; Nigeria to Ethiopia, S to Zambia & Malawi.
EQU: Friis & Vollesen 37 11/1980.

Carduus schimperi Sch.Bip. subsp. *schimperi*

F.T.E.A. Comp. 1: 46 (2000); F.E.E. 4(2): 24 (2004).
Syn: *Carduus chamaecephalus* (Vatke) Oliv. & Hiern – Biol. Skr. 51: 368 (1998).
Syn: *Carduus theodori* R.E.Fr. – F.P.S. 3: 14 (1956).
Perennial herb. Montane grassland.
Distr: EQU; Ethiopia, Kenya, Tanzania.
EQU: Myers 13524 9/1940.

Carthamus nitidus Boiss.

F.Egypt 3: 178 (2002).
Syn: *Carthamus persicus* sensu Andrews, non Willd. – F.P.S. 3: 15 (1956).
Annual herb. Open stony ground.
Distr: RS; Somalia; Egypt, Middle East, Arabia.
RS: Schweinfurth 415 3/1865.

Carthamus tinctorius L.

F.E.E. 4(2): 37 (2004).
Annual herb. Cultivated & occasionally naturalised in disturbed areas.

Distr: NS, KOR, CS; widely cultivated in warm temperate regions.
KOR: <u>Pfund 484</u> 4/1875.

Centaurea aegyptiaca L.

F.Egypt 3: 168 (2002).
Biennial or short-lived perennial herb. Desert wadis.
Distr: RS; Egypt to Arabia.
RS: <u>Drar 295</u> 1933 (n.v.).
Note: the record for this species is from Drar (1936) and the specimen cited is from Gebel Elba in the disputed Hala'ib Triangle. Boulos (l.c.) confirms its presence there.

Centaurea calcitrapa L.

F.P.S. 3: 15 (1956); F.Egypt 3: 170 (2002).
Annual or biennial herb. Weed of disturbed areas & cultivation.
Distr: KOR; Egypt to Iran.
KOR: <u>Kotschy 457</u> 1842.

Centaurea praecox Oliv. & Hiern

F.P.S. 3: 15 (1956); Biol. Skr. 51: 358 (1998); F.T.E.A. Comp. 1: 61 (2000); F.E.E. 4(2): 41 (2004).
Perennial herb. Grassland & wooded grassland.
Distr: EQU; Mali to Ethiopia, S to Zimbabwe & Mozambique.
EQU: <u>Myers 11091</u> 4/1939.

Centaurea senegalensis DC.

F.P.S. 3: 15 (1956); F.J.M.: 132 (1976).
Perennial herb. Grassland & disturbed areas.
Distr: DAR, KOR; Mauritania & Senegal to Chad.
DAR: <u>Wickens 2838</u> 1/1965.

Ceruana praetensis Forssk.

F.P.S. 3: 17 (1956); F.E.E. 4(2): 189 (2004).
Annual herb. Sandy river beds, disturbed areas.
Distr: NS, CS; Senegal to Eritrea & Ethiopia; Egypt.
CS: <u>Jackson 4191</u> 6/1961.

Chrysanthellum indicum DC. var. afroamericanum B.L.Turner

Biol. Skr. 51: 359 (1998); F.E.E. 4(2): 338 (2004); F.T.E.A. Comp. 3: 810 (as subsp.) (2005).
Syn: *Chrysanthellum americanum* sensu auctt., non (L.) Vatke – F.P.S. 3: 17 (1956); F.J.M.: 133 (1976).
Annual herb. Weed of disturbed areas & cultivation.
Distr: NS, DAR, KOR, ES, BAG, UN, EQU; pantropical, originating in South America.
DAR: <u>Wickens 2056</u> 8/1964; **EQU:** <u>Myers 6711</u> 5/1937.

Cineraria deltoidea Sond.

F.P.S. 3: 17 (1956); Biol. Skr. 51: 359 (1998); F.E.E. 4(2): 253 (2004); F.T.E.A. Comp. 3: 563 (2005).
Syn: *Cineraria grandiflora* Vatke – F.P.S. 3: 17 (1956).
Perennial herb. Montane forest, bushland & grassland including secondary scrub.
Distr: EQU; Ethiopia to South Africa.
EQU: <u>Myers 11618</u> 7/1939.

Cirsium buchwaldii O.Hoffm.

Biol. Skr. 51: 359 (1998); F.T.E.A. Comp. 1: 55 (2000).
Syn: *Cirsium* sp. – F.P.S. 3: 18 (1956).
Perennial herb. Upland grassland.

Distr: EQU; Uganda to Zambia.
EQU: <u>Myers 11695</u> 7/1938.

Conyza aegyptiaca (L.) Aiton

F.P.S. 3: 19 (1956); F.J.M.: 133 (1976); Biol. Skr. 51: 359 (1998); F.T.E.A. Comp. 2: 496 (2002); F.E.E. 4(2): 210 (2004).
Annual or biennial herb. A wide variety of open habitats including disturbed areas.
Distr: NS, RS, DAR, KOR, CS, BAG, UN, EQU; palaeotropical.
DAR: <u>Wickens 1349</u> 3/1964; **EQU:** <u>Sillitoe 200</u> 1919.
Note: Andrews (1956) and Drar (1970) both record *C. abyssinica* Sch.Bip. ex A.Rich. from Jebel Marra and Quézel (1969) records it from nearby Jebel Gourgeil. This species is now considered to be restricted to Eritrea and Ethiopia and these Sudanese records are likely to be incorrect (this species certainly wasn't mentioned by Wickens in his Flora of Jebel Marra) but we are unsure as to their true identity.

Conyza bonariensis (L.) Cronquist

Biol. Skr. 51: 360 (1998); F.T.E.A. Comp. 2: 506 (2002); F.E.E. 4(2): 212 (2004).
Syn: *Conyza linifolia* (Willd.) Täckh. – Bot. Exp. Sud.: 34 (1970).
Annual herb. Weed of disturbed areas & grassland.
Distr: NS, EQU; pantropical, originating from S America.
NS: <u>Ahti 16334</u> 10/1962; **EQU:** <u>Jackson 4258</u> 6/1961.

Conyza hochstetteri Sch.Bip. ex A.Rich.

F.P.S. 3: 18 (1956); F.J.M.: 133 (1976); F.T.E.A. Comp. 2: 494 (2002); F.E.E. 4(2): 207 (2004).
Syn: *Conyza gouanii* sensu Friis & Vollesen, non (L.) Willd. – Biol. Skr. 51: 360 (1998).
Perennial herb. Grassland, wooded grassland, open bushland, disturbed areas.
Distr: DAR, CS, EQU; Cameroon to Somalia, S to South Africa; Arabia.
DAR: <u>Wickens 980</u> 1/1964; **EQU:** <u>Myers 10934</u> 4/1939.

Conyza newii Oliv. & Hiern

Biol. Skr. 51: 360 (1998); F.T.E.A. Comp. 2: 497 (2002); F.E.E. 4(2): 205 (2004).
Shrub or perennial herb. Montane forest margins & grassland.
Distr: EQU; Ethiopia to Malawi.
EQU: <u>Friis & Vollesen 980</u> 2/1982.

Conyza pyrrhopappa Sch.Bip. ex A.Rich.

F.J.M.: 133 (1976); Biol. Skr. 51: 360 (1998); F.T.E.A. Comp. 2: 499 (2002); F.E.E. 4(2): 204 (2004).
Syn: *Pluchea crenata* Quézel – F. Darfur Nord-Occ. & J. Gourgeil: 125 (1969).
Shrub or perennial herb. Grassland, bushland, forest margins, disturbed areas.
Distr: DAR, EQU; Nigeria to Somalia, S to Angola & Zambia; Egypt, Arabia.
DAR: <u>Wickens 2826</u> 11/1964; **EQU:** <u>Myers 10984</u> 4/1939.

Conyza ruwenzoriensis (S.Moore) R.E.Fr.

Biol. Skr. 51: 361 (1998); F.T.E.A. Comp. 2: 503 (2002).
Perennial herb or shrub. Montane grassland.

Distr: EQU; D.R. Congo, Uganda, Kenya, Tanzania.
EQU: <u>Friis & Vollesen 203</u> 11/1980.
Note: the identity of the South Sudanese material requires confirmation – in the Kew herbarium it was identified as *C. welwitschii* by Mesfin Tadesse in 1993, but Friis & Vollesen placed it in *C. ruwenzoriensis*; F.T.E.A. records neither species from Sudan.

Conyza steudelii Sch.Bip. ex A.Rich.
Biol. Skr. 51: 361 (1998); F.T.E.A. Comp. 2: 495 (2002); F.E.E. 4(2): 208 (2004).
Annual or short-lived perennial herb. Montane woodland & forest margins.
Distr: EQU; Cameroon to Eritrea, S to D.R. Congo & Tanzania; Arabia.
EQU: <u>Friis & Vollesen 274</u> 11/1980.

Conyza stricta Willd.
F.P.S. 3: 19 (1956); F.J.M.: 133 (1976); F.T.E.A. Comp. 2: 500 (2002); F.E.E. 4(2): 209 (2004).
Syn: *Conyza stricta* Willd. var. *pinnatifida* (D.Don) Kitam. – F.E.E. 4(2): 209 (2004).
Syn: *Conyza schimperi* sensu Wickens, non Sch.Bip. ex A.Rich. – F.J.M.: 133 (1976).
Annual or short-lived perennial herb. Upland grassland, in rocky or disturbed areas.
Distr: RS, DAR, ES, EQU; Guinea to Somalia, S to Angola, Zimbabwe & Mozambique; Madagascar, Arabia to China.
DAR: <u>Pettet 156</u> 12/1962; **EQU:** <u>Andrews 1008</u> 5/1939.
Note: although both F.T.E.A. and F.E.E. list *C. schimperi* as occurring in Sudan, we have seen no authentic material; all the specimens from Jebel Marra identified as such were redetermined by Mesfin Tadesse as *C. stricta* var. *pinnatifida*.

Conyza subscaposa O.Hoffm.
Biol. Skr. 51: 361 pro parte (1998); F.T.E.A. Comp. 2: 510 (2002); F.E.E. 4(2): 215 (2004).
Perennial herb. Montane grassland.
Distr: EQU; Nigeria, Cameroon, Uganda & Kenya, S to Angola, Zimbabwe & Mozambique.
EQU: <u>Myers 11097</u> 4/1939.

Conyza sumatrensis (Retz.) E.Walker
F.E.E. 4(2): 212 (2004).
Annual or short-lived perennial herb. Weed of disturbed areas.
Distr: EQU; pantropical.
EQU: <u>von Ramm 204</u> 2/1956.
Note: in F.T.E.A. this is treated as a synonym of *C. bonariensis* but we follow F.E.E. in maintaining it as distinct.

Conyza tigrensis Oliv. & Hiern
Biol. Skr. 51: 361 (1998); F.T.E.A. Comp. 2: 497 (2002); F.E.E. 4(2): 208 (2004).
Annual or short-lived perennial herb. Montane grassland.
Distr: EQU; Ethiopia to Malawi; Arabia.
EQU: <u>Friis & Vollesen 389</u> 11/1980.

Conyza variegata Sch.Bip. ex A.Rich.
F.T.E.A. Comp. 2: 497 (2002); F.E.E. 4(2): 207 (2004).
Syn: *Conyza subscaposa* sensu Friis & Vollesen pro parte, non O.Hoffm. – Biol. Skr. 51: 361 (1998).

Rhizomatous perennial herb. Montane grassland, bushland & forest margins.
Distr: EQU; Eritrea, Ethiopia, Kenya; Yemen.
EQU: <u>Friis & Vollesen 1004</u> 2/1982.

Cotula abyssinica Sch.Bip. ex A.Rich.
F.P.S. 3: 20 (1956); F.T.E.A. Comp. 2: 523 (2002); F.E.E. 4(2): 227 (2004).
Annual or short-lived perennial herb. Montane grassland, bushland & forest margins, moist ground.
Distr: CS; D.R. Congo, Eritrea, Ethiopia, Uganda, Kenya, Tanzania; Arabia.
CS: <u>Speke & Grant s.n.</u> 3/1863.
Note: this species is usually found in montane habitats but the single Sudanese collection was apparently made from the banks of the Nile at 15°N which is at less than 500 m alt.

Cotula anthemoides L.
F.P.S. 3: 20 (1956); F.T.E.A. Comp. 2: 525 (2002); F.E.E. 4(2): 228 (2004).
Annual herb. Muddy riverbanks and lake shores.
Distr: NS, CS; palaeotropical & warm temperate.
CS: <u>Pettet 132</u> 2/1963.

Cotula kotschyi Benth. & Hook.f.
F.P.S. 3: 20 (1956).
Syn: *Cotula cinerea* sensu Andrews, non Delile – F.P.S. 3: 20 (1956) – see note.
Annual herb. Riverbanks.
Distr: ?NS, ?RS, DAR, ?KOR, CS, BAG; ?Sudan & South Sudan endemic.
IUCN: RD
CS: <u>Kotschy 317</u> 1842; **BAG:** <u>Simpson 7643</u> 3/1930.
Note: the status of this taxon is uncertain; its affinity to *C. cinerea* Delile requires further investigation. Andrews records the latter from Port Sudan and Kordofan, whilst he records *C. kotschyi* from Northern and Central Sudan – we think these refer to a single taxon for which we have only seen material from Darfur, Central Sudan and Bahr el Ghazal.

Crassocephalum bauchiense (Hutch.) Milne-Redh.
Biol. Skr. 51: 361 (1998); F.T.E.A. Comp. 3: 610 (2005).
Annual herb. Wooded grassland.
Distr: EQU; Nigeria, Cameroon, D.R. Congo, Uganda, Zambia.
EQU: <u>Friis & Vollesen 293</u> 11/1980.

Crassocephalum crepidioides (Benth.) S.Moore
F.P.S. 3: 21 (1956); Biol. Skr. 51: 361 (1998); F.E.E. 4(2): 261 (2004); F.T.E.A. Comp. 3: 610 (2005).
Annual herb. Disturbed areas.
Distr: EQU; Senegal to Ethiopia, S to Angola & South Africa; Madagascar, Arabia, naturalised in tropical Asia & Australia.
EQU: <u>Myers 9716</u> 10/1938.

Crassocephalum montuosum (S.Moore) Milne-Redh.
F.P.S. 3: 21 (1956); Biol. Skr. 51: 362 (1998); F.E.E. 4(2): 262 (2004); F.T.E.A. Comp. 3: 602 (2005).

Annual or short-lived perennial herb. Forest margins & clearings.
Distr: EQU; Nigeria to Ethiopia, S to Angola & Zimbabwe; Madagascar.
EQU: Friis & Vollesen 277 11/1980.

Crassocephalum paludum C.Jeffrey

F.T.E.A. Comp. 3: 604 (2005).
Annual herb. Swampy grassland, shallow water.
Distr: UN; D.R. Congo to Kenya, S to Zambia.
UN: J.M. Lock 82/50 8/1982.

Crassocephalum rubens (Juss. ex Jacq.) S.Moore var. rubens

F.P.S. 3: 21 pro parte (1956); F.E.E. 4(2): 263 pro parte (2004); F.T.E.A. Comp. 3: 608 (2005).
Annual herb. Disturbed areas.
Distr: BAG, UN, EQU; Liberia to Ethiopia, S to South Africa; Madagascar, Comoros, Mascarenes.
EQU: Myers 7027 7/1938.

Crassocephalum rubens (Juss. ex Jacq.) S.Moore var. sarcobasis (DC.) C.Jeffrey & Beentje

F.T.E.A. Comp. 3: 609 (2005).
Syn: *Crassocephalum sarcobasis* (DC.) S.Moore – Biol. Skr. 51: 362 (1998).
Syn: *Crassocephalum rubens* sensu auctt., non (Juss. ex Jacq.) S.Moore s.s. – F.P.S. 3: 21 pro parte (1956); Bot. Exp. Sud.: 34 (1970); F.E.E. 4(2): 263 pro parte (2004).
Annual herb. Disturbed areas.
Distr: RS, ?EQU; Ethiopia to South Africa; Madagascar, Comoros, Yemen.
RS: Maffey 20 1928; **EQU:** Kielland-Lund 161 12/1983 (n.v.).
Note: the varietal identity of the Equatoria material cited by Friis & Vollesen requires confirmation since they treated C. rubens as a synonym of C. sarcobasis.

Crassocephalum vitellinum (Benth.) S.Moore

F.P.S. 3: 21 (1956); Biol. Skr. 51: 362 (1998); F.T.E.A. Comp. 3: 606 (2005).
Annual or perennial herb. Forest margins, disturbed grassland & woodland.
Distr: EQU; Nigeria to Uganda, S to Zambia.
EQU: Myers 11710 7/1939.

Crassocephalum × picridifolium (DC.) S.Moore

F.P.S. 3: 21 (1956); F.J.M.: 134 (1976); F.E.E. 4(2): 259 (2004); F.T.E.A. Comp. 3: 604 (2005).
Scrambling perennial herb. Swamps.
Distr: DAR, ?EQU; Gambia to Ethiopia, S to South Africa.
DAR: Wickens 1411 4/1964.
Note: this is considered to be a hybrid between C. paludum and C. vitellinum. Andrews records it from both Jebel Marra and Equatoria but we have only seen material from the former.

Crepis foetida L.

F.T.E.A. Comp. 1: 72 (2000); F.E.E. 4(2): 46 (2004).
Syn: *Crepis schimperi* (Sch.Bip. ex A.Rich.) Schweinf. – F.P.S. 3: 22 (1956); Bot. Exp. Sud.: 34 (1970).
Annual or short-lived perennial herb. Bushland, riverbanks, disturbed areas.

Distr: RS; Eritrea, Ethiopia, Djibouti, Somalia, Kenya; Europe, Middle East to India.
RS: Andrews 3559.

Crepis rueppellii Sch.Bip.

F.P.S. 3: 22 (1956); F.J.M: 134 (1976); Biol. Skr. 51: 363 (1998); F.T.E.A. Comp. 1: 70 (2000); F.E.E. 4(2): 49 (2004).
Syn: *Crepis ugandensis* Babc. – F.P.S. 3: 22 (1956).
Perennial herb. Montane grassland & bushland.
Distr: DAR, EQU; Eritrea to Tanzania; Arabia.
DAR: Wickens 2868 3/1965; **EQU:** Friis & Vollesen 1124 3/1982.

Dichrocephala chrysanthemifolia (Blume) DC. var. chrysanthemifolia

F.P.S. 3: 23 (1956); F.J.M.: 134 (1976); Biol. Skr. 51: 363 (1998); F.T.E.A. Comp. 2: 459 (2002); F.E.E. 4(2): 192 (2004).
Annual or perennial herb. Upland grassland, montane bushland & dry forest margins, disturbed areas.
Distr: RS, DAR, EQU; Cameroon to Eritrea, S to Malawi & Mozambique; Madagascar, Arabia, tropical Asia.
RS: Jackson 2877 4/1953; **EQU:** Friis & Vollesen 306 11/1980.

Dichrocephala integrifolia (L.f.) Kuntze subsp. integrifolia

F.P.S. 3: 23 (1956); Biol. Skr. 51: 363 (1998); F.T.E.A. Comp. 2: 459 (2002); F.E.E. 4(2): 190 (2004).
Annual herb. Disturbed areas, often in wet depressions or swamp margins.
Distr: EQU; Guinea to Eritrea, S to South Africa; Madagascar, Arabia, tropical Asia.
EQU: Myers 11105 4/1939.

Dicoma schimperi (DC.) Baill. ex O.Hoffm. subsp. schimperi

F.E.E. 4(2): 13 (2004).
Syn: *Hochstetteria schimperi* DC. – F.P.S. 3: 10 (1956).
Annual herb. Sandy & stony desert.
Distr: RS; Eritrea, Djibouti, Egypt, Saudi Arabia.
RS: Schweinfurth 385 3/1865.

Dicoma tomentosa Cass.

F.P.S. 3: 23 (1956); F.J.M.: 134 (1976); Biol. Skr. 51: 363 (1998); F.T.E.A. Comp. 1: 14 (2000); F.E.E. 4: 12 (2004).
Annual herb. Semi-desert, woodland & bushland, often on rock outcrops.
Distr: RS, DAR, KOR, CS, EQU; Senegal to Somalia, S to Namibia & South Africa; Egypt to India.
DAR: Wickens 2550 9/1964; **EQU:** Kielland-Lund 236 12/1983 (n.v.).

Doellia bovei (DC.) Anderb.

Willdenowia 25: 21 (1995).
Syn: *Blumea bovei* (DC.) Vatke – Biol. Skr. 51: 357 (1998); F.T.E.A. Comp. 2: 320 (2002); F.E.E. 4(2): 125 (2004).
Perennial herb. Bushland, rock outcrops, often in moist sites.
Distr: RS, EQU; Eritrea, Ethiopia, Somalia; Egypt, Arabia to Afghanistan.
RS: Jackson 3954 4/1959; **EQU:** Kielland-Lund 123 11/1983 (n.v.).

Echinops amplexicaulis Oliv.

F.P.S. 3: 25 (1956); Biol. Skr. 51: 364 (1998); F.T.E.A. Comp. 1: 37 (2000); F.E.E. 4(2): 17 (2004).
Perennial herb. Woodland & wooded grassland.
Distr: EQU; Cameroon to Ethiopia, S to Tanzania.
EQU: Myers 9604 10/1938.

Echinops giganteus A.Rich.

F.P.S. 3: 25 (1956); Biol. Skr. 51: 364 (1998); F.T.E.A. Comp. 1: 39 (2000); F.E.E. 4(2): 19 (2004).
Shrubby perennial herb. Grassland & woodland, often in disturbed areas.
Distr: EQU; Nigeria to Ethiopia, S to Tanzania.
EQU: Friis & Vollesen 646 12/1980.

Echinops gracilis O.Hoffm.

F.P.S. 3: 25 (1956); Bot. Exp. Sud.: 34 (1970); F.T.E.A. Comp. 1: 43˚(2000).
Perennial herb. Grassland.
Distr: ?EQU; Nigeria to Uganda.
EQU: Drar 1235 4/1938 (n.v.).
Note: the Drar specimen cited is sterile and so the identity may not be certain. Andrews also records this species from Equatoria.

Echinops hussonii Boiss.

Bot. Exp. Sud.: 34 (1970); F.Egypt 3: 145 (2002).
Perennial herb. Dry sandy & stony ground.
Distr: RS; Egypt, Saudi Arabia.
RS: Drar 426 3/1938 (n.v.).
Note: the specimen cited is sterile and so identification requires confirmation; Boulos in F.Egypt lists this species from the Gebel Elba area in the Hala'ib Triangle.

Echinops longifolius A.Rich.

F.P.S. 3: 25 (1956); F.J.M.: 134 (1976); Biol. Skr. 51: 364 (1998); F.T.E.A. Comp. 1: 42 (2000); F.E.E. 4(2): 23 (2004).
Perennial herb. Woodland & wooded grassland.
Distr: DAR, KOR, CS, EQU; Guinea to Eritrea, S to Tanzania.
DAR: Wickens 2166 5/1964; **EQU:** Myers 6591 5/1937.

Echinops macrochaetus Fresen.

F.P.S. 3: 25 (1956); F.J.M.: 134 (1976); F.E.E. 4(2): 21 (2004).
Syn: Echinops spinosus sensu Andrews, non L. – F.P.S. 3: 26 (1956).
Perennial herb. Dry open places including disturbed areas & rocky hillslopes.
Distr: RS, DAR, KOR; Eritrea, Ethiopia.
DAR: Robertson 149 12/1957.

Echinops pappii Chiov.

F.J.M.: 134 (1976); F.T.E.A. Comp. 1: 42 (2000); F.E.E. 4(2): 21 (2004).
Syn: Echinops boranensis Lanza – F.J.M.: 134 (1976).
Perennial herb. Bushland & wooded grassland, often on rocky hillslopes.
Distr: DAR, KOR; Mali to Somalia & Kenya; Arabia.
DAR: Macintosh 24 3/1930.

Eclipta prostrata (L.) L.

F.P.S. 3: 26 (1956); F.J.M.: 134 (1976); Biol. Skr. 51: 364 (1998); F.E.E. 4(2): 286 (2004); F.T.E.A. Comp. 3: 728 (2005).
Annual or short-lived perennial herb. Weed of moist ground in a variety of habitats.
Distr: NS, DAR, KOR, CS, ES, UN, EQU; pantropical.
DAR: Wickens 1642 5/1964; **UN:** J.M. Lock 82/45 8/1982.

Elephantopus mollis Kunth

F.P.S. 3: 26 (1956); F.T.E.A. Comp. 1: 283 (2000); F.E.E. 4(2): 109 (2004).
Syn: Elephantopus scaber L. subsp. **plurisetus** sensu Andrews, non (O.Hoffm.) Philipson – F.P.S. 3: 26 (1956).
Perennial herb. Disturbed woodland, wooded grassland & forest clearings.
Distr: UN, EQU; native to the neotropics, now pantropical.
EQU: Friis & Vollesen 476 11/1980.

Emilia abyssinica (Sch.Bip. ex A.Rich.) C.Jeffrey var. abyssinica

Biol. Skr. 51: 365 (1998); F.E.E. 4(2): 254 (2004); F.T.E.A. Comp. 3: 575 (2005).
Syn: Senecio abyssinicus Sch.Bip. ex A.Rich. – F.P.S. 3: 48 (1956); F.J.M.: 137 (1976).
Annual herb. Weed of cultivation, disturbed grassland & woodland.
Distr: RS, DAR, EQU; Nigeria to Eritrea, S to Zimbabwe.
DAR: Wickens 2086 8/1964; **EQU:** Sillitoe 209 1919.

Emilia baberka (Hutch.) C.Jeffrey

Kew Bull. 41: 918 (1986).
Syn: Senecio baberka Hutch. – F.P.S. 3: 49 (1956).
Perennial herb. Grassland with regular burning.
Distr: BAG; Nigeria, Cameroon.
BAG: Hoyle 521 1/1939.

Emilia caespitosa Oliv.

F.E.E. 4(2): 256 (2004); F.T.E.A. Comp. 3: 581 (2005).
Syn: Emilia coccinea sensu auctt., non (Sims) G.Don – F.P.S. 3: 28 pro parte (1956); Biol. Skr. 51: 365 (1998).
Annual herb. Grassland, woodland, disturbed areas.
Distr: EQU; C.A.R. to Ethiopia, S to Angola & Zimbabwe.
EQU: Friis & Vollesen 1254 3/1982.

Emilia coccinea (Sims) G.Don

F.T.E.A. Comp. 3: 583 (2005).
Annual herb. Grassland, woodland, weed of cultivation.
Distr: ?EQU; Uganda & Kenya, S to Angola, Zimbabwe & Mozambique.
EQU: Myers 11138 4/1939.
Note: the single specimen seen and cited was only tentatively identified as E. coccinea by Jeffrey, Beentje & Mesfin in 2004. It seems likely that all the material referred to E. coccinea by Andrews is now placed in either E. caespitosa or E. emilioides.

Emilia discifolia (Oliv.) C.Jeffrey

Biol. Skr. 51: 365 (1998); F.E.E. 4(2): 256 (2004); F.T.E.A. Comp. 3: 576 (2005).
Syn: Senecio discifolius Oliv. – F.P.S. 3: 49 (1956); Bot. Exp. Sud.: 37 (1970).

Annual herb. Grassland, bushland & woodland, disturbed areas.
Distr: EQU; Ethiopia to Zimbabwe.
EQU: Myers 9117 7/1938.

Emilia emilioides (Sch.Bip.) C.Jeffrey
Kew Bull. 41(4): 918 (1986); Biol. Skr. 51: 365 (1998).
Syn: *Emilia coccinea* sensu Andrews pro parte, non (Sims) G.Don – F.P.S. 3: 28 (1956).
Annual herb. Grassland on black clay soils.
Distr: KOR, CS, ES, UN, EQU; C.A.R. (fide Friis & Vollesen).
ES: Beshir 43 8/1951; **UN:** Sherif A.3916 7/1951.

Enydra fluctuans Lour.
F.P.S. 3: 28 (1956); F.E.E. 4(2): 314 (2004); F.T.E.A. Comp. 3: 768 (2005).
Perennial herb. Swamps, lake margins.
Distr: UN, EQU; palaeotropical.
EQU: Andrews 1300 5/1939.

Ethulia conyzoides L.f. subsp. conyzoides
F.P.S. 3: 29 (1956); F.J.M.: 134 (1976); F.T.E.A. Comp. 1: 113 (2000); F.E.E. 4(2): 108 (2004).
Annual herb. Swamps, lake margins.
Distr: DAR, BAG, EQU; palaeotropical.
DAR: Wickens 1111 1/1964; **EQU:** Myers 6990 7/1937.

Ethulia gracilis Delile
Biol. Skr. 51: 366 (1998); F.T.E.A. Comp. 1: 114 (2000); F.E.E. 4(2): 108 (2004).
Annual herb. Woodland, bushland & grassland, disturbed areas.
Distr: DAR, KOR, BAG, UN, EQU; Cameroon, Chad, C.A.R., Eritrea, Ethiopia, Uganda.
DAR: Wickens 2945 4/1965; **UN:** Harrison 296 1/1948.

Felicia abyssinica Sch.Bip. ex A.Rich.
F.T.E.A. Comp. 2: 471 (2002); F.E.E. 4(2): 196 (2004).
Perennial herb. Grassland & wooded grassland, often in disturbed areas.
Distr: RS; Eritrea & Somalia, S to Zambia; Arabia.
RS: Jackson 2890 4/1953.
Note: we here follow F.T.E.A. in treating this as a single variable taxon and so not upholding the infraspecific taxa recognised in F.E.E.

Felicia dentata (A.Rich.) Dandy subsp. nubica Grau
F.P.S. 3: 29 (species only) (1956); F.J.M.: 134 (1976); F.E.E. 4(2): 198 (species only) (2004).
Perennial herb. Montane grassland & rocky hillslopes.
Distr: DAR; Eritrea, Ethiopia; Egypt, Arabia.
DAR: Jackson 3344 12/1954.

Filago abyssinica Sch.Bip. ex A.Rich.
F.E.E. 4(2): 184 (2004).
Annual herb. Cultivated land.
Distr: RS; Eritrea, Ethiopia.
RS: Jackson 2946 4/1953.
Note: a local species but not considered to be threatened.

Filago pyramidata L.
F.Egypt 3: 200 (2002).
Syn: *Filago spathulata* C.Presl. – F. Darfur Nord-Occ. & J. Gourgeil: 124 (1969).
Annual herb. Damp depressions.
Distr: ?DAR; N Africa, E Mediterranean, Arabia to Pakistan.
Note: the record of this species is from Quézel (1969) who records it from the summit of Jebel Gourgeil and notes that it is a remarkable find since this species is absent from Tibesti; we have not seen the specimen on which it is based and the identity requires confirmation.

Flaveria trinervia (Spreng.) C.Mohr
F.P.S. 3: 29 (1956); F.T.E.A. Comp. 3: 725 (2005).
Annual herb. Marshy ground, disturbed areas.
Distr: ES; native to the Americas, now pantropical.
ES: Andrews 47 6/1935.

Galinsoga parviflora Cav.
F.P.S. 3: 30 (1956); F.J.M.: 135 (1976); Biol. Skr. 51: 366 (1998); F.E.E. 4(2): 305 (2004); F.T.E.A. Comp. 3: 763 (2005).
Annual herb. Weed of disturbed areas.
Distr: DAR, KOR, CS, EQU; native to the neotropics, now almost cosmopolitan.
DAR: Wickens 973 1/1964; **EQU:** Myers 11158 4/1939.

Geigeria acaulis Oliv. & Hiern
F.P.S. 3: 31 (1956); F.T.E.A. Comp. 2: 340 (2002); F.E.E. 4(2): 120 (2004).
Perennial herb. Dry grassland & open bushland.
Distr: KOR, DAR; Eritrea, Ethiopia, Uganda, Kenya, Tanzania, Angola, Zimbabwe, Namibia, South Africa.
KOR: Wickens 663 10/1962.

Geigeria alata (Hochst. & Steud. ex DC.) Oliv. & Hiern
F.P.S. 3: 31 (1956); Fl. Jebel Marra 135 (1976); F.T.E.A. Comp. 2: 342 (2002); F.E.E. 4(2): 121 (2004).
Syn: *Geigeria macdougalii* S.Moore – F.P.S. 3: 31 (1956).
Annual herb. Dry bushland & grassland.
Distr: RS, DAR, KOR, CS; Mauritania to Somalia & Kenya, Angola, Namibia; Egypt, Arabia.
KOR: Pfund 84 9/1875.

Gerbera piloselloides (L.) Cass.
F.P.S. 3: 31 (1956); Biol. Skr. 51: 366 (1998); F.T.E.A. Comp. 1: 12 (2000); F.E.E. 4(2): 366 (2004).
Perennial herb. Upland grassland & open woodland, often on rocky hillslopes.
Distr: EQU; Guinea to Eritrea, S to South Africa; Madagascar, Arabia, scattered across Asia.
EQU: Andrews 1991 3/1939.

Gerbera viridifolia (DC.) Sch.Bip.
Biol. Skr. 51: 366 (1998); F.T.E.A. Comp. 1: 9 (2000); F.E.E. 4: 2 (2004).
Perennial herb. Woodland, bushland & grassland.
Distr: EQU; Cameroon, Eritrea to South Africa.
EQU: Myers 10927 4/1939.

Glossocardia bosvallia (L.f.) DC.

Syn: *Bidens minuta* Miré & H.Gillet – F.W.T.A. 2: 234 (1963); F. Darfur Nord-Occ. & J. Gourgeil: 123 (1969).
Annual herb. Dry grassland & degraded areas.
Distr: DAR; Saharan Mts of Niger (Aïr) & Chad (Tibesti & Ennedi).
Note: Quézel (1969) records this species as common in the areas he studied in northern Darfur and Jebel Gourgeil but it doesn't appear to extend to Jebel Marra. We have not seen the specimens on which this record is based.

Gnaphalium polycaulon Pers.

F.T.E.A. Comp. 2: 398 (2002). ˈ
Syn: *Gnaphalium niliacum* Spreng. – F.P.S. 3: 32 (1956).
Annual herb. Seasonally wet open areas.
Distr: CS, BAG; pantropical.
CS: Drar 12 2/1938 (n.v.); **BAG:** Simpson 7638 3/1930.
Note: Andrews records this species as widespread. He may have included specimens collected by Dr. Bromfield from "tropical Nubia" but these were collected from southern Egypt.

Gnaphalium unionis Sch.Bip. ex Oliv. & Hiern var. rubriflorum (Hilliard) Beentje

F.T.E.A. Comp. 2: 396 (2002).
Syn: *Gnaphalium rubriflorum* Hilliard – F.E.E. 4(2): 183 (2004).
Syn: *Helichrysum declinatum* sensu Andrews, non (L.f.) Less. – F.P.S. 3: 36, pro parte? (1956).
Stoloniferous perennial herb. Damp montane grassland & forest margins.
Distr: KOR; Rwanda, Burundi, Ethiopia, Uganda, Kenya, Tanzania.
KOR: Kotschy 27 1842.
Note: Andrews records *Helichrysum declinatum* from Northern Sudan and Kordofan; the latter is certainly referable here, the former almost certainly not.

Gnomophalium pulvinatum (Delile) Greuter

Willdenowia 33: 242 (2003).
Syn: *Gnaphalium pulvinatum* Delile – F.P.S. 3: 32 (1956); Bot. Exp. Sud.: 35 (1970).
Syn: *Homognaphalium pulvinatum* (Delile) Fayed & Zareh – F.Egypt 3: 205 (2002).
Annual herb. Moist ground.
Distr: RS, CS; Egypt, Arabia to India.
RS: Jackson 2692 3/1953.

Goniocaulon indicum (Klein ex Willd.) C.B.Clarke

F.E.E. 4(2): 42 (2004).
Annual herb. Woodland & cultivation on black clay soils.
Distr: CS, ES; Ethiopia; Pakistan, India.
ES: Jackson 3402 1/1955.

Grangea ceruanoides Cass.

Mitt. Bot. Staats., München 15: 462 (1979).
Syn: *Grangea maderaspatana* sensu Andrews pro parte, non (L.) Desf. – F.P.S. 3: 32 (1956).
Annual or perennial herb. Damp ground.
Distr: CS; Mauritania & Senegal to Chad.
CS: Jackson 4173 5/1961.

Grangea maderaspatana (L.) Desf.

F.P.S. 3: 32 pro parte (1956); F.T.E.A. Comp. 2: 466 (2002); F.E.E. 4(2): 190 (2004).
Annual herb. Lake margins, moist depressions, black clay soils.
Distr: BAG, UN; palaeotropical.
BAG: Andrews 648 4/1939.

Guizotia arborescens Friis

Biol. Skr. 51: 367 (1998); F.E.E. 4(2): 314 (2004); F.T.E.A. Comp. 3: 773 (2005).
Shrub. Forest margins, upland bushland.
Distr: EQU; Ethiopia, Uganda.
IUCN: RD
EQU: Friis & Vollesen 391 11/1980.

Guizotia scabra (Vis.) Chiov.

F.P.S. 3: 32 (1956); F.J.M.: 135 (1976); Biol. Skr. 51: 367 (1998); F.E.E. 4(2): 310 (2004); F.T.E.A. Comp. 3: 771 (2005).
Syn: *Guizotia schimperi* sensu Andrews, non Sch. Bip. ex Walp. – F.P.S. 3: 33 (1956).
Annual or perennial herb. Woodland & grassland including swampy ground & disturbed areas.
Distr: DAR, KOR, CS, UN, EQU; Nigeria to Eritrea, S to D.R. Congo & Mozambique; Arabia.
DAR: Wickens 1035 1/1964; **EQU:** Myers 9710 10/1938.
Note: Andrews records also *G. villosa* Sch.Bip. from Fung District; this species is restricted to Eritrea and Ethiopia, and Andrews' record is probably a misidentification, perhaps of *G. scabra*.

Gutenbergia cordifolia Benth. ex Oliv. var. cordifolia

F.P.S. 3: 33 (1956); Biol. Skr. 51: 367 (1998); F.T.E.A. Comp. 1: 124 (2000).
Annual or short-lived perennial herb. Grassland, bushland, rocky hillslopes.
Distr: UN, BAG; D.R. Congo to Kenya, S to Zimbabwe & Mozambique.
BAG: Myers 10103 11/1938.

Gutenbergia rueppellii Sch.Bip. var. rueppellii

F.P.S. 3: 33 (1956); F.J.M.: 135 (1976); Biol. Skr. 51: 368 (1998); F.T.E.A. Comp. 1: 129 (2000); F.E.E. 4(2): 104 (2004).
Syn: *Gutenbergia polycephala* sensu Andrews, non Oliv. & Hiern – F.P.S. 3: 33 (1956).
Annual or short-lived perennial herb. Dry bushland, woodland & grassland.
Distr: DAR, UN, EQU; D.R. Congo, Eritrea, Ethiopia, Somalia, Uganda, Kenya, Tanzania.
DAR: Wickens 2238 8/1964; **UN:** J.M. Lock 81/271 9/1981.

Gynura amplexicaulis Oliv. & Hiern

F.P.S. 3: 34 (1956); F.T.E.A. Comp. 3: 700 (2005).
Rhizomatous perennial herb. Swampy grassland, weed of cultivation.
Distr: EQU; D.R. Congo, Rwanda, Burundi, Uganda, Kenya, Tanzania.
EQU: von Ramm 226 2/1956.

Helichrysum argyranthum O.Hoffm.

F.P.S. 3: 35 (1956); T.S.S.: 415 (1990); Biol. Skr. 51: 368 (1998); F.T.E.A. Comp. 2: 442 (2002); F.E.E. 4(2): 166 (2004).
Shrub. Montane grassland & ericaceous bushland.
Distr: EQU; Eritrea to Tanzania.
EQU: Friis & Vollesen 842 12/1980.

Helichrysum foetidum (L.) Moench

F.P.S. 3: 35 (1956); F.T.E.A. Comp. 2: 434 (2002); F.E.E. 4(2): 176 (2004).
Syn: *Helichrysum foetidum* (L.) Moench var.
microcephalum A.Rich. – Biol. Skr. 51: 368 (1998); F.E.E. 4(2): 177 (2004).
Annual or short-lived perennial herb. Upland grassland, bushland & forest margins.
Distr: EQU; Nigeria to Eritrea, S to Zambia, South Africa; Spain, Macaronesia, Arabia.
EQU: Friis & Vollesen 275 11/1980.

Helichrysum formosissimum (Sch.Bip.) A.Rich. var. formosissimum

F.P.S. 3: 35 (1956); Biol. Skr. 51: 369 (1998); F.T.E.A. Comp. 2: 443 (2002); F.E.E. 4(2):166 (2004).
Perennial herb. Montane grassland & ericaceous bushland.
Distr: EQU; D.R. Congo to Ethiopia, S to Tanzania.
IUCN: DD
EQU: Jackson 1538 6/1950.

Helichrysum forskahlii (J.F.Gmel.) Hilliard & B.L.Burtt var. forskahlii

Biol. Skr. 51: 369 (1998); F.T.E.A. Comp. 2: 415 (2002); F.E.E. 4(2): 170 (2004).
Syn: *Helichrysum fruticosum* Vatke – F.P.S. 3: 36 (1956).
Perennial herb or shrub. Montane grassland & ericaceous bushland.
Distr: EQU; Nigeria to Ethiopia, S to Angola & Zimbabwe; Arabia.
EQU: Friis & Vollesen 1101 3/1982.

Helichrysum globosum Sch.Bip.

Biol. Skr. 51: 370 (1998); F.T.E.A. Comp. 2: 417 (2002); F.E.E. 4(2): 175 (2004).
Syn: *Gnaphalium schultzii* Mendonça – F.J.M.: 135 (1976).
Perennial herb. Upland grassland, ericaceous bushland & forest margins, including swampy areas.
Distr: DAR, EQU; Guinea, Cameroon to Ethiopia, S to Angola & Mozambique; Madagascar.
DAR: Wickens 2629 9/1964; **EQU:** Friis & Vollesen 1102 3/1982.

Helichrysum glumaceum DC.

F.J.M.: 136 (1976); F.T.E.A. Comp. 2: 414 (2002); F.E.E. 4(2): 171 (2004).
Syn: *Achyrocline luzuloides* Sch.Bip. ex Vatke – F. Darfur Nord-Occ. & J. Gourgeil: 123 (1969).
Syn: *Helichrysum luzuloides* (Sch.Bip. ex Vatke) Lanza – F.P.S. 3: 36 (1956).
Annual or perennial herb. Dry grassland & bushland, often on rocky hillslopes.
Distr: RS, DAR; Mauritania to Somalia, S to Angola & Tanzania; Egypt, Arabia, Socotra.
DAR: Wickens 2078 8/1964.

Helichrysum maranguense O.Hoffm.

Biol. Skr. 51: 370 (1998); F.T.E.A. Comp. 2: 412 (2002). Scrambling shrub. Montane grassland, bushland & forest margins.
Distr: EQU; D.R. Congo, Uganda, Kenya, Tanzania.
EQU: Friis & Vollesen 1133 3/1982.

Helichrysum nudifolium (L.) Less. var. nudifolium

Biol. Skr. 51: 370 (1998); F.T.E.A. Comp. 2: 422 (2002); F.E.E. 4(2): 173 (2004).
Syn: *Helichrysum coriaceum* Harv. – Biol. Skr. 51: 368 (1998).
Syn: *Helichrysum gerberifolium* Sch.Bip. ex A.Rich. – Biol. Skr. 51: 370 (1998); F.E.E. 4(2): 173 (2004).
Perennial herb. Montane grassland.
Distr: EQU; Ethiopia to Angola & South Africa.
EQU: Myers 10925 4/1939.

Helichrysum nudifolium (L.) Less. var. oxyphyllum (DC.) Beentje

F.T.E.A. Comp. 2: 423 (2002).
Syn: *Helichrysum oxyphyllum* DC. – Biol. Skr. 51: 371 (1998).
Syn: *Helichrysum undatum* (J.F.Gmel.) Less. – F.P.S. 3: 34 (1956).
Perennial herb. Grassland & ericaceous bushland.
Distr: EQU; Cameroon to Kenya, S to South Africa.
EQU: Myers 6530 5/1937.

Helichrysum odoratissimum (L.) Sweet

F.P.S. 3: 35 (1956); Biol. Skr. 51: 370 (1998); F.T.E.A. Comp. 2: 410 (2002).
Perennial herb. Montane grassland, woodland & bushland, forest margins.
Distr: EQU; D.R. Congo to Kenya, S to South Africa.
EQU: Friis & Vollesen 149 11/1980.

Helichrysum quartinianum A.Rich.

Biol. Skr. 51: 371 (1998); F.T.E.A. Comp. 2: 438 (2002); F.E.E. 4(2): 167 (2004).
Perennial herb. Montane grassland & ericaceous bushland.
Distr: EQU; Nigeria to Ethiopia, S to Angola & Zambia.
EQU: Friis & Vollesen 915 2/1982.

Helichrysum schimperi (Sch.Bip. ex A.Rich.) Moeser

F.P.S. 3: 35 (1956); T.S.S.: 415 (1990); Biol. Skr. 51: 371 (1998); F.T.E.A. Comp. 2: 418 (2002); F.E.E. 4(2): 175 (2004).
Perennial herb or subshrub. Montane grassland, ericaceous bushland & forest margins.
Distr: ?RS, EQU; Ethiopia to Zimbabwe & Mozambique; Arabia.
EQU: Friis & Vollesen 281 11/1980.
Note: Andrews records this species from Has Has in the Red Sea Hills and from the Imatong Mts; we have only seen material from the latter.

Helichrysum splendidum (Thunb.) Less.

Biol. Skr. 51: 372 (1998); F.T.E.A. Comp. 2: 414 (2002); F.E.E. 4(2): 169 (2004).
Perennial herb or shrub. Montane grassland & ericaceous bushland.

Distr: EQU; Ethiopia, Tanzania to South Africa.
EQU: MacDonald 11 12/1938.

Hypericophyllum elatum (O.Hoffm.) N.E.Br.
F.T.E.A. Comp. 3: 720 (2005).
Syn: *Hypericophyllum compositorum* sensu Andrews,
non Steetz – F.P.S. 3: 37 (1956).
Perennial herb. Woodland & grassland.
Distr: BAG; D.R. Congo to Tanzania, S to South Africa.
BAG: Macintosh 37 1932.

Ifloga spicata (Forssk.) Sch.Bip.
F.Egypt 3: 197 (2002).
Annual herb. Sandy desert plains & wadis.
Distr: RS; Spain, Macaronesia, N Africa to SW Asia.
RS: Bent s.n. 1896.
Note: Boulos (l.c.) records two subspecies from the
Gebel Elba area in the Hala'ib Triangle, one of which
is endemic. However, the differences seem very minor.
Therefore, we here treat the species in the broad
sense only.

Inula paniculata (Klatt) Burtt & Davy
Biol. Skr. 51: 372 (1998); F.T.E.A. Comp. 2: 325 (2002);
F.E.E. 4(2): 122 (2004).
Syn: *Inula decipiens* E.A.Bruce – F.P.S. 3: 37 (1956).
Biennial or perennial herb. Grassland & wooded grassland.
Distr: EQU; Ethiopia to South Africa.
EQU: Friis & Vollesen 190 11/1980.

Iphiona scabra DC.
F.P.S. 3: 37 (1956); F.E.E. 4(2): 135 (2004).
Shrub. Desert, wadis.
Distr: RS; Eritrea; Egypt, Palestine, Arabia.
RS: Newberry 153 1/1928.

Kleinia abyssinica (A.Rich.) A.Berger var. abyssinica
F.P.S. 3: 38 (1956); Biol. Skr. 51: 372 (1998); F.E.E. 4(2):
271 (2004); F.T.E.A. Comp. 3: 694 (2005).
Perennial succulent herb. Woodland & grassland, often
amongst rocks.
Distr: EQU; C.A.R. to Ethiopia, S. to Malawi.
EQU: Friis & Vollesen 106 11/1980.

Kleinia picticaulis (P.R.O.Bally) C.Jeffrey
F.E.E. 4(2): 272 (2004); F.T.E.A. Comp. 3: 689 (2005).
Perennial succulent herb. Grassland & dry bushland.
Distr: EQU; Ethiopia, Kenya, Tanzania.
IUCN: RD
EQU: Martin 71 3/1934.

Kleinia schweinfurthii (Oliv. & Hiern) A.Berger
F.P.S. 3: 38 (1956); Biol. Skr. 51: 372 (1998); F.T.E.A.
Comp. 3: 687 (2005).
Perennial succulent herb. Dry bushland.
Distr: BAG; Nigeria, Kenya, Tanzania, Malawi.
BAG: Schweinfurth II.28 1/1871.

Lactuca glandulifera Hook.f.
Biol. Skr. 51: 373 (1998); F.T.E.A. Comp. 1: 85 (2000);
F.E.E. 4(2): 55 (2004).
Perennial herbaceous climber or scrambler. Upland
bushland & grassland, streamsides, forest margins.

Distr: EQU; Sierra Leone to Ivory Coast, Nigeria to
Ethiopia, S to Mozambique.
EQU: Friis & Vollesen 151 11/1980.

Lactuca inermis Forssk.
F.J.M.: 136 (1976); Biol. Skr. 51: 373 (1998); F.T.E.A.
Comp. 1: 83 (2000); F.E.E. 4(2): 55 (2004).
Syn: *Lactuca capensis* Thunb. – F.P.S. 3: 38 (1956); Bot.
Exp. Sud.: 35 (1970); F.J.M.: 136 (1976).
Perennial herb. Grassland, bushland, forest margins,
disturbed areas.
Distr: DAR, UN, EQU; Guinea to Ethiopia, S to Angola &
South Africa; Arabia, Mascarenes.
DAR: Jackson 3390 12/1954; **EQU:** Friis & Vollesen 253
11/1980.

Lactuca schweinfurthii Oliv. & Hiern
F.P.S. 3: 38 (1956); F.T.E.A. Comp. 1: 81 (2000).
Perennial herb. Grassland & woodland.
Distr: EQU; Cameroon to Tanzania, S to Angola &
Zambia.
EQU: Schweinfurth 4015 6/1870.
Note: the single South Sudanese specimen seen was
collected from the Equatoria/Bahr el Ghazal border area.

Lactuca serriola L.
F.E.E. 4(2): 55 (2004).
Annual or perennial herb. Weed of cultivation &
disturbed areas.
Distr: KOR; Eritrea, Ethiopia, Somalia; Mediterranean, W
Asia, introduced in South Africa.
KOR: Kotschy s.n. 1842.
Note: Andrews records *Lactuca virosa* L. from Kordofan;
this is probably based on a misidentification, but the true
identity is unclear.

Laggera crassifolia (A.Rich.) Oliv. & Hiern
F.P.S. 3: 39 (1956); Biol. Skr. 51: 373 (1998); F.E.E. 4(2):
141 (2004).
Perennial herb or subshrub. Grassland & woodland.
Distr: CS, EQU; D.R. Congo, Eritrea, Ethiopia.
CS: Kotschy 564 1837; **EQU:** Friis & Vollesen 333
11/1980.

Laggera crispata (Vahl) Hepper & J.R.I.Wood
Biol. Skr. 51: 373 (1998); F.T.E.A. Comp. 2: 353 (2002);
F.E.E. 4(2): 143 (2004).
Syn: *Laggera alata* (D.Don) Oliv. – F.P.S. 3: 39 (1956).
Syn: *Laggera pterodonta* (DC.) Oliv. – F.P.S. 3: 39
(1956); Bot. Exp. Sud.: 36 (1970); F.J.M.: 136 (1976).
Annual or perennial herb. Woodland, grassland, forest
margins, disturbed areas.
Distr: DAR, CS, BAG, EQU; palaeotropical.
DAR: Wickens 1096 1/1964; **EQU:** Myers 11906
10/1938.

Laggera decurrens (Vahl) Hepper & J.R.I.Wood
Pl. Syst. Evol. 176: 161 (1991).
Syn: *Blumea gariepina* DC. – F.P.S. 3: 14 (1956).
Annual herb. Open dry bushland & woodland.
Distr: DAR; Somalia, Angola, Botswana, South Africa;
Algeria.
DAR: Lynes 406 2/1922.

Laggera elatior R.E.Fr.

F.P.S. 3: 40 (1956); Biol. Skr. 51: 374 (1998); F.T.E.A. Comp. 2: 356 (2002); F.E.E. 4(2): 144 (2004).
Annual or short-lived perennial herb. Montane forest margins, bushland & grassland.
Distr: EQU; Ethiopia to Tanzania & Burundi.
EQU: Friis & Vollesen 979 2/1982.

Launaea brunneri (Webb) Amin ex Boulos

Bot. Not. 115: 59 (1962).
Syn: *Sonchus chevalieri* (O.Hoffm. & Muschl.) Dandy – F.P.S. 3: 51 (1956).
Annual or perennial herb. Dry sandy areas, weed of cultivation.
Distr: KOR; Mauritania & Senegal to Nigeria, ?Chad.
KOR: Kotschy 427 1842.

Launaea capitata (Spreng.) Dandy

F.P.S. 3: 40 (1998); F.E.E. 4(2): 59 (2004).
Perennial herb. Wadis.
Distr: NS, RS; Eritrea; Algeria to Pakistan.
RS: Drummond & Hemsley 942 1/1953.

Launaea cornuta (Hochst. ex Oliv. & Hiern) C.Jeffrey

Biol. Skr. 51: 374 (1998); F.T.E.A. Comp. 1: 104 (2000); F.E.E. 4(2): 59 (2004).
Syn: *Sonchus cornutus* Hochst. ex Oliv. & Hiern – F.P.S. 3: 51 (1956).
Syn: *Sonchus exauriculatus* (Oliv. & Hiern) O.Hoffm. – F.P.S. 3: 51 (1956).
Annual or perennial herb. Grassland including seasonally wet areas, weed of disturbed areas & cultivation.
Distr: DAR, KOR, CS, ES, BAG, UN, EQU; Nigeria to Somalia, S to Zimbabwe & Mozambique.
ES: Schweinfurth 444 11/1865; **EQU:** Myers 6784 5/1937.

Launaea intybacea (Jacq.) Beauverd

F.T.E.A. Comp. 1: 105 (2000); F.E.E. 4(2): 61 (2004).
Syn: *Lactuca pinnatifida* (Lour.) Merr. – F.P.S. 3: 38 (1956).
Annual or biennial herb. Wadis, rocky hillslopes.
Distr: RS, DAR, KOR; pantropical.
RS: Carter 1813 11/1987.

Launaea massavensis (Fresen.) Sch.Bip. ex Kuntze

F.E.E. 4(2): 62 (2000).
Syn: *Heterachaena massavensis* Fresen. – F.P.S. 3: 36 (1956).
Annual herb. Dry grassland, wadis.
Distr: RS; Eritrea, Ethiopia, Somalia; Egypt, Arabia, Socotra, Pakistan.
RS: Andrews 3550.

Launaea mucronata (Forssk.) Muschl. subsp. *mucronata*

F.E.E. 4(2): 63 (2004).
Annual or short-lived perennial herb. Desert plains, wadis.
Distr: RS; Mauritania, Chad, Eritrea; N Africa to Arabia & Iran.
RS: Bally 6900 4/1949.

Launaea mucronata (Forssk.) Muschl. subsp. *cassiniana* Kilian

Englera 17: 409 (1997); F.Egypt 3: 298 (2002).
Short-lived perennial herb. Desert plains, wadis.
Distr: RS; Mauritania, Chad; N Africa to Arabia & Iran.
RS: Bent s.n. 1896.

Launaea nana (Baker) Choiv.

F.J.M.: 136 (1976); Biol. Skr. 51: 375 (1998); F.T.E.A. Comp. 1: 100 (2000).
Syn: *Sonchus elliotianus* Hiern – F.P.S. 3: 53 (1956).
Perennial herb. Grassland & wooded grassland.
Distr: DAR, EQU; Guinea to Kenya, S to Angola & South Africa.
DAR: Wickens 1597 4/1964; **EQU:** Chipp 35 2/1929.

Launaea nudicaulis (L.) Hook.f.

Englera 17: 217 (1997); F.Egypt 3: 291 (2002).
Perennial herb. Dry rocky hillslopes, wadis, desert plains, weed of irrigated land.
Distr: RS; Chad; Spain, Macaronesia, N Africa to Arabia & Iran.
RS: Ahti 16662 10/1962.

Launaea petitiana (A.Rich.) Kilian

F.T.E.A. Comp. 1: 105 (2000); F.E.E. 4(2): 61 (2004).
Perennial herb. Grassland, bushland & wooded grassland.
Distr: RS; Eritrea, Ethiopia, Somalia, Kenya, Tanzania; Yemen.
RS: Jackson 2740 3/1953.

Launaea rarifolia (Oliv. & Hiern) Boulos

F.T.E.A. Comp. 1: 101 (2000); F.E.E. 4(2): 58 (2004).
Syn: *Sonchus welwitschii* (Scott-Elliot) Chiov. – F.P.S. 3: 51 (1956).
Perennial herb. Woodland & wooded grassland.
Distr: EQU; Sierra Leone to Ethiopia, S to Angola & South Africa; Madagascar.
EQU: Wyld 517 5/1939.

Launaea taraxacifolia (Willd.) Amin ex C.Jeffrey

F.T.E.A. Comp. 1: 106 (2000); F.E.E. 4(2): 62 (2004).
Syn: *Lactuca taraxacifolia* (Willd.) Hornem. – F.P.S. 3: 38 (1956).
Annual herb. Dry bushland, weed of cultivation.
Distr: CS, UN, EQU; Senegal to Ethiopia, Tanzania.
CS: Kotschy 337 3/1840; **UN:** Myers 10473 2/1939.

Leysera leyseroides (Desf.) Maire

F.P.S. 3: 40 (1956); F.Egypt 3: 211 (2002).
Annual herb. Semi-desert, wadis.
Distr: RS; N Africa, Palestine to Pakistan.
RS: Bent s.n. 1896.

Litogyne gariepina (DC.) Anderb.

F.T.E.A. Comp. 2: 368 (2002).
Syn: *Epaltes alata* (Sond.) Steetz – F.P.S. 3: 28 (1956); Bot. Exp. Sud.: 35 (1970).
Syn: *Epaltes umbelliiformis* Steetz – F.P.S. 3: 28 (1956); Bot. Exp. Sud.: 35 (1970).
Annual or perennial herb. Riverbanks, floodplains, temporary swamps.

Distr: DAR, KOR, CS; Mali to Kenya, S to Botswana & South Africa.
DAR: Harrison 176 10/1947.

Macledium sessiliflorum (Harv.) S.Ortíz subsp. stenophyllum (G.V.Pope) S.Ortíz
Taxon 50: 742 (2001).
Syn: *Dicoma sessiliflora* Harv. subsp. *stenophylla* G.V.Pope – F.E.E. 4(2): 14 (2004).
Syn: *Dicoma sessiliflora* sensu Andrews, non Harv. s.s. – F.P.S. 3: 23 (1956).
Perennial herb. Woodland.
Distr: BAG; Senegal to Ethiopia.
BAG: Schweinfurth 4282 10/1870.

Melanthera abyssinica (Sch.Bip. ex A.Rich.) Vatke
F.P.S. 3: 41 (1956); F.E.E. 4(2): 298 (2004); F.T.E.A. Comp. 3: 738 (2005).
Perennial herb. Forest margins.
Distr: RS, ES; Sierra Leone to Eritrea, S to Angola & Zimbabwe; Arabia.
RS: Aylmer 180 3/1932.

Melanthera pungens Oliv. & Hiern var. pungens
F.P.S. 3: 43 (1956); F.J.M.: 136 (1976); Biol. Skr. 51: 375 (1998); F.T.E.A. Comp. 3: 741 (2005).
Perennial herb. Grassland, bushland, forest margins.
Distr: DAR, ES, BAG, UN, EQU; Senegal, Mali, D.R. Congo, Uganda, Kenya, Tanzania.
DAR: Wickens 2107 8/1964; **UN:** Sherif A.4029 9/1951.

Melanthera scandens (Schumach. & Thonn.) Roberty subsp. madagascariensis (Baker) Wild
Biol. Skr. 51: 375 (1998); F.E.E. 4(2): 299 (2004); F.T.E.A. Comp. 3: 740 (2005).
Perennial herb. Swamps, river & lake margins, forest margins.
Distr: BAG, UN, EQU; Nigeria to Ethiopia, S to Angola & Botswana; Madagascar.
EQU: Jackson 2003 8/1951.

Melanthera scandens (Schumach. & Thonn.) Roberty subsp. subsimplicifolia Wild
F.E.E. 4(2): 299 (2004); F.T.E.A. Comp. 3: 740 (2005).
Perennial herb. Moist forest margins.
Distr: EQU; Nigeria to Ethiopia, S to Zimbabwe & Mozambique.
EQU: von Ramm 158 4/1956.
Note: this may just be a form of subsp. *madagascariensis* – see note in F.T.E.A.

Micractis bojeri DC.
Biol. Skr. 51: 375 (1998); F.E.E. 4(2): 318 (2004); F.T.E.A. Comp. 3: 776 (2005).
Annual or short-lived perennial herb. Moist upland grassland, streamsides.
Distr: EQU; Nigeria to Ethiopia, S to Malawi; Madagascar.
EQU: Friis & Vollesen 639 11/1980.

Microglossa afzelii O.Hoffm.
F.P.S. 3: 43 (1956); T.S.S.: 416 (1990); F.T.E.A. Comp. 2: 484 (2002).

Scandent shrub. Forest margins & thickets.
Distr: EQU; Guinea to Uganda & Tanzania.
EQU: Wyld 564 7/1939.

Microglossa pyrifolia (Lam.) Kuntze
F.P.S. 3: 43 (1956); T.S.S.: 416 (1990); Biol. Skr. 51: 376 (1998); F.T.E.A. Comp. 2: 484 (2002); F.E.E. 4(2): 195 (2004).
Shrub or scandent shrub. Disturbed areas in grassland, wooded grassland, bushland & forest margins.
Distr: EQU; Senegal to Ethiopia, S to Angola, Zimbabwe & Mozambique; Madagascar, tropical Asia.
EQU: Myers 8201 11/1938.

Mikania chenopodiifolia Willd.
Biol. Skr. 51: 376 (1998); F.T.E.A. Comp. 3: 835 (2005).
Syn: *Mikania capensis* DC. – F.E.E. 4(2): 344 (2004).
Syn: *Mikania cordata* sensu auctt., non (Burm.f.) B.L.Rob. – F.P.S. 3: 43 (1956); T.S.S.: 416 (1990);.
Climbing perennial herb or shrub. Forest margins & thickets.
Distr: EQU; Senegal to Ethiopia, S to Angola & South Africa; Madagascar, Comoros.
EQU: Myers 11 8/1940.
Note: we here follow F.T.E.A. in treating *M. capensis* as a synonym of *M. chenopodiifolia*.

Nicolasia nitens (O.Hoffm.) Eyles var. nitens
F.T.E.A. Comp. 2: 350 (2002).
Syn: *Pluchea nitens* O.Hoffm. – F.J.M.: 137 (1976).
Annual or short-lived perennial herb. Seasonally wet areas in dry bushland, including alkaline soils.
Distr: DAR; Kenya to Namibia & Botswana.
DAR: Wickens 2890 3/1965.

Nidorella spartioides (O.Hoffm.) Cronquist
Biol. Skr. 51: 376 (1998); F.T.E.A. Comp. 2: 486 (2002).
Perennial herb. Grassland, bushland & woodland with regular burning.
Distr: EQU; Nigeria to Uganda, S to Angola & Zambia.
EQU: Friis & Vollesen 943 2/1982.

Ochrocephala imatongensis (Philipson) Dittrich
Biol. Skr. 51: 377 (1998); F.T.E.A. Comp. 1: 36, in notes (2000); F.E.E. 4(2): 33 (2004).
Syn: *Centaurea imatongensis* Philipson – F.P.S. 3: 17 (1956).
Perennial herb or subshrub. Wooded grassland with regular burning.
Distr: EQU; D.R. Congo, Ethiopia, ?Uganda.
IUCN: RD
EQU: Friis & Vollesen 336 11/1980.

Osteospermum vaillantii (Decne.) Norl.
F.P.S. 3: 44 (1956); F.J.M.: 136 (1976); F.T.E.A. Comp. 2: 533 (2002); F.E.E. 4(2): 186 (2004).
Annual or short-lived perennial herb or subshrub. Upland bushland & grassland, disturbed areas.
Distr: NS, RS, DAR; Eritrea, Ethiopia, Djibouti, Somalia, Uganda, Kenya, Tanzania; Egypt, Arabia.
DAR: Wickens 2447 9/1964.
Note: the correct name is given as *Tripteris vaillantii* Decne. in the African Plants Database.

Pegolettia senegalensis Cass.

F.P.S. 3: 44 (1956); F.J.M.: 136 (1976); F.T.E.A. Comp. 2: 332 (2002); F.E.E. 4(2): 128 (2004).
Annual herb. Grassland.
Distr: RS, DAR, KOR; Senegal, Nigeria to Ethiopia, S to Namibia & South Africa; Algeria, Egypt, Arabia, Pakistan, India.
KOR: Wickens 669 10/1962.

Pentanema indicum (L.) Y.Ling

F.T.E.A. Comp. 2: 334 (2002); F.E.E. 4(2): 124 (2004).
Syn: Vicoa leptoclada (Webb) Dandy – F.P.S. 3: 62 (1956); F.J.M.: 139 (1976).
Annual herb. Exposed mud & bare ground.
Distr: DAR, KOR, CS, ES, UN, BAG; Senegal to Eritrea, S to Angola, Zimbabwe & Mozambique; Pakistan, India, Sri Lanka, Myanmar.
KOR: Kotschy 14 4/1837; **BAG:** Schweinfurth 2647 11/1869.

Phagnalon schweinfurthii Sch.Bip. ex Schweinf. var. schweinfurthii

F.P.S. 3: 44 (1956); F.E.E. 4(2): 162 (2004).
Perennial herb. Grassland.
Distr: RS; Eritrea; Egypt.
IUCN: RD
RS: Schweinfurth 365 3/1865.
Note: the second variety, var. androssovii (B.Fedtsch.) Qaiser & Lack is much more widespread, occuring from Egypt to Afghanistan – the two would be better treated as subspecies.

Phagnalon stenolepsis Chiov.

F.E.E. 4(2): 162 (2004).
Syn: Phagnalon scalarum Schweinf. ex Schwartz – F.P.S. 3: 44 (1956); F.J.M.: 137 (1976).
Syn: Phagnalon scalarum Schweinf. ex Schwartz var. **meridionale** (Quézel) Wickens – F.J.M.: 137 (1976).
Syn: Phagnalon sp. – Bot. Exp. Sud.: 37 (1970).
Perennial herb or subshrub. Bushland on rocky hillslopes, upland grassland.
Distr: DAR; Chad, Ethiopia; Arabia.
DAR: Wickens 1233 2/1964.

Pluchea dioscoridis (L.) DC.

F.P.S. 3: 45 (1956); F.J.M.: 137 (1976); T.S.S.: 416 (1990); F.T.E.A. Comp. 2: 362 (2002); F.E.E. 4(2): 144 (2004).
Shrub or subshrub. Moist grassland & wooded grassland, riverine scrub.
Distr: NS, DAR, KOR, CS, ES, UN, EQU; Ethiopia to Namibia & South Africa; Egypt to Arabia.
CS: Jackson 1572 12/1950; **EQU:** Padwa 270 6/1953.

Pluchea ovalis (Pers.) DC.

F.P.S. 3: 45 (1956); T.S.S.: 417 (1990); F.T.E.A. Comp. 2: 359 (2002); F.E.E. 4(2): 146 (2004).
Shrub or perennial herb. Swamps, riverbanks, dry watercourses.
Distr: ?NS, ?CS, BAG, EQU; Mauritania to Somalia, S to Angola & Zambia; Egypt, Arabia.
EQU: Myers 8575 2/1938.
Note: El Amin in T.S.S. records this species from North and Central Sudan but we have only seen specimens from the South.

[Podospermum laciniatum (L.) DC.]

Note: Quézel (1969: 125) recorded Launaea resedifolia (L.) Kuntze from NW Darfur but noted that the identification was unconfirmed since the material lacked mature seeds. L. resedifolia s.s. is a synonym of Podospermum laciniatum which may possibly extend to NW Darfur, having been recorded from the Saharan mountains of Chad, but the name L. resedifolia has also been misapplied in the past. We have therefore decided to omit this record here.

Pseudoconyza viscosa (Mill.) D'Arcy

F.T.E.A. Comp. 2: 357 (2002); F.E.E. 4(2): 139 (2004).
Syn: Blumea aurita (L.f.) DC. – F.P.S. 3: 13 (1956).
Syn: Laggera aurita (L.f.) Benth. ex C.B.Clarke – Bot. Exp. Sud.: 36 (1970).
Annual herb. Riverbanks, grassland, weed of cultivation.
Distr: RS, DAR, KOR, CS, ES, BAG, EQU; widespread in the tropics & subtropics.
CS: Jackson 1597 1/1951; **EQU:** Myers 8505 2/1938.

Pseudognaphalium luteo-album (L.) Hilliard & B.L.Burtt

F.T.E.A. Comp. 2: 399 (2002); F.E.E. 4(2): 180 (2004).
Syn: Gnaphalium luteo-album L. – F.P.S. 3: 31 (1956); Bot. Exp. Sud.: 35 (1970); F.J.M.: 135 (1976); F.J.U. 2: 157 (1999).
Annual or short-lived perennial herb. Montane grassland, streambanks, weed of cultivation.
Distr: NS, DAR, KOR, CS; widespread in the tropics & subtropics.
DAR: Wickens 1661 5/1964.

Pseudognaphalium marranum (Philipson) Hilliard

Bot. J. Linn. Soc. 82: 206 (1981).
Syn: Gnaphalium marranum Philipson – F.P.S. 3: 32 (1956); F.J.M.: 135 (1976).
Annual or perennial herb. Montane grassland.
Distr: DAR; Sudan endemic.
IUCN: RD
DAR: Wickens 1212 2/1964.

Pseudognaphalium oligandrum (DC.) Hilliard & B.L.Burtt

F.T.E.A. Comp. 2: 399 (2002); F.E.E. 4(2): 179 (2004).
Syn: Gnaphalium undulatum L. – F.J.M.: 135 (1976).
Annual herb. Montane grassland & open bushland.
Distr: DAR; Cameroon to Eritrea, S to Namibia & South Africa; Madagascar.
DAR: Kassas 292 (n.v.).
Note: we have not seen any material from Sudan, the identification of the specimen cited follows Wickens (1976).

Pseudognaphalium richardianum (Cufod.) Hilliard & B.L.Burtt

Biol. Skr. 51: 377 (1998); F.E.E. 4(2): 179 (2004).
Annual or short-lived perennial herb. Montane bushland & grassland, upper forest margins.
Distr: EQU; D.R. Congo, Eritrea, Ethiopia.
EQU: Friis & Vollesen 133 11/1980.

Psiadia punctulata (DC.) Vatke

F.T.E.A. Comp. 2: 514 (2002); F.E.E. 4(2): 193 (2004).
Syn: *Psiadia arabica* Jaub. & Spach – F.P.S. 3: 45 (1956);
T.S.S.: 417 (1990).
Shrub or shrubby perennial herb. Woodland, bushland,
grassland, rocky hillslopes.
Distr: RS; Ethiopia & Somalia, S to Angola & South
Africa; Arabia.
RS: Jackson 3949 4/1959.

Pulicaria attenuata Hutch. & B.L.Burtt

F.P.S. 3: 46 (1956); F.E.E. 4(2): 134 (2004).
Annual herb. Grassland including seasonally flooded areas.
Distr: DAR, ES; Eritrea, Somalia.
ES: Andrews 251 11/1936.

Pulicaria grantii Oliv. & Hiern

F.P.S. 3: 45 pro parte (1956); Phanerogam. Monogr. 14:
142 (1981).
Subshrub. Riverbanks.
Distr: CS; Sudan endemic.
IUCN: RD
CS: Speke & Grant 160 4/1863.
Note: Andrews records this species from Gash Delta
and from Khartoum; the former is perhaps based on a
misidentification of *P. attenuata*.

Pulicaria incisa (Lam.) DC.

F.J.M.: 137 (1976); F.J.U. 2: 158 (1999); F.E.E. 4(2): 132
(2004).
Syn: *Pulicaria undulata* sensu Andrews, non (L.)
C.A.Mey. – F.P.S. 3: 46 (1956).
Annual or short-lived perennial herb. Bushland on rocky
hillslopes.
Distr: NS, RS, DAR, KOR, CS, ES; Ghana to Ethiopia &
Djibouti; Egypt.
ES: Carter 1785 11/1987.

Pulicaria petiolaris Jaub. & Spach

F.P.S. 3: 45 (1956); F.T.E.A. Comp. 2: 336 (2002); F.E.E.
4(2): 133 (2004).
Perennial herb or subshrub. Rock crevices in bushland.
Distr: RS, ?CS; Eritrea, Ethiopia, Djibouti, Somalia,
Kenya; Egypt, Arabia.
RS: Jackson 2912 8/1953.

Pulicaria scabra (Thunb.) Druce

F.J.M.: 137 (1976); F.E.E. 4(2): 130 (2004).
Syn: *Pulicaria dysenterica* var. ***stenophylla*** sensu
Andrews, non Boiss. – F.P.S. 3: 46 (1956).
Perennial herb. Stream banks.
Distr: DAR, KOR; Eritrea, Ethiopia, Zambia to Namibia &
South Africa.
DAR: Wickens 1247 2/1964.
Notes: (1) Gamal-Eldin in Phanerogam. Monogr. 14: 125
(1981) records also *P. dysenterica* (L.) Bernh. from Jebel
Marra, based on *Kassas et al.* 526. Whilst we have not
seen the specimen, this record is considered likely to be
based on a misidentification of *P. scabra*. (2) Andrews
records also *P. vulgaris* Gaertn. from Kordofan – we
have not seen the material on which this is based but
specimens of *P. scabra* from Darfur were previously
identified as *P. vulgaris* and it is quite possible that
Andrews also made this mistake.

Pulicaria schimperi DC.

F.E.E. 4(2): 133 (2004).
Annual herb. Dry bushland & semi-desert.
Distr: RS; Eritrea, Ethiopia, Djibouti, Somalia; Arabia.
RS: Jackson 2971 4/1953.

Pulicaria undulata (L.) C.A.Mey. subsp. undulata

F.J.M.: 137 (1976); F.E.E. 4(2): 134 (2004).
Syn: *Pulicaria crispa* (Forssk.) Benth. ex Oliv. – F.P.S. 3:
46 (1956); Bot. Exp. Sud.: 37 (1970); F.J.M.: 137 (1976).
Annual or short-lived perennial herb. Open grassland &
bushland.
Distr: NS, RS, DAR, KOR, CS; Eritrea, Ethiopia, Somalia;
N Africa to Arabia.
DAR: Wickens 1129 2/1964.

Pulicaria undulata (L.) C.A.Mey. subsp. tomentosa (E.Gamal-Eldin) D.J.N.Hind & Boulos

F.E.E. 4(2): 134 (2004).
Annual or short-lived perennial herb. Open grassland,
margins of cultivation.
Distr: NS, CS; Eritrea, Ethiopia, Somalia.
NS: Pettet 81 9/1962.

Reichardia tingitana (L.) Roth

F.P.S. 3: 48 (1956); F.J.M.: 137 (1976); F.T.E.A. Comp. 1:
89 (2000); F.E.E. 4(2): 64 (2004).
Syn: *Reichardia orientalis* (L.) Hochr. – Bot. Exp. Sud.:
37 (1970).
Annual herb. Sparse grassland, weed of cultivation &
disturbed areas.
Distr: RS, DAR; Eritrea, Ethiopia, Somalia, Kenya,
Tanzania; Macaronesia, Mediterranean to Polynesia &
Australia.
DAR: Wickens 1394 4/1964.

Sclerocarpus africanus Jacq. ex Murray

F.P.S. 3: 48 (1956); F.J.M.: 137 (1976); F.E.E. 4(2): 302
(2004); F.T.E.A. Comp. 3: 757 (2005).
Annual herb. Sandy grassland.
Distr: NS, DAR, KOR, CS, ES; Senegal to Somalia, S to
Namibia & South Africa; Arabia, India.
DAR: Wickens 2100 8/1964.

Scolymus maculatus L.

F.P.S. 3: 48 (1956); F.E.E. 4(2): 44 (2004).
Annual herb. Weed of disturbed areas & cultivation.
Distr: KOR; Eritrea, Ethiopia; N Africa, S Europe to
Russia.
KOR: Kotschy 459 1842.

Senecio aegyptius L.

F.E.E. 4(2): 245 (2004).
Syn: *Senecio arabicus* L. – F.P.S. 3: 48 (1956).
Syn: *Senecio aegyptius* L. var. ***discoideus*** Boiss. –
F.E.E. 4(2): 245 (2004).
Annual herb. Riverine black clays.
Distr: NS, KOR, CS; Ethiopia; France, Cyprus, Egypt.
CS: Jackson 4176 5/1961.

Senecio flavus (Decne.) Sch.Bip.

F.Egypt 3: 262 (2002).
Annual herb. Sandy & stony deserts & wadis.
Distr: RS; Spain, Macaronesia, N Africa to Arabia &
Afghanistan.
RS: Drar 222 1933 (n.v.).
Note: the record of this species is from Drar (1936) who
recorded it from Gebel Elba in the disputed Hala'ib
Triangle. Boulos (2002) confirms its presence there.

Senecio hochstetteri Sch.Bip. ex A.Rich.

F.J.M.: 138 (1976); F.E.E. 4(2): 243 (2004); F.T.E.A. Comp.
3: 666 (2005).
Rhizomatous perennial herb. Grassland, open bushland
& woodland.
Distr: DAR; Sierra Leone to Eritrea, S to South Africa.
DAR: Wickens 1684 5/1964.

Senecio inornatus DC.

F.E.E. 4(2): 241 (2004); F.T.E.A. Comp. 3: 633 (2005).
Rhizomatous perennial herb. Montane grassland,
streamsides.
Distr: EQU; D.R. Congo, Ethiopia, Tanzania, S to Angola
& South Africa.
EQU: Jackson 3193 5/1954.

Senecio lyratus Forssk.

F.E.E. 4(2): 246 (2004); F.T.E.A. Comp. 3: 624 (2005).
Shrub or perennial herb. Dry forest.
Distr: RS; Eritrea, Ethiopia, Somalia, Kenya, Tanzania; Yemen.
RS: Jackson 2894 4/1953.

Senecio myriocephalus Sch.Bip. ex A.Rich.

F.P.S. 3: 49 (1956); Biol. Skr. 51: 377 (1998); F.E.E. 4(2):
237 (2004).
Shrub or perennial herb. Montane grassland, ericaceous
bushland, forest margins.
Distr: EQU; Ethiopia.
EQU: Friis & Vollesen 1135 3/1982.
Note: widespread in Ethiopia and so not considered
threatened.

Senecio ragazzii Chiov.

Biol. Skr. 51: 377 (1998); F.E.E. 4(2): 240 (2004).
Rhizomatous perennial herb. Montane bushland & forest
margins.
Distr: EQU; Ethiopia.
EQU: Thomas 1790 12/1935.
Notes: (1) widespread in Ethiopia and so not considered
threatened; (2) S. pachyrhizus O.Hoffm. may also occur
in South Sudan, having been recorded from the Ugandan
side of the Imatong Mts; see Friis & Vollesen (1998: 377).

Senecio ruwenzoriensis S.Moore

F.E.E. 4(2): 243 (2004); F.T.E.A. Comp. 3: 643 (2005).
Rhizomatous perennial herb. Upland grassland &
bushland amongst rocks.
Distr: EQU; Nigeria to Ethiopia, S to South Africa.
EQU: Andrews 1043 5/1939.

Senecio schimperi Sch.Bip. ex A.Rich.

F.E.E. 4(2): 245 (2004).
Annual herb. Forest, open bushland, grassland, disturbed
areas.

Distr: RS; Eritrea, Ethiopia.
RS: Jackson 2886 4/1953.

Senecio subsessilis Oliv. & Hiern

Biol. Skr. 51: 378 (1998); F.E.E. 4(2): 239 (2004); F.T.E.A.
Comp. 3: 627 (2005).
Syn: Senecio trichopterygius Muschl. – F.P.S. 3: 49 (1956).
Perennial herb or shrub. Woodland & forest clearings.
Distr: EQU; D.R. Congo to Ethiopia, S to Tanzania.
EQU: Friis & Vollesen 130 11/1980.
Note: Wickens records Senecio sp. nov. from Jebel Marra
based on his number 2675; we have been unable to
locate this specimen in either Senecio or Crassocephalum
at Kew.

Solanecio angulatus (Vahl) C.Jeffrey

Biol. Skr. 51: 378 (1998); F.E.E. 4(2): 249 (2004); F.T.E.A.
Comp. 3: 676 (2005).
Perennial herb, sometimes climbing. Upland forest
margins, bushland, rocky hillslopes.
Distr: EQU; Cameroon to Eritrea, S to Angola & South
Africa; Madagascar, Comoros, Arabia.
EQU: Andrews 1512 5/1939.

Solanecio mannii (Hook.f.) C.Jeffrey

Biol. Skr. 51: 378 (1998); F.E.E. 4(2): 249 (2004); F.T.E.A.
Comp. 3: 678 (2005).
Syn: Crassocephalum mannii (Hook.f.) Milne-Redh. –
F.P.S. 3: 33 (1956).
Syn: Senecio multicorymbosus Klatt – T.S.S.: 417 (1990).
Shrub or small tree. Montane bushland & forest margins.
Distr: EQU; Nigeria to Ethiopia, S to Angola, Zimbabwe
& Mozambique.
EQU: Myers 11713 7/1939.

Solanecio tuberosus (Sch.Bip. ex A.Rich.) C.Jeffrey var. tuberosus

F.E.E. 4(2): 250 (2004); F.T.E.A. Comp. 3: 671 (2005).
Syn: Senecio tuberosus Sch.Bip. ex A.Rich. – F.P.S. 3:
49 (1956); F.J.M.: 138 (1976).
Perennial herb. Swampy grassland.
Distr: DAR, ES; Eritrea, Ethiopia, Uganda, Kenya.
DAR: Wickens 1816 7/1964.

Sonchus asper (L.) Hill

F.P.S. 3: 50 (1956); F.J.M.: 138 pro parte (1976); F.T.E.A.
Comp. 1: 97 (2000); F.E.E. 4(2): 67 (2004).
Annual or short-lived perennial herb. Weed of cultivation
& disturbed areas.
Distr: DAR; widespread in Africa, Madagascar & Eurasia.
DAR: Wickens 2731 9/1964.

Sonchus bipontini Asch.

Biol. Skr. 51: 379 (1998); F.T.E.A. Comp. 1: 95 (2000);
F.E.E. 4(2): 66 (2004).
Perennial herb. Upland grassland, bushland & forest
margins.
Distr: EQU; D.R. Congo to Ethiopia, S to Malawi.
EQU: Friis & Vollesen 1125 3/1982.

Sonchus gigas Boulos ex Humbert

F.J.M.: 138 (1976); F.E.E. 4(2): 70 (2004).
Annual or short-lived perennial herb. Weed of cultivation
& disturbed areas.

Distr: DAR; Senegal, D.R. Congo, Eritrea, Ethiopia, Zambia to Angola & South Africa; Madagascar.
DAR: Wickens 1703 5/1964.
Note: this name is recorded as a synonym of *S. asper* in the African Plants Database but this appears to be in error.

Sonchus oleraceus L.
F.P.S. 3: 50 (1956); F.T.E.A. Comp. 1: 98 (2000); F.E.E. 4(2): 68 (2004).
Syn: *Sonchus asper* sensu Wickens pro parte, non (L.) Hill – F.J.M.: 138 (1976).
Annual or biennial herb. Weed of disturbed areas & cultivation.
Distr: NS, DAR, CS, EQU; native to Eurasia & N Africa, now widely naturalised elsewhere.
DAR: Wickens 1098 1/1964; **EQU:** Buxton s.n.

Sonchus schweinfurthii Oliv. & Hiern
F.P.S. 3: 51 (1956); F.T.E.A. Comp. 1: 94 (2000); F.E.E. 4(2): 66 (2004).
Syn: *Sonchus angustissimus* sensu Friis & Vollesen, non Hook.f. – Biol. Skr. 51: 379 (1998).
Perennial herb. Wet grassland, riverbanks, weed of cultivation & disturbed areas.
Distr: EQU; Nigeria to Ethiopia, S to Angola & Zimbabwe.
EQU: Friis & Vollesen 213 11/1980.

Sonchus tenerrimus L.
F.E.E. 4(2): 69 (2004).
Annual herb. Montane grassland.
Distr: DAR; Eritrea; Mediterranean, N Africa to Pakistan.
DAR: Lynes 602 8/1920.

Sphaeranthus angustifolius DC.
F.P.S. 3: 53 (1956); F.W.T.A. 2: 267 (1963); F.J.M.: 138 (1976).
Annual herb. Damp depressions.
Distr: DAR, KOR; Senegal to Chad.
DAR: Wickens 1132 2/1964.

Sphaeranthus flexuosus O.Hoffm.
F.P.S. 3: 54 (1956); F.W.T.A. 2: 266 (1963); F.J.M.: 138 (1976).
Annual herb. Upland grassland.
Distr: DAR, KOR; Nigeria to C.A.R. & D.R. Congo.
DAR: Wickens 1468 4/1964.

Sphaeranthus randii S.Moore var. *randii*
F.P.S. 3: 53 (1956); Biol. Skr. 51: 379 (1998); F.T.E.A. Comp. 2: 386 (2002).
Perennial herb. Wet grassland, streamsides.
Distr: EQU; D.R. Congo to Kenya, S to Zimbabwe & Mozambique.
EQU: Myers 11118 4/1939.

Sphaeranthus randii S.Moore var. *bibracteata* Ross-Craig
F.T.E.A. Comp. 2: 387 (2002).
Perennial herb. Swampy grassland.
Distr: BAG, EQU; D.R. Congo, Uganda, Kenya, Tanzania, Zambia.
EQU: Myers 6661 5/1937.

Sphaeranthus steetzii Oliv. & Hiern
F.P.S. 3: 53 (1956); F.T.E.A. Comp. 2: 385 (2002); F.E.E. 4(2): 154 (2004).
Annual or perennial herb or subshrub. Seasonally wet grassland & bushland, wet depressions, riverbanks.
Distr: CS, EQU; Nigeria to Eritrea, S to Zimbabwe.
CS: de Wilde et al. 5843 3/1965; **EQU:** Simpson 7598 2/1930.

Sphaeranthus suaveolens (Forssk.) DC.
F.P.S. 3: 53 (1956); F.T.E.A. Comp. 2: 380 (2002); F.E.E. 4(2): 153 (2004).
Perennial herb. Permanently or seasonally moist sites.
Distr: KOR, CS; D.R. Congo to Eritrea, S to Zambia & Mozambique; Egypt.
CS: Kotschy 463 1/1837.

Sphaeranthus ukambensis Vatke & O.Hoffm.
Biol. Skr. 51: 380 (1998); F.T.E.A. Comp. 2: 378 (2002); F.E.E. 4(2): 150 (2004).
Perennial herb. Seasonally wet grassland, bushland & woodland.
Distr: EQU; Ethiopia, Somalia, Uganda, Kenya, Tanzania.
EQU: Kielland-Lund 428 12/1983 (n.v.).
Note: we have not seen the specimen cited but it is presumably of var. *ukambensis*.

Stomatanthes africanus (Oliv. & Hiern) R.M.King & H.Rob.
Biol. Skr. 51: 380 (1998); F.E.E. 4(2): 349 (2004); F.T.E.A. Comp. 3: 825 (2005).
Syn: *Eupatorium africanum* Oliv. & Hiern – F.P.S. 3: 29 (1956).
Subshrub. Woodland & grassland with regular burning.
Distr: EQU; Guinea to Ethiopia, S to Angola & South Africa.
EQU: Myers 8391 1/1938.

Struchium sparganophora (L.) Kuntze
F.P.S. 3: 54 (1956); F.T.E.A. Comp. 1: 109 (2000).
Annual herb. Riverine & swamp forest.
Distr: EQU; native to the neotropics, widely naturalised in the palaeotropics.
EQU: Schweinfurth 3005 2/1870.

Tagetes minuta L.
Biol. Skr. 51: 380 (1998); F.E.E. 4(2): 281 (2004); F.T.E.A. Comp. 3: 727 (2005).
Annual herb. Weed of cultivation & disturbed areas.
Distr: UN, EQU; native to the neotropics, now largely pantropical.
UN: Sherif A.3994 9/1951.

Tithonia rotundifolia (Mill.) S.F.Blake
F.J.M.: 138 (1976); F.T.E.A. Comp. 3: 759 (2005).
Annual herb. Naturalised from cultivation.
Distr: DAR; native to the neotropics, cultivated & occasionally naturalised elsewhere.
DAR: Wickens 2287 8/1964.

Tolpis capensis (L.) Sch.Bip.
Biol. Skr. 51: 380 (1998); F.T.E.A. Comp. 1: 66 (2000).
Perennial herb. Upland grassland & woodland, including areas with regular burning.

Distr: EQU; D.R. Congo to Kenya, S to Angola & South Africa; Madagascar.
EQU: Myers 11642 7/1939.

Tolpis virgata (Desf.) Bertol.

F.E.E. 4(2): 70 (2004).
Syn: *Tolpis altissima* (Balb.) Pers. – F.P.S. 3: 54 (1956); Bot. Exp. Sud.: 37 (1970).
Perennial herb. Montane grassland & bushland.
Distr: RS, ?DAR, ?EQU; Eritrea, Ethiopia; S Europe, N Africa to Arabia.
RS: Maffey 27 1928; **EQU:** Drar 1383 4/1938 (n.v.).
Note: Drar tentatively lists this species from Amadi, Equatoria and Jebel Marra, Darfur but notes that the material is insufficient. He also lists *T. umbellata* Bertol. from Erkowit but this seems unlikely and it may be a misidentification of *T. virgata*.

Tridax procumbens L.

F.P.S. 3: 55 (1956); F.J.M.: 138 (1976); Biol. Skr. 51: 381 (1998); F.E.E. 4(2): 306 (2004); F.T.E.A. Comp. 3: 764 (2005).
Annual or perennial herb. Disturbed areas.
Distr: DAR, EQU; native to the neotropics, now pantropical.
DAR: Jackson 3293 6/1961; **EQU:** Myers 8717 3/1938.

Urospermum picroides (L.) Scop. ex F.W.Schmidt

F.Egypt 3: 275 (2002).
Annual herb. Weed of cultivation & disturbed areas.
Distr: RS; Mediterranean, N Africa to SW Asia.
RS: Shabetai F.1402 2/1933.
Note: in Sudan, known only from Gebel Elba in the disputed Hala'ib Triangle.

Verbesina encelioides (Cav.) Benth. & Hook.f. ex A.Gray

F.P.S. 3: 55 (1956); F.E.E. 4(2): 297 (2004).
Annual herb. Weed of disturbed areas.
Distr: KOR, CS; native to USA & Mexico, widely introduced elsewhere.
KOR: Pfund 303 11/1875.

Vernonia adoensis Sch.Bip. ex Walp.

F.J.M.: 138 (1976); Biol. Skr. 51: 381 (1998); F.T.E.A. Comp. 1: 245 (2000); F.E.E. 4(2): 98 (2004).
Syn: *Vernonia kotschyana* Sch.Bip. ex Walp. – F.P.S. 3: 61 (1956); Bot. Exp. Sud.: 38 (1970).
Perennial herb or shrub. Grassland & wooded grassland.
Distr: DAR, KOR, CS, UN, EQU; Nigeria to Eritrea, S to South Africa.
DAR: Wickens 2973 5/1965; **EQU:** Friis & Vollesen 263 11/1980.
Note: this species is sometimes separated into varieties, with both var. *adoensis* and var. *kotschyana* (Sch.Bip. ex Walp) G.V.Pope occurring in Sudan.

Vernonia ambigua Kotschy & Peyr.

F.P.S. 3: 57 (1956); F.T.E.A. Comp. 1: 265 (2000); F.E.E. 4(2): 101 (2004).
Annual herb. Wooded grassland.
Distr: BAG, EQU; Senegal to Ethiopia, S to Angola & Zambia.
BAG: Schweinfurth 1293 5/1869.

Vernonia amygdalina Delile

F.P.S. 3: 59 (1956); F.J.M.: 138 (1976); T.S.S.: 419 (1990); Biol. Skr. 51: 381 (1998); F.T.E.A. Comp. 1: 178 (2000); F.E.E. 4(2): 78 (2004).
Shrub or small tree. Forest margins, woodland, bushland & grassland.
Distr: RS, DAR, KOR, CS, BAG, UN, EQU; Guinea to Eritrea, S to Botswana & South Africa; Arabia.
DAR: Wickens 1549 5/1964; **EQU:** Friis & Vollesen 930 2/1982.

Vernonia aschersonii Sch.Bip. ex Schweinf. & Asch.

F.P.S. 3: 58 pro parte (1956); F.E.E. 4(2): 98 (2004).
Perennial herb. Woodland.
Distr: RS, DAR; Chad, Eritrea, Ethiopia, Djibouti.
RS: Robbie 48.

Vernonia biafrae Oliv. & Hiern

Biol. Skr. 51: 382 (1998); F.T.E.A. Comp. 1: 189 (2000); F.E.E. 4(2): 81 (2004).
Climbing shrub. Upland forest margins, swamps, moist grassland.
Distr: EQU; Guinea to Ethiopia, S to Zambia.
EQU: Friis & Vollesen 957 2/1982.

Vernonia chthonocephala O.Hoffm.

F.P.S. 3: 61 (1956); F.T.E.A. Comp. 1: 221 (2000).
Perennial herb. Grassland with regular burning.
Distr: EQU; Sierra Leone to Kenya, S to Angola, Zambia & Mozambique.
EQU: Myers 8321 11/1938.

Vernonia cinerascens Sch.Bip.

F.P.S. 3: 57 (1956); T.S.S.: 419 (1990); F.T.E.A. Comp. 1: 202 (2000); F.E.E. 4(2): 88 (2004).
Perennial herb or shrub. Dry bushland & grassland.
Distr: RS, DAR, EQU; Senegal, Eritrea to Somalia, S to Angola & South Africa; Arabia to India.
RS: Bent s.n. 1896; **EQU:** Myers 14031 5/1939.

Vernonia cinerea (L.) Less. var. *cinerea*

F.P.S. 3: 57 (1956); F.T.E.A. Comp. 1: 209 (2000); F.E.E. 4(2): 90 (2004).
Annual herb. Weed of disturbed areas, cultivation & bare ground.
Distr: ES, EQU; palaeotropical, introduced elsewhere.
ES: Andrews 60 6/1935; **EQU:** Myers 6770 5/1937.

Vernonia conferta Benth.

F.T.E.A. Comp. 1: 198 (2000).
Tree. Forest including swamp forest.
Distr: EQU; Sierra Leone to Uganda, S to Angola.
EQU: Schweinfurth 3095 8/1870.

Vernonia dumicola S.Moore

F.T.E.A. Comp. 1: 242 (2000).
Perennial herb or subshrub. Wet grassland, swamps.
Distr: EQU; D.R. Congo, Rwanda, Uganda, Kenya, Tanzania.
EQU: Sillitoe 196 1919.
Note: this is treated as a synonym of *V. lasiopus* O.Hoffm. var. *lasiopus* in the African Plants Database.

Vernonia galamensis (Cass.) Less. subsp. **galamensis**
F.J.M.: 139 (1976); F.T.E.A. Comp. 1: 227 (2000); F.E.E. 4(2): 93 (2004).
Syn: Vernonia pauciflora (Willd.) Less. – F.P.S. 3: 58 (1956).
Annual herb. Wooded grassland, woodland & dry bushland.
Distr: RS, DAR, KOR, CS, ES, UN; Senegal to Eritrea, S to Tanzania.
KOR: Jackson 4029 9/1959; **UN:** Broun 1712 11/1909.
Note: both var. *galamensis* and var. *petitiana* (A.Rich.) M.G.Gilbert are said to occur in Sudan (see F.E.E.).

Vernonia gerberiformis Oliv. & Hiern var. **gerberiformis**
F.P.S. 3: 58 (1956); F.T.E.A. Comp. 1: 234 (2000).
Perennial herb. Wooded grassland & grassland with seasonal burning.
Distr: BAG; D.R. Congo to Kenya, S to Angola & Zimbabwe.
BAG: Schweinfurth 2688 11/1869.

Vernonia glaberrima Welw. ex O.Hoffm.
Bot. Exp. Sud.: 38 (1970); F.T.E.A. Comp. 1: 199 (2000).
Shrub. Woodland.
Distr: ?EQU; Guinea to Tanzania, S to Angola, Zimbabwe & Mozambique.
EQU: Drar 1841b 4/1938 (n.v.).
Note: we have not seen the speicimen cited, which was identified by Wickens (see Drar 1970).

Vernonia guineensis Benth. var. **guineensis**
F.P.S. 3: 59 (1956); F.W.T.A. 2: 282 (1963); F.E.E. 4(2): 102 (2004).
Perennial herb. Wooded grassland.
Distr: EQU; Guinea to C.A.R., ?Ethiopia.
EQU: Schweinfurth 3153 5/1870.

Vernonia hochstetteri Sch.Bip. var. **hochstetteri**
F.P.S. 3: 61 (1956); T.S.S.: 419 (1990); Biol. Skr. 51: 382 (1998); F.T.E.A. Comp. 1: 212 (2000); F.E.E. 4(2): 90 (2004).
Syn: Vernonia jugalis Oliv. & Hiern – F.P.S. 3: 56 (1956).
Perennial herb, shrub or small tree. Forest margins & secondary bushland.
Distr: EQU; Congo-Brazzaville to Ethiopia, S to Tanzania.
EQU: Sillitoe 197 1919.

Vernonia infundibularis Oliv. & Hiern
F.P.S. 3: 60 (1956); F.T.E.A. Comp. 1: 234 (2000).
Perennial herb. Grassland & wooded grassland.
Distr: EQU; Nigeria to Uganda, S to Tanzania.
EQU: Myers 6994 7/1937.

Vernonia ituriensis Muschl.
Biol. Skr. 51: 382 (1998); F.T.E.A. Comp. 1: 229 (2000); F.E.E. 4(2): 93 (2004).
Annual or short-lived perennial herb. Riverbanks, forest margins, woodland.
Distr: EQU; Cameroon to Ethiopia, S to Tanzania.
EQU: Friis & Vollesen 439 11/1980.

Vernonia karaguensis Oliv.
Biol. Skr. 51: 382 (1998); F.T.E.A. Comp. 1: 200 (2000); F.E.E. 4(2): 85 (2004).
Perennial herb or subshrub. Grassland, woodland, forest margins.
Distr: EQU; Nigeria to Ethiopia, S to Angola & Zimbabwe.
EQU: Friis & Vollesen 207 11/1980.

Vernonia lasiopus O.Hoffm. var. **lasiopus**
Biol. Skr. 51: 383 (1998); F.T.E.A. Comp. 1: 244 (2000); F.E.E. 4(2): 97 (2004).
Perennial herb or shrub. Forest margins, secondary bushland, riverine thickets, disturbed areas.
Distr: EQU; Rwanda, Ethiopia, Uganda, Kenya, Tanzania.
EQU: Shigeta 54 1979 (n.v.).
Note: we have not seen the specimen cited; Friis & Vollesen (1998) report that it is incomplete material and only tentatively identified.

Vernonia myriantha Hook.f.
Biol. Skr. 51: 383 (1998); F.T.E.A. Comp. 1: 195 (2000); F.E.E. 4(2): 82 (2004).
Syn: Vernonia ampla O.Hoffm. – T.S.S.: 419 (1990).
Shrub or small tree. Forest margins, woodland & wooded grassland.
Distr: EQU; Sierra Leone to Ethiopia, S to Angola & South Africa.
EQU: Friis & Vollesen 907 2/1982.

Vernonia perrottetii Sch.Bip.
F.P.S. 3: 57 (1956); F.J.M.: 139 (1976); Biol. Skr. 51: 383 (1998); F.T.E.A. Comp. 1: 256 (2000); F.E.E. 4(2): 98 (2004).
Annual herb. Grassland, bushland, disturbed areas.
Distr: DAR, KOR, UN, EQU; Senegal to Ethiopia, S to Angola & Zambia.
KOR: Wickens 729 10/1962; **EQU:** Myers 10326 11/1938.

Vernonia plumbaginifolia Fenzl ex Oliv. & Hiern var. **plumbaginifolia**
F.P.S. 3: 59 (1956); F.T.E.A. Comp. 1: 260 (2000); F.E.E. 4(2): 100 (2004).
Perennial herb. Habitat unknown.
Distr: CS, BAG; Ethiopia, Uganda.
IUCN: RD
CS: Kotschy 491 1837; **BAG:** Schweinfurth 1434 4/1869.

Vernonia popeana C.Jeffrey
F.T.E.A. Comp. 1: 259 (2000); F.E.E. 4(2): 100 (2004).
Syn: Vernonia aschersonii sensu Andrews pro parte, non Sch.Bip. ex Schweinf. – F.P.S. 3: 58 (1956).
Syn: Vernonia ?sp. nov. sensu Wickens – F.J.M.: 139 (1976).
Perennial herb or subshrub. Bushland, woodland, wooded grassland.
Distr: DAR; Ethiopia, Somalia, Uganda, Kenya, Tanzania.
DAR: Wickens 2909 4/1965.

Vernonia pumila Kotschy & Peyr.
F.P.S. 3: 61 (1956); Biol. Skr. 51: 383 (1998); F.T.E.A. Comp. 1: 249 (2000).
Perennial herb. Woodland & grassland with seasonal burning.

Distr: EQU; Senegal to Uganda & Kenya.
EQU: Friis & Vollesen 1246 3/1982.

Vernonia purpurea Sch.Bip.

F.P.S. 3: 59 (1956); F.J.M.: 139 (1976); Biol. Skr. 51: 384 (1998); F.T.E.A. Comp. 1: 217 (2000); F.E.E. 4(2): 92 (2004).
Syn: *Vernonia inulifolia* Steud. ex Walp. – F.P.S. 3: 61 (1956).
Perennial herb. Upland grassland & woodland.
Distr: DAR, UN, EQU; Senegal to Eritrea, S to Angola, Zambia & Mozambique.
DAR: Wickens 2725 9/1964; **EQU:** Friis & Vollesen 138 11/1980.

Vernonia schweinfurthii Oliv. & Hiern

F.P.S. 3: 60 (1956); Biol. Skr. 51: 384 (1998); F.T.E.A. Comp. 1: 220 (2000).
Perennial herb. Grassland.
Distr: BAG, EQU; Ivory Coast to Kenya, S to Zambia & Malawi.
EQU: Andrews 558 4/1939.

Vernonia smithiana Less.

F.P.S. 3: 58 (1956); Biol. Skr. 51: 384 (1998); F.T.E.A. Comp. 1: 207 (2000); F.E.E. 4(2): 88 (2004).
Perennial herb. Grassland & wooded grassland with regular burning.
Distr: UN, EQU; Guinea to Kenya, S to D.R. Congo & Tanzania.
EQU: Friis & Vollesen 951 2/1982.

Vernonia stellulifera (Benth.) C.Jeffrey

F.T.E.A. Comp. 1: 211 (2000).
Annual herb. Forest clearings.
Distr: ES; Guinea to Uganda, S to Angola & Zambia.
ES: Jackson 4084 7/1960.

Vernonia syringifolia O.Hoffm.

F.P.S. 3: 57 (1956); T.S.S.: 420 (1990); Biol. Skr. 51: 385 (1998); F.T.E.A. Comp. 1: 190 (2000).
Woody herb or shrub, sometimes climbing. Upland forest margins.
Distr: EQU; D.R. Congo to Kenya, S to Zambia & Malawi.
EQU: Thomas 1782 12/1935.

Vernonia theophrastifolia Schweinf. ex Oliv. & Hiern

F.P.S. 3: 62 (1956); Biol. Skr. 51: 385 (1998); F.T.E.A. Comp. 1: 193 (2000); F.E.E. 4(2): 82 (2004).
Syn: *Vernonia richardiana* (Kuntze) Pic.Serm. – F.P.S. 3: 62 (1956); F.J.M.: 139 (1976).
Perennial herb or shrub. Wooded grassland, forest margins.
Distr: DAR, EQU; Togo to Ethiopia, S to D.R. Congo & Kenya.
DAR: Jackson 2598 1/1953; **EQU:** Friis & Vollesen 953 2/1982.

Vernonia thomsoniana Oliv. & Hiern ex Oliv.

F.P.S. 3: 62 (1956); T.S.S.: 420 (1990); Biol. Skr. 51: 385 (1998); F.T.E.A. Comp. 1: 192 (2000); F.E.E. 4(2): 81 (2004).
Perennial herb or shrub. Woodland, grassland & forest margins.

Distr: ?DAR, EQU; Guinea to Ethiopia, S to Angola, Zimbabwe & Mozambique.
EQU: Friis & Vollesen 1270 3/1982.
Note: Andrews and El Amin record this species from both Darfur and Equatoria; we have only seen material from the latter.

Vernonia turbinata Oliv. & Hiern

Biol. Skr. 51: 385 (1998); F.T.E.A. Comp. 1: 201 (2000); F.E.E. 4(2): 87 (2004).
Perennial herb. Wooded grassland.
Distr: EQU; Ethiopia to Kenya.
EQU: Myers 7881 10/1937.

Vernonia undulata Oliv. & Hiern

F.P.S. 3: 58 (1956); F.T.E.A. Comp. 1: 208 (2000).
Perennial herb. Woodland & grassland.
Distr: EQU; Guinea to Kenya, S to Angola & Zambia.
EQU: Andrews 1090 5/1939.

Vernonia unionis Sch.Bip. ex Walp.

F.P.S. 3: 58 (1956); F.E.E. 4(2): 87 (2004).
Perennial herb. Wooded grassland on hillslopes.
Distr: RS, ES; Eritrea, Ethiopia.
IUCN: RD
RS: Jackson 2965 4/1953.

Vernonia wollastonii S.Moore

Biol. Skr. 51: 386 (1998); F.T.E.A. Comp. 1: 208 (2000); F.E.E. 4(2): 89 (2004).
Perennial herb or shrub. Forest margins, wooded grassland.
Distr: EQU; Ethiopia to South Africa.
EQU: Friis & Vollesen 237 11/1980.

Volutaria crupinoides (Desf.) Maire

F.Egypt 3: 177 (2002).
Syn: *Centaurea crupinoides* Desf. – F.P.S. 3: 17 (1956).
Annual herb. Open sandy & stony ground.
Distr: RS; N Africa, Palestine.
RS: Schweinfurth A441 5/1865.

Xanthium strumarium L.

F.E.E. 4(2): 341 (2004); F.T.E.A. Comp. 3: 817 (2005).
Syn: *Xanthium brasilicum* Vell. – F.P.S. 3: 63 (1956); Bot. Exp. Sud.: 38 (1970).
Annual or short-lived perennial herb. Disturbed areas.
Distr: KOR, CS, ES; native to the Americas, now almost cosmopolitan.
CS: Pettet 137 2/1962.

DIPSACALES

Caprifoliaceae

H. Pickering

Dipsacus pinnatifidus Steud. ex A.Rich.

F.P.S. 2: 467 (1952); F.T.E.A. Dipsac.: 2 (1968); Biol. Skr. 51: 350 (1998); F.E.E. 4(1): 286 (2003).
Perennial herb. Grassland, ericaceous bushland.
Distr: EQU; Cameroon, Ethiopia to D.R. Congo & Tanzania.
EQU: Johnston 1511 2/1936.

APIALES

Pittosporaceae

I. Darbyshire

Pittosporum viridiflorum Sims

F.T.E.A. Pittospor.: 1 (1966); F.E.E. 3: 2 (1989); T.S.S.: 54 (1990); Biol. Skr. 51: 115 (1998).
Syn: *Pittosporum mannii* Hook.f. – T.S.S.: 54 (1990).
Syn: *Pittosporum abyssinicum* sensu Andrews, non Delile – F.P.S. 1: 155 (1950).
Shrub, or small tree. Montane forest, often riverine, forest-grassland transition & montane bushland.
Distr: ?RS, ?CS, EQU; Cameroon to Somalia, S to South Africa; Arabia.
EQU: Friis & Vollesen 1078 (fr) 3/1982.
Note: (1) a complex species with a range of variants; our plants fall under the informally recognised form "*mannii*". (2) Andrews' record of *P. abyssinicum* from "White Nile District" is believed to be based on a misidentification of *P. viridiflorum* s.l. El Amin's record of "*P. mannii*" in T.S.S.: 54 is possibly based on Andrews' record of *abyssinicum*. (3) The record for Red Sea region is from El Amin in T.S.S.; no specimen has been seen from Sudan.

Araliaceae

R. Vanderstricht & I. Darbyshire

Cussonia arborea A.Rich.

F.P.S. 2: 356 (1952); F.T.E.A. Aral.: 4 (1968); F.J.M.: 124 (1976); F.E.E. 3: 538 (1989); T.S.S.: 346 (1990); Biol. Skr. 51: 295 (1998).
Syn: *Cussonia hamata* Harms – F.P.S. 2: 356 (1952).
Syn: *Cussonia laciniata* Harms – F.P.S. 2: 356 (1952).
Tree or shrub. Wooded grassland, often in rocky areas.
Distr: DAR, KOR, BAG, EQU; Guinea Bissau to Ethiopia, S to Zimbabwe & Mozambique.
DAR: Wickens 1862 (fr) 7/1964; **EQU:** Friis & Vollesen 1300 (fl) 3/1982.

Cussonia spicata Thunb.

F.T.E.A. Aral.: 3 (1968); T.S.S.: 346 (1990); Biol. Skr. 51: 295 (1998).
Tree. Montane forest.
Distr: EQU; Uganda to South Africa; Comoros.
EQU: Jackson 1095 1/1950.

Hydrocotyle ranunculoides L.f.

F.J.M.: 125 (1976); F.T.E.A. Umbell.: 14 (1989); F.E.E. 4(1): 4 (2003).
Syn: *Hydrocotyle natans* Cirillo – F.P.S. 2: 363 (1952); Bot. Exp. Sud.: 109 (1970).
Aquatic perennial herb. Pools including muddy margins, streams & slow-flowing rivers.
Distr: DAR, BAG, UN; native to the Americas, widely naturalised elsewhere.
DAR: Wickens 1535 (fr) 5/1964; **UN:** Broun 280 3/1903.

Hydrocotyle sibthorpioides Lam.

F.T.E.A. Umbell.: 14 (1989); Biol. Skr. 51: 298 (1998); F.E.E. 4(1): 4 (2003).

Perennial herb. Damp montane grassland, marshes, streamsides.
Distr: EQU; palaeotropical; introduced in the Americas.
EQU: Friis & Vollesen 388 (fr) 11/1980.

Polyscias fulva (Hiern) Harms

F.P.S. 2: 356 (1952); F.T.E.A. Aral.: 12 (1968); F.J.M.: 124 (1976); F.E.E. 3: 537 (1989); T.S.S.: 347 (1990); Biol. Skr. 51: 296 (1998).
Syn: *Polyscias ferruginea* (Hiern) Harms – F.P.S. 2: 357 (1952).
Tree. Forest, particularly as a pioneer in secondary growth & margins.
Distr: DAR, EQU; Guinea to Ethiopia, S to Angola & Mozambique.
DAR: Wickens 1571 (fl fr) 5/1964; **EQU:** Friis & Vollesen 1053 (fr) 2/1982.

Schefflera abyssinica (Hochst. ex. A.Rich.) Harms

F.P.S. 2: 357 (1952); F.T.E.A. Aral.: 20 (1968); F.E.E. 3: 539 (1989); T.S.S.: 347 (1990); Biol. Skr. 51: 296 (1998).
Tree, initially epiphytic. Montane forest.
Distr: EQU; Cameroon to Ethiopia, S to Malawi & Zambia.
EQU: Myers 11144 (fr) 4/1939.
Note: El Amin in T.S.S. records *S. polysciadia* Harms (= *S. myriantha* (Baker) Drake) from the Imatong Mts but Friis & Vollesen (1998) consider this likely to be based on a misidentification.

Apiaceae (Umbelliferae)

I. Darbyshire

Afroligusticum linderi (C.Norman) P.J.D.Winter

Taxon 57: 359 (2008).
Syn: *Peucedanum linderi* C.Norman – F.T.E.A. Umbell.: 105 (1989); Biol. Skr. 51: 300 (1998).
Perennial herb. Montane forest, forest margins, ericaceous scrub & grassland.
Distr: EQU; D.R. Congo & Uganda, S to Zimbabwe & Mozambique.
EQU: Myers 11637 (fl) 8/1939.

Afrosciadium dispersum (C.C.Towns.) P.J.D.Winter

Taxon 57: 360 (2008).
Syn: *Peucedanum dispersum* C.C.Towns. – F.T.E.A. Umbell.: 101 (1989); Biol. Skr. 51: 300 (1998).
Perennial herb. Montane grassland, swamps.
Distr: EQU; Tanzania.
IUCN: RD
EQU: Thomas 1877 12/1935.
Note: the single South Sudanese specimen seen was listed as being from Uganda in F.T.E.A.

Agrocharis incognita (C.Norman) Heywood & Jury

F.T.E.A. Umbell.: 30 (1989); Biol. Skr. 51: 297 (1998); F.E.E. 4(1): 12 (2003).
Syn: *Caucalis incognita* C.Norman – F.P.S. 2: 359 (1952).
Annual or short-lived perennial herb. Forest margins & clearings, montane grassland & bushland.

Distr: ?DAR, EQU; D.R. Congo & Ethiopia, S to
Zimbabwe & Mozambique.
EQU: Friis & Vollesen 1104 (fr) 3/1982.
Note: the record for Darfur is based on Quézel (1969:
120) who noted it as frequent on Jebel Gourgeil; we
have not seen the specimen on which this record was
based and it requires confirmation.

Agrocharis melanantha Hochst.

F.T.E.A. Umbell.: 33 (1989); F.E.E. 4(1): 13 (2003).
Syn: Caucalis melanantha (Hochst.) Hiern – F.J.M.: 125
(1976).
Perennial herb. Montane grassland, wet grassland.
Distr: DAR; Bioko, Cameroon, D.R. Congo, Ethiopia to
Tanzania.
DAR: Wickens 1707 (fl fr) 5/1964.

Agrocharis pedunculata (Baker f.) Heywood & Jury

F.T.E.A. Umbell.: 32 (1989); Biol. Skr. 51: 297 (1998);
F.E.E. 4(1): 12 (2003).
Syn: Caucalis pedunculata Baker f. – F.P.S. 2: 360 (1952).
Perennial herb. Montane grassland & wooded grassland,
forest margins.
Distr: EQU; D.R. Congo, Uganda & Kenya, S to
Zimbabwe & Mozambique.
EQU: Myers 10930 4/1939.

Alepidea peduncularis Steud. ex A.Rich.

F.T.E.A. Umbell.: 89 (1989); Biol. Skr. 51: 297 (1998);
F.E.E. 4(1): 7 (2003).
Syn: Alepidea sp. sensu Andrews – F.P.S. 2: 359 (1952).
Perennial herb. Montane grassland.
Distr: EQU; D.R. Congo to Ethiopia, S to South Africa.
EQU: Friis & Vollesen 833 12/1980.

Anethum graveolens L.

F.J.M.: 124 (1976); F.T.E.A. Umbell.: 89 (1989); F.E.E. 4(1):
35 (2003).
Annual herb. Weed of cultivation, irrigated land.
Distr: NS, DAR, CS; native to Europe & W Asia, widely
cultivated and naturalised elsewhere.
DAR: Lynes 322 (fl fr) 2/1922.
Note: the culinary herb 'dill'.

Berula erecta (Huds.) Coville

F.P.S. 2: 359 (1952); F.J.M.: 125 (1976); F.T.E.A. Umbell.:
73 (1989); F.E.E. 4(1): 31 (2003).
Stoloniferous perennial herb. Streamsides, shallow pools,
marshes.
Distr: DAR; Eritrea to South Africa; widespread in Eurasia
& N America.
DAR: Kamil 1202 (fl) 5/1968.

Centella asiatica (L.) Urb.

F.P.S. 2: 360 (1952); F.J.M.: 125 (1976); F.T.E.A. Umbell.:
15 (1989); Biol. Skr. 51: 298 (1998); F.E.E. 4(1): 5 (2003).
Creeping perennial herb. Damp grassland, riverbanks,
forest margins.
Distr: DAR, UN, EQU; pantropical.
DAR: Wickens 1667 (fr) 5/1964; **UN:** Simpson 7189 (fr)
6/1929.

Coriandrum sativum L.

F.P.S. 2: 360 (1952); F.J.M.: 125 (1976); F.T.E.A. Umbell.:
35 (1989); F.E.E. 4(1): 16 (2003).
Annual herb. Fallow land, escape from cultivation.
Distr: RS, DAR, CS; native of ?SW Asia, widely cultivated
and occasionally naturalised elsewhere.
DAR: Wickens 2587 (fl fr) 9/1964.
Note: the culinary herb 'coriander'.

Daucus carota L.

Bot. Exp. Sud.: 109 (1970); F.E.E. 4(1): 11 (2003).
Syn: Daucus sp. sensu Andrews – F.P.S. 2: 360 (1952).
Biennial herb. Rocky ground.
Distr: RS, DAR; cosmopolitan.
RS: Andrews 2710 (fl fr).
Note: the cited specimen is rather depauperate and its
identification is tentative. We have not seen the material
from Darfur; this record is based on Drar (1970).

Diplolophium africanum Turcz.

F.P.S. 2: 361 (1952); F.J.M.: 125 (1976); F.T.E.A. Umbell.:
81 (1989); Biol. Skr. 51: 298 (1998); F.E.E. 4(1): 33 (2003).
Perennial herb. Grassland & wooded grassland, montane
scrub.
Distr: DAR, EQU; Nigeria to Ethiopia, S to D.R. Congo &
Tanzania.
DAR: Pettet 168 (fl) 12/1962; **EQU:** Myers 6689 (fl)
5/1937.
Note: Andrews records this species from "Central and
Southern Sudan".

Ferula communis L.

F.J.M.: 125 (1976); F.T.E.A. Umbell.: 119 (1989); F.E.E.
4(1): 37 (2003).
Syn: Ferula sp. – F.P.S. 2: 362 (1952); Bot. Exp. Sud.:
109 (1970); F. Darfur Nord-Occ. & J. Gourgeil: 120
(1969).
Perennial herb. Upland scrub & grassland, rocky ground.
Distr: RS, DAR; Eritrea to Tanzania; Mediterranean region
& SW Asia.
DAR: Wickens 1732 (fl) 5/1964.

Foeniculum vulgare Mill.

F.P.S. 2: 362 (1952); F.J.M.: 125 (1976); F.T.E.A. Umbell.:
87 (1989); F.E.E. 4(1): 34 (2003).
Biennial or perennial herb. Naturalised in disturbed areas
or in woodland near cultivation.
Distr: DAR; native of the Mediterranean, widely
cultivated and naturalised elsewhere.
DAR: Wickens 1143 (fl) 2/1964.
Note: the culinary herb 'fennel'.

Heracleum elgonense (H.Wolff) Bullock

F.P.S. 2: 362 (1952); F.T.E.A. Umbell.: 123 (1989); Biol.
Skr. 51: 298 (1998); F.E.E. 4(1): 45 (2003).
Syn: Heracleum abyssinicum sensu Townsend, non
(Boiss.) C.Norman – F.T.E.A. Umbell.: 121 pro parte
(1989).
Perennial herb. Montane grassland, woodland, forest
clearings.
Distr: EQU; Ethiopia, Uganda, Kenya.
EQU: Myers 11661 (fl fr) 7/1939.

Heteromorpha arborescens (Spreng.) Cham. & Schltdl. var. *abyssinica* (Hochst. ex A.Rich.) H.Wolff

F.P.S. 2: 363 (species only) (1952); T.S.S.: 347 (species only) (1990); F.E.E. 4(1): 18 (2003).
Syn: *Heteromorpha trifoliata* (H.L.Wendl.) Eckl. & Zeyh. – F.T.E.A. Umbell.: 38 (1989); Biol. Skr. 51: 299 (1998).
Shrub. Woodland & grassland on rocky slopes, forest margins.
Distr: EQU; Eritrea to Namibia & South Africa.
EQU: Thomas 1769 (fl fr) 12/1935.

Lefebvrea abyssinica A.Rich.

F.T.E.A. Umbell.: 109 (1989); Biol. Skr. 51: 299 (1998); F.E.E. 4(1): 42 (2003).
Perennial herb. Montane grassland & secondary bushland.
Distr: EQU; Cameroon, Ethiopia to Angola, Zimbabwe & Mozambique.
EQU: Friis & Vollesen 95 (fl fr) 11/1980.

Lefebvrea grantii (Hiern) S.Droop

F.T.E.A. Umbell.: 113 (1989).
Syn: *Peucedanum grantii* Hiern – F.P.S. 2: 363 (1952).
Biennial herb. Grassland & open woodland, often in disturbed or heavily grazed areas.
Distr: EQU; Benin to Uganda, S to Angola & South Africa.
EQU: Petherick s.n. (fl fr) 12/1862.

Pimpinella buchananii H.Wolff subsp. buchananii var. *longistyla* C.C.Towns.

F.T.E.A. Umbell.: 64 (1989).
Syn: *Afrosison djurense* H.Wolff – F.P.S. 2: 359 (1952).
Syn: *Afrosison schweinfurthii* H.Wolff – F.P.S. 2: 358 (1952).
Biennial or perennial herb. Woodland including over ironstone.
Distr: BAG, EQU; Tanzania.
EQU: Jackson 4267 (fl) 6/1961.
Note: all material of both *Afrosison djurense* and *A. schweinfurthii* at Kew was redetermined as this taxon by B.-E. van Wyk in 2010. A third species of *Afrosison*, *A. gallabatensis* H.Wolff, was listed by Andrews, based on the type collection by *Schweinfurth* from Matamma near Gallabat. This specimen is believed to have been lost in the Berlin bombing, and no other material has been seen, but this is likely to be a synonym of one of the taxa listed here.

Pimpinella etbaica Schweinf.

F.P.S. 2: 364 (1952); F.E.E. 4(1): 28 (2003).
Annual herb. Rocky ground, hard bare soils.
Distr: RS; Eritrea, Somalia; Egypt.
RS: Andrews 3562 (fl fr).

Pimpinella hirtella A.Rich.

F.T.E.A. Umbell.: 67 (1989); Biol. Skr. 51: 300 (1998); F.E.E. 4(1): 30 (2003).
Syn: *Pimpinella peregrina* sensu Andrews, non L. – F.P.S. 2: 364 (1952).
Biennial or perennial herb. Bamboo forest, montane scrub & grassland.
Distr: KOR, EQU; Guinea, Nigeria, Cameroon, Eritrea & Somalia, S to Tanzania.
KOR: Jackson 2401 (fl) 10/1952; **EQU:** Myers 9720 (fl) 10/1938.

Pimpinella oreophila Hook.f. var. *oreophila*

F.P.S. 2: 363 (1952); F.T.E.A. Umbell.: 66 (1989); Biol. Skr. 51: 300 (1998); F.E.E. 4(1): 28 (2003).
Stoloniferous perennial herb. Montane grassland.
Distr: EQU; Bioko, Cameroon, Ethiopia to Tanzania.
EQU: Friis & Vollesen 832 (fl fr) 12/1980.

Sanicula elata Buch.-Ham. ex D.Don

F.T.E.A. Umbell.: 17 (1989); Biol. Skr. 51: 301 (1998); F.E.E. 4(1): 6 (2003).
Syn: *Sanicula europaea* sensu Andrews, non L. – F.P.S. 2: 364 (1952).
Perennial herb. Forest including stream margins.
Distr: EQU; Bioko to Ethiopia, S to South Africa; subtropical Asia.
EQU: Myers 11761 (fl fr) 8/1939.

Steganotaenia araliacea Hochst.

F.P.S. 2: 366 (1952); F.J.M.: 125 (1976); F.T.E.A. Umbell.: 115 (1989); T.S.S.: 348 (1990); Biol. Skr. 51: 301 (1998); F.E.E. 4(1): 37 (2003).
Small tree. Rocky woodland, gulleys, riverbanks.
Distr: DAR, KOR, CS, ES, EQU; Guinea to Somalia, S to Namibia & South Africa.
KOR: Wickens 945 (fr) 2/1965; **EQU:** Sillitoe 334 (fl) 1919.

Torilis arvensis (Huds.) Link subsp. *heterophylla* (Guss.) Thell.

F.P.S. 2: 366 (species only) (1952); F.J.M.: 126 (species only) (1976); F.T.E.A. Umbell.: 28 (1989); Biol. Skr. 51: 301 (1998).
Annual herb. Dry montane forest & forest margins, secondary scrub, weed of cultivation.
Distr: RS, DAR, EQU; Cameroon to Somalia, S to Angola & South Africa; Mediterranean to Caucasus & SW Asia.
RS: Jackson 2868 (fl fr) 4/1953; **EQU:** Friis & Vollesen 157 (fl fr) 11/1980.

ADDENDUM

During the final proofing of this book, an additional potential record for Sudan came to light that had been previously overlooked:

Ruppiaceae

I. Darbyshire

[*Ruppia maritima* L.]

F.T.E.A. Rupp.: 1 (1989); F.E.E. 6: 24 (1997).
Rhizomatous aquatic herb. Alkaline lakes, brackish coastal waters.

Distr: ?RS, ?DAR; widespread in temperate & tropical regions.

Note: the status of this species in Sudan in unclear. Lye in F.T.E.A. includes Sudan within its distribution and it is likely to occur along the Red Sea coast but we have not seen any specimens from there. *de Wilde et al.* 5605, a sterile specimen from Jebel Marra in Darfur, is tentatively identified as this species but the habitat is described as "in running water in pools at base of fall in small river" which seems wrong for *Ruppia*. Fertile material is required to confirm or refute this record.

Bibliography for the study of plants and their habitats in the Sudan region

This section lists literature cited in the introductory chapters of this book plus a more general botanical bibliography for Sudan and South Sudan, together with particularly useful literature from neighbouring regions. Note that the modern Floras available for neighbouring countries are listed in chapter 4, and since these are multi-volume publications, they are not listed here. We also do not list the large number of more general taxonomic publications that have been consulted during preparation of the checklist, such as global or Africa-wide monographic treatments of genera.

Abdalla, W.S. (1997). *Flora of the Nile Bank in Khartoum State.* Unpubl. M.Sc. thesis, Univ. Khartoum.

Abdelmageed, A.S. (2007). *A study on the flora of Tuti Island in Khartoum.* Unpubl. Ph.D. thesis, Univ. Khartoum.

Ahmed, A.A. (1985). *The grasses of northern Sudan.* Unpubl. M.Sc. thesis, Univ. Khartoum.

Ahti, T., Hämet-Ahti, L. & Pettersson, B. (1973). Flora of the inundated Wadi Halfa reach of the Nile, Sudanese Nubia, with notes on adjacent areas. *Ann. Bot. Fennici* 10: 131–162.

Al Awad, A.A. (1981). *Studies on the family Cucurbitaceae.* Unpubl. M.Sc. thesis, Univ. Khartoum.

Al Awad, A.A. (1995). *Ecotaxonomical studies on the vegetation of the Red Sea State, Sudan.* Unpubl. Ph.D. thesis, Univ. Khartoum.

Andrews, F.W. (1947). *The flora of Erkowit. A. Trees and shrubs.* Dept. Agriculture & Forests, Research Div., Bull. 1. McCorquodale & Co. (Sudan) Ltd., Khartoum.

Andrews, F.W. (1948). The vegetation of the Sudan. pp. 32–61 in: J.D. Tothill (ed.). *Agriculture in the Sudan.* Oxford Univ. Press, London.

Andrews, F.W. (1948). Vernacular names of plants as described in *Flowering plants of the Anglo-Egyptian Sudan. Vol. I.* McCorquodale & Co. (Sudan) Ltd., Khartoum.

Andrews, F.W. (1950). *The flowering plants of the Anglo-Egyptian Sudan. Vol. I (Cycadaceae-Tiliaceae).* T. Buncle & Co. Ltd., Arbroath, Scotland.

Andrews, F.W. (1952). *The flowering plants of the Anglo-Egyptian Sudan. Vol. II (Sterculiaceae-Dipsacaceae).* T. Buncle & Co. Ltd., Arbroath, Scotland.

Andrews, F.W. (1953). Vernacular names of plants as described in *Flowering plants of the Anglo-Egyptian Sudan. Vol. II.* McCorquodale & Co. (Sudan) Ltd., Khartoum.

Andrews, F.W. (1956). *The flowering plants of the Sudan. Vol. III (Compositae-Gramineae).* T. Buncle & Co. Ltd., Arbroath, Scotland.

Andrews, F.W. (1957). Vernacular names of plants as described in *Flowering plants of the Sudan. Vol. III.* McCorquodale & Co. (Sudan) Ltd., Khartoum.

Angiosperm Phylogeny Group [A.P.G.] (2009). An update of the Angiosperm Phylogeny Group classification for the orders and families of flowering plants: APG III. *Bot. J. Linnean Soc.* 161: 105–121.

Badi, K.H., Ahmed, A.E. & Bayoumi, A.A.S. (1989). *The Forests of Sudan.* Ministry of Agriculture, Dept. of Forests, Khartoum.

Baleela, A.R.H. & Kordofani, M.A.Y. (1995). Changing patterns of *Acacia* variations as an indication of climatic change. *Journ. Fac. Sc. Univ. U.A.E.* 8(11): 125–129.

Bebawi, F.F. & Neugebohrn, L. (1991). *A review of plants of northern Sudan with special reference to their uses.* Deutsche Gesellschaft für Technische Zusammenarbeit (GTZ), Eschborn, Germany.

Berry, L. & Whiteman, A.J. (1968). The Nile in the Sudan. *Geogr. Journ.* 134: 1–37.

Braun, M.H., Burgstaller, H., Hamdoun, A.M. & Walter, H. (1991). *Common weeds of Central Sudan.* Verlag Josef Margraf, Weikersheim, Germany.

Broun, A.F. (1906). *Catalogue of Sudan flowering plants.* El-Sudan Printing Press, Khartoum.

Broun, A.F. & Massey, R.E. (1929). *Flora of the Sudan.* The Controller, Sudan Government Office. London.

Brundu, G. & Camarda, I. (2013). The Flora of Chad: a checklist and brief analysis. *Phytokeys* 23: 1–17. doi: 10.3897/phytokeys.23.4752.

Bruneau de Miré, P. & Quézel, P. (1961). Remarques taxonomiques et biogéographiques sur la flore des montagnes de la lisière méridionale du Sahara et plus spécialement du Tibesti et du Djebel Marra. *Journ. Agric. Trop. & Bot. Appl.* 8: 110–133.

Burton, A.N. & Wickens, G.E. (1966). Jebel Marra volcano, Sudan. *Nature* 210: 1146–1147.

Chialvo, N. (1975). *Contribution à l'étude écologique de la vegetation du confluent Atbara-Setit (Rep. Dem. du Soudan)*. Unpubl. thesis, Univ. Scientifique et Médicale Grenoble, France.

Chipp, T.F. (1929). The Imatong Mountains, Sudan. *Bull. Misc. Inf., Kew* 1929: 177–197.

Chipp, T.F. (1930). Forests and plants of the Anglo-Egyptian Sudan. *Geogr. Journ.* 75: 123–141.

Crowfoot, G.M. (1928). *Flowering plants of the northern and central Sudan*. Orphan's Printing Press Ltd., Leominster.

Cufodontis, G. (1962). A preliminary contribution to the knowledge of the botanical exploration of Northeastern Tropical Africa. pp. 233–248 in: A. Fernandes (ed.) *Comptes Rendus de la IVe Réunion Plénière de l'Association pour l'Étude Taxonomique de la Flore d'Afrique Tropicale (AETFAT), Lisbonne et Coïmbra, 16–23 Septembre, 1960*. Junta de Investigações do Ultramar, Lisbon.

Dale, I.R. (1961). The vegetation of Mount Lorienatom. *Sudan Silva* 11: 16–17.

Darbyshire, I. & Kordofani, M.A.Y. (2012). Two new combinations in the South Sudanese flora. *Kew Bull.* 66: 605–607.

Desmond, R. (1977). *Dictionary of British and Irish botanists and horticulturalists, including plant collectors and botanical artists*. Taylor & Francis, London.

Drar, M. (1936). *Enumeration of the plants collected at Gebel Elba during two expeditions*. Government Press, Cairo, Egypt.

Drar, M. (1970). *A botanic expedition to the Sudan in 1938. Edited after author's death with introductory notes by V. Täckholm*. Publications from the Cairo Univ. Herbarium, no. 3.

El Amin, H.M. (1990). *Trees and shrubs of the Sudan*. Ithaca Press. Exeter.

El Daim, A.A. (2000). *The aquatic weeds of Central Sudan*. Unpubl. M.Sc. thesis, Univ. Khartoum.

El Ghazali, G.E.B. (1986). *Medicinal plants of the Sudan. Part 1: medicinal plants of Erkowit*. National Council for Research. Khartoum.

El Hakim, M.S. (2007). *Palyno-taxonomical study on the family Pedaliaceae in Sudan*. Unpubl. M.Sc. thesis, Univ. Khartoum.

El Hakim, M.S. & Kordofani, M.A.Y. (2005). Revision of the genus *Capparis* L. (Family: Capparidaceae) in the Sudan. *Sudan Silva* 11(1): 34–44.

El Hakim, M.S. & Kordofani, M.A.Y. (2006). Tracing evolutionary sequences among the family Pedaliaceae in Sudan. *Sudan Silva* 12(1): 67–78.

El Safori, A.K. (2000). *A study on the flora of El Faw hills (Eastern Sudan)*. Unpubl. M.Sc. thesis, Univ. Khartoum.

El Safori, A.K. (2006). *Eco-taxonomical study on the vegetation of Um Rimmitta area, White Nile State (Central Sudan)*. Unpubl. Ph.D. thesis, Univ. Khartoum.

FAO (1984). *Agroclimatological data for Africa. Vol. 1. Countries north of the equator*. FAO Plant Production and Protection Series No. 22, Rome, Italy.

Farag, I. & Kordofani, M.A.Y. (2012). *Field guide to the wild plants of the Sudanese Coastal Plain*. UNESCO. Sudan Currency Printing Press.

Fluehr-Lobban, C., Lobban, R.A. & Voll, J.O. (1992). *Historical Dictionary of Sudan*. Scarecrow Press, London.

Friis, I. (1992). Forests and forest trees of Northeast Tropical Africa — their natural habitats and distribution patterns in Ethiopia, Djibouti and Somalia. *Kew Bull. Additional Series* 15: i–iv, 1–396.

Friis, I. (1993). Additional observations on forests and forest trees of Northeast Tropical Africa (Ethiopia, Djibouti and Somalia). *Opera Bot.* 121: 119–124.

Friis, I. (1994). Some general features of the afromontane and afrotemperate floras of the Sudan, Ethiopia and Somalia. pp. 958–968 in: J.H. Seyani & A.C. Chikuni (eds), *Proceeding of the XIIIth Plenary Meeting of AETFAT, Zomba, Malawi, 2–11 April 1991*. National Herbarium and Botanic Gardens of Malawi, Zomba, Malawi.

Friis, I., Sebsebe Demissew & van Breugel, P. (2010). *Atlas of the potential vegetation of Ethiopia*. Biologiske Skrifter 58, Det Kongelige Danske Videnskabernes Selskab, Copenhagen.

Friis, I. & Vollesen, K. (1982). New taxa from the Imatong Mountains, South Sudan. *Kew Bull.* 37: 465–479.

Friis, I. & Vollesen, K. (1998). Flora of the Sudan-Uganda border area east of the Nile. I. Catalogue of vascular plants, 1st part. *Biologiske Skrifter* 51:1, Det Kongelige Danske Videnskabernes Selskab, Kommissionær: Munksgaard, Copenhagen.

Friis, I. & Vollesen, K. (1999). *Drimia sudanica*, nom. nov. (Hyacinthaceae), a rare species of the Sudanian grasslands. *Nordic. Journ. Bot.* 19: 209–212.

Friis, I. & Vollesen, K. (2005). Flora of the Sudan-Uganda border area east of the Nile. II. Catalogue of vascular plants, 2nd part, vegetation and phytogeography. *Biologiske Skrifter* 51:2, Det Kongelige Danske Videnskabernes Selskab, Copenhagen.

Frodin, D. (2001). *Guide to standard floras of the world.* 2nd edition. Cambridge Univ. Press, Cambridge.

Gibreel, H.H. (2009). *A taxonomic study on trees and shrubs of El Nour natural forest reserve, Blue Nile State, Sudan.* Unpubl. M.Sc. thesis, Univ. Khartoum.

Good, R. (1924). The geographical affinities of the flora of Jebel Marra. *New Phytologist* 23: 266–281.

Grant, J.A., Oliver, D. & Baker, J.G. (1872–1875). The botany of the Speke and Grant expedition: an enumeration of the plants collected during the journey of the late Captain J.H. Speke and Captain (now Lieut.-Col.) J.A. Grant from Zanzibar to Egypt. The determinations and descriptions by Professor Oliver and others connected with the Herbarium, Royal Gardens, Kew / with an introductory preface, alphabetical list of native names and notes by Colonel Grant. *Trans. Linn. Soc., London* 29: 1–178. Published in three parts.

Greene, H. (1948). Soils of the Anglo-Egyptian Sudan. pp. 144–175 in: J.D. Tothill (ed.) *Agriculture in the Sudan.* Oxford Univ. Press, London.

Gumma, A.G.N. (1988). *The flora of Ingassana Hills, with special reference to khors, S.E. Sudan.* Unpubl. Ph.D. thesis, Univ. Khartoum.

Halwagy, R. (1961). The vegetation of semi-desert North East of Khartoum. *Oikos* 12: 87–110.

Hamada, A.A. (2000). Weeds and weed management in Sudan. *Journ. Weed Science & Technology* 45: 131–136.

Harrison, M.N. & Jackson, J.K. (1958). Ecological classification of the vegetation of the Sudan. *Forests Bull.* No. 2 (New Ser.). Sudan Forests Dept., Ministry of Agriculture, Khartoum.

Hassan, H.M. (1974). *An illustrated guide to the plants of Erkowit.* Khartoum Univ. Press, Khartoum.

Hassan, S.G. (2007). *Chemotaxonomic studies on Sudanese parasitic plants.* Unpubl. M.Sc. thesis, Univ. Khartoum.

Hassan, S.G. & Kordofani, M.A.Y. (2007). Taxonomical studies on Sudanese parasitic plants. *Sudan Silva* Vol. 4(1): 57-69.

Haston, E., Richardson, J.E., Stevens, P.F., Chase, M.W., & Harris, D.J. (2009). The linear Angiosperm Phylogeny Group (LAPG) III: A linear sequence of the families in APG III. *Bot. J. Linnean Soc.* 161: 128–131.

Hedberg, O. (1951). Vegetation belts of the East African mountains. *Svensk. Bot. Tidskr.* 45: 140–202.

Hedberg, O. (1957). Afroalpine vascular plants. A taxonomic revision. *Symb. Bot. Upsal.* 15: 1–411.

Henderson, K.D.D. *'Set under Authority'* Being a portrait of the life of a British District Officer in the Sudan under the Anglo/Egyptian Condominium 1899–1955. Abbey Press.

Heywood, V.H., Brummitt, R.K., Culham, A. & Seberg, O. (2007). *Flowering Plant Families of the World.* Royal Botanic Gardens, Kew.

Hill, R. (1967). *A biographical dictionary of the Sudan.* Second Edition. F. Cass & Co., London.

Hopkins, P.G. (2007). *The Kenana Handbook of Sudan.* Kegan Paul, London.

Howell, P., Lock, M. & Cobb, S. (1988). *The Jonglei Canal.* Cambridge University Press, Cambridge.

Ibrahim, M.A.M. (1996). *A study on the flora of the Gash delta, Eastern Sudan.* Unpubl. M.Sc. thesis, Univ. Khartoum.

Ikram, M.A. (1997). *Eco-taxonomy of El Rawakeeb area, West Omdurman, Khartoum State.* Unpubl. M.Sc. thesis, Univ. Khartoum.

IUCN (2001). *IUCN Red List Categories and Criteria.* Version 3.1. IUCN Species Survival Commission, IUCN, Gland, Switzerland and Cambridge, U.K.

Jackson, J.K. (1950). The Dongotona Hills, Sudan. *Emp. For. Rev.* 29: 139–142.

Jackson, J.K. (1951). Mount Lotuke, Didinga Hills. *Sudan Notes & Rec.* 32: 339–341.

Jackson, J.K. (1956). The vegetation of the Imatong Mountains, Sudan. *Journ. Ecol.* 44: 341–374.

Jackson, J.K. (1957). Changes in the climate and vegetation of the Sudan. *Sudan Notes & Rec.* 38: 47–66.

Kalema, J. & Beentje, H. (2012). *Conservation Checklist of the Trees of Uganda.* Royal Botanic Gardens, Kew.

Kassas, M. (1956a). The mist oasis of Erkowit. *Journ. Ecol.* 44: 180–194.

Kassas, M. (1956b). Landforms and plant cover in the Omdurman desert, Sudan. *Bull. Soc. Géogr., Egypte* 29: 43–58.

Kassas, M. (1957). On the ecology of the Red Sea coastal lowland. *Journ. Ecol.* 45: 187–203.

Keenan, M.L. (2011). *That Hard Hot Land. Botanical collecting expedition in the Egyptian Sudan, 1933–1934.* Self-publ., Parbold, England.

Khawad, M.M. (2006). *Taxonomical studies of some weeds in the Gezira State with special emphasis on Wad Medani and Abu Haraz areas.* Unpubl. M.Sc. thesis, Univ. Khartoum.

Kordofani, M.A.Y. (1985). *Experimental taxonomy of Sudanese bulb bearing species.* Unpubl. M.Sc. thesis, Univ. Khartoum.

Kordofani, M.A.Y. (1989). *Population variation in the genus Acacia Miller in the Northern Sudan.* Unpubl. Ph.D. thesis, Univ. London.

Kordofani, M.A.Y. & Ingrouille, M. (1991). Patterns of morphological variation in the *Acacia* species (Mimosaceae) of the Northern Sudan. *Bot. Journ. Linn. Soc.* 105: 239–256.

Kordofani, M.A.Y. & Ingrouille, M. (1992). Geographical variation in the pollen of *Acacia* (Mimosaceae) in Sudan. *Grana* 31: 113–118.

Kordofani, M.A.Y., Darbyshire, I. & Farag, I. (2013). Two new records for the Sudanese Flora from Erkowit. *Sudan Biota* 1: 5–8.

Kordofani, M.A.Y., El Hakim, M.S.Y., Darbyshire, I., Ali, N.M.M., Farag, I. & Ahmed, H.O. (2013). New medicinal family recorded for Sudan flora (Martyniaceae, species: *Proboscidea parviflora* subsp. *parviflora*). *Journ. Chemical & Pharmaceutical Research* 5: 381–386.

Langdale-Brown, I., Osmaston, H.A. & Wilson, J.G. (1964). *The vegetation of Uganda and its bearing on land-use.* Gov. Uganda, Entebbe & London.

Lebrun, J.-P. & Stork, A. (2003). *Tropical African Flowering Plants. Ecology and Distribution. Vol. 1: Annonaceae-Balanitaceae.* Conservatoire et Jardin botaniques de la Ville de Genève, Switzerland.

Lebrun, J.-P. & Stork, A. (2006). *Tropical African Flowering Plants. Ecology and Distribution. Vol. 2: Euphorbiaceae-Dichapetalaceae.* Conservatoire et Jardin botaniques de la Ville de Genève, Switzerland.

Lebrun, J.-P. & Stork, A. (2008a). *Tropical African Flowering Plants. Ecology and Distribution. Vol. 3: Mimosaceae-Fabaceae.* Conservatoire et Jardin botaniques de la Ville de Genève, Switzerland.

Lebrun, J.-P. & Stork, A. (2008b). *Tropical African Flowering Plants. Ecology and Distribution. Vol. 4: Fabaceae (Desmodium-Zornia).* Conservatoire et Jardin botaniques de la Ville de Genève, Switzerland.

Lebrun, J.-P. & Stork, A. (2010). *Tropical African Flowering Plants. Ecology and Distribution. Vol. 5: Buxaceae-Simaroubaceae.* Conservatoire et Jardin botaniques de la Ville de Genève, Switzerland.

Lebrun, J.-P. & Stork, A. (2011). *Tropical African Flowering plants. Ecology and Distribution. Vol. 6: Burseraceae-Apiaceae and Addendum Vol. 1.* Conservatoire et Jardin botaniques de la Ville de Genève, Switzerland.

Lebrun, J.-P. & Stork, A. (2012). *Tropical African Flowering Plants. Ecology and Distribution. Vol. 7: Monocotyledons I (Limnocharitaceae-Agavaceae).* Conservatoire et Jardin botaniques de la Ville de Genève, Switzerland.

Léonard, J. (1997). Flore et Végétation du Jebel Uweinat (Désert de Libye, Egypte, Sudan). Première partie (Introduction – Des Algues aux Monocotylédones). *Bull. Jard. Bot. Nat. Belg.* 66: 223–340.

Léonard, J. (1999). Flore et Végétation du Jebel Uweinat (Désert de Libye, Egypte, Sudan). Deuxième partie (Dicotylédones: Aizoaceae à Moraceae). *Bull. Jard. Bot. Nat. Belg.* 67: 123–216.

Léonard, J. (1999). Flore et Végétation du Jebel Uweinat (Désert de Libye, Egypte, Sudan). Troisième partie (Dicotylédones: Nyctaginaceae à Zygophyllaceae). *Syst. Geogr. Pl.* 69: 215–264.

Léonard, J. (2000). Flore et Végétation du Jebel Uweinat (Désert de Libye, Egypte, Sudan). Quatrième partie (Considérations generals sur la flore et la végétation). *Syst. Geogr. Pl.* 70: 3–73.

Lester-Garland, L.V. (1921). Some plants from Jebel Marra, Darfur. *Journ. Bot.* 59: 46–48.

Lind, E.M. & Morrison, M.E.S. (1974). *East African Vegetation.* Longman, London.

Lynes, H. (1921). Notes on the natural history of Jebel Marra. *Sudan Notes & Rec.* 4: 119–137.

Macleay, K.N.G. (1953). The ferns and ferns-allies of the Sudan. *Sudan Notes & Rec.* 34: 286–298.

MacLeay, K.N.G. (1955). Geographical relationships of the Pteridophyte flora of the Sudan. *Bull. Jard. Bot. Etat. Brux.* 25: 213–220.

MacLeay, K.N.G. (1955). A preliminary list of Pteridophyta of the Anglo-Egyptian Sudan. *Webbia* 11: 587–606.

MacLeay, K.N.G. (1963). Corrections and emendations to the Flora of the Sudan. *Sudan Notes & Rec.* 44: 140–142.

Malterer, A. (2013). *Merowe Dam Project. Land use and vegetation in the flooding area of a planned hydrodam in Northern Sudan.* Draft Report, Sudan Archaeological Research Society.

Massey, R.E. (1926). *Sudan Grasses.* Dept. Agriculture & Forestry, Sudan Govt., Botanical Series Publ. No 1. McCorquodale & Co. (Sudan) Ltd, Khartoum.

Mohamed, N.M. (2006). *A study of the flora of Fadasi (Central Gezira).* Unpubl. M.Sc. thesis, Sudan Academy of Science, Khartoum.

Musselman, L.J. (1984). Some parasitic angiosperms of Sudan: Hydnoraceae, Orobanchaceae and *Cuscuta* (Convolvulaceae). *Notes Roy. Bot. Gard. Edinburgh* 42: 21–38.

Musselman, L.J. & Hepper, F.N. (1986). The witchweeds (*Striga*, Scrophulariaceae) of the Sudan Republic. *Kew Bull.* 41: 205–221.

Mutasim, M.A. (2012). *Taxonomy of trees and shrubs of Zalingei area, West Darfur state, Sudan.* Unpubl. M.Sc. thesis, Univ. Khartoum.

Norman, C. (1924). Plants from Jebel Marra, Darfur. *Journ. Bot.* 62: 134–139.

Obeid, A. & Mahmoud, A. (1968). The vegetation of Khartoum Province. *Sudan Notes & Rec.* 50: 135–159.

Onana, J.M. (2011). The Vascular Plants of Cameroon. A taxonomic checklist with IUCN assessments. *Flore du Cameroon* Vol. 39, IRAD-Herbier National du Cameroun, Yaoundé.

Osman, M.H. (2012). *Taxonomic revision of the family Boraginaceae in Sudan.* Unpubl. M.Sc. thesis, Univ. Khartoum.

Pfund, J. (1878). *Dr J. Pfund's Reisebriefe aus Kordofan und Dar-Fur 1875–76.* Nach dem Tode des Verfassers herausgegeben von der geographischen Gesellshaft in Hamburg. L. Friederichsen, Hamburg.

Plowes, D.C.H. (1989), The succulents of Sudan. *Excelsa* 14: 57–69.

Polhill, R. & Wiens, D. (1998). *Mistletoes of Africa*. Royal Botanic Gardens, Kew.

Quézel, P. (1968). Premier résultats de l'exploration botanique du Gourgeil (République du Soudan). *Compte Rendu Acad. Sci. Paris Sér. D*, 266: 2061–2063.

Quézel, P. (1969). Flore et végétation des plateaux du Darfur Nord-occidental et du jebel Gourgeil (République du Soudan). Centre Nat. Recherche Sci. Recherche Coop. Prog. No. 45, dossier 5.

Quézel, P. (1970). A preliminary description of the vegetation in the Sahel region of north Darfur. *Sudan Notes & Rec.* 51: 119–125.

Radwanski, S.A. & Wickens, G.E. (1967). The ecology of *Acacia albida* on mantle soils in Zalingei, Jebel Marra, Sudan. *Journ. Appl. Ecol.* 4: 569–579.

Ramsay, D.M. (1958). The forest ecology of central Darfur. *Sudan Govt. For. Bull.* No. 1, Khartoum.

Robertson, P. (2001). Sudan. pp. 877–890 in L.D.C. Fishpool & M.I. Evans (eds) *Important Bird Areas in Africa and associated islands: priority sites for conservation*. Pisces Publications & Birdlife International, Newbury and Cambridge (BirdLife Conservation Series No. 11).

Robertson, V.C. (1965). Jebel Marra. *Geogr. Mag.* 37: 912–925.

Roux, J.P. (2009). Synopsis of the Lycopodiophyta and Pteridophyta of Africa, Madagascar and neighbouring islands. *Strelitzia* 23. South African National Biodiversity Institute, Pretoria.

Sahni, K.C. (1968). *Important Trees of the Northern Sudan*. Khartoum Univ. Press, Sudan.

Schweinfurth, G.A. (1868). *Reliquiae Kotschyanae*. G. Reimer, Berlin.

Schweinfurth, G.A. (1874). *The Heart of Africa. Three years' travels and adventures in the unexplored regions of Central Africa from 1868 to 1871*. Harper & Bros., New York.

Smith, J. (1949). Distribution of tree species in the Sudan in relation to rainfall and soil texture. *Sudan Govt. Agric. Bull.* No. 4, Khartoum.

Sommerlatte, M. (1985). A conservation priority: the tropical forests of the Imatong mountains, southern Sudan. pp. 265–272 in D. Ernst (ed.) *Proc. Seminar on Wildlife Conservation and Management in the Sudan, Khartoum, March 16–21, 1985*. Stubbeman, Hamburg.

Sommerlatte, H. & Sommerlatte, M. (1990). *A Field Guide to the Trees and Shrubs of the Imatong Mountains, Southern Sudan*. Deutsche Gesellschaft fur Technische Zusammenarbeit (GTZ), Nairobi, Kenya.

Sosef, M.S.M., Wieringa, J.J., Jongkind, C.C.H., Achoundong, G., Azizet Issembé, Y., Bedigan, D., van der Berg, R.G., Breteler, F.J., Cheek, M., Degreef, J., Faden, R.B., Goldblatt, P., van der Maesen, L.J.G., Ngok Banak, L., Niangadouma, R., Nzabi, T., Nziengui, B., Rogers, Z.S., Stévart, T., van Valkenburg, J.L.C.H., Walters, G. & de Wilde, J.J.F.E. (2006). Check-list des plantes vasculaires du Gabon / Checklist of Gabonese vascular plants. *Scripta Botanica Belgica* Vol. 35. Jardin Botanique National de Belgique.

Speke, J.H. (1863). *Journal of the Discovery of the the Source of the Nile*. W. Blackwood & Sons, Edinburgh & London.

Sudan/Department of Agriculture Correspondence. (1949–1969). RBG, Kew Archives, QG/0585.

Sudan/Miscellaneous Correspondence. (1954–1987). RBG, Kew Archives QG/0594.

Sudan/Khartoum University Correspondence. (1954–1986). RBG, Kew Archives QG/1144.

Thirakul, S. (1984). *Manual of Dendrology: Bahr el Ghazal and central regions*. Canadian International Development Agency, Quebec, Canada.

Tothill, B.H. (1942). Some extracts of the life and travels of Theodore Kotschy. *Sudan Notes & Rec.* vol 25.

Tothill, J.D. (ed.) (1948). *Agriculture in the Sudan*. Oxford University Press, London.

Traylor, M.A. Jr. & Archer, A.L. (1982). Some results of the Field Museum 1977 Expedition to South Sudan. *Scopus* 6: 5–12.

Tutu, S.O. (2002). *Trees and shrubs of Shambat area: taxonomy, growth, natural regeneration and uses*. Unpubl. M.Sc. thesis, Univ. Khartoum.

Udal, J.O. (1995). John Petherick, First British Resident. *Sudan Studies Society of the United Kingdom*, No. 17.

Vesey-Fitzgerald, D.F. (1955). Vegetation of the Red Sea coast south of Jeddah, Saudi Arabia. *Journ. Ecol.* 43: 477–489.

Vesey-Fitzgerald, D.F. (1957). The vegetation of the Red Sea coast north of Jeddah, Saudi Arabia. *Journ. Ecol.* 45: 547–562.

Vogt, K. (1995). *A field worker's guide to the identification, propagation and uses of common trees and shrubs of dryland Sudan*. SOS Sahel International (UK), London.

White, F. (1965). The savanna woodlands of the Zambesian and Sudanian Domains. An ecological and phytogeographical comparison. *Webbia* 19: 651–681.

White, F. (1981). *UNESCO/AETFAT/UNSO Vegetation Map of Africa. Scale 1:5,000,000, 3 maps and legends*. UNESCO, Paris.

White, F. (1983). *The Vegetation of Africa. A Descriptive Memoir to Accompany the UNESCO/AETFAT/UNSO Vegetation Map of Africa*. UNESCO, Paris.

White, F. (1993). The AETFAT chorological classification of Africa: history, methods and application. *Bull. Jard. Bot. Nat. Belg.* 62: 225–281.

Whitehouse, G.J. (1931). The Langia-Acholi mountain region of the Sudan-Uganda borderland. *Geogr. Journ.* 77: 140–159.

Wickens, G.E. (1968). Some additions and corrections to F.W. Andrews', Flowering Plants of the Sudan. *Sudan For. Bull.* No. 14 (New Series), Khartoum.

Wickens, G.E. (1970). J.D.C. Pfund, a botanist in the Sudan with the Egyptian Military expedition, 1875–76. *Kew Bull.* 24: 191–216.

Wickens, G.E. (1971). Dr. G. Schweinfurth's journeys in the Sudan. *Kew Bull.* 27: 129–146.

Wickens, G.E. (1975).The fossil plants of the Jebel Marra volcanic complex and their palaeoclimatic interpretation. *Palaeogeogr., Palaeoclim., Palaeoecol.* 17: 109–122.

Wickens, G.E. (1975). Changes in the climate and vegetation of the Sudan since 20,000 B.P. Comptes Rendus VIII Réunion A.E.T.F.A.T. Genève 1974. *Boissiera* 24a: 43–65.

Wickens, G.E. (1975). A preliminary note on the status of the oil palm (*Elaeis guinensis* Jacq.) in the Sudan. *Sudan Notes & Rec.* 55: 185–188.

Wickens, G.E. (1976). Speculations on long distance dispersal and the flora of Jebel Marra, Sudan Republic. *Kew Bull.* 31: 105–150.

Wickens, G.E. (1976). The Flora of Jebel Marra (Sudan Republic) and its geographical affinities. *Kew Bull. Additional Series* 5: 1–368.

Wickens, G.E. & Collier, F.W. (1971). Some vegetation patterns in the Republic of the Sudan. *Geoderma* 6: 43–59.

Wilson, R.T. (2012). The Biological Exploration of Darfur, 1799–1998. *Archives of Natural History* 39: 39–58.

Zarb, J.H. (1879). *Rapport sur les spécimens botanique colligés pendant les expéditions égyptiennes au Kordofan et au Darfur 1875 et 1876.* Cairo Nat. Print. Press, Egypt.

Useful websites:

African Plants Database (version 3.4.0). Conservatoire et Jardin botaniques de la Ville de Genève and South African National Biodiversity Institute, Pretoria: http://www.ville-ge.ch/musinfo/bd/cjb/africa/

Stevens, P.F. (2001 onwards). **Angiosperm Phylogeny Website.** Version 12, July 2012 [and more or less continuously updated since]: http://www.mobot.org/MOBOT/research/APweb/

eMonocot (2013). A global online biodiversity information resource for monocot plants. Published on the Internet: http://www.e-monocot.org/

JSTOR Global Plants: http://plants.jstor.org/

The Plant List (2010). Version 1. Published on the Internet: http://www.theplantlist.org/

The Sudan Archive (2010). Durham University: https://www.dur.ac.uk/library/asc/sudan/

World Checklist of Selected Plant Families (2013). Facilitated by the Royal Botanic Gardens, Kew. Published on the Internet: http://apps.kew.org/wcsp/

Index to accepted plant families and genera in the checklist